KT-231-791

GREEN'S FUNCTIONS AND BOUNDARY VALUE PROBLEMS

PURE AND APPLIED MATHEMATICS

A Wiley Series of Texts, Monographs, and Tracts

A complete list of the titles in this series appears at the end of this volume.

GREEN'S FUNCTIONS AND BOUNDARY VALUE PROBLEMS

Third Edition

Ivar Stakgold
Department of Mathematical Sciences
University of Delaware
Newark, Delaware
and
Department of Mathematics
University of California, San Diego
La Jolla, California

Michael Holst
Departments of Mathematics and Physics
University of California, San Diego
La Jolla, California

WILEY

A JOHN WILEY & SONS, INC., PUBLICATION

Published by John Wiley & Sons, Inc., Hoboken, New Jersey.
Published simultaneously in Canada.

Library of Congress Cataloging-in-Publication Data:

Stakgold, Ivar.
Green's functions and boundary value problems / Ivar Stakgold, Michael Holst. — 3rd ed.
p. cm. — (Pure and applied mathematics ; 99)
Includes bibliographical references and index.
ISBN 978-0-470-60970-5 (hardback)
1. Boundary value problems. 2. Green's functions. 3. Mathematical physics. I. Holst, Michael. II. Title.
QA379.S72 2010
515'.35—dc22 2010023290

Printed in the United States of America.

10 9 8 7 6 5 4 3 2 1

To Lainie and Alissa
I.S.

For Mai,
Mason, and Makenna
M.H.

CONTENTS

PREFACE TO THE THIRD EDITION

Why a third edition? The principal reason is to include more material from analysis, approximation theory, partial differential equations, and numerical analysis as needed for understanding modern computational methods that play such a vital role in the solution of boundary value problems. As I am not an expert in computational mathematics, it was essential to find a highly qualified coauthor. When I moved to San Diego in early 2008, I was offered an office at the University of California, San Diego (UCSD), which, luckily, was next to the office of Michael Holst. Here was the perfect coauthor, and it was my good fortune that he agreed to collaborate on the new edition! The most substantial change for the new third edition is a fairly extensive new chapter (Chapter 10), which covers the new material listed above. The sections of the new chapter are:

10.1 Nonlinear Analysis Tools for Banach Spaces
10.2 Best and Near-Best Approximation in Banach Spaces
10.3 Overview of Sobolev and Besov Spaces
10.4 Applications to Nonlinear Elliptic Equations
10.5 Finite Element and Related Discretization Methods
10.6 Iterative Methods for Discretized Linear Equations
10.7 Methods for Nonlinear Equations

To support the inclusion of this new chapter, and to help connect the presentation of the analysis material to standard references, we have added an additional final section to four of the chapters that appeared in the second edition of the book. These completely new sections for the third edition are:

2.6 Weak Derivatives and Sobolev Spaces
4.8 The Hahn-Banach Theorem and Reflexive Banach Spaces
5.9 The Banach-Schauder and Banach-Steinhaus Theorems
8.5 The Lax-Milgram Theorem

We have also added a final subsection on Lebesgue integration at the end of Chapter 0, listing a few of the main concepts and results on Lebesgue integration in $\mathbb{R}^n$. In addition, the titles of a few sections from the second edition have been changed slightly to more clearly bring out the material already contained in the sections, again to help connect the material in the sections to presentations of these topics appearing in standard references. The new section titles are:

4.4 Contractions and the Banach Fixed-Point Theorem
4.5 Hilbert Spaces and the Projection Theorem
4.7 Linear Functionals and the Riesz Representation Theorem
9.1 Introduction and Basic Fixed-Point Techniques

The bibliographies at the end of the chapters in the second edition have also been updated for the third edition, but we have likely left out many outstanding new books and papers that should have been included, and we apologize in advance for all such omissions.

IVAR STAKGOLD

La Jolla, California
November 2010

When Ivar asked me to consider joining him on a third edition of his well-known and popular book, *Green's Functions and Boundary-Value Problems*, I was a bit intimidated; not only had it been a standard reference for me for many years, but it is also used as the main text for the first-year graduate applied analysis sequence in a number of applied mathematics doctoral programs around the country. However, I soon realized it was an opportunity for me to add the material that I feel is often missing from first-year graduate courses in modern applied mathematics, namely, additional foundational material from analysis and approximation theory to support the design, development, and analysis of effective and reliable computational methods for partial differential equations. Although there are some wonderful books covering applied mathematics (such as Ivar's) and some equally strong books on numerical analysis, the bridge between them (built with linear functional analysis, approximation theory, and nonlinear analysis) is often mostly missing in these same books. There are a number of books devoted entirely to building this bridge; however, our goal for the third edition was to add just the right subset of this material so that a course based on this single book, combined with a course based on a strong graduate numerical analysis book, would provide a solid foundation for applied mathematics students in our mathematics doctoral program and in our interdisciplinary Computational Science, Mathematics, and Engineering Graduate Program at UCSD.

After spending substantial time with the second edition of the book over the last year, my appreciation for Ivar's original book has only grown. The second edition is a unique combination of modeling, real analysis, linear functional analysis and operator theory, partial differential equations, integral equations, nonlinear functional analysis, and applications. The book manages to present the topics in a friendly, informal way, and at the same time gives the real theorems, with real proofs, when they are called for. The changes that I recommended we make to the second edition (as Ivar outlined above) were mostly to draw out the existing structure of the book, and also to add in a few results from linear functional analysis to complete the material where it was needed to support the new final chapter of the book. Since those of us who have worked closely with the second edition are very familiar with exactly where to find particular topics, one of my goals for the third edition was to preserve as much of the second edition as possible, right down to theorem, equation, and exercise numbers within the sections of each chapter. This is why I have tried to fit all of the new material into new sections appearing at the end of existing chapters, and into the new final chapter appearing at the end of the book. The index to the second edition also provided finer-grained access to the book than did the table of contents; I always found this a very valuable part of the second edition, so I attempted to preserve the entire second edition index as a subset of the third edition index. My hope is that as a consequence of our efforts, the third edition of the book will be viewed as a useful superset of the second edition, with new material on approximation theory and methods, together with some additional supporting analysis material.

The third edition contains approximately 30% new material not found in the second edition. The longest chapter is now the new final chapter (Chapter 10) on approximation theory and methods. We considered splitting it into two chapters, but it seems to hold together well as a single chapter. In addition to the new material in Chapter 10, we have added material to Chapters 2, 4, 5, and 8 as Ivar outlined above. Chapter 2 now contains an early introduction to Sobolev spaces based on weak differentiation, and Chapter 8 now includes the Lax-Milgram Theorem and some related tools. Chapters 4 and 5 now provide a gentle introduction to many of the central concepts and theorems in linear functional analysis and operator theory, as needed by most first-year graduate students working in applied analysis and applied partial differential equations. Some of the new material in Chapter 10 is a bit more advanced than some of the other sections of the book; however, this material builds only on (old and new) material found in Chapters 2, 4, 5, and 8, with the support of a few new paragraphs added to the end of Chapter 0 (on Lebesgue integration). The only exception is perhaps the last example in Section 10.4, chosen from mathematical physics to illustrate the combined use of several tools from nonlinear analysis and approximation theory; it requires a bit of familiarity with the notation used in differential geometry.

A brief word about the numbering system used in third edition is in order, since we are departing substantially from the convention used in the previous two editions (as outlined in the preface to the first edition). The book is now divided into eleven chapters (beginning with Chapter 0), with the inclusion of a new final chapter (Chapter 10). Each chapter is divided into numbered sections, and equations are numbered by chapter, section, and equation within each section. For example, a reference to equation (8.5.2) is to the second numbered equation appearing in Section 5 of Chapter 8. Similarly, all definitions, theorems, corollaries, lemmas, and the like, as well as exercises, are numbered using the same convention. This convention makes the third edition easier to navigate than the first two editions, with a simple glance at a typical page revealing precisely the section and chapter in which the page appears. However, it also preserves the numbering of items from the second edition; for example, equation (5.2) of Chapter 8 in the second edition is numbered as (8.5.2) in the third edition. Note that some objects remain unnumbered if they were unnumbered in the first two editions (for example, a theorem that is not referred to later in the book). To simplify the presentation without losing the advantages of this numbering convention, we make three consistent exceptions: Figures are numbered only by chapter and figure within the chapter; examples and remarks are numbered only within the section; and the Bibliography continues to consist of a chapter-specific list of references immediately following the chapter, ordered alphabetically. Citations to references are now also numbered within the referring text; for example, a citation to reference [3] occurring within a chapter refers to the third reference appearing in the list of references at the end of the chapter.

I would like to thank my family (Mai, Mason, and Makenna) for their patience during the last few months as I focused on the book. I would also like to thank the faculty in the Center for Computational Mathematics at UCSD, and in particular Randy Bank, Philip Gill, and Jim Bunch, for the support and encouragement they have given me over the last ten years. I am also indebted to the Center for Theoretical Biological Physics, the National Biomedical Computation Resource, the National Science Foundation, the National Institutes of Health, the Department of Energy, and the Department of Defense for their ongoing support of my research. I must express my appreciation for the interactions I have had with Randy Bank, Long Chen, Don Estep, Gabriel Nagy, Gantumur Tsogtgerel, and Jinchao Xu, as each played a role in the development of my understanding of much of the material I wrote for the book. I would also like to thank Ari Stern, Ryan Szypowski, Yunrong Zhu, and Jonny Serencsa for reading the new material carefully and catching mistakes. Finally, I am grateful to my friend and mentor Herb Keller, who greatly influenced my work over the last fifteen years, and this is reflected in the topics that I chose to include in the book. Herb was my postdoctoral advisor at Caltech from 1993 to 1997, and after retiring from Caltech around 2000, he moved to San Diego to join our research group at UCSD. We thoroughly enjoyed the years Herb was with us at the Center (attending the weekly seminars in his biking outfit, after biking down the coast from Leucadia). Unfortunately, Herb passed away just before Ivar joined our research group in 2008; otherwise, we might have had three authors on this new edition of the book.

MICHAEL HOLST

La Jolla, California
November 2010

PREFACE TO THE SECOND EDITION

The field of applied mathematics has evolved considerably in the nearly twenty years since this book's first edition. To incorporate some of these changes, the publishers and I decided to undertake a second edition. Although many fine books on related subjects have appeared in recent years, we believe that the favorable reception accorded the first edition— as measured by adoptions and reviews— justifies the effort involved in a new edition.

My basic purpose is still to prepare the reader to use differential and integral equations to attack significant problems in the physical sciences, engineering, and applied mathematics. Throughout, I try to maintain a balance between sound mathematics and meaningful applications. The principal changes in the second edition are in the areas of modeling, Fourier analysis, fixed-point theorems, inverse problems, asymptotics, and nonlinear methods. The exercises, quite a few of which are new, are rarely routine and occasionally can even be considered extensions of the text. Let me now turn to a chapter-by-chapter list of the major changes.

Chapter 0 [Preliminaries] has assumed a more important role. It is now the starting point for a discussion of the relation among the four alternative formulations of physical problems: integral balance law, boundary value problem, weak form (also known as the principle of virtual work), and variational principle. I have also added new modeling examples in climatology, population dynamics, and fluid flow.

Chapter 1 [Green's functions: intuitive ideas] contains some revisions in exposition, particularly in regard to continuous dependence on the data.

In Chapter 2 [The theory of distributions], the treatment of Fourier analysis has been extended to include Discrete and Fast transforms, band-limited functions, and the sampling theorem using the sinc function.

Chapter 3 [One-dimensional boundary value problems] now includes a more thorough treatment of least-squares solutions and pseudo-inverses. The ideas are introduced through a discussion of unbalanced systems (underdetermined or overdetermined).

Chapter 4 has been retitled "Hilbert and Banach spaces," reflecting an increased emphasis on normed spaces at the expense of general metric spaces. The material on contractions is rewritten from this point of view with some new examples.

Chapter 5 [Operator theory] is virtually unchanged.

Chapter 6 [Integral equations] now includes a treatment of Tychonov regularization for integral equations of the first kind, an important aspect of the study of ill-posed inverse problems. Some new examples of integral equations are presented and there is a short discussion of singular-value decomposition. Part of the material on integrodifferential equations has been deleted.

Chapter 7 [Spectral theory of second-order differential operators] is basically unchanged.

In Chapter 8 [Partial differential equations], I have added a more comprehensive treatment of the spectral properties of the Laplacian, including a discussion of recent results on isospectral problems. The asymptotic behavior of the heat equation is examined. A brief introduction to the finite element method is incorporated in a slightly revised section on variational principles.

Chapter 9 [Nonlinear problems] contains a new subsection comparing the three major fixed-point theorems: the Schauder theorem, the contraction theorem of Chapter 4, and the theorem for order-preserving maps, which is used extensively in the remainder of Chapter 9. I have also included a study of the phenomena of finite-time extinction and blow-up for nonlinear reaction-diffusion problems.

There now remains the pleasant task of acknowledging my debt to the students and teachers who commented on the first edition and diplomatically muted their criticism! I am particularly grateful to my friends Stuart Antman of the University of Maryland, W. Edward Olmstead of Northwestern University, and David Colton and M. Zuhair Nashed of the University of Delaware, who generously provided me with ideas and encouragement. The new material in Chapter 9 owes much to my overseas collaborators, Catherine Bandle (University of Basel) and J. Ildefonso Diaz (Universidad Complutense, Madrid). The TEX preparation of the manuscript was in the highly skilled hands of Linda Kelly and Pamela Haverland.

IVAR STAKGOLD

Newark, Delaware
September 1997

PREFACE TO THE FIRST EDITION

As a result of graduate-level adoptions of my earlier two-volume book, *Boundary Value Problems of Mathematical Physics*, I received many constructive suggestions from users. One frequent recommendation was to consolidate and reorganize the topics into a single volume that could be covered in a one-year course. Another was to place additional emphasis on modeling and to choose examples from a wider variety of physical applications, particularly some emerging ones. In the meantime my own research interests had turned to nonlinear problems, so that, inescapably, some of these would also have to be included in any revision. The only way to incorporate these changes, as well as others, was to write a new book, whose main thrust, however, remains the systematic analysis of boundary value problems. Of course some topics had to be dropped and others curtailed, but I can only hope that your favorite ones are not among them.

My book is aimed at graduate students in the physical sciences, engineering, and applied mathematics who have taken the typical "methods" course that includes vector analysis, elementary complex variables, and an introduction to Fourier series and boundary value problems. Why go beyond this? A glance at modern publications in science and engineering provides the answer. To the lament of some and the delight of others, much of this literature is deeply mathematical. I am referring not only to areas such as mechanics and electromagnetic theory that are traditionally mathematical but also to relative newcomers to mathematization, such as chemical engineering, materials science, soil mechanics, environmental engineering, biomedical engineering, and nuclear engineering. These fields give rise to challenging mathematical problems whose flavor can be sensed from the following short list of examples; integrodifferential equations of neutron transport theory, combined diffusion and reaction in chemical and environmental engineering, phase transitions in metallurgy, free boundary problems for dams in soil mechanics, propagation of impulses along nerves in biology. It would be irresponsible and foolish to claim that readers of my book will become instantaneous experts in these fields, but they will be prepared to tackle many of the mathematical aspects of the relevant literature.

Next, let me say a few words about the numbering system. The book is divided into ten chapters, and each chapter is divided into sections. Equations do *not* carry a chapter designation. A reference to, say, equation 4.32 is to the thirty-second numbered equation in Section 4 of the chapter you happen to be reading. The same system is used for figures and exercises, the latter being found at the end of sections. The exercises, by the way, are rarely routine and, on occasion, contain substantial extensions of the main text. Examples do not carry any section designation and are numbered consecutively within a section, even though there may be separate clusters of examples within the same section. Some theorems have numbers and others do not; those that do are numbered in a sequence within a section— Theorem 1, Theorem 2, and so on.

A brief description of the book's contents follows. No attempt is made to mention all topics covered; only the general thread of the development is indicated.

Chapter 0 presents background material that consists principally of careful derivations of several of the equations of mathematical physics. Among them are the equations of heat conduction, of neutron transport, and of vibrations of rods. In the last-named derivation an effort is made to show how the usual linear equations for beams and strings can be regarded as first approximations to nonlinear problems. There are also two short sections on modes of convergence and on Lebesgue integration.

Many of the principal ideas related to boundary value problems are introduced on an intuitive level in Chapter 1. A boundary value problem (BVP, for short) consists of a differential equation $Lu = f$ with boundary conditions of the form $Bu = h$. The pair (f, h) is known collectively as the data for the problem, and u is the response to be determined. Green's function is the response when f represents a concentrated unit source and $h = 0$. In terms of Green's function, the BVP with arbitrary data can be solved in a form that shows clearly the dependence of the solution on the data. Various examples are given, including some multidimensional ones, some involving interface conditions, and some initial value problems. The useful notion of a well-posed problem is discussed, and a first look is taken at maximum principles for differential equations.

Chapter 2 deals with the theory of distributions, which provides a rigorous mathematical framework for singular sources such as the point charges, dipoles, line charges, and surface layers of electrostatics. The notion of response to such sources is made precise by defining the distributional solution of a differential equation. The related concepts of weak solution, adjoint, and fundamental solution are also introduced. Fourier series and Fourier transforms are presented in both classical and distributional settings.

Chapter 3 returns to a more detailed study of one-dimensional linear boundary value problems. To an equation of order p there are usually associated p independent boundary conditions involving derivatives of order less than p at the endpoints a and b of a bounded interval. If the corresponding BVP with 0 data has only the trivial solution, then the BVP with arbitrary data has one and only one solution which can be expressed in terms of Green's function. If, however, the BVP with 0 data has a nontrivial solution, certain solvability conditions must be satisfied for the BVP with arbitrary data to have a solution. These statements are formulated precisely in an alternative theorem, which recurs throughout the book in various forms. When the BVP with 0 data has a nontrivial solution, Green's function cannot be constructed in the ordinary way, but some of its properties can be salvaged by using a modified Green's function, defined in Section 5.

Chapter 4 begins the study of Hilbert spaces. A Hilbert space is the proper setting for many of the linear problems of applied analysis. Though its elements may be functions or abstract "vectors," a Hilbert space enjoys all the algebraic and geometric properties of ordinary Euclidean space. A Hilbert space is a linear space equipped with an inner product that induces a natural notion of distance between elements, thereby converting it into a metric space which is required to be complete. Some of the important geometric properties of Hilbert spaces are developed, including the projection theorem and the existence of orthonormal bases for separable spaces. Metric spaces can be useful quite apart from any linear structure. A contraction is a transformation on a metric space that uniformly reduces distances between pairs of points. A contraction on a complete metric space has a unique fixed point that can be calculated by iteration from any initial approximation. Examples demonstrate how to use these ideas to prove uniqueness and constructive existence for certain classes of nonlinear differential equations and integral equations.

Chapter 5 examines the theory of linear operators on a separable Hilbert space, particularly integral and differential operators, the latter being unbounded operators. The principal problem of operator theory is the solution of the equation $Au = f$, where A is a linear operator and f an element of the space. A thorough discussion of this problem leads again to adjoint operators, solvability conditions, and alternative theorems. Additional insight is obtained by considering the inversion of the equation $Au - \lambda u = f$, which leads to the idea of the spectrum, a generalization of the more familiar concept of eigenvalue. For compact operators (which include most integral operators) the inversion problem is essentially solved by the Riesz-Schauder theory of Section 7. Section 8 relates the spectrum of symmetric operators to extremal principles for the Rayleigh quotient. Throughout, the theory is illustrated by specific examples.

In Chapter 6 the general ideas of operator theory are specialized to integral equations. Integral equations are particularly important as alternative formulations of boundary value problems. Special emphasis is given to Fredholm equations with symmetric Hilbert-Schmidt kernels. For the corresponding class of operators, the nonzero eigenvalues and associated eigenfunctions can be characterized through successive extremal principles, and it is then possible to give a complete treatment of the inhomogeneous equation. The last section discusses the Ritz procedure for estimating eigenvalues, as well as other approximation methods for eigenvalues and eigenfunctions. There is also a brief introduction to integrodifferential operators in Exercises 5.3 to 5.8.

Chapter 7 extends the Sturm-Liouville theory of second-order ordinary differential equations to the case of singular endpoints. It is shown, beginning with the regular case, how the necessarily discrete spectrum can be constructed from Green's function. A formal extension of this relationship to the singular case makes it possible to calculate the spectrum, which may now be partly continuous. The transition from regular to singular is analyzed rigorously for equations of the first order, but the Weyl classification for second-order equations is given without proof. The eigenfunction expansion in the singular case can lead to integral transforms such as Fourier, Hankel, Mellin, and Weber. It is shown how to use these transforms and their inversion formulas to solve partial differential equations in particular geometries by separation of variables.

Although partial differential equations have appeared frequently as examples in earlier chapters, they are treated more systematically in Chapter 8. Examination of the Cauchy problem— the appropriate generalization of the initial value problem to higher dimensions— gives rise to a natural classification of partial differential equations into hyperbolic, parabolic, and elliptic types. The theory of characteristics for hyperbolic equations is introduced and applied to simple linear and nonlinear examples. In the second and third sections various methods (Green's functions, Laplace transforms, images, etc.) are used to solve BVPs for the wave equation, the heat equation, and Laplace's equation. The simple and double layers of potential theory make it possible to reduce the Dirichlet problem to an integral equation on the boundary of the domain, thereby providing a rather weak existence proof. In Section 4 a stronger existence proof is given, using variational principles. Two-sided bounds for some functionals of physical interest, such as capacity and torsional rigidity, are obtained by introducing complementary principles. Another application involving level-line analysis is also given, and there is a very brief treatment of unilateral constraints and variational inequalities.

Finally, in Chapter 9, a number of methods applicable to nonlinear problems are developed. Section 1 points out some of the features that distinguish nonlinear problems from linear ones and illustrates these differences through some simple examples. In Section 2 the principal qualitative results of branching theory (also known as bifurcation theory) are presented. The phenomenon of bifurcation is understood most easily in terms of the buckling of a rod under compressive thrust. As the thrust is increased beyond a certain critical value, the state of simple compression gives way to the buckled state with its appreciable transverse deflection. Section 3 shows how a variety of linear problems can be handled by perturbation theory (inhomogeneous problems, eigenvalue problems, change in boundary conditions, domain perturbations). These techniques, as well as monotone methods, are then adapted to the solution of nonlinear BVPs. The concluding section discusses the possible loss of stability of the basic steady state when an underlying parameter is allowed to vary.

I have already acknowledged my debt to the students and teachers who were kind enough to comment on my earlier book. There are, however, two colleagues to whom I am particularly grateful: Stuart Antman, who generously contributed the ideas underlying the derivation of the equations for rods in Chapter 0, and W. Edward Olmstead, who suggested some of the examples on contractions in Chapter 4 and on branching in Chapter 9.

IVAR STAKGOLD

Newark, Delaware
September 1979

CHAPTER 0

PRELIMINARIES

As its name and number indicate, this chapter contains background material having no precise place in the subsequent, systematic, mathematical development. Readers already familiar with some of the topics in the present chapter may nevertheless profit from a new presentation; they are particularly urged to read Sections 0.1, 0.5, and 0.6 before proceeding to the later chapters.

The principal purpose here is to give fairly careful derivations of some of the equations of mathematical physics which will be studied more extensively in the rest of the book. The attention paid to modeling in the present chapter could, regrettably, not be sustained in the subsequent ones. Readers who want to further explore aspects of modeling are encouraged to consult the books by Aris [4], Lin and Segel [19], Segel [28], Tayler [32], Keener [16], and Logan [20]. Extensive surveys in mathematical physics, including modern geometric tools, can be found in the recent books of Hassani [13] and of Szekeres [31].

Even when agreement exists on the proper modeling of the physical problem, there are still a number of different possible mathematical descriptions. Although the four formulations we use can be shown to be more or less equivalent (see Section 0.5), each has its distinct advantages. The first, and closest in spirit to the underlying physical law, is the so-called *integral balance* written for a field quantity

Green's Functions and Boundary Value Problems, Third Edition. By I. Stakgold and M. Holst

of interest, such as mass, energy, charge, or momentum. The integral balance is formulated over an arbitrary subregion (control region) of the region in space-time where the physical process takes place. In the absence of external inputs, the integral balance becomes a *conservation law*. The second formulation requires additional regularity assumptions for the inputs and the field quantity; the integral balance can then be transformed into a partial differential equation (PDE) governing the local behavior of the field quantity. Constitutive relations as well as boundary and initial conditions supplement the PDE to yield an *initial boundary value problem* (BVP), which, under normal circumstances, will have one and only one solution. When there is doubt as to the range of validity of the PDE, it is often helpful to return to the integral balance for inspiration and verification.

The third formulation is called the *weak form* of the BVP (also known, in special contexts, as the *variational equation* or the *principle of virtual work*). In many ways this is the most powerful mathematical formulation, as it lends itself to the use of modern techniques of functional analysis and also forms the basis for many numerical methods. As the term *variational equation* suggests, the weak form is often related to a *variational principle* (the fourth formulation), such as a principle of minimum energy. The vanishing of the first variation of the functional being minimized is then just the variational equation or weak form of the BVP.

In Section 0.5 we show, for a simple example, how these various formulations are interconnected. In Sections 0.1 through 0.4 we develop integral balances of energy, mass, and momentum in various physical contexts and show how they lead to the respective BVPs.

The chapter ends with two sections (0.6 and 0.7) of a mathematical nature. Section 0.6 reviews fundamental ideas of convergence and norm which are widely used in the rest of the book. Section 0.7 presents a short treatment of Lebesgue integration. Although only a few essential properties of the Lebesgue integral will be needed, it seemed worthwhile to spend a few pages explaining its construction. These limited goals made it convenient to use Tonnelli's approach (as described, for instance, in Shilov [29]). Another recent approach due to Lax [18] involves defining L^1 as the completion of $C(K)$ in the L^1 norm, where K is a ball in $\mathbb{R}^n$. In this approach, measure is a derived concept.

A few words about terminology are in order. $\mathbb{R}^n$ stands for n-dimensional Euclidean space. The definitions below are given for $\mathbb{R}^3$ but are easily modified for $\mathbb{R}^n$. A point in $\mathbb{R}^3$ is identified by its position vector $\mathbf{x} = (x_1, x_2, x_3)$, where x_1, x_2, x_3 are *Cartesian* coordinates; $|\mathbf{x}| = (x_1^2 + x_2^2 + x_3^2)^{1/2}$, where the nonnegative square root is understood; $d\mathbf{x}$ stands for a volume element $dx_1\, dx_2\, dx_3$. In later chapters the distinguishing notation for vectors is dropped.

An *open ball* of radius a, centered at the origin, is the set of points $\mathbf{x}$ such that $|\mathbf{x}| < a$. The set $|\mathbf{x}| \leqslant a$ is a *closed ball*, and the set $|\mathbf{x}| = a$ is a *sphere*. In $\mathbb{R}^2$ the words *disk* and *circle* are often substituted for ball and sphere, respectively. An *open set* Ω has the property that whenever $\mathbf{x} \in \Omega$, so does some sufficiently small ball with center at $\mathbf{x}$. A point $\mathbf{x}$ belongs to the *boundary* Γ of an open set Ω if $\mathbf{x}$ is not in Ω but if every open ball centered at $\mathbf{x}$ contains a point of Ω. The *closure* $\bar{\Omega}$ of Ω is the union of Ω and Γ. These ideas are best illustrated by an egg with a very thin shell. The interior of the egg is an open set Ω, the shell is Γ, and the egg with shell is $\bar{\Omega}$. An open set Ω is *connected* if each pair of points in Ω can be connected by a curve lying entirely in Ω. A *domain* is an open connected set. Thus an open ball is a domain, but the union of two disjoint open balls is not.

In the definition of the function spaces below, there are some distinctions which are best understood through examples: (a) the function $1/x$ is continuous on $0 < x \leqslant 1$ but cannot be extended continuously to $0 \leqslant x \leqslant 1$; (b) the function $\sqrt{x(1-x)}$ is continuous on $0 \leqslant x \leqslant 1$ with a continuous derivative on $0 < x < 1$ which cannot be extended continuously to $0 \leqslant x \leqslant 1$.

Definition. *Let Ω be a domain in $\mathbb{R}^n$. Then $C^k(\Omega)$ is the set of functions $f(\mathbf{x})$ with continuous derivatives of order 0, 1, 2, ..., k on Ω. (The derivative of order 0 of f is understood to be f itself.) $C^k(\bar{\Omega})$ is the set of functions $f(x) \in C^k(\Omega)$ each of whose derivatives of order 0, 1, 2, ..., k can be extended continuously to $\bar{\Omega}$.*

The sets $C^0(\Omega)$ and $C^0(\bar{\Omega})$ are usually written $C(\Omega)$ and $C(\bar{\Omega})$, respectively.

If Ω is the open interval $a < x < b$ in $\mathbb{R}$, we usually prefer the notation $C^k(a, b)$ for $C^k(\Omega)$ and $C^k[a, b]$ for $C^k(\bar{\Omega})$. Thus the function

$$\sqrt{x(1-x)} \in C[0,1] \cap C^1(0,1) \quad \text{but} \quad \notin C^1[0,1].$$

We shall encounter other function spaces in the sequel (such as the space of piecewise continuous functions, L_p spaces, and Sobolev spaces) with definitions given at the appropriate time.

The symbol $\doteq$ means "set equal to." It is occasionally used to define a new expression. For instance, in writing $D \doteq dS/dx$ we are defining D as dS/dx, which, in turn, is presumably known from earlier discussion.

The terms

$$\inf_{\mathbf{x}\in\Omega} f(\mathbf{x}), \qquad \sup_{\mathbf{x}\in\Omega} f(\mathbf{x})$$

stand for the infimum (greatest lower bound) and supremum (least upper bound) of the real-valued function f on Ω. For instance, if Ω is the open ball in $\mathbb{R}^n$ with radius a and center at the origin, and $f(\mathbf{x}) = |\mathbf{x}|$, then

$$\inf_{\mathbf{x}\in\Omega} f(\mathbf{x}) = 0, \qquad \sup_{\mathbf{x}\in\Omega} f(\mathbf{x}) = a,$$

even though the supremum is not attained for any element $\mathbf{x}$ in Ω.

0.1 HEAT CONDUCTION

Consider heat conduction taking place during a time interval (0, T) in a medium, possibly inhomogeneous, occupying the three-dimensional domain Ω with boundary Γ. There are thus four independent variables x_1, x_2, x_3, t, which we write as $(\mathbf{x}, t)$ since it is convenient to distinguish the time variable from the space variables. The basic domain in space-time is the Cartesian product of Ω and (0, T), written as

$\Omega \times (0, T)$. We now take an energy balance, not over $\Omega \times (0, T)$, but over a subset of the form $R \times (t, t+h)$, where R is an arbitrary portion of Ω and $(t, t+h)$ is an arbitrary interval contained in (0, T). We need flexibility in the choice of the subset $R \times (t, t+h)$ so that we can obtain sufficient information for our purposes. We postulate that only heat energy plays a significant role in the energy budget. The terms which contribute to the heat balance are (a) the change in the heat content of R from time t to time $t+h$, caused by a change in temperature, (b) the heat generated by sources, called *body sources*, in $R \times (t, t+h)$, and (c) the heat flowing in or out through the boundary B of R over the time interval $(t, t+h)$. All of these terms are measured in appropriate units of heat, say calories. The body sources may stem, for instance, from a chemical reaction liberating heat (positive source) or absorbing heat (negative source or sink). Body sources can be of three types: distributed, impulsive, or concentrated. A distributed source is characterized by a density $p(\mathbf{x}, t)$ measured in calories/cm^3 sec, and can generate a finite amount of heat only by acting in a finite volume over a finite time interval. An impulsive source is instantaneous in time and generates a finite amount of heat over an infinitesimal time interval. Concentrated sources are localized in space at points, curves, or surfaces and generate a finite amount of heat over regions of infinitesimal volume.

Whatever the type of source, a heat balance for $R \times (t, t+h)$ gives, in calories,

$$\begin{aligned}&\text{rise in heat content}\\ &\qquad = \text{heat generated by body sources} - \text{outflow of heat through } B.\end{aligned} \tag{0.1.1}$$

With $E_R(t)$ representing the heat content of R at time t, the left side of (0.1.1) becomes $E_R(t+h) - E_R(t)$. Assuming no impulsive sources, we can divide (0.1.1) by h and take the limit as $h \to 0$ to obtain, for any t in (0, T),

$$\frac{dE_R}{dt} = P_R(t) - Q_B(t) \quad \left(\frac{\text{cal}}{\text{sec}}\right), \tag{0.1.2}$$

where $P_R(t)$ is the *rate* of heat generation in R and $Q_B(t)$ is the *rate* of outflow through B. Next, we exclude concentrated sources by expressing P_R and E_R in terms of densities $p(\mathbf{x}, t)$ and $e(\mathbf{x}, t)$ defined on $\Omega \times (0, T)$:

$$P_R(t) = \int_R p(\mathbf{x}, t)\, d\mathbf{x}, \quad E_R(t) = \int_R e(\mathbf{x}, t)\, d\mathbf{x},$$

where e is measured in cal/cm^3 and p in cal/cm^3 sec.

The rate of outflow Q_B is expressed in terms of a *heat flux vector* $\mathbf{J}(\mathbf{x}, t)$ defined on $\Omega \times (0, T)$. The amount of heat flowing per unit time across a surface element of area dS with unit normal $\mathbf{n}$ is given by $\mathbf{J} \cdot \mathbf{n}\, dS$, so that

$$Q_B(t) = \int_B \mathbf{J} \cdot \mathbf{n}\, dS, \tag{0.1.3}$$

where $\mathbf{n}$ is the outward normal to B. Use of the divergence theorem on this term transforms (0.1.2) to

$$\int_R \left(\frac{\partial e}{\partial t} + \operatorname{div} \mathbf{J} - p\right) d\mathbf{x} = 0, \quad 0 < t < T. \tag{0.1.4}$$

If (0.1.4) held only for a particular region R, little information could be extracted, but instead we know that it is true for every subregion R of Ω. We claim that this implies that the integrand vanishes at every $\mathbf{x}$ and t (assuming that the integrand is a continuous function of $\mathbf{x}$ and t). Indeed, suppose that the integrand were positive at $\mathbf{x}$, t; we can then surround $\mathbf{x}$ by a sufficiently small region R in which the integrand is positive, thereby violating (0.1.4). We therefore conclude that

$$\frac{\partial e}{\partial t} + \operatorname{div} \mathbf{J} = p, \quad (\mathbf{x},\ t) \text{ in } \Omega \times (0,\ T). \tag{0.1.5}$$

There are too many unknowns in (0.1.5), but both e and $\mathbf{J}$ are related to the temperature $u(\mathbf{x}, t)$ through the following constitutive relations:

1. When a homogeneous material element of volume $d\mathbf{x}$ is raised from the temperature u to the temperature $u + du$, its heat content is raised by $C\,du\,d\mathbf{x}$, where C is the *specific heat* of the material measured in calories per degree per cm^3. Note that C depends on the material and may also depend on u.

2. Fourier's law of heat conduction for a homogeneous material:

$$\mathbf{J} = -k \text{ grad } u, \tag{0.1.6}$$

where k is the thermal conductivity (which may depend on u) and has units of cal per sec per cm^3 per degree. Thus, the heat flowing across an element of surface per unit time is

$$\mathbf{J} \cdot \mathbf{n}\ dS = -k \text{ grad } u \cdot \mathbf{n}\ dS = -k \frac{\partial u}{\partial n}\, dS,$$

where the minus sign is consistent with the fact that heat flows in the direction of decreasing temperature. Note that if we also wanted to include convection, we would have to modify (0.1.6)—see the remarks below.

Since our medium may be inhomogeneous, both C and k may depend on $\mathbf{x}$ as well as u. Then

$$\frac{\partial e}{\partial t} = C \frac{\partial u}{\partial t} \quad \text{and} \quad \operatorname{div} \mathbf{J} = -\operatorname{div}(k \text{ grad } u),$$

so that (0.1.5) becomes the usual equation of heat conduction:

$$C \frac{\partial u}{\partial t} - \operatorname{div}(k \text{ grad } u) = p, \quad (\mathbf{x},\ t) \in \Omega \times (0,\ T). \tag{0.1.7}$$

If C and k are constants, the equation reduces to

$$\frac{\partial u}{\partial t} - a\,\Delta u = \frac{P}{C}, \tag{0.1.8}$$

where $a = k/C$ is the thermal diffusivity in cm^2/sec, and $\Delta = \text{div grad}$ is the Laplacian operator whose form in Cartesian coordinates is

$$\frac{\partial^2}{\partial x_1^2} + \frac{\partial^2}{\partial x_2^2} + \frac{\partial^2}{\partial x_3^2}.$$

REMARKS

1. The term div $\mathbf{J}$ in (0.1.5) stems from the outflow–inflow of heat through the boundary of a test region—the third term in (0.1.1)—but is now expressed as a point function. We regard div $\mathbf{J}$ as representing *redistribution* of heat through whatever mechanisms are available for that purpose. If the only such mechanism is heat conduction, then Fourier's law applies: $\mathbf{J} = -k \text{ grad } u$. If the medium is a fluid moving with velocity $\mathbf{v}(\mathbf{x}, t)$, then $\mathbf{J} = -k \text{ grad } u + Cu\mathbf{v}$, which incorporates convection. Then $\text{div}\,\mathbf{J} = -\text{div}\,(k \text{ grad } u) + \text{div}\,(Cu\mathbf{v})$; if the fluid is *incompressible*, $\text{div}\,\mathbf{v} = 0$ [see (0.2.9) with constant ρ and $p = 0$] and
$$\text{div } \mathbf{J} = -\text{div}(k \text{ grad } u) + \mathbf{v} \cdot \text{grad } u,$$
so that (0.1.7) becomes
$$C\frac{\partial u}{\partial t} - \text{div}(k \text{ grad } u) + \mathbf{v} \cdot \text{grad } u = p. \tag{0.1.9}$$

2. Suppose that we consider (0.1.8) for a medium covering all of $\mathbb{R}^3$, with $p = 0$, and with initial temperature positive over a small part of $\mathbb{R}^3$ and vanishing elsewhere. Then, we shall see in Chapter 8 that $u(\mathbf{x}, t) > 0$ for all of $\mathbb{R}^3$ when $t > 0$. Of course, $u(\mathbf{x}, t)$ will be small for large $|\mathbf{x}|$, but nevertheless there is something disturbing about our model since a localized initial temperature propagates with infinite velocity to give a positive temperature everywhere for $t > 0$. (See, however, the article by Day [8], who gives a spirited defense of Fourier's law.) One possible remedy is to modify Fourier's law (0.1.6) to
$$\mathbf{J} + \tau\frac{\partial \mathbf{J}}{\partial t} = -k \text{ grad } u,$$
with τ a relaxation time. This leads to an equation of the form
$$\varepsilon\frac{\partial^2 u}{\partial t^2} + \frac{\partial u}{\partial t} - b\,\Delta u = 0,$$
where ε and b are positive constants. The new equation for u has finite propagation velocity, so that a localized initial temperature is only felt within some bounded region which depends on t. Since $\varepsilon \ll 1$, the effect is noticeable only for small times.

3. If the production term p is prescribed as a function of $\mathbf{x}$ and t, (0.1.8) is a linear equation [and so is (0.1.7) if C and k do not depend on u]. There are

many problems, however, where p depends on the unknown temperature u. If, for instance, a chemical reaction takes place which liberates heat, the rate at which heat is released per unit mass of reactant is given by the Arrhenius law $Ae^{-B/u}$, where A and B are known positive constants. We must therefore substitute $p(\mathbf{x}, t) = Ae^{-B/u(\mathbf{x},t)}$ on the right sides of (0.1.7) and (0.1.8) so that these become *nonlinear* partial differential equations of a type known as reaction-diffusion equations (see Section 0.3).

Steady-State Heat Conduction

The temperature $u(\mathbf{x})$ is now independent of time and (0.1.2) reduces to $P_R = Q_B$; on using (0.1.3) and (0.1.6) this becomes

$$-\int_B k\frac{\partial u}{\partial n}\,dS = P_R, \tag{0.1.10}$$

where P_R is the steady heat input per unit time over the region R bounded by B. If these body sources stem from a volume density $p(\mathbf{x})$, we can use reasoning similar to that yielding (0.1.5) to conclude that

$$-\operatorname{div}(k \operatorname{grad} u) = p(\mathbf{x}), \quad \mathbf{x} \in \Omega. \tag{0.1.11}$$

To show that (0.1.10) is indeed more general than (0.1.11), consider the case of a single source concentrated at $\mathbf{x} = \xi$ and generating p calories per second. Then (0.1.10) tells us that

$$-\int_B k\,\frac{\partial u}{\partial n}\,dS = \begin{cases} 0 & \text{if } \xi \text{ is not in } R, \\ p & \text{if } \xi \text{ is in } R. \end{cases} \tag{0.1.12}$$

By a now familiar argument, the first line shows that $-\operatorname{div}(k \operatorname{grad} u) = 0$ for $\mathbf{x} \neq \xi$. By specializing the second line to a small sphere centered at ξ, it is possible to extract precise information on the nature of the singularity in u at ξ (see Section 1.4, for instance).

Whenever there is possible ambiguity in the interpretation of (0.1.11), it is wise to return to the integral formulation (0.1.10) for guidance. As another example, suppose that Ω consists of two media separated by an interface σ. The thermal conductivity k is continuous, with the possible exception of a jump discontinuity across σ. Assuming that there are no prescribed sources or any heat losses (caused by films or imperfect fitting) on the interface, we can apply (0.1.10) to a thin pillbox straddling the interface with its bases parallel to σ. It is permissible to neglect the contribution from the lateral surface of the pillbox to obtain

$$k_+\frac{\partial u}{\partial n_+} = k_-\frac{\partial u}{\partial n_-}, \tag{0.1.13}$$

where the subscripts $+$ and $-$ denote the two sides of σ. Equation (0.1.13) and the continuity of u comprise the interface conditions (see also Section 1.4).

Despite its usefulness, (0.1.10) is not general enough to treat certain important idealized singularities. For instance, a dipole located at ξ would apparently go undetected in (0.1.10) since $P_R = 0$ whether ξ is in R or outside R. We shall see in Chapter 2 that such problems are best handled within the theory of distributions.

Boundary and Initial Conditions

Equation (0.1.7) alone does not determine the temperature $u(\mathbf{x}, t)$. We must in addition give the initial temperature $u(\mathbf{x}, 0)$ and a boundary condition on Γ for $0 < t < T$. This boundary condition is usually one of three types:

$$\left.\begin{array}{ll}\text{(i)} & \text{temperature } u \text{ prescribed on } \Gamma \text{ for } 0 < t < T, \\ \text{(ii)} & \text{heat flow } -k\,\dfrac{\partial u}{\partial n} \text{ prescribed on } \Gamma \text{ for } 0 < t < T, \\ \text{(iii)} & -k\dfrac{\partial u}{\partial n} = \alpha u \text{ on } \Gamma \text{ for } 0 < t < T.\end{array}\right\} \tag{0.1.14}$$

For $\alpha > 0$ the last condition is Newton's law of cooling, which characterizes radiation into a surrounding medium at uniform temperature [which is then taken to be the datum of temperature in (0.1.7)]. Thus heat is lost from the surface of the body at a rate proportional to the difference between the surface temperature and the surrounding temperature. If $\alpha < 0$, condition (iii) states that the larger the boundary temperature, the more heat enters the body, a circumstance which obviously tends to increase the internal temperature. Newton's law of cooling can be regarded as an approximation to Stefan's radiation law, which has a term βu^4 on the right side of (iii). It is, of course, possible to have an inhomogeneous version of (iii) if the surface is simultaneously heated by prescribed heat flow.

One-Dimensional Problems

Equation (0.1.7) can sometimes be reduced to an equation involving derivatives in only one space direction. Let (x_1, x_2, x_3) be Cartesian coordinates; we want to describe classes of problems in which u depends only on x_1 and t.

1. Suppose that Ω is the slab $0 < x_1 < a, -\infty < x_2, x_3 < \infty$. Assume that the source term depends only on x_1 and t, that the boundary conditions on the faces of the slab depend only on t, that the initial temperature depends only on x_1, and that k and C depend only on x_1 and u. On physical grounds it is clear that heat flows only in the x_1 direction, so that u will be a function of x_1 and t alone. The differential equation (0.1.7) then becomes

$$C\frac{\partial u}{\partial t} - \frac{\partial}{\partial x_1}\left(k\frac{\partial u}{\partial x_1}\right) = p(x_1, t), \quad 0 < x_1 < a, \quad 0 < t < T. \tag{0.1.15}$$

The solution $u(x_1, t)$ of the boundary-initial value problem associated with (0.1.15) obviously also satisfies the related boundary-initial value problem associated with

(0.1.7). If we can prove uniqueness in the latter case, we will have shown that $u(x_1, t)$ is, in fact, the desired solution of (0.1.7).

2. Let Ω be a cylindrical rod (of arbitrary cross section) whose axis coincides with the segment $0 < x_1 < a$. In addition to the assumptions in part 1, let us suppose that the *lateral surface of the rod is insulated.* We may imagine the rod as having been punched out normal to the slab of part 1. Since the temperature in the slab is independent of x_2 and x_3, it must satisfy the condition $\partial u/\partial n = 0$ on the lateral surface of the cylinder, which is the criterion of insulation. Thus the flow in the rod is one-dimensional, and (0.1.15) holds as before. In particular, the steady-state equation is

$$-\frac{d}{dx_1}\left(k\frac{du}{dx_1}\right) = \tilde{p}(x_1), \quad 0 < x_1 < a, \tag{0.1.16}$$

where $\tilde{p}(x_1)$ is the *volume density* of sources. If A is the cross-sectional area of the rod, we can write $\tilde{p}(x_1) = p(x_1)/A$, where $p(x_1)$ is the source density per unit length of the rod. Equation (0.1.16) becomes

$$-\frac{d}{dx_1}\left(k\frac{du}{dx_1}\right) = \frac{p(x_1)}{A}, \quad 0 < x_1 < a. \tag{0.1.17}$$

Both sides of (0.1.17) are in units of cal/cm^3 sec. If k is constant, we can nondimensionalize by setting $ax = x_1$ and $u_0 v = u$ to obtain

$$-\frac{d^2v}{dx^2} = \lambda p(ax), \quad 0 < x < 1,$$

where u_0 is some appropriate reference temperature such as the maximum boundary temperature and $\lambda = a^2/Aku_0$. Setting $\lambda p(ax) = f(x)$, we arrive at the nondimensional equation

$$-\frac{d^2v}{dx^2} = f(x), \quad 0 < x < 1. \tag{0.1.18}$$

0.2 DIFFUSION

With a different interpretation of terms, equation (0.1.7) also describes the dispersive behavior of a chemical or of a population. For instance, let $c(\mathbf{x}, t)$ be the *concentration* of a substance diffusing through some underlying medium. The energy balance (0.1.1) is replaced by a mass balance for the substance in question over $R \times (t, t+h)$:

$$\text{mass increase} = \text{mass created by body sources} - \text{outflow of mass through } B. \tag{0.2.1}$$

Passing immediately to rates and densities, we write

$$E_R(t) = \int_R c(\mathbf{x}, t)\, d\mathbf{x} = \text{total mass in } R \text{ at time } t,$$

$$P_R(t) = \int_R p(\mathbf{x}, t)\, d\mathbf{x} = \text{rate of mass production in } R \text{ at time } t,$$

$$Q_B(t) = \int_B \mathbf{J} \cdot \mathbf{n}\; dS = \text{rate of outflow through } B \text{ at time } t.$$

Here $p(\mathbf{x}, t)$ is a density of mass production (grams/sec cm^3) and $\mathbf{J}$ is the mass flux vector. Repeating the steps leading to (0.1.5), we obtain the equation

$$\frac{\partial c}{\partial t} + \text{div}\, \mathbf{J} = p. \tag{0.2.2}$$

The flux $\mathbf{J}$ is usually related to c by *Fick's law* (replacing Fourier's law)

$$\mathbf{J} = -D \text{ grad } c, \tag{0.2.3}$$

where $D > 0$ is the diffusivity. The minus sign in (0.2.3) accounts for the fact that the substance diffuses from regions of higher density to regions of lower density. We are therefore led to the *diffusion equation*

$$\frac{\partial c}{\partial t} - \text{div}(D \text{ grad } c) = p, \tag{0.2.4}$$

and if D is constant,

$$\frac{\partial c}{\partial t} - D\,\Delta c = p. \tag{0.2.5}$$

The term *diffusion equation* is often used even when dealing with (0.1.7) or (0.1.8). A basic property of the equation when $p = 0$ is that it tends to equalize the concentration or the temperature, as the case may be.

The production p may be given as a function of $\mathbf{x}$ and t or may depend on the concentration c. In an absorbing medium it is frequently assumed that the amount of the substance absorbed is proportional to the concentration, so that $p = -q(\mathbf{x})c + n(\mathbf{x}, t)$, where $q(\mathbf{x})$ is a measure of the absorption properties of the medium, which may vary from point to point, and $n(\mathbf{x}, t)$ represents the density of the prescribed sources. As a result, (0.2.4) takes the form

$$\frac{\partial c}{\partial t} - \text{div}(D \text{ grad } c) + qc = n(\mathbf{x}, t) \tag{0.2.6}$$

or, in the steady state,

$$-\text{div}(D \text{ grad } c) + q(\mathbf{x})c = n(\mathbf{x}). \tag{0.2.7}$$

If the medium itself is moving at a velocity $\mathbf{v}(\mathbf{x}, t)$, there is an additional convective contribution

$$\int_B c\mathbf{v} \cdot \mathbf{n}\, dS$$

to the outward mass flow through B. Inserting this term in (0.2.1), with due regard to the minus sign, and again applying the divergence theorem, we find that (0.2.4) becomes

$$\frac{\partial c}{\partial t} - \text{div}(D \text{ grad } c) + \text{div}(\mathbf{v}c) = p. \tag{0.2.8}$$

Porous Medium Equation

In the nondiffusive flow of a compressible fluid with velocity $\mathbf{v}$ and density $\rho(\mathbf{x}, t)$, (0.2.8) becomes

$$\frac{\partial p}{\partial t} + \operatorname{div}(\rho \mathbf{v}) = p. \tag{0.2.9}$$

Problems (0.2.4) and (0.2.5) are linear if both p and D are independent of c. The case where p depends on c will be treated at some length in the next section, but we first consider the case of nonlinear diffusivity, which occurs in a number of physical settings. An important example is that of flow in a porous medium.

Consider a gas flowing through a porous solid. The geometric shape of the pores may be quite complicated, but we assume that there is a fraction ϕ of the volume of the domain which can be occupied by the gas; ϕ is known as the *porosity*. Equation (0.2.9) (with no production term) then takes the form

$$\phi \frac{\partial p}{\partial t} + \operatorname{div}(\rho \mathbf{v}) = 0. \tag{0.2.10}$$

We then introduce *Darcy's Law*, which relates the velocity to the gas pressure P:

$$\mathbf{v} = -\frac{k}{\mu} \operatorname{grad} P, \quad \mu = \text{viscosity}, \quad k = \text{permeability}. \tag{0.2.11}$$

This constitutive law may be regarded as an empirical one, but has been justified by reconciling microscopic and macroscopic behavior through a procedure known as *homogenization* (see, for instance, Hornung and Jaeger [14]). In addition to (0.2.10) and (0.2.11), there is also an *equation of state*, which relates the pressure P and the density ρ:

$$P = P(\rho).$$

For isentropic flow of the gas, $P(\rho) = A\rho^\gamma$, where $\gamma \geqslant 1$. If we use this last relation and Darcy's law in (0.2.10), we obtain

$$\frac{\partial \rho}{\partial t} - \frac{kA\gamma}{\mu\phi} \operatorname{div}(\rho^\gamma \operatorname{grad} \rho) = 0$$

or

$$\frac{\partial \rho}{\partial t} - \frac{kA\gamma}{\mu(\gamma+1)\phi} \Delta \rho^{\gamma+1} = 0.$$

We can rescale the time in either equation to get rid of the constant factor, the last equation becoming

$$\frac{\partial \rho}{\partial t} = \Delta \rho^m, \quad m = 1 + \gamma \geqslant 2. \tag{0.2.12}$$

Equation (0.2.12) is usually known as the *porous medium equation*. For mathematical generality and in view of other possible applications, (0.2.12) is often studied for arbitrary positive values of m.

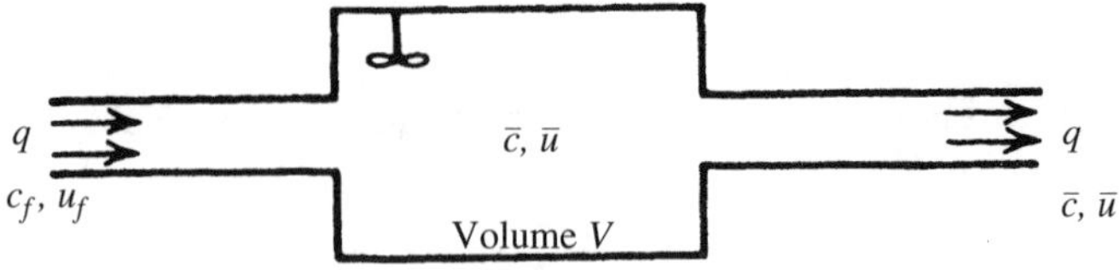

Figure 0.1

0.3 REACTION-DIFFUSION PROBLEMS

We first consider a number of problems of chemical reaction engineering in which a reactant *disappears* at a volume rate dependent on its concentration c and temperature u. In a first-order irreversible reaction this rate is proportional to c and its temperature dependence follows the Arrhenius law with the *production* rate

$$p(c, u) = -Ace^{-B/u}, \tag{0.3.1}$$

where A and B are positive constants. The amount of heat released (or absorbed) in the reaction is proportional to p. For an exothermic reaction, heat is therefore liberated at the rate

$$-hp = hAce^{-B/u}, \tag{0.3.2}$$

where h is a positive constant.

In these equations, u is the absolute temperature, hA the heat of reaction, and B the activation energy. Note that the exponential term increases from 0 to 1 as u increases from 0 to ∞.

Continuous-Flow Stirred Tank Reactor

Here the reaction takes place in a tank of volume V which is stirred continuously to maintain uniform reactant concentration $\bar{c}$ and temperature $\bar{u}$. Of course, $\bar{c}$ and $\bar{u}$ will depend on time $\bar{t}$, and it is this variation that we wish to determine. The tank is fed by a stream that has volume rate q at temperature u_f and in which the concentration of the reactant is c_f. Products of the reaction are removed at the same volume rate q, so that the tank remains full at all times (see Figure 0.1).

A mass balance per unit time on the amount of reactant in the tank gives

$$V\frac{d\bar{c}}{d\bar{t}} = qc_f - q\bar{c} + Vp, \tag{0.3.3}$$

where $p(\bar{c}, \bar{u})$ is given by (0.3.1). We shall assume that the tank is insulated and that no heat is lost in its walls. A heat balance then gives

$$VC\frac{d\bar{u}}{d\bar{t}} = qCu_f - qC\bar{u} - hVp, \tag{0.3.4}$$

where C is the specific heat per unit volume of the reaction mixture, which we take as constant. Introducing the dimensionless variables

$$c = \frac{\bar{c}}{c_f}, \quad u = \frac{\bar{u}}{u_f}, \quad t = \frac{\bar{t}q}{V} \tag{0.3.5}$$

and the dimensionless constants

$$\alpha = \frac{AV}{q}, \quad \beta = \frac{hc_f}{\mu_f C}, \quad \gamma = \frac{B}{u_f}, \tag{0.3.6}$$

we find that (0.3.3) and (0.3.4) become

$$\frac{dc}{dt} = 1 - c - \alpha c e^{-\gamma/u}, \tag{0.3.7}$$

$$\frac{du}{dt} = 1 - u + \beta\alpha c e^{-\gamma/u}, \tag{0.3.8}$$

which form a pair of nonlinear differential equations of the first order for $c(t)$ and $u(t)$. Multiplying (0.3.7) by β and adding to (0.3.8), we obtain

$$\frac{d}{dt}(u + \beta c) = (1 + \beta) - (u + \beta c),$$

so that

$$u + \beta c = 1 + \beta + (\text{constant})e^{-t}. \tag{0.3.9}$$

If we choose the initial value for $u + \beta c$ to be just its steady-state value $1 + \beta$, the system (0.3.7)–(0.3.8) can be reduced to the single nonlinear equation

$$\frac{du}{dt} = 1 - u + \alpha(1 + \beta - u)e^{-\gamma/u} \tag{0.3.10}$$

for the dimensionless temperature $u(t)$. Equation (0.3.10) is discussed in more detail in Example 1, Section 9.5.

Catalyst Pellet

Consider a catalyst pellet occupying the domain Ω in $\mathbb{R}^3$ with boundary Γ. Suppose that a first-order exothermic irreversible reaction is taking place within the catalyst. Let D_c and D_u be effective coefficients for the reactant mass flow and heat flow, respectively. The transport term is negligible so that the usual balances give

$$\frac{\partial c}{dt} = D_c \Delta c + p,$$

$$\frac{\partial u}{dt} = D_u \Delta u - \frac{h}{C}p,$$

where p is given by (0.3.1). One can introduce suitable dimensionless variables to reduce the equations to

$$\begin{aligned}\frac{\partial c}{\partial t} &= \Delta c - \alpha c e^{-\gamma/u},\\ \frac{\partial u}{\partial t} &= \Delta u + \beta \alpha c e^{-\gamma/u},\end{aligned} \tag{0.3.11}$$

where the Lewis number D_c/D_u has been taken equal to 1.

It should be noted that in (0.3.11) both u and c are inherently nonnegative. The reaction is said to be exothermic if $\beta > 0$, isothermal if $\beta = 0$, and endothermic if $\beta < 0$. For a reaction of order m, c is replaced by c^m as the factor preceding the exponential. In Chapter 9 we consider a number of nonlinear problems based on (0.2.8) or on (0.3.11) as well as related problems of extinction or blow-up in finite time.

For various mathematical aspects of chemical engineering, the reader is referred to the books by Aris [3], and by Froment and Bischoff [12].

Climate Modeling

In recent years, considerable interest has been focused on studying the Earth's climate over relatively long time scales by the use of energy balance models (EBMs). These *diagnostic* models, introduced independently by Budyko and Sellers in 1969, are aimed at obtaining qualitative insights such as determining the climate's sensitivity to the variation of critical parameters, rather than at precise weather prediction. EBMs have been used to indicate how small variations in either the solar constant or the amount of greenhouse gases in the atmosphere could drastically alter the Earth's climate. An EBM starts from the premise that the difference between the amount of energy coming from the sun and the infrared radiation escaping into space determines the net energy available to drive the climate system. We shall see later how this appoach can be used to formulate the climate problem mathematically.

By contrast, deterministic weather prediction is possible only for short periods of time (days or weeks) and such predictions are of limited accuracy. These *prognostic* models are known as general circulation models (GCMs) and take into account an array of phenomena and data. Anyone who has examined the maze of interconnections and feedbacks affecting the climate will agree with the statement attributed to Alfonso X (El Sabio) of Castile: "If the Lord Almighty had consulted me before embarking on the Creation, I would have recommended something simpler."

A GCM consists of comprehensive sets of partial differential equations incorporating many of the physical processes for the components of the climate system (atmosphere, oceans, land, etc.). Although some of these equations are derived from first principles (for instance, the various conservation laws), others have to be modeled phenomenologically (radiation and convection, precipitation and evaporation, biological activity, air–ocean–land interactions). Obviously, a rigorous mathematical analysis is out of the question, but even numerical approaches challenge the most powerful computers. Moreover, there is only limited information on oceanic currents and land systems; ocean circulation, cloud formation, and their coupling are not well understood; there is also an intrinsic question as to the limits of predictability of regional climate, uncertainties (stochastic or chaotic) due to astronomical, biological, and human activity.

We refer the reader to the two excellent review papers by North [23] and Drazin and Griffel [10]. The starting point is an energy balance of the atmosphere in the spirit of (0.1.1). The portion of the sun's incoming energy being absorbed by the atmosphere is denoted by R_a, the energy emitted by the Earth and escaping into space by R_e. Thus the production term is $R_a - R_e$ and we therefore have the local balance

$$\text{increase in heat content} = R_a - R_e + D, \tag{0.3.12}$$

where D stands for the redistribution of energy (which we later represent by an equivalent diffusion).

Next we specialize to a one-layer, two-dimensional model with the mean sea-level atmospheric temperature $u(\mathbf{x}, t)$ in degrees Celsius as the only climatic indicator. Horizontal energy transport is simulated by a linear diffusion operator yielding the balance equation [compare with (0.1.7)]

$$c(\mathbf{x})u_t - \operatorname{div}(k(\mathbf{x})\operatorname{grad} u) = R_a - R_e, \tag{0.3.13}$$

where $\mathbf{x}$ is a point on the two-sphere Γ, the Earth's surface (manifold without boundary); $c(\mathbf{x})$ is the heat capacity; $k(\mathbf{x})$ the conductivity; and

$$R_a = QS(\mathbf{x}, t)a(\mathbf{x}, u), \tag{0.3.14}$$

$$R_e = A + Bu. \tag{0.3.15}$$

Here $QS(\mathbf{x}, t)$ represents the insolation with $S(\mathbf{x}, t)$ suitably normalized and Q the solar constant divided by 4, and $a(\mathbf{x}, u)$ is the absorptivity or co-albedo, which is lower over ice-covered regions than over ice-free regions. The constants A and B are obtained by fitting to present observations, but it is understood that they may depend on the amount of greenhouse gases in the atmosphere (alternatively, one could take $R_e = \varepsilon(u)u^4$, where u is in degrees Kelvin and $\varepsilon(u)$ is an emittivity).

Since (0.3.13) holds on Γ, no boundary conditions are required, but an initial temperature $u_0(\mathbf{x})$ must, of course, be specified.

We further simplify (0.3.13)–(0.3.15) by replacing $S(\mathbf{x}, t)$ by its annual average $S(\mathbf{x})$ and by taking the absorptivity a as a function of u alone. This yields the evolution problem

$$c(\mathbf{x})u_t - \operatorname{div}(k(\mathbf{x}) \operatorname{grad} u) = QS(\mathbf{x})a(u) - (A + Bu) \tag{0.3.16}$$

and its steady-state version,

$$-\operatorname{div}(k(\mathbf{x}) \operatorname{grad} u) = QS(\mathbf{x})a(u) - (A + Bu), \tag{0.3.17}$$

or, if k is a constant,

$$-k\,\Delta u = QS(\mathbf{x})a(u) - (A + Bu). \tag{0.3.18}$$

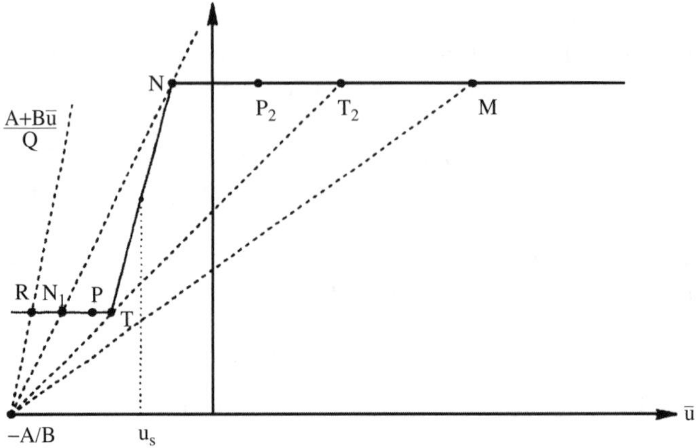

Figure 0.2

The insolation $S(\mathbf{x})$ is *strongly* space dependent, being much larger at the equator than at the poles; $S(\mathbf{x})$ is very nearly even about the equator. The absorptivity $a(u)$ depends strongly on the temperature in a neighborhood of a critical temperature u_s which distinguishes ice-free regions from ice-covered ones. For many purposes, we can take $a(u)$ as a continuous piecewise linear function with a sharp rise near $u = u_s = -10$:

$$a(u) = \begin{cases} a_i, & u + 10 \leqslant -\varepsilon, \\ \dfrac{a_i + a_w}{2} + \dfrac{a_w - a_i}{2\varepsilon}(u + 10), & |u + 10| < \varepsilon, \\ a_w, & u + 10 \geqslant \varepsilon, \end{cases} \tag{0.3.19}$$

where $\varepsilon > 0$ is small.

One of the principal goals of the model is to determine the *free boundary* (where $u = u_s$) separating the ice-free region $u > u_s$ from the ice-covered region $u < u_s$.

A related question of great importance is the sensitivity to changes in parameters such as A, B, and Q. The region $u < u_s$ consists basically of the two polar caps; is the size of these caps appreciably affected by a small change in one of the parameters? In the present section we do not investigate these questions in depth; instead, we confine ourselves to a highly simplified model (the so-called *zero-dimensional model*) which despite its simplicity contains enough of the features of more elaborate models to suggest the possible instability of the Earth's climate. The only climate indicator will be the single number $\bar{u}$, the Earth's surface temperature averaged over a year and over the Earth's surface. The equation for $\bar{u}$ can be obtained by integrating (0.3.18) over Γ:

$$Q\bar{a} - (A + B\bar{u}) = 0, \tag{0.3.20}$$

where the area of the sphere has been normalized, $\bar{a} = \int_\Gamma a(u)S(x)\,dx =$ average absorptivity, and the integral of Δu over Γ vanishes by the divergence theorem (since

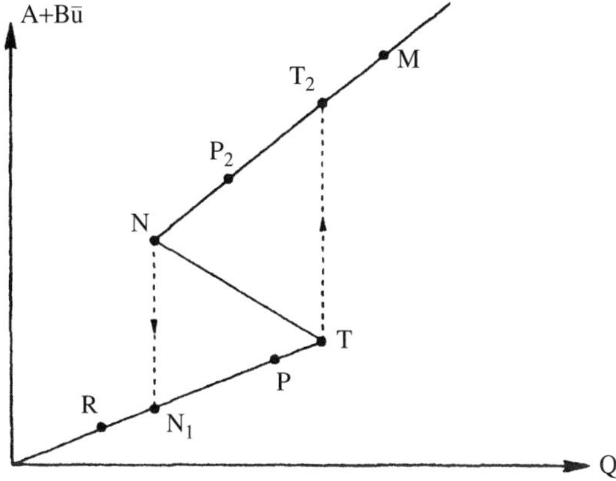

Figure 0.3

Γ has no boundary). We regard $\bar{a}$ as a function of $\bar{u}$ and expect it to have the same general shape as $a(u)$ in (0.3.19). In Figure 0.2 we have sketched $\bar{a}(\bar{u})$, using the explicit form (0.3.19), and $(A + B\bar{u})/Q$ for different values of Q. Intersections of these curves give the solutions of (0.3.20). For small values of Q, there is one intersection at the low temperature R ("frozen" Earth). If Q is large, there is again a single intersection at a high temperature M ("ice-free" Earth). For intermediate values of A there are three intersections, the middle of which is easily seen to be unstable (see Section 9.5). Figure 0.3 shows the dependency of the solution(s) u on Q. Suppose the present conditions are such that we operate at the point P_2. When Q is lowered a little, we move to the left on the upper branch of Figure 0.3 and pretty soon we reach N, where a further decrease in Q leads to a sudden drop of temperature to a frozen Earth at N_1. Conversely, if present conditions are such that the Earth is in state P, a rather small increase in Q brings us to T and we then jump to T_2 (which is interpreted as a partial or full melting of the ice caps).

Our analysis was based on changing the insolation Q while keeping A and B constant, but it is clear that we would encounter similar phenomena if Q is kept constant and the ratio A/B is changed, as might happen if the production of greenhouse gases is increased.

Population Dynamics

This topic is developed in more detail in Chapter 9; here we only derive some simple models of population growth and decay. Strictly speaking, populations are expressed in integers, but it requires no great stretch of the imagination to allow a population (especially if it is large) to take on any nonnegative real value. Consider first an immobile population $u(t)$ which depends only on time. The rate of growth is equal to the difference between the birth and death rates. The simplest model is to take the growth rate proportional to the existing population:

$$\frac{du}{dt} = ku, \tag{0.3.21}$$

which leads to exponential growth if $k > 0$ and exponential decay if $k < 0$.

Since this model does not take into account limitations on food supply or size of the habitat, it is often modified to the logistic model

$$\frac{du}{dt} = ku(u^* - u), \tag{0.3.22}$$

which, for $k > 0$ and $u(0) < u^*$, implies that the population increases asymptotically to the *carrying capacity* u^*. We can easily nondimensionalize (0.3.22) by dividing by u^*; relabeling k and taking u/u^* as the new dependent variable (again called u), we obtain

$$\frac{du}{dt} = ku(1 - u). \tag{0.3.23}$$

Now suppose that the population can move within its habitat Ω (a domain in $\mathbb{R}^2$, say). We let $u(\mathbf{x}, t)$ be the population density so that

$$\int_R u(\mathbf{x}, t)\, d\mathbf{x}$$

gives the population within the subregion R at time t. We keep track of the population as in (0.2.1) to obtain a version of (0.2.2):

$$\frac{\partial u}{\partial t} + \text{div}\mathbf{J} = p, \tag{0.3.24}$$

where p might be chosen as $ku(1 - u)$. Assuming that the species will avoid regions of overcrowding, we may take

$$\mathbf{J} = -D \text{ grad } u.$$

Substituting in (0.3.24) we obtain *Fisher's equation*,

$$\frac{\partial u}{\partial t} - D\,\Delta u = ku(1 - u). \tag{0.3.25}$$

0.4 THE IMPULSE-MOMENTUM LAW: THE MOTION OF RODS AND STRINGS

Motion of Particles

Newton's law of motion for a particle of constant mass m subject to a force $\mathbf{F}(t)$ is $\mathbf{F} = m\mathbf{a}$, where the acceleration $\mathbf{a}(t)$ is the rate of change of the velocity $\mathbf{v}(t)$, which itself is the rate of change of the position vector $\mathbf{x}(t)$. Integrating Newton's law from 0 to t, we obtain the simplest form of the *impulse-momentum law*,

$$\mathbf{I}(t) = \mathbf{p}(t) - \mathbf{p}(0), \tag{0.4.1}$$

where $\mathbf{p}(t) = m\mathbf{v}(t)$ is the (linear) *momentum* of the particle and $\mathbf{I}(t) = \int_0^t \mathbf{F}(\tau)\,d\tau$ is the *linear impulse* on the particle during the time interval (0, t). If $\mathbf{F}(t) \equiv 0$, then momentum is conserved: $\mathbf{p}(t) = \mathbf{p}(0)$. By differentiating (0.4.1) with respect to t we recover Newton's law, but we observe that the impulse-momentum law is somewhat more general, as it can deal with some singular forces. Suppose, for instance, that $\mathbf{F}(t) = Q(t)\mathbf{e}$, where $\mathbf{e}$ is a fixed unit vector, $Q(t) \geqslant 0$ with $Q(t) \equiv 0$ in the interval $J : |t - t_0| \leqslant (\Delta t)/2$, and

$$\int_J Q(t)\,dt = I > 0.$$

Keeping I constant and reducing Δt means that Q gets larger over a smaller time interval while the area under the curve remains the same. As $\Delta t \to 0$, $Q \to \delta(t - t_0)$, a so-called *delta function*, discussed in detail in Chapters 1 and 2. The corresponding force $\mathbf{F}$ becomes an *impulsive force* acting instantaneously at t_0. Such forces are not included in the classical formulation of Newton's law, but (0.4.1) retains a meaning:

$$I\mathbf{e} = \mathbf{p}(t_0+) - \mathbf{p}(t_0-) = \mathbf{p}(t_0+) - \mathbf{p}(0). \tag{0.4.2}$$

The effect of the impulsive force is to increase the velocity instantaneously at time t_0 by the amount $I\mathbf{e}/m$. This corresponds to an infinite acceleration at $t = t_0$.

In preparation for the study of the motion of a continuous body, we formulate the impulse-momentum law for the motion of a set of n particles. Particle $k(1 \leqslant k \leqslant n)$ has mass m_k, position vector $\mathbf{x}_k$, momentum $\mathbf{p}_k = m_k\mathbf{v}_k$, and is subject to a force

$$\mathbf{F}_k = \mathbf{F}_k^{(1)} + \mathbf{F}_k^{(2)},$$

where $\mathbf{F}_k^{(1)}$ is the applied force (including the weight, when appropriate) on particle k and $\mathbf{F}_k^{(2)}$ is the internal force on particle k due to the influence of all other particles in the system. Consider now an *arbitrary subset R* of these n particles. By summing (0.4.1) for all indices in R, we obtain

$$\mathbf{I}_R(t) = \mathbf{p}_R(t) - \mathbf{p}_R(0),$$

where

$$\left.\begin{aligned} \mathbf{p}_R(t) &= \sum_{k\in R} m_k\mathbf{v}_k(t) = \text{linear momentum of } R \text{ at time } t, \\ \mathbf{I}_R(t) &= \int_0^t \mathbf{F}_R(\tau)\,d\tau = \text{impulse on } R \text{ in } (0, t), \\ \mathbf{F}_R(t) &= \sum_{k\in R} \mathbf{F}_k(t) = \text{resultant force on } R \text{ at time } t. \end{aligned}\right\} \tag{0.4.3}$$

The internal forces between a pair of particles in R do not contribute to $\mathbf{F}_R$ since they are equal in magnitude but opposite in direction. (There are cases involving moving charges where Newton's law of action–reaction does not hold, but we shall not consider these.) We can therefore write

$$\mathbf{F}_R = \mathbf{F}_R^{(1)} + \mathbf{F}_R^{(2)},$$

where $\mathbf{F}_R^{(1)}$ is the total applied force on the particles of R and $\mathbf{F}_R^{(2)}$ is the total force on R due to the particles outside R. As we shall see below, $\mathbf{F}_R^{(2)}$ can be considerably simplified for a continuous body when the internal forces are short-range intermolecular forces.

Corresponding to the classical balance of angular momentum, there is also an angular impulse-momentum law. In particle mechanics, this is not an independent principle but follows from the linear impulse-momentum law as long as the internal forces obey the action–reaction law.

Continuum Model

When dealing with liquids or solids, it is sufficient for most engineering purposes to use a *continuum model* where the body is regarded as a continuum of material points rather than as a discrete collection of particles. We identify a material point by its position $\boldsymbol{\xi}$ in a *reference configuration* Ω (often taken as the initial configuration of the body before the deformation under study takes place). The material point $\boldsymbol{\xi}$ will be found at $\mathbf{x} = \mathbf{x}(\boldsymbol{\xi}, t)$ at time t. Our goal is to formulate a boundary value problem for $\mathbf{x}(\boldsymbol{\xi}, t)$. In technical parlance, we are using *Lagrangian* or *material* coordinates, whose major advantage is that most integrations can then be performed in the reference configuration Ω whose geometry is usually simpler than that of the deformed state—for instance, Ω might be a straight rod or a circular plate. Let us point out, however, that material coordinates are less appropriate in fluid flow, where the interest lies in the behavior at a fixed position in space (say, at the outlet of a pipe) rather than in the particular particles being observed.

Consider now the deformation of a solid, whose reference configuration is Ω, under loads consisting of body forces $\mathbf{f}(\boldsymbol{\xi}, t)$ per unit volume and surface forces $\boldsymbol{\sigma}(\boldsymbol{\xi}, t)$ per unit area. Let R be any subregion of Ω such that the boundary B of R has no points in common with the boundary Γ of Ω. The linear momentum of R is now an integral over R instead of the sum (0.4.3):

$$\mathbf{p}_R(t) = \int_R \rho(\boldsymbol{\xi}) \frac{\partial \mathbf{x}}{\partial t}\, d\boldsymbol{\xi}, \tag{0.4.4}$$

where $\rho(\boldsymbol{\xi})$ is the density in the reference configuration. The total applied force $\mathbf{F}_R^{(1)}$ on R is

$$\int_R \mathbf{f}(\boldsymbol{\xi}, t)\, d\boldsymbol{\xi},$$

whereas $\mathbf{F}_R^{(2)}$ is the total force on R exerted by material points outside R. The forces contributing to $\mathbf{F}_R^{(2)}$ are short-range intermolecular forces which are negligible except in a small neighborhood of B, enabling us to write

$$\mathbf{F}_R^{(2)} = \int_B \mathbf{T}(\boldsymbol{\xi}, t, \mathbf{n})\, dS, \tag{0.4.5}$$

where $\mathbf{T}$ is the surface *traction* and $\mathbf{n}$ is the outward unit normal. Our notation displays the dependence of $\mathbf{T}$ on the orientation of the local surface element. For fixed $\boldsymbol{\xi}$ and t, how does $\mathbf{T}$ depend on $\mathbf{n}$? By examining the forces on a small tetrahedron (with three faces in the coordinate planes and the fourth with normal $\mathbf{n}$), we see that $\mathbf{T}$ depends linearly on the direction cosines of $\mathbf{n}$. This is precisely the defining property of a tensor T (the *stress tensor*) which associates with each direction $\mathbf{n}$ the vector $\mathbf{T}(\mathbf{n})$. Using the fixed orthonormal set $(\mathbf{e}_1, \mathbf{e}_2, \mathbf{e}_3)$, we have

$$\mathbf{T}(\mathbf{n}) = \sum_i (\mathbf{n} \cdot \mathbf{e}_i)\mathbf{T}_i \quad \text{where } \mathbf{T}_i = \mathbf{T}(\mathbf{e}_i) = \sum_j T_{ij}\mathbf{e}_j.$$

In T_{ij}, the first index refers to the direction of the normal to the surface and the second index to the component axis. In terms of T_{ij}, we have

$$\mathbf{T}(\mathbf{n}) = \sum_i (\mathbf{n} \cdot \mathbf{e}_i) \sum_j T_{ij}\mathbf{e}_j, \tag{0.4.6}$$

so that the 3×3 matrix (T_{ij}) is a representation of the tensor T, which enables us to calculate the traction $\mathbf{T}$ for any orientation of the surface element. Using tensor notation, we write (0.4.6) as

$$\mathbf{T} = \mathrm{T} \cdot \mathbf{n}.$$

The impulse-momentum law for R then becomes

$$\mathbf{p}_R(t) - \mathbf{p}_R(0) = \int_0^t d\tau \left[\int_R \mathbf{f}(\boldsymbol{\xi}, \tau)\, d\boldsymbol{\xi} + \int_B \mathrm{T}(\boldsymbol{\xi}, \tau) \cdot \mathbf{n}\ dS \right]. \tag{0.4.7}$$

Using (0.4.6) we apply the divergence theorem to obtain

$$\mathbf{p}_R(t) - \mathbf{p}_R(0) = \int_0^t d\tau \left[\int_R (\mathbf{f} + \operatorname{div} \mathrm{T})\, d\boldsymbol{\xi} \right],$$

where the *vector* $\operatorname{div} \mathrm{T}$ is defined in analogy with the ordinary divergence:

$$\operatorname{div} \mathrm{T} = \sum_i \frac{\partial \mathbf{T}_i}{\partial x_i} = \sum_j \mathbf{e}_j \left[\sum_i \frac{\partial T_{ij}}{\partial x_i} \right].$$

Differentiation of (0.4.7) with respect to t gives

$$\int_R \left[\rho(\boldsymbol{\xi}) \frac{\partial^2 \mathbf{x}}{\partial t^2} - \operatorname{div} \mathrm{T} - \mathbf{f} \right] d\boldsymbol{\xi} = 0.$$

By a now familiar argument, the continuity of the integrand and the arbitrariness of R lead to

$$\rho(\boldsymbol{\xi}) \frac{\partial^2 \mathbf{x}}{\partial t^2} - \operatorname{div} \mathrm{T} = \mathbf{f}, \tag{0.4.8}$$

which is a system of three partial differential equations. There are, of course, too many dependent variables in (0.4.8). We need a constitutive law to relate T to the strain tensor, which itself can be expressed in terms of space derivatives of $\mathbf{x}$. Equation (0.4.8) must also be supplemented by initial and boundary conditions. We shall not pursue the general theory further, but switch our attention to the special case of rods and strings. We prefer to rederive the equations in this case to take advantage of the simplified geometry and at the same to introduce the angular impulse-momentum law in this setting.

Equations of Motion for Rods and Strings

What distinguishes a rod from other elastic bodies is its slenderness—length is its only "significant" dimension. More precisely, the elastic behavior of a rod can be described completely by a set of equations having as its only independent variables the time t and the arclength S along a distinguished curve in the reference configuration (usually the centerline of the slender body). The material points along this distinguished curve form the *axis* of the rod. During the motion the axis deforms, and a particle whose arclength coordinate is S in the reference configuration is then located on the deformed axis at a point $\mathbf{x}(S, t)$, so that S now plays the same role as ξ previously.

The surface traction on the cross section is actually distributed over the cross section, but we replace this traction by a resultant force $\mathbf{F}$ and moment M which are statically equivalent to it. It is the slenderness of the rod that justifies the use of this replacement.

We shall be principally concerned with the behavior of initially straight rods, symmetric about the x_1–x_2 plane, and subject to loads lying in this plane of symmetry. Under these assumptions it is reasonable to consider only deformations for which the axis of the rod remains in its original plane. The position vector $\mathbf{x}$ of a particle on the deformed axis will therefore have only x_1 and x_2 components, denoted by $u(S, t)$ and $v(S, t)$, respectively, so that we write $\mathbf{x} = u\mathbf{i} + v\mathbf{j}$. The arclength coordinate along the deformed axis is denoted by s, and the angle between the deformed axis and the horizontal by ϕ.

The nature of the forces that cause the motion of the rod also requires some discussion. Some of these forces are distributed; others may be concentrated. Some may be given explicitly either in the reference configuration or in the deformed state, whereas forces of constraint are known only indirectly from their geometrical effects. Some loads depend on the unknown displacement $\mathbf{x}$ (springlike resistance) or on the unknown velocity $\partial\mathbf{x}/\partial t$ (viscous resistance). Without committing ourselves to any special type of loading, we shall denote by Σ_{AB} the resultant of the external forces applied between the points A and B, and by N_{AB} their moment about the x_3 axis.

The cross-sectional force $\mathbf{F}$ and moment M are defined as the resultant force and moment exerted on the material to the left of the cross section by material on the right (see point B of Figure 0.4). A counterclockwise moment is regarded as positive. It may be convenient to express $\mathbf{F}$ either in terms of its x_1, x_2 components H and V or

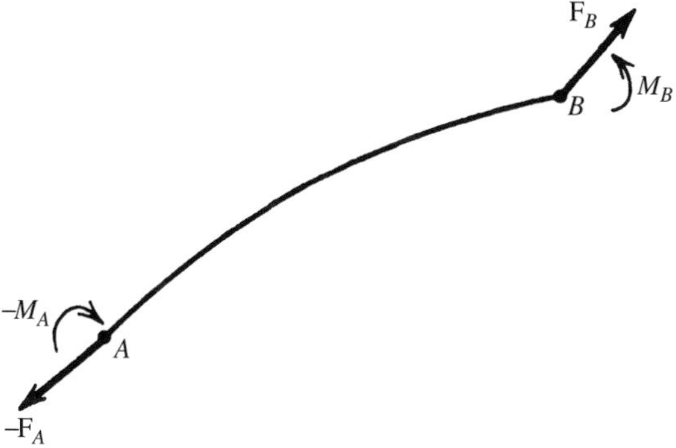

Figure 0.4

in terms of its tangential and normal components T and Q:

$$\text{F} = H\mathbf{i} + V\mathbf{j} = T\mathbf{e} + Q\mathbf{n}, \tag{0.4.9}$$

where **e** $(= d\mathbf{x}/ds)$ is a unit tangential vector and **n** a unit normal vector to the deformed axis. The components T and Q are known as the *tension* and *shear*, respectively. We have the obvious relations

$$Q = -H \sin\phi + V \cos\phi, \quad T = H \cos\phi + V \sin\phi. \tag{0.4.10}$$

The equations of motion for the portion of the rod between A and B (see Figure 0.4) consist of the two-component linear momentum equation

$$\mathbf{F}_B - \mathbf{F}_A + \Sigma_{AB} = \frac{d}{dt}\,(\text{linear momentum}) \tag{0.4.11}$$

and of the one-component angular momentum equation (about the origin)

$$\mathbf{k}(M_B - M_A) + \mathbf{x}_B \times \mathbf{F}_B - \mathbf{x}_A \times \mathbf{F}_A + \mathbf{k}N_{AB} = \frac{d}{dt}\,(\text{angular momentum}). \tag{0.4.12}$$

REMARK. Although these equations are adequate to describe applied forces that are either distributed or concentrated in space, they must be modified to deal with impulsive forces that cause an instantaneous change in momentum. In that case we would have to use an integrated form over time. For instance, (0.4.11) would become: net impulse $=$ change in linear momentum, as discussed earlier in this section.

To derive differential equations from (0.4.11) and (0.4.12) we shall have to express the momenta as functions of a space coordinate and time. Choosing the arclength S in the reference configuration as an independent variable, we find that

$$\text{linear momentum} = \int_{S_A}^{S_B} \rho(S)\frac{\partial \mathbf{x}}{\partial t}(S,t)\,dS,$$
$$\text{angular momentum} = \int_{S_A}^{S_B} \rho(S)\mathbf{x} \times \frac{\partial \mathbf{x}}{\partial t}\,dS,$$

where $\rho(S)$ is the density per unit length of the rod in its reference configuration, and S_A and S_B are the arclength coordinates of A and B in the reference configuration.

If the external forces are distributed, we can write

$$\Sigma_{AB} = \int_{S_A}^{S_B} \boldsymbol{\sigma}(S,t)\,dS, \tag{0.4.13}$$

where $\boldsymbol{\sigma}$ is the density of loading expressed in the reference configuration. Similarly, we find that

$$\mathbf{k}N_{AB} = \int_{S_A}^{S_B} \mathbf{x} \times \boldsymbol{\sigma}\,dS.$$

Let us set $S_B = S_A + dS$ and $S_A = S$, substitute in (0.4.11) and (0.4.12), divide by dS, and let dS tend to 0 to obtain

$$\frac{\partial \mathbf{F}}{\partial S} + \boldsymbol{\sigma} = \rho\frac{\partial^2 \mathbf{x}}{\partial t^2} \tag{0.4.14}$$

and

$$\mathbf{k}\frac{\partial M}{\partial S} + \frac{\partial}{\partial S}(\mathbf{x} \times \mathbf{F}) + \mathbf{x} \times \boldsymbol{\sigma} = \rho\mathbf{x} \times \frac{\partial^2 \mathbf{x}}{\partial t^2},$$

which, in view of (0.4.14) and (0.4.9), reduces to

$$\frac{\partial M}{\partial S} - H\frac{\partial v}{\partial S} + V\frac{\partial u}{\partial S} = 0. \tag{0.4.15}$$

Equation (0.4.15) can also be written as

$$\frac{\partial M}{\partial S} + Q\delta = 0, \quad \delta \doteq \frac{\partial s}{\partial S} = \left[\left(\frac{\partial u}{\partial S}\right)^2 + \left(\frac{\partial v}{\partial S}\right)^2\right]^{1/2}. \tag{0.4.16}$$

We see that (0.4.14) and (0.4.15) are three simultaneous equations for the five unknowns H, V, M, u, v. It is sometimes preferable to use (0.4.14) and (0.4.16) as three simultaneous equations for the alternative set of unknowns T, Q, M, u, v. In any event we must supplement our three equations with two constitutive relations between stress and strain. In the case of a string, one of these constitutive equations is $M \equiv 0$; since δ never vanishes, we find from (0.4.16) that the shear Q also vanishes identically in a string.

Constitutive Relations

We shall try to supplement (0.4.14) and (0.4.16) by two constitutive relations. We shall assume that the local elongation in the rod depends only on the tension T (axial component of the cross-sectional force) and that the local bending depends only on the moment M on the cross section. Natural strain measures consistent with these assumptions are

$$\delta \doteq \frac{\partial s}{\partial S} = (u_S^2 + v_S^2)^{1/2} \text{ (for elongation)} \tag{0.4.17}$$

and

$$\mu \doteq \frac{\partial \phi}{\partial S} = \frac{u_S v_{SS} - v_S u_{SS}}{u_S^2 + v_S^2} \text{ (for bending)}, \tag{0.4.18}$$

where we have used a subscript to denote differentiation, and we have appealed to the relation $\tan\phi = v_S/u_S$. Clearly, δ is a measure of elongation, since it is the ratio of the length of a deformed element to its length in the reference configuration. The strain ε is defined as the change in length of an element divided by its original length: $\varepsilon = \delta - 1$. Whereas δ is always positive, ε is positive for extension but negative for compression. The bending measure μ is in the nature of a curvature, the actual curvature being $K \doteq \partial\phi/\partial s$ (many authors define the curvature with a minus sign). Thus we have

$$\mu = \frac{\partial \phi}{\partial S} = K\frac{\partial s}{\partial S} = K\delta.$$

One of the reasons for preferring μ to K as a measure of bending is that a circular ring which merely expands under applied forces exhibits a change in K but none in μ, and it seems reasonable to say that no additional bending takes place in the process.

The constitutive laws are taken to be of the form

$$\delta = \hat{\delta}(T, S), \quad \mu = \hat{\mu}(M, S), \tag{0.4.19}$$

where $\hat{\delta}$ and $\hat{\mu}$ are specified functions. Substituting the first of (0.4.19) in (0.4.17) gives a relation among T, u, and v, whereas substituting the second of (0.4.19) in (0.4.18) gives a relation among M, u, and v. These two new relations, together with (0.4.14) and (0.4.16), form a set of five equations in the five unknowns T, Q, M, u, v.

Let us now examine the constitutive laws (0.4.19) in a little more detail. For each fixed S, $\hat{\delta}$, and $\hat{\mu}$ have the general appearance shown in Figure 0.5. If the rod is homogeneous, $\hat{\delta}$ and $\hat{\mu}$ will be independent of S. It is clear from Figure 0.5 that we have tacitly assumed that the reference configuration is the stress-free straight rod. For some purposes, such as the vibrations of a taut string, it is more convenient to use a prestressed state as the reference configuration, in which case a change of variables is necessary to express the constitutive relations in terms of the new variables.

The function $\hat{\delta}$ is positive and monotonically increasing in T with $\hat{\delta}(0, S) = 1$ (for the special case of an *inextensible* rod, $\hat{\delta} \equiv 1$). The function $\hat{\mu}$ is odd and increasing in M, so that, by continuity, $\hat{\mu}(0, S) = 0$. For a string the second of

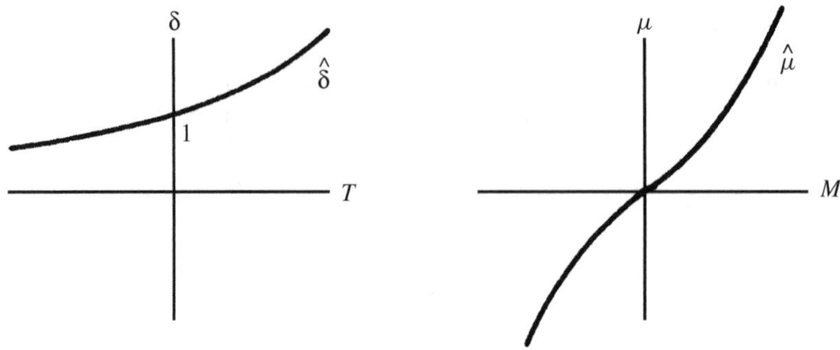

Figure 0.5

(0.4.19) is replaced by $M \equiv 0$, and we then further assume that the first equation can be inverted to give

$$T = \hat{T}(\delta, S), \tag{0.4.20}$$

where $\hat{T}$ is regarded as a *known* function, strictly increasing in δ for each S.

The Vibrating String

We have already noted that both M and Q vanish for a string, so that (0.4.14) becomes

$$(T\mathbf{e})_S + \boldsymbol{\sigma} = \rho\mathbf{x}_{tt}, \tag{0.4.21}$$

where the subscripts denote differentiation. If we take into account the relations

$$\begin{aligned} &\mathbf{x} = u\mathbf{i} + v\mathbf{j}, \quad \delta = (u_S^2 + v_S^2)^{1/2}, \quad T = \hat{T}[\delta(S,t), S], \\ &\mathbf{e} = \mathbf{x}_s = \frac{1}{\delta}\mathbf{x}_S = \frac{1}{\delta}(u_S\mathbf{i} + v_S\mathbf{j}), \end{aligned} \tag{0.4.22}$$

we see that (0.4.21) becomes a pair of nonlinear partial differential equations for $u(S,t)$ and $v(S,t)$ valid for small and large deflections alike. In any specific problem (0.4.21) will be supplemented by appropriate initial and boundary conditions.

We shall consider two problems for a string. The first is the extremely simple problem of taking a string whose length in its natural, stress-free state is L and stretching it between the two points (0, 0) and (l, 0), where $l > L$. With $\boldsymbol{\sigma} = 0$, we are looking for a solution of (0.4.21)–(0.4.22) which has the following properties: (a) it is independent of t and has only an x_1 component, so that we write $\mathbf{x} = \mathbf{x}(S) = u(S)\mathbf{i}$, and (b) $\mathbf{x}(0) = 0$, $\mathbf{x}(L) = l\mathbf{i}$. Then $\mathbf{e} = \mathbf{i}$, and (0.4.21) gives $T_S = 0$. Thus T is a constant along the string, say $T^{(0)}$. When the constitutive law $\delta = \hat{\delta}(T, S)$ is inserted in $u_S = \delta$, we obtain

$$u(S) = \int_0^S \hat{\delta}(T^{(0)}, S^*)\, dS^*, \tag{0.4.23}$$

so that, since $u(L) = l,\ T^{(0)}$ must satisfy

$$l = \int_0^L \hat{\delta}(T^{(0)}, S)\, dS. \tag{0.4.24}$$

If $\hat{\delta}$ is a strictly increasing function of T tending to $+\infty$ with T, then (0.4.24) will have a unique solution $T^{(0)}$ for any given $l > L$. This is the tension required to stretch the string from its natural state to length l. The position of the particles is given by (0.4.23). We shall refer to this state of the string as the state of *simple stretching.*

Let us now turn from this trivial (though nonlinear) problem to the more difficult one of vibrations. We consider a string which is first stretched between (0,0) and (l,0); then with its endpoints fixed it is subject to an initial displacement and velocity and to external forces for $t > 0$. *It will be convenient to use the state of simple stretching as the reference configuration.* To avoid confusion we label particles in the new reference configuration by the letter X; thus $0 \leqslant X \leqslant l$, and we rewrite (0.4.21)–(0.4.22) with X and t as the independent variables:

$$[T\mathbf{e}]_X + \boldsymbol{\sigma} = \rho \mathbf{x}_{tt}, \qquad t > 0, \quad 0 < X < l, \tag{0.4.25}$$

together with

boundary conditions:	$\mathbf{x}(0,t) = 0,$	$\mathbf{x}(l,t) = l\mathbf{i},$	(0.4.26)
initial conditions:	$\mathbf{x}(X,0) = \mathbf{w}(X),$	$\mathbf{x}_t(X,0) = \mathbf{z}(X).$	(0.4.27)

Our measure of elongation is now

$$D \doteq \frac{\partial s}{\partial X} = (u_X^2 + v_X^2)^{1/2}, \tag{0.4.28}$$

and the constitutive relation (0.4.20) can be rewritten in the form

$$T = \tilde{T}(D, X), \tag{0.4.29}$$

where $\tilde{T}(1,X) = T^{(0)}$, the constant tension corresponding to the state of simple stretching. We also have

$$\mathbf{e} = \mathbf{x}_s = \frac{1}{D}\mathbf{x}_X = \frac{1}{D}(u_X \mathbf{i} + v_X \mathbf{j}). \tag{0.4.30}$$

Substitution of (0.4.28) through (0.4.30) in (0.4.25) gives a pair of nonlinear partial differential equations for $u(X,t)$ and $v(X,t)$ which are to be solved under conditions (0.4.26) and (0.4.27). Although some static problems, such as a cable hanging under its own weight, can be solved analytically, this will not be the case for the general dynamic problem. If, however, the forcing terms $\boldsymbol{\sigma}$, $\mathbf{w}$, $\mathbf{z}$ are "small," it is possible to use a linearized theory, which we now develop.

Linearized Problem

To treat the problem of small deflections systematically, we introduce an amplitude parameter α (perturbation parameter) in all forcing terms, writing

$$\boldsymbol{\sigma} = \alpha \mathbf{f}, \quad \mathbf{w} = \alpha \mathbf{W}, \quad \mathbf{z} = \alpha \mathbf{Z}, \tag{0.4.31}$$

where **f**, **W**, **Z** are independent of α.

The solution of (0.4.25) through (0.4.30) will now depend on the parameter α. Thus $\mathbf{x}$, $\mathbf{e}$, D, T will all depend on α. Their behavior for small α is of particular interest. Let us assume that all these quantities can be expanded in power series in α. A typical quantity such as D will have an expansion

$$D(X, t, \alpha) = D^{(0)} + \alpha D^{(1)} + \frac{\alpha^2}{2} D^{(2)} + \cdots, \tag{0.4.32}$$

where

$$D^{(k)}(X, t) = \left. \frac{\partial^k D}{\partial \alpha^k} \right|_{\alpha=0}.$$

We shall be content with characterizing $\mathbf{x}^{(0)}$ and $\mathbf{x}^{(1)}$, the first two terms of the alpha expansion of the solution. On setting $\alpha = 0$ in (0.4.25) through (0.4.31), we are left with the boundary value problem for the state of simple stretch, which in terms of our reference configuration has the solution

$$\begin{gathered} \mathbf{x}^{(0)} = X\mathbf{i}, u^{(0)}(X) = X, \quad v^{(0)}(X) = 0, \quad \mathbf{e}^{(0)} = \mathbf{i}, \\ D^{(0)}(X) = 1, \quad T^{(0)}(X) = T^{(0)}, \end{gathered} \tag{0.4.33}$$

where $T^{(0)}$ is the constant tension in simple stretching.

To formulate a boundary value problem for $\mathbf{x}^{(1)}(X, t)$ requires more work. We differentiate (0.4.25) through (0.4.31) with respect to α and set α equal to 0.

The first step gives

$$(T_\alpha \mathbf{e} + \mathbf{e}_\alpha T)_X + \mathbf{f} = \rho \mathbf{x}_{tt\alpha}; \quad \mathbf{x}_\alpha|_{X=0,l} = 0, \quad \mathbf{x}_\alpha|_{t=0} = \mathbf{W}, \quad \mathbf{x}_{\alpha t}|_{t=0} = \mathbf{Z},$$

and, on setting $\alpha = 0$,

$$\begin{gathered} (T^{(1)}\mathbf{e}^{(0)} + \mathbf{e}^{(1)}T^{(0)})_X + \mathbf{f} = \rho \mathbf{x}_{tt}^{(1)}; \quad \mathbf{x}^{(1)}(0, t) = \mathbf{x}^{(1)}(l, t) = 0, \\ \mathbf{x}^{(1)}(X, 0) = \mathbf{W}, \quad \mathbf{x}_t^{(1)}(X, 0) = \mathbf{Z}. \end{gathered} \tag{0.4.34}$$

To simplify (0.4.34) we note that from (0.4.28) we have

$$D^2 = u_X^2 + v_X^2, \quad DD_\alpha = u_X u_{\alpha X} + v_X v_{\alpha X},$$

and hence

$$D^{(0)} D^{(1)} = u_X^{(0)} u_X^{(1)} + v_X^{(0)} v_X^{(1)},$$

which, by (0.4.33), reduces to

$$D^{(1)} = u_X^{(1)}.$$

A similar calculation starting from (0.4.30) gives

$$\mathbf{e}^{(1)} = v_X^{(1)}\mathbf{j}.$$

Since $\tilde{T} = T(\tilde{D}, X)$, we have $\tilde{T}_\alpha = T_D D_\alpha$ and therefore

$$T^{(1)} = \tilde{T}_D(1,X)D^{(1)} = \tilde{T}_D(1,X)u_X^{(1)}.$$

Substitution in (0.4.34) yields separate *linear* boundary value problems for $u^{(1)}$ and $v^{(1)}$:

$$\begin{gathered}[\tilde{T}_D(1,X)u_X^{(1)}]_X + \mathbf{f}\cdot\mathbf{i} = \rho u_{tt}^{(1)}; \quad u^{(1)}(0,t) = u^{(1)}(l,t) = 0,\\ u^{(1)}(X,0) = \mathbf{W}\cdot\mathbf{i}, \quad u_t^{(1)}(X,0) = \mathbf{Z}\cdot\mathbf{i},\end{gathered} \tag{0.4.35}$$

and

$$\begin{gathered}T^{(0)}v_{XX}^{(1)} + \mathbf{f}\cdot\mathbf{j} = \rho v_{tt}^{(1)}; \quad v^{(1)}(0,t) = v^{(1)}(l,t) = 0,\\ v^{(1)}(X,0) = \mathbf{W}\cdot\mathbf{j}, \quad v_t^{(1)}(X,0) = \mathbf{Z}\cdot\mathbf{j}.\end{gathered} \tag{0.4.36}$$

Both (0.4.35) and (0.4.36) are *wave equations*, which will be useful if **f** and ρ can be expressed in terms of the stretched variable X. In fact, **f** is usually given directly as a function of X, but ρ is normally given in the natural state in terms of S. This presents no difficulty, however, because $X(S)$ is known explicitly from (0.4.23) as $\hat{\delta}(T^{(0)}, S)$. Thus, if $\rho(S)$ is the density of the string in its stress-free state, we find that

$$\rho(X) = \tilde{\rho}(S(X))\frac{dS}{dX}.$$

Equation (0.4.35) describes the small *longitudinal* vibrations of a string and applies equally well to a rod, since bending effects appear only in the transverse equation. Note that the local stiffness of the string or rod is given by $\tilde{T}_D(1,X)$; this plays the role of Young's modulus, which will vary with position when the body is inhomogeneous.

The *transverse* vibrations of a *string* are governed by (0.4.36). Note that the coefficient of $v_{XX}^{(1)}$ is constant even though the string may be inhomogeneous.

For the *transverse* vibrations of a beam we return to (0.4.14), (0.4.16), and the two constitutive laws (0.4.19), using a linearization procedure similar to that for the string, but now about the stress-free state; and recalling (0.4.18), we find that v obeys the *fourth-order linear equation*

$$-[\hat{M}_\mu(0,S)v_{SS}^{(1)}]_{SS} + \mathbf{f}\cdot\mathbf{j} = \rho v_{tt}^{(1)}. \tag{0.4.37}$$

In the usual Euler-Bernoulli theory one takes $\hat{M}_\mu = EI$, where E is Young's modulus and I the moment of inertia about the neutral axis.

In deriving (0.4.37) we have assumed an initially stress-free rod. If there is a tension $T^{(0)}$ in the beam, a term $T^{(0)}v_{SS}^{(1)}$ must be added to the left side of (0.4.37).

In Chapter 9 we consider a fully nonlinear problem: the buckling of an initially straight homogeneous rod under compression.

0.5 ALTERNATIVE FORMULATIONS OF PHYSICAL PROBLEMS

We shall develop the four (nearly) equivalent formulations for steady heat conduction in a cylindrical rod of cross-sectional area A and length 1. The rod is parallel to the x–axis with the end $x = 0$ kept at 0 temperature and the end $x = 1$ insulated. The heat sources along the rod depend only on x. If the lateral surface of the rod is insulated, the steady temperature in the rod will only be a function of x denoted by $u(x)$.

I. *Integral Balance Law.* Consider the portion of the rod lying between $x = a$ and $x = b$. The *integral balance* (0.1.10) reduces to

$$-k(b)u'(b)A + k(a)u'(a)A = P_{ab}, \tag{0.5.1}$$

where P_{ab} is the total heat source in $[a, b]$. Although our interest in this section is in distributed heat sources, we interject a comment about concentrated sources. Suppose that there is a source of strength Q concentrated at $x = x_0$; then (0.5.1) tells us that u' is discontinuous at x_0:

$$-k(x_0)[u'(x_0+) - u'(x_0-)] = Q/A. \tag{0.5.2}$$

This is just the one-dimensional version of (0.1.12). As was pointed out in Section 0.1, the integral law cannot handle more complex singular sources such as a dipole.

In the remainder of this section, we assume that the heat sources are distributed continuously along the rod with a density $f(x)$ per unit volume. Thus, $P_{ab} = \int_a^b f(x)A\,dx$, and (0.5.1) becomes

$$-k(b)u'(b) + k(a)u'(a) = \int_a^b f(x)\,dx. \tag{0.5.3}$$

II. *The Boundary Value Problem (BVP).* Regarding a as fixed and b as variable in (0.5.3), we differentiate with respect to b and then replace b by x. Taking the boundary conditions into account we obtain the *BVP*

$$-[k(x)u'(x)]' = f(x), 0 < x < 1; \quad u(0) = 0, u'(1) = 0. \tag{0.5.4}$$

Here $k \in C^1[0,1]$, $f \in C[0,1]$ and since the left side of the differential equation involves u'', we seek a solution $u \in C^2[0,1]$ which also satisfies both boundary conditions. The theory could be developed under slightly weaker assumptions on f and k, but these will suffice for our present purpose, leading to a straightforward exposition. We shall make use of the following.

Definition.

$$\begin{cases} \phi(x) \in M_1 \textit{ if } \phi \in C^2[0,1] \textit{ and } \phi(0) = 0. \\ \phi(x) \in M_2 \textit{ if } \phi \in M_1 \textit{ and } \phi(1) = 0. \\ \phi(x) \in M_3 \textit{ if } \phi \in M_1 \textit{ and } \phi'(1) = 0. \end{cases}$$

When dealing directly with the BVP (0.5.4), we require that $u \in M_3$. We shall see below, however, that there are other formulations which do not require a priori membership in M_3.

Let us first obtain a simple consequence of (0.5.4) by multiplying the differential equation by $\phi \in M_1$ and integrating by parts:

$$\int_0^1 (ku'\phi' - f\phi)\, dx = 0 \quad \text{for every } \phi \in M_1. \tag{0.5.5}$$

Thus, if u satisfies the BVP, it satisfies (0.5.5).

Before proceeding with a discussion of the weak form of the BVP, we take a short detour.

We shall often be asked to determine a continuous function $h(x)$ when we have indirect information on h. A typical problem is the following: Find $h(x) \in C[0,1]$ given that

$$\langle h, \phi \rangle \doteq \int_0^1 h(x)\phi(x)\, dx = 0 \quad \text{for every } \phi \in \Phi, \tag{0.5.6}$$

where Φ is a given class of *testing functions*. We call $\langle h, \phi \rangle$ the *action* of h on the testing function ϕ.

We would like to conclude from (0.5.6) that $h(x)$ is the zero function. Obviously, we cannot reach this conclusion if Φ is a finite collection of functions. For instance, if Φ contains only the function $\Phi \equiv 1$, then (0.5.6) merely states that the average of $h(x)$ is zero. If Φ is the set of functions $\{\sin \pi x, \sin 2\pi x, \sin 3\pi x\}$, then (0.5.6) tells us only that the first three coefficients in the Fourier sine series of h vanish. If, however, Φ is the countable set $\{\sin n\pi x\}_{n=1}^{n=\infty}$, then (0.5.6) implies that $h(x) \equiv 0$ (see Chapter 2). If we know (0.5.6) for the class Φ of all continuous functions, then by using $\phi = h$ as a testing function we have $\int_0^1 h^2\, dx = 0$, which, since h^2 is continuous and nonnegative, implies that $h(x) \equiv 0, 0 \leqslant x \leqslant 1$. In the applications we have in mind, the class Φ will be somewhat more restricted. Note that if we can conclude that $h \equiv 0$ if (0.5.6) holds for a class Φ, the same conclusion follows if (0.5.6) holds for a larger class Ψ.

Lemma 0.5.1. *Suppose that $h(x) \in C[0,1]$ and $\int_0^1 h(x)\phi(x)\, dx = 0$ for* all $\phi \in M_2$. *Then $h(x) \equiv 0$.*

Proof. We shall show that $h = 0$ at all interior points, so that, by continuity, h will also vanish at the endpoints. Suppose then that $h(x_0) \neq 0$ for some $x_0 \in (0,1)$; we shall take $h(x_0) > 0$, for otherwise we could deal with $-h(x)$ instead. Since $h(x_0) > 0$, continuity implies the existence of an interval $|x - x_0| < \varepsilon$ contained in (0, 1) in which $h(x) > 0$. We can easily find a function $\phi \in C^2[0,1]$ which is positive in $|x - x_0| < \varepsilon$ and vanishes for $|x - x_0| \geqslant \varepsilon$ (see Exercise 0.5.1). This function ϕ belongs to M_2 and for this testing function $\int_0^1 h\phi\, dx > 0$, contradicting the hypothesis. □

We now return to the main line of development.

III. ***The Weak Form of the BVP.*** Let us think of (0.5.5) as characterizing an unknown function $u(x)$:

$$\text{Find } u \in M_1 \text{ such that } \int_0^1 (ku'\phi' - f\phi)\, dx = 0, \qquad \text{for every } \phi \in M_1. \qquad (0.5.7)$$

Note that the function $u(x)$ in (0.5.7) is not required to satisfy $u'(1) = 0$; this will turn out to be a conclusion from (0.5.7). We know a priori that a solution (if any exists) of the BVP (0.5.4) does satisfy (0.5.7); that was shown in (0.5.5). Now we want to show that if a function u satisfies (0.5.7), it must solve the BVP (0.5.4).

Theorem. *If u satisfies (0.5.7), it satisfies (0.5.4).*

Proof. Integrating the first integral by parts we find that, for every $\phi \in M_1$,

$$\int_0^1 \phi[(ku')' + f]\, dx - k(1)u'(1)\phi(1) = 0. \qquad (0.5.8)$$

The fixed unknown function u must satisfy (0.5.8) for all $\phi \in M_1$. As we use different ϕ's in (0.5.8), we may obtain more and more information on u, and this information is cumulative; the fixed u must obey all the conditions imposed by (0.5.8). Let us first consider (0.5.8) for functions $\phi \in M_2$ (a subset of M_1) so that

$$\int_0^1 \phi[(ku')' + f]\, dx = 0 \quad \text{for every } \phi \in M_2.$$

Lemma 0.5.1 then tells us that $(ku')' + f = 0$; hence our function u must satisfy the differential equation in (0.5.4) as well as $u(0) = 0$ since that is contained in the requirement $u \in M_1$. Now that we have found that u satisfies the differential equation, (0.5.8) reduces to

$$-k(1)u'(1)\phi(1) = 0 \quad \text{for every } \phi \in M_1.$$

Taking $\phi \in M_1$ with $\phi(1) \neq 0$, we conclude that $u'(1) = 0$, so that $u \in M_3$ and satisfies the BVP (0.5.4). Thus we have proved the equivalence of (0.5.4) and (0.5.7), but we have not proved the existence of a solution to either problem! □

Let us now turn to the fourth formulation.

IV. ***Variational Principle (Principle of Minimum Energy).*** Define, for $v \in M_1$, the functional

$$\mathbf{J}(v) = \int_0^1 [k(v')^2 - 2fv]\, dx. \qquad (0.5.9)$$

We say that $u \in M_1$ is a *minimizer* of J on M_1 if $J(u) \leqslant J(v)$ for every $v \in M_1$.

Theorem. *A function $u \in M_1$ is a minimizer of J on M_1 if and only if u satisfies (0.5.7).*

Proof. If v and w are two arbitrary elements of M_1, then $\phi = w - v$ belongs to M_1 and

$$J(v) - J(w) = J(w+\phi) - J(w) = \int_0^1 [k(\phi')^2 + 2\,kw'\phi' - 2f\phi]\,dx. \quad (0.5.10)$$

(a) If u satisfies (0.5.7), then upon setting $w = u$, (0.5.10) gives

$$J(v) - J(u) = J(u+\phi) - J(u) = \int_0^1 k(\phi')^2\ dx \geqslant 0,$$

so that u is a minimizer.

(b) Let $u \in M_1$ be a minimizer of (0.5.9); then for each $\varepsilon > 0$ and $\phi \in M_1$, $J(u \pm \varepsilon\phi) \geqslant J(u)$ and (0.5.10) gives

$$0 \leqslant J(u \pm \varepsilon\phi) - J(u) = \int_0^1 [\varepsilon^2 k(\phi')^2 \pm 2\varepsilon(ku'\phi' - f\phi)]\,dx.$$

Dividing by ε and letting $\varepsilon \to 0$, we find that

$$0 \leqslant \pm 2\int_0^1 (ku'\phi' - f\phi)\,dx,$$

from which we conclude that the integral vanishes and, since ϕ is an arbitrary element in M_1, that u obeys (0.5.7). We have thus proved the equivalence of (0.5.7), the minimum principle, and the BVP. Let us emphasize again that we have not proved the existence of a solution to any of these equivalent problems; that will be done in Chapter 1 by constructing the solution to the BVP.

□

REMARKS

1. In the weak formulation (0.5.7) and in seeking a minimizer of (0.5.9) we do not a priori require u to satisfy the BC $u'(1) = 0$, only the condition $u(0) = 0$. This latter is called an *essential* boundary condition, the former a *natural* boundary condition.

2. If $f \in C[0,1]$ and $k \in C^1[0,1]$, the appropriate framework for the BVP (0.5.4) is that $u \in C^2[0,1]$. The weak formulation (0.5.7) and the variational principle involve only the first derivative of u, and it would seem more natural to require only that $u \in C^1[0,1]$. (See Exercise 0.5.3.)

We could have equally well formulated the variational principle by considering the minimum in the class of functions satisfying *both* boundary conditions (and similarly for the weak formulation). Why do we make life apparently more difficult by dropping the natural BC and then showing that the minimizer and the solution of (0.5.7) must satisfy that BC anyhow? The reason is that we want to use the minimum principle as an approximation scheme; we look for a minimum in a restricted class of trial functions (say, polynomials of degree n) and then find the optimal values of the parameters (the coefficients in the polynomial). To impose the requirement that the trial functions have zero derivative at $x = 1$ would cause additional numerical difficulties.

Of course, for the rod problem with both ends at zero temperature, the trial functions are required to satisfy both BCs.

3. *The Principle of Virtual Work (or Virtual Power) in Mechanics. A system is in equilibrium under applied forces if and only if the work done by these applied forces in any small displacement compatible with the constraints is equal to the first-order change in the strain energy.* (Such an additional diplacement is called virtual because it is a mathematical construct rather than an actual physical displacement.)

Let us illustrate the Principle of Virtual Work for the equilibrium deflection $u(x)$ of a string under tension T, subject to a distributed transverse load $f(x)$; we take both ends as fixed: $u(0) = u(1) = 0$. Thus, according to (0.4.36) our BVP is

$$-Tu'' = f, \quad 0 < x < 1, \quad u(0) = u(1) = 0, \tag{0.5.11}$$

where both boundary conditions are essential (see Remark 1).

Now let $\phi(x)$ be an additional virtual displacement compatible with the constraints [that is, $\phi(0) = \phi(1) = 0$]. The work done by the applied forces in this additional displacement is $\int_0^1 \phi(x)f(x)\,dx$. The strain energy in an element dx is proportional to its elongation $(\sqrt{1+(u')^2} - 1)\,dx$ with proportionality factor T. In the linear theory where u' is small, this becomes $T(u')^2/2$. Thus the change in the strain energy in the virtual displacement is

$$T\int_0^1 \frac{(u'+\phi')^2}{2}\,dx - T\int_0^1 \frac{u'^2}{2}\,dx,$$

which for ϕ' small compared to u' becomes approximately

$$T\int_0^1 u'\phi'\,dx.$$

Thus the Principle of Virtual Work states: A function $u(x)$ is the deflection [solution of (0.5.11)] if and only if

$$\int_0^1 f\phi\,dx = T\int_0^1 u'\phi'\,dx \quad \text{for every } \phi(x) \in M_2. \tag{0.5.12}$$

This is just the weak form (0.5.7) for the case when u vanishes at both ends.

4. *The Principle of Minimum Potential Energy. A configuration compatible with the constraints is the equilibrium configuration if and only if its potential energy is a minimum.* What is the potential energy in the case of the string problem? It is the sum of the strain energy and the potential energy of the applied

forces (which is the negative of the work done by the applied forces). Thus, the potential energy of the string in the configuration described by $v(x)$ is

$$\frac{T}{2}\int_0^1 (v')^2\,dx - \int_0^1 f(x)v(x)\,dx,$$

which is just one-half of the quantity

$$J(v) = T\int_0^1 (v')^2\,dx - 2\int_0^1 f(x)v(x)\,dx$$

used in (0.5.9).

EXERCISES

0.5.1 Show that $u = \cos^4(\pi/2)x$ is positive for $|x| < 1$ and that u and its first three derivatives vanish at $x = \pm 1$. Therefore, for any $\varepsilon > 0$, the function

$$\phi(x) = \begin{cases} u\left(\dfrac{x - x_0}{\varepsilon}\right), & |x - x_0| < \varepsilon, \\ 0, & |x - x_0| \geqslant \varepsilon, \end{cases}$$

is in $C^3(-\infty, \infty)$ and if $0 < x_0 < 1$ and $\varepsilon < \min(x_0, 1 - x_0)$, $\phi(0) = \phi(1) = 0$.

0.5.2 Suppose that $h \in C[0,1]$ and $\langle h, \phi' \rangle = 0$ for every $\phi \in M_2$; show that h is a constant without assuming a priori that h is differentiable. [*Hint*: consider the function

$$\phi(x) = \int_0^x (h(t) - \bar{h})\,dt,$$

where $\bar{h}$ is the average of h.]

0.5.3 This exercise expands on Remark 2. Let us define

$$\begin{cases} \phi(x) \in N_1 \text{ if } \phi \in C^1[0,1] \quad \text{and} \quad \phi(0) = 0, \\ \phi(x) \in N_2 \text{ if } \phi \in N_1 \quad \text{and} \quad \phi(1) = 0. \end{cases}$$

Since $N_2 \supset M_2$, Lemma 0.5.1 still holds with $\phi \in N_2$ instead of $\phi \in M_2$. Next consider (0.5.7) with u and ϕ belonging to N_1; show that if u satisfies this new (0.5.7), then u satisfies the BVP (0.5.4). Note that we have not assumed that $u \in C^2[0,1]$; this is a *conclusion* from the new weak formulation! The trick is to integrate the $f\phi$ term by parts. Then use Exercise 0.5.2 (whose result holds for $\phi \in N_2$) to show that u satisfies the differential equation; show also that $u'(1) = 0$. Finally, show that a function $u \in N_1$ is a minimizer of J on N_1 if and only if u satisfies (0.5.7) with M_1 replaced by N_1.

0.5.4 Consider steady heat conduction in a rod of constant thermal conductivity k with the left end at temperature zero and the temperature at the right end obeying Newton's law of cooling [see (0.1.14)(iii)]:

$$-ku'' = f(x), \quad 0 < x < 1, \quad u(0) = 0, \quad -ku'(1) = \alpha u(1), \quad (0.5.13)$$

with α, k positive.

(a) Show that a solution u of the BVP must satisfy

$$\int_0^1 (ku'\phi' - f\phi)\, dx + \alpha u(1)\phi(1) = 0 \quad \text{for every } \phi \in \mathbf{M}_1. \quad (0.5.14)$$

(b) Take (0.5.14) as the basis for the weak formulation: Seek $u \in M_1$ such that (0.5.14) holds for every $\phi \in M_1$. Show that such a function u must satisfy the BVP.

(c) Let

$$J(v) = \int_0^1 [k(v')^2 - 2fv]\, dx + \alpha v^2(1). \quad (0.5.15)$$

Show that a function $u \in M_1$ is a minimizer of J over M_1 if and only if it satisfies (0.5.14).

0.6 NOTES ON CONVERGENCE

A sequence of real numbers $\{s_1, \ldots, s_n, \ldots\}$ is said to *converge to* s (or *have the limit* s) if for each $\varepsilon > 0$ there exists an index N such that

$$|s - s_n| < \varepsilon \quad \text{for all } n > N.$$

This definition states in precise mathematical language the intuitive notion that, for n large enough, s_n can be made arbitrarily close to s. To apply the definition, that is, to exhibit N as a function of ε, we must be able to guess the limit s. The following theorem permits us to determine whether a sequence converges without reference to the actual limit.

Theorem (Cauchy Criterion). *The necessary and sufficient condition for the sequence* $\{s_n\}$ *to converge is that for each* $\varepsilon > 0$, *there exists an index* P *such that*

$$|s_n - s_m| < \varepsilon \quad \textit{for all } m, n > P.$$

Thus, if a sequence converges, all its terms beyond a certain index must be close to each other; conversely, if the terms can be made close to each other by choosing the index large, then there is a number to which the sequence converges. The first part is easy to prove, but the converse relies on a fundamental property of the real number system. These ideas are discussed further in Section 4.3.

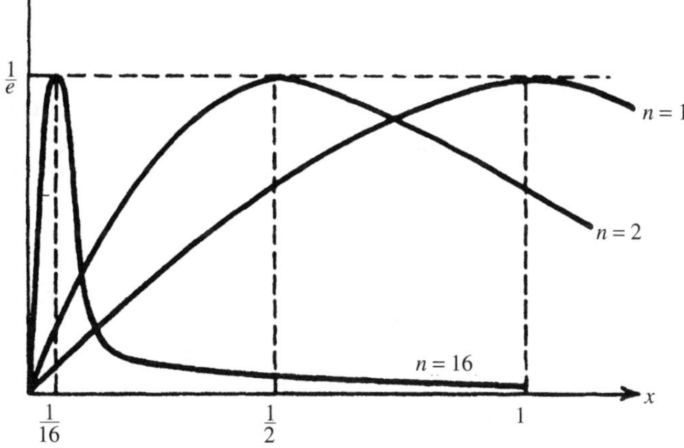

Figure 0.6

For sequences of *functions* there are many different ways of defining convergence; that is, there are various modes of convergence. We shall deal with a fixed interval E (open or closed) on the x axis and with a sequence $\{s_1(x), \ldots, s_n(x), \ldots\}$ of real-valued functions on E.

Perhaps the most natural (but, as we shall see, not the most useful) way of defining convergence for a sequence of functions is to examine each point x in E individually. With x fixed, the sequence $\{s_n(x)\}$ is a sequence of real numbers, and we certainly know what it means for this sequence to converge to the real number $s(x)$. This gives rise to the definition of *pointwise convergence*: For each x in E and for each $\varepsilon > 0$ there exists an index N (depending on both x and ε) such that

$$|s(x) - s_n(x)| < \varepsilon \quad \text{for all } n > N.$$

If we look at this definition graphically, then for each fixed x, the ordinate $s_n(x)$ will be within ε of the ordinate $s(x)$ if n exceeds N. At another point x in E a much larger N may be required to achieve the same degree of approximation. The following example shows some of the deceptive features of pointwise convergence.

Consider the sequence of functions

$$s_n(x) = nxe^{-nx} \quad \text{on } E : 0 \leqslant x \leqslant 1, \tag{0.6.1}$$

a few terms of which are graphed in Figure 0.6. Each function $s_n(x)$ has the same maximum value $1/e$ taken on at $x = 1/n$, so that $s_n(x)$ rises sharply to the height $1/e$ and then dies down quickly. Thus each $s_n(x)$ has a bump of the same height, the bump becoming thinner and thinner and moving closer and closer to $x = 0$ as n becomes larger. If we fix $x > 0$, then, by waiting long enough (that is, taking N sufficiently large), the bump will have moved to the left of the fixed value of x and the ordinate $s_n(x)$ at that fixed point will tend to 0 (of course, the nearer x is to 0, the longer we have to wait to get within a preassigned degree of approximation of the limit). At $x = 0$ each $s_n = 0$, so clearly the limit is 0. Thus, according to the definition of pointwise convergence, which treats each x individually, the sequence $\{s_n(x)\}$ tends to 0 pointwise on $0 \leqslant x \leqslant 1$. Nevertheless, we feel somewhat uncomfortable with the result, since the graph of $s_n(x)$ with its bump of fixed height $1/e$ does not seem like such a good approximation to the limit function, the curve $s \equiv 0$. The situation is even more disturbing for the sequence

$$s_n(x) = n^\alpha x e^{-nx} \quad \text{on } E: 0 \leqslant x \leqslant 1 \tag{0.6.2}$$

when $\alpha > 1$. Here we still have $s_n(x)$ tending to 0 pointwise on $0 \leqslant x \leqslant 1$, but now the bump in $s_n(x)$ increases in height indefinitely as n tends to infinity. The notion of uniform convergence is perhaps more in keeping with our intuition.

The sequence $\{s_n(x)\}$ is said to *converge uniformly to $s(x)$ on E* if for each $\varepsilon > 0$ there exists N (depending only on ε) such that

$$|s_n(x) - s(x)| < \varepsilon \quad \text{for all } n > N \text{ and all } x \text{ in } E.$$

REMARKS

1. If the convergence is uniform and a degree of approximation ε is specified, it is possible to choose a value of N that works for all x in E simultaneously; this means that all the curves $s_n(x)$ with $n > N$ lie entirely within an ε strip of the limit curve $s(x)$. The maximum deviation between the approximating curve $s_n(x)$ and the limit $s(x)$ therefore approaches 0 as n tends to infinity. Thus we have the equivalent definition: $\{s_n(x)\}$ *converges to $s(x)$ uniformly on E* if the numerical sequence

$$\sup_{x \in E} |s_n(x) - s(x)| \tag{0.6.3}$$

tends to 0 as $n \to \infty$. This formulation fits naturally into the framework of metric spaces and normed spaces [see (0.6.6) below and Chapter 4].

2. The Cauchy criterion obviously applies also to uniform convergence: The sequence $\{s_n(x)\}$ converges uniformly on E if and only if for each $\varepsilon > 0$, there exists an index P such that

$$|s_n(x) - s_m(x)| < \varepsilon \quad \text{for all } m, n > P \text{ and all } x \text{ in } E.$$

We shall need two fundamental theorems about uniform convergence.

Theorem 0.6.1. *If the sequence $\{s_n(x)\}$ of continuous functions converges uniformly on E, the limit function $s(x)$ is necessarily continuous.*

Theorem 0.6.2. *If $\{s_n(x)\}$ tends to $s(x)$ uniformly on the* bounded *interval E, and if s_n and s are integrable (say, in the Riemann sense) over E, then*

$$\lim_{n \to \infty} \int_E s_n(x)\, dx = \int_E s(x)\, dx. \tag{0.6.4}$$

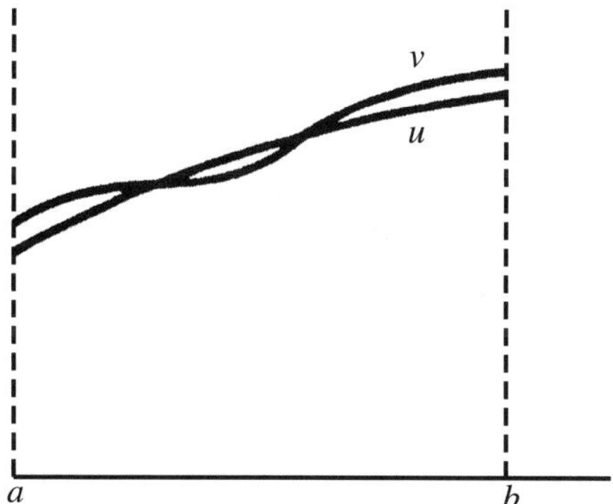

Figure 0.7

The second of these theorems is on term-by-term integration. A much more powerful theorem of this type is the Lebesgue dominated convergence theorem (see Section 0.7). In (0.6.2) the sequence $\{s_n(x)\}$ converges uniformly to 0 on E if $\alpha < 1$. Indeed, the maximum of $s_n(x)$ is $n^{\alpha-1}/e$, which tends to 0 as n approaches infinity if $\alpha < 1$. If $\alpha \geqslant 1$, we still have pointwise convergence to 0, but not uniform convergence. Obviously, $s_n(x)$ is integrable on E, and so is $s(x)$. A direct calculation shows that

$$\int_0^1 s_n(x)\,dx = n^{\alpha-2}\int_0^n xe^{-x}\,dx,$$

and since $\int_0^\infty xe^{-x}\,dx = 1$, we have

$$\lim_{n\to\infty}\int_0^1 s_n(x)\,dx = \begin{cases} 0, & \alpha < 2, \\ 1, & \alpha = 2, \\ \infty, & \alpha > 2. \end{cases} \tag{0.6.5}$$

This explicit result is predicted by Theorem 0.6.2 on uniform convergence only for $\alpha < 1$. However, the Lebesgue dominated convergence theorem will give us the result for the entire range $\alpha < 2$.

It is also worth emphasizing that the boundedness of the interval is needed in Theorem 0.6.2. Consider, for instance, a bounded function $f(x)$ on $0 \leqslant x < \infty$ such that $\int_0^\infty f(x)\,dx = 1$ (for instance, $f = xe^{-x}$ will do). In this case, the sequence $\{s_n(x) \doteq (1/n)f(x/n)\}$ then converges uniformly to 0 on $0 \leqslant x < \infty$, yet

$$\int_0^\infty s_n(x)\,dx = \int_0^\infty \frac{1}{n}f\left(\frac{x}{n}\right)\,dx = \int_0^\infty f(u)\,du = 1,$$

which clearly does not tend to 0 as $n \to \infty$.

Of the many ways of defining the convergence of a sequence of functions to a limiting function on an interval $a \leqslant x \leqslant b$, some of the most useful are based on the concept of a *norm*. A norm gives a numerical measure of the size of a function

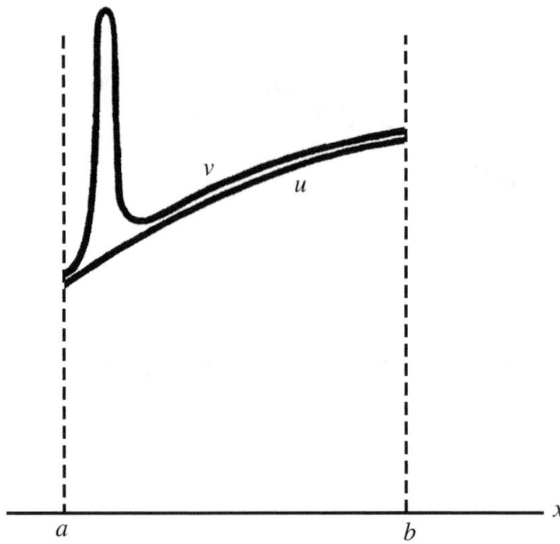

Figure 0.8

$f(x)$ and is denoted by $\|f\|$. To distinguish among different types of norm, we may need to embellish the symbol $\|\cdot\|$ by subscripts or other means of identification. A norm is required to satisfy the conditions (4.3.2). These conditions are imposed so that we can handle $\|f\|$ pretty much the way we would the length of a vector in ordinary three-dimensional space. Once our space of functions is equipped with a norm, we have a natural definition for the convergence of a sequence $s_n(x)$ to a limit $s(x)$: $\lim_{n\to\infty}\|s_n - s\| = 0$. Thus the concept of convergence for a sequence of functions is reduced to the more familiar one for convergence of a sequence of numbers.

Probably the simplest norm is the so-called *sup-norm* $\|\cdot\|_\infty$, defined by

$$\|f\|_\infty = \sup_{a\leqslant x\leqslant b} |f(x)|, \qquad (0.6.6)$$

which is easily seen to meet the requirements of a norm. To say that a sequence $s_n(x)$ converges to $s(x)$ in the sup norm means that

$$\lim_{n\to\infty}\|s_n - s\|_\infty = \lim_{n\to\infty}\sup_{a\leqslant x\leqslant b}|s_n(x) - s(x)| = 0,$$

which is precisely *uniform convergence* of $s_n(x)$ to $s(x)$ as was seen in (0.6.3) with E the interval $a \leqslant x \leqslant b$.

Another useful type of norm is the L_p norm:

$$\|f\|_p = \left[\int_a^b |f(x)|^p\, dx\right]^{1/p}, \quad p > 0. \qquad (0.6.7)$$

Of particular importance are the L_1 and L_2 norms:

$$\|f\|_1 = \int_a^b |f(x)|\,dx, \quad \|f\|_2 = \left[\int_a^b |f(x)|^2\,dx\right]^{1/2}. \tag{0.6.8}$$

In Figure 0.7 we show two functions for which $\|v-u\|_\infty$, $\|v-u\|_1$, and $\|v-u\|_2$ are all small. In Figure 0.8, $\|v-u\|_2$ and $\|v-u\|_1$ are small but $\|v-u\|_\infty$ is not.

To say that $s_n(x)$ converges to $s(x)$ in the L_1 norm (or converges in L_1) means that

$$\lim_{n\to\infty} \|s_n - s\|_1 = \lim_{n\to\infty} \int_a^b |s_n(x) - s(x)|\,dx = 0, \tag{0.6.9}$$

and convergence in the L_2 norm (or convergence in L_2) means that

$$\lim_{n\to\infty} \|s_n - s\|_2 = \lim_{n\to\infty} \left[\int_a^b |s_n(x) - s(x)|^2\,dx\right]^{1/2} = 0. \tag{0.6.10}$$

Of course, $\|s_n - s\|_2 \to 0$ if and only if $\|s_n - s\|_2^2 \to 0$, so that convergence in L_2 can also be stated without the square root:

$$\lim_{n\to\infty} \int_a^b |s_n(x) - s(x)|^2\,dx = 0. \tag{0.6.11}$$

If we consider the sequence (0.6.2) again on the same interval $0 \leqslant x \leqslant 1$ as before, we easily conclude that $s_n(x) \to 0$ in L_1 for any $\alpha < 2$ and that $s_n(x) \to 0$ in L_2 for any $\alpha < 3/2$.

The notation we have introduced for the L_p norm and the sup norm suggests that

$$\lim_{p\to\infty} \|f\|_p = \|f\|_\infty, \tag{0.6.12}$$

and this result is established in Exercise 1.2.6 for continuous functions $f(x)$ on a bounded interval.

0.7 THE LEBESGUE INTEGRAL

No one will dispute the central role of Lebesgue integration in analysis, and this beautiful theory belongs in the repertoire of every aspiring mathematician. For our purposes, however, we shall need only the few facts listed at the end of the section and enough discussion to give them meaning. We can safely omit proofs because the methods involved are not used elsewhere in the book.

The Lebesgue integral may be viewed as an extension of the Riemann integral in the sense that every Riemann-integrable function is also Lebesgue-integrable to the same value and that there exist some functions, such as the Dirichlet function described below, that fail to be Riemann-integrable but are Lebesgue-integrable. The kind of function that is Lebesgue-integrable but not Riemann-integrable rarely, if ever, occurs in practice. Thus it is not for computational reasons that the Lebesgue integral is so important. What the Lebesgue integral does is to give structural unity to analysis. From a philosophical point of view, the relationship of Lebesgue-integrable functions to Riemann-integrable functions is similar to that of real numbers to rational numbers. Concrete calculations require only rational numbers, but mathematics needs irrational numbers. The totality of real numbers (rational plus irrational) has an inner consistency absent from the class of rational numbers alone. It is the completeness (see Chapter 4) of the real number system which makes it powerful. Principally this means that when we apply limiting processes in the class of real numbers we remain within the class. Similarly, we shall find that for most concrete calculations the notion of Riemann integral is adequate, but theorems involving passage to the limit are more easily formulated and proved within the class of Lebesgue-integrable functions.

The difference between these two concepts of integration is illustrated by the following analogy, which, though not strictly apt, has some anecdotal value. A shopkeeper can determine a day's total receipts either by adding the individual transactions (Riemann) or by sorting bills and coins according to their denomination and then adding the respective contributions (Lebesgue). Obviously, the second approach is more efficient!

Consider now a nonnegative real-valued function $f(x)$ defined on the interval $0 \leqslant x \leqslant 1$. In the Riemann scheme one partitions the x interval, then forms the sum $\sum_{k=1}^{n} f(\xi_k)(x_k - x_{k-1})$ for arbitrary ξ_k in $[x_{k-1}, x_k]$, and finally passes to the limit as $n \to \infty$ and the length of the largest subdivision tends to 0. The principal difficulty is proving that the limit exists independent of the choice of ξ_k. In the Lebesgue approach it is the y axis that is partitioned (see Figure 0.9). Let E_i be the set of values of x such that $y_{i-1} \leqslant f(x) < y_i$; in the favorable case shown in the figure, E_i is the union of a finite number of disjoint intervals. We then form the sum $\sum_{i=1}^{n} \eta_i m(E_i)$, where η_i is chosen arbitrarily in $[y_{i-1}, y_i]$ and $m(E_i)$ is the measure of E_i, that is, the sum of the lengths of the disjoint intervals that make up E_i. As the partition is made finer, there is no longer any question as to the existence of the limit of the sum. Indeed, the lower sum $\sum_{i=1}^{n} y_{i-1} m(E_i)$ is monotonically increasing with n and bounded above, and so must converge. The upper sum $\sum_{i=1}^{n} y_i m(E_i)$ differs from the lower sum by less than $\max(y_i - y_{i-1}) \sum_{i=1}^{n} m(E_i)$; since $\sum_{i=1}^{n} m(E_i) = 1$ and $\max(y_i - y_{i-1})$ tends to 0, the upper sum must also converge to the same value as the lower sum. This common value is the Lebesgue integral

$$\int_0^1 f(x)\,dx.$$

In less favorable cases the set E_i may be much more complicated than shown in the figure. A consistent definition for the measure of sets must be given so that the analysis just presented can be suitably adapted.

Rather than developing the measure-oriented approach to the Lebesgue integral, we prefer to follow the method of Tonelli for constructing the Lebesgue integral of a nonnegative real-valued function $f(x)$ on the interval [0, 1]. A set E of points on this interval is said to have *measure less than* ε if E can be contained in a finite or

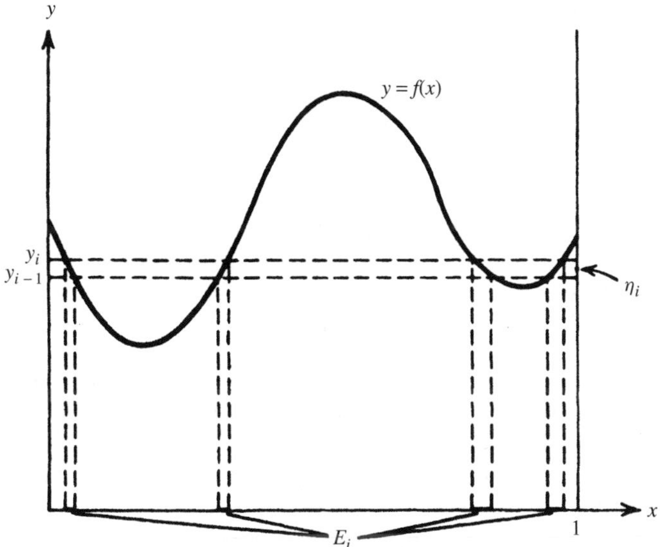

Figure 0.9

countably infinite set of intervals of total length less than ε. The set has *measure* 0 if such a covering can be found for each $\varepsilon > 0$. A function $f(x)$ is *measurable* if, for any $\varepsilon > 0$, we can convert it into a continuous function by changing its values on a set of measure less than ε. Note that f is surely measurable if it can be converted to a continuous function by altering its values on a set of measure 0.

Example 1. Suppose that $f(x)$ is piecewise continuous in [0, 1] with simple jumps at $x_1, \ldots, x_n$. In each of the intervals $[x_i - \varepsilon/2n, x_i + \varepsilon/2n]$ we can replace the function $f(x)$ by a straight line joining the point $(x_i - \varepsilon/2n, f(x_i - \varepsilon/2n))$ and the point $(x_i + \varepsilon/2n, f(x_i + \varepsilon/2n))$. The resulting function is continuous, and we have altered the values of the original function over a set of measure $n(\varepsilon/n) = \varepsilon$.

Example 2. The Dirichlet function $f(x)$ has the value 1 when x is rational and 0 when x is irrational. If we change the value of f from 1 to 0 on the set of rationals, we obtain the continuous function that vanishes identically on [0, 1]. We claim that the set of rationals has measure 0. Since the rational numbers form a countable set, they can be placed into 1-1 correspondence with the positive integers, for instance by ordering them so that they have increasing denominators:

$$0, 1, \tfrac{1}{2}, \tfrac{1}{3}, \tfrac{2}{3}, \tfrac{1}{4}, \tfrac{3}{4}, \tfrac{1}{5}, \tfrac{2}{5}, \tfrac{3}{5}, \tfrac{4}{5}, \tfrac{1}{6}, \ldots .$$

For each $\varepsilon > 0$, we enclose the kth rational number in this list in an interval of length $\varepsilon/2^k$. The total length of the intervals enclosing all rationals is then

$$\sum_{k=1}^{\infty} \frac{\varepsilon}{2^k} = \varepsilon.$$

Thus for each $\varepsilon > 0$ we can enclose the rational numbers in a countably infinite set of intervals whose total length is ε. Since this can be done for each $\varepsilon > 0$, we conclude that the set of rationals has measure 0.

Therefore, the *Dirichlet function is measurable.*

Example 3. Let $f(x) = x^{-\alpha}, 0 < x \leqslant 1$, where $\alpha > 0$; we define $f(0)$ to be 0, but any other value would do as well. Then $f(x)$ is measurable, since it can be converted into a continuous function by replacing it on $0 \leqslant x < \varepsilon$ by the constant function $\varepsilon^{-\alpha}$. Of course, there are many other ways in which such a conversion can be accomplished.

The notion of the Lebesgue integral of a nonnegative measurable function $f(x)$ can now be introduced. Let $\{\varepsilon_n\}$ be a sequence of positive numbers such that $\varepsilon_n \to 0$. For each n construct a nonnegative continuous function $f_n(x)$ which differs from $f(x)$ only on a set of measure less than ε_n. Now $f_n(x)$ is certainly Riemann-integrable on $0 \leqslant x \leqslant 1$. Let us suppose that the functions $f_n(x)$ can be constructed so that *their integrals in the Riemann sense have a common bound.* Then the functions $f_n(x)$ can always be chosen so that their integrals form a convergent sequence. Since there are many possible choices for the sequence $\{f_n\}$, the limit of the sequence of integrals is not uniquely defined by $f(x)$ but can depend on the choice of the sequence $\{f_n\}$. The greatest lower bound of the possible limits is defined as the *Lebesgue integral* of $f(x)$:

$$\int_0^1 f(x)\, dx.$$

It is easily shown that every Riemann-integrable function is Lebesgue-integrable to the same value. If $f(x)$ has an improper Riemann integral, as in Example 3 with $0 < \alpha < 1$, then $f(x)$ is Lebesgue-integrable and the values of the integrals again coincide; if we take $\alpha \geqslant 1$ in Example 3, then neither the improper Riemann integral *nor* the Lebesgue integral exists. Thus the Lebesgue approach does not miraculously reduce infinite areas to finite values. However, the Dirichlet function of Example 2 is Lebesgue-integrable to the value 0 but is not Riemann-integrable (for any partition each subdivision contains both rational and irrational numbers, so that the Riemann sum can be made either 0 or 1 by choice of ξ_k).

If $f(x)$ can take both positive and negative values, we write $f(x) = f_+(x) - f_-(x)$, where f_+ and f_- are both nonnegative functions defined by

$$f_+ = \begin{cases} f, & x \in E_+, \\ 0, & \text{elsewhere}, \end{cases} \qquad f = \begin{cases} -f, & x \in E_-, \\ 0, & \text{elsewhere}, \end{cases}$$

where E_+ and E_- are the sets on the x axis where f is positive and negative, respectively. It is then possible to define

$$\int_0^1 f\,dx = \int_0^1 f_+\,dx - \int_0^1 f_-\,dx.$$

The Lebesgue integral can also be defined for arbitrary finite intervals or for infinite intervals. The integral has the usual linearity property

$$\int_a^b [\alpha f(x) + \beta g(x)]\,dx = \alpha \int_a^b f(x)\,dx + \beta \int_a^b g(x)\,dx.$$

We have already remarked that the importance of the Lebesgue integral lies in the relative impunity with which we can use limiting processes in connection with Lebesgue integration. One of the most important theorems is the following.

Lebesgue Dominated Convergence Theorem. Let $\{s_n(x)\}$ be a sequence of integrable functions over $[a, b]$, which approaches a limit $s(x)$ pointwise except possibly over a set of measure 0. If there exists an integrable function $f(x)$ such that for all sufficiently large n, $|s_n(x)| \leqslant f(x)$, then $s(x)$ is integrable and

$$\lim_{n\to\infty} \int_a^b s_n(x)\,dx = \int_a^b s(x)\,dx.$$

Note that this theorem is much more powerful than Theorem 0.6.2 based on uniform convergence in Section 0.6. Here we need only pointwise convergence (and then only almost everywhere), the interval does not need to be bounded, and the integrability of the limit is guaranteed by the theorem instead of having to be hypothesized.

Let us apply the theorem to sequence (0.6.2). If $\alpha < e$, we can show from elementary calculus that $\alpha \log y < y$ for all $y > 0$. Setting $y = nx$, we find $n^\alpha e^{-nx} < x^{-\alpha}$ or

$$s_n(x) < x^{1-\alpha}, \quad x > 0. \tag{0.7.1}$$

Clearly, $x^{1-\alpha}$ is integrable from 0 to 1 if $\alpha < 2$; since $2 \leqslant e$, (0.7.1) also holds and therefore the Lebesgue theorem yields (0.6.5) for $\alpha < 2$. In fact, we can refine (0.7.1) to show that for $\alpha < 2$, $s_n(x) < f(x), 0 < x < \infty$, where $\int_0^\infty f(x)\,dx$ is finite. The Lebesgue theorem then tells us that

$$\lim_{n\to\infty} \int_0^\infty s_n(x)\,dx = 0, \quad \alpha < 2.$$

The other principal fact we need to know about Lebesgue integration is the completeness of $L_2(a, b)$, the space of real-valued square integrable functions on the interval $a \leqslant x \leqslant b$. Thus $u(x) \in L_2(a, b)$ if and only if

$$\int_a^b u^2(x)\,dx < \infty.$$

In L_2 we use the second of the norms (0.6.8). The significance of the completeness of L_2 will be better appreciated on reading Chapter 4.

Main Theorems on Lebesgue Integration in $\mathbb{R}^n$**.** The need for Lebesgue integration in $\mathbb{R}^n$, and over subsets $\Omega \subset \mathbb{R}^n$, will arise naturally in the study of partial differential equations in Chapter 8 and in the study of Sobolev spaces and approximation methods in Chapter 10 (see also Section 2.6, which gives a brief introduction to Sobolev spaces). Lebesgue integration in $\mathbb{R}^n$ can be defined along the lines of the discussion above, or by first developing measure theory in $\mathbb{R}^n$; classical references following this approach include Kolmogorov and Fomin [17], Royden [25], and Rudin [26]. An interesting recent approach to defining Lebesgue integration is due to Lax [18]; it completely avoids the need to define the concept of measure in order to define the Lebesgue integral. This approach involves defining L^1 directly as the completion (in the sense of metric spaces; see Chapter 4) of $C(K)$ in the L^1 norm, where K is a ball in $\mathbb{R}^n$. In this approach, measure then becomes a derived concept: A set is measurable if its characteristic function is an element of L^1. One can show that measure defined in this way has all of the usual properties developed in measure theory.

In any case, let $L^1(\Omega)$ denote the set of Lebesgue-integrable functions on $\Omega \subset \mathbb{R}^n$, with $\Omega = \mathbb{R}^n$ a possible choice. This set of functions can be given the structure of a Banach space (a complete normed vector space; see Chapter 4), with the norm on L^1 simply the multidimensional analogue of (0.6.8). The Dominated Convergence Theorem for $L^1(a,b)$ was described above; it is considered one of the three big theorems on Lebesgue integration. The multidimensional analogue for $L^1(\Omega)$ takes the following form. (The proof of this theorem, as well as proofs of the theorems to follow, may be found in any standard reference on Lebesgue integration; see, for example, the books Rudin [26], Kolmogorov and Fomin [17], and Royden [25].)

Theorem 0.7.1 (Dominated Convergence Theorem). *Let $\Omega \subset \mathbb{R}^n$ be measurable. Let $\{f_j\}$ be a sequence of measurable functions converging pointwise to a limit on Ω. If there exists $g \in L^1(\Omega)$ such that $|f_j(x)| \leqslant g(x)$, $\forall x \in \Omega$ and for every j, then*

$$\lim_{j\to\infty} \int_\Omega f_j(x)\,dx = \int_\Omega \left(\lim_{j\to\infty} f_j(x)\right) dx.$$

An important related result is the following theorem.

Theorem 0.7.2 (Monotone Convergence Theorem). *Let $\Omega \subset \mathbb{R}^n$ be measurable. Let $\{f_j\}$ be a sequence of measurable functions with $0 \leqslant f_i(x) \leqslant f_{i+1}(x)$ for every $i \geqslant 1$ and $\forall x \in \Omega$. Then*

$$\lim_{j\to\infty} \int_\Omega f_j(x)\,dx = \int_\Omega \left(\lim_{j\to\infty} f_j(x)\right) dx.$$

To understand the third of the three big theorems on Lebesgue integration, we need the concept of *limit inferior* or *lim-inf* of a sequence $\{g_j\}_{j=1}^\infty$:

$$\liminf_{j\to\infty} g_j = \lim_{j\to\infty} \left(\inf_{k\geqslant j} g_k\right).$$

The following result characterizing the relationship between the Lebesgue integral and the lim-inf of a sequence is fundamentally important to the study of variational problems in partial differential equations, among other applications.

Theorem 0.7.3 (Fatou's Lemma). *Let $\Omega \subset \mathbb{R}^n$ be measurable. Let $\{f_j\}$ be a sequence of measurable functions. Then*

$$\int_\Omega \left(\liminf_{j\to\infty} f_j(x) \right) dx \leqslant \liminf_{j\to\infty} \int_\Omega f_j(x)\, dx.$$

The final result on Legesgue integration in $\mathbb{R}^n$ worth mentioning is the following. At times, we will need to freely interchange the order of integration in a multidimensional integral (which of course has no analogue in one dimension). This is justified by

Theorem 0.7.4 (Fubini's Theorem). *Let f be a measurable function on $\mathbb{R}^{m+n}$, and let one or more of the following exist and be finite:*

$$I_1 = \int_{\mathbb{R}^{m+n}} |f(x,y)|\, dxdy, \tag{0.7.2}$$

$$I_2 = \int_{\mathbb{R}^m} \left(\int_{\mathbb{R}^n} |f(x,y)|\, dx \right) dy, \tag{0.7.3}$$

$$I_3 = \int_{\mathbb{R}^n} \left(\int_{\mathbb{R}^m} |f(x,y)|\, dy \right) dx, \tag{0.7.4}$$

with the integrands for the outer integrals in I_2 and I_3 understood in the sense of almost everywhere. Then

(1) $f(\cdot, y) \in L^1(\mathbb{R}^n)$ for almost all $y \in \mathbb{R}^n$.
(2) $f(x, \cdot) \in L^1(\mathbb{R}^m)$ for almost all $x \in \mathbb{R}^m$.
(3) $\int_{\mathbb{R}^n} f(\cdot, y)\, dy \in L^1(\mathbb{R}^n)$.
(4) $\int_{\mathbb{R}^m} f(x, \cdot)\, dx \in L^1(\mathbb{R}^m)$.
(5) $I_1 = I_2 = I_3$.

REFERENCES AND ADDITIONAL READING

1. S. S. Antman, *Nonlinear Problems of Elasticity*, Springer–Verlag, New York, 1995.
2. S. S. Antman and J. E. Osborn, The principle of virtual work and integral laws of motion, *Arch. Rational Mech. Anal.* 69, 1979.
3. R. Aris, *The Mathematical Theory of Diffusion and Reaction in Permeable Catalysts*, Vols. 1 and 2, Oxford University Press, New York, 1975.
4. R. Aris, *Mathematical Modelling Techniques*, Pitman, New York, 1978.
5. D. G. Aronson, The porous medium equation, in *Some Problems in Nonlinear Diffusion* (A. Fasano and M. Primicerio, Eds.), Springer–Verlag, New York, 1986.

6. D. M. Bressoud, *A Radical Approach to Lebesgue's Theory of Integration*, Cambridge University Press, New York, 2008.

7. J. D. Buckmaster and G. S. S. Ludford, *Lectures on Mathematical Combusion*, SIAM, Philadelphia, 1983.

8. W. A. Day, On rates of propagation of heat according to Fourier's theory, *Quart. Appl. Math.*, 55(1), 1997.

9. M. M. Denn, *Process Modeling*, Pitman, New York, 1986.

10. P. G. Drazin and D. H. Griffel, Free boundary problems in climatology, in *Free Boundary Problems: Theory and Application*, Vol. II (A. Fasano and M. Primicerio, Eds.), Pitman, New York, 1983.

11. A. Friedman and W. Littman, *Problems in Industrial Mathematics*, SIAM, Philadelphia, 1993.

12. G. F. Froment and K. B. Bischoff, *Chemical Reactor Analysis and Design*, Wiley, New York, 1990.

13. S. Hassani, *Mathematical Physics*, Springer–Verlag, New York, 2006.

14. U. Hornung and W. Jaeger, Homogenization of reactive transport through porous media, in *International Conference in Differential Equations* (C. Perello et al., Eds.), World Scientific, Hackensack, NJ, 1993.

15. C. Johnson, *Numerical Solutions of Partial Differential Equations by the Finite Element Method*, Cambridge University Press, New York, 1993.

16. J. Keener, *Principles of Applied Mathematics*, 2nd ed., Westview Press, Boulder, CO, 2000.

17. A. N. Kolmogorov and S. V. Fomin, *Introductory Real Analysis*, Dover, New York, 1970.

18. P. D. Lax, *Rethinking the Lebesgue integral, Ameri. Math. Monthly*, Dec. 2009.

19. C. C. Lin and L. A. Segel, *Mathematics Applied to Deterministic Problems in the Natural Sciences*, SIAM, Philadelphia, 1988.

20. J. D. Logan, *Applied Mathematics*, 2nd ed., Wiley, New York, 1997.

21. J. E. Marsden and T. J. R. Hughes, *Mathematical Foundations of Elasticity*, Prentice–Hall, Englewood Cliffs, NJ, 1983.

22. J. D. Murray, *Mathematical Biology*, Springer–Verlag, New York, 1989.

23. G. R. North, Introduction to simple climate models, in *Mathematics, Climate and Environment* (J.-I. Diaz and J.-L. Lions, Eds.), Masson, Paris, 1993.

24. L. A. Peletier, The porous medium equation, in *Applications of Nonlinear Analysis in the Physical Sciences* (H. Amann et al., Eds.), Pitman, New York, 1981.

25. H. L. Royden, *Real Analysis*, Macmillan, New York, 1968.

26. W. Rudin, *Real and Complex Analysis*, McGraw-Hill, New York, 1987.

27. M. H. Santare and M. J. Chajes, *The Mechanics of Solids: History and Evolution*, University of Delaware Press, Cranberry, NJ, 2008.

28. L. Segel, *Mathematics Applied to Continuum Mechanics*, Macmillan, New York, 1977.

29. G. E. Shilov, *Linear Spaces*, Prentice-Hall, Englewood Cliffs, NJ, 1961.

30. I. Stakgold, Free boundary problems in climate modeling, in *Mathematics, Climate and Environment* (J.-I. Diaz and J.-L. Lions, Eds.), Masson, Paris, 1993.

31. P. Szekeres, *A Course in Modern Mathematical Physics*, Cambridge University Press, New York, 2004.

32. A. Tayler, *Mathematical Models in Applied Mechanics*, Oxford University Press, New York, 1986.

33. J. L. Vazquez, Mathematical theory of the porous medium equation, in *Shape Optimization and Free Boundaries* (M. C. Delfour, Ed.), Kluwer, Hingham, MA, 1992.

CHAPTER 1

GREEN'S FUNCTIONS (INTUITIVE IDEAS)

1.1 INTRODUCTION AND GENERAL COMMENTS

For the limited purpose of this section we look mainly at steady heat flow in a homogeneous medium. Consider first the one-dimensional problem of a thin rod occupying the interval (0, a) on the x axis. Using the nondimensional variables of (0.1.18), and replacing v by u, we obtain

$$-\frac{d^2u}{dx^2} = f(x), \quad 0 < x < 1; \quad u(0) = \alpha, \quad u(1) = \beta, \tag{1.1.1}$$

where $f(x)$ is the prescribed source density (per unit *length* of the rod) of heat and α, β are the prescribed end temperatures. The three quantities $\{f(x); \alpha, \beta\}$ are known collectively as the *data* for the problem. The data consists of the *boundary data* α, β and of the *forcing function* $f(x)$.

We shall be concerned not only with solving (1.1.1) for specific data but also with finding a suitable form for the solution that will exhibit its dependence on the data. Thus as we change the data our expression for the solution should remain useful. The feature of (1.1.1) that enables us to achieve this goal is its linearity, as reflected in the *superposition principle*: If $u_1(x)$ is a solution for the data $\{f_1(x); \alpha_1, \beta_1\}$ and

$u_2(x)$ for the data $\{f_2(x); \alpha_2, \beta_2\}$, then $Au_1(x) + Bu_2(x)$ is a solution for the data $\{Af_1(x)+Bf_2(x);\ A\alpha_1+B\alpha_2, A\beta_1+B\beta_2\}$. One can extend this principle in an obvious manner to n solutions corresponding to n sets of data. Under mild restrictions it is even possible to extend the superposition principle to data $\{f(x,\theta); \alpha(\theta), \beta(\theta)\}$ depending on a continuous parameter θ with superposition becoming an integral over θ. In practice, the superposition principle permits us to decompose complicated data into possibly simpler parts, to solve each of the simpler boundary value problems, and then to reassemble these solutions to find the solution of the original problem. One decomposition of the data which is often used is

$$\{f(x); \alpha, \beta\} = \{f(x); 0, 0\} + \{0; \alpha, \beta\}.$$

The problem with data $\{f(x); 0, 0\}$ is an *inhomogeneous equation* with *homogeneous* boundary conditions; the problem with data $\{0; \alpha, \beta\}$ is a *homogeneous equation* with *inhomogeneous* boundary conditions. It should be noted that data $\{0; \alpha, \beta\}$ is itself often split up into $\{0; \alpha, 0\}$ and $\{0; 0, \beta\}$, each of which involves one inhomogeneous and one homogeneous boundary condition.

Later in this section [equation (1.1.12)] and again in Section 1.2 [equations (1.2.9) and (1.2.10)] we show how the superposition principle or other methods lead to the following form for the solution of (1.1.1):

$$u(x) = \int_0^1 g(x,\xi)f(\xi)\,d\xi + (1-x)\alpha + x\beta, \tag{1.1.2}$$

where *Green's function* $g(x, \xi)$ is a function of the real variables x and ξ defined on the square $0 \leqslant x, \xi \leqslant 1$ and is given explicitly by

$$g(x,\xi) = x_<(1-x_>) = \begin{cases} x(1-\xi), & 0 < x < \xi, \\ \xi(1-x), & \xi < x < 1. \end{cases} \tag{1.1.3}$$

Here $x_<$ stands for the lesser of the two quantities x and ξ, and $x_>$ for the greater of x and ξ. We can also write

$$g(x,\xi) = \begin{cases} l(x,\xi), & 0 < x < \xi, \\ r(x,\xi), & \xi < x < 1, \end{cases} \tag{1.1.3a}$$

where $l(x, \xi) = x(1 - \xi)$ is the formula for g when x is to the left of ξ and $r(x, \xi) = \xi(1 - x)$ is the one when x is to the right of ξ. Note that l and r are infinitely differentiable for all x and ξ; it is only by piecing l and r together that we introduce a function g which, though still continuous at $x = \xi$, has a discontinuous first derivative there. We also observe that $l(x, \xi) = r(\xi, x)$, so that $g(x, \xi)$ is itself *symmetric*: $g(x, \xi) = g(\xi, x)$.

Since g does not depend on the data, it is clear that (1.1.2) expresses in a very simple manner the dependence of u on the data $\{f; \alpha, \beta\}$. Symbolically, we can write (1.1.2) as

$$u(x) = F(f, \alpha, \beta),$$

where F is a linear operator transforming the data into the solution. We may regard F as the inverse operator for (1.1.1). If $\alpha = \beta = 0$, this inverse operator is an integral operator of the type studied in Chapters 5 and 6.

For specific f the integration in (1.1.2) can sometimes be performed in closed terms, using elementary integration techniques. One must, however, divide the interval of integration into two parts to take advantage of the simple formulas for g. Since the integration in (1.1.2) is over ξ, we write $\int_0^1 = \int_0^x + \int_x^1$, where in the interval from 0 to x we have $\xi < x$, so that we must use the second line of (1.1.3), whereas in the interval from x to 1 the first line of (1.1.3) applies. The integral term in (1.1.2) therefore becomes

$$(1-x)\int_0^x \xi f(\xi)\,d\xi + x\int_x^1 (1-\xi)f(\xi)\,d\xi.$$

Turning to the three-dimensional problem of heat conduction in a homogeneous medium of unit thermal conductivity, occupying the domain Ω with boundary Γ, we know from (0.1.11) that the steady temperature $u(x)$ satisfies

$$-\Delta u = f(x), \quad x \in \Omega; \quad u = h(x), \quad x \in \Gamma. \tag{1.1.4}$$

Here $x = (x_1, x_2, x_3)$ is a position *vector* in three-dimensional space. The source density per unit volume $f(x)$ is given for $x \in \Omega$, whereas the boundary temperature $h(x)$ is given for x on the surface Γ.

Note that we are no longer using any distinguishing notation for vectors. The context should make it clear whether a quantity is a vector or a scalar. The differential operator Δ appearing in (1.1.4) is the *Laplacian*, which, in *Cartesian coordinates*, takes the form $\partial^2/\partial x_1^2 + \partial^2/\partial x_2^2 + \partial^2/\partial x_3^2$, whereas in other coordinate systems it will look quite different. One of the advantages of the notation of (1.1.4) is that it does not commit us to a particular coordinate system.

In any event the solution of (1.1.4) can be written in terms of Green's function $g(x,\xi)$ (which is now a function of the six real variables $x_1, x_2, x_3, \xi_1, \xi_2, \xi_3$):

$$u(x) = \int_\Omega g(x,\xi)f(\xi)\,d\xi - \int_\Gamma \frac{\partial g}{\partial n}h(\xi)\,dS_\xi, \tag{1.1.5}$$

where $d\xi$ is an element of volume integration ($= d\xi_1\,d\xi_2\,d\xi_3$ if Cartesian coordinates are used), dS_ξ is an element of surface integration at the point ξ on Γ, and $\partial/\partial n$ denotes differentiation with respect to ξ in the outward normal direction on Γ.

Thus (1.1.5) expresses the solution of (1.1.4) in terms of the data $\{f(x); h(x)\}$ with f the forcing function and h the boundary data; again we see that the superposition principle holds. It remains only to confess that the function $g(x,\xi)$ appearing in (1.1.5) is usually not known explicitly (unless the domain Ω is of a very simple type, such as a ball or parallelepiped); nevertheless, one can obtain a great deal of useful information about $g(x,\xi)$. First we point out that $g(x,\xi)$ has a very simple physical interpretation as the temperature at x when the only source is a *concentrated unit source* located at ξ, the boundary being kept at 0 temperature. One can also characterize $g(x, \xi)$ mathematically as the solution of a well-defined boundary value problem; this formulation requires a little delicacy, however, and we shall take up the question in some of the succeeding sections.

The reader may have noticed that in (1.1.1) the differential equation was formulated on the open interval $0 < x < 1$ rather than on the closed interval $0 \leqslant x \leqslant 1$; similarly in (1.1.4) the differential equation held on a *domain* Ω, which, by definition, is an open, connected set (see Section 0.1). Why do we insist that Ω be an open set? The reason is to avoid discussing the differential equation on the boundary. Take (1.1.1), for instance; if we required the differential equation to hold at $x = 1$, we would either have to extend the function u for $x > 1$ (to be able to form the difference quotient at $x = 1$), or would have to use the concept of a one-sided derivative at $x = 1$. For a higher-dimensional problem such as (1.1.4), it is even more awkward to try to use the differential equation on the boundary since this would necessarily require some smoothness for the boundary Γ and the boundary data $h(x)$.

We shall therefore always formulate the differential equation on a domain Ω (*open, connected set*).

How do we relate the boundary values of $u(x)$ to its interior values? The boundary values of u are given, whereas the interior values are obtained by solving a differential equation with its attendant indeterminacy. To see that some clarification is needed, consider (1.1.1) when $f(x) \equiv 0, 0 < x < 1$, and $\alpha = \beta = 0$. We clearly, would like the solution $u(x)$ to be identically 0; we want to rule out ridiculous candidates such as

$$v(x) = \begin{cases} 1, & 0 < x < 1, \\ 0, & x = 0, x = 1. \end{cases}$$

This function $v(x)$ satisfies the differential equation $-d^2v/dx^2 = 0, 0 < x < 1$, and clearly $v(0) = v(1) = 0$; yet $v(x)$ is a spurious solution. We can reject v by requiring that the solution $u(x)$ of (1.1.1) be continuous in the *closed interval* $0 \leqslant x \leqslant 1$ or, equivalently, by requiring that $\lim_{x\to 0+} u(x) = \alpha, \lim_{x\to 1-} u(x) = \beta$.

Similarly, in (1.1.4) *we shall require that the solution* $u(x)$ *be continuous in the closed region* $\bar{\Omega} = \Omega + \Gamma$. [It is, of course, understood that the given boundary data $h(x)$ constitutes a continuous function of position on Γ.]

So far we have said nothing about how to decide whether or not a function $u(x)$ satisfies the differential equation $-u'' = f(x)$ in (1.1.1). At first glance there seems to be little to say: One merely makes sure that $u(x)$ is twice differentiable in $0 < x < 1$ (which implies that u is continuous and has a continuous first derivative) and that the function $-u''(x)$ coincides with the given function $f(x)$ over the entire interval $0 < x < 1$ [in other words, for each x in $0 < x < 1$, the numbers $-u''(x)$ and $f(x)$ should be the same]. This works splendidly if $f(x)$ is continuous, but there are good reasons, both mathematical and physical, for considering forcing functions $f(x)$ that are only piecewise continuous. For instance, one can easily envisage a situation in which the prescribed source density $f(x)$ is a nonzero constant, say 1, in $0 < x < \xi$ and is 0 in $\xi < x < 1$. Note that $f(x)$ is discontinuous at the point $x = \xi$. One often hears the argument that such functions are inadmissible on physical grounds, that the "real" source density is continuous and merely decreases quickly from the value 1 to 0 in a small neighborhood of the point $x = \xi$. Such philosophical arguments are immaterial; all we care about is that the temperature calculated on the basis of the discontinuous source density should be nearly the same (in some suitable sense) as that calculated on the basis of the continuous density (see Exercise 1.2.1, for instance).

We shall return to this question in due time, but now let us try to incorporate piecewise continuous forcing functions into our framework at the cost of slightly reinterpreting the meaning of the differential equation $-u'' = f$. We still want u' to be an integral of f, and of course integrals of piecewise continuous functions are well defined and are necessarily *continuous*; the continuity of u' implies that u is continuous. The new feature is that u'' no longer exists at the points where f has jumps. Let x_0 be such a point, and let us try to calculate $u''(x_0)$ by forming the difference quotient for u':

$$\frac{u'(x_0 + \Delta x) - u'(x_0)}{\Delta x} = \frac{-\int_{x_0}^{x_0+\Delta x} f(x)\,dx}{\Delta x},$$

whose approximate value is $-f(x_0+)$ for $\Delta x > 0$ and $-f(x_0-)$ for $\Delta x < 0$. Thus $u''(x_0)$ cannot exist, no matter how we try to adjust the value of f at x_0, as long as $f(x_0+)$ and $f(x_0-)$ are different. Of course, at points where $f(x)$ is continuous we still require that $u''(x)$ exist and satisfy $-u''(x) = f(x)$. We can easily generalize these ideas to an arbitrary linear differential equation of order p:

$$\begin{aligned} a_p(x)u^{(p)}(x) + a_{p-1}(x)u^{(p-1)}(x) + \cdots + a_1(x)u'(x) \\ + a_0(x)u(x) = f(x), \quad a < x < b. \end{aligned} \tag{1.1.6}$$

Definition. *Let $f(x)$ be piecewise continuous, and let $a_0(x), \ldots, a_p(x)$ be continuous. A* classical solution *of (1.1.6) is a function $u(x)$ belonging to $C^{p-1}(a,b)$—the class of functions with continuous derivatives of order $p-1$ on $a < x < b$—such that at all points of continuity of f, $u^{(p)}(x)$ exists and satisfies the differential equation (1.1.6).*

REMARK. By using the notion of weak solution (see Section 2.5), we can give a reasonable interpretation of (1.1.6) even when f is only integrable. This idea applies also to partial differential equations, where difficulties can arise even if f is continuous.

Let us now solve (1.1.1) for the very simple piecewise continuous, forcing function

$$f(x,a) = H(x-a) = \begin{cases} 0, & 0 < x < a, \\ 1, & a < x < 1, \end{cases} \tag{1.1.7}$$

where $H(x)$ is the usual Heaviside function, which vanishes for $x < 0$ and is equal to 1 for $x > 0$ (its value at $x = 0$ plays no role in the analysis). In (1.1.7) x is the primary variable and a is a parameter. We first solve (1.1.1) when $\alpha = \beta = 0$, that is, for data $\{H(x-a); 0, 0\}$. The solution will be denoted by $u(x,a)$, since it depends not only on x but also on the parameter a. In $0 < x < a$ we have $-d^2u/dx^2 = 0$, whereas in $a < x < 1$ we have $-d^2u/dx^2 = 1$. Integration and use of the boundary conditions gives

$$u = Ax \quad \text{in } (0, a) \qquad \text{and} \qquad u = -\frac{(x-1)^2}{2} + B(1-x) \quad \text{in } (a, 1),$$

where A and B may depend on a but not on x. For u to be a classical solution we must require that u and u' be continuous at $x = a$ (we already have more than enough smoothness in the subintervals $0 < x < a$ and $a < x < 1$). This gives $A = (1-a)^2/2$ and $B = (1-a^2)/2$, so that

$$u(x, a) = \begin{cases} \dfrac{(a-1)^2}{2}x, & 0 < x < a, \\ -\dfrac{(x-1)^2}{2} + \dfrac{1-a^2}{2}(1-x), & a < x < 1, \end{cases} \tag{1.1.8}$$

which is plotted in Figure 1.1. Equation (1.1.17) shows the relation between $u(x, a)$ and $g(x, a)$.

It is of interest to present another approach to (1.1.1) with data $\{f(x); 0, 0\}$, which lends itself to graphical analysis. This method is based on interpreting the problem as the transverse deflection of a taut string with fixed ends. The static version of (0.4.36), with $T^{(0)} = 1, l = 1, X = x, \mathbf{f} \cdot \mathbf{j} = f(x)$, and u instead of v for the transverse deflection, gives us (1.1.1) with $\alpha = \beta = 0$. It then follows that the vertical component of the tension at a point $(x, u(x))$ along the string is just $u'(x)$; thus the reactions at the ends $x = 0$ and $x = 1$ are $-u'(0)$ and $u'(1)$, respectively. By taking moments about the ends of the string, we find that

$$u'(1) + \int_0^1 xf(x)\,dx = 0, \quad -u'(0) + \int_0^1 (1-x)f(x)\,dx = 0, \tag{1.1.9}$$

which could also be derived without recourse to the physical interpretation by multiplying the differential equation in (1.1.1) by x and $1 - x$, respectively, and then integrating from 0 to 1. In any event we have calculated the reactions at the ends and can now find $u'(x)$ at any point from

$$u'(x) = u'(0) - \int_0^x f(\xi)\,d\xi, \tag{1.1.10}$$

which can, of course, be done graphically. Since $u(0) = 0$, we can find $u(x)$ from (1.1.10) by integrating from 0 to x. This again is easy to do graphically; analytically, we find that

$$u(x) = \int_0^x u'(\eta)\,d\eta = u'(0)x - \int_0^x d\eta \int_0^\eta f(\xi)\,d\xi. \tag{1.1.11}$$

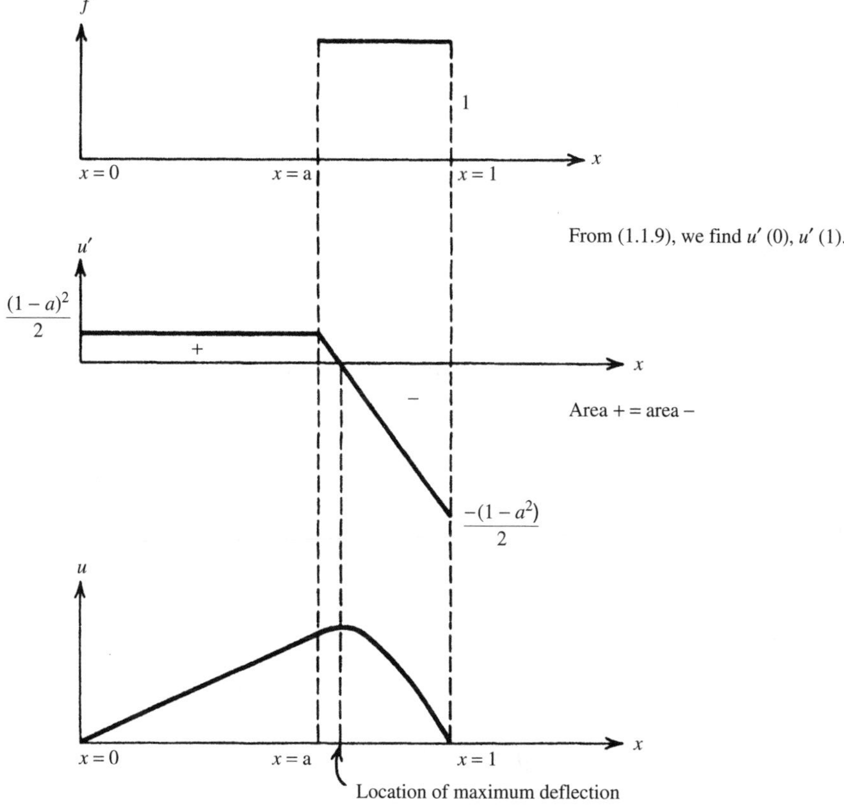

Figure 1.1

The iterated integral can be viewed as a double integral over a triangular region in the ξ–η plane; on changing the order of integration, we obtain

$$\begin{aligned} u(x) &= u'(0)x - \int_0^x (x-\xi)f(\xi)\,d\xi \\ &= x\int_0^1 (1-\xi)f(\xi)\,d\xi - \int_0^x (x-\xi)f(\xi)\,d\xi \\ &= \int_0^1 g(x,\xi)f(\xi)\,d\xi, \end{aligned} \tag{1.1.12}$$

where $g(x, \xi)$ is just Green's function as predicted in (1.1.3). It is then an easy matter to show (1.1.2) holds when the data is $\{f; \alpha, \beta\}$ instead of $\{f; 0, 0\}$. In Figure 1.1 we have illustrated the graphical integration when the data is $\{H(x-a); 0, 0\}$, the corresponding formula for the deflection being (1.1.8).

EXERCISES

1.1.1 Consider the transverse deflection $u(x)$ of a string satisfying

$$-u'' = f(x), \quad 0 < x < 1; \qquad u(0) = 0, \quad u(1) = 0,$$

where

$$f(x) = \begin{cases} x - \frac{1}{4}, & 0 < x < \frac{1}{2}, \\ \frac{1}{4}, & \frac{1}{2} < x < 1. \end{cases}$$

(a) Find u' at one of the ends, and then carry out graphically two successive integrations to obtain the deflection $u(x)$.

(b) Find $u(x)$ using (1.1.2) and (1.1.3). To perform the integration explicitly you must divide the interval into $(0, x)$ and $(x, 1)$; in the first subinterval, x is larger than ξ, so that the second line of (1.1.3) applies. You will then need a further subdivision to handle our specific forcing function. Compare your result with that for part (a).

1.1.2 The small transverse deflection $u(x)$ of a homogeneous beam of unit length subject to a distributed transverse loading $f(x)$ satisfies

$$\frac{d^4u}{dx^4} = f(x), \quad 0 < x < 1, \tag{1.1.13}$$

where we have set $EI = 1$ in (0.4.37). For a beam simply supported at its ends the boundary conditions are

$$u(0) = u''(0) = u(1) = u''(1) = 0. \tag{1.1.14}$$

The shear force V and moment M at a cross section satisfy

$$V = -u'''(x), \qquad M = u''(x),$$

where the choice of signs is in accord with the convention used in Section 0.4. For (1.1.13) subject to (1.1.14), show how to calculate $u'''(0)$. It is therefore straightforward to find $V(x)$ and $M(x)$ by graphical integration. Once $M(x)$ is known, it is easy to calculate $u'(0)$ and hence to proceed in determining $u'(x)$ and $u(x)$ graphically. This problem is *statically determinate* because the reactions at the ends can be found on the basis of statics alone. This would not be the case for a beam clamped at the ends ($u = u' = 0$).

1.1.3 Consider the boundary value problem

$$-\frac{d}{dx}\left[k(x)\frac{du}{dx}\right] = f(x),\ 0 < x < 1; \quad u(0) = u(1) = 0, \tag{1.1.15}$$

where $k(x) > 0$ in $0 \leqslant x \leqslant 1$. Let $K(x)$ be a solution of the homogeneous equation satisfying the boundary condition at $x = 1$. Show how one can calculate $u'(0)$ by multiplying both sides of the differential equation in (1.1.15) by $K(x)$ and integrating from $x = 0$ to $x = 1$. Construct Green's function for the problem.

1.1.4 (a) Consider the transverse deflection of a taut string subject to a concentrated unit load at $x = a$. Proceed as in the last paragraphs of Section 1.1 to find the reactions at the ends. Draw the slope and deflection diagrams. Verify that the deflection is just $g(x, a)$ as given by (1.1.3).

(b) If the notion of a concentrated load is to be useful, the corresponding deflection $g(x, a)$ must be "close" to the one due to a load which approximates the concentrated load. The piecewise continuous function

$$p_\varepsilon(x, a) = \begin{cases} \dfrac{1}{\varepsilon}, & |x - a| < \dfrac{\varepsilon}{2}, \\ 0, & \text{otherwise} \end{cases} \tag{1.1.16}$$

can be regarded, for small positive ε, as a good piecewise continuous approximation to a unit load concentrated at $x = a$. Use both graphical and analytical means to show that the deflection $u_\varepsilon(x, a)$ corresponding to $p_\varepsilon(x, a)$ has the property

$$\lim_{\varepsilon \to 0+} u_\varepsilon(x, a) = g(x, a) \quad \text{uniformly on } 0 \leqslant x \leqslant 1.$$

Note that $u_\varepsilon(x, a)$ can also be expressed as

$$\frac{1}{\varepsilon}\left[u\left(x, a - \frac{\varepsilon}{2}\right) - u\left(x, a + \frac{\varepsilon}{2}\right)\right],$$

where $u(x, a)$ is given by (1.1.8). It therefore follows that

$$g(x, a) = -\frac{d}{da}u(x, a). \tag{1.1.17}$$

The solution of (1.1.1) with data $\{f(x); 0, 0\}$ is given by (1.1.12) or equivalently by

$$u(x) = -\int_0^1 \frac{d}{d\xi}u(x, \xi) f(\xi)\, d\xi.$$

1.1.5 This is a more ambitious attempt than Exercise 1.1.4b to show the closeness of the deflections caused by a unit load concentrated at ξ and by a distributed approximation to it. Choose N so that, for $n \geqslant N$, the interval $(\xi - 1/n, \xi + 1/n)$ belongs to [0, 1]. Let $\{f_n(x)\}$ be a sequence of nonnegative continuous functions with $f_n = 0$ for $|x - \xi| \geqslant 1/n$ and

$$\int_{\xi-1/n}^{\xi+1/n} f_n(x)\, dx = 1.$$

The deflection corresponding to $f_n(x), n \geqslant N$, is called $u_n(x)$. Draw a graph of $u_n'(x)$ for a large value of n. Show that C_n, the constant slope in $0 < x < \xi - 1/n$ approaches $1 - \xi$ as $n \to \infty$. Next, graph $u_n(x)$ and prove that

$$\lim_{n\to\infty} u_n(x) = g(x,\xi) \quad \text{uniformly on } 0 \leqslant x \leqslant 1.$$

1.2 THE FINITE ROD

Construction of Green's Function

We return to the heat conduction problem (1.1.1), repeated below for convenience:

$$-\frac{d^2u}{dx^2} = f(x), \quad 0 < x < 1; \qquad u(0) = \alpha, \quad u(1) = \beta. \tag{1.2.1}$$

We want to solve the problem as compactly as possible for arbitrary data $\{f; \alpha, \beta\}$. The differential operator and the boundary operators appearing on the left sides of the equality signs in (1.2.1) are kept fixed; no one is proposing to solve all differential equations with arbitrary boundary conditions at one stroke!

To solve (1.2.1) for arbitrary data, we introduce an accessory problem where instead of a distributed density of sources, there is only a concentrated source of unit strength at $x = \xi$ and where the *boundary data vanishes* (which means in our case that the temperature is 0 at both ends). Physically, this accessory problem makes sense, and the resulting steady temperature should be well defined; moreover, it is clear that the temperature cannot vanish identically, since there is a steady nonzero heat input from the source. This temperature (solution of the accessory problem) is known as Green's function and is denoted by $g(x, \xi)$. Here ξ is the position of the source, and x is the observation point. We usually regard ξ as a parameter and x as the running variable; but when we are all through we have a function of two real variables, and we are at liberty to forget the original significance of x and ξ. *In any event all differentiations below are with respect to the first variable in g.* Let us see whether we can construct g on the basis of the information available so far. Since there are no sources in $0 < x < \xi$ and in $\xi < x < 1$, we have $-g'' = 0$ in both intervals. Taking into account the fact that g vanishes at $x = 0$ and $x = 1$, we find that

$$g = Ax, \ 0 < x < \xi; \quad g = B(1-x), \ \xi < x < 1. \tag{1.2.2}$$

Here A and B are "constants," that is, independent of x; they may, however, depend on the parameter ξ. If at this stage we demanded the continuity of g and g' at $x = \xi$, we would find $A = B = 0$, so that g would vanish identically—which is nonsense! The requirement that g' be continuous at $x = \xi$ must be *dropped*, although we shall still insist on the continuity of g. The jump of g' at $x = \xi$ is easily calculated if we recall the primary integral formulation of the problem of heat conduction in terms of energy balance [see (0.1.1) and (0.5.2)]. Consider a thin slice of the rod from $\xi - \varepsilon$ to $\xi + \varepsilon$. The one-dimensional character of the problem means that no heat flows

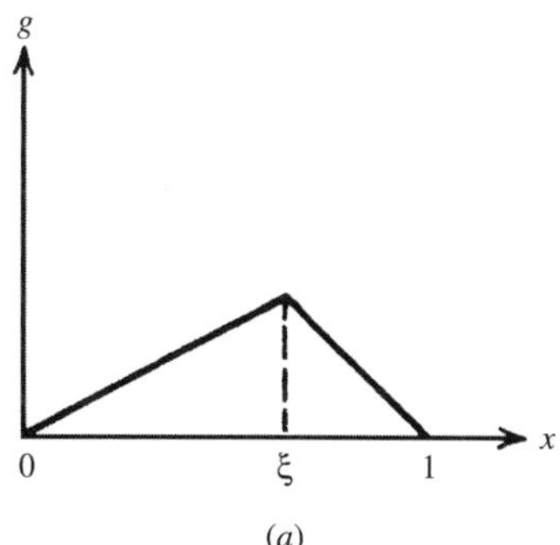

(a)

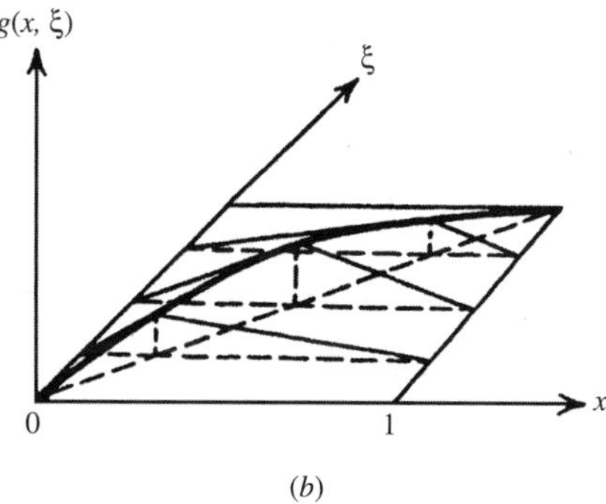

(b)

Figure 1.2

through the lateral surface; the integral balance law then gives

$$-g'|_{x=\xi+\varepsilon} + g'|_{x=\xi-\varepsilon} = 1,$$

which, as ε tends to 0, leads to the *jump condition* for g':

$$g'|_{x=\xi+} - g'|_{x=\xi-} = -1 \quad \text{or} \quad r'(\xi,\xi) - l'(\xi,\xi) = -1. \tag{1.2.3}$$

Condition (1.2.3) and the continuity of g at $x = \xi$ enable us to calculate A and B in (1.2.2) from the simultaneous equations $A = B(1-\xi)$ and $-B - A = -1$. Thus $B = \xi$ and $A = 1 - \xi$, so that

$$g(x,\xi) = \begin{cases} (1-\xi)x, & 0 \leqslant x < \xi, \\ (1-x)\xi, & \xi < x \leqslant 1, \end{cases} \tag{1.2.4}$$

confirming (1.1.3). In Figure 1.2*a* we picture Green's function as a function of x for fixed ξ, and in Figure 1.2*b* as a function of x and ξ. Thus Figure 1.2*a* can be viewed as a cross section of the surface in Figure 1.2*b*.

We have therefore characterized Green's function $g(x,\xi)$ both physically and mathematically. Before proceeding to another characterization, based on the delta function, let us recapitulate what has been done so far.

1. *Physical description.* We chose to describe g in terms of heat conduction in a rod: $g(x,\xi)$ is the temperature at x when the only source is a unit concentrated source at ξ, the ends being at 0 temperature. It is also possible to interpret g as the transverse deflection of a string: $g(x,\xi)$ is the deflection at x when the only load is a unit concentrated force at ξ, the ends being kept fixed on the x axis at $x = 0$ and $x = 1$.

2. *Classical mathematical formulation.* Green's function $g(x, \xi)$ associated with (1.2.1) satisfies

$$\left.\begin{aligned} &-\frac{d^2g}{dx^2} = 0, \quad 0 < x < \xi, \quad \xi < x < 1; \\ &g(0, \xi) = g(1, \xi) = 0; \\ &g \text{ continuous at } x = \xi; \qquad \left.\frac{dg}{dx}\right|_{x=\xi+} - \left.\frac{dg}{dx}\right|_{x=\xi-} = -1. \end{aligned}\right\} \tag{1.2.5}$$

In our third formulation we would like to consider (1.2.5) as a boundary value problem of the form (1.2.1) with specific data. The boundary data for g clearly vanishes, but what is the forcing function? In (1.2.1) the forcing function is a source density (per unit length) rather than the concentrated source of the Green's function problem. How can we describe a concentrated source at the point ξ as a density? This is easy to do symbolically but is not so easy within a consistent mathematical framework. Suppose that we let $\delta(x)$ be the density corresponding to a concentrated source at $x = 0$. We would then need

$$\int_a^b \delta(x)\, dx = \begin{cases} 1, & \text{if}(a, b) \text{ contains the origin,} \\ 0, & \text{otherwise.} \end{cases}$$

Unfortunately, no integrable function satisfies these properties. Nevertheless, we shall use $\delta(x)$ symbolically to represent the source density corresponding to a unit source at the origin. This symbolic function is known as the *Dirac delta function*; $\delta(x - \xi)$ is $\delta(x)$ translated ξ units to the right and so must be the source density for a unit source at $x = \xi$. Perhaps the most natural way to view $\delta(x)$ is as the limit of a sequence of narrow, uniform densities of large magnitude (with total strength unity) such as

$$f_n(x) = \begin{cases} n, & |x| < \dfrac{1}{2n}, \\[2ex] 0, & |x| > \dfrac{1}{2n}. \end{cases}$$

Thus we may think of $\delta(x)$ as the limit as $n \to \infty$ of $f_n(x)$, and $\delta(x - \xi)$ as the limit of $f_n(x - \xi)$. The *sifting property*

$$\int_a^b \delta(x - \xi)\phi(x)\, dx = \begin{cases} \phi(\xi) & \text{if} \quad a < \xi < b, \\ 0 & \text{if} \quad \xi < a \quad \text{or} \quad \xi > b, \end{cases} \tag{1.2.6}$$

where $\phi(x)$ is an arbitrary function continuous at $x = \xi$, then follows by replacing $\delta(x - \xi)$ by $f_n(x - \xi)$ and proceeding to the limit as $n \to \infty$.

3. *Delta function formulation.* Green's function $g(x, \xi)$ assciated with (1.2.1) satisfies

$$-\frac{d^2g}{dx^2} = \delta(x - \xi), \quad 0 < x < 1, \quad 0 < \xi < 1; \quad g(0, \xi) = g(1, \xi) = 0. \tag{1.2.7}$$

At this stage (1.2.7) is nothing but shorthand for (1.2.5), but in Chapter 2 we develop a mathematical framework in which (1.2.7) will have impeccable standing in its own right.

Solution of the Inhomogeneous Equation

The simple physical interpretation for Green's function guides us in constructing the solution of problem (1.2.1) with data $\{f; 0, 0\}$:

$$-u'' = f(x), \quad 0 < x < 1; \quad u(0) = u(1) = 0. \tag{1.2.8}$$

The idea is to decompose the distributed source $f(x)$ into a number of small concentrated sources located at various points along the rod and then add their individual contributions to the temperature to find u. Divide the interval (0, 1) into n equal parts, calling the center of the kth subinterval ξ_k. The length of each subinterval is $\Delta\xi = 1/n$. It is reasonable to suppose that the temperature corresponding to the distributed density $f(x)$ is closely approximated by the temperature corresponding to small concentrated sources $f(\xi_1)\,\Delta\xi, \dots, f(\xi_n)\,\Delta\xi$, located at $\xi_1, \dots, \xi_n$, respectively (see Figure 1.3); that is, the temperature for the data $\{f(x); 0, 0\}$ is close to the temperature for the data $\{\sum_{i=1}^{n} \delta(x - \xi_i) f(\xi_i)\,\Delta\xi; 0, 0\}$. According to the principle of superposition extended to concentrated sources, the temperature at x for all the small concentrated sources is

$$\sum_{i=1}^{n} g(x, \xi_i) f(\xi_i)\,\Delta\xi,$$

which, as $n \to \infty$, tends to

$$u(x) = \int_0^1 g(x, \xi) f(\xi)\, d\xi. \tag{1.2.9}$$

Thus our intuitive (or *heuristic*) argument leads us to believe that (1.2.9) provides a solution to (1.2.8). Observe that this construction will not work directly for (1.2.1) with nonzero boundary data, but the solution is easy to determine. Since $\alpha(1-x) + \beta x$ satisfies (1.2.1) with data $\{0; \alpha, \beta\}$, the superposition principle shows that

$$u(x) = \int_0^1 g(x, \xi) f(\xi)\, d\xi + \alpha(1 - x) + \beta x \tag{1.2.10}$$

satisfies (1.2.1) with data $\{f; \alpha, \beta\}$.

We must now verify that (1.2.10) *actually solves* (1.2.1); *we would also like to show that it is the only solution to the problem and, finally, that* $u(x)$ *depends continuously on the data.*

The rigorous proof below is based, as it must be at this time, on the classical definition 2.5 of g. There will be occasions, however, when we will be satisfied to give merely plausible arguments using the symbolic formulation (1.2.7), together with the sifting property (1.2.6) of the delta function.

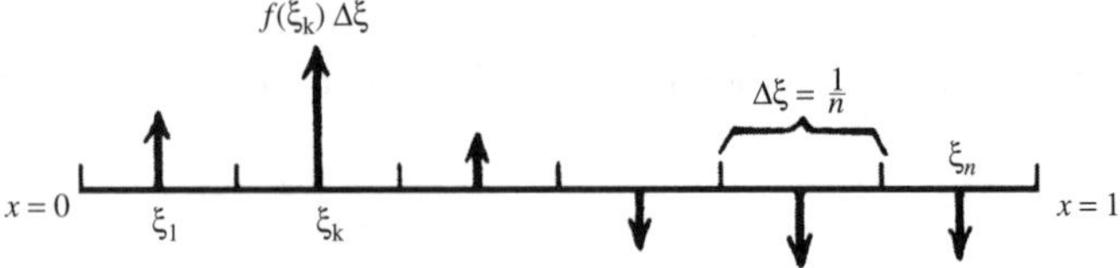

Figure 1.3

Verification of Solution. We confine ourselves here to the case where f is continuous, leaving the more general case to Exercise 1.2.7. Consider first the problem with vanishing boundary data. Clearly, (1.2.9) vanishes at $x = 0$ and 1 because $g(0, \xi) = g(1, \xi) = 0$, so that we only have to show that $-u'' = f$ at each point x, $0 < x < 1$. In view of the discontinuity of g' $(= dg/dx)$ at $x = \xi$, a certain amount of care is required in differentiating expression (1.2.9). Let us split the interval of integration into the parts $(0, x)$ and $(x, 1)$, within each of which g and its derivatives are continuous. Then, using (1.1.3a),

$$\frac{du}{dx} = \frac{d}{dx}\left[\int_0^x r(x,\xi)f(\xi)\,d\xi + \int_x^1 l(x,\xi)f(\xi)\,d\xi\right],$$

and we now appeal to the classical formula for differentiation under the integral sign,

$$\frac{d}{dx}\int_{a(x)}^{b(x)} h(x,\xi)\,d\xi = \int_{a(x)}^{b(x)} \frac{\partial h}{\partial x}\,d\xi + h(x,b(x))\frac{db}{dx} - h(x,a(x))\frac{da}{dx}, \qquad (1.2.11)$$

to obtain

$$\frac{du}{dx} = \int_0^x r'(x,\xi)f(\xi)\,d\xi + \int_x^1 l'(x,\xi)f(\xi)\,d\xi + r(x,x)f(x-) - l(x,x)f(x+).$$

Here the notation $x-, x+$ serves to distinguish between left- and right-hand values at a possible point of discontinuity of a function. Since $r(x,x) = l(x,x)$ and $f(x)$ is continuous, the distinction is unnecessary in the expression for du/dx, in which the last two terms cancel. A further differentiation leads to

$$\begin{aligned}\frac{d^2u}{dx^2} &= \int_0^x r''f(\xi)\,d\xi + \int_x^1 l''f(\xi)\,d\xi + r'(x,x)f(x-) - l'(x,x)f(x+)\\ &= f(x)[r'(x,x) - l'(x,x)] = -f(x).\end{aligned}$$

Hence (1.2.9) is a solution of (1.2.8). Since $\alpha(1-x) + \beta x$ satisfies (1.2.1) with data $\{0; \alpha, \beta\}$, we conclude that (1.2.10) is a solution of (1.2.1) as required.

Uniqueness. Suppose that $u_1(x)$ and $u_2(x)$ satisfy (1.2.1) for the same data, $\{f(x); \alpha, \beta\}$. Then $w(x) = u_1(x) - u_2(x)$ satisfies (1.2.1) with data $\{0; 0, 0\}$. By definition of the concept of a classical solution, u_1' and u_2' must be continuous and

u_1'' and u_2'' exist except at points of discontinuity of f. It follows that w and w' are continuous on $0 \leqslant x \leqslant 1$ and that $w'' = 0$ except possibly at points of discontinuity of f. In each subinterval where f is continuous, w' must be constant; since w' is continuous, the constant is the same in each subinterval and therefore $w = Cx + D$ in $0 < x < 1$. Applying the boundary conditions $w(0) = w(1) = 0$, we find that $w \equiv 0$ in $0 \leqslant x \leqslant 1$.

It is perhaps worth noting that a similar argument shows also that Green's function satisfying (1.2.5) is unique.

Continuity with Respect to the Data. In most experimental situations the data $\{f(x); \alpha, \beta\}$ is not known precisely. It would be comforting to know that the solution of the boundary value problem is not hypersensitive to small changes in the data. We feel that many physical problems should exhibit this kind of stability. We would like to show that a "small" change in the data leads only to a "small" change in the solution. To make this precise we must introduce a notion of "separation" or "distance" between functions (for real numbers there is no problem: the distance between a and b is $|b - a|$).

The most natural way to measure the distance between $f_1(x)$ and $f_2(x)$ is by $\|f_1 - f_2\|$ as defined in Section 0.6. We recall here three norms which will be useful:

$$\|f\|_1 = \int_0^1 |f(x)|\, dx, \qquad \|f\|_2 = \left[\int_0^1 |f|^2\, dx\right]^{1/2}, \qquad \|f\|_\infty = \sup_{0 \leqslant x \leqslant 1} |f(x)|. \tag{1.2.12}$$

These are known as the L_1 norm, the L_2 norm, and the sup norm.

Now let $u_1(x)$ and $u_2(x)$ be the solutions of (1.2.8) corresponding to the respective forcing functions $f_1(x)$ and $f_2(x)$. Our first goal is to show that $\|u_1 - u_2\|$ is small if $\|f_1 - f_2\|$ is small. Because the problem is linear we see that $u_1 - u_2$ is the solution of (1.2.8) with forcing function $f_1 - f_2$; therefore, all we have to show is that the solution u of (1.2.8) has small norm whenever f has small norm.

The solution of (1.2.8) is given by

$$u(x) = \int_0^1 g(x,t) f(t)\, dt,$$

and since $g \geqslant 0$, we have

$$|u(x)| \leqslant \int_0^1 g(x,t)|f(t)|\, dt \leqslant \|f\|_\infty \int_0^1 g(x,t)\, dt = \frac{x(1-x)}{2}\|f\|_\infty, \tag{1.2.13}$$

and taking the maximum with respect to x, we obtain

$$\|u\|_\infty \leqslant \frac{1}{8}\|f\|_\infty \quad \text{or} \quad \|u_2 - u_1\|_\infty \leqslant \frac{1}{8}\|f_2 - f_1\|_\infty. \tag{1.2.14}$$

Thus, if f_1, f_2 are two continuous functions which are close in the sense of maximum deviation (see Figure 0.7), then u_1 and u_2 are also close in the same sense. Note that (1.2.14) is not likely to be useful if f is not continuous.

The first inequality in (1.2.13) also yields

$$\int_0^1 |u(x)|\,dx = \|u\|_1 \leqslant \int_0^1 dx \int_0^1 dt\; g(x,t)|f(t)|$$
$$= \int_0^1 |f(t)| \left\{ \int_0^1 g(x,t)\,dx \right\} dt = \int_0^1 \frac{t(1-t)}{2}|f(t)|\,dt,$$

so that

$$\|u\|_1 \leqslant \tfrac{1}{8}\|f\|_1 \quad \text{or} \quad \|u_2 - u_1\|_1 \leqslant \tfrac{1}{8}\|f_2 - f_1\|_1. \tag{1.2.15}$$

Thus the solution of (1.2.8) depends continuously on the data in both the sup norm and the L_1 norm. Exercise 1.2.1 deals with the L_2 norm and also with a case when different norms are used for the forcing function and the deflection.

For the completely inhomogeneous problem (1.2.1) with u_1, u_2 being the solutions corresponding, respectively, to data $\{f_1(x); \alpha_1, \beta_1\}$ and $\{f_2(x); \alpha_2, \beta_2\}$, a simple calculation shows that

$$\|u_1 - u_2\|_\infty \leqslant \tfrac{1}{8}\|f_1 - f_2\|_\infty + |\alpha_1 - \alpha_2| + |\beta_1 - \beta_2|,$$

so that again we have continuity with respect to the data in the sup norm.

When dealing later with more general boundary value problems (or other equations, such as integral equations), we shall still be faced with these three questions:

1. Is there at least one solution (*existence*)?
2. Is there at most one solution (*uniqueness*)?
3. Does the solution depend continuously on the data?

If the answer to this trio of questions is affirmative, the problem is said to be *well-posed* (otherwise, *ill-posed*). Until recently it was sound dogma to require that every "real" physical problem be well-posed.

Attitudes have changed, however. Insofar as possible we still want to formulate most problems as well-posed, but we no longer fear and reject ill-posed problems, for they occur in practice (particularly in so-called *inverse problems*, where we are trying to infer the data from a partial or full knowledge of the solution). Ill-posed problems need careful physical interpretation, and their mathematical solution may be subtle (see Chapter 6 and the references therein).

Alternative Derivations for the Problem with Nonzero Boundary Data

There is no difficulty in visualizing the role of Green's function in solving a problem with data $\{f(x); 0, 0\}$. We proceed from (1.2.8) to (1.2.9) by straightforward, albeit intuitive arguments. The extra terms in (1.2.10) corresponding to nonzero boundary data were obtained by a different procedure. Could we have used Green's function for this purpose as well? One way of doing this is by translating the problem with data $\{0; \alpha, \beta\}$ into a problem with nonzero f and vanishing boundary data. Consider the boundary value problem

$$-u'' = 0, \quad 0 < x < 1; \quad u(0) = \alpha, \quad u(1) = \beta, \tag{1.2.16}$$

and let $h(x)$ be *any function* (not necessarily satisfying any related differential equation) such that $h(0) = \alpha, h(1) = \beta$. Setting

$$u = h + v,$$

we see that v satisfies

$$-v'' = h'', \quad 0 < x < 1; \quad v(0) = v(1) = 0,$$

whose solution by (1.2.9) is

$$v(x) = \int_0^1 g(x, \xi) h''(\xi) \, d\xi = \int_0^x r(x, \xi) h''(\xi) \, d\xi + \int_x^1 l(x, \xi) h''(\xi) \, d\xi.$$

Integrating by parts twice and using (1.1.3a), we obtain

$$v(x) = -h(x) + \alpha(1 - x) + \beta x = -h(x) + \alpha g'(0, x) - \beta g'(1, x).$$

Since $u = h + v$, we find that

$$u(x) = \alpha g'(0, x) - \beta g'(1, x) = \alpha(1 - x) + \beta x, \tag{1.2.17}$$

in accord with (1.2.10). Observe that, as required, $h(x)$ has disappeared from the final expression (1.2.17) for u.

Another way of arriving at (1.2.17) is to combine the differential equation of (1.2.16) with that for Green's function in the subintervals $(0, \xi)$ and $(\xi, 1)$. Since $g'' = 0$ in each subinterval, we have $ug'' - gu'' = 0$ in $(0, \xi)$ and in $(\xi, 1)$, so that

$$\int_0^\xi (ug'' - gu'') \, dx + \int_\xi^1 (ug'' - gu'') \, dx = 0.$$

The relation

$$ug'' - gu'' = (ug' - gu')', \tag{1.2.18}$$

which is valid classically in each of the subintervals $(0, \xi)$ and $(\xi, 1)$, and the jump condition on g' then yield

$$u(\xi) = u(0) g'(0, \xi) - u(1) g'(1, \xi), \tag{1.2.19}$$

which is the same as (1.2.17).

Both methods used so far are rigorously based on (1.2.5). An alternative to the second of these methods is based formally on the symbolic characterization (1.2.7) and on the sifting property (1.2.6). Multiply (1.2.16) by g and (1.2.7) by u, subtract, and integrate from 0 to 1 to obtain

$$u(\xi) = -\int_0^1 (ug'' - gu'')\, dx.$$

We now use (1.2.18) over the entire interval from 0 to 1; we are entitled to do this because we have accounted for the jump in g' by including the term $\delta(x-\xi)$ in (1.2.7). Thus $u(\xi) = u(0)g'(0,\xi) - u(1)g'(1,\xi)$ as in (1.2.19).

There is a lesson worth remembering here. In the classical approach we use only the subintervals in which all functions are well behaved, the term $u(\xi)$ in (1.2.19) arising from the jump in g' at $x = \xi$. In the symbolic approach we deal with the entire interval at once, the term $u(\xi)$ in (1.2.19) now arising from the fact that there is a delta function on the right side of the differential equation. *Do not mix the two approaches*!

Eigenfunction Expansion

An apparently different approach to (1.2.1) is by way of the associated *eigenproblem*

$$-u'' = \lambda u, \quad 0 < x < 1; \qquad u(0) = u(1) = 0. \tag{1.2.20}$$

Here λ is a complex number regarded as a parameter. Since we are dealing with a homogeneous equation of order 2 with two homogeneous boundary conditions, we might expect that (1.2.20) has only the trivial solution $u \equiv 0, 0 \leqslant x \leqslant 1$. It turns out that this is true for most values of λ, but that there are exceptional values of λ, known as *eigenvalues*, for which the boundary value problem (1.2.20) has nontrivial solutions. These nontrivial solutions are called *eigenfunctions*. Observe that an eigenfunction corresponds to a definite eigenvalue but that to an eigenvalue may be associated more than one independent eigenfunction (it is clear, of course, that any constant multiple of an eigenfunction is again an eigenfunction corresponding to the same eigenvalue; if u_1 and u_2 are eigenfunctions corresponding to the same λ, then $Au_1 + Bu_2$ is also an eigenfunction corresponding to that λ).

For any complex λ we can easily solve the differential equation in (1.2.20); imposition of the boundary conditions then shows that nontrivial solutions are possible only for $\lambda_1 = \pi^2, \lambda_2 = 4\pi^2, \ldots, \lambda_n = n^2\pi^2, \ldots$. To the eigenvalue $\lambda_n = n^2\pi^2$ corresponds essentially one eigenfunction $u_n(x) = \sin n\pi x$ (what this means is that every eigenfunction corresponding to λ_n is necessarily of the form Au_n). We observe that eigenfunctions corresponding to different eigenvalues are orthogonal; that is,

$$\int_0^1 \sin m\pi x \, \sin n\pi x \, dx = 0, \quad m \neq n. \tag{1.2.21}$$

If we now multiply the differential equation in (1.2.1) by $u_n(x)$ and integrate from 0 to 1, we find that

$$-\int_0^1 u'' u_n \, dx = \int_0^1 f u_n \, dx,$$

or, after two integrations by parts and use of the boundary conditions,

$$\lambda_n \int_0^1 u u_n \, dx + n\pi(\beta \cos n\pi - \alpha) = \int_0^1 f u_n \, dx$$

or

$$\int_0^1 u u_n \, dx = \lambda_n^{-1} \left[\int_0^1 f u_n \, dx + n\pi(\alpha - \beta \cos n\pi) \right].$$

The number $\int_0^1 u u_n \, dx$ is just one-half the nth Fourier sine coefficient of $u(x)$, so that we can recover $u(x)$ through

$$u(x) = \sum_{n=1}^{\infty} \frac{2}{n^2\pi^2} \left[\int_0^1 f u_n \, dx + n\pi(\alpha - \beta \cos n\pi) \right] \sin n\pi x. \tag{1.2.22}$$

In particular, for problem (1.2.8) having vanishing boundary data, we find that

$$u(x) = \sum_{n=1}^{\infty} \frac{2}{n^2\pi^2} \left(\int_0^1 f u_n \, dx \right) \sin n\pi x, \tag{1.2.23}$$

which can be considered as an alternative representation to (1.2.9). Comparing the two forms, we deduce the *bilinear series for* Green's function:

$$g(x, \xi) = \sum_{n=1}^{\infty} \frac{2 \sin n\pi x \sin n\pi\xi}{n^2\pi^2}, \tag{1.2.24}$$

which we will study further in Chapters 6 and 7.

We may regard (1.2.20) as a problem of type (1.2.8) with forcing function $\lambda u(x)$; the "solution" is then given by (1.2.9), which becomes

$$u(x) = \lambda \int_0^1 g(x, \xi) u(\xi) \, d\xi, \quad 0 < x < 1. \tag{1.2.25}$$

Since u appears under the integral sign as well as outside, we have not really solved for $u(x)$. Instead, we have shown that (1.2.20) is equivalent to the *integral equation* (1.2.25).

EXERCISES

1.2.1 Consider (1.2.8) with $f(x)$ piecewise continuous.

(a) By using Schwarz's inequality (4.5.14), show that

$$\|u\|_2 \leqslant \left[\int_0^1 \int_0^1 g^2(x, t) \, dx \; dt \right]^{1/2} \|f\|_2,$$

and evaluate the double integral.

(b) Show that

$$\|u\|_\infty \leqslant \tfrac{1}{4}\|f\|_1.$$

The last result shows that if f_1 is piecewise continuous and f_2 is a reasonable continuous approximation to f_1 (such as in the solid and dashed curves of Figure 1.5), the temperatures corresponding to these source functions are *uniformly* close over the entire interval. Thus the statement made in Section 1.1 about replacing certain continuous sources by idealized piecewise continuous ones has been substantiated.

1.2.2 For problem (1.2.8) we have proved the inequalities

$$\|u\|_\infty \leqslant \tfrac{1}{8}\|f\|_\infty \quad \text{and} \quad \|u\|_1 \leqslant \tfrac{1}{8}\|f\|_1,$$

which imply continuity with respect to data in both the sup norm and the L_1 norm. The question remains whether the constant $\frac{1}{8}$ is the best possible. Is there a number c smaller than $\frac{1}{8}$ for which $\|u\|_\infty \leqslant c\|f\|_\infty$ for *all* f? The answer is no. Prove that $\frac{1}{8}$ is the best constant for both inequalities by either exhibiting a function f for which $\|u\| = \frac{1}{8}\|f\|$ or by exhibiting a sequence $\{f_n\}$ for which $\|u_n\|/\|f_n\| \to \frac{1}{8}$. Is the constant obtained in Exercise 1.2.1(a) the best possible?

1.2.3 Let λ be an arbitrary complex number. We shall define the *principal value* of $\sqrt{\lambda}$ as follows. If $\sqrt{\lambda} = 0$, then $\lambda = 0$; if $\lambda \neq 0$, then λ has a unique representation $\lambda = |\lambda|e^{i\theta}$, $0 \leqslant \theta < 2\pi$, and the principal value of $\sqrt{\lambda}$ is defined as $|\lambda|^{1/2}e^{i\theta/2}$, where $|\lambda|^{1/2}$ is the *positive square root* of the positive real number $|\lambda|$. Throughout this exercise $\sqrt{\lambda}$ will stand for the principal value just defined (note that as a function of a complex variable $\sqrt{\lambda}$ has a discontinuity on the positive real axis).

(a) The general solution of $-u'' = \lambda u$ is $u(x) = A + Bx$ if $\lambda = 0$; $u(x) = A\exp(i\sqrt{\lambda}\,x) + B\exp(-i\sqrt{\lambda}\,x)$ [or, alternatively, $u(x) = C\sin\sqrt{\lambda}\,x + D\cos\sqrt{\lambda}\,x$] if $\lambda \neq 0$. Show that only the real values $\lambda_n = n^2\pi^2$ are eigenvalues of (1.2.20).

(b) Find the eigenvalues and eigenfunctions of

$$-u'' = \lambda u, \quad 0 < x < 1; \quad u'(0) = u'(1) = 0.$$

(c) Obtain an expression corresponding to (1.2.23) for the solution of

$$-u'' + qu = f(x), \quad 0 < x < 1; \quad u'(0) = u'(1) = 0,$$

where q is a given positive number. What goes wrong if $q = 0$?

1.2.4 Find Green's function $g(x, \xi)$ satisfying

$$\frac{d^4 g}{dx^4} = \delta(x - \xi), \quad 0 < x, \xi < 1;$$
$$g(0, \xi) = g''(0, \xi) = g(1, \xi) = g''(1, \xi) = 0 \tag{1.2.26}$$

by a graphical method (see Exercise 1.1.2). Note that g is the deflection of a simply supported beam with a concentrated unit load at $x = \xi$. What is the equivalent classical formulation of (1.2.26)? Another method for finding g is to let $h = -g''$. Show that h is given by (1.1.3) and that

$$g(x, \xi) = \int_0^1 h(x, \eta) h(\eta, \xi)\, d\eta.$$

1.2.5 A simply supported beam $(0 < x < l)$ is subject to the distributed transverse loading

$$f(x) = \begin{cases} 0, & |x - \frac{l}{2}| > \varepsilon, \\ p, & \frac{l}{2} < x < \frac{l}{2} + \varepsilon, \\ -p, & \frac{l}{2} - \varepsilon < x < \frac{l}{2}. \end{cases}$$

Find the reactions at the ends; plot shear and moment diagrams. Denote the moment in the beam by $M(x, \varepsilon)$, and calculate $\lim_{\varepsilon \to 0} M(x, \varepsilon)$ in the following cases:

(a) p is fixed.

(b) $p = 1/\varepsilon$.

(c) $p = 1/\varepsilon^2$.

Calculate the limiting deflection corresponding to case (c). What is its physical significance? Formulate the limiting problem as a self-contained mathematical problem without a limiting process.

1.2.6 Recall that in Chapter 0 we introduced the L_p norm

$$\|f\|_p = \left[\int_0^1 |f(x)|^p dx \right]^{1/p}.$$

Show that if $f(x)$ is continuous, then

$$\lim_{p \to \infty} \|f\|_p = \|f\|_\infty,$$

thereby justifying the notation!

(*Hint*: It is easy to show $\|f\|_p \leqslant \|f\|_\infty$ so that $\lim_{p\to\infty} \|f\|_p \leqslant \|f\|_\infty$. To show the opposite inequality, fix $\varepsilon > 0$ and note that

$$\|f\|_p^p = \int_0^1 |f|^p\, dx \geq \int_I |f|^p\, dx,$$

where I is the interval on which $|f| \geq \|f\|_\infty - \varepsilon$. Show then that

$$\lim_{p \to \infty} \|f\|_p \geq \|f_\infty\| - \varepsilon$$

for each ε, and take the limit as $\varepsilon \to 0+$ to obtain the desired result.)

1.2.7 Suppose that $f(x)$ is piecewise continuous on $0 \leqslant x \leqslant 1$. This means that f is continuous except at $a_1, \ldots, a_k$, where f has simple jumps of amounts $J_1, \ldots, J_k$, respectively. We can write

$$f = [f] + \sum_{i=1}^{k} J_i H(x - a_i), \tag{1.2.27}$$

where $[f]$ is continuous on $0 \leqslant x \leqslant 1$ and all the jumps in f are accounted for in the sum of Heaviside functions. We have already proved that if f is continuous, (1.2.9) satisfies (1.2.8). To take care of the piecewise continuous case it is clear [in view of (1.2.27)] that it is enough to treat the special situation where the loading is $H(x - a)$. Show that in this case (1.2.9) satisfies (1.2.8) for all $x \neq a$ and that the deflection has a continuous derivative at $x = a$ (the requirements on a classical solution as defined in Section 1.1 will then be met).

1.3 THE MAXIMUM PRINCIPLE

If $f(x) < 0$ in (1.2.1), we have steady heat conduction with sinks. The temperature $u(x)$ satisfies the *differential inequality*

$$-u'' < 0, \quad 0 < x < 1. \tag{1.3.1}$$

Since heat is removed at every point of the rod, it is physically clear that the maximum temperature must occur on the boundary (which consists of the two points $x = 0$ and $x = 1$) and nowhere else. From the geometrical point of view, u, having a positive second derivative, is strictly convex. This again shows that the maximum is on the boundary. The proof is trivial: If u had even a relative maximum at the interior point x_0, then $u'(x_0) = 0$ and $u''(x_0) \leqslant 0$, contradicting (1.3.1).

If instead of the strict inequality (1.3.1) we know only that

$$-u'' \leqslant 0, \quad 0 < x < 1, \tag{1.3.2}$$

we can still conclude that the maximum of u occurs on the boundary, but now it is also possible for the maximum to be attained in the interior if u is identically constant. We have two versions of the *maximum principle*:

1. *Weak version*. Let u be continuous on $0 \leqslant x \leqslant 1$ and satisfy (1.3.2). Let the maximum of u on the boundary be M. Then $u(x) \leqslant M, 0 \leqslant x \leqslant 1$.

2. *Strong version*. Let u be continuous on $0 \leqslant x \leqslant 1$ and satisfy (1.3.2). Suppose that $u(x) \leqslant M$ in $0 \leqslant x \leqslant 1$ and $u(x_0) = M$ at an interior point x_0; then $u(x) \equiv M$ in $0 \leqslant x \leqslant 1$.

The first version makes no prediction as to whether the maximum can occur at interior points as well as on the boundary; in the strong version this is ruled out unless u is identically constant.

Proof of weak version. For $\varepsilon > 0$ set $v(x) = u(x) + \varepsilon x^2$; then v satisfies the strict inequality $-v'' < 0, 0 < x < 1$, so that the maximum of v is on the boundary and

$$v(x) = u(x) + \varepsilon x^2 \leqslant M + \varepsilon.$$

Thus $u(x) \leqslant M + \varepsilon$ for every $\varepsilon > 0$, and hence $u(x) \leqslant M$. □

Proof of strong version. Suppose that $u(x_1) < M$, where with no loss of generality we can take $x_1 > x_0$. We will show that this leads to a contradiction by constructing a function v satisfying $-v'' < 0$ in $0 < x < x_1, v(0) < M, v(x_1) < M, v(x_0) = M$. Consider the function

$$z = e^{x-x_0} - 1,$$

which is positive for $x > x_0$, negative for $x < x_0$, and vanishes at x_0. Now choose ε so that $0 < \varepsilon < [M - u(x_1)]/z(x_1)$, which is clearly possible since $M > u(x_1)$ and $z(x_1) > 0$. Then the function

$$v(x) = u(x) + \varepsilon z(x)$$

satisfies

$$-v'' = -u'' - \varepsilon z'' \leqslant -\varepsilon z'' = -\varepsilon e^{x-x_0} < 0,$$

which is the strict inequality (1.3.1). But $v(0) < M, v(x_1) < M$, and $v(x_0) = M$, contradicting the fact that v must have its maximum on the boundary of the interval $(0, x_1)$.

□

REMARKS

1. If instead of (1.3.2) we had $-u'' \geqslant 0$, u would be concave and u would satisfy a *minimum* principle. The proof is obtained by noting that $w = -u$ satisfies the maximum principle (in either version).

2. If $u'' = 0$, then both the maximum and minimum principles apply (the result is trivial in one dimension but not in higher dimensions).

The weak version of these principles is easily extended to higher dimensions. Let Ω be a bounded domain with boundary Γ, and let u be continuous on $\bar{\Omega}$. If $-\Delta u \leqslant 0$ in Ω, then max u occurs on Γ; if $-\Delta u \geqslant 0$ in Ω, then min u occurs on Γ; if $\Delta u = 0$ in Ω, then max u and min u occur on Γ.

The strong version of these principles is extended to higher dimensions in Section 8.3.

The weak version by itself leads to the following interesting consequences:

1. *Uniqueness.* The solution of the inhomogeneous problem (1.1.4) is unique. *Proof.* If u_1 and u_2 are two solutions, then $w = u_1 - u_2$ satisfies $-\Delta w = 0$ in Ω with $w = 0$ on Γ. Since the maximum and the minimum of w on the boundary are both 0, w must be identically 0 in the interior.

2. *Comparison*. The data $\{f_1(x); h_1(x)\}$ is said to *dominate* $\{f_2(x); h_2(x)\}$ if $f_1(x) \geqslant f_2(x)$ in Ω and $h_1(x) \geqslant h_2(x)$ on Γ. Suppose that $\{f_1; h_1\}$ dominates $\{f_2; h_2\}$; then the corresponding solutions of (1.1.4) satisfy $u_1(x) \geqslant u_2(x)$. *Proof*. $w = u_2 - u_1$ satisfies $-\Delta w \leqslant 0$ in Ω and $w \leqslant 0$ on Γ. By the maximum principle, max w occurs on Γ; therefore, $w(x) \leqslant 0$ in Ω.

3. *Continuous dependence on data*. We prove this for (1.1.4) with $h = 0$, leaving the case $h \neq 0$ for Exercise 1.3.2. We have to show that there exists a constant A *independent* of f such that $\|u\| \leqslant A\|f\|$. Using the sup norm, we have that $-\|f\|_\infty \leqslant f(x) \leqslant \|f\|_\infty$ so that the comparison principle gives

$$-\|f\|_\infty v(x) \leqslant u(x) \leqslant \|f\|_\infty v(x), \tag{1.3.3}$$

where $v(x)$ is the solution of the Poisson problem

$$-\Delta v = 1, \quad x \in \Omega; \qquad v|_\Gamma = 0. \tag{1.3.4}$$

Thus, we obtain from (1.3.3) that

$$\|u\|_\infty \leqslant \|v\|_\infty \|f\|_\infty, \tag{1.3.5}$$

where $\|v\|_\infty$ is a constant independent of f. We have therefore proved continuous dependence on the data for (1.1.4) with $h = 0$. There is no constant smaller than $\|v\|_\infty$ which will work for all f in (1.3.5), since by taking $f \equiv 1$, the inequality (1.3.5) becomes an equality. Exercise 1.3.2 also asks you to estimate $\|v\|_\infty$.

For a comprehensive, yet accessible treatment of maximum principles, the reader should consult the book by Protter and Weinberger [4].

EXERCISES

1.3.1 The equation $-(ku')' + qu = f$ governs steady diffusion in an absorbing medium. Here $u(x)$ is the *concentration* of the diffusing substance measured relative to some ambient value (so that u can be positive or negative), $-k(x)$ grad u is the diffusion flux vector, $q(x)$ measures the absorption properties, and $f(x)$ is the source density. The effect of the term qu is to try to restore the concentration to its ambient value. The same equation also governs the transverse deflection of a string when there is a springlike resistance (the term qu) to such a deflection. In both cases it is natural to take $k(x) > 0$, $q(x) \geqslant 0$, and we shall do so.

(a) Let $v(x)$ be continuous on $a \leqslant x \leqslant b$ and satisfy the strict inequality

$$-(kv')' + qv < 0, \quad a < x < b. \tag{1.3.6}$$

Show that v cannot have a *positive* (or even nonnegative) relative maximum at an interior point. The example $v = -\cosh x - 1$ on $-1 < x < 1$ satisfies (1.3.6) with $k = q = 1$ and has a negative maximum at the interior point $x = 0$.

(b) Let $u(x)$ be continuous on $a \leqslant x \leqslant b$ and satisfy

$$-\left(ku'\right)' + qu \leqslant 0, \quad a < x < b. \tag{1.3.7}$$

State and prove a weak version of the maximum principle for positive solutions of (1.3.7).

(c) If the inequality in (1.3.7) is reversed, a minimum principle is obtained for negative solutions of the inequality $-(ku')' + qu \geqslant 0$. State appropriate principles for solutions of the equation $-(ku')' + qu = 0$.

(d) Prove uniqueness and continuous dependence on data in the sup norm for $-(ku')' + qu = f; u(0) = \alpha, u(1) = \beta$.

(e) If $q(x) < 0$, no maximum or minimum principle is available. For instance, $u = \sin 2x$ satisfies

$$-u'' - 4u = 0, \quad 0 < x < \pi, \quad u(0) = u(\pi) = 0,$$

yet u has a positive maximum and a negative minimum in the interior of the interval.

1.3.2 (a) Show that for (1.1.4), the following inequality holds:

$$\|u\|_\infty \leqslant \|v\|_\infty \|f\|_\infty + \|h\|_\infty^\Gamma,$$

where v is defined from (1.3.4) and

$$\|h\|_\infty^\Gamma = \sup_{x \in \Gamma} |h(x)|.$$

(b) Solve problem (1.3.4) for a ball of radius a in $\mathbb{R}^n$, and then show that for an arbitrary domain Ω,

$$\|v\|_\infty \leqslant \frac{R^2}{2n},$$

where R is the radius of the smallest ball containing Ω.

1.3.3 Derive a strong maximum principle for solutions of

$$-(ku')' \leqslant 0, \quad 0 < x < 1,$$

where $k(x) > 0$ in $0 \leqslant x \leqslant 1$.

1.3.4 (a) Derive a strong maximum principle for solutions of

$$-u'' + pu' \leqslant 0, \quad 0 < x < 1,$$

where $p(x)$ is an arbitrary continuous function. (*Hint*: First prove the result for solutions of the strict inequality, and then let $v = u + \varepsilon z$, where $z = \exp[\alpha(x - x_0)] - 1$ with α suitably chosen.)

(b) Derive a strong minimum principle for solutions of $-u'' + pu' \geqslant 0$.

(c) State appropriate principles for solutions of $-u'' + pu' = 0$.

1.4 EXAMPLES OF GREEN'S FUNCTIONS

Initial Value Problem

A particle of mass m moves along the u axis under the influence of a force $F(t)$ directed along the axis. The motion of the particle is determined by Newton's law with *initial conditions*

$$m\frac{d^2u}{dt^2} = F(t), \quad t > 0; \qquad u(0) = \alpha, \frac{du}{dt}(0) = \beta. \tag{1.4.1}$$

If the problem were solved over a finite time interval $(0, T)$, T would play no role in the final result. Therefore, we may as well consider the equation on the semi-infinite interval $t > 0$.

Green's function $g(t, \tau)$ associated with (1.4.1) satisfies

$$m\frac{d^2g}{dt^2} = \delta(t-\tau), \quad 0 < t, \tau < \infty; \qquad g(0,\tau) = 0, \quad g'(0,\tau) = 0. \tag{1.4.2}$$

The function $g(t, \tau)$ is the position of a particle initially at rest at the origin and subject to a unit impulse at time τ. As in Section 0.4, we regard the impulse as the limiting case of a very large force $X(t)$ acting over a very short period of time from τ to $\tau + \Delta\tau$ such that

$$\int_\tau^{\tau+\Delta\tau} X(t)\, dt = 1.$$

Such an impulse will cause an instantaneous unit change in the momentum $m(dg/dt)$ of the particle. Thus (1.4.2) can be written in the equivalent form

$$\left.\begin{aligned} &m\frac{d^2g}{dt^2} = 0, \quad 0 < t < \tau, t > \tau; && g(0,\tau) = g'(0,\tau) = 0, \\ &g \text{ continuous at } t = \tau; && m\left.\frac{dg}{dt}\right|_{t=\tau+} - m\left.\frac{dg}{dt}\right|_{t=\tau-} = 1. \end{aligned}\right\} \tag{1.4.3}$$

Since both initial conditions apply to the interval $(0, \tau)$, we find that $g = 0$ until $t = \tau$. The continuity of g and the jump condition on g' give

$$g(t,\tau) = \begin{cases} 0, & 0 \leqslant t < \tau, \\ \dfrac{t-\tau}{m}, & t > \tau. \end{cases}$$

The superposition principle can then be applied to the problem with 0 initial data:

$$mu'' = F(t), \quad t > 0; \quad u(0) = 0, \quad u'(0) = 0$$

with the result

$$u(t) = \int_0^\infty g(t,\tau)F(\tau)\, d\tau = \int_0^t \frac{t-\tau}{m} F(\tau)\, d\tau. \tag{1.4.4}$$

Not surprisingly, the displacement $u(t)$ is independent of the force acting after time t. The solution of the problem with data $\{0; \alpha, \beta\}$ is $\alpha + \beta t$, so that the solution of (1.4.1) is the sum of $\alpha + \beta t$ and (1.4.4). Existence, uniqueness, and continuous dependence on data are easily proved.

Reverting to the x, ξ notation and setting $m = -1$, we see that the function

$$h(x,\xi) = \begin{cases} 0, & 0 < x < \xi, \\ \xi - x, & x > \xi, \end{cases} \tag{1.4.5}$$

satisfies

$$-\frac{d^2h}{dx^2} = \delta(x-\xi), \quad 0 < x, \xi; \quad h(0,\xi) = h'(0,\xi) = 0. \tag{1.4.6}$$

Green's function for an initial value problem is sometimes called a *causal* Green's function. Green's function $g(x, \xi)$ given by (1.2.4) satisfies the same differential equation but with different side conditions. With ξ fixed, $h - g$ satisfies $(h-g)'' = 0$ for all x, and $h - g$ must coincide for all x with a solution of the homogeneous equation, which turns out to be $-(1-\xi)x$. This suggests a method for constructing Green's function for a particular set of boundary conditions: First construct the causal Green's function for the same operator, and then add the appropriate solution of the homogeneous equation to satisfy the original boundary conditions (see the beam problem below, for instance).

Variable Conductivity

Let the thermal conductivity in a rod of unit length be function $k(x)$, which is positive and continuously differentiable. The steady temperature $g(x, \xi)$ in a rod with a concentrated unit source at ξ, with its left end at 0 temperature, and with its right end insulated satisfies

$$\left.\begin{aligned} &-\frac{d}{dx}\left(k(x)\frac{dg}{dx}\right) = \delta(x-\xi), \qquad 0 < x, \xi < 1; \\ &g(0,\xi) = 0, \quad g'(1,\xi) = 0. \end{aligned}\right\} \tag{1.4.7}$$

An equivalent formulation is

$$\left.\begin{aligned} &-(kg')' = 0, \qquad 0 < x < \xi, \quad \xi < x < 1; \\ &g(0,\xi) = 0, \ g'(1,\xi) = 0, \\ &g \text{ continuous at } x = \xi; \\ &k(\xi)[g'(\xi+,\xi) - g'(\xi-,\xi)] = -1, \end{aligned}\right\} \tag{1.4.8}$$

the jump condition on g' stemming from a heat balance for a thin slice of the rod containing the source. The functions

$$u_1(x) = \int_0^x \frac{1}{k(y)}dy \quad \text{and} \quad u_2(x) = 1$$

are solutions of the homogeneous equation satisfying, respectively, the boundary conditions at the left and right endpoints. The matching conditions at $x = \xi$ give

$$g(x,\xi) = \begin{cases} \displaystyle\int_0^x \frac{1}{k(y)}dy, & 0 \leqslant x < \xi, \\ \displaystyle\int_0^\xi \frac{1}{k(y)}dy, & \xi < x \leqslant 1. \end{cases}$$

Simply Supported Beam

Consider a simply supported beam [see (1.1.14)] under a concentrated load at $x = \xi$. The deflection $g(x,\xi)$ satisfies

$$\begin{aligned} \frac{d^4 g}{dx^4} &= \delta(x-\xi), \qquad 0 < x, \xi < 1; \\ g(0,\xi) &= g''(0,\xi) = g(1,\xi) = g''(1,\xi) = 0. \end{aligned} \tag{1.4.9}$$

The shear force $V(x)$ experiences a jump discontinuity from $x = \xi-$ to $\xi+$ to balance the concentrated load:

$$V(\xi+) - V(\xi - 1) = -1.$$

The moment, the slope, and the deflection remain continuous even at $x = \xi$. Since $-V = d^3g/dx^3$, we can write (1.4.9) as

$$\left.\begin{aligned} &\frac{d^4 g}{dx^4} = 0, \quad 0 < x < \xi, \xi < x < 1; \\ &g(0,\xi) = g''(0,\xi) = g(1,\xi) = g''(1,\xi) = 0; \\ &g, g', g'' \text{continuous at } x = \xi; \qquad g'''(\xi+,\xi) - g'''(\xi-,\xi) = 1. \end{aligned}\right\} \tag{1.4.10}$$

Applying the boundary conditions, we find that the solution for $x < \xi$ is $Ax + Bx^3$, while for $x > \xi$ it is $C(1-x) + D(1-x)^3$. It remains to apply the matching conditions at ξ. The conditions on g'' and g''' should be used first to yield $g = Ax - (1-\xi)(x^3/6)$ for $x < \xi$ and $g = C(1-x) - \xi[(1-x)^3/6]$ for $x > \xi$. The continuity of g and g' then gives $A = \frac{1}{6}\xi(1-\xi)(2-\xi)$ and $C = \frac{1}{6}\xi(1-\xi)(1+\xi)$. The same result can, of course, be obtained (perhaps more intuitively) by using the shear and moment diagrams of Exercises 1.1.2 and 1.2.4.

We can also construct g by first finding the causal fundamental solution $h(x,\xi)$ satisfying

$$\frac{d^4 h}{dx^4} = \delta(x-\xi) \quad 0 < x, \xi; \qquad h(0,\xi) = h'(0,\xi) = h''(0,\xi) = h'''(0,\xi) = 0.$$

An easy calculation gives

$$h(x,\xi) = \begin{cases} 0, & x < \xi, \\ \dfrac{(x-\xi)^3}{6}, & x > \xi. \end{cases} \tag{1.4.11}$$

Therefore, g in (1.4.10) must be of the form $h + A + Bx + Cx^2 + Dx^3$. The conditions at the end $x = 0$ give $A = C = 0$. At the right end we have $h''(1, \xi) + 6D = 0$ and $h(1, \xi) + B + D = 0$, that is, $D = -(1-\xi)/6$ and $B = \xi(1-\xi)(2-\xi)/6$, which when substituted in $h + Bx + Dx^3$ confirm the earlier result.

The Infinite Rod with Absorption

The steady-state concentration $u(x)$ of a substance diffusing in a homogeneous absorbing medium satisfies

$$-\frac{d^2u}{dx^2} + q^2u = f(x), \quad -\infty < x < \infty, \tag{1.4.12}$$

where q^2 is a positive constant, $f(x)$ is the source density of the substance, and the process can be considered as taking place in an infinitely long tube, $-\infty < x < \infty$. (The same equation governs the small transverse displacements of a string subject to an applied load and a springlike restoring mechanism.) Green's function corresponding to a steady unit input of the diffusing substance at $x = \xi$ satisfies

$$-\frac{d^2g}{dx^2} + q^2g = \delta(x - \xi), \quad -\infty < x, \xi < \infty \tag{1.4.13}$$

Since the coefficients of the differential equation are constants, it will suffice to find $g(x, 0)$ and then set $g(x, \xi) = g(x - \xi, 0)$. This argument obviously depends also on the fact that we are dealing with the infinite domain $-\infty < x < \infty$. Again we assume that $g(x, 0)$ is continuous; conservation of matter gives $-g'(0+, 0) + g'(0-, 0) = 1$. In keeping with the absorbing nature of the medium, we require that g vanish at $x = \pm\infty$, so that $g(x, 0) = e^{-q|x|}/2q$, and

$$g(x, \xi) = \frac{e^{-q|x-\xi|}}{2q}. \tag{1.4.14}$$

It is perhaps a little surprising that g has no limit as $q \to 0$. The reason is that the nonabsorbing problem cannot obey the condition $g \to 0$ as $|x| \to \infty$. On the other hand, the flux dg/dx obtained from (1.4.14) has the limits $-\frac{1}{2}$ for $x > \xi$ and $+\frac{1}{2}$ for $x < \xi$, so that, by integration, we might suspect that a solution of (1.4.13) for $q = 0$ is $-(|x - \xi|/2) + C$, which is easily confirmed. Although there is no compelling physical argument for doing so, we often set $C = 0$.

Method of Images

Consider (1.4.13) for the semi-infinite interval $0 < x < \infty$. In addition to the condition $g \to 0$ as $x \to \infty$, we now need a boundary condition at $x = 0$, which we will take as $g(0, \xi) = 0$. This means that any of the diffusing substance that reaches $x = 0$ is removed [the boundary condition $g'(0, \xi) = 0$ would model a reflecting wall at $x = 0$]. Thus we wish to solve

$$-\frac{d^2g}{dx^2} + q^2g = \delta(x - \xi), \quad 0 < x, \xi < \infty; \qquad g(0, \xi) = 0, \quad g \to 0 \text{ as } x \to \infty. \tag{1.4.15}$$

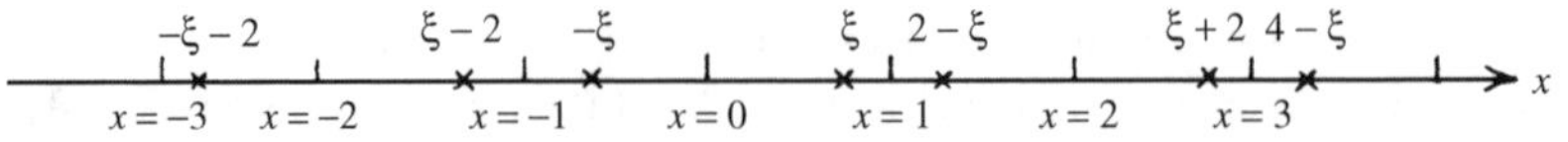

Figure 1.4

Let us look instead at an infinite rod with a unit source at $x = \xi$ and a unit sink at $x = -\xi$. According to (1.4.14), the solution of this problem is

$$\frac{e^{-q|x-\xi|}}{2q} - \frac{e^{-q|x+\xi|}}{2q}. \tag{1.4.16}$$

This function vanishes at $x = 0$ and has only one source singularity in $0 < x < \infty$, namely, the original source at $x = \xi$. The term $e^{-q|x+\xi|}/2q$ arising from the *image* source at $x = -\xi$ satisfies the homogeneous differential equation in $0 < x < \infty$. Thus (1.4.16) is a solution of the boundary value problem (1.4.15).

For Green's function of a finite rod we use a similar idea. If the ends are both *reflecting*, say, the boundary condition is $dg/dx = 0$ at both $x = 0$ and $x = 1$. We consider the related problem of an infinite rod with *positive* unit sources located at the set of points $\xi + 2n$ and $-\xi + 2n$, n ranging through the integers from $-\infty$ to ∞, as in Figure 1.4. The solution of this problem is

$$\sum_{n=-\infty}^{\infty} \frac{e^{-q|x-(\xi+2n)|}}{2q} + \sum_{n=-\infty}^{\infty} \frac{e^{-q|x-(-\xi+2n)|}}{2q}, \tag{1.4.17}$$

which has even symmetry about both $x = 0$ and $x = 1$. Thus this function has a vanishing derivative at $x = 0$ and $x = 1$. It is clear from the figure that of this array of sources the only one in the interval (0, 1) is the original source. Therefore, (1.4.17) is a solution of the problem of the finite tube with reflecting walls.

Steady Diffusion with Absorption in a Three-Dimensional Medium

Let Ω be a bounded or unbounded domain in $\mathbb{R}^3$, and let x be the position vector in $\mathbb{R}^3$. The concentration $u(x)$ of the diffusing substance satisfies the three-dimensional version of (1.4.12):

$$-\Delta u + q^2 u = f(x), \quad x \in \Omega, \tag{1.4.18}$$

where the constant $q^2 \geqslant 0$ is a measure of the absorption of the medium and $f(x)$ is the density of the source. There will, of course, be boundary conditions on Γ, the boundary of Ω. The case $q = 0$ corresponds to diffusion without absorption or to steady heat conduction.

Let us look at the case where Ω is the entire space and there is only a concentrated steady unit source at the origin. A mass balance (or heat balance) shows that the flux through a small sphere about the source must equal the input in the ball, that is,

$$\lim_{\varepsilon \to 0} - \int_{|x|=\varepsilon} \frac{\partial u}{\partial n} dS = 1. \tag{1.4.19}$$

We also expect u to vanish at infinity. The concentration should clearly depend only on the radial coordinate r; since there are no sources for $r \neq 0$, we find, on using the spherical form of Δ, that $u(r)$ satisfies

$$-\frac{1}{r^2}\frac{d}{dr}\left(r^2\frac{du}{dr}\right)+q^2u=0, \quad r>0, \tag{1.4.20}$$

with (1.4.19) becoming

$$-1=\lim_{\varepsilon\to 0} 4\pi\varepsilon^2\left(\frac{du}{dr}\right)_{r=\varepsilon}. \tag{1.4.21}$$

The substitution $u=v/r$ transforms (1.4.20) into $-v''+q^2v=0$, whose general solution is a linear combination of e^{-qr} and e^{qr}. Taking account of the required behavior at $r=\infty$, we obtain $u=Ae^{-qr}/r$. Imposing (1.4.21) gives $A=1/4\pi$, and therefore

$$u=\frac{e^{-qr}}{4\pi r}. \tag{1.4.22}$$

The effect of a source at ξ is obtained from (1.4.22) by translation. The concentration due to such a source is what we call the *free space* Green's function:

$$g=\frac{e^{-q|x-\xi|}}{4\pi|x-\xi|}. \tag{1.4.23}$$

Note that unlike the one-dimensional case, the limit as $q\to 0$ gives the solution $1/4\pi|x-\xi|$ for a nonabsorbing medium. A more important observation is that Green's function is now *singular* at $x=\xi$ (in one dimension g was continuous at $x=\xi$). This is, of course, the free space Green's function for the negative Laplacian:

$$-\Delta\frac{1}{4\pi|x-\xi|}=\delta(x-\xi). \tag{1.4.24}$$

Green's functions for some simple domains (such as a half-space, a quarter-space, a slab, a rectangular parallelepiped) can be found by images when the boundary condition is that the function or its normal derivative vanishes on the boundary. Other methods for constructing Green's functions for partial differential equations are discussed in later chapters, but suppose for the time being that Green's function $g(x,\xi)$ is known for the negative Laplacian in a domain Ω with $g=0$ on the boundary Γ. Then g is the solution of

$$-\Delta g=\delta(x-\xi), \quad x,\xi \text{ in } \Omega; \qquad g=0, \quad x \text{ on } \Gamma. \tag{1.4.25}$$

By the maximum principle, we can show $g>0$ in Ω (see Exercise 8.3.2).

Let $u(x)$ be the solution of the problem with data $\{f;0\}$; that is, $u(x)$ satisfies

$$-\Delta u=f, \quad x \text{ in } \Omega; \qquad u=0, \quad x \text{ on } \Gamma. \tag{1.4.26}$$

By the superposition principle we still expect the solution to be expressible as

$$u(x)=\int_\Omega g(x,\xi)f(\xi)\,d\xi. \tag{1.4.27}$$

Clearly, this function vanishes when x is on Γ because g does. If we formally calculate $-\Delta u$ by differentiating under the integral sign in (1.4.27) and use (1.4.25), we obtain $-\Delta u = f$, as required. The procedure is permissible if f obeys some very mild restrictions.

Next we express the solution of the problem with data $\{0; h\}$ in terms of Green's function. Let $v(x)$ satisfy

$$-\Delta v = 0, \qquad x \text{ in } \Omega; \qquad v = h(x), \quad x \text{ on } \Gamma, \tag{1.4.28}$$

and multiply the differential equation by g, multiply (1.4.25) by v, subtract, and integrate over Ω to obtain

$$v(\xi) = \int_\Omega (g\,\Delta v - v\,\Delta g)\,dx.$$

By using Green's theorem and the fact that g vanishes for x on Γ, we find that

$$v(\xi) = -\int_\Gamma \frac{\partial g(x,\xi)}{\partial n_x} h(x)\,dS_x$$

or, interchanging labels for x and ξ and using the symmetry of g (see Exercise 1.4.9),

$$v(x) = -\int_\Gamma \frac{\partial g(x,\xi)}{\partial n_\xi} h(\xi)\,dS_\xi, \tag{1.4.29}$$

where the subscript indicates the variable of differentition or integration. The solution of the problem with data $\{f; h\}$ is then the sum of (1.4.27) and (1.4.29), as stated in (1.1.5), where the symmetry of $g(x,\xi)$ was also used.

Other problems of interest have sources spread on surfaces in $\mathbb{R}^3$. The forcing function here stands somewhere between an ordinary volume density of sources and the most highly concentrated forcing function, $\delta(x-\xi)$, corresponding to a point source. Suppose, for instance, that a layer of sources whose total strength is unity is spread uniformly over the sphere $|x| = a$. The corresponding solution of (1.4.18) will then depend only on the radial coordinate r measured from the center of the sphere (the differential operator being invariant under rotation). Denoting the solution by $u(r)$, we see, by using the form of Δ appropriate for spherical coordinates, that

$$-\frac{1}{r^2}\frac{d}{dr}\left(r^2\frac{du}{dr}\right) + q^2 u = 0, \quad 0 < r < a, \quad r > a. \tag{1.4.30}$$

We search for a solution which is finite at $r = 0$, vanishes as $r \to \infty$, and represents the appropriate source at $r = a$. The total flux on the sphere $|x| = a + \varepsilon$ minus the flux on $|x| = a - \varepsilon$ must equal the input in the interior of the shell, that is,

$$\lim_{\varepsilon\to 0} -\left[\int_{|x|=a+\varepsilon} \frac{\partial u}{\partial n}\,dS + \int_{|x|=a-\varepsilon} \frac{\partial u}{\partial n} dS\right] = 1.$$

Since u depends only on r, this becomes

$$-1 = 4\pi a^2 \left[\left(\frac{du}{dr}\right)_{r=a+} - \left(\frac{du}{dr}\right)_{r=a-} \right]. \tag{1.4.31}$$

The solution of (1.4.30) must therefore satisfy (1.4.31), vanish at infinity, and be bounded at $r = 0$. We find that

$$u = A\frac{\sinh qr}{r} \quad \text{for } r < a, \qquad u = B\frac{e^{-qr}}{r} \quad \text{for } r > a.$$

Since the problem has been reduced to a one-dimensional problem, it is appropriate to require that u be continuous at $r = a$; this condition, together with (1.4.31), then yields

$$u = \begin{cases} \dfrac{1}{4\pi q}\dfrac{e^{-qa}}{a}\dfrac{\sinh qr}{r}, & r < a, \\[2ex] \dfrac{1}{4\pi q}\dfrac{\sinh qa}{a}\dfrac{e^{-qr}}{r}, & r > a. \end{cases}$$

As $a \to 0$ we should recover the solution $g(x, 0)$ for a unit source at the origin. Taking the limit in the expression valid for $r > a$, we find that

$$u = \frac{1}{4\pi}\frac{e^{-qr}}{r},$$

in agreement with (1.4.22).

In the case of a *line source* of uniform unit density, the response u is independent of the coordinate parallel to the line. It is therefore appropriate to use cylindrical polar coordinates (ρ, ϕ) with the source at $\rho = 0$; the axial symmetry of the problem suggests that u is independent of ϕ and (1.4.18) reduces to

$$-\frac{1}{\rho}\frac{d}{d\rho}\left(\rho\frac{du}{d\rho}\right) + q^2 u = 0, \quad \rho > 0.$$

This is the modified Bessel equation whose independent solutions are $I_0(q\rho)$ and $K_0(q\rho)$. Since I_0 is exponentially large at ∞, we must have $u = AK_0(q\rho)$ with A determined from the unit source condition at $\rho = 0$. This condition has the form

$$-1 = \lim_{\varepsilon \to 0} 2\pi\varepsilon \left(\frac{du}{d\rho}\right)_{\rho=\varepsilon},$$

which in light of the logarithmic singularity of K_0 at the origin gives $A = \frac{1}{2}\pi$ and

$$u = \frac{1}{2\pi}K_0(q\rho). \tag{1.4.32}$$

Note that in three dimensions a concentrated source gives rise to a singularity of order $1/|x|$, a line source to a logarithmic singularity, and a surface source to a simple discontinuity in the normal derivative. Only in the last case is the response continuous across the source.

Interface Problems

Consider steady one-dimensional heat conduction in a rod occupying the interval $-1 < x < 1$. The rod's thermal conductivity is the positive constant k_1 in the interval $-1 < x < 0$ and another positive constant $k_2 > k_1$ in the interval $0 < x < 1$. Such a problem could arise in dealing with a *composite rod* constructed by joining end to end two rods of unit length and of different conductivities or in attempting to idealize a *heterogeneous rod* whose conductivity changes rapidly but continuously from k_1 to k_2. In both interpretations we want to reduce the problem to solving constant conductivity equations in the two halves of the rod, and the question that remains in how to match these solutions at the interface $x = 0$.

For the heterogeneous rod we are considering the limiting case as $\varepsilon \to 0+$ of a problem with a continuously varying positive conductivity $k(x, \varepsilon)$ having the property

$$\lim_{\varepsilon \to 0+} k(x, \varepsilon) = \begin{cases} k_1, & x < 0, \\ k_2, & x > 0. \end{cases} \tag{1.4.33}$$

The limiting conductivity will be denoted $k(x)$; we have $k(x) = k_1 + (k_2 - k_1)H(x)$, where H is the Heaviside function. It turns out that it is not quite sufficient to ask that (1.4.33) hold pointwise. Instead, we will need to require that $1/k(x, \varepsilon)$ tend to its limit in the L_1 sense (see Section 0.7); that is,

$$\lim_{\varepsilon \to 0} \int_{-1}^{1} \left| \frac{1}{k(x, \varepsilon)} - \frac{1}{k(x)} \right| dx = 0, \tag{1.4.34}$$

which means that the area between the curves $1/k(x, \varepsilon)$ and $1/k(x)$ must go to 0 as $\varepsilon \to 0$ (this does not follow from pointwise convergence alone).

This happens, for instance, if $k(x, \varepsilon)$ satisfies (1.4.33) and is an increasing function of x with $k \geqslant k_1$ for $x < 0$, and $k \leqslant k_2$ for $x > 0$, as in the example

$$k(x, \varepsilon) = \frac{k_1 + k_2}{2} + \frac{k_2 - k_1}{\pi} \arctan \frac{x}{\varepsilon},$$

but our results are independent of the particular form of $k(x, \varepsilon)$. The resulting interface conditions are

$$k_2 u'(0+) - k_1 u'(0-) = 0 \tag{1.4.35}$$

and

$$u(0+) - u(0-) = 0. \tag{1.4.36}$$

The first of these conditions is a consequence of the integral formulation of the law of heat conduction, which states in our case that the heat fluxes to the left and right of $x = 0$ must be equal in the absence of concentrated sources at the interface (see Section 0.1). The second condition is nearly obvious but does in fact require (1.4.34), as we shall see in the special case analyzed below.

For a *composite* rod made by joining two rods together, conditions (1.4.35) and (1.4.36) are appropriate only if the unit rods are joined perfectly at $x = 0$ with no film or gap between them.

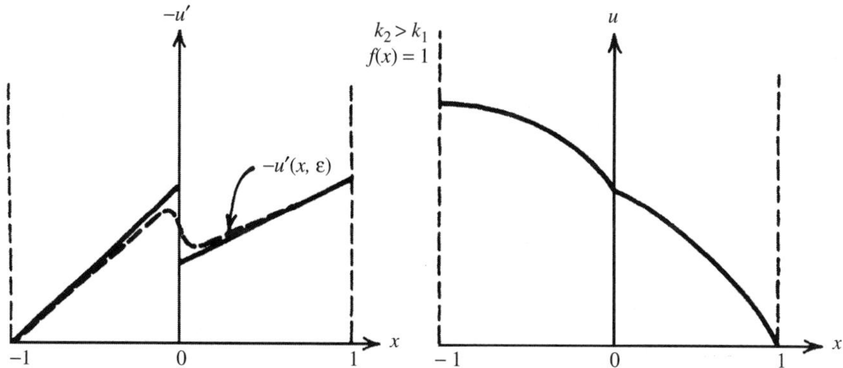

Figure 1.5

Let us now consider the explicitly solvable boundary value problem

$$-\frac{d}{dx}\left(k(x,\varepsilon)\frac{du}{dx}\right) = 1, \quad -1 < x < 1; \qquad u'(-1) = 0 = u(1). \qquad (1.4.37)$$

Here we have a heterogeneous rod subject to a uniform density of sources with its left end insulated and its right end kept at 0 temperature. We are interested in the limiting case of ε tending to 0 for k satisfying (1.4.33) and (1.4.34). In view of (1.4.37) we have $-k(x,\varepsilon)u' = x + 1$, so that, by (1.4.33), the pointwise limit of u' exists, and

$$\lim_{\varepsilon \to 0} u'(x,\varepsilon) = \begin{cases} -\dfrac{x+1}{k_1}, & x < 0, \\ -\dfrac{x+1}{k_2}, & x > 0. \end{cases} \qquad (1.4.38)$$

We shall denote the function on the right of (1.4.38) by $u'(x)$. We observe that $u'(x)$ is discontinuous at $x = 0$ and satisfies (1.4.35) despite the fact that for each $\varepsilon > 0$, $u'(x,\varepsilon)$ is continuous at $x = 0$. The situation is illustrated in Figure 1.5. For a fixed small value of ε there is a very sharp change in $u'(x,\varepsilon)$ in a thin transition layer around the interface; the continuity of $u'(x,\varepsilon)$ at $x = 0$ is deceptive—the useful information is really contained in the discontinuous function $u'(x)$.

Since $u(x,\varepsilon) = -\int_x^1 u'(\eta,\varepsilon)\,d\eta$, we have

$$u(x,\varepsilon) = \int_x^1 \frac{\eta+1}{k(\eta,\varepsilon)}\,d\eta, \qquad (1.4.39)$$

where, in view of (1.4.34), the limit as $\varepsilon \to 0$ may be taken under the integral sign, so that

$$u(x) = \begin{cases} \displaystyle\int_x^1 \frac{\eta+1}{k_2} d\eta = \frac{3}{2k_2} - \frac{1}{k_2}\left(x + \frac{x^2}{2}\right), & x > 0, \\ \displaystyle\int_x^0 \frac{\eta+1}{k_1} d\eta + \int_0^1 \frac{\eta+1}{k_2} d\eta = \frac{3}{2k_2} - \frac{1}{k_1}\left(x + \frac{x^2}{2}\right), & x < 0, \end{cases} \tag{1.4.40}$$

which is certainly continuous at $x = 0$. The continuity of $u(x)$ is guaranteed from the fact that we can pass to the limit under the integral sign in (1.4.39), making $u(x)$ the integral of a piecewise continuous function, and such an integral is, of course, continuous.

Since the interface conditions (1.4.35) and (1.4.36) are deduced by a limiting process from a continuously varying conductivity, we shall call them the *natural interface conditions.*

Let us now calculate Green's function $g(x, \xi)$ for a composite rod with natural interface conditions at $x = 0$ and the same boundary conditions as in (1.4.37). We first place the unit source at ξ in the left half of the rod, so that g satisfies

$$\left.\begin{aligned} &-g'' = 0, && -1 < x < \xi, \xi < x < 0, 0 < x < 1; \\ &g'(-1, \xi) = 0, && g(1, \xi) = 0; \\ &g(\xi+, \xi) = g(\xi-, \xi), && g'(\xi+, \xi) - g'(\xi-, \xi) = -\frac{1}{k_1}; \\ &g(0-, \xi) = g(0+, \xi), && k_1 g'(0-, \xi) = k_2 g'(0+, \xi). \end{aligned}\right\} \tag{1.4.41}$$

Solving the homogeneous equation in each of the three intervals and taking into account the boundary conditions at $x = \pm 1$, we are left with four constants to be determined by the two interface conditions and the two matching conditions at the source. It is often preferable to begin with a solution which already satisfies the source conditions; an obvious candidate is the causal Green's function $h(x, \xi) = H(x - \xi)(\xi - x)/k_1$ for a rod of conductivity k_1. The desired Green's function $g(x, \xi)$ differs from h by a solution of the homogeneous equation in $-1 < x < 0$; in $0 < x < 1$, $g(x, \xi)$ is a solution of the homogeneous equation. In view of the boundary conditions we can write

$$g = \begin{cases} h(x, \xi) + A, & x < 0, \\ B(1 - x), & x > 0. \end{cases} \tag{1.4.42}$$

It remains to apply the interface conditions to obtain

$$\frac{\xi}{k_1} + A = B, \quad -1 = k_2 B,$$

so that

$$B = \frac{1}{k_2}, \quad A = \frac{1}{k_2} - \frac{\xi}{k_1}.$$

We leave as an exercise the calculation of $g(x, \xi)$ when the source is in the right half of the rod.

What are the natural interface conditions for more complicated differential operators? We now present a method which avoids the limiting process described earlier. Suppose that we want to solve the steady diffusion problem

$$-\frac{d}{dx}\left(k(x)\frac{du}{dx}\right) + q(x)u = f(x), \tag{1.4.43}$$

where the source density f is a given piecewise continuous function, $k(x)$ is the diffusion coefficient, and $q(x)$ is a nonnegative absorption coefficient. Normally, one assumes that q is continuous and k smooth and then searches for a classical solution u. However, the left side $-(ku')' + qu$ can be piecewise continuous under weaker conditions on k and q. Suppose, for instance, that k and q are only piecewise continuous; then $-(ku')' + qu$ will be piecewise continuous if (a) ku' is piecewise smooth so that $(ku')'$ is defined as a piecewise continuous function, and (b) qu is piecewise continuous. Now, if ku' is piecewise smooth, it is certainly continuous, and therefore u' is piecewise continuous, so that u is continuous, and the condition on qu is automatically satisfied. In applying these ideas to concrete problems, we usually solve (1.4.43) in the subintervals where both k and q are continuous; this leaves us with constants of integration that are explicitly found by applying interface or matching conditions at the ends of subintervals. Two conditions are needed at each interface x_i:

$$\Delta(ku')_i = 0, \quad \Delta u_i = 0, \tag{1.4.44}$$

where ΔF_i is the jump in F at x_i, that is, $F(x_i+) - F(x_i-)$. If only q is discontinuous at an interface, the matching conditions are

$$\Delta u_i = 0, \quad \Delta u_i' = 0. \tag{1.4.45}$$

In one-dimensional quantum mechanics, the Schrödinger equation has the form

$$u'' + [E - V(x)]u = 0, \tag{1.4.46}$$

where E is a constant and $V(x)$ is the potential. Often, V is only piecewise continuous (as in problems of a rectangular well or a rectangular potential barrier). One then solves for u in the various intervals of continuity of $V(x)$; at the points where V is discontinuous, the matching conditions are (1.4.45).

For a more complicated problem such as

$$(r_2u'')'' + (r_1u')' + r_0u = f,$$

where the coefficients may only be piecewise continuous, one writes the left side as

$$[(r_2u'')' + r_1u']' + r_0u.$$

To make this piecewise continuous, we need the following: (a) r_2u'' piecewise smooth (hence r_2u'' continuous, u'' piecewise continuous, u' and u continuous);

(b) $(r_2u'')' + r_1u'$ piecewise smooth (hence continuous); (c) r_0u piecewise continuous [follows automatically from (a)]. This gives us the four interface conditions

$$\Delta u_i = 0, \quad \Delta u_i' = 0 \quad \Delta(r_2u'')_i = 0, \quad \Delta[(r_2u'')' + r_1u']_i = 0. \tag{1.4.47}$$

As an illustration, consider the small transverse deflection of a beam of constant cross section whose stiffness changes abruptly at $x = 0$. If $E(x)$ is the stiffness of the beam, the deflection satisfies

$$(E(x)u'')'' = f(x),$$

so that, according to (1.4.47), the interface conditions at $x = 0$ are

$$\Delta u_0 = 0, \quad \Delta u_0' = 0, \quad \Delta(Eu'')_0 = 0, \quad \Delta(Eu'')_0' = 0. \tag{1.4.48}$$

These conditions have a very simple interpretation: The deflection, slope, moment, and shear are all continuous at $x = 0$. In particular, if E is the constant E_1 for $x < 0$ and the constant E_2 for $x > 0$, these conditions become

$$u(0+) = u(0-), \quad u'(0+) = u'(0-),$$
$$E_2u''(0+) = E_1u''(0-), \quad E_2u'''(0+) = E_1u'''(0-).$$

In Exercises 1.4.6 and 1.4.7 we give examples of different kinds of problems for composite beams that do *not* lead to natural interface conditions.

EXERCISES

1.4.1 Let q^2 be a positive constant. Find Green's function $g(x, \xi)$ satisfying

$$-g'' + q^2g = \delta(x - \xi), \quad 0 < x, \xi < 1; \qquad g'(0, \xi) = g'(1, \xi) = 0$$

by the direct method of Section 1.2, that is, by starting with two solutions u_1, u_2 of the homogeneous equation satisfying, respectively, the end conditions at $x = 0$ and $x = 1$ and then matching them under the load. Compare your result with the one obtained by images, (1.4.17). Discuss the limit as $q \to 0$ and give a physical reason for the behavior.

1.4.2 (a) Show that the electrostatic potential for a line source of uniform unit density is $u = (1/2\pi)\log(1/\rho)$, where ρ is the cylindrical coordinate measured from the line source (which coincides with the x_3 axis).

(b) Consider the two-dimensional problem

$$\frac{\partial^2 u}{\partial x_1^2} + \frac{\partial^2 u}{\partial x_2^2} = 0, \quad x_2 > 0, -\infty < x_1 < \infty; \qquad u(x_1, 0) = h(x_1), \tag{1.4.49}$$

where $h(x_1)$ is a given function. First find Green's function $g(x, \xi)$ for 0 boundary data by the method of images [using part (a)]. Then write the solution of (1.4.49) by using (1.1.5) as

$$u(x_1, x_2) = \frac{1}{\pi}\int_{-\infty}^{\infty} \frac{x_2}{x_2^2 + (x_1 - \xi_1)^2} h(\xi_1)\, d\xi_1. \tag{1.4.50}$$

1.4.3 An elastic beam is subject to a restoring force proportional to the local displacement and tending to oppose it. If a transverse distributed load $f(x)$ is applied, the appropriate differential equation satisfied by the deflection $u(x)$ is

$$\frac{d^4u}{dx^4} + k^4u = f(x),$$

where k^4 is a positive constant regarded as known.

(a) Find the deflection in an infinite beam $-\infty < x < \infty$, when the applied load is a unit concentrated load at $x = \xi$. This deflection is the free space Green's function.

(b) Find the causal Green's function for the problem.

(c) For a beam simply supported at its ends $x = 0$ and $x = 1$, find Green's function first by using the causal Green's function of part (b) and then by using the method of images.

1.4.4 Consider steady diffusion and absorption of oxygen in tissue occupying the half-space $x > 0$. Oxygen enters the cell wall at $x = 0$ and diffuses into the tissue, where it is absorbed at a rate f. Of course, there is *no further absorption once the concentration of oxygen falls to zero*! Depending on f, the oxygen may either penetrate a finite distance s or an infinite distance in the tissue. The BVP for the concentration $u(x)$ is

$$-u'' = -f, \quad 0 < x < s; \qquad u(0) = 1, \quad u(s) = u'(s) = 0, \qquad (1.4.51)$$

where s and u are both to be found (s could be $+\infty$).

(a) Suppose that f is given as a function of position; show that if

$$\lim_{x\to\infty} f(x) = A > 0,$$

the penetration distance is finite and that (1.4.51) has one and only one solution.

(b) If $f(x) \equiv 1$, find the penetration distance explicitly.

(c) If the absorption is au^p, where $a > 0$ and $p > 0$, show that s is finite if and only if $p < 1$. [*Hint*: Multiply (1.4.51) by u' and note that both sides are perfect derivatives.]

1.4.5 Consider the one-dimensional motion of a mass m attached to a spring whose resistance is proportional to its extension. The displacement $u(t)$ of the mass, measured from the rest position, satisfies

$$mu'' + ku = f(t), \quad t > 0; \qquad u(0) = \alpha, \quad u'(0) = \beta, \qquad (1.4.52)$$

where $f(t)$ is the applied force, k the spring constant, and α, β the initial values of the displacement and velocity, respectively. Find the causal Green's function corresponding to a unit impulse at $t = \tau$ and use it to express the solution of (1.4.52) as

$$u(t) = \int_0^t \frac{1}{\sqrt{km}} \left[\sin \sqrt{\frac{k}{m}}(t-\tau) \right] f(\tau)\, d\tau + \alpha \cos \sqrt{\frac{k}{m}} t + \beta \sqrt{\frac{m}{k}} \sin \sqrt{\frac{k}{m}} t.$$

1.4.6 A simply supported composite beam of constant cross section occupies the interval $0 < x < 2l$. The left half has $EI = 1$, whereas the right half is *rigid* and the two halves are welded together at $x = l$. A concentrated unit transverse force is applied at $x = \xi$, where $0 < \xi < l$. Draw the shear, moment, slope, and deflection diagrams. Express these analytically.

1.4.7 A homogeneous beam of constant cross section is attached to a string. The beam and string are stretched under tension H between the fixed points $x = 0$ and $x = 2l$, the beam occupying the interval $0 < x < l$ and the string the interval $l < x < 2l$. The left end of the beam is simply supported. A transverse concentrated unit force is applied at $x = l/2$. What are the interface conditions at $x = l$? Find the deflection.

1.4.8 Consider the case of a steep potential well or barrier in (1.4.46), which can be ideally represented by a potential $V(x) = \alpha\delta(x)$, where α is a real number. Although such a problem does not fall into the class studied in Section 1.4, it can nevertheless be solved. The principal interest is in the matching (connection) conditions at $x = 0$. If we assume u continuous at the origin, a formal integration of (1.4.46) gives $u'(0+) - u'(0-) = \alpha u(0)$. Show that the same result is obtained by replacing the delta function in (1.4.46) by the sequence

$$f_n(x) = \begin{cases} n, & 0 < x < \frac{1}{n}, \\ 0, & \text{otherwise,} \end{cases}$$

and then proceeding to the limit as $n \to \infty$.

1.4.9 The purpose of this exercise is to show that the Green's function defined in (1.4.25) is symmetric. Write the equations for $g(x, \xi)$ and $g(x, \eta)$; multiply the first equation by $g(x, \eta)$, the second by $g(x, \xi)$; subtract and integrate to obtain the desired result: $g(\xi, \eta) = g(\eta, \xi)$.

REFERENCES AND ADDITIONAL READING

1. G. Barton, *Elements of Green's Functions and Propagation*, Oxford University Press, New York, 1989.
2. J. Keener, *Principles of Applied Mathematics*, 2nd ed., Westview Press, Boulder, CO, 2000.
3. Y. A. Melnikov, *Green's Functions in Applied Mechanics*, CMP Books, San Francisco, 1995.
4. M. H. Protter and H. Weinberger, *Maximum Principles in Differential Equations*, Prentice-Hall, Englewood Cliffs, NJ, 1967.
5. G. Roach, *Green's Function*, Cambridge University Press, New York, 1982.

CHAPTER 2

THE THEORY OF DISTRIBUTIONS

2.1 BASIC IDEAS, DEFINITIONS, AND EXAMPLES

Various examples of Section 1.4 show that one frequently encounters sources that are nearly instantaneous (if time is the independent variable) or almost localized (if a space coordinate is the independent variable). To avoid the cumbersome study of the detailed functional dependence of such sources, we would like to replace them by idealized sources which are truly instantaneous or localized; such idealized sources are said to be *impulsive* or *concentrated* (as opposed to distributed sources). Typical instances of such sources are the concentrated forces and moments of solid mechanics; the heat sources and dipoles in heat conduction; the point masses in the theory of the gravitational potential; the impulsive forces in acoustics and in impact mechanics; the fluid sources and vortices of incompressible fluid mechanics; and the point charges, dipoles, multipoles, line charges, and surface layers in electrostatics.

What do we expect from a mathematical theory of concentrated sources? First, there should be a clear and unambiguous mathematical framework in which such sources have equal standing with distributed sources. Second, a method should be provided for calculating the response to a concentrated source, that is, a means of interpreting and solving a differential equation whose inhomogeneous term is a concentrated source. Third, if a concentrated source is "approximated" by a sequence of distributed sources, the response to the concentrated source should be a suitable limit of the sequence of responses to the distributed sources.

Functions as Linear Functionals

Consider a real-valued continuous function on $\mathbb{R}^n$. The function f is a rule which associates with each point x in $\mathbb{R}^n$ a real number $y = f(x)$, the value of f at x.

Another, indirect description of f is often useful. Instead of giving the value of f at each point x, we give the real number $\int_{\mathbb{R}^n} f(x)\phi(x)\,dx$ for every ϕ belonging to a class K of accessory functions. The role of the class K is secondary; the functions ϕ are merely a vehicle for the description of f. In this alternative characterization we are viewing f as a *functional* on K [that is, f associates with each function ϕ in K the real number $\int_{\mathbb{R}^n} f(x)\phi(x)\,dx$].

Such an indirect description of a point function is common in engineering. A measuring instrument, such as a voltmeter, does not measure the instantaneous value $f(t_0)$ of the voltage at time t_0, but rather a weighted average over a short period $2T$ of time: $(1/2T)\int_{t_0-T}^{t_0+T} f(t)\phi(t)\,dt$, where ϕ is a characteristic of the measuring instrument. The "functional" point of view is also encountered in analytic settings such as in ordinary Fourier series. For instance, if f is a differentiable real-valued function on $0 \leqslant x \leqslant \pi$, then f is equally well characterized by its Fourier sine coefficients $b_n = (2/\pi)\int_0^\pi f(x)\sin nx\,dx$. Here the class K of accessory functions is essentially the set $\sin x$, $\sin 2x, \ldots$. We can, of course, find $f(x)$ from its Fourier sine coefficients by the usual simple Fourier series, but all we care about here is that $f(x)$ is unambiguously determined from the coefficients, with or without an explicit formula. Our purpose in introducing functionals on a space K of accessory functions is not merely to give an alternative description of ordinary functions f; rather, we shall see that the new point of view permits us to enlarge the kinds of functions f that can be described. By choosing the class K of accessory functions ϕ appropriately, we shall be able to characterize as functionals "singular" functions such as the delta function and the dipole function. The most suitable class K of accessory functions will be a class of very smooth functions known as *test functions*. We shall then be able to describe some rather "wild" functions as functionals on this space of very smooth functions. This point of view was introduced by L. Schwartz [12], and the comprehensive theory can be found in his book. (See also Gelfand and Shilov [6], Zemanian [16], Dunford and Schwartz [3], and Haroske and Triebel [7]).

Test Functions

Definition. *A test function* $\phi(x) = \phi(x_1, \ldots, x_n)$ *on* $\mathbb{R}^n$ *is a function which is infinitely differentiable on* $\mathbb{R}^n$ *(this means that every partial derivative to every order exists) and vanishes outside some bounded region (which may vary from test function to test function). The space of all test functions on* $\mathbb{R}^n$ *will be denoted by* $C_0^\infty(\mathbb{R}^n)$.

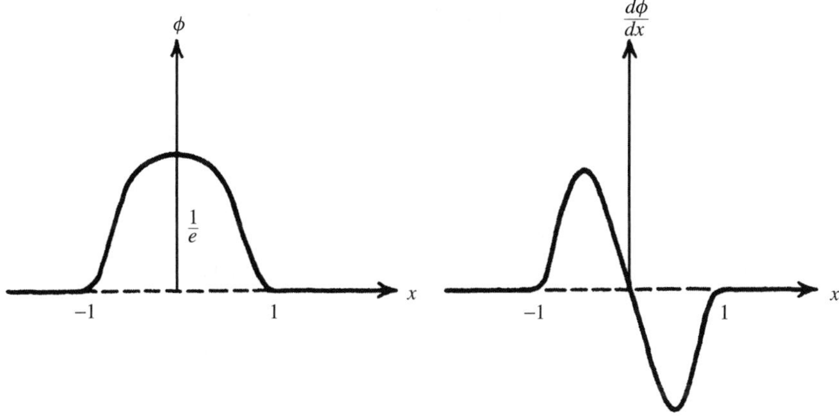

Figure 2.1

REMARKS

1. It is not entirely evident that there exist nontrivial test functions. In $\mathbb{R}$ a test function $\phi(x)$ would vanish identically outside an interval $a < x < b$ but would have to reach nonzero values in the interior, even though every derivative of ϕ is 0 at $x = a$ and $x = b$. At first it is hard to see how a function so "flat" at a and b can differ from the zero function. In fact, if $\phi(x)$ were expandable in a Taylor series of nonzero radius of convergence about $x = a$, then ϕ would have to vanish identically, since all derivatives are 0 at $x = a$. The following example shows that test functions on $\mathbb{R}$ do exist:

$$\phi(x) = \begin{cases} \exp\left(\dfrac{1}{x^2 - 1}\right), & |x| < 1, \\ 0, & |x| \geqslant 1. \end{cases} \tag{2.1.1}$$

This even function is sketched in Figure 2.1. There is no difficulty in seeing that ϕ is infinitely differentiable except possibly at $x = \pm 1$. Because of the evenness of ϕ, it is enough to examine the point $x = 1$. Since

$$\lim_{x \to 1-} \exp\left(\frac{1}{x^2 - 1}\right) = 0,$$

$\phi(x)$ is continuous at $x = 1$. To show that all derivatives of ϕ are 0 at $x = 1$, it is sufficient to note that for any m,

$$\lim_{x \to 1-} \frac{1}{(x^2 - 1)^m} \exp\left(\frac{1}{x^2 - 1}\right) = 0.$$

A spherically symmetric test function on $\mathbb{R}^n$ is given by

$$\phi(x) = \begin{cases} \exp\left(\dfrac{1}{|x|^2 - 1}\right), & |x| < 1, \\ 0, & |x| \geqslant 1, \end{cases} \tag{2.1.2}$$

where $|x| = r$ is the radial coordinate (or distance from the origin).

2. If $\phi_1(x), \phi_2(x)$ are test functions on $\mathbb{R}^n$, so is $c_1\phi_1 + c_2\phi_2$ for any real numbers c_1 and c_2. The space $C_0^\infty(\mathbb{R}^n)$ is therefore a real linear space (see Chapter 4).

3. Let $\phi(x)$ be the test function (2.1.2). Then

$$\phi\left(\frac{x - x_0}{\varepsilon}\right)$$

is also a test function on $\mathbb{R}^n$ which vanishes outside a ball of radius ε with center at x_0.

4. If $\phi(x) \in C_0^\infty(\mathbb{R}^n)$, so does every partial derivative of $\phi(x)$.

5. If $\phi(x) \in C_0^\infty(\mathbb{R}^n)$ and $a(x)$ is infinitely differentiable, then we know that $a(x)\phi(x) \in C_0^\infty(\mathbb{R}^n)$. For instance, if $\phi(x)$ is given by (2.1.1), then $(\sin x)\phi(x)$ and $x^2\phi(x)$ are also test functions on $\mathbb{R}$.

6. If

$$\phi(x_1, \ldots, x_m) \in C_0^\infty(\mathbb{R}^m) \quad \text{and} \quad \psi(x_{m+1}, \ldots, x_n) \in C_0^\infty(\mathbb{R}^{n-m}),$$

then

$$\phi(x_1, \ldots, x_m)\psi(x_{m+1}, \ldots, x_n) \in C_0^\infty(\mathbb{R}^n).$$

Convergence in the Space of Test Functions

We first introduce a concise notation for partial derivatives and differential operators in n independent variables. Let $k_1, \ldots, k_n$ be nonnegative integers; we shall call $k = (k_1, \ldots, k_n)$ a *multi-index of dimension* n. We define

$$|k| = k_1 + \cdots + k_n$$

and

$$D^k = \frac{\partial^{|k|}}{\partial x_1^{k_1} \cdots \partial x_n^{k_n}} = \frac{\partial^{k_1 + \cdots + k_n}}{\partial x_1^{k_1} \cdots \partial x_n^{k_n}},$$

with the understanding that if any component of k is 0, the differentiation with respect to the corresponding variable is omitted. As an example, if $n = 3$ and $k = (2, 0, 5)$, then

$$D^k = \frac{\partial^7}{\partial x_1^2\, \partial x_3^5}.$$

An arbitrary linear differential operator L of order p in n variables can be written as

$$L = \sum_{|k| \leqslant p} a_k(x) D^k, \tag{2.1.3}$$

where the coefficients $a_k(x) = a_{k_1,\ldots,k_n}(x_1,\ldots,x_n)$ are arbitrary functions. The most general differential operator of order 2 in two variables would then be

$$\begin{aligned} L &= \sum_{|k|\leqslant 2} a_k(x)D^k = \sum_{|k|=0} a_k(x)D^k + \sum_{|k|=1} a_k(x)D^k + \sum_{|k|=2} a_k(x)D^k \\ &= a_{0,0} + a_{1,0}\frac{\partial}{\partial x_1} + a_{0,1}\frac{\partial}{\partial x_2} + a_{2,0}\frac{\partial^2}{\partial x_1^2} + a_{1,1}\frac{\partial^2}{\partial x_1 \partial x_2} + a_{0,2}\frac{\partial^2}{\partial x_2^2}. \end{aligned}$$

We also introduce the notion of support of a function $f(x)$.

Definition. *The* support *of $f(x)$ is the closure of the set of points in $\mathbb{R}^n$ on which $f(x) \neq 0$. For instance, the support of the test function (2.1.2) is the closed ball $|x| \leqslant 1$, even though $\phi(x)$ vanishes on the boundary $|x| = 1$. The space of test functions $C_0^\infty(\mathbb{R}^n)$ is also referred to as the space of infinitely differentiable functions with compact support.*

We now introduce a very stringent form of convergence in $C_0^\infty(\mathbb{R}^n)$. We say that the sequence of test functions $\{\phi_1(x), \ldots, \phi_m(x), \ldots\}$ is a *null sequence in* $C_0^\infty(\mathbb{R}^n)$ if and only if the following conditions hold:

1. There exists a common bounded region outside of which all $\phi_m(x)$ vanish (that is, the support of all $\phi_m, m = 1, 2, \ldots,$ is contained within a single, sufficiently large ball).

2. For every multi-index k of dimension n,

$$\lim_{m\to\infty} \max_{x\in\mathbb{R}^n} |D^k \phi_m(x)| = 0.$$

Thus $\{\phi_m(x)\}$ tends *uniformly* to 0 in $\mathbb{R}^n$, and so does the sequence $\{D^k\phi_m(x)\}$. To say that $\{\phi_m\}$ is a null sequence in $C_0^\infty(\mathbb{R}^n)$ therefore means that the approach to 0 is a very strong one: $\{\phi_m\}$ and all its derivatives tend to 0 uniformly in $\mathbb{R}^n$. For example, (a) if $\phi(x)$ is a test function, then $\{(1/m)\phi(x)\}$ is a null sequence in $C_0^\infty(\mathbb{R}^n)$; (b) if $\phi(x)$ is the test function (2.1.1), the sequence of test functions $\{(1/m)\phi(x/m)\}$ fails to meet criterion (1) for a null sequence in $C_0^\infty(\mathbb{R})$, whereas $\{(1/m)\phi(mx)\}$ fails to meet criterion (2). The latter sequence tends to 0 uniformly in $\mathbb{R}$, but the differentiated sequence tends to 0 only pointwise, not uniformly.

Distributions

Definition. *We say that f is a* linear functional *on $C_0^\infty(\mathbb{R}^n)$ if there exists a rule which assigns to each $\phi(x)$ in $C_0^\infty(\mathbb{R}^n)$ a real number [denoted by $\langle f, \phi\rangle$ rather than $f(\phi)$] such that*

$$\langle f, \alpha_1\phi_1 + \alpha_2\phi_2\rangle = \alpha_1\langle f, \phi_1\rangle + \alpha_2\langle f, \phi_2\rangle$$

for all real numbers α_1, α_2, and all ϕ_1, ϕ_2 in $C_0^\infty(\mathbb{R}^n)$.

Note that for any linear functional we have $\langle f, 0\rangle = 0$ and

$$\left\langle f, \sum_{k=1}^{m} \alpha_k \phi_k \right\rangle = \sum_{k=1}^{m} \alpha_k \langle f, \phi_k\rangle.$$

Definition. *A linear functional on $C_0^\infty(\mathbb{R}^n)$ is said to be* continuous *if whenever $\{\phi_m(x)\}$ is a null sequence in $C_0^\infty(\mathbb{R}^n)$, the numerical sequence $\langle f, \phi_m\rangle$ tends to 0 as $m \to \infty$. A continuous linear functional on $C_0^\infty(\mathbb{R}^n)$ is said to be a* distribution *(or an n-*dimensional distribution*). The number $\langle f, \phi\rangle$ is the value of f at ϕ or, perhaps more picturesquely, the* action *of f on ϕ. Two distributions f_1 and f_2 are* equal *if and only if $\langle f_1, \phi\rangle = \langle f_2, \phi\rangle$ for* every *$\phi \in C_0^\infty(\mathbb{R}^n)$. If f_1, f_2 are any distributions and if c_1, c_2 are any real numbers, the distribution $f = c_1 f_1 + c_2 f_2$ is defined in the obvious manner:*

$$\langle f, \phi\rangle = c_1\langle f, \phi\rangle + c_2\langle f_2, \phi\rangle.$$

It is easy to verify that all the conditions for a distribution are satisfied. The space of all n-dimensional distributions is therefore a linear space (see Chapter 4).

Our hope is that the framework of distributions will enable us to incorporate such extraordinary "functions" as the delta function, but first we must make sure that run-of-the-mill functions can be viewed as distributions.

Definition. *A function $f(x)$ on $\mathbb{R}^n$ is said to be* locally integrable *if $\int_\Omega |f| dx$ exists for every bounded domain Ω in $\mathbb{R}^n$.*

Thus, behavior at infinity does not affect local integrability; however, we cannot allow singularities that are too large at any finite point. For instance, $1/|x|^\alpha$ is locally integrable in $\mathbb{R}^n$ if $\alpha < n$ but not if $\alpha \geqslant n$. Of course, continuous functions and piecewise continuous functions are locally integrable.

Theorem 2.1.1. *A locally integrable function $f(x)$ in $\mathbb{R}^n$ defines (generates) an n-dimensional distribution f through the rule*

$$\langle f, \phi\rangle = \int_{\mathbb{R}^n} f(x)\phi(x)\, dx = \int_{-\infty}^{\infty} \cdots \int_{-\infty}^{\infty} f(x_1, \ldots, x_n)\phi(x_1, \ldots, x_n)\, dx_1 \cdots dx_n. \tag{2.1.4}$$

Proof. It is clear that a linear functional on $C_0^\infty(\mathbb{R}^n)$ has been defined. To prove continuity, let $\{\phi_m(x)\}$ be a null sequence all of whose elements vanish outside the finite ball Ω. Then

$$|\langle f, \phi_m\rangle| \leqslant \left(\max_{x\in\Omega}|\phi_m(x)|\right) \int_\Omega |f(x)|\, dx.$$

Since $\{\phi_m\}$ is a null sequence, it certainly follows that $\lim_{m\to\infty} \max |\phi_m(x)| = 0$. The local integrability of f guarantees that $\int_\Omega |f(x)|\, dx$ is finite, so that

$$\lim_{m\to\infty} \langle f, \phi_m\rangle = 0$$

and the functional (2.1.4) is continuous and is therefore a distribution. □

REMARK. By means of (2.1.4) every locally integrable function f can also be regarded as a distribution. As a point function, f has the value $f(x)$ at a point x in $\mathbb{R}^n$; as a distribution, f has an action (value) $\langle f, \phi \rangle$ on a test function $\phi(x)$ belonging to $C_0^\infty(\mathbb{R}^n)$. To what extent does the distribution f determine the point function f? We answer this question below.

If $f_1(x), f_2(x)$ are different continuous functions, they generate different distributions. We must show that for some ϕ in $C_0^\infty(\mathbb{R}^n)$, $\langle f_2, \phi \rangle \neq \langle f_1, \phi \rangle$ or equivalently that $\langle f_2 - f_1, \phi \rangle \neq 0$. Since $f_1(x)$ and $f_2(x)$ are different, there exists x_0 such that $f_2(x_0) \neq f_1(x_0)$, and we may suppose that $f_2(x_0) > f_1(x_0)$. In view of the continuity of these functions, x_0 must have a neighborhood $|x - x_0| < \varepsilon$ in which $f_2(x) - f_1(x) > 0$. We have seen that there is a test function ϕ which is positive in $|x - x_0| < \varepsilon$ and vanishes elsewhere. For this ϕ, rule (2.1.4) shows that $\langle f_2, \phi \rangle$ is larger than $\langle f_1, \phi \rangle$, so that the distributions f_1 and f_2 are different.

If $f_1(x)$ and $f_2(x)$ coincide except at a finite number of points, they generate the same distribution. More generally, two functions are said to be equal almost everywhere if $\int_\Omega |f_1 - f_2|\,dx = 0$ for every bounded domain Ω. Two locally integrable functions that are equal almost everywhere generate the same distribution. If they are not equal almost everywhere, they generate different distributions (proof omitted). *We shall regard functions that are equal almost everywhere as the same function.* (To make all this entirely consistent it is necessary to introduce the notion of equivalence classes, but we shall not do so.)

Definition. *A distribution is* regular *if it can be written in form (2.1.4) with $f(x)$ locally integrable. All other distributions are* singular *(and we shall see that the idea is not vacuous). Even in the latter case we sometimes use formula (2.1.4) symbolically: Given a distribution f, we assign to it a* generalized function *$f(x)$ and write symbolically*

$$\langle f, \phi \rangle = \int_{\mathbb{R}^n} f(x)\phi(x)\,dx. \tag{2.1.5}$$

The symbols f, $f(x)$, and $\langle f, \phi \rangle$ are used interchangeably.

Example 1. Let c be a constant, and consider the functional on $C_0^\infty(\mathbb{R}^n)$ defined by

$$\int_{\mathbb{R}^n} c\phi(x)\,dx.$$

This is clearly a continuous linear functional on $C_0^\infty(\mathbb{R}^n)$, so it is a distribution; in fact, it is a regular distribution generated via (2.1.4) by the constant function $f(x) = c$. We do not distinguish between the number c, the constant point function c, and the distribution c whose value $\langle c, \phi \rangle$ at ϕ is given by $\int_{\mathbb{R}^n} c\phi(x)\,dx$.

Example 2. Let Ω be a domain in $\mathbb{R}^n$, and consider the functional on $C_0^\infty(\mathbb{R}^n)$ defined by

$$\langle I_\Omega, \phi \rangle = \int_\Omega \phi(x)\, dx.$$

This functional is clearly linear and can be written in the form (2.1.4):

$$\int_{\mathbb{R}^n} I_\Omega(x)\phi(x)\, dx,$$

where $I_\Omega(x)$ is the *indicator* or *characteristic* function of Ω; that is, $I_\Omega(x) = 1$, $x \in \Omega$, and $I_\Omega(x) = 0$, otherwise. Since $I_\Omega(x)$ is piecewise continuous, Theorem 2.1.1 tells us that $\langle I_\Omega, \phi \rangle$ is a regular distribution. A particular case of importance occurs in $\mathbb{R}$ with Ω the interval $(0, \infty)$; then $I_\Omega(x)$ is usually denoted by $H(x)$, the *Heaviside function*.

Example 3. Let ξ be a fixed point in $\mathbb{R}^n$. Consider the linear functional δ_ξ, defined from $\langle \delta_\xi, \phi \rangle = \phi(\xi)$, which thus assigns to each test function its value at ξ. If $\{\phi_m(x)\}$ is a null sequence in $C_0^\infty(\mathbb{R}^n)$, then surely the numerical sequence $\{\phi_m(\xi)\}$ tends to 0. Therefore, δ_ξ is a continuous linear functional on $C_0^\infty(\mathbb{R}^n)$, that is, a distribution, known as the *n-dimensional Dirac distribution* (with *pole* at ξ). Let us show that δ_0 (denoted simply by δ) is a *singular* distribution. If δ were regular, there would exist a locally integrable function $f(x)$ such that

$$\int_{\mathbb{R}^n} f(x)\phi(x)\, dx = \phi(0) \quad \text{for every } \phi \in C_0^\infty(\mathbb{R}^n). \tag{2.1.6}$$

Consider the test functions $\psi_a(x) = \phi(x/a)$, where $\phi(x)$ is given by (2.1.2), so that

$$\psi_a(0) = \frac{1}{e}, \quad |\psi_a(x)| \leqslant \frac{1}{e}.$$

Hence

$$\left| \int_{\mathbb{R}^n} f(x)\psi_a(x)\, dx \right| = \left| \int_{|x|<a} f(x) \exp\left(\frac{a^2}{|x|^2 - a^2} \right) dx \right| \leqslant \frac{1}{e} \int_{|x|<a} |f(x)|\, dx.$$

If $f(x)$ is locally integrable, we must have $\lim_{a\to 0} \int_{|x|<a} |f(x)|\, dx = 0$, and hence $\lim_{a\to 0} \int_{\mathbb{R}^n} f(x)\psi_a(x)\, dx = 0$. On the other hand, if (2.1.6) holds, then we have $\int_{\mathbb{R}^n} f(x)\psi_a(x)\, dx = 1/e$, independently of a. This is a contradiction, so that the Dirac distribution δ is singular (the same proof holds if the pole is at ξ instead of at the origin). We visualize $\delta_\xi(x)$ as the symbolic source density for a concentrated unit source at ξ.

Example 4. Translation of a distribution. If the locally integrable function $f(x)$ on $\mathbb{R}^n$ is translated through the vector a, we obtain the locally integrable function

$f(x-a)$, which defines a distribution of its own:

$$\begin{aligned}\langle f(x-a), \phi(x)\rangle &= \int_{\mathbb{R}^n} f(x-a)\phi(x)\,dx \\ &= \int_{\mathbb{R}^n} f(x)\phi(x+a)\,dx = \langle f(x), \phi(x+a)\rangle.\end{aligned}$$

We are therefore led to use the formula

$$\langle f(x-a), \phi(x)\rangle = \langle f(x), \phi(x+a)\rangle \tag{2.1.7}$$

to define the translation through a of any distribution f [note that in (2.1.7) we have used the symbolic notation for generalized functions]. To dot the i's one must show that $\langle f(x), \phi(x+a)\rangle$ really defines a distribution. Since $\phi(x)$ is a test function, so is $\phi(x+a)$; f being a distribution, its action on $\phi(x+a)$ is well defined and linearity is obviously preserved; moreover, if $\{\phi_m(x)\}$ is a null sequence in $C_0^\infty(\mathbb{R}^n)$, so is $\{\phi_m(x+a)\}$, and we have therefore defined a continuous functional.

As a particular case,

$$\langle \delta(x-\xi), \phi(x)\rangle = \langle \delta(x), \phi(x+\xi)\rangle = \phi(\xi),$$

so that we can write

$$\delta_\xi(x) = \delta(x-\xi).$$

In the rest of the work we shall use the notation $\delta(x-\xi)$ for the generalized function corresponding to the Dirac distribution with pole at ξ.

Example 5. Scale expansion and contraction (Similarity transformation). If $f(x)$ is locally integrable, so is $f(\alpha x)$ for any real $\alpha \neq 0$. The distribution corresponding to $f(\alpha x)$ is $\int_{\mathbb{R}^n} f(\alpha x)\phi(x)\,dx$. Setting $\alpha x = y$ and observing that the limits of integration are reversed if α is negative, we find that

$$\langle f(\alpha x),\ \phi(x)\rangle = \frac{1}{|\alpha|^n}\left\langle f(x), \phi\left(\frac{x}{\alpha}\right)\right\rangle. \tag{2.1.8}$$

Even if f is a singular distribution, we use (2.1.8) to define the distribution $f(\alpha x)$.

Example 6. The dipole. Consider in $\mathbb{R}^3$ an electrostatic source configuration consisting of a positive source $1/\varepsilon$ at $\xi + (\varepsilon/2)l$ and a negative source of the same strength at $\xi - (\varepsilon/2)l$, where ξ is a point in $\mathbb{R}^3$ and l is a unit vector (see Figure 2.2). The distribution corresponding to these sources has the action

$$\frac{1}{\varepsilon}\phi\left(\xi + \frac{\varepsilon}{2}l\right) - \frac{1}{\varepsilon}\phi\left(\xi - \frac{\varepsilon}{2}l\right)$$

on any test function in $C_0^\infty(\mathbb{R}^n)$. The *unit dipole with axis l* is obtained as $\varepsilon \to 0$ (note that this brings the charges together, at the same time increasing the magnitude of the charges; otherwise, there would be no effect). The corresponding distribution is

$$\langle f, \phi\rangle = \frac{d\phi}{dl}(\xi), \tag{2.1.9}$$

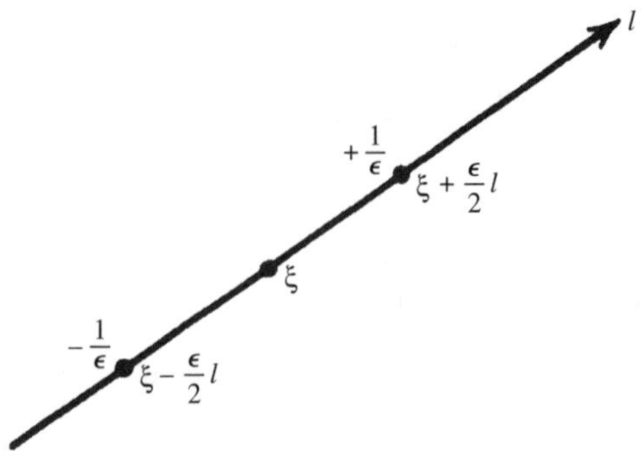

Figure 2.2

which can also be used to define a dipole distribution even if $n \neq 3$. We shall see later that (2.1.9) can be regarded as a distributional derivative of the delta distribution. In any event it is clear that (2.1.9) defines a *singular* distribution.

Example 7. Let Γ be a surface in $\mathbb{R}^3$, and let dS be an area element on Γ. Consider the following functional on $C_0^\infty(\mathbb{R}^3)$:

$$\langle f, \phi \rangle = \int_\Gamma a(x)\phi(x)\, dS,$$

where $a(x)$ is a given, locally integrable function on Γ. This distribution is singular and corresponds to a surface layer of charges (*simple layer*) on Γ with surface density $a(x)$. A special case occurs when Γ is the sphere $|x| = r$ and $a(x) = 1/4\pi r^2$. When acting on a test function $\phi(x)$, our distribution f averages that test function on the sphere:

$$\langle f, \phi \rangle = \frac{1}{4\pi r^2} \int_{|x|=r} \phi(x)\, dS. \tag{2.1.10}$$

Example 8. If $f(x)$ and $a(x)$ are both locally integrable, the product $a(x)f(x)$ need not be locally integrable; and even if af is locally integrable, its action on ϕ may not be related to the individual actions of a and f. If, however, $a(x)$ is infinitely differentiable (but not necessarily of compact support), then $a\phi$ is a test function and we have

$$\langle af, \phi \rangle = \int_{\mathbb{R}^n} a(x)f(x)\phi(x)\, dx = \langle f, a\phi \rangle. \tag{2.1.11}$$

Thus $a(x)f(x)$ generates a distribution through (2.1.11). Obviously, the definition can be used to define the multiplication of any distribution f by an infinitely differentiable function. As a particular case, note that

$$a(x)\delta(x) = a(0)\delta(x). \tag{2.1.11a}$$

Differentiation of Distributions

If $f(x)$ is a differentiable function in $\mathbb{R}$ whose first derivative $f'(x)$ is locally integrable, f' defines its own distribution from (2.1.4):

$$\langle f', \phi\rangle = \int_{-\infty}^{\infty} f'(x)\phi(x)\,dx = -\int_{-\infty}^{\infty} f(x)\phi'(x)\,dx = \langle f, -\phi'\rangle,$$

the middle step being a result of integration by parts combined with the fact that $\phi \equiv 0$ outside a bounded interval. This suggests defining the derivative f' of any distribution f from

$$\langle f', \phi\rangle = \langle f, -\phi'\rangle. \tag{2.1.12}$$

We must check that a distribution is really defined in this way. Since ϕ is a test function, so is $-\phi'$; and f being a distribution, the action of f on $-\phi'$ is defined. Linearity is no problem, but what about continuity? If $\{\phi_m\}$ is a null sequence in $C_0^\infty(\mathbb{R})$, so is $\{-\phi_m'\}$; therefore, $\langle f, -\phi_m'\rangle$ tends to 0 as $m \to \infty$, and (2.1.12) defines a distribution f'.

Similar arguments apply to the partial derivatives of an n-dimensional distribution f. We define

$$\left\langle \frac{\partial f}{\partial x_i}, \phi\right\rangle = \left\langle f, -\frac{\partial \phi}{\partial x_i}\right\rangle, \tag{2.1.13}$$

and by repeated application of this definition we find that

$$\langle D^k f, \phi\rangle = (-1)^{|k|}\langle f, D^k\phi\rangle. \tag{2.1.14}$$

Thus we have the remarkable conclusion that every distribution can be differentiated as often as desired. Distributions can of course be generated by functions which are not differentiable in the ordinary sense, but we now have a way to differentiate such functions in a distributional sense.

Example 9. For a differentiable function such as x^2, the derivative is $2x$ whether x^2 is regarded as a function or as a distribution. What about $(d/dx)|x|$ and $(d/dx)H(x)$, where $H(x)$ is the Heaviside function which vanishes for $x < 0$ and has the value 1 for $x > 0$? Neither $|x|$ nor $H(x)$ is differentiable in the classical sense at $x = 0$, but there should be no harm in thinking of $|x|'$ as sgn x, whereas it is foolish to disregard the *jump* in H and think of H' as the zero function. The reason is that you can recover $|x|$ from sgn x by integration since $\int_0^x \operatorname{sgn} z\, dz = |x|$, but if you regard H' as zero, then integration will not return $H(x)$.

In the distributional sense, (2.1.12) gives

$$\begin{aligned}\langle |x|', \phi\rangle &= -\int_{-\infty}^{\infty} |x|\phi' = -\int_0^{\infty} x\phi'\,dx - \int_{-\infty}^{0} (-x)\phi'\,dx \\ &= \int_0^{\infty} \phi\,dx - \int_{-\infty}^{0} \phi\,dx = \int_{-\infty}^{\infty} (\operatorname{sgn} x)\phi\,dx,\end{aligned}$$

so that $|x|' = \operatorname{sgn} x$. On the other hand,

$$\langle H', \phi \rangle = -\langle H, \phi' \rangle = -\int_0^\infty \phi' \, dx = \phi(0),$$

which shows that H' coincides with δ:

$$H'(x) = \delta(x) \tag{2.1.15}$$

in the sense of distributions. Equation (2.1.15) can at least be formally integrated to give

$$H(x) = \int_{-\infty}^{x} \delta(t) \, dt = \begin{cases} 0, & x < 0, \\ 1, & x > 0. \end{cases}$$

A natural alternative way of defining $H'(x)$ is as the limit of the difference quotient $(1/\varepsilon)[H(x+\varepsilon) - H(x)]$. As a distribution this difference quotient is well defined:

$$\left\langle \frac{1}{\varepsilon}[H(x+\varepsilon) - H(x)], \phi \right\rangle = \frac{1}{\varepsilon}\left(\int_{-\varepsilon}^{\infty} \phi \, dx - \int_0^\infty \phi \, dx \right) = \frac{1}{\varepsilon} \int_{-\varepsilon}^{0} \phi \, dx.$$

Since ϕ is continuous at $x = 0$, the right side has the limit $\phi(0)$ as ε tends to 0, confirming the earlier result. (See also Exercise 2.1.3.)

In $\mathbb{R}^2$, consider the function $f(x_1, x_2) = H(x_1)H(x_2)$, which is unity in the first quadrant and vanishes elsewhere. An easy calculation shows that

$$\frac{\partial^2 f}{\partial x_1 \partial x_2} = \delta(x). \tag{2.1.16}$$

Example 10. If $\delta(x)$ is the Dirac distribution in n dimensions, we can calculate $\partial\delta/\partial x_i$ as follows:

$$\left\langle \frac{\partial \delta}{\partial x_i}, \phi \right\rangle = \left\langle \delta, -\frac{\partial \phi}{\partial x_i} \right\rangle = -\frac{\partial \phi}{\partial x_i}(0),$$

so that $\partial\delta/\partial x_i$ is a unit dipole at the origin with axis in the negative x_i direction [see (2.1.9)].

Example 11. A calculus for functions with jumps. We saw in Example 9 that $H' = \delta$ in the sense of distributions. More generally, if $f(x)$ has simple jumps, its distributional derivative will contain delta contributions from the jumps. Suppose that $f(x)$ is infinitely differentiable on the real line *except* at the points $a_1, \ldots, a_k$, where, however, all left- and right-hand derivatives exist. We denote by $[f'], [f''], \ldots$ the functions obtained by differentiating f without regard to the jumps; for instance, $[H'] = 0$. It is clear that the piecewise continuous function $[f']$ will not usually coincide with the distributional derivative of f. If f has jumps of amounts $\Delta f_1, \ldots, \Delta f_k$ at $a_1, \ldots, a_k$, respectively, the function

$$g = f - \sum_{i=1}^{k} \Delta f_i H(x - a_i)$$

is continuous and has a piecewise continuous derivative g' which coincides with its distributional derivative. Thus, differentiating g in the sense of distributions, we find that

$$g' = f' - \sum_{i=1}^{k} \Delta f_i \delta(x - a_i),$$

and since $g' = [f']$,

$$f' = [f'] + \sum_{i=1}^{k} \Delta f_i \delta(x - a_i). \tag{2.1.17}$$

Therefore each simple jump discontinuity in f contributes a term $\Delta f_i \delta(x - a_i)$ to its distributional derivative.

We can now calculate higher derivatives in this way. Suppose for simplicity that there is only one point, $a_1 = 0$, where discontinuities occur. Let the jump in $[f^{(m)}]$ at $x = 0$ be $\Delta f^{(m)}$. Then

$$\begin{aligned} f' &= [f'] + \Delta f^{(0)} \delta(x), \\ f'' &= [f''] + \Delta f^{(1)} \delta(x) + \Delta f^{(0)} \delta'(x), \\ f^{(m)} &= [f^{(m)}] + \Delta f^{(m-1)} \delta + \Delta f^{(m-2)} \delta' + \cdots + \Delta f^{(0)} \delta^{(m-1)}. \end{aligned}$$

As an illustration let us calculate the successive derivatives of $f = e^{-|x|}$:

$$\begin{aligned} f' &= [f'] = \begin{cases} -e^{-x}, & x > 0, \\ e^{x}, & x < 0, \end{cases} \\ f'' &= [f''] - 2\delta(x) = f - 2\delta(x), \\ f''' &= f' - 2\delta'(x), \\ &\text{etc.} \end{aligned}$$

In particular, we have shown that $f = e^{-|x|}/2$ satisfies $-f'' + f = \delta(x)$ in the sense of distributions, confirming the more intuitive treatment that led to (1.4.14).

Example 12. Let Ω be a bounded domain in $\mathbb{R}^n$ with smooth boundary Γ, and let $I_\Omega(x)$ be the indicator function of Ω (see Example 2). Obviously, I_Ω has a simple jump on Γ, but we are now dealing with a function of more than one variable, so that $\partial I_\Omega / \partial x_i$ cannot be calculated from Example 11. We can, however, go back to definition (2.1.13), which gives

$$\left\langle \frac{\partial I_\Omega}{\partial x_i}, \phi \right\rangle = \left\langle I_\Omega, -\frac{\partial \phi}{\partial x_i} \right\rangle = -\int_\Omega \frac{\partial \phi}{\partial x_i}\, dx = -\int_\Gamma \phi \cos\theta_i \, dS, \tag{2.1.18}$$

where θ_i is the angle between the outward normal to Γ and the x_i axis. Therefore, $\partial I_\Omega / \partial x_i$ is a simple layer of density $-\cos\theta_i$ on the boundary Γ.

Example 13. Let L be the general linear differential operator of order p in n variables $x_1, \ldots, x_n$ [see (2.1.3)]. Assume that the coefficients $\{a_k(x)\}$ are infinitely differentiable. Then Lf is defined for any distribution f by using (2.1.11) and (2.1.14). These give

$$\langle Lf, \phi \rangle = \left\langle \sum_{|k| \leqslant p} a_k \, D^k f, \phi \right\rangle = \left\langle f, \sum_{|k| \leqslant p} (-1)^{|k|} D^k (a_k \phi) \right\rangle . \tag{2.1.19}$$

The pth-order operator appearing on the right side of (2.1.19) is known as the *formal adjoint* of L and is denoted by L^*. We can then write (2.1.19) as

$$\langle Lf, \phi \rangle = \langle f, L^* \phi \rangle, \tag{2.1.20}$$

which defines the distribution Lf in terms of the action of the distribution f on the test function $L^*\phi$. Note that the operator L^* is the one that would appear if we integrated by parts the left side of (2.1.20), treating f as an ordinary p-times-differentiable function. We always have $(L^*)^* = L$. If $L = L^*$, we say that L is *formally self-adjoint.*

Let $L = a_2(x)D^2 + a_1(x)D + a_0(x)$ be the most general second-order operator in one variable, x. Then L^* is defined from

$$L^*\phi = a_2 \phi'' + (2a_2' - a_1)\phi' + (a_2'' - a_1' + a_0)\phi, \tag{2.1.21}$$

and the necessary and sufficient condition for L^* to be formally self-adjoint is

$$a_2' = a_1. \tag{2.1.22}$$

Note that an operator L of any order in any number of variables will be formally self-adjoint if it has constant coefficients and only partial derivatives of even order.

Example 14. We now consider the differentiation of some locally integrable functions whose discontinuities are more severe than the simple jumps covered by (2.1.17).

In $\mathbb{R}$ the locally integrable function $f(x) = H(x) \log x$, where H is the Heaviside function, defines a distribution

$$\langle f, \phi \rangle = \int_0^\infty \phi(x) \log x \, dx.$$

By definition (2.1.12) we have

$$\langle f', \phi \rangle = -\langle f, \phi' \rangle = -\int_0^\infty \phi'(x) \log x \, dx, \tag{2.1.23}$$

which we would like to transform so as to exhibit the function $H(x)(1/x)$, which must somehow be related to $f'(x)$. Since $H(x)/x$ is not locally integrable, it does not define a distribution, so that some qualification is needed. Since the right side of (2.1.23) is a convergent integral, we can write

$$\langle f', \phi \rangle = -\lim_{\varepsilon \to 0} \int_\varepsilon^\infty \phi(x) \log x \, dx,$$

where, performing the integration by parts, we find that

$$\begin{aligned}\langle f', \phi\rangle &= \lim_{\varepsilon\to 0}\left[\int_\varepsilon^\infty \frac{\phi(x)}{x}\,dx + \phi(\varepsilon)\,\log\varepsilon\right]\\ &= \lim_{\varepsilon\to 0}\left[\int_\varepsilon^\infty \frac{\phi(x)}{x}\,dx + \phi(0)\log\varepsilon\right].\end{aligned} \tag{2.1.24}$$

Although the individual terms in brackets will have no limit as $\varepsilon \to 0$ (unless by chance ϕ vanishes at 0), each bracketed term as a whole has a limit, and this limit defines a distribution (no proof is needed since this is merely a restatement, in the present context, of the definition of a distributional derivative).

Since $\log\varepsilon = -\int_\varepsilon^1 (1/x)\,dx$, (2.1.24) can be rewritten in terms of convergent integrals as

$$\langle f', \phi\rangle = \int_0^1 \frac{\phi(x)-\phi(0)}{x}\,dx + \int_1^\infty \frac{\phi(x)}{x}\,dx. \tag{2.1.24a}$$

The right side of (2.1.24a) can be regarded as giving a meaning (*regularization, finite part*) to the divergent integral $\int_0^\infty [\phi(x)/x]\,dx$. In this spirit we define the distribution f' as the *pseudofunction* $H(x)/x$ and write

$$\langle f', \phi\rangle = \left\langle \mathrm{pf}\frac{H(x)}{x}, \phi\right\rangle$$

or

$$\frac{d}{dx}H(x)\log x = \mathrm{pf}\,\frac{H(x)}{x}. \tag{2.1.25}$$

In a similar vein, $H(-x)\log(-x)$ coincides with $\log|x|$ for $x < 0$ and vanishes for $x > 0$; with appropriate definitions mimicking the ones just given, we find that

$$\frac{d}{dx}H(-x)\log(-x) = \mathrm{pf}\,\frac{H(-x)}{x}. \tag{2.1.26}$$

Combining the two formulas, we have

$$\frac{d}{dx}\log|x| = \mathrm{pf}\,\frac{H(x)}{x} + \mathrm{pf}\,\frac{H(-x)}{x} = \mathrm{pf}\,\frac{1}{x}, \tag{2.1.27}$$

where the pseudofunction $1/x$ is defined as the distribution

$$\left\langle \mathrm{pf}\,\frac{1}{x}, \phi\right\rangle = \lim_{\varepsilon\to 0}\left[\int_\varepsilon^\infty \frac{\phi(x)}{x}\,dx + \int_{-\infty}^{-\varepsilon} \frac{\phi(x)}{x}\,dx\right], \tag{2.1.28}$$

which stems from (2.1.24) and a similar formula for the derivative of

$$H(-x)\log(-x),$$

the combination of the integrated terms vanishing as $\varepsilon \to 0$. The right side of (2.1.28) converges as $\varepsilon \to 0$ and defines a distribution, but the individual terms diverge unless $\phi(0) = 0$. Again we can write (2.1.28) in terms of convergent integrals:

$$\left\langle \mathrm{pf}\,\frac{1}{x}, \phi\right\rangle = \int_{-1}^1 \frac{\phi(x)-\phi(0)}{x}\,dx + \int_1^\infty \frac{\phi(x)}{x}\,dx + \int_{-\infty}^{-1} \frac{\phi(x)}{x}\,dx. \tag{2.1.28a}$$

In this way we have assigned a value to the usually divergent integral

$$\int_{-\infty}^{\infty} \frac{\phi(x)}{x}\, dx;$$

this value is known as the *Cauchy principal value.*

Observe that

$$\left\langle x \text{ pf } \frac{1}{x}, \phi \right\rangle = \left\langle \text{pf } \frac{1}{x}, x\phi \right\rangle = \int_{-\infty}^{\infty} \phi(x)\, dx = \langle 1, \phi \rangle,$$

so that

$$x \text{ pf } \frac{1}{x} = 1.$$

Example 15. Consider the locally integrable function $1/|x|$ in $\mathbb{R}^3$. Obviously, this function is not differentiable at the origin in the classical sense. Let us calculate the Laplacian of $1/|x|$ in the distributional sense. Since Δ is formally self-adjoint, we have from (2.1.20)

$$\left\langle \Delta \frac{1}{|x|}, \phi \right\rangle = \left\langle \frac{1}{|x|}, \Delta\phi \right\rangle = \int_{\mathbb{R}^3} \frac{\Delta\phi}{|x|}\, dx.$$

The integral on the right side is convergent because the singularity of $1/|x|$ at the origin is rather weak, so that we may write

$$\int_{\mathbb{R}^3} \frac{\Delta\phi}{|x|}\, dx = \lim_{\varepsilon\to 0} \int_{|x|>\varepsilon} \frac{\Delta\phi}{|x|}\, dx.$$

We calculate the integral on the right by using Green's theorem and the fact that $\phi \equiv 0$ outside a bounded region:

$$\int_{|x|>\varepsilon} \frac{\Delta\phi}{|x|}\, dx = \int_{|x|>\varepsilon} \phi\Delta\left(\frac{1}{|x|}\right) dx + \int_{|x|=\varepsilon} \left[\frac{1}{|x|}\frac{\partial\phi}{\partial n} - \phi\frac{\partial}{\partial n}\left(\frac{1}{|x|}\right)\right] dS,$$

where n is the outward normal to the domain $|x| > \varepsilon$. Setting $|x| = r$, noting that $\partial/\partial n = -(\partial/\partial r)$, and using the fact that $\Delta(1/r) = 0$, $r \neq 0$, we have

$$\int_{|x|>\varepsilon} \frac{\Delta\phi}{|x|}\, dx = -\int_{|x|=\varepsilon} \left(\frac{1}{r}\frac{\partial\phi}{\partial r} + \frac{\phi}{r^2}\right) dS.$$

Since the derivatives of a test function are bounded,

$$\left|\frac{\partial\phi}{\partial r}\right| < M \quad \text{for all } x,$$

and

$$\left|\int_{|x|=\varepsilon} \frac{1}{r}\frac{\partial\phi}{\partial r}\, dS\right| \leqslant \frac{M}{\varepsilon}(4\pi\varepsilon^2) = 4\pi\varepsilon M,$$

which tends to 0 as $\varepsilon \to 0$. We also find that

$$\int_{|x|=\varepsilon} \frac{\phi}{r^2}\, dS = \int_{|x|=\varepsilon} \frac{\phi(0) + [\phi(x) - \phi(0)]}{r^2}\, dS = 4\pi\phi(0) + \int_{|x|=\varepsilon} \frac{\phi(x) - \phi(0)}{r^2}\, dS.$$

Since $\phi(x)$ is continuous at $x = 0$, it is easy to show that the last integral tends to 0 as $\varepsilon \to 0$. It therefore follows that

$$\left\langle \Delta \frac{1}{|x|}, \phi \right\rangle = -4\pi\phi(0),$$

so that, in the distributional sense,

$$\Delta \frac{1}{|x|} = -4\pi\delta(x). \tag{2.1.29}$$

EXERCISES

2.1.1 It is often useful to modify the test function (2.1.2) so that its integral over the unit ball is 1. We therefore define

$$\phi(x) = \begin{cases} k_n \exp\left(\dfrac{1}{|x|^2 - 1}\right), & |x| < 1, \\ 0, & |x| \geqslant 1, \end{cases} \tag{2.1.30}$$

where k_n is chosen so that $\int_{|x|<1} \phi(x)\, dx = 1$. With ϕ given by (2.1.30) and $\varepsilon > 0$, define

$$\phi_\varepsilon(x) = \frac{1}{\varepsilon^n} \phi\left(\frac{x}{\varepsilon}\right), \tag{2.1.31}$$

and note that $\phi_\varepsilon(x) \equiv 0$ for $|x| \geqslant \varepsilon$ and that $\int_{\mathbb{R}^n} \phi_\varepsilon(x)\, dx = 1$. Thus, $\phi_\varepsilon(x)$ is a test function highly peaked about the origin if ε is small. Show that if $f(x)$ is continuous,

$$\lim_{\varepsilon \to 0} \int_{\mathbb{R}^n} \phi_\varepsilon(x - \xi) f(x)\, dx = f(\xi), \tag{2.1.32}$$

so that (2.1.32) recovers the point values of f from its actions on test functions. Prove that if $f \equiv 0$ for $|x| \geqslant R$, then (2.1.32) holds *uniformly* for ξ on $\mathbb{R}^n$.

2.1.2 (a) On $\mathbb{R}$ construct a test function that has the value 1 in $-1 \leqslant x \leqslant 1$. [*Hint*: Look at functions of the type $\int_{-\infty}^{x} \phi(\xi)\, d\xi$, where $\phi(x)$ is the test function on $\mathbb{R}$, given in Exercise 2.1.1.]

(b) If Ω is a bounded domain in $\mathbb{R}^n$ and $\phi_\varepsilon(x)$ is the test function of Exercise 2.1.1, show that the function

$$\psi(x) = \int_\Omega \phi_\varepsilon(x - \xi)\, d\xi$$

is itself a test function and has the value 1 in the part of the interior of Ω lying at a distance greater than ε from the boundary.

2.1.3 We have defined the derivative f' of a one-dimensional distribution f from $\langle f', \phi\rangle = \langle f, -\phi'\rangle$. It is equally reasonable to try instead to define f' using the expression $\lim_{\varepsilon\to 0}[f(x+\varepsilon) - f(x)]/\varepsilon$, where $f(x+\varepsilon)$ is a translated distribution. Adopting this point of view, we find that f' has to satisfy

$$\langle f', \phi\rangle = \langle f, -\phi'\rangle - \lim_{\varepsilon\to 0}\left\langle f(x), \frac{\phi(x)-\phi(x-\varepsilon)}{\varepsilon} - \phi'(x)\right\rangle,$$

which will be in agreement with the earlier definition if

$$\lim_{\varepsilon\to 0}\left\langle f(x), \frac{\phi(x)-\phi(x-\varepsilon)}{\varepsilon} - \phi'(x)\right\rangle = 0$$

for every ϕ in $C_0^\infty(\mathbb{R})$. Since f is a distribution, all we have to show is that $\{[\phi(x)-\phi(x-\varepsilon)]/\varepsilon\} - \phi'(x)$ is a null sequence in $C_0^\infty(\mathbb{R})$. Prove this last statement, thereby reconciling the two definitions.

2.1.4 Show that $g = -(1/2\pi)\log|x|$ satisfies $-\Delta g = \delta(x)$ in the distributional sense (in two dimensions).

2.1.5 Show that $g = e^{-q|x|}/4\pi|x|$ satisfies $-\Delta g + q^2 g = \delta(x)$ in the distributional sense (in three dimensions).

2.1.6 Show that $g = -\frac{1}{2}|x-\xi|$, where ξ is fixed, satisfies $-g'' = \delta(x-\xi)$ in the distributional sense (in one dimension). Show that the same is true for Green's functions defined by (1.1.3) and (1.4.5) (these functions having been suitably extended to the whole real line). Use the method of Example 11 or definition (2.1.20). Also show that $u = H(x)x^3/6$ satisfies $u^{(4)} = \delta(x)$.

2.1.7 (a) Let $\lambda > -1$, then

$$x_+^\lambda = \begin{cases} x^\lambda, & x > 0, \\ 0, & x \leqslant 0, \end{cases}$$

is a locally integrable function in $\mathbb{R}$, and defines a distribution

$$\langle x_+^\lambda, \phi\rangle = \int_0^\infty x^\lambda \phi(x)\,dx, \quad \lambda > -1.$$

Instead of x_+^λ we could have written $H(x)x^\lambda$ as in Example 14, but the present notation has some advantages. As a distribution, x_+^λ is differentiable for $-1 < \lambda < 0$ and its derivative should be related to $\lambda x_+^{\lambda-1}$, thereby serving to define the pseudofunction pf x_+^λ when $-2 < \lambda < -1$. Show (as in Example 14) that for $-1 < \lambda < 0$,

$$\begin{aligned}\langle (x_+^\lambda)', \phi\rangle &= \lim_{\varepsilon\to 0}\left[\int_\varepsilon^\infty \lambda x^{\lambda-1}\phi\,dx + \varepsilon^\lambda \phi(0)\right] \\ &= \int_0^\infty \lambda x^{\lambda-1}[\phi(x)-\phi(0)]\,dx,\end{aligned}$$

where the last integral is convergent and can be used to define the finite part of the divergent integral $\int_0^\infty \lambda x^{\lambda-1}\phi(x)\,dx$ and is written

$$\lambda\langle \mathrm{pf}\, x_+^{\lambda-1}, \phi\rangle.$$

(b) One can also assign a value to $\int_0^\infty x^\mu \phi(x)\,dx$ by analytic continuation in the complex μ-plane. Our integral is well-defined for Re $\mu > -1$ but can be rewritten as

$$\int_0^1 x^\mu [\phi(x) - \phi(0)]\,dx + \int_1^\infty x^\mu \phi(x) + \frac{\phi(0)}{\mu+1}$$

where the first term is defined for Re $\mu > -2$ [because $\phi(x) - \phi(0)$ vanishes at $x = 0$], the second for all μ since ϕ has compact support, and the last for $\mu \neq -1$. Thus all terms are defined for $-2 < \operatorname{Re}\mu < -1$ and in particular for μ real and $-2 < \mu < -1$. Show that the definition is equivalent to the one in part (a).

2.1.8 In $\mathbb{R}^2$ the function $1/r^2$, where $r = |x|$, is *not* locally integrable because the singularity at the origin is too strong. We can, however, find a locally integrable function F such that $\Delta F = 1/r^2$ for $r \neq 0$. In fact, it is easy to see that $F = \frac{1}{2}\log^2 r$ has this property. We then *define* the pseudofunction $1/r^2$ as a distribution from the formula

$$\Delta \tfrac{1}{2}\log^2 r = \mathrm{pf}\frac{1}{r^2}.$$

Carry out the calculations to show that

$$\left\langle \mathrm{pf}\frac{1}{r^2}, \phi\right\rangle = \lim_{\varepsilon\to 0}\left[\int_{r\geqslant\varepsilon} (\phi/r^2)\,dx + 2\pi\phi(0)\log\varepsilon\right].$$

2.1.9 In $\mathbb{R}$ let $a(x)$ be an infinitely differentiable function, and let $f(x)$ be an arbitrary generalized function. Show that

$$[a(x)f(x)]' = a(x)f' + a'(x)f,$$

and therefore

$$a(x)\delta'(x) = a(0)\delta'(x) - a'(0)\delta(x).$$

Thus $x\delta' = -\delta$ and $x^2\delta' = 0$. As pointed out in Example 14, we also have

$$x\,\mathrm{pf}\left(\frac{1}{x}\right) = 1. \tag{2.1.33}$$

2.1.10 Prove (2.1.16).

2.1.11 Let $\phi \in C_0^\infty(\mathbb{R})$; show that

$$\phi_m(x) = \phi\left(x + \frac{1}{m}\right) - \phi(x)$$

is a null sequence in $C_0^\infty(\mathbb{R})$.

2.2 CONVERGENCE OF SEQUENCES AND SERIES OF DISTRIBUTIONS

One of the most natural ways of looking at the delta function is as a limit of ordinary functions (say, locally integrable). For instance, in one dimension we might think of a concentrated unit source at $x = 0$ as the limit as $k \to \infty$ of the sequence $\{s_k(x)\}$ of uniformly distributed sources defined by

$$s_k(x) = \begin{cases} 0, & |x| > \dfrac{1}{2k}, \\ k, & |x| < \dfrac{1}{2k}. \end{cases}$$

In the string problem we would regard a concentrated unit load at $x = 0$ as the limit of distributed loads of increasing density over narrower intervals. The way in which we approximate the concentrated load is not unique; we could, for instance, use distributed loads $s_k(x)$ that are bell-shaped with increasing peak and decreasing base in such a way that the total load is unity. Some one-dimensional examples of this type are

$$s_k(x) = \frac{1}{\pi} \frac{k}{1 + k^2 x^2} \quad \text{and} \quad s_k(x) = \frac{k}{\sqrt{\pi}} e^{-k^2 x^2}.$$

In any event, we realize that we cannot have $\lim_{k \to \infty} s_k(x) = \delta(x)$ in any classical sense; on the other hand, both $s_k(x)$ and $\delta(x)$ are well defined as one-dimensional distributions, and it would be reasonable to say that $\lim_{k \to \infty} s_k = \delta$ in the distributional sense if

$$\lim_{k \to \infty} \langle s_k, \phi \rangle = \langle \delta, \phi \rangle \quad \text{for every } \phi \text{ in } C_0^\infty(\mathbb{R}),$$

or, equivalently,

$$\lim_{k \to \infty} \int_{\mathbb{R}} s_k(x) \phi(x) \, dx = \phi(0) \quad \text{for every } \phi \text{ in } C_0^\infty(\mathbb{R}).$$

Consider a family $\{f_\alpha(x)\}$ of distributions depending on a parameter α that belongs to an index set I (often, this index set will be the set of positive integers, and we then let $\alpha = k$, so that we would be talking about a sequence of distributions); thus, for each α in I, a distribution f_α is defined.

Definition 2.2.1. *Let $\{f_\alpha\}$ be a family of n-dimensional distributions. We say that $\{f_\alpha\}$* converges distributionally *to the distribution f as $\alpha \to \alpha_0$, and write $f_\alpha \to f$ (as $\alpha \to \alpha_0$) if*

$$\lim_{\alpha \to \alpha_0} \langle f_\alpha, \phi \rangle = \langle f, \phi \rangle \quad \textit{for each } \phi \textit{ in } C_0^\infty(\mathbb{R}^n). \tag{2.2.1}$$

Observe that convergence of distributions has been reduced to convergence of numbers.

Often one can ascertain that for each ϕ the left side of (2.2.1) has a limit as $\alpha \to \alpha_0$, but one is not sure that the limiting values are the values of a distribution f. These limiting values clearly are the values of a linear functional f on $C_0^\infty(\mathbb{R}^n)$; it is also true (proof omitted) but not obvious that the functional is continuous. We therefore have the following theorem.

Theorem 2.2.1. *If* $\lim_{\alpha \to \alpha_0} \langle f_\alpha, \phi \rangle$ *exists for each* ϕ *in* $C_0^\infty(\mathbb{R}^n)$, *there exists one and only one distribution* f *such that* $f_\alpha \to f$ *as* $\alpha \to \alpha_0$, *that is,*

$$\lim_{\alpha \to \alpha_0} \langle f_\alpha, \phi \rangle = \langle f, \phi \rangle \quad \textit{for each } \phi \in C_0^\infty(\mathbb{R}^n).$$

As a particular case of the theorem, consider a sequence of distributions $\{f_k\}$ such that $\lim_{k \to \infty} \langle f_k, \phi \rangle$ exists for each ϕ; it then follows that there exists a unique distribution f such that $f_k \to f$ as $k \to \infty$ (that is, $\langle f_k, \phi \rangle \to \langle f, \phi \rangle$ for each ϕ).

We also have an immediate definition for the convergence of a *series* of distributions. Let $u_1, \ldots, u_k, \ldots$ be n-dimensional distributions; we write $\sum_{k=1}^\infty u_k = f$ if and only if the sequence of distributions $f_k = \sum_{j=1}^k u_j$ converges to f.

When dealing with a family $\{f_\alpha\}$ of locally integrable functions on $\mathbb{R}^n$, we may wish to inquire about the relation between distributional convergence and more classical types of convergence. The following theorem is often useful.

Theorem 2.2.2. *Let* $\{f_\alpha(x)\}$ *be a family of locally integrable functions on* $\mathbb{R}^n$. *If* $\lim_{\alpha \to \alpha_0} f_\alpha(x) = f(x)$ *uniformly over every bounded ball in* $\mathbb{R}^n$, *then* $f_\alpha \to f$ *distributionally as* $\alpha \to \alpha_0$.

Proof. It follows from uniformity that $f(x)$ is locally integrable, so that

$$\langle f_\alpha, \phi \rangle = \int_{\mathbb{R}^n} f_\alpha(x)\phi(x)\, dx \to \int_{\mathbb{R}^n} f\phi\, dx = \langle f, \phi \rangle.$$

The middle step is a consequence of the uniform convergence of $f_\alpha \phi$ to $f\phi$ and of the fact that ϕ vanishes outside a bounded ball. □

A more powerful theorem follows from the Lebesgue Dominated Convergence Theorem:

Theorem 2.2.3. *Let* $\{f_\alpha(x)\}$ *be a family of locally integrable real-valued functions on* $\mathbb{R}^n$ *with* $|f_\alpha(x)| \leqslant g(x)$, *where* $g(x)$ *is a locally integrable function. Suppose that as* $\alpha \to \alpha_0$, $f_\alpha(x) \to f(x)$ *pointwise on* $\mathbb{R}$. *Then* $f_\alpha \to f$ *distributionally as* $\alpha \to \alpha_0$.

Next we observe that pointwise convergence is neither necessary nor sufficient for distributional convergence. For instance, the sequence $\{\sin kx\}$ in $\mathbb{R}$ does not converge pointwise to a limiting function on $\mathbb{R}$; yet we can show using integration by parts that

$$\lim_{k \to \infty} \int_{-\infty}^{\infty} \sin kx\, \phi(x)\, dx = 0$$

for each test function ϕ, so that $\sin kx \to 0$ distributionally as $k \to \infty$. On the other hand, the sequence $\{k^\alpha H(x)xe^{-kx}\}$ tends to 0 pointwise on $\mathbb{R}$, but does not converge distributionally if $\alpha > 2$, converges to δ if $\alpha = 2$, and converges to the zero distribution if $\alpha < 2$ (see Exercise 2.2.3).

Let $\{f_\alpha\}$ be a family of arbitrary distributions with the property $f_\alpha \to f$ distributionally as $\alpha \to \alpha_0$. What can be said about the derivatives (which necessarily exist, since everything in sight is a distribution and distributions are differentiable)? We claim that

$$\frac{\partial f_\alpha}{\partial x_i} \to \frac{\partial f}{\partial x_i} \quad \text{distributionally as } \alpha \to \alpha_0.$$

The proof is simple:

$$\left\langle \frac{\partial f_\alpha}{\partial x_i}, \phi \right\rangle = \left\langle f_\alpha, -\frac{\partial \phi}{\partial x_i} \right\rangle \to \left\langle f, -\frac{\partial \phi}{\partial x_i} \right\rangle = \left\langle \frac{\partial f}{\partial x_i}, \phi \right\rangle.$$

It follows immediately that as $\alpha \to \alpha_0$, $D^k f_\alpha \to D^k f$ distributionally for any multi-index k. *Thus every convergent sequence or series of distributions can be differentiated term by term as often as required.* How unlike classical convergence, where term-by-term differentiation is possible only under rather severe restrictions! As a simple one-dimensional example, note that since

$$\lim_{m\to\infty} \sin mx = 0 \quad \text{and} \quad \lim_{m\to\infty} \cos mx = 0,$$

we must also have

$$0 = \lim_{m\to\infty} m^p \sin mx = \lim_{m\to\infty} m^p \cos mx \quad \text{for any integer } p \geqslant 0.$$

In mathematical physics one frequently encounters sequences of functions which fail to converge uniformly on $\mathbb{R}^n$. Some of these sequences converge pointwise; others fail altogether to converge. In either case a distributional interpretation is often possible. Suppose that $\{f_k(x)\}$ is the sequence in question (say on $\mathbb{R}$, for simplicity) and that a positive integer m can be found such that $\{f_k(x)\}$ is the mth derivative of a sequence $\{g_k(x)\}$ which converges uniformly to $g(x)$. Then $g_k \to g$ distributionally; and since we are permitted to differentiate term by term, $g_k^{(m)} \to g^{(m)}$, that is, $f_k \to g^{(m)}$, where $g^{(m)}$ is the mth derivative (in the sense of distributions) of g. This idea is applied to Fourier series in Section 2.3.

Example 1. In $\mathbb{R}$ consider the sequence $\{s_k(x)\}$ of piecewise continuous functions

$$s_k(x) = \begin{cases} 0, & |x| > \dfrac{1}{2k}, \\ k, & |x| < \dfrac{1}{2k}. \end{cases} \tag{2.2.2}$$

Then, denoting the interval $|x| < 1/2k$ by I_k, we have

$$\langle s_k, \phi \rangle = \int_{I_k} k\phi(x)\,dx = \phi(0) + \int_{I_k} k[\phi(x) - \phi(0)]\,dx$$

and

$$\left| k \int_{I_k} [\phi(x) - \phi(0)]\, dx \right| \leqslant k \int_{I_k} |\phi(x) - \phi(0)|\, dx \leqslant \max_{x \in I_k} |\phi(x) - \phi(0)|.$$

If ϕ is continuous at $x = 0$, the maximum tends to 0 as $k \to \infty$. Therefore,

$$\lim_{k \to \infty} \langle s_k, \phi \rangle = \phi(0) \tag{2.2.3}$$

for every ϕ continuous at the origin. This means that (2.2.3) is certainly true for all test functions (with plenty to spare). In the distributional sense we therefore have

$$\lim_{k \to \infty} s_k(x) = \delta(x),$$

but it is important in other applications to know that (2.2.3) holds under much less severe restrictions on ϕ. We should point out that (2.2.3) can hold even if $s_k(0)$ is negative (see Exercise 2.2.16).

Example 2. Let $\{f_\alpha(x)\}$ be a family of locally integrable functions in $\mathbb{R}^n$ with the property

$$\lim_{\alpha \to \alpha_0} \int_{\mathbb{R}^n} f_\alpha(x) \phi(x)\, dx = \phi(0) \quad \text{for each } \phi \text{ in } C_0^\infty(\mathbb{R}^n).$$

We call $\{f_\alpha\}$ an *n-dimensional delta family* (as $\alpha \to \alpha_0$) and subsequently write $\lim_{\alpha \to \alpha_0} f_\alpha(x) = \delta(x)$. (Often, the index α runs through the positive integers k; we then use the name *delta sequence* for $\{f_k\}$.) If $\{f_\alpha(x)\}$ is a delta family, then

$$\lim_{\alpha \to \alpha_0} \int_{\mathbb{R}^n} f_\alpha(x - \xi) \phi(x)\, dx = \phi(\xi), \tag{2.2.4}$$

so that

$$\lim_{\alpha \to \alpha_0} f_\alpha(x - \xi) = \delta(x - \xi).$$

In Example 1, $\{s_k(x)\}$ was a one-dimensional delta sequence. The following theorem shows how easy it is to construct delta sequences by starting with any nonnegative function $f(x)$ and suitably compressing it while increasing its peak.

Theorem 2.2.4. *Let $f(x) = f(x_1, \ldots, x_n)$ be a* nonnegative *locally integrable function on $\mathbb{R}^n$ for which $\int_{\mathbb{R}^n} f(x)\, dx = 1$. With $\alpha > 0$ define*

$$f_\alpha(x) = \frac{1}{\alpha^n} f\left(\frac{x}{\alpha}\right) = \frac{1}{\alpha^n} f\left(\frac{x_1}{\alpha}, \ldots, \frac{x_n}{\alpha}\right); \tag{2.2.5}$$

then $\{f_\alpha(x)\}$ is a delta family as $\alpha \to 0$ [and, setting $\alpha = 1/k$, the sequence $\{s_k(x) \doteq k^n f(kx_1, \ldots, kx_n)\}$ is a delta sequence as $k \to \infty$].

Proof. The substitution $y = x/\alpha$ yields these three properties:

$$\text{(a)} \int_{\mathbb{R}^n} f_\alpha(x)\,dx = 1,$$

$$\text{(b)} \lim_{\alpha \to 0} \int_{|x|>A} f_\alpha(x)\,dx = 0 \quad \text{for each } A > 0,$$

$$\text{(c)} \lim_{\alpha \to 0} \int_{|x|<A} f_\alpha(x)\,dx = 1 \quad \text{for each } A > 0,$$

so that, for small positive α, $f_\alpha(x)$ is highly peaked about $x = 0$ in such a way that the total strength of this distributed source is unity, with most of it near the origin. We must show that for each test function ϕ, $\lim_{\alpha\to 0} \int_{\mathbb{R}^n} f_\alpha \phi\,dx = \phi(0)$; in view of (a) it suffices to show that

$$\lim_{\alpha \to 0} \int_{\mathbb{R}^n} f_\alpha(x)\eta(x)\,dx = 0, \tag{2.2.6}$$

where $\eta(x) = \phi(x) - \phi(0)$.

We divide the region of integration $\mathbb{R}^n$ into the two parts $|x| \leqslant B$, $|x| > B$, so that

$$\left| \int_{\mathbb{R}^n} f_\alpha(x)\eta(x)\,dx \right| \leqslant \left| \int_{|x|\leqslant B} f_\alpha \eta\,dx \right| + \left| \int_{|x|>B} f_\alpha \eta\,dx \right|.$$

Since $\phi(x)$ is bounded, so is $\eta(x)$; thus $|\eta| \leqslant M$ for all x. Setting $p(B) = \max_{|x|\leqslant B} |\eta|$ and using the *nonnegativity* of f and property (a), we have

$$\left| \int_{\mathbb{R}^n} f_\alpha \eta\,dx \right| \leqslant p(B) + M \int_{|x|>B} f_\alpha\,dx.$$

To prove (2.2.6) we must show that for each $\varepsilon > 0$ there exists $\gamma > 0$ such that

$$\left| \int_{\mathbb{R}^n} f_\alpha \eta\,dx \right| < \varepsilon \qquad \text{for all } \alpha, \quad 0 < \alpha < \gamma.$$

Since $\eta(x)$ is continuous and $\eta(0) = 0, \lim_{B\to 0} p(B) = 0$. Therefore, we may choose B (independent of α) such that $p(B) < \varepsilon/2$. With B so chosen, we use property (b) to select γ such that

$$\int_{|x|>B} f_\alpha(x)\,dx < \frac{\varepsilon}{2M} \quad \text{whenever} \quad 0 < \alpha < \gamma.$$

This completes the proof of Theorem 2.2.4. □

REMARK. Examining the proof of Theorem 2.2.4, we see that many of the conditions can be relaxed and the results strengthened. We observe first that it is not essential for the family f_α to be constructed as in (2.2.5) by compression of a single function f; it is sufficient for f_α to satisfy conditions (a), (b), (c) or something similar. Next, we note that (2.2.4) will actually hold for a much larger class of functions $\phi(x)$ than test functions. Exercise 2.2.4 incorporates both of these extensions. Exercise 2.2.5 removes the nonnegativity requirement for compressive families on $\mathbb{R}$, although for general families the requirement cannot be dropped (see Example 5).

Example 3. We now look at some special cases of delta families of the nonnegative type (2.2.5). The first five examples are in $\mathbb{R}$:

(a) $f(x) = \dfrac{1}{\pi(1+x^2)}, \qquad \left\{ f_y(x) = \dfrac{y}{\pi(y^2+x^2)} \right\}$
(delta family as $y \to 0+$),

(b) $f(x) = \dfrac{e^{-x^2/4}}{\sqrt{4\pi}}, \qquad \left\{ f_t(x) = \dfrac{e^{-x^2/4t}}{\sqrt{4\pi t}} \right\}$
(delta family as $t \to 0+$; we have set $\alpha^2 = t$),

(c) $f(t) = H(t)\dfrac{e^{-1/4t}}{\sqrt{4\pi}t^{3/2}}, \qquad \left\{ f_x(t) = H(t)\dfrac{xe^{-x^2/4t}}{\sqrt{4\pi}t^{3/2}} \right\}$
(delta family as $x \to 0+$; we have set $\alpha = x^2$),

(d) $f(x) = \dfrac{\sin^2 x}{\pi x^2}, \qquad \left\{ f_R(x) = \dfrac{\sin^2 Rx}{\pi R x^2} \right\}$
(delta family as $R \to \infty$),

(e) $f(x) = H(x)xe^{-x}, \quad \{f_k(x) = k^2 H(x) x e^{-kx}\}$
(delta sequence as $k \to \infty$).

Case (a) occurs in potential theory; see Exercise 1.4.2. Cases (b) and (c) both occur for the time-dependent equation of heat conduction (see Chapter 8). Case (d) is the Féjer kernel, as it occurs in the theory of Fourier integrals. Case (e) is a delta sequence with $f_k(0) = 0$.

Let us look at some cases in $\mathbb{R}^n$:

(f) $f(x) = \psi(x)$, the test function (2.1.30) renamed ψ; $f_\alpha(x) = \psi(x/\alpha)$ is a delta family as $\alpha \to 0+$. Thus

$$\lim_{\alpha\to 0+} \int_{\mathbb{R}^n} \psi\left(\frac{x}{\alpha}\right)\phi(x)\,dx = \phi(0) \quad \text{for every } \phi \text{ in } C_0^\infty(\mathbb{R}^n).$$

Actually, we showed much more in Exercise 2.1.1, where the present ϕ was denoted by f and was merely assumed continuous.

(g) $f(x)$ depends only on $r = |x|$. The requirement $\int_{\mathbb{R}^n} f(x)\,dx = 1$ becomes $\int_0^\infty r^{n-1} f(x)\,dr = 1/S_n$, where S_n is the area of the unit sphere. (See Exercise

2.2.1.) Case (f) was of this type; another one is the Gauss kernel

$$f(x) = (4\pi)^{-n/2} e^{-r^2/4},$$
$$f_t(x) = (4\pi t)^{-n/2} e^{-r^2/4t} \quad \text{(delta family as } t \to 0+).$$

This family occurs in n-dimensional heat conduction and in probability.

(h$_1$) In $\mathbb{R}^3$ take $f(x) = 1/\pi^2(r^2+1)^2$; then $f_\alpha(x) = \alpha/\pi^2(r^2+\alpha^2)^2$ is a delta family as $\alpha \to 0+$.

(h$_2$) In $\mathbb{R}^2$ take $f(x) = 1/2\pi(r^2+1)^{3/2}$; then $f_\alpha(x) = \alpha/2\pi(r^2+\alpha^2)^{3/2}$ is a delta family as $\alpha \to 0+$.

Both (h$_1$) and (h$_2$) have applications in potential theory.

Example 4. We now give some examples of delta families in $\mathbb{R}$ that are not of type (2.2.5).

(i) The Dirichlet kernel for Fourier integrals,

$$\frac{1}{2\pi}\int_{-R}^{R} e^{i\omega x}\, d\omega = \frac{\sin Rx}{\pi x},$$

is a delta family as $R \to \infty$. Note that we can write $(\sin Rx)/\pi x = Rf(Rx)$, where $f(x) = (\sin x)/\pi x$, $\int_{-\infty}^{\infty} f(x)\, dx = 1$; but $f(x)$ is not $\geqslant 0$. [See Exercise 2.2.5(b).]

(j) Let $I_k = \int_{-1}^{1}(1-x^2)^k\, dx$. The sequence

$$f_k(x) = \begin{cases} (1-x^2)^k/I_k, & |x| < 1, \\ 0, & |x| \geqslant 1, \end{cases}$$

is a delta sequence as $k \to \infty$ and is useful in proving the Weierstrass approximation theorem.

(k) The family

$$f_r(\theta) = \begin{cases} \dfrac{1}{2\pi}\left(\dfrac{1-r^2}{1+r^2-2r\cos\theta}\right), & |\theta| \leqslant \pi, \\ 0, & |\theta| > \pi, \end{cases}$$

is a delta family as $r \to 1-$. It is the so-called *Poisson kernel* that comes up in potential theory (Dirichlet problem for the unit disk).

(l) The sequence

$$f_k(x) = \begin{cases} \displaystyle\sum_{m=-k}^{k} \frac{1}{2\pi} e^{imx} = \frac{\sin(k+\frac{1}{2})x}{2\pi \sin\frac{1}{2}x}, & |x| \leqslant \pi, \\ 0, & |x| > \pi, \end{cases}$$

is a delta sequence as $k \to \infty$. This sequence is the Dirichlet kernel for Fourier series and plays a central role in proving convergence theorems for Fourier series.

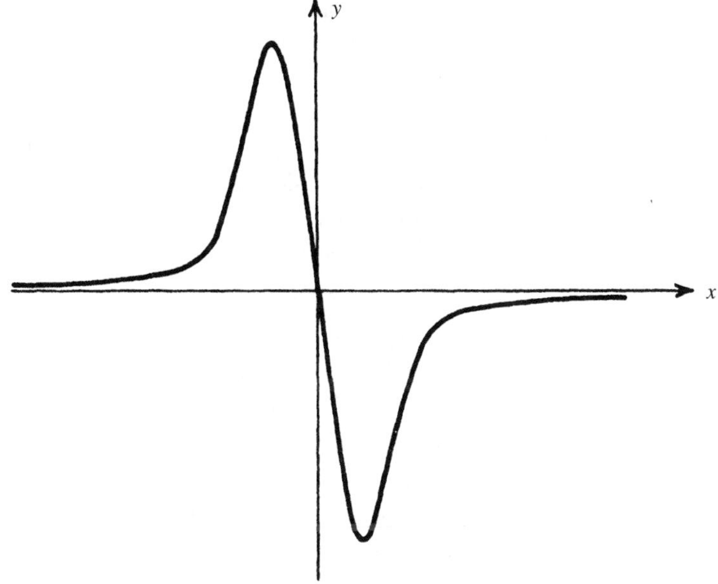

Figure 2.3

Example 5. Consider the family $g_t(x) = -x(e^{-x^2/4t}/4\sqrt{\pi}t^{3/2})$, which is the x derivative of case (b), Example 3 (see also Figure 2.3). In view of the fact that convergent families of distributions can be differentiated termwise, it follows that

$$\lim_{t\to 0+} \int_{-\infty}^{\infty} g_t(x)\phi(x)\,dx = -\phi'(0) \quad \text{for each } \phi \in C_0^\infty(\mathbb{R}).$$

If we examine $g_t(x)$ under the rather dim light of pointwise convergence, we see that $\lim_{t\to 0} g_t(x) = 0$ for every x in $-\infty < x < \infty$. Thus the very large bumps in $g_t(x)$ do not show under pointwise convergence as $t \to 0+$. It is also worth noting that the family

$$f_t(x) + g_t(x) = \frac{e^{-x^2/4t}}{2\sqrt{\pi t}}\left(1 - \frac{x}{2t}\right)$$

satisfies all conditions on a delta family (Exercise 2.2.4) except nonnegativity, yet it is far from a delta family since

$$\lim_{t\to 0+} \int_{-\infty}^{\infty} [f_t(x) + g_t(x)]\phi(x)\,dx = \phi(0) - \phi'(0).$$

Example 6. Differentiation with respect to a parameter. Let $f_\alpha(x)$ be an n-dimensional distribution depending on the continuous parameter α. With α fixed,

$[f_{\alpha+h}(x) - f_\alpha(x)]/h$ is a distribution depending on the parameter h. The limit as $h \to 0$ (*if it exists*) gives us the distribution $df_\alpha/d\alpha$:

$$\left\langle \frac{df_\alpha}{d\alpha}, \phi \right\rangle = \lim_{h\to 0} \frac{1}{h}[\langle f_{\alpha+h}, \phi\rangle - \langle f_\alpha, \phi\rangle]. \tag{2.2.7}$$

Whereas derivatives of f_α with respect to the action variables $x_1, \dots, x_n$ always exist, $df_\alpha/d\alpha$ may fail to exist. For instance, if $f_\alpha(x) = |\alpha| f(x)$, where $f(x)$ is a distribution, then $f_\alpha(x)$ is a distribution whose derivative with respect to the parameter α does not exist at $\alpha = 0$.

(a) As our first illustration of differentiation with respect to a parameter, let $f(x)$ be an n-dimensional distribution, and let $\xi = (\xi_1, \dots, \xi_n)$ be a vector parameter. For each fixed ξ the distribution $f(x-\xi)$ is defined from (2.1.7) as

$$\langle f(x-\xi), \phi(x)\rangle = \langle f(x), \phi(x+\xi)\rangle,$$

so that $f(x-\xi)$ can be regarded as a distribution depending on the parameters $\xi_1, \dots, \xi_n$. We can now calculate the derivative of this distribution with respect to ξ_i:

$$\begin{aligned}\left\langle \frac{\partial}{\partial \xi_i} f(x-\xi), \phi(x)\right\rangle &= \left\langle f(x), \frac{\partial}{\partial \xi_i}\phi(x+\xi)\right\rangle = \left\langle f(x), \frac{\partial}{\partial x_i}\phi(x+\xi)\right\rangle \\ &= -\left\langle \frac{\partial}{\partial x_i} f(x), \phi(x+\xi)\right\rangle \\ &= -\left\langle \frac{\partial}{\partial x_i} f(x-\xi), \phi(x)\right\rangle.\end{aligned}$$

Thus, as expected,

$$\frac{\partial}{\partial \xi_i} f(x-\xi) = -\frac{\partial}{\partial x_i} f(x-\xi), \tag{2.2.8}$$

and, in particular,

$$\frac{\partial}{\partial \xi_i} \delta(x-\xi) = -\frac{\partial}{\partial x_i} \delta(x-\xi), \tag{2.2.9}$$

so that $(\partial/\partial\xi_i)\delta(x-\xi)$ is the volume density arising from a unit dipole at ξ with axis in the positive x_i direction.

(b) Next, let Γ be a surface in $\mathbb{R}^3$ carrying a normally oriented dipole layer of *surface density* $b(x)$. The surface element dS_ξ at ξ carries a dipole whose equivalent volume source density can be expressed by (2.2.9) as

$$b(\xi) dS_\xi \frac{\partial}{\partial n_\xi} \delta(x-\xi)$$

with action

$$b(\xi) \frac{\partial \phi}{\partial n}(\xi)\, dS_\xi.$$

The action of the whole layer is therefore given by

$$\int_{\Gamma} b(\xi)\frac{\partial \phi}{\partial n}(\xi)\,dS_{\xi}, \tag{2.2.10}$$

which we interpret as the value of a functional on $C_0^\infty(\mathbb{R}^3)$ corresponding to a *dipole* (or *double*) *layer* of density $b(x)$ on Γ.

(c) If t is regarded as a parameter, the function $f(x,t) = \frac{1}{2}H(t-|x|)$ defines a one-dimensional distribution. Here we have written $f(x,t)$ instead of $f_t(x)$. Clearly, for $t > 0$,

$$\langle f(x,t), \phi(x)\rangle = \frac{1}{2}\int_{-t}^{t} \phi(x)\,dx, \tag{2.2.11}$$

whereas, for $t \leqslant 0$, f is the zero distribution. Let us calculate both $\partial f/\partial t$ and $\partial^2 f/\partial t^2$ for $t > 0$:

$$\begin{aligned}\left\langle \frac{\partial f}{\partial t}, \phi(x)\right\rangle &= \tfrac{1}{2}\phi(t) + \tfrac{1}{2}\phi(-t),\\ \left\langle \frac{\partial^2 f}{\partial t^2}, \phi(x)\right\rangle &= \tfrac{1}{2}\phi'(t) - \tfrac{1}{2}\phi'(-t).\end{aligned} \tag{2.2.12}$$

Since we also have

$$\left\langle \frac{\partial^2 f}{\partial x^2}, \phi\right\rangle = \left\langle f, \frac{\partial^2 \phi}{\partial x^2}\right\rangle = \frac{1}{2}\int_{-t}^{t} \frac{\partial^2 \phi}{\partial x^2}\,dx = \tfrac{1}{2}\phi'(t) - \tfrac{1}{2}\phi'(-t),$$

we can say that f satisfies the wave equation

$$\frac{\partial^2 f}{\partial t^2} - \frac{\partial^2 f}{\partial x^2} = 0 \quad \text{for}\, t > 0 \tag{2.2.13}$$

with the initial conditions, obtained from (2.2.11) and (2.2.12),

$$\lim_{t\to 0+} f = 0, \qquad \lim_{t\to 0+} \frac{\partial f}{\partial t} = \delta(x). \tag{2.2.14}$$

In (2.2.13) the distributions are one-dimensional, $\partial^2 f/\partial t^2$ representing differentiation with respect to a parameter. If we gave x, t equal status in $f(x,t)$ and treated f as a two-dimensional distribution, we would still find that (2.2.13) is satisfied for $t > 0$ (and, of course, for $t < 0$), but the equation satisfied for $-\infty < t < \infty$ is then $\partial^2 f/\partial t^2 - \partial^2 f/\partial x^2 = \delta(x,t)$. (See Exercise 2.5.4.)

EXERCISES

2.2.1 Let $V_n(r)$ and $S_n(r)$ denote, respectively, the volume and surface area of the n-dimensional ball of radius r. To obtain a formula for $V_n(r)$ we first relate it to the volume of an $(n-1)$-dimensional ball as follows. Let z denote a coordinate on an axis through the origin. If A is a constant smaller than r, the intersection of the hyperplane $z = A$ and the n-dimensional ball is an $(n-1)$-dimensional ball of radius $(r^2 - A^2)^{1/2}$. (The reader should draw appropriate figures for both the three- and two-dimensional cases.) Thus

$$V_n(r) = \int_{-r}^{r} V_{n-1}(\sqrt{r^2 - z^2})\, dz.$$

By induction, show that $V_n(r) = C_n r^n$, where

$$C_{n+1} = 2C_n \int_0^1 (1-u^2)^{n/2}\, du.$$

Show that

$$\int_0^1 (1-u^2)^{n/2}\, du = \frac{\sqrt{\pi}}{2} \frac{(n/2)!}{[(n+1)/2]!}$$

and hence

$$V_n(r) = \frac{\pi^{n/2} r^n}{(n/2)!}, \tag{2.2.15}$$

where $z!$ is defined as $\Gamma(z+1)$. By considering the volume of a thin spherical shell, show that

$$S_n(r) = \frac{n\pi^{n/2} r^{n-1}}{(n/2)!} = \frac{2\pi^{n/2} r^{n-1}}{[(n/2)-1]!}. \tag{2.2.16}$$

Using $\frac{1}{2}! = \sqrt{\pi}/2$, check that (2.2.15) and (2.2.16) are correct for $n = 2$ and $n = 3$.

2.2.2 (a) Show that cases (a) through (e) in Example 3 satisfy the hypotheses in Theorem 2.2.4 and are therefore delta families.

(b) Show that the kernels of Examples 3(g), 3(h_1), and 3(h_2) all generate delta families.

2.2.3 On $\mathbb{R}$ and with fixed $\alpha > 0$, consider the sequence of nonnegative continuous functions (see Figure 0.6)

$$f_k(x) = k^\alpha H(x) x e^{-kx}.$$

Show that

(a) $f_k \to 0$ pointwise as $k \to \infty$ (for any α).

(b) $f_k \to 0$ uniformly on $\mathbb{R}$ if $\alpha < 1$.

(c) $\int_{-\infty}^{\infty} f_k(x)dx = k^{\alpha-2}$, so that $f_k \to 0$ in L_1 for $\alpha < 2$.

(d) $f_k \to 0$ in the sense of distributions if $\alpha < 2$.

(e) $f_k \to \delta$ if $\alpha = 2$.

(f) f_k does not converge in the sense of distributions if $\alpha > 2$. (*Hint*: Estimate $\langle f_k, \phi \rangle$ for a nonnegative test function ϕ with $\phi \equiv 1, 0 \leqslant x \leqslant a$.)

2.2.4 Let $\{f_a(x)\}$ be a family of *nonnegative* locally integrable functions on $\mathbb{R}^n$. Make the following assumptions:

(a) For some $A > 0, \lim_{\alpha \to \alpha_0} \int_{|x|<A} f_\alpha(x)\, dx = 1$.

(b) For every $A > 0, f_\alpha(x) \to 0$ as $\alpha \to \alpha_0$ uniformly on $|x| \geqslant A$.

It follows immediately that (a) holds for every $A > 0$. Show that if $\phi(x)$ is any function continuous at $x = 0$ satisfying $\int_{\mathbb{R}^n} |\phi(x)|\, dx < \infty$, then

$$\lim_{\alpha \to \alpha_0} \int_{\mathbb{R}^n} f_\alpha(x)\phi(x)\, dx = \phi(0).$$

This certainly proves that $\{f_\alpha\}$ is a delta family as $\alpha \to \alpha_0$, but of course also proves a great deal more, since ϕ is not restricted to be a test function. *Hint*: Estimate the terms in

$$\int_{\mathbb{R}^n} f_\alpha \phi\, dx - \phi(0) = \int_{|x| \leqslant B} f_\alpha[\phi - \phi(0)]\, dx + \phi(0) \left[\int_{|x| \leqslant B} f_\alpha\, dx - 1 \right] + \int_{|x|>B} f_\alpha \phi\, dx.$$

2.2.5 (a) On $\mathbb{R}$, let $f(x)$ be a continuous function which may change sign and such that $\int_{-\infty}^{\infty} f(x)\, dx = 1$. Define $f_k(x)$ by compression as $f_k(x) = kf(kx)$ and set $r_k(x) = \int_{-\infty}^{x} f_k(y)\, dy$. Show that $r_k(x)$ tends pointwise to the Heaviside function $H(x)$ and that $|r_k(x)| \leqslant C$, where C is a constant independent of k and x. It therefore follows by the Lebesgue Dominated Convergence Theorem that

$$\lim_{k \to \infty} \int_{-\infty}^{\infty} r_k(x)\phi(x)\, dx = \int_0^{\infty} \phi(x)\, dx \quad \text{for each } \phi \text{ in } C_0^\infty(\mathbb{R}),$$

or, equivalently, $r_k(x) \to H(x)$ distributionally. By differentiation, conclude that f_k is a delta sequence.

(b) The sinc function

$$\text{sinc } x = \frac{\sin x}{x} \tag{2.2.17}$$

plays an important role in Fourier analysis and signal processing (see Stenger [13]). Here sinc $0 = 1$ to be consistent with $\lim_{x \to 0} (\sin x)/x$. By complex variables or otherwise, show that $\int_{-\infty}^{\infty} \text{sinc } x\, dx = \pi$, so that from part (a), (k/π) sinc kx [$= \sin kx/(\pi x)$] is a delta sequence, obtained by compression from the oscillatory function $(1/\pi)$ sinc x, and

$$\lim_{k \to \infty} \int_{-\infty}^{\infty} \frac{\sin kx}{\pi x} \phi(x)\, dx = \phi(0) \quad \text{for each } \phi \text{ in } C_0^\infty(\mathbb{R}). \tag{2.2.18}$$

Formula (2.2.18) will be extended to a larger class of functions $\phi(x)$ in Section 2.4.

2.2.6 Let $\{f_\alpha(x)\}$ be a delta family of the type of Exercise 2.2.4. Show that if in addition to the conditions already imposed there on ϕ we also require that $\phi(x)$ be bounded and *uniformly continuous* on $\mathbb{R}^n$, then

$$\lim_{\alpha\to\alpha_0}\int_{\mathbb{R}^n} f_\alpha(x-\xi)\phi(\xi)\,d\xi = \lim_{\alpha\to\alpha_0}\int_{\mathbb{R}^n}\phi(x-\xi)f_\alpha(\xi)\,d\xi = \phi(x)$$

uniformly on $\mathbb{R}^n$. The proof follows that of Exercise 2.2.4. Write

$$\left|\int_{\mathbb{R}^n}\phi(x-\xi)f_\alpha(\xi)\,d\xi - \phi(x)\right| \leqslant \left|\int_{|\xi|<B} f_\alpha(\xi)[\phi(x-\xi)-\phi(x)]\,d\xi\right| + \left|\phi(x)\left(\int_{|\xi|<B} f_\alpha(\xi)\,d\xi - 1\right)\right| + \left|\int_{|\xi|>B} f_\alpha(\xi)\phi(x-\xi)\,d\xi\right|.$$

Since $\phi(x)$ is uniformly continuous on $\mathbb{R}^n$, we can choose B so small that $\max_{|\xi|\leqslant B}|\phi(x-\xi)-\phi(x)| < \varepsilon$ for all x in $\mathbb{R}^n$. After that it is easy to estimate the remaining terms to obtain the desired result.

2.2.7 *The Weierstrass approximation theorem.* Consider the sequence $\{f_k(x)\}$ of Example 4(j).

(a) Show from Exercise 2.2.1 that $I_k = 2^{2k+1}(k!)^2/(2k+1)!$; by using Stirling's formula, $k! \sim k^k e^{-k}\sqrt{2\pi k}$, valid for large k, show also that $\lim_{k\to\infty}\sqrt{k}\,I_k = \sqrt{\pi}$.

(b) Show the sequence $\{f_k(x)\}$ satisfies the conditions of Exercise 2.2.4. Then let $\phi(x)$ be a continuous function on $\mathbb{R}$ which vanishes for $|x| > \frac{1}{2}$ [this implies that $\phi\left(\frac{1}{2}\right) = \phi\left(-\frac{1}{2}\right) = 0$ and the uniform continuity of $\phi(x)$ in $-\infty < x < \infty$]. Show that $\int_{-\infty}^{\infty} f_n(x-\xi)\phi(\xi)\,d\xi$ reduces to a *polynomial* in x when $-\frac{1}{2} \leqslant x \leqslant \frac{1}{2}$. In view of Exercise 2.2.6 we have therefore approximated $\phi(x)$ uniformly by polynomials on $-\frac{1}{2} \leqslant x \leqslant \frac{1}{2}$.

(c) By a suitable change of variable, show that any function $\phi(x)$ continuous on the finite interval $a \leqslant x \leqslant b$ can be uniformly approximated by polynomials over this interval.

(d) Consider the function $|x|$ on $-1 \leqslant x \leqslant 1$. According to (c), we can approximate this function as closely as we desire by polynomials, yet we cannot expand the function in a power series. Explain the apparent paradox.

2.2.8 Let $f(x)$ be a nonnegative function such that $\int_0^\infty f(x)\,dx = A$. Show that if $\phi(x)$ is an arbitrary, bounded, locally integrable function for which $\phi(0+)$ exists, then

$$\lim_{\alpha\to 0+}\int_0^\infty \frac{1}{\alpha} f\left(\frac{x}{\alpha}\right)\phi(x)\,dx = A\phi(0+). \tag{2.2.19}$$

2.2.9 (a) By setting $e^{i\theta} = z$, transform the real integral

$$J = \int_{-\pi}^{\pi} \frac{1 - r^2}{1 + r^2 - 2r\cos\theta}\, d\theta$$

to a counterclockwise integral on the boundary of the unit circle in the complex z plane, and obtain

$$J = \frac{i(1 - r^2)}{r} \int \frac{dz}{(z - r)(z - 1/r)}.$$

If $0 < r < 1$, there is only the simple pole at $z = r$ in the interior of the circle, so that

$$J = 2\pi.$$

For $r > 1$ show that

$$J = -2\pi.$$

(b) Show that for $0 \leqslant r < 1$,

$$\frac{1 - r^2}{1 + r^2 - 2r\cos\theta} = 2(\tfrac{1}{2} + r\cos\theta + r^2\cos 2\theta + \cdots),$$

and hence derive the first result of part (a) by term-by-term integration.

(c) Show that case (k) in Example 4 is a delta family by using the result of Exercise 2.2.4.

2.2.10 Let $\{f_k(x)\}$ be the Dirichlet kernel of case (1) of Example 4. By using (2.2.18), show that for every test function $\phi(x)$ in $C_0^\infty(\mathbb{R})$,

$$\lim_{k\to\infty} \int_{-\pi}^{\pi} \frac{\sin\left(k + \frac{1}{2}\right)x}{2\pi \sin\frac{1}{2}x}\phi(x)\, dx = \phi(0) \tag{2.2.20}$$

if ϕ is differentiable, $-\pi < x < \pi$. For a stronger theorem, see Section 2.3.

2.2.11 Let $f(x)$ be an n-dimensional distribution. Show that

$$\lim_{\varepsilon\to 0} f(x + \mathbf{e}) = f(x),$$

where $\mathbf{e}$ is a unit vector in some fixed direction. (Use Example 4, Section 2.1 and Exercise 2.1.11.)

2.2.12 It is fortuitous that we were able to calculate the exact value of the integral of the Poisson kernel. All that is necessary in the proof of the delta behavior of the family are the properties of Exercise 2.2.4. Let us now try to establish these properties for the Poisson kernel without reference to the exact value of the integral. (*Hint*: Given $\varepsilon > 0$, choose δ such that

$$1 - (1 + \varepsilon)\frac{\theta^2}{2} \leqslant \cos\theta \leqslant 1 - (1 - \varepsilon)\frac{\theta^2}{2}, \quad |\theta| \leqslant \delta.$$

In this way obtain bounds for $\int_{-\delta}^{\delta} P(r,\theta)\,d\theta$, where $P(r,\theta)$ is the Poisson kernel. Now choose r sufficiently near 1 so that $\int_{-\delta}^{\delta} P(r,\theta)\,d\theta$ is close to 1 and $\int_{|\theta|\geqslant\delta} P(r,\theta)\,d\theta$ is close to 0.)

2.2.13 *Direct product of distributions.* Let $f_1(x_1)$ and $f_2(x_2)$ be locally integrable functions of one variable. Then $f_1(x_1)f_2(x_2)$ is locally integrable in the plane and defines a two-dimensional distribution whose action can be expressed in the equivalent forms

$$\langle f_1 f_2, \phi(x_1, x_2)\rangle = \langle f_1(x_1), \psi(x_1)\rangle = \langle f_2(x_2), \chi(x_2)\rangle, \qquad (2.2.21)$$

where

$$\psi(x_1) \doteq \langle f_2(x_2), \phi(x_1, x_2)\rangle, \quad \chi(x_2) \doteq \langle f_1(x_1), \phi(x_1, x_2)\rangle. \qquad (2.2.22)$$

Clearly, $\psi(x_1)$ and $\chi(x_2)$ belong to $C_0^\infty(\mathbb{R})$, so that the second and third terms of (2.2.21) are well-defined actions of one-dimensional distributions. If f_1 and f_2 are arbitrary one-dimensional distributions, we use (2.2.21) and (2.2.22) to define the *direct product* $f_1 f_2$, which is a two-dimensional distribution since the variables are different (no attempt is made to define the product of two arbitrary distributions in the same variable). Obviously, the same idea can be used to define an n-dimensional distribution as the product of n one-dimensional distributions in n different variables.

(a) Show that the n-dimensional Dirac distribution $\delta(x) = \delta(x_1, \dots, x_n)$ can be written as the direct product of n one-dimensional distributions: $\delta(x_1)\delta(x_2)\cdots\delta(x_n)$.

(b) Define $\delta(x_1)f(x_2, x_3)$ as the direct product of a one-dimensional Dirac distribution with a two-dimensional distribution, and show that the resulting three-dimensional distribution corresponds to a simple layer of sources of surface density $f(x_2, x_3)$ spread on the plane $x_1 = 0$ in $\mathbb{R}^3$.

(c) Show that $f(x_1)\delta(x_2)\delta(x_3)$ represents a line source along the x_1 axis in $\mathbb{R}^3$ with line density $f(x_1)$.

2.2.14 *Coordinate transformations.* A locally integrable function

$$f(x) = f(x_1, \dots, x_n)$$

defines a distribution through

$$\langle f, \phi\rangle = \int_{\mathbb{R}^n} f(x)\phi(x)\,dx.$$

It is sometimes useful to express this distribution in terms of a different coordinate system. Proceeding in a purely formal manner, we let $u_1, \dots, u_n$ be new coordinates that can be obtained from $x_1, \dots, x_n$ by the transformation law $u = u(x)$, that is, $u_1 = u_1(x_1, \ldots, x_n), \ldots, u_n = u_n(x_1, \ldots, x_n)$. If the Jacobian J of x with respect to u is *positive* everywhere,

$$\begin{aligned}\langle f, \phi\rangle &= \int_{\mathbb{R}^n} f(x)\phi(x)\,dx = \int_{u\text{–space}} f(x(u))\phi(x(u))J\;du \\ &= \int_{u\text{–space}} \tilde{f}(u)J\tilde{\phi}(u)\,du,\end{aligned} \tag{2.2.23}$$

where

$$\tilde{\phi}(u) \doteq \phi(x(u)), \quad \tilde{f}(u) \doteq f(x(u)).$$

Note (2.2.23) can be used to interpret coordinate changes for arbitrary distributions. For instance, consider $f = \delta(x - x') = \delta(x_1 - x_1') \cdots \delta(x_n - x_n')$, then $\langle f, \phi\rangle = \phi(x') = \tilde{\phi}(u')$, where $u' = u(x')$. Thus, for (2.2.23) to be consistent, we need

$$\tilde{f}(u)J = \delta(u_1 - u_1') \cdots \delta(u_n - u_n')$$

or, since

$$\tilde{f}(u) = \delta[x(u) - x'],$$

we find that

$$\delta(x - x') = \frac{\delta(u_1 - u_1') \cdots \delta(u_n - u_n')}{J}. \tag{2.2.24}$$

If we know only that $J \geqslant 0$, (2.2.24) can still be used if J does not vanish at u', since only the value of J at u' affects the right side of (2.2.24).

(a) Show that if u_1, u_2, u_3 are the spherical coordinates in $\mathbb{R}^3$ usually denoted by r, θ, ϕ, respectively, then $J = r^2 \sin\theta$ and

$$\delta(x - x') = \frac{1}{r^2 \sin\theta}\delta(r - r')\delta(\theta - \theta')\delta(\phi - \phi'), \quad r' \neq 0, \quad \theta' \neq 0, \pi. \tag{2.2.25}$$

If $r' = 0$ or $\theta' = 0$ or π, the coordinate transformation is singular and J vanishes. For instance, if the source is on the positive x_3 axis (in which case $\theta' = 0, r' > 0$), then (2.2.25) must be modified. One method of attacking the problem is to replace the point source by a uniform ring of sources, of unit total strength, located on the intersection of the sphere $r = r'$ and the cone $\theta = \theta'$, and then to let $\theta' \to 0$. For $\theta' > 0$ the intersection is a circle traced as ϕ ranges from 0 to 2π, say. Since the portion of the ring between ϕ' and $\phi' + \Delta\phi'$ carries a source of strength $\Delta\phi'/2\pi$, its distributional representation is given by (2.2.25) multiplied by $\Delta\phi'/2\pi$. Representation of the entire ring is then obtained by integrating from $\phi' = 0$ to $\phi' = 2\pi$, with the result

$$\frac{\delta(r - r')\delta(\theta - \theta')}{2\pi r^2 \sin\theta}.$$

As $\theta' \to 0$, the configuration tends to a unit point source at $r = r', \theta = 0$, whose distributional representation is therefore

$$\frac{\delta(r - r')\delta(\theta)}{2\pi r^2 \sin\theta}.$$

(b) For the spherical coordinates of part (a), show that a point source at the origin can be represented as

$$\frac{\delta(r)}{4\pi r^2}$$

by considering the limit as $r' \to 0$ of a uniform simple layer of unit strength on the sphere $r = r'$.

(c) For cylindrical coordinates ρ, θ, z, show that

$$\delta(x - x') = \frac{\delta(\rho - \rho')\delta(\theta - \theta')\delta(z - z')}{\rho}, \quad \rho' > 0. \qquad (2.2.26)$$

Show that a uniform ring of sources of unit strength on $\rho = \rho', z = 0$ has the representation

$$\frac{\delta(z)\delta(\rho - \rho')}{2\pi\rho}, \qquad (2.2.27)$$

and that for a unit source at the origin

$$\delta(x) = \frac{\delta(z)\delta(\rho)}{2\pi\rho}. \qquad (2.2.28)$$

Expressions (2.2.26) and (2.2.28) without the z factor remain valid for two-dimensional Dirac distributions in plane polar coordinates.

2.2.15 Let $f \in C^1(\mathbb{R})$ with $f \geqslant 0$ and $\int_{\mathbb{R}} f(x)\,dx = 1$. Show that $\{k^2 f'(kx)\}$ is a dipole sequence; that is,

$$\lim_{k\to\infty} \int_{-\infty}^{\infty} k^2 f'(kx)\phi(x)\,dx = -\phi'(0) \quad \text{for each } \phi \in C_0^\infty(\mathbb{R}).$$

2.2.16 On $\mathbb{R}$, let

$$f(x) = \begin{cases} -1, & |x| < \frac{1}{2}, \\ 2, & \frac{1}{2} < |x| < 1, \\ 0, & |x| > 1. \end{cases}$$

Show that the sequence $f_k(x) = kf(kx)$ is a delta sequence. Note that $f_k(0) \to -\infty$; can you reconcile this behavior with the idea that f_k is an approximation to a concentrated unit load (that is, positive load) at the origin?

2.3 FOURIER SERIES

Review of Classical Results

The term *Fourier series* is used in Chapter 4 to describe an expansion in a general orthonormal set of functions, but in the present section the term is reserved for the special case of expansion in *trigonometric* functions.

Let $f(x)$ be a complex-valued function on the real line $-\infty < x < \infty$, and let f have *period* 2π, that is, $f(x + 2\pi) = f(x)$ for all x. If $f(x)$ is given only on some 2π interval, say $-\pi \leqslant x < \pi$, we can always extend f periodically for all x by the rule $f(x + 2\pi) = f(x)$. Of course, this extension procedure may introduce additional discontinuities into the function; for instance, if we start with $f(x) = x$ in $-\pi \leqslant x < \pi$, the periodic extension will have jump discontinuities at odd integral multiples of π. Instead of looking at a periodic f on $(-\infty, \infty)$, it is sometimes more helpful to view the interval $(-\pi, \pi)$ as wrapped around the unit circle S (whose circumference has length 2π, of course), and in this way the endpoints $-\pi, \pi$ are identified as a single point on the unit circle; for the function x on $-\pi \leqslant x < \pi$, this construction gives a function on S with a single point of discontinuity.

The class of functions f which we are trying to expand will always be a subset of $L_1(-\pi, \pi)$, the set of functions absolutely integrable on $(-\pi, \pi)$; that is,

$$\int_{-\pi}^{\pi} |f(x)|\, dx < \infty.$$

Usually, we shall require $f(x)$ to belong to $L_2(-\pi, \pi)$, the set of square integrable functions. This means that

$$\int_{-\pi}^{\pi} |f^2(x)|\, dx < \infty,$$

and it can be seen that $L_2(-\pi, \pi)$ is a subset of $L_1(-\pi, \pi)$. For certain theorems we will also demand some continuity or smoothness for $f(x)$.

A *trigonometric* series (or, more properly, a trigonometric series in complex-exponential form) is of the type

$$\sum_{k=-\infty}^{\infty} c_k e^{ikx}, \tag{2.3.1}$$

where the coefficients c_k are complex numbers. The sequence $\{s_n(x)\}$ of partial sums of (2.3.1) is defined *symmetrically* by

$$s_n(x) = \sum_{k=-n}^{n} c_k e^{ikx}, \tag{2.3.2}$$

where the index n runs from 0 to infinity. It is possible for (2.3.2) to converge as $n \to \infty$ without the sums from 1 to ∞ and from $-\infty$ to 1 converging separately (see Exercise 2.3.9).

A series of the form

$$\frac{a_0}{2} + \sum_{k=1}^{\infty} (a_k \cos kx + b_k \sin kx) \tag{2.3.3}$$

is completely equivalent to (2.3.1) if the coefficients $\{a_k\}$ and $\{b_k\}$ are related to the $\{c_k\}$ by the formulas

$$a_0 = 2c_0; \quad a_k = c_k + c_{-k} \quad \text{and} \quad b_k = i(c_k - c_{-k}) \quad \text{for} \quad k \geqslant 1. \tag{2.3.4}$$

If (2.3.4) holds, it is easy to see that the nth partial sum

$$(a_0/2) + \sum_{k=1}^{n} (a_k \cos kx + b_k \sin kx)$$

of (2.3.3) is exactly equal to $s_n(x)$ as given by (2.3.2). For most of our purposes it is more convenient to deal with (2.3.1) than with (2.3.3).

All the complex exponentials in (2.3.1) have period 2π, so that the series, if it converges, will also have period 2π. Therefore, the relation between (2.3.1) and any 2π-periodic function it purports to represent needs to be studied only in the fundamental interval $(-\pi, \pi)$ or on the unit circle.

A set $\{e_k(x)\}$ of complex-valued functions is said to be *orthonormal* over (a, b) if

$$\int_a^b e_k(x)\bar{e}_j(x)\, dx = \begin{cases} 0, & k \neq j, \\ 1, & k = j, \end{cases} \tag{2.3.5}$$

where the overbar denotes complex conjugation. The set

$$\left\{ e_k(x) = \frac{e^{ikx}}{\sqrt{2\pi}} \right\}, \tag{2.3.6}$$

where the index k ranges over all integers from $-\infty$ to ∞, is an orthonormal set over any interval of length 2π.

Suppose now that the series (2.3.1) converges uniformly on $-\pi \leqslant x \leqslant \pi$, and let us call the sum $f(x)$. We multiply the equality $f(x) = \sum_{k=-\infty}^{\infty} c_k e^{ikx}$ by e^{-imx} and integrate from $-\pi$ to π. The multiplication by e^{-imx} does not affect the uniform convergence, so that we can perform the integration term by term to obtain

$$c_m = \frac{1}{2\pi} \int_{-\pi}^{\pi} f(x) e^{-imx}\, dx, \quad m = 0, \pm 1, \pm 2, \ldots. \tag{2.3.7}$$

Thus, if the series (2.3.1) converges uniformly on $-\pi \leqslant x \leqslant \pi$ to a function f (which is necessarily continuous, since a uniformly convergent series of continuous functions is continuous), then the coefficients in (2.3.1) are given in terms of f by (2.3.7). If the form (2.3.3) is used instead, the coefficients are

$$\begin{aligned} a_m &= \frac{1}{\pi} \int_{-\pi}^{\pi} f(x) \cos mx\, dx, \quad m = 0, 1, 2, \ldots; \\ b_m &= \frac{1}{\pi} \int_{-\pi}^{\pi} f(x) \sin mx\, dx, \quad m = 1, 2, \ldots. \end{aligned} \tag{2.3.8}$$

Note that c_0 ($= a_0/2$) is the average value of f over a period.

Let us now reverse the process and look at the possible expansion of a given function $f(x)$ on $(-\pi, \pi)$ in the series (2.3.1). Formula (2.3.7) still defines coefficients c_m as long as f is in $L_1(-\pi, \pi)$; these coefficients are the *Fourier coefficients* of f, and the corresponding series (2.3.1) is the *Fourier series* of f. *We no longer know that the series converges to f.* For what functions f does the Fourier series represent f, and in what sense? Obviously, if f_1 and f_2 are equal almost everywhere, they define the same $\{c_m\}$ and hence the same series, so that we cannot in general expect pointwise convergence. Probably the simplest results, and in many ways the most useful, are set in the framework of the Hilbert space $L_2(-\pi, \pi)$. The main theorem (see Chapter 4) states that for each f in $L_2(-\pi, \pi)$ the Fourier series defined by (2.3.1) and (2.3.7) always converges to f in the L_2 sense, that is,

$$\lim_{n\to\infty} \int_{-\pi}^{\pi} \left| f(x) - \sum_{k=-n}^{n} c_k e^{ikx} \right|^2 dx = 0. \tag{2.3.9}$$

We shall not use result (2.3.9) here. Instead, we discuss some aspects of pointwise and uniform convergence of Fourier series. We first note the following lemma.

Lemma. *Let $f \in L_2(-\pi, \pi)$, and let $c_k = (1/2\pi) \int_{-\pi}^{\pi} f(x) e^{-ikx} dx$; then*

$$\lim_{|k|\to\infty} c_k = 0, \tag{2.3.10}$$

$$\sum_{k=-\infty}^{\infty} |c_k|^2 \quad \textit{converges}, \tag{2.3.11}$$

and

$$\sum_{k=-\infty}^{\infty} |c_k|^2 \leqslant \frac{1}{2\pi} \int_{-\pi}^{\pi} |f(x)|^2 \, dx \quad \textit{(Bessel's inequality)}. \tag{2.3.12}$$

Proof. We have

$$\begin{aligned} 0 &\leqslant \int_{-\pi}^{\pi} \left| f - \sum_{k=-n}^{n} c_k e^{ikx} \right|^2 dx \\ &= \int_{-\pi}^{\pi} \left(f - \sum_{k=-n}^{n} c_k e^{ikx} \right) \overline{\left(f - \sum_{k=-n}^{n} c_k e^{ikx} \right)} dx \\ &= \int_{-\pi}^{\pi} |f|^2 \, dx - 2\pi \sum_{k=-n}^{n} |c_k|^2, \end{aligned}$$

so that for every n,

$$\sum_{k=-n}^{n} |c_k|^2 \leqslant \frac{1}{2\pi} \int_{-\pi}^{\pi} |f|^2 \, dx.$$

If we let $\{t_n\}$ be the sequence of symmetric partial sums

$$t_n = \sum_{k=-n}^{n} |c_k|^2,$$

then t_n is monotonically increasing with n and is bounded above by the constant $(1/2\pi)\int_{-\pi}^{\pi} |f|^2\, dx$. Thus t_n converges, so that (2.3.11) and Bessel's inequality obviously hold. It follows that $|c_k|^2$ tends to 0 as $|k| \to \infty$, implying (2.3.10). □

REMARKS

1. Result (2.3.10) holds even if $f \in L_1(-\pi, \pi)$, and it is then known as the *Riemann-Lebesgue lemma.* Suppose that $f \in L_1$; then for any $\varepsilon > 0$ we can write $f = g + h$, where g is bounded and $\int_{-\pi}^{\pi} |h(x)|\, dx < \varepsilon$. It follows that

$$\left| \frac{1}{2\pi} \int_{-\pi}^{\pi} h(x) e^{-ikx}\, dx \right| < \frac{\varepsilon}{2\pi} \quad \text{for all } k,$$

and, by (2.3.10) applied to g, $(1/2\pi)\int_{-\pi}^{\pi} g e^{-ikx}\, dx \to 0$. Therefore,

$$\frac{1}{2\pi} \int_{-\pi}^{\pi} f e^{-ikx}\, dx \to 0,$$

as required. An easy consequence is that for $f \in L_1(a, b)$, where the interval may be finite or infinite,

$$\int_a^b f(x) e^{-ikx}\, dx, \quad \int_a^b f(x) \cos kx\, dx, \quad \int_a^b f(x) \sin kx\ dx, \tag{2.3.13}$$

all tend to 0 as $|k| \to \infty$ (k is not restricted to integral values).

2. Inequality (2.3.12) is actually an equality known as *Parseval's identity*, which holds for all $f \in L_2(-\pi, \pi)$. We shall give a proof in this section for smooth functions and the general proof in Chapter 4. However, (2.3.11) and (2.3.12) do not necessarily hold for L_1 functions.

We shall now try to improve on (2.3.10) for the special classes of functions defined below. It turns out that, the smoother f is (as a function on the unit circle S), the faster the Fourier coefficients vanish. A function f belongs to $C^k(S)$ if it is continuous and has continuous derivatives of orders up to and including k on the unit circle. [If we think of f as a function initially defined on a 2π interval on the real axis, we are requiring that its periodic extension have continuous derivatives of orders up to and including k on the real line. Thus $C^0(S) = C(S)$ is the class of continuous 2π-periodic functions.]

A function f belongs to $D(S)$ (obeys *Dirichlet conditions*) if it is piecewise continuous and if there exists a finite subdivision of S such that f has a continuous derivative within each subinterval and such that the left- and right-hand derivatives exist at the end of the subintervals. The class D is both worse and better than C: In particular, D includes some discontinuous but rather simple functions, and D does not include continuous functions which have cusps (such as $\sqrt{|x|}$). For a function f in D having no jumps, f' has an obvious meaning (see the discussion of $|x|$ in Example 9, Section 2.1).

Theorem 2.3.1. *If f belongs to $C^p(S)$, its Fourier coefficients $\{c_m\}$ satisfy*

$$\lim_{m\to\infty} c_m m^p = 0.$$

Proof. Integration by parts p times gives

$$c_m = \frac{1}{2\pi}\int_{-\pi}^{\pi} f(x)e^{-imx}\,dx = \frac{1}{2\pi(im)^p}\int_{-\pi}^{\pi} f^{(p)}(x)e^{-imx}\,dx,$$

the integrated terms dropping out because of periodicity. Since $f^{(p)}(x)$ is continuous, its Fourier coefficients tend to zero as $m \to \infty$. Therefore, $c_m m^p \to 0$ as $m \to \infty$. □

Theorem 2.3.2. *If $f \in C^{p-1}(S)$ and $f^{(p)} \in D(S)$, the Fourier coefficients $\{c_m\}$ of f satisfy*

$$|c_m m^{p+1}| \leqslant M \quad \textit{for all } m.$$

Proof. Suppose that $p = 0$ (that is, f belongs to D), and that f has a simple jump at only one point, say ξ. Then

$$c_m = \frac{1}{2\pi}\int_{-\pi}^{\xi} fe^{-imx}\,dx + \frac{1}{2\pi}\int_{\xi}^{\pi} fe^{-imx}\,dx,$$

where in each interval, integration by parts is permitted since f is differentiable. It follows that

$$2\pi i m c_m = \left(\int_{-\pi}^{\xi} f'e^{-imx}\,dx + \int_{\xi}^{\pi} f'e^{-imx}\,dx\right) + e^{-im\xi}[f(\xi+) - f(\xi-)].$$

The sum of integrals, being proportional to the mth Fourier coefficient of an L_2 function, certainly tends to 0 as $m \to \infty$. The other term on the right side has nonzero modulus independent of m. Thus mc_m remains bounded as $m \to \infty$ while, for $\varepsilon > 0, m^{1+\varepsilon}c_m$ becomes unbounded as $m \to \infty$.

The proof for $p > 0$ proceeds along similar lines. We have thus established a relation between the smoothness properties of f and the behavior of its Fourier coefficients for large index. □

Example 1. Let $f(x) = x$ in $-\pi \leqslant x < \pi$ and $f(x + 2\pi) = f(x)$. From (2.3.8) it is clear that an odd function such as this has the property that $a_m = 0$ as well as $b_m = (2/\pi)\int_0^{\pi} f(x)\sin\, mx\, dx$. In our case we find that $b_m = (-1)^{m+1}2/m$, which has the behavior for large m expected of a discontinuous function in D.

Example 2. Let $f(x) = |x|$ in $-\pi \leqslant x < \pi$ and $f(x + 2\pi) = f(x)$. This is an even function, so that $b_m = 0$ and $a_m = (2/\pi) \int_0^\pi f(x) \cos mx\, dx$. We find that $a_0 = \pi$, $a_m = 0$ for m even, and $a_m = -(4/\pi m^2)$ for m odd. The coefficients behave like $1/m^2$ for large m, as befits a piecewise smooth function whose discontinuous derivative is in D.

Example 3. For a given real α let $f(x) = \cos \alpha x, -\pi \leqslant x < \pi$ and $f(x + 2\pi) = f(x)$. Note that the 2π-periodic function constructed in this way is *not* the same as cos αx on the whole real line (unless α is an integer). If α is not an integer, a straightforward calculation gives

$$a_m = (-1)^{m+1} \frac{2\alpha}{\pi} \frac{\sin \alpha\pi}{m^2 - \alpha^2}, \qquad m = 0, 1, 2, \ldots,$$

so that $m^2 a_m$ is bounded as $m \to \infty$. The function f is continuous but has a discontinuous derivative at odd multiples of π. If α is an integer, the calculation is even simpler: $a_m = 0$ except for $m = \alpha$, and $a_\alpha = 1$. Thus, as expected, the Fourier series of cos αx is just the single term cos αx *when α is an integer.* Note that in this case $f(x)$ is infinitely differentiable, so that $m^p a_m$ should tend to 0 with m for every p, a condition which is certainly satisfied here since the coefficients are identically 0 past the index α. There are, however, other 2π-periodic, infinitely differentiable functions whose Fourier coefficients do not vanish identically past a certain index. For instance, the temperature $u_t(x)$ at time t in an infinite rod whose initial temperature is the 2π-periodic function of Example 1 is given by (Chapter 8)

$$\sum_{m=1}^{\infty} \frac{2(-1)^{m+1}}{m} e^{-m^2 t} \sin mx, \qquad -\infty < x < \infty.$$

The coefficients are exponentially small as $m \to \infty$, so that they satisfy Theorem 2.3.1 for every p. This implies that $u_t(x)$ is an infinitely differentiable 2π-periodic function.

We now give a version of the principal theorem dealing with pointwise convergence of Fourier series. In anticipation of the result of the theorem, we *redefine* $f(x_0)$ at a point where f has a simple discontinuity (jump):

$$f(x_0) \doteq \frac{f(x_0+) + f(x_0-)}{2}. \tag{2.3.14}$$

The new f has the same jumps as the original f, but the value of the new f at the jump splits the jump. In this notation, the Heaviside function $H(x)$ would satisfy $H(0) = \frac{1}{2}$. Note that the new f and the original f have the same Fourier coefficients (2.3.7) since integrals are unaffected by the value of the integrand at isolated points.

Theorem 2.3.3. *Let $f(x)$ belong to $L_1(S)$ and let x_0 be a point on S at which $f(x_0+)$ and $f(x_0-)$ exist and at which the right- and left-hand derivatives of f also exist. With $\{c_m\}$ defined by (2.3.7), the sequence of partial sums (2.3.2) converges at x_0 to the value of $f(x_0)$ given by (2.3.14).*

REMARKS

1. The local behavior of f at x_0 determines the convergence of the Fourier series at x_0, even though the coefficients (unlike Taylor series coefficients) are given by the global formulas (2.3.7).

2. If $f(x) \in D(S)$, the conditions of the theorem are met at every point x_0 on the circle (or, equivalently, at every point x on the real line for the periodic extension of f). In particular, if $f \in D(S)$ *and also* $f \in C(S)$, the Fourier series converges to f at every point.

Proof. To prove Theorem 2.3.3, we must show that $\lim_{n\to\infty} s_n(x_0) = f(x_0)$. Since rotations leave the circle invariant it is enough to prove this for $x = 0$. Because adding a constant to f merely changes c_0 by that amount without affecting other coefficients, we can confine ourselves to functions $f(x)$ such that $f(0) = 0$. It therefore suffices to show that if $f(0)$ as given by (2.3.14) is zero, then $\lim_{n\to\infty} s_n(0) = 0$.

We note first that

$$\sum_{k=-n}^{n} e^{ikx} = \sum_{k=-n}^{n} e^{-ikx} = -1 + \sum_{k=0}^{n} e^{ikx} + \sum_{k=0}^{n} e^{-ikx}.$$

Since each sum on the right side is a geometric series, we find, after some simplifications, that

$$\frac{1}{2\pi}\sum_{k=-n}^{n} e^{ikx} = \frac{1}{2\pi}\sum_{k=-n}^{n} e^{-ikx} = \frac{\sin\left(n+\frac{1}{2}\right)x}{2\pi\sin(x/2)} \doteq D_n(x), \tag{2.3.15}$$

where D_n is even and

$$\int_{-\pi}^{\pi} D_n(x)\,dx = 1. \tag{2.3.16}$$

The function $D_n(x)$ is known as the *Dirichlet kernel* and has period 2π. The value of D_n at $x = 0$ (and therefore at all multiples of 2π) is understood to be its limiting value as $x \to 0$, that is, $(2n+1)/2\pi$. This understanding is consistent with (2.3.15) at $x = 0$. We have

$$s_n(0) = \sum_{k=-n}^{n} c_k = \sum_{k=-n}^{n} \frac{1}{2\pi}\int_{-\pi}^{\pi} f(x)e^{-ikx}\,dx = \int_{-\pi}^{\pi} D_n(x)F(x)\,dx, \tag{2.3.17}$$

where

$$F(x) = \frac{f(x)+f(-x)}{2} \tag{2.3.18}$$

is an even function, continuous at $x = 0$, with $F(0) = 0$, and whose left- and right-hand derivatives exist at $x = 0$. To prove that $s_n(0) \to 0$, we must essentially show that $D_n(x)$ is a delta sequence on $-\pi < x < \pi$. Writing

$$\frac{F(x)}{\sin(x/2)} = \left[\frac{F(x)}{x}\right]\left[\frac{x}{\sin(x/2)}\right],$$

we observe that the first factor has left- and right-hand limits at $x = 0$ because F has left- and right-hand derivatives at 0; hence F/x belongs to $L_2(S)$. The second factor, $x/[sin(x/2)]$, is bounded so that the product is in $L_2(S)$. From the Riemann-Lebesgue Lemma (2.3.13) we conclude that $s_n(0) \to 0$, as required.

An elegant alternative proof is due to P. Chernoff. Again it suffices to deal with $x = 0$ and to assume that in (2.3.14), $f(0) = 0$. The function F in (2.3.18) has Fourier coefficients F_k related to those of f by

$$F_k = \frac{c_k + c_{-k}}{2}.$$

Now define $g(x) = F(x)/(e^{ix} - 1)$, where the denominator vanishes only at $x = 0$ in $-\pi \leqslant x \leqslant \pi$. In view of the properties of F, g is seen to be bounded near $x = 0$ and so belongs to $L_1(S)$. Denoting the Fourier coefficients of g by g_k, we have

$$F_k = g_{k-1} - g_k;\ s_n(0) = \sum_{k=-n}^{n} c_k = \sum_{k=-n}^{n} F_k = \sum_{k=-n}^{n} (g_{k-1} - g_k) = g_{-n-1} - g_n,$$

so that $\lim_{n\to\infty} s_n(0) = 0$ as was proved. Note that the Dirichlet kernel plays no role in the proof! (See Exercise 2.3.9 for the use of nonsymmetric partial sums.) □

We can now prove uniform convergence of the Fourier series for smooth functions.

Theorem. *If $f(x) \in C^1(S)$, its Fourier series converges uniformly to f.*

Proof. Let $\{c_m\}$ be the Fourier coefficients of f, and $\{d_m\}$ those of f'. Starting with (2.3.7) for d_m and integrating by parts, we find that $d_m = imc_m$, or $c_m = (1/im)d_m$ for $m \neq 0$. Using the Schwarz inequality for sequences, we obtain

$$\sum_{m=-\infty}^{\infty} |c_m| = |c_0| + \sum_{m\neq 0} \frac{|d_m|}{|m|} \leqslant |c_0| + \left(\sum_{m\neq 0} \frac{1}{m^2}\right)^{1/2} \left(\sum_{m\neq 0} |d_m|^2\right)^{1/2}.$$

Since f' is continuous, it is in L_2, so that $\sum |d_m|^2 < \infty$ and, of course,

$$\sum(1/m^2) < \infty.$$

Thus $\sum |c_m| < \infty$, and $\sum c_m e^{imx}$ converges uniformly (by the Weierstrass M-test) to a function, which, according to Theorem 2.3.3, must be f. □

Corollary. *If $f(x) \in C^1(S)$, its Fourier series converges in the L_2 sense to f.*

Proof. We have

$$\int_{-\pi}^{\pi} |f|^2\, dx - 2\pi \sum_{k=-m}^{m} |c_k|^2 = \int_{-\pi}^{\pi} \left| f(x - \sum_{k=-m}^{m} c_k e^{ikx} \right|^2 dx, \qquad (2.3.19)$$

and in view of the uniform convergence of the Fourier series to f, we can pick M so large that

$$\left| f - \sum_{k=-m}^{m} c_k e^{ikx} \right| < \varepsilon, \quad m > M.$$

Hence the right side of (2.3.19) tends to 0 as m approaches infinity, and Parseval's identity holds for $f \in C^1(S)$.

It is now relatively easy to prove that Parseval's identity holds for any L_2 function f by approximating it in the L_2 sense by a function g in $C^1(S)$. Therefore, if $f \in L_2(-\pi, \pi)$, we have

$$\sum_{m=-\infty}^{\infty} |c_m|^2 = \frac{1}{2\pi} \int_{-\pi}^{\pi} |f|^2 \, dx \quad \text{(Parseval's identity)}, \tag{2.3.20}$$

where $c_m = (1/2\pi) \int_{-\pi}^{\pi} f(x) e^{-imx} \, dx$. The proof is given in Chapter 4. □

The Gibbs Phenomenon

The Fourier series of a discontinuous function cannot converge uniformly on an interval enclosing the discontinuity, since a uniformly convergent series of continuous functions is continuous. Let us examine more closely the details of the nonuniformity in the simple case of the function $f(x) = \operatorname{sgn} x [= 2H(x) - 1]$ for $-\pi \leqslant x < \pi$ and $f(x + 2\pi) = f(x)$. Its Fourier series is a sine series, and we find that

$$b_n = 0, \quad n \text{ even}; \qquad b_n = \frac{4}{\pi n}, \quad n \text{ odd}.$$

Thus we can write the partial sums as

$$\begin{aligned}
S_{2n+1}(x) &= \frac{4}{\pi} \sum_{k=0}^{n} \frac{\sin(2k+1)x}{2k+1} = \frac{4}{\pi} \sum_{k=0}^{n} \int_0^x \cos(2k+1)y \, dy \\
&= \frac{2}{\pi} \int_0^x \left[\sum_{k=0}^{n} e^{i(2k+1)y} + \sum_{k=0}^{n} e^{-i(2k+1)y} \right] dy \\
&= \frac{2}{\pi} \int_0^x \left[e^{iy} \frac{1 - e^{i2(n+1)y}}{1 - e^{i2y}} + e^{-iy} \frac{1 - e^{-i2(n+1)y}}{1 - e^{-i2y}} \right] dy \\
&= \frac{2}{\pi} \int_0^x \left[\frac{1 - e^{i2(n+1)y}}{-2i \sin y} + \frac{1 - e^{-i2(n+1)y}}{2i \sin y} \right] dy \\
&= \frac{2}{\pi} \int_0^x \frac{\sin 2(n+1)y}{\sin y} \, dy \\
&= \frac{2}{\pi} \int_0^x \frac{\sin 2(n+1)y}{y} \frac{y}{\sin y} \, dy.
\end{aligned}$$

We shall investigate the behavior near the jump at the origin. Let us choose $x > 0$ and small. Then $y/\sin y$ is close to 1 in the entire interval from 0 to x, and $s_{2n+1}(x)$

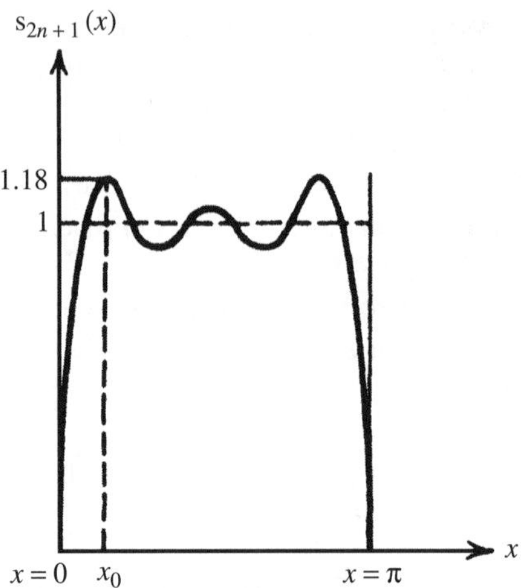

Figure 2.4

is closely approximated by

$$\frac{2}{\pi}\int_0^x \frac{\sin 2(n+1)y}{y}\,dy = \frac{2}{\pi}\int_0^{2(n+1)x} \text{sinc}\,u\;du \doteq \text{Si}[2(n+1)x].$$

Since $\text{Si}(\infty) = 1$, it is easy to see that $\text{Si}(\pi)$ must exceed 1, with an actual value that can be estimated as 1.18. Thus, for large n, there is always a value of x_0 near $0[x_0 \sim \pi/(2n+2)]$ where the partial sum s_{2n+1} takes on the value 1.18 (Figure 2.4). This does not contradict the fact that the Fourier series converges pointwise to 1 in $0 < x < \pi$, but it does show clearly that the convergence is not uniform in an interval including $x = 0$, either in its interior or as an endpoint. At the jump discontinuity the partial sums of high order *overshoot* the limiting function by approximately 9% of the jump. This is known as the *Gibbs phenomenon* and holds for the Fourier series of any function in D with simple jump discontinuities; indeed any such function can be written as the sum of a continuous function in D and terms of the form $J_k H(x - x_k)$, where the x_k's are the points of discontinuity and the J_k's are the numerical values of the jumps. Since $2H(x - x_k) = \text{sgn}(x - x_k) + 1$, the analysis of this section applies almost directly to the terms $J_k H(x - x_k)$; on the other hand, the continuous function in D has Fourier coefficients of order $1/n^2$, and its Fourier series therefore converges uniformly. Thus the entire overshoot is due to the terms $J_k H(x - x_k)$.

In electrical engineering a typical low-pass filter can be regarded as a transformation which cuts off frequencies above a critical frequency ω_c. If the periodic square wave just studied is the input to a low-pass filter, the output will be a *partial sum* of the Fourier series. Any such partial sum exhibits an overshoot at the points where the square wave has jumps, so that the output to the filter looks like Figure 2.4. The phenomenon was once considered mystifying. The overshoot can be avoided by using the Cesaro partial sums (see Exercise 2.3.6).

The overshoot also occurs in expansions in other orthonormal sets, or whenever the expansion is a best least-squares approximation (best fit in the L_2 norm). Foster and Richards [5] discuss the best least-squares approximation of a square wave by a continuous piecewise linear function whose changes in slope occur at equally spaced points. They show that an overshoot occurs (which tends to 13% of the jump in the square wave as the spacing is refined).

Change of Scale

Suppose that $f(x)$ is periodic with period T instead of 2π. We may think of f as defined on a circle whose circumference has length T instead of 2π. Since

$$g(x) \doteq f\left(\frac{T}{2\pi}x\right)$$

has period 2π, the previous results on 2π-periodic functions are easily modified. The Fourier series (in complex exponential form) associated with a T-periodic function $f(x)$ is

$$\sum_{k=-\infty}^{\infty} c_k e^{ik2\pi x/T},$$

where

$$c_k = \frac{1}{T}\int_{-T/2}^{T/2} f(x)e^{-ik2\pi x/T}\,dx = \frac{1}{T}\int_0^T f(x)e^{-ik2\pi x/T}\,dx.$$

In view of the T-periodicity, the integration can be over any interval of the form $(a, a+T)$.

Fourier Series as Distributions

We first extend slightly the notion of distributions to include also generalizations of complex-valued functions (whereas in our earlier treatment we generalized only real-valued functions). We now understand by $C_0^\infty(\mathbb{R})$ the set of all *complex-valued* infinitely differentiable functions $\phi(x)$ vanishing outside a bounded interval. A null sequence in $C_0^\infty(\mathbb{R})$ is defined just as in Section 2.1. A distribution f becomes a continuous, complex-valued, linear functional on $C_0^\infty(\mathbb{R})$; that is, to each test function ϕ in $C_0^\infty(\mathbb{R})$ there is assigned a complex number $\langle f, \phi\rangle$ such that

$$\langle f, \alpha_1\phi_1 + \alpha_2\phi_2\rangle = \alpha_1\langle f, \phi_1\rangle + \alpha_2\langle f, \phi_2\rangle$$

for all *complex* numbers α_1, α_2 and all ϕ_1, ϕ_2 in $C_0^\infty(\mathbb{R})$, and

$$\lim_{n\to\infty}\langle f, \phi_n\rangle = 0 \quad \text{whenever}\{\phi_n\} \text{ is a null sequence in } C_0^\infty(\mathbb{R}).$$

Obviously any complex-valued function f that is locally integrable defines a distribution by (2.1.4). All our earlier work survives with only trivial modifications.

A Fourier series $\sum_{n=-\infty}^{\infty} c_n e^{inx}$ will certainly converge uniformly if $|c_n| \leqslant M/n^2$ for large $|n|$. It therefore also converges in the sense of distributions. How should we interpret $\sum_{n=-\infty}^{\infty} c_n e^{inx}$ if all we know is that for large $|n|$, $|c_n| \leqslant Mn^\alpha$ for some integer α? This class includes series that diverge badly in the pointwise sense. Consider the related series

$$\sum_{n\neq 0} (in)^{-\alpha-2} c_n e^{inx},$$

which clearly converges uniformly, and hence distributionally, to $g(x)$, say. Here the qualification $n \neq 0$ on the summation symbol means that the sum runs from $-\infty$ to ∞, *excluding the term* $n = 0$. On taking the $(\alpha + 2)$th distributional derivative of the equality

$$g(x) = \sum_{n\neq 0} (in)^{-\alpha-2} c_n e^{inx},$$

we find that

$$g^{(\alpha+2)}(x) = \sum_{n\neq 0} c_n e^{inx}$$

and

$$\sum_{n=-\infty}^{\infty} c_n e^{inx} = c_0 + g^{(\alpha+2)}(x),$$

where $g^{(\alpha+2)}$ is the $(\alpha + 2)$th derivative of g in the sense of distributions.

Example 4. Consider the function $f(x) = x - \pi$ in $0 \leqslant x < 2\pi$, with $f(x + 2\pi) = f(x)$ for other values of x. This function is in D and has the Fourier series

$$f(x) = -2\sum_{n=1}^{\infty} \frac{\sin nx}{n} = i\sum_{n\neq 0} \frac{e^{inx}}{|n|}. \tag{2.3.21}$$

The series (2.3.21) does not converge uniformly to f, but it nevertheless converges distributionally to f. That this is the case can be seen by looking at the "integrated" series $2\sum_{n=1}^{\infty}(\cos nx)/n^2$, which converges uniformly, and hence distributionally, to a function $F(x)$ whose derivative is $f(x)$. We can therefore differentiate this integrated series termwise to recover (2.3.21) in the distributional sense. Another differentiation then gives

$$f'(x) = -2\sum_{n=1}^{\infty} \cos nx,$$

where by (2.1.17), $f' = 1 - \sum_{k=-\infty}^{\infty} 2\pi\delta(x - 2k\pi)$, so that

$$\sum_{k=-\infty}^{\infty} \delta(x - 2k\pi) = \frac{1}{2\pi} + \frac{1}{\pi}\sum_{n=1}^{\infty} \cos nx. \tag{2.3.22}$$

In a classical sense the series on the right does not converge, and the left side is totally meaningless. Within the theory of distributions, (2.3.22) means that the action of both sides on a test function ϕ in $C_0^\infty(\mathbb{R})$ is the same. Thus (2.3.22) states that for every test function ϕ,

$$\sum_{k=-\infty}^{\infty} \phi(2k\pi) = \frac{1}{2\pi}\int_{-\infty}^{\infty} \phi(x)\,dx + \frac{1}{\pi}\sum_{n=1}^{\infty}\int_{-\infty}^{\infty} (\cos nx)\phi(x)\,dx.$$

The series on the left is really a finite sum, since ϕ has bounded support. It is less obvious that the series on the right converges, but this is guaranteed by the fact that (2.3.22) has been correctly derived within the theory of distributions.

We can rewrite (2.3.22) as

$$\sum_{k=-\infty}^{\infty} \delta(x-2k\pi) = \frac{1}{2\pi}\sum_{n=-\infty}^{\infty} e^{inx} = \frac{1}{2\pi}\sum_{n=-\infty}^{\infty} e^{-inx} \tag{2.3.23}$$

or, replacing x by $x-\xi$, as

$$\begin{aligned}\sum_{k=-\infty}^{\infty} \delta(x-\xi-2k\pi) &= \frac{1}{2\pi}\sum_{n=-\infty}^{\infty} e^{in(\xi-x)} \\ &= \frac{1}{2\pi} + \frac{1}{\pi}\sum_{n=1}^{\infty}(\cos n\xi\cos nx + \sin n\xi \sin nx).\end{aligned} \tag{2.3.24}$$

Thinking of the parameter ξ as fixed in $(-\pi,\pi)$, let us apply (2.3.24) to a test function whose support is in $(-\pi,\pi)$. Only the delta function with $k=0$ is effective, and we obtain

$$\phi(\xi) = \frac{1}{2\pi}\sum_{n=-\infty}^{\infty} e^{in\xi}\int_{-\pi}^{\pi} e^{-inx}\phi(x)\,dx, \qquad -\pi<\xi<\pi, \tag{2.3.25}$$

which is just the Fourier series expansion formula for ϕ. Of course, in (2.3.25) ϕ has to be a test function; nevertheless, it appears that (2.3.23) contains in a nutshell much of the information needed to generate Fourier series expansions. Equation (2.3.23) is often called a *completeness relation*.

Example 5. The Poisson summation formula. Let $\phi(x)$ be a complex-valued test function on $\mathbb{R}$; if λ, t are real parameters, then $\phi(\lambda x/2\pi)e^{ixt/2\pi}$ is also a test function, to which we apply (2.3.23) to obtain

$$\sum_{k=-\infty}^{\infty} \phi(\lambda k)e^{ikt} = \frac{1}{2\pi}\sum_{n=-\infty}^{\infty}\int_{-\infty}^{\infty} e^{inx}\phi\left(\frac{\lambda x}{2\pi}\right)e^{ixt/2\pi}\,dx.$$

Setting $y=\lambda x/2\pi$, we find that

$$\sum_{k=-\infty}^{\infty} \phi(\lambda k)e^{ikt} = \frac{1}{|\lambda|}\sum_{n=-\infty}^{\infty} \phi^{\wedge}\left(\frac{t+2\pi n}{\lambda}\right), \tag{2.3.26}$$

where

$$\phi^\wedge(\omega) = \int_{-\infty}^{\infty} e^{i\omega x}\phi(x)\,dx$$

is the *Fourier transform* of $\phi(x)$, discussed at greater length in the following section.

If we take successively $t = 0$ and then $\lambda = 1$ in (2.3.26), we obtain

$$\sum_{k=-\infty}^{\infty} \phi(\lambda k) = \frac{1}{|\lambda|}\sum_{n=-\infty}^{\infty} \phi^\wedge\left(\frac{2\pi n}{\lambda}\right) \tag{2.3.27}$$

and

$$\sum_{k=-\infty}^{\infty} \phi(k) = \sum_{n=-\infty}^{\infty} \phi^\wedge(2\pi n). \tag{2.3.28}$$

Identities (2.3.26), (2.3.27), and (2.3.28) are all given the name *Poisson's summation formula*. They can be regarded as transforming one series to another for the purpose of improving the rate of convergence or, in rare instances, carrying out the summation in closed form. Although our formulas have been derived only for test functions, they hold for a much larger class of functions: it suffices for ϕ to be a piecewise smooth function, vanishing at $|x| = \infty$, with $\int_{-\infty}^{\infty} |\phi|\,dx$ finite.

As an illustration of the use of (2.3.27), let $\phi(x) = e^{-x^2}$. An easy calculation gives $\phi^\wedge(\omega) = \sqrt{\pi}e^{-\omega^2/4}$ (see Example 3, Section 2.4). Thus we find that

$$\sum_{k=-\infty}^{\infty} e^{-\lambda^2 k^2} = \frac{\sqrt{\pi}}{|\lambda|}\sum_{n=-\infty}^{\infty} e^{-n^2\pi^2/\lambda^2}. \tag{2.3.29}$$

By examining the ratio of successive terms for the two series in (2.3.29), we see that the series on the left converges rapidly for large $|\lambda|$, and the one on the right for small $|\lambda|$. A similar procedure is often possible when we want to improve the convergence of an infinite series. In fact, (2.3.29) expresses a duality between two forms of the solution of the heat equation with λ^2 proportional to the time [see (8.2.33) and (8.2.35)].

Discrete and Fast Fourier Transforms

A given function on the unit circle can be represented as the Fourier series (2.3.1) with coefficients (2.3.7), or equivalently, as the series (2.3.3) with coefficients (2.3.8). In practice, we may only be given the values f_k at a finite number of points x_k on the unit circle. We say that a function $f(x)$ *interpolates* the data $\{x_k, f_k\}$ if $f(x_k) = f_k$. Suppose that data is given at the $2N+1$ equally spaced points $x_k = 2k\pi/(2N+1)$, where k is an integer ranging from $-N$ to N. There are then many series of the form (2.3.1) which interpolate the data and we may seek one that is particularly simple, namely, a *trigonometric polynomial* not containing high harmonics. Let T_N be the $(2N+1)$-dimensional space of trigonometric polynomials

$$S_N(x) = \sum_{m=-N}^{N} c_m e^{imx} = \frac{a_0}{2} + \sum_{m=1}^{N}(a_m \cos mx + b_m \sin mx).$$

We want to determine the $2N+1$ constants $(c_m)_{-N}^{N}$ from the $2N+1$ equations

$$S_N(x_k) = \sum_{m=-N}^{N} c_m e^{imx_k} = f_k, \quad k = 0, \pm 1, \ldots, \pm N. \tag{2.3.30}$$

Let us denote the vector $(e^{imx_k})_{k=-N}^{k=N}$ by e_m. From (2.3.15), we find that

$$\langle e_m, e_n \rangle = \sum_{k=-N}^{N} e^{imx_k} e^{-inx_k} = \begin{cases} 0, & m \neq n, \\ 2N+1, & m = n. \end{cases}$$

These vectors therefore form an orthogonal set in the sense of Example 2, Section 4.5. In this terminology, the equations (2.3.30) are equivalent to the vector equation

$$f = \sum_{m=-N}^{N} c_m e_m,$$

which has one and only one solution

$$c_m = \frac{\langle f, e_m \rangle}{2N+1} = \frac{1}{2N+1} \sum_{k=-N}^{N} f_k e^{-imx_k}, \quad m = 0, \pm 1, \ldots, \pm N. \tag{2.3.31}$$

Thus, there is one and only one element of T_N which interpolates the data. Its coefficients are given by (2.3.31), known as the *Discrete Fourier Transform* of f. By using (2.3.31) and (2.3.15) we can rewrite

$$\begin{aligned} S_N(x) &= \sum_{m=-N}^{N} \frac{e^{imx}}{2N+1} \left[\sum_{k=-N}^{N} f_k e^{-imx_k} \right] \\ &= \sum_{k=-N}^{N} \frac{f_k}{2N+1} \frac{\sin\left(N+\frac{1}{2}\right)(x-x_k)}{\sin[(x-x_k)/2]} \\ &= h \sum_{k=-N}^{N} f_k D_N(x-x_k), \end{aligned} \tag{2.3.32}$$

where D_N is the Dirichlet kernel in (2.3.15). This representation clearly shows that $S_N(x)$ interpolates the data $\{x_k, f_k\}$ and has similarities to later results for Fourier transforms.

Our problem can be viewed slightly differently. Suppose we know that a function $f(x) \in T_N$ and we sample it at the $2N+1$ points $x_k, k = 0, \pm 1, \ldots, \pm N$. We can then reconstruct $f(x)$ from its values $f(x_k)$ at these points by the formula

$$f(x) = \sum_{m=-N}^{N} c_m e^{imx},$$

where

$$c_m = \frac{1}{2N+1}\sum_{k=-N}^{N} f(x_k)e^{-imx_k}, \quad m = 0, \pm 1, \ldots, \pm N.$$

Note that c_m is just a Riemann sum for the integral

$$\frac{1}{2\pi}\int_{-\pi}^{\pi} f(x)e^{-imx}\,dx,$$

by using a uniform partition with spacing $\Delta x = 2\pi/(2N+1)$. Thus, approximate integration is perfect in this case!

REMARKS

1. Suppose that the data is given at $2N$ equally spaced points on the unit circle. Let these points be $x_k = k\pi/N, k = 0, \pm 1, \ldots, \pm(N-1), -N$; note that x_N and x_{-N} are the same point on the unit circle and only one of them should be included in the summation (2.3.30). We have chosen to include x_{-N} so that (2.3.30) is now a sum from $-N$ to $N-1$, and the rest of the analysis is very similar to that given for the case of $2N+1$ data points.

2. The *Fast Fourier Transform* is a method for calculating (2.3.31) more rapidly by grouping terms in an optimal fashion, recognizing them as discrete transforms over smaller data sets. (See Strang [14].)

3. One can also define discrete sine and cosine transforms. (See Exercise 2.3.3, for instance.)

EXERCISES

2.3.1 Let

$$f(x) = x(\pi - x), \quad 0 \leqslant x \leqslant \pi.$$

(a) Graph $f(x)$ in the interval $(0, \pi)$. Extend $f(x)$ as an odd function in $-\pi \leqslant x \leqslant 0$. What is the formula for this extended function in $(-\pi, 0)$? Extend this new f as a periodic function of period 2π.

(b) Expand the periodic f of (a) into a Fourier series.

(c) Justify the fact that the coefficients behave as $1/n^3$ for large n.

(d) Show that

$$\frac{\pi^3}{32} = 1 - \frac{1}{3^3} + \frac{1}{5^3} - \frac{1}{7^3} + \cdots.$$

(e) Write Parseval's formula for the present case.

2.3.2 We wish to approximate $\sin x$ in $-\pi \leqslant x \leqslant \pi$ by a polynomial $s_n(x) = a_0 + a_1x + \cdots + a_nx^n$.

(a) Plot $s_1(x)$ and $s_3(x)$ when the Maclaurin expansion for $\sin x$ is used $[s_1(x) = x, s_3(x) = x - (x^3/6)]$.

(b) Find the $s_1(x)$ and $s_3(x)$ which give the best L_2 approximation. Plot $s_1(x)$ and $s_3(x)$.

(c) Find the $s_1(x)$ which minimizes the maximum absolute deviation between s_1 and $\sin x$ from $-\pi$ to π; that is, choose $s_1(x)$ so that

$$\max_{-\pi \leqslant x \leqslant \pi} |s_1(x) - \sin x|$$

is as small as possible.

2.3.3 Let $f(x)$ be an unknown, odd, continuous function with period 1. We observe the values $f_1, \ldots, f_N$ of f at the N points $x_k = k/2(N+1), k = 1, 2, \ldots, N$. Show that there is one and only one interpolating function $g(x)$ of the form $\sum_{m=1}^{N} c_m \sin 2\pi m x$ whose value at x_k coincides with f_k, for $k = 1, \ldots, N$. The coefficients c_m are the ones obtained by numerical integration, that is,

$$c_m = \frac{2}{N+1} \sum_{k=1}^{N} f_k \sin 2\pi m x_k.$$

2.3.4 (a) Use the Poisson formula (2.3.27) with $\lambda > 0$ and $\phi(x) = e^{-|x|}$ to show that

$$\sum_{k=-\infty}^{\infty} e^{-\lambda |k|} = \sum_{n=-\infty}^{\infty} \frac{2\lambda}{\lambda^2 + 4\pi^2 n^2}.$$

By noting that the left side is essentially a geometric series, obtain the identity

$$\sum_{n=-\infty}^{\infty} \frac{2\lambda}{\lambda^2 + 4\pi^2 n^2} = \frac{1 + e^{-\lambda}}{1 - e^{-\lambda}}. \tag{2.3.33}$$

(b) Evaluate $\sum_{k=1}^{\infty} 1/(k^4 + a^4)$, which yields a famous formula when $a = 0$.

2.3.5 (a) Consider a 2π-periodic function $f(x)$ whose definition on the fundamental interval $|x| < \pi$ is $f = 1$ for $|x| < \xi$ and $f = 0$ for $\xi < |x| < \pi$ (here ξ is fixed, $0 < \xi < \pi$). Expand $f(x)$ in a cosine series, and differentiate term by term to obtain

$$\frac{2}{\pi} \sum_{n=1}^{\infty} \sin n\xi \sin nx = \sum_{k=-\infty}^{\infty} \delta(x - \xi - 2\pi k) - \sum_{k=-\infty}^{\infty} \delta(x + \xi - 2\pi k). \tag{2.3.34}$$

Apply (2.3.34) to a test function $\phi(x)$ whose support lies in $(0,\pi)$ to derive the sine series expansion

$$\phi(\xi) = \frac{2}{\pi}\sum_{n=1}^{\infty} \sin n\xi \int_0^{\pi} \phi(x)\sin nx\, dx.$$

(b) Combine (2.3.34) and (2.3.24) to find

$$\frac{1}{\pi}+\frac{2}{\pi}\sum_{n=1}^{\infty} \cos n\xi \cos nx = \sum_{k=-\infty}^{\infty} \delta(x-\xi-2\pi k)+\sum_{k=-\infty}^{\infty} \delta(x+\xi-2\pi k).$$

2.3.6 Let $\{s_n(x)\}$ be the sequence of partial sums of the Fourier series for $f(x)$, and define σ_n as the mean of the first n partial sums:

$$\sigma_n(x) = \frac{s_0(x)+\cdots+s_{n-1}(x)}{n}.$$

Using (2.3.17), we can write

$$\sigma_n(0) = \int_{-\pi}^{\pi} \frac{D_0(x)+\cdots+D_{n-1}(x)}{n} f(x)\, dx \doteq int^{\pi}_{-\pi} F_n(x) f(x)\, dx,$$

where $F_n(x)$ is the *Féjer kernel*. Show that

$$F_n(x) = \frac{1}{2\pi n}\frac{\sin^2(nx/2)}{\sin^2(x/2)}$$

and that $\{F_n(x)\}$ is a delta sequence on the unit circle. Prove that for any $f(x)$ in $C(S)$, $\sigma_n(x)$ converges to $f(x)$. We say that the Fourier series is Cesaro summable to $f(x)$.

2.3.7 The function $f(x) = \log(1-\cos x)$ has period 2π and belongs to $L_1(-\pi,\pi)$ despite its singularity at $x=0$. Show that

$$2\,\mathrm{Re}\log(1-e^{ix}) = \log 2 + \log(1-\cos x);$$

use the power series expansion of $\log(1-z)$ to show that

$$\log(1-\cos x) = -\log 2 - 2\sum_{n=1}^{\infty}\frac{\cos nx}{n}, \qquad x \neq 2k\pi. \tag{2.3.35}$$

By differentiating term by term in the distributional sense, obtain a series for $\cot(x/2)$. Since $\cot(x/2)$ is not absolutely integrable, it is to be regarded a pseudofunction.

2.3.8 Let $f(x)$ and $g(x)$ have period 2π, and let their Fourier coefficients be $\{f_n\}$ and $\{g_n\}$, respectively. Express the Fourier coefficients of

$$h(x) \doteq \int_{-\pi}^{\pi} f(x-\xi)g(\xi)\, d\xi = \int_{-\pi}^{\pi} g(x-\xi)f(\xi)\, d\xi$$

in terms of f_n and g_n. The function $h(x)$ is known as the (Fourier series) *convolution* of f and g.

2.3.9 Consider the function $f(x) = \operatorname{sgn} x \quad \text{on} \quad |x| < \pi$, whose Fourier series was calculated when discussing the Gibbs phenomenon. Show that in the form (2.3.2) the separate sums from $-\infty$ to -1 and from 1 to ∞ do not converge. In general, define the nonsymmetric partial sums corresponding to (2.3.2) by

$$s_{m,n}(x) = \sum_{k=-m}^{n} c_k \, e^{ikx}.$$

Consider a function $f(x)$ on S with $f(0) = 0$. Show that

$$\lim_{m,n\to\infty} s_{m,n}(0) = 0$$

if f is continuous at $x = 0$ and has one-sided derivatives there. (Adapt the Chernoff [2] proof of Theorem 2.3.3.)

2.4 FOURIER TRANSFORMS AND INTEGRALS

Classical Results

The Fourier integral expansion is the analog for functions small at $|x| = \infty$ (hence nonperiodic) of the Fourier series expansion of periodic functions. Whereas a periodic function is expanded into harmonics of the same period, the nonperiodic function is a continuous superposition over all frequencies.

Let $f(x)$ be a function vanishing sufficiently fast as $|x| \to \infty$, and let $f_T(x)$ be the T-periodic function coinciding with f on $-T/2 \leqslant x < T/2$. Then, if f_T obeys Dirichlet conditions, we can write for all x,

$$f_T(x) = \sum_{k=-\infty}^{\infty} c_k e^{2\pi i k x/T}, \quad c_k = \frac{1}{T}\int_{-T/2}^{T/2} f(\xi) e^{-2\pi i k \xi/T} \, d\xi,$$

so that for $|x| < T/2$,

$$f(x) = \frac{1}{2\pi} \sum_{k=-\infty}^{\infty} \left[\frac{2\pi}{T} \int_{-T/2}^{T/2} f(\xi) e^{i2\pi k(x-\xi)/T} \, d\xi \right]. \tag{2.4.1}$$

Setting $\omega_k = 2\pi k/T, \Delta\omega = 2\pi/T$, and $g(\omega, x, T) = \int_{-T/2}^{T/2} f(\xi) e^{i\omega(x-\xi)} \, d\xi$, we find that (2.4.1) becomes

$$f(x) = \frac{1}{2\pi} \sum_{k=-\infty}^{\infty} g(\omega_k, x, T) \, \Delta\omega, \quad |x| < T/2, \tag{2.4.2}$$

which, for large T, can be regarded as a Riemann sum over a partition of spacing $2\pi/T$ of the entire ω axis. The sum in (2.4.2) is therefore essentially an integral from $-\infty$ to ∞ in ω. At the same time, since $T \to \infty$, the expression for g also becomes an integral from $-\infty$ to ∞ in ξ [assuming that $\int_{-\infty}^{\infty} f(\xi)e^{-i\omega\xi}\,d\xi$ exists]. Replacing ω by $-\omega$, we are led to the following *Fourier integral expansion* for a function $f(x)$ in $L_1(-\infty,\infty)$:

$$f(x) = \frac{1}{2\pi}\int_{-\infty}^{\infty} f^\wedge(\omega)e^{-i\omega x}\,d\omega, \tag{2.4.3}$$

where

$$f^\wedge(\omega) \doteq \int_{-\infty}^{\infty} f(\xi)e^{i\omega\xi}\,d\xi \left[= \int_{-\infty}^{\infty} f(x)e^{i\omega x}\,dx\right]. \tag{2.4.4}$$

Formula (2.4.4) defines $f^\wedge$ as the *Fourier transform of* f, and (2.4.3) shows how to reconstruct $f(x)$ from its Fourier transform. It turns out that the integral in (2.4.3) has to be interpreted as

$$\lim_{R\to\infty}\frac{1}{2\pi}\int_{-R}^{R} f^\wedge(\omega)e^{-i\omega x}\,d\omega. \tag{2.4.5}$$

[Some authors prefer to define the transform with $e^{-i\omega\xi}$ instead of $e^{i\omega\xi}$; this requires a change in the exponential in (2.4.3) to $e^{i\omega x}$. The factor $1/2\pi$ is sometimes apportioned differently between (2.4.3) and (2.4.4). In some ways the best solution is to use the exponential $e^{i2\pi\omega\xi}$ in definition (2.4.4); this then eliminates all factors of 2π except in the exponent, the drawback being that the differentiation formulas are slightly more cumbersome.]

To prove (2.4.5) we first note that $f^\wedge$ exists in view of the L_1 assumption on f. Substituting the definition of $f^\wedge$, we find that

$$\begin{aligned}\frac{1}{2\pi}\int_{-R}^{R} f^\wedge(\omega)e^{-i\omega x}\,d\omega &= \frac{1}{2\pi}\int_{-R}^{R} e^{-i\omega x}\,d\omega \int_{-\infty}^{\infty} f(\xi)e^{i\omega\xi}\,d\xi \\ &= \int_{-\infty}^{\infty} f(\xi)\frac{\sin R(\xi - x)}{\pi(\xi - x)}\,d\xi \\ &= \int_{-\infty}^{\infty} f(x+y)\frac{R}{\pi}\operatorname{sinc} Ry\,dy.\end{aligned}$$

To calculate the limit as $R \to \infty$ we recall from Exercise 2.2.5 that $(R/\pi)\operatorname{sinc} Ry$ is a delta sequence, so that the limit is $f(x)$ for a fairly large class of functions. Alternatively, we can use a procedure similar to that for Theorem 2.3.3 on Fourier series. We state the result as follows.

Fourier Integral Theorem. Let $f \in L_1(-\infty,\infty)$, and let f satisfy Dirichlet conditions; then $f^\wedge(\omega)$ exists, and the inversion formula (2.4.3) holds in the sense that

$$\frac{f(x+) + f(x-)}{2} = \lim_{R\to\infty}\frac{1}{2\pi}\int_{-R}^{R} f^\wedge(\omega)e^{-i\omega x}\,d\omega. \tag{2.4.6}$$

For further reference we mention two relations, both of which go under the name *Parseval's formula*:

$$\int_{-\infty}^{\infty} |f(x)|^2\,dx = \frac{1}{2\pi}\int_{-\infty}^{\infty} |f^\wedge(\omega)|^2\ d\omega, \tag{2.4.7}$$

$$\int_{-\infty}^{\infty} f^\wedge(x)g(x)\,dx = \int_{-\infty}^{\infty} f(x)g^\wedge(x)\,dx. \tag{2.4.8}$$

The Parseval formulas are easily derived from the definition of the Fourier transform by formally interchanging orders of integration. We shall not, for the present, state precise conditions for the validity of these formulas.

Band-Limited Functions. We have seen at the end of the preceding section how one can interpolate data, given at equally spaced points of a bounded interval, by a trigonometric polynomial. If data is given at equally spaced points a distance h apart on the entire real axis, a different procedure is needed.

Suppose that we are given data f_k at $x_k = kh, k = 0, \pm 1, \ldots$; then the following function (*Whittaker's cardinal function*)

$$C(x) = \sum_{k=-\infty}^{\infty} f_k \operatorname{sinc}\frac{\pi}{h}(x - x_k) \tag{2.4.9}$$

interpolates the data (assuming the series converges, of course). The result follows from the observation that

$$\operatorname{sinc}\frac{\pi}{h}(x_j - x_k) = \begin{cases} 1, & k = j, \\ 0, & k \neq j. \end{cases}$$

For the cardinal function, approximate integration by the trapezoidal rule is exact. Integrating both sides of (2.4.9) from $-\infty$ to ∞ gives

$$\int_{-\infty}^{\infty} C(x)\,dx = h\sum_{k=-\infty}^{\infty} f_k = h\sum_{k=-\infty}^{\infty} C(kh).$$

The right side is just the trapezoidal approximation to the integral on the left and they are equal. So the approximation is perfect! This sort of equality also occurs in related settings: We have already noted such a relation for discrete Fourier transforms, and we now give another instance.

Let $\phi(x)$ be a function for which Poisson's formula (2.3.27) holds. Thinking of x as the time, we shall refer to ϕ as the signal and we shall assume that its Fourier transform $\phi^\wedge(\omega)$ contains no frequencies higher than Ω: $\phi^\wedge(\omega) \equiv 0$ for $|\omega| \geqslant \Omega$; in this case, ϕ is said to be Ω-*band-limited* and Ω is called the *bandwidth*. If we then choose $0 < \lambda \leqslant 2\pi/\Omega$ in (2.3.27), the only nonzero term in the sum on the right side is the one corresponding to $n = 0$, and therefore

$$\sum_{k=-\infty}^{\infty} \lambda\phi(\lambda k) = \int_{-\infty}^{\infty} \phi(x)\,dx, \quad \lambda \leqslant \frac{2\pi}{\Omega}.$$

The left side is the trapezoidal approximation of the right side with spacing λ.

We can now show that if ϕ is Ω-band-limited, then (2.4.9) holds if $h < \pi/\Omega$, that is,

$$\phi(x) = \sum_{k=-\infty}^{\infty} \phi(x_k) \operatorname{sinc} \frac{\phi}{h}(x - x_k), \quad x_k = kh. \tag{2.4.10}$$

Although this could be proved by manipulating Poisson's formula, it is more instructive to use basic principles. Expanding $\phi^\wedge(\omega)$ in a Fourier series for $|\omega| < \Omega$, we have

$$\phi^\wedge(\omega) = \begin{cases} \sum_{k=-\infty}^{\infty} c_k e^{ik\omega\pi/\Omega}, & |\omega| < \Omega, \\ 0, & |\omega| \geqslant \Omega. \end{cases}$$

The Fourier integral representation of $\phi(x)$ is

$$\phi(x) = \frac{1}{2\pi} \int_{-\infty}^{\infty} e^{-i\omega x} \phi^\wedge(\omega)\, dw = \frac{1}{2\pi} \int_{-\Omega}^{\Omega} e^{-i\omega x} \sum_{k=-\infty}^{\infty} c_k e^{ik\omega x/\Omega}\, d\omega,$$

which, after interchanging summation and integration, becomes

$$\phi(x) = \frac{\Omega}{\pi} \sum_{k=-\infty}^{\infty} C_k \operatorname{sinc} \left[\Omega\left(x - \frac{k\pi}{\Omega}\right)\right], \quad x \in \mathbb{R}.$$

At $x = x_m = m\pi/\Omega$, only the term with $k = m$ is nonzero, so that $c_m = (\pi/\Omega)\phi(x_m)$ and

$$\phi(x) = \sum_{k=-\infty}^{\infty} \phi(x_k) \operatorname{sinc} [\Omega(x - x_k)], \quad x_k = \frac{k\pi}{\Omega}, \tag{2.4.10a}$$

or, on setting $h = \pi/\Omega$, we obtain (2.4.10). Of course, if ϕ is Ω-band-limited, it is also Ω'-band-limited with $\Omega' \geqslant \Omega$, so that (2.4.10) and (2.4.10a) hold with a smaller spacing (smaller h or larger Ω). Both equations are known as the *sampling formula*. A band-limited signal can be exactly reconstructed by sampling at the uniform spacing $h = \pi/\Omega$, where $\Omega \geqslant$ bandwidth of the signal. (This should be compared with the result for Fourier series: An element of T_N can be reconstructed by sampling at $2N + 1$ equally spaced points on the unit circle.)

For additional information on the sinc function and its applications, see for example Stenger [13].

Heisenberg's Uncertainty Principle

In its original quantum-mechanical setting the principle gives a quantitative bound on the possible precision of simultaneous measurements of the position and momentum of a particle. Mathematically, the principle reflects a relationship between a function and its Fourier transform. It will be convenient to use the language of electrical engineering and of probability.

Let $f(t)$ be a complex-valued time signal of unit energy: $\int_{-\infty}^{\infty} |f(t)|^2\,dt = 1$. The energy spectrum of the signal is given by $|f^\wedge(\omega)|^2/2\pi$. By Parseval's formula (2.4.7) the total energy in the signal and that in the spectrum are equal, so that

$$\int_{-\infty}^{\infty} |f(t)|^2\,dt = \frac{1}{2\pi}\int_{-\infty}^{\infty} |f^\wedge(\omega)|^2\,d\omega = 1.$$

Both $|f(t)|^2$ and $(1/2\pi)|f^\wedge(\omega)|^2$ can be regarded as densities of probability distribution on their respective axes. A translation in either $f(t)$ or $f^\wedge(\omega)$ merely leads to a phase shift in the other, not affecting its energy distribution. We may therefore assume that the mean of each distribution is 0. The variances of these distributions are then given by

$$\sigma_t^2 \doteq \int_{-\infty}^{\infty} t^2|f(t)|^2\,dt, \quad \sigma_\omega^2 \doteq \frac{1}{2\pi}\int_{-\infty}^{\infty} \omega^2|f^\wedge(\omega)|^2\,d\omega,$$

where σ_t is a measure of the duration of the time signal, and σ_ω a measure of the frequency spread.

Theorem. *If $|t|^{1/2}f$ and $|t|^{1/2}f'$ both tend to 0 as $t \to \infty$, then*

$$\sigma_t\sigma_\omega \geqslant \frac{1}{2}, \tag{2.4.11}$$

and the equality holds if and only if f is the Gaussian signal $Ce^{-\alpha t^2}$, where α is an arbitrary positive number and C is chosen so that its energy is unity.

Proof. For simplicity we shall assume that $f(t)$ is *real*; the reader is invited to supply the appropriate changes for f complex. Since the Fourier transform of $f'(t)$ is $-i\omega f^\wedge(\omega)$, Parseval's formula gives

$$\frac{1}{2\pi}\int_{-\infty}^{\infty} \omega^2|f^\wedge(\omega)|^2\,d\omega = \int_{-\infty}^{\infty} [f'(t)]^2\,dt.$$

On the other hand, we infer from the Schwarz inequality (4.5.14) that

$$\left(\int_{-\infty}^{\infty} \tfrac{1}{2}t\frac{d}{dt}f^2\,dt\right)^2 = \left(\int_{-\infty}^{\infty} tff'\,dt\right)^2 \leqslant \int_{-\infty}^{\infty} t^2f^2\,dt \int_{-\infty}^{\infty} f'^2\,dt = \sigma_t^2\sigma_\omega^2,$$

with equality only if f' is a multiple of tf. Integrating the left side by parts and using the condition on f at infinity, we obtain

$$\sigma_t^2\sigma_\omega^2 \geqslant \tfrac{1}{4}\left(\int_{-\infty}^{\infty} f^2\,dt\right)^2 = \tfrac{1}{4},$$

with equality only if $f' = cft$, that is, only if f is the predicted Gaussian signal. □

Extensions and further applications can be found in Dym and McKean [4] and in Papoulis [11].

Fourier Transform in the Complex Plane

Equations (2.4.4) and (2.4.3) can be generalized to some extent by allowing the transform variable ω to take on complex values. Writing $\omega = u + iv$, where u and v are real, we obtain

$$f^\wedge(\omega) = f^\wedge(u+iv) = \int_{-\infty}^{\infty} e^{i\omega x} f(x)\,dx = \int_{-\infty}^{\infty} e^{iux} e^{-vx} f(x)\,dx. \qquad (2.4.12)$$

Thus $f^\wedge(\omega)$ is just the "old" Fourier transform (2.4.4) (with transform variable u) of the function $e^{-vx}f(x)$. If v is chosen so that $e^{-vx}f(x)$ is in $L_1(-\infty,\infty)$, we deduce from (2.4.5) that

$$e^{-vx}f(x) = \lim_{R\to\infty} \frac{1}{2\pi} \int_{-R}^{R} f^\wedge(u+iv)e^{-iux}\,du.$$

The last integral can be interpreted as an integral in the complex ω plane along a line parallel to the real axis. In fact, we have

$$f(x) = \lim_{R\to\infty} \frac{1}{2\pi} \int_{iv-R}^{iv+R} f^\wedge(\omega)e^{-i\omega x}\,d\omega,$$

or, by a slight abuse of notation,

$$f(x) = \frac{1}{2\pi} \int_{iv-\infty}^{iv+\infty} f^\wedge(\omega)e^{-i\omega x}\,d\omega, \qquad (2.4.13)$$

where v is any real number for which

$$\int_{-\infty}^{\infty} |e^{-vx}f(x)|\,dx < \infty. \qquad (2.4.14)$$

One should observe that the factor e^{-vx} which occurs in (2.4.14) is not necessarily helpful; if $v > 0$, the factor improves convergence at the upper limit but impairs it at the lower limit, and conversely for $v < 0$. Even in the simple case $f(x) = 1$, there is no value of v for which (2.4.14) holds, so that our approach will have to be modified. On the other hand, there are cases where inequality (2.4.14) is satisfied in an entire strip $v_1 < v < v_2$; it can then be shown that $f^\wedge(\omega)$ is an analytic function in that strip.

Example 1. $f(x) = e^{-|x|}$. Then $f(x)e^{-vx}$ is in $L_1(-\infty,\infty)$ for all v such that $-1 < v < 1$. We find that

$$\begin{aligned} f^\wedge(\omega) &= \int_{-\infty}^{\infty} e^{i\omega x}e^{x}\,dx + \int_{0}^{\infty} e^{i\omega x}e^{-x}\,dx \\ &= \frac{1}{1+i\omega} + \frac{1}{1-i\omega} = \frac{2}{1+\omega^2}, \qquad -1 < v < 1. \end{aligned}$$

We now illustrate how (2.4.13), with $v = 0$, can be used to recover $f(x)$ from $f^\wedge(\omega)$ by contour integration. For $x > 0, 2e^{-i\omega x}/(1+\omega^2)$ is exponentially small in the lower half of the ω plane ($v < 0$). Consider the contour C_R, consisting of the entire boundary of a large semicircle of radius R in the lower half-plane (with the diameter on the real axis between $-R$ and R). Then, by Cauchy's theorem,

$$\lim_{R\to\infty} \int_{C_R} \frac{1}{2\pi} \frac{2}{1+\omega^2} e^{-i\omega x}\, d\omega = 2\pi i r,$$

where r is the sum of the residues of $e^{-i\omega x}/\pi(1+\omega^2)$ within C_R. The only singularity is a simple pole at $\omega = -i$, and the corresponding residue is $ie^{-x}/2\pi$. Now, as $R \to \infty$, the contribution from the curved portion of C_R disappears because of the behavior of the integrand. Taking into account the contribution from the diameter, we find that

$$\frac{1}{2\pi} \int_{-\infty}^{\infty} \frac{2}{1+\omega^2} e^{-i\omega x}\, d\omega = e^{-x}, \quad x > 0.$$

By a simple change of variables, one can easily show that the value of the inversion integral is e^x for $x < 0$.

Example 2. $f(x) = -2H(x)\sinh x$, where $H(x)$ is the Heaviside function. Thus

$$f(x) = \begin{cases} -2\sinh x, & x > 0, \\ 0, & x < 0. \end{cases}$$

Since $2\sinh x = e^x - e^{-x}$, we see that $e^{-vx}\sinh x$ is in $L_1(0, \infty)$ for $v > 1$, and therefore $e^{-vx} f(x)$ is in $L_1(-\infty, \infty)$ for $v > 1$. A simple calculation yields

$$f^\wedge(\omega) = \int_0^\infty (e^{-x} - e^x) e^{i\omega x}\, dx = \frac{2}{1+\omega^2}, \quad v > 1.$$

When compared with the result of Example 1, this may at first seem mystifying. Do two different functions have the same Fourier transforms? The answer is that transforms of different functions may have the same functional form valid in *different, nonoverlapping* regions of the ω plane. We can still use the inversion formula (2.4.13), this time with $v > 1$, to recover the original $f(x)$ of our present example. First, for $x < 0$ we use a semicircle in the region $v > 1$ to find $f = 0$ for $x < 0$. Then, for $x > 0$ we use a lower semicircle, which includes both poles of the integrand (at $\omega = +i$ and $\omega = -i$), and obtain $f = -2\sinh x$.

Example 3. $f(x) = e^{-x^2}$. Then fe^{-vx} is in $L_1(-\infty, \infty)$ for all real v, so that $f^\wedge(\omega)$ is analytic in the entire ω plane. We find for $\omega = iv$, v real, that

$$f^\wedge(iv) = \int_{-\infty}^{\infty} e^{-vx} e^{-x^2}\, dx = e^{v^2/4} \int_{-\infty}^{\infty} e^{-(x+v/2)^2}\, dx = \sqrt{\pi}\, e^{v^2/4},$$

so that, by analytic continuation,

$$f^\wedge(\omega) = \sqrt{\pi} e^{-\omega^2/4}.$$

Example 4. If $f(x)$ has bounded support, then $f^\wedge(\omega)$ is an *entire function* (that is, analytic in the entire complex plane). This is illustrated by

$$f_\theta(x) = \begin{cases} \dfrac{1}{2\theta}, & |x| < \theta, \\ 0, & |x| > \theta, \end{cases}$$

where θ is a positive constant.

Clearly, $f_\theta e^{-vx}$ is in $L_1(-\infty, \infty)$ for all real v, so that $f_\theta^\wedge(\omega)$ will be analytic in the entire ω plane. We have

$$f_\theta^\wedge(\omega) = \frac{1}{2\theta}\int_{-\theta}^{\theta} e^{i\omega x}\,dx = \frac{\sin\omega\theta}{\omega\theta} = \operatorname{sinc}\omega\theta,$$

which remains correct at $\omega = 0$. The function so defined is, in fact, analytic in the entire ω plane, including $\omega = 0$. The inversion formula can be used on the real axis of the ω plane, so that

$$f_\theta(x) = \frac{1}{2\pi}\int_{-\infty}^{\infty} (\operatorname{sinc} u\theta)e^{-iux}\,du.$$

Without attention to rigor, let us examine the limit as $\theta \to 0$. Then $f_\theta(x) \to \delta(x)$ and sinc $u\theta \to 1$. Thus we surmise that in some appropriate sense

$$\delta^\wedge = 1, \quad 1^\wedge = 2\pi\delta.$$

These formulas will be shown to hold rigorously in the distributional sense.

One-Sided Functions

A function which vanishes for $x < 0$ is said to be *right-sided* (or causal) and will be denoted by $f_+(x)$; a function which vanishes for $x > 0$ is said to be *left-sided* and will be denoted by $f_-(x)$.

Consider a right-sided function f_+ which is $0(e^{\alpha x})$ at $x = +\infty$, that is, such that there exists a constant C with the property

$$|f_+(x)| < Ce^{\alpha x} \quad \text{for } x \text{ sufficiently large.}$$

Then $f_+(x)e^{-vx}$ is in $L_1(-\infty, \infty)$ for $v > \alpha$, and therefore the Fourier transform $f_+^\wedge(\omega)$ of $f_+(x)$ is analytic in the upper half-plane $v > \alpha$. Formulas (2.4.12) and (2.4.13) then become

$$f_+^\wedge(\omega) = \int_0^\infty f_+(x)e^{i\omega x}\,dx, \quad v > \alpha \tag{2.4.15}$$

and

$$f_+(x) = \frac{1}{2\pi}\int_{iv-\infty}^{iv+\infty} f_+^\wedge(\omega)e^{-i\omega x}\,d\omega, \quad v > \alpha, \tag{2.4.16}$$

respectively. In particular, this implies that the integral in (2.4.16) vanishes identically for $x < 0$.

For a left-sided function f_-, which is $0(e^{\beta x})$ at $x = -\infty$, $f_-(x)e^{-vx}$ belongs to $L_1(-\infty, \infty)$ for $v < \beta$, and the Fourier transform $f_-^\wedge(\omega)$ of $f_-(x)$ is analytic in the lower half-plane $v < \beta$. We therefore have the transform relations

$$f_-^\wedge(\omega) = \int_{-\infty}^{\infty} f_-(x)e^{i\omega x}\, dx, \quad v < \beta \tag{2.4.17}$$

and

$$f_-(x) = \frac{1}{2\pi}\int_{iv-\infty}^{iv+\infty} f_-^\wedge(\omega)e^{-i\omega x}\, d\omega, \quad v < \beta. \tag{2.4.18}$$

The integral in (2.4.18) vanishes identically for $x > 0$.

Now, if $f(x)$ is an arbitrary function on the real line, we can write

$$f(x) = f_+(x) + f_-(x),$$

where

$$f_+(x) = \begin{cases} f(x), & x > 0, \\ 0, & x < 0, \end{cases} \qquad f_-(x) = \begin{cases} 0, & x > 0, \\ f(x), & x < 0. \end{cases}$$

If $f(x)$ is $0(e^{\alpha x})$ at $x = +\infty$ and $0(e^{\beta x})$ at $x = -\infty$, we have, by combining our previous results,

$$f_+^\wedge(\omega) = \int_0^{\infty} f(x)e^{i\omega x}\, dx, \quad v > \alpha, \tag{2.4.19}$$

$$f_-^\wedge(\omega) = \int_{-\infty}^{0} f(x)e^{i\omega x}\, dx, \quad v < \beta, \tag{2.4.20}$$

$$f(x) = \frac{1}{2\pi}\int_{ia-\infty}^{ia+\infty} f_+^\wedge(\omega)e^{-i\omega x}\, d\omega + \frac{1}{2\pi}\int_{ib-\infty}^{ib+\infty} f_-^\wedge(\omega)e^{-i\omega x}\, d\omega, \tag{2.4.21}$$

where $a > \alpha$ and $b < \beta$. These formulas provide a useful generalization of (2.4.12) and (2.4.13). If it happens that $\beta > \alpha$, the Fourier transform $f^\wedge(\omega)$ exists for $\alpha < v < \beta$, and we can choose $a = b$ in this strip to reduce (2.4.21) to (2.4.13).

Functions of Slow Growth

Definition. *A function $f(x)$ on the real line is said to be of* slow growth *if the following conditions hold:*

1. *f is locally integrable; that is, $\int_I |f(x)|\, dx$ is finite for each bounded interval I.*
2. *There exist constants C, n, and R such that*

$$|f(x)| < C|x|^n \quad \textit{for } |x| > R.$$

Thus a function of slow growth is one which grows at infinity more slowly than some polynomial. Of course, a function $f(x)$ of slow growth does not usually have a Fourier transform in the sense of (2.4.4) or (2.4.12); but $f_+(x)e^{-vx}$ is in $L_1(-\infty,\infty)$ for each $v > 0$, and $f_-(x)e^{-vx}$ is in $L_1(-\infty,\infty)$ for each $v < 0$. Hence we can use (2.4.21) for any $a > 0$ and $b < 0$, and therefore

$$f(x) = \frac{1}{2\pi}\lim_{\varepsilon\to 0+}\left[\int_{i\varepsilon-\infty}^{i\varepsilon+\infty} f_+^\wedge(\omega)e^{-i\omega x}\,d\omega + \int_{-i\varepsilon-\infty}^{-i\varepsilon+\infty} f_-^\wedge(\omega)e^{-i\omega x}\,d\omega\right] \tag{2.4.22}$$

or

$$f(x) = \frac{1}{2\pi}\lim_{\varepsilon\to 0+}\int_{-\infty}^{\infty} e^{-iux}[e^{\varepsilon x}f_+^\wedge(u+i\varepsilon) + e^{-\varepsilon x}f_-^\wedge(u-i\varepsilon)]\,du. \tag{2.4.23}$$

One would hope that in some appropriate sense it could be said that $f^\wedge(u)$ exists and that

$$f^\wedge(u) = \lim_{\varepsilon\to 0+}[f_+^\wedge(u+i\varepsilon)+f_-^\wedge(u-i\varepsilon)] = \lim_{\varepsilon\to 0+}\int_{-\infty}^{\infty} f(x)e^{iux}e^{-\varepsilon|x|}\,dx. \tag{2.4.24}$$

Such an interpretation will be shown to be possible in the theory of distributions (see Exercise 2.4.2). At present we content ourselves with a simple example. Let $f(x) = 1, -\infty < x < \infty$; then $f(x)$ is clearly of slow growth, and

$$f_+^\wedge(u+i\varepsilon) = \int_0^\infty e^{i(u+i\varepsilon)x}\,dx = \frac{i}{u+i\varepsilon},$$

$$f_-^\wedge(u-i\varepsilon) = \int_{-\infty}^0 e^{i(u-i\varepsilon)x}\,dx = -\frac{i}{u-i\varepsilon}.$$

Thus

$$f_+^\wedge(u+i\varepsilon) + f_-^\wedge(u-i\varepsilon) = \frac{2\varepsilon}{u^2+\varepsilon^2},$$

and, as was shown in Section 2.2,

$$\lim_{\varepsilon\to 0}\frac{2\varepsilon}{u^2+\varepsilon^2} = 2\pi\delta(u).$$

We are therefore led from (2.4.24) to state that

$$1^\wedge(u) = 2\pi\delta(u),$$

that is, the Fourier transform of 1 is $2\pi\delta(u)$. This confirms the result conjectured in Example 4.

Transforms of Distributions on the Line

In attempting to define the Fourier transform of a distribution f, we would like to use (2.4.4), with the real transform variable u, but unfortunately, e^{iux} is not a test function in $C_0^\infty(\mathbb{R})$, so that the action of f on e^{iux} is not defined. Instead, we try to use Parseval's formula (2.4.8) to define $f^\wedge$ from

$$\langle f^\wedge, \phi\rangle = \langle f, \phi^\wedge\rangle.$$

Again the right side is not defined because $\phi^\wedge$ is not a test function, even though ϕ is. The remedy is to introduce a more suitable class of test functions and correspondingly a new class of distributions. We begin with the one-dimensional case.

Definition. *A complex-valued function $\phi(x)$ of a single real variable is said to belong to $C_\downarrow^\infty(\mathbb{R})$, the space of* test functions of rapid decay, *if the following conditions hold:*

1. *$\phi(x)$ is infinitely differentiable.*

2. *$\phi(x)$, together with all its derivatives, vanishes at $|x| = \infty$ faster than any negative power of x. Thus, for every pair of nonnegative integers k and l,*

$$\lim_{|x|\to\infty} \left| x^k \frac{d^l \phi}{dx^l} \right| = 0. \tag{2.4.25}$$

This class of test functions is larger than the class $C_0^\infty(\mathbb{R})$ introduced in Section 2.1. The test functions in $C_0^\infty(\mathbb{R})$ vanish identically outside a finite interval, whereas those in $C_\downarrow^\infty(\mathbb{R})$ merely decrease rapidly at infinity. Every test function in $C_0^\infty(\mathbb{R})$ also belongs to $C_\downarrow^\infty(\mathbb{R})$, but e^{-x^2} belongs to $C_\downarrow^\infty(\mathbb{R})$ and not to $C_0^\infty(\mathbb{R})$. The test functions in $C_\downarrow^\infty(\mathbb{R})$ form a linear space; moreover, if ϕ is in $C_\downarrow^\infty(\mathbb{R})$, so is $x^k \phi^{(l)}(x)$ for any nonnegative integers k and l.

Convergence in $C_\downarrow^\infty(\mathbb{R})$

A sequence $\{\phi_m(x)\}$ of functions in $C_\downarrow^\infty(\mathbb{R})$ is said to be a *null sequence* in $C_\downarrow^\infty(\mathbb{R})$ if for each pair of nonnegative integers k and l,

$$\lim_{m\to\infty} \max_{-\infty<x<\infty} \left| x^k \frac{d^l \phi_m}{dx^l} \right| = 0.$$

Definition. *A* distribution of slow growth *is a continuous linear functional on the space $C_\downarrow^\infty(\mathbb{R})$. Thus to each ϕ in $C_\downarrow^\infty(\mathbb{R})$ there is assigned a complex number $\langle f, \phi\rangle$ with the properties*

$$\langle f, \alpha_1\phi_1 + \alpha_2\phi_2\rangle = \alpha_1\langle f, \phi_1\rangle + \alpha_2\langle f, \phi_2\rangle,$$
$$\lim_{m\to\infty} \langle f, \phi_m\rangle = 0 \quad \textit{for every null sequence in } C_\downarrow^\infty(\mathbb{R}).$$

Theorem. *Every function $f(x)$ of slow growth generates a distribution of slow growth by the formula*

$$\langle f, \phi\rangle = \int_{-\infty}^{\infty} f(x)\phi(x)\, dx.$$

Proof. The integral converges absolutely by the assumptions on f and ϕ. It is also clear that the functional is linear; we must still prove continuity. Let $\phi_n \to 0$ in $C_\downarrow^\infty(\mathbb{R})$; then

$$\left| \int_{-\infty}^{\infty} f(x)\phi_n(x)\, dx \right| = \left| \int_{-\infty}^{\infty} \frac{f(x)}{(1+x^2)^p} (1+x^2)^p \phi_n(x)\, dx \right|.$$

For p sufficiently large, $f(x)/(1+x^2)^p$ is absolutely integrable from $-\infty$ to ∞, since f is a function of slow growth. With such a value of p, the integral on the right is dominated by

$$\max_{-\infty<x<\infty} [(1+x^2)^p|\phi_n(x)|] \int_{-\infty}^{\infty} \frac{|f(x)|}{(1+x^2)^p}\, dx.$$

Since $\phi_n \to 0$ in $C_\downarrow^\infty(\mathbb{R})$, the maximum which appears also approaches 0. Therefore $\langle f, \phi_n\rangle \to 0$ whenever $\phi_n \to 0$ in $C_\downarrow^\infty(\mathbb{R})$, and $\langle f, \phi\rangle$ is a distribution of slow growth on $C_\downarrow^\infty(\mathbb{R})$. □

Nearly all important distributions on $C_0^\infty(\mathbb{R})$ are also distributions on $C_\downarrow^\infty(\mathbb{R})$. Only the distributions on $C_0^\infty(\mathbb{R})$ which grow too rapidly at infinity cannot be extended to $C_\downarrow^\infty(\mathbb{R})$. Much of the theory of Sections 2.1 and 2.2 can be applied to distributions on $C_\downarrow^\infty(\mathbb{R})$ with only slight modifications. We shall accept this statement and proceed with the new aspects of the theory.

Theorem. *If ϕ is in $C_\downarrow^\infty(\mathbb{R})$, then $\phi^\wedge(u)$ exists and is also in $C_\downarrow^\infty(\mathbb{R})$.*

Proof. The rapid decay of $\phi(x)$ at $|x| = \infty$ implies the absolute convergence of

$$\int_{-\infty}^{\infty} (ix)^k e^{iux}\phi(x)\, dx, \quad k = 0, 1, 2, \ldots.$$

Since this integral is the result of differentiating k times, under the integral sign, the expression for $\phi^\wedge$, it must represent the kth derivative of $\phi^\wedge$. Thus

$$\frac{d^k\phi^\wedge}{du^k}(u) = \int_{-\infty}^{\infty} (ix)^k e^{iux}\phi(x)\, dx, \quad \left|\frac{d^k\phi^\wedge}{du^k}(u)\right| \leqslant \int_{-\infty}^{\infty} |x^k\phi|\, dx,$$

so that the quantity on the left is bounded for all u. Moreover,

$$(iu)^p\frac{d^k\phi^\wedge}{du^k} = \int_{-\infty}^{\infty} (ix)^k\phi(x)\frac{d^p}{dx^p}e^{iux}\, dx,$$

and integration by parts converts the right side to

$$(-1)^p \int_{-\infty}^{\infty} \left[\frac{d^p}{dx^p}(ix)^k\phi(x)\right] e^{iux}\, dx.$$

Since the term multiplying e^{iux} is in $C_\downarrow^\infty(\mathbb{R})$, the integrand is absolutely integrable, and therefore

$$\left|u^p\frac{d^k\phi^\wedge}{du^k}\right|$$

is bounded for all u. Since p and k are arbitrary, it follows that $\phi^\wedge(u)$ is in $C_\downarrow^\infty(\mathbb{R})$. □

The same considerations also apply to the inverse transformation, so that we conclude that every function $\psi(u)$ in $C_\downarrow^\infty(\mathbb{R})$ is the transform of a function $\phi(x)$ in

$C^\infty_\downarrow(\mathbb{R})$. Before proceeding with the principal task of defining the transform of a distribution of slow growth, we list some properties of the transforms of test functions in $C^\infty_\downarrow(\mathbb{R})$. Consider the transform of $d^k\phi/dx^k$; then

$$\int_{-\infty}^{\infty} \frac{d^k\phi}{dx^k} e^{iux}\,dx = (-iu)^k \int_{-\infty}^{\infty} \phi e^{iux}\,dx,$$

by integration by parts. In more compact notation,

$$[\phi^{(k)}]^\wedge(u) = (-iu)^k \phi^\wedge(u). \tag{2.4.26}$$

Also we have

$$\int_{-\infty}^{\infty} (ix)^k \phi(x) e^{iux}\,dx = \frac{d^k}{du^k} \int_{-\infty}^{\infty} \phi(x) e^{iux}\,dx,$$

that is,

$$[(ix)^k \phi]^\wedge(u) = \frac{d^k}{du^k} \phi^\wedge(u). \tag{2.4.27}$$

By the inversion formula for Fourier transforms,

$$\int_{-\infty}^{\infty} e^{ixz} \phi^\wedge(z)\,dz = 2\pi\phi(-x),$$

and therefore

$$\phi^{\wedge\wedge}(x) = 2\pi\phi(-x). \tag{2.4.28}$$

The change of variable $x - a = y$ shows that

$$\int_{-\infty}^{\infty} \phi(x-a) e^{iux}\,dx = e^{iau} \int_{-\infty}^{\infty} \phi(y) e^{iay}\,dy;$$

hence

$$[\phi(x-a)]^\wedge(u) = e^{iau}\phi^\wedge(u). \tag{2.4.29}$$

Definition. *Let f be any distribution of slow growth. Its Fourier transform $f^\wedge$ is the distribution of slow growth defined from*

$$\langle f^\wedge, \phi\rangle = \langle f, \phi^\wedge\rangle. \tag{2.4.30}$$

We must show that we have in fact defined a distribution. Since $\phi^\wedge$ is in $C^\infty_\downarrow(\mathbb{R})$, the action of f on $\phi^\wedge$ is defined, so that $f^\wedge$ is a functional on $C^\infty_\downarrow(\mathbb{R})$. The functional is clearly linear; moreover, it is continuous, since whenever $\phi_m \to 0$ in $C^\infty_\downarrow(\mathbb{R})$, then $\phi_m^\wedge \to 0$ in $C^\infty_\downarrow(\mathbb{R})$ and therefore $\langle f, \phi_m^\wedge\rangle \to 0$.

To show that definition (2.4.30) really does provide an extension of the ordinary Fourier transform, we must show that it coincides with the usual definition when f

is an L_1 function with Fourier transform F. Suppose this to be the case; then by (2.4.30)

$$\begin{aligned}\langle f^\wedge, \phi\rangle = \langle f, \phi^\wedge\rangle &= \int_{-\infty}^{\infty} f(x)\,dx \int_{-\infty}^{\infty} e^{ixy}\phi(y)\,dy \\ &= \int_{-\infty}^{\infty} \phi(y)\,dy \int_{-\infty}^{\infty} f(x)e^{ixy}\,dx \\ &= \int_{-\infty}^{\infty} F(y)\phi(y)\,dy.\end{aligned}$$

Thus the action of $f^\wedge$ on ϕ is the same as the action of F on ϕ. Hence $f^\wedge = F$, and our definition is consistent.

We now show that properties (2.4.26) to (2.4.29) hold for the transform $f^\wedge$ of any distribution. First, consider the Fourier transform of the kth derivative of a distribution f. Then, by definition (2.4.30),

$$\langle [f^{(k)}]^\wedge, \phi\rangle = \langle f^{(k)}(u), \phi^\wedge(u)\rangle = (-1)^k\langle f(u), (\phi^\wedge)^{(k)}(u)\rangle.$$

Now by (2.4.27) we have

$$\begin{aligned}(-1)^k\langle f, (\phi^\wedge)^{(k)}\rangle = \langle f, [(-ix)^k\phi]^\wedge\rangle &= \langle f^\wedge(x), (-ix)^k\phi(x)\rangle \\ &= \langle (-ix)^k f^\wedge(x), \phi(x)\rangle,\end{aligned}$$

where for the last step we have used the definition of multiplication of a distribution by an infinitely differentiable function of slow growth. Thus we find that

$$[f^{(k)}]^\wedge(x) = (-ix)^k f^\wedge(x), \tag{2.4.31}$$

which is just (2.4.26) with a relabeling of the variables.

Turning next to the transform of $(ix)^k f$, we have

$$\langle [(ix)^k f]^\wedge, \phi\rangle = \langle (ix)^k f(x), \phi^\wedge(x)\rangle = (-1)^k\langle f(x), (-ix)^k\phi^\wedge(x)\rangle,$$

and, by using (2.4.26),

$$\begin{aligned}\langle f(x), (-ix)^k\phi^\wedge(x)\rangle = \langle f(x), [\phi^{(k)}]^\wedge(x)\rangle &= \langle f^\wedge(u), \phi^{(k)}(u)\rangle \\ &= (-1)^k\left\langle \frac{d^k f^\wedge}{du^k}(u), \phi(u)\right\rangle.\end{aligned}$$

Consequently,

$$[(ix)^k f]^\wedge(u) = \frac{d^k f^\wedge}{du^k}. \tag{2.4.32}$$

We leave the proofs of the following properties to the reader:

$$f^{\wedge\wedge}(x) = 2\pi f(-x), \tag{2.4.33}$$

$$[f(x-a)]^\wedge(u) = e^{iau} f^\wedge(u), \tag{2.4.34}$$

$$[f(-x)]^\wedge(u) = f^\wedge(-u). \tag{2.4.35}$$

Example 5. Consider the transform of $\delta(x)$. Then

$$\langle \delta^\wedge, \phi \rangle = \langle \delta(x), \phi^\wedge(x) \rangle = \left\langle \delta(x), \int_{-\infty}^{\infty} \phi(y) e^{ixy}\, dy \right\rangle$$
$$= \int_{-\infty}^{\infty} \phi(y)\, dy = \langle 1, \phi \rangle.$$

Therefore,

$$\delta^\wedge = 1. \qquad (2.4.36)$$

Example 6. To find the transform of $f(x) = 1$,

$$\langle 1^\wedge, \phi \rangle = \langle 1, \phi^\wedge(x) \rangle = \int_{-\infty}^{\infty} \phi^\wedge(x)\, dx$$
$$= \left[\int_{-\infty}^{\infty} \phi^\wedge(x) e^{ixy}\, dx \right]_{y=0}.$$

By the inversion formula for $\phi^\wedge$, the last integral is just $2\pi\phi(0)$. Thus

$$1^\wedge = 2\pi\delta. \qquad (2.4.37)$$

The same result can also be obtained by using (2.4.36) and (2.4.33). In fact, from (2.4.36)

$$1^\wedge = \delta^{\wedge\wedge},$$

and from (2.4.33)

$$\delta^{\wedge\wedge}(x) = 2\pi\delta(-x) = 2\pi\delta(x).$$

Example 7. The transform of $\delta(x-a)$ is e^{iau}.

Example 8. We now calculate the transform of the Heaviside function $H(x)$ in three different ways.

(a)

$$\langle H^\wedge, \phi \rangle = \langle H, \phi^\wedge \rangle = \int_0^{\infty} \phi^\wedge(x)\, dx$$
$$= \int_0^{\infty} dx \int_{-\infty}^{\infty} \phi(y) e^{ixy}\, dy$$
$$= \lim_{R\to\infty} \int_{-\infty}^{\infty} \phi(y)\, dy \int_0^R e^{ixy}\, dx$$
$$= \lim_{R\to\infty} \int_{-\infty}^{\infty} \frac{e^{iRy}-1}{iy} \phi(y)\, dy.$$

It can be shown that $\lim_{R\to\infty}(1-\cos Ry)/y = \mathrm{pf}(1/y)$. Moreover, it holds that $\lim_{R\to\infty}\sin Ry/y = \pi\delta(y)$. Therefore,

$$\langle H^\wedge, \phi\rangle = \left\langle i\ \mathrm{pf}\frac{1}{y} + \pi\delta(y), \phi(y)\right\rangle, \quad H^\wedge(y) = i\ \mathrm{pf}\frac{1}{y} + \pi\delta(y). \tag{2.4.38}$$

(b) We have $H'(x) = \delta(x)$. By (2.4.31)

$$(H')^\wedge(x) = -ixH^\wedge(x),$$

and, using (2.4.36),

$$1 = -ixH^\wedge(x).$$

Thus $H^\wedge$ satisfies the distributional equation

$$1 = -ixf(x), \tag{2.4.39}$$

a particular solution of which is [(see (2.1.33)]

$$f(x) = i\ \mathrm{pf}\frac{1}{x}.$$

In fact, substituting in (2.4.39), we find, using (2.1.28), that

$$\begin{aligned}\langle 1, \phi\rangle &= \left\langle x\ \mathrm{pf}\frac{1}{x}, \phi\right\rangle = \left\langle \mathrm{pf}\frac{1}{x}, x\phi\right\rangle \\ &= \lim_{\varepsilon\to 0}\int_{-\infty}^{-\varepsilon} + \int_{\varepsilon}^{\infty}\frac{1}{x}x\phi\ dx = \int_{-\infty}^{\infty}\phi\ dx,\end{aligned}$$

which is an identity.

To find the general solution of (2.4.39) we must add to the particular solution just obtained the general solution of the homogeneous equation $-ixf = 0$, which, by Exercise 2.5.11(c), is $C\delta(x)$. Therefore,

$$H^\wedge(x) = i\ \mathrm{pf}\frac{1}{x} + C\delta(x), \tag{2.4.40}$$

where C is a constant to be determined. The following trick enables us to find C. Consider the equation

$$H(x) + H(-x) = 1,$$

whose transform by (2.4.35) and (2.4.37) yields

$$H^\wedge(u) + H^\wedge(-u) = 2\pi\delta(u).$$

Comparing with (2.4.40) we find that $C = \pi$. Thus

$$H^\wedge(x) = i\ \mathrm{pf}\frac{1}{x} + \pi\delta(x). \tag{2.4.40a}$$

(c) The distribution $H(x)$ may be considered as the limit as $\varepsilon \to 0+$ of $H(x)e^{-\varepsilon x}$ (see Exercise 2.4.2). Therefore, by Exercise 2.4.1,

$$H^\wedge = \lim_{\varepsilon \to 0+} (He^{-\varepsilon x})^\wedge.$$

But $He^{-\varepsilon x}$ has a Fourier transform in the ordinary sense:

$$(He^{-\varepsilon x})^\wedge = \int_0^\infty e^{-\varepsilon x} e^{iux}\, dx = \frac{1}{\varepsilon - iu} = \frac{\varepsilon + iu}{\varepsilon^2 + u^2}.$$

The distributional limit as $\varepsilon \to 0+$ is easily calculated. We have

$$\lim_{\varepsilon \to 0+} \frac{\varepsilon}{\varepsilon^2 + u^2} = \pi\delta(u), \quad \lim_{\varepsilon \to 0} \frac{iu}{\varepsilon^2 + u^2} = i\, \mathrm{pf}\frac{1}{u},$$

which lead again to (2.4.38).

Transforms in More Than One Variable

Definition. *A complex-valued function $\phi(x_1, \ldots, x_n) = \phi(x)$ is said to belong to $C_\downarrow^\infty(\mathbb{R}^n)$, the space of n-*dimensional test functions of rapid decay, *if the following conditions hold:*

1. *$\phi(x)$ is infinitely differentiable, that is, $D^l\phi$ exists for any multi-index l of dimension n.*
2. *For each pair of multi-indices k and l of dimension n,*

$$\lim_{|x| \to \infty} |x^k D^l \phi| = 0,$$

where

$$x^k = x_1^{k_1} \cdots x_n^{k_n}$$

and

$$D^l\phi = \frac{\partial^{l_1 + \cdots + l_n}}{\partial x_1^{l_1} \cdots x_n^{l_n}}.$$

A sequence $\{\phi_m(x)\}$ of functions in $C_\downarrow^\infty(\mathbb{R}^n)$ is said to be a *null sequence* in $C_\downarrow^\infty(\mathbb{R}^n)$ if, for each pair of multi-indices k and l of dimension n,

$$\lim_{m \to \infty} \max_{x \in \mathbb{R}^n} |x^k D^l \phi_m| = 0.$$

Definition. *An n-*dimensional distribution of slow growth *is a continuous linear functional on $C_\downarrow^\infty(\mathbb{R}^n)$. To each test function ϕ in $C_\downarrow^\infty(\mathbb{R}^n)$ there is assigned a complex number $\langle f, \phi \rangle$ with the properties*

$$\langle f, \alpha_1\phi_1 + \alpha_2\phi_2 \rangle = \alpha_1\langle f, \phi_1 \rangle + \alpha_2\langle f, \phi_2 \rangle,$$
$$\lim_{m \to \infty} \langle f, \phi_m \rangle = 0 \quad \textit{for every null sequence in } C_\downarrow^\infty(\mathbb{R}^n).$$

We can now define n-dimensional transforms. First, if ϕ is in $C_\downarrow^\infty(\mathbb{R}^n)$, we define, for real u,

$$\phi^\wedge(u) = \phi^\wedge(u_1, \ldots, u_n) = \int_{-\infty}^{\infty} \cdots \int_{-\infty}^{\infty} \phi(x_1, \ldots, x_n) e^{iu_1x_1} \cdots e^{iu_nx_n}\, dx_1 \cdots dx_n$$
$$= \int_{\mathbb{R}^n} \phi(x) e^{iu\cdot x}\, dx.$$

It is easily seen that $\phi^\wedge(u)$ is an n-dimensional test function of rapid decay. The inversion formula is

$$\phi(x) = \frac{1}{(2\pi)^n} \int_{\mathbb{R}^n} \phi^\wedge(u) e^{-iu\cdot x}\, du.$$

The transform of a distribution of slow growth is then defined by use of Parseval's formula:

$$\langle f^\wedge, \phi \rangle = \langle f, \phi^\wedge \rangle.$$

As examples we observe that

$$1^\wedge = (2\pi)^n \delta, \quad \delta^\wedge = 1,$$

where 1 is the constant function equal to 1 everywhere in $\mathbb{R}^n$, and δ is the n-dimensional Dirac distribution. Formulas similar to (2.4.31) to (2.4.35) are easily derived and are left to the reader.

EXERCISES

2.4.1 Let $\{f_n\}$ be a sequence of distributions on $C_\downarrow^\infty(\mathbb{R})$ such that $f_n \to f$, where f is a distribution on $C_\downarrow^\infty(\mathbb{R})$. Show that $f_n^\wedge \to f^\wedge$.

2.4.2 Show that if $f(x)$ is a function of slow growth on the real line,

$$\lim_{\varepsilon\to 0+} \langle f(x) e^{-\varepsilon|x|}, \phi(x) \rangle = \langle f, \phi \rangle.$$

Thus in the distributional sense

$$f(x) = \lim_{\varepsilon\to 0+} f(x) e^{-\varepsilon|x|} = \lim_{\varepsilon\to 0+} [f_+(x) e^{-\varepsilon x} + f_-(x) e^{\varepsilon x}].$$

Therefore, by Exercise 2.4.1,

$$f^\wedge(u) = \lim_{\varepsilon\to 0+} \left[\int_0^\infty f(x) e^{-\varepsilon x} e^{iux}\, dx + \int_{-\infty}^0 f(x) e^{\varepsilon x} e^{iux}\, dx \right]$$

or

$$f^\wedge(u) = \lim_{\varepsilon\to 0+} [f_+^\wedge(u + i\varepsilon) + f_-^\wedge(u - i\varepsilon)],$$

where $f_+^\wedge$ and $f_-^\wedge$ are the one-sided transforms defined in (2.4.19) and (2.4.20). Thus we have established (2.4.24). In particular, if f vanishes for $x < 0$, we have

$$f^\wedge(u) = \lim_{\varepsilon\to 0+} f_+^\wedge(u + i\varepsilon).$$

2.4.3 Find the Fourier transforms of the following distributions on $C_\downarrow^\infty(\mathbb{R})$:

(a) pf$(1/x)$.

(b) $\operatorname{sgn} x = \begin{cases} 1, & x > 0, \\ -1, & x < 0. \end{cases}$

(c) $\log |x|$.
It will help to recall that $(d/dx) \log |x| = \mathrm{pf}(1/x)$.

2.4.4 Find the Fourier transform of x^k, where k is a positive integer and x is a single real variable.

2.4.5 The *convolution* of f and g is defined as

$$h(x) \doteq \int_{-\infty}^{\infty} f(x-\xi) g(\xi)\, d\xi. \tag{2.4.41}$$

Show in a purely formal manner that

$$h^\wedge(\omega) = f^\wedge(\omega) g^\wedge(\omega) \tag{2.4.42}$$

and that

$$\int_{-\infty}^{\infty} f(x-\xi) g(\xi)\, d\xi = \int_{-\infty}^{\infty} g(x-\xi) f(\xi)\, d\xi.$$

2.4.6 (a) Let $f(t)$ be defined on $0 \leqslant t < \infty$ and be $0(e^{\alpha t})$ at $t = +\infty$. Its *Laplace transform* is defined as

$$\tilde{f}(s) \doteq \int_0^\infty e^{-st} f(t)\, dt, \tag{2.4.43}$$

where s is a complex variable. The integral converges for $\operatorname{Re} s > \alpha$ and is an analytic function of s in that right half-plane. This definition is the same as that of the Fourier transform of a right-sided function (2.4.15), with $s \doteq -i\omega, t \doteq x$. Show that the inversion (2.4.16) then becomes

$$\frac{1}{2\pi i} \int_{a-i\infty}^{a+i\infty} e^{st} \tilde{f}(s)\, ds = \begin{cases} f(t), & t > 0; \\ 0, & t < 0; \end{cases} \quad \operatorname{Re} a > \alpha. \tag{2.4.44}$$

The inversion integral is therefore being taken on a vertical line in the right half-plane of analyticity.

(b) By specializing Exercise 2.4.5 to right-sided functions, show that the convolution of f and g becomes

$$h(t) = \int_0^t f(t-\tau) g(\tau)\, d\tau = \int_0^t g(t-\tau) f(\tau)\, d\tau \tag{2.4.45}$$

and that

$$\tilde{h}(s) = \tilde{f}(s) \tilde{g}(s). \tag{2.4.46}$$

2.5 DIFFERENTIAL EQUATIONS IN DISTRIBUTIONS

Local Properties of Distributions

When a distribution f is generated by a continuous function, one can recover the point values of f from a knowledge of its actions on test functions (see Exercise 2.1.1). If all we know is that the distribution is generated by a locally integrable function f, we cannot obtain complete information about $f(x)$ from $\langle f, \phi \rangle$, since two functions f_1 and f_2 which are equal almost everywhere generate the same distribution. It is possible, however, to determine $f(x)$ up to equality almost everywhere. Suppose, for instance, that $f(x)$ is equal to 0 almost everywhere in an open set Ω; then we have $\langle f, \phi \rangle = 0$ for all test functions ϕ *whose support is contained in* Ω; and vice versa, if a distribution is generated by a locally integrable function f and if $\langle f, \phi \rangle = 0$ for all test functions ϕ whose support is contained in Ω, then $f(x)$ is equal to 0 almost everywhere in Ω. The same ideas are used for an arbitrary distribution.

Definition. *The distribution f is said to* vanish on the open set Ω *if $\langle f, \phi \rangle = 0$ for every test function ϕ with support in Ω. Two distributions f_1 and f_2 are said to be* equal in Ω *if $f_1 - f_2$ vanishes on Ω.*

Examples

(a) Let Ω be the open set consisting of all $\mathbb{R}^n$ with the origin removed. Then δ vanishes on Ω. Indeed, if ϕ has its support in Ω (remember that the support is a closed set), then ϕ must vanish in some neighborhood of the origin and $\langle \delta, \phi \rangle = \phi(0) = 0$ for any such ϕ.

(b) In (2.3.22) we considered the distribution $f(x) = \sum_{k=-\infty}^{\infty} \delta(x - 2k\pi)$, whose action on a test function $\phi(x)$ in $C_0^\infty(\mathbb{R})$ is $\sum_{k=-\infty}^{\infty} \phi(2k\pi)$. Let Ω be the open interval $-\pi < x < \pi$; then, if ϕ has its support in Ω, $\langle f, \phi \rangle = \phi(0)$, so that we can say that $f(x)$ coincides with $\delta(x)$ in $-\pi < x < \pi$.

The Differential Equation $u' = f$ in $\mathbb{R}$

Consider first the homogeneous equation

$$u' = 0, \tag{2.5.1}$$

regarded as an equation for distributions on the real line. By definition this means that we are looking for distributions u such that

$$\langle u, \phi' \rangle = 0 \quad \text{for every } \phi \in C_0^\infty(\mathbb{R}). \tag{2.5.2}$$

Equation (2.5.2) tells us that the action of u is 0 on any test function which is the *derivative* of some other test function. Of course, not every test function has this property. For instance, test function (2.1.1) is not the derivative of a test function. Indeed, any antiderivative $F(x)$ of (2.1.1) will have $F(\infty) \neq F(-\infty)$.

Let M be the subset of $C_0^\infty(\mathbb{R})$ consisting of the elements which are the first derivatives of elements of $C_0^\infty(\mathbb{R})$. Then we have the following lemmas.

Lemma 2.5.1. *Let $\phi \in C_0^\infty(\mathbb{R})$. Then $\phi \in M$ if and only if*

$$\int_{-\infty}^{\infty} \phi\, dx = 0. \tag{2.5.3}$$

Proof.

(a) If $\phi \in M$, then $\phi = \chi'$, where χ belongs to $C_0^\infty(\mathbb{R})$. It follows that $\int_{-\infty}^{\infty} \phi(x)\, dx = \chi]_{-\infty}^{\infty} = 0$, so that (2.5.3) is satisfied.

(b) Let $\phi \in C_0^\infty(\mathbb{R})$, and let $\int_{-\infty}^{\infty} \phi(x)\, dx = 0$. Define $\chi(x) = \int_{-\infty}^{x} \phi(s)\, ds$; then χ is infinitely differentiable, and χ vanishes outside a bounded interval by (2.5.3). Thus χ is a test function; and since $\chi' = \phi$, χ also belongs to M.

□

Lemma 2.5.2. *Let $\phi_0(x)$ be a fixed (but arbitrary) test function such that*

$$\int_{-\infty}^{\infty} \phi_0(x)\, dx = 1.$$

Then for each $\phi(x) \in C_0^\infty(\mathbb{R})$ there are a unique constant a and a unique element ψ in M such that

$$\phi(x) = a\phi_0(x) + \psi(x). \tag{2.5.4}$$

Proof. Choose $a = \langle 1, \phi \rangle = \int_{-\infty}^{\infty} \phi(x)\, dx$, and define $\psi = \phi - a\phi_0$ so that (2.5.4) is clearly satisfied. From the definition of ψ, we see that it is a test function and that $\int_{-\infty}^{\infty} \psi\, dx = 0$. Therefore, $\psi \in M$. The proof of uniqueness is left to the reader. □

We are now ready to solve (2.5.1). If u is any distribution, then, from (2.5.4),

$$\langle u, \phi \rangle = a\langle u, \phi_0 \rangle + \langle u, \psi \rangle,$$

where $\psi \in M$. If u is a solution of (2.5.1), then $\langle u, \psi \rangle = 0$ whenever $\psi \in M$, so that for every $\phi \in C_0^\infty(\mathbb{R})$,

$$\langle u, \phi \rangle = a\langle u, \phi_0 \rangle = \langle u, \phi_0 \rangle \int_{-\infty}^{\infty} \phi\, dx = \langle c, \phi \rangle,$$

where c is the constant $\langle u, \phi_0 \rangle$. We have thus shown that only constant distributions can be solutions of (2.5.1), and it is easy to check that constant distributions do in fact satisfy the equation.

Next we turn to the inhomogeneous equation

$$u' = f, \tag{2.5.5}$$

where f is a given arbitrary distribution. By definition a distribution u satisfies (2.5.5) if and only if

$$\langle u, \phi' \rangle = -\langle f, \phi \rangle \quad \text{for every } \phi \in C_0^\infty(\mathbb{R}).$$

To find the general solution of (2.5.5) we use the decomposition (2.5.4) to write

$$\langle u, \phi \rangle = a\langle u, \phi_0 \rangle + \langle u, \psi \rangle,$$

where $\psi \in M$, say $\psi = \chi', \chi \in C_0^\infty(\mathbb{R})$. The explicit expression for χ in terms of ϕ is

$$\chi = \int_{-\infty}^{x} \psi(s)\, ds = \int_{-\infty}^{x} \phi(s)\, ds - \langle 1, \phi \rangle \int_{-\infty}^{\infty} \phi_0(s)\, ds.$$

Since u is a solution of (2.5.5),

$$\langle u, \psi \rangle = \langle u, \chi' \rangle = -\langle f, \chi \rangle,$$

and therefore

$$\langle u, \phi \rangle = \langle u, \phi_0 \rangle \langle 1, \phi \rangle - \langle f, \chi \rangle.$$

We claim that it is legitimate to define a distribution u_p from

$$\langle u_p, \phi \rangle = -\langle f, \chi \rangle. \tag{2.5.6}$$

Indeed, χ is a test function depending linearly on ϕ, so that (2.5.6) defines a linear functional on the space of test functions $\phi(x)$. If $\{\phi_m\}$ is a null sequence in $C_0^\infty(\mathbb{R})$, so is $\{\psi_m\}$ and hence $\{\chi_m\}$; therefore, the functional defined by (2.5.6) is continuous—in other words, a distribution. Thus every solution of (2.5.5) must be of the form

$$\langle u, \phi \rangle = c\langle 1, \phi \rangle + \langle u_p, \phi \rangle,$$

and it is easily verified that every distribution of this form is indeed a solution. By solving (2.5.5), we have shown that every distribution is integrable (that is, has an antiderivative).

Green's Formula and Lagrange's Identity

We interrupt the distributional treatment to collect some results on linear differential operators.

To fix ideas, consider the ordinary differential operator of order 2 given by

$$L = a_2(x)D^2 + a_1(x)D + a_0(x),$$

where $D = d/dx$ and the coefficients $a_k(x)$ are in $C^2(\mathbb{R})$. Starting from

$$\int_a^b vLu\ dx = \int_a^b (va_2u'' + va_1u' + va_0u)\, dx,$$

where u, v are arbitrary functions in $C^2(\mathbb{R})$, we integrate by parts until all the differentiations are transferred to v. In this way we find that

$$\int_a^b vLu\ dx - \int_a^b uL^*v\ dx = J(u,v)]_a^b, \tag{2.5.7}$$

where the operator L^*, known as the *formal adjoint* of L, is given by

$$L^* = a_2D^2 + (2a_2' - a_1)D + (a_2'' - a_1' + a_0), \tag{2.5.8}$$

and the bilinear form J, the *conjunct* of u and v, is

$$J(u,v) = a_2(vu' - uv') + (a_1 - a_2')uv. \tag{2.5.9}$$

Since (2.5.7) is valid for any upper limit b, we can differentiate with respect to b and then set $b = x$ to obtain

$$vLu - uL^*v = \frac{d}{dx}J(u,v), \tag{2.5.10}$$

which is known as *Lagrange's identity*. The integrated form (2.5.7) is called *Green's formula.*

If the operators L and L^* coincide, we say that L is *formally self-adjoint*; in our case of a second-order ordinary differential operator, L is self-adjoint if and only if

$$a_2' = a_1 \quad \text{so that } Lu = D(a_2Du) + a_0u. \tag{2.5.11}$$

For a formally self-adjoint operator [that is, one satisfying (2.5.11)], we have the simplifications

$$J(u,v) = a_2(vu' - uv'), \tag{2.5.12}$$

$$vLu - uLv = \frac{d}{dx}J(u,v), \tag{2.5.13}$$

$$\int_a^b (vLu - uLv)\,dx = J(u,v)]_a^b. \tag{2.5.14}$$

Of course, if L does not satisfy (2.5.11), we must use the earlier formulas involving L^*.

Let us turn next to an ordinary differential operator of order p,

$$L = a_p(x)D^p + \cdots + a_1(x)D + a_0(x),$$

where the coefficients are in $C^p(\mathbb{R})$. If u and v are in $C^p(\mathbb{R})$, we have

$$D[vD^{m-1}u - v'D^{m-2}u + \cdots + (-1)^{m-1}(D^{m-1}v)u] = vD^mu + (-1)^{m-1}uD^mv,$$

the result following from the observation that most of the terms on the left side cancel out by telescopic action. Thus we have

$$vD^mu = (-1)^m uD^mv + D\sum_{j+k=m-1}(-1)^k(D^kv)(D^ju),$$

where the sum ranges over the $j \geqslant 0$ and $k \geqslant 0$, satisfying $j + k = m - 1$. For $m = 0$ the summation disappears.

Substituting $a_m v$ for v and summing from $m = 0$ to p, we obtain

$$vLu - uL^*v = \frac{d}{dx} J(u, v), \tag{2.5.15}$$

where

$$L^*v = \sum_{m=0}^{p} (-1)^m D^m (a_m v) \tag{2.5.16}$$

and

$$J(u, v) = \sum_{m=1}^{p} \sum_{j+k=m-1} (-1)^k D^k (a_m v)\, D^j u. \tag{2.5.17}$$

The term *Lagrange identity* is used for (2.5.15), while the integrated form

$$\int_a^b (vLu - uL^*v)\, dx = J(u, v)]_a^b \tag{2.5.18}$$

is known as Green's formula. Observe that J contains only derivatives of order up to $p - 1$.

If $L = L^*$, we say that L is formally self-adjoint. This is possible if and only if L is of even order and can be written in the form

$$D^r(b_r D^r) + D^{r-1}(b_{r-1} D^{r-1}) + \cdots + D(b_1 D) + b_0, \tag{2.5.19}$$

where $p = 2r$ and $b_0, \ldots, b_r$ are arbitrary functions.

Partial differential operators pose more of a problem. We take our cue from the Laplacian Δ in $\mathbb{R}^n$, which satisfies

$$v\, \Delta u - u\, \Delta v = \operatorname{div}(v \operatorname{grad} u - u \operatorname{grad} v) \tag{2.5.20}$$

or, in integrated form,

$$\int_\Omega (v\, \Delta u - u\, \Delta v)\, dx = \int_\Gamma n \cdot (v \operatorname{grad} u - u \operatorname{grad} v)\, dS, \tag{2.5.21}$$

which is, of course, the classical Green's formula. For an arbitrary linear operator of order p, as given by (2.1.3), the corresponding forms of (2.5.20) and (2.5.21) are

$$vLu - uL^*v = \operatorname{div} J(u, v) \tag{2.5.22}$$

and

$$\int_\Omega (vLu - uL^*v)\, dx = \int_\Gamma n \cdot J(u, v)\, dS. \tag{2.5.23}$$

The operator L and its formal adjoint L^* are given by

$$Lu = \sum_{|k| \leqslant p} a_k(x) D^k u, \quad L^*v = \sum_{|k| \leqslant p} (-1)^{|k|} D^k (a_k v), \tag{2.5.24}$$

the expression for L^* being obtained by integration by parts as in (2.1.19). If $L = L^*$, we say that L is formally self-adjoint. The expression for the vector bilinear form J is somewhat complicated in general, so that we will merely present explicit expressions for J in specific examples.

Example 1. $x = (x_1, \dots, x_n), p = 2, L = \Delta = \partial^2/\partial x_1^2 + \cdots + \partial^2/\partial x_n^2$. Since the coefficients are constants and there are only terms of even order, Δ is formally self-adjoint. The appropriate formulas are (2.5.20) and (2.5.21).

Example 2. $x = (x_1, \dots, x_n), p = 4, L = \Delta\Delta$. Again, L is formally self-adjoint, and either by direct calculation or by using Exercise 2.5.10 we find that (2.5.22) becomes

$$v\Delta\Delta u - u\Delta\Delta v = \operatorname{div}[v \operatorname{grad} \Delta u - u \operatorname{grad} \Delta v + (\Delta v) \operatorname{grad} u - (\Delta u) \operatorname{grad} v]. \tag{2.5.25}$$

Example 3. The diffusion operator is

$$L = \frac{\partial}{\partial t} - \left(\frac{\partial^2}{\partial x_1^2} + \cdots + \frac{\partial^2}{\partial x_n^2}\right), \tag{2.5.26}$$

where the time variable t has been distinguished from the space variables $x_1, \dots, x_n$. We find that

$$L^* = -\frac{\partial}{\partial t} - \left(\frac{\partial^2}{\partial x_1^2} + \cdots + \frac{\partial^2}{\partial x_n^2}\right) \tag{2.5.27}$$

and

$$vLu - uL^*v = \operatorname{div} J,$$

with

$$J = e_t uv - (v \operatorname{grad}_x u - u \operatorname{grad}_x v),$$

where e_t is a unit vector in the t direction, and the subscript x indicates that differentiation is only with respect to the space variables. Green's formula becomes

$$\int_\Omega (vLu - uL^*v)\, dx\ dt = \int_\Gamma n \cdot (e_t\, uv + u \operatorname{grad}_x v - v \operatorname{grad}_x u)\, dS, \tag{2.5.28}$$

where Ω is a domain in space-time, Γ its boundary, $dx\, dt$ an element of volume in space-time, and dS a hypersurface element on Γ. In most applications of (2.5.28), Ω is a cylinder in space-time having base Ω_x and bounded by the parallel hyperplanes $t = t_1, t = t_2 (> t_1)$. (See Figure 8.8.) The boundary Γ consists of (a) the two bases: $x \in \Omega_x, t = t_1$, and $x \in \Omega_x, t = t_2$, and (b) the lateral surface: $x \in \Gamma_x, t_1 < t < t_2$, where Γ_x is the ordinary bounding surface of the space domain Ω_x. On the base at t_2 the outward normal n is e_t, while on the other base $n = -e_t$; in both cases n is orthogonal to the space directions. On the lateral surface, n is just the outward normal n_x to Ω_x, and n_x is orthogonal to e_t. Thus (2.5.28) reduces to

$$\int_{t_1}^{t_2} dt \int_{\Omega_x} dx(vLu - uL^*v) = \int_{\Omega_x} [uv]_{t_1}^{t_2}\, dx + \int_{t_1}^{t_2} dt \int_{\Gamma_x} dS_x \left(u\frac{\partial v}{\partial n_x} - v\frac{\partial u}{\partial n_x} \right). \tag{2.5.29}$$

Example 4. The wave operator

$$\Box^2 \doteq \frac{\partial^2}{\partial t^2} - \left(\frac{\partial^2}{\partial x_1^2} + \cdots + \frac{\partial^2}{\partial x_n^2} \right) \tag{2.5.30}$$

is formally self-adjoint, and (2.5.22) becomes

$$v\Box^2 u - u\Box^2 v = \operatorname{div}\left[e_t \left(v\frac{\partial u}{\partial t} - u\frac{\partial v}{\partial t} \right) + u \operatorname{grad}_x v - v \operatorname{grad}_x u \right]. \tag{2.5.31}$$

Applying Green's formula to a cylindrical domain as in Example 3, we obtain

$$\int_{t_1}^{t_2} dt \int_{\Omega_x} dx(v\Box^2 u - u\Box^2 v) = \int_{\Omega_x} \left(v\frac{\partial u}{\partial t} - u\frac{\partial v}{\partial t} \right)_{t_1}^{t_2} dx + \int_{t_1}^{t_2} dt \int_{\Gamma_x} dS_x \left(u\frac{\partial v}{\partial n_x} - v\frac{\partial u}{\partial n_x} \right). \tag{2.5.32}$$

Classical, Weak, and Distributional Solutions

Let us look at the ordinary differential equation

$$\frac{du}{dx} = f(x) \quad \text{on the interval } \Omega: \quad a < x < b. \tag{2.5.33}$$

If $f(x)$ is a continuous function, we can define the notion of solution in the classical sense: $u(x)$ is a *classical (or strict) solution* if it has a continuous derivative which satisfies (2.5.33) pointwise on Ω. (An almost trivial extension to the case where f is piecewise continuous was made in Section 1.1.) Denoting the class of test functions with support in Ω by $C_0^\infty(\Omega)$, we find that for any classical solution of (2.5.33),

$$\int_\Omega f\phi\, dx = -\int_\Omega u\frac{d\phi}{dx}\, dx \quad \text{for each } \phi \text{ in } C_0^\infty(\Omega), \tag{2.5.34}$$

the result stemming from integration by parts and the fact that $\phi \equiv 0$ in a neighborhood of the boundary. The two sides of (2.5.34) make sense even if f and u are only locally integrable. This leads to the definition of a *weak solution* of (2.5.33): If f is locally integrable, a locally integrable function u is a weak solution of (2.5.33) if and only if it satisfies (2.5.34) for each ϕ in $C_0^\infty(\Omega)$. If u is a weak solution of (2.5.33), we also say that $du/dx = f$ in the *weak sense*.

Equation (2.5.33) can also be interpreted distributionally. If f is a distribution, we say that a distribution u is a solution of (2.5.33) if and only if

$$-\left\langle u, \frac{d\phi}{dx} \right\rangle = \langle f, \phi \rangle \quad \text{for each } \phi \text{ in } C_0^\infty(\Omega). \tag{2.5.35}$$

Note that the left side is the definition of the distribution u'. If f is a distribution generated by a locally integrable function and if we are looking for solutions u that are functions, (2.5.35) reduces to (2.5.34), that is, to the concept of a weak solution.

We can now extend these ideas to more general operators. Let L be an arbitrary linear differential operator of order p in the n variables $x_1, \ldots, x_n$:

$$L = \sum_{|k| \leqslant p} a_k(x) D^k, \qquad \begin{cases} k = (k_1, \ldots, k_n), \\ |k| = k_1 + k_2 + \cdots + k_n. \end{cases}$$

Assuming that the $a_k(x)$ are infinitely differentiable, the distribution Lu always exists for any distribution u and [see (2.1.19)] is defined by

$$\langle Lu, \phi \rangle = \langle u, L^* \phi \rangle,$$

where L^* is the formal adjoint of L, introduced in (2.5.24),

$$L^* \phi = \sum_{|k| \leqslant p} (-1)^{|k|} D^k (a_k \phi). \tag{2.5.36}$$

We are therefore in a position to give a distributional meaning to the differential equation

$$Lu = f, \quad x \text{ in } \Omega, \tag{2.5.37}$$

where f is a given distribution.

What we require is that the distribution Lu and f coincide in Ω. (See the beginning of Section 2.5.)

Definition. *A distribution u is a solution of (2.5.37)* on Ω *if*

$$\langle u, L^* \phi \rangle = \langle f, \phi \rangle \quad \textit{for each } \phi \textit{ in } C_0^\infty(\Omega). \tag{2.5.38}$$

Definition. *Let f be locally integrable. A locally integrable function u which satisfies (2.5.38) is said to be a* weak solution *of (2.5.37) on Ω.*

REMARK. If $f(x)$ is a continuous function, we can give a classical interpretation to (2.5.37). The function $u(x)$ is a *classical* solution of (2.5.37) if it belongs to $C^p(\Omega)$, the class of functions with continuous derivatives of order up to p, and if it satisfies (2.5.37) at every point of Ω. Since functions can also be interpreted as distributions, we would like to compare the two notions of solutions: weak solutions and classical solutions.

Theorem. *Let $f(x)$ be continuous on Ω. Then (a) a classical solution of (2.5.37) on Ω is also a weak solution, and (b) any weak solution on Ω which has p continuous derivatives is a classical solution.*

Proof.

(a) Let u be a classical solution on Ω, and let $\phi(x)$ be a test function with support in Ω. Then

$$\langle u, L^*\phi\rangle = \int_\Omega uL^*\phi\, dx$$

can be integrated by Green's theorem. The support of ϕ being in Ω, ϕ vanishes in a neighborhood of Γ so that ϕ and all its derivatives vanish on Γ. This guarantees that $J(u, \phi) = 0$ on Γ, and therefore, by (2.5.23),

$$\langle u, L^*\phi\rangle = \int_\Omega \phi Lu\ dx.$$

Since $Lu = f$ at every point of Ω, we obtain

$$\langle u, L^*\phi\rangle = \langle f, \phi\rangle \quad \text{for } \phi \text{ in } C_0^\infty(\Omega).$$

(b) By assumption $\langle u, L^*\phi\rangle = \langle f, \phi\rangle$ for ϕ with support in Ω. Since u has continuous derivatives of order p, we can use Green's theorem on Ω. Again we have $J = 0$ on Γ, so that

$$0 = \int_\Omega \phi q\, dx \quad \text{for } \phi \text{ in } C_0^\infty(\Omega), \tag{2.5.39}$$

where q is the continuous function $Lu - f$. We wish to show that (2.5.39) implies that $q \equiv 0$ in Ω. Suppose that $q(x_0) \neq 0$ at x_0 in Ω. There is no loss of generality in assuming that $q(x_0) > 0$. Since q is continuous, we can find a small ball with center at x_0, lying wholly in Ω, such that $q > 0$ in the ball. We can also construct a test function ϕ positive in the ball and 0 outside. For this ϕ we would have $\int_\Omega \phi q\, dx > 0$, contradicting the hypothesis. Therefore, $q = Lu - f = 0$ at every point in Ω, and u is a classical solution of (2.5.37) in Ω.

□

The question naturally arises as to whether (2.5.37) can have weak solutions that are not classical solutions. By part (b) of the theorem just proved, any such weak solution could not belong to $C^p(\Omega)$. A simple example with $\Omega = \mathbb{R}$ is the first-order equation

$$x\frac{du}{dx} = 0,$$

which has the weak solution $u(x) = H(x)$. Indeed,

$$\langle xH', \phi \rangle = \langle H', x\phi \rangle = \langle \delta, x\phi \rangle = 0,$$

so that $u = H$ is a weak solution which is clearly not a classical solution. In $\mathbb{R}$ a weak solution is associated with the fact that the differential equation has a singular point (in our particular case, at $x = 0$), that is, a point at which the coefficient $a_p(x)$ of the highest derivative $D^p u$ vanishes. The situation is more complicated for partial differential equations. The role formerly played by the coefficient of the highest-order term is now taken over by a matrix of coefficients of the terms of order p. (See also Section 8.1.) Let us look at a few examples.

Example 5. In $\mathbb{R}^2$, with $x = (x_1, x_2)$, consider the equation

$$\frac{\partial u}{\partial x_1} = 0, \tag{2.5.40}$$

whose classical solutions are $u = f(x_2)$ with f differentiable. Since no differentiation with respect to x_2 is involved in (2.5.40), the requirement that $f(x_2)$ be differentiable can be dispensed with; for instance, $u = H(x_2)$ is a weak solution of (2.5.40) since

$$\begin{aligned}\left\langle \frac{\partial H(x_2)}{\partial x_1}, \phi(x_1, x_2) \right\rangle &= \left\langle H(x_2), -\frac{\partial \phi}{\partial x_1} \right\rangle = -\int_0^\infty dx_2 \int_{-\infty}^\infty dx_1 \frac{\partial \phi}{\partial x_1} \\ &= -\int_0^\infty dx_2 [\phi(+\infty, x_2) - \phi(-\infty, x_2)] = 0.\end{aligned}$$

Example 6. In $\mathbb{R}^2$, with $x = (x_1, x_2)$, the equation

$$\frac{\partial^2 u}{\partial x_1 \partial x_2} = 0 \tag{2.5.41}$$

has classical solutions of the form $u = f(x_1) + g(x_2)$, where f and g are twice differentiable. Even if f and g are not differentiable, u of this type is still a solution. Indeed, $u = H(x_1) + H(x_2)$ is a weak solution of (2.5.41) for

$$\left\langle \frac{\partial^2 [H(x_1) + H(x_2)]}{\partial x_1 \partial x_2}, \phi \right\rangle = \left\langle H(x_1) + H(x_2), \frac{\partial^2 \phi}{\partial x_1 \partial x_2} \right\rangle$$

$$= \int_{-\infty}^\infty dx_2 \int_0^\infty \frac{\partial^2 \phi}{\partial x_1 \partial x_2} dx_1 + \int_{-\infty}^\infty dx_1 \int_0^\infty \frac{\partial^2 \phi}{\partial x_1 \partial x_2} dx_2 = 0.$$

Example 7. Consider the homogeneous wave equation in one space dimension x:

$$\frac{\partial^2 u}{\partial t^2} - \frac{\partial^2 u}{\partial x^2} = \Box^2 u = 0. \tag{2.5.42}$$

If f is any function of a real variable with a continuous second derivative, then $u = f(x-t)$ is easily seen to be a solution of (2.5.42). This solution represents a wave traveling to the right with velocity 1. Snapshots taken at times t_1 and t_2 show the same wave form, the one at the later time t_2 being displaced an amount $t_2 - t_1$ to the right with respect to the one taken at time t_1. There seems to be no reason physically to restrict oneself to wave forms that are twice differentiable. Let us show that $u = H(x-t)$ is a weak solution of (2.5.42). Since $\Box^2$ is formally self-adjoint, we must show, according to (2.5.38), that

$$\langle H(x-t), \Box^2\phi(x,t)\rangle = 0 \quad \text{for each test function } \phi(x,t),$$

or, equivalently, that

$$\int\int_{x>t} \left(\frac{\partial^2\phi}{\partial t^2} - \frac{\partial^2\phi}{\partial x^2}\right) dx\, dt = 0 \quad \text{for } \phi \text{ in } C_0^\infty(\mathbb{R}^2). \tag{2.5.43}$$

By changing variables to $x_1 = x-t, x_2 = x+t$, we find that the last integral reduces to an integration of $\partial^2\phi/\partial x_1 \partial x_2$ over the half-plane $x_1 > 0$. Since

$$\int_{-\infty}^{\infty} \frac{\partial^2\phi}{\partial x_1 \partial x_2} dx_2 = 0,$$

we have verified (2.5.43). Weak solutions of (2.5.42) can have jump discontinuities on the *characteristics* $x = t$ and $x = -t$ [the latter arising from solutions of the type $f(x+t)$]. This is discussed further in Section 8.1.

Example 8. *Weak solutions of Laplace's equation are necessarily classical.* We shall only show that a solution of $\Delta u = 0$ in Ω cannot have a simple jump across a hypersurface σ. Suppose that σ divides Ω into the two domains Ω_+ and Ω_-. If u is a weak solution of $\Delta u = 0$ on $\mathbb{R}^n$, we have, for ϕ with support in Ω,

$$0 = \int_\Omega u\, \Delta\phi\, dx = \int_{\Omega+} u\, \Delta\phi\, dx + \int_{\Omega-} u\, \Delta\phi\, dx.$$

If u and its first derivatives have limiting values on both sides of σ, we find by Green's theorem that

$$\int_{\Omega+} u\, \Delta\phi\, dx = \int_{\Omega+} \phi\, \Delta u\, dx + \int_{\sigma+} \left(u\frac{\partial\phi}{\partial n} - \phi\frac{\partial u}{\partial n}\right) dS,$$

where σ_+ is the side of σ bounding Ω_+. There is no contribution from the boundary Γ of Ω, since ϕ vanishes in a neighborhood of Γ. Combining the last equation with a similar one for Ω_- and using the assumption that u is a classical solution within Ω_+ and Ω_-, we obtain

$$0 = \int_\sigma \left[(u_+ - u_-)\frac{\partial\phi}{\partial n} - \left(\frac{\partial u_+}{\partial n} - \frac{\partial u_-}{\partial n}\right)\phi\right] dS, \quad \phi \in C_0^\infty(\Omega), \tag{2.5.44}$$

where n is the outward normal to σ from Ω_+. The factors multiplying $\partial\phi/\partial n$ and ϕ must each vanish on σ. The proof can be adapted from the special case where $\Omega = \mathbb{R}^2$ and σ is the line $x_2 = 0$. Consider the test function $\phi(x_1, x_2) = \phi_1(x_1)\phi_2(x_2)$, where $\phi_2(0) = 0$, $\phi_2'(0) \neq 0$, and $\phi_1(x_1) \geqslant 0$ has its support in a small neighborhood of the point $x_1 = \xi_1$. It then follows from (2.5.44) that $u(\xi_1, 0+) = u(\xi_1, 0-)$; since ξ_1 is arbitrary, we have $u(x_1, 0+) = u(x_1, 0-)$ for any point on σ. Similarly, it can be shown that $(\partial u/\partial x_2)(x_1, 0+) = (\partial u/\partial x_2)(x_1, 0-)$. In the general case the conclusion is $u_+ = u_-$ and $\partial u_+/\partial n = \partial u_-/\partial n$. The first of these relations also implies that tangential derivatives are continuous on crossing σ. Thus u and all its first derivatives must be continuous on crossing σ. More can be shown: u is infinitely differentiable in Ω. We already know this in two dimensions, where a solution of Laplace's equation is the real part of an analytic function of the complex variable $x_1 + ix_2$.

Fundamental Solutions

We consider differential equations on the whole of $\mathbb{R}^n$ (that is, $\Omega = \mathbb{R}^n$). A powerful method for studying such equations is based on the fundamental solution.

Definition. *A fundamental solution for L with pole at ξ is a solution of the equation*

$$Lu = \delta_\xi(x) = \delta(x - \xi), \tag{2.5.45}$$

where ξ is regarded as a parameter.

REMARKS

1. Equation (2.5.45) is to be interpreted in the sense of distributions. A solution of (2.5.45) is denoted by $E(x, \xi)$. It is a distribution in x depending parametrically on ξ. Often (but not always) E will correspond to a locally integrable function of x. In any event, according to (2.5.38), E satisfies (2.5.45) if and only if

$$\langle E, L^*\phi\rangle = \phi(\xi) \quad \text{for each test function } \phi \text{ in } C_0^\infty(\mathbb{R}^n). \tag{2.5.46}$$

2. Equation (2.5.45) will usually have many solutions differing from one another by a solution of the homogeneous equation. For problems in which $\mathbb{R}^n$ can be interpreted as an isotropic geometrical space, we often select a particular solution on grounds of symmetry and behavior at infinity; if, however, one of the coordinates is timelike, we may use causality as the appropriate criterion with respect to that coordinate.

3. If L has *constant coefficients*, it suffices to find the fundamental solution with pole at 0 [that is, $E(x, 0)$] and then translate to obtain the solution with pole at ξ:

$$E(x, \xi) = E(x - \xi, 0).$$

The fundamental solution $E(x, 0)$ will also be denoted by $E(x)$.

There are two parts to determining a fundamental solution. First we must construct, often by intuitive means, a likely candidate, and then we must check that (2.5.46) is in fact satisfied. The first part can itself be divided into two steps: Solve the homogeneous equation for $x \neq \xi$ with proper regard to physical considerations, and then build in the right singularity at $x = \xi$ by using the integrated form of (2.5.45) near $x = \xi$ [for one dimension this involves appropriate matching at ξ of the solutions on either side; for higher dimensions the condition will be a modification of (1.4.19), which is valid for $-\Delta$]. We have already acquired some experience in verifying that a given distribution satisfies a differential equation; see, for instance, Example 15, Section 2.1, and Exercises 2.1.4, 2.1.5, and 2.1.6.

Example 9. Find a fundamental solution E for $-(d^2/dx^2)+q^2$. Since the coefficients are constant, it suffices to find $E(x)$ satisfying

$$-\frac{d^2E}{dx^2} + q^2E = \delta(x), \quad -\infty < x < \infty.$$

On intuitive grounds based on our experience in Chapter 1, we require the continuity of E at $x = 0$ and $E'(0+) - E'(0-) = -1$. If E is to represent the concentration in an absorbing medium, we also demand that E vanish at $|x| = \infty$. This leads to the candidate [see (1.4.14)]

$$E(x) = \frac{e^{-q|x|}}{2q}, \quad E(x, \xi) = \frac{e^{-q|x-\xi|}}{2q}.$$

To check that $E(x)$ is a fundamental solution with pole at 0, we must show that (2.5.46) holds. We have

$$\langle E, L^*\phi\rangle = \int_{-\infty}^{0} \frac{e^{qx}}{2q} L^*\phi\, dx + \int_{0}^{\infty} \frac{e^{-qx}}{2q} L^*\phi\, dx.$$

Using Green's theorem in each interval, we find easily that

$$\langle E, L^*\phi\rangle = \phi(0) \quad \text{for each } \phi \text{ in } C_0^\infty(\mathbb{R}).$$

Alternatively, we can use the result of Example 11, Section 2.1, to differentiate the function $E(x)$ in the sense of distributions.

Example 10. Consider the general ordinary differential operator L of order p. Let us find the *causal* fundamental solution $E(t, \tau)$ which *vanishes* for $t < \tau$ and satisfies

$$LE = a_p \frac{d^pE}{dt^p} + \cdots + a_1 \frac{dE}{dt} + a_0E = \delta(t - \tau), \quad -\infty < t, \tau < \infty. \tag{2.5.47}$$

Proceeding intuitively, we have $E \equiv 0$ for $t < \tau$ and $E, E', \ldots, E^{(p-2)}$ continuous at $t = \tau$ (therefore all are 0 at $\tau+$), and $a_pE^{(p-1)}$ has a unit jump at $t = \tau$, that is,

$$E^{(p-1)}(\tau+, \tau) = \frac{1}{a_p(\tau)}.$$

The last condition was obtained by integrating (2.5.47) from $\tau-$ to $\tau+$. This suggests that, for $t > \tau, E(t,\tau)$ will coincide with the solution $u_\tau(t)$ of the initial value problem

$$\begin{aligned} Lu_\tau(t) = 0, \quad u_\tau(\tau) =& u_\tau'(\tau) = \cdots = u_\tau^{(p-2)}(\tau) = 0, \\ & u_\tau^{(p-1)}(\tau) = \frac{1}{a_p(\tau)}. \end{aligned} \tag{2.5.48}$$

Since (2.5.48) has one and only one solution by the existence and uniqueness theorem for initial value problems (see Chapter 3), we tentatively set

$$E(t,\tau) = H(t-\tau)u_\tau(t). \tag{2.5.49}$$

Let us now check that (2.5.49) satisfies (2.5.46) with $x = t, \xi = \tau$. We have

$$\langle E, L^*\phi\rangle = \int_\tau^\infty u_\tau(t) L^*\phi\, dt,$$

and, using Green's formula (2.5.18),

$$\langle E, L^*\phi\rangle = \int_\tau^\infty \phi L u_\tau(t)\, dt - J(\mu_\tau, \phi)]_{t=\tau}^{t=\infty}.$$

Since $Lu_\tau = 0$, the integral vanishes. The fact that $\phi \equiv 0$ outside a bounded interval shows that $J = 0$ at the upper limit. At the lower limit τ we must look a little more closely at expression (2.5.17) for J. All terms involving derivatives of u_τ of order $p-2$ or less are 0 by (2.5.48); this leaves only a single term with $j = p-1, k = 0, m = p$. Thus, at $t = \tau, J(u_\tau, \phi) = u_\tau^{(p-1)}(\tau)a_p(\tau)\phi(\tau) = \phi(\tau)$, which gives $\langle E, L^*\phi\rangle = \phi(\tau)$, as required. Note that causality removes all arbitrariness from fundamental solutions.

If the coefficients in (2.5.47) are all *constants*, then

$$u_\tau(t) = v(t-\tau),$$

where $v(t)$ is the solution of the initial value problem (2.5.48) with $\tau = 0$. There is a corresponding simplification in (2.5.49). As an illustration, see Exercise 1.4.5.

Example 11. We have already shown in Example 15, Section 2.1, that $E(x) = 1/4\pi|x|$ is a fundamental solution for $-\Delta$ with pole at the origin.

Example 12. Let us find the causal fundamental solution for the equation of diffusion in one space dimension (the term *causal* applies to the time coordinate). With the source at $x = 0, t = 0$, we are searching for the solution $E(x,t)$ of

$$LE = \frac{\partial E}{\partial t} - \frac{\partial^2 E}{\partial x^2} = \delta(x,t), \quad -\infty < t, x < \infty, \tag{2.5.50}$$

with $E \equiv 0, t < 0$, and $E \to 0$ as $|x| \to \infty$. Proceeding without regard to rigor, we take a Fourier transform on the space coordinate; setting

$$E^\wedge = \int_{-\infty}^\infty e^{i\omega x} E\, dx,$$

we obtain, by multiplying (2.5.50) by $e^{i\omega x}$ and integrating,

$$\frac{dE^\wedge}{dt} + \omega^2 E^\wedge = \delta(t), \quad E^\wedge = 0, \quad t < 0.$$

Thus $E^\wedge$ jumps by unity at $t = 0$, and

$$E^\wedge = e^{-\omega^2 t}, \quad t > 0 \quad (\text{of course, } E^\wedge = 0 \quad \text{for } t < 0).$$

The inversion is easily performed:

$$E = \frac{1}{2\pi}\int_{-\infty}^{\infty} e^{-i\omega x} E^\wedge \, d\omega = \frac{e^{-x^2/4t}}{\sqrt{4\pi t}}, \quad t > 0.$$

We are therefore led to believe that

$$E = H(t)\frac{e^{-x^2/4t}}{\sqrt{4\pi t}} \tag{2.5.51}$$

is a fundamental solution of (2.5.50). To verify this we must show that (2.5.51) satisfies

$$\langle E, L^*\phi\rangle = \phi(0,0) \quad \text{for each test function } \phi(x,t).$$

The left side is the convergent integral

$$\int_0^\infty dt \int_{-\infty}^{\infty} dx \frac{e^{-x^2/4t}}{\sqrt{4\pi t}} L^*\phi,$$

which can be written as

$$\lim_{\varepsilon\to 0}\int_\varepsilon^\infty dt \int_{-\infty}^{\infty} dx \frac{e^{-x^2/4t}}{\sqrt{4\pi t}} L^*\phi.$$

Applying Green's formula (2.5.29) to this integral, and noting that $e^{-x^2/4t}/\sqrt{4\pi t}$ satisfies the homogeneous equation for $t > 0$ (that is how it was constructed!), we find that

$$\int_\varepsilon^\infty dt \int_{-\infty}^{\infty} dx \frac{e^{-x^2/4t}}{\sqrt{4\pi t}} L^*\phi = \int_{-\infty}^{\infty} \frac{e^{-x^2/4\varepsilon}}{\sqrt{4\pi\varepsilon}}\phi(x,\varepsilon)\, dx.$$

It is now a simple modification of the argument in case (b), Example 3, Section 2.2, to show that the limit as $\varepsilon \to 0$ is $\phi(0,0)$. An alternative way of describing $E(x,t)$ for $t > 0$ is to state that it satisfies the homogeneous equation with initial value

$$\lim_{t\to 0+} E(x,t) = \delta(x).$$

This point of view is discussed further in Section 8.2.

EXERCISES

2.5.1 Consider the operator in $\mathbb{R}^2$ defined by

$$Lu = \frac{\partial^2 u}{\partial x \partial y} + \gamma^2 u,$$

where γ^2 is a positive constant.

(a) Write Green's formula for L.

(b) Show that $J_0(2\gamma\sqrt{xy})$ is a classical solution of $Lu = 0$ for $x, y > 0$.

(c) Show that $E = H(x)H(y)J_0(2\gamma\sqrt{xy})$ is a fundamental solution for L with pole at the origin.

2.5.2 Consider the wave operator in one space dimension:

$$\Box^2 = \frac{\partial^2}{\partial t^2} - \frac{\partial^2}{\partial x^2}.$$

Write Green's formula for an arbitrary domain Ω with boundary Γ. On Γ introduce a vector q, known as the *transversal* to Γ, defined by

$$q \cdot e_t = n \cdot e_t, \quad q \cdot e_x = -n \cdot e_x,$$

where e_x and e_t are unit vectors in the x and t directions, respectively. Clearly q is a unit vector. Show that Green's formula takes the form

$$\int_\Omega (v\Box^2 u - u\Box^2 v)\, dx\, dt = \int_\Gamma \left(v\frac{\partial u}{\partial q} - u\frac{\partial v}{\partial q} \right) dl, \tag{2.5.52}$$

where dl is an element of arclength along the bounding curve Γ. Note that q is tangent to Γ if and only if Γ is one of the two families of straight lines $x - t =$ constant and $x + t =$ constant. These are the *characteristics* of the differential equation. Suppose we are looking for a curve C across which solutions of $\Box^2 u = 0$ can suffer jump discontinuities. On either side of C we assume that u is a classical solution. Since u is to be a generalized solution, we have $0 = \langle u, \Box^2 \phi \rangle$; using Green's formula (2.5.31) first on one side of C and then on the other, and adding the results, we find that

$$0 = \int_C \left[\frac{\partial \phi}{\partial q}(u_+ - u_-) - \phi \left(\frac{\partial u_+}{\partial q} - \frac{\partial u_-}{\partial q} \right) \right] dl, \tag{2.5.53}$$

where q is the transversal on the positive side of C. If C is nowhere tangent to a characteristic, show that $u_+ - u_- = 0$ and $\partial u_+/\partial q - \partial u_-/\partial q = 0$. If C is a characteristic, then $\partial/\partial q$ is proportional to $\partial/\partial l$ and (2.5.53) reduces to

$$0 = \int_C \left\{ \frac{\partial}{\partial l}[\phi(u_+ - u_-)] - 2\phi\frac{\partial}{\partial l}(u_+ - u_-) \right\} dl.$$

Since ϕ vanishes outside a finite segment of C, the integral of the first term on the right is automatically 0. From the arbitrariness of ϕ, we can, however, conclude that

$$\frac{\partial}{\partial l}(u_+ - u_-) = 0 \quad \text{or} \quad u_+ - u_- = \text{constant on } C. \tag{2.5.54}$$

Thus the only curves that can propagate discontinuities are the characteristics, and across a characteristic the jump in the solution remains constant.

2.5.3 Show that $E = (1/2\pi)K_0(q|x|)$ is a fundamental solution in $\mathbb{R}^2$ for the operator $L = -\Delta + q^2$. Since E is radially symmetric and vanishes at infinity, we can interpret E as the steady concentration in an absorbing medium due to a source placed at the origin of $\mathbb{R}^2$. [See (1.4.32).]

2.5.4 Consider the wave operator of Exercise 2.5.2. Show that $E = \frac{1}{2}H(t - |x|)$ is a causal fundamental solution with pole at $x = 0, t = 0$. Find a fundamental solution which vanishes for $t > 0$.

2.5.5 Show that $E = H(t)(4\pi at)^{-n/2}e^{-r^2/4at}$ is a causal fundamental solution for the diffusion operator $L = (\partial/\partial t) - a\Delta$ in n space dimensions. [See case (g), Example 3, Section 2.2.]

2.5.6 Find a spherically symmetric fundamental solution (with pole at the origin) for the biharmonic operator $L = \Delta^2 = \Delta\Delta$ in two dimensions and in three dimensions. Verify your results distributionally.

2.5.7 (a) Consider one-dimensional unsteady diffusion in an absorbing medium. The causal fundamental solution E with pole at $x = 0, t = 0$ satisfies

$$\frac{\partial E}{\partial t} - \frac{\partial^2 E}{\partial x^2} + q^2 E = \delta(x,t), \quad E \equiv 0, \quad t < 0.$$

Reduce the problem to ordinary diffusion by the transformation $E = e^{-q^2 t}F$ and find E.

(b) What would be the significance of the problem in which $q^2 E$ is replaced by $-q^2 E$? What about the fundamental solution?

2.5.8 In diffusion with *drift*, the causal fundamental solution E satisfies

$$\frac{\partial E}{\partial t} - \frac{\partial^2 E}{\partial x^2} + v\frac{\partial E}{\partial x} = \delta(x,t), \quad E \equiv 0, \quad t < 0,$$

where v is a real constant. Find E. $\left[\text{Hint: Let } E = F\exp\left(vx/2 - v^2t/4\right).\right]$

2.5.9 Let L be the ordinary differential operator of order p:

$$a_pD^p + \cdots + a_1D + a_0,$$

where the coefficients are infinitely differentiable. Let $E(x,\xi)$ be a fundamental solution for L with pole at ξ. Then of course $E, E', \ldots, E^{(p-2)}$ are continuous at ξ, and $E^{(p-1)}$ has a jump $1/a_p(\xi)$. The functions $E^{(p)}, E^{(p+1)}, \ldots,$ are well defined for $x > \xi$ and $x < \xi$. Find the jumps $E^{(p)}(\xi+,\xi) - E^{(p)}(\xi-,\xi)$ and $E^{(p+1)}(\xi+,\xi) - E^{(p+1)}(\xi-,\xi)$, and indicate how jumps in higher derivatives could be found (note that it suffices to consider a causal fundamental solution).

2.5.10 Let L and M be two linear differential operators on $\mathbb{R}^n$, and let $P = LM$. Show that

$$P^* = M^* L^*,$$

and express the right-hand side of Green's theorem for P in terms of the conjuncts J and K corresponding to L and M, respectively.

2.5.11 Let $\phi_0(x)$ be a fixed element of $C_0^\infty(\mathbb{R})$ with $\phi_0(0) = 1$, and let M be the subset of $C_0^\infty(\mathbb{R})$ consisting of elements $\psi(x)$ of the form $\psi = x\chi$, where χ is itself in $C_0^\infty(\mathbb{R})$. Show that the following hold:

(a) A test function ψ belongs to M if and only if $\psi(0) = 0$.

(b) For each ϕ in $C_0^\infty(\mathbb{R})$ there are a unique constant a and a unique element ψ in M such that

$$\phi = a\phi_0 + \psi.$$

(c) The general distributional solution of $xu = 0$ is $C\delta(x)$, where C is an arbitrary constant. What is the general solution of $xu = f$, where f is a given distribution?

2.5.12 For the indicator function I_Ω of Example 2, Section 2.1, show that ΔI_Ω is a dipole distribution on Γ.

2.6 WEAK DERIVATIVES AND SOBOLEV SPACES

Among the most important families of function spaces that we will encounter later in the book, particularly in Chapters 8 and 10, are the *Sobolev spaces*. We will need to assemble some basic analysis tools and concepts (Banach and Hilbert spaces) in Chapters 4 and 5 before we can make a more serious study of Sobolev spaces (which we do in Chapter 10). However, we can make a start here by explaining the connection between distributions and elements of a Sobolev space, through the concept of a *weak derivative*.

We first formalize some of the concepts discussed in the sections above as the *Schwartz theory of distributions* or *generalized functions*. To begin, let $\Omega \subset \mathbb{R}^n$ be a *domain* in $\mathbb{R}^n$, meaning that it is an open subset of $\mathbb{R}^n$, not necessarily bounded. Recall that if the open subset $A \subset \mathbb{R}^n$ is nonempty, the closure of A in $\mathbb{R}^n$ is denoted $\overline{A}$. If $A \subset \Omega$, $\overline{A} \subset \Omega$, and $\overline{A}$ is compact (as a subset of $\mathbb{R}^n$, this is equivalent to being closed and bounded), we denote this as

$$A \subset\subset \Omega.$$

Recall that the *support* of a function $u \colon \Omega \subset \mathbb{R}^n \to \mathbb{R}$ is defined as

$$\operatorname{supp}(u) = \overline{\{ x \in A : u(x) \neq 0 \}}.$$

If $\operatorname{supp}(u) \subset\subset \Omega$, we say that u has *compact support* in Ω.

We now define the vector space $\mathcal{D}(\Omega)$ as the set $C_0^\infty(\Omega)$ of infinitely differentiable functions with compact support in Ω; this set is algebraically closed under linear combinations involving the scalar field $\mathbb{R}$, and therefore has the structure of a real vector space. The vector space $\mathcal{D}(\Omega)$ can be given a *topology* [a collection of open subsets of $\mathcal{D}(\Omega)$ that makes possible the definitions of convergence of sequences and continuity of maps, among other things] either indirectly by defining a norm or metric, or more directly by giving a definition of the convergence of sequences. To this end, given a sequence of functions $\{\phi_j\}$ with $\phi_j \in C_0^\infty(\Omega)$ for all j, we say that the sequence *converges in the sense of the space* $\mathcal{D}(\Omega)$ to $\phi \in C_0^\infty(\Omega)$ if:

(1) The function $(\phi - \phi_j)$ has support in a fixed subset $A \subset\subset \Omega$, for all j.

(2) $\lim_{j\to\infty} D^\alpha \phi_j(x) = D^\alpha \phi(x)$ for each multi-index α, uniformly on the set A.

One can show that with this definition, $\mathcal{D}(\Omega)$ has the structure of a complete (locally convex) topological vector space; we defer to the book of Dunford and Schwartz [3] for a more complete description of this structure that can be placed on a vector space. With this additional structure, $\mathcal{D}(\Omega)$ is referred to as the *space of test functions*.

One can now consider the vector space of linear functionals $T \colon \mathcal{D}(\Omega) \to \mathbb{R}$ (see Section 4.7), and define continuity of a functional T to mean

$$T(\phi_j) \to T(\phi) \quad \text{if and only if} \quad \phi_j \to \phi \quad \text{“in the sense of the space } \mathcal{D}(\Omega)\text{.”}$$

The *dual space* of $\mathcal{D}(\Omega)$, denoted as $\mathcal{D}'(\Omega)$, which consists of the vector space of continuous linear functionals on $\mathcal{D}(\Omega)$ (with continuity defined as above), is called the *space of Schwartz distributions* on Ω.

We recall that the Schwartz distributions have derivatives of all orders in the following sense. Let $u \in C^m(\Omega)$, and consider the integration by parts formula involving any $\phi \in \mathcal{D}(\Omega)$:

$$\int_\Omega (D^\alpha u(x))\phi(x)\, dx = (-1)^{|\alpha|} \int_\Omega u(x)(D^\alpha \phi(x))\, dx,$$

where α is a multi-index with $0 \leqslant |\alpha| \leqslant m$. Note that the boundary terms have vanished due to the compact support of ϕ. This integration by parts formula was the basis for defining the derivative of a distribution in Section 2.1:

$$(D^\alpha T)(\phi) = (-1)^{|\alpha|} T(D^\alpha \phi). \tag{2.6.1}$$

Since $\phi \in \mathcal{D}(\Omega)$, we have also that $D^\alpha \phi \in \mathcal{D}(\Omega)$, so that if $T \in \mathcal{D}'(\Omega)$, then also $D^\alpha T \in \mathcal{D}'(\Omega)$. Therefore, the Schwarz distributions $\mathcal{D}'(\Omega)$ have derivatives of all orders in the sense of (2.6.1).

We have used the notion of a *locally integrable function* throughout the chapter; we now give a name to this vector space of functions:

$$L^1_{\text{loc}}(\Omega) = \{\, u\colon \Omega \to \mathbb{R} \,:\, u \in L^1(A) \text{ for every open set } A \subset\subset \Omega \,\}.$$

The importance of L^1_{loc} was already seen earlier in the chapter; here, we can use it to generate distributions lying in $\mathcal{D}'(\Omega)$: Given any $u \in L^1_{\text{loc}}(\Omega)$,

$$T_u(\phi) = \int_\Omega u(x)\phi(x)\, dx \in \mathcal{D}'(\Omega).$$

The proof that it is continuous follows easily from the properties of ϕ and the integral. We cannot generate every functional in $\mathcal{D}'(\Omega)$ in this way using $L^1_{\text{loc}}(\Omega)$ (for example, the delta function cannot be represented this way), but $L^1_{\text{loc}}(\Omega)$ turns out to be just the right space of functions for making precise the notion of a weak derivative, and subsequently leading to a natural definition of the Sobolev spaces. To this end, let $u \in L^1_{\text{loc}}(\Omega)$. If there exists a $v_\alpha \in L^1_{\text{loc}}(\Omega)$ such that

$$\int_\Omega u(x)(D^\alpha \phi(x))\, dx = (-1)^{|\alpha|} \int_\Omega v_\alpha(x)\phi(x)\, dx, \qquad \forall \phi \in \mathcal{D}(\Omega), \tag{2.6.2}$$

then we say that $D^\alpha u = v_\alpha$ is the *weak* or *distributional* partial derivative of u of order α. One can show that $D^\alpha u$ defined in this way is unique (up to sets of measure zero), and furthermore, if u is sufficiently smooth so that $u \in C^m(\Omega)$ for $m = |\alpha|$, then the classical and weak derivatives are the same object.

It is natural now to consider a vector space of such functions which have weak derivatives in the sense of (2.6.2). This leads to defining

$$W^{m,p}(\Omega) = \{\, u \in L^p(\Omega) \,:\, D^\alpha u \in L^p(\Omega),\ 0 \leqslant |\alpha| \leqslant m \,\},$$

where $D^\alpha u$ is the weak derivative of u in the sense of (2.6.2). The vector space $L^p(\Omega)$, for $1 \leqslant p < \infty$, is defined as

$$L^p(\Omega) = \left\{ u \in \mathcal{M}(\Omega) \,:\, \int_\Omega |u|^p\, dx < \infty \right\},$$

where $\mathcal{M}(\Omega)$ denotes the set of Lebesgue-measurable functions defined over a domain $\Omega \subset \mathbb{R}^n$. The vector spaces $W^{m,p}(\Omega)$ for $m \geqslant 0$ and $1 \leqslant p \leqslant \infty$ are referred to as *Sobolev spaces*, with the case $m = 0$ coinciding with the Lebesgue spaces $L^p(\Omega)$. Both $L^p(\Omega)$ and $W^{m,p}(\Omega)$ can be equipped with norms, turning them into *normed spaces*, and furthermore, both are complete with respect to their norms, giving them the structure of *Banach spaces*. The case $p = 2$ is quite special; the notation

$$H^m(\Omega) = W^{m,2}(\Omega), \qquad m \geqslant 0,$$

is commonly used for this case, with $m = 0$ coinciding with the space $L^2(\Omega)$. In this case, the norm on $H^m(\Omega)$ arises from an inner product, giving it the additional structure of a *Hilbert space*. The case $m = 1$ and $p = 2$, which is then denoted as $H^1(\Omega)$, arises regularly in the study of linear and nonlinear second-order elliptic partial differential equations (see Chapters 8, 9, and 10). We reexamine the Sobolev classes $W^{m,p}(\Omega)$ in more detail in Chapter 10, once we have developed some background in the basic theory of abstract Banach and Hilbert spaces in Chapters 4 and 5.

EXERCISES

2.6.1 Prove that if $u \in L^1_{\text{loc}}(\Omega)$, and if $D^\alpha u$ exists for a multi-index α, then $D^\alpha u$ is uniquely defined (up to a set of measure zero).

2.6.2 Let $u, v, w \in W^{m,p}(\Omega)$, and let α and β be any two multi-indices such that $0 \leqslant |\alpha| + |\beta| \leqslant m$. Prove that if $v = D^\alpha u$ and $w = D^\beta v$, then $w = D^{\alpha+\beta} u$.

2.6.3 Prove that if $u, v \in W^{m,p}(\Omega)$ and $a, b \in \mathbb{R}$, then $au + bv \in W^{m,p}(\Omega)$. In other words, prove that $W^{m,p}(\Omega)$ is a (real) vector space (is algebraically closed).

REFERENCES AND ADDITIONAL READING

1. R. A. Adams, *Sobolev Spaces*, Academic Press, New York, 1978.
2. P. Chernoff, Pointwise convergence of Fourier series, *American Mathematical Monthly* 87, 1980.
3. N. Dunford and J. T. Schwartz, *Linear Operators*, Part II: *Spectral Theory*, Wiley, New York, 1963.
4. H. Dym and H. P. McKean, *Fourier Series and Integrals*, Academic Press, New York, 1972.
5. J. Foster and F. B. Richards, The Gibbs phenomenon for piecewise-linear approximation, *Am. Math. Mon.* 98, 1991.
6. I. M. Gelfand and G. E. Shilov, *Generalized Functions*, Vol. I, Academic Press, New York, 1964.
7. D. D. Haroske and H. Triebel, *Distributions, Sobolev spaces, Elliptic Equations*, EMS Textbooks in Mathematics, European Mathematical Society, 2008.
8. R. P. Kanwal, *Generalized Functions*, Academic Press, New York, 1983.
9. T. W. Körner, *Fourier Analysis*, Cambridge University Press, New York, 1988.
10. P. Kythe, *Fundamental Solutions of Partial Differential Equations and Applications*, Birkhauser, Cambridge, MA, 1996.
11. A. Papoulis, *The Fourier Integral and Its Applications*, McGraw-Hill, New York, 1962.
12. L. Schwartz, *Mathematics for the Physical Sciences*, Hermann, Paris, 1966.
13. F. Stenger, *Numerical Methods based on Sinc and Analytic Functions*, Springer–Verlag, New York, 1993.
14. G. Strang, *Introduction to Applied Mathematics*, Wellesley-Cambridge University Press, New York, 1986.
15. R. Strichartz, *A Guide to Distribution Theory and Fourier Transforms*, CRC Press, Boca Raton, FL, 1993.
16. A. H. Zemanian, *Distribution Theory and Transform Analysis*, McGraw-Hill, New York, 1965.

CHAPTER 3

ONE-DIMENSIONAL BOUNDARY VALUE PROBLEMS

3.1 REVIEW

In this section we deal with classical solutions of the differential equation $Lu = f$ on an open interval I on the x axis. Here L is the general linear ordinary differential operator of order p;

$$L = a_p(x)\frac{d^p}{dx^p} + \cdots + a_1(x)\frac{d}{dx} + a_0(x), \qquad (3.1.1)$$

where the coefficients $\{a_k(x)\}$ are continuous on the closure $\bar{I}$ of I, and *the leading coefficient* $a_p(x)$ *does not vanish anywhere on* $\bar{I}$. A point x at which $a_p(x)$ vanishes is a *singular point*, and the only purpose of introducing I is to avoid such points. If $a_p(x)$ does not vanish anywhere on the real line, we can take I to be the entire line. The inhomogeneous term $f(x)$ is assumed to be piecewise continuous on $\bar{I}$. We recall that a classical solution of $Lu = f$ is a function $u(x)$ with $p-1$ continuous derivatives and a piecewise continuous pth derivative such that the differential equation is satisfied at all points of continuity of f.

From elementary differential equations we know that the general solution of the equation $Lu = f$ involves p arbitrary constants. By imposing p additional con-

Green's Functions and Boundary Value Problems, Third Edition. By I. Stakgold and M. Holst

ditions, known generally as *boundary conditions*, one might hope to single out a specific solution of the differential equation. This conclusion is warranted in the special case of initial conditions, when $u, u', \ldots, D^{p-1}u$ are all given at the same point x_0 in $\bar{I}$, but is *not* always correct for general boundary conditions. We now state the precise form of the existence and uniqueness theorem for initial value problems, with the proof postponed until Chapter 4.

Theorem 3.1.1. *Let L be the operator (3.1.1), and let $a_p(x) \neq 0$ in $\bar{I}$. Let x_0 be a fixed point in $\bar{I}$; let $\gamma_1, \ldots, \gamma_p$ be given numbers, and let $f(x)$ be a given piecewise continuous function on $\bar{I}$. The initial value problem (IVP)*

$$Lu = f, \quad x \text{ in } I; \quad u(x_0) = \gamma_1, \quad u'(x_0) = \gamma_2, \ldots, u^{(p-1)}(x_0) = \gamma_p \qquad (3.1.2)$$

has one and only one classical solution.

REMARKS

1. We refer to (3.1.2) as the IVP with data $\{f(x); \gamma_1, \ldots, \gamma_p\}_{x_0}$. The subscript x_0 serves to remind us of the point where the initial values are specified.

2. The only solution of the completely homogeneous problem

$$Lu = 0, \quad x \text{ in } I; \quad u(x_0) = u'(x_0) = \cdots = u^{(p-1)}(x_0) = 0 \qquad (3.1.3)$$

is $u \equiv 0$; that is, the only solution of the IVP with data $\{0; 0, \ldots, 0\}_{x_0}$ is $u \equiv 0$.

3. The essential nature of the assumption $a_p(x) \neq 0$ is illustrated by the example

$$xu' - 2u = 0 \quad \text{on} -\infty < x < \infty, \quad u(0) = 0.$$

The function $u = Ax^2$ satisfies the condition $u(0) = 0$ and the differential equation for all x, so we clearly do not have uniqueness (in fact, it is even possible to use a function defined as Ax^2 for $x \geqslant 0$ and Bx^2 for $x < 0$ without $A = B$). Another difficulty is illustrated by the example

$$xu' + u = 0, \quad -\infty < x < \infty, \quad u(0) = \gamma \neq 0.$$

The general solution of the differential equation for $x \neq 0$ is $u = A/x$, and no solution can satisfy the initial condition (in fact, the solution is not even continuous at $x = 0$).

Linear Dependence; Wronskians

Consider n continuous functions $f_1(x), \ldots, f_n(x)$ defined on the interval I. These functions are said to be *dependent* over I if there exist constants $c_1, \ldots, c_n$, not all 0, such that

$$c_1 f_1(x) + \cdots + c_n f_n(x) \equiv 0 \quad \text{on } I. \qquad (3.1.4)$$

A set of n functions is therefore dependent if any one of them can be expressed as a linear combination of the others. A set of n functions is independent over I if (3.1.4) can be satisfied only if $c_1 = \cdots = c_n = 0$.

Let $f_1, \ldots, f_n$ have continuous derivatives of orders up to $n-1$. The *Wronskian* of $f_1, \ldots, f_n$ is the function of x defined as the $n \times n$ determinant

$$W(f_1, \ldots, f_n; x) = \begin{vmatrix} f_1 & f_2 & \cdots & f_n \\ f_1' & f_2' & \cdots & f_n' \\ \vdots & & & \\ f_1^{(n-1)} & f_2^{(n-1)} & \cdots & f_n^{(n-1)} \end{vmatrix}. \tag{3.1.5}$$

In particular, the Wronskian of f_1 and f_2 is

$$W(f_1, f_2; x) = f_1(x) f_2'(x) - f_1'(x) f_2(x). \tag{3.1.6}$$

If $f_1, \ldots, f_n$ are dependent over I, their Wronskian vanishes identically over I. The converse is not true, however, as the following example shows. Let I be the interval $-1 < x < 1$, and let $f_1 = x^2, f_2 = x|x|$; then f_1 and f_2 are independent over I, but their Wronskian vanishes identically.

Abel's Formula for the Wronskian

Let $u_1, \ldots, u_p$ be p solutions (independent or not) of the *homogeneous* equation $Lu = 0$. It can be shown (see Exercise 3.1.1) that there exists a constant C such that

$$W(u_1, \ldots, u_p; x) = Ce^{-m(x)}, \quad x \in I, \tag{3.1.7}$$

where $m(x)$ is a particular solution of $m' = a_{p-1}/a_p$. Formula (3.1.7) is known as *Abel's formula* for the Wronskian. If $p = 2$ and L is formally self-adjoint, then $a_2' = a_1$, so that (3.1.7) takes the simple form

$$W(u_1, u_2; x) = \frac{C}{a_2(x)}, \quad x \in I. \tag{3.1.8}$$

In both (3.1.7) and (3.1.8) the constant C is determined by the solutions $\{u_k\}$ used in forming the Wronskian. Different sets of solutions may yield different constants C. In particular, if $u_1, \ldots, u_p$ are dependent, $W \equiv 0$, so that $C = 0$. An immediate conclusion of (3.1.7) is that if $W = 0$ at a single point x_0, then $C = 0$ and hence $W \equiv 0$ on I. We are led to the following theorem.

Theorem 3.1.2. *Let $u_1, \ldots, u_p$ be solutions of $Lu = 0$. The necessary and sufficient condition for these p functions to be dependent is that their Wronskian vanishes at a single point x_0 in I.*

Proof. If $u_1, \ldots, u_p$ are dependent, it is clear that $W \equiv 0$ on I and hence surely vanishes at x_0. If $W(u_1, \ldots, u_p; x_0) = 0$, we have precisely the necessary and sufficient condition for the existence of a nontrivial solution $(c_1, \ldots, c_p)$ to the set of algebraic equations

$$\begin{aligned} c_1 u_1(x_0) \quad &+ \quad c_2 u_2(x_0) \quad + \quad \cdots \quad + \quad c_p u_p(x_0) &= 0, \\ &\vdots \\ c_1 u_1^{(p-1)}(x_0) \quad &+ \quad c_2 u_2^{(p-1)}(x_0) \quad + \quad \cdots \quad + \quad c_p u_p^{(p-1)}(x_0) &= 0. \end{aligned}$$

Let $(c_1, \ldots, c_p)$ be a nontrivial solution to this set of equations, and let $U(x) = c_1 u_1(x) + \cdots + c_p u_p(x)$. Clearly, U is a solution of the homogeneous differential equation satisfying the initial conditions $U(x_0) = U'(x_0) = \cdots = U^{(p-1)}(x_0) = 0$. By Theorem 3.1.1, $U \equiv 0$ and hence the set $\{u_1(x), \ldots, u_p(x)\}$ is dependent. □

Theorem 3.1.3. *Let $u_1(x), u_2(x), \ldots, u_p(x)$ be the respective solutions of the IVPs with data $\{0; 1, 0, \ldots, 0\}_{x_0}, \{0; 0, 1, 0, \ldots, 0\}_{x_0}, \ldots, \{0; 0, 0, \ldots, 0, 1\}_{x_0}$, that is to say, u_k is the solution of the homogeneous equation with 0 initial data at x_0 except that $u^{(k-1)}(x_0) = 1$. Then the set $(u_1, \ldots, u_p)$ is independent over I, and each solution of $Lu = 0$ can be written in the form $u = c_1 u_1 + \cdots + c_p u_p$ for some constants $c_1, \ldots, c_p$.*

Proof. The set $(u_1, \ldots, u_p)$ is independent since its Wronskian is equal to 1 at x_0. If $u(x)$ is any solution of $Lu = 0$, we can write $u(x) = u(x_0)u_1(x) + u'(x_0)u_2(x) + \cdots + u^{(p-1)}(x_0)u_p(x)$, which is just the desired form. □

REMARK. Any set of p independent solutions of $Lu = 0$ is called a *basis* for this equation. In Theorem 3.1.3 we exhibited a special, simple basis, but there are others. Let us look at the equation $u'' + u = 0$, which has the basis $\{\cos x, \sin x\}$ (which is of the type in Theorem 3.1.3 with $x_0 = 0, u_1 = \cos x, u_2 = \sin x$); but the functions $v_1(x) = \cos x + \sin x, v_2(x) = 2\cos x - 3\sin x$ also form a basis. Every solution of $u'' + u = 0$ can be expressed either as $Au_1 + Bu_2$ or as $Cv_1 + Dv_2$.

The Inhomogeneous Equation

Let $a_p(x) \neq 0$ on $-\infty < x < \infty$, and let $u_\xi(x)$ be the solution of the homogeneous equation satisfying the initial conditions

$$u_\xi(\xi) = 0, \quad u'_\xi(\xi) = 0, \ldots, u_\xi^{(p-2)}(\xi) = 0, \quad u_\xi^{(p-1)}(\xi) = 1/a_p(\xi).$$

We have already seen (at least in the case where the coefficients are infinitely differentiable) that $E(x, \xi) = H(x-\xi)u_\xi(x)$ is the causal fundamental solution satisfying, in the distributional sense, $LE = \delta(x - \xi)$ with $E \equiv 0, x < \xi$. Even when the coefficients in L are only continuous, we still define the causal fundamental solution as $H(x - \xi)u_\xi(x)$. By superposition one would then expect that the solution of

$$Lu = f, \quad x > a; \quad u(a) = 0, \ldots, u^{(p-1)}(a) = 0$$

would be given by

$$u(x) = \int_a^\infty H(x-\xi)u_\xi(x)f(\xi)\,d\xi = \int_a^x u_\xi(x)f(\xi)\,d\xi.$$

Let us verify that this is true. We have $u(a) = 0$ and $u'(x) = u_x(x)f(x) + \int_a^x u'_\xi(x)f(\xi)\,d\xi$, so that $u'(a) = 0$. Since $u_x(x) = \cdots = u_x^{(p-2)}(x) = 0$, we find that

$$\begin{aligned} u^{(k)}(x) &= u_x^{(k-1)}(x)f(x) + \int_a^x u_\xi^{(k)}(x)f(\xi)\,d\xi \\ &= \int_a^x u_\xi^{(k)}(x)f(\xi)\,d\xi, \quad k = 1, 2, \ldots, p-1. \end{aligned}$$

Thus $u^{(k)}(a) = 0$ for $k = 1, 2, \ldots, p-1$. We also note that

$$u^{(p)}(x) = u_x^{(p-1)}(x)f(x) + \int_a^x u_\xi^{(p)}(x)f(\xi)\,d\xi = \frac{f(x)}{a_p(x)} + \int_a^x u_\xi^{(p)}(x)f(\xi)\,d\xi.$$

Hence $Lu = a_p(x)u^{(p)} + \cdots + a_0(x)u = f(x) + \int_a^x Lu_\xi(x)f(\xi)d\xi = f(x)$, since $Lu_\xi = 0$. We observe that we have not used the fact that $x > a$. Therefore, we have proved the following theorem.

Theorem 3.1.4. *If $a_p(x) \neq 0$ in $-\infty < x < \infty$, the one and only solution of the IVP with data $\{f; 0, \ldots, 0\}_a$ is*

$$u(x) = \int_a^x u_\xi(x)f(\xi)\,d\xi, \tag{3.1.9}$$

where $u_\xi(x)$ is the solution of the IVP for the homogeneous equation

$$\left.\begin{aligned} &Lu = 0, \quad -\infty < x < \infty; \\ &u(\xi) = 0, \ldots, u^{(p-2)}(\xi) = 0, \\ &u^{(p-1)}(\xi) = \frac{1}{a_p(\xi)}. \end{aligned}\right\} \tag{3.1.10}$$

REMARKS

1. The solution of the problem

$$Lv = f; \quad v(a) = \gamma_1, \ldots, v^{(p-1)}(a) = \gamma_p \tag{3.1.11}$$

can be written as

$$v(x) = \int_a^x u_\xi(x)f(\xi)\,d\xi + \gamma_1 u_1(x) + \cdots + \gamma_p u_p(x), \tag{3.1.12}$$

where $(u_1, \ldots, u_p)$ is the basis of Theorem 3.1.3 (with $x_0 = a$).

2. We have reduced the problem of solving the inhomogeneous differential equation to that of finding the general solution of the homogeneous differential equation. Except for first-order equations and equations with constant coefficients, there is no systematic method for finding explicitly the solutions of the homogeneous equation. If one solution of the homogeneous equation is known, it is possible to reduce the order of the equation by 1. Thus, if we know a solution to a second-order equation, we can find all its solutions.

3. If a_p vanishes at a point c, (3.1.9) still provides the solution of the IVP in the interval $c < x < \infty$ (with $a > c$) and in the interval $-\infty < x < c$ (with $a < c$).

4. If the coefficients in L are constants, then $u_\xi(x) = u_0(x - \xi)$, where $u_0(x)$ is the solution of (3.1.10) with $\xi = 0$. Then (3.1.9) takes the form

$$u(x) = \int_a^x u_0(x - \xi) f(\xi)\, d\xi. \tag{3.1.13}$$

EXERCISES

3.1.1 Prove Abel's formula for the Wronskian. (*Hint*: First show that the derivative of a $p \times p$ determinant is the sum of p determinants, each of which has only one row differentiated.)

3.1.2 Consider the forced wave equation

$$\frac{\partial^2 u}{\partial t^2} - \frac{\partial^2 u}{\partial x^2} = f(x,t), \quad 0 < x < 1, \quad t > 0,$$

with the boundary conditions $u(0,t) = u(1,t) = 0, t > 0$, and the initial conditions $u(x,0) = (\partial u/\partial t)(x,0) = 0, 0 < x < 1$. Write $u(x,t) = \sum_{k=1}^{\infty} u_k(t) \sin k\pi x$, and show that u_k satisfies an IVP for an ordinary differential equation. Using (3.1.9), find $u_k(t)$ and hence $u(x,t)$. Consider the particular case $f(x,t) = \sin \omega t$, and discuss the behavior of the solution for large t (the case where $\omega = k\pi$ has to be treated separately).

3.1.3 The causal fundamental solution $E(t,\tau)$ is known as the *impulse response* in electrical engineering. E satisfies

$$LE = \delta(t - \tau), \quad E \equiv 0, \quad t < \tau.$$

The *step response* $F(t,\tau)$ is the solution of

$$LF = H(t - \tau), \quad F \equiv 0, \quad t < \tau. \tag{3.1.14}$$

(a) Show that $-\partial F/\partial \tau$ is a fundamental solution. Since $\partial F/\partial \tau = 0, t < \tau$, we must have $-\partial F/\partial \tau = E(t,\tau)$.

(b) Show that the solution of

$$Lu = f(t), \quad u(a) = u'(a) = \cdots = u^{(p-1)}(a) = 0$$

can be written in either of the forms

$$\begin{aligned} u(t) &= \int_a^t E(t,\tau) f(\tau)\,d\tau, \\ u(t) &= F(t,a)f(a) + \int_a^t F(t,\tau) f'(\tau)\,d\tau. \end{aligned} \tag{3.1.15}$$

(c) What simplifications are possible when the coefficients in L are constant?

3.2 BOUNDARY VALUE PROBLEMS FOR SECOND-ORDER EQUATIONS

Formulation

In the initial value problem for a second-order equation we specify as accessory conditions the values of u and u' at some point x_0. In a boundary value problem we impose two conditions involving the values of u and u' at the points a and b; we are then interested in solving the differential equation subject to these boundary conditions in the interval $a < x < b$ (which we take to be bounded). Unlike the IVP, the general boundary value problem may not have a solution, or may have more than one solution.

We shall consider the differential equation

$$Lu \doteq a_2(x)u'' + a_1(x)u' + a_0(x)u = f(x), \quad a < x < b, \tag{3.2.1}$$

where the coefficients are continuous in $a \leqslant x \leqslant b, a_2(x) \neq 0$ in $a \leqslant x \leqslant b$, and $f(x)$ is piecewise continuous in $a \leqslant x \leqslant b$. The solution $u(x)$ is required to satisfy the two boundary conditions

$$\begin{aligned} B_1 u &\doteq \alpha_{11}u(a) + \alpha_{12}u'(a) + \beta_{11}u(b) + \beta_{12}u'(b) = \gamma_1, \\ B_2 u &\doteq \alpha_{21}u(a) + \alpha_{22}u'(a) + \beta_{21}u(b) + \beta_{22}u'(b) = \gamma_2, \end{aligned} \tag{3.2.2}$$

where the row vectors $(\alpha_{11}, \alpha_{12}, \beta_{11}, \beta_{12})$ and $(\alpha_{21}, \alpha_{22}, \beta_{21}, \beta_{22})$ are independent (neither row is a multiple of the other).

We refer to B_1 and B_2 as *boundary functionals*, since they assign to each sufficiently smooth function $u(x)$ the numbers $B_1 u$ and $B_2 u$, respectively. The differential equation (3.2.1), together with the boundary conditions (3.2.2), forms a *boundary value problem* (BVP).

REMARKS

1. We regard the coefficients $a_i(x), \alpha_{ij}, \beta_{ij}$ as fixed (that is, L, B_1, B_2 are fixed), and we are interested in studying dependence of the solution on $f(x), \gamma_1, \gamma_2$.

We therefore refer to $\{f; \gamma_1, \gamma_2\}$ as the *data* for the problem. The coefficients and the data consist of *real* functions and *real* numbers.

2. The independence of the row vectors

$$(\alpha_{11}, \alpha_{12}, \beta_{11}, \beta_{12}) \quad \text{and} \quad (\alpha_{21}, \alpha_{22}, \beta_{21}, \beta_{22})$$

guarantees that we really have two distinct boundary conditions. (If the row vectors were dependent, one boundary functional would be a multiple of the other, and the two boundary conditions would be either identical or inconsistent.)

3. If $\beta_{11} = \beta_{12} = \alpha_{21} = \alpha_{22} = 0$, the boundary conditions are said to be *unmixed* (one condition per endpoint):

$$\begin{aligned} B_1 u &= \alpha_{11} u(a) + \alpha_{12} u'(a) = \gamma_1, \\ B_2 u &= \beta_{21} u(b) + \beta_{22} u'(b) = \gamma_2. \end{aligned} \tag{3.2.3}$$

If $\alpha_{12} = \beta_{11} = \beta_{12} = \alpha_{21} = \beta_{21} = \beta_{22} = 0, \alpha_{11} = 1, \alpha_{22} = 1$, we have the *initial conditions* $u(a) = \gamma_1, u'(a) = \gamma_2$.

4. Why do we confine ourselves to boundary conditions of type (3.2.2)? We do so partly because these occur most frequently in applications and partly because a general mathematical theory is most easily developed for conditions of this type. The reason for having two conditions instead of some other number is rather obvious; fewer conditions than the order of the equation would not determine the solution, and more conditions would usually prevent us from having any solution at all. We also exclude in (3.2.2) all derivatives of order greater than or equal to the order of the differential equation. The reason for this can perhaps be seen by studying the example $u'' = f(x)$; clearly, u'' is already determined at the endpoints by the differential equation, so that it would not make sense to specify u'' independently; higher derivatives of u are already known at the endpoints by differentiating the differential equation, so that it would be inappropriate to prescribe such a derivative.

The *superposition principle* applies to solutions of (3.2.1)–(3.2.2). If u satisfies the BVP with data $\{f; \gamma_1, \gamma_2\}$ and U the one with data $\{F; \Gamma_1, \Gamma_2\}$, then $Au + BU$ satisfies the BVP with data $\{Af + BF; A\gamma_1 + B\Gamma_1, A\gamma_2 + B\Gamma_2\}$. If u and v both satisfy the BVP with data $\{f; \gamma_1, \gamma_2\}$, then $u - v$ satisfies the BVP with data $\{0; 0, 0\}$, that is, the completely homogeneous problem. This leads to the important *uniqueness* condition: *If the BVP with data* $\{0; 0, 0\}$ *has only the solution* $u \equiv 0$, *then the BVP with data* $\{f; \gamma_1, \gamma_2\}$ *has at most one solution; if the BVP with data* $\{0; 0, 0\}$ *has a nontrivial solution, then the BVP with data* $\{f; \gamma_1, \gamma_2\}$ *either has no solution or has more than one solution.*

Example 1. The difficulties that can arise are illustrated by the simple BVP

$$-u'' = f(x), \quad 0 < x < 1, \quad u'(0) = \gamma_1, \quad -u'(1) = \gamma_2, \tag{3.2.4}$$

which represents steady one-dimensional heat flow in a rod with prescribed source density along the rod and prescribed heat flux at the ends. By integrating the differential equation from 0 to 1, we see immediately that $f(x), \gamma_1, \gamma_2$ must satisfy the relation

$$\int_0^1 f(x)\,dx = \gamma_2 + \gamma_1, \tag{3.2.5}$$

which merely states that a steady state is possible only if the heat supplied along the rod is removed at the ends. For the data $\{1; 0, 0\}$, (3.2.4) has *no* solution. However, for the data $\{\sin 2\pi x; 0, 0\}$, which satisfies (3.2.5), there are many solutions: $u(x) = A - (x/2\pi) + (1/4\pi^2)\sin 2\pi x$, with A arbitrary.

Example 2. As another illustration consider the BVP

$$u'' + a_0 u = f(x), \quad 0 < x < 1, \qquad u(0) = \gamma_1, \quad u(1) = \gamma_2, \tag{3.2.6}$$

where a_0 is a fixed constant.

(a) If a_0 is *not* one of the numbers $\pi^2, 4\pi^2, 9\pi^2, \ldots$, the BVP with data $\{0; 0, 0\}$ has only the trivial solution and therefore the BVP with data $\{f; \gamma_1, \gamma_2\}$ has at most one solution. We shall show later that there does in fact exist a solution for any data.

(b) If $a_0 = \pi^2$, say, the BVP with data $\{0; 0, 0\}$ has the nontrivial solutions $A \sin \pi x$, where A is an arbitrary constant. The BVP with data $\{1; 0, 0\}$ has *no solution*: Indeed, the general solution of the differential equation is then $u = (1/\pi^2) + A\sin \pi x + B \cos \pi x$, and the boundary conditions give $(1/\pi^2) + B = 0$ and $(1/\pi^2) - B = 0$, which are inconsistent. The BVP with data $\{x - \frac{1}{2}; 0, 0\}$ has *many solutions*: The general solution of the differential equation is $u = (x/\pi^2) - (1/2\pi^2) + A \sin \pi x + B \cos \pi x$, and the boundary conditions give $-(1/2\pi^2) + B = 0$ and $(1/2\pi^2) - B = 0$, so that $B = (1/2\pi^2)$ and A remains arbitrary.

Another way of stating what is happening is in terms of eigenvalues [see (1.2.18)]. If a_0 is not an eigenvalue λ_k of $u'' + \lambda u = 0, 0 < x < 1, u(0) = 0, u(1) = 0$, then (3.2.6) has at most one solution for any data $\{f; \gamma_1, \gamma_2\}$. If a_0 is an eigenvalue, then (3.2.6) either has no solution or has many solutions.

Green's Function and Its Uses

We assume that the completely homogeneous problem (*that is, with data* $\{0; 0, 0\}$) *has only the trivial solution.* We shall then show that problem (3.2.1)–(3.2.2) with data $\{f; \gamma_1, \gamma_2\}$ has one and only one solution. This will be done by an explicit construction using Green's function $g(x, \xi)$, which is the solution corresponding to the specific data $\{\delta(x-\xi); 0, 0\}$, where ξ is a fixed point in $a < x < b$. Of course, the forcing function $\delta(x-\xi)$ is not a piecewise continuous function, so that this problem does not fall directly in the classical category being studied. We regard $g(x,\xi)$ as the fundamental solution that satisfies the boundary conditions $B_1 g = 0, B_2 g = 0$. When the coefficients in L are infinitely differentiable, we found in Chapter 2 that a fundamental solution E satisfies the homogeneous equation for $x < \xi$ and $x > \xi$, that E is continuous at $x = \xi$, and that its first derivative has a jump $1/a_2(\xi)$ at $x = \xi$. Using this experience and the physical examples of Chapter 1 as a guide, it is natural to *define* Green's function $g(x,\xi)$ associated with BVP (3.2.1)–(3.2.2) as the solution of

$$\left.\begin{array}{ll} Lg = 0, \quad a < x < \xi, \xi < x < b; & B_1 g = 0, \quad B_2 g = 0, \\ g \text{ continuous at } x = \xi; & \left.\dfrac{dg}{dx}\right|_{x=\xi+} - \left.\dfrac{dg}{dx}\right|_{x=\xi-} = \dfrac{1}{a_2(\xi)}. \end{array}\right\} \tag{3.2.7}$$

Boundary value problem (3.2.7) can be written more succinctly in the delta function notation:

$$Lg = \delta(x-\xi), \quad a < x, \xi < b; \quad B_1 g = 0, \quad B_2 g = 0. \tag{3.2.8}$$

Thus Green's function is the response, under homogeneous boundary conditions, to a forcing function consisting of a concentrated unit of inhomogeneity at $x = \xi$. Since the difference between two solutions of (3.2.7) has a continuous derivative everywhere, it is a classical solution of the completely homogeneous problem which, by our assumption, necessarily vanishes identically. Therefore, (3.2.7) has at most one solution, which we now construct explicitly in different cases.

1. *Unmixed conditions.* Green's function is to satisfy the boundary conditions $B_1 g = \alpha_{11} g(a,\xi) + \alpha_{12} g'(a,\xi) = 0$ and $B_2 g = \beta_{21} g(b,\xi) + \beta_{22} g'(b,\xi) = 0$. Let $u_1(x)$ be a nontrivial solution of $Lu = 0$ satisfying $B_1 u = 0$. Such a solution must exist. One can, for instance, choose u_1 to be the solution of $Lu = 0$ with the initial conditions $u(a) = \alpha_{12}, u'(a) = -\alpha_{11}$; since α_{12} and α_{11} are not both $0, u_1(x)$ is not identically 0. Similarly, let $u_2(x)$ be a nontrivial solution of $Lu = 0$ satisfying the boundary condition at $x = b(B_2 u = 0)$. *We note that u_1 and u_2 are independent*, since by assumption the completely homogeneous problem has only the trivial solution. Green's function is therefore of the form

$$g(x,\xi) = Au_1(x), \quad a < x < \xi; \qquad g = Bu_2(x), \quad \xi < x < b,$$

where the constants A and B may, of course, depend on ξ. The continuity of g and the jump condition on g' at $x = \xi$ yield

$$\begin{aligned} Au_1(\xi) - Bu_2(\xi) &= 0, \\ -Au_1'(\xi) + Bu_2'(\xi) &= \frac{1}{a_2(\xi)}. \end{aligned}$$

This inhomogeneous system of two algebraic equations in A and B has one and only one solution if and only if the determinant of the coefficients does not vanish. This determinant is just the Wronskian of u_1 and u_2 evaluated at ξ. Since u_1 and u_2 are independent, their Wronskian does not vanish anywhere, so that we can solve for A and B to obtain

$$A = \frac{u_2(\xi)}{a_2(\xi)W(u_1, u_2; \xi)}, \qquad B = \frac{u_1(\xi)}{a_2(\xi)W(u_1, u_2; \xi)},$$

and, hence Green's function is given by

$$g(x, \xi) = \frac{u_1(x_<)u_2(x_>)}{a_2(\xi)W(u_1, u_2; \xi)}, \qquad a < x, \xi < b, \tag{3.2.9}$$

where $x_< = \min(x, \xi), x_> = \max(x, \xi)$. It is easy to verify that g as given by (3.2.9) actually satisfies (3.2.7). One can use Abel's formula (3.1.7) for the Wronskian to cast (3.2.9) in a somewhat simpler form. In the self-adjoint case, C is defined from (3.1.8) and

$$g(x, \xi) = \frac{1}{C} u_1(x_<)u_2(x_>). \tag{3.2.10}$$

2. *Initial conditions.* We have already seen that in this case Green's function (known as the causal fundamental solution) is given by

$$g(x, \xi) = H(x - \xi)u_\xi(x), \tag{3.2.11}$$

where $u_\xi(x)$ is the solution of $Lu = 0$ satisfying the conditions $u(\xi) = 0, u'(\xi) = 1/a_2(\xi)$. If the coefficients are constant, we have $g(x, \xi) = H(x - \xi)u_0(x - \xi)$.

3. *General boundary conditions.* If g is to satisfy mixed conditions $B_1 g = B_2 g = 0$, where B_1 and B_2 are the most general boundary functionals (3.2.2), we write g as the sum of the causal fundamental solution $H(x-\xi)u_\xi(x)$ and of a solution of the homogeneous equation to be determined. We already know that the causal solution satisfies (3.2.7) except for the boundary conditions. Let $u_1(x)$ be a nontrivial solution of $Lu = 0$ satisfying $B_1 u = 0$ (such a solution must exist!), and let $u_2(x)$ be a nontrivial solution of $Lu = 0$ with $B_2 u = 0$. Setting

$$g(x, \xi) = H(x - \xi)u_\xi(x) + Au_1(x) + Bu_2(x), \qquad a < x, \xi < b, \tag{3.2.12}$$

we impose the boundary conditions to find that

$$\begin{aligned} 0 &= \beta_{11} u_\xi(b) + \beta_{12} u'_\xi(b) + B(B_1 u_2), \\ 0 &= \beta_{21} u_\xi(b) + \beta_{22} u'_\xi(b) + A(B_2 u_1), \end{aligned}$$

from which we can solve for A and B (since neither $B_1 u_2$ nor $B_2 u_1$ can vanish).

Green's function just constructed enables us to solve (3.2.1)–(3.2.2). We begin with the case of homogeneous boundary conditions, that is, the data is $\{f(x); 0, 0\}$. We claim that the one and only solution of the BVP

$$Lu = f, \quad a < x < b; \quad B_1 u = 0, \quad B_2 u = 0 \tag{3.2.13}$$

is given by

$$u(x) = \int_a^b g(x, \xi) f(\xi) \, d\xi. \tag{3.2.14}$$

The proof is easy. We note first that u satisfies the homogeneous boundary conditions because g does. Since g can be written in the form (3.2.12), where A and B depend only on ξ, it follows that the right side of (3.2.14) reduces to

$$\int_a^x u_\xi(x) f(\xi) \, d\xi + \theta_1 u_1(x) + \theta_2 u_2(x),$$

where θ_1, θ_2 are constants. The first term satisfies the inhomogeneous differential equation (Theorem 3.1.4), while the remaining terms are solutions of the homogeneous equation. The sum therefore satisfies $Lu = f$. Uniqueness follows from the assumption that the BVP with data $\{0; 0, 0\}$ has only the trivial solution. Thus, when the completely homogeneous BVP has only the trivial solution, (3.2.13) has one and only one solution given by (3.2.14). We have thus "inverted" (3.2.13) through formula (3.2.14), which involves integration. We often write (3.2.14) as

$$u = Gf, \tag{3.2.14a}$$

where G is an integral operator which inverts the differential operator L restricted to the set M of functions in $C^2(a, b)$ satisfying the homogeneous boundary conditions, $B_1 u = B_2 u = 0$. Note that the set M is a *linear manifold*; that is, if $u_1(x)$ and $u_2(x)$ are in M, so is $\alpha u_1(x) + \beta u_2(x)$ for any constants α and β. The operator G is a linear operator (this is just the superposition principle): $G(\alpha f_1 + \beta f_2) = \alpha G f_1 + \beta G f_2$.

Turning next to inhomogeneous boundary conditions we find that the problem with data $\{0; \gamma_1, \gamma_2\}$ has the unique solution

$$v(x) = \frac{\gamma_2}{B_2 u_1} u_1(x) + \frac{\gamma_1}{B_1 u_2} u_2(x),$$

where u_1 is a nontrivial solution of the homogeneous equation satisfying $B_1 u_1 = 0$, and u_2 is a nontrivial solution of the same equation with $B_2 u_2 = 0$. Therefore, by superposition, the problem

$$Lu = f, \quad a < x < b; \quad B_1 u = \gamma_1, \quad B_2 u = \gamma_2 \tag{3.2.15}$$

has the unique solution

$$u = \int_a^b g(x, \xi) f(\xi) \, d\xi + \frac{\gamma_2}{B_2 u_1} u_1(x) + \frac{\gamma_1}{B_1 u_2} u_2(x). \tag{3.2.16}$$

To recapitulate, we have the following theorem.

Theorem 3.2.1. *If the completely homogeneous BVP has only the trivial solution, then the BVP with data $\{f; \gamma_1, \gamma_2\}$ has one and only one solution, given by (3.2.16).*

REMARKS

1. The theorem holds because we have exactly two boundary conditions of type (3.2.2). If we imposed a third condition of the same type, the completely homogeneous BVP would still have only the trivial solution, but then the inhomogeneous problem would usually have no solution.

2. The relation between (3.2.16) and (3.2.15) can also be expressed in terms of direct and inverse operators. Introduce an operator $\tilde{L}$ which transforms any function $u(x)$ in $C^2(a,b)$ into the triple $(Lu; B_1u, B_2u)$ consisting of the function Lu and the two numbers B_1u and B_2u. The BVP (3.2.15) then takes the form $\tilde{L}u = (f(x); \gamma_1, \gamma_2)$ and its solution can be written as $u = \tilde{L}^{-1}(f(x); \gamma_1, \gamma_2)$, where $\tilde{L}^{-1}$ transforms the data $(f(x); \gamma_1, \gamma_2)$ into the element $u(x)$ in $C^2(a,b)$ given by (3.2.16). Note that both $\tilde{L}$ and $\tilde{L}^{-1}$ are linear operators in this description.

The Adjoint Problem

As we saw in Section 2.5, each second-order operator with coefficients in $C^2(a,b)$,

$$L = a_2(x)D^2 + a_1(x)D + a_0(x), \tag{3.2.17}$$

has a well-defined formal adjoint

$$L^* = a_2D^2 + (2a_2' - a_1)D + (a_2'' - a_1' + a_0) \tag{3.2.18}$$

such that for any pair of functions u, v in $C^2(a,b)$,

$$\int_a^b (vLu - uL^*v)\,dx = J(u,v)]_a^b, \tag{3.2.19}$$

where

$$J(u,v) = a_2(vu' - uv') + (a_1 - a_2')uv. \tag{3.2.20}$$

A formally self-adjoint operator is one for which $L = L^*$, which is equivalent to $a_2' = a_1$, so that

$$L = L^* = D(a_2D) + a_0, \quad J = a_2(vu' - uv'), \tag{3.2.21}$$

leading to a simplification of (3.2.19).

We now wish to introduce the notion of adjoint boundary conditions. Given the operator (3.2.17) and a function u satisfying two homogeneous conditions $B_1u = 0$ and $B_2u = 0$ of type (3.2.2), the right side of (3.2.19) will vanish only if v itself satisfies a pair of homogeneous boundary conditions—the so-called *adjoint boundary conditions*. To make matters precise, let M be the set of functions $u(x)$ in $C^2(a,b)$ satisfying $B_1u = 0$ and $B_2u = 0$. As already noted, M is a linear manifold. The set M^* is defined as the set of functions $v(x)$ in $C^2(a,b)$ such that $J(u,v)]_a^b = 0$

for all $u \in M$. It is clear that M^* is a linear manifold. Functions in M^* can be characterized by two homogeneous boundary conditions, $B_1^* v = 0$ and $B_2^* v = 0$, where B_1^* and B_2^* are boundary functionals of type (3.2.2) but usually with different coefficients from those of B_1 and B_2. Although M^* is unambiguously determined from M, the specific functionals B_1^*, B_2^* are not (for instance, we can exchange the order of the boundary functionals, or we could consider other equivalent functionals such as $B_1^{*'} = B_1^* + B_2^*, B_2^{*'} = B_1^* - B_2^*$). This leads to the definition of a self-adjoint problem.

Definition. *We say that a boundary value problem* (L, B_1, B_2) *is* self-adjoint *if* $L^* = L$ *and* $M^* = M$ *(that is, the adjoint boundary conditions define the same set of functions as the homogeneous boundary conditions* $B_1 u = B_2 u = 0$*, so that it is possible to choose* B_1^* *and* B_2^* *with* $B_1^* = B_1, B_2^* = B_2$*).*

We now illustrate these ideas through some examples.

Example 3. $L = D^2, a = 0, b = 1, B_1 u = u'(0) - u(1), B_2 u = u'(1)$. Then $L^* = L$, and Green's formula becomes

$$\int_0^1 (vu'' - uv'')\,dx = v(1)u'(1) - u(1)v'(1) - v(0)u'(0) + u(0)v'(0). \quad (3.2.22)$$

The set M consists of all functions u such that $u(1) = u'(0)$ and $u'(1) = 0$. We want to find the set M^* of all functions v such that the right side of (3.2.22) vanishes whenever u is in M. For u in M the right side of (3.2.22) simplifies to

$$-u(1)[v(0) + v'(1)] + u(0)v'(0),$$

where $u(0)$ and $u(1)$ are arbitrary. Thus the set M^* consists of the functions v for which $v'(0) = 0$ and $v(0) + v'(1) = 0$. The adjoint boundary conditions can be taken as $B_1^* v = v'(0) = 0, B_2^* v = v(0) + v'(1) = 0$. [One could just as well characterize the same set M^* by the conditions $v'(0) + v(0) + v'(1) = 0$, $2v'(0) - v(0) - v'(1) = 0$, or by any pair of conditions obtained by taking independent linear combinations of B_1^* and B_2^*.]

Example 4. Let L be formally self-adjoint, $L = D(a_2 D) + a_0$, with *unmixed* boundary functionals, $B_1 u = \alpha_{11} u(a) + \alpha_{12} u'(a), B_2 u = \beta_{21} u(b) + \beta_{22} u'(b)$. Then $M^* = M$ (see Exercise 3.2.2). This is a typical self-adjoint BVP.

Example 5. $L = a_2 D^2 + a_1 D + a_0, a = 0, b = 1, B_1 u = u(0), B_2 u = u'(1)$. Then M is the set of functions u with $u(0) = u'(1) = 0$. For u in M,

$$J(u, v)]_0^1 = \{[a_1(1) - a_2'(1)]v(1) - a_2(1)v'(1)\}u(1) - a_2(0)v(0)u'(0).$$

Since $u(1)$ and $u'(0)$ are arbitrary, M^* is characterized by

$$v(0) = 0, \quad [a_1(1) - a_2'(1)]v(1) - a_2(1)v'(1) = 0.$$

We see that M^* coincides with M only if $a_2'(1) = a_1(1)$, which will certainly be the case if $L = L^*$.

Example 6. Initial value problem. Whether or not L is formally self-adjoint, we always have $M^* \neq M$. Take $L = a_2 D^2 + a_1 D + a_0$ on an interval $a < x < b$ with $B_1 u = u(a), B_2(u) = u'(a)$. Then M is the set of functions u with $u(a) = u'(a) = 0$. For u in M,

$$J(u, v)]_a^b = \{[a_1(b) - a_2'(b)]v(b) - a_2(b)v'(b)\}u(b) + a_2(b)v(b)u'(b),$$

and, since $u(b)$ and $u'(b)$ are arbitrary, M^* consists of functions v such that $v(b) = v'(b) = 0$. Thus the adjoint boundary conditions never coincide with the original boundary conditions.

A slightly different approach to adjoints is taken in Section 3.3. Let us now look at some consequences of the ideas just introduced.

Consider the BVP

$$Lg(x, \xi) = \delta(x - \xi), \quad a < x, \xi < b; \quad B_1 g = 0, \quad B_2 g = 0, \tag{3.2.23}$$

whose solution will be called the *direct Green's function*. Let $g^*(x, \xi)$ be the *adjoint Green's function* satisfying

$$L^* g^*(x, \xi) = \delta(x - \xi), \quad a < x, \xi < b; \quad B_1^* g^* = 0, \quad B_2^* g^* = 0. \tag{3.2.24}$$

As usual, all differentiations in (3.2.23) and (3.2.24) are with respect to x. We shall show that g and g^* are related by

$$g^*(x, \xi) = g(\xi, x). \tag{3.2.25}$$

Once (3.2.25) has been established, there is no further need to solve (3.2.24). Its solution is just the solution of (3.2.23) with variables interchanged. Another way of saying this is that $g(x, \xi)$ satisfies the direct problem in its first variable and the adjoint problem in its second variable; thus $g(x, \xi)$ satisfies the adjoint boundary conditions in the variable ξ.

To prove (3.2.25), set $\xi = \eta$ in (3.2.24), multiply (3.2.23) by $g^*(x, \eta)$ and (3.2.24) by $g(x, \xi)$, subtract, and integrate from $x = a$ to $x = b$. Using the formal properties of the delta function, we find that

$$\int_a^b [g^*(x, \eta) Lg(x, \xi) - g(x, \xi) L^* g^*(x, \eta)]\, dx = g^*(\xi, \eta) - g(\eta, \xi),$$

or, by Green's formula,

$$J(g^*, g)]_a^b = g^*(\xi, \eta) - g(\eta, \xi).$$

Since g^* satisfies the adjoint homogeneous boundary conditions, $J(g^*, g)]_a^b = 0$ (in fact, that is exactly how adjoint boundary conditions were defined!). This then proves (3.2.25). As an important corollary we note that if (L, B_1, B_2) is *self-adjoint*, then $g(x, \xi)$ is symmetric, that is,

$$g(x, \xi) = g(\xi, x). \tag{3.2.26}$$

Relation (3.2.26) is also known as the *reciprocity principle*: The response at x caused by a unit source at ξ is the same as the response at ξ due to a unit source at x.

One can use the fact that $g(\xi, x)$ satisfies (3.2.24) to write the solution of a problem with nonzero boundary data in terms of Green's function. Consider first

$$Lu = 0, \quad a < x < b; \qquad B_1 u = \gamma_1, \quad B_2 u = \gamma_2. \tag{3.2.27}$$

Multiply (3.2.27) by $g(\xi, x)$ and (3.2.24) by u, subtract, and integrate to obtain

$$-u(\xi) = \int_a^b [g(\xi, x) Lu - uL^* g(\xi, x)]\, dx = J(u(x), g(\xi, x))]_{x=a}^{x=b},$$

or, on relabeling variables,

$$u(x) = -J(u(\xi), g(x, \xi))]_{\xi=a}^{\xi=b}. \tag{3.2.28}$$

The term on the right side of (3.2.28) involves the direct Green's function $g(x, \xi)$, *its derivative with respect to* ξ, and the given boundary data γ_1, γ_2. The last part of this statement is not entirely obvious but is illustrated in the following example and perhaps more convincingly in Section 3.3.

Example 7. Consider the problem

$$u'' = 0, \quad 0 < x < 1; \qquad B_1 u = u'(0) - u(1) = \gamma_1, \quad B_2 u = u'(1) = \gamma_2,$$

which is based on Example 3. We must first calculate $g(x, \xi)$, the solution of $Lg = \delta(x - \xi), B_1 g = 0, B_2 g = 0$. By (3.2.12) we have

$$g(x, \xi) = (x - \xi) H(x - \xi) - x + \xi - 1,$$

so that $g(x, 0) = -1, g(x, 1) = -x, (dg/d\xi)(x, 0) = 0, (dg/d\xi)(x, 1) = 1$, and by (3.2.28),

$$u(x) = -\left[g(x, \xi) u'(\xi) - u(\xi) \frac{dg}{d\xi}(x, \xi) \right]_{\xi=0}^{\xi=1} = x\gamma_2 - \gamma_1.$$

Now, by superposition of (3.2.13) and (3.2.27), we can express the solution of the competely inhomogeneous problem (3.2.15) as the sum of (3.2.14) and (3.2.28):

$$u(x) = \int_a^b g(x, \xi) f(\xi)\, d\xi - J(u(\xi), g(x, \xi))]_{\xi=a}^{\xi=b}, \tag{3.2.29}$$

which should be compared with the earlier form (3.2.16). For ordinary differential equations, (3.2.16) might be regarded as simpler than (3.2.29), but when dealing with partial differential equations, nothing like (3.2.16) is available, whereas (3.2.29) survives with some obvious modifications.

EXERCISES

3.2.1 *Extension of the superposition principle.* Suppose that $Lu = 0$, $a < x < b$, $B_1 u = B_2 u = 0$ has only the trivial solution.

1. Let θ be a real parameter. The BVP

$$Lu = f(x,\theta), \quad a < x < b; \qquad B_1 u = \gamma_1(\theta), \quad B_2 u = \gamma_2(\theta)$$

has one and only one solution $u(x,\theta)$. Show that the solution of

$$Lv = \frac{\partial f}{\partial \theta}, \quad a < x < b; \qquad B_1 v = \frac{d\gamma_1}{d\theta}, \quad B_2 v = \frac{d\gamma_2}{d\theta}$$

is given by $v(x,\theta) = (\partial/\partial\theta)u(x,\theta)$.

2. Show that the solution $h(x,\xi)$ of

$$Lh = \delta'(x-\xi), \quad a < x, \xi < b; \qquad B_1 h = B_2 h = 0,$$

is given by $h = -\partial g/\partial \xi$, where g is Green's function. Find the solution of

$$-h'' = \delta'(x-\xi), \quad 0 < x, \xi < 1, \qquad h(0,\xi) = h(1,\xi) = 0.$$

3.2.2 Let $L = L^*$, and let the boundary functionals be of type (3.2.2). Show that the necessary and sufficient condition for self-adjointness is $a_2(a)P_{34} = a_2(b)P_{12}$, where

$$P_{12} = \begin{vmatrix} \alpha_{11} & \alpha_{12} \\ \alpha_{21} & \alpha_{22} \end{vmatrix}, \quad P_{34} = \begin{vmatrix} \beta_{11} & \beta_{12} \\ \beta_{21} & \beta_{22} \end{vmatrix}.$$

This condition is satisfied for unmixed boundary conditions and for the "periodic" boundary conditions $u(a) = u(b), a_2(a)u'(a) = a_2(b)u'(b)$.

3.2.3 Let $L = D^2 + 4D - 3$. Find L^* and J. If the boundary functionals are $B_1 u = u'(a) + 4u(a)$ and $B_2 u = u'(b) + 4u(b)$, find B_1^* and B_2^*.

3.2.4 Let $0 < a < b$; consider the BVP

$$(xg')' + \left(x - \frac{\nu^2}{x}\right) g = \delta(x-\xi), \quad a < x, \xi < b; \qquad g(a,\xi) = g(b,\xi) = 0.$$

Using information about Bessel functions, determine $g(x,\xi)$. Do the same for the problem

$$(xg')' - \left(x + \frac{\nu^2}{x}\right) g = \delta(x-\xi), \quad a < x, \xi < b; \qquad g(a,\xi) = g(b,\xi) = 0.$$

Are there any values of ν^2 for which either construction fails?

3.2.5 Find Green's function $g(x,\xi)$ of $L = D^2$ subject to the conditions $u'(0) = u(1), u'(1) = 0$. Then $L^* = L$, and the adjoint boundary conditions were found in Example 3. Construct Green's function $g^*(x,\xi)$ for the adjoint problem, and show, by comparing explicit formulas, that $g^*(x,\xi) = g(\xi,x)$.

3.2.6 Let $L = D^2$ on $0 < x < 1$, and consider a problem with only one boundary functional $B_1(u) = u(0)$. Show that the adjoint homogeneous problem has three boundary conditions.

3.2.7 A boundary condition of type (3.2.2) specifies the value of a linear functional of a particular type. One can, of course, consider other kinds of linear functionals, such as $\int_a^b u(x)h(x)\,dx$, where $h(x)$ is a given function. Consider the BVP

$$-u'' = f, \quad -1 < x < 1;$$
$$B_1 u \doteq \int_{-1}^{1} xu(x)\,dx = \gamma_1, \qquad B_2 u \doteq \int_{-1}^{1} u(x)\,dx = \gamma_2. \tag{3.2.30}$$

Find the corresponding Green's function, and then solve (3.2.30).

3.3 BOUNDARY VALUE PROBLEMS FOR EQUATIONS OF ORDER P

Again we let $a \leqslant x \leqslant b$ be a bounded interval on the x axis. The linear operator $L = a_p(x)D^p + \cdots + a_1(x)D + a_0(x)$ has coefficients that belong to C^p on the closed interval $a \leqslant x \leqslant b$, and $a_p(x) \neq 0$ on $a \leqslant x \leqslant b$. We consider the boundary value problem

$$Lu \doteq a_p u^{(p)} + \cdots + a_1 u' + a_0 u = f, \quad a < x < b, \tag{3.3.1}$$

with the p boundary conditions

$$B_1 u \doteq \alpha_{11}u(a) + \cdots + \alpha_{1p}u^{(p-1)}(a) + \beta_{11}u(b) + \cdots + \beta_{1p}u^{(p-1)}(b) = \gamma_1,$$
$$\vdots$$
$$B_p u \doteq \alpha_{p1}u(a) + \cdots + \beta_{pp}u^{(p-1)}(a) + \beta_{p1}u(b) + \cdots + \beta_{pp}u^{(p-1)}(b) = \gamma_p. \tag{3.3.2}$$

The p row vectors $(\alpha_{11},\ldots,\alpha_{1p},\beta_{11},\ldots,\beta_{1p})$, ..., $(\alpha_{p1},\ldots,\alpha_{pp},\beta_{p1},\ldots,\beta_{pp})$ are assumed independent (no one of them is a linear combination of the others; in particular, no row vector has all entries $= 0$). The differential operator L and the boundary functionals $B_1,\ldots,B_p$ are fixed, and we are interested in studying the dependence of the solution u of (3.3.1)–(3.3.2) on the data $\{f;\gamma_1,\ldots,\gamma_p\}$. The problem with $f = 0, \gamma_1 = \cdots = \gamma_p = 0$ is called the completely homogeneous problem. If the completely homogeneous problem has only the trivial solution, (3.3.1)–(3.3.2) has *at most one solution* (and it turns out that a solution actually exists for any data).

If the completely homogeneous problem has nontrivial solutions, (3.3.1)–(3.3.2) either has *no solution* or has *more than one solution.*

Assume next that the completely homogeneous problem

$$Lu = 0, \quad a < x < b; \qquad B_1 u = \cdots = B_p u = 0 \tag{3.3.3}$$

has *only the trivial solution.* We can then construct Green's function $g(x, \xi)$, necessarily unique, associated with $(L, B_1, \ldots, B_p)$. This Green's function satisfies

$$Lg = \delta(x - \xi), \quad a < x, \xi < b; \qquad B_1 g = \cdots = B_p g = 0, \tag{3.3.4}$$

or, equivalently,

$$\left.\begin{array}{l} Lg = 0, \quad a < x < \xi, \quad \xi < x < b; \quad B_1 g = \cdots = B_p g = 0, \\ g, \ldots, g^{(p-2)} \text{continuous}; \quad g^{(p-1)}(\xi+, \xi) - g^{(p-1)}(\xi-, \xi) = \dfrac{1}{a_p(\xi)}. \end{array}\right\} \tag{3.3.5}$$

If u and v are any functions in C^p, Green's formula gives [see (2.5.18)]

$$\int_a^b (vLu - uL^* v)\, dx = J(u, v)]_a^b, \ J = \sum_{m=1}^{p} \sum_{j+k=m-1} (-1)^k D^k (a_m v) D^j u. \tag{3.3.6}$$

We observe that $J(u, v)]_a^b$ can be written as the sum of $2p$ terms

$$u(a) A_{2p} v + \cdots + u^{(p-1)}(a) A_{p+1} v + u(b) A_p v + \cdots + u^{(p-1)}(b) A_1 v, \tag{3.3.7}$$

where each A_k is a linear combination of the $2p$ quantities $v(a), \ldots, v^{(p-1)}(b)$. If we are given p independent boundary functionals $B_1, \ldots, B_p$, we can rewrite (3.3.7) so as to feature the p quantities $B_1 u, \ldots, B_p u$ instead of the $2p$ quantities $u(a), \ldots, u^{(p-1)}(b)$. This suggests introducing p additional boundary (complementary) functionals $B_{p+1}, \ldots, B_{2p}$, so that $B_1, \ldots, B_{2p}$ is a set of $2p$ independent boundary functionals; then (3.3.7) can be rewritten as

$$\begin{aligned} J(u, v)]_a^b &= (B_1 u)(B_{2p}^* v) + \cdots + (B_p u)(B_{p+1}^* v) \\ &\quad + (B_{p+1} u)(B_p^* v) + \cdots + (B_{2p} u)(B_1^* v). \end{aligned} \tag{3.3.8}$$

Although $B_{p+1}, \ldots, B_{2p}$ can be introduced in many different ways, all turn out to be ultimately equivalent for our purposes. Thus we regard $B_{p+1}, \ldots, B_{2p}$ as fixed, and therefore $B_1^*, \ldots, B_{2p}^*$ are unambiguously determined in (3.3.8). The p boundary functionals $B_1^*, \ldots, B_p^*$ are said to be *adjoint* to $B_1, \ldots, B_p$. Given p homogeneous boundary conditions $B_1 u = \cdots = B_p u = 0$, we see from (3.3.8) that $J(u, v)]_a^b$ will vanish if and only if v satisfies the p adjoint conditions $B_1^* v = \cdots = B_p^* v = 0$. If $L = L^*$ and if the adjoint conditions define the same set of functions as $B_1 u = \cdots = B_p u = 0$, then $(L, B_1, \ldots, B_p)$ is said to be *self-adjoint.*

Example 1. Consider the first-order operator $L = D$. A typical problem of form (3.3.1)–(3.3.2) would be

$$\frac{du}{dx} = f, \quad a < x < b; \quad B_1 u \doteq u(a) + \beta u(b) = \gamma.$$

Here $L^* = -D$, and Green's formula for any u, v in C^1 becomes

$$\int_a^b \left(v\frac{du}{dx} + u\frac{dv}{dx} \right) dx = uv]_a^b = J(u,v)]_a^b = u(b)v(b) - u(a)v(a). \tag{3.3.9}$$

With $B_1 u$ defined above, we may introduce the complementary boundary functional $B_2 u \doteq u(b)$, in terms of which we can write

$$u(b)v(b) - u(a)v(a) = [u(a) + \beta u(b)][-v(a)] + u(b)[v(b) + \beta v(a)].$$

Thus $B_2^* v = -v(a), B_1^* v = v(b) + \beta v(a)$. Therefore, the adjoint condition is $v(b)+\beta v(a) = 0$. Thus, if u satisfies $B_1 u = 0$, the necessary and sufficient condition for $J(u,v)]_a^b = 0$ is that v satisfies $B_1^* v = 0$.

Example 2. Consider the fourth-order operator $L = D^2(b_2 D^2) + D(b_1 D) + b_0$, where $b_0(x), b_1(x), b_2(x)$ are arbitrary functions. Then $L = L^*$, and

$$\int_a^b (vLu - uL^*v)\, dx = v(b_2 u'')' - v' b_2 u'' + b_2 v'' u' - u(b_2 v'')' + b_1(vu' - uv')]_a^b.$$

If $B_1 u = u(a), B_2 u = u''(a), B_3 u = u(b), B_4 u = u''(b)$, then (L, B_1, B_2, B_3, B_4) is self-adjoint. This is also true for the case $B_1 u = u(a), B_2 u = u'(a), B_3 u = u(b), B_4 u = u'(b)$. But if we have three conditions at a and one condition at b, the problem is not self-adjoint.

The adjoint Green's function $g^*(x,\xi)$ satisfies

$$L^* g^* = \delta(x - \xi), \quad a < x, \xi < b; \quad B_1^* g^* = \cdots = B_p^* g^* = 0, \tag{3.3.10}$$

and we can once more show that $g^*(x,\xi) = g(\xi, x)$. To find the solution of (3.3.1)–(3.3.2), multiply (3.3.1) by $g(\xi, x)$ and (3.3.10) by $u(x)$, subtract, and integrate to obtain

$$\int_a^b [g(\xi,x)Lu - uL^* g(\xi,x)]\, dx = -u(\xi) + \int_a^b g(\xi,x) f(x)\, dx$$

or, changing the roles of x and ξ,

$$u(x) = \int_a^b g(x,\xi) f(\xi)\, d\xi - J(u(\xi), g(x,\xi))]_{\xi=a}^{\xi=b}, \tag{3.3.11}$$

where [say, by using (3.3.8)] we have that J can be expressed in terms of $\gamma_1, \ldots, \gamma_p$ and $g, \ldots, d^{p-1}g/d\xi^{p-1}$.

The integral term on the right side of (3.3.11) provides the solution of (3.3.1) with homogeneous boundary conditions and can be written as Gf, where G is a linear integral operator, the inverse of the differential operator L on the manifold M of functions satisfying the homogeneous conditions $B_1u = \cdots = B_pu = 0$.

The complete formula (3.3.11) gives the solution of the fully inhomogeneous problem (3.3.1), (3.3.2). It must be emphasized that this solution is derived on the assumption that the completely homogeneous problem (3.3.3) has only the trivial solution and that there are precisely p boundary conditions of type (3.3.2). In the next two sections we discuss what happens when (3.3.3) has nontrivial solutions or when the number of boundary conditions is not the same as the order of the equation.

A complete treatment of BVPs for equations of any order can be found in Coddington and Levinson [2] or in Naimark [4].

EXERCISES

3.3.1 *Adjoints for unbalanced problems.* For an operator of order p there are at most $2p$ independent conditions of type (3.3.2). Consider the problem

$$Lu = f, \quad a < x < b; \quad B_1u = \gamma_1, \ldots, B_mu = \gamma_m,$$

where L is the general differential operator of order p, $B_1, \ldots, B_m$ are independent boundary functionals of type (3.3.2), and $0 \leqslant m \leqslant 2p$. If $m \neq p$, the problem $(L, B_1, \ldots, B_m)$ is unbalanced. We can still define the adjoint of $(L, B_1, \ldots, B_m)$ as $(L^*, B_1^*, \ldots, B_{p-m}^*)$; equivalently, if u satisfies $B_1u = 0, \ldots, B_mu = 0$, the conditions $B_1^*v = 0, \ldots, B_{p-m}^*v = 0$ are just what is required to make $J(u,v)]_a^b$ vanish. As an example, if $L = D$ on $0 < x < 1$ and B_1, B_2 are defined by $B_1u = u(0), B_2u = u(1)$, we clearly have "too many" boundary functionals; the adjoint consists of $L^* = -D$ and *no* boundary functional. If we consider $L = D^2$ on $0 < x < 1$ with only one boundary functional, defined by $B_1u = u(0)$, the adjoint consists of $L^* = D^2$ with B_1^*, B_2^*, B_3^* defined by $B_1^*v = v(1)$, $B_2^*v = v'(1)$, $B_3^*v = v(0)$. Using these ideas, find the adjoint of $L = D^4$ on $0 < x < 1$, with B_1, B_2, B_3, B_4, B_5 defined from $B_1u = u(0)$, $B_2u = u(1)$, $B_3u = u''(0)$, $B_4u = u''(1)$, $B_5u = u'(1) - u'(0)$.

3.3.2 Consider the fourth-order operator $L = D^4$ on $0 < x < 1$, and let the boundary functionals be $B_1u = u(0), B_2u = u^{(3)}(0), B_3(u) = u(1), B_4(u) = u''(1)$. Find the adjoint boundary conditions, and show by explicit calculation that the direct Green's function is not symmetric.

3.3.3 Construct Green's function for $L = D^3$ on $0 < x < 1$ with $g'(0,\xi) = g(1,\xi) = g''(1,\xi) = 0$. Express the solution of

$$u^{(3)} = f(x), \quad 0 < x < 1; \quad u'(0) = u(1) = u''(1) = 0,$$

in terms of Green's function. What are the adjoint BCs?

3.4 ALTERNATIVE THEOREMS

Consider the boundary value problem (3.3.1)–(3.3.2) for a pth-order operator:

$$Lu = f, \quad a < x < b; \quad B_1 u = \gamma_1, \ldots, B_p u = \gamma_p. \tag{3.4.1}$$

For boundary conditions of the type (3.3.2), we saw that (3.4.1) has one and only one solution if the corresponding completely homogeneous problem

$$Lu = 0, \quad a < x < b; \quad B_1 u = 0, \ldots, B_p u = 0 \tag{3.4.2}$$

has only the trivial solution. If (3.4.2) has nontrivial solutions, (3.4.1) usually has no solution unless the data $\{f; \gamma_1, \ldots, \gamma_p\}$ is of a particular type. To characterize the data for which solutions to (3.4.1) exist, we need to consider the related homogeneous adjoint problem

$$L^* v = 0, \quad a < x < b; \quad B_1^* v = 0, \ldots, B_p^* v = 0. \tag{3.4.3}$$

The situation bears considerable resemblance to that encountered for systems of algebraic equations, and we begin our discussion with such systems.

Alternative Theorem for a System of m Equations in n Unknowns

When does a set of m linear equations in n unknowns have a solution? If $m > n$, the system is *overdetermined* and typically there is no solution; if $m < n$, the system is *underdetermined* and one expects many solutions; if $m = n$, the system is *balanced* and usually there is exactly one solution (but if the determinant of the coefficients vanishes there will either be no solution or many solutions). We are interested in studying the exceptional cases as well as the ordinary ones. We shall find a criterion which tells us precisely when solution(s) exist.

We write the set of m equations in n unknowns either in its component form

$$\sum_{j=1}^{n} a_{ij} u_j = f_i, \quad i = 1, \ldots, m, \tag{3.4.4a}$$

or as a single equation

$$Au = f, \tag{3.4.4b}$$

where A is the $m \times n$ matrix of the coefficients $(a_{ij}), f = (f_1, \ldots, f_m)$ is an m-dimensional vector, and $u = (u_1, \ldots, u_n)$ an n-dimensional vector. The entries of A and the components of f are given as *real* numbers and we are seeking a solution u with real components. Thus f is in $\mathbb{R}^m$, u in $\mathbb{R}^n$, and A is a transformation from $\mathbb{R}^n$ to $\mathbb{R}^m$. (We write $A\colon \mathbb{R}^n \to \mathbb{R}^m$ to indicate such a transformation.)

In order to proceed we need to review some linear algebra. A basic notion is that of *inner product* or *dot product*. The inner product of a pair of vectors w and z

belonging to the same p-dimensional space $\mathbb{R}^p$ (p will usually be m or n) is a real number denoted by an angle bracket:

$$\langle w, z\rangle_p = \sum_{i=1}^{p} w_i z_i,$$

where the subscript p may be dropped if there is no confusion about the dimension of the space. Two vectors in $\mathbb{R}^p$ are *orthogonal* if $\langle w, z\rangle = 0$. The quantity $\langle w, w\rangle = w_1^2 + \cdots + w_p^2$ is the square of the length of $w : \langle w, w\rangle = \|w\|^2$. If $\langle w, z\rangle = 0$ for all z, then w is the zero vector (whose components are all zero). If M is a k-dimensional linear manifold in $\mathbb{R}^p (k \leqslant p)$, then the set of vectors orthogonal to M is known as the *orthogonal complement* of M and is denoted by $M^\perp$; a vector u in $\mathbb{R}^p$ can be decomposed in one and only one way, as the sum of a vector in M and a vector in $M^\perp : u = v + w, v \in M, w \in M^\perp$ and, since $\langle v, w\rangle = 0, \|u\|^2 = \|v\|^2 + \|w\|^2$ (Pythagorean theorem). The vector v is the *projection* of u on M.

We now return to the study of (3.4.4a)–(3.4.4b). If $u \in \mathbb{R}^n$, then $Au \in \mathbb{R}^m$ and we may take the inner product of Au with an arbitrary vector v in $\mathbb{R}^m$:

$$\begin{aligned}\langle Au, v\rangle_m &= \sum_{i=1}^{m} v_i \left(\sum_{j=1}^{n} a_{ij} u_j\right) = \sum_{j=1}^{n} u_j \sum_{i=1}^{m} a_{ij} v_i \\ &= \sum_{i=1}^{n} u_i \left(\sum_{j=1}^{m} a_{ji} v_j\right) = \langle u, A^* v\rangle_n,\end{aligned}$$

where A^* is the *adjoint* matrix $(n \times m)$ with entries $a^*_{ij} = a_{ji}$. Thus A^* is obtained from A by transposing its rows and columns. Note that A^* operates on m-dimensional vectors to yield n-dimensional vectors; that is, $A^* \colon \mathbb{R}^m \to \mathbb{R}^n$. If $m = n$ and if $a_{ij} = a_{ji}$, the matrix is *symmetric* and $A = A^*$.

In our discussion of (3.4.4a)–(3.4.4b), two related problems will also play a role:

$$Au = 0 \qquad \text{(direct homogeneous equation)}, \tag{3.4.5}$$

$$A^* v = 0 \qquad \text{(adjoint homogeneous equation)}, \tag{3.4.6}$$

Note that (3.4.5) can be regarded as a special case of (3.4.4a)–(3.4.4b) with $f = 0$. If A is a square symmetric matrix, (3.4.5) and (3.4.6) coincide.

We can now say precisely when (3.4.4a)–(3.4.4b) is solvable.

Theorem (Fredholm Alternative)**.** *The necessary and sufficient condition for equation (3.4.4a)–(3.4.4b) to have solution(s) is that*

$$\langle f, v\rangle_m = 0 \textit{ for every } v \textit{ satisfying } (3.4.6).$$

REMARK. The conditions that f must obey are called *solvability* or *consistency* conditions. The theorem can be rephrased: R_A, the range of A, coincides with the orthogonal complement $N_{A*}^{\perp}$ of the null space of A^*. This point of view is adopted in Section 5.5.

Proof. The proof of necessity is simple. Suppose that (3.4.4a)–(3.4.4b) is solvable; then $f = Au$ for some u and $\langle f, v\rangle_m = \langle Au, v\rangle_m = \langle u, A^*v\rangle_n$, and thus, $\langle f, v\rangle_m = 0$ for every v satisfying (3.4.6). To prove sufficiency, suppose that $\langle f, v\rangle_m = 0$ for every v satisfying (3.4.6); we want to show that f belongs to R_A, the range of A. Since R_A is a linear manifold, we can decompose the vector f as $h + h^{\perp}$, where $h \in R_A$ and $h^{\perp} \in R_A^{\perp}$. The vector $h^{\perp}$ is orthogonal not just to h but to all elements of R_A; hence $0 = \langle Au, h^{\perp}\rangle_m = \langle u, A^*h^{\perp}\rangle_n$ for every $u \in \mathbb{R}^n$ and therefore $A^*h^{\perp} = 0$. Thus, $h^{\perp}$ is a solution of (3.4.6) and, by hypothesis, $f = h + h^{\perp}$ is orthogonal to $h^{\perp}$; since h is orthogonal to $h^{\perp}$, this means that $h^{\perp}$ is orthogonal to itself, so that $h^{\perp} = 0$ and $f \in R_A$.

When the solvability conditions $\langle f, v\rangle_m = 0$ are satisfied, how do we express the solution(s) of (3.4.4a)–(3.4.4b)? The reader may be disappointed to learn that we are not providing new recipes for finding the solution(s). We can, however, characterize the solution space.

If one solution u_p of (3.4.4a)–(3.4.4b) has been found, then every solution of (3.4.4a)–(3.4.4b) can be written as $u = u_p + u_H$, where u_H is a solution of (3.4.5) and, conversely, any u which is of the form $u_p + u_H$ is a solution of (3.4.4a)–(3.4.4b). If (3.4.5) has only the zero solution (and if the solvability condition is satisfied), then (3.4.4a)–(3.4.4b) has one and only one solution. In the balanced case we can say more. □

The Balanced Case ($n \times n$). Consider first the *regular* case when $\det A \neq 0$. Then $\det A^* \neq 0$ and both (3.4.5) and (3.4.6) have only the zero solution. The solvability condition is therefore satisfied automatically for any f and (3.4.4a)–(3.4.4b) has one and only one solution, which can be written $u = A^{-1}f$, where A^{-1} is the *inverse* matrix of A. This inverse can be expressed as a Green's matrix (see Exercise 3.4.4). If $\det A = 0$ (*singular* case), there are interdependencies among the rows of A (say, the third row is the sum of the first two) and (3.4.4a)–(3.4.4b) can be solved only if these interdependencies are reflected in f (in the earlier illustration, we would need $f_3 = f_1 + f_2$). This leads to the solvability conditions, which are automatically satisfied when $f = 0$, that is, in the homogeneous problem (3.4.5). Thus (3.4.5) is always solvable, although in many cases the only solution is the zero solution. If there are $k(\leqslant n)$ independent solutions of (3.4.5), A is said to have a *k-dimensional null space*; it turns out that the same is then true for A^*, although its null space will, in general, be different from that of A. The solvability conditions now consist of k separate conditions $\langle f, v^{(i)}\rangle = 0$, $i = 1, \ldots, k$, where $v^{(1)}, \ldots, v^{(k)}$ is a set of k independent solutions of (3.4.6). If f satisfies these solvability conditions, then the general solution of (3.4.4a)–(3.4.4b) is

$$u = u_p + \sum_{i=1}^{k} c_i u^{(i)},$$

where $u^{(1)}, \ldots, u^{(k)}$ are k independent solutions of (3.4.5) and u_p is a particular solution of (3.4.4a)–(3.4.4b).

Example 1. Consider the 2×2 case where

$$A = \begin{pmatrix} 1 & a \\ 2 & 1 \end{pmatrix}.$$

Then (3.4.4a)–(3.4.4b) becomes

$$u_1 + au_2 = f_1, \quad 2u_1 + u_2 = f_2$$

and

$$A^* = \begin{pmatrix} 1 & 2 \\ a & 1 \end{pmatrix}$$

so that (3.4.6) takes the form

$$v_1 + 2v_2 = 0, \quad av_1 + v_2 = 0.$$

If $a \neq \frac{1}{2}$, then $\det A \neq 0$, $\det A^* \neq 0$ and the solvability condition is automatically satisfied. The one and only solution of (3.4.4a)–(3.4.4b) is given by

$$u = \frac{1}{2a-1}(af_2 - f_1, 2f_1 - f_2).$$

If $a = \frac{1}{2}$, we see by inspection that a solution is possible only if $f_2 = 2f_1$; this also follows from the solvability condition [(3.4.6) has the solution $b(2, -1)$ and the solvability condition is $2f_1 - f_2 = 0$].

Example 2. Consider the system

$$\begin{aligned} u_1 + u_2 + u_3 \quad &= f_1 \\ u_1 + u_2 + au_3 \quad &= f_2, \end{aligned} \qquad Au = f, \qquad A = \begin{pmatrix} 1 & 1 & 1 \\ 1 & 1 & a \end{pmatrix}.$$

Therefore,

$$A^* = \begin{pmatrix} 1 & 1 \\ 1 & 1 \\ 1 & a \end{pmatrix}$$

and (3.4.6) is the system $v_1 + v_2 = 0, v_1 + v_2 = 0, v_1 + av_2 = 0$, which has only the 0 solution when $a \neq 1$, but has the solution $b(1, -1)$ when $a = 1$. Thus $Au = f$ always has solutions if $a \neq 1$, but when $a = 1, Au = f$ has solutions if and only if $f_1 = f_2$. In the case $a \neq 1$, A has a one-dimensional null space and we see that the general solution of $Au = 0$ is $u_H = b(1, -1, 0)$ and the general solution of $Au = f$ is

$$u = u_H + \left(0, \frac{af_1 - f_2}{a-1}, \frac{f_2 - f_1}{a-1}\right).$$

The reader may supply the details when $a = 1$.

Example 3. For the 2×1 matrix

$$A = \begin{pmatrix} 1 \\ 1 \end{pmatrix},$$

the homogeneous equation $Au = 0$ has only the zero solution, but $Au = f$ has no solution unless $f_1 = f_2$.

Example 4. Green's matrix as the inverse matrix. Suppose that A is an $n \times n$ matrix with $\det A \neq 0$. Then $Au = f$ has one and only one solution $u = A^{-1}f$, where the entries of A^{-1} are labeled b_{ij}. We now show how to relate b_{ij} to a *Green's matrix*, playing a role similar to Green's function for BVPs. We obtain the solution of $Au = f$ by superposition, writing $f = (f_1, \ldots, f_n) = \sum_{j=1}^{n} f_j \delta^{(j)}$, where $\delta^{(j)}$ is the vector with jth component equal to 1 and all other components equal to 0. Thus, $\delta^{(j)} = (\delta_1^{(j)}, \ldots, \delta_n^{(j)})$ with $\delta_i^{(j)} = 1$ for $i = j$ and $\delta_i^{(j)} = 0$ for $i \neq j$. Letting $g^{(j)}$ be the solution of $Ag^{(j)} = \delta^{(j)}$, we have $u = \sum_{j=1}^{n} f_j g^{(j)}$ and hence

$$u_i = \sum_{j=1}^{n} g_i^{(j)} f_j = \sum_{j=1}^{n} b_{ij} f_j.$$

Thus $(g_i^{(j)})$ is the inverse matrix of A. The vector $g^{(j)}$ is the jth column in A^{-1}.

Alternative Theorem for Integral Equations

Let $k(x, \xi)$ be a real-valued function defined on the square $a < x < b$, $a < \xi < b$. We consider the inhomogeneous integral equation

$$\int_a^b k(x, \xi) u(\xi)\, d\xi - u(x) = f(x), \quad a < x < b, \tag{3.4.7}$$

the related homogeneous equation

$$\int_a^b k(x, \xi) u(\xi)\, d\xi - u(x) = 0, \quad a < x < b, \tag{3.4.8}$$

and the adjoint homogeneous equation

$$\int_a^b k(\xi, x) v(\xi)\, d\xi - v(x) = 0, \quad a < x < b. \tag{3.4.9}$$

In Chapter 5 we prove the alternative theorem: (a) If (3.4.8) has only the trivial solution, so does (3.4.9), and then (3.4.7) has one and only one solution; (b) if (3.4.8) has nontrivial solutions, so does (3.4.9), and then (3.4.7) will have solutions if and only if

$$\int_a^b f(x) v(x)\, dx = 0$$

for every v which is a solution of (3.4.9).

Alternative Theorem for Boundary Value Problems

Consider first problem (3.4.1) with data $\{f; 0, 0, \ldots, 0\}$,

$$Lu = f, \quad a < x < b; \qquad B_1 u = \cdots = B_p u = 0, \tag{3.4.10}$$

for which the following alternative theorem holds:

(a) If the direct homogeneous problem (3.4.2) has only the trivial solution, so does the adjoint homogeneous problem (3.4.3), and (3.4.10) has one and only one solution.

(b) If (3.4.2) has k independent solutions, (3.4.3) also has k independent solutions [although not necessarily the same as those of (3.4.2)]. Then (3.4.10) has solutions if and only if $\int_a^b fv^{(1)}\,dx = \cdots = \int_a^b fv^{(k)}\,dx = 0$, where $(v^{(1)}, \cdots, v^{(k)})$ is a set of k independent solutions of (3.4.3). If the solvability conditions are satisfied, the general solution of (3.4.10) is of the form $\tilde{u} + \sum_{i=1}^k c_i u^{(i)}$, where $\tilde{u}$ is a particular solution of (3.4.10), the $\{c_i\}$ are arbitrary constants, and $u^{(1)}, \ldots, u^{(k)}$ are k independent solutions of (3.4.2).

The necessity of the solvability conditions is easy to prove. Let u be a solution of (3.4.10), and v a solution of (3.4.3). Multiply (3.4.10) by v and (3.4.3) by u, subtract, and integrate from a to b to obtain

$$\int_a^b fv\,dx = \int_a^b (vLu - uL^*v)\,dx = J(u, v)]_a^b,$$

but $J(u, v)]_a^b = 0$, since u satisfies homogeneous boundary conditions and v satisfies the adjoint boundary conditions (which were designed exactly for the purpose of making $J]_a^b$ vanish). The sufficiency will be proved in Chapter 5.

Example 5.

(a) Consider the problem

$$-u'' = f, \quad 0 < x < 1; \qquad u(1) - u(0) = 0, \quad u'(1) - u'(0) = 0. \tag{3.4.11}$$

The adjoint homogeneous problem is

$$-v'' = 0, \quad 0 < x < 1; \qquad v(1) - v(0) = 0, \quad v'(1) - v'(0) = 0,$$

which has the solutions $v =$ constant. Thus (3.4.11) has solutions if and only if

$$\int_0^1 f\,dx = 0. \tag{3.4.12}$$

There is a simple physical explanation for this condition. We can regard (3.4.11) as governing the steady temperature $u(x)$ in an insulated thin ring of unit perimeter subject to sources of density $f(x)$. The periodic boundary conditions reflect the fact that the points $x = 0$ and $x = 1$ represent the same physical point in the ring, so that the temperature must have the same value at $x = 0$ and $x = 1$ and the gradient of the temperature must also have the same value at $x = 0$ and $x = 1$. Since heat cannot flow through the surface of the ring, conservation of heat shows that a steady state is possible only if the net heat input along the ring is 0, that is, if $\int_0^1 f(x)\,dx = 0$, which is just (3.4.12).

(b) For the problem

$$-u'' - \pi^2 u = f, \quad 0 < x < 1; \qquad u(1) + u(0) = 0, \quad u'(1) + u'(0) = 0$$

there are two solvability conditions:

$$\int_0^1 f \cos \pi x\, dx = 0, \quad \int_0^1 f \sin \pi x dx = 0.$$

Example 6. Consider the inhomogeneous problem

$$u' + u = f, \quad 0 < x < 1; \quad u(0) - eu(1) = 0. \tag{3.4.13}$$

The adjoint homogeneous problem is (see Example 1, Section 3.3),

$$-v' + v = 0, \quad 0 < x < 1; \quad -ev(0) + v(1) = 0,$$

with the nontrivial solution $v = Ae^x$. Therefore, (3.4.13) has solutions if and only if $\int_0^1 fe^x dx = 0$. We can check this by elementary calculations. Solving the differential equation in (3.4.13), we find that $u = \int_0^x f(t)e^{t-x}\,dt + Ce^{-x}$; the boundary condition gives $C - \int_0^1 f(t)e^t\,dt - C = 0$, which is just the predicted solvability condition. If the condition is satisfied, then $u = \int_0^x f(t)e^{t-x} dt + Ce^{-x}$ is the general solution of (3.4.13).

Example 7. Consider next a problem in which there are *too many* boundary conditions:

$$Lu \doteq u' = f, \quad 0 < x < 1; \qquad u(0) = 0, \quad u(1) = 0. \tag{3.4.14}$$

To find the adjoint problem, we proceed as in Example 1, Section 3.3. Clearly, $L^* = -D$, and Green's formula becomes

$$\int_0^1 (vLu - uL^*v)\,dx = uv]_0^1 = J(u,v)]_0^1 = u(1)v(1) - u(0)v(0).$$

Given the boundary conditions $u(0) = 0, u(1) = 0$, we look for homogeneous conditions in v which make $J(u,v)]_0^1$ vanish. But $J]_0^1$ is already 0! Thus, for any v, we have $J]_0^1 = 0$. Hence the adjoint problem is

$$L^*v = -\frac{dv}{dx} = 0, \quad 0 < x < 1,$$

whose general solution is $v =$ constant. Problem (3.4.14) has a solution if and only if f is orthogonal to a constant, that is, $\int_0^1 f(x)\,dx = 0$, a result that could also have been obtained by inspection from (3.4.14).

Next we turn to the completely inhomogeneous problem (3.4.1), repeated here for convenience:

$$Lu = f, \quad a < x < b; \qquad B_1u = \gamma_1, \dots, B_pu = \gamma_p. \tag{3.4.15}$$

We can easily find a necessary condition for (3.4.15) to have solutions. Let v be a solution of the homogeneous adjoint problem (3.4.3). Multiply (3.4.15) by v and (3.4.3) by u, subtract, and integrate from a to b. This yields

$$\int_a^b (vLu - uL^*v)\,dx = J(u,v)]_a^b.$$

The left side reduces to $\int_a^b fv\,dx$, but the right side no longer vanishes because u satisfies inhomogeneous boundary conditions. Using (3.3.8), we can write

$$J(u,v)]_a^b = \gamma_1 B_{2p}^*v + \cdots + \gamma_p B_{p+1}^*v,$$

where the right side is completely known. Thus (3.4.15) can have solutions only if

$$\int_a^b fv\,dx = \gamma_1 B_{2p}^*v + \cdots + \gamma_p B_{p+1}^*v, \tag{3.4.16}$$

a solvability condition which also turns out to be sufficient.

Example 8. Consider the problem

$$u' + u = f, \quad 0 < x < 1; \quad u(0) - eu(1) = \gamma_1. \tag{3.4.17}$$

Then $L^*v = -v' + v$, and

$$\begin{aligned} J(u,v)]_0^1 &= u(1)v(1) - u(0)v(0) \\ &= [u(0) - eu(1)][-v(0)] + u(1)[v(1) - ev(0)] \\ &= (B_1u)(B_2^*v) + (B_2u)(B_1^*v). \end{aligned}$$

Let v be the solution of the homogeneous adjoint problem

$$L^*v = 0, \quad B_1^*v = 0,$$

whose solution (see Example 6) is $v = Ae^x$. With u the solution of (3.4.17) we find that

$$J(u,v)]_0^1 = \gamma_1 B_2^*v = -\gamma_1 v(0) = -\gamma_1 A.$$

Therefore, (3.4.17) has solutions if and only if

$$A\int_0^1 fe^x\,dx = -\gamma_1 A, \quad \text{that is,} \int_0^1 fe^x\,dx = -\gamma_1.$$

Again we can check this easily by a straightforward calculation. The differential equation in (3.4.17) has the general solution

$$u = \int_0^x f(t)e^{t-x}\,dt + Ce^{-x},$$

so that, by imposing the boundary condition

$$C - e\int_0^1 f(t)e^{t-1}\,dt - eCe^{-1} = \gamma_1,$$

we obtain the previously found solvability condition

$$\int_0^1 f(t)e^t\,dt = -\gamma_1.$$

EXERCISES

3.4.1 Refer to Exercise 3.3.1. Find the solvability condition for the problem

$$\begin{aligned}&D^4u = f, \quad 0 < x < 1;\\ &u(0) = u(1) = u''(0) = u''(1) = 0, \quad u'(1) - u'(0) = \gamma.\end{aligned}$$

Check your result by solving the differential equation with the four homogeneous conditions by using Green's function (1.4.10) and then trying to satisfy the last boundary condition.

3.4.2 Find the solvability conditions for

$$\begin{aligned}-u'' - u = f, &\quad -\pi < x < \pi;\\ u(\pi) - u(-\pi) = \gamma_1, &\quad u'(\pi) - u'(-\pi) = \gamma_2.\end{aligned}$$

3.4.3 Consider the fourth-order operator $L = D^4$ on $0 < x < 1$ with the boundary functionals $B_1u = u(0), B_2u = u^{(3)}(0), B_3u = u'(1), B_4u = u^{(3)}(1)$. Find the adjoint BC. What is the solvability condition for the BVP

$$D^4u = f, \quad 0 < x < 1; \qquad B_1u = B_2u = B_3u = B_4u = 0?$$

3.4.4 Part (a) describes the notions of *least-squares solutions* and *pseudoinverse*. The exercise is in the remaining parts.

(a) Let A be an $n \times n$ *real symmetric* matrix with $\det A = 0$. Then $Au = f$ has solutions if and only if $\langle f, v\rangle = 0$ for all v such that $Av = 0$. In other words, R_A and N_A are orthogonal complements. If $f \in R_A$ (that is, if f satisfies the solvability conditions), then $Au = f$ has many solutions, among which we want to choose the one having the smallest norm. We write $u = w + w^{\perp}$, where $w \in R_A$ and $w^{\perp} \in N_A$; hence $Aw^{\perp} = 0$ and $Aw = f, w \in R_A$. The last equation clearly has one and only one solution which is the smallest norm solution of $Au = f$.

Now suppose that f does *not* satisfy the solvability condition (that is, $f \notin R_A$). What meaning can be given to $Au = f$? Since the equation is not solvable, we content ourselves with finding element(s) u in $\mathbb{R}^n$ which *minimize* $\|Au - f\|$ or, equivalently, $\|Au - f\|^2$. Writing $f = h + h^{\perp}$ where $h \in R_A$ (that is, h is the projection of f on R_A) and $h^{\perp} \in R_A^{\perp}(= N_A)$, we have $\|Au - f\|^2 = \|Au - h - h^{\perp}\|^2 = \|Au - h\|^2 + \|h^{\perp}\|^2$, the last equality being a consequence of the fact that $Au - h$ is in R_A and is therefore orthogonal to $h^{\perp}$. The minimum is obtained by choosing u so that $Au = h$. Since $h \in R_A$, the equation is solvable and has, in fact, many solutions, known as *least-squares solutions*. As in the preceding paragraph, we may now choose the solution of smallest norm.

Combining the results of these two paragraphs, we define an $n \times n$ matrix G which has the following properties: Gf is the smallest-norm solution of $Au = h$, where h is the projection of f on R_A. Thus, if $Au = f$ is solvable, Gf is the solution described in the first paragraph, whereas if $Au = f$ is not solvable, then Gf is the least-squares solution of smallest norm. The transformation $G\colon \mathbb{R}^n \to \mathbb{R}^n$ is known as the *pseudoinverse* of A.

(b) Let A be a real, symmetric matrix with det $A \neq 0$. Suppose that A has n distinct eigenvalues $\lambda_1 < \lambda_2 < \cdots < \lambda_n$ with corresponding normalized eigenvectors $e_1, \ldots, e_n$. If λ is not an eigenvalue, solve

$$Au - \lambda u = f$$

by expansion in eigenvectors. How would you handle the case $\lambda = \lambda_k$? Reconcile your results with those of part (a).

(c) Modify the treatment of part (a) if A is not symmetric or is not a square matrix.

(d) Show that u is a least-squares solution of $Au = f$ if and only if it satisfies the equation $A^*Au = A^*f$.

(e) In trying to fit a straight line $y = sx + b$ to m observations (x_i, y_i), $i = 1, \ldots, m$, we are led to a system of m equations in two unknowns:

$$sx_i + b = y_i, \quad i = 1, \ldots, m.$$

This is an overdetermined system for the unknown slope s and intercept b. Use part (d) to find the least-squares solution. Check against the solution obtained by calculus.

3.5 MODIFIED GREEN'S FUNCTIONS

Let us begin with a simple but representative example. The self-adjoint boundary value problem

$$-u'' = f, \quad 0 < x < 1; \quad u'(0) = u'(1) = 0 \tag{3.5.1}$$

describes steady heat conduction in a rod whose lateral surface and *ends* are insulated. Unless the source term $f(x)$ satisfies the solvability condition $\int_0^1 f(x)\,dx = 0$, there is no hope for a solution of (3.5.1). [If one considered the corresponding time-dependent heat conduction equation with a steady source term $f(x)$, the temperature would increase indefinitely with time unless $\int_0^1 f(x)\,dx = 0$.] In particular, if $f = \delta(x-\xi)$, the solvability condition is *not* satisfied and there is no solution to $-g'' = \delta(x-\xi), 0 < x, \xi < 1, g'(0,\xi) = g'(1,\xi) = 0$. We want to construct the nearest thing to a Green's function to enable us to solve (3.5.1) when a solution exists (that is, when $\int_0^1 f\,dx = 0$). This can be done by considering a modified accessory problem with the consistent data $\{\delta(x-\xi) - 1, 0, 0\}$. We have compensated for the concentrated source at ξ by a uniform distribution of sinks along the rod, the net total heat input being 0. Therefore, we examine the problem

$$-g'' = \delta(x-\xi) - 1, \quad 0 < x, \xi < 1; \qquad g'(0,\xi) = g'(1,\xi) = 0, \tag{3.5.2}$$

which is consistent and has many solutions differing from one another by a constant. Since, for $x \neq \xi, -g'' = -1$, we have

$$g(x,\xi) = \begin{cases} A + Bx + \dfrac{x^2}{2}, & 0 < x < \xi, \\ C + Dx + \dfrac{x^2}{2}, & \xi < x < 1. \end{cases}$$

The boundary conditions yield $B = 0, D = -1$; continuity at $x = \xi$ gives $A = C - \xi$, so that

$$g(x,\xi) = \begin{cases} C - \xi + \dfrac{x^2}{2}, & 0 < x < \xi, \\ C - x + \dfrac{x^2}{2}, & \xi < x < 1, \end{cases} \tag{3.5.3}$$

where C can, of course, depend on ξ. At this stage we would normally try to apply the jump condition $g'(\xi+,\xi) - g'(\xi-,\xi) = -1$, but this is now *automatically* satisfied regardless of the value of C. Therefore, (3.5.3) is the general solution of (3.5.2). Moreover, for each position ξ of the source, one could choose a different C. In many calculations it is convenient to single out a particular g by requiring that

$$\int_0^1 g(x,\xi)\,dx = 0 \qquad \text{for all } \xi, \quad 0 < \xi < 1. \tag{3.5.4}$$

Requirement (3.5.4) is *not* a solvability condition, but merely picks out a solution of (3.5.2) that turns out to be symmetric in x and ξ. This symmetric solution of

(3.5.2) will be denoted by $g_M(x,\xi)$ and is known as the *modified Green's function*. By imposing (3.5.4) on (3.5.3), we find explicitly $C = (\xi^2/2) + \frac{1}{3}$ and

$$g_M(x,\xi) = \begin{cases} \dfrac{1}{3} - \xi + \dfrac{x^2+\xi^2}{2}, & 0 \leqslant x \leqslant \xi, \\ \dfrac{1}{3} - x + \dfrac{x^2+\xi^2}{2}, & \xi \leqslant x \leqslant 1. \end{cases} \tag{3.5.5}$$

Note that there are other choices of C in (3.5.3) which make g symmetric, so that (3.5.4) is sufficient but not necessary for symmetry (see Exercise 3.5.1). The principal advantage of our choice is that it gives the g of minimum norm.

We pass now to the general self-adjoint problem of the second order. Assume that the homogeneous problem $Lu = 0, B_1u = B_2u = 0$ has nontrivial solutions of the form $Cu_1(x)$, where $u_1(x)$ is a normalized solution, that is, a solution for which

$$\int_a^b u_1^2(x)\,dx = 1.$$

The problem $Lg = \delta(x-\xi), B_1g = B_2g = 0$ is inconsistent, so that we must introduce the *modified Green's function* g_M, chosen to satisfy

$$Lg_M = \delta(x-\xi) - u_1(x)u_1(\xi), \quad a < x, \xi < b; \qquad B_1g_M = B_2g_M = 0, \tag{3.5.6}$$

$$\int_a^b g_M(x,\xi)u_1(x)\,dx = 0 \quad \text{for every } \xi, a < \xi < b. \tag{3.5.7}$$

The BVP (3.5.6) is consistent, since

$$\int_a^b [\delta(x-\xi) - u_1(x)u_1(\xi)]u_1(x)\,dx = 0.$$

Condition (3.5.7) serves to single out the minimum norm solution of (3.5.6), which turns out to be symmetric in x and ξ (see Exercise 3.5.1). The construction of g_M in any specific case proceeds as for (3.5.2).

Let us now show how the modified Green's function can be used to solve

$$Lu = f, \quad a < x < b; \qquad B_1u = B_2u = 0. \tag{3.5.8}$$

Of course, (3.5.8) is solvable only if $\int_a^b fu_1\,dx = 0$, which we assume holds. Multiply (3.5.6) by the solution u of (3.5.8), multiply (3.5.8) by g_M, subtract, and integrate from a to b. This gives

$$\int_a^b (uLg_M - g_MLu)\,dx = u(\xi) - \int_a^b u_1(x)u_1(\xi)u(x)\,dx - \int_a^b g_M(x,\xi)f(x)\,dx.$$

The left side is equal to $J(u, g_M)]_a^b$, which vanishes by the boundary conditions. Therefore,

$$u(\xi) = Cu_1(\xi) + \int_a^b g_M(x,\xi)f(x)\,dx,$$

and using the symmetry of g_M, we find on relabeling x and ξ that

$$u(x) = Cu_1(x) + \int_a^b g_M(x,\xi) f(\xi)\, d\xi. \tag{3.5.9}$$

This then is the general solution of (3.5.8). If we set $C = 0$ in (3.5.9), we obtain a particular solution of (3.5.8) which is the *minimum norm* solution (see Exercise 3.5.7).

We now discuss the question of least-squares solutions of inconsistent problems by examining an illustrative example. With λ and $f(x)$ arbitrary but fixed, consider the self-adjoint BVP

$$u'' + \lambda u = f(x), \quad 0 < x < 1, \quad u(0) = u(1) = 0, \tag{3.5.10}$$

which was already looked at briefly in Example 2, Section 3.2. The associated eigenproblem [see (1.2.20)] has eigenvalues $\lambda_n = n^2\pi^2, n = 1, 2, \ldots$ and corresponding eigenfunctions $u_n(x) = \sin n\pi x$.

If the given λ in (3.5.10) is not an eigenvalue, then (3.5.10) has one and only one solution. If, on the other hand, $\lambda = \lambda_k$ for some k, then (3.5.10) has solution(s) if and only if the solvability condition

$$\int_0^1 f(x) \sin k\pi x \; dx = 0 \tag{3.5.11}$$

is satisfied. We are interested in finding the solution(s) of solvable problems and in finding least-squares solutions to inconsistent problems. A particularly convenient and useful approach is through a sine series expansion (the eigenfunctions of the homogeneous problem!). For the given $f(x)$ [obeying Dirichlet conditions, that is, belonging to $D(S)$, Section 2.3], we can write

$$f(x) = \sum_{n=1}^{\infty} f_n \sin n\pi x, \quad f_n = 2\int_0^1 f(x) \sin \; n\pi x \; dx.$$

Assuming that a solution $u(x)$ of (3.5.10) exists, we can expand it as

$$\sum_{n=1}^{\infty} b_n \sin n\pi x,$$

where the coefficients are determined by substitution in (3.5.10):

$$b_n(\lambda - \lambda_n) = f_n, \quad n = 1, 2, \ldots \tag{3.5.12}$$

We can then organize our results as follows.

(a) If the given λ in (3.5.10) is not an eigenvalue, then (3.5.10) has the one and only solution

$$u(x) = \sum_{n=1}^{\infty} b_n \sin n\pi x = \sum_{n=1}^{\infty} \frac{f_n}{\lambda - \lambda_n} \sin n\pi x. \tag{3.5.13}$$

(b) If $\lambda = \lambda_k = k^2\pi^2$ for some k, then (3.5.12) gives

$$b_n = \frac{f_n}{\lambda_k - \lambda_n}, \quad n \neq k$$

and

$$b_k = \begin{cases} \text{arbitrary} & \text{if } f_k = 0 \\ \text{does not exist} & \text{if } f_k \neq 0. \end{cases}$$

Thus, if $f_k \neq 0$, (3.5.10) has no solution. [See, however, part (c) below.] If, instead, $f_k = 0$ [that is, condition (3.5.11) is satisfied], then (3.5.10) has the solutions

$$u(x) = b_k \sin k\pi x + \sum_{n \neq k} \frac{f_n}{\lambda_k - \lambda_n} \sin n\pi x, \tag{3.5.14}$$

where b_k is arbitrary, with the *minimum norm* solution of (3.5.10) obtained by taking $b_k = 0$ in (3.5.14).

(c) If $\lambda = \lambda_k$ and f does not satisfy the solvability condition (3.5.11), we can seek a least-squares solution: Find $u(x)$ with $u(0) = u(1) = 0$ so that the L_2 norm

$$\|u'' + \lambda_k u - f\|$$

is least. This is equivalent to finding the element in the range nearest f. Since the range is orthogonal to $\sin k\pi x$, the element closest to f is $f - f_k \sin k\pi x = \sum_{n\neq k} f_n \sin n\pi x$ and therefore the least-squares solutions are given by (3.5.14). By setting $b_k = 0$, we obtain the least-squares solution of minimum norm.

(d) The modified Green's function is a special case of (c). With $\lambda = \lambda_k$ and $f(x) = \delta(x - \xi)$, (3.5.10) is not solvable, but has least-squares solutions. The least-squares solution of minimum norm is $g_M(x, \xi)$. This is also the minimum norm solution of the solvable problem with $\delta(x - \xi)$ replaced by

$$\begin{aligned} \delta(x - \xi) - 2 \left[\int_0^1 \delta(x - \xi) \sin k\pi x \; dx \right] \sin k\pi x \\ = \delta(x - \xi) - 2 \sin k\pi\xi \sin k\pi x. \end{aligned}$$

Note that $g_M(x, \xi)$ is a symmetric function given by

$$\sum_{k \neq n} \frac{2 \sin n\pi\xi \sin n\pi x}{\lambda_k - \lambda_n},$$

and that the least-squares solution of minimum norm of (3.5.10) is given by

$$u = \int_0^1 g_M(x, \xi) f(\xi) \, d\xi = \sum_{n \neq k} \frac{f_n}{\lambda_k - \lambda_n} \sin n\pi x.$$

Therefore, g_M gives rise to an integral operator G_M which inverts (3.5.10) when $\lambda = \lambda_k$ in the following sense: If (3.5.10) is solvable [that is, f satisfies (3.5.11)], then $G_M f$ is the solution of (3.5.10) with minimum norm; if

(3.5.10) is not solvable, then $G_M f$ is the least-squares solution of (3.5.10) with smallest norm. The operator G_M is the *pseudoinverse* of $D^2 + \lambda_k$ subject to the BC $u(0) = u(1) = 0$. (See also Exercise 3.4.4 and a more general treatment in Chapter 9.)

EXERCISES

3.5.1 Show that condition (3.5.7) is sufficient to ensure symmetry of the solution of (3.5.6). It is clear that the condition is *not* necessary; indeed, if C is independent of ξ, the addition of a term $Cu_1(x)u_1(\xi)$ to g_M does not spoil either symmetry or the fact that (3.5.6) is satisfied but (3.5.7) is no longer satisfied.

3.5.2 Let (L, B_1, B_2) be a self-adjoint problem of the second order, and suppose that the completely homogeneous problem $Lu = 0, a < x < b, B_1u = B_2u = 0$ has two independent nontrivial solutions (this means that *every* solution of $Lu = 0$ satisfies the boundary conditions). Develop the theory of the modified Green's function for this case. Find the modified Green's function for $L = (d^2/dx^2) + \pi^2$ with B_1, B_2 defined by $B_1u = u(0) + u(1), B_2u = u'(0) + u'(1)$.

3.5.3 Find the nontrivial solutions of

$$D^4u = 0, \quad 0 < x < 1; \quad u''(0) = u'''(0) = u''(1) = u'''(1) = 0,$$

and give a physical interpretation in beam theory. Show that the problem is self-adjoint. Define and construct the modified Green's function. Solve $D^4u = f$ with the homogeneous boundary conditions above when f satisfies the solvability conditions.

3.5.4 Let L be a second-order differential operator, and suppose that (L, B_1, B_2) is *not* self-adjoint. Assume that the homogeneous problem $Lu = 0, B_1u = B_2u = 0$ has only one independent solution u_1 (so that the homogeneous adjoint problem has only one independent solution v_1). Develop the theory of the modified Green's function in this case. Construct g_M for $L = D^2$ with B_1, B_2 defined by $B_1u = u(0) + u(1), B_2u = u'(0) - u'(1)$.

3.5.5 Show how to use g_M of (3.5.6)–(3.5.7) to solve the completely inhomogeneous problem $Lu = f, B_1u = \gamma_1, B_2u = \gamma_2$ when the appropriate solvability condition is satisfied.

3.5.6 (a) Find the eigenvalues and eigenfunctions of

$$-u'' = \lambda u, \quad -1 < x < 1; \quad u'(1) - u(1) = 0, \quad u'(-1) + u(-1) = 0.$$

Show that there is precisely one negative eigenvalue, that zero is an eigenvalue, and that there are infinitely many positive eigenvalues. Show graphically how the eigenvalues are determined.

(b) Find the modified Green's function when $\lambda = 0$.

3.5.7 Given a fixed interval (a, b), the norm of a function $u(x)$ is defined as

$$\|u\| \doteq \left[\int_a^b u^2(x)\,dx\right]^{1/2}.$$

Two functions u, v are said to be orthogonal if $\int_a^b uv\,dx = 0$. If $u = v + w$, where v and w are orthogonal, the Pythagorean theorem holds: $\|u\|^2 = \|v\|^2 + \|w\|^2$. Show that the two terms on the right side of (3.5.9) are orthogonal and hence that the choice $C = 0$ yields the solution of least norm.

3.5.8 The BVP

$$-u'' = f, \quad 0 < x < 1; \quad u(0) = u(1), \quad u'(0) = u'(1)$$

describes heat flow in a thin ring of unit circumference. Show that the modified Green's function for this problem is

$$g_M(x, \xi) = \tfrac{1}{12} + \frac{(x-\xi)^2}{2} - \tfrac{1}{2}|x - \xi|.$$

3.5.9 We have required least-squares solutions to satisfy the boundary conditions but not necessarily the differential equation. Placing the BCs and the DE on the same footing, we can impose penalties for failing to satisfy either and then seek to minimize the sum of the residuals. Consider, for instance, the problem

$$u'' = f, \quad 0 < x < 1; \quad u'(0) = \alpha, \quad u'(1) = \beta,$$

which is consistent if and only if

$$\int_0^1 f\,dx = \beta - \alpha.$$

Suppose that this condition is *not* satisfied. Find $u(x)$ so that

$$\|u'' - f\| + |u'(0) - \alpha| + |u'(1) - \beta|$$

is least. In the case $\alpha = \beta = 0, f = 1$, show that $u(x) = C$ and $u(x) = x^2/2 + C$ both yield minimum total residual.

3.5.10 The following boundary value problem arises in connection with periodic solutions of the Korteweg–De Vries equation (famous for its solitons):

$$\begin{gathered} -D^3u + a^2Du = f(x), \quad -1 < x < 1; \\ u(-1) = u(1), \quad u'(-1) = u'(1), \quad u''(-1) = u''(1). \end{gathered}$$

Show that the solvability condition is

$$\int_{-1}^1 f(x)\,dx = 0.$$

Assuming that the condition is satisfied, find the solutions and write your answer as

$$u(x) = A + \int_{-1}^{1} g(x, \xi) F(\xi) \, d\xi,$$

where A is an arbitrary constant and

$$F(x) = \int_{-1}^{x} f(t) \, dt; \quad g(x, \xi) = \frac{\cosh a(1 - |x - \xi|)}{2a \sinh a}.$$

REFERENCES AND ADDITIONAL READING

1. A. Ben-Israel and T. N. E. Greville, *Generalized Inverses: Theory and Application*, Wiley-Interscience, New York, 1974.
2. E. A. Coddington and N. Levinson, *Theory of Ordinary Differential Equations*, McGraw-Hill, New York, 1955.
3. C. Lanczos, *Linear Differential Operators*, SIAM, Philadelphia, 1988.
4. M. A. Naimark, *Linear Differential Operators*, Frederick Ungar, New York, 1967.
5. M. Z. Nashed, *Generalized Inverses and Applications*, Academic Press, New York, 1976.

CHAPTER 4

HILBERT AND BANACH SPACES

4.1 FUNCTIONS AND TRANSFORMATIONS

Since the terms *function*, *transformation*, *operator*, and *mapping* are defined in precisely the same manner, a reasonable person might infer that they are used interchangeably, but in practice there are occasional distinctions that will be pointed out as they occur (see, for instance, Example 6 below). The idea of function pervades mathematics, beginning with the familiar real function of a real variable, which can be regarded as a rule f that assigns to each real number x an unambiguous real number $y = f(x)$.

We shall, however, need the concept of function in a more general setting.

Definition. *Let D and E be arbitrary sets. A* function f *is said to be defined* on D *and to assume its values* in E *if to* each *element x belonging to D there is made to correspond exactly one element y in E. We then write $y = f(x)$ and say that f transforms or maps D* into *E; alternatively, we write $f\colon D \to E$.*

REMARKS

1. Of course, y usually varies with x, but to a particular x in D there corresponds one and only one y (known as the *image* of x). We are therefore considering

Green's Functions and Boundary Value Problems, Third Edition. By I. Stakgold and M. Holst

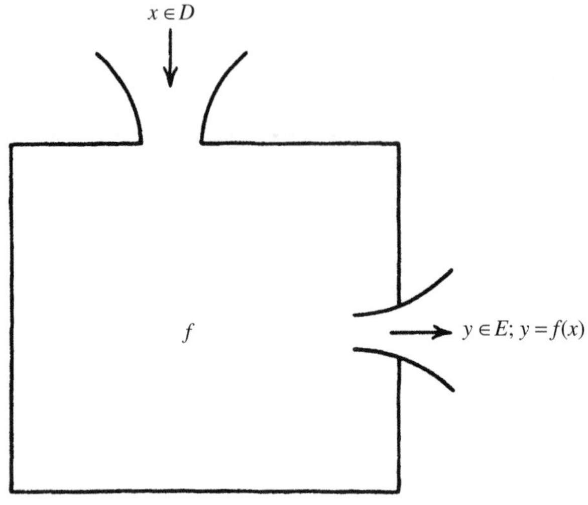

Figure 4.1

only single-valued functions. The element x is also called the *independent variable*, and y the *dependent variable*.

2. A function can be visualized as the black box of Figure 4.1. Every element x in D is an admissible input. The box transforms the input x into a well-defined output y belonging to E.

3. The set D is known as the *domain* of f. The significance of the emphasized word "on" in the definition of function is that every element x in D is admissible as an input, so that f is defined on all of D. On the other hand, not every element of E is necessarily an output. As x traverses D, the set of all outputs y forms a subset of E known as the *range* of f, denoted by R. Thus R consists of all the elements in E which are of the form $f(x)$ for some x in D. The significance of the emphasized words "in" and "into" in the definition of function is that we are not committing ourselves as to whether or not the outputs cover all of E. If, in fact, it happens that $R = E$, we say that f is *onto* (or *surjective*). If we had had the foresight to choose the set E from the start to be just the set R, then f would automatically be onto. The reason why this choice is not usually made is that even for functions of the same type (such as ordinary real-valued functions of a real variable) the range varies from function to function; moreover, the range may be difficult to characterize precisely [see Example 1(d)], so that it is easier to regard R as a subset of a simple, known set E.

4. Strictly speaking, one should distinguish between the function f—the box in Figure 4.1—and the output $f(x)$ corresponding to the input x. We shall, how-

ever, lapse into the traditional time-saving usage of calling both f and $f(x)$ by the name *function*.

5. There is nothing sacred about using the letters x, y, f in their particular roles in the definition. Any other letters will do, but in any specific illustration we must make sure that the same letter is not given two different meanings.

6. The function f is *one-to-one* (or *injective*) if to each y in R there corresponds exactly one x in D such that $y = f(x)$. In other words, f is one-to-one if, whenever $f(x_1) = f(x_2)$, then $x_1 = x_2$. If f is one-to-one, it can be "unraveled" by means of the inverse function f^{-1} (*not* $1/f$), whose domain D' is the range R of f and whose range R' is just the domain D of f. To each element y in D', the function f^{-1} assigns the unambiguous element x in R' for which $y = f(x)$.

Example 1. If D is a subset of the real numbers and E is the set of real numbers, we are dealing with an ordinary real-valued function of a real variable. As specific examples consider the following:

(a) D is the real line, $f(x) = 1/(x^2 + 1)$.
(b) D is the set $0 \leqslant x < \infty$, $f(x) = 1/(x^2 + 1)$.
(c) D is the real line, $f(x) = x^3$.
(d) D is the real line, $f(x) = x^6 + 3x^5 + 2x^4 + x$.

The functions in (a) and (b) are given by the same formula but are defined on different subsets of the real line and should be distinguished. In both cases R is the set $0 < y \leqslant 1$, but only the function in (b) can be inverted. In (c) R is the real line and f is invertible. In (d) it is difficult to determine the range R precisely, but it is of the form $-a \leqslant y < \infty$ for some $a > 0$ and f is not invertible. In (a), (c), (d) we say that f maps the real line *into* itself; only in (c) are we allowed to use the word "onto" in place of "into."

Example 2. If D is a subset of the real numbers and E is the set of complex numbers, f is known as a complex-valued function of a real variable [for instance, $f(x) = e^{ix}$].

Example 3. If D and E are sets of complex numbers, f is known as a function of a complex variable.

Example 4. If D is the set of ordered n-tuples of real numbers $x = (x_1, \ldots, x_n)$ and E is the set of real numbers, f is known as a function of n real variables. We write $y = f(x)$ or $y = f(x_1, \ldots, x_n)$.

Example 5. If D is the set of ordered n-tuples of real numbers and E is the set of ordered m-tuples of real numbers, f is usually known as a *transformation*, *operator*, or *mapping* of D into E, and we tend to use capital letters such as A, F, and T instead of f. Of course, such a transformation can equally well be described by m functions $f_1, \ldots, f_m$ of n real variables.

Example 6. In many problems D and E are themselves sets of functions. This leads to an unavoidable complication in notation and terminology. The rule which assigns to each element in D an element of E is then called a *transformation* or *operator*. This sort of operator is usually denoted by a capital letter such as T. For instance, if D is the set of smooth, real-valued functions $u(t)$ defined on $-\infty < t < \infty$, the differentiation operator T assigns to each $u(t)$ the function $v(t) = du/dt$. In operator notation we would write $v = Tu$ or $v(t) = Tu(t)$. In this book the letters x and y have been avoided, for it is not clear whether x is better used as the input to the box T or as the independent variable within the input function itself. As we proceed, we will have occasion to use x in either capacity. For instance, we might let D be the set of all smooth functions $x(t)$ on $-\infty < t < \infty$, and E the set of all functions $y(t)$ on $-\infty < t < \infty$. The differentiation operator T would then assign to $x(t)$ the function $y(t) = dx/dt$, and we would write $y = Tx$ or $y(t) = Tx(t)$. Alternatively, we might let D be the set of all smooth functions $u(x)$ on $-\infty < x < \infty$, and E the set of all functions $v(x)$ on $-\infty < x < \infty$. The differentiation operator T then assigns to each $u(x)$ in D the element $v(x) = du/dx$ in E, for which we would write $v = Tu$ or $v(x) = Tu(x)$.

Example 7. If D is a set of functions and E a set of real or complex numbers, we use the name *functional* for a rule which assigns to each function in D a number in E. Either capital letters or lowercase letters are used to denote functionals. For instance, if we let D be the set of continuous functions on a bounded interval $a \leqslant x \leqslant b$, then both

$$Tu \doteq \int_a^b u^3(x)\,dx \quad \text{and} \quad Fu \doteq u(b) - u(a)$$

are examples of functionals on D.

Our principal interest is in transformations T defined on function spaces. For further progress restrictions must be imposed on the nature of D, E, and T. The sets D and E, as they occur in most applications, are endowed with both algebraic and metric structures. The algebraic structure will usually be that of a linear space (also known as a vector space); even if D and E are not linear spaces, they will often be subsets of linear spaces. A linear space consists of elements, known as vectors, for which the operations of vector addition and scalar multiplication are defined. Any two vectors may be added to form a vector also in the space. Any vector may be multiplied by any scalar to yield a vector also in the space (the scalar field will always be either the field of real numbers or that of complex numbers). A linear space is therefore said to be *closed* under the operations of vector addition and scalar multiplication. Of course, these operations must obey certain rules which are abstracted from the prototype of all linear spaces, namely, the usual three-dimensional space of directed line segments emanating from a common origin. The sets D and E will also be provided with a metric (or topological) structure. This gives us a way of measuring the distance separating any two elements, the distance satisfying certain axioms similar to the ones for the distance between points in ordinary Euclidean space. In the applications both structures will usually be present simultaneously. In Chapters 9 and 10 and in Section 4.4, we discuss nonlinear transformations, but otherwise we

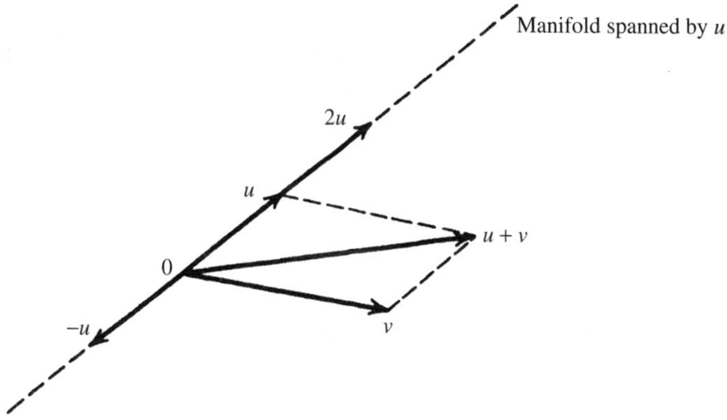

Figure 4.2

deal mostly with *linear transformations*. A linear transformation T satisfies the condition $T(\alpha u_1 + \beta u_2) = \alpha T u_1 + \beta T u_2$ for any vectors u_1, u_2 in D and any scalars α, β.

4.2 LINEAR SPACES

The definition of a linear (or vector) space takes its cue from the vector structure of "ordinary" three-dimensional space V_3. In describing V_3, we shall admittedly use some undefined geometrical terms, but at this stage we must rely on intuition. The elements of V_3 are the directed line segments emanating from a common origin. Scalar multiplication by real numbers and vector addition (based on the parallelogram rule) are illustrated in Figure 4.2. The zero vector is the degenerate line segment coinciding with the origin O. Whether we think of the elements of V_3 as directed line segments or merely as the endpoints of these segments is to some extent a matter of taste, but obviously the terminology has been developed from the first point of view.

In Figure 4.2 the set of vectors αu as α ranges over the real numbers forms a line through the origin; the set of vectors $\alpha u + \beta v$ generates a plane through the origin. The property that characterizes V_3 as a three-dimensional space is the existence of three vectors u, v, w (for instance, u, v in Figure 4.2 and a vector w not coplanar with them) such that each vector in the space can be expressed in one and only one way in the form $\alpha u + \beta v + \gamma w$.

Of course, there are many other properties which we tend to associate automatically with V_3, such as the length of a vector and the angle between a pair of vectors, but we must now dismiss these metric properties from our mind to concentrate on the algebraic structure. The appropriate generalization is contained in the following definition.

Definition. A linear space *(or* vector space*)* X *is a collection of elements (also called vectors)* $u, v, w, \ldots$, *in which each pair of vectors* u, v *can be combined by a process* *known as* vector addition *to yield a well-defined vector* $u+v$ *and in which each vector* u *can be combined with each number* α *by a process known as* scalar multiplication *to yield a well-defined vector* αu. *Vector addition and scalar multiplication satisfy the following axioms:*

1. *(a)* $u + v = v + u$.

   (b) $u + (v + w) = (u + v) + w$.

   (c) There is a unique vector 0 such that $u + 0 = u$ *for every* u *in* X.

   (d) To each vector u *there corresponds a unique vector labeled* $-u$ *such that* $u + (-u) = 0$.

2. *(a)* $\alpha(\beta u) = (\alpha\beta)u$.

   (b) $1u = u$.

3. *(a)* $(\alpha + \beta)u = \alpha u + \beta u$.

   (b) $\alpha(u + v) = \alpha u + \alpha v$.

In these axioms u, v, w *are arbitrary vectors in* X, *and* α, β *are arbitrary numbers.*

REMARKS

1. If the numbers $\alpha, \beta, \ldots$ admitted as scalar multipliers are real numbers, we speak of a *real linear space* $X^{(r)}$; if complex numbers are permitted as multipliers, the space is a *complex linear space* $X^{(c)}$. When no superscript is used, either kind of space is envisaged.

2. No effort has been made to give a minimal set of axioms in the definition of a linear space.

3. The same symbol, 0, is used for both the zero vector and the number zero. The same symbol, +, is used for addition of vectors and addition of numbers. We hope that no serious confusion will arise from this somewhat sloppy practice.

4. Axiom 1(b) shows that the triple sum is independent of order and may therefore be written as $u + v + w$ without ambiguity. The sum of any finite number of vectors $u_1, \ldots, u_k$ is also well defined without parentheses and without regard to order. We have no way at this stage to define the sum of an infinite number of vectors. This requires the notion of convergence (or passage to the limit), which will come later.

5. The vector $u + (-v)$ is also written as $u - v$. It is easy to show that to each pair of vectors u, v there corresponds a unique vector w such that $u = v + w$ and that this vector w is just $u - v$.

6. One can also prove that $0u = 0$ (number zero multipied by vector u is the zero vector) and that $(-1)u = -u$ [number -1 multiplied by vector u is equal to vector $-u$, whose existence is guaranteed by axiom 1(d)].

7. If X contains a nonzero vector u, it automatically contains an infinite number of vectors (those of the form αu as α ranges over all scalars).

Dependence and Independence of Vectors

Definition. *Let $u_1, \dots, u_k$ be vectors in the linear space X. A vector of the form $\alpha_1 u_1 + \dots + \alpha_k u_k$ is said to be a* linear combination *of the vectors $u_1, \dots, u_k$. The combination is said to be* nontrivial *if at least one of the α_i's is nonzero. A set of vectors $u_1, \dots, u_k$ is* dependent *if there exists a nontrivial linear combination of them which is equal to the zero vector; otherwise, the set is independent.*

REMARKS

1. We can restate the above as follows: The vectors $u_1, \dots, u_k$ are dependent if there exist numbers $\alpha_1, \dots, \alpha_k$ *not all* 0 such that

$$\alpha_1 u_1 + \dots + \alpha_k u_k = 0; \tag{4.2.1}$$

   if the only solution of (4.2.1) is $\alpha_1 = \dots = \alpha_k = 0$, the vectors are independent.

2. If $u_1, \dots, u_k$ are dependent (or form a dependent set), at least one vector is a linear combination of the others.

3. If 0 belongs to a set of vectors, the set is dependent.

4. If the set $u_1, \dots, u_k$ is dependent, so is any set which includes all the vectors $u_1, \dots, u_k$.

5. For an infinite set of vectors there are two possible definitions of independence. The first is purely algebraic and states that an infinite set is algebraically independent if every finite subset is independent. Instead, we could demand that the equation $\alpha_1 u_1 + \dots + \alpha_k u_k + \dots = 0$ be satisfied only for $\alpha_1 = \dots = \alpha_k = \dots = 0$. Obviously, the latter definition requires the concept of an infinite series of vectors, which is not yet available (see Sections 4.5 and 4.6).

Dimension of a Linear Space

Definition. *The linear space X is n-*dimensional *if it possesses a set of n independent vectors, but every set of $n + 1$ vectors is a dependent set. The space X is* ∞-dimensional *if for each positive integer k, one can find a set of k independent vectors in X.*

At the risk of stating the obvious, we should point out that even a one-dimensional space contains an infinite number of elements. Next we turn to the concept of a basis in a finite-dimensional space. At this stage we introduce the notion in purely algebraic terms, which we shall not bother to extend to infinite-dimensional spaces (a definition of basis which brings the metric properties into play is given in Section 4.6). The importance of a basis is that it enables us to express all the many vectors in X in terms of a relatively small, fixed set of vectors.

Definition. *A finite set of vectors $h_1, \ldots, h_k$ is said to be a* basis *for the finite-dimensional space X if each vector in X can be represented in* one and only one way *as a linear combination of $h_1, \ldots, h_k$.*

The following are nearly immediate consequences of the definition and the concept of independence. The proofs are left to the reader.

Theorem. *The vectors in a basis are independent.*

Theorem. *In an n-dimensional space, any set of n independent vectors forms a basis.*

Linear Manifolds

In ordinary, three-dimensional space V_3, we can find subsets, known as *linear manifolds*, which are themselves linear spaces under the same definition of vector addition and scalar multiplication used in V_3. The linear manifolds are just the lines and planes containing the origin (of course, the whole space and the set consisting only of the zero vector are also linear manifolds, although not very exciting ones).

Definition. *A set M in a linear space X is a* linear manifold *(in X) if whenever the vectors u and v belong to M, so does $\alpha u + \beta v$ for arbitrary scalars α and β.*

A linear manifold must contain the zero vector and may be regarded as a linear space in its own right with vector addition and scalar multiplication inherited from X. It therefore makes sense to talk of the dimension of a linear manifold. One of the easiest ways to construct linear manifolds is by taking linear combinations of vectors in X.

Definition. *Let U be a set of vectors (finite or infinite) in the linear space X. The* algebraic span *of U, denoted by $S(U)$, is the set of all finite linear combinations of vectors chosen from U.*

Clearly, $S(U)$ is always a linear manifold in X; other names for $S(U)$ are the *linear manifold generated by U* or the *linear hull of U*. If U consists of a single nonzero vector u in X, the algebraic span $S(U)$ is the set of all vectors of the form αu, that is, a one-dimensional linear manifold. If U consists of two independent vectors u and v, the algebraic span of U is the two-dimensional manifold of vectors having the form $\alpha u + \beta v$. If, on the other hand, U is a set of two dependent vectors u and v, not both 0, $S(U)$ is a one-dimensional manifold. Generally, if U is a set of k vectors, its algebraic span $S(U)$ is a linear manifold of dimension $\leqslant k$, with equality if and only if the k vectors are independent. If X is n-dimensional and U is a basis for X (that is, a set of n independent vectors), $S(U) = X$; of course, it is also true that $S(X) = X$, but U has fewer elements than X.

REMARK. If u and v are independent vectors in X, the set P of vectors of the form $u + \alpha v$ is *not* a linear manifold, since P does not contain the origin.

Examples of Linear Spaces

Example 1. We have already encountered the space $\mathbb{R}^n$ of all ordered n-tuples of real numbers. This space appears in a natural way in the solution of n linear equations in n unknowns. Let $u = (\xi_1, \ldots, \xi_n)$ and $v = (\eta_1, \ldots, \eta_n)$ be two arbitrary elements in $\mathbb{R}^n$, and let α be an arbitrary real number. The following definitions clearly make $\mathbb{R}^n$ a linear space:

$$\begin{aligned} u + v &= (\xi_1 + \eta_1, \ldots, \xi_n + \eta_n), \\ \alpha u &= (\alpha\xi_1, \ldots, \alpha\xi_n), \\ 0 &= (0, \ldots, 0), \\ -u &= (-\xi_1, \ldots, -\xi_n). \end{aligned}$$

The set of n vectors $(1, 0, \ldots, 0)$, $(0, 1, 0, \ldots, 0)$, $\ldots$, $(0, 0, \ldots, 0, 1)$ forms a basis in $\mathbb{R}^n$, which is therefore an n-dimensional real linear space. There are, of course, many other bases. Examples of linear manifolds are the set of all vectors of the form $(\xi_1, \xi_2, 0, \ldots, 0)$; the set of all vectors with $\xi_1 = \xi_2 = \cdots = \xi_n$; the set of all solutions of a fixed *homogeneous* system of n linear equations in n unknowns with real coefficients.

Example 2. The space $\mathbb{C}^n$ consists of ordered n-tuples of complex numbers with complex numbers α as scalar multipliers. Otherwise, exactly the same definitions are used as in Example 1; the specific instances of a basis and of linear manifolds shown there carry over to $\mathbb{C}^n$. Thus $\mathbb{C}^n$ is a complex linear space of dimension n. It may at first seem paradoxical that $\mathbb{C}^n$ and $\mathbb{R}^n$ have the same dimension, since $\mathbb{C}^n$ appears much larger than $\mathbb{R}^n$. For instance, $\mathbb{C} = \mathbb{C}^1$ is the set of all complex numbers, which is normally identified with the real plane. However, as a complex linear space, $\mathbb{C}$ has dimension 1.

Example 3. We are now going to create linear spaces whose elements $u, v, \ldots$ are real-valued functions $u(x)$, $v(x)$, $\ldots$ on $a \leqslant x \leqslant b$. All we do is look at ordinary operations on functions, but from a slightly more abstract point of view. The function or curve $u(x)$ is now regarded as a single entity—an element u in a space $X^{(r)}$. How do we add such entities? The element (or vector) $u + v$ is obtained from the elements u and v by adding their ordinates pointwise, as in Figure 4.3. Thus $u + v$ is just the curve or function usually written as $u(x) + v(x)$. The function which vanishes identically in $a \leqslant x \leqslant b$ plays the role of the 0 vector in the space. Scalar multiplication by real numbers is defined in the obvious way. With these definitions we see that the space $X^{(r)}$ of all real-valued functions on $a \leqslant x \leqslant b$ is a real linear space. This space is much too general to be useful, but the following linear manifolds in $X^{(r)}$ will come up in applications [where they are usually regarded as linear spaces in their own right rather than as linear manifolds in $X^{(r)}$]:

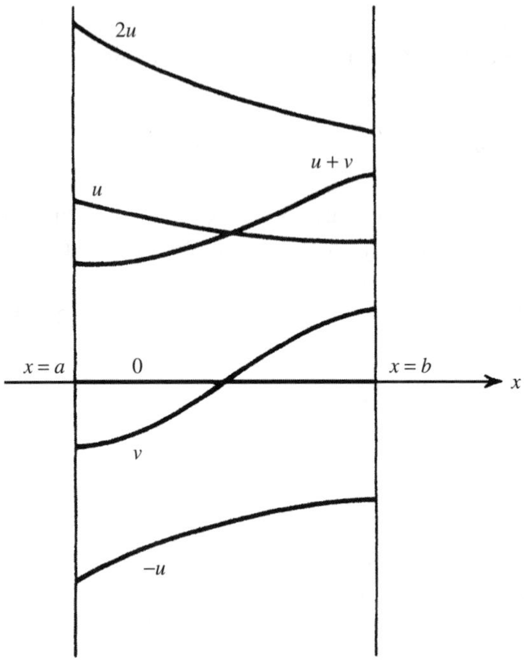

Figure 4.3

(a) P_n, the set of all polynomials of degree *less* than n, is an n-dimensional space. The simplest basis in P_n is $h_1 = 1, h_2 = x, h_3 = x^2, \ldots, h_n = x^{n-1}$, where $h_1 = 1$ is the function equal to 1 on $a \leqslant x \leqslant b$.

(b) The set of all polynomials is an infinite-dimensional space. Any nontrivial polynomial can have only a finite number of zeros, so that the equation

$$\alpha_1 + \alpha_2 x + \alpha_3 x^2 + \cdots + a_n x^{n-1} = 0, \quad a \leqslant x \leqslant b,$$

can only have the solution $\alpha_1 = \alpha_2 = \cdots = \alpha_n = 0$. Thus, whatever n is, the set $1, x, x^2, \ldots, x^{n-1}$ is independent. The set of all polynomials therefore contains independent sets with arbitrarily large numbers of elements. From the definition of dimension, it follows that the space is infinite-dimensional.

(c) The set of all real-valued solutions of $(d^2u/dx^2) - u = 0$, $a \leqslant x \leqslant b$, is a two-dimensional linear space, for which the pair $h_1 = e^x$ and $h_2 = e^{-x}$ is a basis. Another basis is the pair $\sinh x$ and $\cosh x$. Note that the set of all solutions of an inhomogeneous equation does *not* form a linear space, nor does the set of solutions of a nonlinear equation.

(d) The set of all real-valued continuous functions $u(x)$ on $a \leqslant x \leqslant b$ is an infinite-dimensional linear space. The set of functions which are even about the point $x = (a + b)/2$ is an infinite-dimensional linear manifold in this space.

(e) The set of all real-valued functions $u(x)$ on $a \leqslant x \leqslant b$ for which the Riemann integral $\int_a^b u^2(x)\,dx$ is finite is an infinite-dimensional linear space. All that has to be shown is that finite values for $\int_a^b u^2(x)\,dx$ and $\int_a^b v^2(x)\,dx$ imply a finite value for $\int_a^b (u+v)^2\,dx$. This follows from the Schwarz inequality (4.5.14).

Example 4. The set of complex-valued functions $u(x), v(x), \ldots$ on a real interval $a \leqslant x \leqslant b$ forms a complex linear space $X^{(c)}$ under the following definitions. The element $u + v$ is a complex-valued function whose value at x is obtained by adding the complex numbers $u(x)$ and $v(x)$. The element αu, where complex α's are admitted, is the complex-valued function whose value at x is obtained by multiplying the complex numbers α and $u(x)$. Again the function 0 plays the role of the zero vector. The reader should have little difficulty in modifying the real spaces in Example 3 to complex spaces.

Example 5. Let V be a region in the Euclidean space of the k variables $x_1, \ldots, x_k$. The set of all real (or complex)-valued functions $u(x_1, \ldots, x_k)$ on V is an infinite-dimensional linear space.

EXERCISES

4.2.1 Prove the statements in Remarks 5 and 6 following the definition of a linear space.

4.2.2 Define a linear space as follows: Elements are n-tuples of complex numbers, but only scalar multiplication by real numbers is allowed; otherwise, the same definitions as in Example 1 of linear spaces hold. Verify that this is indeed a *real* linear space and that its dimension is $2n$.

4.2.3 Show from the definition of a linear space that $\alpha 0 = 0$; $u + v = u + w$ implies that $v = w$; $\alpha u = \alpha v$ and $\alpha \neq 0$ imply that $u = v$; $\alpha u = \beta u$ and $u \neq 0$ imply that $\alpha = \beta$.

4.2.4 Let C^* be the set of all real-valued continuous functions on $-\infty < x < \infty$ which fail to have a first derivative at the fixed point $x = 0$. The operations are defined as usual for a function space. Is C^* a linear space?

4.2.5 Consider the set P_n^0 of all polynomials of degree less than n that have the value 0 at some fixed point, say $x = \xi$. With operations as in P_n, show that P_n^0 is a linear space. What is its dimension? Find a basis for P_n^0. Show that the set of polynomials that take on the value 1 at $x = \xi$ is not a linear space.

4.2.6 Consider the differential equation $d^2u/dx^2 = 0$ on $0 < x < 1$. What is the dimension of the linear space of solutions satisfying the boundary condition $u(0) = u(1)$; the boundary conditions $u(0) = u(1) = 0$; the boundary conditions $u'(0) = u'(1) = 0$?

4.2.7 Another way of showing that $1, x, \ldots, x^{n-1}$ are independent on $a \leqslant x \leqslant b$ is to consider the equation $\alpha_1 + \alpha_2 x + \cdots + \alpha_n x^{n-1} = 0$ on $a \leqslant x \leqslant b$, and to differentiate the equation $n-1$ times. We then have n homogeneous equations in the n unknowns $\alpha_1, \dots, \alpha_n$. Since the determinant of the coefficients is easily seen to be different from 0, the required independence follows. Let $c_1, \dots, c_k$ be arbitrary *unequal* real numbers. Show that $x^{c_1}, \dots, x^{c_k}$ are independent on any interval $0 < a \leqslant x \leqslant b$.

4.3 METRIC SPACES, NORMED LINEAR SPACES, AND BANACH SPACES

The vectors encountered so far possess algebraic properties similar to those of ordinary three-dimensional vectors—they can be added and multiplied by scalars subject to the rules of a linear space—but the equally familiar notions of length and direction have not yet made their appearance. We wish to endow linear spaces of higher dimension with a structure as rich as the one of three-dimensional vector analysis. On the algebraic structure already introduced, we must superimpose a metric structure. The metric structure can be studied without reference to any coexisting algebraic structure and leads immediately to the definition of a convergent sequence. A *metric space* is a set of elements (often called *points* to emphasize their geometric interpretation) endowed with a suitable *metric* $d(u, v)$ measuring the distance between pairs of points. The metric is required to satisfy the conditions (4.3.1), which certainly agree with our experience. A sequence of points $\{u_k\}$ is then said to converge to the point u if the *numerical* sequence $d(u_k, u)$ converges to zero. If, as in most applications, the elements of the metric space also form a linear space, they can be regarded interchangeably as vectors emanating from a common origin or as the endpoints of these vectors. It then seems natural to derive the metric from the more primitive notion of length (or *norm*) of a vector, the norm $\|u\|$ being required to satisfy conditions (4.3.2) reflecting familiar properties of Euclidean space; $\|u\|$ can be viewed as either the length of u or as the distance of the point u from the origin; in this spirit, we then define $d(u, v) = \|u - v\|$ and easily check that conditions (4.3.1) are satisfied. Thus, any linear space provided with a norm automatically becomes a metric space by setting $d(u, v) = \|u - v\|$. It is worth emphasizing that in a general metric space, a function of two variables is needed to determine the distance between two points, whereas in a normed linear space a function of a single variable suffices.

Let us now give the appropriate definitions.

Definition. *A set X of* elements *(or* points*) $u, v, w, \dots$ is said to be a* metric space *if to each pair of elements u, v there is associated a real number $d(u, v)$ the* distance *between u and v, satisfying*

$$\left.\begin{array}{ll} d(u,v) > 0 \textit{ for } u, v \textit{ distinct}, & \\ d(u,u) = 0, & \\ d(u,v) = d(v,u), & \\ d(u,w) \leqslant d(u,v) + d(v,w) & \textit{triangle inequality}. \end{array}\right\} \tag{4.3.1}$$

The function d is known as the metric.

Definition. *A* normed linear space *is a linear space (real or complex) in which a real-valued function* $\|u\|$ *(known as the* norm *of* u*) is defined, with the properties*

$$\left.\begin{array}{ll}(a)\ \|u\|>0, \quad u\neq 0, & \\ (b)\ \|0\|=0, & \\ (c)\ \|\alpha u\|=|\alpha|\,\|u\|, & \\ (d)\ \|u+v\|\leqslant\|u\|+\|v\| & \textit{triangle inequality}.\end{array}\right\} \tag{4.3.2}$$

REMARKS

1. Here $\|u\|$ plays the same role as the length of u in ordinary three-dimensional space. Property (c) states that the length of a scalar multiple of a vector u is just the appropriate multiple of the length of the vector u (note that the vector $-3u$ should have and has three times the length of u). Property (d) is a form of the triangle inequality. Clearly, all the properties (4.3.2) hold for the usual three-dimensional vectors.

2. A normed linear space is *automatically* a metric space with the metric defined by
$$d(u,v)=\|u-v\|.$$
Let us see that (4.3.1) now follows from (4.3.2). Clearly, $d(u,v)>0$ for u, v distinct and $d(u,u)=0$. Since, by (4.3.2)(c), $\|u-v\|=\|v-u\|$, we have $d(u,v)=d(v,u)$. In (4.3.2)(d) we substitute $u-v$ for u and $v-w$ for v to obtain $\|u-w\|\leqslant\|u-v\|+\|v-w\|$ or $d(u,w)\leqslant d(u,v)+d(v,w)$. Thus $\|u-v\|$ defines a suitable metric known as the *natural metric generated by the norm*. Observe that we can recover the norm from d by
$$\|u\|=d(u,0).$$
The metric generates an automatic notion of convergence: A sequence of points $\{u_k\}$ converges to the point u if and only if the sequence of *real numbers* $\{d(u,u_k)\}$ converges to 0.

Definition. *We write* $\lim_{k\to\infty}u_k=u$ *or* $u_k\to u$*, and say that* $\{u_k\}$ converges *to* u*, or that* $\{u_k\}$ *has the* limit u*, if for each* $\varepsilon>0$ *there exists an index* N *such that*

$$d(u,u_k)\leqslant\varepsilon \quad \textit{whenever}\ k>N.$$

It is a consequence of the triangle inequality that a sequence can have but one limit.

We cannot speak of series in a general metric space, as there is no notion of addition. In normed spaces, however, we can add elements and introduce the concept of convergence of a series.

Definition. *In a normed space, the* series

$$u_1+u_2+\cdots+u_n+\cdots$$

is said to converge to u if the sequence of partial sums converges to u, that is, if to each $\varepsilon > 0$, there exists $N > 0$ such that

$$\left\| \sum_{k=1}^{n} u_k - u \right\| < \varepsilon, \quad n > N.$$

We then write $\sum_{k=1}^{\infty} u_k = u$.

To use the definition of convergence of a sequence, we need to know or guess the limit of the sequence. Can we determine if a sequence converges without information on the limit? The answer is yes in many cases. We shall see that in a complete space it suffices to show that a sequence is a Cauchy sequence to guarantee its convergence.

Definition. *A sequence $\{u_k\}$ is a* Cauchy *(or* fundamental*) sequence if for each $\varepsilon > 0$ there exists N such that*

$$d(u_m, u_p) \leqslant \varepsilon \quad \text{whenever}\, m, p > N.$$

Thus, a Cauchy sequence is one where the points beyond a certain index are close to each other. We easily prove the following:

Theorem. *If a sequence $\{u_k\}$ converges, it is fundamental.*

Proof. Let u be the limit of $\{u_k\}$. From the triangle inequality, it follows that $d(u_m, u_p) \leqslant d(u_m, u) + d(u, u_p)$. Since $u_k \to u$, there exists an integer N such that $d(u_n, u) \leqslant \varepsilon/2$ whenever $n > N$. If p and m are chosen larger than N, then $d(u_m, u_p) \leqslant \varepsilon$, as required. □

It might appear at first that the converse is also true. If all the points of a sequence beyond a certain index are close to each other, they should be close to some definite point u. This will be true if the point u has not been carelessly left out of the space.

Definition. *A metric space X is* complete *if every Cauchy sequence of points from X converges to a limit in X.*

Definition. *A normed linear space which is complete in its natural metric $d(u, v) = \|u - v\|$ is a* Banach space.

In a *complete* metric space we can ascertain that a sequence converges by showing that it is a Cauchy sequence—a much simpler task, since the actual limit is not involved. The reverse side of the coin is that this approach does not help us find the limit of the sequence!

The importance of completeness of a space lies in our ability to carry out limiting processes within the space; this, in turn, is the glue needed to build a sound mathematical framework to support calculations.

The Set of Rational Numbers Is Incomplete

The space *Rat* of all rational numbers $u, v, \ldots$ with the usual metric $|u - v|$ is not complete and must be enlarged to the real number system so that simple equations with rational coefficients such as $u^2 = 2$ can be solved. We now exhibit a Cauchy sequence in *Rat* which does not converge in *Rat*:

$$u_1 = 1, u_2 = 1 + \frac{1}{1!}, \ldots, u_n = 1 + \frac{1}{1!} + \frac{1}{2!} + \cdots + \frac{1}{(n-1)!}, \ldots.$$

For $p > m \geqslant 1$ we have

$$\begin{aligned} |u_p - u_m| &= \frac{1}{m!} + \frac{1}{(m+1)!} + \cdots + \frac{1}{(p-1)!} \\ &= \frac{1}{m!}\left[1 + \frac{1}{(m+1)} + \cdots + \frac{1}{(m+1)(m+2)\cdots(p-1)}\right] \\ &\leqslant \frac{1}{m!}\left(1 + \frac{1}{2} + \frac{1}{4} + \cdots\right) = \frac{2}{m!}, \end{aligned}$$

so that for all m, p sufficiently large we can make $|u_p - u_m|$ as small as we wish. The sequence is therefore a fundamental sequence in *Rat*, but we now show that it does not have a limit in *Rat*. In fact, suppose that $u_k \to p/q$, where p and q are integers and q is chosen larger than 2 (which can obviously always be done). From the definition of convergence there exists N such that

$$\left|\frac{p}{q} - \left(1 + \cdots + \frac{1}{k!}\right)\right| < \frac{1}{4q!} \quad \text{whenever } k > N.$$

Since the sequence $\{u_k\}$ is strictly increasing, this means that

$$0 < \frac{p}{q} - \left(1 + \cdots + \frac{1}{k!}\right) < \frac{1}{4q!}, \quad k > N.$$

Multiplying by $q!$ and rearranging, we obtain

$$0 < \text{integer} < \frac{1}{4} + \frac{q!}{(q+1)!} + \cdots + \frac{q!}{k!},$$

and since $q > 2$,

$$\begin{aligned} \frac{q!}{(q+1)!} + \cdots + \frac{q!}{k!} &< \frac{1}{q+1} + \cdots + \frac{1}{(q+1)(q+2)\cdots(k)} \\ &< \frac{1}{3} + \frac{1}{3^2} + \cdots = \frac{1}{2}. \end{aligned}$$

The earlier inequality then yields $0 <$ integer $< \frac{3}{4}$, which is a contradiction. Thus $\{u_k\}$ cannot converge in *Rat*. If, however, the sequence $\{u_k\}$ is viewed as a sequence in $\mathbb{R}$, the space of real numbers with the metric $|u - v|$, then $\{u_k\}$ converges to the real number e, which is not a rational number.

The difficulty in this example lies in the space of rational numbers, which, unfortunately, is full of gaps invisible to the naked eye. We fill in the gaps in the space *Rat* as follows. If a Cauchy sequence of rational numbers does not have a rational limit, we associate with this sequence an abstract element which plays the role of the limit. These abstract elements, known as *irrational numbers*, are used to enlarge the space of rational numbers, the larger space being called the *space of real numbers*. Is this space complete? The answer is yes, either by adopting this as an axiom of the real number system or by showing that it follows from some other "equivalent" axiom. In any event the completeness of the real number system is the rock on which analysis is based. Once the completeness of the real number system is accepted, it becomes relatively easy to show that every incomplete metric space can be completed (see Exercise 4.3.1).

Examples of Metric Spaces

Most of the metric spaces below are normed linear spaces endowed with the natural metric $d(u, v) = \|u - v\|$.

Example 1. On the real line $\mathbb{R}$, set $\|u\| = |u|$ to obtain a normed linear space with natural metric $d(u, v) = |u - v|$. As stated above, this space is complete—a Banach space. For other metrics on $\mathbb{R}$, see Exercise 4.3.2.

Example 2. The set of all complex numbers $z = x + iy$ is a Banach space under the norm

$$\|z\| = |z| = [x^2 + y^2]^{1/2}.$$

Example 3. Consider the set $\mathbb{R}^n$ of all ordered n-tuples, $u = (\xi_1, \dots, \xi_n), v = (\eta_1, \dots \eta_n), \dots,$ of real numbers. Various metrics can be defined, yielding different metric spaces. The ones here are derived from norms. If $p \geqslant 1$, we introduce the class of norms

$$\|u\|_p = (|\xi_1|^p + \cdots + |\xi_n|^p)^{1/p}.$$

Of particular interest are the cases $p = 1, p = 2$:

$$\|u\|_1 = |\xi_1| + \cdots + |\xi_n| \tag{4.3.3}$$

$$\|u\|_2 = (|\xi_1|^2 + \cdots + |\xi_n|^2)^{1/2}, \quad \text{the } Euclidean\ norm. \tag{4.3.4}$$

Another important norm is the *sup norm*

$$\|u\|_\infty = \max_k |\xi_k|. \tag{4.3.5}$$

The first three properties in (4.3.2) are easily verified in all cases but note that the exponent $1/p$ is needed for condition (4.3.2)(c). Condition (4.3.2)(d) is straightforward for (4.3.3) and (4.3.5).

For other p, you can use Minkowski's inequality (see Exercise 4.3.7). The special case $p = 2$ is easy to treat separately by first deriving the Schwarz inequality (4.3.8). Exercise 4.3.8 shows that the index notation for the norms is consistent.

All these spaces are complete; let us prove this for $\mathbb{R}^n$ with the norm $\|u\|_p$. Let $\{u_k = (\xi_1^{(k)}, \ldots, \xi_n^{(k)})\}$ be a fundamental sequence; then for each $\varepsilon > 0$, there exists N such that

$$\|u_k - u_m\|_p = [|\xi_1^{(k)} - \xi_1^{(m)}|^p + \cdots + |\xi_n^{(k)} - \xi_n^{(m)}|^p]^{1/p} < \varepsilon$$

whenever $k, m > N$. This implies that

$$|\xi_1^{(k)} - \xi_1^{(m)}| \leqslant \varepsilon, \ldots, |\xi_n^{(k)} - \xi_n^{(m)}| \leqslant \varepsilon, \quad k, m > N.$$

From the completeness of the real number system, we conclude that each of the n sequences $\{\xi_1^{(k)}\}, \ldots, \{\xi_n^{(k)}\}$ converges as $k \to \infty$. Letting $\lim_{k\to\infty} \xi_i^{(k)} = \xi_i$, we see that $\lim_{k\to\infty} u_k = (\xi_1, \ldots, \xi_n)$. Thus, $\mathbb{R}^n$ with the norm $\|u\|_p$ is a Banach space for each $p \geqslant 1$. It is also easy to show that $\mathbb{R}^n$ with the sup norm is a Banach space.

Example 4. For the set of ordered n-tuples of complex numbers, all of the special cases and proofs of Example 3 carry over without additional difficulty.

Example 5. Consider the set of all real-valued continuous functions $u(x)$ defined on $a \leqslant x \leqslant b$ with norm

$$\|u\|_\infty = \max_{a\leqslant x\leqslant b} |u(x)|. \tag{4.3.6}$$

Properties (4.3.2) are easily verified, so that we have a normed linear space denoted by $C[a, b]$. Let $\{\dot{u}_k(x)\}$ be a Cauchy sequence in $C[a, b]$; then for each $\varepsilon > 0$ there exists N such that

$$\max_{a\leqslant x\leqslant b} |u_k(x) - u_m(x)| < \varepsilon, \quad m, k > N.$$

But this is just the Cauchy criterion for uniform convergence; hence there exists a function $u(x)$, necessarily continuous, to which the sequence $\{u_k(x)\}$ converges uniformly. The space $C[a, b]$ is therefore complete. The metric (4.3.6) is known as the *uniform norm* or *sup norm*.

Example 6. Consider, as in Example 5, the set of all real-valued continuous functions $u(x)$ on the bounded interval $a \leqslant x \leqslant b$ but now with the L_2 norm:

$$\|u\|_2 = \left[\int_a^b |u(x)|^2 \, dx\right]^{1/2}. \tag{4.3.7}$$

This is easily seen to meet conditions (4.3.2), the triangle inequality again being a consequence of the Schwarz inequality, this time in the form (4.5.14). The space is, however, *not complete*. In fact, we have already seen in Chapter 1 that functions with jump discontinuities can be approximated in the L_2 sense (that is, in the metric $d_2 = \|u - v\|_2$) by continuous functions. The sequence of continuous functions

$$u_k(x) = \frac{1}{2} + \frac{1}{\pi} \arctan kx, \quad -1 \leqslant x \leqslant 1,$$

is a Cauchy sequence in the metric d_2, which also converges pointwise to the discontinuous function $u(x) = \frac{1}{2} + \frac{1}{2} \operatorname{sgn} x$. Moreover, $\lim_{k\to\infty} d_2(u_k, u) = 0$; that is, u_k approaches u in the L_2 sense. There is no continuous function $v(x)$ for which $d_2(u, v) = 0$, so that there cannot exist a continuous function v for which $\lim_{k\to\infty} d_2(u_k, v) = 0$.

Example 7. The space $L_2^{(r)}(a, b)$, probably the most important in applications, is the *completion* of the set of continuous functions in the metric d_2 of Example 6. The difficulty with Example 6 is that the space of continuous functions is not extensive enough to accommodate the metric d_2, just as the space of rational numbers is not extensive enough for the metric $|u - v|$. We fill the "gaps" in the set of continuous functions by adjoining to the set abstract elements associated with nonconvergent Cauchy sequences. The procedure is described in Exercise 4.3.1. The resulting space is the space $L_2^{(r)}(a, b)$ of real-valued functions square integrable in the Lebesgue sense (see Chapter 0).

Starting instead with the set of complex-valued functions on $a \leqslant x \leqslant b$ and completing it with respect to the metric d_2, we obtain the space $L_2^{(c)}(a, b)$ of complex-valued functions square integrable in the Lebesgue sense.

Similarly, we can consider the completion of the set of real- or complex-valued functions $u(x) = u(x_1, \ldots, x_k)$ defined on a k-dimensional domain D in the variables $x_1, \ldots, x_k$ under the metric

$$d_2(u, v) = \|u - v\|_2 = \left[\int_D |u(x) - v(x)|^2 \, dx_1 \cdots dx_k \right]^{1/2}.$$

Example 8. The following metric space appears in coding theory. A message is an ordered n-tuple of binary digits such as $(0, 1, 1, 0, \ldots, 0)$. Let $x = (\xi_1, \ldots, \xi_n)$ and $y = (\eta_1, \ldots, \eta_n)$ be two messages. We define $d(x, y) = |\xi_1 - \eta_1| + \cdots + |\xi_n - \eta_n|$; in other words, d is just the total number of places in which the two messages differ. It is easy enough to verify that all the axioms for a metric space are satisfied and that the space is complete. Obviously, this is not a linear space!

Additional Topological Concepts

There are some additional properties of metric spaces that will be needed in Section 4.4. We shall present these in the context of a *Banach space* X (complete normed linear space).

The *closure* of S (in X) is the set $\bar{S}$ consisting of the limits of all sequences that can be constructed from S. $\bar{S}$ must contain all points u in S, since the sequence

$u, u, u, \ldots$ clearly has limit u. $\bar{S}$ may in addition contain points of X not in S. If S is the set $0 < u < 1$ on the real line X with the usual distance between points, $\bar{S}$ is the set $0 \leqslant u \leqslant 1$. If S is the set $x^2 + y^2 + z^2 < 1$ in the usual three-dimensional space, $\bar{S}$ is the set $x^2 + y^2 + z^2 \leqslant 1$.

A set S is *closed* (or closed in X if the underlying space is in doubt) if $\bar{S} = S$. Thus limiting processes do not take us out of a closed set. The set $\bar{S}$ is closed whether or not S is. A set containing only a finite number of elements is closed. A set S is *open* if whenever u is in S, so is some neighborhood of u. Precisely: For each u in S there exists $a > 0$ such that all points v in X satisfying $\|u - v\| < a$ are also in S. On the real line the set $0 < u < 1$ is open; in three-dimensional space the set $x^2 + y^2 + z^2 < 1$ is open. A set does not have to be either open or closed.

Let $S \subset T$ be two subsets of X. S is *dense* in T if for each element u in T and each $\varepsilon > 0$, there exists an element v of S such that $\|u - v\| < \varepsilon$. Thus every element of T can be approximated to arbitrary precision by elements of S. A set is always dense in itself and in its closure. If S is dense in T and T is dense in U, then S is dense in U. The set of rational numbers is dense on the real line. The set of all polynomials is dense in the set of continuous functions under the norm of Example 5; this is just the content of the Weierstrass approximation theorem proved in Exercise 2.2.7. The set of polynomials is also dense in the set of continuous functions under the norm of Example 6. In fact, if $u(x)$ is a continuous function and $p(x)$ a polynomial, we have

$$\|u - p\|_2^2 = d_2^2(u, p) = \int_a^b |u - p|^2\, dx \leqslant (b - a) \max_{a \leqslant x \leqslant b} |u - p|^2,$$

and since by the Weierstrass theorem we can find p such that $\max |u - p|$ is arbitrarily small, we can make d_2 arbitrarily small. Under the metric d_2 it can also be shown that continuous functions are dense in $L_2(a, b)$. Thus polynomials are dense in $L_2(a, b)$, that is, *any square-integrable function can be approximated in* L_2 *by polynomials.* The statement remains valid for complex-valued functions.

A set S is *bounded* if there exists a number M such that $\|u - v\| \leqslant M$ for all u, v in S.

A set is *compact* if each sequence of points in S contains a subsequence which converges to a point in S. (Given a sequence $u_1, u_2, \ldots$ and a sequence of positive integers $k_1 < k_2 < \cdots$, the sequence $u_{k_1}, u_{k_2}, \ldots$ is said to be a subsequence of $u_1, u_2, \ldots$. Thus $1, 5, 9, 13, \ldots$ is a subsequence of $1, 3, 5, 7, \ldots$, but $1, 1, 1, \ldots$ is not.) Note that a compact subset of X is necessarily closed in X and bounded. In the n-dimensional spaces of Example 3, any bounded closed set is compact, but this will not be the case in some infinite-dimensional spaces. For instance, in the Banach space $C[0, 1]$ of Example 5 (continuous functions on $0 \leqslant x \leqslant 1$ with uniform norm), the set S of continuous functions satisfying $|u(x)| \leqslant 1$ is a bounded closed set that is *not* compact, because the sequence $1, x, x^2, \ldots$ chosen from S has no converging subsequence in the uniform norm. An infinite orthonormal set in $L_2(0, 1)$ is another example of a bounded closed set which is not compact. A set S in X is *relatively compact* if $\bar{S}$ is compact. In other words, each sequence from S contains a subsequence which converges to an element of X (not necessarily in S). Thus on the real line any bounded set (closed, open, or neither) is relatively compact.

On a general metric space it makes no sense to talk about linear and nonlinear transformations since there is no algebraic structure. For a normed linear space X, however, we define a linear transformation L to have the property

$$L(\alpha u + \beta v) = \alpha L u + \beta L v$$

for all u, v in X and all real (or complex) α and β. In the next section we discuss transformations on normed linear spaces, but these transformations will usually *not* be linear.

EXERCISES

4.3.1 *Completion of a metric space X with metric d*

(a) Two Cauchy sequences $\{u_k\}$ and $\{u_k'\}$ in X are equivalent if and only if $d(u_k, u_k') \to 0$ as $k \to \infty$. We write $\{u_k\} \sim \{u_k'\}$. Show that if $\{u_k\} \sim \{u_k'\}$ and $\{u_k'\} \sim \{u_k''\}$, then $\{u_k\} \sim \{u_k''\}$. The space of all Cauchy sequences in X is therefore partitioned into so-called *equivalence classes*. Each equivalence class consists of all Cauchy sequences equivalent to one another. These equivalence classes are denoted by capital letters such as U and V. By a representative of U we mean any Cauchy sequence $\{u_k\}$ in U.

(b) Let Y be the space of equivalence classes of Cauchy sequences defined in (a). We tentatively define a metric δ on Y by

$$\delta(U, V) = \lim_{k \to \infty} d(u_k, v_k),$$

where $\{u_k\}$ and $\{v_k\}$ are representatives of U and V, respectively. Show that this definition makes sense, that is, that the right side exists and is independent of the representatives used. Next, show that δ satisfies the axioms of a metric.

(c) Show that Y is complete in the metric δ.

(d) Each element u in X can be regarded as an element of Y by identifying u with the equivalence class of all Cauchy sequences that converge to u (one such sequence always exists: $u, u, u, \ldots$). If u, v are elements in X, and U, V are their respective equivalence classes, show that $d(u, v) = \delta(U, V)$. Thus X can be regarded as a subset of Y, and Y is therefore a completion of X. Show that X is dense in Y. In this sense the completion is unique.

4.3.2 (a) Is $d(u, v) = (u - v)^2$ an admissible metric on the real line? $\sqrt{|u - v|}$?

(b) Is $d(u, v) = |\xi_1 - \eta_1|$ an admissible metric on the set of all ordered n-tuples of real numbers $(\xi_1, \ldots, \xi_n)$?

(c) For what functions $f(u)$ is $f(|u-v|)$ a metric on $\mathbb{R}$?

4.3.3 (a) Two metrics d, δ on the same set X are said to be *equivalent* if convergence to a limit in either metric entails convergence to the same limit in the other. Show that d and δ are equivalent if there exist positive constants α and β independent of u, v such that

$$ad(u,v) \leqslant \delta(u,v) \leqslant \beta\, d(u,v).$$

(b) Let X be the set of continuously differentiable real functions $u(x)$ on the interval $0 \leqslant x \leqslant 1$, satisfying $u(0)=0$. For $u, v \in X$, define

$$d^2(u,v) = \int_0^1 (u-v)^2\,dx, \quad \delta^2(u,v) = \int_0^1 (u'-v')^2\,dx,$$
$$D^2(u,v) = d^2 + \delta^2,$$

and show that d, δ, D are all admissible metrics. Show that δ and D are equivalent. Of course, δ and d are *not* equivalent, as is seen by examining elements of the form $\sin \alpha x$. What about d and D?

4.3.4 (a) Show that

$$d(u,w) \leqslant d(u,v_1) + d(v_1,v_2) + d(v_2,v_3) + \cdots + d(v_n,w)$$

for any u, w, $v_1, \ldots, v_n$. The geometric interpretation is that the length of a side of a polygon does not exceed the sum of the lengths of the other sides.

(b) Show that the metric is a continuous function; that is,

$$\lim_{\substack{k\to\infty \\ j\to\infty}} d(u_k, v_j) = d(u,v) \quad \text{whenever } u_n \to u \text{ and } v_n \to v.$$

(c) In a normed linear space show that:

If $u_n \to u$, $\|u_n\| \to \|u\|$.
If $\alpha_n \to \alpha$, $\alpha_n u \to \alpha u$.
If $u_n \to u$ and $v_n \to v$, $\alpha u_n + \beta v_n \to \alpha u + \beta v$.
For any u and v, $\quad |\,\|u\| - \|v\|\,| \leqslant \|u+v\|$.

4.3.5 *The Schwarz inequality in* $\mathbb{R}^n$. Let $u = (\xi_1, \ldots, \xi_n), v = (\eta_1, \ldots, \eta_n)$; in the norm (4.3.4), we have

$$\|u\|_2^2 = \xi_1^2 + \cdots + \xi_n^2 \qquad \text{and} \qquad \|v\|_2^2 = \eta_1^2 + \cdots + \eta_n^2.$$

We want to prove the *Schwarz inequality*

$$\left(\sum_{i=1}^{n} \xi_i \eta_i\right)^2 \leqslant \|u\|_2^2 \|v\|_2^2. \tag{4.3.8}$$

Observe that for each real α,

$$0 \leqslant \|u - \alpha v\|_2^2 = \|u\|_2^2 + \alpha^2 \|v\|_2^2 - 2\alpha \sum_{i=1}^{n} \xi_i \eta_i.$$

Assume that $\|v\|_2 \neq 0$ [otherwise, (4.3.8) is trivially true], and take

$$\alpha = \left(\sum_{i=1}^{n} \xi_i \eta_i\right) \Big/ \|v\|_2^2$$

to obtain

$$0 \leqslant \|u\|_2^2 - \left(\sum_{i=1}^{n} \xi_i \eta_i\right)^2 \Big/ \|v\|_2^2,$$

which immediately yields (4.3.8). Show that (4.3.2)(d) follows at once. For the Schwarz inequality in a general inner product space, see (4.5.3).

4.3.6 We have seen that polynomials and continuous functions are dense in $L_2(a,b)$. Thus the rather "wild" functions that are in L_2 can be approximated in the L_2 norm by "well-behaved" functions. Here are some similar results, for which the reader is invited to supply the proofs:

(a) The set of differentiable functions is dense in $L_2(a,b)$.

(b) The set of twice-differentiable functions which vanish at a and b is dense in $L_2(a,b)$.

(c) The set of test functions $C_0^\infty(a,b)$ is dense in $L_2(a,b)$ (see Exercise 2.1.1). Thus the completion in the metric d_2 of any of the sets (a), (b), (c) yields $L_2(a,b)$.

4.3.7 (a) Let $a, b \geqslant 0; p, q > 1; 1/p + 1/q = 1$. Show that $ab \leqslant a^p/p + b^q/q$ by considering the function $\phi(x) = 1/q + x/p - x^{1/p}$ on (0, ∞), noting that its minimum occurs at $x = 1$ and comparing $\phi(1)$ with $\phi(a^p/b^q)$.

(b) Prove Hölder's inequality in $\mathbb{R}^n$ or $\mathbb{C}^n$,

$$\sum_{k=1}^{n} |\xi_k \eta_k| \leqslant \|u\|_p \|v\|_q,$$

by setting $a_k = |\xi_k|/\|u\|_p$, $b_k = |\eta_k|/\|v\|_q$ in (a) and summing over k. Note that this reduces to Schwarz's inequality if $p = q = 2$.

(c) Prove Minkowski's inequality

$$\|u + v\|_p \leqslant \|u\|_p + \|v\|_p, \quad p \geqslant 1.$$

For another method of proof, see Maligranda [10].

4.3.8 Show that on $\mathbb{R}^n$, $\lim_{p\to\infty} \|u\|_p = \|u\|_\infty$, where these norms are defined in Example 3.

4.3.9 Consider the sequence of rational numbers

$$s_1 = \frac{1}{10}, s_2 = \frac{1}{10} + \frac{1}{10^4}, \ldots, s_k = \sum_{j=1}^{k} \frac{1}{10^{j^2}}, \ldots .$$

Show that this is a Cauchy sequence which does not converge to a rational number.

4.3.10 The unit sphere in a Banach space is the set of elements satisfying $\|u\| = 1$. In the discussion immediately preceding this set of exercises, we showed that the unit sphere in $C[0, 1]$ is not compact. This can also be proved by exhibiting a sequence of continuous functions of unit norm with nonoverlapping supports. Let $\phi(x)$ be the test function (2.1.1) and consider the sequence

$$\phi_n(x) = e\phi\left(\frac{x - x_n}{\varepsilon_n}\right), \quad n = 1, 2, \ldots,$$

where $\varepsilon_n = 1/2^{n+1}, x_1 = 1/4, x_{n+1} - x_n = \varepsilon_{n+1} + \varepsilon_n$. Show for $i \neq j$ that $\|\phi_i - \phi_j\| = 1$, and that since $\|\phi_n\| = 1$, the unit sphere is not compact.

4.4 CONTRACTIONS AND THE BANACH FIXED-POINT THEOREM

Many functional, differential, and integral equations can be recast in the form

$$u = Tu, \tag{4.4.1}$$

where T is a transformation of a metric space into itself. From this point of view, solutions of (4.4.1) are *fixed points* of the transformation, that is, elements u which are invariant under T. A popular—and often successful—method for solving (4.4.1) is the so-called *method of successive approximations* (or method of iteration). Starting with some "initial approximation" u_0, one defines successive approximations $u_1 = Tu_0, u_2 = Tu_1, \ldots$, in the hope that for large n, u_n will be a good approximation to a fixed point u of T. One of our goals is to find conditions under which the procedure just outlined is justified. A particular type of transformation, known as a *contraction*, will play a significant role.

In our problems, the underlying space X will be a normed linear space, often a complete space (Banach space). The transformation T will not always be defined on the whole space X, but rather on a subset $S \subset X$ and will map S into itself. Of course, S is automatically a metric space with the metric $\|u - v\|$ inherited from X, but S does not have to be a linear manifold in X. For instance, if X is $C[a, b]$ with the sup norm, as in Example 5 of Section 4.3, S might be taken as the unit ball $\|u\|_\infty \leqslant 1$. We begin with some simple definitions.

Definition. *Let $T\colon S \to S$. T is* continuous *at $u \in S$ if, whenever $\{u_n\}$ is a sequence in S converging to u, then $Tu_n \to Tu$. If T is continuous at every u in S we say that T is* continuous on S.

Definition. *Let $T\colon S \to S$. T is* Lipschitz on S *if there exists a constant ρ, independent of u and v, such that*

$$\|Tu - Tv\| \leqslant \rho\|u - v\| \tag{4.4.2}$$

for all $u, v \in S$. If (4.4.2) holds with $\rho < 1$, then T is called a contraction on S *(or, for emphasis, a* strict contraction*).*

REMARKS

1. If S is known from the context, the qualifier "on S" may be dropped in these definitions.

2. It is clear that if T is Lipschitz, it is continuous; hence, a contraction is continuous.

3. Let $S = X = \mathbb{R}$ with $\|u\| = |u|$. Then the four transformations $T_1 u = u$, $T_2 u = u/2, T_3 u = |u|, T_4 u = \sqrt{|u|}$ are all continuous, the first three are Lipschitz, but only the second is a contraction.

4. A contraction brings points *uniformly* closer together, shrinking the distance between any pair of points by at least the fixed scale factor ρ. Let us contrast this with a *weak contraction* T which shrinks distances but not necessarily uniformly: $\|Tu - Tv\| < \|u - v\|$ for all $u, v \in S, u \neq v$. The transformation in Exercise 4.4.3 is a weak contraction which is not a contraction.

Theorem (The Contraction Mapping or Banach Fixed-Point Theorem). *Let S be a closed set in the Banach space X and let $T\colon S \to S$ be a contraction. Then T has one and only one fixed point in S. This fixed point can be obtained from any initial element u_0 in S as the limit of the iterative sequence $u_{k+1} = Tu_k, k = 0, 1, 2, \ldots$.*

Proof.

(a) *Uniqueness.* Suppose that u and v are two fixed points of T. Then there exists $\rho < 1$ such that

$$\|u - v\| = \|Tu - Tv\| \leqslant \rho\|u - v\|,$$

which for $u \neq v$ can be divided by $\|u - v\|$ to yield the contradiction $\rho \geqslant 1$. Thus, $u = v$ and T has at most one fixed point.

(b) *The sequence $\{u_k\}$ is fundamental.* We have

$$\|u_{k+1} - u_k\| = \|Tu_k - Tu_{k-1}\| \leqslant \rho\|u_k - u_{k-1}\| \leqslant \cdots \leqslant \rho^k\|u_1 - u_0\|,$$

and hence for $p > m$,

$$\begin{aligned}
\|u_p - u_m\| &= \|(u_p - u_{p-1}) + (u_{p-1} - u_{p-2}) + \cdots + (u_{m+1} - u_m)\| \\
&\leqslant \|u_p - u_{p-1}\| + \|u_{p-1} - u_{p-2}\| + \cdots + \|u_{m+1} - u_m\| \\
&\leqslant (\rho^{p-1} + \cdots + \rho^m)\|u_1 - u_0\| \\
&\leqslant \rho^m(1 + \rho + \rho^2 + \cdots)\|u_1 - u_0\| = \frac{\rho^m}{1-\rho}\|u_1 - u_0\|.
\end{aligned}$$

The first inequality in the chain stems from the polygonal inequality of Exercise 4.3.4. Thus for all p, m,

$$\|u_p - u_m\| \leqslant \frac{\|u_1 - u_0\|}{1-\rho} \rho^{\min(m,p)},$$

and since the right side can be made arbitrarily small for m, p sufficiently large, $\{u_k\}$ is a fundamental sequence.

(c) $\{u_k\}$ *converges to u belonging to S*. Indeed, since $\{u_k\}$ is a fundamental sequence in X and X is complete, $u_k \to u$ where $u \in X$. The elements of the convergent sequence $\{u_k\}$ also belong to S and S is closed so that the limit u must belong to S.

(d) *u is the fixed point of T*. We have $u_{k+1} = Tu_k$ with T continuous; passing to the limit, we obtain $u = Tu$, as required.

□

REMARK. We need all the hypotheses in the theorem! If S is not closed, there may not be a fixed point in S (for instance, if S is the open interval $0 < u < 1$ in $\mathbb{R}$ and $Tu = u/2$, obviously a contraction on S, T has no fixed point in S). If S is closed but T is not a contraction, there may be no fixed point or there may be more than one. The identity transformation satisfies (4.4.2) with $\rho = 1$ and has every point of S as a fixed point. If $S = X = \mathbb{R}^2$, a translation satisfies (4.4.2) with $\rho = 1$ and has no fixed point; if S is a closed annulus in $\mathbb{R}^2$, then a rotation through an angle less than 2π has no fixed point. Even if the transformation T on S is a weak contraction ($\|Tu - Tv\| < \|u - v\|$), there may not be a fixed point: Consider $Tu = u + 1/u$ on S, where S is the closed set $1 \leqslant u < \infty$ on the real line; then $Tu \in S$ when $u \in S$ and is easily seen to be a weak contraction but has no fixed point. However, a weak contraction on a compact set has a fixed point (see Exercise 4.4.2).

Example of the Contraction Theorem. Suppose that we have two maps of the United States, map A being larger than B. Map B is placed on top of map A, so that the United States in B fits entirely within the United States in A (see Figure 4.4). Our theorem states that there is one and only one place in the United States whose position on the two maps will coincide (that is, there is one and only one place which will be found on map B directly above its position on map A). The transformation T is from A into itself: If u is a point in A, say Cincinnati, find Cincinnati on map B and look directly below on map A to find the point v which defines $v = Tu$. The set A can be viewed as a closed subset of $\mathbb{R}^2$ equipped with the Euclidean norm. Since T is clearly a contraction on A, it has one and only one fixed point—the coincidence point referred to earlier. This point can be found by iteration from any initial point in A.

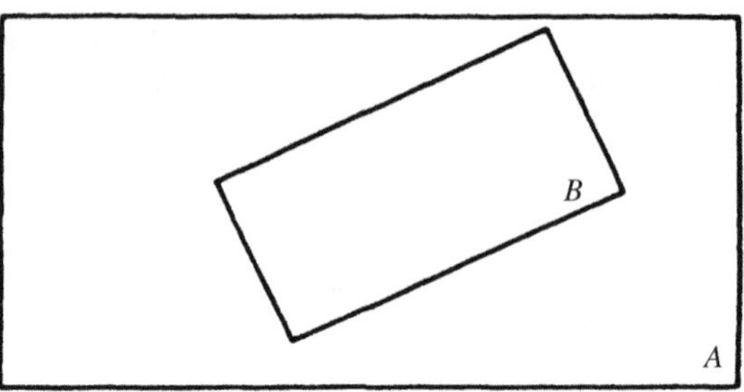

Figure 4.4

Corollary 4.4.1. *If a is a fixed point of a transformation T, it is also a fixed point of T^k for every positive integer k. If, furthermore, T is a contraction, then so is T^k, and both T and T^k have the same fixed point a.*

Corollary 4.4.2. *Let T be a transformation, not assumed to be continuous, on a closed subset of a Banach space, and let T^k, for some positive integer k, be a contraction with fixed point a. Then T has the fixed point a and no other.*

Proof of Corollary 4.4.2. Since $T^k a = a, T^{k+1}a = Ta$ and hence $T^k(Ta) = Ta$, so that Ta is also a fixed point of T^k. Since T^k is a contraction, it has only one fixed point; therefore, $Ta = a$ and a is a fixed point of T. By Corollary 4.4.1, any other fixed point of T would also be a fixed point of T^k, but T^k, being a contraction, has a unique fixed point. □

Example 1. Let $f(u)$ be a real-valued continuous function of a real variable which transforms the closed interval $[a, b]$ into itself (thus f is defined on $a \leqslant u \leqslant b$, and its range is contained in $[a, b]$ as in Figure 4.5). The equation $u = f(u)$ then determines the points where the curve $f(u)$ crosses the diagonal. If $f(u)$ is a contraction [that is, if there exists $\rho < 1$ such that $|f(u) - f(v)| \leqslant \rho|u - v|$ for all u, v], there is exactly one such crossing point, and this point can be found by iteration from any initial try such as u_0 or $\tilde{u}_0$. The contraction condition will be satisfied if $|f'(u)| \leqslant \rho < 1, a \leqslant u \leqslant b$. If $f(u)$ is increasing, as in the figure, the iterations will approach the fixed point monotonically; but if $f(u)$ is decreasing, the iterations will alternate about the fixed point. (See, however, Exercise 4.4.11.)

If f is not a contraction but just a continuous transformation of $[a, b]$ into itself, it will have at least one fixed point but may have many (this is a form of Brouwer's fixed-point theorem; see Chapter 9).

Root-finding problems of the form $g(u) = 0$ occur frequently and can be translated into $u = f(u)$ with $f(u) = u - \lambda g(u)$, $\lambda \neq 0$. Suppose that $g(a) < 0$, $g(b) > 0$, and $0 < \alpha \leqslant g'(u) \leqslant \beta$ on $a \leqslant u \leqslant b$. There is one and only one root of $g(u)$ in (a, b). Setting $\lambda = 1/\beta$, and using the mean value theorem, we see that $f(u)$ maps $[a, b]$ into itself and is a contraction. Starting with an initial approximation u_0,

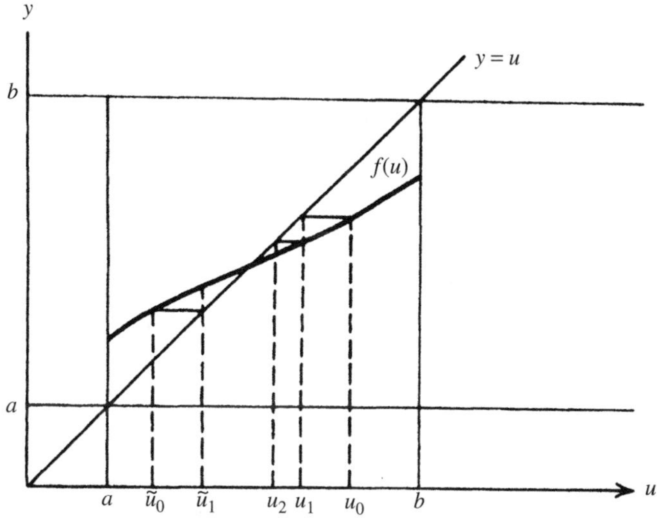

Figure 4.5

the iterates $u_n = f(u_{n-1}) = u_{n-1} - \lambda g(u_{n-1})$ then converge to the root of $g(u)$. Of course, numerical analysts employ a wider variety of iterative procedures; in particular, λ is often changed from step to step as in Newton's method. Many of these schemes can also be shown to have fixed points, but we are not going to pursue this subject further.

Example 2. Contractions are also useful for proving the existence and uniqueness of solutions of simultaneous equations (Exercise 4.4.1).

Example 3.

(a) Let $u \in C[a, b]$, the space of real-valued continuous functions on the bounded closed interval $[a, b]$, and let $k(x, y, z)$ be a given real-valued continuous function on $a \leqslant x \leqslant b, a \leqslant y \leqslant b, -\infty < z < \infty$, satisfying the *Lipschitz condition*

$$|k(x, y, z_2) - k(x, y, z_1)| \leqslant M|z_1 - z_2|, \tag{4.4.3}$$

where M is a positive constant independent of x, y, z_1, z_2. Substitution of $u(y)$ for z in $k(x, y, z)$ gives a continuous function of x and y for which the integral $\int_a^b k(x, y, u(y))\, dy$ is in $C[a, b]$. The *nonlinear Fredholm integral equation*

$$u(x) = \int_a^b k(x, y, u(y))\, dy + f(x), \quad a \leqslant x \leqslant b, \tag{4.4.4}$$

where $f \in C[a, b]$ can therefore be written as $u = Tu$, where T is the transformation of $C[a, b]$ into itself defined by the right side of (4.4.4). Let us determine whether T is a contraction on the space $C[a, b]$ equipped with the uniform norm (4.3.6). For any two continuous functions u and v we have

$$|(Tu)(x) - (Tv)(x)| \leqslant \int_a^b |k(x, y, u(y)) - k(x, y, v(y))| \, dy,$$

so that, by (4.4.3),

$$|Tu - Tv| \leqslant M \int_a^b |u(y) - v(y)| \, dy \leqslant M(b - a)\|u - v\|_\infty.$$

Thus

$$\|Tu - Tv\|_\infty \leqslant M(b - a)\|u - v\|_\infty,$$

and T is a contraction if

$$M < \frac{1}{b - a}, \tag{4.4.5}$$

as will happen if either M or $b - a$ is sufficiently small. If this condition is satisfied, any initial element $u_0(x)$ in $C[a, b]$ generates an iterative sequence through

$$u_n(x) = f(x) + \int_a^b k(x, y, u_{n-1}(y)) \, dy, \tag{4.4.6}$$

which tends uniformly as $n \to \infty$ to the one and only continuous solution of (4.4.4).

(b) A particular case of (4.4.4) results from setting $k(x, y, z) = \mu k(x, y)z$, where μ is a real number and $k(x, y)$ is continuous on the square $a \leqslant x, y \leqslant b$. Then (4.4.3) is satisfied for $M = |\mu|N$, where N is the maximum of $|k|$ on $a \leqslant x, y \leqslant b$. Equation (4.4.4) becomes the *linear integral equation*

$$u = \mu Ku + f, \quad a \leqslant x \leqslant b, \tag{4.4.7}$$

where

$$Ku = \int_a^b k(x, y)u(y) \, dy, \tag{4.4.8}$$

and K is known as an *integral operator* with *kernel* $k(x, y)$. From (4.4.5) we conclude that (4.4.7) has one and only one solution for

$$|\mu| < \frac{1}{N(b - a)}; \tag{4.4.9}$$

in particular, the equation $u = \mu Ku$ has only the trivial solution $u \equiv 0$ when (4.4.9) is satisfied. In the notation of operator theory, we write $\mu = 1/\lambda$, so that $\lambda u = Ku$ and we can then say that all eigenvalues of K must be less than $N(b - a)$ in absolute value. The iterative scheme (4.4.6) applied to (4.4.7) becomes

$$\begin{aligned} u_n &= f + \mu K u_{n-1} = f + \mu K(f + \mu K u_{n-2}) \\ &= \cdots = f + \mu K f + \mu^2 K^2 f + \cdots + \mu^{n-1} K^{n-1} f + \mu^n K^n u_0. \end{aligned}$$

By the contraction property, u_n is known to converge uniformly to the solution of (4.4.7) when (4.4.9) is satisfied. We can also show easily that the term $\mu^n K^n u_0$ approaches 0 uniformly as $n \to \infty$, so that the solution of (4.4.7) is given by the uniformly convergent *Neumann series*

$$u(x) = f + \sum_{j=1}^{\infty} \mu^j K^j f, \quad |\mu| < \frac{1}{N(b-a)}. \tag{4.4.10}$$

If (4.4.7) is taken as an equation on $L_2(a, b)$, many results remain true. (See Chapter 6.)

Example 4.

(a) Consider now the *Volterra* equation obtained by substituting x for the upper limit b in (4.4.4):

$$u(x) = \int_a^x k(x, y, u(y))\, dy + f(x), \quad a \leqslant x \leqslant b. \tag{4.4.11}$$

The function $k(x, y, z)$ needs to be defined only for $y \leqslant x$. We shall assume that k is continuous on G: $a \leqslant x \leqslant b, a \leqslant y \leqslant x, -\infty < z < \infty$, and that, on G, k satisfies the Lipschitz condition

$$|k(x, y, z_2) - k(x, y, z_1)| \leqslant M|z_1 - z_2|. \tag{4.4.12}$$

Equation (4.4.11) is of the form $u = Tu$, where

$$Tu = f(x) + \int_a^x k(x, y, u(y))\, dy$$

defines a transformation of $C[a, b]$ into itself. We have

$$|(Tu)(x) - (Tv)(x)| \leqslant \int_a^x |k(x, y, u(y)) - k(x, y, v(y))|\, dy,$$

and using (4.4.12), we obtain

$$\begin{aligned} |(Tu)(x) - (Tv)(x)| &\leqslant M \int_a^x |u(y) - v(y)|\, dy \\ &\leqslant M(x-a) \max_{a \leqslant y \leqslant x} |u(y) - v(y)| \\ &\leqslant M(x-a)\|u - v\|_\infty. \end{aligned}$$

Therefore,

$$\begin{aligned} |(T^2u)(x) - (T^2v)(x)| &\leqslant M \int_a^x |(Tu)(y) - (Tv)(y)|\, dy \\ &\leqslant \frac{M^2(x-a)^2}{2}\|u - v\|_\infty, \end{aligned}$$

and proceeding step by step, we find that

$$|(T^n u)(x) - (T^n v)(x)| \leqslant \frac{M^n (x-a)^n}{n!} \|u - v\|_\infty, \tag{4.4.13}$$

so that

$$\|T^n u - T^n v\|_\infty \leqslant \frac{M^n (b-a)^n}{n!} \|u - v\|_\infty. \tag{4.4.14}$$

By choosing n sufficiently large, we can ensure that $M^n(b-a)^n/n! < 1$ and therefore T^n is a contraction. Corollary 4.4.2, then tells us that T has one and only one fixed point. A consequence of Corollary 4.4.2 (see Exercise 4.4.8) is that the fixed point of T can be obtained by the usual iterative scheme, which in our case has the form (4.4.6) with x instead of b at the upper limit. Note that these results hold for arbitrarily large b as long as the Lipschitz condition (4.4.12) remains valid.

(b) If $k(x, y, z) = \mu k(x, y) z$, with k continuous on the triangular region given by $a \leqslant x \leqslant b, a \leqslant y \leqslant x$, then (4.4.11) becomes the linear Volterra equation

$$u(x) = \mu \int_a^x k(x, y) u(y)\, dy + f(x), \quad a \leqslant x \leqslant b. \tag{4.4.15}$$

The Lipschitz condition (4.4.12) is automatically satisfied, and we can therefore assert that (4.4.15) has one and only one solution and that, in particular, the homogeneous equation ($f = 0$) has only the trivial solution $u \equiv 0$, so that the Volterra equation has no eigenvalues.

Example 5. We shall consider the initial value problem for a (nonlinear) ordinary differential equation of the first order. Let $f(t, z)$ be a real-valued continuous function on the plane; we look for solutions $u(t)$, necessarily continuous, of

$$\frac{du}{dt} = f(t, u(t)), \quad u(t_0) = z_0. \tag{4.4.16}$$

With only the continuity assumption on f, it can be shown that (4.4.16) has at least one solution in some two-sided t interval about t_0. If, in addition, $f(t, z)$ satisfies a Lipschitz condition in z, there is *exactly* one solution in some two-sided t interval about t_0. We shall give a proof of this statement, but first let us see through simple examples that the conditions being imposed are not superfluous. Consider the IVP

$$\frac{du}{dt} = 2\sqrt{|u|}, \quad u(0) = 0. \tag{4.4.17}$$

Here $f(t, z) = 2\sqrt{|z|}$ is independent of the variable t (a so-called *autonomous* equation). Clearly, f is continuous in the t–z plane, but f does *not* satisfy a Lipschitz condition in any neighborhood of $z = 0$. Problem (4.4.17) has the two solutions

$$u_1(t) \equiv 0, \quad u_2(t) = \begin{cases} t^2, & t \geqslant 0, \\ -t^2, & t \leqslant 0. \end{cases}$$

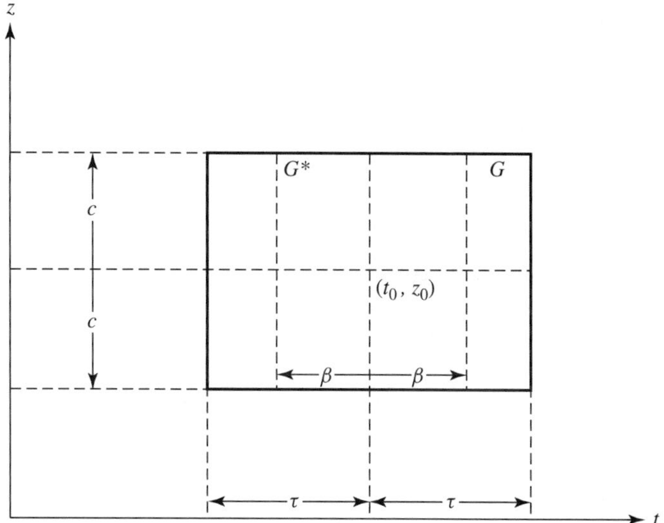

Figure 4.6

Note that $u_2(t)$ is continuous and differentiable everywhere (even at $t = 0$); and that $u_2'(t) = 2\sqrt{|u_2|}$ for all t. In fact, for each $a > 0$,

$$u_a(t) = \begin{cases} 0, & t \leqslant a, \\ (t-a)^2, & t \geqslant a, \end{cases}$$

is also a solution of (4.4.17).

Next consider the IVP

$$\frac{du}{dt} = u^2, \quad u(1) = 1. \tag{4.4.18}$$

It is easily seen that this problem has one and only one solution, given by

$$u(t) = \frac{1}{2-t}, \quad t < 2.$$

Note, however, that the solution becomes infinite as $t \to 2$, so that the IVP (4.4.18) has one and only one solution, but only in the interval $-\infty < t < 2$. Here $f(t, z) = z^2$ satisfies a Lipschitz condition for any *bounded* z interval but not for $-\infty < z < \infty$. This is enough to prevent the solution of (4.4.18) from being defined for all t; if the Lipschitz condition had held for $-\infty < z < \infty$, the IVP would have had one and only one solution for all t.

We now return to the general problem (4.4.16). We shall assume that $f(t, z)$ satisfies the Lipschitz condition

$$|f(t, z_2) - f(t, z_1)| \leqslant M|z_2 - z_1|, \tag{4.4.19}$$

for all t, z_1, z_2 in the Lipschitz region G (large rectangle in Figure 4.6), defined by

$$G : |t - t_0| \leqslant \tau, \quad |z - z_0| \leqslant c. \tag{4.4.20}$$

Problem (4.4.16) is clearly equivalent to the Volterra integral equation

$$u(t) = z_0 + \int_{t_o}^{t} f(y, u(y))\, dy \doteq Tu, \tag{4.4.21}$$

where t_0 is even allowed to be larger than t. This equation is a simple form of (4.4.11) with an obvious relabeling of some of the variables. As we saw in Example 4(a), (4.4.21) will have one and only one solution in $t_0 \leqslant t \leqslant t_0 + \tau$ if condition (4.4.19) holds for the region $t_0 \leqslant t \leqslant t_0 + \tau$ and *all* z_1, z_2. If (4.4.19) is also valid for $t_0 - \tau \leqslant t \leqslant \tau_0$ and all z_1, z_2, a trivial modification of the argument in Example 4(a) shows that we also have a unique solution for $t_0 - \tau \leqslant t \leqslant t_0$. *Thus* (4.4.16) *has one and only one solution in* $|t - t_0| < \tau$ *if* (4.4.19) *holds for* $|t - t_0| < \tau$ *and all* z_1, z_2. In particular, if the Lipschitz condition holds for all z_1, z_2 and for every finite τ, (4.4.16) has one and only one solution in $-\infty < t < \infty$; this will be true, for instance, if df/dz is bounded in each strip $|t - t_0| \leqslant \tau, -\infty < z < \infty$.

Unfortunately, however, a more frequent case is one where (4.4.19) holds only for the rectangle G defined by (4.4.20). Our proof of existence and uniqueness for (4.4.11) needs repair. Result (4.4.13) is a consequence of the fact, no longer true, that whenever $u(t)$ is in the Lipschitz region, so is Tu. We now give a modified proof for (4.4.21).

Let $\max_{(t,z)\in G} |f(t, z)| = p$, and let $\beta = \min(\tau, c/p)$. The rectangle G^* defined by $|t - t_0| \leqslant \beta, |z - z_0| \leqslant c$ has the same height as G but is perhaps not as wide (see Figure 4.6). Consider the set C^* of continuous functions $u(t)$ defined on $|t - t_0| \leqslant \beta$ and such that $|u(t) - z_0| \leqslant c$. We say, for short, that C^* is the set of continuous functions lying in G^*. We claim that if u is in C^*, so is Tu. Indeed, it is clear that Tu is continuous and

$$\begin{aligned}\|(Tu)(t) - z_0\| &= \left|\int_{t_0}^{t} f(y, u(y))\, dy\right| \leqslant \left|\left[\int_{t_0}^{t} |f(y, u(y))|\, dy\right]\right| \\ &\leqslant p|t - t_0| \leqslant p\beta \leqslant c,\end{aligned}$$

so that $Tu \in C^*$ (observe that the repeated absolute value signs in the first inequality are needed to cover the case where the upper limit t is less than t_0). Since G^* is part of the Lipschitz region G, it follows that for $u, v \in C^*$, we have

$$\begin{aligned}|(Tu)(t) - (Tv)(t)| &\leqslant \left|\left[\int_{t_0}^{t} |f(y, u(y)) - f(y, v(y))|\, dy\right]\right| \\ &\leqslant M\left|\left[\int_{t_0}^{t} |u - v|\, dy\right]\right| \\ &\leqslant M|t - t_0|\|u - v\|_\infty \leqslant M\beta\, d\|u - v\|_\infty,\end{aligned}$$

where $\|u - v\|_\infty$ is the uniform norm on $|t - t_0| \leqslant \beta$.

Since Tu and Tv are in the Lipschitz region, we have

$$
\begin{aligned}
|(T^2u)(t) - (T^2v)(t)| &\leqslant M\left|\left[\int_{t_0}^{t} |Tu - Tv|\, dy\right]\right| \\
&\leqslant M^2\|u - v\|_\infty \left|\left[\int_{t_0}^{t} |y - t_0|\, dy\right]\right| \\
&\leqslant \frac{|t - t_0|^2}{2} M^2 \|u - v\|_\infty \leqslant \frac{M^2}{2}\beta^2\|u - v\|_\infty
\end{aligned}
$$

and

$$
|T^n u - T^n u| \leqslant \frac{M^n}{n!}\beta^n \|u - v\|_\infty, \quad \|T^n u - T^n v\|_\infty \leqslant \frac{M^n \beta^n}{n!}\|u - v\|_\infty .
$$

Thus T^n is a contraction of C^* for n sufficiently large. Since C^* is a closed subset of the Banach space $C[t_0 - \beta, t_0 + \beta]$, we can apply the Contraction Mapping Theorem. Therefore, T^n has a unique fixed point, and by Corollary 4.4.2 so does T. We have therefore established that (4.4.16) has one and only one solution in some two-sided interval about $t = t_0$. The solution can be obtained as follows: Start with any initial approximation $u_0(t)$ in C^* [it is not necessary but is nevertheless desirable to have $u_0(t_0) = z_0$; a commonly used initial choice is $u_0(t) \equiv z_0$], and define successively for $n = 1, 2, \ldots$

$$
u_n(t) = z_0 + \int_{t_0}^{t} f(y, u_{n-1}(y))\, dy.
$$

Of course, for $n \geqslant 1, u_n(t_0) = z_0$; as $n \to \infty$, the sequence $\{u_n(t)\}$ converges uniformly to the solution of (4.4.16), at least on $|t - t_0| \leqslant \beta$.

In Chapter 9, we prove another existence theorem for (4.4.16) using the Schauder fixed-point theorem.

Example 6. A satellite web coupling can be idealized as a thin sheet connecting two cylindrical satellites. A cross section of the arrangement is shown in Figure 4.7. Both satellites remain at the same constant absolute temperature U while heat is radiated from the surface of the web into space at 0 absolute temperature. There is no temperature variation in the direction perpendicular to the paper. Since the web is relatively thin, we may also assume that the temperature is constant across the thickness of the web. Therefore, in the *steady state*, the temperature in the web will depend only on the x coordinate.

A heat balance for a web section one unit deep between x and $x + dx$ yields $-kh(d^2u/dx^2) + 2q = 0$, where k is the thermal conductivity, h the thickness, and q the heat radiated per unit area of the surface (the factor 2 being required because there is radiation from both the top and bottom surfaces). According to the Stefan-Boltzmann law, $q = \alpha u^4$, where α is a positive constant describing the radiation properties of the surface of the web. This leads to the nonlinear BVP

$$
-khu'' = -2\alpha u^4, \quad 0 < x < l; \quad u(0) = u(l) = U.
$$

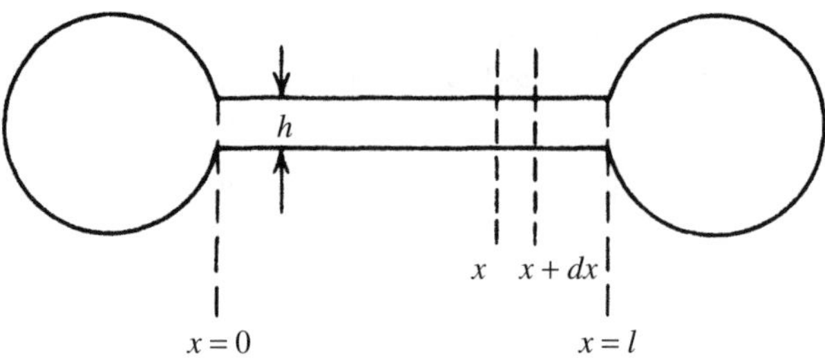

Figure 4.7

In terms of the dimensionless variables

$$x^* = \frac{x}{l}, \quad u^* = \frac{u}{U},$$

the BVP becomes, after *dropping the asterisks* for convenience,

$$-u'' = -\lambda u^4, \quad 0 < x < 1, \quad u(0) = u(1) = 1, \tag{4.4.22}$$

where the *nondimensional positive* constant λ is given by

$$\lambda = \frac{2\alpha l^2 U^3}{kh}. \tag{4.4.23}$$

We can translate (4.4.22) into an integral equation by using Green's function $g(x, \xi)$ of $-d^2/dx^2$ with *vanishing* boundary conditions at $x = 0$ and $x = 1$. The explicit expression $g(x, \xi) = x_<(1 - x_>)$ was derived in Chapter 1. Writing $u = 1 - v$, we observe that v satisfies

$$-v'' = \lambda u^4, \quad 0 < x < 1, \quad v(0) = v(1) = 0,$$

so that $v(x) = \lambda \int_0^1 g(x, y) u^4(y)\, dy$ and therefore

$$u(x) = Tu \doteq 1 - \lambda \int_0^1 g(x, y) u^4(y)\, dy, \quad 0 < x < 1, \tag{4.4.24}$$

which is equivalent to the BVP (4.4.22).

Our goal is to solve (4.4.24), and at first an attempt might be made to view the problem as a special case of the nonlinear Fredholm equation of Example 3(a). In our case $k(x, y, z) = \lambda g(x, y) z^4$, and the Lipschitz condition (4.4.3) is not satisfied for all z. Fortunately, it turns out that we will need a Lipschitz condition to hold only for $0 \leqslant z \leqslant 1$, since the desired solution $u(x)$ of (4.4.22) or (4.4.24) is confined to that range; indeed, an absolute temperature must be *nonnegative*, and the maximum principle applied to (4.4.22) shows that $u(x) \leqslant 1$ (as is even more obvious from the integral equation for u). This suggests that instead of dealing with the space

$C[0,1]$ we should consider the subset S of $C[0,1]$ consisting of functions satisfying $0 \leqslant u(x) \leqslant 1$. Clearly, S is a closed subspace of $C[0,1]$ when equipped with the uniform norm. We want to show that the transformation T defined by the right side of (4.4.24) is a *contraction* on S. First we must show that T transforms elements of S into elements of S.

If $u \in S$, we have

$$(Tu)(x) = 1 - \lambda \int_0^1 g(x,y)u^4(y)\,dy \geqslant 1 - \lambda \int_0^1 g(x,y)\,dy \geqslant 1 - \frac{\lambda}{8},$$

so that $Tu \geqslant 0$ if $\lambda \leqslant 8$, and obviously $Tu \leqslant 1$. Thus, if $\lambda \leqslant 8$, T is a transformation on S. Is T a contraction?

If u and v are arbitrary elements in S, then

$$|v^4(y) - u^4(y)| \leqslant |v - u|(v^3 + uv^2 + vu^2 + u^3) \leqslant 4\|u - v\|_\infty,$$

so that

$$\begin{aligned}|(Tv)(x) - (Tu)(x)| &\leqslant \lambda \int_0^1 g(x,y)|v^4(y) - u^4(y)|\,dy \\ &\leqslant 4\lambda\|u - v\|_\infty \int_0^1 g(x,y)\,dy.\end{aligned}$$

Since the last integral is equal to $x(1-x)/2$, we find that

$$\|Tu - Tv\|_\infty \leq \frac{\lambda}{2}\|u - v\|_\infty,$$

and T is a contraction if $\lambda < 2$. Hence, for $\lambda < 2$, (4.4.24) or, equivalently, (4.4.22) has one and only one solution in the range $0 \leqslant u(x) \leqslant 1$. The iteration scheme

$$-u_n'' = -\lambda u_{n-1}^4, 0 < x < 1; \quad u_n(0) = u_n(1) = 0,$$

converges uniformly to the desired solution as long as the initial element u_0 satisfies $0 \leqslant u_0(x) \leqslant 1$. The sequence $\{u_n\}$ is not monotone, but in Chapter 9 we modify the procedure to generate a monotone sequence and further prove that a unique solution in S exists for all $\lambda > 0$.

EXERCISES

4.4.1 Consider the set of n nonlinear equations

$$Su = u, \tag{4.4.25}$$

where $u = (u_1, \ldots, u_n)$ is an n-tuple of real numbers and Su is the n-tuple $(S_1(u_1, \ldots, u_n), \ldots, S_n(u_1, \ldots, u_n))$, where each S_i is a real-valued function of the real variables $u_1, \ldots, u_n$.

We shall assume the Lipschitz condition, for $i = 1, \ldots, n$,

$$|S_i(u_1, \ldots, u_n) - S_i(v_1, \ldots, v_n)| \leqslant M_{i1}|u_1 - v_1| + \cdots + M_{in}|u_n - v_n|. \tag{4.4.26}$$

Using, as appropriate, the norms (4.3.3), (4.3.4), (4.3.5), show that (4.4.25) has one and only one solution if *any* of the following conditions hold:

$$\max_j \sum_{j=1}^{n} M_{ij} < 1, \max_i \sum_{j=1}^{n} M_{ij} < 1, \sum_{i,j=1}^{n} M_{ij}^2 < 1. \tag{4.4.27}$$

Next, consider the linear problem

$$Au = b,$$

where $A = [a_{ij}]$ is an $n \times n$ matrix and b is a given vector in $\mathbb{R}^n$. Prove that there is one and only one solution if the matrix is *diagonally dominant*, that is, if

$$\sum_{j \neq i} |a_{ij}| < |a_{ii}|, \quad i = 1, \ldots, n.$$

4.4.2 Let S be a *bounded closed interval* on the real line and let $T\colon S \to S$ be a weak contraction, that is,

$$|Tu - Tv| < |u - v| \qquad \text{for every pair } u \neq v \in S.$$

Show that T has one and only fixed point. (*Hint*: Look at $\min |Tu - u|$.) As an example, $Tu = \sin u$ is a weak contraction on $0 \leqslant u \leqslant \pi/2$ and has a unique fixed point. The same proof holds for a compact set S in an arbitrary Banach space X.

4.4.3 On $\mathbb{R}$, consider the transformation $Tu = \ln(1 + e^u)$. Show that T is a weak contraction and has no fixed point. Why does this not contradict Exercise 4.4.2?

4.4.4 Consider the Volterra equation (4.4.11) under the following conditions:

(a) $|k(x, y, z_2) - k(x, y, z_1)| \leqslant M|z_2 - z_1|$ for all x, y, z_1, z_2 in G, defined by $a \leqslant x \leqslant b$, $a \leqslant y \leqslant x$, $|z - \eta| \leqslant c$.

(b) $|f(x) - \eta| \leqslant c_1 < c$ for $a \leqslant x \leqslant b$.

(c) $\max_G |k(x, y, z)| \leqslant (c - c_1)/(b - a)$.

Show that (4.4.11) has one and only one continuous solution in $a \leqslant x \leqslant b$.

4.4.5 Consider the sets of all n-tuples of real-valued continuous functions $u(t) = (u_1(t), \ldots, u_n(t))$ on a fixed bounded interval $a \leqslant t \leqslant b$. Show that this set becomes a Banach space under any of the following norms:

$$\|u\|_1 = \sum_{i=1}^{n} \max_{a \leqslant t \leqslant b} |u_i(t)|,$$

$$\|u\|_2 = \left\{ \sum_{i=1}^{n} \max_{a \leqslant t \leqslant b} |u_i(t)|^2 \right\}^{1/2},$$

$$\|u\|_\infty = \max_i \max_{a \leqslant t \leqslant b} |u_i(t)|.$$

4.4.6 Let $f_1(t, z_1, \ldots, z_n), \ldots, f_n(t, z_1, \ldots, z_n)$ be n real-valued continuous functions of the real variables $t, z_1, \ldots, z_n$. Consider the first-order *system* of n (nonlinear) ordinary differential equations

$$\frac{du_1}{dt} = f_1(t, u_1(t), \ldots, u_n(t)), \ldots, \frac{du_n}{dt} = f_n(t, u_1(t), \ldots, u_n(t)),$$

with the initial conditions

$$u_1(t_0) = \eta_1, \ldots, u_n(t_0) = \eta_n.$$

In vector notation the IVP takes the form

$$\frac{du}{dt} = f(t, u(t)), \quad u(t_0) = \eta, \tag{4.4.28}$$

where $u(t) = (u_1(t), \ldots, u_n(t)), \eta = (\eta_1, \ldots, \eta_n)$, and

$$f(t, z) = (f_1(t, z_1, \ldots, z_n), \ldots, f_n(t, x_1, \ldots, z_n)),$$

Note that a single differential equation of order n can always be transformed to a first-order system of n equations. Translate (4.4.28) into a vector Volterra integral equation. Assume that f satisfies the Lipschitz condition, for $i = 1, \ldots, n$,

$$|f_i(t, z) - f_i(t, w)| \leqslant M_{i1}|z_1 - w_1| \cdots + M_{in}|z_n - w_n|, \tag{4.4.29}$$

which holds for *all* z and w and for $|t - t_0| \leqslant \tau$. Show that (4.4.28) has one and only one solution for all t on $|t - t_0| \leqslant \tau$. It is easiest to use the norm $\|u\|_\infty$ of Exercise 4.4.5 and to note that (4.4.29) implies that there exists M such that

$$|f_i(t, z) - f_i(t, w)| \leqslant M \max_j |z_j - w_j|, \quad i = 1, \ldots, n.$$

4.4.7 (a) Consider the linear system of n equations

$$\frac{du}{dt} = A(t)u(t) + q(t), \qquad u(t_0) = \eta, \tag{4.4.30}$$

which is of the form (4.4.28) with

$$f_i(t, u_1, \ldots, u_n) = \sum_{j=1}^{n} a_{ij}(t)u_j + q_i(t),$$

where the coefficients $a_{ij}(t)$ and the forcing terms $q_i(t)$ are assumed continuous. Show that the Lipschitz condition (4.4.29) is satisfied for any bounded interval whatever, so that (4.4.30) has one and only one solution for any bounded interval and hence for $-\infty < t < \infty$.

(b) Consider the IVP for a single linear differential equation of order n for an unknown scalar function $y(t)$. Set $u_1 = y, u_2 = y', \ldots, u_n = y^{(n-1)}$ to translate the differential equation into a first-order linear system for the vector $u = (u_1, \ldots, u_n)$, which will be of the form (4.4.30) if $a_n(t) \neq 0$, where $a_n(t)$ is the leading coefficient in the given ordinary differential equation. This proves the existence and uniqueness theorem for the IVP for a single scalar equation of any order.

4.4.8 Suppose that T^n is a contraction on a complete metric space X with fixed point a. Then Corollary 4.4.2 guarantees that a is the one and only fixed point of T. Show that if u_0 is any element in X, the sequence $u_0, u_1 = Tu_0, u_2 = Tu_1, \ldots$ tends to a.

4.4.9 As we have seen, a contraction T on a nonclosed set U may not have a fixed point. By imposing additional conditions, we can recover the fixed point property. Let U contain a closed ball $B : \|u - u_0\| \leqslant a$ such that $\|Tu_0 - u_0\| < (1 - \rho)a$, where ρ is the contraction constant of T. Then T has a unique fixed point u^* in B. To prove this, show first by induction that the sequence $u_{n+1} = Tu_n, n = 0, 1, 2, \ldots$ satisfies $\|u_n - u_0\| \leqslant (1 - \rho^n)a$, so that the sequence $\{u_n\}$ belongs to B. Next show that the sequence is fundamental and hence converges to u^* (which belongs to B, since B is closed). Then show that $Tu^* = u^*$ and that u^* is the only fixed point of T in B.

4.4.10 Consider the nonlinear BVP

$$-u'' = \alpha \sin u, \quad 0 < x < 1, \quad u'(0) = u(1) = 0, \tag{4.4.31}$$

which governs the buckling of a suitably supported rod under a compressive thrust to which the parameter α is proportional. By introducing an appropriate Green's function, translate (4.4.31) into an integral equation

$$u(x) = Tu \doteq \alpha \int_0^1 g(x, \xi) \sin u(\xi)\, d\xi,$$

and show that T is a contraction in a suitably defined metric space when $\alpha < 2$. Thus for $\alpha < 2$ the only solution of (4.4.31) is $u = 0$, so that the critical load must exceed 2 (in fact, we show in Chapter 9 that the critical load is $\pi^2/4$).

4.4.11 Let $f(u)$ be a differentiable function transforming $0 \leqslant u \leqslant 1$ into itself with $-M \leqslant f'(u) \leqslant \rho < 1$. Show that the fixed points of f are the same as those of $h(u) \doteq [Mu + f(u)]/(1 + M)$. Note that h is a contraction on [0, 1] and therefore has a single fixed point, and so must $f(u)$. The iteration leading to the fixed point of h is monotonic.

4.4.12 *Implicit function theorem (simple form)*. Let f be a real-valued function on the strip in $\mathbb{R}^2 : a \leqslant x \leqslant b, -\infty < y < \infty$. We ask if the equation

$$f(x, y) = 0 \tag{4.4.32}$$

can be solved for y as a function of x; in other words, does (4.4.32) define y implicitly as a function of x? Prove that if f is continuous with continuous partial derivatives of the first order and

$$0 < m \leqslant \frac{\partial f}{\partial y} \leqslant M,$$

then there exists a unique function $y = u(x)$ so that $f(x, u(x)) = 0$. [*Hint*: Consider $C[a, b]$ equipped with the sup norm, and define the transformation $T\colon C[a, b] \to C[a, b]$ by

$$(Tu)(x) = u(x) - \frac{2}{m + M} f(x, u(x))$$

and show that T is a contraction.]

4.4.13 (a) Consider the IVP

$$\frac{du}{dt} = u^2, \quad t > 0; \quad u(0) = z_0 > 0.$$

Show that $|u^2 - v^2| \leqslant 2(z_0 + c)|u - v|$ in the strip $|u - z_0| < c$ (corresponding to $\tau = \infty$ in Example 5 and Figure 4.6). Show that $\beta = c/(z_0 + c)^2$ with largest value $1/4z_0$. Thus a solution exists at least in the time interval $0 < t < 1/4z_0$. By solving the IVP in closed form show that a solution actually exists until $t = 1/z_0$.

(b) Generalize part (a) by replacing u^2 by $f(u)$, where f is an even, convex function with $f(0) \geqslant 0$. (See also Chapter 9.) Show that the exact lifetime of the solution is given by

$$T = \int_{z_0}^{\infty} \frac{ds}{f(s)}.$$

4.5 HILBERT SPACES AND THE PROJECTION THEOREM

In a normed linear space a vector has a length. We want to refine the structure further so that the angle between vectors is also defined. In particular, we want a criterion for determining whether or not two vectors are perpendicular. As in three-dimensional space, these notions are most easily derived from an inner product (or dot product or scalar product) between vectors.

Definition. *An* inner product $\langle u, v\rangle$ *on a* real *linear space* $X^{(r)}$ *is a* real-*valued function of ordered pairs of vectors* u, v *with the properties*

$$\left.\begin{aligned}&\langle u, v\rangle = \langle v, u\rangle,\\ &\langle \alpha u, v\rangle = \alpha\langle u, v\rangle,\\ &\langle u+v, w\rangle = \langle u, w\rangle + \langle v, w\rangle,\\ &\langle u, u\rangle > 0 \quad \textit{for}\, u \neq 0,\end{aligned}\right\} \tag{4.5.1}$$

where u, v, w *are arbitrary vectors and* α *is an arbitrary real number. Such a linear space is called a* real inner product space.

REMARK. It follows that $\langle u, 0\rangle = 0$ for every u. Although $\langle u, u\rangle$ is nonnegative, $\langle u, v\rangle$ may be positive, negative, or 0. The requirement on $\langle u, u\rangle$ is made so that it will have a positive square root which can act as a norm.

The definition of an inner product on a complex vector space is somewhat different. Since the inner product will still be used to generate a norm, the property $\langle u, u\rangle > 0$ for $u \neq 0$ must be preserved. This forces some changes in (4.5.1); otherwise, $\langle iv, iv\rangle = i\langle v, iv\rangle = i\langle iv, v\rangle = i^2\langle v, v\rangle = -\langle v, v\rangle$, and it would be impossible to have $\langle u, u\rangle$ positive for all $u \neq 0$. The remedy is to change the first property in (4.5.1) to $\langle u, v\rangle = \overline{\langle v, u\rangle}$.

Definition. *An* inner product $\langle u, v\rangle$ *on a* complex *linear space* $X^{(c)}$ *is a* complex-valued *function of ordered pairs* u, v *with the properties*

$$\left.\begin{aligned}&\langle u, v\rangle = \overline{\langle v, u\rangle} \quad (\textit{this implies } \langle u, u\rangle \textit{ real}),\\ &\langle \alpha u, v\rangle = \alpha\langle u, v\rangle,\\ &\langle u+v, w\rangle = \langle u, w\rangle + \langle v, w\rangle,\\ &\langle u, u\rangle > 0 \quad \textit{for } u \neq 0,\end{aligned}\right\} \tag{4.5.2}$$

where u, v, w *are arbitrary vectors in* $X^{(c)}$ *and* α *is an arbitrary complex number. The linear space* $X^{(c)}$ *provided with an inner product is called a* complex inner product space.

An immediate consequence of the definition is that $\langle u, \alpha v\rangle = \bar{\alpha}\langle u, v\rangle$, where the appearance of the complex conjugate means that the inner product is not quite linear in the second term. The inner product is an instance of a bilinear form (see Exercise 4.5.7). We also note that $\langle u, 0\rangle = \langle 0, v\rangle = 0$. Observe also that if α is restricted to real values, properties (4.5.2) reduce to (4.5.1). Unless otherwise stated, the results below are valid for both real and complex inner products; the derivations will, however, be given only for the more difficult complex case.

Theorem (Schwarz Inequality). *For any two vectors* u, v *in an inner product space,*

$$|\langle u, v\rangle|^2 \leqslant \langle u, u\rangle\langle v, v\rangle, \tag{4.5.3}$$

with equality if and only if u and v are dependent.

REMARK. Remember that $\langle u, v\rangle$ is a complex number, so that the absolute value sign in (4.5.3) is not wasted; without it we could not guarantee that the left side is real! On a real space the absolute value sign can be dropped.

Proof. For any two vectors u, w, the inequality $\langle u-w, u-w\rangle \geqslant 0$ yields

$$\langle u, u\rangle \geqslant \langle w, u\rangle + \langle u, w\rangle - \langle w, w\rangle = 2\,\mathrm{Re}\,\langle w, u\rangle - \langle w, w\rangle, \tag{4.5.4}$$

with equality if and only if $w = u$.

If $v = 0$, then (4.5.3) is trivially true with the equality sign; if $v \neq 0$, set

$$w = \frac{\langle u, v\rangle}{\langle v, v\rangle} v$$

in (4.5.4) to obtain the Schwarz inequality. Since $w = u$ if and only if u is a multiple of v, we have equality in (4.5.3) if and only if either $v = 0$ or u is a multiple of v. Thus equality in (4.5.3) occurs if and only if u and v are dependent. □

Corollary.

$$\langle u+v, u+v\rangle^{1/2} \leqslant \langle u, u\rangle^{1/2} + \langle v, v\rangle^{1/2}. \tag{4.5.5}$$

Proof. Expand the real nonnegative quantity $\langle u+v, u+v\rangle$, and use (4.5.3). □

It is therefore clear that $\langle u, u\rangle^{1/2}$ is an admissible norm on our linear space, since

$$\|u\| = \langle u, u\rangle^{1/2}$$

satisfies conditions (4.3.2). Thus an inner product space is *automatically a normed linear space* and hence a metric space with

$$d(u, v) = \|u - v\| = \langle u-v, u-v\rangle^{1/2},$$

and the Schwarz inequality (4.5.3) takes on the more attractive form

$$|\langle u, v\rangle| \leqslant \|u\|\,\|v\|. \tag{4.5.6}$$

With a metric at our disposal we have the usual definition of convergence, which we repeat for convenience: A sequence $\{u_k\}$ is said to converge to u if for each $\varepsilon > 0$ there exists $N > 0$ such that $\|u_k - u\| < \varepsilon, k > N$. Moreover, since a linear structure exists together with the metric structure, we can use the notion of convergence of series introduced in Section 4.3.

It follows from the Schwarz inequality that the *inner product is a continuous function of its arguments*. Thus $\lim_{n\to\infty}\langle u_n, v_n\rangle = \langle u, v\rangle$ whenever $u_n \to u$ and $v_n \to v$. In particular, $\|u_n\| \to \|u\|$ if $u_n \to u$. If $\sum_{k=1}^{\infty} u_k = u$, the series $\sum_{k=1}^{\infty}\langle u_k, v\rangle$ automatically converges to $\langle u, v\rangle$.

Definition. *An inner product space which is complete in its natural norm is called a* Hilbert space.

To distinguish between real and complex spaces, we sometimes use the superscript r or c.

Definition. *An n-dimensional inner product space is known as a* Euclidean space E_n.

We shall see later that a Euclidean space is automatically complete; hence it is a Hilbert space. Of course, many of the inner product spaces of interest are infinite-dimensional; such spaces are not automatically Hilbert spaces and may require the completion procedure described in Exercise 4.3.1 to convert them into Hilbert spaces. Before proceeding to a more systematic investigation of Hilbert spaces and operators defined on them, we try to substantiate an earlier claim that an inner product provides additional structure to a linear space.

First, we note the *parallelogram law*

$$\|u+v\|^2 + \|u-v\|^2 = 2\|u\|^2 + 2\|v\|^2, \tag{4.5.7}$$

which states that the sum of the squares of the diagonals of the parallelogram built from u and v is equal to the sum of the squares of the sides. Relation (4.5.7) holds, not for arbitrary normed linear spaces, but only for those which can be provided with an inner product (see Exercise 4.5.4).

In an inner product space there is also a natural definition of perpendicularity.

Definition. *Two vectors are* orthogonal *(or perpendicular) if $\langle u, v\rangle = 0$. A set of vectors, each pair of which is orthogonal, is called an* orthogonal set *(the set is* proper *if all vectors in it $\neq 0$); if, in addition, every vector in the set has unit norm, the set is* orthonormal.

The terminology is justified by observing that $\langle u, v\rangle = 0$ implies that $\|u+v\|^2 = \langle u+v, u+v\rangle = \|u\|^2 + \|v\|^2$, which is just the Pythagorean theorem. It follows that if the sum of two orthogonal vectors vanishes, each vector must be the zero vector.

Let H be a Hilbert space and M a linear manifold in H, that is, a set of elements in H such that, whenever $u, v \in M$, then $\alpha u + \beta v \in M$ for all scalars α, β. At first we consider only the simplest linear manifold, a line through the origin.

Definition. *The set of vectors αv_0, where $v_0 \neq 0$ and α runs through the scalars (real or complex numbers, depending on the space), is called the* line *generated by* v_0.

In the real case the line generated by v_0 contains only two unit vectors: $v_0/\|v_0\|$ and $-v_0/\|v_0\|$, but in the complex case the line contains all unit vectors of the form $e^{i\beta}v_0/\|v_0\|$ with β real.

Definition. *The (orthogonal)* projection *of u on the line M generated by v_0 is the vector $v = \langle u, e\rangle e$, where e is a unit vector on the line (the various possible choices for e all yield the* same *vector v).*

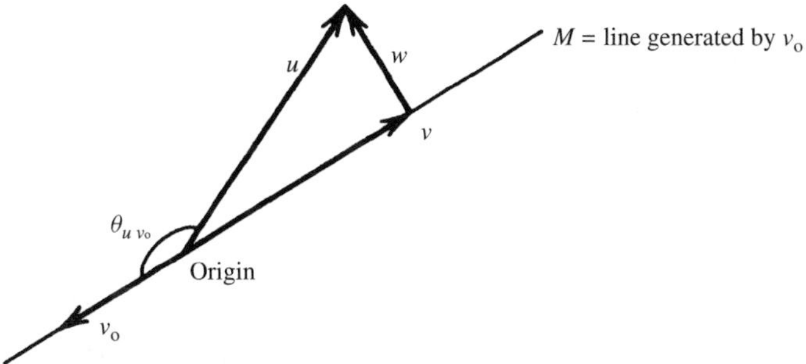

Figure 4.8

Without reference to any unit vector, we can write the projection of v on the line generated by v_0 as

$$v = \frac{\langle u, v_0 \rangle}{\|v_0\|^2} v_0. \tag{4.5.8}$$

(In mathematical terminology a projection is always a vector, although in engineering it sometimes means the magnitude of v.) The definition agrees with intuition, for we can easily check that $u - v$ is orthogonal to all vectors in M. Thus u has been written as a sum of two vectors, one in M and the other perpendicular to M. Moreover, such a decomposition is unique, for suppose that we had $u = v + w = v' + w'$, where both v and v' are in M and both w and w' are perpendicular to M; then by subtraction we find that $0 = (v - v') + (w - w')$, where $(v - v')$ and $(w - w')$ are orthogonal. But if the sum of two orthogonal vectors vanishes, they must both vanish, so that $v = v'$ and $w = w'$. The vector v can also be unambiguously characterized as the vector in M which lies closest to u. These ideas are illustrated in Figure 4.8 for the real case, where we can also define the angle θ_{uv_0} between u and v_0 by

$$\cos \theta_{uv_0} = \frac{\langle u, v_0 \rangle}{\|u\| \|v_0\|}. \tag{4.5.9}$$

The Schwarz inequality shows that the right side is between -1 and 1, so that a real value of the angle between 0 and π is indeed defined.

These notions can be extended to closed linear manifolds in H. We recall that a closed linear manifold M has the property that whenever $u_1, u_2, \ldots$ is a sequence in M which converges in H, the limit u actually belongs to M. Example 3(b) below shows that linear manifolds in H are not necessarily closed. There are, however, classes of manifolds that are always closed: finite-dimensional manifolds and manifolds orthogonal to a given manifold.

Definition. *Let M be a linear manifold (closed or not) in H. The set $M^\perp$* (linear manifold orthogonal to M) *is defined as follows: u belongs to $M^\perp$ if and only if it is orthogonal to every vector in M.*

The set $M^\perp$ is seen to be a linear manifold in fact as well as in name. Moreover, $M^\perp$ is closed. Indeed, let $u_1, u_2, \ldots$ be a sequence in $M^\perp$ with limit u in H; since $\langle u_k, f\rangle = 0$ for every f in M, continuity of the inner product shows that $\langle u, f\rangle = 0$, so that u is in $M^\perp$ and hence $M^\perp$ is closed.

We are now in a position to state the projection theorem, which plays a central role in both the theory and applications of Hilbert spaces. Two proofs are given. The one below is quite general, whereas the one in the next section is valid only for separable spaces (the definition of this concept will be given in due time).

Theorem (Projection Theorem). *Let M be a closed linear manifold in H. Each vector u in H can be written in one and only one way as the sum of a vector in M and a vector in $M^\perp$. The vector in M is known as the (orthogonal)* projection *of u on M and can also be characterized as the unique vector in M which is closest to u.*

REMARK. Figure 4.8 illustrates the theorem when M is one-dimensional. It is also easy to visualize the meaning of the theorem when M is a two-dimensional linear manifold (that is, a plane through the origin) in a three-dimensional Hilbert space H. Let u be a fixed, arbitrary vector emanating form the origin, and drop a perpendicular from u to M. The theorem asserts the obvious geometric fact that u is the sum of its projection v lying in M and of the vector $w = u - v$, which is orthogonal to M. If we consider $\|u - z\|$ for all z lying in M with the same origin as u, then $z = v$ gives the smallest possible value of $\|u - z\|$; identifying vectors with their endpoints, we can say that v is the point (or vector) in M closest to u.

Proof.

(a) We first show that there exists at least one vector in M which is closest to u. As z ranges through M, the set of real numbers $\|u - z\|$ is bounded below (since these numbers are nonnegative), so that we can define the nonnegative number

$$d = \inf_{z \in M} \|u - z\|, \tag{4.5.10}$$

and we wish to prove that the infimum is attained. By definition there exists a minimizing sequence $\{z_n\}$ in M such that $\|u - z_n\| \to d$. We apply the parallelogram law (4.5.7) with $u - z_m$ and $u - z_n$ playing the roles of u and v, respectively, to obtain

$$\|z_n - z_m\|^2 = 2\|u - z_m\|^2 + 2\|u - z_n\|^2 - 4\|u - (z_n + z_m)/2\|^2.$$

Since M is linear, the vector $(z_n + z_m)/2$ belongs to M and

$$\|u - (z_n + z_m)/2\|^2$$

must exceed d^2. Therefore,

$$\|z_n - z_m\|^2 \leqslant 2\|u - z_m\|^2 + 2\|u - z_n\|^2 - 4d^2,$$

and as m, n tend to infinity, the right side tends to 0 and $\{z_n\}$ is a Cauchy sequence. Since H is complete, $z_n \to v$ in H; moreover, $v \in M$ because M is closed.

(b) Let v be the limit of the minimizing sequence of part (a), and let w be the vector $u - v$. We want to show that w is orthogonal to M, or, equivalently, that $\langle w, z\rangle = 0$ for every z in M. This statement is obviously true for $z = 0$, so let us take z to be an arbitrary nonzero element of M. Then $v + \alpha z$ belongs to M for all complex α, and by the definition of v we have

$$\|w - \alpha z\|^2 = \|u - (v + \alpha z)\|^2 \geqslant \|u - v\|^2 = \|w\|^2,$$

so that $\|w - \alpha z\|^2$ has its minimum at $\alpha = 0$. Taking α real, we can write

$$\|w - \alpha z\|^2 = \|w\|^2 + \alpha^2\|z\|^2 - \alpha\langle w, z\rangle - \alpha\langle z, w\rangle,$$

and setting the derivative with respect to α equal to 0 at $\alpha = 0$, we find that $\text{Re}(w, z) = 0$ for every $z \in M$. Since iz also belongs to M, we have also that $\text{Re}(w, iz) = 0$ or $\text{Im}(w, z) = 0$. We can then conclude that $\langle w, z\rangle = 0$.

(c) We showed in part (b) that $u = v + w, v \in M, w \in M^\perp$. Such a decomposition is unique. Suppose that we also had $u = v' + w', v' \in M, w' \in M^\perp$; then $0 = (v - v') + (w - w')$ with $v - v'$ in M and $w - w'$ in $M^\perp$. If the sum of two orthogonal vectors vanishes, each of the vectors must vanish. It follows therefore that the vector v of part (a) is unique.

□

REMARKS ON THE PROOF

1. If M is not closed, the infimum may not be attained. Take, for instance, $H = L_2(a, b)$ and $M =$ the set of all polynomials. If u is an element in L_2 but is not a polynomial, there exists for each $\varepsilon > 0$ an element v_ε in M for which $\|u - v_\varepsilon\| \leqslant \varepsilon$. Thus $\inf_{z\in M}\|u - z\| = 0$, but there is no element v in M for which $\|u - v\| = 0$.

2. If M is a closed linear manifold in a Banach space B, the infimum may not be attained because of the failure of the parallelogram law. The set M of continuous functions on $0 \leqslant x \leqslant 1$ vanishing at $x = 0$ is a closed linear manifold of $C[0, 1]$ equipped with the sup norm, yet the function $u(x) \equiv 1$ does not have a closest element in M.

3. To show that the minimizing sequence $\{z_n\}$ in part (a) was Cauchy, we only needed that $(z_m + z_n)/2$ be in M whenever z_m and z_n were. This is exactly the defining property of a *convex* set (if the midpoint of the line segment joining two points in the set is also in the set, the entire line segment is in the set). A linear manifold is a convex set, but the class of convex sets is much larger.

One can prove (see Exercise 4.5.8) that given $u \in H$ and a *closed* convex set C, there exists a unique element in C which is closest to u.

Corollary. *Let M be a closed linear manifold in H, with $M \neq H$. Then there exists a nonzero element in $M^{\perp}$.*

Proof. Let $u \neq 0$ be in H but not in M. Let the projection of u on M be v; then $w = u - v$ is a nonzero vector in $M^{\perp}$. □

The Gram-Schmidt Orthogonalization Process

It is often more convenient to deal with orthonormal sets of vectors rather than just independent sets. The Gram-Schmidt process converts an independent set into an orthonormal set with the same span.

Let $h_1, h_2, \ldots$ be a finite or infinite sequence of vectors in H; for each $k \geqslant 1$ (but not exceeding the number of elements in the sequence) let $h_1, \ldots, h_k$ be *independent*. We shall construct an orthonormal set $e_1, e_2, \ldots$ such that for each k, e_k is a linear combination of $h_1, \ldots, h_k$. Note that if the original sequence has only a finite number of elements, the orthonormal sequence has the same number of elements; if the original sequence is infinite, so is the orthonormal sequence. The construction is as follows. The unit vector e_1 is defined as $h_1/\|h_1\|$. To obtain e_2 we proceed in two steps. First, we construct g_2 by subtracting from h_2 its projection on e_1, that is, $g_2 = h_2 - \langle h_2, e_1 \rangle e_1$; g_2 is clearly orthogonal to e_1 and is not the zero vector, since it is a linear combination of h_2 and h_1 with the coefficient of h_2 not 0; normalization then gives $e_2 = g_2/\|g_2\|$. Continuing in this way, we construct successively

$$g_k = h_k - \sum_{j=1}^{k-1} \langle h_k, e_j \rangle e_j; \quad e_k = \frac{g_k}{\|g_k\|}. \tag{4.5.11}$$

The construction goes on ad infinitum if the original sequence is infinite; otherwise, it stops when the finite number of vectors in the original sequence has been exhausted. From (4.5.11) we see that e_k is a linear combination of $h_1, \ldots, h_k$, so that $e_1, \ldots, e_k$ all lie in the span M_k of $h_1, \ldots, h_k$. Now M_k is a k-dimensional manifold containing the k independent vectors $e_1, \ldots, e_k$, which must therefore span M_k.

An illustration of the process is given in Example 3(c) below.

Corollary. *Every Euclidean space E_n has an orthonormal basis.*

Proof. Use the Gram-Schmidt process on any basis for the space. □

By introducing orthonormal bases in n-dimensional spaces, we can often simplify proofs that might otherwise be awkward.

Theorem. *Every Euclidean space is complete.*

Proof. Let $\{u_k\}$ be a Cauchy sequence of elements in E_n, the index k ranging from 1 to ∞. With $e_1, \dots, e_n$ an orthonormal basis for E_n, we can write

$$u_k = \sum_{j=1}^{n} \alpha_{k,j} e_j,$$

and taking the inner product with e_i, we find that

$$\alpha_{k,i} = \langle u_k, e_j \rangle,$$

and by Schwarz's inequality,

$$|\alpha_{k,i} - \alpha_{p,i}| = |\langle u_k - u_p, e_i \rangle| \leqslant \|u_k - u_p\|.$$

Thus, for i fixed, $\{\alpha_{k,i}\}$ is a Cauchy sequence of numbers (real or complex, depending on whether E_n is real or complex) and must therefore converge to a number, say α_i. It follows that

$$\lim_{k\to\infty} u_k = \sum_{i=1}^{n} \alpha_i e_i,$$

which is a vector in E_n, so that E_n is complete. The proof for the concrete space of n-tuples of real numbers was already given in Example 3 of Section 4.3. □

Corollary. *Let M be a finite-dimensional manifold in a Hilbert space H (finite- or infinite-dimensional). Then M is closed in H.*

Example 1. A concrete example of a real n-dimensional Euclidean space is the space of n-tuples of real numbers, which becomes a real inner product space with the definition

$$\langle u, v \rangle = \xi_1\eta_1 + \cdots + \xi_n\eta_n,$$

generating the norm

$$\|u\| = (\xi_1^2 + \cdots + \xi_n^2)^{1/2}$$

and the metric

$$d_2(u, v) = \|u - v\| = [(\xi_1 - \eta_1)^2 + \cdots + (\xi_n - \eta_n)^2]^{1/2}.$$

The Schwarz inequality (4.5.3) becomes

$$\left[\sum_{k=1}^{n} \xi_k\eta_k\right]^2 \leqslant \left(\sum_{k=1}^{n} \xi_k^2\right)\left(\sum_{k=1}^{n} \eta_k^2\right). \tag{4.5.12}$$

This is just the space already encountered in Example 3, Section 4.3, with the norm (4.3.4). A direct proof of (4.5.12) is given in Exercise 4.3.5. We showed there that the space is complete. Note that the other norms considered in that example cannot be generated from an inner product. These spaces are complete normed spaces (that is, Banach spaces) but *not* Hilbert spaces (see Exercise 4.5.4).

Example 2. A concrete example of a complex Euclidean space is the space of n-tuples of complex numbers with the inner product defined as

$$\langle u, v\rangle = \sum_{k=1}^{n} \xi_k \bar{\eta}_k.$$

Conditions (4.5.2) are easily verified. The norm and metric are given by

$$\|u\| = \left[\sum_{k=1}^{n} |\xi_k|^2\right]^{1/2} ; \quad d_2(u, v) = \|u - v\| = \left[\sum_{k=1}^{n} |\xi_k - \eta_k|^2\right]^{1/2},$$

while the Schwarz inequality (4.5.3) becomes

$$\left|\sum_{k=1}^{n} \xi_k \bar{\eta}_k\right|^2 \leqslant \left(\sum_{k=1}^{n} |\xi_k|^2\right)\left(\sum_{k=1}^{n} |\eta_k|^2\right). \tag{4.5.13}$$

Example 3.

(a) Consider the space $L_2^{(r)}(a, b)$ of real-valued functions $u(x)$ on $a \leqslant x \leqslant b$ for which the Lebesgue integral $\int_a^b u^2(x)\, dx$ exists and is finite. The inner product is defined by

$$\langle u, v\rangle = \int_a^b u(x)v(x)\, dx,$$

which meets all the conditions (4.5.1). The corresponding norm and metric are

$$\|u\| = \left[\int_a^b u^2\, dx\right]^{1/2} ; \quad d_2(u, v) = \|u - v\| = \left[\int_a^b (u - v)^2\, dx\right]^{1/2},$$

and the Schwarz inequality becomes

$$\left|\int_a^b uv\, dx\right| \leqslant \left[\int_a^b u^2\, dx\right]^{1/2} \left[\int_a^b v^2\, dx\right]^{1/2}. \tag{4.5.14}$$

As was pointed out in Example 7, Section 4.3, $L_2^{(r)}(a, b)$ is complete with this metric. Therefore, $L_2^{(r)}(a, b)$ is a Hilbert space.

(b) Let us look at some linear manifolds in $H = L_2^{(r)}(a, b)$. Any such manifold has an inner product structure inherited from H. The set M_1 of continuous functions on $a \leqslant x \leqslant b$ is a linear manifold in H; M_1 is *not* closed, but $\bar{M}_1 = H$. The set M_2 of polynomials on $a \leqslant x \leqslant b$ is a linear manifold which is not closed, but again $\bar{M}_2 = H$. The set M_3 of the elements in H which are even about $x = (a + b)/2$ is a closed linear manifold with $M_3 \neq H$. The set M_4 of polynomials even about $x = (a + b)/2$ is a linear manifold in H with $M_4 \neq$

$\bar{M}_4 \neq H$. All these linear manifolds are infinite-dimensional. The set M_5 of all polynomials of degree < 17 is a finite-dimensional linear manifold in H and as such is automatically closed.

(c) Consider the set $(1, x, x^2, \ldots)$ in $L_2^{(r)}(-1,1)$. We use a slight variant of the Gram-Schmidt process to convert the set into an orthogonal set. Instead of normalizing g_k in (4.5.11), it is customary to require that $g_k(1) = 1$. This leads to the orthogonal set of *Legendre polynomials*, whose first few elements are

$$\psi_1(x) = 1, \;\; \psi_2(x) = x, \;\; \psi_3(x) = \tfrac{1}{2}(3x^2 - 1), \;\; \psi_4(x) = \tfrac{1}{2}(5x^3 - 3x), \ldots .$$

Example 4.

(a) Let $L_2^{(c)}(a,b)$ be the set of complex-valued functions $u(x)$ defined on $a \leqslant x \leqslant b$ for which the Lebesgue integral $\int_a^b |u|^2 \, dx$ exists and is finite. It is easily seen that

$$\langle u, v \rangle = \int_a^b u(x)\bar{v}(x)\, dx$$

is an admissible inner product, yielding the norm and metric

$$\|u\| = \left[\int_a^b |u|^2\right]^{1/2}; \quad d_2(u,v) = \|u - v\| = \left[\int_a^b |u - v|^2 \, dx\right]^{1/2}.$$

The Schwarz inequality takes the form

$$\left|\int_a^b u\bar{v} \, dx\right| \leqslant \left[\int_a^b |u|^2 \, dx\right]^{1/2} \left[\int_a^b |v|^2 \, dx\right]^{1/2}. \tag{4.5.15}$$

It is a consequence of the properties of the Lebesgue integral that the space is complete. Thus $L_2^{(c)}(a,b)$ is a Hilbert space.

(b) Consider the set $L_2^{(c)}(D)$ of all complex-valued functions $u(x_1, \ldots, x_k)$ on a domain D in the k variables $x_1, \ldots, x_k$ for which the k-dimensional Lebesgue integral

$$\int_D |u(x_1, \ldots, x_k)|^2 dx_1 \cdots dx_k$$

exists and is finite. This space is a Hilbert space with the inner product defined as

$$\langle u, v \rangle = \int_D u\bar{v} \, dx_1 \cdots dx_k.$$

Example 5. The space $C[a, b]$ of continuous functions (real- or complex-valued) on a bounded interval $a \leqslant x \leqslant b$ is a normed space under the definition

$$|u| = \max_{a \leqslant x \leqslant b} |u(x)|.$$

This norm generates the natural metric

$$d_\infty(u, v) = \|u - v\| = \max_{a \leqslant x \leqslant b} |u(x) - v(x)|,$$

under which $C[a, b]$ was shown to be complete in Example 5, Section 4.3. Thus $C[a, b]$ is a Banach space, but Exercise 4.5.5 shows that the norm *cannot* be derived from an inner product.

Example 6. Sobolev spaces. Let Ω be a bounded domain in $\mathbb{R}^n$ with boundary Γ. Consider the set S of real-valued functions which are continuous and have a continuous gradient on $\bar{\Omega}$. The bilinear form

$$\langle u, v \rangle \doteq \int_\Omega (\text{grad}\, u \cdot \text{grad}\, v + uv)\, dx$$

is clearly an admissible inner product on S. The completion of S in the norm generated by this inner product is known as the Sobolev space $H^1(\Omega)$. A second distinct definition of the space $H^1(\Omega)$ was given in Section 2.6, as the space of functions in $L^2(\Omega)$ with first weak derivative also in $L^2(\Omega)$, where the weak derivative was defined through distribution theory. Under reasonable assumptions about the domain Ω, it can be shown that these two characterizations lead to precisely the same function space (the Meyers-Serin Theorem). The Hilbert space $H^1(\Omega)$ and the generalizations $W^{m,p}(\Omega)$ described in Section 2.6 play an important role in the theory of partial differential equations (see Section 8.4) and in the development of modern numerical methods for these equations. Chapter 10 contains a fairly complete overview of the properties of Sobolev spaces on bounded Lipschitz domains $\Omega \subset \mathbb{R}^n$.

EXERCISES

4.5.1 Refer to Example 3, Section 4.3. Show that only by taking $p = 2$ does the norm satisfy the parallelogram law (4.5.7). Therefore, only then is the space a Hilbert space, according to Exercise 4.5.4.

4.5.2 Consider a linear space provided with a metric satisfying

$$d(\alpha u, \alpha v) = |\alpha|\, d(u, v)$$

for all vectors u, v and all scalars α. Show that $d(u, 0)$ is an admissible norm on the space.

4.5.3 Show that on any inner product space

$$\langle u, v \rangle + \langle v, u \rangle = \tfrac{1}{2}[\|u + v\|^2 - \|u - v\|^2],$$

and that for a complex inner product space

$$\langle u, v\rangle - \langle v, u\rangle = \tfrac{i}{2}[\|u+iv\|^2 - \|u-iv\|^2].$$

Consequently, in a complex inner product space

$$\langle u, v\rangle = \tfrac{1}{4}[\|u+v\|^2 - \|u-v\|^2 + i\|u+iv\|^2 - i\|u-iv\|^2], \qquad (4.5.16)$$

whereas in a real inner product space

$$\langle u, v\rangle = \tfrac{1}{4}[\|u+v\|^2 - \|u-v\|^2]. \qquad (4.5.17)$$

Thus in an inner product space, knowledge of the norm of every element determines the inner product unambiguously. Therefore, in any normed linear space *there can exist at most one inner product which generates the norm.*

4.5.4 *The Jordan–Von Neumann theorem.* Consider a *real* Banach space whose norm satisfies the parallelogram law (4.5.7). Define the real-valued function $f(u, v)$ from

$$f(u, v) = \tfrac{1}{4}[\|u+v\|^2 - \|u-v\|^2].$$

(a) Show that f is a continuous function of v.

(b) Show that for any integer k, $f(ku, v) = kf(u, v)$, and hence that for any real α, $f(\alpha u, v) = \alpha f(u, v)$.

(c) Show that $f(u_1 + u_2, v) = f(u_1, v) + f(u_2, v)$.

In the light of these properties $f(u, v)$ is an admissible inner product with $f(u, u) = \|u\|^2$. By the final remark in Exercise 4.5.3, no other inner product can generate the given norm. Thus *the necessary and sufficient condition for a real Banach space to be a Hilbert space is that the norm satisfies the parallelogram law* (4.5.7). If the law is satisfied, the inner product can be calculated from the norm by (4.5.17).

4.5.5 Use the result of Exercise 4.5.4 to show that $C[a, b]$ with the uniform metric is not a Hilbert space even though it is a Banach space.

4.5.6 Consider the Hilbert space $H = L_2^{(r)}(-1, 1)$.

(a) Let M be the linear manifold of even functions in H. What is $M^\perp$? What is the unique decomposition predicted by the projection theorem?

(b) Answer the same question if M is the linear manifold of all functions in H such that $\int_{-1}^{1} u\, dx = 0$.

(c) Is the set of functions such that $\int_{-1}^{1} u\, dx = 1$ a linear manifold?

4.5.7 A *bilinear form* $b(u, v)$ on a complex linear space X is a complex-valued function of pairs of elements u and v satisfying

$$b(\alpha u_1 + \beta u_2, v) = \alpha b(u_1, v) + \beta b(u_2, v),$$
$$b(u, \alpha v_1 + \beta v_2) = \bar{\alpha} b(u, v_1) + \bar{\beta} b(u, v_2).$$

It follows that $b(0, v) = b(u, 0) = 0$. The bilinear form $b(u, v)$ generates a quadratic functional $b(u, u) = B(u)$, known as the *associated quadratic form.* The bilinear form is said to be symmetric if $b(u, v) = \bar{b}(v, u)$, in which case the associated quadratic form is *real*. If, moreover, $B(u) \geqslant 0$ for all u, we say that both the quadratic form and the bilinear form are *nonnegative*. If $B(u) > 0$ for all $u \neq 0$, the forms are *positive*. The simplest example of a positive, symmetric bilinear form is the inner product satisfying (4.5.2). Show that any bilinear form satisfies the *polar identity*

$$b(u, v) = \tfrac{1}{4}\{B(u + v) - B(u - v) + iB(u + iv) - iB(u - iv)\}, \quad (4.5.18)$$

which should be compared with (4.5.16). If $b(u, v)$ is *nonnegative*, prove the Schwarz-like inequality

$$|b(u, v)|^2 \leqslant B(u)B(v).$$

Show that any positive bilinear form on X generates an admissible inner product $\langle u, v\rangle_b \doteq b(u, v)$.

4.5.8 Let C be a closed convex set in a Hilbert space H, and let u be an arbitrary vector in H. Show that there exists one and only one element in C which is closest to u. (See Remark 3 following the projection theorem.)

4.5.9 In a Banach space show that $\|u\| - \|v\| \leqslant \|u - v\|$ and therefore, by interchanging u and v, that

$$|\,\|u\| - \|v\|\,| \leqslant \|u - v\|. \quad (4.5.19)$$

4.5.10 Let Ω be a bounded domain in $\mathbb{R}^3$ with boundary Γ. Consider the linear space X of vector functions $\overrightarrow{w} = (w_1(x), w_2(x), w_3(x))$ defined on $\bar{\Omega}$, where the components w_1, w_2, w_3 are real and have continuous partial derivatives of order 2 in $\bar{\Omega}$. With the definition

$$\langle \overrightarrow{w}, \overrightarrow{z}\rangle = \int_\Omega \overrightarrow{w} \cdot \overrightarrow{z}\, dx,$$

where dx is a volume element in $\mathbb{R}^3$, X becomes an inner product space. We introduce two linear manifolds in X: M_1 is the set of *irrotational* vectors [that is, $\overrightarrow{u} \in M_1$ if there exists a scalar function $p(x)$ such that $\overrightarrow{u} = \operatorname{grad} p$]; M_2 is the set of *solenoidal* vectors $\overrightarrow{v}$ (that is, such that div $\overrightarrow{v} = 0$) with $\overrightarrow{v} \cdot \overrightarrow{n} = 0$ on Γ.

(a) Prove that M_1 and M_2 are orthogonal.

(b) For $\overrightarrow{w}$ in X, write

$$\overrightarrow{w} = \overrightarrow{u} + \overrightarrow{v}, \quad \overrightarrow{u} \in M_1, \overrightarrow{v} \in M_2.$$

Show that if such a decomposition exists, it is unique.

(c) Existence of the decomposition is proved by exhibiting the required $\overrightarrow{u}$ and $\overrightarrow{v}$: We see that p satisfies

$$\Delta p = \operatorname{div} \overrightarrow{w}, \qquad \left.\frac{\partial p}{\partial n}\right|_{\Gamma} = \overrightarrow{w} \cdot \overrightarrow{n},$$

which is a Neumann BVP for $p(x)$ since the right sides are given. In Chapter 8 it is shown that the problem has a solution p determined only up to an additive constant. Thus $\overrightarrow{u}$ is uniquely determined as grad p; setting $\overrightarrow{v} = \overrightarrow{w} - \overrightarrow{u}$, we see that $\overrightarrow{v}$ belongs to M_2.

4.5.11 Let T be a Lipschitz operator [see (4.4.2)] defined on the real Hilbert space H. We say that T is *strongly monotone* if there exists $c > 0$ such that

$$\langle Tu - Tv, u - v \rangle \geqslant c\|u - v\|^2 \quad \text{for all } u, v \in H.$$

Solving the equation $Tu = f$ is equivalent to finding the fixed points of the operator B, where $Bu = u - t(Tu - f)$ and $t > 0$ is a parameter. Show that B becomes a contraction on H for a suitable choice of t. Hence $Tu = f$ has one and only one solution.

4.6 SEPARABLE HILBERT SPACES AND ORTHONORMAL BASES

We recall that a Hilbert space H is a linear space on which an *inner product* is defined, the space being *complete* in the metric generated by the inner product. Our principal interest in this section is in infinite-dimensional spaces, particularly in $L_2(a, b)$, but the results we shall derive for the infinite-dimensional case will apply—with obvious simplifications—to Euclidean spaces also. This is hardly surprising, since much of the infinite-dimensional theory is motivated by analogy with the finite-dimensional case. We shall deal with separable Hilbert spaces (the condition of separability defined below ensures that our infinite-dimensional space is not too "large"), of which L_2 is a particular instance. The development will be presented for the general, abstract case because the notation is simpler and the geometrical analogy more vivid. In specific applications to L_2 we shall revert to functional notation with Examples 3 and 4 of Section 4.5 serving as a dictionary of translation.

Let U be a set of vectors in H. The algebraic span $S(U)$ has been defined as the set of all finite linear combinations of vectors drawn from U. In general, the set $S(U)$ will not be closed in H; its closure $\bar{S}(U)$ is known as the *closed span* of U. If U happens to contain only a finite number of independent vectors, $S(U)$ is a finite-dimensional manifold and is therefore closed. In other cases $S(U)$ and $\bar{S}(U)$ may differ markedly; for instance, if U is the set $(1, x, x^2, \ldots)$ in $L_2(a, b)$, $S(U)$ is the set of all polynomials, whereas $\bar{S}(U)$ is the whole of $L_2(a, b)$ (see Section 4.3).

If the closed span of U coincides with H, we say that U is a *spanning set*; alternatively, we may say that the algebraic span of U is dense in H. By taking U to be the whole of H, it is trivially true that $S(U) = \bar{S}(U) = H$, but the goal is to find a relatively small set U which will be a spanning set for H.

Definition. *The Hilbert space H is* separable *if it contains a* countable *spanning set U.*

We recall that a set is countable if its elements can be placed into one-to-one correspondence with the positive integers or some subset of the positive integers (a set containing only a finite number of elements is countable; the rational numbers form a countable set, but the real numbers do not). Without the crucial word *countable* in the definition of separability, H itself could be used as the spanning set. The existence of a countable spanning set means that H is not too "large," although it will always contain an uncountable number of elements.

A finite-dimensional space is separable since any basis serves as a finite spanning set. Obviously, an infinite-dimensional space cannot have a finite spanning set. Thus a separable, infinite-dimensional Hilbert space must contain a countably infinite set $U : \{u_1, u_2, \ldots\}$ such that each vector u in H can be approximated to any desired accuracy by a linear combination of a finite number of elements of U; in other words, for each u in H and each $\varepsilon > 0$, there exist an index N and scalars $\alpha_1, \ldots, \alpha_N$ (*which usually depend on* ε) such that

$$\left\| u - \sum_{k=1}^{N} \alpha_k u_k \right\| < \varepsilon. \tag{4.6.1}$$

A separable space H contains an algebraically independent spanning set (all that is necessary is to eliminate from the spanning set U the elements u_k, which are linear combinations of elements with lower indices). By the Gram-Schmidt procedure one can convert this spanning set into an orthonormal set with the same algebraic span and hence with the same closed span. *Thus a separable space H contains an orthonormal spanning set.*

It should be pointed out the condition (4.6.1) does not guarantee that u can be expanded in an infinite series $\sum_{k=1}^{\infty} \alpha_k u_k$. For (4.6.1) to be equivalent to this series expansion it must be possible to choose the coefficients $\{\alpha_k\}$ in (4.6.1) independently of ε, and this cannot always be done. The space $L_2(a, b)$ is separable since the countable set $\{1, x, x^2, \ldots\}$ is a spanning set, but there are few elements of $L_2(a, b)$ that can be expanded in a power series $\sum_{k=1}^{\infty} \alpha_k x^{k-1}$; such a function would, at the very least, have to be infinitely differentiable, and most L_2 functions are not that smooth. This distinction, discussed in Exercise 2.2.7(d), *fortunately disappears for orthonormal spanning sets.* (See Theorem 4.6.1.)

In addition to $L_2(a, b)$, the spaces $L_2(a, \infty)$, $L_2(-\infty, b)$, $L_2(-\infty, \infty)$ are all separable, but since polynomials do not belong to these spaces, the earlier argument must be modified (see Exercise 4.6.6).

From here on we confine ourselves to separable Hilbert spaces. We begin with the notion of a basis for such a space.

Definition. *A set of vectors $\{h_1, \ldots, h_n, \ldots\}$ is said to be a* basis *(a* Schauder basis*) for H if each vector u in H can be represented in* one and only one way *as*

$$u = \sum_{k=1}^{\infty} \alpha_k h_k.$$

The numerical coefficients α_k *are the* coordinates *of* u *in the basis.*

REMARKS

1. We have not yet shown that H has a basis.

2. The coordinates α_k are easily calculated if the basis is orthonormal. Suppose, in fact, that $\{e_1, \ldots, e_n, \ldots\}$ is an orthonormal basis, so that $u = \sum_{k=1}^{\infty} \alpha_k e_k$. Let us calculate α_i. Choose $n \geqslant i$, and let $s_n = \sum_{k=1}^{n} \alpha_k e_k$. Then $\langle s_n, e_i \rangle = \langle \sum_{k=1}^{n} \alpha_k e_k, e_i \rangle = \alpha_i$. Since $s_n \to u$ and the inner product is continuous, we have
$$\langle u, e_i \rangle = \lim_{n\to\infty} \langle s_n, e_i \rangle = \alpha_i. \tag{4.6.2}$$
An orthonormal basis is also known as a *complete orthonormal set.* Note that if the orthonormal set is not a basis but u happens to be expressible as $\sum_{k=1}^{\infty} \alpha_k e_k$, it is still true that $\alpha_i = \langle u, e_i \rangle$. In the section on linear functionals we show how to calculate the coordinates in an oblique basis.

3. If $\{h_1, \ldots, h_n, \ldots\}$ is a Schauder basis, the equation $\alpha_1 h_1 + \cdots + \alpha_n h_n + \cdots = 0$ has only the solution $\alpha_1 = \cdots = \alpha_n = \cdots = 0$. Thus the set satisfies a stronger "independence" condition than the algebraic independence previously introduced. It is, of course, clear that any finite subset of the basis is independent, so that the basis is certainly algebraically independent.

4. In an infinite-dimensional space a basis necessarily consists of an infinite number of independent vectors, but not every infinite independent set is a basis. (For instance, removing a single element from a basis leaves an infinite independent set which is not a basis.)

In the next section we prove the existence of an orthonormal basis for H, but we stop here to prove the following preliminary theorem.

Theorem (Riesz-Fischer Theorem). *Let* $\{e_k\}$ *be an orthonormal set (not necessarily a basis) in* H, *and let* $\{\alpha_k\}$ *be a sequence of complex numbers. Then* $\sum_{k=1}^{\infty} \alpha_k e_k$ *and* $\sum_{k=1}^{\infty} |\alpha_k|^2$ *converge or diverge together.*

Proof. Let $\{s_n\}$ be the sequence of partial sums for the series $\sum_{k=1}^{\infty} \alpha_k e_k$, and let $\{t_n\}$ be the sequence of partial sums for the series $\sum_{k=1}^{\infty} |\alpha_k|^2$. Note that $\{s_n\}$ is a sequence of elements in H, whereas $\{t_n\}$ is a sequence of nonnegative real numbers. These sequences are related by
$$\|s_p - s_m\|^2 = \left\| \sum_{k=m+1}^{p} \alpha_k e_k \right\|^2 = \sum_{k=m+1}^{p} |\alpha_k|^2 = (t_p - t_m), \quad p > m. \tag{4.6.3}$$
If $\sum \alpha_k e_k$ coverges, $\{s_n\}$ is a Cauchy sequence in H and (4.6.3) shows that $\{t_n\}$ is a Cauchy sequence of numbers and must therefore converge. Conversely, if $\sum |\alpha_k|^2$ converges, $\{t_n\}$ is a Cauchy sequence of numbers and (4.6.3) now shows that $\{s_n\}$ is a Cauchy sequence in H. Since H is complete, $\{s_n\}$ must converge. □

Problem of Best Approximation

We wish to approximate an arbitrary vector u in H by a linear combination of the given independent set $\{u_1, \ldots, u_k\}$; that is, we seek the element $\sum_{i=1}^{k} \alpha_i u_i$ closest to u in the sense of the norm of H. Since the set of linear combinations of $\{u_1, \ldots, u_k\}$ is a k-dimensional linear manifold M_k (and hence necessarily closed), the problem is a special case of that considered in the projection theorem. Our purpose here is to study the question more deeply and concretely when H is separable. By the Gram-Schmidt procedure we can construct from $\{u_1, \ldots, u_k\}$ an orthonormal basis $\{e_1, \ldots, e_k\}$ for M_k. We are then interested in finding the element $\sum_{i=1}^{k} \beta_i e_i$ which is closest to u. The advantage of using an orthonormal basis rather than an oblique one is that the calculation of the coefficients is much simpler. For any $\beta_1, \ldots, \beta_k$ we have

$$\begin{aligned} \left\| u - \sum_{i=1}^{k} \beta_i e_i \right\|^2 &= \|u\|^2 + \sum_{i=1}^{k} \beta_i \bar{\beta}_i - \sum_{i=1}^{k} \beta_i \langle e_i, u \rangle - \sum_{i=1}^{k} \bar{\beta}_i \langle u, e_i \rangle \\ &= \|u\|^2 + \sum_{i=1}^{k} |\langle u, e_i \rangle - \beta_i|^2 - \sum_{i=1}^{k} |\langle u, e_i \rangle|^2. \end{aligned} \tag{4.6.4}$$

It is clear that the minimum is attained for $\beta_i = \langle u, e_i \rangle$. These optimal values of the coefficients are known as the *Fourier coefficients* of u with respect to the set $\{e_1, \ldots, e_k\}$ and will be permanently denoted by γ_i; thus $\gamma_i = \langle u, e_i \rangle, i = 1, \ldots, k$. The uniquely determined best approximation to u in M_k is the *Fourier sum* $v_k = \sum_{i=1}^{k} \gamma_i e_i = \sum_{i=1}^{k} \langle u, e_i \rangle e_i$. Geometrically, the single term $\langle u, e_i \rangle e_i$ is the orthogonal projection of u on the line generated by e_i. The Fourier sum is the orthogonal projection of u on the linear manifold M_k. We easily confirm by explicit calculation that $w_k = u - v_k$ is orthogonal to M_k, as predicted by the projection theorem.

Our best approximation to u will usually be far from perfect. In fact, the square of the distance between u and the Fourier sum v_k is

$$\left\| u - \sum_{i=1}^{k} \langle u, e_i \rangle e_i \right\|^2 = \|u\|^2 - \sum_{i=1}^{k} |\langle u, e_i \rangle|^2,$$

from which we deduce that

$$\sum_{i=1}^{k} |\langle u, e_i \rangle|^2 \leqslant \|u\|^2. \tag{4.6.5}$$

Naturally, the best approximation to u improves as the orthonormal set is taken larger. Suppose that we include another vector e_{k+1} in our orthonormal set of vectors; the new best approximation is the Fourier sum $\sum_{i=1}^{k+1} \langle u, e_i \rangle e_i$, which is just the previous one with the added term $\langle u, e_{k+1} \rangle e_{k+1}$ corresponding to the extra axis e_{k+1}. *What is crucial is that previously calculated coefficients do not have to be changed.* This suggests the possibility of using infinite orthonormal sets, with the best approximation to u being an infinite series in $\{e_k\}$. If we had used independent, but not orthonormal, vectors $u_1, \ldots, u_k$ to approximate u, we would find that the inclusion of an additional vector u_{k+1} in the approximating pool usually leads to entirely new coefficients in the best approximation.

Before proceeding to consider the problem of best approximation when the set $\{e_k\}$ is infinite, we pause to show that an orthonormal spanning set for H is necessarily a basis.

Theorem 4.6.1. *Let $\{e_1, \ldots, e_n, \ldots\}$ be an orthonormal spanning set for H. Then the set is a basis for H.*

Proof. We must show that for each u in H, $u = \sum_{k=1}^{\infty} \langle u, e_k \rangle e_k$, that is, we must show for each u and each $\varepsilon > 0$ there exists N such that $\|u - \sum_{k=1}^{n} \langle u, e_k \rangle e_k\| < \varepsilon$ whenever $n > N$. In order to establish this, choose N and $\alpha_1, \ldots, \alpha_N$ such that $\|u - \sum_{k=1}^{N} \alpha_k e_k\| < \varepsilon$, which we know can be done since $\{e_i\}$ is a spanning set. Now $\sum_{k=1}^{N} \alpha_k e_k$ lies in M_N, the linear manifold generated by $e_1, \ldots, e_N$. The best approximation to u in M_N is given by $\sum_{k=1}^{N} \langle u, e_k \rangle e_k$, so that

$$\left\| u - \sum_{k=1}^{N} \langle u, e_k \rangle e_k \right\| \leqslant \left\| u - \sum_{k=1}^{N} \alpha_k e_k \right\| < \varepsilon.$$

If $n > N$, we have

$$\left\| u - \sum_{k=1}^{n} \langle u, e_k \rangle e_k \right\| \leqslant \left\| u - \sum_{k=1}^{N} \langle u, e_k \rangle e_k \right\|,$$

and the theorem has therefore been proved. □

Corollary 4.6.1. *Let $\{e_1, \ldots, e_n, \ldots\}$ be an orthonormal set which is not necessarily a spanning set for H. Let the closed span of this set be denoted by M. Then the set is a basis for M.*

We now return to the problem of best approximation when the approximating set is a countably infinite orthonormal set $\{e_1, \ldots, e_n, \ldots\}$. Then, according to Corollary 4.6.1, each element in the closed span M of the set can be expressed as an infinite series $\sum \beta_k e_k$. Let u be an element of H. We want to find the vector in M which lies closest to u. We suspect that this best approximation is the *Fourier series*

$$v = \sum_{i=1}^{\infty} \langle u, e_i \rangle e_i = \sum_{i=1}^{\infty} \gamma_i e_i. \tag{4.6.6}$$

To show that this series converges we first note that the sequence of partial sums for the series $\sum_{i=1}^{\infty} |\gamma_i|^2$ is monotonically increasing and, by (4.6.5), is bounded above by the fixed number $\|u\|^2$. Thus the series $\sum_{i=1}^{\infty} |\gamma_i|^2$ converges, and so does

$\sum_{i=1}^{\infty} \gamma_i e_i$ (by the Riesz-Fischer theorem). Moreover, we note *Bessel's inequality* [compare with (2.3.12)]

$$\sum_{i=1}^{\infty} |\gamma_i|^2 = \sum_{i=1}^{\infty} |\langle u, e_i\rangle|^2 \leqslant \|u\|^2 \tag{4.6.7}$$

and its trivial consequence

$$\lim_{i\to\infty} \langle u, e_i\rangle = 0. \tag{4.6.8}$$

That v in (4.6.6) gives the best approximation to u among vectors in M now follows by letting k tend to infinity in (4.6.4). It is also easy to see that the vector $w = u - v$ is orthogonal to M.

What if the closed manifold M in H is given to us without the a priori knowledge that it is spanned by a countable orthonormal set? Since M is a subset of a separable space, it can be shown that it is itself separable and is therefore spanned by a countable set which can be made orthonormal by the Gram-Schmidt process. Thus M is the span of an orthonormal set which, by the corollary, must be a basis for M. We are therefore in a position to state a refined version of the projection theorem.

Theorem (Projection Theorem for Separable Spaces). *Let H be a separable space and M a closed linear manifold in H. Then M has an orthonormal basis $\{e_k\}$, and each vector u in H can be decomposed in one and only one way as the sum of a vector v in M and a vector w in $M^{\perp}$. The vector v is the one and only vector in M lying closest to u; moreover, v has the explicit form (4.6.6).*

Characterization of an Orthonormal Basis

Let H be a separable space. By definition an orthonormal set $\{e_k\}$ is a basis for H if each vector u in H has a unique representation $u = \sum_{k=1}^{\infty} \alpha_k e_k$. From this it follows that $\alpha_k = \langle u, e_k\rangle$, so that $\{e_k\}$ is an orthonormal basis if and only if each u in H can be written as $u = \sum_{k=1}^{\infty} \langle u, e_k\rangle e_k$. We now give a number of equivalent criteria for an orthonormal set to be a basis for H.

Theorem 4.6.2. *If* any one *of the following criteria is met, the orthonormal set $\{e_k\}$ is a basis for H.*

1. *For each u in H, $u = \sum_{k=1}^{\infty} \langle u, e_k\rangle e_k$. This is just a direct consequence of the definition of an orthonormal basis. It says that each u is equal to its Fourier series or, equivalently, that the best approximation to u in terms of the $\{e_k\}$ is perfect.*

2. *The set $\{e_k\}$ is maximal. (An orthonormal set $\{e_1, \ldots, e_n, \ldots\}$ is* maximal *if there exists no element e in H such that $\{e, e_1, \ldots, e_n, \ldots\}$ is an orthonormal set.)*

3. *The only element u in H for which* all *the Fourier coefficients $\langle u, e_1\rangle, \ldots$ are 0 is the element $u = 0$.*

4. Parseval's identity. *For each u in H,*

$$\|u\|^2 = \sum_{k=1}^{\infty} |\langle u, e_k\rangle|^2. \tag{4.6.9}$$

REMARK. If H is finite-dimensional of dimension n, there is a very simple way of characterizing an orthonormal basis: it must have exactly n elements. Note, however, that any one of criteria 1 through 4 also characterizes an orthonormal basis in n-space; each can be regarded as an indirect way of saying that there are n elements in the orthonormal set. Unfortunately, counting the elements is not good enough in an infinite-dimensional space H; admittedly, an orthonormal set must have an infinite number of elements to be a basis, but that is not sufficient. On the other hand, criteria 1 through 4 carry over to the infinite-dimensional case.

Proof.

(a) Criterion 2 implies 3. Assuming criterion 3 is false, there would exist $\psi \neq 0$ such that $\langle \psi, e_i\rangle = 0$ for all i. Then $\psi/\|\psi\|$ would be a unit element orthogonal to $e_1, \ldots, e_n, \ldots$ and could therefore be used to enlarge the orthonormal set, thereby violating criterion 2.

(b) Criterion 3 implies 1. We have previously shown that $\sum_{k=1}^{\infty} \langle u, e_k\rangle e_k$ converges to some element v in H; we must prove that $v = u$ if criterion 3 holds. Now it is clear that v and u have the same Fourier series, so that $\langle u - v, e_k\rangle = 0$ for all k; therefore, criterion 3 guarantees that $u - v = 0$ or $u = v$.

(c) Criterion 1 implies 4. This is automatic by passing to the limit as $k \to \infty$ in the identity $\|u - \sum_{i=1}^{k} \langle u, e_i\rangle e_i\|^2 = \|u\|^2 - \sum_{i=1}^{k} |\langle u, e_i\rangle|^2$.

(d) Criterion 4 implies 2. Assuming criterion 2 false, there exists e with $\|e\| = 1$ and $\langle e, e_k\rangle = 0$ for all k. But then criterion 4 would tell us that $\|e\|^2 = \sum_{k=1}^{\infty} |\langle e, e_k\rangle|^2$, an evident contradiction.

□

REMARK. In view of Theorem 4.6.1, it is clear that criterion 1 is equivalent to the following: For each u in H and each $\varepsilon > 0$ there exist $N > 0$ and $c_1, \ldots, c_N$ such that $\|u - \sum_{k=1}^{N} c_k e_k\| < \varepsilon$.

Orthonormal Bases on Concrete Functional Spaces

In $L_2^{(c)}(a,b)$ the elements u are complex-valued functions $u(x)$ that are Lebesgue square integrable on the interval $a \leqslant x \leqslant b$. We revert to functional notation in this section. Let $\{e_1(x), \ldots, e_n(x), \ldots\}$ be an orthonormal basis for $L_2^{(c)}(a,b)$. The criterion for orthonormality takes the form

$$\int_a^b e_i(x)\bar{e}_j(x)\,dx = \begin{cases} 0, & i \neq j, \\ 1, & i = j. \end{cases} \tag{4.6.10}$$

Since $\{e_i\}$ is an orthonormal basis, each $u(x)$ in $L_2^{(c)}(a,b)$ can be expanded as a Fourier series

$$u(x) = \sum_{k=1}^{\infty} \gamma_k e_k(x), \quad \gamma_k = \int_a^b u(x)\bar{e}_k(x)\,dx, \tag{4.6.11}$$

where convergence in (4.6.11) stands for mean-square convergence (that is, L_2 convergence)

$$\lim_{n\to\infty} \int_a^b \left| u(x) - \sum_{k=1}^{\infty} \gamma_k e_k(x) \right|^2 dx = 0. \tag{4.6.12}$$

If we want (4.6.11) *to hold in some other mode such as pointwise convergence or uniform convergence, it is not enough to know that* $\{e_i(x)\}$ *is a basis* (see Section 0.6 for relations among various modes of convergence). Such questions require delicate analysis, and the desired type of convergence may be achieved only if $u(x)$ belongs to some class of functions such as differentiable functions or functions of bounded variation. Nevertheless, the methods used to show that $\{e_i(x)\}$ is a basis can often be adapted—at the cost of additional effort—to prove that (4.6.11) holds in some stronger sense. Since mean-square convergence is by far the most important type of convergence in applications, we will usually be satisfied to show that an orthonormal set $\{e_i(x)\}$ is a basis, which will often not be easy to do. We remark that it is enough to show that (4.6.11) holds in the sense of (4.6.12) for any subset M of functions $u(x)$ that is dense in L_2. In fact, it is enough to show that each u in M can be approximated (in the L_2 norm) to arbitrarily prescribed accuracy by finite linear combinations of the set $\{e_i(x)\}$. The following three methods are particularly important for showing that an orthonormal set is a basis:

1. Adapt the Weierstrass approximation theorem (see Exercises 4.6.3 and 4.6.4, where trigonometric orthonormal sets are treated in this way).

2. Use the general theorem on eigenfunctions of compact self-adjoint operators (Theorem 6.3.5).

3. Refer to the theory of delta sequences in Section 2.2.

Let us explain what is meant by the third method. Consider the sequence $\{t_n(x)\}$ of partial sums of (4.6.11). We have

$$t_n(x) = \sum_{k=1}^{n} \gamma_k e_k(x) = \sum_{k=1}^{n} \left(\int_a^b u(y)\bar{e}_k(y)\,dy \right) e_k(x),$$

which, since we are dealing with finite sums, can be rewritten as

$$t_n(x) = \int_a^b S_n(x,y)u(y)\,dy, \tag{4.6.13}$$

where

$$S_n(x,y) \doteq \sum_{k=1}^{n} e_k(x)\bar{e}_k(y)$$

is known as the kernel of (4.6.13). We must then show that as $n \to \infty$ (4.6.13) approaches $u(x)$. Admittedly, we only want to prove this in the L_2 sense, but we could try from (4.6.13) to analyze the possible convergence of t_n in some other mode, although we shall not do so. What is needed roughly is to show that as $n \to \infty$, $S_n(x,y) \to \delta(x-y)$, where δ is the Dirac delta function of Chapter 2. We shall use this approach to show later that the set (4.6.15) is in fact an orthonormal basis.

REMARKS

1. In some problems [see, for instance, (4.6.17)] it is more convenient to index the orthonormal set starting from $k=0$. On occasion [see (4.6.15)] the index is even allowed to run from $-\infty$ to ∞, the infinite series $\sum_{k=-\infty}^{\infty} \gamma_k e_k(x)$ being interpreted as $\lim_{n\to\infty} \sum_{k=-n}^{n} \gamma_k e_k(x)$.

2. If $\{e_k(x)\}$ is an orthonormal basis, so is $\{\bar{e}_k(x)\}$.

3. If $\{e_k(x)\}$ is an orthonormal basis and $\{\alpha_k\}$ is a sequence of nonzero complex numbers, $\{\alpha_k e_k(x)\}$ is an orthogonal basis. Note that when we expand an element u in this new basis, the coordinates of u may not tend to 0 as the index tends to infinity.

4. If $\{e_k(x)\}$ is an orthonormal basis for $L_2(-1,1)$, then

$$\left\{ \left(\frac{2}{b-a}\right)^{1/2} e_k\left[\frac{2}{b-a}\left(x - \frac{a+b}{2}\right)\right] \right\}$$

is an orthonormal basis for $L_2(a,b)$.

5. In $L_2^{(r)}(a,b)$ all functions are real-valued, and a basis consists of real functions. Formulas (4.6.10) through (4.6.12) are still valid, but the complex conjugates may be omitted. A basis for $L_2^{(r)}(a,b)$ is also a basis for $L_2^{(c)}(a,b)$, and any real basis for $L_2^{(c)}(a,b)$ is automatically a basis for $L_2^{(r)}(a,b)$.

6. In applications to differential equations (see Chapter 7), one frequently encounters sets of functions that are *orthogonal with weight* $s(x)$, that is, such that
$$\int_a^b s(x)e_k(x)\bar{e}_j(x)\,dx = \begin{cases} 0, & k \neq j, \\ 1, & k = j, \end{cases} \tag{4.6.14}$$
where $s(x) > 0$ in $a < x < b$. Although the set $\{s^{1/2}e_k\}$ is orthonormal in the sense of (4.6.10), it is often preferable to introduce a new inner product $\langle u, v\rangle_s = \int_a^b s(x)u(x)\bar{v}(x)\,dx$ and its corresponding norm, $\|u\|_s = [\int_a^b s(x)|u(x)|^2\,dx]^{1/2}$. With these definitions the space of functions for which $\|u\|_s$ exists and is finite is a Hilbert space, and our general analysis can be applied. If $\{e_k(x)\}$ is an orthonormal basis in the sense of (4.6.14), then every u with $\|u\|_s$ finite can be expanded in the form
$$u(x) = \sum_{k=1}^{\infty} \gamma_k e_k(x), \quad \gamma_k = \langle u, e_k\rangle_s = \int_a^b s(x)u(x)\bar{e}_k(x)\,dx,$$
where the series converges in the sense of
$$\lim_{n\to\infty} \int_a^b s(x)|u(x) - \sum_{k=1}^{n} \gamma_k e_k(x)|^2\,dx = 0.$$

Example 1. We have already encountered the Legendre polynomials (Example 3, Section 4.5) as the outcome of the Gram-Schmidt procedure of orthogonalization on the monomials $1, x, x^2, \ldots$ on $-1 \leqslant x \leqslant 1$. Since the monomials form a spanning set, the Legendre polynomials form an orthogonal basis.

Example 2. The following sets of trigonometric functions are orthonormal bases on the respective intervals shown:

(a) On any interval of length 2π:
$$\left(\frac{1}{2\pi}\right)^{1/2}, \left(\frac{1}{2\pi}\right)^{1/2} e^{ix}, \left(\frac{1}{2\pi}\right)^{1/2} e^{-ix}, \left(\frac{1}{2\pi}\right)^{1/2} e^{2ix}, \left(\frac{1}{2\pi}\right)^{1/2} e^{-2ix}, \ldots. \tag{4.6.15}$$

(b) On any interval of length 2π:
$$\left(\frac{1}{2\pi}\right)^{1/2}, \left(\frac{1}{\pi}\right)^{1/2} \cos x, \left(\frac{1}{\pi}\right)^{1/2} \sin x, \left(\frac{1}{\pi}\right)^{1/2} \cos 2x, \left(\frac{1}{\pi}\right)^{1/2} \sin 2x, \ldots. \tag{4.6.16}$$

(c) On $0 \leqslant x \leqslant \pi$:
$$\left(\frac{1}{\pi}\right)^{1/2}, \left(\frac{2}{\pi}\right)^{1/2} \cos x, \left(\frac{2}{\pi}\right)^{1/2} \cos 2x, \ldots. \tag{4.6.17}$$

(d) On $0 \leqslant x \leqslant \pi$:

$$\left(\frac{2}{\pi}\right)^{1/2} \sin x, \ \left(\frac{2}{\pi}\right)^{1/2} \sin 2x, \ldots . \tag{4.6.18}$$

(a_1) On any interval of length $2L$:

$$\left\{\left(\frac{1}{2L}\right)^{1/2} e^{ik\pi x/L}\right\}, \ k = 0, \pm 1, \pm 2, \ldots . \tag{4.6.19}$$

(b_1) On any interval of length $2L$:

$$\left(\frac{1}{2L}\right)^{1/2}, \ \left\{\left(\frac{1}{L}\right)^{1/2} \cos \frac{k\pi x}{L}\right\}, \ \left\{\left(\frac{1}{L}\right)^{1/2} \sin \frac{k\pi x}{L}\right\}, \ k = 1, 2, \ldots . \tag{4.6.20}$$

(c_1) On $0 \leqslant x \leqslant L$:

$$\left(\frac{1}{L}\right)^{1/2}, \ \left\{\left(\frac{2}{L}\right)^{1/2} \cos \frac{k\pi x}{L}\right\}, \ k = 1, 2, \ldots . \tag{4.6.21}$$

(d_1) On $0 \leqslant x \leqslant L$:

$$\left\{\left(\frac{2}{L}\right)^{1/2} \sin \frac{k\pi x}{L}\right\}, \ k = 1, 2, \ldots . \tag{4.6.22}$$

Expansions using the functions in (a_1) or (b_1) are known as *full-range* expansions, whereas those using (c_1) or (d_1) are *half-range* expansions. It is a simple exercise in integration to prove that each set is orthonormal on the stated interval. We shall prove that (4.6.15) is a basis by showing that the kernel in (4.6.13) is a delta sequence. It then follows from Exercise 4.6.5 that all other listed sets are also bases.

Since set (4.6.15) is indexed from $n = -\infty$ to $n = \infty$, we shall want to show that

$$\lim_{n \to \infty} \left\| u(x) - \sum_{k=-n}^{n} \gamma_k \frac{e^{ikx}}{(2\pi)^{1/2}} \right\| = 0,$$

where

$$\gamma_k = \frac{1}{(2\pi)^{1/2}} \int_{-\pi}^{\pi} u(y) e^{-iky}\, dy.$$

The sequence of partial sums $\{t_n(y)\}$ is defined from

$$t_n(x) = \sum_{k=-n}^{n} \gamma_k \frac{e^{ikx}}{(2\pi)^{1/2}} = \frac{1}{2\pi} \int_{-\pi}^{\pi} s_n(x - y) u(y)\, dy,$$

where

$$s_n(x) = \sum_{k=-n}^{n} e^{ikx}.$$

We showed in Theorem 2.3.3 that $\{s_n(x)\}$ is a delta sequence in $-\pi \leqslant x < \pi$. In fact, if the 2π-periodic extension of u is continuously differentiable, $t_n(x) \to u(x)$ uniformly on $(-\infty, \infty)$ and hence certainly on $-\pi \leqslant x < \pi$. Therefore, not only is set (4.6.15) a basis but we also have results on uniform convergence of trigonometric Fourier series.

Example 3. By combining orthonormal bases in one dimension, we can construct orthonormal bases in higher dimensions. Suppose that $\{e_k(x)\}$ is an orthonormal basis for $L_2^{(c)}(a,b)$ and $\{f_k(y)\}$ is an orthonormal basis for $L_2^{(c)}(c,d)$; then the set $\{e_1(x)f_1(y), e_1(x)f_2(y), e_2(x)f_1(y), e_2(x)f_2(y), \ldots\}$, consisting of *all products* of pairs from the original bases, is an orthonormal basis for $L_2^{(c)}(D)$, where D is the rectangle $a \leqslant x \leqslant b, c \leqslant y \leqslant d$. The proof is omitted. Note that the diagonal set $\{e_1(x)f_1(y), e_2(x)f_2(y), \ldots\}$ is orthonormal but is not a basis for $L_2^{(c)}(D)$.

As an example of this construction, we deduce from (4.6.22) that

$$\left\{ \frac{2}{ab} \sin \frac{j\pi x}{a} \sin \frac{k\pi y}{b} \right\}$$

is an orthonormal basis for the rectangle $0 \leqslant x \leqslant a, 0 \leqslant y \leqslant b$. Repeated application of the procedure enables us to construct bases in higher dimensions.

Example 4. Other examples of orthonormal bases such as Hermite polynomials and Bessel functions are given in Chapter 7.

EXERCISES

4.6.1 Let D be a dense subset of H, and suppose that every element of D is in the closed span of an orthonormal set. Show that the orthonormal set is a basis for H. As an illustration of this theorem, the following procedure is often used to prove that an orthonormal set $\{e_k(x)\}$ is a basis for $H \doteq L_2(a,b)$. The set D of continuous functions is known to be dense in H. Suppose that elements of D can be uniformly approximated by linear combinations of elements of $\{e_k\}$. This means that for each g in D and each $\varepsilon > 0$, we can find a linear combination h of a finite number of elements of $\{e_k\}$ such that $\|g-h\|_\infty < \varepsilon$, using the uniform norm of Example 5, Section 4.3. It then follows that in the L_2 norm, $\|g - h\| < \varepsilon(b-a)^{1/2}$ so that every continuous function is in the closed span of $\{e_k\}$. Hence $\{e_k\}$ is a basis for H. Thus the only thing that has to be proved for the set $\{e_k\}$ is that continuous functions can be uniformly approximated by finite linear combinations of $\{e_k\}$.

4.6.2 Show that a necessary and sufficient condition for the orthonormal set $\{e_i\}$ to be a basis for H is that

$$\langle u, v\rangle = \sum_{i=1}^{\infty} \langle u, e_i\rangle \langle e_i, v\rangle \quad \text{for all } u, v \text{ in } H.$$

4.6.3 (a) Show that $\cos^k x$ is a linear combination of $1, \cos x, \ldots, \cos kx$.

(b) Let $f(x)$ be a complex-valued continuous function on $0 \leqslant x \leqslant \pi$. Make the admissible change of variables $y = \cos x$, and use the Weierstrass approximation theorem on $-1 \leqslant y \leqslant 1$ to show that $f(x)$ can be uniformly approximated by linear combinations of $1, \cos x, \ldots, \cos kx$. It then follows from Exercise 4.6.1 that set (4.6.17) is an orthonormal spanning set for $L_2(0, \pi)$ and therefore a basis.

4.6.4 (a) Show that $\sin^k x$ is a linear combination of $1, \sin x, \cos x, \ldots, \sin kx$, $\cos kx$.

(b) Let $f(x, y)$ be a complex-valued continuous function defined on the disk $x^2 + y^2 \leqslant 1$. An extension of the Weierstrass theorem shows that $f(x, y)$ can be uniformly approximated by linear combinations of $x^m y^k$. Introducing polar coordinates, we see that $f(\cos\theta, \sin\theta)$, which is the value of $f(x, y)$ on the circumference of the disk, can be uniformly approximated by linear combinations of $\cos^m\theta \sin^k\theta$ on $0 \leqslant \theta \leqslant 2\pi$. Use Exercise 4.6.3(a) and 4.6.4(a) to show that (4.6.16) is an orthonormal basis over the stated interval.

4.6.5 Show that sets (4.6.15) and (4.6.16) lead to the same Fourier series by giving explicit relations between the coefficients of the two series

$$u(x) = \sum_{k=-\infty}^{\infty} \gamma_k \frac{e^{ikx}}{(2\pi)^{1/2}}$$

and

$$u(x) = a_0 \left(\frac{1}{2\pi}\right)^{1/2} + \sum_{k=1}^{\infty} \left[a_k \frac{\cos kx}{(\pi)^{1/2}} + b_k \frac{\sin kx}{(\pi)^{1/2}}\right].$$

These relations are

$$a_0 = \gamma_0, \quad a_k = \frac{1}{\sqrt{2}}(\gamma_k + \gamma_{-k}), \quad b_k = \frac{i}{\sqrt{2}}(\gamma_k - \gamma_{-k})$$

and, therefore,

$$\gamma_0 = a_0, \quad \gamma_k = \frac{1}{\sqrt{2}}(a_k - ib_k) \quad \text{for } k > 0,$$

$$\gamma_k = \frac{1}{\sqrt{2}}(a_{-k} + ib_{-k}) \quad \text{for } k < 0.$$

Show that these results are in agreement with (2.3.4). Since set (4.6.15) has been shown to be a basis, it follows that the orthonormal set (4.6.16) is also a basis. Since the sets have period 2π, they are bases on any 2π interval. If $u(x)$ is an odd function, that is, $u(-x) = -u(x)$, then all the $\{a_k\}$ vanish. Since u is arbitrary in $(0, \pi)$, we have shown that set (4.6.18) is a basis on $(0, \pi)$. Now let $u(x)$ be even, so that $u(x) = u(-x)$. This leads to $b_k = 0$ for all k. Hence set (4.6.17) is a basis on $(0, \pi)$. A simple change of variables lets us extend the results to the remaining sets, (4.6.19) through (4.6.22).

4.6.6 We wish to show that $L_2^{(c)}(0, \infty)$ is separable. Recall that a function $u(x)$ belongs to $L_2^{(c)}(0, \infty)$ if $\int_0^\infty |u(x)|^2 dx < \infty$. [Note that this does not imply that $\lim_{x \to \infty} u(x) = 0$.] Show that this space is separable by proving that the countable set $H(k - x)x^m$, where $k = 1, 2, \ldots, m = 0, 1, 2, \ldots$, and H is the Heaviside function, is dense in $L_2^{(c)}(0, \infty)$.

4.6.7 Let $\{e_1, \ldots, e_n, \ldots\}$ be an orthonormal basis for H. The following scheme constructs a related spanning set for H which is *not* a basis. The spanning set in question is $\{h_1, \ldots, h_n, \ldots\}$, where $h_n = e_1 + e_{n+1}/(n + 1)$.

(a) Show that the set $\{h_i\}$ is algebraically independent.

(b) Show that the sets $\{h_i\}$ and $\{e_i\}$ have the same closed span (observe that e_1 is in the span of $\{h_i\}$ since $\lim_{n \to \infty} h_n = e_1$).

(c) Show that e_1 cannot be expanded in a series in the set $\{h_i\}$. Assume that $e_1 = \alpha_1 h_1 + \cdots$, and take the inner product with respect to each of the e_k's to arrive at a contradiction.

4.7 LINEAR FUNCTIONALS AND THE RIESZ REPRESENTATION THEOREM

Let M be a linear manifold in the separable complex Hilbert space H. If to each u in M there corresponds a complex number, denoted by $l(u)$, satisfying the condition

$$l(\alpha u + \beta v) = \alpha l(u) + \beta l(v) \quad \text{for all } u, v \text{ in } M \text{ and all } \alpha, \beta \in \mathbb{C}, \qquad (4.7.1)$$

we say that l is a *linear functional* on M. The domain M will often be the entire Hilbert space. Note that a linear functional satisfies $l(0) = 0$ and $l(\sum_{i=1}^k \alpha_i u_i) = \sum_{i=1}^k \alpha_i l(u_i)$.

REMARK. On a real Hilbert space a linear functional is real-valued and satisfies (4.7.1) for $\alpha, \beta \in \mathbb{R}$.

Example 1. The functional whose value at u is $\|u\|$ is not linear.

Example 2. Let f be a fixed vector in H, and associate with each u in H the complex number $l(u) = \langle u, f \rangle$. Then l is a linear functional—in fact, the prototype of linear functionals, as we shall see below.

Example 3. The functionals defined on H by $l(u) = \langle u, f \rangle + \alpha$ and $l(u) = \langle f, u \rangle$ are not linear.

A linear functional is *bounded* on its domain M if there exists a constant c such that *for all* u in M, $|l(u)| \leqslant c\|u\|$. The smallest constant c for which the inequality holds for all u in M is known as the *norm* of l and is denoted by $\|l\|$.

To say that l is continuous at a point u in its domain means roughly that $l(v)$ is near $l(u)$ whenever v is near u. Precisely: For each $\varepsilon > 0$ there exists $\delta > 0$ such that $|l(v) - l(u)| < \varepsilon$ whenever v is in M and $\|u - v\| < \delta$. An equivalent definition is the following: The functional l is *continuous* at u if, whenever $\{u_n\}$ is a sequence in M with limit u, then $l(u_n) \to l(u)$. A functional continuous at every point of M is said to be continuous on M.

Theorem. *A linear functional continuous at the origin is continuous on its entire domain of definition M.*

Proof. Let $\{u_n\}$ be a sequence in M with limit u in M. Then $u_n - u \to 0$ and hence, by hypothesis, $l(u_n - u) \to l(0) = 0$. By the linearity of l, $l(u_n - u) = l(u_n) - l(u)$, and therefore $l(u_n) \to l(u)$, as desired. □

Theorem. *Boundedness and continuity are equivalent for linear functionals.*

Proof.

(a) Let l be bounded; then $|l(u_n) - l(u)| = |l(u_n - u)| \leqslant \|l\| \, \|u_n - u\|$. As $u_n \to u$, we therefore have $l(u_n) \to l(u)$ and l is continuous.

(b) Let l be continuous. If l is unbounded, there exists a sequence $\{u_n\}$ of nonzero elements in M such that $|l(u_n)| \geqslant n\|u_n\|$; hence the sequence $v_n = u_n/n\|u_n\|$, which tends to 0, has the property $|l(v_n)| \geqslant 1$, which violates continuity at the origin.

□

We now give an example of an unbounded linear functional on a dense subset of an infinite-dimensional Hilbert space. Let $H = L_2(-1, 1)$, and let M be the subset of continuous functions. The functional $l(u) = u(0)$ which assigns to each continuous function its value at the origin is clearly linear. Now let $f(x)$ be a real continuous function such that $f(-1) = f(1) = 0$, and $f > 0$ in $-1 < x < 1$. Define the sequence $u_k(x) = f(kx)$ when $|x| \leqslant k^{-1}$ and $u_k(x) = 0$ elsewhere. Since $\int_{-1}^{1} u_k^2(x)\,dx = (1/k)\int_{-1}^{1} f^2(x)\,dx$, we see that $u_k(x) \to 0$ in the L_2 norm. We have, however, $u_k(0) = f(0)$, so that the numerical sequence $u_k(0)$ does not tend to 0 as $k \to \infty$. Thus $l(u)$ is not continuous at 0 and is therefore unbounded. On the other hand, every linear functional defined on an n-dimensional manifold in H is bounded. This is an immediate consequence of the following result.

Theorem. *Every linear functional on E_n is bounded.*

Proof. It suffices to show that $l(u_k) \to 0$ whenever $u_k \to 0$. In terms of the orthonormal basis $\{e_1, \ldots, e_n\}$, $u_k = \sum_{i=1}^{n} \alpha_{k,i} e_i$, where $\alpha_{k,i} = \langle u_k, e_i \rangle$. The continuity of the inner product shows that if $u_k \to 0$, then $\lim_{k\to\infty} \alpha_{k,i} = 0$ for $i = 1, \ldots, n$. Hence $\lim_{k\to\infty} l(u_k) = \lim_{k\to\infty} \sum_{i=1}^{n} \alpha_{k,i} l(e_i) = 0$. □

One of the fundamental theorems of Hilbert space (akin in importance to the projection theorem, Section 4.5, to which it is closely related) is the Riesz representation theorem, which tells us that every bounded linear functional on H is the inner product with respect to some fixed vector in H.

Theorem (Riesz Representation Theorem). *To each continuous linear functional l defined on the whole of H corresponds an unambiguously defined vector f such that*

$$l(u) = \langle u, f\rangle \quad \textit{for every } u \textit{ in } H. \tag{4.7.2}$$

Proof. The set of vectors N for which $l(u) = 0$ is easily seen to be a linear manifold. If N coincides with H, there is no problem since $f = 0$ clearly satisfies (4.7.2). Suppose then that N is a proper subset of H. According to the corollary following the projection theorem, there exists a nonzero vector in $N^\perp$; by multiplication by an appropriate scalar one can choose a vector f_0 in $N^\perp$ such that $\|f_0\| = 1$ and $l(f_0)$ is *real*. Indeed, for any u the vector $l(u)f_0 - l(f_0)u$ is clearly in N and therefore orthogonal to f_0. Hence $l(u)\langle f_0, f_0\rangle - l(f_0)\langle u, f_0\rangle = 0$, so that, since $l(f_0)$ is real, $l(u) = \langle u, l(f_0)f_0\rangle$. The required f in (4.7.2) is then $l(f_0)f_0$. To prove uniqueness, suppose that f and g satisfy (4.7.2) for all u. Then $\langle u, f - g\rangle = 0$ for all u, and by choosing $u = f - g$, we find that $\|f - g\| = 0$ or $f = g$. □

REMARK. Two other proofs of this theorem are outlined in Exercises 4.7.2 and 4.7.3. Observe that every bounded linear functional on the space $L_2^{(c)}(a, b)$ is of the form $\int_a^b u(x)\bar{f}(x)\,dx$ for some f.

Representation of Bounded Linear Functionals: Dual Bases

A linear functional l on E_n can be characterized by its values on a basis. If such a basis is $\{h_1, \dots, h_n\}$, the n complex numbers $l(h_1), \dots, l(h_n)$ enable us to calculate $l(u)$ for each l in E_n by the formula

$$l(u) = l\left(\sum \alpha_i h_i\right) = \sum l(\alpha_i h_i) = \sum \alpha_i l(h_i). \tag{4.7.3}$$

Conversely, if we are given n complex numbers $l_1, \dots, l_n$, there exists one and only one linear functional l on E_n such that $l(h_1) = l_1, \dots, l(h_n) = l_n$. To see this, all that is necessary is to read (4.7.3) from right to left: Express u in the basis $\{h_i\}$ as $\sum \alpha_i h_i$, and then set $l(u) = \sum \alpha_i l_i$ to define the linear functional l.

The situation is more delicate in an infinite-dimensional space since now continuity of the linear functional is required for the second equality in (4.7.3). Thus a *continuous* linear functional is characterized completely by its values on a basis. Given a sequence of complex numbers $\{l_i\}$, when is there a bounded linear functional such that $l(h_i) = l_i$? A sufficient condition is $\sum |l_i|^2 < \infty$, for we can then define $l(\sum \alpha_i h_i)$ as $\sum \alpha_i l_i$, this series being convergent by the Schwarz inequality for series; moreover, l is uniquely determined.

We have already frequently noted that the coordinates of a vector u expanded in an orthonormal basis $\{e_i\}$ are easily calculated by $\alpha_i = \langle u, e_i\rangle$. If the basis is an oblique one, say $\{h_i\}$, the calculation of the coordinates of u is more difficult unless we introduce the so-called *dual* or *reciprocal* basis $\{h_i^*\}$ with the properties

$$\langle h_i, h_j^* \rangle = \begin{cases} 0, & i \neq j, \\ 1, & i = j. \end{cases} \tag{4.7.4}$$

Let us now prove the existence of such a basis. We can treat the h_j^* one at a time. For j fixed, the preceding discussion shows that there exists a unique bounded linear functional l_j with the property $l_j(h_i) = 0, i \neq j$, and $l_j(h_j) = 1$. By the Riesz representation theorem there must exist a unique vector h_j^* such that $l_j(u) = \langle u, h_j^* \rangle$. Clearly, h_j^* has the properties $\langle h_i, h_j^* \rangle = 0, i \neq j, \langle h_j, h_j^* \rangle = 1$. Obviously, we can perform this construction for every index j, so that (4.7.4) is indeed satisfied. Now let $\{h_i\}$ be a basis in H, and let $\{\alpha_i\}$ be the coordinates of a vector u in this basis, that is, $u = \sum \alpha_i h_i$. Taking the inner product with respect to h_j^* and using the continuity of the inner product, we find this simple expression for the coordinates:

$$\alpha_j = \langle u, h_j^* \rangle. \tag{4.7.5}$$

EXERCISES

4.7.1 Let l be a bounded linear functional on H; does there exist an element f^* such that $l(u) = \langle f^*, u \rangle$? [*Hint*: Look at $l(\alpha u)$.]

4.7.2 Prove the Riesz representation theorem by the following alternative method. The underlying idea is that if $l(u)$ were really of the form $\langle u, f \rangle$, then it would hold that the max $|l(u)|/\|u\|$ would occur for $u = \alpha f$.

(a) Since l is bounded, let $\|l\|$ = supremum over nonzero u of $|l(u)|/\|u\|$. Everything is trivial if $\|l\| = 0$, so assume that $\|l\| > 0$. Show, by using the parallelogram law, that there exists an element u_0 with $\|u_0\| = 1$ for which $l(u_0) = \|l\|$.

(b) Now show that the vector f in Riesz's theorem is $\|l\|u_0$. This means proving that $l(u) - \langle u, \|l\|u_0 \rangle = 0\ \forall u$ or $\langle z, \|l\|u_0 \rangle = 0$, where z is defined as $z = l(u)u_0/\|l\| - u$. Since $l(z) = 0$, it suffices to show that $l(z) = 0$ implies that $\langle z, \|l\|u_0 \rangle = 0$. Prove this in a manner similar to the way in which orthogonality is proved in the projection theorem.

The uniqueness proof is the same as that in the text.

4.7.3 Another proof of the Riesz representation theorem consists of exhibiting the element f explicitly in terms of an orthonormal basis $\{e_i\}$. For each u in H we have $u = \sum \langle u, e_i \rangle e_i$, and $l(u) = \sum \langle u, e_i \rangle l(e_i)$. On the other hand, for any pair of vectors u, f, we have

$$\langle u, f \rangle = \left\langle \sum \langle u, e_i \rangle e_i, \sum \langle f, e_i \rangle e_i \right\rangle = \sum \langle u, e_i \rangle \overline{\langle f, e_i \rangle}.$$

Thus the desired f has the property that $\bar{l}(e_i) = \langle f, e_i \rangle$, and therefore $f = \sum_i \bar{l}(e_i)e_i$. It remains only to show that this definition makes sense [that is, that $\|f\|$ is finite or, equivalently, that $\sum |l(e_i)|^2$ converges; assume the contrary and arrive at a contradiction by considering elements of the form $\sum_{k=1}^{N} \bar{l}(e_k)e_k$].

4.8 THE HAHN-BANACH THEOREM AND REFLEXIVE BANACH SPACES

Having an inner product is in fact not necessary to consider linear functionals on sets; the only structure really needed is the algebraic structure of a vector space (although one loses the Riesz Representation Theorem). The most useful case is when the vector space is equipped with a norm (a normed space), and in some cases it will be useful if that normed space is complete (so it is a Banach space).

Let X be a Banach space, which we recall means three things:

(1) The space X is a linear (vector) space built from a set of vectors X and an associated field of scalars $\mathbb{K}$, which together obey the vector space axioms with respect to scalar-vector multiplication "$\cdot$" and vector-vector addition "$+$" (meaning that it is *algebraically closed*; see Section 4.2). It is fairly common to "overload" the symbol X to mean both the set of vectors, as well as the vector space itself; we will sometimes indicate the structure on X as

$$X = \langle X, \mathbb{K}, \cdot, + \rangle.$$

(2) The space X is equipped with a norm satisfying the three norm axioms (see Section 4.3); we think about this as an additional structure imposed on the existing vector space structure of X, and denote this as

$$X = \langle \langle X, \mathbb{K}, \cdot, + \rangle, \| \cdot \|_X \rangle.$$

(3) It is complete with respect to this norm (meaning that it is *topologically closed*; see Section 4.3).

Our interest in this section is primarily a *real* Banach space X, so we will assume that the scalar field $\mathbb{K}$ associated with the vector space structure of X is in fact $\mathbb{R}$ rather than $\mathbb{C}$. The *dual space* X' of bounded linear functionals on the Banach space X naturally forms a linear (vector) space itself, since it is obviously (*algebraically*) closed under the natural vector space operation of linear combinations of functionals: For $f, g \in X'$, it holds that $h \in X'$, where

$$h(u) = \alpha f(u) + \beta g(u), \qquad \alpha, \beta \in \mathbb{R}, \quad \forall u \in X.$$

The dual space X' becomes a normed space when equipped with the operator norm induced by X, namely

$$\|f\|_{X'} = \sup_{0 \neq u \in X} \frac{|f(u)|}{\|u\|_X}. \tag{4.8.1}$$

We note that this definition implies the following inequality for any member of the dual space:

$$|f(u)| \leqslant L\|u\|_X, \qquad \forall u \in X,$$

where L is any upper bound for $\|f\|_{X'}$. That X' is in fact also *topologically closed*, meaning that it is complete as a metric space with respect to the metric induced by the dual norm, is something we will examine shortly.

While the Riesz Representation Theorem and the Projection Theorem are two of the most important fundamental results on Hilbert spaces, the three basic principles of linear analysis in Banach spaces are

I: The Hahn-Banach Theorem

II: The Banach-Schauder (Open Mapping) Theorem

III: The Banach-Steinhaus Theorem (Principle of Uniform Boundedness)

We will come to the latter two principles in Chapter 5 when we look at operator theory in Banach spaces; the first principle (the Hahn-Banach Theorem) concerns only linear functionals on vector spaces with no other structure, so we can discuss it completely here. In order to state the theorem, we need a couple of simple concepts. A *sublinear* functional $p(u)\colon X \to \mathbb{R}$ on a real vector space X satisfies

$$\begin{aligned} p(u+v) &\leqslant p(u) + p(v), \quad \forall u, v \in X. \\ p(\alpha u) &= \alpha p(u), \quad \forall u \in X, \quad \alpha > 0. \end{aligned}$$

Note that an example of a sublinear functional on a real normed space X is the norm that X comes equipped with: $p(u) = \|u\|_X$. Recall that a subspace $U \subset X$ of a vector space X is by definition algebraically closed with respect to the vector space structure inherited from X, meaning that

$$U = \langle U, \mathbb{R}, \cdot, + \rangle$$

is a self-contained vector space. Note that U is not necessarily *topologically* closed with respect to the norm on X; if U is also topologically closed, we refer to U as a *closed subspace* of X.

Theorem 4.8.1 (Hahn-Banach Theorem). *Let X be a real vector space and let $p(u)$ be a sublinear functional on X. Let $U \subset X$ be a subspace of X, and let $f\colon U \to \mathbb{R}$ be a linear functional on U such that*

$$f(u) \leqslant p(u), \quad \forall u \in U.$$

Then there exists a linear functional $F\colon X \to \mathbb{R}$, extending f to all of X, such that

$$\begin{aligned} F(u) &= f(u), \quad \forall u \in U, \\ F(u) &\leqslant p(u), \quad \forall u \in X. \end{aligned}$$

Proof. See, for example, the books of Dunford and Schwartz [5] or Yosida [15]. □

The importance of this fairly simple theorem in functional analysis is quite remarkable; three immediate consequences of the Hahn-Banach Theorem are the following.

Theorem 4.8.2. *Let X be a normed vector space and let $0 \neq u \in X$ be arbitrary. Then there exists a bounded linear functional $f \in X'$ such that*

$$\|f\|_{X'} = 1, \qquad f(u) = \|u\|_X.$$

Proof. A proof may be found in the book of Schechter [14]. □

Theorem 4.8.3. *If $u \in X$ such that $f(u) = 0 \ \ \forall f \in X'$, then $u = 0$.*

Proof. See Schechter [14]. □

The result above allows one to connect weak and strong formulations of operator equations, and will be used repeatedly in Chapter 10.

Theorem 4.8.4. *Let X be a normed space. Then X' is a Banach space.*

Proof. See, for example, Dunford and Schwartz [5] or Schechter [14]. □

This last result states that the space X' of bounded linear functionals on a normed space X is always both algebraically and topologically closed, even if the base normed space is not complete; this is due to the fact that the *range space* (in this case $\mathbb{R}$) is complete, which is all that is needed.

Since the dual space X' of any normed space X is always itself a Banach space, we can consider *its* dual space, namely $X'' = (X')'$, the *second* or *double dual* of X, which is the space of bounded linear functionals on X'. Again, by the result above, the double dual X'' is always a Banach space. There is a natural linear map from X to X'' given by

$$J \colon X \to X'', \qquad J_u(f) = f(u), \quad \forall u \in X, \quad \forall f \in X'.$$

In other words, for a fixed $u \in X$, $J_u(f)$ is a linear functional on X', defined by evaluating f at the point u. Both the linearity and the range of the map $J_u(\cdot)$ being $\mathbb{R}$ are inherited from f. Moreover, the Hahn–Banach Theorem ensures that J_u is bounded, and in fact preserves the norm on X (is isometric), so that for each given $u \in X$, we have $J_u \in X''$. The map $J \colon X \to X''$ being an isometric map implies that it is injective (one-to-one). If it is also surjective (onto all of X''), then we say that the space X is a *reflexive Banach space.*

Here are some standard useful facts about normed spaces, Banach spaces, and subspaces that can be established after some work following the discussion above:

(1) All reflexive normed spaces X are Banach spaces (X is isometric to the complete space X'').
(2) If X is reflexive, so is X'.
(3) If a Banach space X is isomorphic to a reflexive Banach space Y, then X is also reflexive.

(4) All Hilbert spaces are reflexive (by the Riesz Representation Theorem).
(5) A reflexive Banach space X is separable if and only if its dual X' is separable.
(6) Every closed subspace U of a reflexive space X is reflexive.
(7) Every finite-dimensional subspace U of a normed space X is closed.
(8) All finite-dimensional normed spaces X are reflexive.
(9) All finite-dimensional normed spaces X are Banach spaces.

In the list, by *separable* we mean precisely the same property introduced in Section 4.6 for Hilbert spaces (namely, that the space X has a countable basis).

Nearly all of the function spaces we have run into so far are examples of reflexive Banach spaces; for example, the Banach spaces $L^p(\Omega)$ (Lebesgue spaces) and the Banach spaces $W^{m,p}(\Omega)$ (Sobolev spaces) introduced in Section 2.6 can be shown to be reflexive for $1 < p < \infty$ as well as separable for $1 \leqslant p < \infty$ (see Chapter 10). One of the most useful properties of a reflexive Banach space is given in Theorem 4.8.8. This result makes it possible to apply some of the most powerful tools from nonlinear functional analysis to solve nonlinear partial differential equations posed in Sobolev spaces. To understand the statement of Theorem 4.8.8, we first recall the definitions of strong, weak, and weak-$*$ convergence along with some simple results relating them.

(1) A sequence $\{u_j\}_{j=1}^{\infty}$ in a Banach space X is *strongly convergent* (or simply *convergent*) to $u \in X$ if $\lim_{j\to\infty} \|u - u_j\| = 0$. This is denoted $u_j \to u$.
(2) A sequence $\{u_j\}_{j=1}^{\infty}$ in a Banach space X is *weakly convergent* to $u \in X$ if $\lim_{j\to\infty} f(u) = f(u_j), \forall f \in X'$. This is denoted $u_j \rightharpoonup u$.
(3) A sequence $\{f_j\}_{j=1}^{\infty}$ in the dual space X' of a Banach space X is *weak-$*$ convergent* to $f \in X'$ if $\lim_{j\to\infty} f_j(u) = f(u), \forall u \in X$. This is denoted $f_j \overset{*}{\rightharpoonup} f$.

There are several distinct notions of the *rate of strong convergence* of a sequence $u_j \to u \in X$.

(1) *Q-linear convergence:* There exists $c \in [0, 1)$ and $N \geqslant 0$ such that for $j \geqslant N$, $\|u - u_{j+1}\| \leqslant c\|u - u_j\|$.
(2) *Q-superlinear convergence:* There exists an auxillary sequence $\{c_j\}$ such that $c_j \to 0$ and $\|u - u_{j+1}\| \leqslant c_j\|u - u_j\|$.
(3) *Q-order(p) convergence:* There exists $p > 1$, $c \geqslant 0$, and $N \geqslant 0$ such that for $j \geqslant N$, $\|u - u_{j+1}\| \leqslant c\|u - u_j\|^p$.
(4) *R-order(p) convergence:* There exists an auxillary sequence $\{v_j\}$ converging with rate Q-order(p), and $N \geqslant 0$, such that for $j \geqslant N$, $\|u - u_j\| \leqslant \|v - v_j\|$.

Two easy theorems relating the weak and strong convergence are:

Theorem 4.8.5. *A strongly convergent sequence in a Banach space is also weakly convergent.*

Proof. Let $u_j \to u \in X$, so that $\lim_{j\to\infty} \|u - u_j\|_X = 0$. By the definition of the dual norm (4.8.1), for any $f \in X'$ we have

$$|f(u - u_j)| \leqslant \|u - u_j\|_X \|f\|_{X'}.$$

We have then

$$\lim_{j\to\infty} |f(u-u_j)| \leqslant \lim_{j\to\infty} \|u-u_j\|_X \|f\|_{X'} = 0, \qquad \forall f \in X'.$$

Therefore, $u_j \rightharpoonup u \in X$. □

Theorem 4.8.6. *In a finite-dimensional Banach space, strong, weak, and weak-∗ convergence are equivalent.*

Proof. We have just shown that strong convergence implies weak convergence in any Banach space X; we now show that when X is finite-dimensional, then weak convergence implies strong convergence. Let $\dim(X) = n < \infty$, and let $u_j \rightharpoonup u \in X$. Let $\{e^1, \ldots, e^n\}$ be a basis for X, so that $u = \sum_{i=1}^n c_i e^i$ and $u_j = \sum_{i=1}^n c_i^j e^i$. Now pick a very specific set of linear functionals from X':

$$f_i(e^j) = \delta_{ij}.$$

Since $u_j \rightharpoonup u \in X$, and since $f_i \in X'$, $1 \leqslant i \leqslant n$, we have $\lim_{j\to\infty} f_i(u_j) = f_i(u)$. Since $f_i(u_j) = c_i^j$ and $f_i(u) = c_i$, we have $c_i^j \to c_i$, for $1 \leqslant i \leqslant n$. Now let $\epsilon > 0$ be given, set $M = \max_i \|e^i\|_X$, and take $\overline{\epsilon} = \epsilon/(Mn)$. Since $c_i^j \to c_i$, we know there exists $N > 0$ such that $|c_i - c_i^j| < \overline{\epsilon}$ whenever $j > N$. Consider now

$$\|u-u_j\|_X = \|\sum_{i=1}^n (c_i - c_i^j) e^i\|_X \leqslant \sum_{i=1}^n |c_i - c_i^j| \, \|e^i\|_X \leqslant M \sum_{i=1}^n \overline{\epsilon} < \epsilon,$$

giving that $u_j \to u \in X$. Equivalence of weak-∗ convergence to weak convergence can be shown similarly. □

Another important result is

Theorem 4.8.7. *Weakly convergent sequences in a Banach space are bounded.*

Proof. The proof follows from the Principle of Uniform Boundedness (see Chapter 5 or Dunford and Schwartz [5]). □

Before we can state the result we are after for use in Chapter 10, recall that a subset U of a normed space X is called *sequentially compact* if every sequence $\{u_j\}$ from U contains a subsequence that converges (strongly in X) to a point $u \in U$. Recall that compactness of U (every open cover of U contains a finite subcover) and *sequential* compactness of U are equivalent in metric spaces (normed spaces being one example). We know that any compact subset $U \subset X$ is closed and bounded; conversely, if $U \subset X$ is closed and bounded *and if* X is finite-dimensional, then U is compact (the Heine-Borel and Bolzano-Weierstrass Theorems). A subset U of a normed space X is called *weakly sequentially compact* if every sequence $\{u_j\}$ from U contains a subsequence that converges weakly (in X) to a point $u \in U$. A fundamentally important property of reflexive Banach spaces is that bounded subsets are always weakly sequentially compact.

Theorem 4.8.8 (Eberlein-Shmulian Theorem). *A Banach space X is reflexive if and only if the closed unit ball in X is weakly sequentially compact.*

Proof. A proof may be found in the book of Dunford and Schwartz [5] or that of Yosida [15]. □

An equivalent formulation of the theorem is

Theorem 4.8.9. *A Banach space X is reflexive if and only if every bounded sequence in X contains a subsequence that converges weakly to an element of X.*

Proof. A proof may be found in the book of Dunford and Schwartz [5]. □

EXERCISES

4.8.1 Let H be a Hilbert space. Prove that for any $x \in H$,

$$\|x\| = \sup_{\substack{y \in H \\ \|y\|=1}} |(x, y)|.$$

4.8.2 Prove that any Hilbert space is reflexive. (*Hint:* Use the Riesz Representation Theorem.)

4.8.3 Prove that if M is any subset of a Hilbert space H, then $M^{\perp}$ (the orthogonal complement of M in H) is a closed subspace of H.

4.8.4 Prove that any finite-dimensional subspace of a normed space is closed.

4.8.5 Prove that all finite-dimensional normed spaces are reflexive Banach spaces.

REFERENCES AND ADDITIONAL READING

1. G. Auchmuty, Potential representations of incompressible vector fields, in *Nonlinear Problems in Applied Mathematics* (T. S. Angell, L. P. Cook, R. E. Kleinman, and W. E. Olmstead, Eds.), SIAM, Philadelphia, 1996.
2. B. Bollobas, *Linear Analysis*, Cambridge University Press, New York, 1990.
3. L. Debnath and P. Mikusinski, *Introduction to Hilbert Spaces with Applications*, Academic Press, San Diego, 1990.
4. J. Dieudonné, *Foundations of Modern Analysis*, Academic Press, New York, 1960.
5. N. Dunford and J. T. Schwartz, *Linear Operators*, Part I: *General Theory*, Wiley, New York, 1957.
6. C. W. Groetsch, *Elements of Applicable Functional Analysis*, Marcel Dekker, New York 1980.
7. S. Kesavan, *Topics in Functional Analysis and Applications*, Wiley, New York, 1989.
8. A. N. Kolmogorov and S. V. Fomin, *Functional Analysis*, Vols. I and II, Graylock, 1957, 1961.

9. E. Kreyszig, *Introductory Functional Analysis with Applications*, Wiley, New York, 1990.

10. L. Maligranda, A simple proof of the Hölder and the Minkowski inequality, *American Mathematical Monthly* 102, 1995.

11. D. Mitrovic and D. Zubrinic, *Fundamentals of Applied Functional Analysis*, Pitman Monographs and Surveys in Pure and Applied Mathematics, Wiley, New York, 1998.

12. M. Reed and B. Simon, *Methods of Modern Mathematical Physics*, 4 vols., Academic Press, New York, 1972–1979.

13. F. Riesz and B. Sz.-Nagy, *Functional Analysis*, Frederick Ungar, New York, 1955.

14. M. Schechter, *Principles of Functional Analysis*, Academic Press, New York, 1971.

15. K. Yosida, *Functional Analysis*, Springer–Verlag, New York, 1980.

16. N. Young, *An Introduction to Hilbert Space*, Cambridge University Press, New York, 1988.

17. E. Zeidler, *Applied Functional Analysis*, Springer–Verlag, New York, 1995.

18. E. Zeidler, *Nonlinear Functional Analysis and its Applications*, I: *Fixed-point theorems*, Springer–Verlag, New York, 1991.

CHAPTER 5

OPERATOR THEORY

5.1 BASIC IDEAS AND EXAMPLES

Whatever linear problem we are trying to solve, whether a set of algebraic equations, a differential equation, or an integral equation, there are advantages of clarity, generality, and geometric visualization in setting the problem in a suitable Hilbert space framework. To do so does, however, require some introductory definitions, and the impatient reader may find it refreshing to refer from time to time to the set of examples that follow shortly. Throughout we shall be dealing with a separable Hilbert space.

A (linear) *transformation* or *operator* A from the Hilbert space H into itself is a correspondence which assigns to each element u in H a well-defined element v, written Au, also belonging to H with the property

$$A(\alpha u + \beta v) = \alpha Au + \beta Av \quad \text{for all } u, v \text{ in } H \text{ and all scalars } \alpha, \beta. \tag{5.1.1}$$

Note that a transformation takes vectors into vectors, whereas a functional maps vectors into scalars. A slightly more general definition of a linear transformation is obtained if we allow A to be defined only on a linear manifold D_A in H. Here D_A is the *domain* of A; often $D_A = H$, but it is not always possible to define an operator on the whole of H. (For instance, if H is L_2, a differential operator can be defined only on part of H.) A linear operator always satisfies $A0 = 0$, and the linearity property (5.1.1) can be extended to *finite* sums. The set of all images (that is, the set of all vectors of the form Au for some u in D_A) is labeled R_A and is known as the *range* of A. The set N_A of all vectors for which $Au = 0$ is called the *null space* of A (the null space is the set of all solutions of the homogeneous equation associated with A). The sets N_A and R_A are always linear manifolds.

A transformation A is *bounded* on its domain if, for all u in D_A, there exists a constant c such that $\|Au\| \leqslant c\|u\|$. Thus the ratio of the "output" norm to the "input" norm is bounded above. The smallest number c which satisfies the inequality for all u in D_A is the *norm* of A, written as $\|A\|$. We see with no difficulty that

$$\|A\| = \sup_{\|u\| \neq 0} \frac{\|Au\|}{\|u\|} = \sup_{\|u\|=1} \|Au\|. \tag{5.1.2}$$

Even if A is bounded, the supremum may not be attained for any element u (see Example 4). If $\|A\| = 0$, then A is the zero operator. To prove that a number m is the norm of an operator A, one must show that $\|Au\| \leqslant m\|u\|$ for all $u \in D_A$ *and* either that there exists $u_0 \neq 0$ in D_A such that $\|Au_0\| = m\|u_0\|$ or that there exists a sequence $\{u_n\} \in D_A$ with $\|u_n\| \neq 0$ and $\|Au_n\| - m\|u_n\| \to 0$. To prove that A is unbounded, one must exhibit a sequence $\{u_n\}$ with $\|u_n\| \neq 0$ such that $\|Au_n\|/\|u_n\| \to \infty$.

A transformation A is *continuous at the point* u in D_A if whenever $\{u_n\}$ is a sequence in D_A with limit u, then $Au_n \to Au$. A transformation is continuous (on its domain) if it is continuous at every point in D_A. The following theorems have proofs similar to those of corresponding theorems for functionals.

Theorem. *If A is continuous at the origin, it is continuous on all of D_A.*

Theorem. *A is continuous if and only if it is bounded.*

Theorem. *In E_n all linear transformations are bounded.*

Representation of Bounded Linear Operators; Matrices

A linear transformation A from $E_n^{(c)}$ into itself is completely characterized by its (vector) values on a fixed basis $h_1, \ldots, h_n$. Indeed, if u is a vector in $E_n^{(c)}$, we can write $u = \sum_{j=1}^n \alpha_j h_j$ and $v \doteq Au = \sum_{j=1}^n \alpha_j Ah_j$, so that knowledge of the n vectors $Ah_1, \ldots, Ah_n$ enables us to calculate Au. We can go further: Each vector Ah_j is itself in $E_n^{(c)}$ and can in turn be written as

$$Ah_j = \sum_{i=1}^n a_{ij} h_i \qquad \text{(summation on } \textit{first index} \text{ in } a_{ij}\text{)}, \tag{5.1.3}$$

where the complex numbers a_{ij} are defined unambiguously from (5.1.3) and can be calculated from

$$a_{ij} = \langle Ah_j, h_i^* \rangle \quad (= \langle Ae_j, e_i \rangle \text{ in an orthonormal basis}), \tag{5.1.4}$$

in which $\{h_i^*\}$ is the dual basis introduced in (4.7.4). Thus

$$v = Au = \sum_{j=1}^{n} \alpha_j Ah_j = \sum_{i=1}^{n} \left(\sum_{j=1}^{n} a_{ij}\alpha_j \right) h_i,$$

and if we denote the ith coordinate of v in the $\{h_i\}$ basis by β_i, we have

$$\beta_i = \sum_{j=1}^{n} a_{ij}\alpha_j \quad \text{(summation on } \textit{second index} \text{ in } a_{ij}). \tag{5.1.5}$$

The set of n^2 complex numbers a_{ij} completely characterizes the transformation A. These numbers are usually displayed as a square array:

$$\begin{bmatrix} a_{11} & \dots & a_{1n} \\ \vdots & & \vdots \\ a_{n1} & \dots & a_{nn} \end{bmatrix},$$

known as the *matrix* of A in the basis $\{h_i\}$. The shorter notation $[a_{ij}]$ will often be used. By tradition the first index in a_{ij} gives the *row* location; the second, the *column* location. Note that the matrix is used in different ways in (5.1.3) and (5.1.5). To obtain the vector Ah_j, we "multiply" the jth *column* of A by $(h_1, \dots, h_n)$; whereas to find the ith coordinate of the image vector v in terms of the coordinates of u, we "multiply" the ith *row* of A by $(\alpha_1, \dots, \alpha_n)$.

The relation between a transformation and a matrix is as follows.

Relative to a given basis, each linear transformation A determines a unique matrix $[a_{ij}]$ whose entries can be calculated from (5.1.3), and, conversely, each matrix (that is, each square array of n^2 numbers) generates a unique linear transformation via (5.1.3).

Everything said holds also for transformations on $E_n^{(r)}$, but then the entries in the matrix are real.

When dealing with an infinite-dimensional separable Hilbert space H, we find, just as we did for linear functionals, that the analogy with E_n can be carried out successfully only for bounded operators. Let A be a *bounded* linear operator defined on the whole of H, and, for simplicity, let $\{e_i\}$ be an *orthonormal* basis. We can then write $u = \sum_j \alpha_j e_j$ and $v \doteq Au = A(\sum_j \alpha_j e_j) = \sum_j \alpha_j Ae_j$, where the continuity of A has been used in the last equality. Each vector Ae_j can be written as $\sum_i a_{ij} e_i$, where $a_{ij} = \langle Ae_j, e_i \rangle$. It follows that $\sum_i |a_{ij}|^2 < \infty$ (and it can also be shown that $\sum_j |a_{ij}|^2 < \infty$). Also $\beta_i = \langle v, e_i \rangle = \langle Au, e_i \rangle = \langle A \sum_j \alpha_j e_j, e_i \rangle = \sum_j \alpha_j \langle Ae_j, e_i \rangle = \sum_j a_{ij}\alpha_j$. Thus, relative to a fixed basis, each bounded linear transformation generates a unique infinite matrix $[a_{ij}]$ each of whose rows and columns is square summable. These conditions are, however, not sufficient for a set of numbers $[a_{ij}]$ to be the matrix of a bounded linear transformation (see Exercise 5.1.1).

Examples of Linear Operators

The following examples deserve careful examination; they will be referred to frequently as we take up new material.

Example 1.

(a) The zero transformation 0 is defined by $v \doteq 0u \doteq 0$ for every u in H. The transformation is clearly linear and bounded and has norm 0. The range consists of the single element 0, and the null space is H.

(b) The identity I is a transformation defined by $v \doteq Iu \doteq u$ for each u in H. The transformation is linear, $\|I\| = 1, R_I = H, N_I = \{0\}$.

Example 2. Let M be a closed linear manifold in H, and for each u in H define $v = Pu$ as the (orthogonal) projection of u on M. P is linear, and $\|v\| \leqslant \|u\|$ with equality if and only if u is in M. We have $\|P\| = 1, R_P = M, N_P = M^{\perp}$.

Example 3.

(a) Let A be a linear transformation from $E_n^{(c)}$ into itself. We have already shown that A is necessarily bounded, but it is not always easy to calculate $\|A\|$. Suppose that A has the matrix $[a_{ij}]$ in the *orthonormal* basis $\{e_1, \ldots, e_n\}$. Then by (5.1.5) and (4.5.13), we have

$$\|v\|^2 = \sum_{i=1}^{n} |\beta_i|^2 = \sum_{i=1}^{n} \left| \sum_{j=1}^{n} a_{ij}\alpha_j \right|^2 \leqslant \sum_{i,j=1}^{n} |a_{ij}|^2 \|u\|^2.$$

It follows that

$$\|A\| \leqslant \left[\sum_{i,j=1}^{n} |a_{ij}|^2 \right]^{1/2}, \tag{5.1.6}$$

but the upper bound can be grossly conservative. If A happens to have *diagonal* form in the orthonormal basis, life becomes much simpler (which is a strong incentive for studying the possibility of choosing a basis in which A is diagonal). Then $a_{ii} = m_i$ and $a_{ij} = 0, i \neq j$, so that

$$\|v\|^2 = \sum_{i=1}^{n} |m_i\alpha_i|^2 \leqslant m^2 \|u\|^2, \qquad m = \max |m_i|.$$

This shows that $\|A\| \leqslant m$, but since $\|Ae_i\| = |m_i|$, we see that there is a unit vector for which $\|Ae\| = m$. Thus $\|A\| = m$. Although we have carefully refrained from using the word *eigenvalue*, it is clear that $\|A\|$ is closely related to the eigenvalues of A. This question and the characterization of N_A and R_A will be taken up later. We now look at two simple cases of linear transformations on E_2.

(b) On the real plane $E_2^{(r)}$, consider the transformation A which rotates each vector counterclockwise through the angle θ. To find the matrix of A relative to a right-handed orthonormal basis (e_1, e_2) we have to draw a sketch from which we conclude that

$$Ae_1 = (\cos\theta)e_1 + (\sin\theta)e_2, \quad Ae_2 = (-\sin\theta)e_1 + (\cos\theta)e_2,$$

and therefore

$$A = \begin{bmatrix} \cos\theta & -\sin\theta \\ \sin\theta & \cos\theta \end{bmatrix}.$$

The coordinates (β_1, β_2) of the rotated vector are related to the coordinates (α_1, α_2) of the original vector by

$$\beta_1 = \alpha_1 \cos\theta - \alpha_2 \sin\theta, \quad \beta_2 = \alpha_1 \sin\theta + \alpha_2 \cos\theta.$$

We could have considered a linear transformation on the complex space $E_2^{(c)}$ with the same matrix as above relative to some orthonormal basis. The formulas would remain the same, but the geometric interpretation would be lost. In either case, $N_A = \{0\}, R_A = E_2, \|A\| = 1$.

(c) On $E_2^{(c)}$ or $E_2^{(r)}$ consider the transformation A whose matrix, relative to an orthonormal basis (e_1, e_2), is

$$\begin{bmatrix} 0 & 1 \\ 0 & 0 \end{bmatrix}.$$

The coordinates of the image vector are related to those of the original by

$$\beta_1 = \alpha_2, \quad \beta_2 = 0.$$

Both N_A and R_A consist of all vectors proportional to e_1; also, $\|A\| = 1$.

Example 4. On $L_2^{(c)}(0,1)$ consider the transformation corresponding to multiplication by the independent variable x. Thus $v(x) \doteq Au \doteq xu(x)$. The transformation is defined for all elements of $L_2^{(c)}(0,1)$ and is linear. From

$$\|Au\|^2 = \int_0^1 x^2|u|^2\,dx \leqslant \int_0^1 |u|^2\,dx = \|u\|^2$$

we infer that $\|A\| \leqslant 1$ and that A is therefore bounded. Next we show that $\|A\| = 1$ by exhibiting a family $u_\varepsilon(x)$ with the property $\|Au_\varepsilon\|/\|u_\varepsilon\| \to 1$ as $\varepsilon \to 0$. Let $u_\varepsilon(x)$ be 0 except on the small interval $1 - \varepsilon \leqslant x \leqslant 1$, where $u_\varepsilon(x) = 1$. Then $\|u_\varepsilon\|^2 = \varepsilon$ and $\|Au_\varepsilon\|^2 = \varepsilon[1 - \varepsilon + (\varepsilon^2/3)]$, which shows that u_ε has the desired property. Although $\|A\| = 1$, there is no nonzero element in $L_2^{(c)}(0,1)$ for which $\|Au\| = \|u\|$. Note that $N_A = \{0\}$ but that R_A is *not* all of $L_2^{(c)}(0,1)$.

Example 5. In the interval $0 < x < 1$ we want to consider the transformation defined by $v(x) \doteq Au \doteq (1/x)u(x)$. Since $1/x$ becomes infinite at $x = 0$, the transformation is not defined for all u in $L_2^{(c)}(0,1)$. On the other hand, if u vanishes sufficiently fast at $x = 0$, then u/x will be in $L_2^{(c)}(0,1)$. Thus, the natural domain of definition for this operator is the linear manifold D_A of functions $u(x)$ in $L_2^{(c)}(0,1)$ for which u/x is also in $L_2^{(c)}(0,1)$. D_A clearly includes the functions u in $L_2^{(c)}(0,1)$ which vanish identically in a neighborhood of $x = 0$. Since such functions are dense in $L_2^{(c)}(0,1)$, D_A is certainly dense in $L_2^{(c)}(0,1)$. We now show that A is *unbounded* on D_A. Consider the function $u_\varepsilon(x)$ in D_A defined by $u_\varepsilon = 0$, $0 < x < \varepsilon$; $u_\varepsilon = 1$, $\varepsilon < x < 1$. Then $\|u_\varepsilon\|^2 = 1 - \varepsilon$ and $\|Au_\varepsilon\|^2 = (1-\varepsilon)/\varepsilon$. As $\varepsilon \to 0$, the ratio $\|Au_\varepsilon\|/\|u_\varepsilon\|$ becomes arbitrarily large, so that A is unbounded. We have $N_A = \{0\}$, $R_A = L_2^{(c)}(0,1)$.

Example 6. On $L_2^{(c)}(a,b)$ consider the *integral operator* defined by

$$v(x) \doteq Ku \doteq \int_a^b k(x,y)u(y)\,dy. \tag{5.1.7}$$

Here $k(x,y)$ is a given function defined over the square $a \leqslant x, y \leqslant b$. The function $k(x,y)$ is known as the *kernel* of the operator. We have already considered such operators on the space of continuous functions in Section 4.4. Although a function of two variables is involved in an intermediate stage, the operator K takes a function of one variable into another function of one variable. If $u(x)$ is in $L_2^{(c)}(a,b)$, so will be $v(x)$ as long as $k(x,y)$ satisfies some very mild conditions. If, for instance,

$$\int_a^b \int_a^b |k^2(x,y)|\,dx\,dy < \infty, \tag{5.1.8}$$

then $k(x,y)$ is said to be a *Hilbert-Schmidt kernel*, and the operator K maps the space $L_2^{(c)}(a,b)$ into itself and is bounded. In fact, by the Schwarz inequality,

$$\begin{aligned} |v(x)|^2 &= \left| \int_a^b k(x,y)u(y)\,dy \right|^2 \\ &\leqslant \int_a^b |k(x,y)|^2\,dy \int_a^b |u(y)|^2\,dy = \|u\|^2 \int_a^b |k(x,y)|^2\,dy. \end{aligned}$$

Therefore,

$$\|v\|^2 \leqslant \|u\|^2 \left[\int_a^b \int_a^b |k^2(x,y)|\,dx\,dy \right],$$

so that K is bounded:

$$\|K\| \leqslant \left[\int_a^b \int_a^b |k^2(x,y)|\,dx\,dy \right]^{1/2}. \tag{5.1.9}$$

Example 7. On $L_2^{(c)}(0,1)$ consider the operator defined by

$$v(x) \doteq Au \doteq \int_0^x u(y)\,dy. \tag{5.1.10}$$

The operator is defined for all u in $L_2^{(c)}(0,1)$ since square integrability implies that u is integrable over any subinterval of (0, 1). This operator is linear and can be regarded as a special case of Example 6 with $a = 0, b = 1$ and $k(x,y) = H(x-y)$, where $H(x)$ is the usual Heaviside function. From (5.1.9) we find that

$$\|A\| \leqslant \left[\int_0^1 \int_0^1 H^2(x-y)\,dx\,dy\right]^{1/2} = \left(\frac{1}{2}\right)^{1/2}.$$

The exact expression for $\|A\|$ is $2/\pi$. (See Exercise 5.1.4.) We observe that $N_A = \{0\}$ and that R_A consists of differentiable functions vanishing at $x = 0$ with the derivative in $L_2^{(c)}(0,1)$.

Example 8. We make a preliminary attempt to define a differentiation operator A for functions on $0 \leqslant x \leqslant 1$. Since functions in $L_2(0,1)$ may not even be continuous, we cannot define the operator for the whole of L_2. As a first try, let D_A be the set of functions in L_2 with a continuous derivative on $0 \leqslant x \leqslant 1$, and define

$$v(x) \doteq Au \doteq \frac{du}{dx}.$$

Both D_A and R_A are subsets of L_2, and D_A is a dense linear manifold in $L_2(0,1)$. The operator A is *unbounded* on D_A: Let $u_n = \sin n\pi x$; then $\|u_n\|^2 = \frac{1}{2}$ and $\|Au_n\|^2 = n^2\pi^2/2$, so that $\|Au_n\|/\|u_n\|$ is unbounded as $n \to \infty$. Note that N_A is the set of constant functions and that R_A (by the way we have defined A) is the set of continuous functions on $0 \leqslant x \leqslant 1$.

When faced with solving the inhomogeneous equation $du/dx = f, u \in D_A$, we note two unpleasant features. First, the solution, when it exists, is not unique since we can add a constant function and still have a solution. This difficulty is easily overcome by restricting the domain of the operator to functions satisfying an appropriate "boundary condition" such as $u(0) = 0$. The second, and more serious, trouble is that we are able to solve the inhomogeneous equation only for continuous f rather than for every f in L_2. Admittedly, this is, in part, a self-imposed restriction since we have defined D_A so that the range contains only continuous functions. We shall see in Example 1, Section 5.3 (after the topic of closed operators) how to reformulate the problem.

Example 9. The shift operator. Let H be an infinite-dimensional separable space, and let $\{e_1, e_2, \ldots, e_n, \ldots\}$ be an orthonormal basis in H. Define the operator A by giving its effect on the basis: A transforms each unit vector in the next one on the list (*right shift*); that is, $Ae_i = e_{i+1}$. The image of $u = \sum_{i=1}^{\infty} \alpha_i e_i$ is therefore $Au = \sum_{i=1}^{\infty} \alpha_i e_{i+1}$. It is clear that A is linear and is defined on the whole of H and that $\|Au\| = \|u\|$ for every u; hence $\|A\| = 1$. The range of A is the closed manifold orthogonal to e_1; that is, R_A consists of all vectors whose first coordinate is 0. R_A is closed but not dense in H. The null space of A consists only of the element 0. The matrix of A consists of 0's except for the diagonal directly below the main diagonal, which consists of 1's.

Example 10. A modified shift. Again $\{e_i\}$ is an orthonormal basis, but now define $Be_k = (1/k^2)e_{k+1}$. Then $\|B\| = 1, N_B = \{0\}$, and R_B consists of all vectors $\sum_{i=1}^{\infty} \beta_i e_i$ with $\beta_1 = 0$ and $\sum_{k=1}^{\infty} k^4 |\beta_{k+1}|^2 < \infty$. We observe that R_B is not closed, but that $\bar{R}_B = M_1^{\perp}$, where M_1 is the linear manifold generated by e_1. *Proof.* If $u \in \bar{R}_B$, there exists $\{u_n\}$ in R_B with $u_n \to u$. Since $\langle u_n, e_1 \rangle = 0$, it follows by continuity of the inner product that $\langle u, e_1 \rangle = 0$, so that $\bar{R}_B \subset M_1^{\perp}$. For $u \in M_1^{\perp}$, consider the truncated vector $[u]_n$ obtained from u by setting equal to 0 all coordinates with indices larger than n. Then $[u]_n$ is in R_B, and since $[u]_n \to u$, u must lie in $\bar{R}_B$. Thus $\bar{R}_B \supset M_1^{\perp}$, which, together with the reverse inclusion, shows that $\bar{R}_B = M_1^{\perp}$.

EXERCISES

5.1.1 Let A be a bounded linear transformation on H, and let $\{e_k\}$ be an orthonormal basis. If $[a_{ij}]$ is the matrix of A with respect to this basis, we have shown that $\sum_i |a_{ij}|^2 < \infty$ for each j; that is, each column is square summable. By considering the adjoint (Section 5.4), it can also be shown that each row is square summable. Show that these two conditions, even taken together, are not sufficient for A to be bounded. (*Hint*: Take a diagonal matrix whose elements increase indefinitely along the diagonal.) Show that a sufficient (but not necessary) condition for A to be bounded is that $\sum_{ij} |a_{ij}|^2 < \infty$.

5.1.2 On $L_2^{(c)}(0,1)$ consider the transformation A defined by $Au = f(x)u(x)$, where $f(x)$ is a fixed function. Show that if f is continuous on $0 \leqslant x \leqslant 1$, then $\|A\| = \max_{0 \leqslant x \leqslant 1} |f(x)|$.

5.1.3 *Holmgren kernels.* Consider the integral operator of Example 6, but instead of the Hilbert-Schmidt condition (5.1.8), assume that

$$\max_{a \leqslant z \leqslant b} \int_a^b \int_a^b |k(x,y)k(x,z)|\, dx\, dy \doteq M < \infty. \tag{5.1.11}$$

Show that K is a bounded operator and that $\|K\| \leqslant M^{1/2}$. On $L_2(-\infty, \infty)$ the example $k(x,y) = h(x-y)$, where $\int_{-\infty}^{\infty} |h(x)| < \infty$, shows that there are Holmgren kernels that are not Hilbert-Schmidt.

5.1.4 Consider the operator of Example 7.

(a) Show that the set

$$\left\{ e_n(x) = \sqrt{2} \cos \frac{2n-1}{2} \pi x \right\}, \quad n = 1, 2, \ldots,$$

is an orthonormal basis in $L_2^{(c)}(0,1)$. [*Hint*: The set $\cos(k\pi x/2)$, for $k = 0, 1, 2, \ldots$, is a basis for $L_2(0,2)$; any function in (0, 1) can be extended to (0, 2) with the property $f(x+1) = -f(1-x)$.]

(b) Expand u in the set $\{e_n(x)\}$ of part (a) to show that $\|Au\| \leqslant (2/\pi)\|u\|$.

(c) Show $\|A\| = 2/\pi$ by finding a function u such that $\|Au\| = (2/\pi)\|u\|$.

5.1.5 Obtain the results corresponding to those of Example 9 for the left-shift operator defined by $Ae_i = e_{i-1}, i = 2, \ldots; Ae_1 = 0$.

5.1.6 Consider the two-sided basis

$$\left\{e_n(x) = \frac{1}{\sqrt{2}}e^{2n\pi i x}\right\}, \quad n = 0, \pm 1, \pm 2, \ldots, \quad \text{on } L_2^{(c)}(-1,1).$$

Define a linear operator by $Ae_n = e_{n+1}$. Show that this operator can be represented as a multiplication by a function. Find $\|A\|$, N_A, and R_A.

5.2 CLOSED OPERATORS

Example 8 of Section 5.1 shows that one encounters difficulties in finding a suitable mathematical framework for differential operators. There are three accepted methods for handling these problems: (a) One can deal with the inverse operator, usually an integral operator that is not only bounded but also compact (see Section 5.7); (b) one can introduce the idea of a closed operator as we shall do in the present section; or (c) one can change the norm in the domain to include appropriate derivatives, an approach equivalent to introducing Sobolev spaces (see Example 6, Section 4.5).

Definition. *The linear operator B is said to be an* extension *of the linear operator A if $D_B \supset D_A$ and $Bu = Au$ for each u in D_A.*

Bounded linear operators defined on a linear manifold (closed or not) in H can always be extended to the whole of H without affecting continuity or changing the norm.

(a) If D_A is closed and $D_A \neq H$, we construct the extension by defining the new operator B to be 0 on $(D_A)^\perp$ and to coincide with A on D_A, letting linearity take care of everything else. Precisely: If $u \in H$, write $u = v + w, v \in D_A, w \in D_A^\perp$; then Bu is defined as Av. The operator B is easily seen to be linear on H and $\|B\| = \|A\|$.

(b) If D_A is not closed, we first extend A to $\bar{D}_A$ by "continuity." Let $\{u_n\}$ be a sequence in D_A with limit u in $\bar{D}_A$. Since A is bounded, the sequence $\{Au_n\}$ is Cauchy and has a limit, say f. It is natural to set Au equal to f. To see that this definition depends only on u and not on $\{u_n\}$, let $\{v_n\}$ be another sequence in D_A with limit u. Then $u_n - v_n \to 0$, and by continuity, $A(u_n - v_n) \to 0$, so that $\{Au_n\}$ and $\{Av_n\}$ have the same limit, f. In this way we extend A to $\bar{D}_A$ and then by (a) to H.

Thus, if in a problem we are permitted to change the domain of a given bounded operator A, we may as well take the operator as a bounded operator *defined on the whole of* H.

Suppose now that A is an *unbounded* linear operator on a linear manifold D_A in H. Let $\{u_n\}$ be a sequence of elements in D_A for which $\lim_{n\to\infty} u_n = 0$; such a sequence will be called a *null sequence in* D_A. We cannot have $Au_n \to 0$ for *all* null sequences in D_A, for otherwise A would be continuous at 0 and hence bounded. (On the other hand, there will always be some null sequences for which $Au_n \to 0$. For instance, let $u \neq 0$ be in D_A and $\{\alpha_n\}$ be a sequence of scalars such that $\alpha_n \to 0$; then $\{u_n \doteq \alpha_n u\}$ is a null sequence in D_A for which $Au_n \to 0$.) Thus there must exist some null sequences in D_A for which $Au_n \to f \neq 0$ or for which Au_n has no limit. The first possibility leads to very pathological operators which are *never encountered* in the study of differential or integral equations. We may therefore confine ourselves to the following class of operators.

Definition. *A linear operator is* closable *if for every null sequence* $\{u_n\}$ *in* D_A, *either* $Au_n \to 0$ *or* Au_n *has no limit.*

Note the definition includes all bounded operators and presumably unbounded operators such as those of Examples 5 and 8, Section 5.1. Let us examine the differentiation operator of Example 8. For any null sequence $\{u_n\}$ in D_A we are to show that if $Au_n \to f$ (that is, if $u_n' \to f$), then $f = 0$. If $u_n' \to f$, then $\int_0^1 u_n'\bar{z}\,dx \to \langle f, z\rangle$ for every z in L_2, so certainly for every z in the class M of continuously differentiable functions with $z(0) = z(1) = 0$; hence for z in M we obtain, through integration by parts,

$$-\int_0^1 u_n\bar{z}'\,dx \to \langle f, z\rangle,$$

which, by the continuity of the inner product, gives $\langle f, z\rangle = 0$ for every z in M. Since M is dense in H, we have $f = 0$, so that A on D_A is *closable*. This means that if $\{u_n(x)\}$ is a sequence of functions in D_A with $u_n \to 0$, then either $u_n' \to 0$ or u_n' has no limit [the latter possibility occurs if $u_n = (\sin nx)/n$, for instance].

The advantage of a closable operator is that it can usefully be extended by the following simple procedure. Let $\{u_n\} \in D_A$ with $u_n \to u$ (which may or may not be in D_A), and suppose that $Au_n \to f$. Let $\{v_n\}$ be another sequence in D_A which approaches the same limit u; then $u_n - v_n$ is a null sequence in D_A, so that either $A(u_n - v_n) \to 0$ or this sequence has no limit. Thus either $Av_n \to f$ or Av_n has no limit. We therefore yield gracefully to the temptation to include u in the domain of the operator and to let f be the image of u. If we do this for all convergent sequences $\{u_n\}$ in D_A for which $\{Au_n\}$ has a limit, we obtain a new operator known as the *closure* of A and denoted by $\tilde{A}$. It is clear that $\tilde{A}$ is a linear operator defined on a domain $D_{\tilde{A}}$ which includes D_A and that $\tilde{A}u = Au$ whenever $u \in D_A$; thus $\tilde{A}$ is an extension of A. The operator $\tilde{A}$ belongs to the important class of closed operators.

Definition. *Let* A *be a linear operator on the linear manifold* D_A. *We say that* A *is* closed *if it has the following property: Whenever* $\{u_n\}$ *is in* D_A *and* $u_n \to u$, *and* $Au_n \to f$, *then* u *is in* D_A *and* $Au = f$.

REMARKS

1. If A is closable, $\tilde{A}$ is closed.

2. A bounded operator defined on the whole of H (or even defined on a closed set) is closed.

3. *A closed operator on a closed domain is bounded.* We shall not prove this, but we analyze some of its consequences. It means first that a closed, *unbounded* operator can never be defined on the whole of H. Differential operators are the most important among closed, unbounded operators, and they are usually defined on domains *dense in* H. The only unbounded operators that can be defined on the whole of H are "highly discontinuous" and never occur in the study of differential or integral equations.

4. Since a closed operator does not necessarily have a closed domain or a closed range, the reader may wonder how it got its name. The graph of an operator A is the set of ordered pairs (u, Au) with u in D_A. The graph can be viewed as a subset of a new normed space consisting of pairs (u, v) with $u, v \in H$ and $\|(u, v)\| \doteq \|u\| + \|v\|$. It then turns out that the original operator A is closed if and only if its graph is a closed set in the new normed space.

5. The null space of a closed operator is a closed set.

Now let us examine again the differentiation operator A of Example 8, Section 5.1. This operator was defined on the domain D_A of continuously differentiable functions on $0 \leqslant x \leqslant 1$. We have shown that the operator is closable, but it is easy to see that it is not closed. In Exercise 5.2.1 we exhibit a sequence of very smooth functions $\{u_n(x)\}$ such that u_n tends in the L_2 sense to a function u that is only piecewise differentiable, while at the same time u_n' tends to a limit v. Thus u belongs to $D_{\tilde{A}}$ and $\tilde{A}u = v$, but u does not belong to D_A. It is negligent of us to leave out piecewise smooth functions from the original domain, but even then the differentiation operator would not be closed. To see exactly what the significance of $\tilde{A}$ is, let $\{u_n\}$ be a sequence in D_A such that, simultaneously,

$$\|u_n - u\| \to 0 \quad \text{and} \quad \|u_n' - v\| \to 0. \tag{5.2.1}$$

Then by definition $u(x)$ belongs to the domain of $\tilde{A}$ and $\tilde{A}u = v$. Of course such a definition is appealing only if $\tilde{A}$ can be characterized concretely as a generalization of differentiation with v the derivative of u.

Since $u_n'(x)$ is continuous, we have

$$u_n(x) = \int_0^x u_n'(t)\, dt + u_n(0). \tag{5.2.2}$$

The L_2 convergence of u_n' to v implies that $\int_0^x u_n'\, dt$ converges in L_2 to $\int_0^x v\, dt$. Equation (5.2.2) then shows that the sequence of constants $\{u_n(0)\}$ must converge in L_2; obviously, the limit is equivalent to a constant, say α, so that (5.2.2) yields

$$u(x) = \int_0^x v(t)\,dt + \alpha \quad \text{almost everywhere.} \tag{5.2.3}$$

Since the right side of (5.2.3) is continuous, $u(x)$ can be redefined, if necessary, on a set of measure 0 to make the equation valid for all x in $0 \leqslant x \leqslant 1$. As an immediate consequence we see that $\alpha = u(0)$ and that $u(x)$ and $v(x)$ are related by

$$u(x) = \int_0^x v(t)\,dt + u(0), \quad 0 \leqslant x \leqslant 1. \tag{5.2.4}$$

If v is continuous, u is differentiable everywhere and $u' = v$, but even if v is only integrable, (5.2.4) provides a useful extension of the notion of differentiability. It follows from (5.2.4) that u' exists almost everywhere and that $u' = v$ almost everywhere; in addition, u is the integral of its derivative. [Everyone should know Cantor's example of a nonconstant continuous function whose derivative is equal to 0 almost everywhere. Obviously, such a function does not satisfy (5.2.4).] A function satisfying (5.2.4) is said to be *absolutely continuous* (for which we often use the abbreviation a.c.).

Therefore, the closure $\tilde{A}$ of the differentiation operator has a domain $D_{\tilde{A}}$ consisting of all a.c. functions with first derivative in L_2. For a differential operator A of order p the domain of the closure $\tilde{A}$ consists of functions $u(x)$ whose first $(p-1)$ derivatives are continuous and such that $u^{(p-1)}$ is a.c. with $u^{(p)}$ in L_2. This exact delineation of the smoothness of the functions in the domain is not of great importance to us; usually it suffices to know that the domain is the largest class of functions for which the differential operator makes sense and the image is in L_2.

We shall see in the next section how boundary conditions behave under closure.

EXERCISES

5.2.1 On $L_2^{(c)}(0, 1)$, let $u(x) = |x - \frac{1}{2}|$, and let $\{u_n(x)\}$ be the sequence of partial sums of the Fourier cosine series of $u(x)$. Show that $\|u_n - u\| \to 0$ and that $\|u_n' - v\| \to 0$, where $v = \operatorname{sgn}(x - \frac{1}{2})$. Since u is not continuously differentiable, it does not belong to the domain of the differentiation operator A as originally defined. Of course, u belongs to $D_{\tilde{A}}$ and $\tilde{A}u = v$. This exercise shows that piecewise smooth functions belong to $D_{\tilde{A}}$.

5.2.2 In (2.5.34), we gave another generalization of the notion of differentiation. We said that a function u satisfies $u' = f$ on (0, 1) if, for each test function $\phi(x)$ with support in (0, 1),

$$\int_0^1 f\phi\ dx = -\int_0^1 u\phi'\ dx.$$

Show that this definition is equivalent to saying that u is absolutely continuous on $0 \leqslant x \leqslant 1$.

5.3 INVERTIBILITY: THE STATE OF AN OPERATOR

Suppose A is a given linear operator (not necessarily bounded) on a linear manifold D_A in H. The central problem of operator theory is the solution ("inversion") of the inhomogeneous equation

$$Au = f, \tag{5.3.1}$$

where f is an arbitrary, given element in H and we are looking for solution(s) u lying in D_A. In a perfect world, (5.3.1) would have one and only one solution for each f in H, and the solution u would depend continuously on the "data" f. In the language of Chapter 1, such a problem would be well-posed. If this were always true, there might result a sharp drop in the employment of mathematicians, so that perhaps we should accept as a partial blessing the difficulties we are about to encounter in the analysis of (5.3.1).

We can divide the problem into two overlapping parts: the question of invertibility, and the question of characterizing the range of A. We take up invertibility first. If the mapping A is one-to-one from its domain to its range, then for each $f \in R_A$ there exists one and only one solution u (in D_A) of (5.3.1). This correspondence enables us to define the inverse operator A^{-1} by $u = A^{-1}f$, and A^{-1} is clearly linear. We are also interested in whether or not A^{-1} is bounded (if A^{-1} is bounded, it is continuous and therefore the solution u depends continuously on the "data" f). On E_n the inverse, if it exists, must be bounded, but on an infinite-dimensional space *the inverse of a bounded operator is not necessarily bounded.* The two very simple theorems that follow characterize, respectively, the existence of an inverse and the existence of a bounded inverse. Throughout, the linear operator A is defined on a fixed domain D_A, a linear manifold in H.

Theorem 5.3.1. *A^{-1} exists (A has an inverse, A is one-to-one) if and only if the homogeneous equation $Au = 0$ has only the zero solution (that is, the null space N_A contains only the zero element).*

Proof. If A is one-to-one, then $Au = 0$ can have at most one solution, and since $u = 0$ is obviously a solution, it is the only one. If $Au = 0$ has only the zero solution, we must show that $Av = f$ can have at most one solution; if $Av_1 = f$ and $Av_2 = f$, the linearity of A and of its domain implies that $A(v_1 - v_2) = 0$, so that $v_1 - v_2 = 0$ or $v_1 = v_2$. □

Definition. *The operator A (on D_A) is* bounded away from 0 *if there exists $c > 0$ such that $\|Au\| \geqslant c\|u\|$ for all $u \in D_A$.*

REMARK. Obviously, if A is bounded away from 0, then $Au = 0$ has only the zero solution and therefore A^{-1} exists. Moreover, A^{-1} must be bounded, for if we set $v = Au$, then $\|v\| = \|Au\| \geqslant c\|u\|$, so that $\|u\| \leqslant c^{-1}\|v\|$, and since $u = A^{-1}v$, we find that $\|A^{-1}\| \leqslant c^{-1}$. This leads easily to the following theorem.

Theorem 5.3.2. *The operator A (on D_A) has a bounded inverse if and only if it is bounded away from 0.*

REMARK. Thus, if A fails to have a bounded inverse, there exists a sequence $\{u_n\}$ of unit elements such that $Au_n \to 0$.

We now turn to the characterization of R_A. By definition of the range, (5.3.1) can be solved, although perhaps not uniquely, if and only if $f \in R_A$. It would be nice if R_A were all of H, but that is not the case in many of the examples we have just given. It turns out that it is often difficult to describe the range precisely (see, for instance, Examples 4, 6, 7, and 10, in Section 5.1), but easier to characterize its closure $\bar{R}_A$. Either this closure is all of H, or it is a proper subset of H (in which case there exist nonzero elements in H orthogonal to $\bar{R}_A$ and hence to R_A). Often, R_A and H are both infinite-dimensional, so that it is easier to talk about the dimension of $R_A^{\perp}$, the so-called *codimension* of R_A.

We are now ready to present at least a coarse-grained classification of operators. The *state* of an operator will be represented by a Roman numeral (I, II, or III), followed by an Arabic numeral (1 or 2). The Roman numeral describes the invertibility properties of A, whereas the Arabic numeral tells us whether or not $\bar{R}_A = H$.

$$\left.\begin{array}{rl} \text{I.} & \text{Bounded inverse.} \\ \text{II.} & A^{-1} \text{ exists but is unbounded.} \\ \text{III.} & A^{-1} \text{ does not exist } (Au = 0 \text{ has a nontrivial solution}). \\ 1. & \bar{R}_A = H. \\ 2. & \bar{R}_A \neq H. \end{array}\right\} \tag{5.3.2}$$

A further refinement in the classification will sometimes be made. The subscript c or n on the Arabic numeral tells us that the range is closed or not closed, respectively. Only state (I,1_c) represents the ideal operator with a bounded inverse and $R_A = H$. Such an operator is said to be *regular*. The other states characterize operators with various assortments of ills. Often one tries to adjust the definition of A and D_A (say, by closure) so that state (I,1_c) is achieved (see Example 1, for instance).

The possible states of an operator are illustrated in Figure 5.1 without regard to subscripts. In the interior of the left circle something is wrong with the inverse, whereas in the interior of the right circle $\bar{R}_A \neq H$. The table below shows the states for the examples of Section 5.1.

Example	State	Example	State
1(a)	(III,2_c)	5	(I,1_c)
1(b)	(I,1_c)	7	(II,1_n)
2	(III,2_c)	8	(III,1_n)
3(b)	(I,1_c)	9	(I,2_c)
4	(II,1_n)	10	(II,2_n)

(5.3.3)

Taking subscripts into account, there are 12 possible states in symbols if not in fact. Four of these can be eliminated for closed operators. We need the following theorems, the first of which (unproved) was discussed in Remark 3 of Section 5.2.

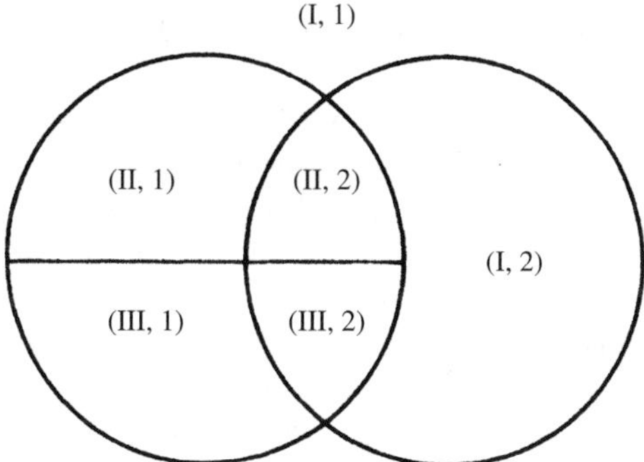

Figure 5.1 Possible states of operators as given in (5.3.2). The diagram can also be interpreted as showing how for a fixed A, the state of $A - \lambda I$ may vary with λ (without suggesting that every region is necessarily traversed as λ varies): II is the continuous spectrum, III the point spectrum, 2 the compression spectrum, II$\cup$III the approximate spectrum, and (I,1) the resolvent set.

Theorem 5.3.3. *A closed operator on a closed domain is bounded.*

Theorem 5.3.4. *If A is closed and A^{-1} exists, then A^{-1} is closed.*

Proof. Let $\{f_n\}$ be a sequence in R_A with $f_n \to f$ and $u_n \doteq A^{-1} f_n \to u$; we must show that f is in R_A and $A^{-1} f = u$. Since $Au_n = f_n, f_n \to f$, and $u_n \to u$, it follows from the fact that A is closed that u must be in the domain of A and that $Au = f$; therefore, f is in R_A and $u = A^{-1} f$. □

Theorem 5.3.5. *Let A be closed, and let A^{-1} exist; then R_A is closed if and only if A^{-1} is bounded.*

Proof.

(a) Let A^{-1} be bounded, and let $\{f_n\}$ be a sequence in R_A with $f_n \to f$. The boundedness of A^{-1} shows that $u_n \doteq A^{-1} f_n$ is a Cauchy sequence which must therefore have a limit, say u, in H. Thus $\{u_n\} \in D_A, u_n \to u$, and $Au_n \to f$; hence, since A is closed, $u \in D_A$ and $Au = f$. This shows that f is in R_A and therefore R_A is closed.

(b) By Theorem 5.3.4, A^{-1} is closed; if also R_A is closed, then, by Theorem 5.3.3, A^{-1} is bounded.

□

This means that states (I,1_n), (I,2_n), (II,1_c), and (II,2_c) cannot occur for a closed operator, leaving us with eight states: (I,1_c), (I,2_c), (II,1_n), (II,2_n), and all four states associated with III. States (III,1_c) and (III,2_n) are possible for closed operators even though they are not included in (5.3.3). State (III,1_c) is obtained when we close the operator in Example 8, Section 5.1, and state (III,2_n) occurs in Exercises 5.3.1 and 5.3.3.

There are important classes of operators on infinite-dimensional spaces for which much more can be said. We must postpone that discussion until we have studied adjoints in the next section. For an operator on $E_n^{(c)}$, however, the situation is exceedingly simple. First of all, R_A is always a closed set. If the homogeneous equation has only the trivial solution, then $R_A = H$ and $Au = f$ has the solution $u = A^{-1}f$, where A^{-1} is itself a linear operator on $E_n^{(c)}$ and hence bounded. If the homogeneous equation has a nontrivial solution, we know that $R_A \neq H$ (in fact, the codimension of R_A is equal to the dimension of N_A). Thus an operator on $E_n^{(c)}$ can only be in state (I,1_c) or in state (III,2_c); see Theorem 5.5.3.

Let us now look at the differentiation operator d/dx on various domains. We shall be particularly interested in the closed operators associated with various boundary conditions.

Example 1.

(a) Let D_1 be the subset of $L_2(0,1)$ consisting of continuously differentiable functions $u(x)$ on $0 \leqslant x \leqslant 1$ with $u(0) = 0$. Define $A_1 u = du/dx$ for $u \in D_1$. This operator is closable, and its *closure* T_1 is an extension of A_1 with domain $D_{T_1} \supset D_{A_1}$. We claim that the boundary condition $u(0) = 0$ survives in D_{T_1}. In fact, suppose that $\{u_n\}$ is in D_{A_1}, $u_n \to u$, and $u_n' \to f$; then by definition of the closure we have $u \in D_{T_1}$ and $T_1 u = f$. We now show that $u(0) = 0$; indeed, since $u_n(0) = 0$, we have $u_n(x) = \int_0^x u_n'(y)\,dy$, so that by taking limits $u(x) = \int_0^x f(y)\,dy$ and $u(0) = 0$. It is also clear that R_{T_1} consists of all elements in L_2. In fact, T_1^{-1} is closed because T_1 is; furthermore, it follows from the inversion formula $u(x) = \int_0^x f(y)\,dy$ and Example 7, Section 5.1, that T_1^{-1} is bounded, so that Theorem 5.3.5 guarantees that R_{T_1} is closed. Since R_{A_1} is already dense in L_2, R_{T_1} must also be dense in L_2 and $R_{T_1} = L_2$. Thus T_1 is in state (I,1_c); that is, T_1 is regular.

(b) If instead of the boundary condition in (a) we impose $u(0) = au(1)$, where a is a fixed real number, we have a new operator A_2 which is also closable and whose closure is denoted by T_2. The boundary condition survives in the process of closure. If $a \neq 1$, the one and only solution of $T_2 u = f$ is $u = \int_0^1 g(x,y) f(y)\,dy$, where $g = a/(1-a)$ for $x < y$, and $g = 1/(1-a)$ for $x > y$. Obviously, g is just Green's function for the problem. We see that $R_{T_2} = L_2$ and that T_2 is regular. If $a = 1$, then $u = C$ is a solution of the homogeneous equation and T_2 is in state (III,2_c).

(c) If instead of the boundary condition in (a) we impose the two boundary conditions $u(0) = 0$ *and* $u(1) = 0$, we obtain an operator A_3 which is closable, with closure T_3. Again, boundary conditions survive the process of closure, but now $R_{T_3} \neq L_2$. The unique solution of $T_3 u = f$ is as in (a), but a solution is now possible only if f satisfies the solvability condition $\int_0^1 f(x)\,dx = 0$. Thus T_3 is in state (I,2_c).

(d) If we impose the boundary condition $u'(0) = 0$ instead of $u(0) = 0$, we discover a new phenomenon. The operator A_4 is closable, but the boundary condition $u'(0) = 0$ *disappears* in the process of closure, so that the closure T_4 of A_4 is just the same as the closure of the differentiation operator without boundary conditions. Thus T_4 is in state (III,1_c).

We have the following general principle (stated somewhat loosely) for linear differential operators of order n: Boundary conditions involving derivatives of order less than n survive closure, but those involving derivatives of order greater than or equal to n disappear in closure.

Example 2. Let Ω be an open set in $\mathbb{R}^n$, and let Γ be its boundary. We are interested in the following boundary value problem:

$$-\Delta u = -\left(\frac{\partial^2 u}{\partial x_1^2} + \cdots + \frac{\partial^2 u}{\partial x_n^2}\right) = f(x), \quad x = (x_1, \ldots, x_n) \in \Omega, \quad u|_{x \text{ on } \Gamma} = 0.$$

This is just the Dirichlet problem for Laplace's equation (actually, we have, for convenience, written the equation with the negative Laplacian on the left side; obviously, the change $f \to -f$ restores the equation to the form $\Delta u = f$). To treat this problem in a suitable functional framework, we first define an operator $A = -\Delta$ on the domain D_A of functions having continuous partial derivatives of order 2 on $\bar{\Omega}$, and vanishing on Γ. The closure T of this operator is defined on a domain $D_T \supset D_A$. Although it is possible to develop the theory in this manner, the modern theory of partial differential equations uses a different approach (see Section 8.4).

EXERCISES

5.3.1 Let $\{e_i\}$ be an orthonormal basis for $L_2(a, b)$, and let the operator A be defined by

$$Ae_1 = 0; \quad Ae_k = \frac{e_k}{k}, \quad k \geqslant 2.$$

Show that A is in state (III,2_n).

5.3.2 Let A be a linear transformation from E_n to E_m. Describe all possible states of A.

5.3.3 On $L_2(0, \pi)$ consider the integral operator K with kernel

$$k(x, y) = \sum_{n=1}^{\infty} \frac{\cos nx \cos ny}{n^2}.$$

Show that K is in state (III,2_n).

5.3.4 On $L_2(0, 1)$ consider the operator A defined by

$$Au = e^x u - au,$$

where a is a real constant. Discuss the state of the operator for different values of a.

5.4 ADJOINT OPERATORS

Consider first a bounded operator A defined on the whole Hilbert space H. With v a fixed element in H, we can regard $\langle Au, v\rangle$ as a complex number which varies with u; it is clear that $\langle Au, v\rangle$ is a bounded linear functional in u, and so, by the Riesz representation theorem of Chapter 4, there must exist a well-defined element g in H such that

$$\langle Au, v\rangle = \langle u, g\rangle \quad \text{for all } u \text{ in } H. \tag{5.4.1}$$

The element g depends on v, and we write $g = A^*v$, where A^* is seen to be a linear operator defined on the whole of H. The operator A^* is known as the *adjoint* of A; A^* is bounded and $\|A^*\| = \|A\|$ (see Theorem 5.8.1). If u, v are any two elements in H, we have

$$\langle Au, v\rangle = \langle u, A^*v\rangle. \tag{5.4.2}$$

The matrix representation $[a_{ij}^*]$ of A^* in an orthonormal basis $\{e_i\}$ is easily calculated in terms of the matrix $[a_{ij}]$ of A in the same basis. From (5.1.4) we have

$$a_{ij} = \langle Ae_j, e_i\rangle = \langle e_j, A^*e_i\rangle = \overline{\langle A^*e_i, e_j\rangle},$$

so that

$$a_{ij} = \overline{a_{ji}^*} \quad (\text{or } a_{ij}^* = \bar{a}_{ji}). \tag{5.4.3}$$

If we are dealing with a real Hilbert space, the matrix entries are real and the complex conjugate may be omitted in (5.4.3). The matrix representation is a satisfactory way to describe adjoints in finite-dimensional spaces and in some infinite-dimensional problems like Example 9, Section 5.1, where an easy calculation shows that A^* is the *left* shift defined by $A^*e_i = e_{i-1}, i > 1; A^*e_1 = 0$. In other problems a more striking form for the adjoint is found by using (5.4.2) directly. Turning to Example 6, Section 5.1, we have, for any $u,\ z$,

$$\begin{aligned}\langle Ku, z\rangle &= \int_a^b dx\ \bar{z}(x) \int_a^b k(x, y)u(y)\, dy = \int_a^b u(y)\, dy \int_a^b k(x, y)\bar{z}(x)\, dx \\ &= \int_a^b u(x)\, dx \int_a^b k(y, x)\bar{z}(y)\, dy,\end{aligned}$$

where the second equality results from a change in the order of integration, whereas the last equality is merely a convenient relabeling of variables enabling us to identify the last term as $\int_a^b u(x)\overline{K^*z}\, dx$, where

$$K^*z = \int_a^b \bar{k}(y, x)z(y)\, dy. \tag{5.4.4}$$

Thus K^* is also an integral operator on $L_2(a, b)$ with kernel

$$k^*(x, y) = \bar{k}(y, x), \tag{5.4.5}$$

a formula which bears comparison with (5.4.3) for matrices.

As a specific illustration of these formulas, consider the operator A of Example 7, Section 5.1, which can be regarded as an integral operator on $L_2(0,1)$ with kernel $k(x,y) = H(x-y)$. Thus $k^*(x,y) = H(y-x)$, and (5.4.4) becomes

$$A^*z = \int_0^1 H(y-x)z(y)\,dy = \int_x^1 z(y)\,dy.$$

We now turn to unbounded operators and confine ourselves to closed operators defined on domains dense in H. Let A on D_A be such an operator, and, as before, let v be a fixed element in H. As u varies over D_A, $\langle Au, v\rangle$ takes on various numerical values, so that $\langle Au, v\rangle$ is a functional (possibly unbounded) in u. Although no longer guaranteed by the Riesz theorem, it may happen that for some elements v we can write

$$\langle Au, v\rangle = \langle u, g\rangle \quad \text{for all } u \text{ in } D_A. \tag{5.4.6}$$

Any pair (v, g) for which (5.4.6) holds for all u in D_A will be called an *admissible pair*. The pair (0, 0) is admissible, but it is not clear that there are other admissible pairs. It is true, however, that g is unambiguously determined from v: If (5.4.6) were true for the same v and for different g_1 and g_2, we would have $\langle u, g_1 - g_2\rangle = 0$ for all u in D_A, and since D_A is dense in H, $g_1 - g_2 = 0$. Thus, for a given v, there is at most one g for which (5.4.6) holds. Considering *all* admissible pairs, we see that g depends on v, and therefore we can write $g = A^*v$. The operator A^* is the *adjoint* of A; it has a well-defined domain D_{A^*} which includes at least the element 0, and $A^*0 = 0$. The operator A^* is linear, and we have

$$\langle Au, v\rangle = \langle u, A^*u\rangle \quad \text{for all } u \text{ in } D_A, \quad v \text{ in } D_{A^*}. \tag{5.4.7}$$

It can also be shown that A^* is closed and that its domain D_{A^*} is dense in H.

We say that A on D_A is *self-adjoint* if $D_{A^*} = D_A$ and $A^* = A$ on their common domain of definition. If A is bounded, $D_{A^*} = D_A = H$, and therefore A is self-adjoint if and only if

$$\langle Au, v\rangle = \langle u, Av\rangle \quad \text{for all } u, v \in H. \tag{5.4.8}$$

An operator is *symmetric* on its domain of definition if

$$\langle Au, v\rangle = \langle u, Av\rangle \quad \text{for all } u, v \in D_A. \tag{5.4.9}$$

Every self-adjoint operator is symmetric and every *bounded* symmetric operator is self-adjoint, but there are unbounded symmetric operators which are *not* self-adjoint. (See Example 3 below.) A bounded operator will be self-adjoint if its matrix representation in an orthonormal basis satisfies $a_{ij} = \bar{a}_{ji}$. For a bounded integral operator K the condition of self-adjointness is that the kernel satisfies $k(x,y) = \bar{k}(y,x)$.

If α is a scalar, we define αA and $A + \alpha I$ in the obvious way: $(\alpha A)(u) = \alpha Au$, $(A + \alpha I)(u) = Au + \alpha u$; both operators are defined on the same domain D_A as A. We then have

$$(\alpha A)^* = \bar{\alpha}A^*, \quad (A + \alpha I)^* = A^* + \bar{\alpha}I, \tag{5.4.10}$$

with $(\alpha A)^*$ and $(A + \alpha I)^*$ both having the same domain D_{A^*} as A^*.

Properties of the Adjoint

1. If A is closed and D_A dense in H, A^* is closed and D_{A^*} is dense in H. Moreover, $A^{**} = A$.

2. If A is closable, both A and its closure have the same adjoint (therefore, if A is closable on a dense domain D_A, we can find the closure of A by calculating A^{**} instead).

Example 1. We look at the closed operator T_1 of Example 1(a), Section 5.3, where $T_1 u = du/dx$ and D_{T_1} is the linear manifold in $L_2^{(c)}(0,1)$ consisting of absolutely continuous functions $u(x)$ with $u'(x)$ in $L_2^{(c)}(0,1)$ and $u(0) = 0$. The important thing is that D_{T_1} contains all sufficiently smooth functions with $u(0) = 0$. We want to calculate T_1^*. We must find *all* admissible pairs (v, g) satisfying

$$\langle T_1 u, v\rangle = \langle u, g\rangle \quad \text{for all } u \text{ in } D_{T_1};$$

that is,

$$\int_0^1 \bar{v}u' \, dx = \int_0^1 u\bar{g} \, dx \quad \text{for all } u \text{ in } D_{T_1}. \tag{5.4.11}$$

We give two methods (one heuristic, the other precise) for finding the adjoint. The types of argument are similar to those used in variational approaches (see Section 8.4).

Heuristic Method

We integrate by parts on the left side of (5.4.11) (this assumes v absolutely continuous with v' in L_2) to obtain, using the fact that $u(0) = 0$,

$$\int_0^1 (\bar{v}' + \bar{g})u \, dx - \bar{v}(1)u(1) = 0 \quad \text{for all } u \text{ in } D_{T_1}. \tag{5.4.12}$$

Let us examine the consequences of this equation as u varies in D_{T_1}. If we consider first the subset M of functions on D_{T_1} that also vanish at $x = 1$, we see that the integral in (5.4.12) vanishes for $u \in M$. This implies that $\bar{v}' + \bar{g} = 0$, so that the integral term must vanish identically; if we then turn to the subset of functions u in D_{T_1} for which $u(1) \neq 0$, we find that $v(1) = 0$. Thus (v, g) is admissible if $g = -v'$ *and* $v(1) = 0$ and v is absolutely continuous with v' in L_2. We therefore suspect that

$$T_1^* v = -\frac{dv}{dx}, \quad D_{T_1^*} : v(1) = 0, \quad v \text{ a.c., } v' \in L_2. \tag{5.4.13}$$

Of course, it is not clear that we have found *all* admissible pairs in this way.

Precise Method

We start from (5.4.11), but do not assume anything about v. We integrate by parts on the *right* side of (5.4.11); with $G(x) = -\int_x^1 g(y)\,dy$, we see that G' exists, $G' = g$, and $G(1) = 0$, so that, since $u(0) = 0$,

$$\int_0^1 u'[\bar{v} + \bar{G}]\,dx = 0 \quad \text{for all } u \text{ in} D_{T_1}. \tag{5.4.14}$$

It is perhaps obvious that this implies that $\bar{v} + \bar{G} = 0$, but we can prove it as follows. Let $u = \int_0^x [v(y) + G(y)]\,dy$; then $u \in D_{T_1}$ and $u' = v + G$; substituting this particular u in (5.4.14), we see that $\int_0^1 |v + G|^2\,dx = 0$ and therefore

$$v + G = 0 \quad \text{or} \quad v = \int_x^1 g(y)\,dy.$$

Thus v is necessarily differentiable with $v' = -g$ and $v(1) = 0$, showing that (5.4.13) does in fact define the adjoint T_1^*. In other examples we shall be satisfied with the heuristic method for determining the adjoint. Note that T_1 is not self-adjoint for two reasons: first, the formal differentiation in T_1^* is $-(d/dx)$ rather than d/dx, and second, the boundary condition on D_{T_1} is transferred to the other end of the interval in $D_{T_1}^*$. For the operator $S_1 = -i(d/dx) = -iT_1$, defined on the same domain as T_1, we obtain, using (5.4.10),

$$S_1^* v = -i\frac{dv}{dx}, \quad D_{S_1^*} : v(1) = 0, \quad v \text{ a.c.}, \; v' \in L_2.$$

Since the boundary condition is not the same as for S_1, we still find that S_1 is not self-adjoint, although at least S_1^* consists of the same formal differentiation as S_1.

Example 2. In the calculation of the adjoint for the closed operator T_2 of Example 1(b), Section 5.3, with the boundary condition $u(0) = au(1)$, we see that (5.4.11) is to hold for all u in D_{T_2}. From this we conclude that

$$\int_0^1 (\bar{v}' + \bar{g})u\;dx - u(1)[\bar{v}(1) - a\bar{v}(0)] = 0 \quad \text{for all } u \text{ in } D_{T_2}.$$

Therefore, admissible pairs (v, g) satisfy

$$g = -v', \quad v(1) = av(0).$$

Thus

$$T_2^* v = -\frac{dv}{dx}, \quad D_{T_2^*} : v(1) = av(0), \quad v \text{ a.c.}, \; v' \in L_2.$$

If we let $S_2 = -iT_2$, then

$$S_2^* v = -i\frac{dv}{dx}, \quad D_{S_2^*} : v(1) = av(0), \quad v \text{ a.c}, \; v' \in L_2.$$

Therefore, S_2 is *self-adjoint* if and only if $a = 1$ or $a = -1$.

Example 3. Consider the adjoint for the operator T_3 of Example 1(c), Section 5.3, where "too many" boundary conditions have been imposed. We still have (5.4.11), but now for $u \in D_{T_3}$. Integration by parts yields

$$\int_0^1 (\bar{v}' + \bar{g})u \, dx = 0 \quad \text{for all } u \text{ in } D_{T_3},$$

from which we conclude that $v' = -g$, but that no boundary conditions are needed on v. Thus

$$T_3^* v = -\frac{dv}{dx}, \quad D_{T_3^*} : v \text{ a.c., } v' \in L_2.$$

For the operator $S_3 = -iT_3$, we have

$$S_3 u = -i\frac{du}{dx}, \quad D_{S_3} : u(0) = u(1) = 0, \quad u \text{ a.c., } u' \in L_2,$$

$$S_3^* v = -i\frac{dv}{dx}, \quad D_{S_3^*} : v \text{ a.c., } v' \in L_2.$$

The operator S_3 is an example of a symmetric operator which is not self-adjoint. It is clear that if $u, v \in D_{S_3}$, then $\langle S_3 u, v\rangle = \langle u, S_3 v\rangle$, but $S_3^* \neq S_3$. In fact, S_3^* is an extension of S_3. For any symmetric operator A it is always true that A^* is an extension of A.

EXERCISES

5.4.1 If A, B are bounded operators on H, show that

$$(AB)^* = B^* A^*.$$

Even if A, B are both self-adjoint, the product AB may not be.

5.4.2 Let A be a symmetric operator on $E_n^{(c)}$, and let $\{h_k\}$ be a *nonorthogonal* basis in which A has the matrix $[a_{ij}]$. What is the matrix of A^* in the same basis?

5.4.3 Let $H = L_2^{(c)}(0, 1)$, and let D be the linear manifold of elements $u(x)$ with u' absolutely continuous and u'' in L_2. Consider the operator $A = -(d^2/dx^2)$ on the domain D_A of functions in D satisfying the boundary conditions $u(0) = u(1) = u'(0) = 0$. Is A symmetric? What is the adjoint of A?

5.4.4 Let $a(x)$ be continuous but *not* differentiable on $0 \leqslant x \leqslant 1$. Let $A = a(x)(d/dx)$ be an operator defined on the domain D_A consisting of elements $u(x)$ in $L_2^{(c)}(0, 1)$ that are absolutely continuous with u' in L_2. Find the adjoint operator A^*.

5.4.5 Consider the differentiation operator D on $L_2(-\infty, \infty)$ restricted to functions that are differentiable with derivative in $L_2(-\infty, \infty)$. Prove that D is unbounded. For what $f \in L_2(-\infty, \infty)$ is the equation $Du = f$ solvable?

5.4.6 Consider the translation operator T_a defined by $T_a[u(x)] = u(x - a)$ on $L_2(-\infty, \infty)$. What is its adjoint? Its inverse?

5.5 SOLVABILITY CONDITIONS

Let A be a linear operator on a domain D_A dense in H. We want to characterize the range of A: For what forcing terms f can we solve the inhomogeneous equation

$$Au = f, \quad u \in D_A? \tag{5.5.1}$$

The statement $u \in D_A$ is solely for emphasis; it is already incorporated in the equation $Au = f$ since A is defined only on D_A. It is convenient to study at the same time the adjoint homogeneous equation

$$A^*v = 0, \quad v \in D_{A^*}, \tag{5.5.2}$$

where again the statement $v \in D_{A^*}$ is for emphasis only. Observe that (5.5.2) always has the solution $v = 0$ and perhaps some others.

If u is a solution of (5.5.1) and v a solution of (5.5.2), we have, by taking inner products,

$$\langle Au, v\rangle = \langle f, v\rangle.$$

From (5.4.7) and (5.5.2) it follows that $\langle Au, v\rangle = \langle u, A^*v\rangle = 0$, so that a *necessary* condition for (5.5.1) to have solution(s) is that

$$\langle f, v\rangle = 0 \quad \text{for all } v \text{ satisfying } (5.5.2). \tag{5.5.3}$$

Of course, the *solvability condition* (5.5.3) has content only if (5.5.2) has nontrivial solutions. Some questions immediately come to mind. Is (5.5.3) sufficient for solvability of (5.5.1)? How many nontrivial solutions does (5.5.2) have? If (5.5.1) has solutions, how many are there?

We shall establish the sufficiency of (5.5.3) for particular classes of operators, but first we prove a general result which characterizes $(R_A)^\perp$ instead of R_A.

Theorem. *The orthogonal complement of the range of A is the null space of the adjoint:*

$$(R_A)^\perp = N_{A^*}. \tag{5.5.4}$$

Proof. (a) Let $z \in N_{A^*}$, so that $A^*z = 0$. Then $\langle u, A^*z\rangle = 0$ for every u in H, and hence surely for every u in D_A; hence $\langle Au, z\rangle = 0$ for every $u \in D_A$, so that z is in $(R_A)^\perp$.

(b) Let $z \in R_A^\perp$; then $\langle Au, z\rangle = 0 = \langle u, 0\rangle$ for every u in D_A. Thus $(z, 0)$ is an admissible pair, which means that $z \in D_{A^*}$ and $A^*z = 0$. Therefore, $z \in N_{A^*}$.

Combining (a) and (b), we see that the theorem has been proved.

□

By taking orthogonal complements we find that $(R_A)^{\perp\perp} = (N_{A^*})^\perp$. But also $(R_A)^{\perp\perp} = \bar{R}_A$, so that

$$\bar{R}_A = (N_{A^*})^\perp. \tag{5.5.5}$$

This result could have been proved directly by a slight modification of the proof of the Fredholm alternative for linear systems of algebraic equations in Section 3.4 (see Exercise 5.5.1). Unfortunately, (5.5.5) does not characterize R_A but only $\bar{R}_A$, but it is the best that can be done in general. If, however, R_A is a closed set, $R_A = \bar{R}_A$ and we have the following necessary and sufficient solvability condition.

Theorem 5.5.1 (Solvability for operators with closed range)**.** *If A has a closed range, $Au = f$ has solution(s) if and only if f is orthogonal to every solution of the adjoint homogeneous equation; that is,*

$$R_A = (N_{A^*})^{\perp}. \tag{5.5.6}$$

When does A have a closed range? Clearly, if A is an operator on E_n or, more generally, an operator with a finite-dimensional range, A obviously has a closed range. Theorem 5.3.5 gives an important criterion for an operator to have a closed range. This theorem can be generalized as follows.

Theorem 5.5.2. *Let A be closed, and let A be bounded away from 0 on $N_A^{\perp} \cap D_A$. Then R_A is closed.*

REMARK. The hypothesis tells us there exists $c > 0$ such that $\|Au\| \geqslant c\|u\|$ for all $u \in N_A^{\perp} \cap D_A$. This means that the operator A *restricted to* $N_A^{\perp} \cap D_A$ has a bounded inverse; in particular, if N_A is $\{0\}$, the criterion is just that A on D_A has a bounded inverse.

Proof. Let $f_n \in R_A$ with $f_n \to f$. Let $Av_n = f_n$, and define $u_n = v_n - Pv_n$, where Pv_n is the projection of v_n on N_A; then u_n is in $D_A \cap N_A^{\perp}$ and $Au_n = f_n$. Now $\|u_n - u_m\| \leqslant (1/c)\|Au_n - Au_m\| = (1/c)\|f_n - f_m\|$, so that $\{u_n\}$ is a Cauchy sequence in D_A and therefore $u_n \to u, u$ in H. Therefore, we have $Au_n \to f$, $u_n \to u$; by the definition of a closed operator this means that $u \in D_A, Au = f$, so that R_A is closed. □

We are now in a position to prove the complete Fredholm alternative for various classes of operators. We do so for operators A on E_n; the proofs given are not the simplest possible but are the ones most easily extended to wider classes of operators (see Section 5.7). The theorem deals with the relationship among the three equations

$$Au = 0, \quad Au = f, \quad A^*v = 0.$$

Theorem 5.5.3 (Fredholm Alternative)**.** *Let A be an operator on $E_n = H$. Then* either *of the following holds:*

(a) $Au = 0$ has only the zero solution, in which case $A^*v = 0$ has only the zero solution and $Au = f$ has precisely one solution for each f in H.

(b) N_A has dimension k, in which case N_{A^*} also has dimension k and $Au = f$ has solutions if and only if f is orthogonal to N_{A^*}.

REMARK. The difference between this version of the Fredholm alternative and the one in Section 3.4 is that we also show here that the dimensions of N_A and N_{A^*} are equal for a balanced system. We can elaborate here a little on part (b). Let $\{u_1, \ldots, u_k\}$ be a basis for N_A, and let $\{v_1, \ldots, v_k\}$ be a basis for N_{A^*}. Then the solvability conditions for the equation $Au = f$ consist of the k equations $\langle f, v_1\rangle = \cdots = \langle f, v_k\rangle = 0$, which guarantee that f is orthogonal to N_{A^*}. *If these conditions are satisfied*, $Au = f$ has many solutions, all of which are of the form

$$u = \tilde{u} + c_1 u_1 + \cdots + c_k u_k,$$

where $c_1, \ldots, c_k$ are arbitrary constants and $\tilde{u}$ is any particular solution of $Au = f$.

Proof. (Theorem 5.5.3) The proof consists of three steps.

1. $Au = f$ has solution(s) if and only if $\langle f, v\rangle = 0$ for every solution of $A^*v = 0$, that is, $R_A = (N_{A^*})^\perp$. This is just Theorem 5.5.1, which requires only that the range be closed, a trivial fact for operators on E_n.

2. If $R_A = H$, then $N_A = \{0\}$, and vice versa. We already know from step 1 that $N_{A^*} = \{0\}$ is necessary and sufficient for $R_A = H$. Thus, if we manage to prove step 2, we will also have shown that

$$N_A = \{0\} \Leftrightarrow N_{A^*} = \{0\}.$$

We now give the proof.

(a) Suppose that $R_A = H$. We must show that $N_A = \{0\}$. Assuming the contrary, there exists $u_1 \neq 0$ with $Au_1 = 0$. Consider successively the equations $Au_2 = u_1, Au_3 = u_2, \ldots, Au_p = u_{p-1}, \ldots$. By hypothesis each equation is solvable, and $u_p \neq 0$, $A^{p-1}u_p = u_1, A^p u_p = 0$. Thus u_p belongs to the null space N_p of A^p. Obviously, $N_p \supset N_{p-1}$, and the inclusion is strict since u_p belongs to N_p but not to N_{p-1}. We are thus led to an infinite sequence of spaces $N_1, N_2, \ldots$ of increasing dimension, a circumstance which violates the fact that H is finite-dimensional. Therefore, we must have $N_A = \{0\}$.

(b) Assume that $N_A = \{0\}$. We must show that $R_A = H$. The hypothesis implies that $N_A^\perp = H$. Reversing the roles of A and A^* in step 1 (which is permissible since $A^{**} = A$), we have $N_A^\perp = R_{A^*} = H$. Applying part (a) above to A^*, we find that $N_{A^*} = \{0\}$.

3. N_A and N_{A^*} have the same dimension. By step 2 we know that if one of these has dimension 0, so does the other. Thus we need to consider only the case where both dimensions are positive. We shall proceed by contradiction. Let ν, ν^* be the respective dimensions, and assume, without loss of generality, that $\nu^* > \nu$. Choose orthonormal bases $\{u_1, \ldots, u_\nu\}$ and $\{v_1, \ldots, v_{\nu^*}\}$ for N_A and N_{A^*}, respectively. Consider the operator B defined by

$$Bu = Au - \sum_{j=1}^{\nu} \langle u_j, u\rangle v_j.$$

Then $\langle Bu, v_k\rangle = -\langle u_k, u\rangle$ for $k \leqslant \nu$, and $\langle Bu, v_k\rangle = 0$ for $k > \nu$. If $Bu = 0$, then $\langle u_k, u\rangle = 0$ for $k \leqslant \nu$ and therefore $Au = 0$ for $u \in N_A$; since u is also orthogonal to a basis for N_A, it follows that $u = 0$. Thus $Bu = 0$ implies that $u = 0$, but by step 2 we can then solve $Bu = v_{\nu+1}$, which yields $\|v_{\nu+1}\|^2 = \langle Bu, v_{\nu+1}\rangle = 0$, a contradiction.

These three steps together prove the Fredholm alternative.

□

REMARKS

1. Theorem 5.5.3 can be phrased in the language of systems of algebraic equations (see also Section 3.4). We consider the relationship among these three systems:

$$\sum_{j=1}^{n} a_{ij}u_j = 0, \quad i = 1, \ldots, n, \tag{5.5.7}$$

$$\sum_{j=1}^{n} a_{ij}u_j = f_i, \quad i = 1, \ldots, n, \tag{5.5.8}$$

$$\sum_{j=1}^{n} \bar{a}_{ji}v_j = 0, \quad i = 1, \ldots, n. \tag{5.5.9}$$

The theorem states that *either* of the following holds:

(a) The homogeneous system (5.5.7) has only the trivial solution

$$u_1 = \cdots = u_n = 0,$$

in which case (5.5.9) has only the trivial solution and (5.5.8) has precisely one solution.

(b) The homogeneous system (5.5.7) has k independent nontrivial solutions, say $(h_1^{(1)}, \ldots, h_n^{(1)}), \ldots, (h_1^{(k)}, \ldots, h_n^{(k)})$, in which case (5.5.9) also has k independent nontrivial solutions, say $(p_1^{(1)}, \ldots, p_n^{(1)}), \ldots, (p_1^{(k)}, \ldots, p_n^{(k)})$, and (5.5.8) has solutions if and only if $\sum_{i=1}^{n} f_i\bar{p}_i^{(j)} = 0, j = 1, \ldots, k$. If these solvability conditions are satisfied, the general solution of (5.5.8) is of the form $u = u_p + \sum_{i=1}^{k} c_i h^{(i)}$, where u_p is any particular solution of (5.5.8).

As we know from elementary considerations, alternative (a) occurs if the determinant of $[a_{ij}]$ does *not* vanish. Alternative (b) occurs if the determinant vanishes. The number k of independent solutions of (5.5.9) is then $n - r$, where r is the *rank* of $[a_{ij}]$.

2. In Section 5.7 the theorem will be extended to operators A of the form $K-\lambda I$, where K is compact and $\lambda \neq 0$.

EXERCISES

5.5.1 Prove (5.5.5) directly by adapting the proof of the Fredholm alternative in Section 3.4.

5.5.2 Consider the right-shift operator A of Example 9, Section 5.1. Find A^*, N_A, N_{A^*}, R_A, R_{A^*}; verify (5.5.6) and the relation $R_{A^*} = (N_A)^\perp$.

5.5.3 With apologies for the inelegant notation, let P_n be the space of real-valued polynomials of degree less than n. Thus, an element of P_n is of the form

$$\xi_1 + \xi_2 x + \cdots + \xi_n x^{n-1},$$

where $(\xi_1, \xi_2, \ldots, \xi_n)$ is an n-tuple of real numbers. In what follows, we identify P_n with $\mathbb{R}^n$, the space of n-tuples of real numbers, for which the natural basis is $(1, 0, \ldots, 0), \ldots, (0, \ldots, 0, 1)$. Consider the differentiation operator D on P_n. Characterize N_D and R_D. In the natural basis, what is the matrix of D of D^*? Check that R_D and N_{D^*} are orthogonal complements. Solve the equation $Du = f$ when $f \in R_D$ and write the minimum norm solution as $u = D^{-1}f$, where D^{-1} is the pseudoinverse introduced in Exercise 3.4.4.

Note that D^{-1} is defined on all of P_n and can be used to provide an approximate solution to $Du = f$ even when $f \notin R_D$. Then $D^{-1}f$ is the least-squares solution of minimum norm. We have found an element u which has minimum norm among those for which the residual $\|Au - f\|$ is least. Note that D^{-1} depends on the norm used and will change if, for instance, we regard the space P_n as a subset of $L_2(-1, 1)$ with its attendant inner product and norm.

5.5.4 (a) The notion of pseudoinverse can be extended to bounded operators with closed range on infinite-dimensional Hilbert spaces. Let A be a bounded operator on the Hilbert space H. Then A^* is also bounded and the sets N_A, N_{A^*} are closed. *Suppose that, in addition, R_A is a closed set.* It follows from (5.5.5) that R_A and N_{A^*} are orthogonal complements. Consider now the inhomogeneous problem

$$Au = f. \tag{5.5.10}$$

If $f \in R_A$, (5.5.10) has solution(s) which can be written as $u = w + w^\perp$, where w is an arbitrary element in N_A and $w^\perp \in N_A^\perp$ and is uniquely determined. Indeed, $w^\perp$ is the minimum norm solution of (5.5.10) and can be written as Gf, where G takes R_A into $N_A^\perp$ and is one-to-one. Thus, if A is restricted to $N_A^\perp$, G is its inverse. Even if f is not in R_A we can find an approximate solution of (5.5.10) by projecting f on the closed linear manifold R_A. Writing $f = h + h^\perp$ where $h \in R_A$ and $h^\perp \in N_{A^*}$, we let $u = Gh$ be the desired approximate solution (having both least residual and minimum norm). The operator G is defined on all of H and is the *pseudoinverse* of A.

(b) Use the ideas of part (a) for the operator on $L_2^{(r)}(-1,1)$ defined by

$$Au = -\frac{2}{3}u(x) + \int_{-1}^{1} y^2 u(y)\, dy.$$

Determine $A^*, N_A, R_A, N_{A^*}, R_{A^*}$ and verify (5.5.4) and (5.5.6). Find the solutions of $Au = f$ when f is in R_A and determine the minimum norm solution. Find the pseudoinverse G and use it to "solve" $Au = f$ even when $f \notin R_A$.

5.6 THE SPECTRUM OF AN OPERATOR

The principal purpose in studying the spectrum is to make precise and to generalize, so far as possible, the method of eigenfunction expansion. The program can be successfully completed for self-adjoint operators, culminating in the spectral theorem for such operators. In applications it is not enough, however, to stop at the spectral theorem; the important thing is to construct explicitly the eigenfunctions (or the generalization referred to earlier). Although we give some restricted forms of the spectral theorem, we devote greater energy in Chapter 7 to methods for constructing the spectrum and the appropriate series or integral expansions needed in specific problems.

Let A be a closed operator whose domain is a linear manifold D_A dense in H. Our attention will be focused on the operator $A-\lambda I$, where λ is an arbitrary *complex* number and I is the identity operator. For any λ we may and shall take the domain of $A-\lambda I$ to be the same as that of A. This domain will be denoted simply by D. The range of $A-\lambda I$ will, however, usually vary with λ, and we shall therefore denote the range by $R(\lambda)$. Since $A-\lambda I$ is closed and defined on a dense set, its adjoint $A^*-\bar{\lambda}I$ is defined on a domain D^* which is dense in H. We now classify all points λ in the complex plane according to whether or not $A-\lambda I$ is regular.

Definition. *The values of λ for which $A-\lambda I$ is regular, that is, $A-\lambda I$ has a bounded inverse and $R(\lambda) = H$, form the* resolvent set *of A. All other values of λ comprise the* spectrum *of A. It can be shown that the resolvent set is* open *and therefore the spectrum is* closed.

Perhaps it is time to revise our earlier attitude, which implied that being in the resolvent set was marvelous and being in the spectrum was bad. If there were no spectrum at all, we could not develop anything resembling an eigenfunction expansion!

There are many competing classification schemes for further division of the spectrum, all of them based on the states of closed operators as given in Section 5.3. The one used here is a slight modification of the scheme proposed by Halmos [4] and seems intuitively the simplest. Its disadvantage is that it divides the spectrum into sets which are *not* necessarily disjoint.

Definition. *Values of λ in the spectrum are classified as follows:*

1. *λ belongs to the* point spectrum *if $A-\lambda I$ is in state III [that is, $(A-\lambda I)u=0$ has nontrivial solutions; in other words, λ is an* eigenvalue *of A and any corresponding nontrivial solution u is an* eigenvector *of A belonging to λ; the (geometric) multiplicity of λ is the dimension of the null space $N(\lambda)$ of $A-\lambda I$].*

2. *λ belongs to the* continuous spectrum *if $A-\lambda I$ is in state II [that is, $(A-\lambda I)^{-1}$ exists but is unbounded].*

3. *λ belongs to the* compression spectrum *if $A-\lambda I$ is in state 2 [that is, $\bar{R}(\lambda) \neq H$; the range has been compressed; thus $[R(\lambda)]^{\perp}$ has a nonzero dimension, the* deficiency *of λ, which is just the codimension of $R(\lambda)$].*

The point spectrum and the continuous spectrum are disjoint sets in the λ plane. Their union is the *approximate point spectrum.* For λ to be in the approximate spectrum, $A-\lambda I$ must fail to have a bounded inverse or, equivalently, $A-\lambda I$ must not be bounded away from 0. Thus λ is in the approximate spectrum if and only if there is a sequence of unit elements $\{u_n\}$ such that $Au_n - \lambda u_n \to 0$. (This criterion is automatically satisfied if there exists $u \neq 0$ such that $Au - \lambda u = 0$; just choose $u_n = u/\|u\|$.) The approximate point spectrum often overlaps the compression spectrum (see Figure 5.1). In fact, for transformations on $E_n^{(c)}$ we have shown (Theorem 5.5.3) that if λ is an eigenvalue of multiplicity m, it also has deficiency m. Thus for a transformation on $E_n^{(c)}$, either λ is in the resolvent set or it belongs *simultaneously* to the point and compression spectrum. On an arbitrary Hilbert space, we can prove the following theorem.

Theorem 5.6.1. *If λ has deficiency m in the compression spectrum of A, then $\bar{\lambda}$ is an eigenvalue of A^* with multiplicity m, and conversely.*

Proof. We have shown [see (5.5.4)] that the orthogonal complement of the range is equal to the null space of the adjoint. Apply this theorem to the operator $A-\lambda I$. □

Theorem 5.6.2. *If A is symmetric, the following hold:*

(a) $\langle Au, u\rangle$ is real for all $u \in D$.

(b) The approximate spectrum (which consists of the eigenvalues and of the continuous spectrum) is real.

(c) Eigenvectors corresponding to different eigenvalues are orthogonal.

Proof.

(a) For u in D, $\langle Au, u\rangle = \langle u, Au\rangle = \overline{\langle Au, u\rangle}$, the first equality from symmetry and the second from the definition of inner product. Thus $\langle Au, u\rangle$ is real.

(b) If $Au = \lambda u$, then $\langle Au, u\rangle = \lambda\|u\|^2$, and for $\|u\| \neq 0$ we have λ real. With $\lambda = \xi + i\eta, \xi$ and η real, we find that

$$\|(A-\lambda I)u\|^2 = \langle(Au-\xi u)-i\eta u, (Au-\xi u)-i\eta u\rangle = \|Au-\xi u\|^2+\eta^2\|u\|^2,$$

where the symmetry of $A - \xi I$ has been used to eliminate cross terms. Thus $\|(A-\lambda I)u\|^2 \geqslant \eta^2\|u\|^2$, and $A - \lambda I$ is bounded away from 0 for $\eta \neq 0$. By Theorem 5.3.2, $A - \lambda I$ has a bounded inverse and the approximate spectrum is therefore confined to the real axis.

(c) Let $\lambda \neq \mu$ and $Au = \lambda u, Av = \mu v$. Then $\langle Au, v\rangle - \langle u, Av\rangle = (\lambda - \bar{\mu})\langle u, v\rangle$. The left side vanishes by symmetry. Since μ is real, $\lambda \neq \bar{\mu}$ and therefore $\langle u, v\rangle = 0$.

□

For a self-adjoint operator we can say more.

Theorem 5.6.3. *If A is self-adjoint, every λ with Im $\lambda \neq 0$ is in the resolvent set. The point spectrum (necessarily real) coincides with the compression spectrum, and the multiplicity of any eigenvalue is equal to its deficiency.*

Proof. Since $A = A^*$, A is certainly symmetric. By Theorem 5.6.2, λ with Im $\lambda \neq 0$ is either in the resolvent set or the compression spectrum. The latter would imply that A^* (hence A) has $\bar{\lambda}$ as an eigenvalue, which is a contradiction. The whole spectrum is therefore real. By Theorem 5.6.1 the point and compression spectra coincide, and the multiplicity of an eigenvalue is equal to its deficiency. We may therefore omit reference to the compression spectrum. □

In what follows, we refer to our standard examples, Examples 1 through 10 of Section 5.1.

Example 1. See Example 1, Section 5.1.

(a) A is the zero operator. The only point in the spectrum is $\lambda = 0$, which is an eigenvalue of infinite multiplicity.

(b) A is the identity. The only point in the spectrum is $\lambda = 1$, an eigenvalue of infinite multiplicity.

Example 2. See Example 2, Section 5.1. Let P be the operator defined by $v = Pu$, where Pu is the (orthogonal) projection on a closed manifold M in H. The operator P is easily seen to be self-adjoint. The spectrum of P consists of the two eigenvalues $\lambda = 0$ and $\lambda = 1$. The null space of $\lambda = 1$ consists of all vectors in M, and the null space of $\lambda = 0$ consists of all vectors in $M^\perp$. If λ is not an eigenvalue, the solution of $(P - \lambda I)u = f$ is $Pf/(1-\lambda) - (f - Pf)/\lambda$, so that $P - \lambda I$ has a bounded inverse and λ is in the resolvent set.

Example 3. See Example 3(b), Section 5.1. With $\theta \neq k\pi$, the eigenvalue problem leads to the equations

$$(\cos\theta - \lambda)u_1 - (\sin\theta)u_2 = 0, \quad (\sin\theta)u_1 + (\cos\theta - \lambda)u_2 = 0,$$

so that nontrivial solutions are possible if and only if

$$(\cos\theta - \lambda)^2 + \sin^2\theta = 0 \quad \text{or} \quad \lambda = e^{i\theta}, \quad e^{-i\theta}.$$

When this operator is viewed in $E_2^{(r)}$, its spectrum is empty; on $E_2^{(c)}$, however, there are two eigenvalues, $\lambda_1 = e^{i\theta}$, and $\lambda_2 = e^{-i\theta}$, with respective eigenvectors $e_1 - ie_2$ and $e_1 + ie_2$.

Example 4. See Example 3(c), Section 5.1. This example has many striking features. The eigenvalue problem reduces to the pair of equations

$$u_2 = \lambda u_1, \quad 0 = \lambda u_2,$$

from which we conclude that $\lambda = 0$ is the only eigenvalue; the corresponding null space has dimension 1 and is spanned by the eigenvector whose coordinates are $(1, 0)$. The range $R(0)$ consists of all vectors with vanishing second coordinate, so that $R(0) = N(0)$, a surprising result [on E_n we always have $R_A = (N_{A^*})^\perp$, and if A is self-adjoint, $R_A = N_A^\perp$; the case $R_A = N_A$ therefore cannot occur for self-adjoint transformations].

Example 5. See Example 4, Section 5.1. On $L_2^{(c)}(0, 1)$, let $Au \doteq xu(x)$. This operator is clearly self-adjoint. The homogeneous equation $xu(x) = \lambda u(x)$ has only the trivial solution, so that A has *no eigenvalues*. For f in L_2 the inhomogeneous equation

$$xu - \lambda u = f$$

has the formal solution $u = f/(x - \lambda)$, which is a genuine L_2 solution whenever λ is *not* in the real interval $0 \leqslant \lambda \leqslant 1$. If λ is in that interval, the solution belongs to L_2 only if f vanishes to sufficiently high order at $x = \lambda$; the inverse of $A - \lambda I$ is seen to be unbounded, and $R(\lambda)$ is dense in L_2. Thus every value of λ in $0 \leqslant \lambda \leqslant 1$ is in the continuous spectrum—state (II,1_n)—and every other value of λ is in the resolvent set.

REMARK. For some purposes it is useful to view $xu = \lambda u$ as a distributional equation. For λ in [0, 1] the equation then admits the solution $c\delta(x - \lambda)$, which can be regarded as a "singular" eigenfunction.

Example 6. On $L_2^{(c)}(0, 1)$ consider the bounded self-adjoint operator defined by

$$Au \doteq xu(x) + \int_0^1 u(x)\,dx. \tag{5.6.1}$$

Since A is self-adjoint, its spectrum is real. Let us search first for eigenvalues and eigenfunctions by examining the equation

$$xu + \int_0^1 u(x)\,dx = \lambda u, \quad 0 < x < 1. \tag{5.6.2}$$

If $\int_0^1 u(x)\,dx = 0$, we find that $u \equiv 0$; we therefore assume that $\int_0^1 u\,dx \neq 0$. In view of the fact that we are dealing with a linear homogeneous equation we may as well normalize u, so that

$$\int_0^1 u\ dx = 1, \tag{5.6.3}$$

and then (5.6.2) gives

$$u = \frac{1}{\lambda - x}, \tag{5.6.4}$$

which belongs to L_2 if λ is outside $0 \leqslant \lambda \leqslant 1$. To satisfy (5.6.3) we need $\lambda > 1$, and performing the required integration, we obtain

$$\frac{\lambda}{\lambda - 1} = e, \tag{5.6.5}$$

which has the one and only solution

$$\lambda_0 = \frac{e}{e-1}, \tag{5.6.6}$$

the corresponding eigenfunction being

$$u(x) = \frac{1}{\lambda_0 - x}. \tag{5.6.7}$$

Next we show that the segment $0 \leqslant \lambda \leqslant 1$ is in the continuous spectrum. It suffices to show that $A - \lambda I$ is not bounded away from 0. We must exhibit a sequence $\{u_n\}$ such that $\|Au_n - \lambda u_n\|/\|u_n\|$ tends to 0 as $n \to \infty$. Let λ be fixed, $0 < \lambda < 1$, and for n sufficiently large take

$$u_n(x) = \begin{cases} 1, & \lambda < x < \lambda + \dfrac{1}{n}, \\ -1, & \lambda - \dfrac{1}{n} < x < \lambda, \\ 0, & |x - \lambda| \geqslant \dfrac{1}{n}. \end{cases}$$

Then $Au_n - \lambda u_n = (x - \lambda)u_n$, and

$$\|Au_n - \lambda u_n\|^2 = 2\int_\lambda^{\lambda+1/n} (x-\lambda)^2\,dx = \frac{2}{3n^3},$$

$$\|u_n\|^2 = \frac{2}{n},$$

so that

$$\lim_{n\to\infty}\left(\frac{\|Au_n - \lambda u_n\|}{\|u_n\|}\right) = 0.$$

A slight modification is needed for $\lambda = 0$ and $\lambda = 1$.

As in Example 5, it is sometimes useful to interpret (5.6.2) distributionally. Setting $\int_0^1 u(x)\,dx = 1$, we have

$$(x-\lambda)u = -1, \quad 0 \leqslant x \leqslant 1,$$

an equation which has already been discussed [(2.4.39)], and whose solution is (if $0 \leqslant \lambda \leqslant 1$)

$$u = C\delta(x-\lambda) - \text{pf}\frac{1}{x-\lambda},$$

where $C(\lambda)$ can be calculated from the normalization (5.6.3).

Example 7. See Example 7, Section 5.1. Let $Au = \int_0^x u(y)\,dy$ on $L_2^{(c)}(0,1)$. The equation $Au = \lambda u$ implies for $\lambda \neq 0$ that $\lambda u' = u, u(0) = 0$, and therefore $u \equiv 0$; for $\lambda = 0, Au = 0$ implies that $u = 0$. Thus A has *no* eigenvalues. Consider the equation $Au - \lambda u = f$; setting $z = \lambda u + f$, we see that z is differentiable and $z(0) = 0$. Differentiation gives $z' = u$, or, for $\lambda \neq 0, z' = (z-f)/\lambda$ and

$$z = -\frac{e^{x/\lambda}}{\lambda}\int_0^x f(y)e^{-y/\lambda}\,dy,$$

$$u = \frac{z-f}{\lambda} = -\frac{f(x)}{\lambda} - \frac{e^{x/\lambda}}{\lambda^2}\int_0^x f(y)e^{-y/\lambda}\,dy = (A-\lambda I)^{-1}f. \tag{5.6.8}$$

One can now verify that this function u satisfies the original equation $Au - \lambda u = f$, and it is the only solution. Furthermore, (5.6.8) shows clearly that $(A - \lambda I)^{-1}$ is bounded. For $\lambda = 0$ the equation $Au = f$ has solutions only if f is differentiable and $f(0) = 0$; this set is dense in $L_2(0,1)$, and the inversion formula is $u = f'$. Thus A^{-1} is unbounded. Recapitulating: Every value of $\lambda \neq 0$ is in the resolvent set, and $\lambda = 0$ is in the continuous spectrum [state (II,1_n)].

Example 8. See also Example 9 in Section 5.1. For the right-shift operator, if $u = \sum_{k=1}^{\infty}\alpha_k e_k$, then $Au = \sum_{k=1}^{\infty}\alpha_k e_{k+1}$. The eigenvalue problem is therefore equivalent to the system $0 = \lambda\alpha_1, \alpha_1 = \lambda\alpha_2, \alpha_2 = \lambda\alpha_3, \ldots$. If $\lambda \neq 0$, then $\alpha_1 = 0$ and therefore all $\alpha_k = 0$. If $\lambda = 0$, all α_k are again 0. Therefore, A has no eigenvalues. Let us now look at the inhomogeneous equation $Au - \lambda u = f$, which is explicitly $-\lambda\alpha_1 e_1 + \sum_{k=2}^{\infty}(\alpha_{k-1} - \lambda\alpha_k)e_k = \sum_{k=1}^{\infty} f_k e_k$. If $\lambda = 0$ *and* $f_1 = 0$, there is one and only one solution, given by $\alpha_1 = f_2, \alpha_2 = f_3, \ldots$. Thus $\lambda = 0$ is in the compression spectrum and has deficiency 1, and A^{-1} is bounded. If $\lambda \neq 0$, we obtain

$$\alpha_1 = -\frac{f_1}{\lambda}$$

and the recurrence relations $\alpha_k = (\alpha_{k-1} - f_k)/\lambda$.

If the series $\sum \alpha_k e_k$ obtained in this way converges, that is, if $\sum |\alpha_k|^2 < \infty$, it provides a solution to the inhomogeneous equation. Rather than ascertaining the convergence directly, we obtain useful information by studying the adjoint A^*, which is the left shift $A^* e_1 = 0, A^* e_{k+1} = e_k, k = 1, 2, \ldots$. Since A has no eigenvalues, the compression spectrum of A^* is empty. The eigenvalue problem for A^* is $\sum_{k=1}^{\infty} \alpha_{k+1} e_k = \lambda \sum_{k=1}^{\infty} \alpha_k e_k$, or $\lambda \alpha_k = \alpha_{k+1}, k = 1, 2, \ldots$. In terms of α_1, these equations have the solution $\alpha_k = \lambda^{k-1} \alpha_1$, and these will form the coordinates of a vector in H if and only if $\sum |\alpha_k|^2 < \infty$, that is, if and only if $|\lambda| < 1$, which is then the point spectrum of A^* and the compression spectrum of A. The circle $|\lambda| = 1$ must be in the spectrum (since the spectrum is a closed set) for both A and A^* but cannot be in the compression spectrum and so must be in the continuous spectrum. For $|\lambda| < 1$ we have $\|Af - \lambda f\| \geqslant |\|Af\| - \|\lambda f\|| = (1 - |\lambda|)\|f\|$, so that $A - \lambda I$ is bounded away from 0 and λ cannot be in the continuous spectrum.

Example 9. See Example 1, Section 5.3.

(a) Let

$$T_1 u = \frac{du}{dx}, \quad D_{T_1} : u \text{ a.c.}, \quad u' \in L_2(0, 1), u(0) = 0.$$

Then

$$T_1^* v = -\frac{dv}{dx}, \quad D_{T_1^*} : v \text{ a.c.}, \quad v' \in L_2(0, 1), v(1) = 0.$$

The only solution of $T_1 u = \lambda u, u \in D_{T_1}$, is $u \equiv 0$, so that the point spectrum of T_1 is empty. The inhomogeneous equation $T_1 u - \lambda u = f, u \in D_{T_1}$, has the unique solution

$$u(x) = e^{\lambda x} \int_0^x e^{-\lambda y} f(y)\, dy,$$

so that $T_1 - \lambda I$ has a bounded inverse for each λ, and $R(\lambda) = H$. *Therefore, the spectrum of T_1 is empty.*

(b) Let

$$S_2 u = -i\frac{du}{dx}, \quad D_{S_2} : u \text{ a.c.}, \quad u' \in L_2(0, 1), u(0) = au(1),$$

where a is a given real constant $\neq 0$. Then

$$S_2^* v = -i\frac{dv}{dx}, \quad D_{S_2^*} : v \text{ a.c.}, \quad v' \in L_2(0, 1), v(1) = av(0).$$

Thus S_2 is self-adjoint if $a = 1$ or $a = -1$. Consider the case $a = 1$; then the eigenvalue problem is $-iu' = \lambda u$, $u(0) = u(1)$, from which we conclude that $\lambda = 2n\pi$, where n is an integer $-\infty < n < \infty$. Each eigenvalue has multiplicity 1, the null space being spanned by the eigenfunction $e^{2n\pi ix}$. Note that the *eigenfunctions form an orthonormal basis for* $L_2(0, 1)$. Consider the inhomogeneous equation $S_2 u - \lambda u = f$, which has the explicit form $-iu' - \lambda u = f$, $u(0) = u(1)$. If $\lambda \neq 2n\pi$, the one and only solution can be written in terms of a Green's function, and the corresponding integral representation of the solution shows that $(S_2 - \lambda I)^{-1}$ is bounded. Thus $\lambda \neq 2n\pi$ is in the resolvent set. The case $a = -1$ proceeds along similar lines.

(c) Let

$$S_3 u = -i\frac{du}{dx}, \quad D_{S_3} : u \text{ a.c.}, \quad u' \in L_2(0,1), u(0) = u(1) = 0.$$

Then

$$S_3^* v = -i\frac{dv}{dx}, \quad D_{S_3^*} : u \text{ a.c.}, \quad u' \in L_2(0,1).$$

Note that S_3 is symmetric and has no point spectrum. Every value of λ is in the compression spectrum of S_3 and is an eigenvalue of S_3^*.

Spectral Theory on $E_n^{(c)}$

For an operator on $E_n^{(c)}$ there are only two possibilities: Either λ is in the resolvent set, or λ simultaneously is an eigenvalue and is in the compression spectrum. We may therefore omit further reference to the compression spectrum.

We can easily show that every operator A on $E_n^{(c)}$ must have at least one eigenvalue. Let $\{h_k\}$ be a basis in $E_n^{(c)}$. If the coordinates of u in this basis are $\{\alpha_i\}$, then, according to (5.1.5), those of Au are $\left\{\beta_i = \sum_j a_{ij}\alpha_j\right\}$, where $[a_{ij}]$ is the matrix of A in the basis $\{h_i\}$. Thus the eigenvalue problem $Au - \lambda u = 0$ is equivalent to the homogeneous system

$$\begin{aligned} (a_{11} - \lambda)\alpha_1 + a_{12}\alpha_2 + \cdots + a_{1n}\alpha_n &= 0 \\ &\vdots \\ a_{n1}\alpha_1 + a_{n2}\alpha_2 + \cdots + (a_{nn} - \lambda)\alpha_n &= 0, \end{aligned} \tag{5.6.9}$$

which has a nontrivial solution if and only if

$$\det(A - \lambda I) \doteq \begin{vmatrix} a_{11} - \lambda & a_{12} & \cdots & a_{1n} \\ \vdots & \vdots & & \vdots \\ a_{n1} & a_{n2} & \cdots & a_{nn} - \lambda \end{vmatrix} = 0. \tag{5.6.10}$$

Clearly, $\det(A - \lambda I)$ is a polynomial of degree n in λ which must therefore have n zeros (not necessarily distinct), which we label $\lambda_1, \ldots, \lambda_n$. Each distinct λ_k, when substituted in (5.6.9), gives rise to a nontrivial vector solution $(\alpha_1^{(k)}, \ldots, \alpha_n^{(k)})$. We are therefore entitled to call this λ_k an eigenvalue, and we have shown that *every operator on $E_n^{(c)}$ has at least one eigenvalue*. [An operator on $E_n^{(r)}$ does not necessarily have an eigenvalue, as illustrated by the rotation operator of Example 3 with $\theta \neq k\pi$.] To a k-fold zero of (5.6.10) may correspond as few as one or as many as k independent eigenvectors. For instance, the identity operator has $\lambda = 1$ as an n-fold zero, and since every vector in $E_n^{(c)}$ is a corresponding eigenvector, we can obviously choose n independent ones. On the other hand, Example 4 has only the eigenvalue $\lambda = 0$, which is a double zero of (5.6.10) but has geometric multiplicity equal to 1.

It is important to know when A has enough eigenvectors to form a basis (that is, when A has n independent eigenvectors). Exercise 5.6.7 asks the reader to show that if A has n *distinct* eigenvalues, the eigenvectors form a basis. More important for us is the fact that the eigenvectors of any *symmetric* operator can be chosen to form an orthonormal basis even if not all the eigenvalues are distinct. We now prove this assertion.

Let A be a symmetric operator on $E_n^{(c)}$. We have already established that A has at least one eigenvalue, say λ_1. Since A is symmetric, λ_1 is real. Let e_1 be a unit eigenvector corresponding to λ_1, and let M_1 be the one-dimensional linear manifold generated by e_1. Every vector u in $E_n^{(c)}$ can be decomposed in one and only one way as a sum $v + w$, where $v \in M_1$ and $w \in M_1^\perp$. Clearly, $Av = \lambda_1 v$, so that every vector in M_1 remains in M_1 after the transformation A. We say that M_1 is *invariant* under A. $M_1^\perp$ is also invariant under A since $\langle Aw, v\rangle = \langle w, Av\rangle = \lambda_1\langle w, e_1\rangle = 0$. We can therefore consider A as an operator on the $(n-1)$-dimensional space $M_1^\perp$. This operator is clearly symmetric and has therefore at least one eigenvalue λ_2 (not necessarily different from λ_1). Let e_2 be a unit eigenvector corresponding to λ_2; since e_2 is in $M_1^\perp, e_2$ is orthogonal to e_1. Let M_2 be the two-dimensional linear manifold generated by e_1 and e_2. We see that M_2 and $M_2^\perp$ are invariant under A and that A can therefore be regarded as a symmetric operator on M_2 with at least one real eigenvalue λ_3 and a unit eigenvector e_3. We continue this procedure for n steps until $M_n^\perp$ is empty.

We have therefore generated in this way n eigenvalues $\lambda_1, \dots, \lambda_n$, which are *not necessarily distinct*, but to which correspond n orthonormal eigenvectors $e_1, \dots, e_n$. Such a set of vectors is of course a basis for $E_n^{(c)}$. We have

$$Ae_k = \lambda_k e_k,\ k = 1, \dots, n,\ \langle e_i, e_j\rangle = 0,\ i \neq j;\ \langle e_i, e_i\rangle = 1. \tag{5.6.11}$$

The matrix of A in this basis has diagonal form: The eigenvalues appear along the principal diagonal, and all off-diagonal elements vanish.

The set of eigenvectors of (5.6.11) provides a convenient way to solve the inhomogeneous equation $Au = f$ or, even more generally, the equation

$$Au - \lambda u = f, \tag{5.6.12}$$

where A is a symmetric operator on $E_n^{(c)}$, f is a given element in $E_n^{(c)}$, and λ is a given complex number. Since the eigenvectors (5.6.11) of A form a basis, we can write

$$f = \sum_{k=1}^{n} \beta_k e_k, \text{ where } \beta_k = \langle f, e_k\rangle.$$

If (5.6.12) has a solution u, we can express u as

$$u = \sum_{k=1}^{n} \alpha_k e_k,$$

and, on substitution in (5.6.12), we find that

$$\sum_{k=1}^{n} [(\lambda_k - \lambda)\alpha_k - \beta_k]e_k = 0,$$

from which we conclude that

$$(\lambda_k - \lambda)\alpha_k = \beta_k, \quad k = 1, 2, \ldots, n. \tag{5.6.13}$$

If λ is *different* from $\lambda_1, \ldots, \lambda_n$, (5.6.13) can be solved for each α_k, and we obtain the prospective solution

$$u = \sum_{k=1}^{n} \frac{\beta_k}{\lambda_k - \lambda} e_k, \tag{5.6.14}$$

which is easily checked to be a solution of (5.6.12) in fact as well as in form. Of course, if λ is equal to some eigenvalue λ_m (say, nondegenerate), then there is no solution of (5.6.13) for α_m unless $\beta_m = 0$. If $\beta_m = 0$, then α_m is arbitrary. For $k \neq m$, we still have $\alpha_k = \beta_k/(\lambda_k - \lambda)$. Thus a necessary and sufficient condition for (5.6.12) to have solutions is that $\langle f, e_m \rangle = 0$; and assuming that this solvability condition is satisfied, all solutions of (5.6.12) are of the form

$$u = \sum_{k \neq m} \frac{\beta_k}{\lambda_k - \lambda} e_k + c e_m, \tag{5.6.15}$$

where c is arbitrary. A slight but obvious modification is needed if λ is equal to a degenerate eigenvalue.

EXERCISES

5.6.1 Let P be the orthogonal projection on a closed manifold M in the Hilbert space H. We saw in Example 2 that the only points in the spectrum are $\lambda = 0$ and $\lambda = 1$, which are eigenvalues. What are the multiplicities of these eigenvalues? What are the solvability conditions for the equations $Pu = f$ and $Pu - u = f$? Find the corresponding solutions if the solvability conditions are satisfied.

5.6.2 Consider the problem

$$u' = f, \; 0 < x < 1; \quad u(0) = 0, u'(0) = 0.$$

Under what conditions on f is the problem solvable? What is the adjoint homogeneous problem? Why does the Fredholm alternative not apply?

5.6.3 Carry out the analysis in Example 9(b) for the case $a = -1$. Construct Green's function when λ is not in the point spectrum.

5.6.4 On $L_2^{(c)}(0, 1)$, let $Au = a(x)u(x)$, where $a(x)$ is a fixed continuous function. Describe the spectrum.

5.6.5 On $L_2^{(c)}(-1,1)$ consider the operator

$$Au \doteq xu(x) + \theta \int_{-1}^{1} u(x)\, dx,$$

where θ is a given real number. Describe completely the spectrum of A.

5.6.6 On $E_n^{(c)}$ consider the nearest thing to a shift operator. Let A be defined by its effect on an orthonormal basis $\{e_1, \dots, e_n\}$ as

$$Ae_i = e_{i+1}, i = 1, \dots, n-1; Ae_n = e_1.$$

Find the spectrum and discuss the limit as $n \to \infty$ (compare with Example 8).

5.6.7 Prove that an operator with n distinct eigenvalues on $E_n^{(c)}$ has n independent eigenvectors.

5.6.8 Consider the differentiation operator D on P_n (see Exercise 5.5.3). Find the spectrum of D and D^*. Discuss the solution of $Du - \lambda u = f$, where $f \in P_n$ and λ is a real number.

5.6.9 On the real line consider the multiplication operator A defined by $Au = xu$. The appropriate Hilbert space H is $L_2^{(c)}(\mathbb{R})$, but in order to have the image belong to the space, the domain of A must be restricted to elements for which $xu \in L_2^{(c)}(\mathbb{R})$. D_A is dense in H since $C_0^\infty(\mathbb{R}) \in D_A$ and is itself dense in H. Show that A is self-adjoint and find its spectrum. Compare with Example 5.

5.7 COMPACT OPERATORS

Definition. *An operator K on H is said to be* compact *(or* completely continuous*) if K transforms bounded sets into relatively compact sets (see Section 4.3), that is, whenever $\{u_n\}$ is a sequence in H with $\|u_n\| < M$, then $\{Ku_n\}$ contains a subsequence which converges to some point in H.*

A compact operator is necessarily bounded, for otherwise there would exist a sequence with $\|u_n\| = 1$ and $\|Ku_n\| \to \infty$; by eliminating superfluous elements if necessary, we can suppose that $\|Ku_{n+1}\| > \|Ku_n\|$, and clearly $\{Ku_n\}$ does not contain a convergent subsequence.

Every bounded set in a finite-dimensional space is relatively compact (this is the Bolzano-Weierstrass theorem), so that every bounded operator on E_n is compact and every bounded operator on H having finite-dimensional range is compact. Arbitrary bounded operators on infinite-dimensional spaces are not necessarily compact. In fact, the identity operator is clearly bounded but is not compact since it transforms the infinite orthonormal set $\{e_i\}$ into itself and $\{e_1, e_2, \dots\}$ is a bounded sequence containing no convergent subsequence. The most important compact operators on infinite-dimensional spaces are integral operators of the Hilbert-Schmidt type (see Example 6, Section 5.1).

Compact operators seem to be very obliging since they transform sequences that are merely bounded into "nicer" sequences having convergent subsequences. This is a two-edged sword, however, since unavoidably there will be trouble with the inverse. An operator which strikes a better balance between itself and its inverse is $I + K$, where I is the identity. Here one can think of the operator K as being a perturbation of I. Operators of the form $I + K$ (or, more generally, of the form $K - \lambda I$, where λ is a complex number $\neq 0$) will have invertibility properties similar to those of operators on finite-dimensional spaces.

Theorem 5.7.1. *If K is compact and $\{e_n\}$ is an infinite orthonormal sequence in H, then* $\lim_{n\to\infty} Ke_n = 0$.

Proof. If the contrary were true, there would exist a subsequence $\{f_n\}$ of $\{e_n\}$ such that $\|Kf_n\| > \varepsilon$ for all sufficiently large n. Since K is compact, a subsequence $\{g_n\}$ can be extracted from $\{f_n\}$ such that Kg_n converges, say to u. This element u is not 0 since $\|Kg_n\| > \varepsilon$ for n large. The continuity of the inner product shows that $\langle Kg_n, u\rangle \to \|u\|^2 \neq 0$; also $\langle Kg_n, u\rangle = \langle g_n, K^*u\rangle$, which tends to 0 as $n \to \infty$ by (4.6.8). This contradiction proves the theorem. □

Corollary. *If K is a compact, invertible operator on an infinite-dimensional space, its inverse is* unbounded.

Proof. If $\{e_n\}$ is an infinite orthonormal set, we have $Ke_n \to 0$ by Theorem 5.6.1, and, since $\|e_n\| = 1$, K is not bounded away from 0. Therefore, if K is invertible, its inverse is unbounded. □

How can we tell whether an operator is compact? It is not always easy to apply the definition directly. The following theorem tells us that K is compact if it can be approximated in norm by compact operators (which, in applications, will often be operators with finite-dimensional ranges).

Theorem 5.7.2. *The operator A is compact if there exists a sequence of compact operators $\{K_n\}$ such that $\|A - K_n\| \to 0$ as $n \to \infty$.*

Proof. It suffices to show that each bounded sequence $\{u_k\}$ contains a subsequence $\{v_k\}$ for which Av_k converges. Since K_1 is compact, there exists a subsequence of $\{u_k\}$, say $\{u_k^{(1)}\}$, for which $K_1u_k^{(1)}$ converges. In the bounded sequence $\{u_k^{(1)}\}$ we can find a subsequence $\{u_k^{(2)}\}$ such that $K_2u_k^{(2)}$ converges; proceeding in this manner, we find a subsequence $\{u_k^{(n)}\}$ of $\{u_k^{(n-1)}\}$ for which $K_nu_k^{(n)}$ converges as $k \to \infty$. Consider now the diagonal sequence

$$v_1 = u_1^{(1)}, v_2 = u_2^{(2)}, \ldots, v_n = u_n^{(n)}, \ldots,$$

which is a subsequence of $\{u_k\}$ transformed into a convergent sequence by each of the operators $K_1, K_2, \ldots$. We claim that A also transform this diagonal sequence into a convergent sequence. Indeed, we have

$$\begin{aligned}\|Av_n - Av_m\| &\leqslant \|Av_n - K_kv_n\| + \|K_kv_n - K_kv_m\| + \|K_kv_m - Av_m\| \\ &\leqslant \|A - K_k\|(\|v_n\| + \|v_m\|) + \|K_k(v_n - v_m)\|.\end{aligned}$$

Since $\{u_k\}$ is bounded, so is $\{v_n\}$, and there exists c such that $\|v_n\| + \|v_m\| \leqslant c$ for all m, n. By hypothesis we can choose k so large that $\|A - K_k\| \leqslant \varepsilon/2c$. With k fixed in this way we choose m and n so large that $\|K_k v_n - K_k v_m\| \leqslant \varepsilon/2$; this can be done since the sequence $K_k v_n$ converges as $n \to \infty$. Therefore, for m and n sufficiently large we have

$$\|Av_n - Av_m\| \leqslant \varepsilon,$$

so that Av_n is a Cauchy sequence and hence converges, completing the proof of the theorem. □

Theorem 5.7.2 will be applied when we take up the theory of integral equations. We now turn to the Fredholm Alternative for operators of the form $A = K - \lambda I, K$ compact, $\lambda \neq 0$.

Theorem (Fredholm Alternative). *Let H be a separable Hilbert space, K a compact operator on H, and λ a complex number $\neq 0$; let $A = K - \lambda I (A^* = K^* - \bar{\lambda} I)$. Then* either *the following alternatives hold (compare with Theorem 5.5.3 for E_n):*

(a) $Au = 0$ has only the zero solution (λ is not an eigenvalue of K), in which case $A^* v$ has only the zero solution ($\bar{\lambda}$ is not an eigenvalue of K^*) and $Au = f$ has precisely one solution for each f in H.

(b) N_A has finite dimension k (λ is an eigenvalue of K with multiplicity k), in which case N_{A^*} also has dimension k ($\bar{\lambda}$ is an eigenvalue of K^* with multiplicity k) and $Au = f$ has solutions if and only if f is orthogonal to N_{A^*}. If $\{u_1, \ldots, u_k\}$ is a basis for N_A (that is, a set of independent eigenvectors of K corresponding to the eigenvalue λ), and if $\{v_1, \ldots, v_k\}$ is a basis for N_{A^*} (that is, a set of independent eigenvectors of K^* corresponding to the eigenvalue $\bar{\lambda}$), the solvability conditions are $\langle f, v_1 \rangle = \cdots = \langle f, v_k \rangle = 0$; and if these conditions are satisfied, the general solution of $Au = f$ is $u = \tilde{u} + c_1 u_1 + \cdots + c_k u_k$, where $c_1, \ldots, c_k$ are arbitrary constants and $\tilde{u}$ is any particular solution of $Au = f$.

We indicate how the steps of the proof of Theorem 5.5.3 have to be carried out.

1. We must show that R_A is a closed set, so that we will have $R_A = N_{A^*}^{\perp}$. Since A is closed and, moreover, $D_A = H$, it suffices by Theorem 5.5.2 to show that $\|Au\| \geqslant c\|u\|$ for some fixed $c > 0$, and all $u \in N_A^{\perp}$. If this were not true, there would exist $\{u_i\} \in N_A^{\perp}, \|u_i\| = 1, \|Au_i\| \to 0$; that is, $Ku_i - \lambda u_i \to 0$. But since K is compact, we can restrict ourselves to a sequence $\{u_i\}$ such that Ku_i converges; then $Ku_i - \lambda u_i \to 0$ implies that λu_i converges, and since $\lambda \neq 0, u_i$ converges, say to u. It then follows that Ku_i converges to Ku and therefore $Ku - \lambda u = 0$ and $u \in N_A$; u is also the limit of a sequence of elements of the closed set $N_A^{\perp}$, so that $u \in N_A^{\perp}$; hence $u = 0$. However, since u is the limit of a sequence of unit elements, $\|u\| = 1$. Therefore, A must be bounded away from 0 on $N_A^{\perp}$, and R_A is closed and $R_A = N_{A^*}^{\perp}$.

2. If $R_A = H$, then $N_A = \{0\}$, and vice versa. We already know from step 1 that $N_{A^*} = \{0\}$ is necessary and sufficient for $R_A = H$.

(a) Suppose that $R_A = H$. We proceed as in Section 5.5, but now it is not obvious that the strict inclusion $N_p \supset N_{p-1}$ for all p leads to a contradiction. The proof is left for Exercise 5.7.1 and shows that $N_A = \{0\}$.

(b) This is exactly the same as in Section 5.5.

3. N_A and N_{A^*} have the same *finite* dimension. First we show that N_A has a finite dimension. Since N_A is closed, let $\{e_i\}$ be an orthonormal basis for N_A; then if the dimension of N_A were infinite, we would have $Ke_i \to 0$ by Theorem 5.7.1. Now $Ke_i = \lambda e_i$, so that $\lambda e_i \to 0$, which is impossible for $\lambda \neq 0$. The rest of the proof is exactly as in Section 5.5.

REMARK. In the alternative theorem for E_n, the common dimension of N_A and N_{A^*} is an integer which obviously cannot exceed n. In the theorem just proved, k can be any finite integer.

EXERCISES

5.7.1 This exercise fills in the missing step in 2(a) of the proof of the Fredholm alternative. Let $A = K - \lambda I$, where $\lambda \neq 0$ and K is a compact operator on H. We want to show that if $R_A = H$, then N_A contains only the zero element. As in Theorem 5.5.3, we assume that there exists $u_1 \neq 0$ with $Au_1 = 0$ and construct successively $u_2, u_3, \ldots$, where $Au_p = u_{p-1}$. Each $u_p \neq 0$ and $A^p u_p = 0$, so that u_p is a nonzero element in the null space N_p of A^p. It is easy to see that $N_p \supset N_{p-1}$ and that the inclusion is strict; it is therefore possible to choose $u_p \in N_p$ with $\|u_p\| = 1$ and $u_p \perp N_{p-1}$. For $n > m$ consider $\|Ku_n - Ku_m\| = \|\lambda u_n + g\|$, where $g = Au_n - Au_m - \lambda u_m$. Show that $g \in N_{n-1}$ and $\|Ku_n - Ku_m\|^2 \geqslant |\lambda| > 0$, which contradicts compactness.

5.7.2 Let K be a compact operator. Then we know from the Fredholm alternative that the multiplicity of any nonzero eigenvalue is finite. More can be said: Let M_ε be the linear manifold consisting of the eigenvectors corresponding to *all* eigenvalues λ with $|\lambda| \geqslant \varepsilon$, where ε is any positive number; then M_ε is finite-dimensional. Prove the theorem for the case when K is self-adjoint.

5.7.3 We say that u_n *converges weakly* to u if $\langle u_n, h\rangle \to \langle u, h\rangle$ for every h in H. If $\{e_n\}$ is an infinite orthonormal set in H, then $e_n \to 0$ *weakly* (but of course e_n does not converge). It is easy to see that convergence entails weak convergence (see Section 4.8). Prove that if K is compact, it maps every weakly convergent sequence into a convergent sequence.

5.8 EXTREMAL PROPERTIES OF OPERATORS

Let A be an operator, possibly unbounded, defined on a domain D_A dense in the separable Hilbert space H. We are interested in finding bounds for the spectrum of

A, particularly its point spectrum, if any. If (λ, u) is an eigenpair, that is, if λ is an eigenvalue with corresponding nontrivial eigenvector u, then

$$Au = \lambda u, \tag{5.8.1}$$

and hence

$$|\lambda| = \frac{\|Au\|}{\|u\|}, \quad \lambda = \frac{\langle Au, u\rangle}{\|u\|^2}. \tag{5.8.2}$$

For any nonzero element v in D_A we define the *Rayleigh quotient*

$$R(v) \doteq \frac{\langle Av, v\rangle}{\|v\|^2}. \tag{5.8.3}$$

In general, $R(v)$ may take on complex values, but if A is symmetric, $R(v)$ is real for all v in D_A. If (λ, u) is an eigenpair, we have $R(u) = \lambda$. The set of all possible values of $R(v)$ with $v \neq 0$ (or, equivalently, the set of all possible values of $\langle Av, v\rangle$ with $\|v\| = 1$) is known as the *numerical range* of A, denoted by $W(A)$.

Bounded Operators

We may take the domain of a bounded operator A to be the entire Hilbert space H. We recall the definition of a norm in (5.1.2):

$$\|A\| = \sup_{\|v\|=1} \|Av\| = \sup_{v\neq 0} \frac{\|Av\|}{\|v\|} \tag{5.8.4}$$

and introduce the new definition

$$M_A \doteq \sup_{\|v\|=1} |\langle Av, v\rangle| = \sup_{v\neq 0} |R(v)|, \tag{5.8.5}$$

where in both definitions we must use "sup" instead of "max" since the supremum is not necessarily attained. We shall see in Theorem 5.8.3 that M_A is necessarily finite. It follows from (5.8.2) that every eigenvalue λ satisfies

$$|\lambda| \leqslant \|A\|, \quad |\lambda| \leqslant M_A.$$

One of our goals is to find out whether these inequalities can be transformed into equalities for the eigenvalues of largest modulus. We shall need a sequence of theorems.

Theorem 5.8.1. $\|A^*\| = \|A\|$.

Proof. For each v in H, the Schwarz inequality gives

$$0 \leqslant \|Av\|^2 = \langle Av, Av\rangle = \langle v, A^*Av\rangle \leqslant \|v\| \, \|A^*Av\| \leqslant \|v\| \, \|A^*\| \, \|Av\|.$$

Thus $\|Av\| \leqslant \|A^*\| \, \|v\|$, and by reversing the roles of A and A^*, we also find that $\|A^*v\| \leqslant \|A\| \, \|v\|$. Together, these inequalities prove the theorem. □

Theorem 5.8.2. *If λ is in the spectrum of A, then $|\lambda| \leqslant \|A\|$.*

Proof. The result has already been proved for eigenvalues. If λ is in the compression spectrum of A, $\bar{\lambda}$ is an eigenvalue of A^*; hence $|\lambda| = |\bar{\lambda}| \leqslant \|A^*\| = \|A\|$. If λ is in the continuous spectrum of A, there exists a sequence $\{v_n\}$ with $\|v_n\| = 1$ such that $w_n \doteq Av_n - \lambda v_n \to 0$. Hence

$$|\lambda| = \|Av_n - w_n\| \leqslant \|Av_n\| + \|w_n\| \leqslant \|A\| + \|w_n\|,$$

and since $\|w_n\| \to 0$, we have $|\lambda| \leqslant \|A\|$. □

Theorem 5.8.3. $M_A \leqslant \|A\|$.

Proof. By the Schwarz inequality,

$$|\langle Av, v\rangle| \leqslant \|Av\|\,\|v\| \leqslant \|A\|\,\|v\|^2.$$

□

For an arbitrary bounded operator the numbers M_A and $\|A\|$ are quite different. For instance, the operator A describing a rotation through an angle $\pi/2$ [see Example 3(b), Section 5.1] has $\langle Av, v\rangle = 0$ for every v and $\|A\| = 1$. However, we have the following theorem.

Theorem 5.8.4. *If A is* symmetric, $M_A = \|A\|$.

Proof. It suffices to show that $M_A \geqslant \|A\|$. The symmetry of A guarantees that $R(v)$ is real. Together with the definition of M_A, this implies that for all v, w,

$$\langle A(v+w), v+w\rangle \leqslant M_A\|v+w\|^2, \quad \langle A(v-w), v-w\rangle \geqslant -M_A\|v-w\|^2,$$

from which we obtain

$$\langle A(v+w), v+w\rangle - \langle A(v-w), v-w\rangle \leqslant M_A(\|v+w\|^2 + \|v-w\|^2).$$

The left side is equal to $2\langle Av, w\rangle + 2\langle Aw, v\rangle$. Using the parallelogram law on the right side, we find that

$$\langle Av, w\rangle + \langle Aw, v\rangle \leqslant M_A(\|v\|^2 + \|w\|^2).$$

If $Av \neq 0$, we may substitute $w = Av(\|v\|/\|Av\|)$ to obtain

$$\langle Av, Av\rangle + \langle AAv, v\rangle \leqslant 2M_A\|v\|\,\|Av\|.$$

Since A is symmetric, the left side is $2\|Av\|^2$, yielding $\|Av\| \leqslant M_A\|v\|$, an equality which remains valid if $Av = 0$. We have thus proved Theorem 5.8.4. □

We now confine ourselves to symmetric operators. It turns out that parts of the theory are just as easily developed for unbounded operators that are either bounded below or bounded above.

Symmetric Operators

Let A be a symmetric operator defined on a domain D_A dense in H. Since $R(v)$ is now real for all v in D_A, it makes sense to look at sup $R(v)$ and inf $R(v)$ rather than sup $|R(v)|$. Of course, it is possible that sup $R(v) = +\infty$ or inf $R(v) = -\infty$. If both are finite, the operator is bounded; if only one is finite, the operator is said to be *bounded from one side.*

Definition. *The symmetric operator A is* bounded below *if there exists a constant c such that*

$$\langle Av, v\rangle \geqslant c\|v\|^2 \quad \text{for all } v \in D_A. \tag{5.8.6}$$

We say that A is nonnegative *if c can be chosen $\geqslant 0$ in (5.8.6) and* strongly positive *(or* coercive*) if c can be chosen > 0. An intermediate concept is that of a* positive *operator: $\langle Av, v\rangle > 0$ for all $v \neq 0$ in D_A. The operator A is* bounded above *if there exists a constant C such that*

$$\langle Av, v\rangle \leqslant C\|v\|^2 \quad \text{for all } v \in D_A,$$

with similar definitions for nonpositive, negative, and strongly negative operators.

We note that an operator is bounded if and only if it is bounded both above and below.

Example. Let A be the operator $-d^2/dx^2$ defined on the domain D_A of functions $u(x)$ with two continuous derivatives on $0 \leqslant x \leqslant 1$ and satisfying $u(0) = u(1) = 0$. Then A is a strongly positive, unbounded operator. Indeed, we have

$$\langle Au, u\rangle = -\int_0^1 u''\bar{u}\;dx = \int_0^1 |u'|^2\,dx \geqslant 0,$$

which shows that A is nonegative. Moreover, since $u(0) = 0$,

$$|u(x)|^2 = \left|\int_0^x u'(t)\,dt\right|^2 \leqslant \int_0^x 1^2\;dt \int_0^x |u'(t)|^2\;dt \leqslant x\int_0^1 |u'(t)|^2\;dt,$$

and therefore

$$\int_0^1 |u'(t)|^2\;dt \geqslant 2\|u\|^2,$$

which proves that A is strongly positive. The eigenvalues of A are $\lambda_n = n^2\pi^2$, $n = 1, 2, \ldots$. Since $\lambda_n \to \infty$ as $n \to \infty$, (5.8.2) shows that the operator is unbounded.

For operators that are bounded from one side, we can only hope to find extremal principles for the eigenvalues at the bounded end of the spectrum.

Definition.

$$L_A \doteq \inf_{\substack{v\in D_A\\ v\neq 0}} R(v) = \inf_{\substack{v\in D_A\\ \|v\|=1}} \langle Av, v\rangle, \tag{5.8.7}$$

$$U_A \doteq \sup_{\substack{v\in D_A\\ v\neq 0}} R(v) = \sup_{\substack{v\in D_A\\ \|v\|=1}} \langle Av, v\rangle. \tag{5.8.8}$$

REMARK. If A is bounded below, L_A is finite; if A is bounded above, U_A is finite. For a bounded operator both L_A and U_A are finite, and

$$M_A = \|A\| = \max(|U_A|, |L_A|). \tag{5.8.9}$$

Thus the larger of U_A and $-L_A$ is equal to $\|A\|$.

Theorem 5.8.5. *Let A be symmetric and bounded below. If there is an element u in D_A for which the infimum in (5.8.7) is attained, (L_A, u) is an eigenpair and L_A is the lowest eigenvalue of A.*

Proof. By assumption $R(v)$ is a minimum for $v = u$, and therefore, for any $\eta \in D_A$ and any real number ε, we have

$$R(u + \varepsilon\eta) \geqslant R(u) = L_A.$$

It follows that

$$\left[\frac{d}{d\varepsilon} R(u + \varepsilon\eta)\right]_{\varepsilon=0} = 0.$$

Performing the required calculation, we find that

$$\langle \eta, Au\rangle + \langle Au, \eta\rangle - L_A[\langle \eta, u\rangle + \langle u, \eta\rangle] = 0, \quad \eta \in D_A.$$

Since $i\eta$ also belongs to D_A, we may substitute $i\eta$ for η to obtain

$$\langle \eta, Au\rangle - \langle Au, \eta\rangle - L_A[\langle \eta, u\rangle - \langle u, \eta\rangle] = 0.$$

Adding this equation to the preceding one, we have

$$\langle \eta, Au\rangle - L_A\langle \eta, u\rangle = 0,$$

that is,

$$\langle \eta, Au - L_A u\rangle = 0 \quad \text{for every } \eta \in D_A.$$

Since D_A is dense in H, it follows that $Au - L_A u = 0$, so that (L_A, u) is an eigenpair. It remains only to prove that A has no eigenvalue smaller than L_A. Let (λ, ϕ) be any eigenpair. Then, from (5.8.2), $\lambda = R(\phi)$. Since the minimum of R is L_A, we have $\lambda \geqslant L_A$, thereby completing the proof. □

Theorem 5.8.6. *Let A be symmetric and bounded above. If there is an element u in D_A for which the supremum in (5.8.8) is attained, (U_A, u) is an eigenpair and U_A is the largest eigenvalue of A.*

Proof. The operator $-A$ is symmetric and bounded below. By applying Theorem 5.8.5, we find that $-U_A$ is the lowest eigenvalue of $-A$ and hence U_A is the largest eigenvalue of A. □

In applications it is relatively easy to determine whether A is bounded above or below or both. The difficulty arises in trying to establish whether or not the supremum or infimum is actually attained. Take, for instance, the multiplication operator on $L_2^{(c)}(0,1)$ defined by $Av = xv(x)$. Then

$$R(v) = \int_0^1 x|v|^2\,dx \Big/ \int_0^1 |v|^2\ dx,$$

and clearly $0 \leqslant R(v) \leqslant 1$. In this case it can be seen that $U_A = 1$ and $L_A = 0$ by choosing sequences with support near $x = 1$ and $x = 0$, respectively. It can also be shown that neither the supremum nor the infimum of R is achieved for a nonzero element in D_A, so that the fact that A has no eigenvalues does not contradict Theorems 5.8.5 and 5.8.6. We note that in this last example, L_A and U_A belong to the continuous spectrum. More generally, we have the following theorem.

Theorem 5.8.7. *If A is bounded and symmetric, L_A and U_A are both in the approximate spectrum.*

Proof. To show that U_A belongs to the approximate spectrum, we must exhibit a sequence $\{v_k\}$ of unit elements such that $Av_k - U_A v_k \to 0$. From the definition of U_A there exists a sequence of unit elements $\{v_k\}$ such that $\langle Av_k, v_k\rangle \to U_A$. Consider first the case where A is a positive operator. We then have $U_A = \|A\|$ and

$$\begin{aligned}\|Av_k - U_A v_k\|^2 &= \|Av_k\|^2 - 2U_A\langle Av_k, v_k\rangle + U_A^2\\ &\leqslant \|A\|^2 - 2U_A\langle Av_k, v_k\rangle + U_A^2\\ &= 2U_A^2 - 2U_A\langle Av_k, v_k\rangle,\end{aligned}$$

which yields the desired result,

$$Av_k - U_A v_k \to 0. \tag{5.8.10}$$

If A is not positive, we choose a positive number α so that $B \doteq A + \alpha I$ is positive. Clearly,

$$\sup_{\|v\|=1} \langle Bv, v\rangle = U_A + \alpha,$$

and the maximizing sequence $\{v_k\}$ for A is then also a maximizing sequence for B: $\langle Bv_k, v_k\rangle \to U_A + \alpha$. Hence we can apply (5.8.10) to B to obtain

$$Bv_k - (U_A + \alpha)v_k \to 0$$

or

$$Av_k - U_A v_k \to 0.$$

We have shown that U_A belongs to the approximate spectrum. The proof for L_A is similar. □

If the symmetric operator A is not only bounded but also *compact*, the supremum (5.8.7), if nonzero, is actually attained. The same is true for the infimum (5.8.8).

Theorem 5.8.8. *If A is a symmetric compact operator, then if $U_A \neq 0$, it is an eigenvalue, and if $L_A \neq 0$, it is also an eigenvalue.*

Proof. Since A is bounded and symmetric, U_A is in the approximate spectrum by Theorem 5.8.7. There exists a sequence $\{v_k\}$ of unit elements such that $Av_k - U_A v_k \to 0$. Because A is compact, we can redefine $\{v_k\}$ by eliminating superfluous elements of the original sequence, so that $\{Av_k\}$ converges. Thus $U_A v_k$ converges, and if $U_A \neq 0$, v_k converges to a unit element v which, by the continuity of A, satisfies

$$Av - U_A v = 0.$$

A similar proof holds for L_A if $L_A \neq 0$. □

Corollary. *If A is a nontrivial symmetric, compact operator, at least one of the numbers $\|A\|, -\|A\|$ is an eigenvalue.*

More detailed information on the spectrum of compact, symmetric operators will be presented in the next chapter.

We end this section with two important results relating to the inhomogeneous equation

$$Au = f \tag{5.8.11}$$

when A is a bounded, symmetric, strongly positive operator on H [see (5.8.6)].

Theorem 5.8.9. *If A is bounded, symmetric, and strongly positive, (5.8.11) has one and only one solution.*

Proof. This follows from Exercise 4.5.11, but let us give the proof in the present setting. By assumption, there exists $c > 0$ such that

$$c\|u\|^2 \leqslant \langle Au, u\rangle \leqslant \|A\|\ \|u\|^2$$

for every $u \in H$. To solve $Au = f$ is equivalent to finding a fixed point of the operator B defined by

$$Bu = u - \frac{c}{\|A\|^2}(Au - f).$$

We show next that B is a contraction:

$$\begin{aligned}\|Bu - Bv\|^2 &= \left\langle u - v - \frac{c}{\|A\|^2} A(u-v), u - v - \frac{c}{\|A\|^2} A(u-v) \right\rangle \\ &= \|u-v\|^2 + \frac{c^2}{\|A\|^4}\|A(u-v)\|^2 - \frac{2c}{\|A\|^2}\langle A(u-v), u-v\rangle \\ &\leqslant \|u-v\|^2 + \frac{c^2}{\|A\|^2}\|u-v\|^2 - \frac{2c^2}{\|A\|^2}\|u-v\|^2 \\ &= \left(1 - \frac{c^2}{\|A\|^2}\right)\|u-v\|^2.\end{aligned}$$

Thus B is a contraction on H and therefore has a unique fixed point, the one and only solution of $Au = f$. □

Our next theorem shows that the solution of (5.8.11) can be characterized by a variational principle which forms the basis for numerical methods (see Chapters 6 and 8).

Definition. *Let*

$$J(v) = \langle Av, v\rangle - 2\langle f, v\rangle; \tag{5.8.12}$$

we define a minimizer u *of* J *to be any element in* H *such that*

$$J(u) \leqslant J(v) \quad \textit{for all } v \in H.$$

Theorem 5.8.10. *Let A be bounded, symmetric, and strongly positive; then the solution of (5.8.11) is also the unique minimizer of (5.8.12).*

Proof. Let u be the unique solution of (5.8.11) and let v be any other element of H. Then $v = u + h$ where $h \neq 0$. A simple calculation gives

$$J(u+h) - J(u) = 2\langle Au - f, h\rangle + \langle Ah, h\rangle = \langle Ah, h\rangle.$$

Since $\langle Ah, h\rangle > 0$, u is a minimizer of J. It is also clear that the minimizer is unique. □

EXERCISES

5.8.1 On $L_2^{(r)}(0,1)$ consider the compact symmetric operator A defined by

$$Au = \int_0^1 xyu(y)\,dy.$$

Find U_A and L_A.

5.8.2 On $L_2^{(r)}(0,1)$ consider the compact symmetric operator K defined by

$$Ku = \int_0^1 k(x,y)u(y)\,dy,$$

where

$$k(x,y) = \begin{cases} x(1-y), & x < y, \\ y(1-x), & x > y. \end{cases}$$

Find U_K, L_K. Note that $U_K = \|K\|$ is the largest eigenvalue of K. Compare with the estimate (5.1.9).

5.9 THE BANACH-SCHAUDER AND BANACH-STEINHAUS THEOREMS

One of the remarks made about closed operators in Section 5.2 is that *a closed operator on a closed domain is bounded.* This result is known more formally as:

Theorem 5.9.1 (Closed Graph Theorem)**.** *Let X and Y be Banach spaces. Any closed linear operator $A\colon X \to Y$ defined on the whole space X is bounded.*

Proof. A proof may be found in any classic reference on linear functional analysis, such as Dunford and Schwartz [1], Yosida [11], or Kreyszig [6]. □

Recall now that a bijective map is one that is both injective (one-to-one) and surjective (onto). An important result that we frequently use is the

Theorem 5.9.2 (Bounded Inverse Theorem)**.** *Let X and Y be Banach spaces. If the bounded linear operator $A\colon X \to Y$ is a bijection, then A^{-1} exists as a bounded linear operator.*

Proof. See again Dunford and Schwartz [1], Yosida [11], or Kreyszig [6]. □

As it turns out, both of these results are consequences of the Open Mapping Theorem, and in fact the three results are equivalent. The *Open Mapping Theorem* is regarded (as noted in Section 4.8) as being one of the three basic principles of linear functional analysis:

I: The Hahn-Banach Theorem

II: The Banach-Schauder (Open Mapping) Theorem

III: The Banach-Steinhaus Theorem (Principle of Uniform Boundedness)

We examined the Hahn-Banach Theorem in Section 4.8. Now that we are familiar with some of the basic definitions and properties of linear operators on Banach spaces, we can look briefly at the other two principles. First recall that a map $A\colon X \to Y$ from a Banach space X to a Banach space Y is called an *open map* if whenever $U \subset X$ is an open set in X, then $A(U) \subset Y$ is an open set in Y.

Theorem 5.9.3 (Banach-Schauder or Open Mapping Theorem)**.** *If X and Y are Banach spaces and $A\colon X \to Y$ is a surjective (onto all of Y) bounded linear operator, then A is an open map.*

Proof. See Volume I of Dunford and Schwartz [1]. □

The third principle states that for a family of bounded linear operators whose domain is a Banach space, pointwise boundedness is equivalent to uniform boundedness in the operator norm:

Theorem 5.9.4 (Banach-Steinhaus Theorem or Principle of Uniform Boundedness). *Let X be a Banach space, let Y be a normed space, and let F be a collection of bounded linear operators from X to Y. If for all $x \in X$,*

$$\sup_{A \in F} \|A(x)\|_Y < \infty,$$

then

$$\sup_{A \in F} \|A\|_{\mathcal{L}(X,Y)} < \infty.$$

Proof. A classic reference is again Volume I of Dunford and Schwartz [1]. □

In the theorem, the operator norm is just the usual operator norm induced by the spaces X and Y:

$$\|A\|_{\mathcal{L}(X,Y)} = \sup_{x \in X} \frac{\|Ax\|_Y}{\|x\|_X}.$$

This result is useful in the study of solutions to linear and nonlinear operator equations in Banach spaces, and we make use of it in Chapter 10.

We finish the discussion with a key tool we will need in Chapter 10; it is often referred to as either the *Banach Lemma* or the *Operator Perturbation Lemma*.

Lemma 5.9.1 (Banach Lemma). *Let X be a Banach space and let the bounded linear operator $A\colon X \to X$ have operator norm $\|A\| = \|A\|_{\mathcal{L}(X,X)} < 1$. Then the operator $[I - A]$ is invertible, and*

(1) $[I - A]^{-1} = \sum_{k=0}^{\infty} A^k$,

(2) $\|[I - A]^{-1}\| \leqslant \dfrac{1}{1 - \|A\|} < \infty$,

where the series in (1) converges in the operator norm on $\mathcal{L}(X, X)$.

Proof. The operator norm induced by the underlying Banach norm on X has the property $\|A^k\| \leqslant \|A\|^k$, $k \geqslant 0$, which together with the triangle inequality gives

$$\left\| \sum_{k=0}^{\infty} A^k \right\| \leqslant \sum_{k=0}^{\infty} \|A\|^k = \frac{1}{1 - \|A\|}.$$

This is due to the series of norms converging as a geometric series in $\mathbb{R}$ as a result of $\|A\| < 1$. This shows that the operator series (1) is absolutely convergent. Since the range space X of A is complete, the space of bounded linear operators $\mathcal{L}(X, X)$ is also complete and hence is a Banach space with respect to the operator norm. (See Section 10.1 for a discussion.) An absolutely convergent series in a Banach space is always convergent (see the book of Kreyszig [6]), hence the series in (1) is convergent.

It remains to show that the series representation in (1) is in fact the inverse operator to $[I - A]$. To this end, we form the telescoping sum

$$(I - A)(I + A + \cdots + A^n) = (I + A + \cdots + A^n)(I - A) = I - A^{n+1}.$$

Since $\|A\| < 1$, taking the limit as $n \to \infty$ ensures $A^{n+1} \to 0$. With $B = \sum_{k=0}^{\infty} A^k$, we thus have $(I - A)B = B(I - A) = I$, giving that $B = [I - A]^{-1}$. □

EXERCISES

5.9.1 Let X and Y be Banach spaces, and equip the vector space $X \times Y$ with the graph norm

$$\|(u, v)\|_{X;Y} = \|u\|_X + \|v\|_Y, \quad (u, v) \in X \times Y.$$

Prove that $X \times Y$ equipped with the graph norm is a Banach space.

5.9.2 Let X and Y be normed spaces with Y compact. Prove that if $A\colon X \to Y$ is a closed linear operator, then A is bounded.

5.9.3 Let X and Y be normed spaces with X compact. Prove that if $A\colon X \to Y$ is a closed linear operator that is also bijective, then A^{-1} is a bounded linear operator.

5.9.4 Let X and Y be Banach spaces, and let $A\colon X \to Y$ be a bounded linear operator that is injective (one-to-one). Prove that $A^{-1}\colon \mathcal{R}(A) \to X$ is a bounded linear operator if and only if $\mathcal{R}(A)$ is closed in Y.

5.9.5 Prove the following corollary to the Banach-Steinhaus Theorem: Let X and Y be Banach spaces, and let $\{A_j\}$ be a sequence of bounded linear operators $A_j\colon X \to Y$. If the sequence converges pointwise, so that $\lim_{j\to\infty} A_j u \in Y$ exists for all $u \in X$, then the pointwise limits define a bounded linear operator $A\colon X \to Y$.

5.9.6 Use the Banach-Steinhaus Theorem to prove that any weakly bounded subset $U \subset X$ is bounded.

REFERENCES AND ADDITIONAL READING

1. N. Dunford and J. T. Schwartz, *Linear Operators*, Part I: *General Theory*, Wiley, New York, 1957.
2. N. Dunford and J. T. Schwartz, *Linear Operators*, Part II: *Spectral Theory*, Wiley, New York, 1963.
3. N. Dunford and J. T. Schwartz, *Linear Operators*, Part III: *Spectral Operators*, Wiley, New York, 1971.
4. P. R. Halmos, *A Hilbert Space Problem Book*, Van Nostrand, New York, 1967.
5. S. Kesavan, *Topics in Functional Analysis and Applications*, Wiley, New York, 1989.

6. E. Kreyszig, *Introductory Functional Analysis with Applications*, Wiley, New York, 1990.
7. E. R. Lorch, *Spectral Theory*, Oxford University Press, New York, 1962.
8. A. W. Naylor and G. R. Sell, *Linear Operator Theory in Engineering and Science*, Holt, Rinehart and Winston, New York, 1971.
9. J. R. Retherford, *Hilbert Space: Compact Operators and The Trace Theorem*, Cambridge University Press, New York, 1993.
10. F. Riesz and B. Sz.-Nagy, *Functional Analysis*, Frederick Ungar, New York, 1955.
11. K. Yosida, *Functional Analysis*, Springer–Verlag, New York, 1980.
12. E. Zeidler, *Applied Functional Analysis*, Springer–Verlag, New York, 1995.
13. E. Zeidler, *Nonlinear Functional Analysis and Its Applications*, I: *Fixed-Point Theorems*, Springer–Verlag, New York, 1991.

CHAPTER 6

INTEGRAL EQUATIONS

6.1 INTRODUCTION

Integral equations arise directly in various mathematical and applied settings, but for our purposes their principal function is to provide alternative formulations of boundary value problems. Some of the advantages of the integral equation approach are as follows:

1. The integral operator appearing in the equation is a bounded operator and often compact, whereas the differential operator is unbounded.

2. The boundary conditions are incorporated in the integral equation through its kernel, which is a Green's function.

3. Associated with the integral equation are variational principles and approximation schemes that complement those arising from the formulation as a differential equation.

4. Some BVPs for partial differential equations can be translated into integral equations of lower dimensionality.

The following examples of integral equations give some idea of the scope of the present chapter; in each case the unknown function appears under the integral sign.

Example 1. Abel's integral equation

$$\int_0^y \frac{u(\eta)}{\sqrt{y-\eta}}\,d\eta = f(y), \quad y > 0, \tag{6.1.1}$$

where $f(y)$ is a *prescribed* function for $y > 0$, and $u(y)$ is to be found for $y > 0$. The problem arises in finding a curve C lying in the first quadrant, terminating at the origin, and having the following property: A particle starting from rest at the elevation y slides down C under the influence of gravity in the prescribed time $f(y)$. By using conservation of energy, we can then derive (6.1.1) with $u(y) = (2g)^{-1/2}ds/dy$, where g is the acceleration of gravity and $s(y)$ is the arclength along C measured from $y = 0$. If (6.1.1) is viewed purely mathematically, $f(y)$ may have to satisfy some conditions for a solution to exist. The physical interpretation imposes the further restriction that the solution u should generate a feasible arclength. This means that $u \geqslant (2g)^{-1/2}$ and therefore, from (6.1.1), that $f(y) \geqslant (2y/g)^{1/2}$, which states the obvious fact that the prescribed time of descent along a curve must exceed the time of free fall from the same elevation.

Example 2. The Fourier transform of $f(x)$ is defined as

$$f^\wedge(\omega) = \int_{-\infty}^{\infty} e^{i\omega x} f(x)\,dx, \quad -\infty < \omega < \infty. \tag{6.1.2}$$

To recover $f(x)$ from $f^\wedge(\omega)$ we must solve the integral equation (6.1.2). The answer is given by the inversion formula (2.4.3):

$$f(x) = \frac{1}{2\pi}\int_{-\infty}^{\infty} e^{-i\omega x} f^\wedge(\omega)\,d\omega. \tag{6.1.3}$$

Example 3. Let $H = L_2^{(c)}(-\infty, \infty)$, and let S_Ω be the linear manifold in H consisting of Ω-band-limited functions; thus $f(t) \in S_\Omega$ if $\int_{-\infty}^{\infty} |f(t)|^2\,dt < \infty$ and if its Fourier transform $f^\wedge(\omega)$ vanishes for $|\omega| \geqslant \Omega$. (See Section 2.4.) For functions of *unit energy* in S_Ω, Parseval's formula gives $(1/2\pi)\int_{-\Omega}^{\Omega} |f^\wedge(\omega)|^2\,d\omega = 1 = \int_{-\infty}^{\infty} |f(t)|^2\,dt$. We then pose the problem of finding the function of unit energy in S_Ω that maximizes the energy $\int_{-T}^{T} |f(t)|^2\,dt$ in a fixed, preassigned time interval $|t| \leqslant T$. The optimal function $f(t)$ is the eigenfunction corresponding to the largest eigenvalue of

$$\lambda f(t) = \frac{1}{\pi}\int_{-T}^{T} [\operatorname{sinc} \Omega(t-s)] f(s)\,ds, \quad -\infty < t < \infty, \tag{6.1.4}$$

where the sinc function is defined in (2.2.17).

It suffices to regard (6.1.4) as an equation on $-T < t < T$ to determine all eigenpairs $(\lambda_n, f_n(t))$ on $-T < t < T$. With f_n known for $|t| < T$, the right side of (6.1.4) has unambiguous meaning for all t, and we take this to be $\lambda_n f_n(t)$ for $|t| \geqslant T$.

Example 4. The normal modes $u(x)$ and normal frequencies λ of a vibrating membrane whose edge is fixed satisfy

$$-\Delta u = \lambda u, \quad x \in \Omega; \quad u|_\Gamma = 0, \tag{6.1.5}$$

where Ω is the plane domain covered by the membrane and Γ is its boundary. The equivalent integral equation is

$$u(x) = \lambda \int_\Omega g(x, \xi) u(\xi)\, d\xi, \quad x \in \Omega, \tag{6.1.6}$$

where $g(x, \xi)$ is Green's function satisfying (1.4.25). Note that x and ξ are position vectors in the plane, so that (6.1.6) is an integral equation involving functions of two independent variables. If Ω is a unit disk and polar coordinates r, ϕ are introduced, (6.1.6) becomes

$$u(r, \phi) = \lambda \int_0^1 r'\, dr' \int_0^{2x} g(r, \phi; r', \phi') u(r', \phi')\, d\phi',$$

where $g(r, \phi; r', \phi')$ is given explicitly by (8.3.13).

Example 5. The concentration $u(x)$ of a substance diffusing in an absorbing medium between parallel walls satisfies

$$-u'' + q(x)u = f(x), \quad 0 < x < 1; \quad u(0) = \gamma_1, \quad u(1) = \gamma_2, \tag{6.1.7}$$

where γ_1, γ_2 are the prescribed concentrations at the walls, $f(x)$ is the given source density, and $q(x)$ is the known absorption coefficient. If we knew Green's function for the differential operator in (6.1.7) we could write the solution u explicitly in the form (3.2.16). Usually, this Green's function is not known, however, so that we try to formulate the problem in terms of the simpler Green's function $g(x, \xi)$ satisfying

$$-g'' = \delta(x - \xi), \quad 0 < x, \xi < 1; \quad g(0, \xi) = g(1, \xi) = 0. \tag{6.1.8}$$

This function $g(x, \xi)$ is given explicitly by (1.1.3). Regarding $q(x)u$ as an additional inhomogeneous term, we can translate (6.1.7) into the integral equation

$$u(x) = \gamma_1 + (\gamma_2 - \gamma_1)x + \int_0^1 g(x, \xi)[f(\xi) - q(\xi)u(\xi)]\, d\xi, \quad 0 < x < 1. \tag{6.1.9}$$

If $q(x)$ is small, (6.1.9) can be used as the basis of a regular perturbation scheme (see Chapter 9). If, on the other hand, q is very large, which is equivalent to a small diffusion constant, the techniques of singular perturbation will be needed.

Example 6. Tomography. One of the most important medical applications of mathematics is the noninvasive detection of tumors. Such names as CAT (computer-aided tomography), MRI (magnetic resonance imaging), PET (positron emission tomography), and EIT (electric impedance tomography) have entered the popular lexicon.

As an electromagnetic wave travels in a straight line through a portion of the human body, it is absorbed at different rates by healthy and diseased tissues. By observing waves emanating from different directions as they have passed through the body, it may be possible to detect if and where tumors exist. For illustration we shall consider a much simplified two-dimensional model in which the tumor lies within a circular region (say, the cross section of a skull) of radius a with center at the origin of the xy plane. An electromagnetic beam traveling in the y direction is emitted at $(x, -a)$ and is detected at (x, a). The absorption coefficient $f(x, y)$ varies throughout the circular skull but vanishes outside. The intensity of the beam satisfies the differential equation

$$\frac{dI}{dy} = -fI,$$

so that

$$I(x, a) = I_0 \exp\left(-\int_{-\sqrt{a^2-x^2}}^{\sqrt{a^2-x^2}} f(x, y)\, dy\right),$$

where I_0 is the initial intensity of the emitted beam. From a knowledge of $I(x, a)$ for $|x| \leqslant a$, we would like to determine $f(x, y)$, but this is not possible, in general, by just considering beams in the y–direction. If, however, we confine ourselves to the case where f depends only on the distance r from the origin, then the preceding equation can be rewritten

$$p(x) = \int_x^a \frac{2r}{\sqrt{r^2 - x^2}} f(r)\, dr, \tag{6.1.10}$$

where

$$p(x) = \log[I_0 / I(x, a)].$$

The change of variables $\xi = a^2 - r^2$, $\eta = a^2 - x^2$ reduces (6.1.10) to Abel's integral equation of Example 1.

Example 7. The steady temperature $u(x)$ in a homogeneous medium occupying the domain $\Omega \subset \mathbb{R}^3$ with prescribed temperature $f(x)$ on the boundary Γ satisfies

$$-\Delta u = 0, \quad x \in \Omega; \quad u|_\Gamma = f(x). \tag{6.1.11}$$

This BVP is equivalent to the integral equation

$$\int_\Gamma \frac{1}{4\pi|x - \xi|} I(\xi)\, dS_\xi = f(x), \quad x \in \Gamma, \tag{6.1.12}$$

where once $I(x)$ has been found on Γ, we can calculate $u(x)$ in Ω from

$$u(x) = \int_\Gamma \frac{1}{4\pi|x - \xi|} I(\xi)\, dS_\xi. \tag{6.1.13}$$

Note that (6.1.12) is an integral equation on the two-dimensional surface Γ, whereas the original problem (6.1.11) is three-dimensional.

Example 8. A nonlinear integral equation. As we saw in Example 6, Section 4.4, the problem of radiation from the web coupling between two satellites leads to the nonlinear differential equation for the temperature $u(x)$:

$$-u'' - \lambda u^4 = 0, \quad 0 < x < 1; \quad u(0) = u(1) = 1. \tag{6.1.14}$$

By means of Green's function (6.1.8), the problem can be translated into the nonlinear integral equation

$$u(x) = 1 - \lambda \int_0^1 g(x, \xi) u^4(\xi)\, d\xi, \tag{6.1.15}$$

which was solved by successive approximations for a range of values of λ.

Integral Operators

The principal ingredient in a linear integral equation is a linear integral operator K defined by

$$z(x) = Ku = \int_\Omega k(x, \xi) u(\xi)\, d\xi, \quad x \in \Omega. \tag{6.1.16}$$

Here Ω is a region in $\mathbb{R}^n$ which is possibly unbounded; $x = (x_1, \ldots, x_n)$ and $\xi = (\xi_1, \ldots, \xi_n)$ are points in Ω; and $d\xi = d\xi_1 \cdots d\xi_n$ is an n-dimensional volume element. The complex-valued functions $u(x)$ and $z(x)$, defined on Ω, are regarded as elements of a linear (function) space X to be specified later. The transformation K maps X into itself and is clearly linear:

$$K(\alpha u_1 + \beta u_2) = \alpha K u_1 + \beta K u_2$$

for all scalars α, β and all elements u_1 and u_2 in X. The transformation takes place through the intermediary of a *kernel* $k(x, \xi)$, a complex-valued function defined on the $2n$-dimensional Cartesian product $\Omega \times \Omega$. In many problems $k(x, \xi)$ has a natural definition for x outside Ω, so that K can be viewed as mapping functions u defined on Ω into functions z defined on some other region. The drawback to this point of view comes when we try to invert K. If z is assigned on a region Ω' larger than Ω, (6.1.16) will usually not have any solution, whereas if z is specified on a region smaller than Ω, the data is insufficient to determine u (see, however, Example 3).

For simplicity, but with no real loss of generality, we confine ourselves to problems where Ω is an interval, possibly unbounded, on the real axis. Thus our integral operator K transforms functions defined on $a \leqslant x \leqslant b$ into functions defined on the same interval through the formula

$$z(x) = Ku = \int_a^b k(x, \xi) u(\xi)\, d\xi, \quad a \leqslant x \leqslant b, \tag{6.1.17}$$

where $k(x,\xi)$ is a given function on the square $a \leqslant x, \xi \leqslant b$. We shall try to develop an L_2 theory of integral equations, so that we can take advantage of all the structure of a Hilbert space. Thus the domain of K will be $L_2^{(c)}(a,b)$, the set of all complex-valued functions $u(x)$ with finite L_2 norm, that is,

$$\|u\|^2 = \int_a^b |u(x)|^2 \, dx < \infty.$$

This norm is generated by the inner product

$$\langle u, v \rangle = \int_a^b u(x)\overline{v}(x) \, dx.$$

If K operates on an arbitrary element u of $L_2^{(c)}(a,b)$, how can we guarantee that $z = Ku$ is again in $L_2^{(c)}(a,b)$? This places only very mild restrictions on the kernel $k(x,\xi)$. Hilbert-Schmidt kernels (Example 6, Section 5.1) generate integral operators that are not only bounded but will be shown to be compact; the Holmgren kernels of Exercise 5.1.3, also generate bounded operators. We recapitulate some of these definitions below and also extend the theory.

Recall that an operator K on $L_2^{(c)}(a,b)$ is bounded if there exists a constant M such that $\|Ku\| \leqslant M\|u\|$ for every u in $L_2^{(c)}(a,b)$. The norm of K is then defined as

$$\|K\| = \sup_{u \neq 0} \frac{\|Ku\|}{\|u\|}.$$

The operator K is compact if it transforms bounded sets into relatively compact sets.

Definition. *A kernel of the form $\sum_{i=1}^n p_i(x)\bar{q}_i(\xi)$ with n finite and*

$$\int_a^b |p_i(x)|^2 \, dx < \infty, \qquad \int_a^b |q_i(x)|^2 \, dx < \infty$$

is said to be separable *(or* degenerate*), and so is the corresponding integral operator defined by (6.1.17).*

Definition. *A kernel $k(x,\xi)$ for which*

$$\int_a^b \int_a^b |k(x,\xi)|^2 \, dx \, d\xi < \infty \tag{6.1.18}$$

is a Hilbert-Schmidt *(or* H.-S.*)* kernel*, and the corresponding integral operator is known as an* H.-S. operator.

Definition. *A kernel $k(x,\xi)$ for which*

$$\sup_{a \leqslant \eta \leqslant b} \int_a^b \int_a^b |k(x,\xi)| \, |k(x,\eta)| \, dx \, d\xi < \infty \tag{6.1.19}$$

is said to be a Holmgren kernel.

It is clear that a separable kernel is Hilbert-Schmidt. We showed in (5.1.9) that for an H.-S. operator the following estimate holds:

$$\|K\| \leqslant \left[\int_a^b \int_a^b |k(x,\xi)|^2\,dx\ d\xi\right]^{1/2}. \tag{6.1.20}$$

A Holmgren kernel also generates a bounded operator (see Exercise 5.1.3). We now come to the principal feature of H.-S. operators.

Theorem. *An H.-S. operator is compact.*

Proof.

(a) We first show that a separable operator S is compact. For each f in $L_2^{(c)}(a,b)$ we have

$$Sf = \int_a^b \sum_{i=1}^n p_i(x)\bar{q}_i(\xi)f(\xi)\,d\xi = \sum_{i=1}^n c_i p_i(x).$$

The range of S is finite-dimensional and S is clearly bounded, so that it is compact (see Section 5.7).

(b) It can be shown that if $\{e_i(x)\}$ is an orthonormal basis for $L_2^{(c)}(a,b)$, then $\{e_i(x)\bar{e}_j(\xi)\}$ is an orthonormal basis for $L_2^{(c)}(D)$, where D is defined as the square $a \leqslant x,\xi \leqslant b$.

(c) Let $k(x,\xi)$ be an H.-S. kernel; we can write $k(x,\xi) = \sum_{i,j=1}^{\infty} k_{ij}e_i(x)\bar{e}_j(\xi)$, where $k_{ij} = \int_a^b \int_a^b k(x,\xi)\bar{e}_i(x)e_j(\xi)\,dx\,d\xi$. By Parseval's identity

$$\sum_{i,j=1}^{\infty} |k_{ij}|^2 = \int_a^b \int_a^b |k(x,\xi)|^2\,dx\ d\xi,$$

and therefore

$$\sum_{i,j=n+1}^{\infty} |k_{ij}|^2 = \int_a^b \int_a^b \left|k(x,\xi) - \sum_{i,j=1}^{n} k_{ij}e_i(x)\bar{e}_j(\xi)\right|^2 dx\ d\xi.$$

By choosing n sufficiently large the last double integral can be made smaller than ε. With n so chosen let S_ε be the separable operator generated by the kernel $s_\varepsilon(x,\xi) = \sum_{i,j=1}^{n} k_{ij}e_i(x)\bar{e}_j(\xi)$. Clearly $K - S_\varepsilon$ is Hilbert-Schmidt and is generated by $k(x,\xi) - s_\varepsilon(x,\xi)$. Therefore, by (6.1.20), we have

$$\|K - S_\varepsilon\|^2 \leqslant \int_a^b \int_a^b |k(x,\xi) - s_\varepsilon(x,\xi)|^2\,dx\ d\xi < \varepsilon.$$

Thus K can be approximated in norm by compact operators and must itself be compact by Theorem 5.7.2.

□

An H.-S. kernel $k(x,\xi)$ always generates a *compact operator*, even though $k(x,\xi)$ may be unbounded as a function of x and ξ. For instance, let $a = 0, b = 1, k(x,\xi) = 1/|x-\xi|^\alpha$ with $0 < \alpha < \frac{1}{2}$; then (6.1.18) is satisfied, so that k generates an H.-S. operator; if, however, $\frac{1}{2} \leqslant \alpha < 1$, the operator is no longer Hilbert-Schmidt but is nevertheless compact (see Exercise 6.1.2). The kernel $e^{-|x-\xi|}$ on $-\infty < x,\xi < \infty$ is not Hilbert-Schmidt and generates an operator that is *not* compact; the operator is bounded, however, because it is a Holmgren operator (see Exercise 5.1.3). The kernel $e^{-x^2-\xi^2}$ on $-\infty < x,\xi < \infty$ is Hilbert-Schmidt.

If K is a bounded operator, its adjoint K^* is defined from

$$\langle Ku, v\rangle = \langle u, K^*v\rangle \quad \text{for all } u,v \in L_2^{(c)}(a,b). \tag{6.1.21}$$

By Theorem 5.8.1 we see that K^* is also bounded and that

$$\|K^*\| = \|K\|. \tag{6.1.22}$$

If K is an integral operator with kernel $k(x,\xi)$, then K^* is also an integral operator whose kernel $k^*(x,\xi)$ satisfies

$$k^*(x,\xi) = \overline{k}(\xi,x). \tag{6.1.23}$$

The operator K is *self-adjoint* (or *symmetric*, the two words being synonymous for bounded operators) if $K = K^*$. The kernel of a self-adjoint operator satisfies

$$k(x,\xi) = \overline{k}(\xi,x). \tag{6.1.24}$$

The operator $K^2 = KK$ is defined from

$$\begin{aligned} K^2u &= \int_a^b k(x,\xi_1)\,d\xi_1 \int_a^b k(\xi_1,\xi)u(\xi)\,d\xi \\ &= \int_a^b k_2(x,\xi)u(\xi)\,d\xi, \end{aligned}$$

where

$$k_2(x,\xi) = \int_a^b k(x,\xi_1)k(\xi_1,\xi)\,d\xi_1. \tag{6.1.25}$$

Thus K^2 is itself an integral operator whose kernel $k_2(x,\xi)$ can be obtained from k by (6.1.25). Similarly, K^m is an integral operator with kernel

$$k_m(x,\xi) = \int_a^b \cdots \int_a^b k(x,\xi_1)k(\xi_1,\xi_2)\cdots k(\xi_{m-1},\xi)\,d\xi_1\cdots d\xi_{m-1}, \tag{6.1.26}$$

requiring $m-1$ integrations. The kernels (6.1.25) and (6.1.26) are known as *iterated kernels*. We note that $k_m(x,\xi)$ can also be written as

$$k_m(x,\xi) = \int_a^b k(x,\xi_1)k_{m-1}(\xi_1,\xi)\,d\xi_1. \tag{6.1.27}$$

The iterated kernels will be symmetric if $k(x,\xi)$ is symmetric. In any event we find that

$$\|K^m\| \leqslant \|K\|^m.$$

EXERCISES

6.1.1 Consider the kernel $k(x,y) = 1/|x-y|^\alpha, a \leqslant x, y \leqslant b$, where a and b are finite and $\alpha < 1$.

(a) The corresponding integral operator K is Hilbert-Schmidt for $\alpha < \frac{1}{2}$. Show that K is compact for $\alpha < 1$. [*Hint*: Use Theorem 5.7.2, which states that K is compact if for each $\varepsilon > 0$ one can write K as the sum of a compact operator H and a bounded operator L with $\|L\| < \varepsilon$. The splitting required is of the type

$$k(x,y) = h(x,y) + l(x,y),$$

where l coincides with k on $|x-y| < \eta$ and *vanishes* for $|x-y| > \eta$.]

(b) Show that the iterated kernel $k_2(x,y)$ is bounded for $\alpha < \frac{1}{2}$ and satisfies $|k_2| \leqslant c/|x-y|^{2\alpha-1}$ for $\frac{1}{2} < \alpha < 1$.

6.1.2 Let D be a bounded domain in $\mathbb{R}^n$, $n \geqslant 3$, and let

$$k(x,\xi) = \frac{a(x,\xi)}{|x-\xi|^{n-2}}, \quad x,\xi \in D,$$

where $a(x,\xi)$ is continuous for x and ξ in $\bar{D}$.

(a) Show that for $n=3$ the operator K is Hilbert-Schmidt.

(b) Show that for $n \geqslant 3$ the operator K is compact (use the hint in Exercise 6.1.1).

6.2 FREDHOLM INTEGRAL EQUATIONS

We shall consider linear integral equations of the form

$$Ku - \lambda u = f, \quad \text{that is,} \int_a^b k(x,\xi)u(\xi)\,d\xi - \lambda u(x) = f(x), \quad a \leqslant x \leqslant b. \tag{6.2.1}$$

Here the unknown function $u(x)$ always appears under the integral sign and, unless $\lambda = 0$, also appears outside the integral. The kernel $k(x, \xi)$, the inhomogeneous term $f(x)$, and the complex parameter λ are regarded as given, although we shall want to study the dependence of the solution(s) on f and λ.

For $\lambda \neq 0$, (6.2.1) is known as a *Fredholm equation of the second kind* and, for $\lambda = 0$, as a *Fredholm equation of the first kind.* If $f(x) \equiv 0$, we have the *eigenvalue problem*

$$Ku - \lambda u = 0 \quad \text{or} \int_a^b k(x, \xi)u(\xi)\, d\xi = \lambda u(x), \quad a \leqslant x \leqslant b, \tag{6.2.2}$$

which for most values of λ can be expected to have the unique solution $u(x) \equiv 0$; the interest is in finding the exceptional values of λ (*eigenvalues*) for which (6.2.2) has nontrivial solutions $u(x)$ (*eigenfunctions*). Since K is a linear operator, the set of all eigenfunctions corresponding to the *same* eigenvalue forms a linear manifold. The dimension of this manifold is the *multiplicity* of the eigenvalue.

Our theory will be developed principally for the case where K is a compact integral operator (say, generated by an H.-S. kernel), but we will occasionally consider problems where K is only a bounded operator.

Before trying to solve (6.2.1) in a more or less explicit fashion, let us recall the alternative theorem in Section 5.7, which tells us about existence, uniqueness, and solvability conditions. As we know, an important role is played by the adjoint homogeneous problem

$$K^*v - \bar{\lambda}v = 0 \quad \text{or} \int_a^b \bar{k}(\xi, x)v(\xi)\, d\xi = \bar{\lambda}v(x), \quad a \leqslant x \leqslant b; \tag{6.2.3}$$

this features the operator $K^* - \bar{\lambda}I$, which is the adjoint of $K - \lambda I$ appearing in (6.2.1) and (6.2.2).

Theorem 6.2.1. *Let λ be fixed, $\lambda \neq 0$, and let K be* compact. *Then* either *of the following alternatives holds:*

(a) The number λ is not an eigenvalue of K and the number $\bar{\lambda}$ is not an eigenvalue of K^*. For each f, (6.2.1) has one and only one solution.

(b) The number λ is an eigenvalue of K having finite multiplicity k, say. The number $\bar{\lambda}$ is an eigenvalue of K^* having the same multiplicity. Equation (6.2.1) has solution(s) if and only if f is orthogonal to all k independent solutions of (6.2.3).

REMARKS

1. The case $\lambda = 0$ must be treated separately. The equation $Ku = f$ will have solutions only if f is orthogonal to the null space of K^*, but this condition will usually not be sufficient because the range of K may not be closed.
2. $\lambda = 0$ can be an eigenvalue of infinite multiplicity (see Example 1 below).
3. If K is bounded but not compact, even a nonzero eigenvalue can have infinite multiplicity.

Example 1. Consider the separable kernel

$$k(x, \xi) = \sum_{j=1}^{n} p_j(x)\bar{q}_j(\xi), \quad 0 \leqslant x, \xi \leqslant 1, \tag{6.2.4}$$

where each of the two sets $\{p_j\}$ and $\{q_j\}$ is assumed to be an independent set. (This is no restriction, for otherwise the kernel could be rewritten in this form with a smaller n.) We denote by M_P and M_Q the n-dimensional linear manifolds spanned by $\{p_j\}$ and $\{q_j\}$, respectively. If u is any vector in H, we have

$$Ku = \int_0^1 k(x, \xi)u(\xi)\, d\xi = \sum_{j=1}^{n} \langle u, q_j \rangle p_j(x). \tag{6.2.5}$$

(a) *The eigenvalue problem.* We see that (6.2.2) becomes

$$\sum_{j=1}^{n} \langle u, q_j \rangle p_j(x) = \lambda u(x), \quad 0 \leqslant x \leqslant 1. \tag{6.2.6}$$

Obviously, $\lambda = 0$ is an eigenvalue. In view of the independence of the set $\{p_j\}$, the corresponding eigenfunctions must satisfy

$$\langle u, q_1 \rangle = \cdots = \langle u, q_n \rangle = 0, \qquad \text{that is, } u \in M_Q^{\perp}.$$

For λ to be a nonzero eigenvalue of (6.2.6) we must have

$$u(x) = \sum_{j=1}^{n} c_i p_i(x),$$

which, on substitution in (6.2.6), leads to

$$\sum_{j=1}^{n} \langle p_j, q_i \rangle c_j = \lambda c_i, \quad i = 1, \ldots, n, \tag{6.2.7}$$

which is the eigenvalue problem for the $n \times n$ matrix $\alpha_{ij} = \langle p_j, q_i \rangle$. Thus any nonzero eigenvalue of (6.2.6) must be a nonzero eigenvalue of (6.2.7). Conversely, we can check that any nonzero eigenvalue λ of (6.2.7) is an eigenvalue of (6.2.6).

Let us look at some special cases with $n = 2$. If all four matrix elements vanish, (6.2.7) has no nonzero eigenvalues. This is the case for the kernel $\sin \pi x \sin 2\pi\xi + \sin 3\pi x \sin 4\pi\xi$, which therefore has only the zero eigenvalue with eigenfunctions u such that

$$\int_0^1 u(x) \sin 2\pi x\, dx = \int_0^1 u(x) \sin 4\pi x\, dx = 0.$$

If $k(x,\xi) = \sin\pi x \sin\pi\xi + \theta\cos\pi x\cos\pi\xi$, given $\theta \neq 0$, then (6.2.7) becomes the pair of equations

$$\tfrac{1}{2}c_1 = \lambda c_1, \quad \frac{1}{2}\theta c_2 = \lambda c_2,$$

from which we find (if $\theta \neq 1$) that there are two eigenvalues, $\lambda_1 = \frac{1}{2}$, $\lambda_2 = \theta/2$, with corresponding eigenfunctions $u_1 = A_1 \sin\pi x$, $u_2 = A_2\cos\pi x$, where A_1 and A_2 are arbitrary. If, however, $\theta = 1$, then $\lambda = \frac{1}{2}$ is a double eigenvalue with the eigenfunctions $A_1\sin\pi x + A_2\cos\pi x$. Regardless of the value of $\theta(\neq 0)$, we have $\lambda = 0$ as an eigenvalue with eigenfunctions satisfying

$$\int_0^1 u(x)\sin\pi x\,dx = \int_0^1 u(x)\cos\pi x\,dx = 0.$$

(b) *The inhomogeneous equation.* If $\lambda = 0$, (6.2.1) becomes

$$\sum_{j=1}^{n}\langle u, q_j\rangle p_j(x) = f(x), \quad 0 \leqslant x \leqslant 1. \tag{6.2.8}$$

A solution is possible *only if f is in M_P*. If this is the case, write $u = v + w$, with $v \in M_Q$, $w \in M_Q^{\perp}$. Equation (6.2.8) then states that

$$\sum_{j=1}^{n}\langle v, q_j\rangle p_j = f, \quad f \in M_P. \tag{6.2.9}$$

For each f in M_P there is therefore a unique v in M_Ω satisfying (6.2.9). In fact, if $f = \sum_{j=1}^{n}\xi_j p_j$, we need only find $v \in M_Q$ such that $\langle v, q_j\rangle = \xi_j$, $j = 1,\dots,n$, which has the unique solution $v = \sum_{j=1}^{n}\xi_j q_j^*$, where $\{q_j^*\}$ is the dual basis (for M_Q) of $\{q_j\}$. Thus the general solution of (6.2.8) is $u = v + w$ with an arbitrary $w \in M_Q^{\perp}$.

If $\lambda \neq 0$, (6.2.1) becomes

$$\sum_{j=1}^{n}\langle v, q_j\rangle p(x) - f(x) = \lambda u(x), \quad 0 \leqslant x \leqslant 1, \tag{6.2.10}$$

so that u must be of the form $\sum_{i=1}^{n} c_i p_i - f/\lambda$. Substituting in (6.2.10), we obtain the inhomogeneous algebraic system

$$\sum_{j=1}^{n}\langle p_j, q_i\rangle c_j - \lambda c_i = \frac{1}{\lambda}\langle f, q_i\rangle, \quad i = 1,\dots,n. \tag{6.2.11}$$

The corresponding homogeneous system is (6.2.7), and the usual theory for algebraic systems holds. In particular, if λ is not an eigenvalue of (6.2.7), (6.2.11) has a unique solution for each f. If this solution is $(c_1,\dots,c_n)$, then (6.2.10) has the unique solution $u = \sum_{i=1}^{n} c_i p_i(x) - f(x)/\lambda$.

Example 2. Consider the H.-S. kernel

$$k(x,\xi) = \begin{cases} 1-\xi, & 0 \leqslant x \leqslant \xi, \\ 1-x, & \xi \leqslant x \leqslant 1, \end{cases} \tag{6.2.12}$$

which is a continuous function on the square $0 \leqslant x,\ \xi \leqslant 1$ but has a discontinuous gradient along the diagonal $x = \xi$. This behavior suggests that k might be Green's function for a differential operator of the second order. If ξ is regarded as a fixed parameter, a simple calculation shows that

$$-\frac{d^2k}{dx^2} = \delta(x-\xi), \quad 0 < x, \xi < 1; \quad \frac{dk}{dx}(0,\xi) = 0, \quad k(1,\xi) = 0; \tag{6.2.13}$$

and k is indeed a Green's function. Note that (6.2.12) is the one and only solution of (6.2.13). From our previous work on differential equations, we can therefore conclude that the one and only solution $v(x)$ of

$$-v'' = f, \quad 0 < x < 1; \quad v'(0) = 0, \quad v(1) = 0 \tag{6.2.14}$$

is

$$v(x) = \int_0^1 k(x,\xi) f(\xi)\, d\xi \doteq Kf, \tag{6.2.15}$$

where K is the integral operator generated by the kernel $k(x,\xi)$. Thus K is the *inverse* of the differential operator $-(d^2/dx^2)$ defined on a domain of suitably smooth (precisely: v' absolutely continuous, v'' square integrable) functions $v(x)$ satisfying $v'(0) = v(1) = 0$. Note that the null space N_K consists only of the zero element. Indeed, if $Kf = 0$, then $v = 0$ and $-v'' = 0 = f$. The operator K in (6.2.15) is defined for all f in L_2, but the range R_K consists only of functions v which are sufficiently smooth and satisfy $v'(0) = v(1) = 0$. It is true, however, that $\bar{R}_K = L_2^{(c)}(0,1)$.

We shall now look at integral equations involving K. In the analysis we often transform the problem to an equivalent differential equation which we then solve. This may seem to deflate our earlier claim that integral equations are supposed to shed light on differential equations, rather than the other way around. For certain specific problems it may be simpler to solve the differential equation, but it is often easier to prove general theorems within the theory of integral equations.

(a) Consider first the eigenvalue problem

$$Ku - \lambda u = 0. \tag{6.2.16}$$

For $\lambda \neq 0$, any solution of (6.2.16) must be in R_K and is therefore twice differentiable. From (6.2.14) we have $(-Ku)'' = u$, so that

$$-u'' = \frac{1}{\lambda} u, \quad 0 < x < 1; \qquad u'(0) = u(1) = 0,$$

which we write (with $\theta = 1/\lambda$) in the standard eigenvalue form

$$-u'' = \theta u, \quad 0 < x < 1; \qquad u'(0) = u(1) = 0. \tag{6.2.17}$$

Thus the nonzero eigenvalues of (6.2.16) and (6.2.17) are *reciprocals* of each other, while the corresponding eigenfunctions are the same. It is straightforward to solve (6.2.17) explicitly. We find that

$$\theta_n = \frac{(2n-1)^2\pi^2}{4}, \qquad u_n(x) = c\,\cos(2n-1)\frac{\pi x}{2}.$$

Using Green's function (6.2.12), we see that any solution of (6.2.17) solves (6.2.16) with $\lambda = 1/\theta$. Therefore, the nonzero eigenvalues and normalized eigenfunctions of (6.2.16) are

$$\lambda_n = \frac{4}{(2n-1)^2\pi^2}, \qquad e_n(x) = \sqrt{2}\cos(2n-1)\frac{\pi x}{2}, \qquad n = 1, 2, \ldots. \tag{6.2.18}$$

We have already seen that N_K consists only of the zero element, so that $\lambda = 0$ is *not* an eigenvalue of (6.2.16). It is worth noting, however, that 0 is a limit point of the eigenvalues λ_n and the set $\{e_n(x)\}$ is an orthonormal basis for $L_2^{(c)}(0,1)$. These properties are shared by all symmetric compact operators for which 0 is not an eigenvalue.

(b) Let us now investigate the inhomogeneous equation of the second kind,

$$Ku - \lambda u = f, \quad \lambda \neq 0. \tag{6.2.19}$$

Set $z = \lambda u + f$, and note that $z \in R_K$. Thus $-z'' = u$, and

$$-z'' - \frac{z}{\lambda} = -\frac{f}{\lambda}, \quad 0 < x < 1; \qquad z'(0) = z(1) = 0. \tag{6.2.20}$$

Through the relation $z = \lambda u + f$, any solution u of (6.2.19) generates a solution z of (6.2.20), and vice versa. If $\lambda \neq \lambda_n$ [see (6.2.18)], then (6.2.20) has a unique solution which can be obtained by using Green's function $g(x, \xi; \lambda)$ satisfying

$$-g'' - \frac{g}{\lambda} = \delta(x-\xi), \quad 0 < x, \xi < 1; \qquad g'|_{x=0} = g|_{x=1} = 0. \tag{6.2.21}$$

If necessary, g can be calculated explicitly from (3.2.10). In any event, we have

$$u = \lambda^{-1}(z-f) = -\lambda^{-1}\left[\lambda^{-1}\int_0^1 g(x,\xi;\lambda)f(\xi)\,d\xi + f(x)\right]. \tag{6.2.22}$$

The solution can also be expressed in another form by expanding (6.2.19) or (6.2.20) in terms of the eigenfunctions $\{e_n\}$, a procedure which we carry out systematically in Section 6.4; see (6.4.8) and (6.4.10).

If λ in (6.2.19) coincides with an eigenvalue, say λ_k, then (6.2.20) and (6.2.19) can be solved only if $\langle f, e_k\rangle = 0$. The solution is then not unique. Problem

(6.2.21) is inconsistent, and a modified Green's function must be used (see Section 3.5). If $\lambda = 0$ in (6.2.19), we have an equation of the first kind. We have already seen that R_K consists only of twice-differentiable functions $v(x)$ satisfying $v'(0) = v(1) = 0$. If $f(x)$ is such a function, $Ku = f$ has the one and only solution $u = -f''$.

Example 3. Consider the kernel

$$k(x, \xi) = \frac{e^{-|x-\xi|}}{2} \quad \text{on the plane} \quad -\infty < x, \xi < \infty.$$

Since

$$\int_{-\infty}^{\infty} \int_{-\infty}^{\infty} |k(x, \xi)|^2 \, dx \, d\xi = \int_{-\infty}^{\infty} dx \int_{-\infty}^{\infty} \frac{e^{-2|y|}}{4} \, dy = \infty,$$

we do *not* have an H.S. kernel. However, the criterion of Exercise 5.1.3, for Holmgren kernels is satisfied so that the integral operator K generated by the kernel k is bounded, and $\|K\| \leqslant 1$. Exercise 6.2.2 shows that $\|K\| = 1$. Since K is also symmetric, it is self-adjoint and its spectrum is confined to the real axis.

(a) The eigenvalue problem $Ku - \lambda u = 0$ becomes

$$\int_{-\infty}^{\infty} \frac{e^{-|x-\xi|}}{2} u(\xi) \, d\xi - \lambda u(x) = 0, \quad -\infty < x < \infty. \tag{6.2.23}$$

The integral term is the convolution of the functions $e^{-|x|}/2$ and $u(x)$, so that Fourier transforms are indicated. Using the results of Chapter 2 (Example 1, Section 2.4, and Exercise 2.4.5), we find, with ω the real transform variable, that

$$u^{\wedge}(\omega) \left[\frac{1}{1 + \omega^2} - \lambda \right] = 0. \tag{6.2.24}$$

If λ is not in $0 < \lambda \leqslant 1$, the bracketed term does not vanish. If $\lambda = 1$, the bracketed term vanishes at $\omega = 0$, and, if $0 < \lambda < 1$, it vanishes at $\omega = \omega_1$ and $\omega = \omega_2$, where

$$\omega_{1,2} = \pm \left(\frac{1}{\lambda} - 1 \right)^{1/2}. \tag{6.2.25}$$

In any event the only L_2 solution of (6.2.24) is $u^{\wedge} = 0$, so that (6.2.23) has *no eigenvalues*. If for the moment we permit distributional solutions, then, for $0 < \lambda < 1$, we can have

$$u^{\wedge}(\omega) = c_1 \delta(\omega - \omega_1) + c_2 \delta(\omega - \omega_2),$$

which, on inversion, gives

$$u(x) = c_1 e^{-i\omega_1 x} + c_2 e^{-i\omega_2 x}. \tag{6.2.26}$$

This function is not in L_2 but nevertheless satisfies (6.2.23) in the classical sense. Thus one can view the function $e^{-i\omega_1 x}$ and $e^{-i\omega_2 x}$ as "pseudoeigenfunctions" corresponding to the value $\lambda(0 < \lambda < 1)$. A more satisfactory description is that the interval $0 \leqslant \lambda \leqslant 1$ is in the *continuous spectrum* of K. Since eigenvalues have already been ruled out, it suffices to show that there exists a sequence $\{u_n\}$ such that $\|Ku_n - \lambda u_n\|/\|u_n\|$ tends to 0 as $n \to \infty$. Now

$$(Ku_n)^\wedge = \frac{1}{1+\omega^2} u_n^\wedge,$$

and by Parseval's formula

$$\frac{\|Ku_n - \lambda u_n\|}{\|u_n\|} = \frac{\|(Ku_n)^\wedge - \lambda u_n^\wedge\|}{\|u_n^\wedge\|} = \frac{\|[1/(1+\omega^2) - \lambda]u_n^\wedge(\omega)\|}{\|u_n^\wedge(\omega)\|}.$$

It remains only to show that the interval $0 \leqslant \lambda \leqslant 1$ is in the continuous spectrum for the multiplication operator $1/(1+\omega^2)$ on $L_2^{(c)}(-\infty, \infty)$. This is essentially the content of Example 5, Section 5.6, and of Exercise 5.6.4. All one has to do is to take for $u_n^\wedge(\omega)$ the function which has the value 1 in a small interval about a point where $1/(1+\omega^2) - \lambda = 0$, and has the value 0 elsewhere [for the case $\lambda = 0$ this means that $u_n^\wedge(\omega)$ is taken to be 1 outside a large interval on the ω axis].

An alternative approach is to translate (6.2.23) into a differential equation. It is easy to see that $k(x, \xi)$ is a Green's function:

$$-\frac{d^2k}{dx^2} + k = \delta(x - \xi), \quad -\infty < x, \xi < \infty; \quad k(\pm\infty, \xi) = 0. \qquad (6.2.27)$$

Thus (6.2.23) is equivalent to $u - \lambda(-u'' + u) = 0$. If $\lambda = 0, u = 0$, and for $\lambda \neq 0$ we have

$$-u'' + u = \theta u, \quad \theta = \frac{1}{\lambda}, \quad -\infty < x < \infty. \qquad (6.2.28)$$

For $\theta \neq 1$ the general solution of (6.2.28) is

$$A\exp(\sqrt{1-\theta}\,x) + B\exp(-\sqrt{1-\theta}\,x),$$

which is never in $L_2^{(c)}(-\infty, \infty)$ for any value of θ in the complex plane (unless $A = B = 0$). For $\theta = 1$ the solution is $A + Bx$, which is not in L_2 unless $A = B = 0$. Thus we see that (6.2.28) has no eigenvalues, and neither does (6.2.23).

(b) The inhomogeneous equation $Ku - \lambda u = f$ becomes

$$\int_{-\infty}^{\infty} \frac{e^{-|x-\xi|}}{2} u(\xi)\,d\xi - \lambda u(x) = f(x), \quad -\infty < x < \infty. \qquad (6.2.29)$$

Given the element f in L_2 and the complex number λ, we want to find the L_2 solution(s) u of (6.2.29). In the absence of eigenvalues the solution, if any, is necessarily unique. Taking Fourier transforms, we obtain

$$u^\wedge\left[\frac{1}{1+\omega^2}-\lambda\right]=f^\wedge \quad \text{or} \quad u^\wedge=-\frac{f^\wedge}{\lambda}\left[\frac{\omega^2+1}{\omega^2+1-(1/\lambda)}\right], \qquad \lambda\neq 0.$$

In the second equality the term in brackets is further decomposed as

$$1+\frac{1}{\lambda}\frac{1}{\omega^2+1-(1/\lambda)},$$

where the second term now vanishes at $\omega=\pm\infty$. Thus we can write

$$u^\wedge=-\frac{f^\wedge}{\lambda}-\frac{f^\wedge}{\lambda^2}\left[\frac{1}{\omega^2+1-(1/\lambda)}\right]. \tag{6.2.30}$$

If λ is real but *not* in $0\leqslant\lambda\leqslant 1$, then $\omega^2+1-(1/\lambda)$ does not vanish on the real axis. In fact, we have

$$\omega^2+1-\frac{1}{\omega}=(\omega-\beta_1)(\omega-\beta_2), \quad \beta_{1,2}\doteq\pm i\left(1-\frac{1}{\lambda}\right)^{1/2}.$$

We are therefore entitled to set

$$\frac{1}{\omega^2+1-(1/\lambda)}\doteq b^\wedge(\omega),$$

where $b^\wedge$ is the Fourier transform of the L_2 function

$$b(x)=\frac{1}{2\pi}\int_{-\infty}^{\infty}e^{-i\omega x}b^\wedge(\omega)\,d\omega,$$

which can be evaluated as in Section 2.4 by using a contour in the lower half-plane when $x>0$ and one in the upper half-plane when $x<0$. A straightforward calculation gives

$$b(x)=\frac{1}{2[1-(1/\lambda)]^{1/2}}e^{-|x|[1-(1/\lambda)]^{1/2}},$$

and after use of the convolution theorem, (6.2.30) yields

$$u(x)=-\frac{f(x)}{\lambda}-\frac{1}{\lambda^2}\int_{-\infty}^{\infty}b(x-\xi)f(\xi)\,d\xi, \tag{6.2.31}$$

a result which can also be obtained by using a Green's function technique on the BVP equivalent to (6.2.29). Even if λ is complex, $\lambda\notin[0,1]$, (6.2.31) provides the one and only solution of (6.2.29) as long as $[1-(1/\lambda)]^{1/2}$ in the definition of $b(x)$ is chosen to have a positive real part. If λ is in the continuous spectrum, that is, if $\lambda\in[0,1]$, then (6.2.29) will not have an L_2 solution for every f in L_2. In fact, the bracketed term in (6.2.30) then has singularities on the real axis at $\omega_{1,2}$ given by (6.2.25) and is no longer an L_2 function of ω; however, if its product with f is in L_2 (as happens if f vanishes rapidly enough at $\omega_{1,2}$), then (6.2.31) can still be used. Such questions are examined systematically in Chapter 7.

Example 4. Let us take twice the kernel of Example 3 but over the finite square $-1 \leqslant x, \xi \leqslant 1$. The kernel is then Hilbert-Schmidt and again satisfies the differential equation (6.2.27) but with boundary conditions at the endpoints. For $x > \xi$, $k = e^{-(x-\xi)}$ and $k' = -e^{-(x-\xi)}$; therefore we have both $k(1, \xi) = e^{-(1-\xi)}$ and $k'(1, \xi) = -e^{-(1-\xi)}$. For $x < \xi$, $k = e^{x-\xi}$ and $k' = e^{x-\xi}$, so $k(-1, \xi) = e^{-1-\xi}$ and $k'(-1, \xi) = e^{-1-\xi}$. The boundary conditions can then be taken as

$$k'(1, \xi) + k(1, \xi) = 0, \quad k'(-1, \xi) - k(-1, \xi) = 0.$$

The integral equation

$$\int_{-1}^{1} e^{-|x-\xi|} u(\xi)\, d\xi = \lambda u(x), \quad -1 < x < 1, \tag{6.2.32}$$

is therefore equivalent to

$$-u'' + u = \theta u, \quad -1 < x < 1; \quad u'(1) + u(1) = u'(-1) - u(-1) = 0, \tag{6.2.33}$$

where $\theta = 2/\lambda$. Problem (6.2.33) yields an infinite set of eigenvalues with eigenfunctions forming an orthonormal basis.

Example 5. Consider the Fourier transform as an integral operator on the space $L_2^{(c)}(-\infty, \infty)$. Instead of denoting this operator by a capital letter in front of the function on which it acts, we use the caret ($^\wedge$) as a superscript following the function. Thus

$$f^\wedge(x) = \int_{-\infty}^{\infty} e^{ixy} f(y)\, dy \tag{6.2.34}$$

is regarded as defining an integral operator $\wedge$ from $L_2^{(c)}(-\infty, \infty)$ into itself. It follows from the inversion formula for Fourier transforms that

$$f^{\wedge\wedge}(x) = 2\pi f(-x) \quad \text{and} \quad f^{\wedge\wedge\wedge\wedge}(x) = 4\pi^2 f(x). \tag{6.2.35}$$

The eigenvalue problem for the $^\wedge$ operator is

$$u^\wedge(x) = \lambda u(x), \tag{6.2.36}$$

which implies that $u^{\wedge\wedge\wedge\wedge}(x) = \lambda^4 u(x)$. In view of (6.2.35), the only possible eigenvalues of (6.2.36) are

$$\lambda = \pm\sqrt{2\pi}, \quad \pm i\sqrt{2\pi}. \tag{6.2.37}$$

To find the corresponding eigenfunctions, let $f(x)$ be an even function in the space $L_2^{(c)}(-\infty, \infty)$. Then $f^\wedge(x)$ is also an even function of x. Let

$$u = f + \alpha f^\wedge,$$

so that

$$\begin{aligned} u^\wedge(x) &= f^\wedge(x) + \alpha f^{\wedge\wedge}(x) = f^\wedge(x) + 2\pi\alpha f(-x) = f^\wedge(x) + 2\pi\alpha f(x) \\ &= 2\pi\alpha\left[f(x) + \frac{1}{2\pi\alpha} f^\wedge(x)\right]. \end{aligned}$$

Thus $u^\wedge$ will satisfy (6.2.36) if $1/2\pi\alpha = \alpha$; that is, $\alpha = \pm(1/\sqrt{2\pi})$. The eigenvalues corresponding to these choices of α are $\pm\sqrt{2\pi}$. Clearly, each of these eigenvalues has infinite multiplicity. Start with any even function, add $1/\sqrt{2\pi}$ times its transform, and the resulting function will satisfy (6.2.36) with $\lambda = \sqrt{2\pi}$. For example,

$$u = e^{-a|x|} + \frac{1}{\sqrt{2\pi}} \frac{2a}{a^2 + x^2}$$

satisfies (6.2.36) with $\lambda = \sqrt{2\pi}$ for *any* value of $a > 0$. These functions are independent, so that $\sqrt{2\pi}$ has infinite multiplicity. Similarly,

$$u = e^{-ax^2} - \frac{1}{\sqrt{2a}} e^{-x^2/4a}$$

satisfies (6.2.36) with $\lambda = -\sqrt{2\pi}$. It can be shown that the eigenfunctions can be expressed in a natural way in terms of an orthonormal basis of Hermite functions (see Dym and McKean [2]).

EXERCISES

6.2.1 Solve (6.2.19) by expanding in the orthonormal basis (6.2.18). Make sure to treat the special cases $\lambda = \lambda_n$ and $\lambda = 0$.

6.2.2 Consider the integral operator K generated by a difference kernel:

$$Ku = \int_{-\infty}^{\infty} k(x - \xi) u(\xi)\, d\xi.$$

Observe that the kernel depends only on $x - \xi$ and stems from a function $k(x)$ of a single variable. Let $v = Ku$, and let $k^\wedge, u^\wedge, v^\wedge$ denote the Fourier transforms of $k(x), u(x), v(x)$, respectively. In the transform variable we have

$$v^\wedge(\omega) = k^\wedge(\omega) u^\wedge(\omega),$$

so that the integral operator is reduced to multiplication by the function $k^\wedge(\omega)$. Show that

$$\|K\| = \sup_{-\infty<\omega<\infty} |k^\wedge(\omega)|,$$

which applied to Example 3 gives $\|K\| = 1$. Another way of obtaining $\|K\| = 1$ for Example 3 is to show first that $\|K\| \leqslant 1$ (as was done in the text) and then to exhibit an L_2 sequence $\{u_n(x)\}$ such that

$$\lim_{n\to\infty} \frac{\|Ku_n\|}{\|u_n\|} = 1.$$

Show that the sequence

$$u_n(x) = \begin{cases} 1, & |x| < n, \\ 0, & |x| > n, \end{cases}$$

has the required property. This sequence was suggested by the fact that $u(x) = 1$ is a pseudoeigenfunction associated with $\lambda = 1$, the largest value in the spectrum.

6.2.3 Find all the eigenvalues and eigenfunctions in Example 4, and discuss the inhomogeneous equation.

6.2.4 Consider the eigenvalue problem $Ku - \lambda u = 0$ on $L_2^{(c)}(-1,1)$, where K is generated by the kernel $k(x,\xi) = 1 - |x - \xi|$. Translate into a differential equation with boundary conditions to find the eigenvalues and eigenfunctions.

6.2.5 In Example 5 exhibit eigenfunctions corresponding to the eigenvalues $i\sqrt{2\pi}$ and $-i\sqrt{2\pi}$.

6.2.6 Consider the kernel

$$k(x,\xi) = \sum_{n=1}^{\infty} \frac{[\sin(n+1)x] \sin n\xi}{n^2}, \quad 0 \leqslant x, \xi \leqslant \pi.$$

Show that this is an unsymmetric H.-S. kernel and that the corresponding integral operator K has *no* eigenvalue. Compute $\|K\|$. Show that every value of λ except $\lambda = 0$ is in the *resolvent* set [that is, $Ku - \lambda u = f$ has a unique solution for each f, and $(K - \lambda I)^{-1}$ is bounded]. Show that $\lambda = 0$ is in the compression spectrum (compare with the modified shift of Example 10, Section 5.1) and in the continuous spectrum.

6.2.7 Give an example of a bounded self-adjoint integral operator with a nonzero eigenvalue of infinite multiplicity. (*Hint*: Modify Example 5 by using a cosine or sine transform instead of a complex-exponential transform.)

6.3 THE SPECTRUM OF A SELF-ADJOINT COMPACT OPERATOR

On a finite-dimensional space a nontrivial operator has at least one eigenvalue; if the operator is also self-adjoint, the set of all eigenvectors forms an orthonormal basis. Neither of these results holds for infinite-dimensional spaces such as $L_2^{(c)}(a,b)$ even if we confine ourselves to bounded integral operators. Indeed, the Volterra operator of Example 7, Section 5.6, and the modified shift operator of Exercise 6.2.6 are not only bounded but also compact; neither of these unsymmetric operators has eigenvalues. The integral operator of Example 3, Section 6.2, is bounded (but not compact), is self-adjoint, and has no eigenvalues. However, if we restrict ourselves—as we shall in this section—to compact self-adjoint integral operators, much of the theory for finite-dimensional spaces can be extended.

Let $k(x,\xi)$ be a symmetric kernel on $a \leqslant x, \xi \leqslant b$, that is, $k(x,\xi) = \bar{k}(\xi, x)$; then k generates a self-adjoint operator K on $L_2^{(c)}(a,b)$. If k satisfies the Hilbert-Schmidt condition (6.1.18), K will be a compact operator. The eigenvalues are real, and eigenfunctions corresponding to different eigenvalues are orthogonal. To exclude the trivial operator we shall assume that $\|K\| \neq 0$. The existence of a nonzero eigenvalue then follows from the extremal principle of Theorem 5.8.8 and from its corollary. The eigenvalue λ_1 so characterized is the one of largest modulus, and $|\lambda_1| = \|K\|$. We restate the principle as follows.

Theorem 6.3.1. *Let K be a compact self-adjoint operator. Then*

$$\sup_{\|v\|=1} |\langle Kv, v\rangle| \tag{6.3.1}$$

is attained for a normalized eigenfunction $e_1(x)$ of K, corresponding to an eigenvalue λ_1 whose absolute value is just the maximum in question. No eigenvalue of K can have a larger modulus. We have $|\lambda_1| = \|K\|$ and therefore $\|Kv\| \leqslant |\lambda_1|\|v\|$ for every v in H.

Eigenvectors corresponding to other eigenvalues are orthogonal to e_1. This suggests maximizing $|\langle Kv, v\rangle|$ over unit functions orthogonal to e_1 and thus obtaining another eigenpair (λ_2, e_2) with $|\lambda_2| \leqslant |\lambda_1|$. Let M_1 be the manifold (line) generated by e_1, and let $M_1^\perp$ be the manifold (necessarily closed) of elements orthogonal to e_1. We then have the following theorem.

Theorem 6.3.2.

$$\sup_{\substack{\|v\|=1 \\ v \in M_1^\perp}} |\langle Kv, v\rangle| \tag{6.3.2}$$

is attained for a normalized eigenfunction $e_2(x)$ of K corresponding to an eigenvalue λ_2 whose modulus is the maximum in question. Moreover, $|\lambda_2| \leqslant |\lambda_1|$ and

$$\|Kv\| \leqslant |\lambda_2|\|v\| \quad \textit{for every } v \textit{ in } M_1^\perp. \tag{6.3.3}$$

REMARK. It is clear that the supremum (6.3.1) cannot be less than the supremum (6.3.2), since the functionals are the same but the latter supremum is over a subspace of the former. It is possible, however, for the suprema to be equal if the eigenvalue of largest modulus is degenerate or if both $\|K\|$ and $-\|K\|$ are eigenvalues.

Proof. $M_1^\perp$ is itself a Hilbert space. We claim that K can be regarded as an operator on $M_1^\perp$, that is, K transforms elements of $M_1^\perp$ into elements of $M_1^\perp$. Indeed, if $v \in M_1^\perp$, we have $\langle v, e_1\rangle = 0$ and

$$\langle Kv, e_1\rangle = \langle v, Ke_1\rangle = \lambda_1\langle v, e_1\rangle = 0.$$

Moreover, it is clear that K is compact and self-adjoint on $M_1^\perp$. Therefore, we can apply Theorem 6.3.1 to the operator K on the Hilbert space $M_1^\perp$ to obtain Theorem 6.3.2. □

We can now continue in this way to characterize successive eigenpairs (λ_1, e_1), ..., (λ_n, e_n), where $|\lambda_1| \geqslant |\lambda_2| \geqslant \cdots \geqslant |\lambda_n|$. The set $(e_1, \ldots, e_n)$ is an orthonormal set. Let M_n be the manifold generated by $(e_1, \ldots, e_n)$, and let $M_n^\perp$ be the orthogonal manifold, that is, the closed subspace of H consisting of functions orthogonal to all the elements $e_1(x), \ldots, e_n(x)$. We can then characterize the next eigenpair as follows.

Theorem 6.3.3.

$$\sup_{\substack{\|v\|=1 \\ v \in M_n^\perp}} |\langle Kv, v\rangle| \tag{6.3.4}$$

is attained for normalized eigenfunction $e_{n+1}(x)$ of K corresponding to an eigenvalue λ_{n+1} whose modulus is the maximum in question. Moreover, $|\lambda_{n+1}| \leqslant |\lambda_n|$, and

$$\|Kv\| \leqslant |\lambda_{n+1}| \|v\| \quad \textit{for every } v \in M_n^\perp. \tag{6.3.5}$$

Proof. The proof of Theorem 6.3.3 is similar to that of Theorem 6.3.2. □

There are two possible outcomes of the succession of maximum principles.

1. At some finite stage (say, $n+1$) we have for the first time that the supremum is 0. Thus $|\lambda_n| > 0$, but the supremum (6.3.4) is 0. All successive suprema obviously also vanish. Thus we have n nonzero eigenvalues (counted with multiplicity); clearly, $\lambda = 0$ is an eigenvalue of infinite multiplicity since every element in $M_n^\perp$ is a corresponding eigenfunction.

2. At no finite stage do we get a vanishing supremum. We therefore generate an infinite sequence of eigenvalues of positive modulus

$$|\lambda_1| \geqslant |\lambda_2| \geqslant \cdots \geqslant |\lambda_n| \geqslant \cdots > 0 \tag{6.3.6}$$

with a corresponding orthonormal set of eigenfunctions

$$e_1(x), \quad e_2(x), \ldots, \quad e_n(x), \ldots. \tag{6.3.7}$$

In case 1, the operator K is *separable*. Indeed, let z be any element in H, and decompose z by the projection theorem as the sum of an element in M_n and one in $M_n^\perp$:

$$z = \langle z, e_1\rangle e_1 + \cdots + \langle z, e_n\rangle e_n + v,$$

where $v \in M_n^\perp$. Since the supremum (6.3.4) is 0, (6.3.5) shows that $Kv = 0$. Thus

$$Kz = \sum_{j=1}^{n} \lambda_j \langle z, e_j\rangle e_j = \sum_{j=1}^{n} \lambda_j e_j(x) \int_a^b z(\xi)\bar{e}_j(\xi)\, d\xi = \int_a^b k(x,\xi) z(\xi)\, d\xi, \tag{6.3.8}$$

where

$$k(x, \xi) = \sum_{j=1}^{n} \lambda_j e_j(x) \bar{e}_j(\xi). \tag{6.3.9}$$

In case 2 we have an infinite sequence of nonzero eigenvalues of decreasing modulus. We show first that $\lambda_n \to 0$. If not, the sequence $\{e_n/\lambda_n\}$ would be bounded and $K(e_n/\lambda_n) = e_n$; by the compactness of K, $\{e_n\}$ would have to contain a convergent subsequence, which is clearly false. Thus $\lambda_n \to 0$.

The question immediately arises whether the orthonormal set $\{e_j\}$ of eigenfunctions corresponding to nonzero eigenvalues forms a basis. Obviously, it does not in case 1; even in case 2, where the set of eigenfunctions is infinite, it need not form a basis. We shall show, however, that the eigenfunctions corresponding to nonzero eigenvalues always form a basis for the *range* of K. This is an immediate conclusion from (6.3.8) for case 1. We need therefore consider only case 2. Let f be an arbitrary element of H; we want to show that

$$\lim_{n\to\infty} \left\| Kf - \sum_{k=1}^{n} \langle Kf, e_k \rangle e_k \right\| = 0. \tag{6.3.10}$$

An easy calculation gives

$$Kf - \sum_{k=1}^{n} \langle Kf, e_k \rangle e_k = K\left(f - \sum_{k=1}^{n} \langle f, e_k \rangle e_k \right).$$

The element $v \doteq f - \sum_{k=1}^{n} \langle f, e_k \rangle e_k$ belongs to $M_n^{\perp}$, so that (6.3.5) yields

$$\|Kv\| \leqslant |\lambda_{n+1}| \left\| f - \sum_{k=1}^{n} \langle f, e_k \rangle e_k \right\| \leqslant |\lambda_{n+1}| \|f\|.$$

Since $\lambda_n \to 0$ as $n \to \infty$, we have proved (6.3.10). The result is contained in the following theorem.

Theorem 6.3.4 (Expansion theorem). *Any function in the range of K can be expanded in a Fourier series in the eigenfunctions of K corresponding to nonzero eigenvalues. These eigenfunctions $\{e_k\}$ must therefore form an orthonormal basis for R_K, but not necessarily for H. Thus, for every f in H, we have*

$$Kf = \sum \langle Kf, e_k \rangle e_k = \sum \langle f, Ke_k \rangle e_k = \sum \lambda_k \langle f, e_k \rangle e_k, \tag{6.3.11}$$

even though the series $\sum \langle f, e_k \rangle e_k$ may not represent f. All equalities are of course understood in the L_2 sense.

We can now show that the extremal principles yield all the nonzero eigenvalues and their corresponding eigenfunctions. Suppose that there existed an eigenfunction v with eigenvalue $\lambda \neq 0$ not listed among the λ_k's. Then $Kv = \lambda v$, and by Theorem 6.3.4,

$$Kv - \sum_{k=1}^{\infty} \langle Kv, e_k \rangle e_k = \lambda \sum_{k=1}^{\infty} \langle v, e_k \rangle e_k.$$

But v is orthogonal to the e_k's so that $Kv = 0$, contradicting the assumption that v is an eigenfunction for a nonzero eigenvalue. Thus the sequence (6.3.6) contains all nonzero eigenvalues. Since $\lambda_n \to 0$, the multiplicity of every nonzero eigenvalue is finite (see also Section 5.7).

Suppose next that f is an arbitrary function not necessarily in R_K. The Riesz-Fischer theorem (Section 4.6) guarantees that $\sum \langle f, e_k \rangle e_k$ converges to some element, say g, in H. Elements Kf and Kg have the same expansion $\sum \lambda_k \langle f, e_k \rangle e_k$, so that the element $h = g - f$ is in the null space of K. Thus an arbitrary function f can be decomposed in one and only one way as

$$f = h + \sum \langle f, e_k \rangle e_k, \quad \text{where } Kh = 0. \tag{6.3.12}$$

Of course, h is just the projection of f on the null space of K. Thus h is an eigenfunction of K corresponding to the zero eigenvalue and is orthogonal to all the e_k's. We can restate the content of (6.3.12) as follows.

Theorem 6.3.5. *The set of all eigenfunctions of K (including those corresponding to the zero eigenvalue) forms a basis for H. The set $\{e_n\}$ of eigenfunctions corresponding to nonzero eigenvalues forms a basis for H if and only if $\lambda = 0$ is* not *an eigenvalue of K.*

This theorem has an important application to self-adjoint boundary value problems. Let L be a formally self-adjoint operator of order p with real coefficients (p has to be even), and let $B_1, \ldots, B_p$ be p boundary functionals of the form (3.3.2), so that $(L, B_1, \ldots, B_p)$ is self-adjoint. Consider the eigenvalue problem

$$Lu = \theta u, \quad a < x < b; \qquad B_1 u = \cdots = B_p u = 0. \tag{6.3.13}$$

We may assume without loss of generality that $\theta = 0$ is not an eigenvalue (otherwise we merely add a suitable term $\theta' u$ to both sides of $Lu = \theta u$ and relabel the operator and the eigenvalue parameter). We can then construct Green's function $g(x, \xi)$ satisfying

$$Lg = \delta(x - \xi), \quad a < x, \xi < b; \qquad B_1 g = \cdots = B_p g = 0.$$

The function $g(x, \xi)$ is real and symmetric, and (6.3.13) can be translated into the integral equation

$$u(x) = \theta \int_a^b g(x, \xi) u(\xi)\, d\xi, \qquad a < x < b,$$

which is of the form $Gu = \lambda u, \lambda = 1/\theta$. Since G is a symmetric, compact integral operator, Theorem 6.3.5 applies. Now the eigenfunctions of (6.3.13) are just the eigenfunctions of G corresponding to $\lambda \neq 0$. We claim that $\lambda = 0$ is not an eigenvalue of G. Indeed, the BVP

$$Lz = u, \quad a < x < b; \qquad B_1 z = \cdots = B_p z = 0$$

has the one and only solution

$$z = Gu,$$

and if $z = 0$, then $Lz = 0$ and $u = 0$.

Thus the eigenfunctions of G corresponding to $\lambda \neq 0$ and hence the eigenfunctions of (6.3.13) form an orthonormal basis.

Theorem 6.3.6. *The eigenfunctions of a self-adjoint boundary value problem form an orthonormal basis.*

Definite and Indefinite Operators

Let K be a *compact symmetric operator* defined on the whole of H. We recall from Section 5.8 that K is *nonnegative* if $\langle Kv, v\rangle \geqslant 0$ for all v and *positive* if $\langle Kv, v\rangle > 0$ for all $v \neq 0$. A compact operator cannot be coercive in the sense of Section 5.8. The fact that an operator is positive does *not* mean that its kernel is a positive function of x and ξ (See, however, Exercise 6.4.5). All eigenvalues of a positive operator are positive, and all eigenvalues of a nonnegative operator are nonnegative. A compact symmetric operator K, all of whose eigenvalues are positive, is a positive operator (the statement remains true if "nonnegative" is substituted for "positive" throughout). The reader can supply the definitions of nonpositive and negative operators and the obvious consequences. If, however, $\langle Kv, v\rangle$ is positive for some elements v and negative for others, K is said to be *indefinite*.

If K is a positive (or nonnegative) compact operator, we can reformulate the extremal principles as follows:

$$\lambda_1 = \sup_{\|v\|=1} \langle Kv, v\rangle, \ldots, \lambda_{n+1} = \sup_{\substack{\|v\|=1 \\ v\in M_n^\perp}} \langle Kv, v\rangle, \ldots, \tag{6.3.14}$$

where M_n is the manifold spanned by the eigenfunctions $e_1(x), \ldots, e_n(x)$. Similarly, for a nonpositive operator K, we have

$$\lambda_1 = \inf_{\|v\|=1} \langle Kv, v\rangle, \ldots, \lambda_{n+1} = \inf_{\substack{\|v\|=1 \\ v\in M_n^\perp}} \langle Kv, v\rangle, \ldots.$$

Even if K is an indefinite compact operator, we can develop separate principles for its positive and negative eigenvalues. Instead of listing all nonzero eigenvalues by decreasing modulus as in (6.3.6), we define two sequences of eigenvalues: $\lambda_1^+, \lambda_2^+, \ldots$ are the positive eigenvalues in decreasing order, and $\lambda_1^-, \lambda_2^-, \ldots$ are the negative eigenvalues in increasing order. The corresponding eigenfunctions are denoted by $\{e_n^+\}$ and $\{e_n^-\}$, respectively. By the expansion theorem (Theorem 6.3.4) we have, for any v in H,

$$\langle Kv, v\rangle = \sum \lambda_k |\langle v, e_k\rangle|^2 = \sum \lambda_k^+ |\langle v, e_k^+\rangle|^2 + \sum \lambda_k^- |\langle v, e_k^-\rangle|^2.$$

It therefore follows that

$$\lambda_1^- \|v\|^2 \leqslant \langle Kv, v\rangle \leqslant \lambda_1^+ \|v\|^2.$$

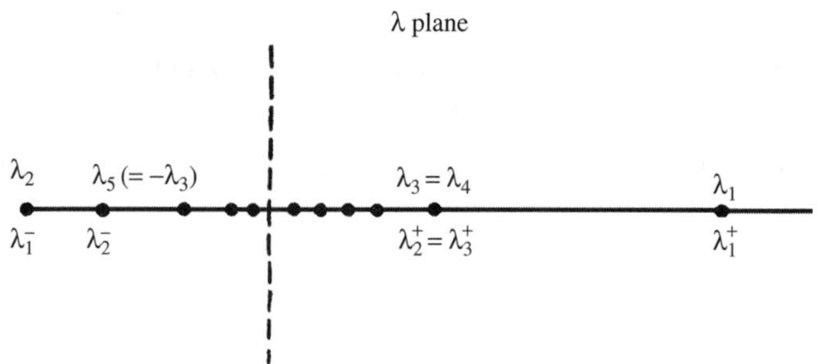

Figure 6.1 If there are only finitely many nonzero eigenvalues, $\lambda = 0$ is an eigenvalue (of infinite multiplicity). If there is an infinite number of nonzero eigenvalues, $\lambda = 0$ is an eigenvalue (of finite or infinite multiplicity) or is in the continuous spectrum.

One easily establishes the principles

$$\sup_{\substack{\|v\|=1 \\ v\in(M_n^+)^\perp}} \langle Kv, v\rangle = \lambda_{n+1}^+, \qquad \inf_{\substack{\|v\|=1 \\ v\in(M_n^+)^\perp}} \langle Kv, v\rangle = \lambda_{n+1}^-, \tag{6.3.15}$$

where M_n^+ is the manifold spanned by $(e_1^+, \ldots, e_n^+)$, and M_n^- the one spanned by $(e_1^-, \ldots, e_n^-)$.

In the general case of an indefinite, symmetric, compact operator, both $\{\lambda_n^-\}$ and $\{\lambda_n^+\}$ will be infinite sequences; the number 0 may or may not be an eigenvalue. In special cases either or both of the sequences $\{\lambda_n^-\}$, $\{\lambda_n^+\}$ may contain only a finite number of terms or be empty. Of course, if both sequences are finite, then 0 must be an eigenvalue of infinite multiplicity. Figure 6.1 illustrates the two different methods of indexing the eigenvalues of a compact self-adjoint operator.

Bilinear Expansions

Let K be the self-adjoint integral operator on $L_2^{(c)}(a, b)$, generated by the symmetric H.-S. kernel $k(x, \xi)$. The nonzero eigenvalues and corresponding orthonormal eigenfunctions are indexed as in (6.3.6) and (6.3.7). Thus we have

$$Ke_n = \lambda_n e_n \quad \text{and} \quad K^m e_n = \lambda_n^m e_n. \tag{6.3.16}$$

The self-adjoint integral operator K^m is generated by the symmetric iterated kernel $k_m(x, \xi)$ given by (6.1.26) or (6.1.27). Let us consider for fixed ξ the Fourier series of $k_m(x, \xi)$ in the orthonormal set $\{e_n(x)\}$:

$$k_m(x, \xi) \sim \sum_n \left(\int_a^b k_m(x, \xi)\bar{e}_n(x)\, dx \right) e_n(x).$$

Since $k_m(x,\xi) = \bar{k}_m(\xi,x)$ and $\int_a^b k_m(\xi,x)e_n(x)\,dx = \lambda_n^m e_n(\xi)$, we have

$$k_m(x,\xi) \sim \sum_n \lambda_n^m \bar{e}_n(\xi)e_n(x).$$

For $m \geqslant 2$, (6.1.27) shows that $k_m(x,\xi)$ is in R_K for each fixed ξ, so that the series converges to $k_m(x,\xi)$ in the mean square in the variable x for each fixed ξ. Obviously, the same is true in the variable ξ for each fixed x. Moreover, one can also show that if $k_2(x,\xi)$ is continuous, the following stronger theorem holds.

Theorem 6.3.7. *Let $k_2(x,\xi)$ be continuous on $a \leqslant x,\xi \leqslant b$; then the m-th iterated kernel has the* bilinear expansion

$$k_m(x,\xi) = \sum_n \lambda_n^m e_n(x)\bar{e}_n(\xi), \tag{6.3.17}$$

and the convergence is uniform on the square $a \leqslant x,\xi \leqslant b$.

REMARK. Even if $k(x,\xi)$ is unbounded, it may happen that $k_2(x,\xi)$ is continuous, as, for instance, in Exercise 6.1.2. For $m = 1$, (6.3.17) does not necessarily hold in the sense of uniform convergence even if $k(x,\xi)$ is continuous. We do, however, have the following theorem.

Theorem 6.3.8 (Mercer's Theorem). *If $k(x,\xi)$ is continuous and if all but a finite number of eigenvalues are of one sign, then*

$$k(x,\xi) = \sum \lambda_n e_n(x)\bar{e}_n(\xi) \quad \textit{uniformly on } a \leqslant x,\xi \leqslant b. \tag{6.3.18}$$

From (6.3.17) we conclude that

$$\sum_n \lambda_n^m = \int_a^b k_m(x,x)\,dx,$$

and if K is a *positive* operator, we obtain the *trace inequality*

$$\lambda_1 \leqslant \left[\int_a^b k_m(x,x)\,dx\right]^{1/m}, \tag{6.3.19}$$

which is valid for $m = 1$ if k is continuous.

EXERCISES

6.3.1 A standard method for constructing examples of integral operators with special properties is as follows. Begin with an infinite orthonormal set $\{e_n(x)\}$ on $L_2^{(c)}(a,b)$ which may or may not be a basis, and with a sequence of real numbers $\{\lambda_n\}$ of decreasing modulus such that $\sum |\lambda_n|^2 < \infty$. Construct the function

$$k(x,\xi) = \sum_{n=1}^{\infty} \lambda_n e_n(x)\overline{e}_n(\xi),$$

which is clearly a symmetric kernel.

(a) Show that the corresponding integral operator K has eigenpairs (λ_n, e_n) and $(0, u)$, where u is any element orthogonal to all the $\{e_n\}$. In particular, if $\{e_n\}$ is a basis, 0 is not an eigenvalue.

(b) Construct a symmetric kernel which has infinitely many nonzero eigenvalues and has 0 as an eigenvalue of multiplicity k (do both the case where k is finite and the case where k is infinite).

(c) Construct a symmetric kernel that has infinitely many positive and negative eigenvalues and for which 0 is not an eigenvalue. Show nevertheless that there are nontrivial elements u for which $\langle Ku, u\rangle = 0$.

6.3.2 The expansion theorem (6.3.11) no longer holds if K is not symmetric (in fact, K may have no eigenfunctions, yet $\overline{R}_K = H$; this is the case for the Volterra operator of Example 7, Section 5.6, which becomes a Fredholm operator by setting $k(x, y) = 0$ for $x < y < 1$). Something can be rescued, however, by introducing the left and right iterates: $L = K^*K$, $R = KK^*$.

(a) Show that these are both symmetric, nonnegative integral operators and that $N_L = N_K, N_R = N_{K^*}$.

(b) Show that L and R have the same positive eigenvalues, λ_n^2 (indexed in decreasing order with due respect to multiplicity), with $\lambda_n > 0$. If $\{v_n\}$ is a corresponding orthonormal set of eigenfunctions for L, show that $\{u_n = Kv_n/\lambda_n\}$ is an orthonormal set of eigenfunctions for R. If $\{u_n\}$ is an orthonormal set of eigenfunctions for R, show that $\{v_n = K^*u_n/\lambda_n\}$ is an orthonormal set of eigenfunctions for L. The numbers λ_n are known as the *singular values* of K and $\{\lambda_n, u_n, v_n\}$ is called a *singular system* for K.

From the theory of symmetric operators we know that $\{v_n\}$ is an orthonormal basis for $\overline{R}_L$ and hence for $(N_L)^\perp = (N_K)^\perp = \overline{R}_{K^*}$ [see (5.5.5)]. Similarly, $\{u_n\}$ is a basis for $\overline{R}_K$. Thus, for any $f \in \overline{R}_K$, $f = \sum\langle f, u_n\rangle u_n$; and, for any $f \in \overline{R}_{K^*}$, $f = \sum\langle f, v_n\rangle v_n$. Show that this implies that for any $f \in H$, the following expansions hold:

$$f = f_0 + \sum\langle f, v_n\rangle v_n, \quad (f_0 \in N_K), \tag{6.3.20}$$

$$f = f_0^* + \sum\langle f, u_n\rangle u_n, \quad (f_0^* \in N_{K^*}). \tag{6.3.21}$$

We now obtain a representation for nonsymmetric K somewhat similar to (6.3.11) for symmetric K. Since $Kf \in R_K$, we can write

$$Kf = \sum \langle Kf, u_n \rangle u_n = \sum \langle f, K^* u_n \rangle u_n = \sum \lambda_n \langle f, v_n \rangle u_n. \tag{6.3.22}$$

This is known as the *singular value decomposition* of K.

6.3.3 If K is a symmetric, positive H.-S. operator, show that

$$\lambda_1 = \lim_{m \to \infty} \left[\int_a^b k_m(x, x)\, dx \right]^{1/m},$$

where $k_m(x, \xi)$ is the mth iterated kernel of K.

6.3.4 As an illustration of Exercise 6.3.2, consider the Volterra operator of (5.1.10) (see also Example 7, Section 5.6). Find L and R. Show that the eigenvalue problem for L is equivalent to a boundary value problem whose eigenfunctions form the orthonormal basis of Exercise 5.1.4. Write explicitly the singular value decomposition of the Volterra operator.

6.4 THE INHOMOGENEOUS EQUATION

Neumann Series

The equation $Ku - \lambda u = f$ can be viewed as a fixed-point problem in a suitable metric space. With $\lambda \neq 0$ the equation can be rewritten as

$$Tu = u \quad \text{with } Tu \doteq \frac{1}{\lambda} Ku - \frac{f}{\lambda}. \tag{6.4.1}$$

In Chapter 4 we showed that for λ sufficiently large, T was a contraction mapping on the space of continuous functions with the uniform norm [on the assumption that the kernel $k(x, \xi)$ was a continuous function of x and ξ]. It is easy to reformulate the problem in an L_2 setting.

Assume that $k(x, \xi)$ is a kernel that generates a bounded integral operator K on $L_2^{(c)}(a, b)$. From

$$\|Tu - Tv\| \leqslant \frac{\|K\|}{|\lambda|} \|u - v\|,$$

it follows that T is a contraction on $L_2^{(c)}(a, b)$ for $|\lambda| > \|K\|$. Thus (6.4.1) has one and only one solution whenever $|\lambda| > \|K\|$. (This is, of course, already known from the fact that the spectrum of K is confined to $|\lambda| \leqslant \|K\|$.) Moreover, the solution u can be determined by iteration from an arbitrary initial element v_0:

$$u = \lim_{n \to \infty} v_n, \quad v_n = T^n v_0.$$

Thus

$$\begin{aligned} u &= \lim_{n\to\infty}\left[\frac{1}{\lambda^n}K^n v_0 - \sum_{m=1}^{n-1}\frac{K^m f}{\lambda^{m+1}} - \frac{f}{\lambda}\right] \\ &= -\frac{f}{\lambda} - \sum_{m=1}^{\infty}\frac{K^m f}{\lambda^{m+1}} = -\sum_{m=0}^{\infty}\frac{K^m f}{\lambda^{m+1}}, \end{aligned} \tag{6.4.2}$$

where K^0 is the identity operator. The operator K^m in (6.4.2) is an integral operator whose kernel is the iterated kernel (6.1.26). These iterated kernels are not always easy to calculate explicitly. Although we have shown only that the solution (6.4.2) is valid for $|\lambda| > \|K\|$, it is actually valid for $|\lambda| > |\lambda_c|$, where λ_c is the complex number of largest modulus in the spectrum of K.

Example 1. Consider the case of a one-term separable kernel

$$k(x,\xi) = p(x)\bar{q}(\xi), \quad 0 \leqslant x, \xi \leqslant 1,$$

for which

$$v \doteq Ku \doteq \int_0^1 p(x)\bar{q}(\xi)u(\xi)\,d\xi,$$

so that

$$\|v\|^2 = \|p\|^2|\langle u, q\rangle|^2 \leqslant \|p\|^2\|q\|^2\|u\|^2,$$

which becomes an equality if u is proportional to q. Thus

$$\|K\| = \|p\|\|q\|.$$

The iterated kernel k_2 is

$$\begin{aligned} k_2(x,\xi) &= \int_0^1 p(x)\bar{q}(\xi_1)p(\xi_1)\bar{q}(\xi)\,d\xi_1 \\ &= p(x)\bar{q}(\xi)\langle p, q\rangle, \end{aligned}$$

and similarly,

$$k_n(x,\xi) = \langle p, q\rangle^{n-1}p(x)\bar{q}(\xi).$$

Thus (6.4.2) becomes

$$u = -\frac{f}{\lambda} - \frac{1}{\lambda^2}p(x)\langle f, q\rangle\sum_{k=1}^{\infty}\left(\frac{\langle p, q\rangle}{\lambda}\right)^{k-1}. \tag{6.4.3}$$

The series converges for $|\lambda| > |\langle p, q\rangle|$, and, using the formula for a geometric series, we find that

$$u = -\frac{f(x)}{\lambda} - \frac{\langle f, q\rangle}{\lambda^2}\left(1 - \frac{\langle p, q\rangle}{\lambda}\right)^{-1} p(x). \tag{6.4.4}$$

We observe that (6.4.3) is the one and only solution of (6.4.1) for $|\lambda| > |\langle p, q\rangle|$. Since $\langle p, q\rangle$ is the only nonzero eigenvalue of K, we have confirmed that the Neumann series converges not only for $|\lambda| > \|K\|$ but even for $|\lambda| > |\lambda_c|$, where λ_c is the point in the spectrum having the largest modulus (of course, if K is self-adjoint, $|\lambda_c| = \|K\|$, as is easily checked in this particular problem since then $p = q$ and $|\langle p, q\rangle| = \|p\|\,\|q\| = \|K\|$). Expression (6.4.4) provides the unique solution of $Ku - \lambda u = f$ whenever $\lambda \neq 0$ and $\lambda \neq \langle p, q\rangle$, but, in general, it is not possible to perform explicitly the analytic continuation which enabled us to go from (6.4.3) to (6.4.4) in our very simple special case.

Solution by Eigenfunction Expansion

Let the nonzero eigenvalues of the self-adjoint compact operator K be indexed with due regard to multiplicity as

$$|\lambda_1| \geqslant |\lambda_2| \geqslant \cdots \geqslant |\lambda_n| \geqslant \cdots,$$

the corresponding *orthonormal* set of eigenfunctions being denoted by $e_1(x)$, $e_2(x)$, ..., $e_n(x)$, ... Although occasionally there may be only a finite number of such eigenvalues (if the kernel is separable), in most cases of interest the sequence $\{\lambda_k\}$ will be infinite and, as was shown earlier, $\lim_{k\to\infty} \lambda_k = 0$.

Consider the inhomogeneous equation

$$Ku - \lambda u = f, \tag{6.4.5}$$

where λ is a given complex number and f is an element in H. For a solution to exist, the element $z \doteq f + \lambda u$ must be in the range of K and so can be represented by its Fourier series in the set $\{e_k\}$:

$$z = Ku = \sum_k \gamma_k e_k, \qquad \gamma_k = \langle z, e_k\rangle = \langle Ku, e_k\rangle.$$

Since $\langle z, e_k\rangle = \langle f, e_k\rangle + \lambda\langle u, e_k\rangle$, and

$$\langle Ku, e_k\rangle = \langle u, Ke_k\rangle = \lambda_k\langle u, e_k\rangle,$$

we find that a solution u of (6.4.5) must satisfy

$$(\lambda_k - \lambda)\langle u, e_k\rangle = \langle f, e_k\rangle, \quad k = 1, 2, \ldots. \tag{6.4.6}$$

We must now consider a number of different cases.

1. If $\lambda \neq 0$ and λ is *not* one of the eigenvalues $\lambda_1, \lambda_2, \ldots$, then (6.4.6) gives

$$\langle u, e_k\rangle = \frac{\langle f, e_k\rangle}{\lambda_k - \lambda}, \quad \gamma_k = \lambda_k \frac{\langle f, e_k\rangle}{\lambda_k - \lambda}, \tag{6.4.7}$$

so that

$$u = \frac{z - f}{\lambda} = -\frac{f}{\lambda} + \sum_k \frac{\lambda_k}{\lambda(\lambda_k - \lambda)}\langle f, e_k\rangle e_k, \tag{6.4.8}$$

which is our candidate for the solution of (6.4.5). The coefficients of the series appearing in (6.4.8) tend to 0 more rapidly than $\langle f, e_k \rangle$ since the only possible limit point of $\{\lambda_k\}$ is 0. In view of the convergence of $\sum_k |\langle f, e_k \rangle|^2$, the series

$$\sum_k \left| \frac{\lambda_k}{\lambda(\lambda_k - \lambda)} \right|^2 |\langle f, e_k \rangle|^2$$

must also converge—in fact, more rapidly—and hence the series in (6.4.8) converges in L_2. One can therefore insert (6.4.8) in the integral equation and apply K term by term to verify that the equation is satisfied. Thus (6.4.8) is the one and only solution of $Ku - \lambda u = f$. By setting $f = \sum \langle f, e_k \rangle e_k + f_0$, where f_0 is the projection of f on N_K, we can rewrite (6.4.8) as

$$u = -\frac{f_0}{\lambda} + \sum_k \frac{1}{\lambda_k - \lambda} \langle f, e_k \rangle e_k, \tag{6.4.9}$$

which, despite converging more slowly than (6.4.8), has a simple appearance and consists only of orthogonal terms. By interchanging integration and summation, (6.4.8) can also be put in the form

$$u = -\frac{f}{\lambda} + \frac{1}{\lambda} \int_a^b R(x, \xi, \lambda) f(\xi)\, d\xi, \tag{6.4.10}$$

where

$$R(x, \xi, \lambda) = \sum_k \frac{\lambda_k}{\lambda_k - \lambda} e_k(x) \bar{e}_k(\xi) \tag{6.4.11}$$

is known as the *resolvent kernel* of K.

2. Let $\lambda = \lambda_m$, one of the nonzero eigenvalues of K. If λ_m is not degenerate, the first factor in (6.4.6) vanishes only for $k = m$. If $\langle f, e_m \rangle \neq 0$, (6.4.6) cannot be solved for $\langle u, e_m \rangle$ and $Ku - \lambda_m u = f$ has *no* solution. If $\langle f, e_m \rangle = 0$, $\langle u, e_m \rangle$ is arbitrary, but for $k \neq m$, $\langle u, e_k \rangle$ is still given by (6.4.7), so that $Ku - \lambda_m u = f$ has the infinite number of solutions

$$u = -\frac{f}{\lambda_m} + \sum_{k \neq m} \frac{\lambda_k}{\lambda_m(\lambda_k - \lambda_m)} \langle f, e_k \rangle e_k + c e_m, \quad \langle f, e_m \rangle = 0. \tag{6.4.12}$$

If λ_m is degenerate, a number of eigenvalues with successive indices will be equal to λ_m. Let these resonance indices be $m_1, \ldots, m_j$, where j is the multiplicity of λ_m; one of these indices will, of course, be m. Then, unless $\langle f, e_{m_i} \rangle = 0$ for each resonance index, $Ku - \lambda_m u = f$ will have *no* solution. If, however, these j *solvability conditions* are satisfied, the solutions of the inhomogeneous equation are given by

$$u = -\frac{f}{\lambda_m} + \sum_{k \neq m_i} \frac{\lambda_k}{\lambda_m(\lambda_k - \lambda_m)} \langle f, e_k \rangle e_k + \sum_{i=1}^{j} c_i e_{m_i}, \qquad \langle f, e_{m_i} \rangle = 0, \quad i = 1, \ldots, j, \tag{6.4.13}$$

which can also be written [see (6.4.9)] as

$$u = -\frac{f_0}{\lambda_m} + \sum_{k \neq m_i} \frac{1}{\lambda_k - \lambda_m} \langle f, e_k \rangle e_k + \sum_{i=1}^{j} c_i e_{m_i}. \tag{6.4.13a}$$

Note that the solution of minimum norm is obtained by setting all the c_i's equal to zero. If f does not satisfy the orthogonality conditions, then (6.4.13a) still furnishes least-squares solutions, and the one with minimum norm has $c_1 = \cdots = c_j = 0$. Thus the pseudoinverse of $K - \lambda_m I$ is the operator T defined by

$$Tf = -\frac{f_0}{\lambda_m} + \sum_{k \neq m_i} \frac{1}{\lambda_k - \lambda_m} \langle f, e_k \rangle e_k; \quad f \in H. \tag{6.4.13b}$$

3. If $\lambda = 0$, the inhomogeneous equation becomes $Ku = f$, and we need to characterize R_K. With u an arbitrary element of H, we have

$$Ku = \sum_k \lambda_k \langle u, e_k \rangle e_k.$$

Since the set $\{e_k\}$ is orthogonal to N_K, clearly f will have to be orthogonal to N_K. Then $f = \sum \langle f, e_k \rangle e_k$, and $Ku = f$ reduces to

$$\langle u, e_k \rangle = \frac{1}{\lambda_k} \langle f, e_k \rangle, \quad k = 1, 2, \ldots.$$

The series $\sum_k \langle u, e_k \rangle e_k$ will converge if and only if

$$\sum_k \frac{1}{|\lambda_k|^2} |\langle f, e_k \rangle|^2 < \infty. \tag{6.4.14}$$

To recapitulate, f is in R_K if and only if f is orthogonal to N_K and satisfies (6.4.14). The general solution of $Ku = f$ will then be

$$u = u_0 + \sum_k \frac{1}{\lambda_k} \langle f, e_k \rangle e_k, \tag{6.4.15}$$

where u_0 is an arbitrary element in N_K (that is, an arbitrary eigenfunction corresponding to $\lambda = 0$).

If the sequence $\{\lambda_k\}$ is in fact an infinite sequence, condition (6.4.14) implies that R_K is not closed but that its closure is the set of all elements orthogonal to N_K; in particular, $\bar{R}_K = H$ if 0 is not an eigenvalue. In this latter case a solution (necessarily unique) of $Ku = f$ exists if and only if (6.4.14) is satisfied; this solution is given by (6.4.15) without the u_0 term. The solution does not, however, depend continuously on the data. Let $f \in R_K$ and let us perturb f to $\tilde{f} = f + \varepsilon e_N$, where $\varepsilon > 0$ is small. The element $\tilde{f}$ still satisfies (6.4.14) and therefore belongs to R_K. The corresponding solution $\tilde{u}$ of the integral equation has been changed by $e_N\varepsilon/\lambda_N$, so that

$$\|\tilde{u} - u\| = \left|\frac{\varepsilon}{\lambda_N}\right|,$$

which since $\lambda_N \to 0$ as $N \to \infty$ can be made arbitrarily large by choosing N sufficiently large. Thus K^{-1} is clearly unbounded. The lack of continuous dependence on data (or *instability*) is characteristic of integral equations of the first kind. Such problems are said to be ill-posed or improperly posed, but they occur frequently in applications and are discussed at the end of the present section.

Example 2. Let

$$k(x, \xi) = p(x)\bar{p}(\xi), \quad 0 \leqslant x, \xi \leqslant 1.$$

Then there is only one nonzero eigenvalue $\lambda_1 = \int_0^1 |p|^2\, dx = \|p\|^2$ with normalized eigenfunction $e_1 = p(x)/\|p\|$.

If $\lambda \neq 0$, $\lambda \neq \lambda_1$, the one and only solution of $Ku - \lambda u = f$ is given by (6.4.8):

$$u = -\frac{f(x)}{\lambda} + \frac{\lambda_1 \langle f, p\rangle}{\lambda(\lambda_1 - \lambda)\|p\|^2} p(x) = -\frac{f(x)}{\lambda} + \frac{\langle f, p\rangle}{\lambda(\|p\|^2 - \lambda)} p(x),$$

which coincides with (6.4.4) when $p = q$.

The cases $\lambda = 0$ and $\lambda = \lambda_1$ are left as exercises.

Example 3. Consider the integral operator with difference kernel

$$Ku \doteq \int_{-\pi}^{\pi} k(\theta - \phi)u(\phi)\, d\phi, \quad -\pi \leqslant \theta \leqslant \pi. \tag{6.4.16}$$

In order for the function Ku to be defined on $-\pi \leqslant \theta \leqslant \pi$, values of $k(x)$ will be required for $-2\pi \leqslant x \leqslant 2\pi$. Not much can be done for general k, but *if k has period* 2π, the operator K simplifies considerably by using the convolution formula for Fourier series (see Exercise 2.3.8). Writing

$$u(\theta) = \sum \gamma_n e^{in\theta}, \quad k(\theta) = \sum k_n e^{in\theta},$$

we find that

$$Ku = 2\pi \sum k_n \gamma_n e^{in\theta}.$$

The eigenvalue problem $Ku - \lambda u = 0$ reduces to

$$(2\pi k_n - \lambda)\gamma_n = 0, \tag{6.4.17}$$

which must be satisfied for all integers, $-\infty < n < \infty$. This, in turn, requires that λ be one of the numbers $2\pi k_n$, the corresponding eigenfunction being proportional to $e^{in\theta}$. Of course, it may happen that some number $2\pi k_m$ occurs for a set of indices $m_1, \ldots, m_p$, which means that the eigenvalue $2\pi k_m$ has multiplicity p with eigenfunctions $c_1 e^{im_1\theta} + \cdots + c_p e^{im_p\theta}$.

As a particular case, consider the Poisson kernel that occurs in potential theory:

$$K(\theta) = \frac{1}{2\pi}\frac{1-r^2}{1+r^2-2r\cos\theta}, \qquad \text{where } r \text{ is fixed}, \quad 0<r<1.$$

We have already studied this kernel in Section 2.2, particularly in Exercise 2.2.9, where we obtained the Fourier expansion

$$k(\theta) = \frac{1}{2\pi} + \frac{1}{\pi}\sum_{n=1}^{\infty} r^n \cos n\theta = \frac{1}{2\pi}\sum_{n=-\infty}^{\infty} r^{|n|} e^{in\theta}.$$

The eigenvalues of K are therefore

$$\lambda_n = r^n, \quad n = 0, 1, 2, \ldots.$$

The eigenvalue $\lambda_0 = 1$ has the eigenfunction $e_0(\theta) = A_0$, whereas each of the other eigenvalues λ_n is degenerate with independent eigenfunctions $A_n \cos n\theta$, $B_n \sin n\theta$. For $\lambda \neq r^n$ the inhomogeneous equation $Ku - \lambda u = f$ has the one and only solution

$$u(\theta) = -\frac{f(\theta)}{\lambda} + \sum_{n=-\infty}^{\infty} \frac{r^{|n|}e^{in\theta}}{2\pi\lambda(r^{|n|}-\lambda)}\int_{-\pi}^{\pi} f(\theta)e^{-in\theta}\,d\theta.$$

If $\lambda = r^n$, a solution is possible only if $0 = \int_{-\pi}^{\pi} f(\theta)\cos n\theta\,d\theta = \int_{-\pi}^{\pi} f(\theta)\sin n\theta\,d\theta$.

Example 4. Consider again Example 3, Section 6.1. We are searching for an Ω-band-limited function $f(t)$ of unit energy which maximizes the energy in the time interval $|t| \leqslant T$:

$$E_T = \int_{-T}^{T} |f(t)|^2\,dt. \tag{6.4.18}$$

Of course, the maximum value of E_T is $\leqslant 1$, but we shall see below that it is strictly smaller than 1.

We shall set the problem in the functional framework of the Hilbert space $H = L_2^{(c)}(-\infty,\infty)$, on which we define the two closed linear manifolds:

M_T: the set of elements $f(t)$ vanishing for $|t| \geqslant T$,

S_Ω: the set of elements $f(t)$ which are Ω-band-limited.

The operator P_T, defined by

$$P_T f = \begin{cases} f(t), & |t| \leqslant T, \\ 0, & |t| > T, \end{cases} \tag{6.4.19}$$

is clearly the orthogonal projection on M_T. The operator Q_Ω, defined by

$$\Omega_\Omega f = \frac{1}{2\pi}\int_{-\Omega}^{\Omega} f^\wedge(\omega)e^{-i\omega t}\,d\omega, \tag{6.4.20}$$

is the orthogonal projection on S_Ω [as can be seen by using a slight modification of Parseval's formula (2.4.8)]. Both P_T and Q_Ω are symmetric nonnegative operators on H.

An easy calculation gives, for each f in H,

$$\begin{aligned} Q_\Omega P_T f &= \frac{1}{\pi}\int_{-T}^{T}[\text{sinc}\,\Omega(t-s)]f(s)\,ds \\ &\doteq \int_{-\infty}^{\infty} k(t,s)f(s)\,ds, \quad -\infty < t < \infty, \end{aligned} \tag{6.4.21}$$

where

$$k(t,s) = \begin{cases} \dfrac{1}{\pi}\text{sinc}\,\Omega(t-s), & |s| \leqslant T, \\ 0, & |s| > T, \end{cases} \tag{6.4.22}$$

For f in S_Ω, expression (6.4.18) for E_T becomes

$$E_T = \langle P_T f, f\rangle = \langle P_T f, Q_\Omega f\rangle = \langle Q_\Omega P_T f, f\rangle. \tag{6.4.23}$$

Since $Q_\Omega P_T$ transforms elements of S_Ω into elements of S_Ω, it can be viewed as an operator on the infinite-dimensional Hilbert space S_Ω. On S_Ω the operator $Q_\Omega P_T$ is symmetric and positive. Indeed, for f and g in S_Ω, we have

$$\langle Q_\Omega P_T f, g\rangle = \langle P_T f, Q_\Omega g\rangle = \langle P_T f, g\rangle = \langle f, P_T g\rangle = \langle Q_\Omega f, P_T g\rangle = \langle f, Q_\Omega P_T g\rangle.$$

Clearly, (6.4.23) shows that $Q_\Omega P_T$ is a nonnegative operator on S_Ω; it is also known that a nontrivial band-limited function cannot vanish on any time interval, so that $Q_\Omega P_T$ is actually a positive operator.

When $Q_\Omega P_T$ is regarded as an operator on $L_2^{(c)}(-\infty,\infty)$, it is an integral operator generated by the kernel (6.4.22). The calculation [see the Féjer kernel of Example 3(d), Section 2.2]

$$\int_{-\infty}^{\infty}\int_{-\infty}^{\infty} |k|^2\,ds\,dt = \int_{-T}^{T} ds \int_{-\infty}^{\infty} \frac{\sin^2 \Omega w}{\pi^2 w^2}\,dw = \frac{2T\Omega}{\pi} \tag{6.4.24}$$

shows that k is Hilbert-Schmidt and hence that $Q_\Omega P_T$ is compact on H and, a fortiori, on S_Ω. Thus $Q_\Omega P_T$ is a positive compact operator on S_Ω, therefore generating a sequence of eigenpairs (λ_n, e_n) with $\lambda_n > 0$ and $\{e_n\}$ an orthonormal basis. By the theory of Section 6.3 [see (6.3.14)], the largest eigenvalue λ_1 is the maximum of $\langle Q_\Omega P_T f, f\rangle$ over unit elements f in S_Ω. We see from (6.4.23) that λ_1 *is therefore just the desired maximal value of* E_T *and* $e_1(t)$ *is the optimal signal* $f(t)$. The pair (λ_1, e_1) therefore satisfies the integral equation

$$\lambda_1 e_1(t) = \int_{-\infty}^{\infty} k(t,s)e_1(s)\,ds, \quad -\infty < t < \infty,$$

that is,

$$\lambda_1 e_1(t) = \int_{-T}^{T} \frac{1}{\pi}[\text{sinc}\,\Omega(t-s)]e_1(s)\,ds, \quad -\infty < t < \infty. \tag{6.4.25}$$

Note that λ_1 and $P_T e_1$ are determined by looking at (6.4.25) as an integral equation on $-T \leqslant t \leqslant T$ (more precisely: $P_T e_1$ is determined up to a multiplicative constant). The right side then has meaning for all t and when divided by λ_1 provides the values of $e_1(t)$ for $|t| > T$; of course, $e_1(t)$ is determined up to a multiplicative constant, which is in turn calculated by the normalization requirement

$$\int_{-\infty}^{\infty} |e_1(t)|^2 \, dt = 1.$$

Let us calculate some rough bound for λ_1. By the trace inequality (6.3.19) with $m = 1$, we find that

$$\lambda_1 \leqslant \int_{-T}^{T} k(t,t) \, dt = \frac{2T\Omega}{\pi}.$$

Of course, this upper bound yields new information only if $\Omega T < \pi/2$. A lower bound is obtained from

$$\lambda_1 \geqslant \int_{-T}^{T} |f(t)|^2 \, dt, \quad \|f\| = 1, \quad f \in S_\Omega.$$

The simplest trial function is the function $f(t)$ of unit energy whose Fourier transform is constant for $|\omega| \leqslant \Omega$ and vanishes for $|\omega| > \Omega$. Then

$$f^\wedge(\omega) = \begin{cases} \left(\dfrac{\pi}{\Omega}\right)^{1/2}, & |\omega| \leqslant \Omega, \\ 0, & |\omega| > \Omega, \end{cases}$$

and we find that

$$f(t) = \frac{1}{2(\pi\Omega)^{1/2}} \int_{-\Omega}^{\Omega} e^{-i\omega t} \, d\omega = \frac{\sin \Omega t}{(\pi\Omega)^{1/2} t},$$

so that

$$\lambda_1 \geqslant \int_{-T}^{T} |f|^2 \, dt = \int_{-\Omega T}^{\Omega T} \frac{\sin^2 z}{\pi z^2} \, dz.$$

We note that most of these results are due to Landau, Pollak, and Slepian (see Dym and McKean [2]).

The Volterra Equation

Consider the integral equation

$$\int_{a}^{x} k(x,\xi) u(\xi) \, d\xi = \lambda u(x) + f(x), \quad a \leqslant x \leqslant b, \tag{6.4.26}$$

where the upper limit of integration is x instead of b. Such an equation can be considered as a special case of a Fredholm equation on $a \leqslant x \leqslant b$ with kernel

$$\tilde{k}(x,\xi) = \begin{cases} 0, & x < \xi, \\ k(x,\xi), & \xi < x. \end{cases}$$

From this point of view $\tilde{k}$ will *never* be symmetric, and therefore the results on the existence of eigenfunctions cannot be applied. However, the situation is quite a bit simpler than that for Fredholm equations, and it is easier to proceed directly from (6.4.26). Problem (6.4.26) has already been discussed in Example 4(b), Section 4.4, in the functional framework of $C[a,b]$ with the uniform norm. We now merely assume that

$$\int_a^b \int_a^b |\tilde{k}^2|\, dx\; d\xi = M < \infty$$

and that f is in $L_2^{(c)}(a,b)$. We then look for solutions u in $L_2^{(c)}(a,b)$. Using the same fixed-point techniques as in Chapter 4, we can show that no value of $\lambda \neq 0$ can be an eigenvalue; if $k(x,x)$ vanishes for some x, 0 can be an eigenvalue. For instance, $u = x^2$ is a solution of $\int_0^x (x - \frac{4}{3}\xi)u(\xi)\, d\xi = 0$, a phenomenon related to singular points of differential equations.

Since the initial value problem for a differential equation of order p with $a_p(x) \neq 0$ can be transformed into a Volterra equation, we now have the existence and uniqueness of solution for such an IVP in an L_2 framework.

Regularization for Equations of the First Kind

Earlier in this section, we noted that if a compact operator is invertible, then K^{-1} is unbounded, so that the solution u of the integral equation of the first kind $Ku = f$ does not depend continuously on the data f.

To focus on the core of the difficulty and to develop explicit remedies, we confine ourselves to the case when K is compact, symmetric, and positive (so that N_K consists only of the zero element). Then R_K is *not* closed, but $\bar{R}_K = H$. A typical operator of this type is the one of Example 2, Section 6.2, all of whose eigenvalues are indeed positive. The equation $Ku = f$ is then equivalent to

$$-f'' = u, \quad 0 < x < 1; \quad f'(0) = f(1) = 0. \tag{6.4.27}$$

Thus, we are seeking the distributed source $u(x)$ generating the given temperature $f(x)$ in a unit rod whose left end is insulated and whose right end is kept at zero temperature. Usually, we are not so fortunate as to be able to solve the integral equation $Ku = f$ by merely differentiating f, but let us take advantage of our good fortune to develop insight into the problem.

It is clear from (6.4.27) that $Ku = f$ is solvable only if f is twice differentiable and satisfies the boundary conditions, restrictions which are not apparent from the integral equation. The instability is obvious from (6.4.27): A small change in f, say by $\varepsilon e_N(x)$, with $\varepsilon > 0$ and ε_N given by (6.2.18) changes u by $\varepsilon(2N-1)^2(\pi^2/4)e_N(x)$, which is large if $N > 1/\varepsilon$. Continuity with respect to data or the lack of it must be stated in a particular norm; we are using the L_2 norm because of the Hilbert space setting, but the instability in this example also occurs in the supnorm: A small-norm, high-frequency error in measuring the temperature f yields a large error in u. Moreover, how are we to deal with a measured temperature which is not twice differentiable or fails to satisfy the boundary conditions exactly? We can no longer fall back on the notion of least-squares solutions since if $f \notin R_K$, it has no closest element in R_K but to each $\delta > 0$ there is an element in R_K which is within δ of f.

Let us now leave our special example and consider a compact symmetric positive integral operator K. The eigenvalues $\{\lambda_k\}$ of K then form an infinite sequence with

$$\lambda_k > 0, \quad \lambda_{k+1} \leqslant \lambda_k, \quad \lim_{k\to\infty} \lambda_k = 0.$$

Since zero is not an eigenvalue of K, (6.4.14) shows that f belongs to R_K if and only if

$$\sum_{k=1}^{\infty} \frac{|\langle f, e_k\rangle|^2}{\lambda_k^2} < \infty. \tag{6.4.28}$$

If (6.4.28) is satisfied, the solution of $Ku = f$ is found from (6.4.15) as

$$u = \sum_{k=1}^{\infty} \frac{1}{\lambda_k}\langle f, e_k\rangle e_k. \tag{6.4.29}$$

As was noted in the discussion of (6.4.15), the solution u does not depend continuously on f in the L_2-norm. In other words, the problem $Ku = f$ is ill-posed. How shall we stabilize the problem? The idea is as follows. Let the "true" data f belong to R_K with $Ku = f$; suppose that f is subject to a small measurement error so that the measured data is f^ε where $\|f - f^\varepsilon\| \leqslant \varepsilon$. The element f^ε belongs to H but may or may not belong to R_K. We then seek an "inversion" scheme which when applied to f^ε yields a solution $\tilde{u}$ with the property

$$\lim_{\varepsilon\to\infty} \tilde{u} = u \quad \text{whenever } f^\varepsilon \in H \quad \text{and} \quad \lim_{\varepsilon\to 0} f^\varepsilon = f. \tag{6.4.30}$$

Note that the correct inversion scheme (6.4.29) does not satisfy (6.4.30) as can be seen by letting

$$f^\varepsilon = f + \varepsilon e_{N(\varepsilon)},$$

where $N(\varepsilon)$ is chosen so that $\lambda_{N(\varepsilon)} \leqslant \varepsilon$; clearly, $f^\varepsilon \to f$, but

$$\tilde{u} = u + \frac{\varepsilon}{\lambda_{N(\varepsilon)}} e_N,$$

so that $\|\tilde{u} - u\| \geqslant 1$.

To regularize $Ku = f$, we proceed in two steps:

STEP 1. Construct a family T_α of *bounded* operators depending on a parameter $\alpha > 0$ so that as α tends to a distinguished value α_0:

$$\lim_{\alpha\to\alpha_0} T_\alpha K u = u \quad \text{for every } u \in H, \tag{6.4.31}$$

which implies that

$$\lim_{\alpha\to\alpha_0} T_\alpha f = K^{-1} f \quad \text{for every } f \in R_K. \tag{6.4.32}$$

Thus, for α near α_0, T_α and $T_\alpha K$ try to approximate K^{-1} and the identity, respectively. The nature of the approximation is clarified in Exercise 6.4.10, but it is clear that (6.4.32) requires $\|T_\alpha\| \to \infty$ as $\alpha \to \alpha_0$.

Now let f^ε be an element of H near the "true" $f(\|f - f^\varepsilon\| \leqslant \varepsilon)$. Our candidate for an approximation to u is u_α^ε corresponding to the approximate inverse T_α applied to the perturbed data f^ε:

$$u_\alpha^\varepsilon \doteq T_\alpha f^\varepsilon. \tag{6.4.33}$$

The error between u_α^ε and the true solution $u = K^{-1}f$ is

$$\begin{aligned}\|u_\alpha^\varepsilon - u\| = \|T_\alpha f^\varepsilon - u\| &\leqslant \|T_\alpha f^\varepsilon - T_\alpha f\| + \|T_\alpha f - u\| \\ &\leqslant \|T_\alpha\|\varepsilon + \|T_\alpha f - u\| = \|T_\alpha\|\varepsilon + \|T_\alpha K u - u\|.\end{aligned} \tag{6.4.34}$$

For u_α^ε to be a good approximation to u, the right-hand side should be small. The parameter α is at our disposal but the two terms pull in opposite directions. In view of (6.4.32) we need α near α_0 for the second term to be small, but $\|T_\alpha\| \to \infty$ as $\alpha \to \alpha_0$ so that α cannot be too close to α_0 in order for the first term to be small. This is the recurrent dilemma between *accuracy* ($T_\alpha f$ near u) and *stability* ($\|T_\alpha\|$ not too large).

STEP 2. To resolve the dilemma we choose α to *depend* on ε and to satisfy

$$\lim_{\varepsilon \to 0} \alpha(\varepsilon) = \alpha_0, \quad \lim_{\varepsilon \to 0} \varepsilon \|T_{\alpha(\varepsilon)}\| = 0. \tag{6.4.35}$$

The trick is to pick $\alpha(\varepsilon) \to \alpha_0$ so that $\|T_{\alpha(\varepsilon)}\|$ is $o(\varepsilon^{-1})$. Once this is done, $u_{\alpha(\varepsilon)}^\varepsilon$ can play the role of $\tilde{u}$ in (6.4.30).

Next, we illustrate steps 1 and 2 for two of the most popular regularization methods: (I) spectral cutoff and (II) Tychonov regularization.

I. Spectral Cutoff

The instability in (6.4.29) arises from the values of λ_k near zero (that is, corresponding to large k). This suggests chopping off all terms beyond N and replacing the infinite series by the finite sum

$$T_N f = \sum_{k=1}^{N} \frac{1}{\lambda_k} \langle f, e_k \rangle e_k, \tag{6.4.36}$$

which is defined for all f in H. The parameter α is N and $\alpha_0 = \infty$.

From (6.4.36) it follows that

$$T_N K u = \sum_{k=1}^{N} \frac{1}{\lambda_k} \langle K u, e_k \rangle e_k = \sum_{k=1}^{N} \frac{1}{\lambda_k} \langle u, K e_k \rangle e_k = \sum_{k=1}^{N} \langle u, e_k \rangle e_k,$$

so that

$$\lim_{N \to \infty} T_N K u = u.$$

and hence (6.4.31) and (6.4.32) are satisfied.

We also find that

$$\|T_N z\|^2 = \sum_{k=1}^{N} \frac{1}{\lambda_k^2} |\langle z, e_k \rangle|^2 \leqslant \frac{1}{\lambda_N^2} \|z\|^2,$$

with equality for $z = e_N$, so that

$$\|T_N\| = \frac{1}{\lambda_N}, \tag{6.4.37}$$

and (6.4.34) becomes, with $u_N^\varepsilon = T_N f^\varepsilon$,

$$\|u_N^\varepsilon - u\| \leqslant \frac{\varepsilon}{\lambda_N} + \|T_N f - u\| = \frac{\varepsilon}{\lambda_N} + \left\| \sum_{k=N+1}^{\infty} \langle u, e_k \rangle e_k \right\|. \tag{6.4.38}$$

Now we wish to choose $N(\varepsilon)$ so the right side tends to zero as $\varepsilon \to 0$. Of course, we need $N \to \infty$ to have the second term approach zero. Since $\lambda_k \to 0$ as $k \to \infty$, we can choose $N(\varepsilon) \to \infty$ *and* $\varepsilon/(\lambda_{N(\varepsilon)}) \to 0$ as $\varepsilon \to 0$ [say, by picking $N(\varepsilon)$ so that $\lambda_{N+1} \leqslant \sqrt{\varepsilon} \leqslant \lambda_N$].

The requirements (6.4.35) are therefore met and (6.4.38) then gives

$$\lim_{\varepsilon \to 0} \|u_{N(\varepsilon)}^\varepsilon - u\| = 0.$$

In Example 2, Section 6.2, where $\lambda_k = 4/\pi^2(2k-1)^2$, an acceptable choice for $N(\varepsilon)$ is the smallest integer exceeding $(\pi\varepsilon^{1/4})^{-1}$.

II. Tychonov Regularization

This procedure is based on replacing $Ku = f$ by an integral equation of the second kind

$$(K + \alpha I)u = f, \tag{6.4.39}$$

where $\alpha > 0$. Here we take $\alpha_0 = 0$ in (6.4.31). Because K is positive and $\alpha > 0$, the operator $K + \alpha I$ has a bounded inverse and by (6.4.9) the unique solution of (6.4.39) is

$$T_\alpha f = \sum_{k=1}^{\infty} \frac{1}{\lambda_k + \alpha} \langle f, e_k \rangle e_k, \tag{6.4.40}$$

which is valid for every $f \in H$.

It follows that

$$T_\alpha K u = \sum_{k=1}^{\infty} \frac{\lambda_k}{\lambda_k + \alpha} \langle u, e_k \rangle e_k$$

and

$$\|u - T_\alpha K u\|^2 = \sum_{k=1}^{\infty} \left(\frac{\alpha}{\lambda_k + \alpha} \right)^2 |\langle u, e_k \rangle|^2.$$

Since this last series is dominated by $\|u\|^2$ for all $\alpha > 0$, we may interchange summation and limit as $\alpha \to 0$ to obtain

$$\lim_{\alpha \to 0} T_\alpha K u = u, \tag{6.4.41}$$

which proves (6.4.31). Note that

$$\|T_\alpha z\|^2 = \sum_{k=1}^{\infty} \frac{1}{(\lambda_k + \alpha)^2} |\langle z, e_k \rangle|^2 \leqslant \frac{1}{\alpha^2} \sum_{k=1}^{\infty} |\langle z, e_k \rangle|^2 = \frac{\|z\|^2}{\alpha^2}.$$

Thus, we find $\|T_\alpha\| \leqslant 1/\alpha$ so that T_α is indeed a bounded operator. For the sequence of unit elements e_n, we find that

$$\|T_\alpha e_n\| = \frac{1}{\lambda_n + \alpha},$$

which tends to $1/\alpha$ as $n \to \infty$ so that, combining this with the previous bound on $\|T_\alpha\|$, we have

$$\|T_\alpha\| = \frac{1}{\alpha}. \tag{6.4.42}$$

Defining $u_\alpha^\varepsilon = T_\alpha f^\varepsilon$, we see that (6.4.34) becomes

$$\|u_\alpha^\varepsilon - u\| \leqslant \frac{\varepsilon}{\alpha} + \|T_\alpha K u - u\|.$$

The second term tends to zero as $\alpha \to 0$ by (6.4.41). By choosing $\alpha = \sqrt{\varepsilon}$ for instance, we ensure that

$$\lim_{\varepsilon \to 0} \|u_{\alpha(\varepsilon)}^\varepsilon - u\| = 0,$$

which completes the regularization procedure.

Extensions of the regularization methods to the case where K is not symmetric may be found in the books by Groetsch [3] and Kress [8].

EXERCISES

6.4.1 The function $k(x) = \log(1 - \cos x)$ has period 2π and mild singularities at $x = 2n\pi$. Equation (2.3.35) shows that its Fourier series is

$$-\log 2 - 2 \sum_{n=1}^{\infty} \frac{\cos nx}{n}.$$

Consider the integral operator on $L_2^{(c)}(-\pi, \pi)$:

$$Ku \doteq \int_{-\pi}^{\pi} k(x - \xi) u(\xi)\, d\xi,$$

and show that K is Hilbert-Schmidt and self-adjoint. Find all eigenfunctions and eigenvalues of K. Discuss the equation $Ku = f$.

6.4.2 Let $k(x)$ be an even, real function of period 2π such that $\int_{-\pi}^{\pi} k^2(x)\,dx < \infty$. The function k has the Fourier cosine series

$$k(x) = \frac{1}{2}k_0 + \sum_{n=1}^{\infty} k_n \cos nx.$$

Consider the integral operator on $L_2^{(c)}(-\pi, \pi)$:

$$Ku \doteq \int_{-\pi}^{\pi} k(x+\xi)u(\xi)\,d\xi$$

and find all its eigenvalues and eigenfunctions. Compare with Example 3.

6.4.3 Let $k(x)$ be an odd, real function of period 2π such that $\int_{\pi}^{\pi} k^2(x)\,dx < \infty$. Find the eigenvalues and eigenfunctions of the integral equation

$$\int_{-\pi}^{\pi} k(x+\xi)u(\xi)\,d\xi = \lambda u(x), \quad -\pi \leqslant x \leqslant \pi.$$

(Note that $\lambda = 0$ is always an eigenvalue in Exercise 6.4.3 but not necessarily in 6.4.2.)

6.4.4 For heat conduction (without sources or sinks) in a homogeneous infinite rod, the temperature $U(x,t)$ at time $t > 0$ is related to the initial temperature $u(x)$ by

$$U(x,t) = \frac{1}{\sqrt{4\pi t}} \int_{-\infty}^{\infty} e^{-(x-\xi)^2/4t} u(\xi)\,d\xi \doteq, Ku. \tag{6.4.43}$$

Suppose that we want to determine the initial temperature from observations on the temperature U at some given time $t > 0$, say $t = 1$ for definiteness. We then view (6.4.43) as an integral equation of the first kind for $u(x)$ with data $U(x,1)$ given on $-\infty < x < \infty$. Our kernel is not Hilbert-Schmidt (see Example 3, Section 6.2 for a similar phenomenon) but we can apply a Fourier transform on the space variable to try to solve (6.4.43). As in other ill-posed problems, a solution is possible only if the data satisfies certain restrictions. In particular, show that a solution is possible if $U(x,1)$ is band-limited (that is, $U^\wedge \equiv 0$ for $|\omega| > \omega_0$). Show that K^{-1} is unbounded but that $\bar{R}_K = L_2^{(c)}(-\infty, \infty)$.

6.4.5 Let $k(x,\xi)$ be a continuous symmetric H.-S. kernel that generates a nonnegative operator K. Show that $k(x,x)$ is pointwise nonnegative. (Note that k can be negative off the diagonal.)

6.4.6 Let A be a bounded operator on H. Show that if $\langle Au, u\rangle$ is real for all u, then A is symmetric. [Use the polar identity (4.5.18) with $b(u,v) \doteq \langle Au, v\rangle$.]

6.4.7 Consider a Volterra equation with a difference kernel:

$$\int_0^x k(x-y)u(y)\,dy - \lambda u(x) = f(x), \quad x > 0.$$

Apply a Laplace transform to both sides, and express u as an inversion integral. Specialize to the Abel equation (6.1.1), and by bringing the Laplace transform of f' into the picture, show that

$$u(x) = \frac{1}{\pi} \int_0^x \frac{f'(y)}{\sqrt{x-y}}\, dy,$$

which can be rewritten as

$$u(x) = \frac{1}{\pi} \frac{d}{dx} \int_0^x \frac{f(y)}{\sqrt{x-y}}\, dy.$$

What restrictions are placed on f in these two formulas?

6.4.8 (a) (Continuation of Exercise 6.3.2) Let K be a compact, nonsymmetric integral operator. Show that $Ku = f$ can have solutions if and only if *Picard's solvability conditions* are fulfilled:

$$f \in (N_{K^*})^{\perp} \quad \text{and} \quad \sum_n \frac{1}{\lambda_n^2} |\langle f, u_n \rangle|^2 < \infty. \tag{6.4.44}$$

Show that the corresponding solution(s) are given by

$$u = u_0 + \sum_n \frac{\langle f, u_n \rangle}{\lambda_n} v_n, \tag{6.4.45}$$

where u_0 is an arbitrary element of N_K.

(b) Apply part (a) to Example 7, Section 5.1.

6.4.9 Show that for a regularization procedure defined by (6.4.31), T_α is *not* uniformly bounded in α, $\lim_{\alpha \to \alpha_0} \|T_\alpha u - K^{-1}u\| = 0$ for each fixed u, but the convergence is not uniform with respect to u on the set $\|u\| = 1$, so

$$\lim_{\alpha \to \alpha_0} \|T_\alpha - K^{-1}\| \neq 0.$$

6.4.10 Consider spectral cutoff regularization: In (6.4.38), the choice of $N(\varepsilon)$ such that $\lambda_{N+1} \leqslant \sqrt{\varepsilon} \leqslant \lambda_N$ makes the term ε/λ_N of order $\sqrt{\varepsilon}$. What about the other term? We cannot say much about it, in general, except that it tends to zero as $\varepsilon \to 0$. Suppose, however, f is in the range not just of K but of K^2; then show that

$$\|T_N f - u\| \leqslant \lambda_{N+1} \|v\|,$$

where $K^2 v = f$. Thus, show that the choice of $N(\varepsilon)$ above will make the entire right-hand side of (6.4.38) of order $\sqrt{\varepsilon}$.

6.4.11 Tychonov regularization admits a variational formulation which is particularly simple in our case of positive K. Then $K + \alpha I$ is strongly positive for $\alpha > 0$ and Theorem 5.8.10 applies. The minimizer of

$$J_\alpha(v) = \langle Kv, v \rangle - 2\langle v, f^\varepsilon \rangle + \alpha \|v\|^2$$

is precisely the solution of (6.4.39) with $f = f^{\varepsilon}$. The advantage of the variational approach is that trial functions v with a few unknown parameters can be substituted in J_α and the optimal values of the parameters can be calculated by minimizing J_α. There is then no need to find the eigenfunctions $\{e_k\}$, which could be a laborious task.

6.5 VARIATIONAL PRINCIPLES AND RELATED APPROXIMATION METHODS

We want to characterize the eigenvalues of a self-adjoint operator by extremal principles which can be used for approximate calculations. Our interest in this chapter is in compact operators, but the method applies with slight modifications to bounded operators (or even operators that are only bounded below or bounded above) as long as either the upper or lower part of the spectrum consists of eigenvalues.

To avoid the notational difficulties associated with the general case, let us confine the analysis to a *self-adjoint, nonnegative, compact* operator K. We will make remarks later about possible extensions. The spectrum of K consists of positive eigenvalues

$$\lambda_1 \geqslant \lambda_2 \geqslant \cdots \geqslant \lambda_n \geqslant \cdots > 0 \tag{6.5.1}$$

listed with due regard to multiplicity, and of $\lambda = 0$, which either is an eigenvalue or is in the continuous spectrum. We shall be concerned only with the positive eigenvalues of K and their corresponding orthonormal eigenfunctions $e_1, \dots, e_n, \dots$. The span of $\{e_1, \dots, e_n\}$ is denoted by M_n (thus M_n is an n-dimensional manifold in H). The set (6.5.1) of eigenvalues can be finite or infinite; *in the former case we agree to define* $\lambda_n = 0$ *for* $n > m$, *where m is the number of nonzero eigenvalues*. In this way we always have an infinite sequence $\{\lambda_n\}$, which proves convenient in the statement of theorems. If $\lambda_n = 0$, the corresponding e_n is understood to be an element of N_K.

The *Rayleigh quotient* is defined as

$$R(u) = \frac{\langle Ku, u\rangle}{\|u\|^2}, \tag{6.5.2}$$

where it is *always* assumed that $\|u\| \neq 0$. Theorem 6.5.3 can be restated as follows.

Theorem 6.5.1.

$$\max_{u \in M_n^{\perp}} R(u) = \lambda_{n+1}. \tag{6.5.3}$$

Proof. By (6.3.11) we have, for u in H,

$$\langle Ku, u\rangle = \sum_k \lambda_k |\langle u, e_k\rangle|^2.$$

If $u \in M_n^{\perp}$, $\langle u, e_1\rangle = \cdots = \langle u, e_n\rangle = 0$ and

$$\langle Ku, u\rangle = \sum_{k>n} \lambda_k |\langle u, e_k\rangle|^2 \leqslant \lambda_{n+1} \sum_{k>n} |\langle u, e_k\rangle|^2 \leqslant \lambda_{n+1}\|u\|^2,$$

so that $R(u) \leqslant \lambda_{n+1}$. On the other hand, $e_{n+1} \in M_n^\perp$ and $R(e_{n+1}) = \lambda_{n+1}$, which proves the theorem. (If $n = 0$, the condition $u \in M_n^\perp$ is understood to be absent and the proof is the same.) □

It is possible to characterize λ_{n+1} by an extremal principle that does not refer to the eigenfunctions $e_1, \ldots, e_n$, which, after all, are unknown. The idea is to first maximize $R(u)$ subject to orthogonalization with respect to a "wrong" set of functions $\{v_1, \ldots, v_n\}$ whose span is E_n rather than M_n. Next we choose E_n to minimize this maximum. We find that the minimax is just λ_{n+1}.

Theorem 6.5.2 (Weyl-Courant Minimax Theorem). *Set*

$$\nu(E_n) = \max_{u \in E_n^\perp} R(u),$$

where, as indicated, ν will depend on the choice of E_n. Then

$$\lambda_{n+1} = \min_{\substack{\text{over all} \\ \text{choices of } E_n}} \nu(E_n),$$

or, combining the two statements, we obtain

$$\lambda_{n+1} = \min_{E_n \in S_n} \max_{u \in E_n^\perp} R(u),$$

where S_n is the set of all n-dimensional manifolds in H.

Proof. First we note that by Theorem 6.5.1 of Chapter 6, $\nu(M_n) = \lambda_{n+1}$, so that clearly, $\min \nu(E_n) \leqslant \lambda_{n+1}$. To prove the reverse inequality, and hence the theorem, it is sufficient to exhibit for each choice of E_n an element w in $E_n^\perp$ such that $R(w) \geqslant \lambda_{n+1}$. Let us try an element w of the form

$$w = c_1 e_1 + \cdots + c_{n+1} e_{n+1},$$

with $c_1, \ldots, c_{n+1}$ chosen so that $\|w\| \neq 0$ and $0 = \langle w, v_1 \rangle = \cdots = \langle w, v_n \rangle$, where $\{v_1, \ldots, v_n\}$ is a basis for E_n. Such a choice is always possible since these conditions reduce to finding a nontrivial solution to a homogeneous system of n equations in $n+1$ unknowns. Now

$$R(w) = \frac{\langle Kw, w \rangle}{\|w\|^2} = \frac{\sum_{k=1}^{n+1} \lambda_k |c_k|^2}{\sum_{k=1}^{n+1} |c_k|^2} \geqslant \lambda_{n+1},$$

which completes the proof. □

Ritz-Rayleigh Procedure

The simplest practical procedure for estimating eigenvalues starts from taking $n = 0$ in Theorem 6.5.1:

$$\lambda_1 = \max R(u), \tag{6.5.4}$$

and instead of taking the maximum of R over all u we take the maximum only for elements of the form $c_1v_1 + \cdots + c_kv_k$, where $v_1, \ldots, v_k$ is a fixed, judiciously chosen set of independent functions in H. Denoting the span of $\{v_1, \ldots, v_k\}$ by E_k, we then try to calculate

$$\max_{u \in E_k} R(u) = \max_{c_1, \ldots, c_k} R(c_1v_1 + \cdots + c_kv_k). \tag{6.5.5}$$

Since the functional in (6.5.5) is the same as that in (6.5.4) but the maximum is taken over a subspace rather than the whole of H, it follows that (6.5.5) yields a value that cannot exceed λ_1. Let us try to calculate the maximum of (6.5.5) as explicitly as possible. We have

$$R(c_1v_1 + \cdots + c_kv_k) = \frac{\left\langle \sum_{i=1}^{k} c_iKv_i, \sum_{j=1}^{k} c_jv_j \right\rangle}{\left\langle \sum_{i=1}^{k} c_iv_i, \sum_{j=1}^{k} c_jv_j \right\rangle} = \frac{\sum_{i,j=1}^{k} c_i\bar{c}_jk_{ij}}{\sum_{i,j=1}^{k} c_i\bar{c}_j\alpha_{ij}}, \tag{6.5.6}$$

where

$$\alpha_{ij} = \langle v_i, v_j \rangle, \quad k_{ij} = \langle Kv_i, v_j \rangle$$

are regarded as known quantities that can be calculated before any maximization procedure. Note that $\alpha_{ij} = \bar{\alpha}_{ji}, k_{ij} = \bar{k}_{ji}$, so that both matrices are symmetric. We obtain greater geometrical insight by rewriting (6.5.6) in terms of the operator P (defined on H) which associates with each u in H its orthogonal projection Pu on E_k. If u is in E_k, Ku will not usually be in E_k but PKu will be; for u in E_k we have $u = Pu$ and, by the symmetry of P,

$$R(u) = \frac{\langle Ku, u \rangle}{\|u\|^2} = \frac{\langle Ku, Pu \rangle}{\|u\|^2} = \frac{\langle PKu, u \rangle}{\|u\|^2}, \quad u \in E_k. \tag{6.5.7}$$

Since PK transforms elements of E_k into elements of E_k, it can be regarded as an operator on E_k; PK is known as the *part of* K *in* E_k and is easily seen to be symmetric and nonnegative. Thus (6.5.6) is just the Rayleigh quotient for the nonnegative symmetric PK on E_k.

Since such an operator is certainly compact, the usual extremal principles apply and the maximum of (6.5.7) and (6.5.6) is the largest eigenvalue Λ_1 of the algebraic eigenvalue problem

$$PKw - \Lambda w = 0, \quad w \in E_k, \tag{6.5.8}$$

which we can easily write in coordinate form. Since $\{v_1, \ldots, v_k\}$ is a basis for E_k, (6.5.8) will hold if and only if $\langle PKw - \Lambda w, v_j \rangle = 0$, $j = 1, \ldots, k$. By using

$\langle PKw, v_j\rangle = \langle Kw, Pv_j\rangle = \langle Kw, v_j\rangle$, we find that

$$\langle Kw, v_j\rangle = \Lambda\langle w, v_j\rangle, \quad j = 1, \dots, k,$$

and setting $w = \sum_{i=1}^{k} c_i v_i$, we obtain

$$\sum_{i=1}^{k}\langle Kv_i, v_j\rangle c_i = \Lambda\sum_{i=1}^{k}\langle v_i, v_j\rangle c_i, \quad j = 1, \dots, k, \tag{6.5.9}$$

which is just the set of algebraic equations that would be obtained by maximizing (6.5.6) using the calculus (keeping in mind, however, that the c_i's might be complex).

Equation (6.5.8) or its equivalent coordinate form (6.5.9) is known as the *Galerkin equation*. Although we were led to the equation from an extremal principle, the equation has a simple, intrinsic, geometrical meaning independent of any variational principle. To see this let us examine again the original problem

$$Ku - \lambda u = 0, \quad u \in H, \tag{6.5.10}$$

for which we try to find approximate solutions lying in E_k. Such an approximate solution would be of the form $\sum_{i=1}^{k} c_i v_i$. Substitution in (6.5.10) would lead to an inconsistent equation since $K\left(\sum_{i=1}^{k} c_i v_i\right)$ is not usually in E_k. Setting our sights lower, we merely require that the projection of the left side of (6.5.10) on E_k vanish. This gives the Galerkin equation (6.5.8) or (6.5.9).

REMARK. Even if a variational principle is not available (for instance, if K is not symmetric or even if K is not linear), we can regard (6.5.8) and (6.5.9) as approximations to (6.5.10). The advantage of also having a variational principle is that we can make more precise statements about the relationship between the eigenvalues of (6.5.8) and (6.5.10).

For the class of problems we are considering here, PK is a symmetric nonnegative operator on E_k. Therefore, PK has k nonnegative eigenvalues (not necessarily distinct):

$$\Lambda_1 \geqslant \Lambda_2 \geqslant \cdots \geqslant \Lambda_k \geqslant 0. \tag{6.5.11}$$

The corresponding eigenvectors $\{w_1, \dots, w_k\}$ are chosen to form an orthonormal basis for E_k. Of course, in dealing with finite-dimensional problems the largest eigenvalue Λ_1 is characterized by a maximum principle, whereas the lowest eigenvalue Λ_k is characterized by the minimum principle,

$$\Lambda_k = \min_{u\in E_k}\frac{\langle PKu, u\rangle}{\|u\|^2} = \min_{u\in E_k} R(u). \tag{6.5.12}$$

We now compare the eigenvalues (6.5.11) of (6.5.8) with those of (6.5.10).

Theorem 6.5.3 (Poincaré)**.**

$$\Lambda_1 \leqslant \lambda_1, \dots, \Lambda_k \leqslant \lambda_k. \tag{6.5.13}$$

Proof. Let us show that $\Lambda_j \leqslant \lambda_j$ by applying Theorem 6.5.1 for $n = j-1$. Our trial element w is a linear combination $d_1 w_1 + \cdots + d_j w_j$ of the first j eigenvectors of PK. We require that $\|w\| \neq 0$ and that w satisfy the $j-1$ orthogonality conditions $\langle w, e_1\rangle = \cdots = \langle w, e_{j-1}\rangle = 0$, where $e_1, \ldots, e_{j-1}$ are the familiar eigenfunctions of K. Since these orthogonality conditions reduce to $j-1$ homogeneous equations in j unknowns, it is always possible to construct the desired element w. From Theorem 6.5.1 it follows that $R(w) \leqslant \lambda_j$, and from (6.5.7)

$$R(w) = \frac{\langle PKw, w\rangle}{\|w\|^2} = \frac{\sum\limits_{i=1}^{j} \Lambda_i |d_i|^2}{\sum\limits_{i=1}^{j} |d_i|^2} \geqslant \Lambda_j,$$

so that $\Lambda_j \leqslant \lambda_j$ as claimed. □

Thus the Galerkin equation provides lower bounds to the largest k eigenvalues of K.

An immediate consequence of Theorem 6.5.3 is the maximin theorem below. Like Theorem 6.5.2, it gives a characterization of the eigenvalues of K without reference to eigenfunctions of lower index, but has the advantage of not requiring an extremum over an infinite-dimensional subspace such as $E_n^{\perp}$.

Theorem 6.5.4 (Poincaré Maximin Theorem). *Let S_k be the set of all linear manifolds of dimension k lying in H, and let E_k be a particular member of S_k. If we define*

$$\mu(E_k) = \min_{u \in E_k} R(u), \tag{6.5.14}$$

then

$$\max_{E_k \in S_k} \mu(E_k) = \lambda_k, \tag{6.5.15}$$

or combining these statements,

$$\max_{E_k \in S_k} \min_{u \in E_k} R(u) = \lambda_k. \tag{6.5.16}$$

Proof. From (6.5.12) we see that $\mu(E_k) = \Lambda_k$, the lowest eigenvalue of the part of K in E_k. By Theorem 6.5.3, $\Lambda_k \leqslant \lambda_k$, and therefore the maximum of Λ_k over all possible choices of E_k does not exceed λ_k. On the other hand, $\mu(M_k) = \lambda_k$, so Theorem 6.5.4 is proved. □

Remarks on the Ritz-Rayleigh Procedure

1. If K is nonpositive, Theorems 6.5.1, 6.5.2, 6.5.3, and 6.5.4 hold in revised form with the words *maximum* and *minimum* interchanged and all inequalities reversed. If K is indefinite, the original theorems hold for the upper end of the spectrum and the revised theorems for the lower end. In general, if A is merely bounded

from one side and it is known that the spectrum at the bounded end is discrete, there are appropriate extremal principles for the discrete end of the spectrum.

2. With a limited set of trial functions $v_1(x), \ldots, v_k(x)$ one cannot expect a very good approximation to the eigenfunction $e_1(x)$, but the approximation to λ_1 is much better. The reason is that the functional $R(u)$ is stationary (has a maximum) at $u = e_1$ and that the maximum value is λ_1; roughly speaking, a first-order change in u about e_1 leads to a second-order change in R.

3. With k fixed, the approximation Λ_1 to λ_1 is usually better than the approximation Λ_2 to λ_2, and so on.

4. If we increase k, we improve the approximations. In theory we could start with a basis $\{v_1, \ldots, v_n, \ldots\}$ for H and define E_k as the span of $\{v_1, \ldots, v_k\}$; as $k \to \infty$, the eigenvalues of (6.5.8) would tend to those of K, since the compact operator PK tends to K in the operator norm.

5. Equation (6.5.8) provides us with lower bounds to the eigenvalues of K. How do we obtain complementary bounds (that is, upper bounds)? The trace inequality (6.3.19) is one such bound. Another, the Kohn-Kato method, is presented in the following section. A third, due to Weinstein and Arondszajn and later modified by Bazley and Fox, is described by Weinstein and Stenger [13].

6. For numerical purposes we must avoid ill-conditioned matrices. This means that the set $v_1, \ldots, v_n$ used as a basis for E_n should be either orthogonal or nearly so (in a real space the cosine of the angle between v_i and v_j is $\langle v_i, v_j \rangle / \|v_i\| \|v_j\|$, and we require that this number not be close to $+1$ or -1). In particular, if we are dealing with a problem originally in $L_2(-1, 1)$, the set $v_k(x) = x^{k-1}$ is unsuitable and should first be orthogonalized by the Gram-Schmidt process. For many purposes it is useful to employ a set $v_1, \ldots, v_n$ with narrow support. These functions are usually constructed by piecing together polynomials and requiring some continuity and perhaps smoothness at the junctions. A simple special example is the set of *roof* functions, one of which is shown in Figure 6.2. We divide the interval $0 \leqslant x \leqslant 1$ into n equal parts, and for $j = 1, \ldots, n-1$ we let

$$v_j(x) = \begin{cases} 0, & x < \dfrac{j-1}{n}, \quad x > \dfrac{j+1}{n}, \\[2ex] nx - (j-1), & \dfrac{j-1}{n} < x < \dfrac{j}{n}, \\[2ex] -nx + (j+1), & \dfrac{j}{n} < x < \dfrac{j+1}{n}. \end{cases}$$

Clearly v_j and v_k are orthogonal if $|j-k| \geqslant 2$, but v_j is not orthogonal to v_{j-1} and v_{j+1}. These functions play a role in the finite element method discussed in Chapters 8 and 10.

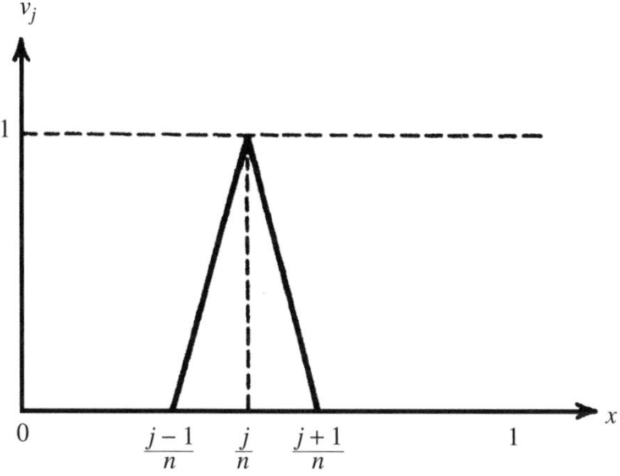

Figure 6.2

Eigenvalue Estimation Based on Spectral Theory

Although the methods described below are applicable for more general symmetric operators, we shall confine ourselves for simplicity to *positive compact operators*. Let K be such an operator. Its eigenvalues $\{\lambda_n\}$ are all positive, and the corresponding eigenfunctions $\{e_n\}$ form an orthonormal basis. Let z be an arbitrary unit vector which we call a trial element. We associate with z a mass distribution on the λ axis by placing at each point λ_n the concentrated mass $|\langle z, e_n\rangle|^2$, which is, of course, nonnegative. The total mass is $\sum_n |\langle z, e_n\rangle|^2 = \|z\|^2 = 1$, so that this distribution of mass can also be regarded as the probability distribution of a discrete random variable X. If S is a set on the λ axis, the mass in S is the probability $P(X \in S)$. As we change z, we merely redistribute our unit of mass over the same points $\{\lambda_n\}$.

The mean or center of mass is given by

$$m = E(X) = \sum \lambda_n |\langle z, e_n\rangle|^2 = \langle Kz, z\rangle. \tag{6.5.17}$$

If $f(X)$ is any function of the random variable X, we have

$$E(f(X)) = \sum f(\lambda_n) |\langle z, e_n\rangle|^2, \tag{6.5.18}$$

and, in particular,

$$E(X^m) = \sum \lambda_n^m |\langle z, e_n\rangle|^2 = \langle K^m z, z\rangle$$

and

$$\sigma^2 = E[(X-m)^2] = \sum (\lambda_n - m)^2 |\langle z, e_n\rangle|^2 = \|Kz - mz\|^2, \tag{6.5.19}$$

where σ^2 is the *variance* of X (that is, the moment of inertia of the mass distribution about its center of mass). Note that m and σ^2 can be calculated directly from the trial element z by integration. Of course, higher moments require iterated integrals, which may be hard to compute. Our goal is to estimate an eigenvalue in terms of m and σ alone.

Since our mass is distributed on the part of the λ axis between $\lambda = 0$ and $\lambda = \lambda_1$, it is clear that the center of mass also lies in that interval, so that

$$m = \langle Kz, z\rangle \leqslant \lambda_1, \tag{6.5.20}$$

which is just the familiar upper bound for the Rayleigh quotient with $\|z\| = 1$. The inequality will be useful if we can cleverly choose z so that most of the associated unit mass is at λ_1.

If we can show that a certain interval contains mass, then there must be at least one eigenvalue in that interval. We have immediately the following theorem.

Theorem 6.5.5. *The closed interval* $[m - \sigma, m + \sigma]$ *contains an eigenvalue.*

Proof. If all the mass is outside the interval in question, $(\lambda_n - m)^2 > \sigma^2$ for each n having $\langle z, e_n\rangle \neq 0$, and (6.5.19) gives an immediate contradiction. □

REMARK. The theorem is nothing more than the simplest application of the famous Chebyshev inequality of probability theory.

To use the theorem to estimate, say, the largest eigenvalue λ_1, we must know that the interval $[m - \sigma, m + \sigma]$ contains only the eigenvalue λ_1. This will be the case if $m - \sigma > \lambda_2$. Obviously, this requires that the trial element z be chosen judiciously enough so that its mass center is larger than λ_2 by at least one standard deviation. If the initial trial element does not do the trick, the Schwarz iteration described in Exercise 6.5.2 will enable us to construct a suitable z. We then have

$$m - \sigma \leqslant \lambda_1 \leqslant m + \sigma \tag{6.5.21}$$

or, in view of (6.5.20),

$$m \leqslant \lambda_1 \leqslant m + \sigma, \tag{6.5.22}$$

where $m = \langle Kz, z\rangle, \sigma = \|Kz - mz\|$.

We can improve on these bounds by the following procedure, based on Chebyshev inequalities (see Marshall and Olkin [9]). Let S be an interval on the λ axis which we want to contain mass, and let T be the complement of S. We would like to estimate $P(X \in S)$ and to show that it is positive or, equivalently, that $P(X \in T) < 1$. Let $f(\lambda)$ be a real-valued function satisfying

$$f(\lambda) \leqslant 0 \quad \text{for all } \lambda; \qquad f(\lambda) \geqslant 1, \quad \lambda \in T. \tag{6.5.23}$$

We would like to choose $f(\lambda)$ as close as possible to the indicator function (see Section 2.1) of the set T. We have

$$P(X \in T) = \sum_{\lambda_n \in T} |\langle z, e_n\rangle|^2 \leqslant \sum_{\lambda_n \in T} f(\lambda_n)|\langle z, e_n\rangle|^2$$
$$\leqslant \sum_{n=1}^{\infty} f(\lambda_n)|\langle z, e_n\rangle|^2 = E(f(X)),$$

and therefore

$$P(X \in S) \geqslant 1 - E(f(X)), \tag{6.5.24}$$

which will be positive if

$$E(f(X)) < 1. \tag{6.5.25}$$

We shall consider functions f, which also depend on certain parameters adjusted so that (6.5.23) and (6.5.25) are satisfied and the interval S containing mass is as narrow as possible. Suppose, for instance, that $m > A$ and we want to find the narrowest interval $(A, A + \Delta)$ which can be guaranteed to contain mass (note there must be some mass for $\lambda \geqslant m$ since m is the center of mass). We would like to determine this interval by using only the first two moments of X. Let $h > A$, and consider the function

$$f(\lambda, h) = \frac{(\lambda - h)^2}{(A - h)^2},$$

which satisfies (6.5.23) when $|\lambda - h| \geqslant |A - h|$. Thus we take S as the interval $(A, 2h - A)$, of which h is the midpoint.

A simple calculation gives

$$E(f(X, h)) = \frac{(h - m)^2 + \sigma^2}{(h - A)^2},$$

which satisfies (6.5.25) if

$$h > \frac{m + A}{2} + \frac{\sigma^2}{2(m - A)}.$$

Therefore, the open interval

$$\left(A, m + \frac{\sigma^2}{m - A} + \varepsilon\right)$$

contains mass for every $\varepsilon > 0$. In view of the discrete nature of the spectrum and the fact that $m + \sigma^2/(m - A) > 0$, we conclude that the half-closed interval

$$\left(A, m + \frac{\sigma^2}{m - A}\right]$$

must contain mass. Similarly, if $m < B$, the interval

$$\left[m - \frac{\sigma^2}{B - m}, B\right)$$

contains mass.

Now suppose that we are trying to estimate some particular eigenvalue λ^* in the spectrum and that we know (perhaps by previous rough calculations) that the interval (A, B) contains only the eigenvalue λ^*. In this case we find the *Kohn-Kato inclusion interval*

$$m - \frac{\sigma^2}{B - m} \leqslant \lambda^* \leqslant m + \frac{\sigma^2}{m - A} \tag{6.5.26}$$

if $A < m < B$ (as should be the case if z is a reasonable trial element). In fact, for a reasonable z, both $\sigma/(m - A)$ and $\sigma/(B - m)$ will be much smaller than 1 and the inclusion interval (6.5.26) will be narrower than the one given by Theorem 6.5.5. If we apply (6.5.26) to estimate the kth eigenvalue λ_k, we need to set A equal to an upper bound U_{k+1} to λ_{k+1} and B to a lower bound L_{k-1} to λ_{k-1}. Then (6.5.26) gives, for a trial element z such that $U_{k+1} < m < L_{k-1}$,

$$m - \frac{\sigma^2}{L_{k-1} - m} \leqslant \lambda_k \leqslant m + \frac{\sigma^2}{m - U_{k+1}}. \tag{6.5.27}$$

The method gives an improved upper bound to λ_1:

$$\lambda_1 \leqslant m + \frac{\sigma^2}{m - U_2}, \tag{6.5.28}$$

where U_2 is an upper bound (perhaps fairly crude) to λ_2 and $m > U_2$.

REMARK. If an eigenvalue is degenerate or if some eigenvalues are clustered together, the method can still be applied but gives information only about the overall location of the cluster.

Eigenfunction Approximation

Consider a trial element z intended to approximate the eigenfunction e^* corresponding to the simple eigenvalue λ^*. Note that $-e^*$ and, more generally, $e^* \exp(i\psi)$ for real ψ are also unit eigenfunctions corresponding to λ^*. We are perfectly satisfied if z is close to any of these unit eigenvectors. Obviously, z will be a good approximation to a unit eigenvector on the one-dimensional manifold generated by e^* if the magnitude of the projection $\langle z, e^* \rangle e^*$ is close to 1. By the projection theorem we can decompose z as

$$z = u + v = \langle z, e^* \rangle e^* + v, \quad \langle v, e^* \rangle = 0. \tag{6.5.29}$$

We will have $|\langle z, e^* \rangle|$ near 1 if

$$\|v\|^2 = 1 - |\langle z, e^* \rangle|^2 \tag{6.5.30}$$

is near 0.

If z is supposed to be an approximation to e_1, we can easily find a bound for $1 - |\langle z, e_1\rangle|^2$. Indeed, we have

$$m = \langle Kz, z\rangle = \sum_k \lambda_k |\langle z, e_k\rangle|^2 \leqslant \lambda_1 |\langle z, e_1\rangle|^2 + \lambda_2 \sum_{k\geqslant 2} |\langle z, e_k\rangle|^2$$

or

$$m \leqslant \lambda_1 |\langle z, e_1\rangle|^2 + \lambda_2(1 - |\langle z, e_1\rangle|^2), \tag{6.5.31}$$

so that

$$1 - |\langle z, e_1\rangle|^2 \leqslant \frac{\lambda_1 - m}{\lambda_1 - \lambda_2}. \tag{6.5.32}$$

Of course, neither λ_1 nor λ_2 is generally known. However, we may use upper and lower bounds for λ_1, λ_2 in (6.5.32) to obtain

$$1 - |\langle z, e_1\rangle|^2 \leqslant \frac{U_1 - m}{L_1 - U_2}. \tag{6.5.33}$$

A somewhat better bound for $1 - |\langle z, e^*\rangle|^2$ can be found by using the decomposition (6.5.29). For any real c we have

$$Kz - cz = Ku - cu + Kv - cv,$$

where $Ku - cu$ is seen to be proportional to e^* and $Kv - cv$ is orthogonal to e^*. Therefore, we find that

$$\|Kv - cv\|^2 \leqslant \|Kz - cz\|^2 = \sigma^2 + (m - c)^2,$$

so that

$$\|v\|^2 \leqslant \frac{\sigma^2 + (m-c)^2}{\|Kv - cv\|^2/\|v\|^2}.$$

Since

$$\min_{\langle v, e^*\rangle = 0} \frac{\|Kv - cv\|^2}{\|v\|^2} = (\tilde{\lambda} - c)^2,$$

where $\tilde{\lambda}$ is the eigenvalue different from λ^* nearest c, we have

$$\|v\|^2 \leqslant \frac{\sigma^2 + (m-c)^2}{(\tilde{\lambda} - c)^2}.$$

If the interval (A, B) contains only the eigenvalue λ^* and $A < m < B$, we find, by choosing $c = m$, that

$$\begin{aligned} \|v\|^2 &\leqslant \frac{\sigma^2}{(B-m)^2}, \quad m \geqslant \frac{A+B}{2}, \\ \|v\|^2 &\leqslant \frac{\sigma^2}{(m-A)^2}, \quad m \leqslant \frac{A+B}{2}. \end{aligned} \tag{6.5.34}$$

The optimal choice of c is discussed in Exercise 6.5.9. If U_2 is an upper bound to λ_2 and $m > U_2$, we find from (6.5.34) that

$$1 - |\langle z, e_1 \rangle|^2 \leqslant \frac{\sigma^2}{(m - U_2)^2}, \tag{6.5.35}$$

which is usually better than (6.5.33).

EXERCISES

6.5.1 Let A be a symmetric (perhaps indefinite) operator on $E_n^{(c)}$ whose eigenvalues listed in decreasing order are: $\lambda_1 \geqslant \lambda_2 \geqslant \cdots \geqslant \lambda_n$. Show that λ_k can be characterized by two theorems of the Weyl-Courant type (see Theorem 6.5.2), one requiring orthogonality to a $(k-1)$-dimensional space and the other to an $(n-k)$-dimensional space. The second of these is a maximin theorem, but is it the same as the Poincaré maximin theorem?

6.5.2 Let K be a real, nonnegative, symmetric, H.-S. operator defined on $L_2(a,b)$. Starting with an arbitrary real function f_0 such that $Kf_0 \neq 0$, let us define

$$f_n = Kf_{n-1} = \cdots = K^n f_0$$

and

$$a_k = \langle f_i, f_{k-i} \rangle,$$

where, as the notation suggests, the definition is independent of i. By expanding in the eigenfunctions of K, show that $a_n > 0$ and that the sequence

$$\theta_{k+1} \doteq \frac{a_{k+1}}{a_k}$$

is monotonically increasing and bounded above by the largest eigenvalue λ_1 of K. Show that if f_0 is *not* orthogonal to the fundamental eigenfunction e_1, then $\theta_k \to \lambda_1$.

6.5.3 *An integrodifferential operator.* Let A be the operator defined by

$$Au = -\frac{d^2u}{dx^2} + \int_0^1 xyu(y)\,dy = -u''(x) + x\langle x, u \rangle, \tag{6.5.36}$$

with domain D_A consisting of all functions $u(x)$ on $0 < x < 1$ with a continuous second derivative and satisfying the boundary conditions

$$u(0) = 0, \quad u'(1) = 0. \tag{6.5.37}$$

(a) Show that A on D_A is a symmetric positive operator. Since A is real, we can (and shall) restrict the domain of A to real-valued functions.

(b) Consider the inhomogeneous integrodifferential equation

$$-u'' + \int_0^1 xyu(y)\,dy = f(x), \quad 0 < x < 1; \quad u(0) = u'(1) = 0. \tag{6.5.38}$$

A function $u(x)$ satisfies (6.5.38) if and only if $u(x)$ and α satisfy simultaneously

$$-u'' = f - \alpha x, \quad 0 < x < 1; \quad u(0) = u'(1) = 0;$$

and

$$\alpha = \int_0^1 xu(x)\, dx = \langle x, u \rangle.$$

Using Green's function $g(x, \xi)$ for $-D^2$ with the boundary conditions (6.5.37), and letting G be the corresponding integral operator, show that

$$\alpha = \frac{\langle f, Gx \rangle}{1 + \langle Gx, x \rangle}$$

and

$$u(x) = \int_0^1 \left[g(x,\xi) - \frac{5}{204}(3x - x^3)(3\xi - \xi^3) \right] f(\xi)\, d\xi. \qquad (6.5.39)$$

6.5.4 Consider the eigenvalue problem $Au = \lambda u$, where A is the operator of Exercise 6.5.3 with the boundary conditions (6.5.37). Since A on D_A is symmetric and positive, all eigenvalues are positive and eigenfunctions corresponding to different eigenvalues are orthogonal. In view of the fact that A is a real operator, we may restrict ourselves to real-valued eigenfunctions.

(a) Show that the problem can be reduced to an eigenvalue problem for a pure integral operator; hence show that the eigenfunctions form a basis.

(b) By observing that the problem $Au = \lambda u$ has the form $u'' + \lambda u = cx$, show that the eigenvalues $\lambda = \alpha^2$ are obtained from the positive solutions of

$$\tan \alpha = \alpha + \frac{\alpha^3}{3} - \alpha^5.$$

Sketch the functions $\tan \alpha$ and $\alpha + (\alpha^3/3) - \alpha^5$. Compute an approximate value for the smallest eigenvalue, λ_1.

(c) In the extremal principle

$$\lambda_1 = \min_{v \in D_A} \frac{\int_0^1 (v')^2\, dx + \left(\int_0^1 xv\, dx \right)^2}{\int_0^1 v^2\, dx},$$

use the trial element $v = x(2 - x)$ to find an upper bound to λ_1. Use the trace inequality for k_2 [the iterate of the kernel of part (a)] to find a lower bound to λ_1.

6.5.5 Show that the BVP

$$-u'' + 4\pi^2 \int_0^1 u(x)\, dx = \lambda u, \ 0 < x < 1; \ u(0) = u(1), \ u'(0) = u'(1) \qquad (6.5.40)$$

has $\lambda = 4\pi^2$ as an eigenvalue of *multiplicity 3.*

6.5.6 Find all eigenvalues and eigenfunctions of (6.5.40). Does the set of eigenfunctions form a basis?

6.5.7 For $A < c < B$ the inequality preceding (6.5.34) yields

$$\|v\|^2 \leqslant \frac{\sigma^2 + (m-c)^2}{\alpha^2},$$

where $\alpha = \min(c - A, B - c)$. Find the optimal value of c if $m \leqslant (A+B)/2$.

REFERENCES AND ADDITIONAL READING

1. D. Colton and R. Kress, *Inverse Acoustic and Electromagnetic Scattering Theory*, Springer–Verlag, New York, 1992.
2. H. Dym and H. P. McKean, *Fourier Series and Integrals*, Academic Press, New York, 1972.
3. C. Groetsch, *Inverse Problems in the Mathematical Sciences*, Vieweg, Wiesbaden, Germany, 1993.
4. W. Hackbusch, *Integral Equations*, Birkhauser, Cambridge, MA, 1995.
5. H. Hochstadt, *Integral Equations*, Wiley-Interscience, New York, 1973.
6. G. C. Hsiao and W. L. Wendland, *Boundary Integral Equations*, Volume 164, Applied Mathematical Sciences Series, Springer–Verlag, New York, 2008.
7. A. Kirsch, *An Introduction to the Mathematical Theory of Inverse Problems*, Springer–Verlag, New York, 1996.
8. R. Kress, *Linear Integral Equations*, 2nd ed., Springer–Verlag, New York, 1999.
9. A. W. Marshall and I. Olkin, *Inequalities: Theory of Majorization and Its Applications*, Academic Press, New York, 1979.
10. B. Noble, *The Wiener-Hopf Technique*, Pergamon Press, Elmsford, NY, 1958.
11. L. E. Payne, *Improperly Posed Problems in Partial Differential Equations*, SIAM, Philadelphia, 1975.
12. I. Stakgold, *Boundary Value Problems of Mathematical Physics*, Vols. I and II, Macmillan, 1967. Reprinted as Vo. 29 in SIAM Classics in Applied Mathematics, 2000.
13. A. Weinstein and W. Stenger, *Methods of Intermediate Problems for Eigenvalues*, Academic Press, New York, 1972.

CHAPTER 7

SPECTRAL THEORY OF SECOND-ORDER DIFFERENTIAL OPERATORS

7.1 INTRODUCTION; THE REGULAR PROBLEM

The main purpose of the chapter is to study the relationship between two methods for solving boundary value problems, the one based on eigenfunction expansion and the other on Green's function. We first deal with operators whose spectrum is discrete; in this case the eigenfunction expansion is an infinite series, and we show how Green's function can be expressed as a series of eigenfunctions and how, conversely, the eigenfunctions can be generated from the closed-form expression for Green's function. If the spectrum is continuous, an integral expansion over a continuous parameter plays the role of the infinite series, but it is still possible to generate this expansion from a knowledge of Green's function.

We shall develop the theory for self-adjoint differential operators of the second order. The corresponding differential equations, which usually arise when separating variables in curvilinear coordinates for Laplace's equation or related equations, have the form

$$-\frac{d}{dx}\left(p(x)\frac{du}{dx}\right)+q(x)u-\lambda s(x)u=0, \quad a<x<b, \tag{7.1.1}$$

Green's Functions and Boundary Value Problems, Third Edition. By I. Stakgold and M. Holst

where λ, which started life as a separation constant, is now viewed as an eigenvalue parameter. We can write (7.1.1) as

$$Lu - \lambda u = 0, \tag{7.1.2}$$

where L is the operator defined by

$$Lu \doteq \frac{1}{s}[-(pu')' + qu]. \tag{7.1.3}$$

Note that we have placed a $1/s$ factor in L, so that (7.1.2) will have the standard eigenvalue form. This causes the slight complication that L is no longer formally self-adjoint in the usual inner product $\langle u, v\rangle = \int_s^b u\bar{v}\,dx$. As we shall see, an appropriate remedy is to introduce the new inner product

$$\langle u, v\rangle_s = \int_a^b su\bar{v}\,dx.$$

Since $s(x) > 0$ in $a < x < b$, $\langle u, u\rangle_s$ is positive for $u \neq 0$ and can be used to define a norm

$$\|u\|_s = \langle u, u\rangle_s^{1/2} = \left(\int_a^b s|u|^2\,dx\right)^{1/2}.$$

Most of our work will take place in the Hilbert space H_s, consisting of all functions $u(x)$ for which $\int_a^b s|u|^2\,dx$ is finite. Orthogonality in H_s means that $\langle u, v\rangle_s = 0$, that is, $\int_a^b su\bar{v}\,dx = 0$. Let $\{e_n\}$ be an orthonormal basis in H_s; then, for each f in H_s,

$$f = \sum_n \langle f, e_n\rangle_s e_n = \sum_n \left[\int_a^b s(x)f(x)\bar{e}_n(x)\,dx\right] e_n(x). \tag{7.1.4}$$

If f is in $L_2(a, b)$, then f/s is in H_s and

$$\frac{f}{s} = \sum_n \left\langle \frac{f}{s}, e_n\right\rangle_s e_n = \sum_n \langle f, e_n\rangle e_n = \sum_n \left(\int_a^b f\bar{e}_n\,dx\right) e_n(x). \tag{7.1.5}$$

We make the following assumptions throughout the chapter regarding the coefficients in (7.1.1): p, q, s are *real-valued* functions on $a < x < b$; p, p', q, s are *continuous* on $a < x < b$; p and s are *positive* on $a < x < b$. If the interval is *finite* and all assumptions on the coefficients hold for the *closed* interval $a \leqslant x \leqslant b$, the problem is said to be *regular*; otherwise it is *singular*. In a singular problem solutions of (7.1.1) need not lie in H_s, and a more delicate analysis is required. In the regular case we associate with (7.1.1) two homogeneous boundary conditions of the *unmixed* type [see (3.2.3)]:

$$\begin{aligned} 0 &= B_a u \doteq u(a)\cos\alpha - u'(a)\sin\alpha, \\ 0 &= B_b u \doteq u(b)\cos\beta + u'(b)\sin\beta, \end{aligned} \tag{7.1.6}$$

where α and β are given real numbers, $0 \leqslant \alpha < \pi, 0 \leqslant \beta < \pi$. As β, for instance, takes on different values, the boundary condition at b specifies the ratio $u'(b)/u(b)$ as any positive or negative number [the cases $\beta = 0$ and $\beta = \pi/2$ correspond to $u(b) = 0$ and $u'(b) = 0$]. A similar statement holds for the endpoint a. The signs in (7.1.6) have been chosen so that the eigenvalues decrease as α or β increases.

We note the following properties of L and the boundary operators. Let u and v be arbitrary twice-differentiable functions:

(a)
$$L\bar{u} = \overline{(Lu)}, \quad B_a\bar{u} = \overline{(B_a u)}, \quad B_b\bar{u} = \overline{(B_b u)}. \tag{7.1.7}$$

(b)
$$\bar{v}Lu - uL\bar{v} = \frac{1}{s}[p(u\bar{v}' - u'\bar{v}]' = \frac{1}{s}[p(x)W(u, \bar{v}; x)]', \tag{7.1.8}$$

where $W(u, \bar{v}; x) = u\bar{v}' - \bar{v}u'$ is the Wronskian of u and $\bar{v}$ [see (3.1.6)].

(c)
$$\langle Lu, v\rangle_s - \langle u, Lv\rangle_s = p(b)W(u, \bar{v}; b) - p(a)W(u, \bar{v}; a). \tag{7.1.9}$$

(d) If u and v *both satisfy* (7.1.6), then $W(u, \bar{v}; b) = W(u, \bar{v}; a) = 0$, so that
$$\langle Lu, v\rangle_s = \langle u, Lu\rangle_s \tag{7.1.10}$$
and
$$\langle Lu, u\rangle_s \quad \text{is real.} \tag{7.1.11}$$

Thus, if we consider L as an operator whose domain D consists of functions having a continuous second derivative and satisfying (7.1.6), then L is symmetric. If we are willing to enlarge D slightly by relaxing the smoothness condition (to include functions whose first derivative is absolutely continuous and whose second derivative is in H_s), then L is actually self-adjoint on this new domain D_L.

We now investigate some of the properties of the eigenvalue problem (7.1.1)–(7.1.6):

$$-(pu')' + qu - \lambda su = 0, \quad a < x < b; \tag{7.1.12a}$$
$$u(a)\cos\alpha - u'(a)\sin\alpha = 0, \quad u(b)\cos\beta + u'(b)\sin\beta = 0,$$

or, equivalently, using definition (7.1.3),

$$Lu - \lambda u = 0, \quad a < x < b; \quad B_a u = B_b u = 0. \tag{7.1.12b}$$

1. *The eigenvalues are real.* Multiply (7.1.12b) by $s\bar{u}$ and integrate to obtain

$$\langle Lu, u\rangle_s = \lambda\langle u, u\rangle_s = \lambda\int_a^b s(x)|u(x)|^2\,dx.$$

Since u is an eigenfunction, $\int_s^b s|u|^2\,dx > 0$, and it follows from (7.1.11) that λ is real.

2. *Eigenfunctions corresponding to different eigenvalues are orthogonal in H_s.* Let u and v satisfy (7.1.12b) with respective eigenvalues λ and μ. Then

$$\langle Lu, v\rangle_s = \lambda\langle u, v\rangle_s = \lambda \int_a^b su\bar{v}\,dx,$$

$$\langle u, Lv\rangle_s = \langle u, \mu v\rangle_s = \bar{\mu} \int_a^b su\bar{v}\,dx = \mu \int_a^b su\bar{v}\,dx.$$

By (7.1.10), $\langle Lu, v\rangle_s = \langle u, Lv\rangle_s$; hence, for $\lambda \neq \mu$,

$$0 = \int_a^b su\bar{v}\,dx = \langle u, v\rangle_s, \tag{7.1.13}$$

so that u and v are orthogonal in H_s [other ways of expressing this property; u and v are orthogonal with weight s; $\sqrt{s}\,u$ and $\sqrt{s}\,v$ are orthogonal in $L_2(a, b)$].

3. The eigenvalues of (7.1.12b) are simple and can be listed as the sequence

$$\lambda_1 < \lambda_2 < \lambda_3 < \cdots < \lambda_n < \cdots,$$

with $\lim_{n\to\infty} \lambda_n = +\infty$ *(thus there are at most finitely many negative eigenvalues). The corresponding normalized eigenfunctions $\{u_n\}$ together form an orthonormal basis in H_s.* Assume without loss of generality that $\lambda = 0$ is not an eigenvalue of (7.1.12b), and introduce Green's function g_0 satisfying $-(pg_0')' + qg_0 = \delta(x - \xi)$, $B_a g_0 = B_b g_0 = 0$. Then (7.1.12b) is equivalent to the integral equation

$$u(x) = \lambda \int_a^b g_0(x, \xi)s(\xi)u(\xi)\,d\xi,$$

or, setting $k(x, \xi) = \sqrt{s(x)}\,g_0(x, \xi)\sqrt{s(\xi)}$ and $\sqrt{s(\xi)}\,u(x) = v(x)$,

$$v(x) = \lambda \int_a^b k(x, \xi)v(\xi)\,d\xi \doteq \lambda K v. \tag{7.1.14}$$

Since $g_0(x, \xi)$ is symmetric and Hilbert-Schmidt, so is $k(x, \xi)$ and the theory of self-adjoint H.-S. operators applies. Thus the eigenvalues of (7.1.12b) are the *reciprocals* of the eigenvalues of K (recall that an eigenvalue of K is a number γ such that $Kv = \gamma v$). Since $Kv = 0$ implies that $v = 0$, the eigenfunctions $\{v_n\}$ form an orthonormal basis; now $v_n = \sqrt{s}\,u_n$, where u_n is the corresponding eigenfunction of (7.1.12b); thus $\{u_n\}$ is an orthonormal basis in H_s. The differential equation (7.1.12b) has two independent solutions, so that no eigenvalue can have multiplicity greater than 2. In fact, it can be shown that for unmixed boundary conditions the eigenvalues are simple. The theory of compact operators tells us that 0 is the only limit point of the eigenvalues of K. Therefore, $|\lambda_n| \to \infty$; in addition, one can show (Exercise 7.1.1) that there are only finitely many negative eigenvalues, and hence it is possible to index the eigenvalues λ_n so that $\lambda_1 < \lambda_2 < \cdots$ and $\lambda_n \to +\infty$.

Relation between Green's Function and the Eigenfunctions

We now study the relation between the eigenfunctions of (7.1.12b) and Green's function $g(x, \xi; \lambda)$ satisfying

$$-(pg')' + qg - \lambda sg = \delta(x - \xi), \quad a < x.\xi < b; \quad B_a g = B_b g = 0. \tag{7.1.15}$$

It is an easy matter (and not our primary concern) to determine the Fourier expansion of $g(x, \xi; \lambda)$ in the eigenfunctions $u_n(x)$ of (7.1.12b). We have, by (7.1.4),

$$g(x, \xi; \lambda) = \sum_n g_n(\xi, \lambda) u_n(x), \quad g_n = \langle g, u_n \rangle_s = \int_a^b s(x) g(x, \xi; \lambda) \bar{u}_n(x)\, dx.$$

To find g_n, multiply (7.1.15) by $\bar{u}_n(x)$ and integrate from a to b to obtain

$$\langle Lg, u_n \rangle_s - \lambda \langle g, u_n \rangle_s = \bar{u}_n(\xi),$$

or, using (7.1.10),

$$\langle g, u_n \rangle_s (\lambda_n - \lambda) = \bar{u}_n(\xi),$$

so that

$$g(x, \xi; \lambda) = \sum_n \frac{u_n(x) \bar{u}_n(\xi)}{\lambda_n - \lambda}, \tag{7.1.16}$$

which is the *bilinear series* for g. As expected, g has singularities at $\lambda = \lambda_n$; in any event, g can be constructed if $\{u_n\}$ and $\{\lambda_n\}$ are known. We can also find the Fourier expansion of the solution $w(x, \lambda)$ of the inhomogeneous equation

$$-(pw')' + qw - \lambda sw = f, \quad a < x < b; \quad B_a w = B_b w = 0, \tag{7.1.17}$$

either by imitating the steps leading to (7.1.16) or by recalling that

$$\begin{aligned} w(x, \lambda) &= \int_a^b g(x, \xi; \lambda) f(\xi)\, d\xi \\ &= \sum_n u_n(x) \frac{\int_a^b f(\xi) \bar{u}_n(\xi)\, d\xi}{\lambda_n - \lambda} = \sum_n \frac{\langle f, u_n \rangle}{\lambda_n - \lambda} u_n. \end{aligned} \tag{7.1.18}$$

We now ask the question that concerns us most in the present chapter: Given an explicit formula for $g(x, \xi, \lambda)$ as obtained, say, by the methods of Section 3.2, how do we determine the eigenfunctions $\{u_n\}$ and eigenvalues $\{\lambda_n\}$ of (7.1.12b)? Representation (7.1.16) shows that as a function of the complex parameter λ, $g(x, \xi; \lambda)$ has simple poles at the real points $\lambda = \lambda_n$ with corresponding residues $-u_n(x)\bar{u}_n(\xi)$. Thus all we have to do is to examine $g(x, \xi; \lambda)$ in the complex λ plane and to pick out the singularities (which will be the eigenvalues) and the residues at these singularities (which are related to the eigenfunctions). This information is formally contained in the compact formula

$$\frac{1}{2\pi i} \int_{C_\infty} g(x, \xi; \lambda)\, d\lambda = -\sum_n u_n(x) \bar{u}_n(\xi), \tag{7.1.19}$$

where the integral in the λ plane is taken counterclockwise around the infinitely large circle C_∞, thus containing all the poles of g in (7.1.16). Admittedly, the series in (7.1.19) fails to converge in the ordinary sense but has a distributional meaning (see Section 2.3); in fact, using (7.1.5) with $f = \delta(x-\xi)$ and $e_n = u_n$, we have the completeness relation

$$\frac{\delta(x-\xi)}{s(x)} = \sum_n u_n(x)\bar{u}_n(\xi). \tag{7.1.20}$$

If one feels uncomfortable with the series in (7.1.19), it is possible instead to focus attention on (7.1.18). By integrating around C_∞, we find that

$$\frac{1}{2\pi i}\int_{C_\infty} w(x,\lambda)\,d\lambda = -\sum_n \langle f, u_n\rangle u_n = -\frac{f(x)}{s(x)}, \tag{7.1.21}$$

which can be used as a basis for a more rigorous theory; we prefer, however, to use (7.1.19) to generate the eigenfunctions. Even in problems with a continuous spectrum it will be possible to use (7.1.19) with the summation replaced by an integral over a continuous parameter (Green's function then has a branch instead of poles).

The first step in using (7.1.19) is to construct $g(x,\xi;\lambda)$ in such a way that we can keep track of its dependence on λ. For the regular problem (7.1.15) we know that for $x < \xi$, g is a solution of the homogeneous equation satisfying $0 = B_a g = (g\cos\alpha - g'\sin\alpha)_{x=a}$. Therefore, g is just a constant multiple of the solution $v(x,\lambda)$ of an appropriate initial value problem for the homogeneous equation; for instance, we can require v to satisfy $v(a,\lambda) = \sin\alpha, v'(a,\lambda) = \cos\alpha$. Then v obviously satisfies $B_a v = 0$. The advantage of proceeding in this way is that v is an *analytic* function of λ in the whole λ plane (a so-called *entire function*). This follows from the fact that v satisfies *initial* conditions independent of λ and a differential equation where λ appears analytically. Such a problem can be translated into a Volterra integral equation whose Neumann series converges in the whole λ plane (see Example 4, Section 4.4). Similarly, let $z(x,\lambda)$ be the unique solution of the homogeneous equation with the initial conditions $z(b,\lambda) = \sin\beta, z'(b,\lambda) = -\cos\beta$. Then $B_b z = 0$ and g is proportional to z for $x > \xi$. Thus we have, after ensuring continuity at $x = \xi$,

$$g(x,\xi;\lambda) = Av(x_<,\lambda)z(x_>,\lambda),$$

where $x_< = \min(x,\xi)$ and $x_> = \max(x,\xi)$. The jump condition on g' is

$$\left.\frac{dg}{dx}\right|_{x=\xi+} - \left.\frac{dg}{dx}\right|_{x=\xi-} = -\frac{1}{p(\xi)},$$

which becomes

$$A[v(\xi,\lambda)z'(\xi,\lambda) - v'(\xi,\lambda)z(\xi,\lambda)] = -\frac{1}{p(\xi)}.$$

The quantity in brackets is the Wronskian of v and z, which, by (3.1.8), is of the form $C/p(\xi)$, where C is independent of ξ but may depend on λ. Therefore,

$$g(x,\xi;\lambda) = -\frac{v(x_<,\lambda)z(x_>,\lambda)}{C(\lambda)}, \tag{7.1.22}$$

where $C(\lambda)$ is unambiguously determined from

$$W[v(x,\lambda), z(x,\lambda); x] = vz' - zv' = \frac{C(\lambda)}{p(x)}. \tag{7.1.23}$$

Since v and z are entire functions, so are v', z', W, and $C(\lambda)$. Let $\lambda = \mu$ be a zero of C; that is, $C(\mu) = 0$. Then the Wronskian of $v(x,\mu)$ and $z(x,\mu)$ vanishes, and these functions are linearly dependent. In view of their initial values, neither function can vanish identically in x. Therefore, $v(x,\mu)$ is a nontrivial constant multiple of $z(x,\mu)$, and both functions satisfy the two boundary conditions and the differential equation in (7.1.12b). Thus μ is an eigenvalue of (7.1.12b) with eigenfunction (not normalized) $v(x,\mu)$. From (7.1.16) it is clear that at an eigenvalue g has a singularity, and therefore C must vanish. We conclude that the zeros of $C(\lambda)$ coincide with the eigenvalues of (7.1.12b); we therefore label these zeros (or eigenvalues) $\lambda_1 < \lambda_2 < \cdots < \lambda_n < \cdots$, with $\lambda_n \to \infty$, and

$$v(x,\lambda_n) = k_n z(x,\lambda_n), \tag{7.1.24}$$

where k_n is a real nonzero constant. We also know from (7.1.16) that g has only simple poles, so that the zeros of C are likewise simple. To simplify the notation a little, we let

$$v_n(x) \doteq v(x,\lambda_n), \quad z_n(x) \doteq z(x,\lambda_n).$$

Thus $v_n(x)$ (or z_n) is a real eigenfunction corresponding to the simple eigenvalue λ_n and $v_n(x) = k_n z_n(x)$. Neither v_n nor z_n is normalized.

The residue of g at $\lambda = \lambda_n$ is

$$-\frac{v_n(x_<)z_n(x_>)}{C'(\lambda_n)} = -k_n\frac{z_n(x_<)z_n(x_>)}{C'(\lambda_n)} = -\frac{v_n(x_<)v_n(x_>)}{k_n C'(\lambda_n)}.$$

The quantity $z_n(x_<)z_n(x_>)$ is equal to $z_n(x)z_n(\xi)$ whether $x < \xi$ or $x > \xi$. It is remarkable that the discontinuity in the first derivative of g at $x = \xi$ has left no trace. The residue of g at $\lambda = \lambda_n$ becomes

$$-k_n\frac{z_n(x)z_n(\xi)}{C'(\lambda_n)} = -\frac{v_n(x)v_n(\xi)}{k_n C'(\lambda_n)} = -u_n(x)\bar{u}_n(\xi),$$

from which we recognize that the *real normalized eigenfunction* $u_n(x)$ is given by

$$u_n(x) = \pm\frac{v_n(x)}{[k_n C'(\lambda_n)]^{1/2}} = \pm\left[\frac{k_n}{C'(\lambda_n)}\right]^{1/2} z_n(x). \tag{7.1.25}$$

Another method of obtaining the normalization constant is described in Exercise 7.1.2.

Example 1. Consider one-dimensional heat conduction without sources in a rod $0 < x < 1$. The initial temperature is given, the left end is kept at temperature 0, and at the right end the temperature gradient is proportional to the temperature. The temperature $\Theta(x, t)$ in the rod satisfies

$$\frac{\partial \Theta}{\partial t} - \frac{\partial^2 \Theta}{\partial x^2} = 0, \quad 0 < x < 1, t > 0; \quad \Theta(x, 0) = f(x), \tag{7.1.26}$$
$$\Theta(0, t) = 0, \quad \Theta(1, t) \cos \beta + \frac{\partial \Theta}{\partial x}(1, t) \sin \beta = 0,$$

where β is a given real number in $[0, \pi]$. One tries to construct the solution of this problem by using as building blocks functions that are separable in x and t [that is, of the form $u(x)T(t)$]. These functions are chosen to satisfy the differential equation and the boundary conditions but *not* the arbitrary initial condition, which is ultimately satisfied by an appropriate sum of such separable functions. The steps in the procedure are purely formal, and one must check afterward that the proposed solution actually meets all the requirements of (7.1.26).

Substituting $\Theta = u(x)T(t)$ in the differential equation, we find that

$$\frac{1}{T} \frac{dT}{dt} = \frac{1}{u} \frac{d^2 u}{dx^2} = -\lambda,$$

where λ is an arbitrary complex parameter whose values will be determined from the x equation with its boundary conditions:

$$-u'' - \lambda u = 0, \quad 0 < x < 1; \quad u(0) = 0, u(1) \cos \beta + u'(1) \sin \beta = 0. \tag{7.1.27}$$

Let the eigenvalues of (7.1.27) be arranged in increasing order $\lambda_1 < \lambda_2 < \cdots$, with the corresponding eigenfunctions $u_1(x), \ldots, u_n(x), \ldots$ forming an orthonormal basis. For $\lambda = \lambda_k$ the solution of the t equation is $T = e^{-\lambda_k t}$ and $\Theta = e^{-\lambda_k t} u_k(x)$. The sum $\sum_{k=1}^{\infty} c_k e^{-\lambda_k t} u_k(x)$ formally satisfies the partial differential equation and the boundary conditions; it will also take on the correct initial value if the $\{c_k\}$ are chosen so that $f(x) = \sum_{k=1}^{\infty} c_k u_k$. Thus c_k must be chosen to be the Fourier coefficient $\langle f, u_k \rangle$.

For large values of time the behavior of the temperature is controlled by λ_1. According as $\lambda_1 > 0$ or $\lambda_1 < 0$, the temperature decays exponentially at the rate $e^{-\lambda_1 t}$ or blows up exponentially at the rate $e^{|\lambda_1| t}$ if $\langle f, u_1 \rangle \neq 0$. In the case $\lambda_1 < 0$ we would say that the zero solution of the completely homogeneous problem ($f = 0$) is *unstable*, for a slight change in the initial condition would create a disturbance that would grow in time.

Before analyzing (7.1.27) in detail we note that the eigenvalues $\{\lambda_n\}$ decrease as β increases in $0 \leqslant \beta < \pi$. Mathematically, this is a consequence of the Weyl-Courant variational principle (see Exercise 7.1.3). Physically, as β increases, less and less heat is removed at the right end; $\beta = 0$ means that the right end is kept at 0 temperature; if $0 < \beta < \pi/2$, we have radiation into a surrounding medium at 0 temperature, heat being removed from the right end of the rod at a rate proportional to the temperature (the proportionality constant decreasing with increasing β); if

$\beta = \pi/2$, the right end is insulated; if $\pi/2 < \beta < \pi$, heat is being fed into the rod from its right end at a rate proportional to the temperature. Obviously, the last case is one of potential instability, but since heat is removed from the left end (to keep it at 0 temperature), it is not clear how large β has to be before a negative eigenvalue λ_1 of (7.1.27) appears. In any event it seems physically obvious that at least the lowest eigenvalue $\lambda_1(\beta)$ is a decreasing function of β.

Green's function $g(x, \xi; \lambda)$ corresponding to (7.1.27) satisfies

$$\begin{aligned} -g'' - \lambda g &= \delta(x - \xi), \quad 0 < x, \xi < 1; \\ g|_{x=0} &= 0, \quad (g \cos \beta + g' \sin \beta)_{x=1} = 0. \end{aligned} \tag{7.1.28}$$

Let $v(x, \lambda), z(x, \lambda)$ be the solutions of the homogeneous equation satisfying the initial conditions

$$v(0, \lambda) = 0, \quad v'(0, \lambda) = 1; \qquad z(1, \lambda) = \sin \beta, \quad z'(1, \lambda) = -\cos \beta.$$

Then

$$\begin{aligned} v(x, \lambda) &= \frac{\sin \sqrt{\lambda} x}{\sqrt{\lambda}}, \\ z(x, \lambda) &= (\sin \beta) \cos \sqrt{\lambda}(x - 1) - \frac{\cos \beta}{\sqrt{\lambda}} \sin \sqrt{\lambda}\,(x - 1), \end{aligned} \tag{7.1.29}$$

where for definiteness we have chosen $\sqrt{\lambda}$ unambiguously as follows: For $\lambda = 0$, $\sqrt{\lambda} = 0$; for $\lambda \neq 0$, λ has the unique polar representation $\lambda = |\lambda| e^{i\theta}, 0 \leqslant \theta < 2\pi$, and we define

$$\sqrt{\lambda} = |\lambda|^{1/2} e^{i\theta/2}, \quad 0 \leqslant \theta < 2\pi, \tag{7.1.30}$$

where $|\lambda|^{1/2}$ is the positive square root of the positive number $|\lambda|$. In this way each complex number λ has a well-defined square root $\sqrt{\lambda}$, which is analytic in the complex plane except on the positive real axis (for λ directly above the positive real axis, $\sqrt{\lambda} = |\lambda|^{1/2}$, whereas directly below, $\sqrt{\lambda} = -|\lambda|^{1/2}$). This is just the *principal value* of $\sqrt{\lambda}$ as defined in Exercise 1.2.3. Note that $\sqrt{\lambda}$ has a *positive* imaginary part as long as λ is not on the positive real axis. Despite the branch singularity of $\sqrt{\lambda}$, the functions defined by (7.1.29) are analytic in the whole λ plane (both v and z are even functions of $\sqrt{\lambda}$, so that whereas $\sqrt{\lambda}$ abruptly changes sign across the positive real axis, v and z do not; note also that there is no singularity at $\lambda = 0$ if we use the limiting value of (7.1.29) as $\lambda \to 0$).

Next we calculate, from (7.1.23) with $p = 1$ and (7.1.29),

$$C(\lambda) = vz' - zv' = -(\sin \beta) \cos \sqrt{\lambda} - \frac{\cos \beta}{\sqrt{\lambda} \sin \sqrt{\lambda}}, \tag{7.1.31}$$

which is also an analytic function of λ. The eigenvalues of (7.1.27), which are necessarily real, are the zeros of $C(\lambda)$.

We first look for negative eigenvalues. Setting $\lambda = -r^2, r > 0$, we find that the equation $C(\lambda) = 0$ becomes (see Figure 7.1)

$$\tanh r = -r \tan \beta, \quad r > 0, \tag{7.1.32}$$

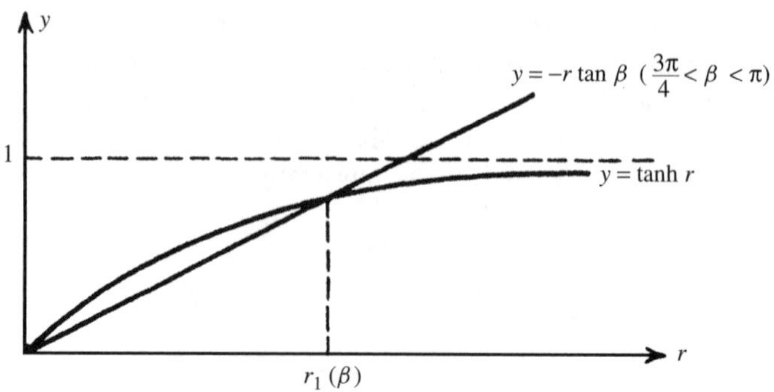

Figure 7.1

which has a single root $r_1(\beta)$ if and only if $3\pi/4 < \beta < \pi$. The corresponding negative eigenvalue $\lambda_1(\beta) = -r_1^2(\beta)$ decreases from

$$\lambda_1\left(\frac{3\pi}{4}+\right) = 0 \quad \text{to} \quad \lambda_1(\pi-) = -\infty$$

as β increases from $(3\pi/4)+$ to $\pi-$.

If $\beta = 3\pi/4$, then $\lambda_1 = 0$ is an eigenvalue with an eigenfunction $u_1(x)$ proportional to x.

Turning next to positive eigenvalues, we set $\lambda = r^2, r > 0$. The equation $C(\lambda) = 0$ now becomes

$$\tan r = -r \tan \beta, \quad r > 0. \tag{7.1.33}$$

The values of r at the intersections of the curve $y = \tan r$ with the straight line $y = -r \tan \beta$ then give the square roots of the desired positive eigenvalues. Figure 7.2 shows these intersections for different values of β. For a fixed β there are infinitely many intersections. Since for β in $0 \leqslant \beta < 3\pi/4$ all eigenvalues are positive, we label the r values at the intersections as $r_1(\beta), r_2(\beta), \ldots$; then $\lambda_k(\beta) = r_k^2(\beta)$ gives the sequence of eigenvalues in increasing order. In particular, we have $r_k(0) = k\pi$ and $\lambda_k(0) = k^2\pi^2$. As β increases in $0 \leqslant \beta < 3\pi/4$, it is clear that $r_k(\beta)$ decreases and hence so does $\lambda_k(\beta)$. When β is slightly smaller than $3\pi/4$, the first intersection occurs close to $r = 0$; when $\beta = 3\pi/4$, the line $y = -r \tan \beta$ (that is, $y = r$) is tangent to the curve $y = \tan r$ at $r = 0$ and 0 now becomes an eigenvalue (as we saw earlier). For $3\pi/4 < \beta < \pi$ the smallest eigenvalue λ_1 is the negative eigenvalue $-r_1^2(\beta)$, found from Figure 7.1; the intersections on Figure 7.2 are then labeled $r_2(\beta), r_3(\beta), \ldots$ in increasing order, and the corresponding eigenvalues are $\lambda_2(\beta) = r_2^2(\beta), \ldots$.

With this notation we see that, as β increases from 0 to $\pi-$,

$$\lambda_1(\beta) \text{ decreases from } \pi^2 \text{ to } -\infty,$$
$$\lambda_k(\beta) \text{ decreases from } k^2\pi^2 \text{ to } (k-1)^2\pi^2, k \geqslant 2.$$

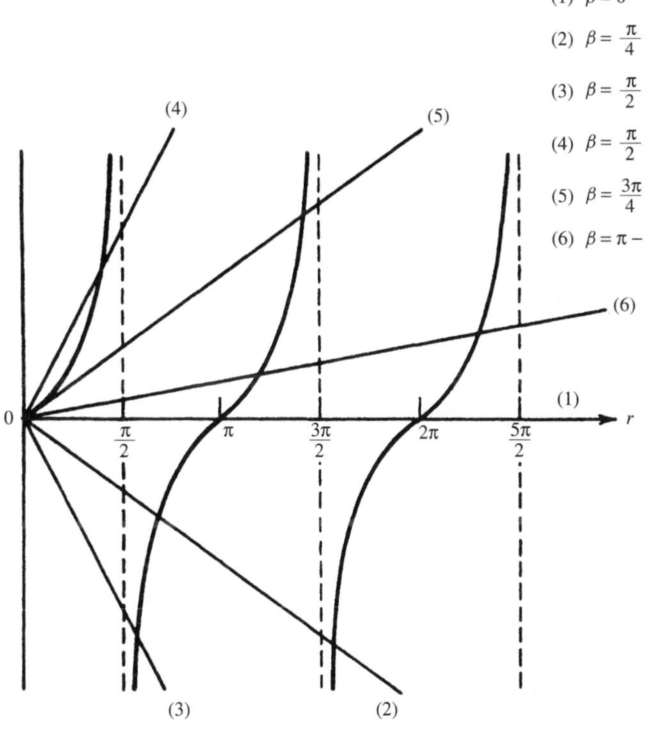

Figure 7.2

In all cases the dependence on β is continuous.

We now use (7.1.25) to find the normalized eigenfunctions (for the case $\beta < 3\pi/4$). We have, with $\lambda_n = r_n^2$,

$$v_n(x) = \frac{\sin r_n x}{r_n}, \quad z_n(x) = -\frac{\cos\beta}{r_n \cos r_n} \sin r_n x,$$

where we have made use of (7.1.33), which is satisfied for $r = r_n$. From (7.1.24) we see that

$$k_n = \frac{-\cos r_n}{\cos\beta}$$

and from (7.1.31),

$$C'(\lambda_n) = \frac{1}{2r_n}\left((\sin\beta)\sin r_n - \frac{\cos\beta}{r_n}\cos r_n + \cos\beta\frac{\sin r_n}{r_n^2}\right).$$

A straightforward calculation now gives

$$k_n C'(\lambda_n) = \frac{1}{r_n^2}\left(\frac{1}{2} - \frac{\sin 2r_n}{4r_n}\right).$$

Therefore, the normalized eigenfunctions can be found from (7.1.25):

$$u_n(x) = \frac{1}{N_n} \sin \sqrt{\lambda_n}\, x, \quad N_n^2 = \frac{1}{2} - \frac{\sin 2\sqrt{\lambda_n}}{4\sqrt{\lambda_n}}.$$

Of course, in our case the normalization could have been calculated by elementary integration of $\sin^2 \sqrt{\lambda_n}\, x$, but this will usually be impossible in more difficult cases. Alternatively, the method of Exercise 7.1.2 can be used.

Example 2. Any circularly symmetric natural mode u and the corresponding natural frequency λ of an annular membrane with fixed boundary satisfy

$$-(xu')' = \lambda x u, \quad 0 < a < x < b; \qquad u(a) = u(b) = 0, \tag{7.1.34}$$

where x is the *radial* coordinate measured from the center of the annulus.

Green's function $g(x, \xi; \lambda)$ is the solution of

$$-(xg')' - \lambda x g = \delta(x - \xi), \quad a < x, \xi < b; \quad g = 0 \quad \text{at}\, x = a, b. \tag{7.1.35}$$

The homogeneous equation has the independent solutions $J_0(\sqrt{\lambda}\, x), N_0(\sqrt{\lambda}\, x)$. The functions

$$\begin{aligned} v(x, \lambda) &= J_0(\sqrt{\lambda}\, a) N_0(\sqrt{\lambda}\, x) - J_0(\sqrt{\lambda}\, x) N_0(\sqrt{\lambda}\, a), \\ z(x, \lambda) &= J_0(\sqrt{\lambda}\, b) N_0(\sqrt{\lambda}\, x) - J_0(\sqrt{\lambda}\, x) N_0(\sqrt{\lambda}\, b) \end{aligned} \tag{7.1.36}$$

satisfy the homogeneous equation and the respective initial conditions

$$\begin{aligned} v(a, \lambda) = 0, \quad v'(a, \lambda) &= \sqrt{\lambda}\, W(J_0, N_0; \sqrt{\lambda}\, a) = \frac{2}{\pi a}, \\ z(b, \lambda) = 0, \quad z'(b, \lambda) &= \sqrt{\lambda}\, W(J_0, N_0; \sqrt{\lambda}\, b) = \frac{2}{\pi b}, \end{aligned}$$

where use has been made of the Wronskian relation

$$W(J_0, N_0; x) = \frac{2}{\pi x}. \tag{7.1.37}$$

Clearly, $v(x, \lambda)$ and $z(x, \lambda)$ are entire functions, and

$$g = A v(x_<, \lambda) z(x_>, \lambda).$$

The jump condition on dg/dx is $g'|_{\xi-}^{\xi+} = -(1/\xi)$, so that

$$A W(v, z; \xi) = -\frac{1}{\xi}.$$

From (7.1.36) and (7.1.37) we have

$$W(v, z; x) = \frac{2}{\pi x}[J_0(\sqrt{\lambda}\, a) N_0(\sqrt{\lambda}\, b) - J_0(\sqrt{\lambda}\, b) N_0(\sqrt{\lambda}\, a)],$$

so that

$$g = -\frac{\pi}{2}\frac{v(x_<, \lambda)z(x_>, \lambda)}{J_0(\sqrt{\lambda}\, a)N_0(\sqrt{\lambda}\, b) - J_0(\sqrt{\lambda}\, b)N_0(\sqrt{\lambda}\, a)}. \tag{7.1.38}$$

The eigenvalues of (7.1.34) are the zeros of the denominator $D(\lambda)$ in (7.1.38). These are known from (7.1.34) to be real and positive. Setting $\lambda = r^2$, we find that the eigenvalues $\lambda_n = r_n^2$ are determined from

$$D(r^2) \doteq J_0(ra)N_0(rb) - J_0(rb)N_0(ra) = 0 \quad \text{or} \quad \frac{J_0(ra)}{J_0(rb)} = \frac{N_0(ra)}{N_0(rb)}. \tag{7.1.39}$$

This equation gives rise to a sequence of positive simple roots

$$r_1 < r_2 < \dots,$$

and we set

$$R_n = \frac{J_0(r_n a)}{J_0(r_n b)} = \frac{N_0(r_n a)}{N_0(r_n b)}. \tag{7.1.40}$$

We then find that

$$\begin{aligned} v_n(x) &\doteq v(x, r_n^2) = J_0(r_n a)N_0(r_n x) - J_0(r_n x)N_0(r_n a) \\ &= R_n J_0(r_n b)N_0(r_n x) - R_n N_0(r_n b)J_0(r_n x) = R_n z(x, r_n^2) \\ &\doteq R_n z_n(x). \end{aligned}$$

The residue of g at $\lambda = \lambda_n = r_n^2$ is

$$-\frac{\pi}{2}\frac{v_n(x_<)z_n(x_>)}{D'(\lambda_n)} = -\frac{\pi}{2}\left(\frac{2r_n}{R_n}\right)\frac{v_n(x)v_n(\xi)}{[dD(r^2)/dr]_{r=r_n}},$$

and on using (7.1.39) and (7.1.40), we obtain

$$\left.\frac{dD(r^2)}{dr}\right|_{r=r_n} = -\frac{a}{R_n}\frac{2}{\pi r_n a} + bR_n\frac{2}{\pi r_n b} = \frac{2}{\pi r_n}\left(\frac{R_n^2 - 1}{R_n}\right).$$

The normalized eigenfunctions are therefore given by

$$u_n(x) = \frac{\pi r_n}{\sqrt{2}\sqrt{R_n^2 - 1}}[J_0(r_n a)N_0(r_n x) - J_0(r_n x)N_0(r_n a)], \tag{7.1.41}$$

where r_n is determined from (7.1.39) and R_n from (7.1.40).

Example 3. (See Eastham [5])

(a) In propagation through periodic lattices one encounters the equation

$$-u'' + Q(x)u = 0, \quad -\infty < x < \infty, \tag{7.1.42}$$

where $Q(x)$ is periodic in x. Such an equation does not necessarily have a nontrivial periodic solution (for instance, $-u'' + u = 0$), but any periodic solution

u has a period which coincides with one of the periods of Q. Suppose, without loss of generality, that

$$Q(x+1) = Q(x), \tag{7.1.43}$$

and let us look not only for solutions of (7.1.42) with period 1 but also, more generally, for solutions satisfying

$$u(x+1) = \rho u(x), \tag{7.1.44}$$

where ρ is an undetermined constant known as a *multiplier*.

Let $v(x)$ and $z(x)$ be the solutions of the homogeneous equation (7.1.42) satisfying, respectively, the initial conditions

$$v(0) = 1, \quad v'(0) = 0; \quad z(0) = 0, \quad z'(0) = 1. \tag{7.1.45}$$

In view of (7.1.43) the functions $v(x+1)$ and $z(x+1)$ are also solutions of (7.1.42); moreover, these functions are independent. We can therefore write $v(x+1)$ and $z(x+1)$ as linear combinations of $v(x)$ and $z(x)$:

$$\begin{aligned} v(x+1) &= v(1)v(x) + v'(1)z(x), \\ z(x+1) &= z(1)v(x) + z'(1)z(x). \end{aligned} \tag{7.1.46}$$

The general solution of (7.1.42) is $u = Av(x) + Bz(x)$; requiring (7.1.44) and using (7.1.46), we find that nontrivial solutions are possible if and only if the multiplier ρ satisfies the quadratic equation

$$\rho^2 - \rho[v(1) + z'(1)] + v(1)z'(1) - z(1)v'(1) = 0.$$

The usual Wronskian relation gives the relation

$$v(1)z'(1) - z(1)v'(1) = v(0)z'(0) - z(0)v'(0) = 1,$$

so that

$$\rho^2 - \rho[v(1) + z'(1)] + 1 = 0. \tag{7.1.47}$$

If (7.1.47) has $\rho = 1$ as a solution, (7.1.42) will have solution(s) with period 1. If (7.1.47) has $\rho = -1$ as a solution, (7.1.42) will have solution(s) with period 2.

In many applications, $Q(x) = q(x) - \lambda$, where $q(x)$ has period 1 and λ is an eigenvalue parameter. We look for values of λ which give rise to periodic solutions of (7.1.42). Then the corresponding functions v and z depend on λ, and so does the coefficient of ρ in (7.1.47). We see below how to generate periodic solutions by considering eigenvalue problems on the basic interval $0 < x < 1$.

(b) We look at two related problems.

(i) Given q such that $q(x+1) = q(x)$, find λ and $u(x)$ with $u(x+1) = u(x)$ satisfying

$$-u'' + q(x)u - \lambda u = 0, -\infty < x < \infty.$$

(ii) Find the eigenvalues and eigenfunctions of

$$-u'' + q(x)u - \lambda u = 0, 0 < x < 1; \quad u(0) = u(1), u'(0) = u'(1).$$

If (λ, u) is a solution of (i), then (λ, u) is clearly an eigenpair of (ii). If (λ, u) is an eigenpair of (ii), we can, by virtue of the boundary conditions, extend $u(x)$ to $-\infty < x < \infty$ as a continuously differentiable function with period 1. On $-\infty < x < \infty$ this periodic extension satisfies problem (i), where $q(x)$ stands for the periodic extension of the $q(x)$ that appears in (ii). *Thus problems (i) and (ii) are equivalent.*

If instead of the boundary conditions in (ii) we take $u(0) = -u(1), u'(0) = -u'(1)$, another eigenvalue problem results. Any eigenfunction of this new problem can be extended as a continuously differentiable function in $1 \leqslant x \leqslant 2$ by setting $u(x+1) = -u(x)$. This new function satisfies $u(0) = u(2), u'(0) = u'(2)$ and can therefore be extended as a function $u(x)$ *with period* 2 on $-\infty < x < \infty$. The extended u satisfies $-u'' + qu - \lambda u = 0$, where q is the periodic extension referred to earlier: $q(x+1) = q(x)$. Let us now examine problem (ii) more closely.

Nontrivial solutions of (ii) are possible if and only if (7.1.47) is satisfied for $\rho = 1$. Since v and z are now functions of x and λ, this means that λ is determined from

$$2 - v(1,\lambda) - z'(1,\lambda) = 0. \tag{7.1.48}$$

Simple roots of (7.1.48) lead to simple eigenvalues of (ii), and double roots to eigenvalues of multiplicity 2. If the conditions $v'(1,\lambda) = 0$ and $z(1,\lambda) = 0$ hold simultaneously in addition to (7.1.48), the eigenvalue is of multiplicity 2. Indeed, from the Wronskian relation we then have $v(1,\lambda)z'(1,\lambda) = 1$, which together with (7.1.48) gives $v(1,\lambda) = z'(1,\lambda) = 1$. Thus both v and z satisfy the boundary conditions in (ii) and are therefore (independent) eigenfunctions corresponding to that value of λ. Since the differential equation is of the second order, every nontrivial solution is then an eigenfunction.

(c) We consider a special simple case of problem (ii):

$$-u'' - \lambda u = 0, \quad 0 < x < 1; \quad u(0) = u(1), \quad u'(0) = u'(1). \tag{7.1.49}$$

It is easy to find the eigenvalues and eigenfunctions explicitly. To this end, let $v(x,\lambda), z(x,\lambda)$ be the solutions of the homogeneous equation satisfying, respectively, the initial conditions

$$v(0,\lambda) = 1, v'(0,\lambda) = 0; \quad z(0,\lambda) = 0, \quad z'(0,\lambda) = 1.$$

We find that

$$v(x,\lambda) = \cos\sqrt{\lambda}\,x, \quad z(x,\lambda) = \frac{\sin\sqrt{\lambda}\,x}{\sqrt{\lambda}},$$

which are correct even for $\lambda = 0$ if the limiting value is used when needed.

Therefore, (7.1.48) becomes

$$\cos\sqrt{\lambda} = 1. \tag{7.1.50}$$

Equation (7.1.50) has a simple zero at $\lambda = 0$ and double zeros at $\lambda = 4n^2\pi^2$, $n = 1, 2, \ldots$. Only $\lambda = 0$ is a simple eigenvalue of (7.1.49); all the others have multiplicity 2. To index all eigenvalues with regard to multiplicity, we set

$$\lambda_0 = 0, \quad \lambda_{2n-1} = \lambda_{2n} = 4n^2\pi^2, n = 1, 2, \ldots. \tag{7.1.51}$$

The corresponding orthonormal basis of eigenfunctions is

$$u_0 = 1, \quad u_{2n-1} = \sqrt{2}\cos n\pi x, \quad u_{2n} = \sqrt{2}\sin n\pi x. \tag{7.1.52}$$

Let us now see how we can generate the spectrum (7.1.51)–(7.1.52) from the residues at the singularities of Green's function $g(x, \xi; \lambda)$ satisfying

$$-g'' - \lambda g = \delta(x - \xi), 0 < x, \xi < 1; \quad g|_{x=0} = g|_{x=1}, g'|_{x=0} = g'|_{x=1}.$$

We can write

$$g(x, \xi; \lambda) = h(x, \xi; \lambda) + Av(x, \lambda) + Bz(x, \lambda), \tag{7.1.53}$$

where h is the causal Green's function given by

$$h(x, \xi; \lambda) = -\frac{\sin\sqrt{\lambda}\,(x - \xi)}{\sqrt{\lambda}} H(x - \xi).$$

On applying the boundary conditions, we find that

$$\begin{aligned} A(1 - \cos\sqrt{\lambda}) - B\frac{\sin\sqrt{\lambda}}{\sqrt{\lambda}} &= -\frac{\sin\sqrt{\lambda}\,(1 - \xi)}{\sqrt{\lambda}}, \\ A\sqrt{\lambda}\sin\sqrt{\lambda} + B(1 - \cos\sqrt{\lambda}) &= -\cos\sqrt{\lambda}\,(1 - \xi), \end{aligned}$$

from which follow

$$A = -\frac{\sin\sqrt{\lambda}\,(1 - \xi)}{2\sqrt{\lambda}} - \frac{\sin\sqrt{\lambda}}{2\sqrt{\lambda}(1 - \cos\sqrt{\lambda})}\cos\sqrt{\lambda}\,(1 - \xi), \tag{7.1.54}$$

$$B = -\frac{\cos\sqrt{\lambda}\,(1 - \xi)}{2} + \frac{\sin\sqrt{\lambda}}{2(1 - \cos\sqrt{\lambda})}\sin\sqrt{\lambda}\,(1 - \xi). \tag{7.1.55}$$

Expansion (7.1.16) is still valid, and we need only to keep track of the singularities of g. Since h, v, z, and the first term on the right of each of (7.1.54) and (7.1.55) are all analytic, the residues all stem from the singular part of $Av(x, \lambda) + Bz(x, \lambda)$, that is, from

$$\begin{aligned} R &= -\frac{\sin\sqrt{\lambda}\cos\sqrt{\lambda}\,(1 - \xi)\cos\sqrt{\lambda}\,x}{2(1 - \cos\sqrt{\lambda})} + \frac{\sin\sqrt{\lambda}\sin\sqrt{\lambda}\,(1 - \xi)\sin\sqrt{\lambda}\,x}{2(1 - \cos\sqrt{\lambda})\sqrt{\lambda}} \\ &= -\frac{\cos\sqrt{\lambda}\,(x - \xi + 1)}{2\sqrt{\lambda}}\left(\frac{\sin\sqrt{\lambda}}{1 - \cos\sqrt{\lambda}}\right) \\ &= -\frac{\cos\sqrt{\lambda}\,(x - \xi + 1)}{2\sqrt{\lambda}}\frac{\cos(\sqrt{\lambda}/2)}{\sin(\sqrt{\lambda}/2)}. \end{aligned}$$

At $\lambda = 0$ we have a simple pole from the simple zero of $1 - \cos\sqrt{\lambda}$; this gives the residue

$$-\frac{1}{2}\frac{1}{[(d/d\lambda)(1-\cos\sqrt{\lambda})]_{\lambda=0}} = -1.$$

At $\lambda = 4n^2\pi^2$, $1 - \cos\sqrt{\lambda}$ has a double zero but $\sin\sqrt{\lambda}$ has a simple zero, so that we again have a simple pole whose residue is

$$-\frac{\cos 2n\pi(x-\xi+1)}{4n\pi}\frac{\cos n\pi}{[(d/d\lambda)\sin(\sqrt{\lambda}/2)]_{\lambda=2n\pi}} = -2\cos 2n\pi(x-\xi+1)$$
$$= -2\cos 2n\pi(x-\xi) = -2(\cos 2n\pi x\cos 2n\pi\xi + \sin 2n\pi x\sin 2n\pi\xi).$$

Using (7.1.19), we can set the sum of these residues equal to $-\sum_n u_n(x)\bar{u}_n(\xi)$, and we recover (7.1.52). Note that (7.1.20) becomes the familiar [see (2.3.24)]

$$\delta(x-\xi) = 1 + 2\sum_{n=1}^{\infty}(\cos 2n\pi x\cos 2n\pi\xi + \sin 2n\pi x\sin 2n\pi\xi), \qquad (7.1.56)$$
$$0 < x, \xi < 1.$$

Transition to a Singular Problem for First-Order Equations

Consider the eigenvalue problem

$$u' = i\lambda s(x)u, \quad 0 < x < b; \quad u(b) = u(0), \qquad (7.1.57)$$

where $s(x) > 0$ for $x \geqslant 0$. Letting $H_s(0,b)$ be the space of complex-valued functions such that $\int_s^b s|u|^2\,dx < \infty$, we can rewrite our problem in the form

$$Lu = \lambda u, \qquad (7.1.58)$$

where

$$Lu \doteq -\frac{i}{s(x)}u',$$

and the domain D_L consists of absolutely continuous functions with Lu in $H_s(0,b)$ satisfying $u(b) = u(0)$. We proved in Chapter 5 that L is self-adjoint. Problem (7.1.57) or (7.1.58) is regular for finite b, and we shall be interested in the limit as $b \to \infty$ when the problem becomes singular.

For finite b we can solve (7.1.57) explicitly. Setting $S(x) = \int_0^x s(t)\,dt$, we see that the general solution of the differential equation is $A\phi(x,\lambda)$, where

$$\phi(x,\lambda) = e^{i\lambda S(x)}. \qquad (7.1.59)$$

Applying the boundary condition, we find that

$$e^{i\lambda S(b)} = 1,$$

so that the eigenvalues are given by

$$\lambda_n = \frac{2n\pi}{S(b)}, \quad n = 0, \pm 1, \pm 2, \ldots. \tag{7.1.60}$$

Note that the eigenvalues are uniformly spaced along the entire real axis.

The eigenfunctions are $Ae^{i\lambda_n S}$, which we now proceed to normalize. Setting $\phi_n = \phi(x, \lambda_n)$, we find that

$$\langle \phi_n, \phi_m \rangle_s = \int_0^b s(x) e^{i(\lambda_n - \lambda_m)S(x)}\, dx.$$

Since S is increasing and differentiable, we may make the change of variable

$$y = \frac{2\pi S(x)}{S(b)}$$

to obtain

$$\langle \phi_n, \phi_m \rangle_s = \frac{S(b)}{2\pi} \int_0^{2\pi} e^{i(n-m)y}\, dy = \begin{cases} 0, & n \neq m, \\ S(b), & n = m. \end{cases}$$

Thus (7.1.57) leads to the orthonormal basis of eigenfunctions in H_s,

$$u_n(x) = [S(b)]^{-1/2} e^{i\lambda_n S(x)}. \tag{7.1.61}$$

It now becomes easy to analyze the behavior as $b \to \infty$. Since $S(x)$ is an increasing continuous function, there are only two possibilities: (1) $\lim_{b\to\infty} S(b)$ is a finite number denoted by $S(\infty)$, or (2) $S(b) \to +\infty$ as $b \to +\infty$. If $s(x) = 1$, for instance, case 2 obtains; but if $s(x) = 1/(1 + x^2)$, we have case 1.

In case 1 everything is pretty much the same as for the bounded interval. We see from (7.1.60) that the spectrum remains discrete. It suffices to replace $S(b)$ by $S(\infty)$ in (7.1.60) and (7.1.61) to obtain the normalized eigenpairs for the interval $0 < x < \infty$. In fact, even the boundary condition in (7.1.57) is satisfied in the form $u(\infty) = u(0)$. We should also note that the general solution of the differential equation (7.1.57) without boundary condition is in H_s for all λ in the complex plane. Indeed, we have, from (7.1.59)

$$\|\phi\|_s^2 = \int_0^\infty s e^{-2\beta S}\, dx = \int_0^{S(\infty)} e^{-2\beta z}\, dz, \tag{7.1.62}$$

where $\beta = \text{Im}\, \lambda$. Since $S(\infty)$ is finite, so is the integral in (7.1.62) for all β.

In case 2, $\lim_{b\to\infty} S(b) = \infty$, so that the eigenvalues (7.1.60) coalesce into the entire real axis. The solution $\phi(x, \lambda)$ is no longer in H_s when λ is real. In fact, (7.1.62) shows that $\phi(x, \lambda)$ is in H_s if and only if $\text{Im}\, \lambda > 0$. Thus our limiting spectral problem has no eigenvalues. What happens is that the real λ axis is now in the continuous spectrum.

Additional information can be obtained from a study of the Green's function $g_b(x, s; \lambda)$ satisfying

$$-ig_b' - \lambda s g_b = \delta(x - \xi), \quad 0 < x, \xi < b; \qquad g_b|_{x=b} = g_b|_{x=0}. \tag{7.1.63}$$

The bilinear series (7.1.16), as well as (7.1.19) and (7.1.20), remain valid for regular self-adjoint boundary value problems involving equations of the first order. The functions $\{u_n\}$ are now given explicitly by (7.1.61). Of course Green's function can also be constructed in closed form from (7.1.63) when λ is not one of the eigenvalues (7.1.60). Certainly, this construction is valid for Im $\lambda \neq 0$. The function g_b is proportional to ϕ for $x < \xi$ and for $x > \xi$ but with different proportionality constants. At $x = \xi, g_b$ satisfies the jump condition

$$g_b(\xi+, \xi; \lambda) - g_b(\xi-, \xi; \lambda) = i.$$

A straightforward calculation gives

$$g_b(x, \xi; \lambda) = ie^{i\lambda[S(x)-S(\xi)]}\left[\frac{1}{e^{-i\lambda S(b)} - 1} + H(x - \xi)\right], \tag{7.1.64}$$

where H is the Heaviside function.

As expected, g_b has simple poles at the eigenvalues λ_n given by (7.1.60), and the residue of g_b at $\lambda = \lambda_n$ is easily verified to be

$$\frac{-e^{i\lambda_n S(x)}e^{-i\lambda_n S(\xi)}}{S(b)},$$

which agrees with $-u_n(x)\bar{u}_n(\xi)$ from (7.1.61), thereby confirming (7.1.19).

We can also determine the behavior of g_b as $b \to \infty$. In case 1 when $S(b)$ tends to the finite limit $S(\infty), g_b$ itself has a limiting value with simple poles at $\lambda_n = 2\pi/S(\infty)$. The spectrum therefore remains discrete. If, however, $S(b) \to \infty$, the simple poles in (7.1.64) disappear. Indeed, we have, on setting $\lambda = \alpha + i\beta$,

$$\frac{1}{e^{-i\lambda S(b)} - 1} = \frac{1}{e^{-i\alpha S(b)}e^{\beta S(b)} - 1},$$

which tends, as $b \to \infty$, to 0 for $\beta > 0$ and to -1 for $\beta < 0$ (there is no limit for $\beta = 0$). The limiting Green's function g therefore has different analytic expressions in the upper and lower halves of the λ plane. We have the explicit forms

$$\begin{aligned} g(x, \xi; \lambda) &= ie^{i\lambda[S(x)-S(\xi)]}H(x - \xi) && \text{for Im } \lambda > 0, \\ g(x, \xi; \lambda) &= ie^{i\lambda[S(x)-S(\xi)]}[-1 + H(x - \xi)] && \text{for Im } \lambda < 0. \end{aligned} \tag{7.1.65}$$

If we had used the more general self-adjoint boundary condition $u(b) = e^{i\gamma}u(0)$ in (7.1.63), the same limiting form for g would result.

Since g is analytic in both half-planes, we reduce the integral over a large circle C_∞ in the complex plane to an integral along the real axis:

$$\int_{C\infty} g(x, \xi; \lambda)\, d\lambda = -\int_{-\infty}^{\infty} [g(x, \xi; \lambda_+) - g(x, \xi; \lambda_-)]\, d\lambda, \tag{7.1.66}$$

where λ_+ and λ_- denote the values just above the real axis and just below, respectively. Using (7.1.65), we find that

$$\frac{1}{2\pi i}\int_{C\infty} g(x,\xi;\lambda)\,d\lambda = -\frac{1}{2\pi}\int_{-\infty}^{\infty} e^{i\lambda[S(x)-S(\xi)]}\,d\lambda. \tag{7.1.67}$$

Thus the integral over the real λ axis now plays the role of the sum appearing in (7.1.19). The equivalent form of (7.1.20) is then

$$\frac{\delta(x-\xi)}{s(x)} = \frac{1}{2\pi}\int_{-\infty}^{\infty} e^{i\lambda[S(x)-S(\xi)]}\,d\lambda, \tag{7.1.68}$$

which contains, for the case $s(x) = 1$, the famous Fourier integral formula

$$\delta(x-\xi) = \frac{1}{2\pi}\int_{-\infty}^{\infty} e^{i\lambda(x-\xi)}\,d\lambda, \tag{7.1.69}$$

which has been derived for $0 < x, \xi < \infty$, but is clearly valid for all x, ξ, since $x-\xi$ takes on all possible real values as x and ξ vary between 0 and infinity.

We shall pursue the transition to singular problems for second-order BVPs in Sections 7.2 and 7.3.

EXERCISES

7.1.1 A symmetric operator L defined on a dense domain D in H_s is said to be *bounded below* [see (5.8.6)] if there exists a real c such that

$$\langle Lu, u\rangle_s \geqslant c\|u\|_s^2 \quad \text{for all } u \in D. \tag{7.1.70}$$

Show that the operator L given by (7.1.3) on the domain D of functions having a continuous second derivative and satisfying (7.1.6) is bounded below. (It follows that L on D has only a finite number of negative eigenvalues since we have shown elsewhere that the eigenvalues cannot have a finite limit point.) [*Hint:* Follow these steps with $\|u\|^2 = \int_a^b |u|^2\,dx$:

(a) There exists x_0 such that $\|u\|^2 = (b-a)|u(x_0)|^2$.

(b) $\big||u(x_0)|^2 - |u(a)|^2\big| = \left|\int_a^{x_0}(d/dx)(u\bar{u})\,dx\right| \leq 2\|u\|\,\|u'\|$.

(c) $|u(a)|^2 \leq [\|u\|^2/(b-a)] + 2\|u\|\,\|u'\|$. This is also true for $|u(b)|^2$.

(d) $\langle Lu, u\rangle_s \geq m_1\|u'\|^2 + m_2\|u\|^2 + m_3\|u\|\,\|u'\|, m_1 > 0$.

(e) $\langle Lu, u\rangle_s \geq d\|u\|^2$.

(f) $\langle Lu, u\rangle_s \geq c\|u\|_s^2$.]

7.1.2 Consider the eigenvalue problem

$$-(pu')' + qu - \lambda su = 0, \quad a < x < b; \qquad B_a u = 0, \quad B_b u = 0,$$

where B_a and B_b are defined by (7.1.6). Let $v(x, \lambda)$ be the solution of the homogeneous differential equation satisfying $v(a, \lambda) = \sin \alpha, v'(a, \lambda) = \cos \alpha$. Note that $v(x, \lambda)$ is real for λ real. The nontrivial function $v(x, \lambda)$ satisfies the boundary condition at the left end; an eigenvalue (necessarily real) λ_n is a number such that $v(x, \lambda_n)$ also satisfies the boundary condition at the right end. The real function $v_n(x) = v(x, \lambda_n)$ is the corresponding eigenfunction but is not normalized. We would like to calculate $\int_a^b s v_n^2(x)\, dx$. From the respective BVPs satisfied by $v_n(x)$ and $v(x, \lambda)$, where λ is a real number near λ_n, show that

$$(\lambda - \lambda_n) \int_a^b s v_n v \, dx = p(b) W(v, v_n; b),$$

and by taking the limit as $\lambda \to \lambda_n$,

$$\int_a^b s v_n^2 \, dx = p(b) \left[v_n'(b) \frac{dv(b, \lambda)}{d\lambda} \bigg|_{\lambda = \lambda_n} - v_n(b) \frac{dv'(b, \lambda)}{d\lambda} \bigg|_{\lambda = \lambda_n} \right].$$

Prove the equivalence of this result and (7.1.25).

7.1.3 Consider the eigenvalue problem

$$\begin{aligned} -(pu')' + qu - \lambda s u = 0, \quad & a < x < b; \\ u(a) \cos \alpha - u'(a) \sin \alpha = 0, \quad & u(b) \cos \beta + u'(b) \sin \beta = 0, \end{aligned} \tag{7.1.71}$$

where $0 \leqslant \alpha < \pi, 0 \leqslant \beta < \pi$. The eigenvalues (which depend on α, β) are indexed in the usual order $\lambda_1 < \lambda_2 < \cdots < \lambda_n < \cdots$. Show that $\lambda_n(\alpha, \beta)$ is a decreasing function of α and of β. [*Hint:* Use the Weyl-Courant principle,

$$\lambda_{n+1}(\alpha, \beta) = \max_{E_n \in S_n} \min_{\substack{u \in E_n \\ u \in D(\alpha, \beta)}} R(u),$$

where

$$R(u) = \frac{\int_a^b (pu'^2 + qu^2)\, dx + p(b) u^2(b) \cot \beta + p(a) u^2(a) \cot \alpha}{\int_a^b s u^2 \, dx},$$

$D(\alpha, \beta)$ is the set of piecewise differentiable real functions $u(x)$ (if α or $\beta = 0$, u is also required to vanish at the corresponding endpoint, and the term involving α or β in R is omitted), and S_n is the class of all n-dimensional subspaces of $D(\alpha, \beta)$.]

7.1.4 Consider the eigenvalue problem (7.1.71) and the related eigenvalue problem obtained by replacing $p(x)$ by $P(x)$ and $q(x)$ by $Q(x)$, but keeping s, α, β the same. Denote the eigenvalues of the related problem by Λ_n. Show that if $P \geqslant p, Q \geqslant q$, then

$$\Lambda_n \geqslant \lambda_n, \quad n = 1, 2, \ldots.$$

7.1.5 Show that the eigenvalues of

$$-(x^2u')' - \lambda u = 0, \quad 1 < x < e; \quad u(1) = u(e) = 0$$

are positive. By trying solutions of the form x^μ, find the general solution of the differential equation. Show that the eigenvalues are $\lambda_n = n^2\pi^2 + \frac{1}{4}$, $n = 1, 2, \ldots$, with corresponding (nonnormalized) eigenfunctions

$$v_n(x) = x^{-1/2} \sin(n\pi \log x).$$

By considering the differential equations satisfied by $v_n(x)$ and also by $v = x^{-1/2} \sin(\alpha \log x)$, show that

$$\int_1^e v_n^2(x)\, dx = \frac{1}{2},$$

a result that could have also been obtained by elementary integration.

7.1.6 Consider the special case of (7.1.71)

$$\begin{aligned} &-u'' - \lambda u = 0, \quad -1 < x < 1; \\ &u'(-1) = u(-1) \cot \alpha, \quad u'(1) = -u(1) \cot \beta, \end{aligned}$$

where $0 \leqslant \alpha < \pi, 0 \leqslant \beta < \pi$, with the understanding that the cases $\alpha = 0$, $\beta = 0$ are to be interpreted as $u(-1) = 0, u(1) = 0$, respectively. Find Green's function $g(x, \xi; \lambda)$, and use (7.1.19) to generate the normalized eigenfunctions. Show that there exists a negative eigenvalue if and only if $\cot \alpha + \cot \beta < 0$.

7.1.7 Follow the procedure of Example 3(c) to find the eigenvalues and normalized eigenfunctions of

$$-u'' - \lambda u = 0, \quad 0 < x < 1; \quad u(0) = -u(1), \quad u'(0) = -u'(1).$$

7.1.8 Consider the eigenvalue problem

$$-(pu')' - \lambda u = 0, -1 < x < 1; \quad u(-1) = u(1) = 0, \qquad (7.1.72)$$

where we are interested in the limiting case $p(x) = p_1$, $x < 0$, $p(x) = p_2$, $x > 0$, where p_1 and p_2 are positive constants (for a discussion of such interface problems, see Section 1.4). Even though p is discontinuous at $x = 0, pu'$ and u remain continuous at $x = 0$, as noted in (1.4.35) and (1.4.36). One can therefore carry out an analysis similar to the one leading to (7.1.22). Let $v(x, \lambda)$ be the solution of the homogeneous equation satisfying $v(-1, \lambda) = 0$, $v'(-1, \lambda) = 1$. Show that the eigenvalues of (7.1.72) are the zeros of $v(1, \lambda)$, where

$$v(1, \lambda) = \sqrt{\frac{p_1}{\lambda}} \left(\sin \sqrt{\frac{\lambda}{p_1}} \cos \sqrt{\frac{\lambda}{p_2}} + \sqrt{\frac{p_1}{p_2}} \cos \sqrt{\frac{\lambda}{p_1}} \sin \sqrt{\frac{\lambda}{p_2}} \right).$$

7.1.9 Let $g(x)$ have period 1. Show that the function

$$u(x) = e^{G(x)},$$

where $G(x) = \int_0^x g(y)\, dy$, is a solution of the equation with periodic coefficient

$$-u'' + (g' + g^2)u = 0.$$

What is the multiplier [see (7.1.44)] in this case? When is this solution periodic?

7.1.10 Consider the eigenvalue problem

$$u' = i\lambda s(x)u, \quad a < x < b; \quad u(b) = e^{i\gamma}u(a),$$

where $s > 0$ on the real axis and γ is a real constant. Find the eigenvalues and normalized eigenfunctions. Discuss the transition to a singular problem as $b \to \infty, a \to -\infty$. Construct Green's function for the regular problem, and calculate the limit as $b \to \infty, a \to -\infty$. Find the corresponding forms of (7.1.67) and (7.1.68).

7.1.11 Consider the equation

$$-iu' + qu - \lambda su = 0, \tag{7.1.73}$$

where $q(x)$, $s(x)$ are real and $s(x) > 0$ for $x \geqslant 0$. Define

$$S(x) \doteq \int_0^x s(t)\, dt, \qquad Q(x) \doteq \int_0^x q(t)\, dt, \qquad P(x) = \lambda S - Q.$$

Show that, if $S(\infty)$ is finite, the solutions of (7.1.73) are in $H_s(0, \infty)$ for all λ, whereas if $S(\infty) = \infty$, the nontrivial solutions are in $H_s(0, \infty)$ only for $\operatorname{Im} \lambda > 0$.

Consider next (7.1.73) with the general boundary condition $u(b) = e^{i\gamma}u(0)$, where γ is a real number. Find all eigenvalues of this self-adjoint problem.

Show that Green's function $g_b(x, \xi; \lambda)$ satisfying

$$-ig_b' + qg_b - \lambda sg_b = \delta(x - \xi), \quad 0 < x, \xi < b; \quad g_b|_{x=b} = e^{i\gamma}g_b|_{x=0}$$

is given by

$$g_b(x, \xi; \lambda) = ie^{i[P(x) - P(\xi)]}[H(x - \xi) + C], \tag{7.1.74}$$

where

$$C = [e^{i\gamma}e^{-iP(b)} - 1]^{-1}.$$

Show that for fixed b and β the locus of C is a circle in the complex C plane. For β fixed and b increasing, each circle is enclosed in the preceding one, so that as $b \to \infty$ the locus of C is either a point (*limit point*) or a circle (*limit circle*). If $S(\infty) = \infty$, show that the limit point case obtains and that

$g_b \to g, g$ being given by (7.1.65) with $\lambda S(x)$ replaced by $P(x)$. If $S(\infty)$ is finite, we have the limit circle case. Points on the limit circle correspond to boundary conditions of the form $|u(\infty)| = |u(0)|$. Then $g_b \to g, g$ being given by (7.1.74) with

$$C = [e^{i\theta} e^{-i\lambda S(\infty)} - 1]^{-1},$$

where θ is a real constant.

7.2 WEYL'S CLASSIFICATION OF SINGULAR PROBLEMS

Suppose that (7.1.1) has *one* endpoint singular; that is, either the interval is semi-infinite or, if finite, $p(x)$ vanishes at one of the endpoints (this may also be accompanied by the unboundedness of q or s at the same endpoint). The solutions of the differential equation are no longer necessarily in the space H_s of functions such that $\int_a^b s|u|^2\, dx < \infty$. The following theorem of Weyl enables us to classify the singular endpoint according to the number of solutions in H_s.

Theorem (Weyl's Theorem). *Consider*

$$-(pu')' + qu - \lambda su = 0, \quad a < x < b, \tag{7.2.1}$$

when one of the endpoints is regular and the other singular. The coefficients are fixed except for the parameter λ. Then the following hold:

1. *If for some particular value of λ every solution of (7.2.1) is in H_s, for any other value of λ every solution is again in H_s.*

2. *For every λ with Im $\lambda \neq 0$, there exists at least one solution of (7.2.1) in H_s.*

REMARKS

1. The theorem (whose proof can be found, for instance, in Hajmirzaahmad and Krall [7]) tells us that the singular point must fall into one of the following two mutually exclusive categories:

   Limit circle case. All solutions are in H_s for all λ.

   Limit point case. For λ with Im $\lambda \neq 0$ there is exactly one (independent) solution in H_s, and then for Im $\lambda = 0$ there may be one or no solution in H_s. To determine which case applies it suffices to examine a *single* value of λ.

2. If both endpoints are singular, we introduce an intermediate point $l, a < l < b$, and then classify a according to the behavior of solutions in $a < x < l$ and classify b according to the behavior of solutions in $l < x < b$ (the classification is clearly independent of l).

Example 1. Consider $-u'' - \lambda u = 0$ on three different intervals:

(a) The interval $a < x < \infty$, where a is finite. Then $s = p = 1$ and $H_s = L_2^{(c)}(a, \infty)$. The point a is regular, but the right endpoint $b = \infty$ is singular. If $\lambda = 0$, the general solution is $Ax + B$, so that *no* solution is in $L_2^{(c)}(a, \infty)$, which means that we have the limit point case at the singular point $b = \infty$. It is easy to check the prediction, contained in Weyl's theorem, that there is exactly one solution in $L_2^{(c)}(a, \infty)$ for Im $\lambda \neq 0$. Indeed, the general solution of $-u'' - \lambda u = 0$ is $A \exp(i\sqrt{\lambda}\, x) + B \exp(-i\sqrt{\lambda}\, x)$, where, as usual, $\sqrt{\lambda}$ is the principal value (7.1.30). We denote the real interval $0 \leqslant \lambda < \infty$ by $[0, \infty)$. If $\lambda \notin [0, \infty)$, $\sqrt{\lambda}$ has a positive imaginary part and $\exp(i\sqrt{\lambda}\, x)$ is in $L_2^{(c)}(a, \infty)$ while $\exp(-i\sqrt{\lambda}\, x)$ is not. If $\lambda \in [0, \infty)$, no solution is in $L_2^{(c)}(a, \infty)$. Thus when Im $\lambda \neq 0$ there is exactly one solution in H_s, whereas when Im $\lambda = 0$ there is one solution in H_s if $\lambda < 0$ and no solution in H_s if $\lambda \geqslant 0$.

(b) The interval $-\infty < x < b$ with b finite. Again we have the limit point case at the singular endpoint (which is now the left endpoint $a = -\infty$).

(c) The interval $-\infty < x < \infty$. By Remark 2 and the results of (a) and (b) it follows that both endpoints are in the limit point case.

Example 2. Bessel's equation of fixed order $\nu \geqslant 0$ with parameter λ

$$-(xu')' + \frac{\nu^2}{x} u - \lambda x u = 0. \tag{7.2.2}$$

(a) On $0 < x < b$, with b finite. Then 0 is a singular point since $p(0) = 0$ [and also $q(0) = \infty$ if $\nu \neq 0$]. The space H_s consists of functions u for which $\int_0^b x|u|^2\, dx < \infty$. For $\lambda = 0$, $\nu \neq 0$, (7.2.2) degenerates into a much simpler equation with the independent solutions x^ν and $x^{-\nu}$, which are both in H_s if $\nu < 1$; if $\nu \geqslant 1$, only $x^{-\nu}$ is in H_s. For $\lambda = 0$ and $\nu = 0$, both the independent solutions 1 and $\log x$ are in H_s. Thus if $\nu < 1$, we have the *limit circle* case at $x = 0$, whereas for $\nu \geqslant 1$ we have the *limit point* case at $x = 0$. We can easily confirm the prediction of Weyl's theorem that for $\nu < 1$ all solutions are in H_s for all λ. The functions $J_\nu(\sqrt{\lambda}\, x)$ and $J_{-\nu}(\sqrt{\lambda}\, x)$ are independent solutions of (7.2.2) for $\lambda \neq 0$ and $\nu \neq 0$. J_ν is finite at $x = 0$ and $J_{-\nu}$ has a singularity of order $x^{-\nu}$; thus J_ν and $J_{-\nu}$ are both in H_s. If $\nu = 0$, the independent solutions can be taken as $J_0(\sqrt{\lambda}\, x)$ and $N_0(\sqrt{\lambda}\, x)$, the first of which is finite at $x = 0$ while the second has a logarithmic singularity at $x = 0$; again both solutions are in H_s.

(b) On $a < x < \infty$, where $a > 0$. Then the right endpoint is singular, but the left endpoint is regular. The space H_s: $\int_a^\infty x|u|^2\, dx < \infty$. Taking $\lambda = 0$, we see that neither of the independent solutions $x^\nu, x^{-\nu} (\nu > 0)$ is in H_s; the same is true for $\nu = 0$ when 1 and $\log x$ are independent solutions. Thus we have the *limit point* case at ∞ for all ν.

(c) On $0 < x < \infty$ with both endpoints singular. Here we combine the results of parts (a) and (b).

Example 3. The Hermite equation. The quantum-mechanical problem of a particle in a quadratic potential field (the so-called *harmonic oscillator*) is described by the equation

$$-u'' + x^2u - \lambda u = 0, \quad -\infty < x < \infty. \tag{7.2.3}$$

Each of the endpoints is singular. The substitution $u = ze^{x^2/2}$ transforms the differential equation into

$$-z'' - 2xz' - z = \lambda z,$$

which for $\lambda = -1$ has the independent solutions 1 and $\int_0^x e^{-t^2}\,dt$. Thus, for $\lambda = -1$, (7.2.3) has the solutions

$$u_1(x) = e^{x^2/2}, \quad u_2(x) = e^{x^2/2}\int_0^x e^{-t^2}\,dt. \tag{7.2.4}$$

Since $u_1(x)$ is of infinite norm in both $(-\infty, l)$ and (l, ∞), both endpoints are in the *limit point* case.

Example 4. The Legendre equation

$$-[(1-x^2)u']' + \lambda u = 0, \quad -1 < x < 1; \quad p(x) = 1 - x^2, \quad s(x) = 1 \tag{7.2.5}$$

has both of the endpoints -1 and 1 as singular points since $p(-1) = p(1) = 0$. For $\lambda = 0$ the equation has the independent solutions $u_1(x) = 1$ and $u_2(x) = \log(1+x) - \log(1-x)$. For any l, $-1 < l < 1$, each of the integrals $\int_{-1}^l |u_1|^2\,dx$, $\int_l^1 |u_1|^2\,dx$, $\int_{-1}^l |u_2|^2\,dx$, $\int_l^1 |u_2|^2\,dx$ is finite. Thus we have the *limit circle* case at both endpoints.

Construction of Green's Function in the Limit Point Case

If both endpoints in (7.2.1) are regular, then, after associating the boundary conditions (7.1.6), Green's function can be constructed whenever λ is not an eigenvalue. Since these eigenvalues form a discrete set of real numbers $\{\lambda_n\}$ with $\lim_{n\to\infty}\lambda_n = \infty$, we know a priori that $g(x, \xi; \lambda)$ exists whenever Im $\lambda \neq 0$ and even when λ is real negative with sufficiently large modulus. If the endpoint $x = a$ is regular but b is singular in the *limit point* case, we can proceed in the same way for Im $\lambda \neq 0$ by attaching a condition of the type (7.1.6) at $x = a$ but *no* condition at b (it suffices to require that $g \in H_s$). The condition at a selects a solution of the homogeneous equation for $a < x < \xi$, and the requirement g in H_s picks out a solution for $x > \xi$. The arbitrary multiplicative constants are then determined by the continuity of g and the jump condition in dg/dx at $x = \xi$. Consider, for instance, Green's function for Example 1(a), where we have set $a = 0$,

$$-\frac{d^2g}{dx^2} - \lambda g = \delta(x-\xi), \quad 0 < x, \xi < \infty; \quad g|_{x=0} = 0, \quad g \in L_2^{(c)}(0, \infty). \tag{7.2.6}$$

For $x < \xi$ we have $g = A \sin \sqrt{\lambda}\, x$; for $x > \xi$ the only solution in $L_2^{(c)}(0, \infty)$ is $B \exp(i\sqrt{\lambda}\, x)$, an observation which is true not only when Im $\lambda \neq 0$ but also when λ is real and negative. The matching conditions then give

$$g(x, \xi, \lambda) = \frac{1}{\sqrt{\lambda}} \sin \sqrt{\lambda}\, x_< \exp(i\sqrt{\lambda}\, x_>), \quad \lambda \notin [0, \infty), \tag{7.2.7}$$

Regarded as a function of the complex variable λ, g has no poles (not even at $\lambda = 0$) but has a branch on the positive real axis. In fact, letting $[g]$ stand for the jump in g across the positive real axis, we find that

$$\begin{aligned}[g] &= g(x, \xi; |\lambda| e^{io+}) - g(x, \xi; |\lambda| e^{i2\pi-}) \\ &= \frac{2i \sin |\lambda|^{1/2} x_< \sin |\lambda|^{1/2} x_>}{|\lambda|^{1/2}} \\ &= \frac{2i}{|\lambda|^{1/2}} \sin |\lambda|^{1/2} x \sin |\lambda|^{1/2} \xi.\end{aligned} \tag{7.2.8}$$

It is perhaps worth interrupting the discussion to compare (7.2.7) with Green's function g_l for the regular problem

$$-\frac{d^2 g_l}{dx^2} - \lambda g_l = \delta(x - \xi), \quad 0 < x, \xi < l; \quad g_l|_{x=0} = g_l|_{x=l} = 0. \tag{7.2.9}$$

An easy calculation gives

$$g_l(x, \xi; \lambda) = \frac{1}{\sqrt{\lambda} \sin \sqrt{\lambda}\, l} [\sin \sqrt{\lambda}\, x_< \sin \sqrt{\lambda}\, (l - x_>)], \quad 0 < x, \xi < l, \tag{7.2.10}$$

which has simple poles at the eigenvalues $\lambda_n = n^2\pi^2/l^2, n = 1, 2, \ldots,$ of the problem $-u_l'' - \lambda u_l = 0, u_l(0) = u_l(l) = 0$. Setting $\sqrt{\lambda} = \alpha + i\beta (\beta \geqslant 0)$, we find that

$$g_l = \frac{\sin \sqrt{\lambda}\, x_<}{\sqrt{\lambda}} \left[\frac{e^{i\sqrt{\lambda}\, x_>} - e^{2i\sqrt{\lambda}\, l} e^{-i\sqrt{\lambda}\, x_>}}{1 - e^{2i\sqrt{\lambda}\, l}} \right].$$

When $\lambda \notin [0, \infty)$, we have $\beta > 0$ and $|e^{2i\sqrt{\lambda}\, l}| = e^{-2\beta l}$, which tends to zero as $l \to \infty$. Therefore, if $\lambda \notin [0, \infty)$,

$$\lim_{l \to \infty} g_l(x, \xi; \lambda) = \frac{\sin \sqrt{\lambda}\, x_< \exp(i\sqrt{\lambda}\, x_>)}{\sqrt{\lambda}},$$

which is just (7.2.7). We have shown explicitly how the simple poles of (7.2.10) coalesce into the branch singularity in (7.2.7). Even if we had used a different boundary condition of type (7.1.6) at $x = l$ to define g, we would still find that g_l tends to (7.2.7) as $l \to \infty$. Therefore, (7.2.7) is the common limit as $l \to \infty$ of solutions of (7.2.9), whatever boundary condition is imposed at $x = l$.

The inhomogeneous problem related to (7.2.9) is

$$-w_l'' - \lambda w_l = f(x), 0 < x < l; \quad w_l|_{x=0} = w_l|_{x=l} = 0. \tag{7.2.11}$$

We know that if $\lambda \neq n^2\pi^2/l^2$, (7.2.11) has one and only one solution, given by

$$w_l(x) = \int_0^1 g_l(x, \xi; \lambda) f(\xi)\, d\xi.$$

If λ is an eigenvalue, g_l does not exist and (7.2.11) has solutions only if f satisfies a solvability condition. What is the corresponding statement for the inhomogeneous equation associated with (7.2.6)? Consider the problem

$$-w'' - \lambda w = f(x), 0 < x < \infty; \quad w(0) = 0, \tag{7.2.12}$$

where $f(x) \in L_2^{(c)}(0, \infty)$, and we are also looking for solutions in $L_2^{(c)}(0, \infty)$. Then, if $\lambda \notin [0, \infty)$, (7.2.12) has one and only one solution in $L_2^{(c)}(0, \infty)$, given by

$$w(x) = \int_0^\infty g(x, \xi; \lambda) f(\xi)\, d\xi, \tag{7.2.13}$$

where g is Green's function (7.2.7). The proof is simple: For fixed $\lambda, \lambda \notin [0, \infty)$, let G be the integral operator with kernel g. Then G is a Holmgren operator; indeed, we have the estimate

$$\int_0^\infty |g(x, \xi; \lambda)|\, d\xi = \int_0^\infty |g(\xi, x; \lambda)|\, d\xi \leqslant \frac{2}{\beta|\lambda|^{1/2}}; \quad \sqrt{\lambda} = \alpha + i\beta, \beta > 0,$$

which implies condition (5.1.11), characterizing Holmgren kernels. Thus G is a bounded operator on $L_2^{(c)}(0, \infty)$, although G is, of course, not compact. It is now easy to verify by differentiation that (7.2.13) satisfies (7.2.12). Let us state the results in spectral terminology. Consider the operator $L = -(d^2/dx^2)$ defined on a domain D_L consisting of complex-valued functions $u(x)$ in $L_2^{(c)}(0, \infty)$, with u continuous, u' absolutely continuous, u'' in $L_2^{(c)}(0, \infty)$, and $u(0) = 0$. Then every value of λ not in $[0, \infty)$ is in the resolvent set. Moreover, in Exercises 7.2.1 and 7.2.2 we show that every value of λ in $[0, \infty)$ is in the continuous spectrum.

Next let us consider a case where *both endpoints* are singular in the limit point case. It is still easy to construct $g(x, \xi; \lambda)$ for Im $\lambda \neq 0$. Take, for instance, the specific problem

$$-\frac{d^2 g}{dx^2} - \lambda g = \delta(x - \xi), \quad -\infty < x, \xi < \infty, \tag{7.2.14}$$

where both endpoints are in the limit point case [$\exp(i\sqrt{\lambda}\, x)$ is in $L_2^{(c)}(l, \infty)$, and $\exp(-i\sqrt{\lambda}\, x)$ is in $L_2^{(c)}(-\infty, l)$]. We impose no boundary condition at either endpoint, but merely require square integrability. Thus we have the expression $g = A \exp(-i\sqrt{\lambda}\, x_<) \exp(i\sqrt{\lambda}\, x_>)$, and an easy calculation gives $A = i/2\sqrt{\lambda}$, so that

$$g(x, \xi; \lambda) = \frac{i}{2\sqrt{\lambda}} \exp(i\sqrt{\lambda}\, |x - \xi|). \tag{7.2.15}$$

We note that $g \in L_2^{(c)}(-\infty, \infty)$ for $\lambda \notin [0, \infty)$. The equation

$$-w'' - \lambda w = f, \quad -\infty < x < \infty; \quad f \in L_2^{(c)}(-\infty, \infty) \tag{7.2.16}$$

has one and only one solution in $L_2^{(c)}(-\infty, \infty)$ for $\lambda \notin [0, \infty)$. That solution is given by

$$w(x) = \int_{-\infty}^{\infty} \frac{1}{2\sqrt{\lambda}} \exp(i\sqrt{\lambda}\,|x - \xi|) f(\xi)\, d\xi.$$

There are instances in applications when one actually wants to calculate g (or w) when λ is in the continuous spectrum, that is, when λ is real and $0 \leqslant \lambda < \infty$. There is no longer a natural mathematical criterion to select the solutions of (7.2.14) for $x < \xi$ and $x > \xi$ since no solution is square integrable at either $+\infty$ or $-\infty$. Obviously, this means that an L_2 theory is really insufficient for our purposes. Fortunately, however, there will usually be a physical criterion to guide us. The case where λ is real positive often arises in nondissipative wave propagation, and the natural requirement is that the problem be considered as the limit of vanishing dissipation. It turns out that small dissipation means that λ has a small, positive imaginary part; thus the limit of vanishing dissipation is obtained by letting λ approach real values from above the real axis. The corresponding limit for g is

$$\frac{i}{2|\lambda|^{1/2}} e^{i|\lambda|^{1/2}|x-\xi|},$$

which will then be the appropriate Green's function for a nondissipative problem with $0 < \lambda < \infty$.

Green's Function in the Limit Circle Case

Suppose that $x = a$ is a regular point and $x = b$ is a singular point in the limit circle case. With the regular point we associate a boundary condition of the form $0 = B_a u \doteq u(a) \cos\alpha - u'(a) \sin\alpha$ with *real* coefficients. In constructing Green's function this boundary condition determines the form of g for $x < \xi$; for $x > \xi$ the criterion of belonging to H_s is not sufficient to specify g since all solutions of the homogeneous equation are in H_s. Thus we will have to *impose a boundary condition* at the singular point $x = b$. In trying to imitate the regular case we encounter the difficulty that $u(b)$ and $u'(b)$ may be infinite, so that the usual condition relating $u(b)$ and $u'(b)$ may not make sense. We shall, however, reformulate below the boundary condition in the regular case to allow its extension to the singular limit circle problem.

If u and v have continuous second derivatives, and $a < b_0 < b$, then, the point b_0 being regular, we have from Green's theorem,

$$\int_a^{b_0} s(\bar{v}Lu - uL\bar{v})\, dx = p(b_0)W(u, \bar{v}; b_0) - p(a)W(u, \bar{v}; a). \tag{7.2.17}$$

Let D be the class of functions u such that both u and Lu are in H_s (in other words, $\int_a^b s|u|^2\,dx < \infty$ and $\int_a^b s|Lu|^2\,dx < \infty$). If u and v both belong to D, the left side of (7.2.17) has a limit as $b_0 \to b$ and therefore

$$\lim_{b_0 \to b} p(b_0)W(u, \bar{v}; b_0) \doteq [u, \bar{v}]_b \tag{7.2.18}$$

exists. At a regular point we can, of course, use the same notation for $p(a)W(u, \bar{v}; a)$, and (7.2.17) becomes

$$\int_a^b s(\bar{v}Lu - uL\bar{v})\,dx = [u, \bar{v}]_b - [u, \bar{v}]_a, \quad u, v \in D. \tag{7.2.19}$$

If u is also a solution of $Lu - \lambda u = 0$, then $L\bar{u} - \bar{\lambda}\bar{u} = 0$ because the coefficients in L are real, and therefore (7.2.19) with $v = u$ gives

$$2i(\operatorname{Im}\lambda) \int_a^b s|u|^2\,dx = [u, \bar{u}]_b - [u, \bar{u}]_a. \tag{7.2.20}$$

Now if at the regular point $x = a$, u satisfies a condition of the type $u(a)\cos\alpha - u'(a)\sin\alpha = 0$ with *real* coefficients, *then*

$$[u, \bar{u}]_a = \{u(a)\bar{u}'(a) - \bar{u}(a)u'(a)\}p(a) = 0;$$

and conversely, if $[u, \bar{u}]_a = 0$, then u satisfies $u(a)\cos\alpha - u'(a)\sin\alpha = 0$ for some real α. At the singular point b, $[u, \bar{u}]_b$ exists if $u \in D$; it is therefore natural to regard

$$[u, \bar{u}]_b = 0 \tag{7.2.21}$$

as a *boundary condition* at b equivalent to the boundary condition with real coefficients that applies in the regular case. It follows from (7.2.20) that no solution of $Lu - \lambda u = 0$ except the trivial solution can satisfy both $[u, \bar{u}]_a = 0$ and $[u, \bar{u}]_b = 0$ when $\operatorname{Im}\lambda \neq 0$.

Another important feature of the regular case is that the boundary condition is independent of the parameter λ; this is not taken into account by (7.2.21), which merely makes sure that the coefficients in the boundary condition are real. To see how to deal with the other question, let us begin with Green's function $g(x, \xi; \lambda_0)$ for a particular $\lambda = \lambda_0$ with $\operatorname{Im}\lambda_0 \neq 0$:

$$Lg - \lambda_0 g = \frac{\delta(x-\xi)}{s(x)}, \quad a < x, \xi < b; \quad B_a g = 0, \quad [g, \bar{g}]_b = 0. \tag{7.2.22}$$

For $x < \xi$ let $v(x, \lambda_0)$ be the solution of the homogeneous equation satisfying $B_a v = 0$; that is, $(\cos\alpha)v(a, \lambda_0) - (\sin\alpha)v'(a, \lambda_0) = 0$ for a particular α. Thus v is determined up to a multiplicative constant, and $[v, \bar{v}]_a = 0$. For $x > \xi$ let $z(x, \lambda_0)$ be a solution of the homogeneous equation satisfying $[z, \bar{z}]_b = 0$, which is regarded as a boundary condition at the singular point b. Unlike v, z is not determined up to a multiplicative constant; nevertheless, the functions z and v are independent since there is no nontrivial solution of the homogeneous equation satisfying both $[u, \bar{u}]_a = 0$ and $[u, \bar{u}]_b = 0$. Thus, if we choose the multiplicative constant in v so that

$$p(x)W(v, z; x) = -1, \tag{7.2.23}$$

then

$$g(x, \xi; \lambda_0) = v(x_<, \lambda_0)z(x_>, \lambda_0). \tag{7.2.24}$$

Obviously, g as a function of either x or ξ is in H_s, but we have the even more powerful result (not always true in the limit point case)

$$\int_a^b \int_a^b s(x)s(\xi)|g(x, \xi; \lambda_0)|^2 \, dx \, d\xi < \infty, \tag{7.2.25}$$

which will pave the way for application of the Hilbert-Schmidt theory. To prove (7.2.25), we observe from (7.2.24) that

$$\begin{aligned}\int_a^b |g|^2 s(\xi) \, d\xi &= |z(x)|^2 \int_a^x s(\xi)|v(\xi)|^2 \, d\xi + |v(x)|^2 \int_x^b s(\xi)|z(\xi)|^2 \, d\xi \\ &\leqslant |z(x)|^2 \|v\|_s^2 + |v(x)|^2 \|z\|_s^2,\end{aligned}$$

where we have suppressed the dependence on λ_0. Hence

$$\int_a^b \int_a^b |g(x, \xi; \lambda_0)|^2 s(x)s(\xi) \, dx \, d\xi \leqslant 2\|v\|_s^2 \|z\|_s^2,$$

which is finite since we are in the limit circle case. If $\tilde{z}$ is a solution of $L\tilde{z} - \lambda_0 \tilde{z} = 0$ which satisfies $[\tilde{z}, \bar{z}]_b = 0$ and is not a multiple of z, we will obtain a different Green's function in (7.2.24) corresponding to a "different" boundary condition at the singular point b.

Consider next the inhomogeneous problem

$$Lw - \lambda_0 w = f, \quad a < x < b; \quad B_a w = 0, \tag{7.2.26}$$

where we have not yet specified the boundary condition at the singular point b. We assume that $f \in H_s$, and we look for solutions in H_s. Green's function (7.2.24) enables us to write one solution of (7.2.26) as

$$\begin{aligned}w(x, \lambda_0) &= \int_a^b g(x, \xi; \lambda_0)s(\xi)f(\xi) \, d\xi \\ &= z(x, \lambda_0) \int_a^x v(\xi, \lambda_0)s(\xi)f(\xi) \, d\xi + v(x, \lambda_0) \int_x^b z(\xi, \lambda_0)s(\xi)f(\xi) \, d\xi,\end{aligned} \tag{7.2.27}$$

from which follow

$$w'(x, \lambda_0) = z'(x, \lambda_0) \int_a^x v(\xi, \lambda_0)s(\xi)f(\xi) \, d\xi + v'(x, \lambda_0) \int_x^b z(\xi, \lambda_0)s(\xi)f(\xi) \, d\xi,$$

$$[w, \bar{z}]_x = [z, \bar{z}]_x \int_a^x v(\xi, \lambda_0)s(\xi)f(\xi) \, d\xi + [v, \bar{z}]_x \int_x^b z(\xi, \lambda_0)s(\xi)f(\xi) \, d\xi.$$

Now let $x \to b$, and use the fact that $[z, \bar{z}]_b = 0$ to obtain

$$[w, \bar{z}]_b = 0. \tag{7.2.28}$$

From (7.2.27) we find that

$$\|w\|_s^2 \leqslant \left[\int_a^b \int_a^b |g(x, \xi; \lambda_0)|^2 s(x)s(\xi)\,dx\,d\xi\right] \|f\|_s^2. \tag{7.2.29}$$

Thus the BVP

$$Lw - \lambda_0 w = f, \quad a < x < b; \quad B_a w = 0, \quad [w, \bar{z}]_b = 0 \tag{7.2.30}$$

has one and only one solution, given by (7.2.27); moreover, from (7.2.29) together with (7.2.25) we conclude that the inverse operator is bounded.

We now formulate a spectral problem for the operator L:

$$Lu - \lambda u = 0, \quad a < x < b; \quad B_a u = 0, \quad [u, \bar{z}]_b = 0, \tag{7.2.31}$$

where the same auxiliary function $z(x, \lambda_0)$ satisfying $Lz - \lambda_0 z = 0$ and $[z, \bar{z}]_b = 0$ is used for all λ. We rewrite the differential equation in (7.2.31) as $Lu - \lambda_0 u = (\lambda - \lambda_0)u$; from (7.2.27), (7.2.28), and (7.2.30) it then follows that the BVP (7.2.31) is equivalent to the integral equation

$$u(x, \lambda) = (\lambda - \lambda_0) \int_a^b g(x, \xi; \lambda_0)s(\xi)u(\xi, \lambda)\,d\xi, \tag{7.2.32}$$

which can be written as

$$\mu y(x) = \int_a^b k(x, \xi)y(\xi)\,d\xi \quad \text{or} \quad \mu y = Ky, \tag{7.2.33}$$

where

$$\mu = \frac{1}{\lambda - \lambda_0}, \quad y = \sqrt{s}\,u, \quad k = \sqrt{s(x)}\sqrt{s(\xi)}g(x, \xi; \lambda_0).$$

From (7.2.25), k is a Hilbert-Schmidt kernel, so that (7.2.33) has at most denumerably many eigenvalues and there must therefore exist a *real* number θ which is *not* an eigenvalue of (7.2.31). If we then repeat our derivation using $\lambda_0 = \theta$, k will be a *real symmetric* H.-S. kernel. The eigenfunctions of (7.2.31) must therefore form an orthonormal basis in H_s.

We can therefore conclude that problems with singular points in the limit circle case (at one end or both) are quite similar to regular problems. The boundary condition at the singular point(s) is specified in a somewhat indirect way as in (7.2.31), but the problem then generates a discrete spectrum and the usual type of eigenfunction. The only difference from the regular case is that the eigenvalues λ_n need only satisfy $|\lambda_n| \to \infty$ rather than $\lambda_n \to \infty$.

Example 5. Setting $\nu = 0$ in (7.2.2), we have Bessel's equation of order 0:

$$-(xu')' - \lambda xu = 0. \tag{7.2.34}$$

We shall consider spectral problems for this equation on the interval $0 < x < 1$. At the regular endpoint $x = 1$ we impose the boundary condition $u(1) = 0$. At the singular (limit circle) endpoint $x = 0$, we impose, in keeping with (7.2.31), a boundary condition of the form $[u, \bar{z}]_0 = 0$, where the auxiliary function z remains to be specified. From our previous theory we know that z can be chosen to be a solution of $Lz - \lambda_0 z = 0$ with $\text{Im}\lambda_0 \neq 0$ and $[z, \bar{z}]_0 = 0$. It is permissible, however, to use instead a real value of λ_0 as long as the solution z is chosen independent of the solution v of the same equation satisfying the boundary condition at the regular endpoint.

With $\lambda_0 = 0$, (7.2.34) has independent solutions 1 and $\log x$. Thus $v(x) = \log x$, and every independent solution is a multiple of $z = -1 + A \log x$, where A must be *real* to satisfy $[z, \bar{z}]_0 = 0$. The particular form chosen for z has the property $[\bar{z}, v]_x = x(\bar{z}v' - v\bar{z}') = -1$. The condition $[u, \bar{z}]_0 = 0$, with A fixed, becomes

$$\lim_{x \to 0} Au - (A \log x - 1)xu' = 0.$$

For each real A we can consider the spectral problem

$$-(xu')' - \lambda xu = 0,\ 0 < x < 1;\ u(1) = 0,\ \lim_{x \to 0} Au - (A \log x - 1)xu' = 0, \tag{7.2.35}$$

and each problem yields a different set of eigenvalues and eigenfunctions. By far the most important problem in applications occurs for $A = 0$, when the boundary condition at $x = 0$ is

$$\lim_{x \to 0} xu' = 0. \tag{7.2.36}$$

The independent solutions of (7.2.34) for $\lambda \neq 0$ are $J_0(\sqrt{\lambda}\, x)$ and $N_0(\sqrt{\lambda}\, x)$, of which only $J_0(\sqrt{\lambda}\, x)$ satisfies (7.2.36). In applications to heat conduction, x is the radial cylindrical coordinate, and (7.2.36) states that the total heat flux through a small circle surrounding the origin vanishes, that is, there is no heat source at the origin. It is usual to replace (7.2.36) by the condition that the solution be *finite* at the origin, which has the same effect as far as selecting $J_0(\sqrt{\lambda}\, x)$ as the admissible solution. One would argue, however, that it is both mathematically *and* physically more natural to impose condition (7.2.36).

Let us consider in more detail problem (7.2.35) with $A = 0$ [that is, with boundary condition (7.2.36)]. Then $J_0(\sqrt{\lambda}\, x)$ is the only solution satisfying (7.2.36). The boundary condition at $x = 1$ implies that $J_0(\sqrt{\lambda}) = 0$, which gives rise to a sequence $\{\lambda_k\}$ of positive eigenvalues. The corresponding, nonnormalized eigenfunctions are $J_0(\sqrt{\lambda_k}\, x)$. It is instructive to calculate the normalization factor by the method of Section 7.1 [see, for instance, (7.1.25)]. We first need Green's function $g(x, \xi; \lambda)$ satisfying, for $\text{Im}\ \lambda \neq 0$,

$$-(xg')' - \lambda xg = \delta(x - \xi), \quad 0 < x, \xi < 1; \qquad xg'|_{x=0} = 0, \quad g|_{x=1} = 0.$$

We find, on using the Wronskian relationship (7.1.37), that

$$g = J_0(\sqrt{\lambda}\, x_<)v(x_>, \lambda),$$

where

$$v(x, \lambda) = \frac{\pi}{2J_0(\sqrt{\lambda})}[J_0(\sqrt{\lambda}\, x)N_0(\sqrt{\lambda}) - J_0(\sqrt{\lambda})N_0(\sqrt{\lambda}\, x)].$$

Thus g has simple poles at $\lambda = \lambda_k$, where $J_0(\sqrt{\lambda_k}) = 0$. The residue at $\lambda = \lambda_k$ is

$$\frac{\pi\sqrt{\lambda_k}}{J_0'(\sqrt{\lambda_k})} J_0(\sqrt{\lambda_k}\, x)J_0(\sqrt{\lambda_k}\, \xi)N_0(\sqrt{\lambda_k}).$$

From the Wronskian relation we have $N_0(\sqrt{\lambda_k})J_0'(\sqrt{\lambda_k}) = -2/(\pi\sqrt{\lambda_k})$, and the residue can be rewritten as

$$-\frac{2}{[J_0'(\sqrt{\lambda_k})]^2} J_0(\sqrt{\lambda_k}\, x)J_0(\sqrt{\lambda_k}\, \xi),$$

so that from (7.1.19) we find that

$$u_k(x) = \frac{\sqrt{2}}{J_0'(\sqrt{\lambda_k})} J_0(\sqrt{\lambda_k}\, x) \tag{7.2.37}$$

is an orthonormal basis of eigenfunctions in H_s; that is,

$$\int_0^1 x u_k(x)\bar{u}_j(x)\, dx = \begin{cases} 0, & k \neq j, \\ 1, & k = j. \end{cases}$$

The completeness relation (7.1.20) takes the form

$$\frac{\delta(x - \xi)}{x} = \sum_{k=1}^{\infty} \frac{2}{[J_0'(\sqrt{\lambda_k})]^2} J_0(\sqrt{\lambda_k}\, x)J_0(\sqrt{\lambda_k}\, \xi), \tag{7.2.38}$$

which gives rise to the usual Fourier-Bessel series:

$$F(\xi) = 2\sum_{k=1}^{\infty} \frac{J_0(\sqrt{\lambda_k}\, \xi)}{[J_0'(\sqrt{\lambda_k})]^2} \int_0^1 x J_0(\sqrt{\lambda_k}\, x) f(x)\, dx. \tag{7.2.39}$$

The question remains open whether (7.2.35) might occur in a physical setting with $A \neq 0$. The eigenfunctions would then contain a combination of a source term $N_0(\sqrt{\lambda}\, x)$ and a term $J_0(\sqrt{\lambda}\, x)$. Only the ratio of J_0 and N_0 would be determined. No simple physical interpretation seems to be available.

EXERCISES

In Exercises 7.2.1 and 7.2.2 we complete the discussion of the spectral properties of the operator $L \doteq -d^2/dx^2$ defined on the domain D_L of complex-valued functions $u(x)$ in $L_2^{(c)}(0,\infty)$, with u continuous, u' absolutely continuous, u'' in $L_2^{(c)}(0,\infty)$, and $u(0)=0$. We want to prove that every value of λ in $[0,\infty)$ is in the continuous spectrum. Setting $\lambda = \alpha^2, \alpha \geqslant 0$, we must do two things

1. Show that

$$-u'' - \alpha^2 u = f(x), \quad 0 < x < \infty; \quad u(0) = 0 \tag{7.2.40}$$

has an L_2 solution for $f \in M$, where M is a dense set in L_2. This means that the range of $L - \alpha^2 I$ is dense in L_2.

2. Show that the inverse of $L - \alpha^2 I$ is unbounded; that is, for each α exhibit a sequence $\{u_n \neq 0\}$ such that

$$\frac{\|Lu_n - \alpha^2 u_n\|}{\|u_n\|} \to 0. \tag{7.2.41}$$

7.2.1 Clearly, there are some $f \in L_2$ for which (7.2.40) has an L_2 solution: Start with any function U of *compact support* satisfying $U(0) = 0$ and having a piecewise continuous second derivative. Then $f \doteq -U'' - \alpha^2 U$ will be in L_2, so that for that particular f, (7.2.40) has the solution U.

Show that if f is any L_2 function with compact support, (7.2.40) has a solution in L_2 if and only if

$$\int_0^\infty (\sin \alpha x) f(x)\, dx = 0, \quad \alpha > 0; \quad \int_0^\infty x f(x)\, dx = 0, \quad \alpha = 0, \tag{7.2.42}$$

and determine the corresponding solutions. It can also be shown that the set M_α of L_2 functions with compact support satisfying condition (7.2.42) for fixed α is dense in L_2.

7.2.2 To exhibit a sequence with property (7.2.41), we note first that $\sin \alpha x$ satisfies the homogeneous equation $Lu - \alpha^2 u = 0, u(0) = 0$ but fails to be in $L_2^{(c)}(0,\infty)$. Nevertheless, $\sin \alpha x$ is "nearly" an eigenfunction and suitably modified into an L_2 function will become the desired u_n. Let $F(x)$ be a fixed twice-differentiable function on $0 \leqslant x \leqslant 1$, satisfying $F(0) = 0, F'(0) = 1, F(1) = 0, F'(1) = 0$. Using the definition

$$u_n(x) = \begin{cases} \sin \alpha x, & 0 \leqslant, x \leqslant l_n, \\ F(x - l_n), & l_n \leqslant x \leqslant l_n + 1, \\ 0, & x \geqslant l_n + 1, \end{cases}$$

where $l_n = 2n\pi/\alpha$, show that $\{u_n\}$ satisfies (7.2.41).

7.2.3 Consider the spectral problem (7.2.3). Construct Green's function for the case $\lambda = -1$ and translate the problem into an integral equation. Show that the kernel is symmetric and H.S. with eigenfunctions forming an orthonormal basis. The *Hermite polynomials* are defined by

$$H_n(x) = (-1)^n \exp(x^2)\frac{d^n}{dx^n}\exp(-x^2), \quad n = 0, 1, 2, \ldots.$$

Set $u_n(x) = \exp(-x^2/2)H_n(x)$ and show that $u_n(x)$ is an eigenfunction of the Hermite equation (7.2.3) corresponding to $\lambda_n = 2n + 1$.

7.3 SPECTRAL PROBLEMS WITH A CONTINUOUS SPECTRUM

In regular problems (or problems in the limit circle case) we can use (7.1.19) to generate the orthonormal eigenfunctions. Relation (7.1.20) tells us further that the eigenfunctions form a basis; indeed, multiplying (7.1.20) by $f(\xi)s(\xi)$ and integrating from $\xi = a$ to $\xi = b$, we obtain

$$f(x) = \sum_n \langle f, u_n\rangle_s u_n(x), \tag{7.3.1}$$

which is the Fourier expansion of f in the set $\{u_n\}$. Of course, this approach only guarantees the validity of (7.3.1) for test functions f with compact support in (a, b), but this is enough to tell us that the set $\{u_n\}$ is a basis.

If we have a singular problem with at least one endpoint in the *limit point* case, it is possible (but not necessary) for the spectrum to be in part continuous. This continuous portion arises as a consequence of a branch in $g(x, \xi; \lambda)$, as we saw in (7.2.7), for instance. We shall still accept the validity of the formula

$$-\frac{\delta(x-\xi)}{s(x)} = \frac{1}{2\pi i}\int_{C_\infty} g(x, \xi; \lambda)\, d\lambda, \quad a < x, \xi < b. \tag{7.3.2}$$

The integral around the large circle in the complex plane can now be reduced to a sum of residues (the point spectrum) plus a branch-cut integral over a portion of the real axis (the continuous spectrum). Instead of (7.1.19) and (7.1.20), one then obtains for $a < x, \xi < b$

$$\begin{aligned} -\frac{\delta(x-\xi)}{s(x)} &= \frac{1}{2\pi i}\int_{C_\infty} g(x, \xi; \lambda)\, d\lambda \\ &= -\sum_n u_n(x)\overline{u}_n(\xi) - \int u_\nu(x)\overline{u}_\nu(\xi)\, d\nu. \end{aligned} \tag{7.3.3}$$

If the sum is not present, one has an expansion over a continuous index, leading to transform pairs rather than series expansions. The ideas will perhaps be clearer through examples.

Sine Transform

Consider the spectral problem

$$-u'' - \lambda u = 0, \quad 0 < x < \infty; \quad u(0) = 0, \quad u \in L_2^{(c)}(0, \infty). \tag{7.3.4}$$

To apply (7.3.3) we must first construct Green's function $g(x, \xi; \lambda)$. This was done in (7.2.7), and we observed there that g was analytic in the λ plane except for a branch on the positive real axis. It therefore follows from Cauchy's integral theorem that

$$\frac{1}{2\pi i}\int_{C_\infty} g(x, \xi; \lambda)\, d\lambda = -\frac{1}{2\pi i}\int_0^\infty [g]\, d|\lambda|,$$

where $[g]$ stands for the jump of g across the positive real axis. Thus we can write, since $s(x) = 1$,

$$\delta(x - \xi) = \frac{1}{2\pi i}\int_0^\infty [g]\, d|\lambda|, \quad 0 < x, \xi < \infty, \tag{7.3.5}$$

where, from (7.2.8),

$$[g] = \frac{2i}{|\lambda|^{1/2}} \sin |\lambda|^{1/2} x \sin |\lambda|^{1/2} \xi.$$

Letting $\lambda = \nu^2$, we find that (7.3.5) becomes

$$\delta(x - \xi) = \frac{2}{\pi}\int_0^\infty \sin \nu x \sin \nu \xi \, d\nu, \quad 0 < x, \xi < \infty. \tag{7.3.6}$$

The completeness relation (7.3.6) contains the formula for the Fourier sine transform. Indeed, if we multiply (7.3.6) by $f(x)$ and integrate from $x = 0$ to ∞, we find that

$$f(\xi) = \frac{2}{\pi}\int_0^\infty \sin \nu \xi F_s(\nu)\, d\nu, \quad 0 < \xi < \infty, \tag{7.3.7}$$

where

$$F_s(\nu) \doteq \int_0^\infty \sin \nu x \; f(x)\, dx, \quad 0 < \nu < \infty. \tag{7.3.8}$$

Equation (7.3.8) defines the Fourier sine transform of $f(x)$, and (7.3.7) is an inversion integral or, if one prefers, an expansion of f in the set $\sin \nu x$ over the continuous parameter ν.

We can now use (7.3.7) and (7.3.8) to solve the inhomogeneous problem

$$-w'' - \lambda w = f(x), \quad 0 < x < \infty; \quad w(0) = \alpha, \quad w \in L_2^{(c)}(0, \infty) \tag{7.3.9}$$

on the assumption that f is given in $L_2^{(c)}(0, \infty)$ and λ is not in the spectrum, that is, $\lambda \notin [0, \infty)$. We multiply (7.3.9) by $\sin \nu x$ and integrate from 0 to ∞ to obtain

$$-\int_0^\infty w'' \sin \nu x \, dx - \lambda W_s(\nu) = F_s(\nu),$$

where F_s and W_s are the sine transforms of f and w, respectively. We integrate by parts to find

$$-\int_0^\infty w'' \sin \nu x \, dx = \nu^2 \int_0^\infty w \sin \nu x \, dx + (w\nu \cos \nu x - w' \sin \nu x)_0^\infty,$$

where the contribution from 0 is $-\alpha\nu$, while the terms from infinity are dropped (at this stage it usually does not pay to argue that w and w' vanish at infinity; instead proceed as if they did and then check that the solution obtained actually satisfies the original problem). Thus we have

$$(\nu^2 - \lambda)W_s(\nu) = \alpha\nu + F_s(\nu), \quad W_s(\nu) = \frac{\alpha\nu}{\nu^2 - \lambda} + \frac{F_s(\nu)}{\nu^2 - \lambda}$$

and

$$\begin{aligned} w(x) &= \frac{2}{\pi} \int_0^\infty \frac{\alpha\nu}{\nu^2 - \lambda} \sin \nu x \, d\nu + \frac{2}{\pi} \int_0^\infty \frac{F_s(\nu)}{\nu^2 - \lambda} \sin \nu x \, d\nu \\ &= \frac{2i}{2\pi} \int_{-\infty}^\infty e^{-i\nu x} \left[\frac{\alpha\nu}{\nu^2 - \lambda} + \frac{F_s(\nu)}{\nu^2 - \lambda} \right] d\nu, \end{aligned}$$

where we have used (7.3.8) to define $F_s(\nu)$ for negative ν. The problem has been reduced to ordinary Fourier inversion. The first integral is easily calculated by residues and gives (for $x > 0$) $\alpha \exp(i\sqrt{\lambda}x)$. The second integral is the convolution of the Fourier originals of $2iF_s(\nu)$ and $1/(\nu^2 - \lambda)$, which are, respectively, $f_0(x)$ and $i \exp(i\sqrt{\lambda}\,|x|)/2\sqrt{\lambda}$, where $f_0(x)$ is the odd extension of $f(x)$. Thus, for $x > 0$,

$$\begin{aligned} w(x) &= \alpha \exp(i\sqrt{\lambda}\,x) + i \int_{-\infty}^\infty \frac{\exp(i\sqrt{\lambda}\,|x - \xi|)}{2\sqrt{\lambda}} f_0(\xi) \, d\xi \\ &= \alpha \exp(i\sqrt{\lambda}\,x) + i \int_0^\infty \frac{\exp(i\sqrt{\lambda}\,|x - \xi|) - \exp(i\sqrt{\lambda}\,|x + \xi|)}{2\sqrt{\lambda}} f(\xi) \, d\xi, \end{aligned} \tag{7.3.10}$$

which is just the same solution that would be obtained by using Green's function (7.2.7), which in turn can be deduced from (7.2.15) by images.

As an application of the Fourier sine transform in partial differential equations, consider the problem of finding Green's function (with Dirichlet boundary conditions) for the negative Laplacian in a two-dimensional semi-infinite strip. Let x, y be Cartesian coordinates in the plane, and let the strip occupy the domain $\Omega : 0 < x < \pi, 0 < y < \infty$, with the source at $Q = (\xi, \eta)$. Green's function $g(P, Q) = g(x, y; \xi, \eta)$ satisfies

$$-\Delta g = \delta(P - Q) = \delta(x - \xi)\delta(y - \eta), \quad P, Q \in \Omega; \quad g(P, Q)|_{P \in \Gamma} = 0, \tag{7.3.11}$$

where Γ is the boundary of the strip. The relation between delta functions may be found in Exercise 2.2.13.

Multiply both sides of (7.3.11) by $\sin \nu y$, and integrate from $y = 0$ to $y = \infty$ to obtain (after integration by parts of the term $\sin \nu y\, \partial^2 g/\partial y^2$)

$$-\frac{\partial^2 G_s}{\partial x} + \nu^2 G_s = \sin \nu\eta\, \delta(x-\xi),\ 0 < x, \xi < \pi; \quad G_s|_{x=0} = G_s|_{x=\pi} = 0, \tag{7.3.12}$$

where

$$G_s = G_s(x, \nu; Q) = \int_0^\infty g(x, y; Q) \sin \nu y\, dy, \tag{7.3.13}$$
$$g(x, y; Q) = \frac{2}{\pi} \int_0^\infty G_s(x, \nu; Q) \sin \nu y\, d\nu.$$

In deriving (7.3.12), we have dropped integrated terms of the form $g \cos \nu y$ and $g' \sin \nu y$ evaluated at $y = \infty$. This seems reasonable since g and g' can be expected to vanish at $y = \infty$, but full justification can come only after careful examination of the final solution (a step we leave to the conscientious reader).

The solution of (7.3.12) is easily obtained by the methods of Chapter 1. We find that

$$G_s = A \sinh \nu x_< \sinh \nu(x_> - \pi),$$

where A is determined from the jump condition

$$\frac{dG_s}{dx}(\xi+, \nu; Q) - \frac{dG_s}{dx}(\xi-, \nu; Q) = -\sin \nu\eta,$$

which gives $A = -\sin \nu\eta/(\nu \sinh \nu\pi)$, so that

$$G_s = -\frac{\sin \nu\eta}{\nu \sinh \nu\pi} \sinh \nu x_< \sinh \nu(x_> - \pi). \tag{7.3.14}$$

We can calculate $g(P, Q)$ from (7.3.13), but the integration can be performed numerically with reasonable effort only if G_s decreases exponentially for large ν. Analysis of (7.3.14) shows that this behavior occurs as long as $|x - \xi|$ is not small. Thus the form (7.3.13) is useful whenever the x coordinate of the observation point is not too close to the x coordinate of the source.

Alternatively, we can find an expression for g that is particularly useful if $|y - \eta|$ is not small. This requires an expansion in the x direction rather than in the y direction. On separating variables, the x problem is $-u'' - \lambda u = 0$ with $u(0) = u(\pi) = 0$ giving rise to the eigenfunctions $\sin nx$. Starting with (7.3.11), we multiply by $(2/\pi) \sin nx$ and integrate from 0 to π to obtain

$$n^2 g_n - \frac{\partial^2 g_n}{\partial y^2} = \frac{2}{\pi} \sin n\xi\, \delta(y - \eta), \quad 0 < y, \eta < \infty; \quad g_n|_{y=0} = 0, \tag{7.3.15}$$

where

$$g_n = g_n(y; Q) = \frac{2}{\pi} \int_0^\pi g(x, y; Q) \sin nx\, dx, \tag{7.3.16}$$
$$g(x, y; Q) = \sum_{n=1}^\infty g_n(y; Q) \sin nx.$$

The solution of (7.3.15) which vanishes at $y = \infty$ is

$$g_n = \frac{2}{\pi n} \sin n\xi \sinh ny_< \exp(-ny_>), \tag{7.3.17}$$

where $y_< = \min(y, \eta), y_> = \max(y, \eta)$. We then substitute g_n in (7.3.16) to find g. The series for g converges rapidly if g_n tends to 0 exponentially as $n \to \infty$, as turns out to be the case whenever $|y - \eta|$ is not small.

By using contour integration or the Poisson sum formula it is possible to pass directly from (7.3.13)–(7.3.14) to (7.3.16)–(7.3.17) without reference to the differential equation. The solution of (7.3.11) could also have been found by images or by conformal mapping.

For another illustration of the use of the Fourier sine transform, consider heat flow in a semi-infinite rod with 0 initial temperature and temperature 1 at the end. The temperature $u(x, t)$ satisfies

$$\frac{\partial u}{\partial t} - \frac{\partial^2 u}{\partial x^2} = 0, \quad 0 < x < \infty, t > 0; \quad u(x, 0) = 0, \quad u(0, t) = 1. \tag{7.3.18}$$

Apply a sine transform to the x coordinate to obtain

$$\frac{\partial U_s}{\partial t} + \nu^2 U_s = \nu, \quad t > 0; \quad U_s|_{t=0} = 0, \tag{7.3.19}$$

where

$$U_s(\nu, t) = \int_0^\infty u(x, t) \sin \nu x \, dx, \tag{7.3.20}$$

$$u(x, t) = \frac{2}{\pi} \int_0^\infty U_s(\nu, t) \sin \nu x \, d\nu.$$

Again we dropped some integrated terms at infinity in the derivation of (7.3.19), whose solution is

$$U_s(\nu, t) = \frac{1}{\nu}(1 - e^{-\nu^2 t}), \tag{7.3.21}$$

so that

$$u(x, t) = \frac{2}{\pi} \int_0^\infty \frac{\sin \nu x}{\nu} \, d\nu - \frac{2}{\pi} \int_0^\infty \frac{\sin \nu x}{\nu} e^{-\nu^2 t} \, d\nu.$$

The first term on the right is equal to 1 for all $x > 0$. To evaluate the second integral $I(x, t)$ note that $\partial I/\partial x = \int_0^\infty \cos \nu x \, e^{-\nu^2 t} \, d\nu = \frac{1}{2}(\pi/t)^{1/2} e^{-x^2/4t}$. Since $I(0, t) = 0$, we have

$$I(x, t) = \int_0^x \frac{1}{2} \sqrt{\frac{\pi}{t}} e^{-y^2/4t} \, dy = \int_0^{x/2\sqrt{t}} \sqrt{\pi} \, e^{-z^2} \, dz,$$

and hence

$$u(x, t) = 1 - \frac{2}{\sqrt{\pi}} \int_0^{x/2\sqrt{t}} e^{-z^2} \, dz = 1 - \operatorname{erf}\left(\frac{x}{2\sqrt{t}}\right), \tag{7.3.22}$$

where "erf" stands for the error function

$$\operatorname{erf} x = \frac{2}{\sqrt{\pi}} \int_0^x e^{-z^2}\, dz.$$

The more general problem with boundary temperature $h(t)$ instead of 1 can be solved by superposition. First we note that if the boundary temperature in (7.3.18) is 0 up to time τ and is then suddenly increased to 1, the solution will be 0 up to time τ and then $u(x, t-\tau)$, where $u(x,t)$ is given by (7.3.22). In other words, the solution for the displaced step temperature $H(t-\tau)$ is $H(t-\tau)u(x,t-\tau)$. Now we can write, for $t > 0$,

$$\begin{aligned} h(t) &= \int_0^\infty \delta(t-\tau)h(\tau)\,d\tau = -\int_0^\infty \frac{dH(t-\tau)}{d\tau} h(\tau)\,d\tau \\ &= \int_0^\infty H(t-\tau)h'(\tau)\,d\tau + h(0). \end{aligned}$$

Therefore, by superposition,

$$\begin{aligned} w(x,t) &\doteq \int_0^\infty H(t-\tau)u(x,t-\tau)h'(\tau)\,d\tau + h(0)u(x,t) \\ &= \int_0^t u(x,t-\tau)h'(\tau)\,d\tau + h(0)u(x,t) \\ &= -\int_0^t \left[\frac{\partial}{\partial\tau}u(x,t-\tau)\right] h(\tau)\,d\tau = \int_0^t \frac{d}{dt}u(x,t-\tau)h(\tau)\,d\tau \qquad (7.3.23) \end{aligned}$$

is the solution of

$$\frac{\partial w}{\partial t} - \frac{\partial^2 w}{\partial x^2} = 0, \quad 0 < x < \infty, t > 0; \quad w(x,0) = 0, \quad w(0,t) = h(t). \quad (7.3.24)$$

The form (7.3.23) of the solution is known as *Duhamel's formula* (see also Exercise 3.1.3). In view of (7.3.22) we can write explicitly

$$w(x,t) = \int_0^t \frac{x}{2\sqrt{\pi}} \frac{e^{-x^2/4(t-\tau)}}{(t-\tau)^{3/2}} h(\tau)\,d\tau. \qquad (7.3.25)$$

The general initial value problem for the semi-infinite rod is best attacked by finding Green's function $g(x,t;\xi,0)$ corresponding to an initial temperature $\delta(x-\xi)$ and a vanishing boundary temperature

$$\frac{\partial g}{\partial t} - \frac{\partial^2 g}{\partial x^2} = 0, \quad 0 < x < \infty, \quad t > 0; \quad g|_{x=0} = 0, \quad g|_{t=0} = \delta(x-\xi). \quad (7.3.26)$$

Applying a Fourier sine transform on x, we obtain

$$\frac{\partial}{\partial t}G_s + \nu^2 G_s = 0, \quad t > 0; \quad G_s|_{t=0} = \sin\nu\xi, \qquad (7.3.27)$$

$$G_s = \int_0^\infty g(x,t;\xi,0)\sin\nu x\,dx, \quad g = \frac{2}{\pi}\int_0^\infty G_s \sin\nu x\,d\nu.$$

Solving (7.3.27) and using the inversion formula, we find that

$$g = \frac{2}{\pi}\int_0^\infty e^{-\nu^2 t}\sin\nu\xi\sin\nu x\,d\nu = \frac{1}{\pi}\int_0^\infty e^{-\nu^2 t}[\cos\nu(x-\xi) - \cos\nu(x+\xi)]\,d\nu,$$

an integral already calculated just above (7.3.22). We conclude that

$$g(x,t;\xi,0) = \frac{1}{\sqrt{4\pi t}}[e^{-(x-\xi)^2/4t} - e^{-(x+\xi)^2/4t}], \tag{7.3.28}$$

which confirms the result obtained by the method of images.

The IVP

$$\frac{\partial v}{\partial t} - \frac{\partial^2 v}{\partial x^2} = 0, \quad 0 < x < \infty, \quad t > 0; \quad v(0,t) = 0, \quad v(x,0) = f(x), \tag{7.3.29}$$

has the solution

$$v(x,t) = \int_0^\infty g(x,t;\xi,0)f(\xi)\,d\xi. \tag{7.3.30}$$

The more general problem involving both a nonzero initial value and a nonzero boundary value at $x = 0$ can be solved by superposition of (7.3.25) and (7.3.30). If a constant diffusion coefficient D multiplies the term $\partial^2/\partial x^2$, we can divide by D and rescale the time variable by $\bar{t} = Dt$, of if we prefer, we can use a new space variable $\bar{x} = x/D^{1/2}$. For further discussion of this and related examples for the heat equation, see Section 8.2 (particularly Exercise 8.2.6).

Hankel Transform of Order 0 and Applications

Consider Bessel's equation of order 0 on $0 < x < \infty$:

$$-(xu')' - \lambda xu = 0, \quad 0 < x < \infty. \tag{7.3.31}$$

Here both endpoints are singular, and we saw in Example 2, Section 7.2, that the left end $x = 0$ is in the limit circle case and the right end is in the limit point case. We shall therefore require no boundary condition at infinity; at $x = 0$ we shall ask that the boundary condition $\lim_{x\to 0} xu' = 0$ be satisfied. As we saw in (7.2.36), this is an appropriate condition for (7.3.31) at $x = 0$ and has the effect of selecting the solution which is bounded at the origin. We proceed with the construction of Green's function satisfying

$$-(xg')' - \lambda xg = \delta(x-\xi), \quad 0 < x, \xi < \infty; \tag{7.3.32}$$
$$(xg')_{x=0+} = 0, \quad \int_0^\infty x|g|^2\,dx < \infty.$$

Whenever $\lambda \notin [0,\infty)$, we can actually meet these requirements and we find that

$$g(x,\xi;\lambda) = \frac{\pi i}{2} J_0(\sqrt{\lambda}\,x_<)H_0^{(1)}(\sqrt{\lambda}\,x_>), \tag{7.3.33}$$

which has a branch on the positive real axis and no other singularity. Applying (7.3.2), we obtain for $0 < x, \xi < \infty$,

$$\frac{\delta(x-\xi)}{x} = -\frac{1}{2\pi i}\int_{C_\infty} g(x,\xi;\lambda)\,d\lambda = \frac{1}{2\pi i}\int_0^\infty [g]\,d|\lambda|, \tag{7.3.34}$$

where

$$[g] = \frac{\pi i}{2}\{J_0(|\lambda|^{1/2}x_<)H_0^{(1)}(|\lambda|^{1/2}x_>) - J_0(-|\lambda|^{1/2}x_<)H_0^{(1)}(-|\lambda|^{1/2}x_>)\} \tag{7.3.35}$$

is the jump in g across the real axis.

By using the formulas

$$J_0(-a) = J_0(a),\, H_0^{(1)}(a) - H_0^{(1)}(-a) = 2J_0(a), \tag{7.3.36}$$

we can simplify (7.3.35), and after setting $|\lambda| = \gamma^2$, (7.3.34) becomes

$$\frac{\delta(x-\xi)}{x} = \int_0^\infty \gamma J_0(\gamma x)J_0(\gamma\xi)\,d\gamma, \qquad 0 < x, \xi < \infty. \tag{7.3.37}$$

The completeness relation (7.3.37) contains the pair of transform formulas

$$F_H(\gamma) = \int_0^\infty xJ_0(\gamma x)f(x)\,dx, \qquad 0 < \gamma < \infty, \tag{7.3.38}$$

$$f(x) = \int_0^\infty \gamma J_0(\gamma x)F_H(\gamma)\,d\gamma, \qquad 0 < x < \infty, \tag{7.3.39}$$

where (7.3.38) defines the Hankel transform of order 0 of f, and (7.3.39) is an inversion formula enabling us to recover f from its Hankel transform.

We now consider a number of related applications of the Hankel transform of order 0. All of these problems are axisymmetric, with the z axis as the axis of symmetry. First, in free space let us place a ring of sources on the circle $r = a, z = 0$; here r is the radial cylindrical coordinate (that is, $r = \sqrt{x^2+y^2}$). The total strength of this ring is taken to be unity, so that the source density is $\delta(r-a)\delta(z)/2\pi a = \delta(r-a)\delta(z)/2\pi r$. We shall suppose that we are dealing with steady heat conduction with absorption [see (1.4.18)]. The corresponding temperature u satisfies

$$-\Delta u + k^2u = \frac{\delta(r-a)\delta(z)}{2\pi r}, \qquad 0 < r, \qquad -\infty < z < \infty,$$

or

$$-\frac{\partial}{\partial r}\left(r\frac{\partial u}{\partial r}\right) - r\frac{\partial^2 u}{\partial z^2} + k^2ru = \frac{\delta(r-a)\delta(z)}{2\pi}. \tag{7.3.40}$$

If we separated variables for the homogeneous equation, the radial part would satisfy (7.3.31), so that a Hankel transform (with r playing the role of x) of order 0 should help in solving (7.3.40). Multiply the equation by $J_0(\gamma r)$ and integrate from $r = 0$ to ∞ to obtain, after integration by parts of the first term,

$$-\frac{\partial^2 U_H}{\partial z^2} + (k^2 + \gamma^2)U_H = \frac{\delta(z)J_0(\gamma a)}{2\pi}, \tag{7.3.41}$$

where U_H is the Hankel transform of u:

$$U_H(\gamma, z) = \int_0^\infty rJ_0(\gamma r)u(r, z)\, dr;\ u(r, z) = \int_0^\infty \gamma J_0(\gamma r)U_H(\gamma, z)\, d\gamma. \tag{7.3.42}$$

We solve the ordinary differential equation in the usual way, requiring that at $z = \pm\infty$ the temperature tend to 0. This gives

$$U_H(\gamma, z) = \frac{\exp(-\sqrt{k^2 + \gamma^2}\,|z|)}{4\pi\sqrt{k^2 + \gamma^2}} J_0(\gamma a), \tag{7.3.43}$$

$$u(r, z) = \frac{1}{4\pi}\int_0^\infty \gamma J_0(\gamma a)J_0(\gamma r)\frac{\exp(-\sqrt{k^2 + \gamma^2}\,|z|)}{\sqrt{k^2 + \gamma^2}}\, d\gamma.$$

If $a \to 0$, the ring degenerates into a unit point source at the origin; the corresponding solution of (7.3.40) depends only on the distance from the origin and $u = (1/4\pi)\exp(-k\sqrt{r^2 + z^2})/\sqrt{r^2 + z^2}$. Comparing with (7.3.43) for $a = 0$, we find that

$$\frac{\exp(-k\sqrt{r^2 + z^2})}{\sqrt{r^2 + z^2}} = \int_0^\infty \gamma J_0(\gamma r)\frac{\exp(-\sqrt{k^2 + \gamma^2}|z|)}{\sqrt{k^2 + \gamma^2}}\, d\gamma, \tag{7.3.44}$$

an interesting relation due essentially to Sommerfeld, who studied the corresponding acoustical problem. For $k = 0$, (7.3.44) remains valid and reduces to

$$\frac{1}{\sqrt{r^2 + z^2}} = \int_0^\infty J_0(\gamma r)e^{-\gamma|z|}\, d\gamma. \tag{7.3.45}$$

Another expression for (7.3.43) can be obtained by applying a Fourier transform in the z direction to (7.3.40). Letting

$$u^\wedge(r, \alpha) = \int_{-\infty}^\infty e^{i\alpha z}u(r, z)\, dz,$$

we find that

$$-\frac{d}{dr}\left(r\frac{du^\wedge}{dr}\right) + (k^2 + \alpha^2)ru^\wedge = \frac{\delta(r - a)}{2\pi}, \qquad 0 < r < \infty.$$

The corresponding homogeneous equation is the modified Bessel equation with two independent solutions $I_0(\sqrt{k^2 + \alpha^2}\, r)$ and $K_0(\sqrt{k^2 + \alpha^2}\, r)$, the first of which is bounded at $r = 0$, and the second square integrable at $r = \infty$. Thus, after using the Wronskian relations for I_0, K_0, we have

$$u^\wedge(r, \alpha) = \frac{1}{2\pi}I_0(\sqrt{k^2 + \alpha^2}\, r_<)K_0(\sqrt{k^2 + \alpha^2}\, r_>),$$

where $r_< = \min(r,a), r_> = \max(r,a)$. The inversion formula then gives

$$u(r,z) = \frac{1}{4\pi^2}\int_{-\infty}^{\infty} e^{-i\alpha z} I_0(\sqrt{k^2+\alpha^2}\, r_<) K_0(\sqrt{k^2+\alpha^2}\, r_>)\, d\alpha,$$

which reduces, as $a \to 0$, to

$$\frac{\exp(-k\sqrt{r^2+z^2})}{\sqrt{r^2+z^2}} = \frac{1}{\pi}\int_{-\infty}^{\infty} e^{-i\alpha z} K_0(\sqrt{k^2+\alpha^2}\, r)\, d\alpha$$

or

$$K_0(\sqrt{k^2+\alpha^2}\, r) = \frac{1}{2}\int_{-\infty}^{\infty} e^{i\alpha z} \frac{\exp(-k\sqrt{r^2+z^2})}{\sqrt{r^2+z^2}}\, dz.$$

On setting $\alpha = 0$, we obtain the integral representation of the Macdonald function:

$$K_0(r) = \frac{1}{2}\int_{-\infty}^{\infty} \frac{\exp(-\sqrt{r^2+z^2})}{\sqrt{r^2+z^2}}\, dz. \tag{7.3.46}$$

EXERCISES

7.3.1 *Fourier transform*

(a) Consider the spectral problem

$$-u'' - \lambda u = 0, \quad -\infty < x < \infty,$$

where both endpoints are in the limit point case. Show that when the eigenvalue $\lambda \notin [0,\infty)$, Green's function is given by (7.2.15). By using (7.3.2) obtain the completeness relation

$$\delta(x-\xi) = \frac{1}{2\pi}\int_{-\infty}^{\infty} e^{i\nu x} e^{-i\nu\xi}\, d\nu, \quad -\infty < x, \xi < \infty, \tag{7.3.47}$$

from which the usual Fourier transform and inversion follow.

(b) Consider the Dirichlet problem for a strip:

$$\begin{aligned} -\Delta u = 0, \quad & -\infty < x < \infty, \quad 0 < y < b; \\ & u(x,0) = 0, \quad u(x,b) = f(x), \end{aligned} \tag{7.3.48}$$

where $f(x)$ is a given function. Apply a Fourier transform on the x coordinate to obtain

$$u^\wedge(\omega,y) = f^\wedge(\omega)\frac{\sinh \omega y}{\sinh \omega b}.$$

By the convolution theorem for Fourier transforms we then have

$$u(x,y) = \int_{-\infty}^{\infty} I(x-\xi,y) f(\xi)\, d\xi, \tag{7.3.49}$$

where $I(x, y)$ is the inverse transform of $(\sinh \omega y)/(\sinh \omega b)$.

(c) Show by the calculus of residues that

$$I(x, y) = \frac{1}{2b} \frac{\sin(\pi y/b)}{\cos(\pi y/b) + \cosh(\pi x/b)}. \tag{7.3.50}$$

7.3.2 The following problem can have a spectrum which is partly discrete and partly continuous:

$$-u'' - \lambda u = 0, \quad 0 < x < \infty; \quad u'(0) = u(0) \cot \alpha,$$

where α is a real number $0 \leqslant \alpha < \pi$. The physical significance of this boundary condition was explained in Example 1, Section 7.1 (the change in sign in the boundary condition is a result of applying the physical condition at the left endpoint instead of the right). See also Exercise 7.1.6.

(a) By elementary methods show that an eigenvalue is possible if and only if $\pi/2 < \alpha < \pi$. Then there is exactly one eigenvalue, given by $\lambda = -\cot^2 \alpha$ with normalized eigenfunction $\sqrt{-2 \cot \alpha}\, e^{x \cot \alpha}$.

(b) Show that Green's function for Im $\lambda \neq 0$ is given by

$$g(x, \xi; \lambda) = \frac{i[\exp(i\sqrt{\lambda}\, x_>)]}{\sqrt{\lambda} \sin \alpha + i \cos \alpha} \cdot \left((\sin \alpha) \cos \sqrt{\lambda}\, x_< + \frac{(\cos \alpha) \sin \sqrt{\lambda}\, x_<}{\sqrt{\lambda}} \right).$$

For $0 \leqslant \alpha \leqslant \pi/2$ the spectrum is purely continuous and consists of the real interval $0 \leqslant \lambda < \infty$; for $\pi/2 < \alpha < \pi$ the spectrum consists of the continuous part $0 \leqslant \lambda < \infty$ and of the single eigenvalue $\lambda = -\cot^2 \alpha$. Derive the completeness relations

$$\delta(x - \xi) = \frac{2}{\pi} \int_0^\infty u_\nu(x) u_\nu(\xi)\, d\nu, \quad 0 \leqslant \alpha \leqslant \frac{\pi}{2},$$

$$\delta(x - \xi) = \frac{2}{\pi} \int_0^\infty u_\nu(x) u_\nu(\xi)\, d\nu - 2(\cot \alpha) e^{(x+\xi) \cot \alpha}, \quad \frac{\pi}{2} < \alpha < \pi,$$

where

$$u_\nu(x) = \frac{\nu}{(\cos^2 \alpha + \nu^2 \sin^2 \alpha)^{1/2}} \left[(\sin \alpha) \cos \nu x + \frac{\cos \alpha}{\nu} \sin \nu x \right].$$

(c) Show that the first of these completeness relations leads to the transform pair

$$F(\nu) = \int_0^\infty u_\nu(x) f(x)\, dx, \quad f(x) = \frac{2}{\pi} \int_0^\infty u_\nu(x) F(\nu)\, d\nu.$$

These reduce to the ordinary sine and cosine transforms for $\alpha = 0$ and $\alpha = \pi/2$, respectively.

(d) Solve the problem

$$\frac{\partial u}{\partial t} - \frac{\partial^2 u}{\partial x^2} = 0, \quad t > 0, \quad x > 0; \quad u(x,0) = f(x),$$
$$(\cos\alpha)u(0,t) - (\sin\alpha)\frac{\partial u}{\partial x}(0,t) = 0,$$

by applying the transform in (c) when $0 \leqslant \alpha \leqslant \pi/2$.

7.3.3 *The Weber transform*

(a) Consider the spectral problem for Bessel's equation:

$$-(xu')' - \lambda x u = 0, \quad a < x < \infty; \qquad u(a) = 0,$$

where a is a given positive number. Clearly the left endpoint is regular, and the right endpoint is in the limit point case. Show that for Im $\lambda \neq 0$

$$g(x,\xi;\lambda) = \frac{\pi i}{2H_0^{(1)}(\sqrt{\lambda}\,a)} v(x_<,\lambda)H_0^{(1)}(\sqrt{\lambda}\,x_>),$$

where

$$v(x,\lambda) = J_0(\sqrt{\lambda}\,x)H_0^{(1)}(\sqrt{\lambda}\,a) - J_0(\sqrt{\lambda}\,a)H_0^{(1)}(\sqrt{\lambda}\,x)$$

is the solution of the homogeneous equation satisfying $v(a,\lambda) = 0$, $v'(a,\lambda) = -2i/\pi a$. Show that the spectrum is purely continuous and consists of the real values $0 \leqslant \lambda < \infty$. Obtain the expansions

$$\frac{\delta(x-\xi)}{x} = -\int_0^\infty \frac{v(x,\mu^2)v(\xi,\mu^2)}{J_0^2(a\mu) + N_0^2(a\mu)}\mu\, d\mu, \quad a < x, \xi < \infty,$$
$$\frac{\delta(x-\xi)}{x} = \int_0^\infty \frac{w(x,\mu^2)w(\xi,\mu^2)}{J_0^2(a\mu) + N_0^2(a\mu)}\mu\, d\mu, \quad a < x, \xi < \infty,$$

where

$$v(x,\mu^2) = J_0(\mu x)H_0^{(1)}(\mu a) - J_0(\mu a)H_0^{(1)}(\mu x),$$
$$w(x,\mu^2) = J_0(\mu x)N_0(\mu a) - J_0(\mu a)N_0(\mu x).$$

Formulas (7.3.36) may be useful in the derivation. The transform pair corresponding to the second completeness relation is known as the *Weber transform* pair.

(b) Use a Weber transform to find the solution of the heat equation problem in two space dimensions (cylindrical hole):

$$\frac{\partial u}{\partial t} - \Delta u = \frac{\partial u}{\partial t} - \frac{1}{r}\frac{\partial}{\partial r}\left(r\frac{\partial u}{\partial r}\right) = 0, \quad r > a, \quad t > 0;$$
$$u(a,t) = 1, \quad u(r,0) = 0.$$

7.3.4 *The Mellin transform.* Consider the spectral problem

$$-(rR')' - \frac{\lambda}{r}R = 0, \quad 0 < r < \infty; \quad p(r) = r, \quad s(r) = \frac{1}{r},$$

where the independent variable has been called r because the problem arises in polar coordinates in the *plane*.

(a) Show that the limit point case holds at both singular points, $r = 0$ and $r = \infty$.

(b) For $\lambda \notin [0, \infty)$ construct Green's function

$$g(r, r_0) = \frac{i}{2\sqrt{\lambda}} \, r_<^{-i\sqrt{\lambda}} \, r_>^{i\sqrt{\lambda}},$$

where $r_< = \min(r, r_0), \quad r_> = \max(r, r_0)$.

(c) Obtain formally the completeness relation

$$r\delta(r - r_0) = \frac{1}{2\pi} \int_{-\infty}^{\infty} r_0^{i\nu} r^{-i\nu} \, d\nu, \quad 0 < r, r_0 < \infty, \tag{7.3.51}$$

and the *Mellin transform pair*

$$F_{\mathrm{M}}(\nu) \doteq \int_0^{\infty} f(r) r^{-i\nu - 1} \, dr \quad \text{(Mellin transform)}, \tag{7.3.52}$$

$$f(r) = \frac{1}{2\pi} \int_{-\infty}^{\infty} r^{i\nu} F_{\mathrm{M}}(\nu) \, d\nu. \tag{7.3.53}$$

(d) Solve the BVP for the Laplacian in a wedge:

$$\Delta u = 0, \quad 0 < r < \infty, \ 0 < \phi < \alpha; \ u(r, 0) = 0, \ u(r, \alpha) = h(r),$$

where $h(r)$ is a prescribed function. Proceed formally by taking the Mellin transform to obtain the solution in the form of an inverse transform.

REFERENCES AND ADDITIONAL READING

1. F. V. Atkinson, *Discrete and Continuous Boundary Problems*, Academic Press, New York, 1964.
2. N. Dunford and J. T. Schwartz, *Linear Operators*, Part I: *General Theory*, Wiley, New York, 1957.
3. N. Dunford and J. T. Schwartz, *Linear Operators*, Part II: *Spectral Theory*, Wiley, New York, 1963.

4. N. Dunford and J. T. Schwartz, *Linear Operators*, Part III: *Spectral Operators*, Wiley, New York, 1971.

5. M. S. P. Eastham, *The Spectral Theory of Periodic Differential Equations*, Scottish Academic Press, Edinbergh, UK, 1973.

6. C. T. Fulton and S. A. Pruess, Eigenvalue and eigenfunction asymptotics for regular Sturm-Liouville problems, *J. Math. Anal. Appl.* 188, 1994.

7. M. Hajmirzaahmad and A. M. Krall, Singular second-order operators, *SIAM Rev.* 34, 1992.

8. P. R. Halmos, *A Hilbert Space Problem Book*, Van Nostrand, New York, 1967.

9. S. Kesavan, *Topics in Functional Analysis and Applications*, Wiley, New York, 1989.

10. M. A. Naimark, *Linear Differential Operators*, Frederick Ungar, New York, 1967.

11. E. C. Titchmarsh, *Eigenfunction Expansions Associated with Second-Order Differential Equations*, Oxford University Press, New York, 1946.

12. K. Yosida, *Lectures on Differential and Integral Equations*, Interscience, New York, 1960.

13. K. Yosida, *Functional analysis*, Springer–Verlag, New York, 1980.

CHAPTER 8

PARTIAL DIFFERENTIAL EQUATIONS

8.1 CLASSIFICATION OF PARTIAL DIFFERENTIAL EQUATIONS

The only difficulty that can arise for the initial value problem for an ordinary linear differential equation is the presence of a singular point (see Remark 3 following Theorem 3.1.1). The equivalent concept for partial differential equations is more subtle. Not surprisingly, one must first recast the one-dimensional problem before finding a useful generalization to higher dimensions.

In the IVP for an ordinary linear differential equation of order p, we look for a solution $u(x)$ of

$$Lu \doteq a_p(x)\frac{d^p u}{dx^p} + \cdots + a_1(x)\frac{du}{dx} + a_0(x)u = f(x), \tag{8.1.1}$$

with given initial data $\{u(x_0), u'(x_0), \ldots, u^{(p-1)}(x_0)\}$. We can rewrite (8.1.1) as

$$a_p(x)u^{(p)}(x) = f(x) - Mu, \tag{8.1.2}$$

where M is a differential operator of order $p-1$.

If the coefficients $\{a_m(x)\}$ and $f(x)$ are continuous and if $a_p(x_0) \neq 0$, the IVP has one and only one solution in a neighborhood of x_0. If the leading coefficient

$a_p(x)$ vanishes at x_0, one can try dividing by $a_p(x)$, but this will usually only cause some of the other coefficients to become unbounded at x_0.

In the case $a_p(x_0) \neq 0$, the initial data *together* with the differential equation enable us to construct the solution in a neighborhood of x_0 by a stepwise numerical procedure. As a first and crucial step we are able to calculate $u^{(p)}(x_0)$ unambiguously from the data and the differential equation since the right side of (8.1.2) is known at x_0 from the initial data, and division by $a_p(x_0)$ gives $u^{(p)}(x_0)$. Armed with this additional information, we can calculate new initial data $\{u(x_1), \ldots, u^{(p-1)}(x_1)\}$ at the neighboring point $x_1 = x_0 + \Delta x$ through the approximate formulas

$$u(x_1) = u(x_0) + u'(x_0)\,\Delta x, \ldots, u^{(p-1)}(x_1) = u^{(p-1)}(x_0) + u^{(p)}(x_0)\,\Delta x.$$

If x_1 is sufficiently close to x_0, $a_p(x_1)$ will not vanish and we can use the differential equation again to find $u^{(p)}(x_1)$. We now repeat the procedure to construct the solution of the original problem on an interval (x_0, b) as long as $a_p(x)$ does not vanish anywhere in $x_0 \leqslant x \leqslant b$.

If, however, $a_p(x_0) = 0$, we cannot calculate $u^{(p)}(x_0)$ unambiguously from (8.1.2). The right side is known from the initial data, and the left side vanishes; thus either $u^{(p)}(x_0)$ does not exist or it is indeterminate. In any event the earlier numerical construction fails, and so does the existence and uniqueness theorem. Another way of stating the difficulty is that Lu itself (rather than just Mu) is known at x_0 from the initial data alone.

Cauchy Problem

Consider a linear partial differential equation of order p in the n variables $x_1, \ldots, x_n$:

$$Lu = f(x), \tag{8.1.3}$$

where, in the usual notation of (2.1.3),

$$L = \sum_{|k| \leqslant p} a_k(x) D^k, \tag{8.1.4}$$

with

$$k = (k_1, \ldots, k_n), \quad |k| = k_1 + \cdots + k_n, \quad D^k = D_1^{k_1} \cdots D_n^{k_n}, \quad D_i = \frac{\partial}{\partial x_i}.$$

Anticipating that many of the qualitative properties of the solution of (8.1.3) will depend only on the terms of highest order in L, we define the *principal part* of the operator (8.1.4) as

$$L_p = \sum_{|k|=p} a_k(x) D^k = \sum_{|k|=p} a_k(x) D_1^{k_1} \ldots D_n^{k_n}. \tag{8.1.5}$$

It is often convenient to regard L_p, for fixed x, as a p-th-degree polynomial in the n variables $D_1, \ldots, D_n$, the components of the n-dimensional vector D. When this point of view is taken, we shall write $L_p(x, D)$ for (8.1.5). On occasion we shall use P rather than x to denote a point in $\mathbb{R}^n$. The coefficients are then $a_k(P)$, and $L_p(x, D)$ becomes $L_p(P, D)$.

The coefficients and the forcing term in (8.1.3) are assumed continuous, and a solution is sought in a specified domain in $\mathbb{R}^n$ (often this domain is the whole of $\mathbb{R}^n$). We now wish to associate initial data with (8.1.3). The appropriate generalization of the one-dimensional case is to assign the values of u and its *normal* derivatives of order $\leqslant p-1$ on a smooth hypersurface Γ (a hypersurface being a manifold of dimension $n-1$). This type of initial data is known as *Cauchy data*, and the resulting initial value problem as the *Cauchy problem* for L. As we shall see, the Cauchy data actually determines on Γ all derivatives (with respect to $x_1, \dots, x_n$, say) of order $\leqslant p-1$.

Let P be a point on Γ. In view of the smoothness of Γ we can introduce an $(n-1)$-dimensional coordinate system $\xi_2, \dots, \xi_n$ to label points on Γ in some neighborhood of P which is taken as the origin. This can be done, for instance, by assigning to each point on Γ the Cartesian coordinates of its projection on the tangent plane to Γ at P. In any event we shall refer to $\xi_2, \dots, \xi_n$ as *tangential* coordinates. Now let $\nu = \xi_1$ stand for a coordinate along the normal to Γ with origin on Γ. Then $(\nu, \xi_2, \dots, \xi_n)$ is a normal-tangential coordinate system for all points in $\mathbb{R}^n$ sufficiently close to P. The coordinate surface $\nu = 0$ is Γ, and the point $(0, \dots, 0)$ is P.

From the given Cauchy data on Γ we know the quantities

$$u(0, \xi_2, \dots, \xi_n), \quad \frac{\partial u}{\partial \nu}(0, \xi_2, \dots, \xi_n), \dots, \quad \frac{\partial^{p-1} u}{\partial \nu^{p-1}}(0, \xi_2, \dots, \xi_n).$$

Provided that there is enough smoothness, we can therefore calculate on Γ derivatives of any order with respect to the tangential coordinates and of order $\leqslant p-1$ with respect to the normal coordinate. In the coordinates $(\nu, \xi_2, \dots, \xi_n)$ the only derivative of order p on Γ which is not known from the Cauchy data is $\partial^p u/\partial \nu^p$. To return to the original coordinates $x_1, \dots, x_n$, we use the transformation law relating the two coordinate systems. We have

$$\frac{\partial u}{\partial x_i} = \sum_{j=1}^{n} \frac{\partial u}{\partial \xi_j} \frac{\partial \xi_j}{\partial x_i}, \qquad \nu \doteq \xi_1,$$

so that $\partial u/\partial x_i$ is a linear combination of $\partial u/\partial \xi_j$ with variable coefficients. Thus any mth derivative with respect to the x coordinates will involve only derivatives of order $\leqslant m$ with respect to the ξ coordinates. Since derivatives of order $\leqslant p-1$ with respect to the ξ coordinates are known from the Cauchy data, so are derivatives of order $\leqslant p-1$ with respect to the x coordinates. In fact, we even have partial information about derivatives of order p.

To construct the solution numerically for points near P but away from Γ we will need to be able to calculate $\partial^p u/\partial \nu^p$ at P (or, equivalently, to calculate unambiguously the set of all pth derivatives with respect to $x_1, \dots, x_n$ at P). This calculation will *not* be possible if Lu at P can be determined from the Cauchy data alone. We are therefore led to the following definition.

Definition. *If Lu can be evaluated at a point P on Γ from the Cauchy data alone, the surface Γ is said* to be characteristic *(*for L*) at P. If the surface Γ is characteristic (for L) at every point P on Γ, we say that Γ is a* characteristic surface *(*for L*).*

REMARK. Since the operator L is usually fixed throughout the discussion, the qualifier "for L" may be omitted.

If Γ is characteristic at P, then when we express L in terms of the normal tangential coordinates $(\nu, \xi_2, \ldots, \xi_n)$, the coefficient of $\partial^p/\partial\nu^p$ will vanish at $P = (0, \ldots, 0)$. In that sense the leading coefficient "vanishes" as for singular points of ordinary differential equations. We must note, however, that the new notion depends not only on the point P but also on Γ (or, more precisely, on the normal direction to Γ at P). This leads to the following theorem and corollaries, whose proofs we omit.

Theorem 8.1.1. *Let L be a given operator of order p, and let P be a point on the smooth hypersurface Γ. The necessary and sufficient condition for Γ to be characteristic at P is that the coefficient of $\partial^p/\partial\nu^p$ vanishes when L is expressed in the coordinate system $(\nu, \xi_2, \ldots, \xi_n)$.*

Corollary 8.1.1. *Γ is characteristic at the point P if and only if*

$$L_p(P, \nu) = 0, \tag{8.1.6}$$

where ν is the normal vector to Γ at P.

Corollary 8.1.2. *Γ is* not *characteristic at the point P if and only if all p-th-order derivatives of u with respect to $x_1, \ldots, x_n$ are unambiguously determined at P from the Cauchy data on Γ together with the differential equation.*

The results so far are essentially negative: If Γ is characteristic at P, there is no hope that the Cauchy problem is well-posed, even in a neighborhood of P. Is the Cauchy problem well-posed if Γ is nowhere characteristic? We shall find out that this is so for certain classes of partial differential equations but not for others.

Equations of the First Order in Two Independent Variables

Let x, y be Cartesian coordinates in $\mathbb{R}^2$, and let Γ be a smooth *curve* in the plane. Since L is to be an operator of the first order, the Cauchy data will consist of specifying u on Γ.

We begin with the operator $L = \partial/\partial x$ and the simple Cauchy problem

$$\frac{\partial u}{\partial x} = 0, \quad u \text{ given on } \Gamma. \tag{8.1.7}$$

Let A be a point on Γ. The curve Γ will be characteristic at A if $\partial u/\partial x$ can be calculated at A from the Cauchy data alone. Obviously, this is possible if and only if the tangent to Γ at A is in the x direction. The characteristic curves are therefore the

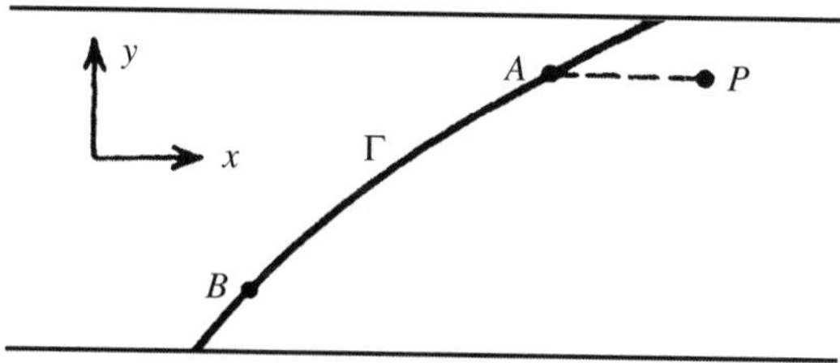

Figure 8.1

straight lines $y =$ constant. The general solution of (8.1.7) is $u = F(y)$, where F is arbitrary. Thus u is constant on any horizontal line. If the initial curve Γ on which the Cauchy data is given is *nowhere characteristic*, as in Figure 8.1, the solution will be unambiguously determined in the entire strip by the formula $u_P = u_A$. [Note that even if the Cauchy data has a discontinuity a some point B on Γ, we can still define the solution in the horizontal strip by $u_P = u_A$, but the solution is then discontinuous along the entire horizontal line through B. We showed in Example 5, Section 2.5, that a discontinuous function $F(y)$ still satisfies (8.1.7) in the sense of distributions.] If the initial curve is characteristic even at a single point A, as in Figure 8.2, the Cauchy data will usually be incompatible with the differential equation; since u is constant along a horizontal line, a solution is possible only if the given values at points such as A_1 and A_2 are the same.

Let us now look at the equation of the first order:

$$a(x,y)\frac{\partial u}{\partial x} + b(x,y)\frac{\partial u}{\partial y} + c(x,y,u) = 0, \tag{8.1.8}$$

which is a linear equation if $c(x,y,u)$ is of the form $c(x,y)u + d(x,y)$ and is said to be *semilinear* if $c(x,y,u)$ depends nonlinearly on u (note that in a semilinear equation the highest derivatives appear linearly). Since the theory of characteristics applies just as easily to semilinear equations, we may as well consider this more general case. Let Γ be a smooth curve in $\mathbb{R}^2$, and let P be a point on Γ. The value of u is given on Γ. By Corollary 8.1.2, only if we can calculate $\partial u/\partial x$ and $\partial u/\partial y$ unambiguously at P from the data and the differential equation will Γ not be characteristic at P. Consider two neighboring points $P = (x,y)$ and $Q = (x + dx, y + dy)$ located on Γ. Let us try to calculate $\partial u/\partial x$ and $\partial u/\partial y$ at P. The following relations hold:

$$\begin{aligned} u(Q) - u(P) &= dx\frac{\partial u}{\partial x}(P) + dy\frac{\partial u}{\partial y}(P), \\ c(P, u(P)) &= a(P)\frac{\partial u}{\partial x}(P) + b(P)\frac{\partial u}{\partial y}(P). \end{aligned}$$

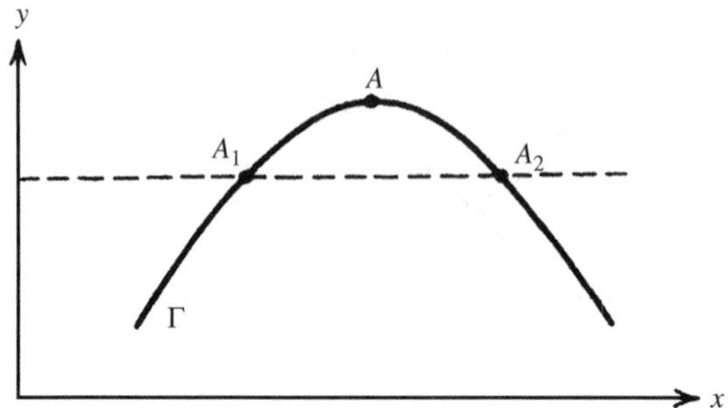

Figure 8.2

These simultaneous linear equations for $\partial u/\partial x$ and $\partial u/\partial y$ at P will have a unique solution if and only if

$$\begin{vmatrix} dx & dy \\ a(P) & b(P) \end{vmatrix} \neq 0 \quad \text{or} \quad b(P)\,dx - a(P)\,dy \neq 0. \tag{8.1.9}$$

The curve Γ is characteristic at $P = (x, y)$ if and only if

$$b(x, y)\,dx - a(x, y)\,dy = 0. \tag{8.1.10}$$

Thus at every point in the plane a unique characteristic direction is defined by (8.1.9) as long as a and b do not vanish simultaneously. If the ordinary differential equation (8.1.10) is integrated, a "one-parameter" family of curves is obtained—the characteristic curves for (8.1.8). The same result could have been obtained by using (8.1.6). We have $p = 1$ and

$$L_1(P, \nu) = a(P)\nu_1 + b(P)\nu_2.$$

On setting $L_1 = 0$, we find that Γ is characteristic at P if its normal is in the direction $(-b(P), a(P))$, which means that its slope must be $b(P)/a(P)$, confirming (8.1.10).

We state without proof for the linear version of (8.1.8) a theorem similar to that obtained for $\partial u/\partial x = 0$: If Γ is *nowhere characteristic*, the solution of (8.1.8) exists and is unique in the characteristic "strip" bounded by the characteristic curves through the endpoints of Γ. Moreover, the solution depends continuously on the data. If the given value of u on Γ has a discontinuity at a point P, the discontinuity is propagated along the characteristic curve through P. The solution is then a weak solution in the sense of Section 2.5.

As a simple illustration of (8.1.8), consider the equation

$$\frac{\partial u}{\partial x} + \frac{\partial u}{\partial t} + \sigma u = 0, \tag{8.1.11}$$

where σ is a nonnegative constant, x is a space coordinate, and t is the time (corresponding to y in the earlier notation). The case $\sigma = 0$ governs the plug flow of a fluid [see (0.2.8) with $D = p = 0$], whereas the case $\sigma > 0$ describes an absorptive flow. Since $a = b = 1$, the characteristics of (8.1.11) are the straight lines $x - t =$ constant. The usual Cauchy problem associated with (8.1.11) involves data given on the initial curve $t = 0$, which is *nowhere characteristic*. We then want to solve (8.1.11) for $t > 0$ and $-\infty < x < \infty$, subject to the initial condition $u(x, 0) = f(x)$, where $f(x)$ is arbitrary. By making the change of variables $\alpha = x - t, \beta = x + t$, we reduce (8.1.11) to

$$\frac{\partial u}{\partial \beta} + \frac{\sigma}{2} u = 0,$$

so that $u(x, t) = F(x - t)e^{-\sigma(x+t)/2}$, and on imposing the initial condition,

$$u(x, t) = f(x - t)e^{-\sigma t}. \tag{8.1.12}$$

If $\sigma = 0$, the solution is $f(x - t)$, which for fixed $t > 0$ is just the initial function $f(x)$ displaced t units to the right along the x axis. We can therefore view $f(x-t)$ as a wave traveling to the right with unit velocity. If $\sigma > 0$, we also have an attenuation factor from the exponential in (8.1.12). This is in agreement with our intuitive ideas of absorption.

Equations of the Second Order in Two Variables

Consider the equation of the second order in the two independent variables x and y:

$$a(x, y)\frac{\partial^2 u}{\partial x^2} + 2b(x, y)\frac{\partial^2 u}{\partial x\, \partial y} + c(x, y)\frac{\partial^2 u}{\partial y^2} + F\left(x, y, u, \frac{\partial u}{\partial x}, \frac{\partial u}{\partial y}\right) = 0. \tag{8.1.13}$$

The equation will be *linear* if

$$F\left(x, y, u, \frac{\partial u}{\partial x}, \frac{\partial u}{\partial y}\right) = d\frac{\partial u}{\partial x} + e\frac{\partial u}{\partial y} + fu + g,$$

where d, e, f, g are arbitrary functions of x and y alone, and otherwise *semilinear*. We are given u and its normal derivative on a smooth curve Γ in the x–y plane. Thus both $\partial u/\partial x$ and $\partial u/\partial y$ are known on Γ.

If we can calculate all second derivatives at a point P on Γ from the data together with the differential equation, Γ is not characteristic at P. Again let $P = (x, y)$ and $Q = (x + dx, y + dy)$ be two neighboring points on Γ. Then the following simultaneous linear equations for $\partial^2 u/\partial x^2$, $\partial^2 u/\partial y^2$, $\partial^2 u/\partial x\, \partial y$, all evaluated at P, must hold:

$$\frac{\partial u}{\partial x}(Q) - \frac{\partial u}{\partial x}(P) = dx\frac{\partial^2 u}{\partial x^2}(P) + dy\frac{\partial^2 u}{\partial x\, \partial y}(P),$$

$$\frac{\partial u}{\partial y}(Q) - \frac{\partial u}{\partial y}(P) = dx\frac{\partial^2 u}{\partial x\, \partial y}(P) + dy\frac{\partial^2 u}{\partial y^2}(P),$$

$$-F_P = a(P)\frac{\partial^2 u}{\partial x^2}(P) + 2b(P)\frac{\partial^2 u}{\partial x\, \partial y}(P) + c(P)\frac{\partial^2 u}{\partial y^2}(P),$$

where $F_P \doteq F(P, u(P), \partial u/\partial x(P), \partial u/\partial y(P))$.

The left sides are known from the Cauchy data, so that a necessary and sufficient condition for a unique solution is

$$\begin{vmatrix} dx & dy & 0 \\ 0 & dx & dy \\ a(P) & 2b(P) & c(P) \end{vmatrix} \neq 0 \quad \text{or} \quad a\,dy^2 - 2b\,dx\,dy + c\,dx^2 \neq 0. \tag{8.1.14}$$

If $a(P) \neq 0, \Gamma$ will be characteristic at $P = (x, y)$ if and only if the slope of Γ at P satisfies

$$\frac{dy}{dx} = \frac{b \pm \sqrt{b^2 - ac}}{a}. \tag{8.1.15}$$

We distinguish among three cases:

If $b^2 - ac < 0$ at P, no real curve Γ can satisfy (8.1.15) and no curve is characteristic at P. Equation (8.1.13) is said to be *elliptic* at P. For the Laplace equation $\partial^2 u/\partial x^2 + \partial^2 u/\partial y^2 = 0$, we have $b^2 - ac = -1$, so that the equation is elliptic in the entire plane. The Laplace equation is the prototype of elliptic equations.

If $b^2 - ac > 0$ at P, there are two characteristic directions at P and the equation is said to be *hyperbolic* at P. The equation $\partial^2 u/\partial x_1\,\partial x_2 = 0$ and the wave equation $\partial^2 u/\partial t^2 - c^2 \partial^2 u/\partial x^2 = 0$ are hyperbolic in the entire x_1–x_2 and x–t planes, respectively. The first equation can be obtained from the second by the simple transformation $x_1 = x - ct, x_2 = x + ct$.

If $b^2 - ac = 0$ at P, there is only one characteristic direction at P and the equation is *parabolic* at P. The diffusion equation $\partial u/\partial t - a\,\partial^2 u/\partial x^2 = 0$ is parabolic in the entire x–t plane.

REMARKS

1. If the coefficients are not constant, the equation may be of different types in various parts of the plane. For instance, $\partial^2 u/\partial x^2 + x\,\partial^2 u/\partial y^2 = 0$ is elliptic in the half-plane $x > 0$ and hyperbolic in the half-plane $x < 0$.

2. If the equation is hyperbolic in the entire plane, we can integrate (8.1.15) to find two families of characteristic curves, whereas for the parabolic case we can find only one such family.

3. Characteristics can be a blessing or a curse. If the equation is hyperbolic and Cauchy data is given on a curve Γ, the problem is well-posed when Γ is not a characteristic but ill-posed if Γ is a characteristic. One might therefore think that in the elliptic case—when characteristics are totally absent—the Cauchy problem would be well-posed. We shall see, however, that this is not the case and that, instead, boundary value problems are usually well-posed for elliptic equations (but not for hyperbolic ones).

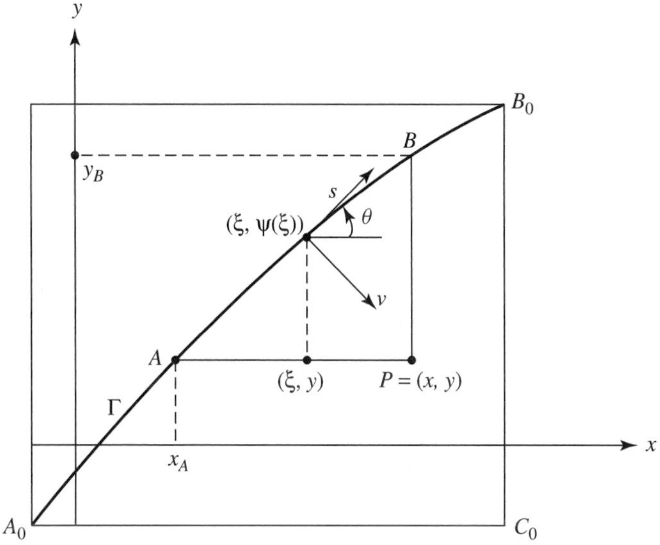

Figure 8.3

To show how characteristics can be used, we shall investigate a few problems related to the hyperbolic equation

$$\frac{\partial^2 u}{\partial x\, \partial y} = f(x, y, u), \tag{8.1.16}$$

where f is a continuous function of its arguments. If $f = 0$, we have the simple equation

$$\frac{\partial^2 u}{\partial x\, \partial y} = 0, \tag{8.1.17}$$

which we study first. Since it holds that $(\partial/\partial x)(\partial u/\partial y) = 0$, $\partial u/\partial y = g(y)$ and $u = F(x) + G(y)$ is the general solution of (8.1.17). The characteristics for (8.1.17) as well as for (8.1.16) are the straight lines $x = \text{constant}, y = \text{constant}$. Let us consider a few initial value problems for (8.1.17).

1. *The Cauchy problem.* Let Γ be a curve that is nowhere characteristic, as in Figure 8.3. We shall show how the solution can be determined within the large rectangle from the Cauchy data on Γ. Indeed, let $P = (x, y)$ be a point in the rectangle. Integrating $\partial u/\partial x$ from A to P, we obtain

$$u(P) = u(A) + \int_{x_A}^{x} \frac{\partial u}{\partial x}(\xi, y)\, d\xi.$$

Here x_A is the value of x at the point A, that is, $x_A = \psi^{-1}(y)$, where $\psi(x)$ is the equation of the curve Γ. The differential equation tells us that $\partial u/\partial x$ is independent of y, so that

$$\frac{\partial u}{\partial x}(\xi, y) = \frac{\partial u}{\partial x}(\xi, \psi(\xi)).$$

Thus we obtain

$$u(P) = u(A) + \int_{x_A}^{x} \frac{\partial u}{\partial x}(\xi, \psi(\xi))\, d\xi, \tag{8.1.18}$$

which expresses $u(x, y)$ in terms of known quantities on Γ. One can easily verify that (8.1.18) actually solves the Cauchy problem for (8.1.17). By integrating $\partial u/\partial y$ from P to B, one also finds that

$$u(P) = u(B) - \int_{y}^{y_B} \frac{\partial u}{\partial x}(\psi^{-1}(\eta), \eta)\, d\eta. \tag{8.1.19}$$

We can express all integrals in terms of the arclength s along Γ by using the relations $d\xi = \cos\theta\, ds, d\eta = \sin\theta\, ds$. Adding (8.1.18) and (8.1.19), we then find that

$$u(P) = \frac{u(A)}{2} + \frac{u(B)}{2} + \frac{1}{2}\int_{s_A}^{s_B} \left(\frac{\partial u}{\partial x}\cos\theta - \frac{\partial u}{\partial y}\sin\theta\right) ds.$$

Introducing the unit vector $q = (\cos\theta, -\sin\theta)$, known as the *transversal* to Γ, we can simplify the last formula to

$$u(P) = \frac{u(A)}{2} + \frac{u(B)}{2} + \frac{1}{2}\int_{s_A}^{s_B} \frac{\partial u}{\partial q}\, ds, \tag{8.1.20}$$

where $\partial u/\partial q$ is known from the Cauchy data. The transversal will coincide with the unit normal $\nu = (\sin\theta, -\cos\theta)$ if $\theta = \pi/4$. Thus, if Γ is a 45° line, we have the useful result

$$u(P) = \frac{u(A)}{2} + \frac{u(B)}{2} + \frac{1}{2}\int_{A}^{B} \frac{\partial u}{\partial \nu}\, ds. \tag{8.1.21}$$

Formulas such as (8.1.18), (8.1.19), and (8.1.20) establish the existence of a solution and its continuity with respect to the data. Uniqueness is easy. Suppose that we have two solutions u_1 and u_2 corresponding to the same Cauchy data. Then the difference $u = u_1 - u_2$ satisfies (8.1.17) with vanishing Cauchy data on Γ. Since the general solution of (8.1.17) is $u = F(x) + G(y)$, it follows that

$$0 = F(x_A) + G(y), 0 = F(x) + G(y_B),$$

so that $u(x, y)$ is constant in the large rectangle of Figure 8.3. Since u vanishes on Γ, that constant is 0.

2. If Γ is characteristic at P, the Cauchy data and the differential equations are generally incompatible. This is clearly seen if Γ is a portion of the characteristic

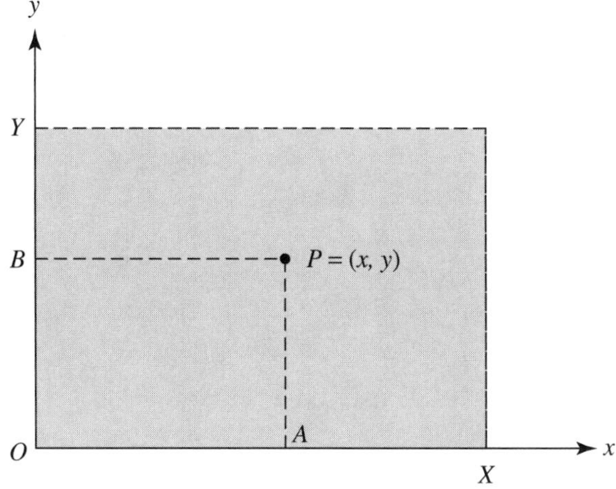

Figure 8.4

curve $y = $ constant. The Cauchy problem consists of assigning u and $\partial u/\partial y$ arbitrarily on Γ, but the differential equation tells us that $\partial u/\partial y$ is independent of x, so that $\partial u/\partial y$ must be constant on Γ.

3. There are sensible problems that can be formulated with data on characteristics, but these are not strict Cauchy problems. Consider two intersecting characteristics which, for simplicity, we have taken along the coordinate axes, as in Figure 8.4. If we are given u *alone* on the segments OX and OY, we can easily show that the solution is uniquely determined in the shaded rectangle by the formula

$$u(P) = u(A) + u(B) - u(O), \tag{8.1.22}$$

where $u(O)$ is the value of u at the origin (we have assumed that the given values of u on the segments OX and OY coincide at O). With enough smoothness in the data, it is easy to verify that (8.1.22) is a classical solution. Even if the data is discontinuous, we find that (8.1.22) still provides a weak solution to the Cauchy problem. Discontinuities in the data are propagated along characteristics.

4. Another type of problem occurs when u is given on a characteristic segment and on an intersecting curve nowhere characteristic, as in Figure 8.5. We then find the solution in the shaded region as

$$u(P) = u(A) + u(B) - u(C). \tag{8.1.23}$$

Let us now turn to the more difficult nonlinear problem (8.1.16) with Cauchy data on the noncharacteristic curve Γ. We still refer to Figure 8.3. Integrating $\partial u/\partial x$

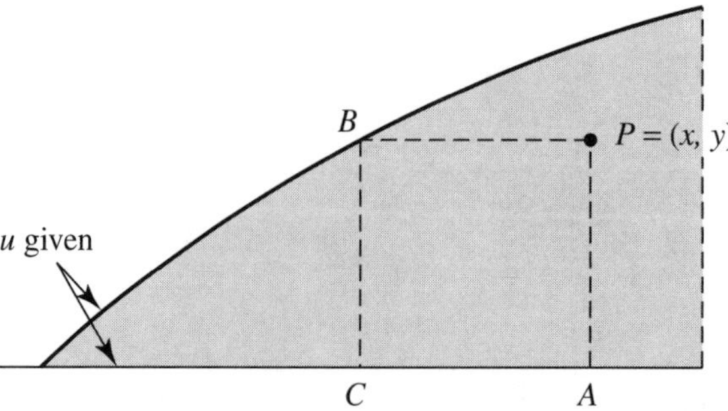

Figure 8.5

from A to P, we have

$$u(P) = u(A) + \int_{x_A}^{x} \frac{\partial u}{\partial x}(\xi, y)\, d\xi,$$

and from the differential equations,

$$\frac{\partial u}{\partial x}(\xi, \psi(\xi)) - \frac{\partial u}{\partial x}(\xi, y) = \int_{y}^{\psi(\xi)} f(\xi, \eta, u(\xi, \eta))\, d\eta,$$

so that

$$u(P) = u(A) + \int_{x_A}^{x} \frac{\partial u}{\partial x}(\xi, \psi(\xi))\, d\xi - \iint\limits_{D(x,y)} f(\xi, \eta, u(\xi, \eta))\, d\xi\, d\eta, \qquad (8.1.24)$$

where $D(x, y)$ is the triangular region bounded by AP, PB, and Γ. Although the first two terms on the right side of (8.1.24) are known from the Cauchy data, the last term is not. Thus (8.1.24) is an integral equation (nonlinear) for the unknown $u(x, y)$, completely equivalent to the Cauchy problem for (8.1.16).

By subtracting from the solution of the Cauchy problem for (8.1.16) the solution (previously calculated) of (8.1.17) with the same Cauchy data, we obtain a problem of type (8.1.16) with vanishing Cauchy data but different f. It is therefore with no loss of generality that we may analyze (8.1.24) with vanishing Cauchy data:

$$u(x, y) = -\iint\limits_{D(x,y)} f(\xi, \eta, u(\xi, \eta))\, d\xi\, d\eta. \qquad (8.1.25)$$

We would like to show that (8.1.25) has one and only one solution, at least in some sufficiently small region to the right of Γ. We shall use the ideas developed in Section 4.4, on contraction transformations. First we must make some restriction on the nature of the nonlinearity in f. We shall require that the Lipschitz condition

$$|f(x, y, u_2 - u_1)| \leqslant C|u_2 - u_1| \qquad (8.1.26)$$

holds for all u_1, u_2 and for all x, y in the closed region R bounded by the segments A_0C_0, C_0B_0, and Γ (see Figure 8.3).

Let M_R be the space of continuous functions on R with the uniform norm

$$\|u_1 - u_2\|_\infty = \max_R |u_1 - u_2|.$$

The right side of (8.1.25) defines a transformation of M_R into itself. We shall denote this transformation by T. A solution of (8.1.25) is then just a fixed point of T, and vice versa. Let u_1 and u_2 be arbitrary elements of M_R, and let v_1 and v_2 be their respective images under T. Then

$$\begin{aligned}|v_1(x,y) - v_2(x,y)| = |Tu_1 - Tu_2| &\leqslant \iint\limits_{D(x,y)} |f(\xi,\eta,u_2) - f(\xi,\eta,u_1)|\,d\xi\,d\eta \\ &\leqslant CS(x,y)\|u_1 - u_2\|_\infty,\end{aligned}$$

where $S(x,y)$ is the area of $D(x,y)$. Therefore,

$$\|v_1 - v_2\|_\infty \leqslant CS\|u_1 - u_2\|_\infty,$$

where S is the area of R. Clearly, if the constant C in (8.1.26) is smaller than $1/S$, then T is a contraction on M_R and therefore has a unique fixed point which is the required solution of (8.1.25). In any event we can always choose a smaller region R (of the same triangular form), so that $CS < 1$; then, within this smaller region, (8.1.25) and the equivalent Cauchy problem for (8.1.16) will have one and only one solution.

REMARKS

1. Existence and uniqueness for the Cauchy problem (8.1.16) can be extended to the case where f depends on first derivatives, but again a growth condition is needed.

2. It may appear that we have studied only a very special hyperbolic equation. However, by a change of coordinates, any hyperbolic equation of the second order in two variables can be transformed into (8.1.16) with f also depending on $\partial u/\partial x$ and $\partial u/\partial y$. For instance, the linear wave equation

$$\frac{\partial^2 u}{\partial t^2} - c^2\frac{\partial^2 u}{\partial x^2} = f(x,t)$$

takes the form

$$\frac{\partial^2 u}{\partial \alpha\,\partial \beta} = \tilde{f}(\alpha,\beta)$$

with $\alpha = x - ct, \beta = x + ct$, and

$$\tilde{f}(\alpha,\beta) = -\frac{1}{4c^2} f\left(\frac{\alpha+\beta}{2}, \frac{\beta-\alpha}{2c}\right).$$

EXERCISES

8.1.1 The Cauchy problem for elliptic equations is ill-posed despite the absence of characteristics. Consider Laplace's equation $\Delta u = 0$ in the half-plane $x > 0$, $-\infty < y < \infty$ with the Cauchy data $u(0, y) = 0, \partial u/\partial x = \varepsilon \sin \alpha y$, where ε and α are constants. A solution of this problem is $u(x, y) = \varepsilon \sin \alpha y \sinh \alpha x$, which reduces—as it should—to $u \equiv 0$ when $\varepsilon = 0$. If $|\varepsilon|$ is small, the data is small in the uniform norm, independent of α, that is,

$$\|u(0, y)\|_\infty \leqslant \varepsilon \quad \text{and} \quad \left\| \frac{\partial u}{\partial x}(0, y) \right\|_\infty \leqslant \varepsilon.$$

Show that for any $\varepsilon \neq 0$ and any preassigned $x > 0$, the solution $u(x, y)$ can be made arbitrarily large by choosing α large enough. Thus the solution does not depend continuously on the data.

8.1.2 The pure BVP with data on a noncharacteristic rectangle is ill-posed for hyperbolic equations. Consider the homogeneous wave equation

$$\frac{\partial^2 u}{\partial t^2} - \frac{\partial^2 u}{\partial x^2} = 0$$

on the rectangle $0 < x < \pi, 0 < t < \tau$ with $u(x, 0) = u(0, t) = u(\pi, t) = 0$ and $u(x, \tau) = \varepsilon f(x)$, where $f(x)$ is a given function in $C^2[(0, \pi)]$ that satisfies $f(0) = f(\pi) = 0$ and all of whose Fourier sine coefficients are nonvanishing. Show that if τ is rational, the solution obtained by separation of variables cannot be made to satisfy the condition at $t = \tau$; if τ is irrational, the solution does not depend continuously on the data.

8.2 WELL-POSED PROBLEMS FOR HYPERBOLIC AND PARABOLIC EQUATIONS

Hyperbolic Equations

Consider the small, transverse vibrations of a very long, taut string subject to a pressure $q(x, t)$ for $t > 0$. At $t = 0$ we give the initial "state" of the string, that is, the initial displacement $f_1(x)$ and the initial velocity $f_2(x)$. Our boundary value problem is

$$\frac{\partial^2 u}{\partial t^2} - \frac{\partial^2 u}{\partial x^2} = q(x, t), \quad t > 0, -\infty < x < \infty; \tag{8.2.1}$$
$$u(x, 0) = f_1(x), \quad \frac{\partial u}{\partial t}(x, 0) = f_2(x).$$

The positive constant which originally multiplied the term $\partial^2 u/\partial x^2$ has been incorporated in the time variable.

We first consider the *homogeneous* equation ($q \equiv 0$) whose general solution is

$$F(x-t) + G(x+t), \tag{8.2.2}$$

where F and G are arbitrary. As we have already seen in Section 8.1 and in Example 7, Section 2.5, $F(x-t)$ can be regarded as a wave traveling to the right with unit velocity. Similarly, $G(x+t)$ is a wave traveling to the left with unit velocity. To solve the initial value problem for the homogeneous equation it suffices to express F and G in terms of f_1 and f_2. This is easily done for the infinite string. Although it is no trouble to handle both initial conditions simultaneously, we prefer, with an eye to future applications, to treat them separately. We begin with the case where $f_1 = 0$, that is, with the BVP

$$\frac{\partial^2 u}{\partial t^2} - \frac{\partial^2 u}{\partial x^2} = 0, \quad t > 0, -\infty < x < \infty; \quad u(x,0) = 0, \frac{\partial u}{\partial t}(x,0) = f(x). \tag{8.2.3}$$

Since $u(x,t)$ can be written as $F(x-t) + G(x+t)$, we immediately find that

$$F(x) + G(x) = 0, \quad -F'(x) + G'(x) = f(x).$$

The last equation gives

$$-F(x) + G(x) = \int_0^x f(\xi)\, d\xi + A,$$

so that together with $F + G = 0$, we obtain

$$G(x) = \frac{A}{2} + \frac{1}{2}\int_0^x f(\xi)\, d\xi, \quad F(x) = -\frac{1}{2}\int_0^x f(\xi)\, d\xi - \frac{A}{2},$$

and

$$u(x,t) = F(x-t) + G(x-t) = \frac{1}{2}\int_{x-t}^{x+t} f(\xi)\, d\xi, \tag{8.2.4}$$

which is the solution of (8.2.3).

If the initial displacement is $f(x)$ and the initial velocity 0, the corresponding solution of (8.2.3) is just the time derivative of (8.2.4). This is contained in the following theorem.

Theorem. *If $u(x,t)$ is a solution of (8.2.3), then $v(x,t) \doteq (\partial u/\partial t)(x,t)$ is a solution of*

$$\frac{\partial^2 v}{\partial t^2} - \frac{\partial^2 v}{\partial x^2} = 0, \quad t > 0, -\infty < x < \infty; \qquad v(x,0) = f(x), \quad \frac{\partial v}{\partial t} = 0. \tag{8.2.5}$$

Proof. Since u is a solution of a homogeneous linear differential equation with constant coefficients, any derivative of u is also a solution of the same equation. Also,

$$v(x,0+) = f(x) \quad \text{and} \quad \frac{\partial v}{\partial t}(x,0+) = \frac{\partial^2 u}{\partial t^2}(x,0+) = \frac{\partial^2 u}{\partial x^2}(x,0+) = 0,$$

since $u(x, 0+) = 0$. □

It therefore follows from (8.2.4) that the solution of (8.2.5) is

$$v(x,t) = \tfrac{1}{2}f(x+t) + \tfrac{1}{2}f(x-t). \tag{8.2.6}$$

If the initial displacement is $\delta(x)$ and the initial velocity is 0 (that is, we are plucking the string at $x = 0$), the corresponding solution (8.2.6) is $\frac{1}{2}\delta(x+t) + \frac{1}{2}\delta(x-t)$, which means that the initial displacement splits into two waves, one traveling to the left and the other to the right with unit velocity. If the initial displacement is 0 and the initial velocity is $\delta(x)$ (that is, we are striking the string at $x = 0$), a well of constant depth forms under the blow and the well front spreads outward with unit velocity.

In terms of the original time variable, the propagation velocity is c, where c^2 is the coefficient of $\partial^2/\partial x^2$ in the original version of (8.2.1).

The solution of (8.2.3) with initial velocity $= \delta(x)$ coincides for $t > 0$ with the causal fundamental solution with pole at $x = 0, t = 0$. A similar relationship has already been shown for ordinary differential equations. The causal fundamental solution $E(t, 0)$ for the operator

$$L = a_p(t)\frac{d^p}{dt^p} + \cdots + a_1(t)\frac{d}{dt} + a_0(t)$$

satisfies

$$LE = \delta(t), \quad -\infty < t < \infty; \quad E \equiv 0, \quad t < 0.$$

In (2.5.48) we showed that E coincides for $t > 0$ with the solution of the IVP (for the homogeneous equation)

$$Lu = 0, \quad t > 0; \quad u(0) = u'(0) = \cdots = u^{(p-2)}(0) = 0, \quad u^{(p-1)}(0) = \frac{1}{a_p(0)}.$$

For our partial differential equation we want a fundamental solution that is causal with respect to time; thus $E(x, t; 0, 0)$ satisfies

$$\frac{\partial^2 E}{\partial t^2} - \frac{\partial^2 E}{\partial x^2} = \delta(x)\delta(t), \quad -\infty < x, t < \infty; \quad E \equiv 0, \quad t < 0. \tag{8.2.7}$$

By analogy with ordinary differential equations we can equally well characterize E for $t > 0$ as the solution of the IVP

$$\frac{\partial^2 u}{\partial t^2} - \frac{\partial^2 u}{\partial x^2} = 0, \quad -\infty < x < \infty,\ t > 0; \quad u(x,0) = 0,\ \frac{\partial u}{\partial t}(x,0) = \delta(x). \tag{8.2.8}$$

As a check we recall that (8.2.7) was solved in Exercise 2.5.4. The solution was found to be

$$E(x,t) = \tfrac{1}{2}H(t - |x|) = \tfrac{1}{2}H(t)[H(x+t) - H(x-t)]. \tag{8.2.9}$$

On the other hand, the solution of (8.2.8) follows from (8.2.4) with $f(x) = \delta(x)$,

$$u(x,t) = \frac{1}{2}\int_{x-t}^{x+t} \delta(\xi)\,d\xi = \frac{1}{2}[H(x+t) - H(x-t)], \quad t > 0, \tag{8.2.10}$$

and (8.2.10) coincides with (8.2.9) for $t > 0$. We note that the function

$$\tfrac{1}{2}[H(x+t) - H(x-t)]$$

is a solution of the homogeneous wave equation for $-\infty < x, t < \infty$. This is true despite the discontinuities along the characteristics $x = t$ and $x = -t$. Function (8.2.9), however, has a source at $x = 0, t = 0$ created by piecing together two different solutions of the homogeneous equation: The zero solution for $t < 0$ and (8.2.10) for $t > 0$.

If the source is introduced at $x = \xi, t = \tau$, the corresponding causal fundamental solution is

$$E(x,t;\xi,\tau) = \tfrac{1}{2}H(t - \tau - |x - \xi|). \tag{8.2.11}$$

Consider now the inhomogeneous equation with vanishing initial data

$$\frac{\partial^2 w}{\partial t^2} - \frac{\partial^2 w}{\partial x^2} = q(x,t),\ -\infty < x < \infty,\ t > 0;\ w(x,0) = \frac{\partial w}{\partial t}(x,0) = 0. \tag{8.2.12}$$

We can immediately write the solution in terms of the fundamental solution (8.2.11) as

$$w(x,t) = \int_0^t d\tau \int_{-\infty}^{\infty} d\xi\ E(x,t;\xi,\tau)q(\xi,\tau).$$

Since $E = \frac{1}{2}$ for $x - (t-\tau) < \xi < x + (t-\tau)$ and vanishes elsewhere, we have

$$w(x,t) = \frac{1}{2}\int_0^t d\tau \int_{x-(t-\tau)}^{x+(t-\tau)} q(\xi,\tau)\,d\xi = \frac{1}{2}\iint\limits_D q(\xi,\tau)\,d\xi\,d\tau,$$

where D is the triangular domain in Figure 8.6.

The solution of the general problem (8.2.1) is therefore

$$u(x,t) = \frac{1}{2}\iint\limits_D q(\xi,\tau)\,d\xi\,d\tau + \frac{1}{2}f_1(x-t) + \frac{1}{2}f_1(x+t) + \frac{1}{2}\int_{x-t}^{x+t} f_2(\xi)\,d\xi. \tag{8.2.13}$$

It is clear that the solution depends continuously on the data $\{q, f_1, f_2\}$ in the sup-norm, but not in the norm involving the derivative. Note that the solution at (x,t) depends only on the data in $\bar{D}$, which is called the *domain of dependence* of (x,t).

Let us now turn to the semi-infinite string. If the left end $x = 0$ is kept fixed, the causal Green's function $g(x,t;\xi,\tau)$ satisfies

$$\begin{gathered}\frac{\partial^2 g}{\partial t^2} - \frac{\partial^2 g}{\partial x^2} = \delta(x-\xi)\delta(t-\tau), \quad 0 < x, \xi < \infty, \quad -\infty < t, \tau < \infty; \\ g|_{x=0} = 0; \quad g \equiv 0, \quad t < \tau.\end{gathered} \tag{8.2.14}$$

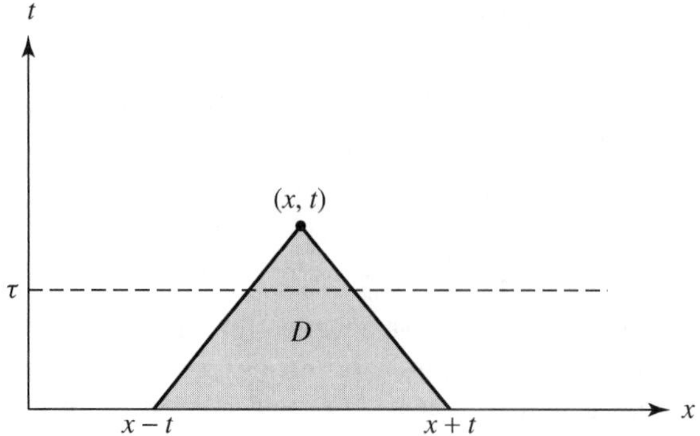

Figure 8.6

It obviously suffices to find $g(x, t; \xi, 0)$ and set $g(x, t; \xi, \tau) = g(x, t - \tau; \xi, 0)$. Of course we cannot handle the space part in the same way because the geometry spoils the invariance under translations in space. To find $g(x, t; \xi, 0)$ for $t > 0$ we may choose to solve instead the IVP

$$\begin{aligned} \frac{\partial^2 g}{\partial t^2} - \frac{\partial^2 g}{\partial x^2} &= 0, \quad 0 < x < \infty, \quad t > 0; \\ g|_{x=0} = 0, \quad g|_{t=0} = 0, \quad \left.\frac{\partial g}{\partial t}\right|_{t=0} &= \delta(x - \xi). \end{aligned} \tag{8.2.15}$$

Whether we deal with (8.2.14) or (8.2.15), the solution is most easily found by images as

$$g(x, t; \xi, \tau) = \tfrac{1}{2} H(t - \tau - |x - \xi|) - \tfrac{1}{2} H(t - \tau - |x + \xi|). \tag{8.2.16}$$

The problem

$$\begin{aligned} \frac{\partial^2 u}{\partial t^2} - \frac{\partial^2 u}{\partial x^2} &= q(x, t), \quad 0 < x < \infty, \quad t > 0 \\ u(x, 0) = f_1(x), \quad \frac{\partial u}{\partial t}(x, 0) &= f_2(x), \quad u(0, t) = 0 \end{aligned} \tag{8.2.17}$$

has the solution

$$\begin{aligned} u(x, t) = &\int_0^t d\tau \int_0^\infty d\xi \; g(x, t; \xi, \tau) q(\xi, \tau) + \int_0^\infty g(x, t; \xi, 0) f_2(\xi) \, d\xi \\ &+ \frac{d}{dt} \int_0^\infty g(x, t; \xi, 0) f_1(\xi) \, d\xi. \end{aligned}$$

Alternatively, we can extend $q(x, t), f_1(x), f_2(x)$ as *odd functions* for $x < 0$ and then use (8.2.13).

A more interesting feature of the semi-infinite string is the possibility of an inhomogeneous boundary condition at $x = 0$. Consider, for instance, the problem

$$\begin{aligned} \frac{\partial^2 v}{\partial t^2} - \frac{\partial^2 v}{\partial x^2} &= 0, \quad 0 < x < \infty, t > 0; \\ v(x,0) = \frac{\partial v}{\partial t}(x,0) &= 0, \quad v(0,t) = h(t). \end{aligned} \tag{8.2.18}$$

There are many interrelated methods for solving a problem of this type: reduction to a problem with a homogeneous boundary condition, Green's function, Duhamel's formula, Fourier sine transform on space, Laplace transform on time. We shall illustrate the last method and the method of characteristics.

First let us look at the method of characteristics. The solution for $x > t$ is determined from the Cauchy data on the noncharacteristic initial curve $t = 0, x > 0$, and therefore $v \equiv 0, x > t$. For $x < t$ we have a problem in which the value of v is given on the characteristic $x = t$ and on the noncharacteristic $x = 0$. Since the jump along a characteristic remains constant for this particular hyperbolic equation [see (2.5.54)], we have $v(x, x+) = h(0+)$. Therefore, $v(x,t) = v(0, t-x) = h(t-x), x < t$. Thus the solution of (8.2.18) is

$$v(x,t) = H(t-x)h(t-x). \tag{8.2.19}$$

Let us illustrate how to apply the Laplace transform to (8.2.18). We multiply both sides by e^{-st} and integrate from $t = 0$ to ∞. Integrating the first term by parts twice, and using the initial conditions and the fact that e^{-st} is exponentially small at $t = \infty$ for Re $s > 0$, we find that

$$-\frac{\partial^2 \tilde{v}}{\partial x^2} + s^2 \tilde{v} = 0, \quad 0 < x < \infty; \quad \tilde{v}(0,s) = \tilde{h}(s), \tag{8.2.20}$$

where

$$\tilde{v}(x,s) = \int_0^\infty e^{-st} v(x,t)\, dt, \quad \tilde{h}(s) = \int_0^\infty e^{-st} h(t)\, dt$$

are the Laplace transforms of v and h, respectively.

If we solve (8.2.20) under the requirement that

$$\lim_{x \to \infty} \tilde{v}(x,s) = 0, \quad \text{Re } s > 0,$$

we obtain

$$\tilde{v}(x,s) = \tilde{h}(s) e^{-sx}, \tag{8.2.21}$$

which we want to invert for fixed $x > 0$. The original of e^{-sx} is $\delta(t-x)$, so that by the convolution theorem for Laplace transforms [see (2.4.46)]

$$v(x,t) = \int_0^t \delta(t - \tau - x) h(\tau)\, d\tau = \begin{cases} h(t-x), & t > x, \\ 0, & t < x, \end{cases}$$

in agreement with (8.2.19). If this treatment using delta functions is deemed a bit cavalier, (8.2.21) can be written instead as

$$\tilde{v}(x,s) = [s\tilde{h} - h(0)]\frac{e^{-sx}}{s} + h(0)\frac{e^{-sx}}{s}.$$

The original of $s\tilde{h} - h(0)$ is $h'(t)$, and that of e^{-sx}/s is $H(t-x)$. Therefore,

$$v(x,t) = \int_0^t H(t-\tau-x)h'(\tau)\,d\tau + h(0)H(t-x), \qquad (8.2.22)$$

which is easily reduced to (8.2.19). Note that (8.2.22) is just the version of Duhamel's formula (7.3.23), appropriate for our problem since $H(t-x)$ is the solution of (8.2.18) with $h(t) \equiv 1$.

Parabolic Equations

The one-dimensional heat operator $\partial/\partial t - \partial^2/\partial x^2$ has characteristics $t =$ constant. Here these characteristics do not play the same significant role as they do for hyperbolic equations. A typical well-posed problem for heat conduction in an infinite rod is

$$\frac{\partial u}{\partial t} - \frac{\partial^2 u}{\partial x^2} = q(x,t), \quad -\infty < x < \infty, t > 0; \quad u(x,0) = f(x), \qquad (8.2.23)$$

where only the initial value for u is required since the equation is first order in time. The constant diffusivity that would normally multiply the term $\partial^2 u/\partial x^2$ has been absorbed in the time variable. The causal Green's function $g(x,t;0,0)$ satisfies

$$\frac{\partial g}{\partial t} - \frac{\partial^2 g}{\partial x^2} = \delta(x)\delta(t), \quad -\infty < x, t < \infty; \quad g \equiv 0, \quad t < 0 \qquad (8.2.24)$$

or, equivalently (for $t > 0$),

$$\frac{\partial g}{\partial t} - \frac{\partial^2 g}{\partial x^2} = 0, \quad -\infty < x < \infty, t > 0; \quad g|_{t=0} = \delta(x). \qquad (8.2.25)$$

In the first of these characterizations we view g as a distribution on $C_0^\infty(\mathbb{R}^2)$ with the properties

$$\phi(0,0) = \langle g, L^*\phi\rangle = \left\langle g, -\frac{\partial \phi}{\partial t} - \frac{\partial^2 \phi}{\partial x^2}\right\rangle \quad \text{for each } \phi(x,t) \in C_0^\infty(\mathbb{R}^2),$$

$$\langle g, \phi\rangle = 0 \quad \text{for each } \phi \in C_0^\infty(\mathbb{R}^2) \text{ with support in } t < 0.$$

In (8.2.25) we regard g as a one-dimensional distribution with t appearing as a parameter. Thus

$$\frac{\partial}{\partial t}\langle g, \phi\rangle - \left\langle g, \frac{\partial^2 \phi}{\partial x^2}\right\rangle = 0 \quad \text{for } t > 0 \text{ and each } \phi(x) \in C_0^\infty(\mathbb{R}),$$

$$\lim_{t\to 0+}\langle g, \phi\rangle = \phi(0) \quad \text{for each } \phi \in C_0^\infty(\mathbb{R}).$$

We saw in (2.5.51) that

$$g(x,t;0,0) = H(t)\frac{e^{-x^2/4t}}{\sqrt{4\pi t}}, \tag{8.2.26}$$

so that

$$g(x,t;\xi,\tau) = H(t-\tau)\frac{e^{-(x-\xi)^2/4(t-\tau)}}{\sqrt{4\pi(t-\tau)}}, \tag{8.2.27}$$

and the solution of (8.2.23) is

$$\begin{aligned} u(x,t) = &\int_0^t d\tau \int_{-\infty}^{\infty} d\xi \frac{e^{-(x-\xi)^2/4(t-\tau)}}{\sqrt{4\pi(t-\tau)}} q(\xi,\tau) \\ &+ \int_{-\infty}^{\infty} d\xi \frac{e^{-(x-\xi)^2/4t}}{\sqrt{4\pi t}} f(\xi). \end{aligned} \tag{8.2.28}$$

It is clear that u depends continuously on the data.

An unexpected consequence of (8.2.26) is that for $t > 0$, g is positive for all x: A disturbance initially concentrated at $x = 0$ is felt immediately afterward at every point on the real line! This infinite speed of propagation is a counterintuitive property of the model and is not attributable to the singular nature of the initial disturbance; indeed, even if $q = 0$ and f is a continuous function positive over a bounded interval and vanishing elsewhere, (8.2.28) shows that $u(x,t)$ is positive for all x when $t > 0$ (instantaneous loss of compact support). Returning to (8.2.26), we observe that g is in fact very small for large $|x|$ so that the expression may still be useful in most practical settings (see Day [8]). In Section 0.1 we introduced a modified model for heat conduction leading to a finite speed of propagation. In Chapter 9 we show that the standard heat conduction model, when combined with strong absorption, leads to a finite propagation speed.

In an IVP such as (8.2.23) we have not yet specified precisely how the initial values are to be taken on by the solution. A natural requirement would be

$$\lim_{t\to 0+} u(x,t) = f(x) \quad \text{for each } x, \quad -\infty < x < \infty, \tag{8.2.29}$$

but as we shall see, the use of a pointwise limit to obtain uniqueness is unsatisfactory both mathematically and physically. To see what goes wrong, let us look at the function

$$h(x,t) = \frac{x}{4\pi^{1/2}t^{3/2}} e^{-x^2/4t}, \quad t > 0, \tag{8.2.30}$$

which is just $-(\partial/\partial x)g(x,t;0,0)$. Since $g(x,t;0,0)$ satisfies the homogeneous heat equation for $t > 0$, so does any partial derivative of g (because the coefficients in the differential equation are constant). The function $h(x,t)$ therefore satisfies $\partial h/\partial t - \partial^2 h/\partial x^2 = 0, t > 0$; moreover,

$$\lim_{t\to 0+} h(x,t) = 0 \quad \text{for each } x, \quad \infty < x < \infty.$$

Of course, we expect the trivial solution to be the only solution of the homogeneous heat equation with vanishing initial data. The paradox is only apparent, for h really does not correspond to vanishing initial data. Let us see what $\lim_{t\to 0+} h(x,t)$ is in the distribution sense. We think of h as a distribution in x with parameter t so that

$$\langle h, \phi(x)\rangle = \int_{-\infty}^{\infty} h(x,t)\phi(x)\,dx, \quad t > 0.$$

Hence

$$\langle h, \phi\rangle = \int_{-\infty}^{\infty} -\frac{\partial g}{\partial x}\phi\,dx = \int_{-\infty}^{\infty} g\frac{d\phi}{dx}\,dx = \int_{-\infty}^{\infty} \frac{e^{-x^2/4t}}{\sqrt{4\pi t}}\frac{d\phi}{dx}\,dx$$

and

$$\lim_{t\to 0+}\langle h, \phi\rangle = \phi'(0),$$

which is not the zero distribution. In fact, $h(x,t)$ is the temperature corresponding to an initial unit dipole at $x = 0$.

Requirement (8.2.29) is not sufficient for a proper formulation of the IVP. There are two ways of remedying the situation. The first works adequately if $f(x)$ and $q(x,t)$ are continuous functions. We then look for a solution $u(x,t)$ of (8.2.23) for $t > 0$ which, together with the initial values $f(x)$, defines a continuous function in $t \geqslant 0$. Note that the function $h(x,t)$ for $t > 0$, together with the value $f = 0$ for $t = 0$, is not a continuous function in $t \geqslant 0$. In fact, if we approach the origin along the curve $x = 2\sqrt{t}$, h tends to $+\infty$ (and along $x = -2\sqrt{t}$ to $-\infty$).

Another formulation is more suitable when f or g may be discontinuous (or perhaps distributions). Then we require that the initial values be taken in the distributional sense:

$$\lim_{t\to 0+}\langle u, \phi\rangle = \langle f, \phi\rangle \quad \text{for each } \phi(x) \in C_0^\infty(\mathbb{R}). \tag{8.2.31}$$

The first formulation tells us only that $h(x,t)$ in (8.2.30) is not a solution which assumes zero initial values. The second formulation gives much more information: the initial value for $h(x,t)$ is $\delta'(x)$.

Let us turn next to some heat conduction problems for finite rods. Consider first a thin ring constructed by bending a rod of unit length so that its ends meet perfectly. If x measures position along the centerline of the ring, we may assume that the ring extends from $x = -\frac{1}{2}$ to $x = \frac{1}{2}$, these two values of x corresponding to the same *physical* point in the ring. If the ring is insulated, the flow of heat is essentially one-dimensional along the centerline of the ring. The novel feature is that the boundary conditions state that the temperature and any derivative of the temperature at $x = \frac{1}{2}$ must be equal to the corresponding quantity at $x = -\frac{1}{2}$. It actually suffices to require that the temperature and its first derivative have, respectively, the same value at $x = \pm\frac{1}{2}$; the relations for the higher derivatives then follow from successive differentiations of the differential equation.

The causal fundamental solution $E(x,t;\xi,\tau)$ for this ring is the temperature when a unit of heat is suddenly introduced at time τ at the point ξ when the ring was at 0 temperature up to time τ. It is clear that E depends only on $x - \xi$ and $t - \tau$, so that we need only to calculate $E(x,t) \doteq E(x,t;0,0)$ satisfying

$$\begin{gathered} \frac{\partial E}{\partial t} - \frac{\partial^2 E}{\partial x^2} = \delta(x)\delta(t), \quad -\frac{1}{2} < x < \frac{1}{2}, \quad -\infty < t < \infty; \\ E \equiv 0,\ t < 0;\ E\left(-\frac{1}{2}, t\right) = E\left(\frac{1}{2}, t\right), \ \frac{\partial E}{\partial x}\left(-\frac{1}{2}, t\right) = \frac{\partial E}{\partial x}\left(\frac{1}{2}, t\right). \end{gathered} \tag{8.2.32}$$

The solution is most easily obtained by images. Consider a fictitious infinite straight rod with positive unit sources at all integers. The resulting temperature is cearly periodic in x with period 1, so that the boundary conditions in (8.2.32) are satisfied; moreover, the part of the rod between $-\frac{1}{2}$ and $\frac{1}{2}$ contains only the single source at $x = 0$. If we believe that the solution of (8.2.32) is unique, it must coincide with the temperature in the infinite rod between $-\frac{1}{2}$ and $\frac{1}{2}$. The temperature in the infinite rod is the *theta function*

$$\theta(x,t) \doteq \frac{1}{\sqrt{4\pi t}} \sum_{n=-\infty}^{\infty} e^{-(x-n)^2/4t}, \quad t > 0, \tag{8.2.33}$$

and therefore $E(x,t) = \theta(x,t)$, $-\frac{1}{2} < x < \frac{1}{2}$, $t > 0$. Formula (8.2.33) is useful for small t, and only the term corresponding to $n = 0$ contributes appreciably to the temperature in the ring. Physically, the interpretation is that the finiteness of the ring is not felt for small t, and it is as if we were dealing with a source in an infinite rod.

Another way of determining E is by an expansion in a series of spatial eigenfunctions $e^{i2n\pi x}$. We multiply (8.2.32) by $e^{-i2n\pi x}$ and integrate from $x = -\frac{1}{2}$ to $x = \frac{1}{2}$ to obtain

$$\frac{\partial E_n}{\partial t} + 4n^2\pi^2 E_n = \delta(t); \qquad E_n = 0, \quad t < 0, \tag{8.2.34}$$

where

$$E(x,t) = \sum_{n=-\infty}^{\infty} E_n(t) e^{i2n\pi x}, \quad E_n(t) = \int_{-1/2}^{1/2} E(x,t) e^{-i2n\pi x}\, dx.$$

From (8.2.34) we find that $E_n = e^{-4n^2\pi^2 t}$ and

$$E = \sum_{n=-\infty}^{\infty} e^{-4n^2\pi^2 t} e^{i2n\pi x} = 1 + 2\sum_{n=1}^{\infty} e^{-4n^2\pi^2 t} \cos 2n\pi x. \tag{8.2.35}$$

The cosine series in (8.2.35) converges rapidly for large t since the ratio of two successive terms is exponentially small. It is possible to pass from (8.2.33) to (8.2.35) directly by using the Poisson summation formula (2.3.26). Note that $E \to 1$ as $t \to \infty$ so that the temperature is uniformly distributed in the ring. This is, of course, the expected steady state for the insulated ring.

We shall now use the results for the ring to deal with the finite rod $0 < x < l$ with vanishing end temperatures. The corresponding causal Green's function $g(x,t;\xi,0)$

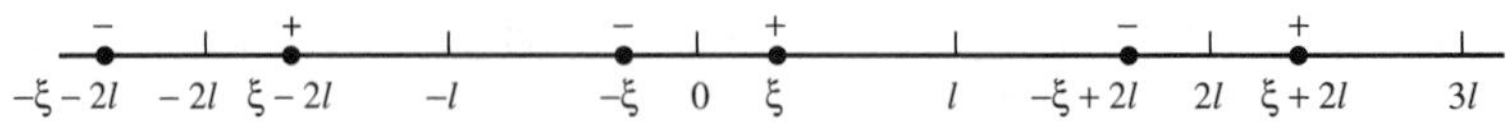

− + − + − +

$-\xi-2l$ $-2l$ $\xi-2l$ $-l$ $-\xi$ 0 ξ l $-\xi+2l$ $2l$ $\xi+2l$ $3l$

Figure 8.7

satisfies

$$\begin{aligned}&\frac{\partial g}{\partial t}-\frac{\partial^2 g}{\partial x^2}=\delta(x-\xi)\delta(t), \quad 0<x,\xi<l, \quad \infty<t<\infty;\\&g\equiv 0, \quad t<0; \quad g|_{x=0}=g|_{x=l}=0, \quad t>0.\end{aligned} \tag{8.2.36}$$

Consider an infinite rod with the array of unit sources and sinks shown in Figure 8.7. We have placed a unit source at each of the points $\xi+2nl$ and a unit sink at each of the points $-\xi+2nl$, where n is an arbitrary integer ranging from $-\infty$ to ∞. The resulting temperature is an odd function about $x=0$ and about $x=l$, so that the temperatures at these points must vanish for $t>0$. Since there is only a unit source in the part of the rod between $x=0$ and $x=l$, the temperature for the infinite rod will satisfy in $0<x<l$ the BVP (8.2.36). We have, explicitly,

$$\begin{aligned}g(x,t;\xi,0)&=\sum_{n=-\infty}^{\infty}\frac{1}{\sqrt{4\pi t}}[e^{-(x-\xi-2nl)^2/4t}-e^{-(x+\xi-2nl)^2/4t}]\\&=\frac{1}{2l}\left[\theta\left(\frac{x-\xi}{2l},\frac{t}{4l^2}\right)-\theta\left(\frac{x+\xi}{2l},\frac{t}{4l^2}\right)\right],\end{aligned}$$

where θ is the theta function (8.2.33).

Either by using relation (8.2.35) for theta functions or by expanding (8.2.36) directly in the space eigenfunctions $\sin(n\pi x/l)$, we find the alternative expression

$$g(x,t;\xi,0)=\frac{2}{l}\sum_{n=1}^{\infty}\sin\frac{n\pi x}{l}\sin\frac{n\pi\xi}{l}e^{-n^2\pi^2t/l^2}. \tag{8.2.36a}$$

Using Green's function $g(x,t;\xi,\tau)=g(x,t-\tau;\xi,0)$, it is easy to solve the initial-boundary value problem

$$\frac{\partial u}{\partial t}-\frac{\partial^2 u}{\partial x^2}=q(x,t), \quad 0<x<l,\ t>0; \quad u(x,0)=f(x),\ u(0,t)=u(l,t)=0,$$

in the form

$$u(x,t)=\int_0^t d\tau\int_0^l d\xi\ g(x,t;\xi,\tau)q(\xi,\tau)+\int_0^l g(x,t;\xi,0)f(\xi)\,d\xi, \tag{8.2.37}$$

which invites comparison with (8.2.28).

Continuous dependence on the data is an immediate consequence of (8.2.37). For the problem where the boundary values depend on t, one can use either a Green's

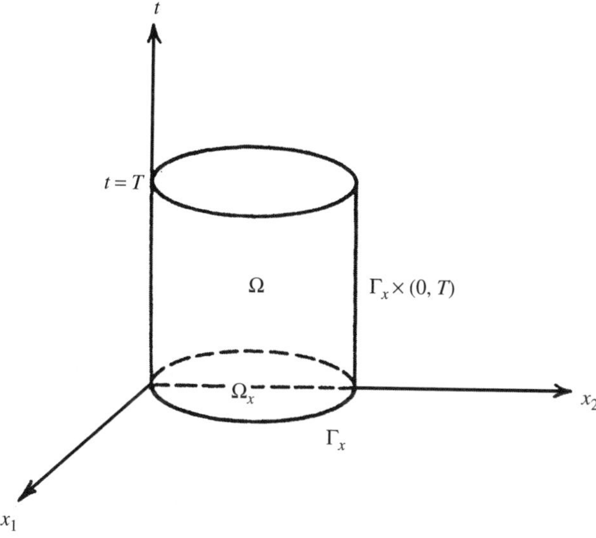

Figure 8.8

function method or one of the many techniques mentioned in the discussion of the wave equation. Again, continuity with respect to data follows from the explicit expression for the solution.

We turn now to questions of uniqueness and continuous dependence on the data when simple explicit forms for the solution are not available (say, for problems with more than one space dimension). A typical well-posed problem for the heat equation is a combined initial value and boundary value problem. Let Ω_x be a *bounded* domain in the space variables $x = (x_1, \ldots, x_n)$, and let its boundary be Γ_x. We shall want to solve the heat equation with given source density $q(x,t)$ in the cylindrical space-time domain Ω, the Cartesian product of Ω_x and the segment of the time axis $0 < t < T$. (See Figure 8.8.) Here T is a large value of time used for convenience instead of $t = \infty$. The boundary Γ of Ω consists of three portions: the bases $(x \in \Omega_x, t = 0)$ and $(x \in \Omega_x, t = T)$, and the lateral surface $(x \in \Gamma_x, 0 < t < T)$. Initial data is given on the lower base, and boundary data on the lateral surface. We shall denote the union of the lower base and the lateral surface by Γ_d, known as the *parabolic boundary*. We assume that the data is continuous on Γ_d. Our BVP can be formulated as follows: to find $u(x,t)$ continuous in $\bar{\Omega}$, satisfying

$$\begin{gathered}\frac{\partial u}{\partial t} - \Delta u = q(x,t), \quad (x,t) \in \Omega; \qquad u(x,0) = f(x), \quad x \in \Omega_x; \\ u(x,t) = h(x,t), \quad x \in \Gamma_x, \quad 0 < t < T\end{gathered} \tag{8.2.38}$$

(these two conditions mean that u is given on Γ_d). We are, of course, taking all data to be real-valued.

As always, the question of existence is delicate, and we shall concentrate on uniqueness and continuous dependence on the data. If (8.2.38) has two solutions u_1 and u_2, their difference must satisfy the completely homogeneous problem

$$\frac{\partial u}{\partial t} - \Delta u = 0, \quad (x,t) \in \Omega; \qquad u = 0 \text{ on } \Gamma_d. \tag{8.2.39}$$

We now show that the only solution of (8.2.39) is the zero solution, so that (8.2.38) has at most one solution (uniqueness). Multiply the differential equation in (8.2.39) by $u(x,t)$, and integrate over the space domain Ω_x to obtain

$$0 = \int_{\Omega_x} \left(u\frac{\partial u}{\partial t} - u\,\Delta u \right) dx = \frac{1}{2}\frac{d}{dt}\int_{\Omega_x} |u|^2\,dx + \int_{\Omega_x} |\text{grad } u|^2\,dx - \int_{\Gamma_x} u\frac{\partial u}{\partial \nu}\,dS.$$

Since u vanishes on Γ_x for $t < T$, we find that

$$0 = \frac{1}{2}\frac{d}{dt}\int_{\Omega_x} |u|^2\,dx + \int_{\Omega_x} |\text{grad } u|^2\,dx \geqslant \frac{1}{2}\frac{d}{dt}\int_{\Omega_x} |u|^2\,dx. \tag{8.2.40}$$

The initial condition implies that $\int_{\Omega_x} |u|^2\,dx = 0$ at $t = 0$, and therefore (8.2.40) shows that $\int_{\Omega_x} |u|^2\,dx = 0$ for all $t < T$. The continuity of u then yields $u(x,t) \equiv 0$, $(x,t) \in \bar{\Omega}$.

A more powerful method of proving uniqueness (and also continuous dependence on data) utilizes the *maximum principle* for solutions of the homogeneous heat equation $(\partial u/\partial t) - \Delta u = 0$. Since any solution of this homogeneous equation also obeys a *minimum principle* of equal importance, it is somewhat misleading to give special status to the maximum principle, but we bow to tradition. When we deal with differential inequalities, however, the distinction between principles is essential.

Theorem 8.2.1 (Maximum Principle). *Let $u(x,t)$ be continuous in $\bar{\Omega}$ and satisfy the differential inequality*

$$\frac{\partial u}{\partial t} - \Delta u \leqslant 0, \quad (x,t) \in \Omega. \tag{8.2.41}$$

If $u \leqslant M$ on Γ_d, then $u(x,t) \leqslant M$ on $\bar{\Omega}$.

The physical interpretation is simple. Inequality (8.2.41) tells us that no heat is being added in the interior. It therefore follows that the interior temperature must be less than the maximum of the temperature on the parabolic boundary.

Corollary (Minimum Principle). *Suppose that u is continuous in $\bar{\Omega}$ and*

$$\frac{\partial u}{\partial t} - \Delta u \geqslant 0, \quad (x,t) \in \Omega.$$

If $u \geqslant m$ on Γ_d, then $u(x,t) \geqslant m$ in $\bar{\Omega}$.

The proof of the corollary follows by setting $u = -v$ and applying Theorem 8.2.1. To prove Theorem 8.2.1 we consider first the simpler case where there is a sink at every point in Ω [strict inequality in (8.2.41)].

Lemma. *Let v be continuous in $\bar{\Omega}$ and satisfy*

$$\frac{\partial v}{\partial t} - \Delta v < 0 \textit{ in } \Omega; \tag{8.2.42}$$

then the maximum of v occurs on Γ_d.

Proof. (Lemma) A maximum for v occurring at an interior point (x, t) would imply that $\Delta v \leqslant 0$ and $\partial v/\partial t = 0$ at that point, which contradicts (8.2.42). If the maximum occurred at a point in Ω_x at time T, we would have $\Delta v \leqslant 0$ and $\partial v/\partial t \geqslant 0$, which again contradicts (8.2.42). □

Proof. (Theorem 8.2.1) Since Ω_x is bounded, it can be enclosed in an n-ball of radius a with center at the origin. Let $u(x,t)$ satisfy the hypothesis in Theorem 8.2.1, and define $v(x,t) = u(x,t) + \varepsilon|x|^2$. Then

$$\frac{\partial v}{\partial t} - \Delta v \leqslant -2n\varepsilon < 0,$$

so that v satisfies the inequality of the lemma. Thus

$$u(x,t) \leqslant v(x,t) \leqslant \max_{(x,t)\in\Gamma_d} v \leqslant [\varepsilon a^2 + \max_{(x,t)\in\Gamma_d} u] \leqslant \varepsilon a^2 + M,$$

and, by letting $\varepsilon \to 0$, we find that $u \leqslant M$. □

For the *homogeneous* heat equation, both Theorem 8.2.1 and its corollary apply, so that we have the following theorem.

Theorem 8.2.2. *Let $u(x,t)$ be a continuous function on $\bar{\Omega}$ satisfying the equation $(\partial u/\partial t) - \Delta u = 0$. If, on Γ_d, we have $m \leqslant u \leqslant M$, then, everywhere in Ω, $m \leqslant u(x,t) \leqslant M$.*

REMARK. Our proof does not exclude the possibility that at an interior point the temperature may equal the maximum of the initial and the boundary temperature. If, for instance, $f(x) = 1$ and $h(x,t) = 1$, $0 < t < t_0$, $h(x,t) < 1$, $t > t_0$, then $u(x,t) \equiv 1$ for $t < t_0$ but starts decreasing in the interior thereafter. The only way that we can have an interior maximum or minimum is by having constant initial and boundary temperature's up to the time at which the interior extremum occurs.

There are many important consequences of the maximum and minimum principles. First let us consider the inhomogeneous problem (8.2.38) for two different sets of data, $\{q_1, f_1, h_1\}$ and $\{q_2, f_2, h_2\}$. Assume that the second set *dominates* the first; that is, $q_2 \geqslant q_1, f_2 \geqslant f_1, h_2 \geqslant h_1$. Physically, this means that the sources and boundary data in the second set are larger than those in the first set, so that the solution should also be larger.

Theorem 8.2.3 (Comparison Theorem). *If the data $\{q_2, f_2, h_2\}$ dominates the data $\{q_1, f_1, h_1\}$, then $u_2(x,t) \geqslant u_1(x,t)$.*

Proof. The function $v = u_1 - u_2$ satisfies $(\partial v/\partial t) - \Delta v \leqslant 0$ in $\Omega, v \leqslant 0$ on Γ_d. By Theorem 8.2.1, $v \leqslant 0$ in $\bar{\Omega}$. □

Theorem 8.2.4 (Uniqueness). *Problem (8.2.38) has at most one solution.*

Proof. Let u_1 and u_2 satisfy (8.2.38) for the *same* data. Then $\{q_2, f_2, h_2\}$ dominates $\{q_1, f_1, h_1\}$, and vice versa. Therefore, by Theorem 8.2.3, $u_1 \leqslant u_2$ and $u_2 \leqslant u_1$. □

Theorem 8.2.5 (Continuous Dependence on Data). *Let u_1 and u_2 be solutions of (8.2.38) corresponding, respectively, to data $\{q_1, f_1, h_1\}$ and $\{q_2, f_2, h_2\}$. Let*

$$|q_1 - q_2| \leqslant \alpha, \quad (x,t) \in \bar{\Omega},$$
$$|u_1 - u_2| \leqslant \beta, \quad (x,t) \in \Gamma_d;$$

then

$$|u_1(x,t) - u_2(x,t)| \leqslant \alpha T + \beta, \quad (x,t) \in \bar{\Omega}.$$

Proof. The function $u \doteq u_1 - u_2$ satisfies the heat equation with data lying between $\{-\alpha, -\beta, -\beta\}$ and $\{\alpha, \beta, \beta\}$. Write $u = v + w$, where v has a vanishing inhomogeneous term and w has vanishing initial and boundary data. Then, by Theorem 8.2.2, $|v| \leqslant \beta$. The functions $w_1 = \alpha t, w_2 = -\alpha t$ satisfy the heat equation with inhomogeneous terms $+\alpha$ and $-\alpha$, respectively; on $\Gamma_d, w_1 \geqslant 0$ and $w_2 \leqslant 0$. Thus the data for w_1 dominates that for w; and the data for w dominates that for w_2. From Theorem 8.2.3 it follows that $w_2 \leqslant w \leqslant w_1$ and hence $-\alpha T \leqslant w \leqslant \alpha T$. Together with $|v| \leqslant \beta$, this gives the required result. □

It is possible to extend these results with some modifications to unbounded domains and to more general parabolic equations (see Exercise 8.2.2, for instance).

EXERCISES

8.2.1 The concentration $C(x,t)$ of a pollutant subject to diffusion and convection along the x axis satisfies

$$\frac{\partial C}{\partial t} + U(t)\frac{\partial C}{\partial x} - D\frac{\partial^2 C}{\partial x^2} = 0, \tag{8.2.43}$$

where D is the constant diffusivity and $U(t)$ is the wind velocity in the positive x direction. Both D and U are regarded as known. Make a transformation to an appropriate moving coordinate system $t' = t, x' = x - V(t)$ to reduce (8.2.43) to

$$\frac{\partial C}{\partial t'} - D\frac{\partial^2 C}{\partial x'^2} = 0.$$

Find the free space causal fundamental solution for (8.2.43) with source at $x = 0, t = 0$. The curves $x - V(t) =$ constant are called *drift curves*. Let Γ

be the drift curve through the origin. Solve the initial-boundary value problem for (8.2.43) with $C(x,0)$ given for $x > 0$ and C given on Γ for $t > 0$.

8.2.2 Let Ω be defined as in Figure 8.8. Suppose that u is continuous in $\bar{\Omega}$ and satisfies

$$\frac{\partial u}{\partial t} - \Delta u + cu \leqslant 0, \quad (x,t) \in \Omega; \quad c(x) \geqslant 0.$$

Show that if on Γ_d, $u \leqslant M$, where $M \geqslant 0$, then $u \leqslant M$ in $\bar{\Omega}$. What is the corresponding principle when the sign is reversed in the differential inequality? What is the appropriate principle if u satisfies

$$\frac{\partial u}{\partial t} - \Delta u + cu = 0?$$

8.2.3 For *c constant*, solve by separation of variables

$$\frac{\partial u}{\partial t} - \Delta u + cu = 0, \quad (x,t) \in \Omega; \quad u(x,0) = f(x); \quad u|_{\Gamma_x} = 0, \quad t > 0.$$

When can you guarantee that the solution tends to the steady state $u \equiv 0$ as $t \to \infty$? (See Section 8.3.)

8.2.4 Consider the wave equation on the domain of Figure 8.8:

$$\left.\begin{aligned} &\frac{\partial^2 u}{\partial t^2} - \Delta u = q(x,t), && (x,t) \in \Omega, \\ &u(x,0) = f(x), \frac{\partial u}{\partial t}(x,0) = g(x), && x \in \Omega_x, \\ &u(x,t) = h(x,t), && x \in \Gamma_x, 0 < t < T. \end{aligned}\right\} \tag{8.2.44}$$

To prove uniqueness, consider the completely homogeneous problem satisfied by the difference v of two solutions u_1 and u_2 of (8.2.44). Multiply the homogeneous differential equation by $\partial v/\partial t$, and integrate over Ω_x to show that $v \equiv 0$ in Ω.

8.2.5 Use the Poisson summation formula of Chapter 2 to show the equality of (8.2.33) and (8.2.35).

8.2.6 Consider the heat conduction problem for a semi-infinite rod:

$$\frac{\partial w}{\partial t} - \frac{\partial^2 w}{\partial x^2} = 0, \quad 0 < x < \infty, \quad t > 0; \quad w(x,0) = 0, \quad w(0,t) = 1. \tag{8.2.45}$$

[We shall see in Exercise 8.2.7 that $w(x,t)$ enables us to solve the problem in which an arbitrary temperature is imposed on the left-hand end.] We now solve (8.2.45) in three ways:

(a) Write $w = 1 - z(x,t)$, and find $z(x,t)$ by using Green's function (obtained by images) for the semi-infinite rod with vanishing temperature at $x = 0$.

(b) Use a Laplace transform on the time variable to show that $\tilde{w}(x, s) = e^{-x\sqrt{s}}/s$, and invert.

(c) Use the method of similarity solution: Show first that the equation and boundary conditions in (8.2.45) are invariant under the transformation $x^* = \alpha x$, $t^* = \alpha^2 t$, so that $w(\alpha x, \alpha^2 t) = w(x, t)$ for all $\alpha > 0$. With x, t fixed, choose $\alpha = 1/2\sqrt{t}$ to show that $w(x, t) = w\left(x/2\sqrt{t}, \frac{1}{4}\right)$. Thus all points in the x–t plane with the same value of $x/2\sqrt{t}$ have the same temperature; hence

$$w(x, t) = h\left(\frac{x}{2\sqrt{t}}\right).$$

Substitution in (8.2.45) shows that $h(\xi)$ satisfies the ordinary differential equation

$$h'' + 2\xi h' = 0$$

with the boundary conditions $h(0) = 1, h(\infty) = 0$, whose solution is

$$h = 1 - \operatorname{erf} \xi, \quad \operatorname{erf} \xi \doteq \frac{2}{\sqrt{\pi}} \int_0^\xi e^{-\eta^2}\, d\eta.$$

This proves that

$$w(x, t) = 1 - \operatorname{erf} \frac{x}{2\sqrt{t}}, \tag{8.2.46}$$

as was shown by still another method in (7.3.22).

8.2.7 Consider the heat conduction problem

$$\frac{\partial u}{\partial t} - \frac{\partial^2 u}{\partial x^2} = 0, \quad 0 < x < \infty, \quad t > 0; \quad u(x, 0) = 0, \quad u(0, t) = h(t).$$

Take a Laplace transform on time, and use the convolution theorem and Exercise 8.2.6 to show that

$$u(x, t) = \int_0^t w(x, t - \tau) h'(\tau)\, d\tau + h(0)\, w(x, t),$$

which can be reduced to

$$u(x, t) = -\int_0^t \frac{\partial w}{\partial \tau}(x, t - \tau) h(\tau)\, d\tau.$$

8.2.8 (a) Find a similarity solution (see Exercise 8.2.6) to the problem

$$\frac{\partial u}{\partial t} - \frac{\partial^2 u}{\partial x^2} = 0, \quad -\infty < x < \infty, \quad t > 0; \qquad u(x, 0) = H(x).$$

(b) Find Green's function $g(x, t; 0, 0)$ for an infinite rod by noting that its initial value is dH/dx.

8.2.9 Find a similarity solution to the nonlinear heat conduction problem

$$\frac{\partial u}{\partial t} - \frac{\partial}{\partial x}\left(k(u)\frac{\partial u}{\partial x}\right) = 0, \quad 0 < x < \infty, t > 0; \;\; u(x,0) = 0, \;\; u(0,t) = 1.$$

8.2.10 Consider the wave equation in three space dimensions. In analogy with (8.2.8), the causal fundamental solution $E(x,t)$ with pole at $x = 0, t = 0$, satisfies

$$\frac{\partial^2 E}{\partial t^2} - \Delta E = 0, \quad x \in \mathbb{R}^3, \; t > 0; \; E(x,0) = 0, \; \frac{\partial E}{\partial t}(x,0) = \delta(x). \tag{8.2.47}$$

Regarding E as a three-dimensional distribution with t a parameter, show that

$$E = \frac{\delta(t - |x|)}{4\pi|x|}$$

is a solution of (8.2.47). This means that you must prove that for each $\varphi \in C_0^\infty(\mathbb{R}^3)$,

$$\frac{\partial^2}{\partial t^2}\langle E, \varphi\rangle = \langle E, \Delta\varphi\rangle \quad \text{for } t > 0,$$

$$\lim_{t\to 0+}\langle E, \varphi\rangle = 0, \quad \lim_{t\to 0+}\frac{\partial}{\partial t}\langle E, \varphi\rangle = \varphi(0).$$

8.2.11 *Hadamard's method of descent for the wave equation.* A two-dimensional point source at (x_1, x_2) in the plane can also be regarded as a line of sources along the x_3 axis in three dimensions. Each element of this line produces a response (see Exercise 8.2.10)

$$\frac{\delta(t - r)}{4\pi r}\, dx_3 \qquad [r = (x_1^2 + x_2^2 + x_3^2)^{1/2}],$$

so that the total field can be calculated at $(x_1, x_2, 0)$ and identified with the causal fundamental solution E for the plane. Show in this way that

$$E = \begin{cases} \dfrac{1}{2\pi}\dfrac{1}{\sqrt{t^2 - \rho^2}}, & t > \rho, \\ 0, & t < \rho, \end{cases}$$

where $\rho = (x_1^2 + x_2^2)^{1/2}$.

8.3 ELLIPTIC EQUATIONS

Introduction

Since elliptic equations have no characteristics, it might seem at first that they are ideally suited to the Cauchy problem. However, we saw in Exercise 8.1.1 that this is *not* the case. In fact, one obtains existence and uniqueness only in the small and in situations when everything in sight is analytic; morever, there is no continuous dependence on the data. It turns out that the pure boundary value problem is well-posed for elliptic equations (a certain amount of preliminary grooming may be necessary). For instance, if Ω is a bounded domain in $\mathbb{R}^n$, the BVP

$$-\Delta u = q(x), \quad x \in \Omega; \qquad u = f, \quad x \in \Gamma \tag{8.3.1a}$$

is well-posed if appropriate conditions (much less severe than analyticity) are placed on the nature of the boundary Γ and the data $\{q(x); f(x)\}$; its solution can be obtained by superposition of the solutions of the problems with data $\{q; 0\}$ and $\{0; f\}$. If $q = 0$, the equation is Laplace's equation, whereas if $q \neq 0$ it is known as Poisson's equation. We refer to (8.3.1a) as a Dirichlet problem because u is given on Γ; if, instead, the normal derivative is given, we have a Neumann problem (see Exercise 8.3.6).

We shall now consider (8.3.1a) for continuous data: f is continuous on Γ and q is continuous on $\bar{\Omega}$. It seems natural (see, however, Exercise 8.3.8, for some troublesome exceptions) to seek a solution u in $C^2(\bar{\Omega})$. There are then two other, equivalent, formulations for the BVP (8.3.1a): the weak formulation (also known as the *variational equation*) and the Dirichlet principle (principle of minimum energy). Such formulations were introduced in Section 0.5, for a one-dimensional problem. In our case we find it convenient to first define the sets of functions R and M:

$$\begin{aligned} &w \in R \quad \text{if } w \in C^2(\bar{\Omega}) \qquad \text{and} \qquad w = f \quad \text{on } \Gamma, \\ &v \in M \quad \text{if } v \in C^2(\bar{\Omega}) \qquad \text{and} \qquad v = 0 \quad \text{on } \Gamma. \end{aligned}$$

The BVP (8.3.1a) then seeks a function $u \in R$ such that $-\Delta u = q$ in Ω.

The *weak form* of (8.3.1a) is: Find $u \in R$ such that

$$\int_\Omega \operatorname{grad} u \cdot \operatorname{grad} v \, dx = \int_\Omega qv \, dx \quad \text{for every } v \in M. \tag{8.3.1b}$$

The *Dirichlet principle* states: Find $u \in R$ such that for every $w \in R$,

$$J(u) \leqslant J(w), \quad \text{where } J(w) \doteq \int_\Omega (|\operatorname{grad} w|^2 - 2qw) \, dx. \tag{8.3.1c}$$

If such a u exists, it is called a *minimizer*.

The equivalence of these three formulations is shown in Exercise 8.3.9 by arguments similar to those in Section 0.5. Of course, this equivalence does not prove existence of a solution. Indeed, there are cases for Laplace's equation with f continuous when there is no solution to the BVP or when the solution is not in $C^2(\bar{\Omega})$. We now give three such examples.

Examples.

1. Exercise 8.3.8 shows that for the unit disk (as simple a domain in $\mathbb{R}^2$ as one could want) it is possible for $|\operatorname{grad} u|$ to tend to infinity as x approaches the circular boundary. Thus, the solution is in $C^2(\Omega) \cap C(\bar{\Omega})$ but is not in $C^2(\bar{\Omega})$.

2. Let Ω be the punctured unit disk $0 < |x| < 1$; then Γ consists of the unit circle together with the isolated point $x = 0$. The Dirichlet problem for Laplace's equation in Ω with $f = 0$ on the unit circle $|x| = 1$ and $f = 1$ at $x = 0$ has no solution.

3. If Ω is a three-dimensional domain with a very sharp inner spike, Laplace's equation in Ω may not have a solution for some continuous functions f. An extreme case of this kind occurs when Ω consists of the open unit ball from which a radial line has been removed.

For a physical interpretation of the nonexistence phenomenon, consider the capacity problem [see (8.4.95)] when the charged conductor Γ is the unit sphere with a protruding sharp spike (an inner spike for the domain Ω_e exterior to Γ). The nonexistence of a solution can be interpreted as charge leaking out through the spike.

To ensure existence, restrictions need to be made on Γ. Let Ω be a bounded domain with boundary Γ. The *exterior sphere condition* is satisfied at $x_0 \in \Gamma$ if there exists an open ball B lying outside Ω whose boundary ∂B touches Γ at x_0 and only at x_0. If such a condition holds at every point of Γ, we say that Γ satisfies the exterior sphere condition; then the Dirichlet problem for Laplace's equation with f continuous on Γ has a solution in $C^2(\Omega) \cap C(\bar{\Omega})$. For further discussion of sufficient conditions on Γ, f and q for existence of a solution to (8.3.1a), see, for instance, DiBenedetto [10]. These questions can also be studied from the point of view of Sobolev spaces (see Section 8.4). In the remainder of this section we confine ourselves to sufficiently well-behaved Γ, f, and q.

Interior Dirichlet Problem for the Unit Disk

A function which satisfies Laplace's equation $\Delta u = 0$ in a domain Ω is said to be *harmonic* in Ω. Let Ω be the unit disk; we wish to construct a function u harmonic in Ω and continuous in $\bar{\Omega}$, which takes on prescribed continuous boundary values on the unit circle Γ.

For purposes of calculations it will be advantageous to introduce in Ω polar coordinates (r, ϕ) with $0 \leqslant r < 1, -\pi \leqslant \phi \leqslant \pi$. Before proceeding, we should pause to reflect on the implications of using this coordinate system. The coordinate endpoints $r = 0, \phi = \pi, \phi = -\pi$ do *not* correspond to any physical boundaries; $r = 0$ is the center of the disk, whereas $\phi = \pi$ and $\phi = -\pi$ both represent the same radial line, a line which has no special distinction in the original formulation of the problem and which should ultimately be returned to the anonymity it deserves. We must therefore make sure that any proposed solution $u(r, \phi)$ behaves properly at $r = 0$ and on the radial line whose equation is both $\phi = \pi$ and $\phi = -\pi$. With this warning we now express the Laplacian in polar coordinates, so that the problem has the form

$$0 = \Delta u = \frac{1}{r}\frac{\partial}{\partial r}\left(r\frac{\partial u}{\partial r}\right) + \frac{1}{r^2}\frac{\partial^2 u}{\partial \phi^2}, \quad (r, \phi) \in \Omega; \quad u(1, \phi) = f(\phi), \tag{8.3.2}$$

where $f(\phi)$ is continuous, $-\pi \leqslant \phi \leqslant \pi$, and $f(-\pi) = f(\pi)$.

The simplest way of solving (8.3.2) is by separation of variables. We shall construct u from basic building blocks u^*, which are harmonic functions of the special form $R(r)\Phi(\phi)$. Substitution in $\Delta u = 0$ yields, after division by $R\Phi$,

$$\frac{r}{R}\frac{d}{dr}\left(r\frac{dR}{dr}\right) = -\frac{1}{\Phi}\frac{d^2\Phi}{d\phi^2}.$$

Each of the two sides depends on a single coordinate. Since these coordinates are independent, the two sides must be equal to the same constant, labeled λ and known as a *separation constant*. We are thus led to the two ordinary differential equations

$$-\frac{d^2\Phi}{d\phi^2} = \lambda\Phi, \quad -\pi < \phi < \pi, \tag{8.3.3}$$

$$r\frac{d}{dr}\left(r\frac{dR}{dr}\right) = \lambda R, \quad 0 < r < 1. \tag{8.3.4}$$

For $u^* = R\Phi$ to be harmonic in Ω, its second derivatives must exist, so that both u^* and grad u^* certainly need to be continuous. To ensure the continuity of u^* and grad u^* on the radial line whose equation is both $\phi = \pi$ and $\phi = -\pi$, we must associate with (8.3.3) the so-called *periodic boundary conditions*

$$\Phi(-\pi) = \Phi(\pi), \quad \Phi'(-\pi) = \Phi'(\pi). \tag{8.3.5}$$

For the radial equation (8.3.4) we require that $R(0)$ be finite (or, equivalently, that $\lim_{r\to 0} rR' = 0$, which tells us that there is no source at $r = 0$). Equation (8.3.3) subject to (8.3.5) is an eigenvalue problem with eigenvalues $n^2, n = 0, 1, 2, \ldots$, and eigenfunctions $\Phi_n = C\sin n\phi + D\cos n\phi$ (or $Ae^{in\phi} + Be^{-in\phi}$) for $n > 0$ and $\Phi_0 = A$. For our purposes it is slightly more convenient to let n also assume negative values and to regard $e^{in\phi}$ as the single eigenfunction corresponding to n.

Setting $\lambda = n^2$, we obtain from (8.3.4)

$$R_n(r) = Cr^n + Dr^{-n}, \quad n \neq 0,$$
$$R_0(r) = E + F\log r,$$

where we set $F = 0$ and either D or $C = 0$ according to whether $n > 0$ or $n < 0$. This guarantees that R is bounded near the origin. In all cases we can write

$$R_n(r) = Cr^{|n|}.$$

For any integer n the function $u_n^*(r,\phi) = r^{|n|}e^{in\phi}$ is harmonic in the entire plane, and so is any finite linear combination of such functions. To satisfy the boundary condition at $r = 1$ it will be necessary, however, to consider an infinite linear combination

$$u(r,\phi) = \sum_{n=-\infty}^{\infty} a_n r^{|n|} e^{in\phi}, \quad r < 1, \tag{8.3.6}$$

which must reduce to $f(\phi)$ at $r = 1$, so that

$$f(\phi) = \sum_{n=-\infty}^{\infty} a_n e^{in\phi} \quad \text{and} \quad a_n = \frac{1}{2\pi}\int_{-\pi}^{\pi} f(\phi)e^{-in\phi}\,d\phi. \tag{8.3.7}$$

With these values for a_n substituted in (8.3.6) we presumably have the solution of (8.3.2). Rather than verifying this directly, we first transform (8.3.6) into an expression which exhibits more clearly the dependence of u on the boundary data. From (8.3.6) and the expression for a_n, we have

$$u(r,\phi) = \int_{-\pi}^{\pi}\left[\frac{1}{2\pi}\sum_{n=-\infty}^{\infty} r^{|n|}e^{in(\phi-\psi)}\right] f(\psi)\,d\psi, \quad r < 1.$$

Now, for any complex z with $|z| < 1$ and for any real a,

$$\begin{aligned}\sum_{n=-\infty}^{\infty} z^{|n|}e^{ina} &= 1 + \sum_{n=1}^{\infty} z^n e^{ina} + \sum_{n=1}^{\infty} z^n e^{-ina}\\ &= 1 + \frac{ze^{ia}}{1 - ze^{ia}} + \frac{ze^{-ia}}{1 - ze^{-ia}} = \frac{1 - z^2}{1 + z^2 - 2z\cos a},\end{aligned}$$

so that we obtain

$$u(r,\phi) = \frac{1}{2\pi}\int_{-\pi}^{\pi} \frac{1 - r^2}{1 + r^2 - 2r\cos(\phi - \psi)} f(\psi)\,d\psi, \quad r < 1. \tag{8.3.8}$$

The function

$$k(r,\phi) = \frac{1}{2\pi}\frac{1 - r^2}{1 + r^2 - 2r\cos\phi} \tag{8.3.9}$$

is known as *Poisson's kernel* and can be interpreted as a boundary influence function. Thus (8.3.9) is the temperature at the interior point (r, ϕ) when the boundary temperature is a delta function concentrated at $\phi = 0$.

In view of the fact that (8.3.9) is harmonic for $r < 1$ it is easy to see that (8.3.8) represents a harmonic function for $r < 1$. Moreover, we showed in Exercise 2.2.9, that

$$\lim_{r\to 1-} k(r,\phi) = \delta(\phi), \quad |\phi| \leqslant \pi.$$

Application of the methods of Chapter 2 shows that $u(r, \phi)$ tends to $f(\phi)$ uniformly in $|\phi| \leqslant \pi$ and that the boundary values are therefore assumed in the sense of continuity. Continuous dependence on the boundary data is a fairly easy consequence of the explicit formula (8.3.8).

We now turn to the inhomogeneous equation (Poisson's equation) in the unit disk. It suffices to consider only 0 boundary data since we can always add the solution of a problem of type (8.3.2). We shall therefore consider

$$-\Delta v = q(r,\phi), \quad (r,\phi) \in \Omega; \quad v(1,\phi) = 0. \tag{8.3.10}$$

Although we can solve (8.3.10) directly by a Fourier expansion in ϕ, it is perhaps more in keeping with the spirit of our work to find Green's function and then use superposition as in (1.4.27). Green's function g for a unit source at $\phi = 0$, $r = r_0$ satisfies

$$-\frac{1}{r}\frac{\partial}{\partial r}\left(r\frac{\partial g}{\partial r}\right) - \frac{1}{r^2}\frac{\partial^2 g}{\partial \phi^2} = \frac{\delta(r-r_0)\delta(\phi)}{r}, \quad r < 1; \quad g|_{r=1} = 0. \tag{8.3.11}$$

The right side of the differential equation is the delta function expressed in polar coordinates; see (2.2.26). Although the solution depends on r, ϕ, and r_0, we shall sometimes suppress the dependence on r_0 for notational convenience.

We multiply (8.3.11) by $(1/2\pi)e^{-in\phi}$ and integrate from $\phi = -\pi$ to $\phi = \pi$. Integrating the second term by parts and using the fact that both g and $e^{-in\phi}$ satisfy (8.3.5), we find that

$$-\frac{d}{dr}\left(r\frac{dg_n}{dr}\right) + \frac{n^2}{r}g_n = \frac{1}{2\pi}\delta(r-r_0), \quad 0 < r, r_0 < 1; \quad g_n|_{r=1} = 0, \tag{8.3.12}$$

where

$$g_n(r, r_0) = \frac{1}{2\pi}\int_{-\pi}^{\pi} g e^{-in\phi}\, d\phi, \quad g = \sum_{n=-\infty}^{\infty} g_n(r, r_0)e^{in\phi}.$$

The general solution of the homogeneous equation corresponding to (8.3.12) is $Ar^n + Br^{-n}$ for $n \neq 0$ and $A + B\log r$ for $n = 0$. We require that g_n be bounded at $r = 0$ and vanish at $r = 1$. On applying the usual matching conditions at $r = r_0$, we find that

$$g_n(r, r_0) = -\frac{1}{4\pi|n|}(r_<^{|n|})(r_>^{|n|} - r_>^{-|n|}), \quad n \neq 0,$$

$$g_0(r, r_0) = -\frac{1}{2\pi}\log r_>,$$

where $r_> = \max(r, r_0)$, $r_< = \min(r, r_0)$. We then conclude that

$$g = -\frac{1}{2\pi}\log r_> - \frac{1}{4\pi}\sum_{\substack{n=-\infty \\ n\neq 0}}^{\infty} \frac{e^{in\phi}}{|n|}(r_<^{|n|})(r_>^{|n|} - r_>^{-|n|})$$

$$= -\frac{1}{2\pi}\log r_> - \frac{1}{2\pi}\sum_{n=1}^{\infty}\frac{\cos n\phi}{n}(r_<^n)(r_>^n - r_>^{-n}).$$

If the source is at $\phi = \phi_0$, we have, instead,

$$g(r, \phi; r_0, \phi_0) = -\frac{1}{2\pi}\log r_> - \frac{1}{2\pi}\sum_{n=1}^{\infty}\frac{\cos n(\phi - \phi_0)}{n}(r_<^n)(r_>^n - r_>^{-n}). \tag{8.3.13}$$

The solution of (8.3.10) can therefore be written as

$$v(r, \phi) = \int_{-\pi}^{\pi} d\phi_0 \int_0^1 r_0\, dr_0\; g(r, \phi; r_0, \phi_0) q(r_0, \phi_0), \tag{8.3.14}$$

from which one can deduce the continuous dependence of v on the forcing term q.

If we prefer to express the solution as a series in trigonometric exponentials, we set

$$q(r,\phi) = \sum_{n=-\infty}^{\infty} q_n(r)e^{in\phi}, \quad v(r,\phi) = \sum_{n=-\infty}^{\infty} v_n(r)e^{in\phi},$$

and, by substitution in either (8.3.14) or (8.3.10), we obtain

$$v_n(r) = 2\pi \int_0^1 r_0 q_n(r_0) g_n(r, r_0)\, dr_0.$$

The solution of the Dirichlet problem (8.3.2) can also be expressed in terms of Green's function as

$$u(r,\phi) = -\int_{-\pi}^{\pi} \frac{\partial g}{\partial r_0}(r,\phi; r_0,\phi_0) f(\phi_0) r_0\, d\phi_0, \tag{8.3.15}$$

which, on substitution of (8.3.13), reduces to (8.3.6) and hence to (8.3.8).

Of course, for most domains, we cannot find an explicit formula for Green's function. We must then resort to other techniques, such as that of integral equations, described later in the section.

Mean Value Theorem and The Maximum Principle

An immediate consequence of (8.3.8) or (8.3.6) is that the *value of a harmonic function at the center of a disk is the average of its values on the boundary* (the extension to a disk of radius a instead of 1 is easily accomplished by a change of variables). By splitting the disk into concentric annuli, we conclude that the value at the center is also the average over the area of the disk. Similar properties hold in n dimensions. Thus, if a function is harmonic in a ball Ω (and continuous in $\bar{\Omega}$), its value at the center is equal to its average on any sphere or any ball having the same center and lying within $\bar{\Omega}$. This mean value theorem enables us to prove a strong form of the maximum principle for harmonic functions.

Theorem. *Let u be harmonic in the bounded domain Ω and continuous in $\bar{\Omega}$. If $m \leqslant u \leqslant M$ on the boundary Γ of Ω, then $m < u < M$ in Ω unless $m = M$, in which case u is constant in $\bar{\Omega}$.*

Proof. Suppose that u has a maximum at the interior point x_0. Then $u(x) \leqslant u(x_0)$ for all $x \in \bar{\Omega}$. Let B_0 be the largest ball, with center at x_0, that lies entirely in $\bar{\Omega}$. If at some point of B_0 we had $u(x) < u(x_0)$, the continuity of u and the fact that $u(x) \leqslant u(x_0)$ for all x in B_0 would imply that $u(x_0)$ is greater than its average on B_0. Since this contradicts the mean value theorem, we must have $u(x) \equiv u(x_0)$ in B_0. We now show that u must have this same value everywhere in Ω (and hence in $\bar{\Omega}$, by continuity). Let ξ be an arbitrary point in Ω, and connect ξ with x_0 by a curve lying in Ω. The intersection of this curve and B_0 is denoted by x_1; next, we construct the largest ball B_1, with center at x_1, lying entirely in $\bar{\Omega}$. Repeating the previous argument, we find that $u(x) \equiv u(x_0)$ in B_1. Continuing in this manner, we finally cover the point ξ by some ball B_k and therefore $u(\xi) = u(x_0)$. We conclude that if u is harmonic in Ω, continuous in $\bar{\Omega}$, and not identically constant, its maximum occurs on the boundary and not in the interior. The same reasoning applied to the harmonic function $-u(x)$ shows that the minimum of u occurs on the boundary. □

REMARK. A weaker version of the maximum principle was proved in Section 1.3. We were then able to prove uniqueness and continuous dependence on the data for (8.3.1a).

An alternative method for proving the uniqueness of (8.3.1a) is based on "energy" integrals. Suppose that u_1 and u_2 are two solutions of (8.3.1a). Their difference $v(x)$ satisfies the homogeneous problem $-\Delta v = 0, x \in \Omega; v = 0$ on Γ. Therefore,

$$0 = \int_\Omega v\,\Delta v\,dx = -\int_\Omega |\text{grad}\, v|^2\,dx + \int_\Gamma v\frac{\partial v}{\partial \nu}\,dS = -\int_\Omega |\text{grad}\, v|^2\,dx,$$

and $v =$ constant in Ω. Since $v = 0$ on Γ, this constant must be 0. Thus $v \equiv 0$ and $u_1 = u_2$.

Surface Layers

It will be convenient in this section to use the language of electrostatics, although the results, being of a purely mathematical nature, can be applied whenever the physical laws lead to potential theory.

If a unit positive charge is located at the point ξ in $\mathbb{R}^3$, the corresponding potential (free space fundamental solution) $E(x, \xi)$ satisfies

$$-\Delta E = \delta(x - \xi), \quad x, \xi \in \mathbb{R}^3. \tag{8.3.16}$$

The solution is determined only up to an arbitrary solution of the homogeneous equation. If we also require that E be spherically symmetric about the source and vanish at infinity, the unique solution of (8.3.16) becomes

$$E(x, \xi) = \frac{1}{4\pi|x - \xi|}, \tag{8.3.17}$$

as was shown in Chapter 2. Equation (8.3.16) is interpreted in its distributional sense:

$$\langle E, \Delta\phi\rangle = -\phi(\xi) \quad \text{for each } \phi \in C_0^\infty(\mathbb{R}^3).$$

Now suppose that charge is distributed in free space with density $q(x)$, where, for simplicity, we shall assume that $q \equiv 0$ outside some bounded ball. The element of volume $d\xi$ at ξ carries a charge $q(\xi)\,d\xi$ whose potential is $(1/4\pi|x - \xi|)q(\xi)\,d\xi$. By superposition the potential $u(x)$ of the entire charge distribution is

$$u(x) = \int_{\mathbb{R}^3} \frac{1}{4\pi|x - \xi|} q(\xi)\,d\xi = \int_\Omega \frac{1}{4\pi|x - \xi|} q(\xi)\,d\xi, \tag{8.3.18}$$

where Ω is the domain where q does not vanish. If x is outside Ω, the integrand is not singular as ξ traverses Ω. We can therefore calculate derivatives with respect to x by differentiating under the integral sign, and we see that $\Delta u = 0$. If x is in the charge region Ω, the integrand in (8.3.18) has a very mild singularity which causes little difficulty. The integral (8.3.18) still converges, as can easily be seen by passing to spherical coordinates with origin at x, the element of volume $d\xi$ furnishing a factor proportional to $|x-\xi|^2$ which more than cancels the denominator. Differentiating formally under the integral sign, we obtain

$$-\Delta u = \int_{\mathbb{R}^3} \delta(x-\xi)q(\xi)\,d\xi = q(x), \tag{8.3.19}$$

a result which is rigorous in the sense of the theory of distributions (and also holds classically if we assume that q is slightly better than just continuous—Hölder continuity suffices; differentiability is more than enough).

A unit dipole at ξ with axis l is a singular charge distribution obtained by taking the limit as $h \to 0$ of charges $-(1/h)$ and $1/h$ located at ξ and $\xi + hl$, respectively, where l is a given unit vector. The corresponding potential $D_l(x,\xi)$ satisfies (see also Sections 2.1 and 2.2)

$$-\Delta D_l = \lim_{h\to 0} \frac{1}{h}\{\delta[x-(\xi+hl)] - \delta(x-\xi)\} = \frac{d}{dl_\xi}\delta(x-\xi),$$

where d/dl_ξ refers to differentiation with respect to ξ in the l direction. By the principle of superposition, we can find D_l by performing the differentiation on the solution of $-\Delta E = \delta(x-\xi)$. A simple calculation then gives

$$D_l(x,\xi) = \frac{d}{dl_\xi}\frac{1}{4\pi|x-\xi|} = \frac{\cos(x-\xi,l)}{4\pi|x-\xi|^2}, \tag{8.3.20}$$

where $(x-\xi,l)$ is the angle between the vectors $x-\xi$ and l.

In (8.3.18) we considered the potential of a volume distribution of charges with density q. The potential $u(x)$ not only is continuous within the charge region but even satisfies the differential equation $-\Delta u = q$. We now turn to surface distributions of charge. In the hierarchy of charge configurations, a surface distribution is less singular than a point charge but not as regular as a volume distribution. (A line charge is less singular than a point charge but more so than a surface charge.) Let the surface Γ carry a charge with surface density $a(x)$. Such a charge configuration is known as a *simple layer*. This means that the portion Γ' of Γ has the charge $\int_{\Gamma'} a(\xi)\,dS_\xi$. One can think of the surface element dS_ξ at ξ as carrying the small concentrated charge $a(\xi)\,dS_\xi$. The potential of the simple layer is

$$u(x) = \int_\Gamma \frac{1}{4\pi|x-\xi|}a(\xi)\,dS_\xi, \quad x \in \mathbb{R}^3. \tag{8.3.21}$$

Consider next a surface distribution of dipoles of density $b(x)$ with axis *normal* to Γ. This configuration is known as a *double layer*. The element of surface dS_ξ at ξ

carries a dipole of strength $b(\xi)\,dS_\xi$ with axis n, where n is the normal to Γ at ξ. The potential $v(x)$ of the double layer follows from (8.3.20) by superposition:

$$v(x) = \int_\Gamma \frac{\cos(x-\xi, n)}{4\pi|x-\xi|^2} b(\xi)\,dS_\xi = \int_\Gamma \frac{\partial}{\partial n_\xi} \frac{1}{4\pi|x-\xi|} b(\xi)\,dS_\xi. \tag{8.3.22}$$

The integrals in (8.3.21) and (8.3.22) are obviously well defined for x not on Γ. Even if x is a point on Γ, the integral in (8.3.21) converges. Any difficulty would arise from points ξ near x, and then the corresponding integration is over a flat portion of Γ; introducing polar coordinates with origin at x, we see that dS_ξ contributes a factor proportional to $|x-\xi|$ which cancels the denominator. Similarly, it can be shown that (8.3.22) converges when x is on Γ because $\cos(x-\xi, n)$ tends to 0 as ξ approaches x. Of course, the fact that (8.3.21) and (8.3.22) make sense even when x is on Γ does not tell us how these integrals, evaluated for x near Γ, are related to the values on Γ. We study this question next.

The potential of a point source located at the origin is

$$E(x_1, x_2, x_3) = \frac{1}{4\pi(x_1^2 + x_2^2 + x_3^2)^{1/2}}. \tag{8.3.23}$$

By regarding x_3 as a parameter, we would like to relate the behavior of E as x_3 tends to 0 with its behavior at $x_3 = 0$. In effect, we have selected a set of parallel planes (x_3 = constant) on which to study the potential (8.3.22). For each $x_3 \neq 0$, E is a continuous function of x_1, x_2 and so defines a distribution on the space of test functions $\phi(x_1, x_2)$ of compact support in $\mathbb{R}^2$. When $x_3 = 0$, E becomes unbounded at $x_1 = x_2 = 0$ but remains locally integrable in the x_1–x_2 plane and therefore still defines a distribution. Setting $\rho^2 = x_1^2 + x_2^2$, we have the following theorem.

Theorem 8.3.1. *In the sense of distributions,*

$$\lim_{x_3 \to 0\pm} E(x_1, x_2, x_3) = \lim_{x_3 \to 0\pm} \frac{1}{4\pi(\rho^2 + x_3^2)} = \frac{1}{4\pi\rho} = E(x_1, x_2, 0), \tag{8.3.24}$$

$$\lim_{x_3 \to 0\pm} \frac{\partial E}{\partial x_3}(x_1, x_2, x_3) = \lim_{x_3 \to 0\pm} \frac{-x_3}{4\pi(\rho^2 + x_3^2)^{3/2}} = \mp\frac{1}{2}\delta(x_1)\delta(x_2). \tag{8.3.25}$$

Proof. For all x_3, $1/4\pi(\rho^2 + x_3^2)^{1/2}$ is dominated by the locally integrable function $1/4\pi\rho$. In the pointwise sense, as $x_3 \to 0$, $1/4\pi(\rho^2 + x_3^2)^{1/2}$ tends to $1/4\pi\rho$ everywhere except at $\rho = 0$. It follows from the Lebesgue dominated convergence theorem (see Section 0.7) that

$$\lim_{x_3 \to 0\pm} \int_{\mathbb{R}^2} \frac{\phi(x_1, x_2)}{4\pi(\rho^2 + x_3^2)^{1/2}} dx_1\,dx_2 = \int_{\mathbb{R}^2} \frac{\phi(x_1, x_2)}{4\pi\rho} dx_1\,dx_2 \tag{8.3.26}$$

for every integrable function ϕ (hence certainly for test functions).

Next we turn to $\partial E/\partial x_3$, which is the derivative normal to the planes x_3 = constant. For $x_3 \neq 0$ we have

$$\frac{\partial E}{\partial x_3}(x_1, x_2, x_3) = -\frac{x_3}{4\pi(\rho^2 + x_3^2)^{3/2}}, \tag{8.3.27}$$

whereas on $x_3 = 0, \partial E/\partial x_3 = 0$ except at $\rho = 0$. The Lebesgue theorem is not applicable because we cannot find an integrable function which dominates (8.3.27) in a neighborhood of $x_3 = 0$. However, (8.3.25) follows from Example 3(h_2), Section 2.2. Thus, if ϕ is a test function, or even just a function continuous at the origin and integrable, we have

$$\lim_{x_3 \to 0\pm} \int_{\mathbb{R}^2} \phi(x_1, x_2) \frac{\partial E}{\partial x_3}(x_1, x_2, x_3)\, dx_1\, dx_2 = \mp \frac{1}{2}\phi(0,0). \tag{8.3.28}$$

□

REMARK. Results (8.3.26) and (8.3.28) remain valid even if the integral is taken only over a portion σ of the plane $x_3 = 0$. Indeed, it suffices to extend ϕ to be identically 0 outside σ and then apply (8.3.26) and (8.3.28), which hold for piecewise continuous integrands.

Now suppose that Γ is a *smooth* surface, one side of which is taken as the positive side. By definition the *normal to* Γ is the normal on the positive side. If Γ is a closed surface without boundary, such as a sphere, it is customary to label the exterior side as positive. We shall investigate the behavior of potentials of layers in the neighborhood of a *fixed point* s on Γ. The normal at s will be called ν; of course, the direction ν is then also determined by parallelism at every point in space.

Let $u(x)$ be the potential of a simple layer of surface density $a(x)$ spread on Γ; then

$$u(x) = \int_\Gamma \frac{a(\xi)}{4\pi|x-\xi|} dS_\xi, \tag{8.3.29}$$

which is well defined not only when x is outside Γ but also when x is on Γ. Furthermore, if we imitate the arguments that led to (8.3.26), we can show that

$$\lim_{x_3 \to s\pm} u(x) = \int_\Gamma \frac{a(\xi)}{4\pi|s-\xi|} dS_\xi, \tag{8.3.30}$$

so that the limits from either side coincide with the potential right on the surface. Let us now turn to the derivative $\partial u/\partial \nu$ in the direction ν, which coincides with the normal at s. If x is not on Γ, we can calculate $\partial u/\partial \nu$ by differentiating (8.3.29) under the integral sign to obtain

$$\frac{\partial u}{\partial \nu}(x) = \int_\Gamma a(\xi) \frac{\partial}{\partial \nu} \frac{1}{4\pi|x-\xi|} dS_\xi = \int_\Gamma a(\xi) \frac{\cos(\xi - x, \nu)}{4\pi|x-\xi|^2} dS_\xi.$$

To analyze the behavior as $x \to s$, we divide Γ into the complementary parts Γ_ε and $\Gamma - \Gamma_\varepsilon$, where Γ_ε is the portion of Γ within a ball of radius ε centered at s. Thus we have

$$\frac{\partial u}{\partial \nu}(x) = \int_{\Gamma - \Gamma_\varepsilon} a(\xi) \frac{\cos(\xi - x, \nu)}{4\pi|x-\xi|^2} dS_\xi + \int_{\Gamma_\varepsilon} a(\xi) \frac{\cos(\xi - x, \nu)}{4\pi|x-\xi|^2} dS_\xi. \tag{8.3.31}$$

The first integral is clearly continuous at any point x not lying on $\Gamma - \Gamma_\varepsilon$, and will therefore be continuous as $x \to s$. For ε sufficiently small, Γ_ε is nearly a flat surface and we can regard Γ_ε as lying in the plane tangent to Γ at s. We introduce a Cartesian coordinate system (x_1, x_2, x_3) with origin at s; x_1, x_2 lie in the tangent plane, and the positive x_3 axis is in the ν direction. If we let x lie on the normal to Γ at s, the second integral in (8.3.31) may be rewritten as

$$\int_{\Gamma_\varepsilon} a(\xi_1, \xi_2) \left[\frac{-x_3}{4\pi(\xi_1^2 + \xi_2^2 + x_3^2)^{3/2}} \right] d\xi_1 \, d\xi_2.$$

To let $x \to s\pm$ is the same as allowing x_3 to tend to $0\pm$. Hence from (8.3.28) and the remark following Theorem 8.2.1, the integral on Γ_ε tends to $\mp\frac{1}{2}a(s)$. The remaining integral in (8.3.31) approaches the integral

$$\int_{\Gamma-\Gamma_\varepsilon} a(\xi) \frac{\cos(\xi - s, \nu)}{4\pi|s - \xi|^2} dS_\xi,$$

which in the limit as $\varepsilon \to 0$ is equal to the convergent integral

$$\int_{\Gamma} a(\xi) \frac{\cos(\xi - s, \nu)}{4\pi|s - \xi|^2} dS_\xi.$$

We therefore conclude that

$$\lim_{x \to s\pm} \frac{\partial u}{\partial \nu}(x) = \mp\frac{1}{2}a(s) + \int_{\Gamma} a(\xi) \frac{\cos(\xi - s, \nu)}{4\pi|s - \xi|^2} dS_\xi, \tag{8.3.32}$$

which contains the well-known result of electrostatics that at a charged surface, the normal component of the electric field [that is, $-(\partial u/\partial \nu)$] jumps by an amount equal to the surface charge density. This result is usually derived by elementary arguments from Gauss's theorem, but (8.3.32) also yields deeper information about one-sided limits.

The potential (8.3.32) of a normally oriented *double layer* of surface density $b(x)$ spread on Γ can be written as

$$v(x) = \int_{\Gamma-\Gamma_\varepsilon} b(\xi) \frac{\cos(x - \xi, n)}{4\pi|x - \xi|^2} dS_\xi + \int_{\Gamma_\varepsilon} b(\xi) \frac{\cos(x - \xi, n)}{4\pi|x - \xi|^2} dS_\xi,$$

where n *is the normal to* Γ *at* ξ. The first integral is continuous as $x \to s$, and the second is over a nearly flat surface if ε is sufficiently small. Using the same Cartesian coordinate system introduced earlier, and with x lying on the normal to Γ at s, we have

$$\frac{\cos(x - \xi, n)}{4\pi|x - \xi|^2} = \frac{x_3}{4\pi(\xi_1^2 + \xi_2^2 + x_3^2)^{3/2}},$$

and hence

$$\lim_{x \to s\pm} v(x) = \pm\frac{1}{2}b(s) + \int_{\Gamma} b(\xi) \frac{\cos(s - \xi, n)}{4\pi|s - \xi|^2} dS_\xi. \tag{8.3.33}$$

Thus the potential of a dipole layer is *discontinuous* on crossing the layer. We shall not need to investigate the behavior of the normal derivative of the potential of a double layer.

Interior Dirichlet Problem: Integral Equation of the Second Kind

Let Ω be a bounded domain in $\mathbb{R}^3$ with smooth boundary Γ. Consider the Dirichlet problem for Laplace's equation

$$\Delta w = 0, \quad x \in \Omega; \quad w = f, \quad x \in \Gamma, \tag{8.3.34}$$

where we assume f continuous on Γ, and we are looking for a solution w in $C^2(\bar{\Omega})$.

The classical approach, which at the same time provides an existence proof, attempts to find the solution of (8.3.34) as a double layer on Γ. Thus our candidate for the solution of (8.3.34) is

$$w(x) = \int_\Gamma b(\xi) \frac{\cos(x - \xi, n)}{4\pi |x - \xi|^2} \, dS_\xi, \tag{8.3.35}$$

where b is to be determined from

$$\lim_{x \to s-} w(x) = f(s) \quad \text{for every } s \text{ on } \Gamma. \tag{8.3.36}$$

In view of (8.3.33), condition (8.3.36) becomes the Fredholm integral equation of the second kind:

$$f(s) = -\frac{1}{2} b(s) + \int_\Gamma k(s, \xi) b(\xi) \, dS_\xi, \tag{8.3.37}$$

where the kernel

$$k(s, \xi) = \frac{\cos(s - \xi, n)}{4\pi |s - \xi|^2} \tag{8.3.38}$$

is usually not symmetric.

The singularity of the kernel at $x = \xi$ is only of order $1/|x - \xi|$ because of the mitigating effect of the numerator. The corresponding integral operator K is Hilbert-Schmidt (the proof is similar to Exercise 6.1.2). It can be shown that $\frac{1}{2}$ is not an eigenvalue of K, so that (8.3.37) has one and only one solution b for each f. The corresponding double layer potential (8.3.35) is then a solution of Laplace's equation in Ω satisfying (8.3.36). With a little more work it can be shown that the boundary values are taken on in the sense of continuity. We have therefore shown the existence of a solution to (8.3.34). Existence under less restrictive conditions on Γ and f will be obtained in Section 8.4 by variational methods. Uniqueness was proved earlier.

Interior and Exterior Dirichlet Problems by an Integral Equation of the First Kind

Correct formulation of the Dirichlet problem for an unbounded domain requires a boundary condition at infinity. Consider, for instance, the domain exterior to the unit ball in $\mathbb{R}^3$. The functions $u_1(x) = 1$ and $u_2(x) = 1/|x|$ are both harmonic for $|x| > 1$ and take on the same value on the unit sphere. In many (but not all) physical applications the appropriate requirement is that u vanish as $|x| \to \infty$.

Definition. *Let* Ω_e *be the exterior of a bounded domain* Ω_i *in* $\mathbb{R}^3$. *The common boundary of* Ω_i *and* Ω_e *is denoted by* Γ. *The* exterior Dirichlet problem *is the BVP*

$$\Delta u_e = 0, \quad u \in \Omega_e; \quad u_e|_\Gamma = f; \quad \lim_{|x|\to\infty} u_e = 0. \tag{8.3.39}$$

Expansion in a series of spherical harmonics shows that the solution of (8.3.39) has the properties

$$u_e = 0\left(\frac{1}{|x|}\right), \quad |\text{grad}\, u_e| = 0\left(\frac{1}{|x|^2}\right) \quad \text{as } |x| \to \infty.$$

An immediate consequence is that

$$\lim_{R\to\infty} \int_{\Gamma_R} \left(E\frac{\partial u_e}{\partial R} - u_e \frac{\partial E}{\partial R}\right) dS_x = 0, \tag{8.3.40}$$

where $E = 1/4\pi|x-\xi|$, ξ is a fixed point and Γ_R is a sphere of radius R with center at ξ.

Simultaneously with (8.3.39) we consider the *interior* Dirichlet problem with the same boundary values on Γ:

$$\Delta u_i = 0, \quad x \in \Omega_i; \qquad u_i|_\Gamma = f. \tag{8.3.41}$$

Note that the solutions of the interior and exterior problems are not related in any obvious manner. For instance, if $f = 1$, then $u_i \equiv 1$ but u_e can be quite complicated (if Ω_i is the unit ball, $u_e = 1/|x|$). Nevertheless, we shall be able to reduce both problems to a single integral equation on Γ whose solution can then be used to calculate both $u_i(x)$ and $u_e(x)$.

Recall that $E(x,\xi) = 1/4\pi|x-\xi|$ satisfies

$$-\Delta E = \delta(x-\xi), \quad x, \xi \in \mathbb{R}^3. \tag{8.3.42}$$

We regard ξ as a fixed point in either Ω_i or Ω_e. Multiply (8.3.41) by E, (8.3.42) by u_i, add, and integrate over Ω_i to obtain

$$\int_{\Omega_i} (E\,\Delta u_i - u_i\,\Delta E)\,dx = \begin{cases} u_i(\xi), & \xi \in \Omega_i, \\ 0, & \xi \in \Omega_e. \end{cases}$$

On applying Green's theorem, this becomes

$$\int_\Gamma \left(E\frac{\partial u_i}{\partial \nu} - u_i\frac{\partial E}{\partial \nu}\right) dS_\xi = \begin{cases} u_i(\xi), & \xi \in \Omega_i, \\ 0, & \xi \in \Omega_e, \end{cases} \tag{8.3.43}$$

where ν is the outward normal to Γ (that is, pointing away from Ω_i into Ω_e).

Similarly, we multiply (8.3.39) by E, (8.3.42) by u_e, add, and integrate over the domain bounded internally by Γ and externally by a sphere Γ_R of large radius R with center at ξ (such a large sphere will contain Γ if R is large enough). After applying Green's theorem, we note that the contribution from Γ_R tends to 0 as $R \to \infty$ by

(8.3.40). With respect to Ω_e the outward normal on Γ is the negative of the previous ν. Therefore, we obtain

$$\int_\Gamma \left(-E\frac{\partial u_e}{\partial \nu} + u_e\frac{\partial E}{\partial \nu}\right) dS_x = \begin{cases} 0, & \xi \in \Omega_i, \\ u_e(\xi), & \xi \in \Omega_e. \end{cases} \tag{8.3.44}$$

Adding (8.3.43) and (8.3.44), and noting that $u_i = u_e = f$ on Γ, we find that

$$\int_\Gamma E(x,\xi)\left[\frac{\partial u_i}{\partial \nu}(x) - \frac{\partial u_e}{\partial \nu}(x)\right] dS_x = \begin{cases} u_i(\xi), & \xi \in \Omega_i, \\ u_e(\xi), & \xi \in \Omega_e. \end{cases}$$

Interchanging the labels of the variables and invoking the symmetry of E, we obtain

$$\int_\Gamma E(x,\xi)I(\xi)dS_\xi = \begin{cases} u_i(x), & x \in \Omega_i, \\ u_e(x), & x \in \Omega_e, \end{cases} \tag{8.3.45}$$

where

$$I(\xi) \doteq \frac{\partial u_i}{\partial \nu}(\xi) - \frac{\partial u_e}{\partial \nu}(\xi), \quad \xi \in \Gamma. \tag{8.3.46}$$

We have succeeded in expressing u_i and u_e within their respective domains as the potential of the same simple layer with unknown surface density $I(\xi)$. Letting x tend to a point s on Γ, we obtain the integral equation for I:

$$\int_\Gamma \frac{1}{4\pi|s-\xi|}I(\xi)\, dS_\xi = f(s), \quad s \in \Gamma, \tag{8.3.47}$$

where we have used the continuity of the simple layer potential. After solving (8.3.47) for I, we would substitute in (8.3.45) to find both $u_i(x)$ and $u_e(x)$.

Equation (8.3.47) is a Fredholm equation of the first kind with a symmetric H.-S. kernel. We know that 0 is in the continuous spectrum of the corresponding integral operator, so that (8.3.47) cannot have an L_2 solution for each $f \in L_2(\Gamma)$. Nevertheless, even in these cases one can interpret the "solution" in a distributional sense, and although the formal eigenfunction expansion for I may diverge, the corresponding expressions for u_i and u_e obtained from (8.3.45) will be well behaved (see Hsiao and Wendland [24], and the references quoted there for more complete results on this problem).

Another advantage of (8.3.47) is that the corresponding integral operator is *positive*. To show this let $I(\xi)$ be an arbitrary function in $L_2(\Gamma)$, and *define* $f(s)$ from (8.3.47). What we wish to prove is that

$$\int_\Gamma f(s)I(s)\, dS > 0 \quad \text{for } I \not\equiv 0. \tag{8.3.48}$$

Let us define the simple layer potential

$$u(x) = \int_\Gamma E(x,\xi)I(\xi)\, dS_\xi,$$

which is harmonic in both Ω_i and Ω_e; moreover, u vanishes at infinity. Thus we obtain

$$0 = \int_{\Omega_i} u\,\Delta u\,dx = \int_\Gamma u(s-)\frac{\partial u}{\partial \nu}(s-)\,dS - \int_{\Omega_i} |\text{grad}\,u|^2\,dx,$$
$$0 = \int_{\Omega_e} u\,\Delta u\,dx = -\int_\Gamma u(s+)\frac{\partial u}{\partial \nu}(s+)\,dS - \int_{\Omega_e} |\text{grad}\,u|^2\,dx.$$

Adding these equations and recalling the simple layer properties

$$u(s-) = u(s+) = f(s), \quad \frac{\partial u}{\partial \nu}(s-) - \frac{\partial u}{\partial \nu}(s+) = I(s),$$

we find that

$$0 = \int_\Gamma f(s)I(s)\,dS - \int_{\mathbb{R}^3} |\text{grad}\,u|^2\,dx.$$

It follows that $\int_\Gamma fI\,dS > 0$ unless u is constant, in which case the behavior at infinity tells us that $u \equiv 0$ and therefore $I \equiv 0$. Hence we have proved (8.3.48).

Eigenvalue Problem for the Negative Laplacian

The three principal (but related) methods for solving the inhomogeneous problem (8.3.1a) are Green's function, eigenfunction expansion, and integral equation. The BVP (8.3.1a) features the operator $-\Delta$ rather than Δ. This is natural both physically and mathematically; indeed, as written, q is a source term and an increase in q leads to an increase in the response u. A related mathematical property is that $-\Delta$ subject to vanishing boundary conditions is a positive operator. We shall find that the eigenvalues of this operator are positive. The purpose of the present section is to study the eigenvalue problem and show some of its uses.

Let Ω be a bounded domain in $\mathbb{R}^n$ and consider the eigenvalue problem

$$-\Delta u = \lambda u, \quad x \in \Omega; \qquad u = 0, \quad x \in \Gamma. \tag{8.3.49}$$

For our present purposes, it suffices to consider solutions in $C^2(\bar{\Omega})$. In operator form, we rewrite (8.3.49) as

$$Au = \lambda u, \quad u \in D_A, \tag{8.3.50}$$

where $A = -\Delta$ and D_A consists of all $C^2(\bar{\Omega})$ functions vanishing on Γ. Using the standard definition of the inner product, we find that for any $u, v \in D_A$,

$$\langle Au, v\rangle = -\int_\Omega \bar{v}\,\Delta u\,dx = -\int_\Omega u\,\Delta\bar{v}\,dx = \langle u, Av\rangle,$$

so that A is symmetric and its eigenvalues must be real. Furthermore, for $u \in D_A$, we have

$$\langle -\Delta u, u\rangle = -\int_\Omega \bar{u}\,\Delta u\,dx = \int_\Omega |\text{grad}\,u|^2\,dx \geqslant 0.$$

The right side vanishes only if $u \equiv C$; but the only constant in D_A is the zero constant. Thus, we have

$$\langle Au, u\rangle > 0 \quad \text{for every } u \neq 0 \text{ in } D_A. \tag{8.3.51}$$

This is precisely the definition of a positive operator given in Section 5.8. In fact, we shall see in Section 8.4 that the operator A is actually strongly positive.

On account of (8.3.51), the eigenvalues of A must be positive. As we have seen in Chapter 5, eigenfunctions corresponding to different eigenvalues are orthogonal.

By introducing Green's function $g(x, \xi)$, which satisfies

$$-\Delta g = \delta(x-\xi), \quad x, \xi \in \Omega; \quad g = 0, \quad x \in \Gamma, \tag{8.3.52}$$

we can translate (8.3.49) into the integral equation

$$u(x) = \lambda \int_\Omega g(x,\xi)u(\xi)\,d\xi \doteq \lambda G u, \tag{8.3.53}$$

where G is the integral operator with kernel $g(x, \xi)$. Since the solution of $Au = q$, $u \in D_A$ can be written as $u = Gq$, G can be interpreted as the inverse of A. Clearly, $Gq = 0$ implies that $q = 0$, so that zero is not an eigenvalue of the operator G whose eigenvalues are defined from

$$Gu = \mu u, \quad \mu = 1/\lambda. \tag{8.3.54}$$

Thus, the eigenvalues of G are precisely the reciprocals of those of A and the two operators have the same eigenfunctions. The operator G is symmetric, positive, and *compact* (see Exercise 8.3.2). By Theorem 6.3.5 the eigenfunctions of G (and hence, of A) form an orthonormal basis for $H = L_2^{(c)}(\Omega)$. The eigenvalues (indexed with due regard to multiplicity) form a sequence $\{\mu_k\}$ with $\lim_{k\to\infty}\mu_k = 0$ so that the eigenvalues λ_k of A have the property $\lim_{k\to\infty}\lambda_k = \infty$. The corresponding orthonormal basis of eigenfunctions is denoted by $\{e_k\}$. In Chapter 6 we also saw that the eigenpairs of G can be characterized by a variational principle; in particular, the largest eigenvalue μ_1 [whose reciprocal λ_1 is the lowest eigenvalue of (8.3.49)] can be characterized by

$$\mu_1 = \frac{1}{\lambda_1} = \max_{\|u\|=1} \langle Gu, u\rangle; \tag{8.3.55a}$$

or

$$\lambda_1 = \min_{\|u\|=1} \langle Gu, u\rangle^{-1}. \tag{8.3.55b}$$

In either form the element u providing the extremum is ce_1, where $|c| = 1$. Lower eigenvalues of G can also be expressed by a variational principle as in Section 6.3. There is also a direct variational principle for λ_1:

$$\lambda_1 = \min_{u\in D_A, \|u\|=1} \int_\Omega |\text{grad } u|^2\,dx, \tag{8.3.56}$$

with the minimizer $u = ce_1$, $|c| = 1$. To prove (8.3.56) note that for any $u \in D_A$,

$$\int_\Omega |\text{grad } u|^2 \, dx = -\int_\Omega u \, \Delta u \, dx = \sum \lambda_k |\langle u, e_k \rangle|^2;$$

since λ_k increases with k and $\|u\| = 1$,

$$\sum \lambda_k |\langle u, e_k \rangle|^2 \geqslant \lambda_1 \sum |\langle u, e_k \rangle|^2 = \lambda_1.$$

Thus, for any $u \in D_A$ with $\|u\| = 1$, we have

$$\lambda_1 \leqslant \int_\Omega |\text{grad } u|^2 \, dx,$$

and equality is achieved for $u = ce_1$ with $|c| = 1$. Exercise 8.3.11 shows that e_1 does not change sign in Ω. We can, as in Chapter 6, derive variational principles for higher eigenvalues:

$$\lambda_{n+1} = \min_{u \in D_A \cap M_n^\perp, \|u\|=1} \int_\Omega |\text{grad } u|^2 \, dx, \tag{8.3.56a}$$

where M_n is the span of $\{e_1, \ldots, e_n\}$.

Next we consider how the eigenpairs of (8.3.49) can be used to solve various inhomogeneous problems.

Solution of BVPs by Eigenfunction Expansion

Consider first the problem

$$-\Delta u = q, \quad x \in \Omega; \quad u = 0, \quad x \in \Gamma. \tag{8.3.57}$$

Writing $q = \sum q_k e_k(x)$ and $u = \sum u_k e_k(x)$, where the index k is understood to run from 1 to ∞, our goal is to find the coefficients $\{u_k = \langle u, e_k \rangle\}$ in terms of the coefficients $\{q_k = \langle q, e_k \rangle\}$, which are regarded—somewhat optimistically—as known. Since q, unlike the eigenfunctions, is not required to vanish on the boundary, the series for q may converge rather slowly. [Think of the one-dimensional example $\Omega = (0, 1)$: The expansion of $q = 1$ in terms of the eigenfunctions $e_k = (2/\pi)^{1/2} \sin k\pi x$ yields the series $\sum_{k \text{ odd}} (4 \sin k\pi x)/k\pi$, whose coefficients decrease slowly to zero, the series converging only because it is essentially an alternating series—as indicated by its behavior at $x = 1/2$.] On the other hand, the solution u satisfies the same boundary conditions as the eigenfunctions, so we can expect more rapid convergence. Since it is not clear a priori that the series for u can be differentiated twice term–by–term, we shall not substitute the series in (8.3.57). Instead, we multiply both sides of the PDE by $\bar{e}_k(x)$ and integrate over Ω to obtain

$$\langle q, e_k \rangle = -\langle \Delta u, e_k \rangle = -\langle u, \Delta e_k \rangle = \bar{\lambda}_k \langle u, e_k \rangle = \lambda_k \langle u, e_k \rangle.$$

We can therefore write the solution of (8.3.57) as

$$u(x) = \sum \langle u, e_k \rangle e_k(x) = \sum_k \frac{\langle q, e_k \rangle}{\lambda_k} e_k(x). \tag{8.3.58}$$

Since $\lambda_k \to \infty$ as $k \to \infty$, the coefficients in the series for u tend to zero more rapidly than those for q. A particular case of (8.3.58) is the Green's function $g(x, \xi)$ corresponding to $q = \delta(x - \xi)$ and $q_k = \bar{e}_k(\xi)$. This yields the so-called *bilinear series*

$$g(x, \xi) = \sum_k \frac{e_k(x)\bar{e}_k(\xi)}{\lambda_k}. \tag{8.3.59}$$

We turn next to the Dirichlet problem for Laplace's equation:

$$\Delta v = 0, \quad x \in \Omega; \quad v = f, \quad x \in \Gamma. \tag{8.3.60}$$

We could again use an eigenfunction expansion for v despite the inhomogeneous boundary condition. Multiply both sides of the PDE in (8.3.60) by $\bar{e}_k(x)$ and integrate over Ω to obtain

$$0 = \langle \Delta v, e_k \rangle = \langle v, \Delta e_k \rangle - \int_\Gamma v \frac{\partial \bar{e}_k}{\partial \nu} \, dS,$$

yielding

$$\langle v, e_k \rangle = -\frac{1}{\lambda_k} \int_\Gamma f \frac{\partial \bar{e}_k}{\partial \nu} \, dS. \tag{8.3.61}$$

The expansion for the completely inhomogeneous problem (8.3.1a) can then be found by superposition. Since the series $\sum \langle v, e_k \rangle e_k$ formed from the coefficients (8.3.61) usually converges slowly, other methods are often preferable for solving (8.3.60). One way is to reduce (8.3.60) to a problem of the form (8.3.57) by writing

$$v = w + F, \tag{8.3.62}$$

where F is any smooth function on Ω with $F(x) = f(x)$ for $x \in \Gamma$; then w satisfies

$$-\Delta w = \Delta F, \quad x \in \Omega; \quad w = 0, \quad x \in \Gamma,$$

whose solution by (8.3.58) is

$$w = \sum \frac{\langle \Delta F, e_k \rangle}{\lambda_k} e_k(x).$$

The representation (8.3.62) is often more useful than the series $\sum \langle v, e_k \rangle e_k$ with $\langle v, e_k \rangle$ given by (8.3.61).

Both the heat equation and the wave equation can be formally solved by an expansion in the eigenfunctions of (8.3.49). For illustration, we confine ourselves to the case of vanishing boundary conditions on Γ. Consider, for instance, the BVP for the heat equation

$$\begin{aligned} &\frac{\partial u}{\partial u} - \Delta u = q(x,t), \quad (x,t) \in \Omega \times (0,\infty), \\ &u(x,0) = f(x); \qquad u(x,t) = 0 \text{ for } (x,t) \in \Gamma \times (0,\infty). \end{aligned} \tag{8.3.63}$$

We write

$$f(x) = \sum f_k e_k(x), \quad q(x,t) = \sum q_k(t) e_k(x), \quad u(x,t) = \sum u_k(t) e_k(x),$$

and proceed to express $u_k(t)$ in terms of f_k and $q_k(t)$. For $t > 0$, multiply the PDE in (8.3.63) by $\bar{e}_k(x)$ and integrate over Ω to obtain

$$\frac{d}{dt} u_k(t) + \lambda_k u_k(t) = q_k(t), \quad t > 0; \quad u_k(0) = f_k,$$

whose solution is

$$u_k(t) = f_k e^{-\lambda_k t} + \int_0^t e^{-\lambda_k (t-\tau)} q_k(\tau)\, d\tau.$$

Green's function can be calculated by choosing *either* $q = 0$ and $f(x) = \delta(x - \xi)$ or $q = \delta(x - \xi)\delta(t)$ and $f(x) = 0$ [as in (8.2.24) and (8.2.25)]. We find that

$$g(x,t;\xi,0) = \sum e_k(x) \bar{e}_k(\xi) e^{-\lambda_k t}, \tag{8.3.64}$$

which generalizes (8.2.36a). Similar expansions for the wave equation are left as Exercise 8.3.4.

A well-known particular case of (8.3.63) is that of $q = 0$:

$$u_t - \Delta u = 0, \quad u(x,0) = f(x), \quad u(\Gamma,t) = 0, \tag{8.3.65}$$

whose solution (which could also be obtained by separation of variables) is

$$u(x,t) = \sum f_k e^{-\lambda_k t} e_k(x) = \sum \langle f, e_k \rangle e^{-\lambda_k t} e_k(x). \tag{8.3.66}$$

Note that $\lim_{t \to 0} u(x,t) = 0$, uniformly on $\bar{\Omega}$. Thus, $u \equiv 0$ is the steady-state temperature after all "transients" due to the initial temperature have disappeared. (See also Exercise 8.3.10.)

Let us now list some properties of the solution of (8.3.65) whose explicit form is given by (8.3.66):

1. If $f_1(x) \geqslant f_2(x)$, the corresponding solutions satisfy $u_1(x,t) \geqslant u_2(x,t)$. This follows from the Comparison Theorem (Theorem 8.2.3).

2. If $f(x) \geqslant 0$, then $\max u(x,t)$ is a *monotonically decreasing function of t*. Indeed, let $\max_x f(x) = \|f\|$; then the constant $\|f\|$ is a solution of the heat equation

with data $\{0, \|f\|, \|f\|\}$ which dominates $\{0, f, 0\}$, so that $u(x,t) \leqslant \|f\|$. A similar argument applied to $t > T$ shows that $u(x,t) \leqslant \beta \doteq \max_x u(x,T)$, which is the desired result.

3. If $f(x) \geqslant 0$ and $\Delta f < 0$ in Ω, then for each $x \in \Omega$, $u(x,t)$ decreases monotonically in t. The proof consists of first noting that

$$u_t(x, 0+) = \Delta u(x, 0+) = \Delta f,$$

so u_t is initially negative where $\Delta f < 0$, but is initially positive where $\Delta f > 0$. Thus, where f is convex, u initially increases. If, however, $\Delta f < 0$ for all x in Ω, then $u(x,t) \leqslant f(x)$ for small t, say $t \leqslant \tau$. Setting $v(x,t) = u(x, t+\tau)$, we see that $v(x,t)$ satisfies (8.3.65) with smaller data so that $v(x,t) \leqslant u(x,t)$ and hence $u(x, t+\tau) \leqslant u(x,t)$. Since this inequality holds for all sufficiently small τ, $u(x,t)$ decreases monotonically in time.

4. (a) If $\langle f, e_1 \rangle \neq 0$, then for large t, $u(x,t)$ behaves like $\langle f, e_1 \rangle e^{-\lambda_1 t} e_1(x)$, and since $e_1(x) > 0$ in Ω, the approach to the zero steady state is entirely from one side. (Thus, even if f changes sign but $\langle f, e_1 \rangle > 0$, u is strictly positive in Ω for large t.)

   (b) If $f = e_2$, then $u(x,t) = e^{-\lambda_2 t} e_2(x)$, and since $\langle e_1, e_2 \rangle = 0, e_2$ must change sign on Ω and so does u.

5. By integrating (8.3.65) with respect to t, we obtain

$$\eta_t - \Delta \eta = f(x), \quad \eta(x, 0) = \eta(\Gamma, t) = 0$$

where $\eta = \int_0^t u(x,\tau)\, d\tau$. If $f(x) > 0, \eta(x,t)$ increases monotonically in time and tends, as $t \to \infty$, to the solution of the steady-state problem

$$-\Delta \eta_\infty = f(x), \quad x \in \Omega; \quad \eta_\infty(x) = 0, \quad x \in \Gamma.$$

Thus, we find that

$$\int_0^\infty u(x,\tau)\, d\tau = \eta_\infty(x), \tag{8.3.67}$$

a result which could also have been obtained by integrating (8.3.66) term–by–term and using the eigenfunction expansion for the Poisson equation. Equation (8.3.67) has a useful interpretation: For fixed x the curve $u(x,t)$ decreases from $f(x)$ to 0; the area under the curve is $\eta_\infty(x)$, which can be regarded as a representative decay time for the approach to the steady state.

Let us now discuss the relationship between the eigenvalues of (8.3.49) and the domain Ω, when $\Omega \subset \mathbb{R}^2$. The eigenvalues are just the natural frequencies of a membrane (or drum) stretched over Ω and fixed at the boundary Γ. The eigenvalue λ_1 is the fundamental frequency and the corresponding fundamental mode $e_1(x)$ is of one sign (taken as positive); the higher eigenvalues are known as *harmonics*. Except for the simplest geometries, such as the rectangle and the circle (see Exercise 8.3.3), there is little hope of calculating the λ_k's explicitly. Of course, we can use variational methods [such as (8.3.56) and (8.3.56a)] or perturbation methods (see Section 9.3) to approximate the eigenvalues. In a different direction, Exercise 8.4.14 shows that the circular membrane possesses the following property: Of all membranes of equal area, the circular one has the lowest fundamental frequency.

It is also possible to obtain information on the asymptotic behavior of the eigenvalues of (8.3.49) by studying the bilinear formula (8.3.64) for Green's function of the diffusion equation! Setting $\xi = x$ and integrating over Ω, we find that

$$\int_\Omega g(x,t;x,0)\,dx = \sum_{k=1}^{\infty} e^{-\lambda_k t} = t\int_0^\infty N(\lambda)e^{-\lambda t}\,d\lambda, \tag{8.3.68}$$

where $N(\lambda) = \sum_{k=1}^{\infty} H(\lambda - \lambda_k)$ is the number of eigenvalues less than λ and H is the usual Heaviside function. We interpret $g(x,t;\xi,0)$ as the concentration due to an interior unit source at ξ in the presence of an absorbing boundary. It seems plausible that for small t, g has not yet felt the presence of the boundary and must be nearly indistinguishable from the free space fundamental solution (see Exercise 2.5.5)

$$E(x,t;\xi,0) = \frac{1}{4\pi t}e^{-|x-\xi|^2/4t}, \quad x,\xi \in \mathbb{R}^2.$$

Replacing g by E on the left side of (8.3.68), we find that

$$\int_0^\infty N(\lambda)e^{-\lambda t}\,d\lambda \sim \frac{|\Omega|}{4\pi t^2} \quad (\text{small } t), \tag{8.3.69}$$

where $|\Omega|$ is the area of Ω. The left side of (8.3.69) is a Laplace transform whose behavior for small values of the transform variable t is related to the behavior of N for large values of the original variable λ. We can then conclude that

$$N(\lambda) \sim \frac{\lambda|\Omega|}{4\pi} \quad \text{for large } \lambda.$$

A somewhat more delicate analysis (see, for example, Stakgold [40]) taking account of the boundary (but neglecting its curvature) yields

$$N(\lambda) \sim \frac{\lambda|\Omega|}{4\pi} - \frac{|\Gamma|}{4\pi}\sqrt{\lambda} \quad \text{for large } \lambda, \tag{8.3.70}$$

where $|\Gamma|$ is the length of the boundary.

From just the asymptotic behavior of the eigenvalues of a drum we can find both its area and perimeter. It would appear—and was long thought to be true—that the exact knowledge of all the eigenvalues $\{\lambda_k\}$ would surely determine Ω unambiguously (modulo translations and rotations, of course). So it was quite a surprise when in 1992, Carolyn Gordon, David Webb, and Scott Wolpert exhibited two noncongruent polygons with the same eigenvalues! Such domains are called *isospectral*. Since these polygons are not convex, there remains an open question as to whether there exist two isospectral convex domains. (See Chapman [6] or Driscoll [11] for an accessible discussion of the question.)

EXERCISES

8.3.1 Green's function for the unit disk Ω can also be obtained by images. We set

$$g(x,\xi) = \frac{1}{2\pi}\log\frac{1}{|x-\xi|} + v(x,\xi), \quad \xi \in \Omega, \tag{8.3.71}$$

where $v(x,\xi)$ is required to be harmonic in Ω and g must vanish on Γ. The idea of images is to express v as the potential due to sources lying outside Ω. Let ξ^* be the inverse point of ξ with respect to the unit circle, that is, the polar coordinates of ξ^* are $(1/r, \phi)$ if those of ξ are (r, ϕ).

(a) Show that

$$g(x,\xi) = \frac{1}{2\pi}\log\frac{1}{|x-\xi|} - \frac{1}{2\pi}\log\frac{1}{|x-\xi^*|} + \frac{1}{2\pi}\log|\xi|. \tag{8.3.72}$$

(b) Show that using (8.3.72) in (8.3.15) reduces the latter to (8.3.8).

8.3.2 Let Ω be an arbitrary bounded domain in $\mathbb{R}^3$, and let Γ be its boundary. Green's function $g(x,\xi)$ satisfies (8.3.52). It will sometimes be useful to write

$$g = \frac{1}{4\pi|x-\xi|} + v(x,\xi),$$

where v is harmonic in Ω. Demonstrate the following properties of g:

(a) g exists and is unique.

(b) $g(x,\xi) = g(\xi,x)$.

(c) g is positive in Ω. (*Hint*: Draw a small ball about ξ, and use the maximum principle outside the ball.)

(d) $g < 1/4\pi|x-\xi|, \quad x,\xi \in \Omega$.

(e) The integral operator G defined by (8.3.53) is an H.-S. operator on the space $L_2(\Omega)$.

8.3.3 (a) Consider (8.3.49) on the rectangle $0 < x_1 < a$, $0 < x_2 < b$. Show that the eigenvalues are

$$\lambda_{m,n} = \frac{m^2\pi^2}{a^2} + \frac{n^2\pi^2}{b^2}, \quad m = 1,2,\ldots; \quad n = 1,2,\ldots$$

and the eigenfunctions (orthonormal over the rectangle) are

$$e_{m,n} = \left(\frac{4}{ab}\right)^{1/2} \sin\frac{m\pi x_1}{a} \sin\frac{n\pi x_2}{b}.$$

Obviously, a double-index notation is more convenient in this problem.

(b) Consider (8.3.49) on the unit disk. Use polar coordinates and separation of variables to show that the eigenvalues are

$$\lambda_{n,k} = [\beta_k^{(n)}]^2; \quad n = 0, \pm 1, \pm 2, \ldots; \quad k = 1, 2, \ldots,$$

where $\beta_k^{(n)}$ is the kth positive root of $J_n(x)$. Since J_{-n} is proportional to J_n, $\lambda_{n,k} = \lambda_{-n,k}$. The corresponding eigenfunctions are proportional to

$$e^{in\phi} J_n(\beta_k^{(n)} r).$$

The normalization factor can be found by the method of Example 5, Section 7.2, yielding the orthonormal basis

$$e_{n,k} = \frac{e^{in\phi}}{(\pi)^{1/2} J_n'(\beta_k^{(n)})} J_n(\beta_k^{(n)} r).$$

8.3.4 Let $\{e_i\}$ be the orthonormal basis of eigenfunctions of (8.3.49). Write the solution of

$$\frac{\partial^2 u}{\partial t^2} - \Delta u = q(x,t), \quad x \in \Omega, \quad t > 0;$$

$$u(x,0) = f(x), \quad \frac{\partial u}{\partial t}(x,0) = g(x); \qquad u(x,t) = 0, \quad x \in \Gamma, \quad t > 0,$$

as an eigenfunction expansion.

8.3.5 Let a unit source in free space in $\mathbb{R}^2$ be located at the point with polar coordinates $(r_0, 0)$. By taking a Mellin transform on the radial coordinate, show that

$$g(r,\phi; r_0, 0) = \frac{1}{2\pi} \int_{-\infty}^{\infty} \frac{r^{i\nu} r_0^{-i\nu}}{2\nu \sinh \nu\pi} \cosh \nu(\pi - |\phi|)\, d\nu.$$

8.3.6 (a) Show that the Neumann problem

$$-\Delta u = q(x), \quad x \text{ in } \Omega; \qquad \frac{\partial u}{\partial n} = f, \quad x \in \Gamma, \tag{8.3.73}$$

can have a solution only if f and q satisfy the solvability condition

$$\int_\Omega q(x)\, dx = -\int_\Gamma f(x)\, dS_x, \tag{8.3.74}$$

and give a physical interpretation.

(b) The modified Green's function for the Neumann problem satisfies

$$-\Delta g_M = \delta(x-\xi) - \frac{1}{V}, \quad x, \xi \in \Omega; \qquad \frac{\partial g_M}{\partial n} = 0, \quad x \in \Gamma, \tag{8.3.75}$$

where V is the volume of Ω. Construct g_M when Ω is the unit disk in $\mathbb{R}^2$ by a Fourier expansion in the angle ϕ. Show that

$$g_M(r, \phi; r_0, 0) = A + \frac{r^2}{4\pi} - \frac{1}{2\pi} \log r_> + \sum_{n=1}^{\infty} \frac{\cos n\phi}{2\pi n} r_<^n (r_>^n + r_>^{-n}),$$

where the notation is the same as in (8.3.13).

8.3.7 Calculate explicitly the potentials of the following charge distributions in $\mathbb{R}^3$:

(a) Line density 1 along the segment $-1 \leqslant z \leqslant 1$.

(b) Surface density 1 on the unit disk $z = 0, r \leqslant 1$.

(c) A normally oriented double layer of uniform density 1 on the unit disk $z = 0, r \leqslant 1$.

Show in (a) that the potential itself becomes singular as the observation point approaches the charge distribution. For (b) verify (8.3.30) and (8.3.32) at the center of the charge distribution. For (c) verify (8.3.33) at the center of the dipole layer.

8.3.8 Consider (8.3.2) when $f(\phi) = \sum_{n=1}^{\infty} b_n \sin n\phi$. Show that (8.3.6) now takes the form

$$u(r, \phi) = \sum_{n=1}^{\infty} b_n r^n \sin n\phi.$$

Show that

$$\frac{\partial u}{\partial \phi}(r, 0) = \sum_{n=1}^{\infty} n b_n r^n$$

and that for $a < 1$,

$$J_a(u) \doteq \int_0^a r\, dr \int_{-\pi}^{\pi} |\text{grad } u|^2 \, d\phi = \pi \sum_{n=1}^{\infty} n b_n^2 a^{2n}.$$

Now consider the special case

$$f(\phi) = \sum_{k=1}^{\infty} \frac{\sin k^4 \phi}{k^2},$$

which is clearly a continuous function on the unit circle. Show that

$$\frac{\partial u}{\partial \phi}(r, 0) \to \infty$$

as $r \to 1-$ and that $\lim_{a \to 1} J_a(u) = \infty$.

8.3.9 This exercise shows the equivalence of formulations (8.3.1a), (8.3.1b), and (8.3.1c). The arguments follow the ideas of Section 0.5.

(a) If u is a solution of (8.3.1a), it satisfies (8.3.1b).

(b) If u satisfies (8.3.1b), it solves (8.3.1a). This requires a generalization of Lemma 0.5.1 to $\mathbb{R}^n$.

(c) If u is a minimizer of J, then u satisfies (8.3.1b).

(d) If u satisfies (8.3.1b), it is a minimizer of J.

8.3.10 Consider the evolution problem

$$\frac{\partial u}{\partial t} - \Delta u = q(x), \quad (x,t) \in \Omega \times (0,\infty);$$
$$u(x,0) = f(x); \qquad u(x,t) = h(x), \quad x \in \Gamma,$$

where the source term q and the boundary temperature h may depend on position but not on time. Writing $U(x,t) = U(x) + v(x,t)$, where $U(x)$ is the solution of the elliptic problem

$$-\Delta U = q(x), \quad x \in \Omega; \qquad U(x) = h(x), \quad x \in \Gamma,$$

show that $v(x,t) \to 0$ as $t \to \infty$ and therefore $U(x)$ is the steady-state temperature.

8.3.11 The goal is to show that the fundamental eigenfunction $e_1(x)$ of (8.3.49) is of one sign in Ω. Suppose that e_1 changed sign. Let $e_1 > 0$ on Ω^+ and $e_1 \leqslant 0$ on Ω^-. Clearly, $|e_1|$ is also a minimizer of (8.3.56) reformulated for $u \in H_0^{(1)}(\Omega)$. Thus $|e_1|$ and $e_1^+(x) = \max[e_1(x), 0]$ are also fundamental eigenfunctions of (8.3.49). Show that this contradicts the strong maximum principle: If $-\Delta u \geqslant 0$ in Ω and $u(x)$ assumes its minimum in Ω, then u is identically constant on $\bar{\Omega}$.

8.4 VARIATIONAL PRINCIPLES FOR INHOMOGENEOUS PROBLEMS

As already pointed out in the preceding section and in Section 0.5, the solution $u(x)$ of a boundary value problem can also be characterized as the element for which a related functional J is stationary. Such a stationary characterization is a *variational principle* whose Euler-Lagrange equation is the differential equation of the original BVP. The functional J will usually have physical significance as an energy or equivalent quantity such as capacity or torsional rigidity. In many instances the stationary value of J turns out to be a minimum or maximum rather than a saddle-point. When the BVP is linear, the corresponding J is quadratic.

The equivalence between the BVP and the variational principle can be exploited numerically and theoretically. The variational principle is the basis for the Ritz-Rayleigh method for finding an approximation, say $\tilde{u}$, to the exact solution u of the BVP, and the corresponding approximation $J(\tilde{u})$ to the stationary value $J(u)$. Since J is "flat" at $\tilde{u}$, it is not surprising that $J(\tilde{u})$ furnishes a better estimate of $J(u)$ than

$\tilde{u}$ does of u. The new aspect of the present section is that the variational principle will be used to prove existence for the BVP (at least in some generalized sense).

We shall analyze the case of an inhomogeneous linear differential equation (ordinary or partial) subject to homogeneous boundary conditions. The appropriate modifications for inhomogeneous boundary conditions will be taken up later.

To fix ideas, let us examine the concrete BVP

$$-u'' + ku = f(x), \quad 0 < x < 1; \qquad u(0) = u(1) = 0, \tag{8.4.1}$$

where $f(x)$ is a given real-valued function and k a *nonnegative* constant. Here $f(x)$ denotes the source term in the differential equation (whereas in Section 8.3 it stood for the boundary data). To translate this problem into an operator equation we must specify the domain of the differential operator on the left of (8.4.1). The simplest approach is to define the operator A by

$$Au \doteq -u'' + ku \tag{8.4.2}$$

on the domain D_A consisting of all functions $u(x)$ in $C^2[0,1]$ with $u(0) = u(1) = 0$. Here we shall view D_A as a linear manifold in $H = L_2^{(r)}(0,1)$; clearly, D_A is dense in H. Thus (8.4.1) will be replaced by the operator equation

$$Au = f, \quad u \in D_A, \tag{8.4.3}$$

where we realize that we can hope for solutions only if $f(x)$ is a continuous function. The reader who has thoroughly mastered Chapter 5 might start more ambitiously with (8.4.2) defined on the larger domain consisting of functions $u(x)$ satisfying the boundary conditions and such that u' is absolutely continuous with u'' in L_2, but we shall not take this as a starting point.

One physical interpretation of (8.4.1) is that $u(x)$ is the transverse deflection of a taut string, fixed at its ends, under an applied, transverse, distributed load $f(x)$ and subject to a restoring force proportional to the displacement u (so that k is a "spring constant"). The three terms

$$\frac{1}{2}\int_0^1 (u')^2\,dx, \quad \frac{k}{2}\int_0^1 u^2\,dx, \quad \int_0^1 fu\,dx$$

represent, respectively, the strain energy of the string, the energy stored in the spring, and the work done by the applied load to take the string from its reference configuration to its deflected state. We shall find it convenient to deal with *twice* the total potential energy in the deflected state:

$$\int_0^1 (u')^2\,dx + k\int_0^1 u^2\,dx - 2\int_0^1 fu\,dx. \tag{8.4.4}$$

If u is replaced in (8.4.4) by an arbitrary element v in D_A, a functional J is defined on D_A:

$$J(v) = \int_0^1 (v')^2\,dx + k\int_0^1 v^2\,dx - 2\int_0^1 fv\,dx, \quad v \in D_A. \tag{8.4.5}$$

We can regard $J(v)$ as the total potential energy associated with the virtual deflection v, whereas $J(u)$ is the energy associated with the correct deflection u. For any element $v \in D_A$ we have

$$\langle Av, v\rangle = \int_0^1 (-v'' + kv)v\,dx = \int_0^1 [(v')^2 + kv^2]\,dx, \tag{8.4.6}$$

so that we can rewrite J as

$$J(v) = \langle Av, v\rangle - 2\langle f, v\rangle, \quad v \in D_A. \tag{8.4.7}$$

We note that if u is a solution of (8.4.3), then $\langle Au, u\rangle = \langle f, u\rangle$, so that

$$J(u) = -\langle f, u\rangle = \langle Au, u\rangle. \tag{8.4.8}$$

We now consider the problem of finding the minimum of $J(v)$ among all functions $v \in D_A$. We write this minimal problem in the schematic form

$$J(v) \to \min, \quad v \in D_A. \tag{8.4.9}$$

Of course, it is not evident that J assumes a minimum on D_A.

The following theorem of *minimum potential energy* shows that problems (8.4.3) and (8.4.9) are equivalent.

Theorem.

(a) Let u be a solution of (8.4.3); then $J(u) \leqslant J(v)$ for all $v \in D_A$, so that u is a solution of the minimum problem (8.4.9).

(b) Let u be a solution of the minimum problem (8.4.9); then u satisfies (8.4.3).

Proof. Let v and w be arbitrary elements of D_A, and set $v = w + h$. Then from (8.4.7) we find that

$$J(v) - J(w) = \langle A(w+h), w+h\rangle - \langle Aw, w\rangle - 2\langle f, h\rangle = 2\langle Aw - f, h\rangle + \langle Ah, h\rangle. \tag{8.4.10}$$

Now let (8.4.1) have a solution u, and set $w = u$ in (8.4.10). Then, since $\langle Ah, h\rangle \geqslant 0$ for $h \in D_A$, we conclude that $J(v) \geqslant J(u)$, which proves part (a).

Next, assume that there exists an element $u \in D_A$ for which the minimum in (8.4.9) is attained. Then, from (8.4.10) with $w = u$, we find that

$$2\langle Au - f, h\rangle + \langle Ah, h\rangle \geqslant 0 \quad \text{for every } h \in D_A.$$

Considering elements h of the form $\varepsilon\eta$, where ε is a real number and η is an element of D_A, we obtain

$$\begin{aligned} 2\langle Au - f, \eta\rangle + \varepsilon\langle A\eta, \eta\rangle &\geqslant 0, \qquad \varepsilon > 0, \quad \eta \in D_A,\\ 2\langle Au - f, \eta\rangle + \varepsilon\langle A\eta, \eta\rangle &\leqslant 0, \qquad \varepsilon < 0, \quad \eta \in D_A. \end{aligned}$$

By taking the limit as $\varepsilon \to 0$, we conclude that $\langle Au - f, \eta \rangle = 0$ for all $\eta \in D_A$. Since the set D_A is dense in H, it follows that $Au - f = 0$, completing the proof of part (b). □

REMARK. We have *not* proved existence in (8.4.3) or (8.4.9), but have proved only that a solution to either problem is necessarily a solution to the other. If we are willing to accept (or prove independently) that (8.4.3) has a solution, that solution is also the minimizing function in (8.4.9). If, however, we want to use variational methods to establish that (8.4.3) or (8.4.1) has a solution, we must prove directly the existence of a minimum for J.

We shall give the existence proof for the general case shortly, but first let us outline the steps as related to the present problem (8.4.1) or (8.4.3). Our starting point is $J(v)$ as given by (8.4.5). It is relatively easy to show that J is bounded below on D_A, so that J has an infimum, say d, on D_A. If there is an element $u \in D_A$ for which $J(u) = d$, the infimum is attained on D_A, so that the minimum problem for J on D_A has the solution u. There are two reasons for considering the minimum problem for J on a larger domain M (as yet unspecified). First, J may not have a minimum on D_A; if f is only piecewise continuous, the solution of (8.4.1) is not C^2, so that (8.4.3) has no solution and hence, by the theorem of minimum potential energy, J does not have a minimum on D_A. The trouble here is simply that D_A is not a suitable domain on which to analyze the BVP when f is not continuous. The second reason is more subtle: even if J has a minimum on D_A (and therefore the BVP has a classical solution in D_A), it may be hard to prove this fact directly; it turns out to be easier to prove the existence of a minimum on the larger domain M and *then* to show that the minimizing element lies in D_A.

Thus we shall try to enlarge D_A so that a minimum for J can be guaranteed on the new domain. It is reasonable to hope for such an extension since the expression for J involves only the first derivative of v, whereas an element of D_A has a continuous derivative of order 2.

We write (8.4.5) as

$$J(v) = a(v, v) - 2l(v),$$

where $a(v, w)$ is the bilinear form

$$a(v, w) \doteq \int_0^1 (v'w' + kvw)\, dx,$$

and $l(v)$ is the linear functional

$$l(v) \doteq \int_0^1 fv\, dx.$$

Since $a(v, v) > 0$ for all nonzero elements v in D_A, it is clear that $a(v, w)$ is an acceptable inner product on D_A, known as the *energy inner product*. The corresponding *energy norm* $\|\cdot\|_a$ is given by

$$\|v\|_a^2 = a(v, v) = \int_0^1 [(v')^2 + kv^2]\, dx. \tag{8.4.11}$$

We now complete the set D_A in the energy norm. This gives us the larger domain M, on which $J(v)$ given by (8.4.11) can be guaranteed to have a minimum. The corresponding minimizing function $u \in M$ satisfies the equivalent variational equation

$$a(u, \eta) = l(\eta) \quad \text{for every } \eta \in M. \tag{8.4.12}$$

Equation (8.4.12) is a weak form of the original BVP (8.4.1) and is the integral form of the Euler-Lagrange equation for the variational principle. It is worth noting that (8.4.12) is known in mechanics as the principle of virtual work (see Section 0.5). If f is given as a continuous function, one can show that the solution of (8.4.12) is a classical solution of (8.4.1). When f is only in L_2, the solution of (8.4.12) is regarded as a generalized solution of (8.4.1). In our problem, which involves an ordinary differential equation, this generalized solution is in the class of functions u with u' absolutely continuous and u'' in L_2. However, for partial differential equations the generalized solution is not characterized so easily.

To avoid too much repetition, we now proceed to the general case, after which we shall return to our special BVP (8.4.1) and to other examples.

Quadratic Functionals

Let H be a real Hilbert space with inner product $\langle \cdot, \cdot \rangle$ and norm $\| \cdot \|$. We warn the reader that a second inner product will be introduced later. Let J be a real-valued functional defined on a linear manifold D in H. We say that J is *stationary* at $u \in D$ if

$$\left[\frac{d}{d\varepsilon} J(u + \varepsilon \eta) \right]_{\varepsilon=0} = 0 \quad \text{for every } \eta \in D, \tag{8.4.13}$$

so that, at u, the directional derivative of J vanishes in every direction lying in D. Similar ideas, for operators as well as functionals, are discussed in Chapter 9.

Since our interest is in linear BVPs, we shall need to consider only quadratic functionals as in (8.4.5). Throughout we take

$$J(v) = a(v, v) - 2l(v), \tag{8.4.14}$$

where l is a linear functional and $a(v, v)$ is the quadratic form associated with a bilinear form $a(v, w)$ defined for $v, w \in D$. The following definitions are given in order of increasing stringency.

Definition. *The bilinear form $a(v, w)$ defined on D (that is, for $v, w \in D$) is said to be*

$$\left.\begin{array}{ll} \textit{symmetric:} & \textit{if } a(v,w) = a(w,v) \textit{ for all } v, w \in D, \\ \textit{nonnegative:} & \textit{if symmetric and } a(v,v) \geqslant 0 \textit{ for all } v \in D, \\ \textit{positive:} & \textit{if symmetric and } a(v,v) > 0 \textit{ for all } v \in D, v \neq 0, \\ \textit{coercive:} & \textit{if symmetric and there exists a positive constant } c \\ & \textit{such that } a(v,v) \geqslant c\|v\|^2 \textit{ for all } v \in D. \end{array}\right\} \tag{8.4.15}$$

The same terminology is used for an operator A on D if the bilinear form it generates, namely $a(v,w) \doteq \langle Av, w\rangle$, has the corresponding property (see also Section 5.8). A coercive form is also called *strongly positive*.

If the form is nonnegative, we have the Schwarz inequality

$$a^2(v,w) \leqslant a(v,v)a(w,w), \quad v,w \in D. \tag{8.4.16a}$$

If the form is positive, $a(v,w)$ can serve as an *energy inner product* with corresponding *energy norm* given by

$$\|v\|_a^2 = a(v,v),$$

and the Schwarz inequality can be rewritten as

$$|a(v,w)| \leqslant \|v\|_a \|w\|_a, \quad v,w \in D. \tag{8.4.16b}$$

Assuming only that the bilinear form $a(v,w)$ is symmetric, we note from (8.4.14) that

$$J(v+h) - J(v) = 2a(v,h) - 2l(h) + a(h,h) \quad \text{for all } v,h \in D, \tag{8.4.17}$$

so that

$$\left[\frac{d}{d\varepsilon}J(v+\varepsilon\eta)\right]_{\varepsilon=0} = 2a(v,\eta) - 2l(\eta) \quad \text{for all } v,\eta \in D. \tag{8.4.18}$$

We are therefore led to the following theorem.

Theorem 8.4.1. *Let $a(v,w)$ be a symmetric bilinear form for v, $w \in D$. Then the necessary and sufficient condition for the problem*

$$J(v) \to \textit{stationary}, \quad v \in D, \tag{8.4.19}$$

to have a solution $u \in D$ is that this element satisfy the variational equation

$$a(u,\eta) = l(\eta) \quad \textit{for every } \eta \in D. \tag{8.4.20}$$

For any solution u of these equivalent problems we have the reciprocity relation

$$a(u,u) = l(u), \tag{8.4.21}$$

so that

$$J(u) = -a(u,u) = -l(u). \tag{8.4.22}$$

Proof. If u is an element of D for which the variational equation (8.4.20) is satisfied, (8.4.18) shows that

$$\left[\frac{d}{d\varepsilon}J(u+\varepsilon\eta)\right]_{\varepsilon=0} = 0,$$

so that, by definition (8.4.13), J is stationary at u. If J is stationary at u, (8.4.18) and the definition of stationarity show that (8.4.20) is satisfied. Thus the equivalence of problems (8.4.19) and (8.4.20) has been proved. If u is a solution of these equivalent problems, we may substitute $\eta = u$ in (8.4.20) to obtain (8.4.21) and (8.4.22). □

For positive bilinear forms we can replace "stationary" by "minimum," and we can also guarantee uniqueness (but, of course, not existence as yet).

Theorem 8.4.2. *Let $a(v, w)$ be a positive bilinear form on D. Then the problem*

$$J(v) \to \min, \quad v \in D, \tag{8.4.23}$$

is equivalent to the variational equation (8.4.20). There is at most one solution to these equivalent problems.

Proof. If J has a minimum at u, J is certainly stationary at u, so that (8.4.20) holds. If (8.4.20) holds, then (8.4.17) shows that $J(u + h) - J(u) = a(h, h)$. Since $a(h, h) > 0$ for $h \neq 0$, J clearly has an absolute minimum at u and there is at most one minimizing element. □

For an existence proof we need two additional hypotheses: l must be bounded in the energy norm, and D must be complete in energy. The first hypothesis postulates the existence of a constant c such that

$$|l(v)| \leqslant c\|v\|_a \quad \text{for every } v \in D, \tag{8.4.24}$$

and the second means whenever a sequence $\{v_n\}$ in D satisfies $\|v_n - v_m\|_a \to 0$, there exists $v \in D$ such that $\|v_n - v\|_a \to 0$.

Theorem 8.4.3. *Let $a(v, w)$ be a positive bilinear form on D which is complete in energy. Let l satisfy (8.4.24). Then the minimum in (8.4.23) is actually attained for an element (necessarily unique) in D.*

Proof. Since D is complete in energy, it is a Hilbert space in its own right under the energy inner product. The boundedness of l allows us to use the Riesz representation theorem [see (4.7.2)] to guarantee the existence of a unique element $u \in D$ such that

$$l(v) = a(v, u) \quad \text{for all } v \in D. \tag{8.4.25}$$

We can therefore write

$$J(v) = \|v\|_a^2 - 2a(v, u) = \|v - u\|_a^2 - \|u\|_a^2, \tag{8.4.26}$$

so that it is clear that J attains its minimum at the element u and nowhere else. Hence u is the solution of the minimum problem and of the variational equation. Moreover, the minimal value of J is

$$J(u) = -\|u\|_a^2 = -l(u). \tag{8.4.27}$$

□

Theorem 8.4.3 still begs the question since the original domain D on which $a(v, w)$ is defined may not satisfy the additional hypotheses. Condition (8.4.24) is usually easy to verify in applications, but we are still left with the problem of constructing a manifold that is complete in energy.

We now show how to add elements to D to make it complete in energy. *We shall require that $a(v, w)$ be coercive* [*see* (8.4.15)] *on D*. Let $\{v_n\}$ be a sequence in D which is fundamental in energy. Thus $\|v_n - v_m\|_a \to 0$, and because of coercivity we also have $\|v_n - v_m\| \to 0$, so that $\{v_n\}$ is also fundamental in the ordinary norm. Since H is complete in the ordinary norm, there exists a unique element $v \in H$ such that $\|v_n - v\| \to 0$. The set of all such elements in H is the space of elements of finite energy.

Definition. *The space M of* elements of finite energy *is the set of elements $v \in H$ with the property that there exists a sequence $\{v_n\} \in D$ such that, simultaneously,*

$$\|v_n - v_m\|_a \to 0, \quad \|v_n - v\| \to 0. \tag{8.4.28}$$

The sequence $\{v_n\}$ is said to be representative *for v.*

We note that all elements of D belong to M since we can take $v_n = v$ in the definition. The energy norm and inner product can be extended to the whole of M by the obvious definitions

$$\|v\|_a = \lim_{n\to\infty} \|v_n\|_a, \quad a(v, w) = \lim_{n\to\infty} a(v_n, w_n),$$

where $\{v_n\}$ and $\{w_n\}$ are representative for v and w, respectively. One must, of course, show that the limits in these definitions exist and are independent of the representative sequences used, but we omit the proofs. Both (8.4.15) and (8.4.24) are then easily extended to the whole of M. It can be shown that M is complete in energy, so that the representative sequence $\{v_n\}$ in (8.4.28) also has the property

$$\|v_n - v\|_a \to 0. \tag{8.4.29}$$

Much of what we have done is related to the standard completion procedure described in Exercise 4.3.1.

Theorem 8.4.4. *Let $a(v, w)$ be coercive on D, and let M be the set of elements of finite energy (that is, the completion of D in energy). Then the problem*

$$J(v) \to \min, \quad v \in M, \tag{8.4.30}$$

has one and only one solution $u \in M$, which is also the solution of the variational equation

$$a(u, \eta) = l(\eta) \quad \textit{for every } \eta \in M. \tag{8.4.31}$$

Proof. Since $a(v, w)$ is coercive on M and (8.4.24) holds for $\eta \in M$, we can apply Theorem 8.4.3. □

In most applications we are interested in the inhomogeneous problem

$$Au = f, \quad u \in D_A, \tag{8.4.32}$$

where A is a symmetric operator on the linear manifold D_A dense in H. We associate with A the symmetric bilinear form

$$a(v, w) \doteq \langle Av, w \rangle, \quad v, w \in D_A. \tag{8.4.33}$$

The linear functional l is defined from

$$l(v) \doteq \langle f, v \rangle, \tag{8.4.34}$$

so that

$$J(v) = \langle Av, v \rangle - 2\langle f, v \rangle, \quad v \in D_A. \tag{8.4.35}$$

The variational equation on D_A becomes

$$\langle Au - f, \eta \rangle = 0 \quad \text{for all } \eta \in D_A,$$

which is equivalent (since D_A is dense in H) to

$$Au = f, \quad u \in D_A. \tag{8.4.36}$$

Theorems 8.4.1 and 8.4.2 now show the equivalence between a variational principle for (8.4.35) and problem (8.4.36), whereas Theorem 8.4.3 gives existence when D_A is complete in energy. Of course when we construct M by completing D_A in energy, relation (8.4.33) between a and A is lost, so that Theorem 8.4.4 cannot be stated in terms of A. However, if it happens that the solution u of (8.4.31), and hence of (8.4.30), lies in the subset D_A of M, then u satisfies $Au = f$; if not, we view u as a *generalized* solution of $Au = f$.

Suppose that A is coercive and that (8.4.32) has a solution u (necessarily unique) in D_A. We can then obtain bounds for $J(u)$ *from both sides* in terms of an arbitrary element $v \in D_A$. With $h = v - u$, we have

$$J(u) = J(v) - \langle Ah, h \rangle.$$

Since $\langle Ah, h \rangle > 0$ for $h \neq 0$, we have the usual upper bound $J(u) \leqslant J(v)$. To find a lower bound we first recall that, from coercivity,

$$\langle Ah, h \rangle \geqslant L_A \|h\|^2, \quad \text{where } L_A > 0.$$

Schwarz's inequality then shows that

$$\|Ah\| \geqslant L_A \|h\|.$$

It follows that

$$\langle Ah, h \rangle \leqslant \|Ah\| \, \|h\| \leqslant \frac{1}{L_A} \|Ah\|^2 = \frac{1}{L_A} \|Av - f\|^2,$$

which gives the explicit lower bound

$$J(u) \geqslant J(v) - \frac{1}{L_A}\|Av - f\|^2. \tag{8.4.37}$$

An improved upper bound for $J(u)$ can be obtained on the additional assumption that A is bounded above:

$$\langle Ah, h\rangle \leqslant U_A\|h\|^2.$$

It is known that there exists an unambiguous positive symmetric operator $A^{1/2}$ (the *square root* of A) with the property $A^{1/2}A^{1/2} = A$. We then find that

$$\begin{aligned}\|Ah\|^2 = \langle Ah, Ah\rangle &= \langle A(A^{1/2}h), A^{1/2}h\rangle \leqslant U_A\|A^{1/2}h\|^2\\ &= U_A\langle A^{1/2}h, A^{1/2}h\rangle = U_A\langle Ah, h\rangle,\end{aligned}$$

so that

$$\langle Ah, h\rangle \geqslant \frac{1}{U_A}\|Ah\|^2 = \frac{1}{U_A}\|Av - f\|^2,$$

and, together with (8.4.37), we have

$$J(v) - \frac{1}{L_A}\|Av - f\|^2 \leqslant J(u) \leqslant J(v) - \frac{1}{U_A}\|Av - f\|^2. \tag{8.4.38}$$

A few examples may serve to clarify some of the ideas just presented, particularly the construction of the space M complete in energy.

Example 1. For purposes of comparison we shall simultaneously study the BVP (8.4.1) and a second BVP obtained from (8.4.1) by changing the boundary condition at the right end to $u'(1) = 0$. In operator form we have

$$A_1u = f, \quad u \in D_1, \tag{8.4.39}$$

$$A_2u = f, \quad u \in D_2, \tag{8.4.40}$$

where $A_1u = A_2u = -u'' + ku$, D_1 is the domain consisting of all functions $u(x)$ in $C^2[0, 1]$ with $u(0) = u(1) = 0$, and D_2 is the domain of all functions $u(x)$ in $C^2[0, 1]$ with $u(0) = u'(1) = 0$. Both D_1 and D_2 are linear manifolds dense in $H = L_2(0, 1)$.

If v, $w \in D_1$, we find, on integration by parts, that

$$\langle A_1v, w\rangle = \int_0^1 (v'w' + kvw)\,dx.$$

If v, $w \in D_2$, we obtain exactly the same expression for $\langle A_2v, w\rangle$, so that for both problems we define the same bilinear form:

$$a(v, w) \doteq \int_0^1 (v'w' + kvw)\,dx$$

with the associated quadratic form

$$a(v, v) = \int_0^1 [(v')^2 + kv^2]\, dx.$$

It is easy to see that $a(v, v) > 0$ for all nontrivial v in D_1 and D_2. Thus $a(v, w)$ can serve as the energy inner product with corresponding energy norm given by $\|v\|_a^2 = a(v, v)$. Note that the energy norm can also be written in terms of the L_2 norms of v and its first derivative,

$$\|v\|_a^2 = a(v, v) = \|v'\|^2 + k\|v\|^2. \tag{8.4.41}$$

It is clear that for $k > 0$ the form is coercive [it suffices to set $c = k$ in (8.4.15)]. Consider, therefore, the case $k = 0$ [see also the Example following (5.8.6)]. Then, if $v \in D_1$ or D_2, we have $v(0) = 0$ and hence $v(x) = \int_0^x v'(t)\, dt$, so that by the Schwarz inequality,

$$v^2(x) \leqslant \int_0^x 1^2\, dt \int_0^x (v')^2\, dt \leqslant x \int_0^1 (v')^2\ dt,$$

from which we infer the desired inequality

$$\int_0^1 (v')^2\ dt \geqslant 2\|v\|^2, \qquad v \in D_1 \quad \text{or} \quad v \in D_2.$$

The constant 2 that appears on the right side can be improved by appealing to the minimum characterization of eigenvalues. For instance, the best possible inequality for $v \in D_1$ is

$$\int_0^1 (v')^2\, dt \geqslant \pi^2\|v\|^2.$$

Let M_1 and M_2 be the sets of functions of finite energy corresponding to the BVPs (8.4.39) and (8.4.40), respectively. A function $v \in M_1$ if there exists a sequence $\{v_n\}$ in D_1 such that

$$\|v_n - v_m\|_a \to 0 \quad \text{and} \quad \|v_n - v\| \to 0. \tag{8.4.42}$$

The same definition applies to M_2 except that the sequence $\{v_n\}$ must then be chosen from D_2. In either case we see from (8.4.41) that conditions (8.4.42) are equivalent to

$$\|v_n - v\| \to 0 \quad \text{and} \quad \|v_n' - v_m'\| \to 0,$$

from the second of which we infer that v_n' converges in L_2 to some element w. Conditions (8.4.42) are therefore equivalent to the existence of elements $v, w \in L_2$, for which, simultaneously,

$$\|v_n - v\| \to 0 \quad \text{and} \quad \|v_n' - w\| \to 0. \tag{8.4.43}$$

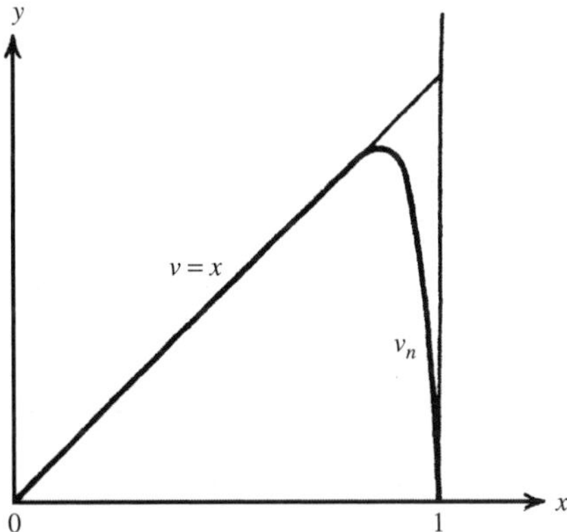

Figure 8.9

Whenever (8.4.43) holds with a sequence $\{v_n\}$ in $C^1[0, 1]$, we refer to w as the *strong* L_2 *derivative* of v and write $w = v'$. In one dimension this notion is identical with the following property: $v(x)$ is absolutely continuous with the first derivative in L_2 (see Chapter 5 for the definition and discussion of absolute continuity). The space of elements v having a strong L_2 derivative is known as the *Sobolev space* $H^1(0, 1)$. (See also Example 6, Section 4.5.) Thus both M_1 and M_2 are subsets of H^1, M_1 being formed by sequences $\{v_n\}$ in D_1 and M_2 being formed by sequences $\{v_n\}$ in D_2.

An immediate consequence of (8.4.43) is that $v_n(x)$ tends to $v(x)$ *uniformly* on $0 \leqslant x \leqslant 1$. We can verify this easily for elements in M_1 and M_2. From Schwarz's inequality we have

$$\left| \int_0^x v_n' \, dt - \int_0^x w \, dt \right|^2 \leqslant \int_0^x 1^2 \, dt \int_0^x |v_n' - w|^2 \, dt \leqslant \|v_n' - w\|^2,$$

so that $\int_0^x v_n' \, dt$ converges uniformly to $\int_0^x w \, dt$ on $0 \leqslant x \leqslant 1$. Since $v_n(0) = 0$, we find that $v_n(x)$ converges uniformly to $\int_0^x w \, dt$, which must therefore coincide with $v(x)$. Hence

$$v(x) = \int_0^x w \, dt,$$

which confirms our earlier statement that $v(x)$ is absolutely continuous with its derivative in L_2. For elements in M_1 the same procedure, using integrals from x to 1 instead of from 0 to x, yields $v(x) = -\int_x^1 w \, dt$. Thus elements in M_1 satisfy the two boundary conditions $v(0) = v(1) = 0$, but all we can show for elements in M_2 is that $v(0) = 0$.

One can also show geometrically that M_1 and M_2 must be different. We claim that the function $v = x$ belongs to M_2 but not to M_1. If v were to belong to M_1, there would be a sequence $\{v_n\}$ in D_1 with the simultaneous properties $\|v_n - x\| \to 0$,

$\|v_n' - 1\| \to 0$. In Figure 8.9 we have drawn a candidate for such a sequence $\{v_n\}$. The sequence satisfies $\|v_n - x\| \to 0$ but not $\|v_n' - 1\| \to 0$. The latter condition is violated because v_n' becomes infinite too rapidly near $x = 1$; more precisely, if we had $\|v_n' - 1\| \to 0$ and $\|v_n - x\| \to 0$, then v_n would converge uniformly to x on $0 \leqslant x \leqslant 1$, but this is impossible as long as v_n vanishes at $x = 1$ while v does not. On the other hand, $v \in M_2$; the sequence $\{z_n\}$ of Figure 8.10 satisfies $\|z_n - x\| \to 0$ and $\|z_n' - 1\| \to 0$. To prove the second condition, note that

$$\begin{aligned} &|z_n' - 1| \leqslant 1, \quad && x_n \leqslant x \leqslant 1, \\ &z_n' = 1, && 0 \leqslant x < x_n. \end{aligned}$$

Since $x_n \to 1$ as $n \to \infty$, $\|z_n' - 1\| \to 0$, as claimed.

To recapitulate: M_1 is the set of elements $v(x)$ in $H^1(0,1)$ satisfying $v(0) = v(1) = 0$. M_2 is the set of elements $v(x)$ in $H^1(0,1)$ satisfying $v(0) = 0$.

Theorem 8.2.4 then gives the following results. Setting

$$J(v) = a(v,v) - 2l(v) = \int_0^1 [(v')^2 + kv^2]\, dx - 2\int_0^1 fv\, dx,$$

we find that the problem

$$J(v) \to \min, \quad v \in M_1,$$

has a unique solution u which satisfies

$$a(u,\eta) = l(\eta) \quad \text{for every } \eta \in M_1. \tag{8.4.44}$$

The problem

$$J(v) \to \min, \quad v \in M_2,$$

has a unique solution u which satisfies

$$a(u,\eta) = l(\eta) \quad \text{for every } \eta \in M_2. \tag{8.4.45}$$

We would now like to show that when f is continuous, the solutions of (8.4.44) and (8.4.45) satisfy (8.4.39) and (8.4.40), respectively. This means that we must show that our minimizing functions are twice differentiable (functions in M_1 and M_2 need have only one derivative) and that in (8.4.45) the minimizing function satisfies the boundary condition $u'(1) = 0$. Any boundary condition in the BVP which need not be imposed on the set of admissible functions in the variational principle is said to be a *natural boundary condition*. Other boundary conditions are *essential*. (See also Section 0.5, keeping in mind the change in the definitions of M_1 and M_2.) Thus the boundary condition $u(0) = 0$ is essential, as is $u(1) = 0$, but the boundary condition $u'(1) = 0$ is natural. We shall carry out the proofs for the more interesting case, that of (8.4.45).

Equation (8.4.45) characterizes a function $u \in M_2$ such that

$$\int_0^1 (u'\eta' + ku\eta - f\eta)\, dx = 0 \quad \text{for every } \eta \in M_2. \tag{8.4.46}$$

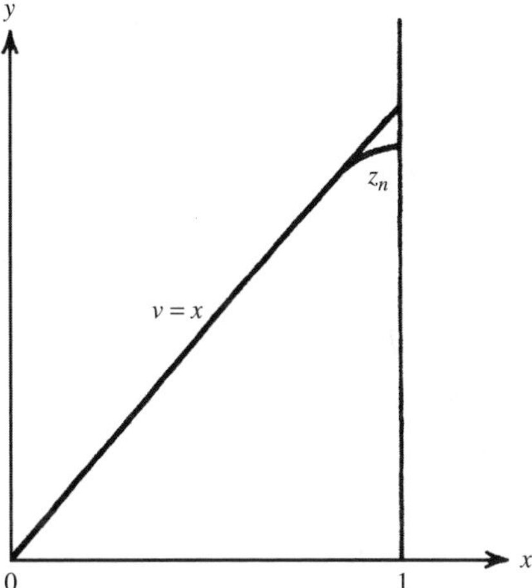

Figure 8.10

Since we are not allowed to assume that u'' exists, we shall use integration by parts on the term $(ku - f)\eta$. Setting

$$G(x) = \int_0^x (ku - f)\, dt, \tag{8.4.47}$$

we obtain from (8.4.46)

$$\int_0^1 \eta'(u' - G)\, dx + \eta G]_0^1 = 0 \quad \text{for every } \eta \in M_2. \tag{8.4.48}$$

Since (8.4.48) holds for all $\eta \in M_2$, it certainly holds for those $\eta \in M_2$ that also satisfy $\eta(1) = 0$. This happens to be the set $\eta \in M_1$. Thus we find that

$$\int_0^1 \eta'(u' - G)\, dx = 0 \quad \text{for every } \eta \in M_1,$$

from which we conclude (see Exercise 8.4.1) that

$$u'(x) = G(x) + C.$$

Hence from (8.4.47) we see that u'' exists and $u'' = ku - f$. Now that the existence of u'' has been established, we return to (8.4.46) and integrate the first term by parts to obtain

$$0 = \int_0^1 \eta(-u'' + ku - f)\, dx + \eta(1)u'(1) - \eta(0)u'(0), \quad \eta \in M_2.$$

Since we have shown that $-u'' + ku - f = 0$, and since $\eta(0) = 0$, we have

$$\eta(1)u'(1) = 0 \quad \text{for every } \eta \in M_2.$$

But when $\eta \in M_2$, the value of $\eta(1)$ is arbitrary, so that $u'(1) = 0$. We have shown that the solution of (8.4.40) or, equivalently, the function that minimizes J in M_2 is the solution of

$$-u'' + ku = f, \quad 0 < x < 1; \quad u(0) = u'(1) = 0.$$

Example 2. Let Ω be a bounded domain in $\mathbb{R}^n$ with boundary Γ, and consider the BVP

$$-\Delta u + ku = f, \quad x \in \Omega; \quad u|_\Gamma = 0, \tag{8.4.49}$$

where k is a nonnegative constant and f is a given element of the Hilbert space $H = L_2(\Omega)$, whose inner product and norm are denoted by $\langle \cdot, \cdot \rangle$ and $\| \cdot \|$, respectively. We translate the BVP (8.4.49) into the operator equation

$$Au = f, \quad u \in D_A,$$

where $A = -\Delta + k$ and D_A is the set of functions $u(x)$ in $C^2(\bar{\Omega})$ which vanish on Γ. Clearly, D_A is dense in H.

For v, $w \in D_A$ we have

$$\langle Av, w \rangle = \int_\Omega (-\Delta v + kv)w \, dx = \int_\Omega (\text{grad } v \cdot \text{grad } w + kvw) \, dx;$$

this suggests introducing the bilinear form

$$a(v, w) \doteq \int_\Omega (\text{grad } v \cdot \text{grad } w + kvw) \, dx,$$

which is clearly symmetric. We observe that

$$a(v, v) = \int_\Omega (|\text{grad } v|^2 + kv^2) \, dx \tag{8.4.50}$$

is positive for every nontrivial v in D_A. The statement is obvious for $k > 0$; if $k = 0$, then $a(v, v) = 0$ implies that $v =$ constant, but since elements in D_A vanish on Γ, the constant must be 0.

Hence $a(v, w)$ can serve as an energy inner product with $a(v, v)$ the square of the energy norm. Coercivity follows immediately from (8.4.50) for $k > 0$. If $k = 0$, it is necessary to prove the existence of a positive constant c such that

$$\int_\Omega |\text{grad } v|^2 \, dx \geqslant c \int_\Omega v^2 \, dx \quad \text{for every } v \in D_A.$$

The inequality is known in the literature as Friedrichs' inequality. We shall omit its proof, but we note that the best possible value of c is just the lowest eigenvalue of $-\Delta u = \lambda u$ in Ω, with u vanishing on Γ.

The completion of D_A in the energy norm is the Sobolev space $H_0^1(\Omega)$, defined independently as the closure of $C_0^\infty(\Omega)$ in the Sobolev norm $\| \cdot \|_1$, given by

$$\|u\|_1^2 = \|u\|^2 + \|\text{grad } u\|^2.$$

Thus $M = H_0^1(\Omega)$.

Our minimum principle then reads as follows:

$$J(v) \doteq \int_\Omega (|\text{grad } v|^2 + kv^2)\, dx - 2\int_\Omega fv\, dx \to \min, \quad v \in H_0^1(\Omega), \tag{8.4.51}$$

has a unique solution which is also the solution of the variational equation $a(u, v) = \langle f, v\rangle$. By fiat, this function $u(x)$ is declared to be the generalized solution of (8.4.49). Even if there is some physical validity to this assertion, it is essential to know when $u(x)$ is an ordinary solution of (8.4.49).

We summarize the results for the case $k = 0$ when (8.4.49) becomes

$$-\Delta u = f, \quad x \in \Omega; \quad u(x) = 0, \quad x \in \Gamma, \tag{8.4.52}$$

where we take $f \in C(\bar{\Omega})$ but make no particular restrictions on the smoothness of Γ. A solution $u(x)$ is said to be *classical* if it belongs to $C^2(\bar{\Omega})$. In contrast with the one-dimensional case, such a solution does *not* always exist! By reformulating (8.4.52) as a minimum principle and allowing functions in $H_0^1(\Omega)$ as trial functions, we established existence of a solution $u(x)$ to the modified problem. This function also solves the variational equation

$$\int_\Omega \text{grad } u \cdot \text{grad } v\, dx = \int_\Omega fv\, dx \quad \text{for every } v \in H_0^1(\Omega). \tag{8.4.52a}$$

When will $u(x)$ be a classical solution of (8.4.52)? We need more than continuity of f, and Γ must not be too wild: If $f \in C^1(\bar{\Omega})$ and Γ satisfies the exterior sphere condition of Section 8.3, the solution of the modified problem will be a classical solution of (8.4.52).

We conclude with a few remarks on the related problem

$$\Delta v = 0, \quad x \in \Omega; \quad v(x) = h(x), \quad x \in \Gamma, \tag{8.4.53}$$

where $h(x) \in C(\Gamma)$. If h can be extended to the interior of Ω as a $C^2(\bar{\Omega})$ function, then $u \doteq v - h$ satisfies (8.4.52) with $f = \Delta h$. We are then back to our previously analyzed problem. We have seen in the introduction to Section 8.3 that $h \in C(\Gamma)$ cannot always be extended to a $C^2(\bar{\Omega})$ function.

Example 3. Let A be a bounded, symmetric operator defined on the whole Hilbert space H. We shall also assume that A is coercive, that is, there exists a positive constant L_A such that

$$\langle Au, u\rangle \geqslant L_A \|u\|^2 \quad \text{for every } u \in H.$$

The bilinear form

$$a(v, w) \doteq \langle Av, w\rangle$$

is clearly symmetric, positive, and coercive. Thus $a(v, w)$ can serve as an energy inner product with energy norm

$$\|v\|_a^2 = a(v, v) = \langle Av, v\rangle.$$

Since A is both bounded and coercive, we have

$$L_A\|v\|^2 \leqslant \|v\|_a^2 \leqslant \|A\|\|v\|^2,$$

so that the ordinary norm and the energy norm are equivalent. Thus the completion of H in the energy norm is just H, and any linear functional $\langle f, v\rangle$ is also bounded in energy. We therefore have the following minimum principle:

$$J(v) \doteq \langle Av, v\rangle - 2\langle f, v\rangle, \quad v \in H, \tag{8.4.54}$$

has one and only one minimizing element u, which is the solution of the inhomogeneous equation

$$Au = f. \tag{8.4.55}$$

This completes the proof of the existence of a solution to (8.4.55). This result should be compared with Theorem 5.8.9.

One particular case of importance occurs when $A = K - \mu I$, where K is an integral operator on $L_2(a, b)$:

$$Ku \doteq \int_a^b k(x, \xi)u(\xi)\, d\xi.$$

We assume that K is symmetric, nonnegative, and Hilbert-Schmidt and that the real number μ is negative. Our previous results apply, and the problem

$$J(v) \doteq \int_a^b \int_a^b k(x, \xi)v(x)v(\xi)\, dx\, d\xi - \mu \int_a^b v^2\, dx - 2\int_a^b fv\, dx \to \min$$

has one and only one solution $u(x)$, which is also the solution of the integral equation

$$\int_a^b k(x, \xi)u(\xi)\, d\xi - \mu u(x) = f(x), \quad a < x < b.$$

Example 4. Schwinger-Levine Principle. Let u be the solution of the problem

$$J(v) \doteq a(v, v) - 2l(v) \to \min, \quad v \in M.$$

Then the problem

$$R(v) \doteq -\frac{l^2(v)}{a(v, v)} \to \min, \quad v \in M, \quad v \neq 0, \tag{8.4.56}$$

has the same minimum value as J, and the minimum is attained for any element of the form cu, $c \neq 0$. The proof is easy. First we note by the reciprocity relation (8.4.21) that

$$R(cu) = -\frac{l^2(u)}{a(u, u)} = J(u),$$

so that $\inf R(v) \leqslant \min J(v)$. If $\inf R < \min J$, there exists an element $z \in M$ such that $R(z) < \min J$. With $c = l(z)/a(z, z)$ we find that $J(cz) = R(z)$, which is a contradiction. Therefore, $\inf R = \min J$. Clearly, the infimum is achieved for elements of the form cu and for no others.

The Ritz-Rayleigh Approximation

We want to approximate the solution u of the minimum problem

$$J(v) \doteq a(v, v) - 2l(v), \quad v \in M,$$

by minimizing J on an n-dimensional subspace E_n of M. We then have

$$\min_{v \in E_n} J(v) \geqslant \min_{v \in M} J(v) = J(u).$$

Since J is bounded below, it has an infimum on E_n and that infimum must be attained because E_n is finite-dimensional. We recall that if $v \in M$, we can write, from (8.4.26),

$$J(v) = \|v - u\|_a^2 - \|u\|_a^2,$$

so that since u is a fixed (though unknown) element,

$$\min_{v \in E_n} J(v) = -\|u\|_a^2 + \min_{v \in E_n} \|v - u\|_a^2. \tag{8.4.57}$$

The minimum appearing on the right side characterizes the element in E_n which is nearest in energy to the fixed element u, that is, the orthogonal projection (in the energy inner product) of u on E_n. We shall denote this projection by $u^{(n)}$, the superscript anticipating that the dimension of E_n may vary. Introducing a basis $\{h_i\}$ in E_n, we can write

$$u^{(n)} = \sum_{j=1}^{n} \alpha_j^{(n)} h_j, \tag{8.4.58}$$

and since $u^{(n)} - u$ is orthogonal to E_n in energy,

$$a(u^{(n)} - u, h_i) = 0, \quad i = 1, \ldots, n. \tag{8.4.59}$$

A consequence of the variational equation (8.4.31) satisfied by u is that $a(u, h_i) = l(h_i), i = 1, \ldots, n$. Thus we can write (8.4.59) as

$$a(u^{(n)}, h_i) = l(h_i), \quad i = 1, \ldots, n, \tag{8.4.60}$$

or as the set of n linear equations for $\alpha_1^{(n)}, \ldots, \alpha_n^{(n)}$,

$$\sum_{j=1}^{n} a_{ij} \alpha_j^{(n)} = l(h_i), \quad i = 1, \ldots, n, \tag{8.4.61}$$

where

$$a_{ij} \doteq a(h_i, h_j). \tag{8.4.62}$$

Equations (8.4.61) are known as the *Galerkin equations*, which determine unambiguously the *Ritz-Rayleigh approximation* $u^{(n)}$.

If we multiply (8.4.60) by $\alpha_i^{(n)}$ and sum over i, we note that our Ritz-Rayleigh approximation $u^{(n)}$ satisfies the reciprocity principle

$$a(u^{(n)}, u^{(n)}) = l(u^{(n)}). \tag{8.4.63}$$

In view of the orthogonality of $u^{(n)}$ and u, we see from (8.4.57) that

$$J(u^{(n)}) = -\|u^{(n)}\|_a^2 = -a(u^{(n)}, u^{(n)}) = -l(u^{(n)}), \tag{8.4.64}$$

which is an upper bound to $J(u)$.

If the elements in the basis for E_n are chosen to be orthonormal in energy, some simplifications take place. Denoting the elements of the basis by $\{e_i\}$ instead of $\{h_i\}$, we have

$$a(e_i, e_j) = \begin{cases} 0, & i \neq j, \\ 1, & i = j. \end{cases}$$

Then (8.4.61) immediately gives

$$\alpha_i^{(n)} = l(e_i), \tag{8.4.65}$$

the Ritz-Rayleigh approximation is

$$u^{(n)} = \sum_{j=1}^{n} l(e_j) e_j, \tag{8.4.66}$$

and from (8.4.64) the corresponding value of J is

$$J(u^{(n)}) = -\sum_{j=1}^{n} l^2(e_j), \tag{8.4.67}$$

which provides an explicit upper bound to $J(u)$.

Now suppose that the set $\{e_i\}$ is an infinite orthonormal set (in energy) which is also a basis (in energy) for M. Then every element $v \in M$ can be written as

$$v = \sum_{j=1}^{\infty} a(v, e_j) e_j,$$

and in particular, the solution u of the original minimum problem is given by

$$u = \sum_{j=1}^{\infty} a(u, e_j) e_j = \sum_{j=1}^{\infty} l(e_j) e_j, \tag{8.4.68}$$

so that (8.4.66) converges in energy to u as $n \to \infty$. Since $a(v, w)$ is a coercive form, convergence in the ordinary norm follows from energy convergence, so that (8.4.68) also holds in the norm of H. From (8.4.67) we can conclude that

$$J(u) = -\sum_{j=1}^{\infty} l^2(e_j). \tag{8.4.69}$$

Although (8.4.68) and (8.4.69) give explicit expressions for the solution of the minimum problem, they are not as useful as they appear, for it may be numerically awkward to construct (say, by the Gram-Schmidt procedure) an orthonormal energy basis.

In many applications, such as those of Example 1, (8.4.39)–(8.4.40), the minimum problem stems from the operator equation

$$Au = f, \quad u \in D_A, \tag{8.4.70}$$

where A is a coercive operator on D_A. In that case, as we have seen in (8.4.33) and (8.4.34), we let

$$l(v) \doteq \langle v, f \rangle, \quad v \in D_A, \tag{8.4.71}$$

$$a(v, w) \doteq \langle Av, w \rangle \quad \text{for } v, w \in D_A. \tag{8.4.72}$$

If f is such that (8.4.70) has a solution u in D_A, we can characterize u as the minimum of $J(v) \doteq a(v,v) - 2l(v)$ either on D_A or on M. Even in such a favorable case it is important to study the problem on M because the Ritz-Rayleigh procedure can then use approximating elements $h_1, \ldots, h_n$ which lie in M rather than D_A. Often, calculations become simpler if the elements $\{h_i\}$ are only piecewise smooth (say, piecewise linear, such as the roof functions used in the finite element method) rather than being C^2 (as will usually be the case for elements of D_A). Another advantage of looking at the approximation problem on M is that the approximating elements $\{h_i\}$ need only satisfy the essential boundary conditions. The Ritz-Rayleigh approximations (8.4.66) and (8.4.67) then become

$$u^{(n)} = \sum_{j=1}^{n} \langle f, e_j \rangle e_j, \quad J(u^{(n)}) = -\sum_{j=1}^{n} \langle f, e_j \rangle^2, \tag{8.4.73}$$

and if $\{e_i\}$ is a basis for M, then

$$u = \sum_{j=1}^{\infty} \langle f, e_j \rangle e_j, \quad J(u) = -\sum_{j=1}^{\infty} \langle f, e_j \rangle^2. \tag{8.4.74}$$

If the $\{e_i\}$ are the eigenvectors of A, we know that they form an orthogonal set in the original inner product $\langle \cdot, \cdot \rangle$; they are also orthogonal in energy since

$$\langle e_i, e_j \rangle_a = a(e_i, e_j) = \langle Ae_i, e_j \rangle = \lambda_i \langle e_i, e_j \rangle.$$

Note, however, that the normalization factors differ: $\|e_i\|_a^2 = \lambda_i \|e_i\|^2$. This explains the apparent discrepancy between (8.4.74) and the solution obtained in (8.3.58). The expression (8.4.73) for $u^{(n)}$ is just the nth partial sum of the Fourier expansion of the exact solution of (8.4.70) in terms of the eigenvectors.

From both the theoretical and practical points of view, it is important to know how fast the approximation $u^{(n)}$ of (8.4.58) converges to u as $n \to \infty$. Although we do not attempt a comprehensive discussion of this question until Chapter 10, we observe here that we can find an a posteriori estimate for the error between $u^{(n)}$ and u. Equations (8.4.26) and (8.4.27) give

$$J(u^{(n)}) = J(u) + \|u_n - u\|_a^2,$$

so that

$$\|u^{(n)} - u\|_a^2 = J(u^{(n)}) - J(u) \leqslant J(u^{(n)}) - m,$$

where m is any lower bound to $J(u)$. Using the lower bound (8.4.37), we obtain

$$\|u^{(n)} - u\|_a \leqslant (L_A)^{-1/2}\|Au^{(n)} - f\|.$$

Once we have calculated $u^{(n)}$, we can compute the right side of the inequality to estimate the energy error between the approximation $u^{(n)}$ and the exact (but unknown) solution u.

We now illustrate the Ritz-Rayleigh method for the simple BVP (8.4.75) by using two different n-dimensional spaces of approximating functions $\{h_i\}$. The first of these (Example 5) corresponds to a polynomial approximation, whereas the second (Example 6) provides an introduction to the finite element method.

Example 5. Consider the BVP

$$Au \doteq -u'' + u = 1, \quad 0 < x < 1; \quad u(0) = u'(1) = 0, \tag{8.4.75}$$

which is a particular case of (8.4.40) corresponding to $k = 1$ and the constant forcing function $f(x) = 1$. According to Example 1, the appropriate formulations of the minimum principle and the variational equation are (8.4.30) and (8.4.31) with $M = M_2$ (see Example 1) and

$$\begin{aligned} a(v, w) &= \int_0^1 (v'w' + vw)\, dx, \quad l(v) = \langle 1, v\rangle = \int_0^1 v\, dx, \\ J(v) &= a(v, v) - 2l(v) = \int_0^1 (v'^2 + v^2 - 2v)\, dx. \end{aligned} \tag{8.4.76}$$

As our approximating functions we shall use $h_j = x^j, j = 1, \ldots, n$. These functions span the n-dimensional space P_n^* of polynomials of degree n vanishing at $x = 0$. P_n^* is clearly a subspace of M_2; each element of P_n^* satisfies the essential boundary condition at $x = 0$. An easy calculation from (8.4.76) gives

$$l(h_j) = \frac{1}{j+1}, \quad a_{ij} = a(h_i, h_j) = \frac{ij}{i+j-1} + \frac{1}{i+j+1}.$$

We use these values in (8.4.61) to solve for $\alpha_j^{(n)}$ and then substitute in (8.4.58) to get our Ritz-Rayleigh approximation $u^{(n)}$. Some drawbacks are apparent. The matrix $\{a_{ij}\}$ is a full matrix with no zero entries, making the solution of (8.4.61) quite laborious. Of course, we could have used the Gram-Schmidt procedure on the $\{h_j\}$ to transform them to polynomials orthogonal in energy, but that is also computationally lengthy. Even after we have found $\alpha_j^{(n)}$, (8.4.58) still involves summing many nonzero terms (if n is large, as required for accuracy), making it awkward to determine $u^{(n)}(x)$ at prescribed points of interest on [0, 1]. We shall see in the next example how the finite element method circumvents these difficulties. Before passing to that example, let us note that $J(u^{(n)})$ is a much better approximation to $J(u)$ than $u^{(n)}$ is to u. Taking $n = 2$ and using the earlier calculations, we find that

$$\alpha_1^{(2)} = .726, \quad \alpha_2^{(2)} = -.375, \quad u^{(2)}(x) = (.726)x - (.375)x^2,$$

$$Au^{(2)} = .75 + (.726)x - (.375)x^2, \quad Au^{(2)}|_{x=1} = 1.101, \frac{du^{(2)}}{dx}(1) = -.024,$$

$$J(u^{(2)}) = -l(u^{(2)}) = -.238.$$

For purposes of comparison, we now conveniently recall that (8.4.75) has the closed-form solution

$$u(x) = 1 - \frac{\cosh(x-1)}{\cosh(1)}, \quad J(u) = -1 + \tanh(1) = -.238.$$

We observe that $u^{(2)}(x)$ is not a spectacular approximation to $u(x)$, as can be seen from the fact that the derivative of $u^{(2)}$ at $x = 1$ does not vanish and that $Au^{(2)}$ differs appreciably from the correct value 1. On the other hand, $J(u^{(2)})$ agrees with $J(u)$ to three decimal places!

Example 6. When the Ritz-Rayleigh method uses functions $\{h_i\}$ which are piecewise polynomials with narrow support (see, for instance, the roof functions of Figure 6.2) it is called a *finite element method.* Although the finite element method's full power only manifests itself for BVPs of dimensions larger than 1, and particularly those on irregularly shaped domains, we content ourselves with illustrating the main ideas for the one-dimensional BVP (8.4.75).

We divide the interval [0, 1] into n equal parts of width $1/n$ (the so-called *mesh size*) with partition points (*nodes*) $x_k = k/n, k = 0, 1, \dots, n$. Consider the space U_n of functions $h(x) \in C[0,1]$ with $h(0) = 0$, and piecewise linear (that is, linear in each subinterval $[x_k, x_{k+1}]$). Clearly, U_n is a linear space; a function in U_n is completely characterized by its values at the nodes. We next introduce a basis in U_n as follows: For $j = 1, \dots, n$, let $h_j(x) \in U_n$ be defined by

$$h_j(x_j) = 1, \quad h_j(x_k) = 0 \quad \text{for } k \neq j. \tag{8.4.77}$$

Note that $h_j(x)$ is just the roof function $v_j(x)$ shown in Figure 6.2. Such a function is a special example of a "finite element." An arbitrary element $h(x)$ in U_n has the unique representation

$$h(x) = \sum_{j=1}^{n} \xi_j h_j(x) \quad \text{with } \xi_j = h(x_j). \tag{8.4.78}$$

We have $\langle h_j, h_k \rangle = 0$ whenever $|k-j| \geqslant 2$. From the definitions (8.4.76), it follows that

$$\begin{gathered} l(h_j) = \frac{1}{n}, \quad j = 1, \ldots, n-1; \quad l(h_n) = \frac{1}{2n}. \\ a_{jk} = a(h_j, h_k) = 0 \quad \text{for } |k-j| \geqslant 2. \end{gathered} \tag{8.4.79}$$

Since U_n is a subspace of M_2, all ingredients are in place for the Ritz-Rayleigh method. Our next step is to solve the algebraic system (8.4.61) for $\alpha_j^{(n)}$. The coefficient matrix is now sparse: Only the main diagonal and the two adjacent ones contain nonzero elements. Such *tridiagonal* systems are relatively easy to solve numerically. Once the $\alpha_j^{(n)}$ have been found, (8.4.78) tells us that $u^{(n)}(x_j) = \alpha_j^{(n)}$ so that we know the approximate solution of (8.4.75) at the nodes and (8.4.58) provides the (linear) interpolation between the nodes.

Estimates of the convergence of $u^{(n)}$ to u can be obtained without great difficulty (see, for instance, Johnson [26] or Zeidler [50], or refer to Chapter 10 for a more complete discussion).

Complementary Variational Principles

Consider a linear BVP, possibly with inhomogeneous boundary conditions, for a function $u(x)$ on a domain Ω. Existence is not in question here, as we shall assume that the BVP has a solution. It is sometimes tedious to obtain even an approximate solution that is valid throughout the domain Ω. We may, however, be willing to settle for an approximate value of a single numerical quantity having physical significance as some sort of "average" of the solution. Such examples as capacity, torsional rigidity, and scattering cross section immediately come to mind. In each case a single number gives a great deal of information about a phenomenon whose details are governed by a BVP of some complexity. The number in question can usually be expressed as a functional of the solution $u(x)$ of the BVP. We shall try to obtain upper and lower bounds for this functional in terms of comparison functions $v(x)$ that may be required to satisfy certain constraints.

The proper setting for our work is a Hilbert space H with inner product $\langle \cdot, \cdot \rangle$ and norm $\| \cdot \|$. Let $a(v, w)$ be a *symmetric, nonnegative* bilinear form on the linear manifold D in H. Then $a(v, w)$ does not quite serve as a new inner product because the vanishing of $a(v, v)$ does not guarantee that v is the zero element. Nevertheless, we shall see that the Schwarz inequality and the Pythagorean theorem remain true, and we shall therefore still refer to $a(v, w)$ as an energy inner product. The following simple inequalities suffice for our purposes:

$$a(u, u) \geqslant 2a(u, v) - a(v, v), \tag{8.4.80}$$

$$a(u, u) \geqslant \frac{a^2(u, v)}{a(v, v)} \quad \text{if } a(v, v) \neq 0, \tag{8.4.81}$$

$$a(u, u) \leqslant a(v, v) \quad \text{if } a(v-u, u) = 0. \tag{8.4.82}$$

The first of these inequalities follows from $a(u-v,u-v) \geqslant 0$, the second from replacing v by $va(u,v)/a(v,v)$ in the first, and the third from

$$\begin{aligned} a(v,v) &= a(u+v-u,u+v-u) = a(u,u) + 2a(u,v-u) + a(v-u,v-u) \\ &= a(u,u) + a(v-u,v-u) \geqslant a(u,u). \end{aligned}$$

Inequality (8.4.82) has a simple geometrical meaning. Since $v-u$ and u are orthogonal in the energy inner product, we may view v as the hypotenuse of a right triangle whose other two sides are u and $v-u$, so that the length of u cannot exceed that of v.

Our first application of these principles is to a BVP for the Poisson equation:

$$-\Delta u = f(x), \quad x \in \Omega; \quad u|_\Gamma = 0, \tag{8.4.83}$$

where Ω is a bounded domain in $\mathbb{R}^n$ with boundary Γ, and f is a given function. For simplicity we shall assume that f is smooth enough so that (8.4.83) has a classical solution. When $f(x) = 2$ and $n = 2$, (8.4.83) governs the torsion of a cylinder of cross section Ω. The function $u(x) = u(x_1, x_2)$ is then the stress function. The constant of proportionality relating the twisting moment and the angle of twist per unit length of the cylinder is known as the *torsional rigidity* and is denoted by T. If, for convenience, we set the shear modulus equal to unity, it can be shown that

$$T = \int_\Omega 2u(x)\,dx = \int_\Omega |\text{grad } u|^2\,dx,$$

where u is the solution of (8.4.83) for $f(x) \equiv 2$.

Since it is just as easy to deal with the case of a general function in (8.4.83), we shall consider the problem of estimating

$$a(u,u) \doteq \int_\Omega |\text{grad } u|^2\,dx = \int_\Omega fu\,dx, \tag{8.4.84}$$

where u is the solution of (8.4.83). The corresponding bilinear form is

$$a(v,w) \doteq \int_\Omega \text{grad } v \cdot \text{grad } w\,dx, \tag{8.4.85}$$

defined on the set D of piecewise smooth functions on $\bar{\Omega}$.

We want to estimate (8.4.84) without solving (8.4.83) for u. To use (8.4.80) we will need to find comparison functions v for which $a(u,v)$ can be calculated without a knowledge of u. We note that

$$a(u,v) = \int_\Omega (\text{grad } u \cdot \text{grad } v)\,dx = -\int_\Omega v\,\Delta u\,dx + \int_\Gamma v\frac{\partial u}{\partial n}\,dS,$$

so that since $-\Delta u = f$,

$$a(u,v) = \int_\Omega fv\,dx \quad \text{if } v|_\Gamma = 0, \quad v \in D.$$

Thus we obtain

$$a(u,u) \geqslant 2\int_\Omega fv\,dx - \int_\Omega |\text{grad } v|^2\,dx \quad \text{if } v|_\Gamma = 0, \quad v \in D, \tag{8.4.86}$$

which is just the minimum principle (8.4.51) since $a(u,u) = -J(u)$. Applying (8.4.81), we find that

$$a(u,u) \geqslant \frac{\left(\int_\Omega fv\,dx\right)^2}{\int_\Omega |\text{grad } v|^2\,dx} \quad \text{if } v|_\Gamma = 0, \quad v \in D, \quad v \not\equiv 0, \tag{8.4.87}$$

which is the Schwinger-Levine principle (8.4.56) with a change of sign. Of course, in (8.4.87) the maximum is achieved not just for the solution of (8.4.83) but also for any nonzero multiple of it.

A bound from the opposite direction can be obtained from (8.4.82) by finding comparison functions v which satisfy $a(v-u,u) = 0$, where u is the unknown solution of (8.4.83). A simple calculation gives

$$a(v-u,u) = -\int_\Omega u\,\Delta(v-u)\,dx + \int_\Omega u\frac{\partial}{\partial n}(v-u)\,dS.$$

Since u is 0 on Γ, it suffices that $-\Delta v = f$ for $a(v-u,u)$ to vanish. We therefore find from (8.4.82) that

$$a(u,u) \leqslant \int_\Omega |\text{grad } v|^2\,dx \quad \text{for every } v \text{ satisfying } -\Delta v = f. \tag{8.4.88}$$

Because the comparison function v is required to satisfy Poisson's equation (without the boundary condition, however), it is clear that v must be restricted to a subset of D containing sufficiently smooth functions.

In the particular case of torsional rigidity, we have the bounds

$$\frac{4\left(\int_\Omega w\,dx\right)^2}{\int_\Omega |\text{grad } w|^2\,dx} \leqslant T \leqslant \int_\Omega |\text{grad } v|^2\,dx, \tag{8.4.89}$$

where v satisfies $-\Delta v = 2$, and w vanishes on Γ.

We can easily use the inequalities in (8.4.89) to obtain explicit bounds for T in terms of simpler domain functionals. For an arbitrary choice of the origin, the function $v = -(x_1^2 + x_2^2)/2$ satisfies $-\Delta v = 2$, and therefore

$$T \leqslant \int_\Omega (x_1^2 + x_2^2)\,dx_1\,dx_2.$$

The right side is smallest if the origin is chosen at the center of gravity of the cross section Ω. The integral then becomes the polar moment of inertia I_p, so that

$$T \leqslant I_p. \tag{8.4.90}$$

An improvement of this upper bound can be found in Exercise 8.4.4.

A different type of upper bound can be obtained by using level line coordinates. The solution u of the torsion problem

$$-\Delta u = 2, \quad x \in \Omega; \quad u|_\Gamma = 0 \tag{8.4.91}$$

is positive in Ω by the maximum principle. Let us introduce a coordinate t such that $0 \leqslant t \leqslant u_m$, where u_m is the unknown maximum value of the solution. Let $A(t)$ be the area enclosed within the level curve $u = t$:

$$A(t) \doteq \int_{u>t} dx, \quad \text{so that } A(u_m) = 0, \quad A(0) = \text{area of } \Omega.$$

By considering the annular domain between $u = t$ and $u = t + dt$, we find that

$$-A'(t) = \int_{u=t} \frac{dl}{|\text{grad } u|},$$

known as the *coarea formula*. Integrating the differential equation (8.4.91), we obtain

$$\int_{u=t} |\text{grad } u| \, dl = 2A(t).$$

Multiplying the expressions for $2A$ and for $-A'$, we obtain, after using Schwarz's inequality,

$$-2AA' = \int_{u=t} \frac{dl}{|\text{grad } u|} \int_{u=t} |\text{grad } u| \, dl \geqslant L^2(t),$$

where $L(t)$ is the length of the curve $u = t$. Of course, a curve of length L cannot enclose more area than a circle of the same length. Therefore,

$$L^2(t) \geqslant 4\pi A(t) \quad \text{and} \quad -\frac{d}{dt}A^2 \geqslant 4\pi A.$$

We integrate the latter inequality from $t = 0$ to $t = u_m$ to obtain

$$A^2(0) \geqslant 4\pi \int_0^{u_m} A \, dt = -4\pi \int_0^{u_m} tA' \, dt = -4\pi \int_0^{u_m} \psi' \, dt,$$

where

$$\psi(t) \doteq \int_{u>t} u \, dx, \quad \psi(0) = \frac{T}{2}.$$

We therefore recover Pólya's isoperimetric inequality:

$$T \leqslant \frac{A^2}{2\pi}, \tag{8.4.92}$$

the term *isoperimetric* referring to the fact that equality is achieved for a particular domain—in this case, a circular disk.

To find a lower bound we shall use a special class of comparison functions w whose level lines are similar to the curve Γ. We assume that Γ is star-shaped with respect to an origin O within Ω; this means that every ray emanating from O intersects

Γ in only one point whose polar coordinates are (r, θ) with $r = f(\theta)$, $0 \leqslant \theta < 2\pi$. Let $\phi(t)$ be a nonnegative function on $0 \leqslant t \leqslant 1$ with $\phi(0) = 1$, $\phi(1) = 0$. Choosing $w = \phi(r/f(\theta))$, we see that w is constant on the curves $r = cf(\theta)$, $0 \leqslant c \leqslant 1$, which are similar to the curve Γ, whose equation is $r = f(\theta)$. A simple calculation gives

$$\int_\Omega w\,dx = \int_0^{2\pi} d\theta \int_0^f \phi\left(\frac{r}{f}\right) r\,dr$$
$$= \int_0^{2\pi} f^2(\theta)\,d\theta \int_0^1 c\phi(c)\,dc = 2A \int_0^1 c\phi(c)\,dc,$$

where A is the area of Ω.

We also obtain

$$\int_\Omega |\text{grad } w|^2\,dx = \int_0^{2\pi} d\theta \int_0^f \left\{ \left[\frac{\partial}{\partial r}\phi\left(\frac{r}{f}\right)\right]^2 + \frac{1}{r^2}\left[\frac{\partial}{\partial \theta}\phi\left(\frac{r}{f}\right)\right]^2 \right\} r\,dr$$
$$= \int_0^{2\pi} \left[1 + \frac{(f')^2}{f^2}\right] d\theta \int_0^1 c[\phi'(c)]^2\,dc.$$

The quantity $[1 + (f')^2/f^2]d\theta$ can be written as $dl/h(\theta)$, where dl is an element of length along the boundary and $h(\theta)$ is the perpendicular distance from the origin to the line tangent to the curve at the point whose angular coordinate is θ. We therefore find from (8.4.89) that

$$T \geqslant \frac{16A^2 \left[\displaystyle\int_0^1 c\phi(c)\,dc\right]^2}{\displaystyle\int_0^{2\pi} [dl/h(\theta)] \int_0^1 c[\phi'(c)]^2\,dc}; \quad \phi(0) = 1, \quad \phi(1) = 0. \tag{8.4.93}$$

Let us now choose $\phi(c)$ to be proportional to the actual stress function for a circular cylinder of radius R. The exact solution of (8.4.91) for a disk of radius R is

$$\tfrac{1}{2}(R^2 - r^2),$$

so that $\phi(c) = (1 - c^2)$ and (8.4.93) gives

$$T \geqslant \frac{A^2}{\displaystyle\int_0^{2\pi} dl/h(\theta)}. \tag{8.4.94}$$

Next we consider a problem from electrostatics. A thin metallic conductor coincides with the closed surface Γ in $\mathbb{R}^3$. We let Ω_e stand for the unbounded domain exterior to Γ, and Ω_i for the bounded interior domain. If a charge is placed on Γ so that the potential on Γ is unity while the potential at large distances tends to 0, the electrostatic potential satisfies the exterior Dirichlet problem [see (8.3.39)]

$$\Delta u = 0, \quad x \in \Omega_e; \quad u|_\Gamma = 1, \quad \lim_{|x| \to \infty} |x|u = 0, \quad \lim_{|x| \to \infty} |x|^2|\text{grad } u| = 0. \tag{8.4.95}$$

The capacity C of the conductor is the total charge on Γ, that is,

$$C = -\int_\Gamma \frac{\partial u}{\partial \nu}\, dS,$$

where ν is the normal to Γ outward from Ω_i. By integrating $u\,\Delta u$ over Ω_e and using the limiting behavior of u and $|\text{grad } u|$ at infinity, we find that

$$C = -\int_\Gamma \frac{\partial u}{\partial \nu}\, dS = \int_{\Omega_e} |\text{grad } u|^2 \; dx,$$

where the latter equality shows that the capacity is also equal to the electrostatic energy stored in the field.

To apply our variational principles, we introduce the inner product

$$a(v,w) = \int_{\Omega_e} \text{grad } v \cdot \text{grad } w \;\, dx,$$

which is defined on the set D of functions v with continuous second derivatives in Ω_e and such that $|x|v$ and $|x|^2|\text{grad } v|$ are bounded at infinity. Then the capacity C can be expressed as

$$C = a(u,u).$$

Since

$$a(v-u,u) = \int_{\Omega_e} \text{grad } u \cdot \text{grad}(v-u)\, dx = -\int_\Gamma (v-u)\frac{\partial u}{\partial \nu}\, dS,$$

we have $a(v-u,u) = 0$ if $v = 1$ on Γ. Therefore,

$$C \leqslant \int_{\Omega_e} |\text{grad } v|^2\, dx \quad \text{for } v \in D, \quad v|_\Gamma = 1. \tag{8.4.96}$$

It is clear that the minimum of the right side is achieved for $v = u$. A simple calculation gives

$$a(u,v) = -\int_{\Omega_e} u\,\Delta v\, dx - \int_\Gamma u \frac{\partial v}{\partial \nu}\, dS,$$

so that

$$a(u,v) = -\int_\Gamma \frac{\partial v}{\partial \nu}\, dS \quad \text{if } v \in D, \quad \Delta v = 0 \text{ in } \Omega_e.$$

From (8.4.81) we then obtain the lower bound

$$C \geqslant \frac{\left[\displaystyle\int_\Gamma \frac{\partial v}{\partial \nu}\, dS\right]^2}{\displaystyle\int_{\Omega_e} |\text{grad } v|^2\, dx}, \quad v \in D, \quad v \not\equiv 0, \quad \Delta v = 0 \text{ in } \Omega_e. \tag{8.4.97}$$

The capacity problem can also be formulated as an integral equation on Γ. In fact, we can use (8.3.47) with $u_i(x) \equiv 1, u_e(x) = u(x), \partial u_i/\partial \nu = 0$, so that

$$1 = \int_\Gamma \frac{1}{4\pi|s-\xi|} I(\xi)\, dS_\xi, \qquad s \in \Gamma, \tag{8.4.98}$$

where $I(\xi) = -\partial u/\partial \nu$ is just the charge density on Γ.

Once (8.4.98) has been solved for I, the potential $u(x)$ is given, as in (8.3.45), by

$$u(x) = \int_\Gamma \frac{1}{4\pi|s-\xi|} I(\xi)\, dS_\xi, \qquad x \in \Omega_e.$$

Rather than solving (8.4.98), we try to estimate the capacity. Since

$$C = \int_\Gamma I(x)\, dS_x,$$

we also obtain from (8.4.98)

$$C = \int_\Gamma \int_\Gamma \frac{1}{4\pi|x-\xi|} I(x) I(\xi)\, dS_x\, dS_\xi.$$

Since at the end of Section 8.3 the integral operator appearing in (8.4.98) was shown to be positive, we can introduce the inner product

$$a(p,q) = \int_\Gamma \int_\Gamma \frac{1}{4\pi|x-\xi|} p(x) q(\xi)\, dS_x\, dS_\xi,$$

which is certainly well defined for continuous functions on Γ. Then, applying the inequality (8.4.81), we find that

$$C \geqslant \frac{\left[\int_\Gamma p(x)\, dS_x\right]^2}{\int_\Gamma \int_\Gamma dS_x\, dS_\xi \left[\dfrac{p(x)p(\xi)}{4\pi|x-\xi|}\right]}, \qquad p(x) \in C(\Gamma), \quad p \not\equiv 0.$$

Unilateral Constraints

Boundary value problems with inequality constraints arise frequently in applications, particularly in solid mechanics, where they often go under the name of *contact problems*. As the barest introduction to the subject, we shall study only two simple examples in beam theory. Although we use the linearized differential equation for transverse deflections, the problems are nevertheless intrinsically nonlinear because of the nature of the constraints. As evidence of this nonlinearity, the general form of the superposition principle does not apply (see Dundurs [12] for a beautiful treatment of this question).

In our first problem the unilateral constraint is on the boundary. Consider a beam $0 < x < 1$ subject to a transverse distributed load $f(x)$. The left end of the beam

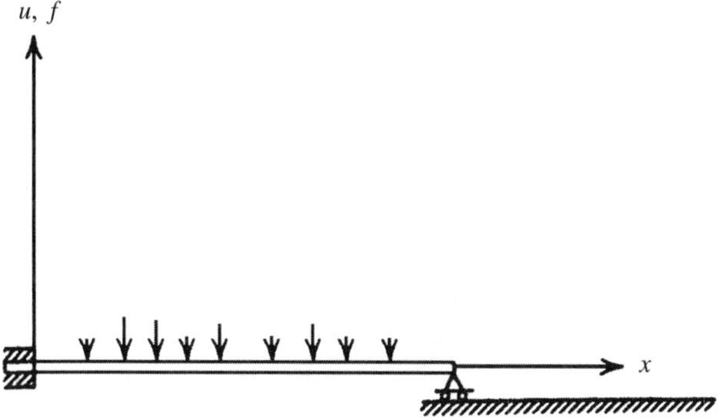

Figure 8.11

is built in, but its right end rests on weightless rollers near the edge of a smooth table, as in Figure 8.11. Depending on the nature of $f(x)$, the beam may remain in contact with the table or may be lifted from it. For instance, if $f = -1$, contact is maintained, but if $f = 1$, the right end is raised from its support. Of course, if $f = 0$, the deflection vanishes and is therefore not equal to the sum of the deflections corresponding to the respective loads $f = 1$ and $f = -1$. For an arbitrary load $f(x)$ which changes sign along the beam, it is not clear at the outset whether or not contact is maintained, so that both possibilities must be taken into account in the formulation of the boundary conditions at the right end. In either case the moment vanishes at the right end, and therefore $u''(1) = 0$. It would seem at first that the only other information about the right endpoint is that $u(1) \geqslant 0$. However, if $u(1) = 0$, the reaction of the table on the beam must be upward, so that $u'''(1) \leqslant 0$; also, if $u(1) > 0$ the end is free and $u'''(1) = 0$. A suitable formulation of the BVP is therefore

$$\frac{d^4u}{dx^4} = f(x), \quad 0 < x < 1; \tag{8.4.99}$$
$$u(0) = u'(0) = u''(1) = 0, \quad u(1)u'''(1) = 0, \quad u(1) \geqslant 0, \quad u'''(1) \leqslant 0.$$

We would like to characterize the solution of (8.4.99) as the minimizing function of a variational principle. Our goal here is not to prove existence theorems but merely to show the equivalence between the BVP and the variational principle. The potential energy of the deflected beam is given by

$$\int_0^1 (u'')^2\,dx - 2\int_0^1 fu\,dx.$$

This suggests that we consider the minimum, over a suitable set K of admissible functions, of the functional

$$J(v) \doteq a(v,v) - 2l(v), \tag{8.4.100}$$

where

$$a(v,w) \doteq \int_0^1 v''w''\,dx, \quad l(v) \doteq \int_0^1 fv\,dx.$$

We see easily that $a(v,w)$ is a positive bilinear form on the subset D of functions in $C^2[0,1]$ satisfying the conditions $v(0) = v'(0) = 0$. Thus we can use $a(v,w)$ as an energy inner product on D. Leaving aside questions of smoothness, let us analyze the boundary conditions to be satisfied by the elements v of K. Clearly, v must satisfy the two essential boundary conditions $v(0) = v'(0) = 0$ at the left end. Since $u''(1) = 0$ is a natural boundary condition, a function v in K does not have to satisfy that condition. As for the other conditions at the right end, it suffices to require that $v(1) \geqslant 0$, the other conditions being automatically satisfied by the solution of the variational problem.

Theorem 8.4.5. *Let K be the set of functions v in $C^2[0,1]$ satisfying the conditions $v(0) = v'(0) = 0$ and $v(1) \geqslant 0$. Then the necessary and sufficient condition for the problem*

$$J(v) \to \min, \quad v \in K, \tag{8.4.101}$$

to have a solution $u \in K$ is that this element satisfies the variational inequality

$$a(u, v-u) - l(v-u) \geqslant 0 \quad \textit{for all } v \in K. \tag{8.4.102}$$

REMARK. This variational inequality should be contrasted with variational equation (8.4.20). The reason for the difference is that the set K is not a linear manifold but a *convex* set. A set K is said to be convex if it contains the straight-line segments joining any two points in the set. Thus, if $v, w \in K$, then

$$v + t(w-v) \in K, \quad 0 \leqslant t \leqslant 1.$$

Of course, any linear manifold is a convex set, but not vice versa.

Proof. Let u be an element of K that provides the minimum in (8.4.101). Then, if $v \in K$, so does $u + t(v-u)$ for $0 \leqslant t \leqslant 1$. Therefore,

$$J[u + t(v-u)] \geqslant J(u), \quad 0 \leqslant t \leqslant 1,$$

and hence

$$\left\{ \frac{d}{dt} J[u + t(v-u)] \right\}_{t=0} \geqslant 0,$$

from which (8.4.102) follows. Now suppose that $u \in K$ satisfies (8.4.102). Then, if $v \in K$, we have

$$\begin{aligned} J(v) &= J(u+v-u) = a(u+v-u, u+v-u) - 2l(u+v-u) \\ &= J(u) + 2[a(u,v-u) - l(v-u)] + a(v-u, v-u) \geqslant J(u), \end{aligned}$$

so that u furnishes the minimum in (8.4.101). □

Furthermore, we note that $a(v, v) > 0$ for all $v \not\equiv 0$ satisfying the boundary conditions $v(0) = v'(0) = 0$. This guarantees that the solution of the equivalent problems is unique (assuming existence, of course). If instead of K we use its completion in the energy norm, we can also prove existence (see Exercise 8.4.5).

Next we must show that the solution u of the variational inequality (8.4.102) also satisfies the BVP (8.4.99). The solution u of (8.4.102) is an element of K for which

$$\int_0^1 u''(v'' - u'')\,dx - \int_0^1 f(v - u)\,dx \geqslant 0 \quad \text{for every } v \in K.$$

Assuming we have shown that u actually belongs to C^4, we find, using integration by parts, that

$$\int_0^1 (u^{(4)} - f)(v - u)\,dx - u'''(1)[v(1) - u(1)] + u''(1)[v'(1) - u'(1)] \geqslant 0. \quad (8.4.103)$$

If ϕ is any sufficiently smooth function for which $\phi(0) = \phi'(0) = \phi(1) = \phi'(1) = 0$, then $v \doteq u + \phi$ belongs to K and (8.4.103) shows that $\int_0^1 (u^{(4)} - f)\phi\,dx = 0$ for every such ϕ and therefore certainly for every test function ϕ with compact support in $(0, 1)$. Hence u is a weak solution of the differential equation $u^{(4)} = f$. Such a weak solution is a classical solution if f is continuous or piecewise continuous. What boundary conditions does u satisfy? Since $u^{(4)} - f = 0$, we find from (8.4.103) that

$$-u'''(1)[v(1) - u(1)] + u''(1)[v'(1) - u'(1)] \geqslant 0 \quad \text{for every } v \in K.$$

Now let ϕ satisfy $\phi(0) = 0, \phi'(0) = 0, \phi(1) = 0, \phi'(1) < 0$. Then $v = u + \phi \in K$, $v(1) - u(1) = 0, v'(1) - u'(1) < 0$, so that $u''(1) = 0$, which is the first of the boundary conditions at the right end. We finally have

$$u'''(1)[v(1) - u(1)] \leqslant 0 \quad \text{for every } v \in K.$$

If $u(1) = 0$, we choose $v(1) > 0$ to find that $u'''(1) \leqslant 0$, If $u(1) > 0$, we choose $v(1) > u(1)$ to conclude that $u'''(1) = 0$. Thus we have shown that the solution of (8.4.102) satisfies (8.4.99).

In our second example the unilateral constraint is active throughout the domain rather than on the boundary alone. A beam whose ends are clamped is subject to a transverse loading $f(x)$ and rests on an impenetrable obstacle whose upper boundary is the curve $y = h(x)$, where $h(x) \leqslant 0, 0 \leqslant x \leqslant 1$ (see Figure 8.12). Under the load the beam may be in contact with the obstacle over some unknown portion of its length.

The potential energy of the deflected beam is then

$$\int_0^1 (u'')^2\,dx - 2\int_0^1 fu\,dx,$$

and we again introduce the functional $J(v)$ as in (8.4.100). Let K be the set of sufficiently smooth functions $v(x)$ satisfying $v(0) = v'(0) = v(1) = v'(1) = 0$ and

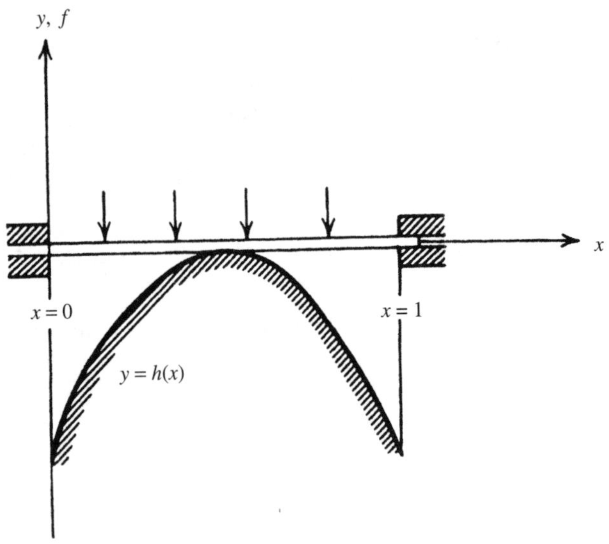

Figure 8.12

$v(x) \geqslant h(x)$. Then K is a convex set, and by the same reasoning as was used in Theorem 8.2.5, we have that the necessary and sufficient condition for the problem

$$J(v) \to \min, \quad v \in K,$$

to have a solution $u \in K$ is that this element satisfies the variational inequality

$$a(u, v-u) - l(v-u) \geqslant 0 \quad \text{for all } v \in K.$$

Thus u is characterized as the element in K for which

$$\int_0^1 u''(v''-u'')\,dx - \int_0^1 f(v-u)\,dx \geqslant 0 \quad \text{for every } v \in K. \tag{8.4.104}$$

What consequences can we derive from (8.4.104)? First, let $\phi(x)$ be a nonnegative function in $C_0^\infty(0,1)$; then $v = u + \phi$ belongs to K and $\int_0^1 (u''\phi'' - f\phi)\,dx \geqslant 0$, which is a weak form of the differential inequality $u^{(4)} \geqslant f$ on $0 < x < 1$.

At any point x where $u(x) > h(x)$, we can take variations of either sign about u so that the differential equation $u^{(4)} = f$ must be satisfied. In any interval where $u(x) = h(x)$, the inequality $u^{(4)} \geqslant f$ must hold, so that the additional reaction force due to the obstacle is in the positive u direction, in keeping with physical requirements.

EXERCISES

8.4.1 Suppose that w is a continuous function on $0 \leqslant x \leqslant 1$ such that

$$\int_0^1 w(x)\phi(x)\,dx = 0$$

for every $\phi \in C_0^\infty(0,1)$ satisfying $\int_0^1 \phi\,dx = 0$; then show that w is constant on $0 \leqslant x \leqslant 1$.

8.4.2 Let A be a linear operator for which the inhomogeneous equation

$$Au = f, \quad u \in D_A,$$

has one and only one solution. Determine the element which has the form $u^* = \sum_{i=1}^n c_i v_i$, where the v_i's are independent elements, which minimizes $\|Au^* - f\|^2$. Show that the coefficients $\{c_i\}$ satisfy

$$\langle f, Av_j\rangle = \sum_{k=1}^{n} c_k \langle Av_k, Av_j\rangle, \quad j = l, \ldots, n. \tag{8.4.105}$$

The method just described is known as the *method of least squares*. (See also Exercise 3.4.4.)

8.4.3 Let J be the functional (8.4.14), where $a(v, w)$ is coercive and $l(v)$ is bounded in energy [see (8.4.24)]. Then J is bounded below on D and has an infimum d. On the space M of elements of finite energy, J has a minimum. Show that this minimum coincides with d.

8.4.4 By considering trial elements of the form

$$v = -\frac{x_1^2 + x_2^2}{2} + A(x_1^2 - x_2^2),$$

show that (8.4.90) can be improved to

$$T \leqslant \frac{4I_1 I_2}{I_p},$$

where I_1 and I_2 are the moments of inertia about the x_1 and x_2 axes, respectively.

8.4.5 *Existence for unilateral problem.* Let $l(v)$ be bounded in energy, and let $a(v, w)$ be coercive on the manifold M, where M is complete in energy. Let K be a convex subset of M that is closed in energy (thus K is itself a metric space complete in energy). Show that the minimum problem

$$J(v) \to \min, \quad v \in K,$$

has one and only one solution. (*Hint*: Since J is bounded below on K, construct a minimizing sequence $\{v_n\}$ in K and show that v_n converges to an element of K.)

8.4.6 Show that the natural boundary conditions associated with the functional

$$J(u) \doteq \int_0^1 (u'')^2\,dx - 2\int_0^1 fu\,dx$$

are $u''(0) = u''(1) = u'''(0) = u'''(1) = 0$. Give the physical interpretation in beam theory.

8.4.7 Let Ω be a bounded domain in $\mathbb{R}^n$ with smooth boundary Γ. Consider the BVP

$$-\Delta u = f, \quad x \in \Omega; \quad -\frac{\partial u}{\partial n} = k(x)u, \quad x \in \Gamma, \tag{8.4.106}$$

where f is a given continuous function in $\bar{\Omega}$ and $k(x)$ is a positive function on Γ. Assuming that (8.4.106) has a classical solution $u(x)$, show that $u(x)$ can also be characterized as the element which minimizes

$$J(v) \doteq \int_\Omega |\text{grad } v|^2\,dx - 2\int_\Omega fv\,dx + \int_\Gamma kv^2\,dS \tag{8.4.107}$$

among functions which are sufficiently smooth but are not required to satisfy any boundary condition on Γ. Thus the boundary condition in (8.4.106) is natural for (8.4.107).

8.4.8 Let Ω be a domain in $\mathbb{R}^2$ with boundary Γ. Consider the *biharmonic* equation

$$\Delta\Delta u = f, \quad x \in \Omega; \quad u = \frac{\partial u}{\partial \nu} = 0, \quad x \in \Gamma, \tag{8.4.108}$$

which we also write as

$$Au = f, \quad u \in D_A,$$

where $A = \Delta\Delta$ and D_A is the set of functions in $C^4(\bar{\Omega})$ satisfying the condition $u = \partial u/\partial \nu = 0$ on Γ.

(a) Show that the operator A on D_A is positive and that its eigenvalues must therefore also be positive.

(b) Introduce the bilinear form

$$a(v, w) = \int_\Omega (\Delta v)(\Delta w)\,dx$$

and obtain two-sided bounds for

$$\alpha \doteq \int_\Omega fu\,dx.$$

(c) Despite the positivity of A on D_A, there is no straightforward maximum principle. If $\Delta\Delta u \geqslant 0$ in Ω for $u \in D_A$, it does not follow that $u \geqslant 0$. This, in turn, means that Green's function for (8.4.108) is not necessarily positive at all points in Ω (as first pointed out by Duffin and Garabedian [16]). A simple example of the failure of the maximum principle is given in the article by Shapiro and Tegmark [37].

8.4.9 (a) Consider the integrodifferential equation (6.5.38), repeated here:

$$-u'' + \int_0^1 xyu(y)\,dy = f(x), \quad 0 < x < 1; \quad u(0) = u'(1) = 0,$$

which has one and only one solution $u(x)$. Show that

$$\int_0^1 fu\;dx = \max_{v \in D_A} \left[2\int_0^1 fv\;dx - \int_0^1 (v')^2\;dx - \left(\int_0^1 xv\;dx \right)^2 \right],$$

where D_A is the set of functions in $C^2[0,1]$ satisfying $v(0) = v'(1) = 0$.

(b) Show further that the boundary condition $v'(1) = 0$ need not be imposed in the variational principle.

8.4.10 Consider the beam problem (8.4.99) when the applied load is $A \sin 2\pi x$. For what values of A is contact maintained? Find the deflection of the beam.

8.4.11 A beam with clamped ends is subject to a uniform load $f(x) = -1$. The beam lies a distance δ above a flat, impenetrable foundation. Find the deflection of the beam. (*Hint*: If δ is sufficiently small, the beam will be in contact with the foundation along part of its length; there may be reaction forces at the end of the contact portion.)

8.4.12 Consider the beam of Figure 8.12 when the obstacle has the equation $h(x) = -\alpha\left(x - \frac{1}{2}\right)^4, \alpha > 0$. Show that if $\alpha > \frac{1}{24}$, contact can occur only at isolated points along the beam and no contact can occur at $x = \frac{1}{2}$. What happens? Find the deflection.

8.4.13 Let Ω be a bounded plane domain with boundary Γ, and let $u(x)$ be positive on Ω and vanish on Γ. We introduce the level curves $u = t$, and, as in the discussion following (8.4.91), we define

$$A(t) \doteq \int_{u>t} dx, \quad J(t) \doteq \int_{u>t} |\text{grad } u|^2\,dx.$$

Show that

$$-A' = \int_{u=t} \frac{dl}{|\text{grad } u|}, \quad -J' = \int_{u=t} |\text{grad } u|\,dl, \quad 4\pi A \leqslant A'J',$$

and therefore

$$\int_\Omega |\text{grad } u|^2\,dx \geqslant -4\pi \int_0^{u_m} \frac{A(t)}{A'(t)}\,dt, \tag{8.4.109}$$

where equality holds if and only if Ω is a disk and $u(x)$ depends only on the distance from the center. For $u(x)$ given on Ω, we construct a circularly symmetrized function $u^*(x)$ on a disk Ω^* of area equal to Ω as follows:

$$u^* = t \quad \text{on the circle } |x| = r \quad \text{where } \pi r^2 = A(t).$$

Clearly, u^* depends only on the radial coordinate, and the function $A^*(t) = \int_{u^*>t} dx$ is identically equal to $A(t)$. From (8.4.109) it follows that

$$\int_{\Omega} |\text{grad } u|^2 \, dx \geqslant \int_{\Omega^*} |\text{grad } u^*|^2 \, dx, \tag{8.4.110}$$

so that circular symmetrization decreases the Dirichlet integral. If h is an arbitrary function, we also find that

$$\int_{\Omega} h(u(x)) \, dx = \int_{\Omega^*} h(u^*(x)) \, dx. \tag{8.4.111}$$

8.4.14 (a) Let $u(x)$ be the solution of (8.4.91), and let u^* be the circularly symmetrized function (see Exercise 8.4.13). Show that if T and T^* are the respective torsional rigidities for Ω and Ω^*, then

$$T = \frac{4\left(\int_{\Omega} u \; dx\right)^2}{\int_{\Omega} |\text{grad } u|^2 \; dx} \leqslant \frac{4\left(\int_{\Omega^*} u^* \; dx\right)^2}{\int_{\Omega} |\text{grad } u^*|^2 \; dx} \leqslant T^*,$$

where the first inequality follows from (8.4.110) and (8.4.111), and the second from the variational characterization of the torsional rigidity.

(b) Let $u(x)$ be the fundamental eigenfunction (chosen positive) of

$$-\Delta u = \lambda u, \quad x \in \Omega; \quad u = 0, \quad x \in \Gamma, \tag{8.4.112}$$

and let $u^*(x)$ be the circularly symmetrized function. Let λ^* be the fundamental eigenvalue for the disk Ω^*, and show that $\lambda^* \leqslant \lambda$. Thus, of all domains of equal area, the disk has the highest torsional rigidity and the lowest fundamental frequency.

8.4.15 Let $u(x)$ be the fundamental eigenfunction of (8.4.112). Define the function $I(t) = \int_{u>t} u \, dx$. Show that

$$-\lambda I I' \geqslant 4\pi t A,$$

and therefore obtain the Payne-Rayner inequality

$$\left(\int_{\Omega} u \; dx\right)^2 \geqslant \frac{4\pi}{\lambda} \int_{\Omega} u^2 \; dx.$$

8.4.16 For the torsion problem (8.4.91), $|\text{grad } u|$ is proportional to the stress. Define $M = |\text{grad } u|^2$ and show that $\Delta M \geqslant 0$ so that the maximum stress must occur on the boundary (confirming physical intuition!).

8.5 THE LAX-MILGRAM THEOREM

We now consider two results involving linear functionals, bilinear forms, and linear operators which will be particularly important in Chapter 10. Both results can be viewed as generalizations or extensions of the Riesz Representation Theorem from Chapter 4. These are similar to results we established in Section 8.4 in the context of variational methods, under the assumption that the underlying bilinear form is symmetric. However, symmetry is not required for the following generalizations. Since only a single Hilbert space H is involved, we will denote the norm and inner product simply as $\|\cdot\|$ and $(\cdot,\cdot)$, respectively.

Theorem 8.5.1 (Bounded Operator Theorem). *Let H be a real Hilbert space, and let $a(u,v)$ be a bounded bilinear form on $H \times H$. Then there exists a unique bounded linear operator $A \in H \to H$ such that*

$$a(u,v) = (Au,v), \quad \forall u,v \in H. \tag{8.5.1}$$

Proof. For fixed $u \in H$, the form $a(u,v)$ is a bounded linear functional of v, hence by the Riesz Representation Theorem, there exists a unique $Au \in H$ such that $a(u,v) = (Au,v)$, $\forall v \in H$. The mapping $A\colon u \to Au$ defined implicitly by the Riesz Theorem for all $u \in H$ is clearly linear, by linearity of $a(\cdot,\cdot)$ in both arguments. Boundedness of A also follows from that of $a(\cdot,\cdot)$, since

$$|(Au,v)| = |a(u,v)| \leqslant M\|u\|\|v\|, \quad \forall u,v \in H,\ M > 0.$$

Taking $v = Au$ in particular gives $\|Au\|^2 \leqslant M\|u\|\|Au\|$. If $Au \neq 0$, this gives $\|Au\| \leqslant M\|u\|$, so A is bounded. Therefore, we have proven the existence of a bounded linear operator $A\colon H \to H$ satisfying (8.5.1).

It remains to show that such an A is unique. To this end, assume that there exists a second operator B that also satisfies (8.5.1). Then $(Au,v)-(Bu,v) = 0, \forall u,v \in H$, or $(Au - Bu, v) = 0$, $\forall u,v \in H$. Taking $u \neq 0$ and $v = (A-B)u$, we have $\|Au - Bu\|^2 = 0$, which holds for all $u \in H$ if and only of $Au = Bu$, $\forall u \in H$. This is true only if $A = B$, and therefore A is unique. □

Theorem 8.5.2 (Lax-Milgram Theorem). *Let H be a real Hilbert space, let the bilinear form $a(u,v)$ be bounded and coercive on $H \times H$, and let $f(u)$ be a bounded linear functional on H. Then there exists a unique solution to the problem:*

$$\text{Find } u \in H \text{ such that } a(u,v) = f(v), \quad \forall v \in H. \tag{8.5.2}$$

Proof. We give a proof based on use of the Banach Fixed-Point Theorem, which we have proven in Section 4.4. The assumptions on a and f are as follows:

(1) Boundedness of a: $a(u,v) \leqslant M\|u\|\|v\|$, $\forall u,v \in H$.

(2) Coerciveness of a: $a(u,u) \geqslant m\|u\|^2$, $\forall u \in H$.

(3) Boundedness of f: $f(v) \leqslant L\|v\|$, $\forall v \in H$.

By the Riesz Representation Theorem, there exists an $F \in H$ such that

$$f(v) = (F, v), \quad \forall v \in H.$$

Similarly, by Theorem 8.5.1, there exists a bounded linear operator $A\colon H \to H$ such that

$$a(u, v) = (Au, v), \quad \forall u, v \in H.$$

Therefore, our problem is equivalent to:

$$\text{Find } u \in H \text{ such that } (Au, v) = (F, v), \quad \forall v \in H.$$

This is equivalent to $(Au - F, v) = 0$, $\forall v \in H$. Taking $v = Au - F$, we have $\|Au - F\|^2 = 0$, or simply $Au = F$. Therefore, our problem is, in fact, equivalent to:

$$\text{Find } u \in H \text{ such that } Au = F,$$

where $A\colon H \to H$, and $F \in H$. If $Au = F$, then for any $\rho > 0$, we also have the identity $u = u - \rho(Au - F)$. This defines a fixed point map:

$$u = u - \rho(Au - F) = T(u).$$

Let us examine the contraction properties of T:

$$\begin{aligned}\|T(u) - T(v)\| &= \|[(I - \rho A)u + \rho F] - [(I - \rho A)v + \rho F]\| \\ &= \|(I - \rho A)(u - v)\| \\ &\leqslant \|(I - \rho A)\|\|u - v\|.\end{aligned}$$

Note that the operator norm is

$$\|(I - \rho A)\| = \sup_{u \in H} \frac{\|(I - \rho A)u\|}{\|u\|}.$$

Therefore, for T to be a contraction, we must construct an inequality of the form $\|(I - \rho A)u\| \leqslant \gamma\|u\|$, for some $\gamma \in [0, 1)$. Consider then

$$\begin{aligned}\|(I - \rho A)u\|^2 &= \|u\|^2 - 2\rho(Au, u) + \rho^2\|Au\|^2 \\ &\leqslant \|u\|^2 - 2\rho m\|u\|^2 + \rho^2 M^2\|u\|^2 \\ &= (1 - 2\rho m + \rho^2 M^2)\|u\|^2,\end{aligned}$$

which gives $\|(I - \rho A)u\| \leqslant (1 - 2\rho m + \rho^2 M^2)^{1/2}\|u\|$. Therefore, T is a contraction if $(1 - 2\rho m + \rho^2 M^2) < 1$, which holds when $\rho < 2m/M^2$. By the Banach Fixed-Point Theorem, if $\rho \in (0, 2m/M^2)$, there exists a unique fixed point to $u = T(u)$, and hence a unique solution to the problem (8.5.2). □

EXERCISES

8.5.1 Give an alternative proof of the Lax-Milgram Theorem that does not involve the Banach Fixed-Point Theorem.

8.5.2 Extend the proof of the Lax-Milgram Theorem to establish the following result for variational inequalities, known as the *Lions-Stampacchia Theorem*:

Theorem. *Let H be a Hilbert space, and let $U \subset H$ be a closed convex subset. Let $a(u, v)$ be a bounded and coercive bilinear form on $H \times H$, and let $f(v)$ be a bounded linear functional on H. Then there exists a unique solution to the variational inequality:*

$$\textit{Find } u \in U \textit{ such that } a(u, v-u) \geqslant f(v-u), \quad \forall v \in U.$$

8.5.3 Prove that when $U = H$ in the Lions-Stampacchia Theorem in Exercise 8.5.2, the Lions-Stampacchia Theorem reduces to the Lax-Milgram Theorem.

8.5.4 Use the Lax-Milgram Theorem to prove that the following elliptic partial differential equation is well-posed:

$$-\nabla \cdot (a\nabla u) + bu = f, \text{ in } \Omega \subset \mathbb{R}^2, \qquad u = 0, \text{ on } \partial\Omega,$$

where $a, b, f \in C^\infty(\Omega; \mathbb{R})$, $a(x) \geqslant 1$, $b(x) \geqslant 1$, $\forall x \in \Omega$, and where Ω is a bounded simply connected open set with a smooth boundary.

8.5.5 Prove the following semilinear extension of the Lax-Milgram Theorem.

Theorem. *Let H be a Hilbert space, let $a\colon H \times H \to \mathbb{R}$ be a bounded and coercive bilinear form on $H \times H$, and let $b\colon H \to H'$ be a continuous map from H to H' having the monotonicity property:*

$$(b(u) - b(v), u - v)_H \geqslant 0, \quad \forall u, v \in H.$$

Then the following problem has a unique solution:

$$\textit{Find } u \in H \textit{ such that } a(u, v) + (b(u), v) = 0, \quad \forall v \in H.$$

REFERENCES AND ADDITIONAL READING

1. S. Agmon, *Lectures on Elliptic Boundary Value Problems*, Van Nostrand, New York, 1965.
2. M. S. Ashbaugh and R. D. Benguria, Proof of the Payne-Polya-Weinberger conjecture, *Bull. Amer. Math. Soc.* 25, 1991.
3. C. Bandle, *Isoperimetric Inequalities and Their Applications*, Pitman, New York, 1980.

4. S. Bergman and M. Schiffer, *Kernel Functions and Elliptic Differential Equations in Mathematical Physics*, Academic Press, New York, 1953.

5. J. R. Cannon, *The One-Dimensional Heat Equation*, Addison-Wesley, Reading, MA, 1984.

6. S. J. Chapman, Drums that sound the same, *Amer. Math. Monthly* 102, 1995.

7. R. Courant and D. Hilbert, *Methods of Mathematical Physics*, Vols. I and II, Interscience, New York, 1953, 1962.

8. W. A. Day, On rates of propagation of heat according to Fourier's theory, *Quart. Appl. Math.* 55(1), 1997.

9. M. C. Delfour, Shape derivatives and differentiability of min max, in *Shape Optimization and Free Boundaries* (M. C. Delfour, Ed.), Kluwer, Hingham, MA, 1992.

10. E. DiBenedetto, *Partial Differential Equations*, Birkhauser, Cambridge, MA, 1995.

11. T. A. Driscoll, Eigenmodes of isospectral drums, *SIAM Rev.* 39, 1997.

12. J. Dundurs, Properties of elastic bodies in contact, in *The Mechanics of the Contact Between Deformable Bodies* (A. D. de Pater and J. J. Kalker, Eds.), Delft University Press, Delft, The Netherlands, 1975.

13. G. Duvaut and J. L. Lions, *Inequalities in Mechanics and Physics*, Springer–Verlag, New York, 1976.

14. L. C. Evans, *Partial Differential Equations*, Graduate Studies in Mathematics, Vol. 19, American Mathematical Society, Providence, RI, 2010.

15. A. Friedman, *Partial Differential Equations*, Holt, Rinehart and Winston, New York, 1969.

16. P. R. Garabedian, *Partial Differential Equations*, Wiley, New York, 1964.

17. D. Gilbarg and N. S. Trudinger, *Elliptic Partial Differential Equations of Second Order*, 2nd ed., Springer–Verlag, New York, 1983.

18. C. Gordon, D. Webb, and S. Wolpert, Isospectral plane domains and surfaces via Riemannian orbifolds, *Invent. Math.* 110, 1992.

19. P. Grisvard, *Elliptic Problems in Nonsmooth Domains*, Pitman, New York, 1985.

20. R. B. Guenther and J. W. Lee, *Partial Differential Equations of Mathematical Physics and Integral Equations*, Dover, Mineola, NY, 1996.

21. K. E. Gustafson, *Partial Differential Equations*, Wiley, New York, 1980.

22. W. Hackbusch, *Elliptic Differential Equations*, Springer–Verlag, New York, 1992.

23. Q. Han and F. Lin, *Elliptic Partial Differential Equations*, American Mathematical Society, Providence, RI, 2000.

24. G. C. Hsiao and W. L. Wendland, A finite element method for some integral equations of the first kind, *J. Math. Anal. Appl.* 58(3), 1977.

25. F. John, *Partial Differential Equations*, Springer–Verlag, New York, 1982.

26. C. Johnson, *Numerical Solutions of Partial Differential Equations by the Finite Element Method*, Cambridge University Press, New York, 1993.

27. M. Kac, Can one hear the shape of a drum?, *Amer. Math. Monthly* 73, 1966.

28. J. Keener, *Principles of Applied Mathematics*, 2nd ed., Westview Press, Boulder, CO, 2000.

29. J. Kervorkian, *Partial Differential Equations*, Wadsworth & Brooks/Cole, Belmont, CA, 1990.

30. S. Kesavan, *Topics in Functional Analysis and Applications*, Wiley, New York, 1989.

31. R. Leis, *Initial Boundary Value Problems in Mathematical Physics*, Wiley, New York, 1986.

32. J. L. Lions and E. Magenes, *Non-homogeneous Boundary Value Problems and Applications*, Vol. I, Springer–Verlag, New York, 1972.

33. J. D. Logan, *Applied Mathematics*, 2nd ed., Wiley, New York, 1997.

34. R. McOwen, *Partial Differential Equations*, Prentice Hall, Upper Saddle River, NY, 1995.

35. M. Renardy and R. C. Rogers, *An Introduction to Partial Differential Equations*, Springer–Verlag, New York, 1993.

36. M. J. Sewell, *Maximum and Minimum Principles*, Cambridge University Press, New York, 1987.

37. H. S. Shapiro and M. Tegmark, An elementary proof that the biharmonic Green function of an eccentric ellipse changes sign, *SIAM Rev.* 36, 1994.

38. R. E. Showalter, *Hilbert Space Methods for Partial Differential Equations*, Pitman, New York, 1977.

39. R. Sperb, *Maximum Principles and Their Applications*, Academic Press, New York, 1981.

40. I. Stakgold, *Boundary Value Problems of Mathematical Physics*, Vols. I and II, Macmillan, 1967. Reprinted as Volume 29 in SIAM Classics in Applied Mathematics, 2000.

41. W. A. Strauss, *Partial Differential Equations*, Wiley, New York, 1992.

42. M. Taylor, *Partial Differential Equations*, Vol I: *Basic Theory*, Springer–Verlag, New York, 1996.

43. M. Taylor, *Partial Differential Equations*, Vol II: *Qualitative studies of linear equations*, Springer–Verlag, New York, 1996.

44. M. Taylor, *Partial Differential Equations*, Vol III: *Nonlinear equations*, Springer–Verlag, New York, 1996.

45. J. L. Troutman, *Variational Calculus with Elementary Convexity*, Springer–Verlag, New York, 1983.

46. H. F. Weinberger, *Variational Methods for Eigenvalue Approximation*, SIAM, Philadelphia, 1974.

47. J. Wloka, *Partial Differential Equations*, Cambridge University Press, New York, 1992.

48. E. Zauderer, *Partial Differential Equations of Applied Mathematics*, Wiley, New York, 1989.

49. E. Zeidler, *Nonlinear Functional Analysis and its Applications*, Vo. IV: *Applications to Mathematical Physics*, Springer–Verlag, New York, 1991.

50. E. Zeidler, *Applied Functional Analysis*, Springer–Verlag, New York, 1995.

CHAPTER 9

NONLINEAR PROBLEMS

9.1 INTRODUCTION AND BASIC FIXED-POINT TECHNIQUES

Fixed-Point Theorems

We have already used the power of the Contraction Theorem (Section 4.4) to prove existence and uniqueness for both linear and nonlinear differential and integral equations. The Contraction Theorem is one of the three principal types of *fixed-point theorems of analysis*, the other two being the *Schauder fixed-point theorem* and the *fixed-point theorem for order-preserving maps*. In the latter two theorems, we discard the contraction (distance-shortening) assumption and replace it by compactness plus convexity in the Schauder case or by an appropriate notion of order and monotonicity in the other case.

Since order-preserving maps will be featured in the rest of the chapter, we use the present section to make some remarks on the Schauder theorem and its applications. The Schauder theorem is itself an extension to infinite dimensions of the Brouwer fixed-point theorem in $\mathbb{R}^n$:

Theorem (Brouwer's Theorem). *Let $S \subset \mathbb{R}^n$ be compact and convex and let $T : S \to S$ be continuous. Then T has a fixed point.*

We omit the proof (see, for example, Franklin [22]) but make a few observations:

1. In $\mathbb{R}^n$ a compact set is a set which is bounded and closed. Both of these conditions are needed: If $S = \mathbb{R}$, a translation T maps S into itself but has no fixed point; if $S \subset \mathbb{R}$ is the open interval $0 < u < 1$, then T defined from $Tu = u/2$ transforms S into itself but has no fixed point.

2. Convexity (or something akin to it) is also required: In $\mathbb{R}^2$, let S be the closed annulus $0 < a \leqslant |u| \leqslant b$; then a rotation T through an angle $\theta \neq 2n\pi$ has no fixed point. Instead of convexity it is enough to have S homeomorphic to a convex set. We say that S_1 and S_2 are homeomorphic if there is a one-to-one continuous transformation (with continuous inverse) from S_1 onto S_2. Thus, a disk with a protuberance or indentation is homeomorphic to the undeformed disk, but an annulus is not.

3. Giving up the contraction property means giving up uniqueness. As an example, let S be the interval $0 \leqslant u \leqslant 1$. Then T defined from $Tu = u + (\sin 2\pi u)/2$ transforms S into itself but has three fixed points. Note also that the Contraction Theorem does not require the set S to be convex. Consider, for instance, the example in Chapter 4 corresponding to Figure 4.4, but with a country in the shape of a thick annulus. Then the smaller-scale map B might fit into the annulus of map A. There would then be a unique fixed point which can be obtained by iteration.

4. The Contraction Theorem is constructive, but Brouwer's Theorem is not.

5. As an application of Brouwer's Theorem, let us prove the *Perron-Frobenius theorem*: A matrix whose entries are all nonnegative has a nonnegative eigenvalue with an eigenvector whose coordinates are all nonnegative. The proof is simple: Let A be a linear operator on $\mathbb{R}^n$ with matrix (a_{ij}) whose elements a_{ij} are $\geqslant 0$. We write $u \geqslant 0$ if all its coordinates $\xi_1, \ldots, \xi_n$ are themselves nonnegative. We shall use the L_1-norm on $\mathbb{R}^n$: $\|u\| = |\xi_1| + \cdots + |\xi_n|$. Let S be the subset of $\mathbb{R}^n$ with $u \geqslant 0$ and $\|u\| = 1$ so that $\|u\| = \xi_1 + \cdots + \xi_n$. S is convex and compact. For $u \in S$, we have $Au \geqslant 0$. If $Au = 0$, then u is a unit eigenvector with nonnegative components corresponding to the eigenvalue 0, and our theorem is proved; so let us assume that $Au \neq 0$. Then $\|Au\| \geqslant c$ for some $c > 0$ and $Au/\|Au\|$ is a continuous transformation from S into S. By Brouwer's Theorem we must have $Au = \|Au\|u$ for some $u \in S$, which proves the Perron-Frobenius theorem. Observe that if we had assumed that all $a_{ij} > 0$, we could have inferred the existence of a positive eigenvalue.

We now state without proof two versions of the *Schauder fixed-point theorem* on a Banach space B:

(a) Let $S \subset B$ be compact and convex and let T be a continuous transformation from S into S. Then there exists an element $u \in S$ with $u = Tu$.

(b) Let $S \subset B$ be closed and convex and let T be a compact transformation from S into S. Then there exists $u \in S$ with $u = Tu$.

Version (a) is a consequence of (b) since a continuous transformation on a compact set is compact. Recall that we gave the definition of a *compact linear operator* at the beginning of Section 5.7. For nonlinear transformations the definition is practically the same: T is compact if it is continuous and transforms bounded sets into relatively compact sets.

As an application of the Schauder theorem, let us revisit the ODE initial-value problem (4.4.16). This IVP was translated into the nonlinear Volterra integral equation (4.4.21), and under the Lipschitz condition (4.4.19) we were able to prove local existence and uniqueness by using the Contraction Theorem. If we give up the Lipschitz condition, we can no longer expect uniqueness [see (4.4.17)], but what about existence? This is where Schauder comes to the rescue! We assume only that $f(t,z)$ is continuous in a neighborhood of (t_0, z_0), which implies continuity in some rectangle R defined by

$$R : |t - t_0| \leqslant \tau, |z - z_0| \leqslant c.$$

Theorem. *There exists* $\alpha > 0$ *such that (4.4.16) has a solution* $u(t)$ *in* $|t - t_0| \leqslant \alpha$.

Proof. Since f is continuous on the closed rectangle R, $|f|$ is bounded on R. Choose $M > \max\{1, c/\tau, |f|\}$ and define $\alpha = c/M$. Let C^* be the set of continuous functions $u(t)$ on the interval $|t - t_0| \leqslant \alpha$ and satisfying

$$|u(t) - z_0| \leqslant c, \qquad |u(t_2) - u(t_1)| \leqslant M|t_2 - t_1|.$$

The first condition means that C^* is a bounded set and the second guarantees the equicontinuity needed for the Arzela-Ascoli theorem (see, for instance, Rudin [50] or Marsden and Hoffman [46]). Thus the set C^* is compact. The operator T defined by

$$Tu = z_0 + \int_{t_0}^{t} f(y, u(y))\, dy$$

is continuous and can be seen to transform C^* into itself. Thus by version (a) of the Schauder theorem, T has a fixed point and (4.4.16), has a solution $u(t)$ on the interval $|t - t_0| \leqslant \alpha$. □

In applications, it is more usual to employ version (b) of the Schauder theorem (see, for instance, Exercise 9.4.14).

One-Dimensional Equations

Although our principal interest is in nonlinear differential equations and integral equations whose appropriate setting is an infinite-dimensional Hilbert or Banach space, we begin modestly with a single nonlinear equation in one independent variable. Let A be a real-valued function of a real variable ($A\colon \mathbb{R} \to \mathbb{R}$), and consider either of the nonlinear equations $A(u) = 0$ and $A(u) = u$, where it is clear that one form can be obtained from the other by a suitable change in the definition of the function A.

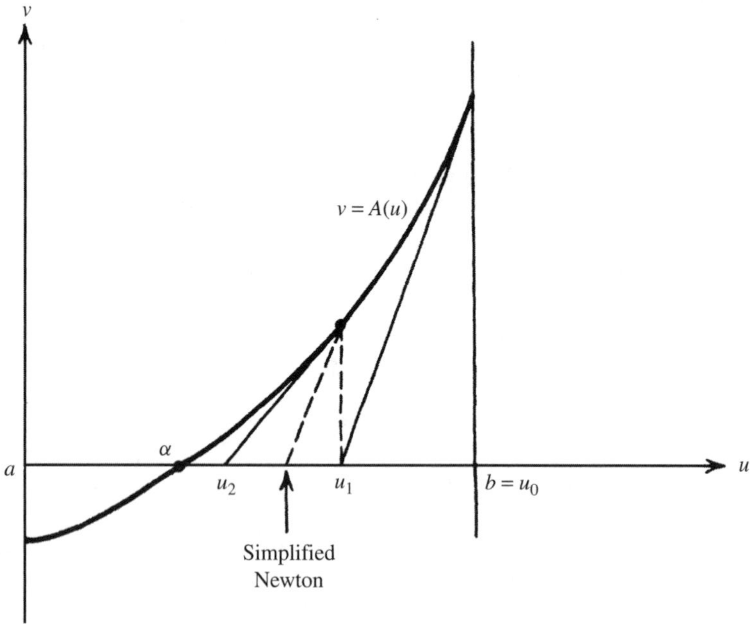

Figure 9.1

A popular method for solving $A(u) = 0$ is *Newton's method* (or the method of tangents), illustrated in Figure 9.1. To find a root α known to be located in $a \leqslant u \leqslant b$, we start with the initial approximation $u_0 = b$ and define the next approximation u_1 as the intersection of the u axis and the tangent line drawn to $A(u)$ at u_0. Proceeding in this way, we obtain a sequence

$$u_n = u_{n-1} - \frac{A(u_{n-1})}{A'(u_{n-1})}, \tag{9.1.1}$$

which, under favorable circumstances, will converge to α. The important feature of Newton's method is that at each step in the approximation procedure the curve is replaced by its tangent line (linearization). The calculations are greatly reduced if we use $A'(u_0)$ instead of $A'(u_{n-1})$ in (9.1.1). Geometrically, we are then drawing a line *parallel* to the original tangent instead of drawing the new tangent. This simplified Newton's method is equivalent to the method of successive substitutions (or successive approximations) described below.

We now consider our nonlinear equation in the fixed-point form

$$u = A(u). \tag{9.1.2}$$

To solve this equation means to find the intersections of the straight line $y = u$ with the curve $y = A(u)$. If A is a contraction on the interval $a \leqslant u \leqslant b$, we know

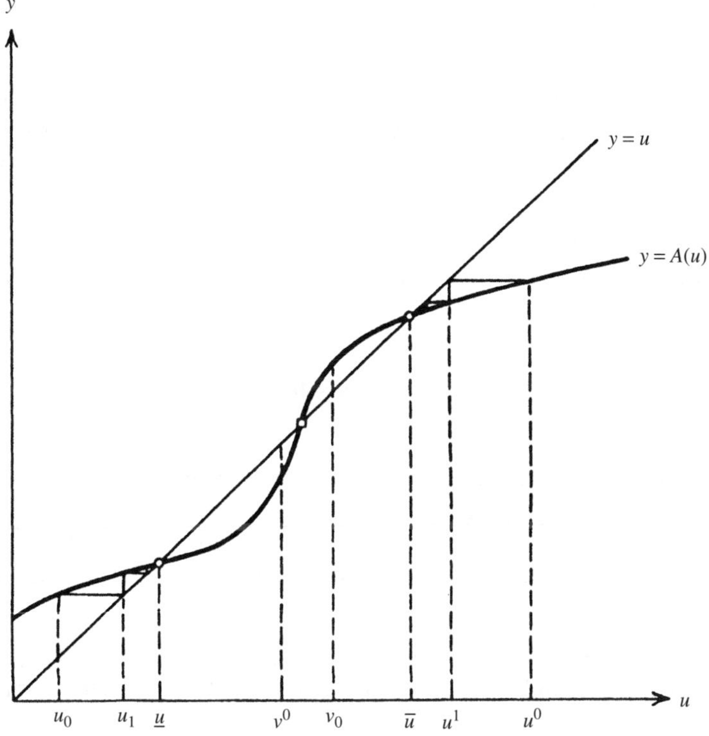

Figure 9.2

from Section 4.4, that (9.1.2) has one and only one solution, which is the limit of the iterative sequence $u_n = A(u_{n-1})$ independently of how the initial element u_0 is chosen in $[a, b]$. We now want to discard the contraction assumption and concentrate instead on the possible monotonicity of the iterative sequence. In monotone iteration we construct a pair of sequences, one converging to the solution from above and the other from below. The method we outline works for the solutions marked by a circle in Figure 9.2 but not for the one marked by a square.

Definition. *The number* u_0 *is said to be a* lower solution *to (9.1.2) if*

$$u_0 \leqslant A(u_0), \tag{9.1.3}$$

and u^0 *is an* upper solution *to (9.1.2) if*

$$u^0 \geqslant A(u^0). \tag{9.1.4}$$

REMARK. A value of the abscissa for which the straight line lies below the curve is a lower solution, and one for which the straight line lies above the curve is an upper solution. Thus in Figure 9.2, u_0 and v_0 are lower solutions, whereas u^0 and v^0 are upper solutions.

The following result seems obvious from a glance at Figure 9.2 and will be proved in a more general setting later. Let the lower solution u_0 and the upper solution u^0 satisfy $u_0 \leqslant u^0$, and let A be *increasing* on the closed interval $[u_0, u^0]$. The sequence

$$u_n = A(u_{n-1}), \quad n = 1, 2, \ldots \tag{9.1.5}$$

is increasing and converges to the minimal solution u of (9.1.2) on $[u_0, u^0]$. The sequence

$$u^n = A(u^{n-1}), \quad n = 1, 2, \ldots \tag{9.1.6}$$

is decreasing and converges to the maximal solution $\overline{u}$ of (9.1.2) on $[u_0, u^0]$.

REMARKS

1. If we had started iteration (9.1.6) with the upper solution v^0 instead of u^0, the iterates would have converged downward to the one and only solution of (9.1.2) in the interval $[u_0, v^0]$. Observe that A is not a contraction on either $[u_0, u^0]$ or $[u_0, v^0]$. In any event, uniqueness can be guaranteed only by making additional assumptions on A.

2. The upper solution v^0 is smaller than the lower solution v_0. Although there is a solution of (9.1.2) in $[v^0, v_0]$, it cannot be obtained by monotone iteration. Such a solution is termed *unstable*.

3. If A is not increasing, the simple iteration schemes (9.1.5) and (9.1.6) may diverge or only converge in an alternating manner. However, addition of a large linear term to both sides of (9.1.2) reduces the problem to the case of increasing A. Indeed, suppose that we have
$$u = S(u),$$
where all we know is that $u_0 \leqslant S(u_0), u^0 \geqslant S(u^0)$, and $u_0 \leqslant u^0$. Let the constant M be chosen large enough so that $S'(u) + M \geqslant 0$ in $[u_0, u^0]$. Then $S(u) + Mu$ is increasing on $[u_0, u^0]$, and $u = S(u)$ can be rewritten as
$$u = \frac{S(u) + Mu}{M + 1} \doteq A(u),$$
where A is now increasing on $[u_0, u^0]$ and $u_0 \leqslant A(u_0), u^0 \geqslant A(u^0)$.

   This procedure should remind you somewhat of the one used for contraction mappings (see Exercise 4.4.11), but since we are not restricting the derivative, we will not have uniqueness.

Another type of method for solving a nonlinear equation introduces an additional real parameter λ to provide a continuous transition between the desired problem and a simpler, *base problem* whose solution is *explicitly known*. Suppose that the equation to be solved is $A(u) = 0$ and that a solution u_0 is known for the base problem $B(u) = 0$. We construct a function $F(\lambda, u)$ depending smoothly on λ and u, such that $F(0, u) = B(u)$ and $F(1, u) = A(u)$. We then focus our attention on the equation

$$F(\lambda, u) = 0 \tag{9.1.7}$$

with the hope that (9.1.7) will determine implicitly a function $u(\lambda)$ on $0 \leqslant \lambda \leqslant 1$ with the property $u(0) = u_0$. If this is the case, $u(\lambda)$ is said to be the branch of (9.1.7) passing through $(0, u_0)$, and $u(1)$ will be a solution of the original problem, $A(u) = 0$. Two remarks should be made at this time. First, (9.1.7) often occurs in its own right, with λ having a natural significance as a perturbation parameter. Second, the choice of the interval $0 \leqslant \lambda \leqslant 1$ is arbitrary and sometimes inconvenient; we could as well have λ vary between λ_0 and λ_1.

In view of these remarks we look at the following slightly more general problem: Given a solution (λ_0, u_0) of (9.1.7), that is, a point in the solution set of (9.1.7), can we construct a branch of solutions passing through this point? The implicit function theorem tells us that if grad F is continuous and if $(\partial F/\partial u)(\lambda_0, u_0) \neq 0$, there will exist a branch through (λ_0, u_0) having the functional form $u = u(\lambda)$, but this branch's existence is guaranteed only for a small λ interval around λ_0. In the *method of continuity* (also known as an *embedding method*) we extend $u(\lambda)$ by solving a succession of initial value problems obtained by differentiating (9.1.7) with respect to λ. Denoting partial derivatives by subscripts, we find that

$$F_u(\lambda, u)u_\lambda + F_\lambda(\lambda, u) = 0, \tag{9.1.8}$$

and if $F_u \neq 0$,

$$u_\lambda = -\frac{F_\lambda(\lambda, u)}{F_u(\lambda, u)}. \tag{9.1.9}$$

At $\lambda = \lambda_0$ we have $u = u_0$, and the right side of (9.1.9) is therefore known; hence we have $u_\lambda(\lambda_0)$. For $\Delta\lambda$ sufficiently small we can write

$$u(\lambda_0 + \Delta\lambda) = u(\lambda_0) + u_\lambda(\lambda_0)\,\Delta\lambda, \tag{9.1.10}$$

and we can then use (9.1.9) again to find $u_\lambda(\lambda_0 + \Delta\lambda)$. We can then repeat the procedure to find $u(\lambda_0 + 2\Delta\lambda)$, and so on. In this way we hope to construct the desired branch away from the initial point. Note that in (9.1.10) $\Delta\lambda$ can be either positive or negative, so that we can move forward or backward in λ. Assuming that grad F is well behaved throughout the λ–u plane, the construction can run into trouble only if $F_u = 0$ somewhere along the branch. What goes wrong geometrically if $F_u = 0$?

The following two simple examples are typical of the principal kinds of difficulties encountered:

(a) $u^2 - \lambda^2 = 0$.

(b) $u^2 - 1 + \lambda^2 = 0$.

For (a) the solution set consists of the two intersecting straight lines $u = \lambda$ and $u = -\lambda$. Pretending to be unaware of this but armed with the information that the point $P = (\lambda_0, u_0) \doteq (-1, 1)$ is a solution, we try to construct the branch through P (see Figure 9.3a). Equation (9.1.9) gives

$$u_\lambda = \frac{\lambda}{u} \tag{9.1.11}$$

and $u_\lambda(\lambda_0) = -1$. No difficulty is encountered in integrating (9.1.11) numerically for $\lambda < 0$. At $\lambda = 0$ a new branch of solutions $(u = \lambda)$ intersects the branch on which P lies $(u = -\lambda)$. This phenomenon is signaled by the fact that the denominator on the right side of (9.1.11) vanishes. Admittedly, the numerator also vanishes, so that careful numerical integration might enable one to cross 0 and continue along the original branch. It is certainly not clear how one would go about picking up the new branch at the origin.

For (b) the solution set is the unit circle in the λ–u plane shown in Figure 9.3b. Suppose that we start from the solution $P = (0, -1) \doteq (\lambda_0, u_0)$ and try to construct the branch through P. Equation (9.1.9) becomes

$$u_\lambda = -\frac{\lambda}{u} \tag{9.1.12}$$

and $u_\lambda(\lambda_0) = 0$. We have no difficulty in integrating (9.1.12) numerically until we reach a neighborhood of Q where the denominator of the right side of (9.1.12) vanishes. This time the numerator does *not* vanish, and u_λ becomes positively infinite. Obviously, the branch becomes vertical and is no longer suitably parametrized by λ. This difficulty could be circumvented by making a change of variables (say, using u as the independent variable), but all we are interested in at this time is to characterize the nature of the difficulty.

To recapitulate, we have found that the vanishing of F_u along a branch may signal (a) intersection with a new branch, or (b) the fact that the branch becomes vertical and may then turn back so that further extension in λ is impossible. A third possibility is that $u(\lambda)$ tends to infinity as we approach the point where $F_u = 0$ (say, for $F = \lambda u - 1$ at $\lambda = 0$).

A closely related method for constructing the branch through a solution (λ_0, u_0) is the *perturbation method*. Here we attempt to calculate higher derivatives of u at the initial point by further differentiation of (9.1.7) or (9.1.8) with respect to λ. For instance, $u_{\lambda\lambda}$ is determined from

$$u_{\lambda\lambda} F_u + u_\lambda(2F_{u\lambda} + u_\lambda F_{uu}) + F_{\lambda\lambda} = 0. \tag{9.1.13}$$

We then write

$$u(\lambda) = u(\lambda_0) + (\lambda - \lambda_0)u_\lambda(\lambda_0) + \frac{(\lambda - \lambda_0)^2}{2} u_{\lambda\lambda}(\lambda_0) + \dots, \tag{9.1.14}$$

where the three dots can be interpreted in two ways. Either one hopes that the infinite series converges, or one substitutes a remainder for which an estimate has to be found. The perturbation method is more difficult to justify mathematically and numerically since it is not clear that information pertaining only to the immediate

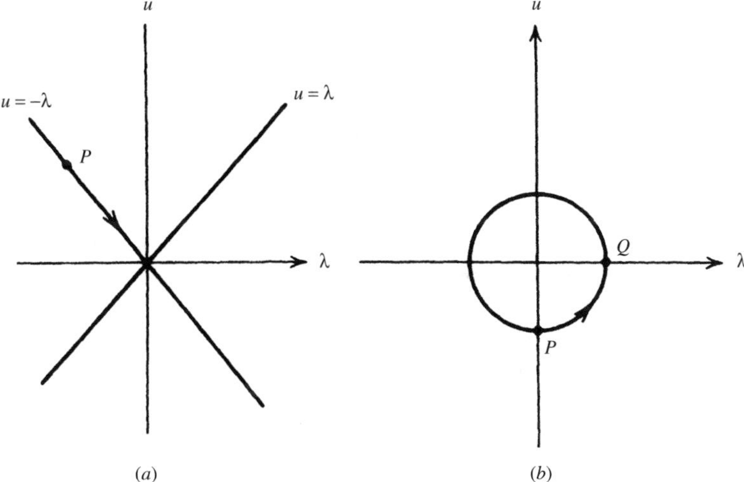

Figure 9.3

neighborhood of the initial point on the branch can predict the behavior at values of λ appreciably distant. The compensating advantage is that the differentiations, however complicated, are all performed at the same point (λ_0, u_0). We shall illustrate these various methods when treating nonlinear problems in function space.

Linear Versus Nonlinear: Buckling of a Rod

The phenomenon of buckling is a familiar one. When a flexible ruler is compressed, it retains its straight shape until the compressive force reaches a critical value which causes the ruler to bend (buckle) rather suddenly with appreciable transverse deflection. The phenomenon is essentially nonlinear since the transverse deflection is not proportional to the applied compressive load P, whereas in a linear theory such as elementary beam theory the deflection is in fact proportional to the applied transverse load.

Figure 9.4 shows a homogeneous thin rod whose ends are pinned, the left end being fixed and the right end free to move along the x axis. In its unloaded state (reference configuration) the axis of the rod coincides with the portion of the x axis between 0 and l. Under a compressive load P a possible state for the rod is that of pure compression, but experience shows that for sufficiently large P, transverse deflections occur. Assuming that this buckling takes place in the x–y plane, we investigate the equilibrium of forces on a portion of the rod including its left end. The free-body diagram shown in Figure 9.5 uses the same sign convention as in Figure 0.4, so that T and M must turn out to be negative. Note that by taking a free-body diagram of the entire rod, we conclude that the only reaction at the left end is the force P shown.

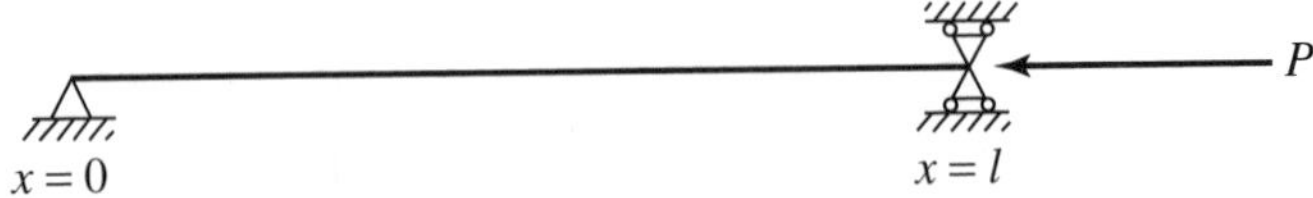

Figure 9.4

A particle occupying the position $(S, 0)$ before loading has moved to the position $(u(S), v(S))$. We let ϕ be the angle between the tangent to the buckled rod and the x axis, and let s be the arclength measured from the left end. Although we could proceed from (0.4.14) and (0.4.15), it is simpler to start afresh for our particular problem. The equilibrium equations are

$$M = -Pv, \tag{9.1.15}$$

$$T = -P\cos\phi, \tag{9.1.16}$$

$$Q = P\sin\phi, \tag{9.1.17}$$

whose form suggests the use of v and ϕ as dependent variables rather than v and u. The strain measures δ and μ of (0.4.17) and (0.4.18), can be expressed in terms of ϕ and v:

$$\mu = \frac{d\phi}{dS}, \quad \delta\sin\phi = \frac{ds}{dS}\frac{dv}{ds} = \frac{dv}{dS}. \tag{9.1.18}$$

Since the rod is homogeneous, the constitutive laws of (0.4.19) take on the simpler form

$$\delta = \hat{\delta}(T), \quad \mu = \hat{\mu}(M), \tag{9.1.19}$$

where $\hat{\delta}$ and $\hat{\mu}$ are prescribed functions having the general properties shown in Figure 0.5. The function $\hat{\delta}$ is positive and monotonically decreasing, with $\hat{\delta}(0) = 1$. The function $\hat{\mu}$ is odd and increasing, with $\hat{\mu}(0) = 0$. Using (9.1.15), (9.1.16), (9.1.18), and (9.1.19), we obtain the system of two nonlinear differential equations in the unknown functions $\phi(S)$ and $v(S)$:

$$\frac{d\phi}{dS} = \hat{\mu}(-Pv), \quad \frac{dv}{dS} = \hat{\delta}(-P\cos\phi)\sin\phi. \tag{9.1.20}$$

The two boundary conditions associated with (9.1.20) can be taken either as $v(0) = v(l) = 0$ or as $\phi'(0) = \phi'(l) = 0$. The fact that the coordinate S refers to the undeformed rod is the reason why these boundary conditions are so easy to impose.

A particularly simple (yet important) case occurs when the rod is *inextensible* and obeys the *Euler-Bernoulli* bending law. In our notation this means that

$$\hat{\delta}(T) \equiv 1, \quad \hat{\mu}(M) = \frac{M}{EI}, \tag{9.1.21}$$

where the constants E and I are, respectively, Young's modulus and the moment of inertia about the neutral axis (see Chapter 0). Then $s = S$, and we could use s as an independent variable but shall not do so. The system (9.1.20) becomes

$$-\phi' = \alpha v, \quad v' = \sin\phi, \quad 0 < S < l; \qquad v(0) = v(l) = (0), \tag{9.1.22}$$

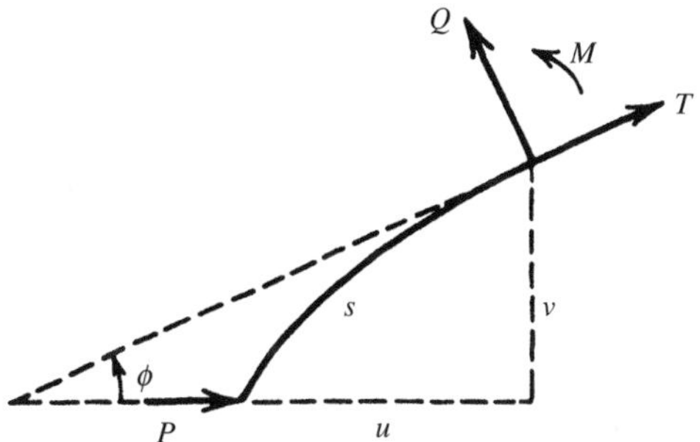

Figure 9.5

where $\alpha \doteq P/EI$. The boundary conditions could also have been written as $\phi'(0) = \phi'(l) = 0$.

Note that (9.1.22) is still *nonlinear* even though linear constitutive relations have been used. The nonlinearity stems from the fact that we have used the exact expression for the curvature. In the analysis of (9.1.22) an important role is played by the *linearized system* derived on the assumption that $|\phi|$ is small. Either by using the linearization procedure of Chapter 0 or, more simply, by replacing $\sin\phi$ by ϕ, we obtain

$$-\phi' = \alpha v, \quad v' = \phi,$$

or the single equation

$$v'' + \alpha v = 0, \quad 0 < S < l; \qquad v(0) = v(l) = 0, \tag{9.1.23}$$

which could have been derived directly by using the approximate form v'' for the curvature $d\phi/dS$.

According to (9.1.23), a nontrivial transverse deflection is possible only for certain discrete values of α (that is, for discrete values of the compressive load). These eigenvalues are

$$\alpha_n = \frac{n^2\pi^2}{l^2}, \quad n = 1, 2, \ldots. \tag{9.1.24}$$

with corresponding deflection $A\sin(n\pi S/l)$. Thus only the shape, not the size, of the deflection is determined. This is depicted in Figure 9.6, where we have chosen $v'(0)$ as a measure of the deflection. The indeterminacy in $v'(0)$ at $\alpha = \alpha_n$ reflects the indeterminacy in the deflection $v(S)$. These results are physically unrealistic. The deflection should be determinate, and the rod should not return to its undeflected form when the load is slightly increased beyond the critical level. We shall see how the *nonlinear system* (9.1.22) resolves these inconsistencies.

A solution of (9.1.22) is a pair of functions (v, ϕ) satisfying the coupled differential equations and the boundary conditions. If (v, ϕ) is a solution, so is $(v, \phi + 2n\pi)$ for any integer n, but these new solutions give rise to the same physical deflection v. If (v, ϕ) is a solution, so is $(-v, -\phi)$, which corresponds to a deflection that is

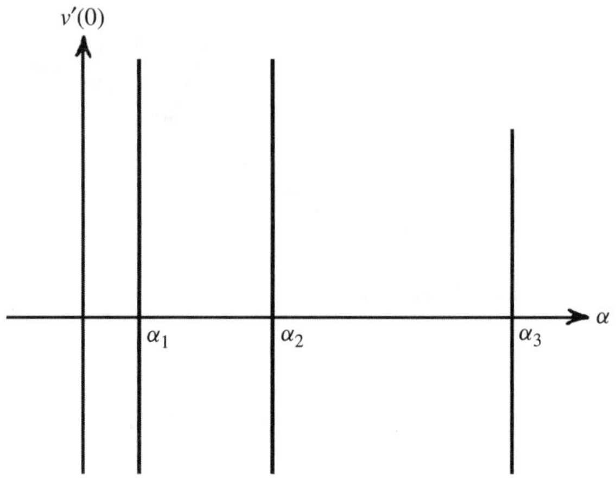

Figure 9.6

the mirror image of the deflection v about the undeformed axis of the rod. System (9.1.22) always has the solutions $(0, 2n\pi)$, which all yield the trivial physical solution $v \equiv 0$, easily seen to be the only solution for $\alpha = 0$. To obtain a nontrivial deflection v, we must have $\phi(0) \neq 2n\pi$ [otherwise, we would have $v(0) = 0$ and $\phi(0) = 2n\pi$, so that, by the uniqueness theorem for the initial value problem, the solution would be $v(S) \equiv 0$, $\phi(S) \equiv 2n\pi$, which corresponds to a trivial deflection]. Thus any solution of (9.1.22) that yields a nontrivial deflection must satisfy the IVP

$$-\phi' = \alpha v, \quad v' = \sin\phi; \quad v(0) = 0, \quad \phi(0) = \phi_0 \neq 2n\pi. \tag{9.1.25}$$

With α and ϕ_0 given, the IVP in equation (9.1.25) has one and only one solution $(v(S, \alpha, \phi_0), \phi(S, \alpha, \phi_0))$ because the nonlinear term satisfies the Lipschitz condition of Exercise 4.4.6. This unique solution will usually *not* satisfy the condition $v(l) = 0$ unless α and ϕ_0 are suitably related. In any event our previous discussion shows that it suffices to *consider only initial values satisfying* $0 < \phi_0 < \pi$.

For some purposes it is convenient to reduce system (9.1.25) to a single differential equation of the second order for v or ϕ. The simpler of these is

$$\phi'' + \alpha \sin\phi = 0, \quad \phi(0) = \phi_0, \quad \phi'(0) = 0, \tag{9.1.26}$$

which, for $\alpha > 0$, also describes the motion of a simple pendulum (here S is the time, and ϕ the angle between the pendulum and the downward vertical; initially, the bob is displaced by an angle ϕ_0 and released from rest). For $\alpha > 0$ we shall see that the solution of (9.1.25) [or (9.1.26)] is periodic in S, tracing out the orbit in the v–ϕ plane shown in Figure 9.7. The equation of this orbit is obtained from (9.1.25) by adding the results of multiplying the first equation by v' and the second by ϕ':

$$\frac{\alpha}{2}\frac{d}{dS}v^2 - \frac{d}{dS}\cos\phi = 0,$$

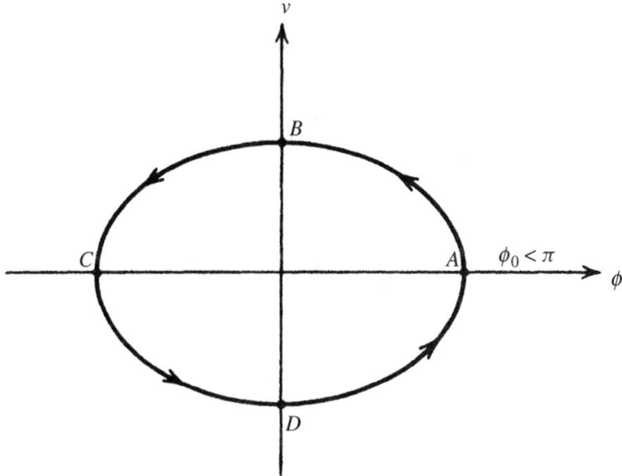

Figure 9.7

or, using the initial conditions,

$$v^2 = \frac{2}{\alpha}(\cos\phi - \cos\phi_0). \tag{9.1.27}$$

From (9.1.27) we have $|\phi| \leqslant \phi_0$. At the beginning of the motion we are at A with $v = 0, \sin\phi > 0$, and hence $v'(0) > 0$. For small positive S we can therefore infer from (9.1.25) that v is positive and ϕ' negative; these properties continue to hold until we reach B. At that point $\phi = 0$ and $v > 0$, which we may regard as new initial values for the next "time" interval. Since the transformation $S \to -S$, $\phi \to -\phi$, $v \to v$ keeps the equations invariant, the path starting at B will be the mirror image about the v axis of the path from A to B. Similar arguments show that the path CDA is the reflection about the ϕ axis of ABC. We have returned to A with the same initial values of ϕ and v, so that we will just keep going around the same orbit. If $\alpha < 0$, it can easily be seen that the motion is *not* periodic.

The period of the orbit (the time it takes to go around the orbit once) depends on α and ϕ_0 and will be denoted by $T(\alpha, \phi_0)$. For the linearized problem

$$-\phi' = \alpha v, \quad v' = \phi; \quad v(0) = 0, \quad \phi(0) = \phi_0, \tag{9.1.28}$$

the explicit solution is

$$\phi = \phi_0 \cos\sqrt{\alpha}\, S, \quad v = \frac{\phi_0}{\sqrt{\alpha}} \sin\sqrt{\alpha}\, S,$$

with period $2\pi/\sqrt{\alpha}$ independent of ϕ_0. For the nonlinear problem with small ϕ_0 the period will tend to $2\pi/\sqrt{\alpha}$, but, as we increase ϕ_0, the period increases. This can be seen by comparing (9.1.26) with the corresponding linear problem and noting that $\sin\phi < \phi$ for $0 < \phi < \pi$.

With this information we can analyze the BVP (9.1.22) qualitatively. A solution of (9.1.25) will satisfy $v(l) = 0$ if and only if l is an integral multiple of a half-period $T/2$, that is,

$$\frac{2l}{n} = T(\alpha, \phi_0). \tag{9.1.29}$$

Since $T(\alpha, \phi_0)$ is increasing as a function of ϕ_0, with $T(\alpha, 0+) = 2\pi/\sqrt{\alpha}$ and $T(\alpha, \pi-) = \infty$, nontrivial solutions will be possible for a *given* l only if $\alpha > \pi^2/l^2$. If

$$\frac{n^2\pi^2}{l^2} < \alpha < \frac{(n+1)^2\pi^2}{l^2}, \tag{9.1.30}$$

there will be exactly n solutions of (9.1.22), each having a different value of ϕ_0 between 0 and π. If we also take into account the possibility of deflections which are mirror images about the x axis, we have established the existence of n pairs of solutions of (9.1.22) in the interval (9.1.30). All of this is confirmed by writing T as a complete elliptic integral. From (9.1.27) we find, for $\alpha > 0$,

$$(\phi')^2 = 4\alpha\left(\sin^2\frac{\phi_0}{2} - \sin^2\frac{\phi}{2}\right).$$

Along the path AB of Figure 9.7, ϕ decreases from ϕ_0 to 0, while S increases from 0 to $T/4$, so that

$$2\sqrt{\alpha}\frac{dS}{d\phi} = -\left(\sin^2\frac{\phi_0}{2} - \sin^2\frac{\phi}{2}\right)^{-1/2}$$

and, on integration from $S = 0$ to $S = T/4$,

$$T\frac{\sqrt{\alpha}}{2} = \int_0^{\phi_0}\left(\sin^2\frac{\phi_0}{2} - \sin^2\frac{\phi}{2}\right)^{-1/2} d\phi.$$

The permissible change of variables

$$\sin z = \frac{\sin(\phi/2)}{\sin(\phi_0/2)}$$

then gives

$$T = \frac{4}{\sqrt{\alpha}}\int_0^{\pi/2}\frac{dz}{\sqrt{1 - p^2\sin^2 z}}, \qquad p = \sin\frac{\phi_0}{2}, \tag{9.1.31}$$

from which it is apparent that $T(\alpha, \phi_0)$ increases with ϕ_0 and that $T(\alpha, 0+) = 2\pi/\sqrt{\alpha},\ T(\alpha, \pi-) = \infty$.

The *branching diagram* (solid lines) in Figure 9.8 summarizes the results of the nonlinear analysis and contrasts them with those for the linearized problem of Figure 9.6. There are still some important unanswered questions, the principal one being as follows: Which of the mathematical buckled states does the rod actually choose? This is a problem of *stability* and can be answered only by studying the dynamical

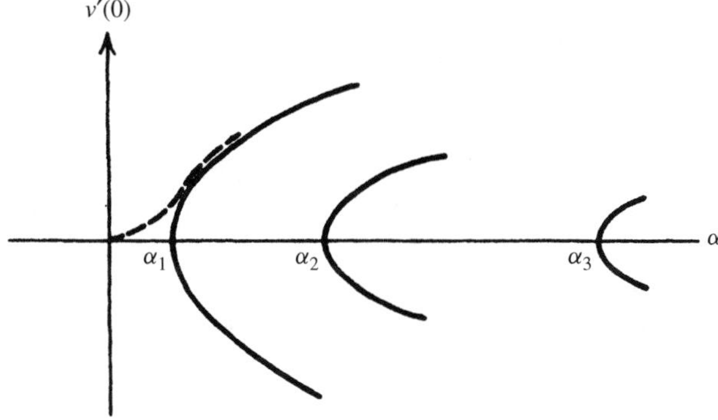

Figure 9.8

equations. Sometimes arguments involving potential energy can be used to discuss the stability of the various states, but such apparently "static" arguments work only because a suitable connection has been made between potential energy and stability. What is found in our particular problem is that the zero solution is stable for $\alpha < \alpha_1$ but not for $\alpha > \alpha_1$. For $\alpha > \alpha_1$ only the branches emanating from $\alpha = \alpha_1$ are stable. This still leaves the minor ambiguity of sign in the stable buckled state. No model based on a perfectly straight homogeneous rod, centrally compressed, can resolve this point. If, however, we assume that the load P is applied with a slight eccentricity or that the column is slightly imperfect, the ambiguity is removed. A typical load-deflection curve for such a case is shown as the dashed line in Figure 9.8.

Nonlinear Operators

We shall deal with nonlinear operators on a *real* Hilbert space H (with an occasional excursion into a Banach space of continuous functions). The class of nonlinear operators is understood to include linear operators as a subset. Some of the ideas below have already been encountered in Section 4.4.

Examples.

(a) $H = \mathbb{R}$. An arbitrary real-valued function of a real variable maps $\mathbb{R}$ into itself and is also said to be a nonlinear operator A on H. The value of A at u is written as Au or $A(u)$. A linear operator is one whose value at every u is obtained by multiplying u by a fixed, real constant.

(b) $H = \mathbb{R}^n$. A nonlinear operator A maps $\mathbb{R}^n$ into itself. Thus A operates on a vector $u = (u_1, \dots, u_n)$ to give an image vector $Au \doteq v = (v_1, \dots, v_n)$. Each component v_i of Au is the value at u of a real-valued function A_i of the n variables $u_1, \dots, u_n : v_i = A_i u = A_i(u_1, \dots, u_n)$. For a linear operator L each A_i is linear, so that $A_i(u) = \sum_j a_{ij} u_j$; the real $n \times n$ matrix (a_{ij}) then completely specifies the linear operator L.

(c) $H = L_2^{(r)}(0,1)$. The elements u of H are functions of the real variable x, $0 < x < 1$. The operator A maps functions into functions. Even simple nonlinear operators such as $Au = u^2$ are not defined on the whole of H [if $u(x)$ is square integrable, $u^2(x)$ may not be]. This is a minor inconvenience which we dismiss by restricting the domain of A either explicitly or implicitly to elements of H for which Au is also in H. Perhaps a more satisfactory approach would be to consider operators on the Banach space $C[0, 1]$ of continuous functions with the uniform norm, but we usually prefer to enjoy the comparative familiarity of Hilbert spaces. A more interesting example of a nonlinear operator is the *Hammerstein integral operator*, defined as follows:

$$v(x) = \int_0^1 k(x,\xi) f(u(\xi))\, d\xi \quad \text{or} \quad v = Tu, \tag{9.1.32}$$

where f is a real-valued function of a real variable and k is a real-valued function on $\mathbb{R}^2$. Note that

$$T = KF, \tag{9.1.33}$$

where F is the nonlinear operator mapping the element u into the element $f(u)$, and K is the real linear integral operator with kernel $k(x,\xi)$. T can be considered as an operator on $C[0, 1]$ or on $L_2(0,1)$.

(d) $H = L_2^{(r)}(\Omega)$, where Ω is a domain in $\mathbb{R}^n$. Consider the set of sufficiently smooth functions $u(x)$ which vanish on the boundary Γ of Ω, and set

$$v(x) = -\Delta u(x) + f(u(x)) \doteq Au, \tag{9.1.34}$$

where Δ is the Laplacian in $\mathbb{R}^n$. This operator A occurs in problems of combined diffusion and reaction.

(e) Consider the space of pairs of real-valued functions $(v(x), \phi(x))$ square integrable on $0 \leqslant x \leqslant l$. This is a real Hilbert space H under the inner product

$$\langle (v_1, \phi_1), (v_2, \phi_2) \rangle = \int_0^1 (v_1 v_2 + \phi_1 \phi_2)\, dx.$$

If v and ϕ are continuously differentiable, we can define a nonlinear operator A on H by

$$A(v,\phi) = (-\phi' - \alpha v, v' - \sin\phi).$$

Except for the boundary conditions, this is the operator that appears in (9.1.22).

Linearization

Let us begin by paraphrasing the great French mathematician Jean Dieudonné: A closely guarded secret in elementary calculus is that the basic idea of differential calculus is the local approximation of a nonlinear function by a linear one. This idea is obscured by the accidental fact that in a one-dimensional vector space there is a one-to-one correspondence between numbers and linear transformations.

What does this mean? Suppose that A is a real-valued function of the real variable u and we want to calculate $A(u)$ for all u near a fixed u_0, where $A(u_0)$ is regarded as known. If A is differentiable, we can write

$$A(u) - A(u_0) = A'(u_0)[u - u_0] + r, \qquad \lim_{u \to u_0} \frac{r}{u - u_0} = 0,$$

the number $A'(u_0)$ being the derivative of A at u_0. But the number $A'(u_0)$ can also be regarded as a linear operator on the increment $h = u - u_0$; this is the point of view that generalizes to $\mathbb{R}^n$ and to infinite-dimensional spaces:

$$A(u) - A(u_0) = Lh + r, \quad \lim_{h \to 0} \frac{r}{h} = 0. \tag{9.1.35}$$

Thus, the change in A from u to u_0 is the sum of a principal part linear in the increment $h = u - u_0$ and of a remainder r small compared to the linear term as the increment tends to zero. (Our notation has suppressed the dependence of r on h.) Although the operator L is defined for all h, the linearized approximation

$$A(u) - A(u_0) = Lh$$

is useful only when h is small.

These ideas are easily extended to the case of an operator A on an arbitrary Hilbert or Banach space (see Chapter 10 for just such extensions).

Definition. *The operator A is* linearizable *at u_0 if there exists a* bounded linear operator *L such that*

$$Au - Au_0 = Lh + r, \quad \lim_{h \to 0} \frac{\|r\|}{\|h\|} = 0, \tag{9.1.36}$$

where $h = u - u_0$.

REMARKS

1. If A is linearizable at u_0, the operator L is uniquely determined and is known as the *Fréchet derivative* of A at u_0. We sometimes write $L = A'(u_0)$. If A is linearizable at many (or all) points u_0, the dependence of L on u_0 will usually *not* be linear. The remainder r in (9.1.36) depends on h and u_0. When we wish to emphasize the dependence of r on h, we write $r = Rh$, where R is a nonlinear operator.

2. If A is a bounded linear operator, $A'(u_0) = A$ for all u_0.

3. The notion of linearization can be extended to operators from one Banach space to another (hence to functionals on either a Banach or a Hilbert space; these ideas are pivotal in the calculus of variations, as we have seen in Section 8.4). See Chapter 10 for a more extensive discussion.

4. If A is linearizable at u_0, the following method is often useful in calculating $L = A'(u_0)$. We write $h = \varepsilon\eta$, where ε is a real number and η is a fixed element in H. Then (9.1.36) implies that

$$\lim_{\varepsilon \to 0} \frac{A(u_0 + \varepsilon\eta) - Au_0}{\varepsilon} = L\eta \quad \text{for each } \eta \in H. \tag{9.1.37}$$

Examples.

(a) Let A be an operator on $\mathbb{R}^n$. Let $\tilde{u} = (\tilde{u}_1, \ldots, \tilde{u}_n)$ be a fixed point in $\mathbb{R}^n$ (the terminology u_0 would be confusing), and let $u = (u_1, \ldots, u_n)$ be an arbitrary point, $h = (h_1, \ldots, h_n) = u - \tilde{u}$. The value of A at u can be expressed as

$$Au \doteq v = (v_1(u), \ldots, v_n(u)),$$

where all $v_i(u)$ are assumed to be continuously differentiable. Then, by Taylor's theorem in n variables, we find that

$$v_i(u) - v_i(\tilde{u}) = \sum_{j=1}^{n} \frac{\partial v_i}{\partial u_j}(\tilde{u})h_j + r_i, \qquad \lim_{h \to 0} \frac{r_i}{\|h\|} = 0.$$

Hence

$$Au - A\tilde{u} = Lh + r, \qquad \lim_{h \to 0} \frac{\|r\|}{\|h\|} = 0,$$

where L is the *Jacobian* matrix with entries $(\partial v_i/\partial u_j)(\tilde{u})$ and r is the vector $(r_1, \ldots, r_n)$.

(b) To linearize the integral operator (9.1.32) at u_0, we use (9.1.37):

$$\begin{aligned} L\eta &= \lim_{\varepsilon \to 0} \int_0^1 k(x,\xi) \left[\frac{f(u_0(\xi) + \varepsilon\eta(\xi)) - f(u_0(\xi))}{\varepsilon}\right] d\xi \\ &= \int_0^1 k(x,\xi) f_u(u_0(\xi))\eta(\xi)\, d\xi. \end{aligned} \tag{9.1.38}$$

Thus L is a linear integral operator with kernel $k(x,\xi)f_u(u_0(\xi))$.

(c) Let f be a real-valued nonlinear functional on H. We say that f is linearizable at u_0 if there exists a bounded linear functional l such that

$$f(u) - f(u_0) = l(h) + r, \qquad \lim_{\|h\| \to 0} \frac{|r|}{\|h\|} = 0. \tag{9.1.39}$$

We also write $l = f'(u_0)$. A *critical point* for f is a point u_0 at which $f' = 0$, that is, $l = 0$, and then (9.1.39) shows that $f(u_0 + h) - f(u_0)$ is small in comparison with h for h near 0. A necessary condition for u_0 to be a critical point of f is condition (8.4.13), with J replaced by f and u by u_0. If f is linearizable at u_0, we can use the Riesz representation theorem to write $l(h) = \langle h, z\rangle$, where z is uniquely determined from l. If f is linearizable at every u_0, the element z will depend on u_0 and we will have

$$z = Au_0,$$

where A is a *nonlinear operator* known as the *gradient* of f. It is useful to proceed in the opposite direction. Suppose that A is an operator for which there happens to exist a functional f such that

$$f(u_0 + h) - f(u_0) = \langle h, Au_0\rangle + r, \quad \lim_{h\to 0} \frac{|r|}{\|h\|} = 0. \tag{9.1.40}$$

We then say that A is a *gradient operator* (and, of course, A is the gradient of f) and call f a (scalar) *potential* of A. The potential is determined up to an arbitrary additive constant element. In particular, the potential which vanishes at $u = 0$ is

$$f(u) = \int_0^1 \langle u, A(tu)\rangle\, dt. \tag{9.1.41}$$

If A is linear and symmetric, it is a gradient and its potential is $\frac{1}{2}\langle Au, u\rangle$. If A is a nonlinear, continuously differentiable operator with $A'(u_0)$ symmetric, then A is a gradient. Thus, if

$$\langle A'(u_0)h_1, h_2\rangle = \langle h_1, A'(u_0)h_2\rangle \quad \text{for all } u_0, h_1, h_2,$$

then A is a gradient.

Monotone Iteration on $C(R)$: Order-Preserving Transformations

The idea of monotone iteration introduced earlier in this section can easily be extended to function spaces. To be specific, consider the space $C(R)$ of real-valued continuous functions $u(x)$ on the closed region R in $\mathbb{R}^n$. We order the elements of $C(R)$ in the natural way: $u \leqslant v$ if $u(x) \leqslant v(x)$ for every x in R. A similar definition holds for $u \geqslant v$. If $u \leqslant v$, we denote by $[u, v]$ the set of all elements w such that $u \leqslant w \leqslant v$, and we refer to $[u, v]$ as an *order interval*. Of course, not all elements of $C(R)$ are comparable; there exist u, v for which neither $u \leqslant v$ nor $u \geqslant v$ holds. The space $C(R)$ is said to be partially ordered.

A typical nonlinear boundary value problem can often be translated into an equation of the form

$$u = Tu, \tag{9.1.42}$$

where T is a nonlinear integral operator (usually, of the Hammerstein type) on $C(R)$. Thus solving the BVP is equivalent to finding the fixed point(s) of T. Often, we shall look for fixed points in the order interval $[u_0, u^0]$. The operator T is said to be increasing on $[u_0, u^0]$ if, whenever $u \leqslant v$ and $u,\ v \in [u_0, u^0]$, then $Tu \leqslant Tv$, which means that T is order-preserving. In all our applications T will be a *compact* operator, that is, continuous and mapping bounded sets into relatively compact sets (see the beginning of the present section and Section 5.7). Since our norm is the uniform norm, convergence in $C(R)$ means uniform convergence on R. A fixed point u of T is said to be *maximal* in $[u_0, u^0]$ if for any fixed point v of T in $[u_0, u^0]$ we have $u \geqslant v$. A similar definition is used for *minimal* fixed point.

Theorem 9.1.1. *Let $u_0 \leqslant Tu_0, u^0 \geqslant Tu^0$, and $u_0 \leqslant u^0$. If T is compact and increasing on $[u_0, u^0]$, the sequence $u_0, u_1, \ldots$, where*

$$u_n = Tu_{n-1}, \quad n = 1, 2, 3, \ldots \tag{9.1.43}$$

is increasing and converges to the minimal solution $\underline{u}$ of (9.1.42) on $[u_0, u^0]$. The sequence $u^0, u^1, \ldots$, where

$$u^n = Tu^{n-1}, \quad n = 1, 2, 3, \ldots \tag{9.1.44}$$

is decreasing and converges to the maximal solution $\overline{u}$ of (9.1.42) on $[u_0, u^0]$.

Proof. If $u_0 \leqslant u \leqslant u^0$, then $Tu_0 \leqslant Tu \leqslant Tu^0$, so that $u_0 \leqslant Tu \leqslant u^0$ and Tu lies in the order interval $[u_0, u^0]$. Thus T maps $[u_0, u^0]$ into itself, and the sequences $\{u_n\}$ and $\{u^n\}$ are well defined. Since $u_0 \leqslant u_1$, we have $Tu_0 \leqslant Tu_1$, that is, $u_1 \leqslant u_2$. Hence $\{u_n\}$ is increasing, and, similarly, $\{u^n\}$ is decreasing. Moreover, $u_0 \leqslant u^0$ implies that $u_1 \leqslant u^1$ and, recursively, $u_n \leqslant u^n$. Thus the two sequences can be ordered as

$$u_0 \leqslant u_1 \leqslant \cdots \leqslant u_n \leqslant \cdots \leqslant u^n \leqslant \cdots \leqslant u^1 \leqslant u^0.$$

The sequence $\{u_n\}$ is bounded, so that $\{Tu_n\}$ contains a converging subsequence. Since $Tu_n = u_{n+1}$, the sequence $\{u_n\}$ contains a converging subsequence. Because $\{u_n\}$ is increasing, the entire sequence actually converges, to $\underline{u}$, say. Since T is continuous and $Tu_n = u_{n+1}$, we pass to the limit to obtain $T\underline{u} = \underline{u}$. Similarly, it can be shown that u^n converges to $\overline{u}$, where $T\overline{u} = \overline{u}$. Clearly, $\underline{u} \leqslant \overline{u}$.

Let us now show that $\underline{u}$ is the minimal solution of (9.1.42) in $[u_0, u^0]$. Let v be any fixed point of T in $[u_0, u^0]$. Then $u_0 \leqslant v$ and $Tu_0 \leqslant Tv$ or, since $Tv = v, u_1 \leqslant v$. Continuing in this way, we find that $u_n \leqslant v$ and hence $\underline{u} \leqslant v$, so that $\underline{u}$ is the minimal solution in $[u_0, u^0]$. Similarly, it can be shown that $\overline{u}$ is the maximal solution in the order interval.

Extensive treatments of monotone methods and related techniques may be found in the works of Amann [1], Sattinger [51], and Pao [48]. □

9.2 BRANCHING THEORY

Guided by our limited experience of Section 9.1, we shall now investigate the equation

$$F(\lambda, u) = 0, \tag{9.2.1}$$

where λ is a real number, but u is now an *element of a real Hilbert space* H (in most cases of interest H will be a function space and hence infinite-dimensional). The operator F is an arbitrary nonlinear operator mapping ordered pairs (λ, u) in $\mathbb{R} \times H$ into elements of H. Thus, despite its simple appearance, (9.2.1) may disguise a difficult nonlinear partial differential equation or integral equation. Strictly speaking, solutions of (9.2.1) are ordered pairs (λ, u), but we shall often refer to an element u as a solution corresponding to a particular λ. Throughout we work in λ–u space, that is, in $\mathbb{R} \times H$; we unabashedly use geometrical language and visualization that would be appropriate if H were one-dimensional instead of infinite-dimensional. Some care will have to be exercised in translating the conceptual image into numerical calculations.

Often, we shall deal with a slightly simpler version of (9.2.1):

$$Au - \lambda u = 0, \tag{9.2.2}$$

where A is a nonlinear operator mapping H into itself. Note that (9.2.2) includes as special cases both linear homogeneous and linear inhomogeneous problems (in the first instance, set $A = L$; in the second, let $Au = Lu + f$). We can imitate this distinction when A is nonlinear by defining the problem to be *unforced* if $A0 = 0$, and *forced* if $A0 \neq 0$. If the problem is unforced, we always have the *basic solution branch* $(\lambda, 0)$ extending from $\lambda = -\infty$ to $\lambda = \infty$.

The problem of finding all solutions of (9.2.1) or (9.2.2) is much too difficult to entertain in any kind of generality (although there will be particular examples where the complete solution set can be exhibited). *Branching theory*, also known as *bifurcation theory*, deals with some particular aspects of (9.2.1) and (9.2.2). Given a solution (λ_0, u_0), can we use the method of continuity (or some other method) to construct a branch of solutions through (λ_0, u_0)? If so, when does the extension run into difficulties? Does the branch sprout twigs? Does the branch turn around at some limiting value of λ? As a special case, consider the *unforced* problem (9.2.2). We then have at our disposal the basic branch $(\lambda, 0)$, which can be viewed as the infinitely long trunk of a tree. There is no problem of extending this solution since it already goes from $-\infty$ to ∞, nor does the solution turn around at some value of λ. Therefore, the principal remaining task is to locate the branch points at which other branches join the trunk and to obtain qualitative information about the shape of these branches near their intersections with the trunk.

We shall also make some observations about a problem which is not strictly within the province of branching theory. There are good physical grounds for trying to find *positive* solutions of (9.2.1) or (9.2.2) (obviously, we are thinking of H as a space where this notion makes sense). Positive solutions can often be constructed by monotone methods (see Section 9.4).

Without completely abandoning our arboreal analogy, we turn now to more mathematical forms of expression.

Definition. *Consider (9.2.2) with* $A0 = 0$. *We say that* $\lambda = \lambda^0$ *is a* branch point *(of the basic solution) if in every neighborhood of* $(\lambda^0, 0)$ *in* $\mathbb{R} \times H$ *there exists a solution* (λ, u) *of (9.2.2) with* $\|u\| \neq 0$.

REMARKS

1. The definition places the burden on small neighborhoods of $(\lambda^0, 0)$. Observe that each of the numbers $n^2\pi^2/l^2$ is a branch point in both the linear and nonlinear problems in Figures 9.6 and 9.8. In this case the branch points of the nonlinear problem coincide with those values of λ (eigenvalues) for which the linearized problem has nontrivial solutions.

2. Our definition is clearly equivalent to the existence of a sequence (λ^n, u^n) of solutions of (9.2.2) with $\|u^n\| \neq 0$ and $(\lambda^n, u^n) \to (\lambda^0, 0)$.

3. The definition does not guarantee the existence of a *continuous* branch of solutions. Such a continuous branch will be formed if the boundary of each sufficiently small, open ball in $\mathbb{R} \times H$ with center at $(\lambda^0, 0)$ contains a solution (λ, u) with $\|u\| \neq 0$.

We expect the branch points of the nonlinear problem to be related to the spectrum of the linearized problem. Suppose that we consider (9.2.2) in the *unforced* case, and let us look for the branch points of the basic solution. The Fréchet derivative of $A - \lambda I$ at $u = 0$ is the operator $L - \lambda I$, where $L = A'(0)$. In one dimension the vanishing of the number $L - \lambda I$ alerted us to the possibility of a branch point; in a general Hilbert space the equivalent condition is that the operator $L - \lambda I$ fails to have a bounded inverse (that is, 0 is in the spectrum of $L - \lambda I$, or, in other words, λ is in the spectrum of L).

Theorem 9.2.1. *The number λ^0 can be a branch point of the basic solution of (9.2.2) only if it is in the spectrum of $L = A'(0)$.*

Proof. Let λ^0 be a branch point. There must therefore exist a sequence $(\lambda^n, u^n) \to (\lambda^0, 0)$ with $\|u^n\| \neq 0$ and $Au^n - \lambda^n u^n = 0$. Using (9.1.36), we can write the last equation as

$$Lu^n - \lambda^0 u^n = (\lambda^n - \lambda^0)u^n - Ru^n, \qquad \lim_{n\to\infty} \frac{\|Ru^n\|}{\|u^n\|} = 0.$$

Setting $z^n = u^n/\|u^n\|$, we find that $Lz^n - \lambda^0 z^n = f^n$, where $\lim_{n\to\infty} \|f^n\| = 0$ and $\|z^n\| = 1$. Thus $L - \lambda^0 I$ is not bounded away from 0 and hence cannot have a bounded inverse. Therefore, λ^0 must be in the spectrum of L. □

REMARK. Theorem 9.2.1 tells us that branch points of the nonlinear problem are to be found in the spectrum of the linearized problem. In the problems we shall consider, L will be either a compact self-adjoint operator or its inverse. Thus the spectrum of L will consist solely of eigenvalues (except perhaps for $\lambda = 0$). Even so, we cannot guarantee that an eigenvalue of L will be a branch point of the nonlinear problem. In Example 4(a) below there is no branching from an eigenvalue of the linearized problem. Theorems 9.2.2 and 9.2.3, however, give some sufficient conditions for an eigenvalue of the linearized problem to be a branch point of the nonlinear problem.

Theorem 9.2.2 (Leray-Schauder-Krasnoselskii). *If A is compact and $\lambda^0 \neq 0$ is an eigenvalue of odd multiplicity for L, then λ^0 is a branch point of the basic solution of (9.2.2).*

REMARK. The proof is based on topological degree theory and is omitted. The multiplicity referred to in the theorem is the algebraic multiplicity. For a self-adjoint operator this is the same as the geometric multiplicity, which is the dimension of the eigenspace corresponding to that eigenvalue. In applying Theorem 9.2.2 one is limited by the need for predicting the multiplicity of the eigenvalues of a linear problem. The only available theorems are those which state that certain eigenvalues are simple. This is the case, for instance, for Sturm-Liouville problems of ordinary differential equations, for certain special kinds of matrices and integral operators, and for the lowest eigenvalue of elliptic BVPs. Rabinowitz later proved that the branch emanating from λ^0 must either terminate at another eigenvalue of $A'(0)$ or must tend to infinity.

Theorem 9.2.3 (Krasnoselskii). *Let A be a compact operator which is the gradient of a uniformly differentiable functional, and let A have a second derivative at the origin. Then any nonzero eigenvalue of L is a branch point.*

REMARK. The proof, which is based on the category theory of Liusternik and Schnirelmann, is omitted. We shall not clarify the hypotheses of uniform differentiability and existence of a second derivative; suffice it to say that the hypotheses are met in all but pathological cases. Note that if A is a gradient, problem (9.2.2) is equivalent to finding the critical points of $f(u) - \lambda \|u\|^2/2$, where f is the potential of A.

We now take up some simple examples. The reader may be disappointed that these examples are chosen from finite-dimensional spaces. Some of the qualitative features of branching, however, are already present in these examples. We turn to physically more important applications in Section 9.4.

Example 1. With $H = \mathbb{R}$, consider the nonlinear operator

$$A(u) = Lu + cu^2, \tag{9.2.3}$$

where L and c are given real numbers, $c \neq 0$. Of course Lu is just the value at u of a linear operator on $\mathbb{R}$. Equation (9.2.2) becomes

$$Lu + cu^2 = \lambda u, \tag{9.2.4}$$

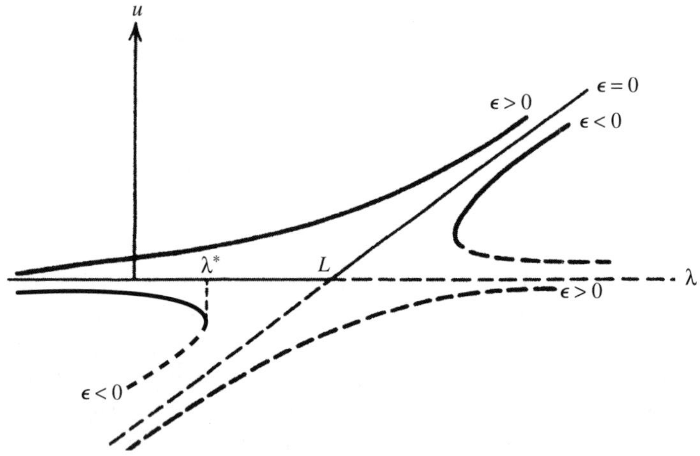

Figure 9.9

which always admits the solution $u = 0$. For $\lambda \neq L$ we also have the solution

$$u = \frac{\lambda - L}{c}.$$

The branching diagram consists of the intersecting straight lines appearing in Figure 9.9 (dashed lines represent unstable solutions; see Section 9.5), where c and L have been chosen positive. The only branch point of the basic solution is at $\lambda = L$. The linearized problem about $u = 0$ is $Lu = \lambda u$, which has $\lambda = L$ as its only eigenvalue.

Let us now alter (9.2.3) slightly to generate a forced problem:

$$A(u, \varepsilon) = Lu + cu^2 - \varepsilon, \quad \varepsilon \neq 0.$$

Equation (9.2.2) now has the form

$$Lu + cu^2 - \varepsilon = \lambda u \quad \text{or} \quad cu^2 - \varepsilon = (\lambda - L)u, \tag{9.2.5}$$

whose solutions are easily found graphically by plotting the parabola $cu^2 - \varepsilon$ and the straight line $(\lambda - L)u$. The branching diagrams for $\varepsilon > 0$ and $\varepsilon < 0$ are both shown in Figure 9.9. In the case $\varepsilon > 0$ there is no branch point along either solution branch and no limiting value of λ; it is easy to check that along both branches

$$\frac{d}{du}[A(u, \varepsilon) - \lambda u] = A_u - \lambda \neq 0.$$

The case $\varepsilon < 0$ exhibits two solution branches, each of which has a limiting value of λ where $A_u(u, \varepsilon) - \lambda$ vanishes.

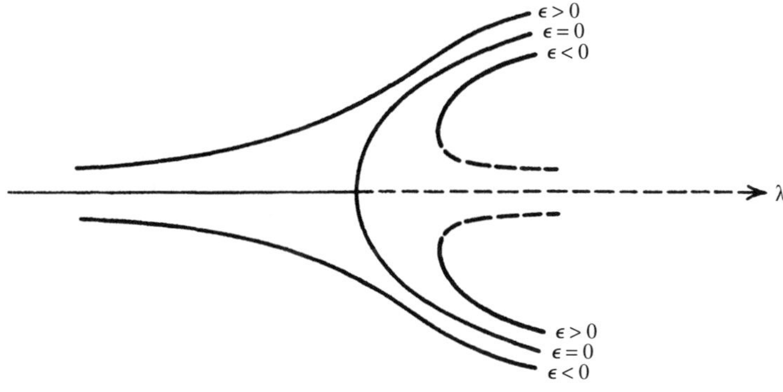

Figure 9.10

If $c < 0$, there is no need to redraw the diagram. It is essentially enough to reverse the λ direction (that is, to regard λ as increasing from right to left) to obtain the appropriate diagrams.

It turns out that branching diagrams similar to those of Figures 9.9 and 9.10 occur in many stability problems in elasticity. Here λ is a loading parameter, ε an imperfection parameter, and u a measure of the deflection. Thus the case $\varepsilon = 0$ corresponds to a perfect system (no eccentricity in the loading or imperfection in the structure), whereas $\varepsilon \neq 0$ corresponds to an imperfect system. A dangerous feature of the slightly imperfect system with $\varepsilon < 0$ in Figure 9.9 is that a critical load (limit load) λ^* is reached much earlier than expected for the perfect system. If λ is increased beyond this critical value, the system usually snaps into another static solution [not shown in our diagram but presumably present if a more realistic model than (9.2.5) is used] or else experiences large oscillations.

Example 2. With $H = \mathbb{R}$ we now consider the second canonical nonlinear operator

$$A(u) = Lu + cu^3, \quad c \neq 0, \tag{9.2.6}$$

which differs from (9.2.3) in that the nonlinear term is cubic rather than quadratic. Equation (9.2.2) becomes

$$Lu + cu^3 = \lambda u,$$

which has the solution $u = 0$ for all λ. In addition, if $c > 0$, we find two nontrivial solutions for $\lambda > L$, given by

$$u = \pm\sqrt{(\lambda - L)/c}. \tag{9.2.7}$$

The branching diagram is shown in Figure 9.10. Again,, $\lambda = L$ is the only branch point, and it is also the only eigenvalue of the linearized problem $Lu = \lambda u$. If c were negative, the branching would be to the left (again it suffices to reverse the λ direction).

For the forced problem

$$Lu + cu^3 - \varepsilon = \lambda u, \quad \varepsilon \neq 0, \tag{9.2.8}$$

the branching diagram is also shown in Figure 9.10. The situation is now quite symmetric in ε. The solution for $\varepsilon > 0$ consists of two branches, one of which increases smoothly from $\lambda = -\infty$ to $+\infty$, whereas the other exhibits a turning point.

Example 3. With $H = \mathbb{R}^2$ the problem is two-dimensional, and hence the linearized problem can have at most two eigenvalues. *Suppose that these eigenvalues are distinct*, say $\lambda_1 < \lambda_2$. Each eigenvalue is simple, so Theorem 9.2.2 guarantees that λ_1 and λ_2 are actually branch points of the basic solution (we are considering an *unforced problem*). The structure of the solution set can now be more complicated, even for relatively simple-looking problems. All the cases we shall study are of the form

$$Au = Lu + Ru, \tag{9.2.9}$$

where

$$Lu = (\lambda_1 u_1, \lambda_2 u_2), \quad \lambda_1 < \lambda_2, \tag{9.2.10}$$

and R, to be specified later, is always *homogeneous of the third degree*:

$$R(\alpha u) = \alpha^3 Ru. \tag{9.2.11}$$

We see that L is the linearization of A at $u = 0$ and that branching will occur at the eigenvalues λ_1, λ_2 of L.

(a) If

$$Ru = (c_1 u_1^3, c_2 u_2^3), \tag{9.2.12}$$

then (9.2.2) becomes the pair of uncoupled equations

$$\lambda_1 u_1 + c_1 u_1^3 = \lambda u_1, \quad \lambda_2 u_2 + c_2 u_2^3 = \lambda u_2. \tag{9.2.13}$$

Thus, as in Example 2, a parabolic branch lying in the λ–u_1 plane emanates from λ_1, and a parabolic branch lying in the λ–u_2 plane emanates from $\lambda = \lambda_2$. If $c_1 < 0, c_2 > 0$, the first parabola points to the left and the other to the right; there are then no other solutions of (9.2.13). If, however, c_1 and c_2 are of the same sign (positive, say), (9.2.13) yields additional solutions with neither u_1 nor $u_2 = 0$: the intersection of the two parabolic cylinders $\lambda - \lambda_1 = c_1 u_1^2, \lambda - \lambda_2 = c_2 u_2^2$. As is easily seen, the intersection consists of two curves which represent secondary branching from the parabolic branch that lies in the λ–u_1 plane. Exercise 9.2.1 gives an even more picturesque example of secondary branching.

(b) Example 4(a) below gives an instance where there is no branching from a double eigenvalue ($\lambda = 1$) of the linearized problem. A slight perturbation splits the double eigenvalue into two simple eigenvalues $\lambda_1 = 1 - \varepsilon$ and $\lambda_2 = 1 + \varepsilon$. By

Theorem 9.2.2, we must now have branching from both these eigenvalues. How do the branches disappear as $\varepsilon \to 0$? In (9.2.9) we take

$$Ru = (u_2^3, -u_1^3), \tag{9.2.14}$$

so that (9.2.2) becomes

$$u_2^3 = [\lambda - (1-\varepsilon)]u_1, \quad -u_1^3 = [\lambda - (1+\varepsilon)]u_2.$$

Nontrivial solutions are possible only for $1 - \varepsilon < \lambda < 1 + \varepsilon$. It can be seen that this solution is a closed loop connecting the branch points $\lambda_1 = 1 - \varepsilon$ and $\lambda_2 = 1 + \varepsilon$. The loop does not wander far from the λ axis since the value of $u_1^2 + u_2^2$ does not exceed 2ε. As ε tends to 0, all dimensions of the loop tend to 0. This gives a geometrical picture of what can happen when two simple eigenvalues coalesce into a multiple eigenvalue for which no branching occurs.

Example 4. With $H = \mathbb{R}^2$ we consider branching from a double eigenvalue of the linearized problem. We still use operators satisfying (9.2.9), (9.2.10), and (9.2.11), but now we take $\lambda_1 = \lambda_2 = 1$. We confine ourselves to two extreme examples showing the variety of phenomena that can now occur.

(a) R is given by (9.2.14), so that (9.2.2) becomes

$$u_2^3 = (\lambda - 1)u_1, \quad -u_1^3 = (\lambda - 1)u_2.$$

Multiply these equations by u_2 and u_1, respectively, and subtract to obtain

$$u_1^4 + u_2^4 = 0,$$

whose only solution is $u_1 = u_2 = 0$. Thus there is only the basic solution branch with no branching at $\lambda = 1$ (even though this is an eigenvalue of the linearized system). This example does not violate Theorem 9.2.2, because $\lambda = 1$ has even multiplicity, nor Theorem 9.2.3, because A is not a gradient operator.

(b) Taking

$$Ru = (u_1\|u\|^2, u_2\|u\|^2),$$

we find that (9.2.2) has the form

$$(\lambda - 1)u_1 = u_1(u_1^2 + u_2^2), \quad (\lambda - 1)u_2 = u_2(u_1^2 + u_2^2),$$

which is equivalent to the single equation $u_1^2 + u_2^2 = \lambda - 1$. This equation describes a paraboloid of revolution about the λ axis; thus, an entire surface of solutions branches off from the basic solution at $\lambda = 1$.

Before attempting a study of branching problems in function spaces, we develop in the next section some perturbation theory for linear problems.

EXERCISES

9.2.1 Consider the problem $Au - \lambda u = 0$, with A given by (9.2.9), (9.2.10), (9.2.11), and

$$Ru = (c_1 u_1 \|u\|^2, c_2 u_2 \|u\|^2), \quad c_1 > c_2 > 0.$$

Find the entire solution set, and show that there are two parabolic branches joined by a circular hoop.

9.2.2 In $\mathbb{R}^2$ consider the equation $Au - \lambda u = 0$, with

$$Au = (u_1 + 2u_1 u_2, u_2 + u_1^2 + 2u_2^2).$$

Show that $\lambda = 1$ is a double eigenvalue of the linearized problem but that there is only a single branch emanating from the basic solution at $\lambda = 1$.

9.2.3 In $\mathbb{R}^2$ let $Au = (u_1^3 + u_2, u_1^2 u_2 - u_1^3)$, and consider the equation $Au = \lambda u$. Show that the linearized problem about $u = 0$ has only one eigenvalue and that this eigenvalue has geometric multiplicity 1 and algebraic multiplicity 2. Show that there is no branching from the basic solution.

9.2.4 Construct integral equations with separable kernels that lead to the branching diagrams of Examples 1 and 2.

9.2.5 Consider the problem

$$\phi(x) - \lambda \int_0^1 \phi^2(y)\, dy = f(x)$$

when (a) $f(x) = 0$ and (b) $f(x) = \varepsilon > 0$. Draw the bifurcation diagrams and obtain the explicit solution(s). Compare the behaviors.

9.3 PERTURBATION THEORY FOR LINEAR PROBLEMS

The methods used in this section can be loosely grouped under the heading of perturbation theory. With an eye to the nonlinear applications in Sections 9.4 and 9.5, we first employ these approximate methods on a variety of linear problems. Our purpose is not to carry out detailed calculations but rather to show how different types of problems can be treated in a fairly systematic manner. We deal successively with inhomogeneous problems, eigenvalue perturbations, change in boundary conditions, and domain perturbations.

Inhomogeneous Problems

For a given continuous $f(x)$, we consider the class of problems

$$P(\varepsilon): -u'' + (1 + \varepsilon x^2)u = f, \quad 0 < x < 1; \quad u(0) = u(1) = 1. \qquad (9.3.1)$$

We may be interested in solving the problem for a particular value of $\varepsilon \neq 0$ or in studying the dependence of the solution on ε. In any event we shall show that for a suitable range of ε, the problem has one and only one solution, denoted by $u(x, \varepsilon)$. The *base problem* $P(0)$ has the solution

$$u(x, 0) = w_0(x) + F(x), \tag{9.3.2}$$

where

$$w_0(x) = \frac{\sinh x + \sinh(1-x)}{\sinh 1}, \quad F(x) = \int_0^1 g(x,\xi) f(\xi)\, d\xi,$$

and $g(x,\xi)$ is Green's function for $-D^2 + 1$ with vanishing boundary conditions

$$g(x,\xi) = \frac{\sinh x_< \sinh(1 - x_>)}{\sinh 1}; \quad x_< \doteq \min(x,\xi), \quad x_> \doteq \max(x,\xi). \tag{9.3.3}$$

One way of determining $u(x,\varepsilon)$ approximately is by expanding u in a Taylor series about $\varepsilon = 0$. The coefficients in this expansion are derivatives of u with respect to ε evaluated at $\varepsilon = 0$. These coefficients satisfy BVPs which can be obtained by differentiating (9.3.1) with respect to ε. Denoting ε derivatives by subscripts and x derivatives by primes, we find that

$$-u_\varepsilon'' + (1 + \varepsilon x^2) u_\varepsilon = -x^2 u, \quad 0 < x < 1; \quad u_\varepsilon|_{x=0} = u_\varepsilon|_{x=1} = 0. \tag{9.3.4}$$

On setting $\varepsilon = 0$ and using (9.3.3), we obtain

$$u_\varepsilon(x, 0) = -\int_0^1 g(x,\xi)\xi^2 u(\xi, 0)\, d\xi,$$

which, together with (9.3.2), gives the approximation

$$\begin{aligned} u(x,\varepsilon) \sim u(x,0) + \varepsilon u_\varepsilon(x,0) &= w_0(x) + F(x) \\ &\quad - \varepsilon \int_0^1 g(x,\xi)\xi^2 [w_0(\xi) + F(\xi)]\, d\xi. \end{aligned} \tag{9.3.5}$$

Alternatively, by moving the term $\varepsilon x^2 u$ to the right side, (9.3.1) can be written as the integral equation

$$u(x,\varepsilon) = w_0(x) + F(x) - \varepsilon \int_0^1 g(x,\xi)\xi^2 u(\xi,\varepsilon)\, d\xi. \tag{9.3.6}$$

With ε fixed, (9.3.6) can be regarded as a fixed-point problem $u = Tu$ in the Banach space $C[0, 1]$. With $\|\|_\infty$ denoting the uniform norm, we find that

$$\|Tu_1 - Tu_2\|_\infty \leqslant |\varepsilon|\, \|u_1 - u_2\|_\infty \max_{0 \leqslant x \leqslant 1} \int_0^1 \xi^2 g(x,\xi)\, d\xi,$$

where no absolute value sign is needed since the integrand is positive. Using (9.3.3), one can show that the mapping T is a contraction for $|\varepsilon| \leqslant 6$. Thus (9.3.1) has one and only one solution $u(x, \varepsilon)$ for $|\varepsilon| \leqslant 6$. We can find $u(x, \varepsilon)$ by successive substitutions:

$$u(x, \varepsilon) = \lim_{n \to \infty} u_n(x, \varepsilon), \quad u_n(x, \varepsilon) \doteq T u_{n-1}(x, \varepsilon),$$

where $u_0(x, \varepsilon)$ is arbitrary. If we use for $u_0(x, \varepsilon)$ the solution (9.3.2) of the base problem, then $u_1(x, \varepsilon) \doteq T u_0(x, \varepsilon)$ is just (9.3.5).

The problem $P(\varepsilon)$ can also be treated by monotone methods, even though their main efficacy is for nonlinear problems. Since the technique does not rely on small ε, we shall consider the particular value $\varepsilon = 1$ and drop further reference to ε. The BVP under consideration is

$$-u'' + (1 + x^2)u = f(x), \quad 0 < x < 1; \quad u(0) = u(1) = 1. \tag{9.3.7}$$

To apply Theorem 9.2.1, we rewrite (9.3.7) as

$$-u'' + 2u = f + (1 - x^2)u, \quad 0 < x < 1; \quad u(0) = u(1) = 1, \tag{9.3.8}$$

where now the right side of the differential equation is an increasing function of u for each x in $0 \leqslant x \leqslant 1$. The basis for the iteration scheme is the equation

$$-w'' + 2w = f + (1 - x^2)v, \quad 0 < x < 1; \quad w(0) = w(1) = 1, \tag{9.3.9}$$

whose one and only solution can be written as

$$w = Tv \doteq 1 + \int_0^1 g(x, \xi)[f(\xi) - 2 + (1 - \xi^2)v(\xi)]\, d\xi, \tag{9.3.10}$$

where g is Green's function for $-D^2 + 2$ with vanishing boundary conditions. The BVP (9.3.7) is then equivalent to the fixed-point problem $u = Tu$.

To verify the hypotheses of Theorem 9.1.1 we shall use the following maximum principle (see also Exercise 1.3.1): If $z(x)$ satisfies

$$-z'' + 2z \geqslant 0, \quad 0 < x < 1; \quad z(0) \geqslant 0, \quad z(1) \geqslant 0, \tag{9.3.11}$$

then $z(x) \geqslant 0$ for $0 \leqslant x \leqslant 1$. Indeed, if $z < 0$ at an interior point, there is an interior negative minimum where $z < 0$ and $z'' \geqslant 0$; but these conditions violate the differential inequality. We can now show easily that T is an increasing operator; that is, $v_1 \leqslant v_2$ implies that $Tv_1 \leqslant Tv_2$ [we write $v_1 \leqslant v_2$ if $v_1(x) \leqslant v_2(x)$ on $0 \leqslant x \leqslant 1$]. It suffices to set $z = Tv_2 - Tv_1$ and use (9.3.9) to obtain

$$-z'' + 2z = (1 - x^2)(v_2 - v_1) \geqslant 0, \quad 0 < x < 1; \qquad z(0) \geqslant 0, \quad z(1) \geqslant 0,$$

so that the maximum principle gives $z(x) \geqslant 0$ in $0 \leqslant x \leqslant 1$, that is, $Tv_1 \leqslant Tv_2$. This result could have been derived directly from (9.3.10) by using the fact that $g(x, \xi) \geqslant 0$.

Starting with elements $u_0(x), u^0(x)$ satisfying $u_0 \leqslant u^0$ and

$$-u_0'' + (1+x^2)u_0 \leqslant f(x), \quad 0 < x < 1; \quad u_0(0) \leqslant 1, \quad u_0(1) \leqslant 1, \tag{9.3.12}$$

$$-(u^0)'' + (1+x^2)u^0 \geqslant f(x), \quad 0 < x < 1; \quad u^0(0) \geqslant 1, \quad u^0(1) \geqslant 1, \tag{9.3.13}$$

we use (9.3.10) or, equivalently, (9.3.9) to obtain the sequences $\{u_n\}$ and $\{u^n\}$ needed in Theorem 9.1.1. We must show that $u_0 \leqslant Tu_0$ and $u^0 \geqslant Tu^0$. Let us prove the first of these, the second yielding to a similar argument. Since $Tu_0 = u_1$, we have

$$-u_1'' + 2u_1 = f + (1-x^2)u_0, \quad 0 < x < 1; \qquad u_1(0) = u_1(1) = 1,$$

and, combining with (9.3.12), we find that

$$-(u_1-u_0)''+2(u_1-u_0) \geqslant 0, \quad 0 < x < 1; \, u_1(0)-u_0(0) \geqslant 0, \, u_1(1)-u_0(1) \geqslant 0,$$

so that $u_1(x) - u_0(x)$ satisfies our maximum principle and hence we have $u_1 \geqslant u_0$ or $Tu_0 \geqslant u_0$.

Theorem 9.1.1 then tells us that there is at least one solution of (9.3.7) in the order interval $[u_0, u^0]$. That there exist elements that satisfy (9.3.12) and (9.3.13) is easily verified. We can choose u_0 and u^0 to be the *constants* given by

$$u_0 = \min\left(1, \frac{m}{2}, m\right), \qquad u^0 = \max\left(1, \frac{M}{2}, M\right),$$

where $m \leqslant f(x) \leqslant M$ on $0 \leqslant x \leqslant 1$. Better choices can often be made for u_0 and u^0, but this need not concern us now. Elementary arguments (unrelated to monotone methods) show that the solution of (9.3.7) is unique. This unique solution is sandwiched between the iterative sequences $\{u_n\}$ and $\{u^n\}$.

Eigenvalue Perturbation

For each ε satisfying $|\varepsilon| < a$, suppose that $L(\varepsilon)$ is a self-adjoint compact operator on the real Hilbert space H. We shall assume that $L(\varepsilon)$ depends continuously on ε, that is, $\lim_{\Delta\varepsilon\to 0} \|L(\varepsilon+\Delta\varepsilon) - L(\varepsilon)\| = 0$. The eigenvalue problem for $L(\varepsilon)$ is

$$L(\varepsilon)u = \lambda u, \tag{9.3.14}$$

and we would like to relate the spectrum of the *perturbed operator* $L(\varepsilon)$ to the presumably known spectrum of the *base operator* $L \doteq L(0)$.

Theorem. *Let $\lambda_n \neq 0$ be a simple eigenvalue of L with normalized eigenvector e_n. Then, in some neighborhood of $\varepsilon = 0$, there exists an eigenpair $(\lambda(\varepsilon), e(\varepsilon))$ of (9.3.14) with the properties*

$$\lim_{\varepsilon\to 0} \lambda(\varepsilon) = \lambda_n, \quad \lim_{\varepsilon\to 0} e(\varepsilon) = e_n.$$

Proof. We now sketch the proof of this theorem by a method which can be adapted to nonlinear problems. We rewrite (9.3.14) as

$$Lu - \lambda_n u = \delta u - R(\varepsilon)u, \quad \delta \doteq \lambda - \lambda_n, \qquad R(\varepsilon) \doteq L(\varepsilon) - L. \tag{9.3.15}$$

Setting $f = \delta u - R(\varepsilon)u$, we see that (9.3.15) has the form $Lu - \lambda_n u = f$. From Chapter 6 we know that this equation has solutions if and only if $\langle f, e_n \rangle = 0$. With this solvability condition satisfied, the equation has many solutions, differing from one another by a multiple of e_n, but there is only one solution orthogonal to e_n. This particular solution is denoted by Tf, where T is known as the *pseudoinverse* of $L - \lambda_n I$. Since λ_n is simple, (6.4.13b) gives the explicit expansion

$$Tf = -\frac{Pf}{\lambda_n} + \sum_{k \neq n} \frac{1}{\lambda_k - \lambda_n} \langle f, e_k \rangle e_k, \tag{9.3.16}$$

where P is the projection on the null space of L. Note that T is defined as a bounded operator on all of H, but Tf solves the equation $Lu - \lambda_n u = f$ only if $\langle f, e_n \rangle = 0$.

Thus, if u is to satisfy (9.3.15), we must have simultaneously

$$\langle \delta u - R(\varepsilon)u, e_n \rangle = 0 \tag{9.3.17}$$

and

$$u = ce_n + T[\delta u - R(\varepsilon)u]. \tag{9.3.18}$$

It is also clear that any solution of (9.3.17) and (9.3.18) will satisfy (9.3.15). We can proceed further if $|\delta|$ and $|\varepsilon|$ are small ($|\varepsilon|$ small means that the perturbation is small, and $|\delta|$ small gives the eigenvalue branch passing through λ_n). Our goal is to obtain a relation between δ and ε. With δ, ε, c fixed and $|\delta|, |\varepsilon|$ small, we shall show that (9.3.18) has a unique solution which depends linearly on c and can therefore be written as $u = c\overline{u}(\delta, \varepsilon)$. We then substitute in (9.3.17), canceling c since we are looking for nontrivial solutions, to find

$$\delta = \langle R(\varepsilon)\overline{u}(\delta, \varepsilon), e_n \rangle = \langle \overline{u}(\delta, \varepsilon), R(\varepsilon)e_n \rangle, \tag{9.3.19}$$

which is the desired relation between δ and ε.

Let us carry out these steps. The right side of (9.3.18) can be regarded as mapping an element $u \in H$ into the element $v \doteq Bu$ given by

$$v = ce_n + T[\delta u - R(\varepsilon)u].$$

Fixed points of B coincide with the solutions of (9.3.17)–(9.3.18). If u_1, u_2 are arbitrary elements of H,

$$\|Bu_2 - Bu_1\| \leqslant \|T\|\{|\delta| + \|R(\varepsilon)\|\}\|u_2 - u_1\|,$$

so that B is a contraction whenever

$$|\delta| + \|R(\varepsilon)\| < \frac{1}{\|T\|}. \tag{9.3.20}$$

Assuming $|\delta|, |\varepsilon|$ small enough so that (9.3.20) is satisfied, we find that B has a unique fixed point given by

$$u = c\sum_{k=0}^{\infty}\{T[\delta I - R(\varepsilon)]\}^k e_n \doteq cS(\delta, \varepsilon)e_n, \tag{9.3.21}$$

where S is a linear operator depending nonlinearly on the parameters δ and ε, and having the property $S(\delta, 0) = I$. Substitution in (9.3.17) gives (9.3.19), or

$$\delta = \langle S(\delta, \varepsilon)e_n, R(\varepsilon)e_n\rangle, \tag{9.3.22}$$

which is a *single nonlinear equation in two real variables.*

Writing (9.3.22) as $F(\delta, \varepsilon) = 0$, we observe that $F(\delta, 0) = \delta$, so that (0, 0) is a solution of (9.3.22) and $\partial F/\partial\delta = 1$ at the origin. Under very mild restrictions on $R(\varepsilon)$ we can show that there is a neighborhood of (0, 0) where grad F is continuous and $\partial F/\partial\delta \neq 0$. The implicit function theorem then guarantees the existence of a unique solution of (9.3.22) in the form $\delta = \delta(\varepsilon)$, at least for $|\delta|$ and $|\varepsilon|$ sufficiently small. The first-order change in λ_n is obtained by using $S(\delta, 0) = I$ in (9.3.22):

$$\lambda(\varepsilon) - \lambda_n \sim \langle e_n, R(\varepsilon)e_n\rangle, \tag{9.3.23}$$

a well-known formula that could have been derived in other ways. The eigenvector of $L(\varepsilon)$ corresponding to $\lambda(\varepsilon) = \lambda_n + \delta(\varepsilon)$ is $\overline{u}(\delta(\varepsilon), \varepsilon)$, where the normalization $\langle\overline{u}, e_n\rangle = 1$ (that is, $c = 1$) has been used. To first order, we have

$$\overline{u} \sim e_n + T(\delta e_n) - TR(\varepsilon)e_n = e_n - TR(\varepsilon)e_n. \tag{9.3.24}$$

Substitution of (9.3.16) in (9.3.24) gives

$$\overline{u} \sim e_n + \frac{PR(\varepsilon)e_n}{\lambda_n} + \sum_{k\neq n}\frac{1}{\lambda_k - \lambda_n}\langle R(\varepsilon)e_n, e_k\rangle e_k. \tag{9.3.25}$$

□

Once the theory has been established by the rigorous method outlined above, simpler ways can be found to perform the calculations. Let us see how the method of continuity (parametric differentiation) can be used in the BVP

$$-u'' + (1 + \varepsilon x^2)u = \mu u, \quad 0 < x < 1; \qquad u(0) = u(1) = 0. \tag{9.3.26}$$

For $\varepsilon = 0$ the eigenvalues are all simple: $\mu_n = n^2\pi^2 + 1, n = 1, 2, \ldots$, with corresponding normalized eigenfunctions $e_n(x) = \sqrt{2}\sin n\pi x$, where we have reverted to functional notation. To apply the earlier results we would have to translate (9.3.26) into an integral equation by using an appropriate Green's function. For $\varepsilon = 0$ the integral equation has the eigenvalues $\lambda_n = 1/\mu_n$ and the eigenfunctions $e_n(x)$. From our treatment of compact operators we know that there is a branch of eigenpairs $(\lambda(\varepsilon), u(x, \varepsilon))$ of the perturbed problem with the property

$\lambda(\varepsilon) \to \lambda_n, u(x,\varepsilon) \to e_n(x)$ as $\varepsilon \to 0$. Therefore, for (9.3.26), there is a branch $(\mu(\varepsilon), u(x,\varepsilon))$ tending to $(\mu_n, e_n(x))$ as $\varepsilon \to 0$. To calculate the perturbed eigenpair, we assume that we can differentiate (9.3.26) with respect to ε to obtain

$$-u_\varepsilon'' + (1+\varepsilon x^2)u_\varepsilon - \mu u_\varepsilon = \mu_\varepsilon u - x^2 u; \qquad u_\varepsilon = 0 \quad \text{at } x = 0, 1. \quad (9.3.27)$$

Equation (9.3.27) should be viewed as an inhomogeneous equation for $u_\varepsilon(x,\varepsilon)$ with a right side that involves $\mu_\varepsilon(\varepsilon)$ and $u(x,\varepsilon)$. Setting $\varepsilon = 0$ gives an inhomogeneous equation for $u_\varepsilon(x,0)$:

$$-u_\varepsilon'' + u_\varepsilon - \mu_n u_\varepsilon = \mu_\varepsilon(0)e_n(x) - x^2 e_n(x); \qquad u_\varepsilon = 0 \quad \text{at } x = 0, 1. \quad (9.3.28)$$

At first we are troubled by the presence of the unknown $\mu_\varepsilon(0)$, which makes it appear that we cannot solve (9.3.28) for $u_\varepsilon(x,0)$. Note, however, that the corresponding homogeneous problem has the nontrivial solution $e_n(x)$, so that (9.3.28) must obey the solvability condition

$$\langle \mu_\varepsilon(0)e_n(x) - x^2 e_n(x), e_n(x)\rangle = 0, \qquad (9.3.29)$$

which serves to determine

$$\mu_\varepsilon(0) = \int_0^1 x^2 e_n^2(x)\,dx, \quad \mu(\varepsilon) \sim \mu_n + \varepsilon \int_0^1 x^2 e_n^2(x)\,dx. \qquad (9.3.30)$$

Just as in (9.3.23), the first-order change in the eigenvalue can be calculated independent of the change in the eigenfunction. Exercise 9.3.2 compares (9.3.23) and (9.3.30).

Now that (9.3.28) is consistent, we can solve for $u_\varepsilon(x,0)$. Of course, the solution will contain an additive term proportional to $e_n(x)$. There is, however, a unique solution $w(x)$ that is orthogonal to $e_n(x)$:

$$w(x) = Q[u_\varepsilon(0)e_n(x) - x^2 e_n(x)],$$

where Q is the pseudoinverse of the differential operator in (9.3.28) with the corresponding boundary conditions. This pseudoinverse is just the integral operator whose kernel is the modified Green's function of Section 3.5. For any f in H we have

$$Qf = \sum_{m\neq n} \frac{\langle f, e_m\rangle}{\mu_m - \mu_n} e_m(x),$$

so that

$$w(x) = -\sum_{m\neq n} \frac{\langle x^2 e_n(x), e_m(x)\rangle}{(m^2-n^2)\pi^2} e_m(x), \quad e_m(x) = \sqrt{2}\sin m\pi x. \qquad (9.3.31)$$

Therefore, we have the first-order change in the eigenfunction

$$u_n(x,\varepsilon) - e_n(x) = \varepsilon w(x), \qquad (9.3.32)$$

where the normalization $\langle u_n, e_n\rangle = 1$ has been used.

Change in Boundary Conditions

Consider the class of eigenvalue problems

$$-u'' - \lambda u = 0, \quad 0 < x < 1; \quad u(0) = 0, \quad \varepsilon u'(1) + u(1) = 0, \tag{9.3.33}$$

where the parameter ε occurs in the boundary condition. [The problem has already been discussed fully; see (7.1.27).] For $\varepsilon = 0$ we have a base problem whose eigenvalues are $\lambda_n = n^2\pi^2$ with normalized eigenfunctions $e_n(x) = \sqrt{2}\sin n\pi x$. We want to calculate the perturbed eigenpair $(\lambda(\varepsilon), u(x, \varepsilon))$, which tends to $(\lambda_n, e_n(x))$ as ε approaches 0. As usual, we differentiate (9.3.33) with respect to ε to obtain

$$-u_\varepsilon'' - \lambda u_\varepsilon = \lambda_\varepsilon u; \quad u_\varepsilon(0, \varepsilon) = 0, \quad \varepsilon u_\varepsilon'(1, \varepsilon) + u_\varepsilon(1, \varepsilon) = -u'(1, \varepsilon), \tag{9.3.34}$$

where note should be made of the additional contribution in the boundary term [stemming from the presence of ε in the boundary condition at $x = 1$ in (9.3.33)]. Setting $\varepsilon = 0$ gives the following BVP for $u_\varepsilon(x, 0)$:

$$-u_\varepsilon'' - \lambda_n u_\varepsilon = \lambda_\varepsilon(0) e_n(x); \quad u_\varepsilon(0, 0) = 0, \quad u_\varepsilon(1, 0) = -e_n'(1, 0). \tag{9.3.35}$$

Since the homogeneous problem has a nontrivial solution, there will be a solvability condition to be satisfied. In view of the inhomogeneous boundary condition in (9.3.35), the solvability condition is found as follows. We multiply the equation for e_n by $u_\varepsilon(x, 0)$ and (9.3.35) by $e_n(x)$, subtract, and integrate from $x = 0$ to 1. This gives

$$\lambda_\varepsilon(0) = \int_0^1 e_n^2(x)\, dx = -[e_n'(1)]^2 = -2n^2\pi^2,$$

and to first order,

$$\lambda_n(\varepsilon) \sim n^2\pi^2 - 2\varepsilon n^2\pi^2, \tag{9.3.36}$$

in general agreement with Figure 7.2, where $\varepsilon = 0$ corresponds to $\beta = 0$, $\varepsilon = 0+$ to $\beta = 0+$, $\varepsilon = 0-$ to $\beta = \pi-$. Note that approximation (9.3.36) becomes progressively worse as n increases; in fact, Figure 9.2 shows that $|\sqrt{\lambda_n(\varepsilon)} - \sqrt{\lambda_n}|$ cannot exceed $\pi/2$, whereas (9.3.36) gives $|(1 - \sqrt{1 - 2\varepsilon})n\pi|$, which, for fixed $\varepsilon \neq 0$, tends to ∞ with n.

Another interesting feature of the problem is that the perturbed eigenfunctions obtained in this way do not form a basis for $\varepsilon < 0$. We saw in Chapter 7 that for $\varepsilon = 0-$ (which corresponds to $\beta = \pi-$) there is a large negative eigenvalue (not arising from one of our perturbation branches) whose eigenfunction is therefore not obtainable by perturbing the $\{e_n(x)\}$ of the base problem. This is not surprising when we observe that the physical problem is drastically changed as ε changes from 0 to $0-$ (this corresponds to β going from the value 0 to a value just below π, which is a totally different problem, as explained in Chapter 7). If, however, we had considered instead the boundary condition $u'(1) + \tau u(1) = 0$, the problem for $|\tau|$ small is close to the one for $\tau = 0$.

Domain Perturbations

Consider first the problem of finding the natural frequencies of a vibrating membrane of elliptical shape. The base problem is the eigenvalue problem for a disk. Let $\Omega(\varepsilon)$ represent a continuously varying domain depending on ε, with $\Omega(0)$ being the unit disk and $\Omega(\varepsilon_0)$ the desired domain [we could, for instance, let $\Omega(\varepsilon)$ be an ellipse with semiaxes $1-\varepsilon$ and 1]. The eigenvalue problem is

$$\Delta u + \lambda u = 0, \quad x \in \Omega(\varepsilon); \quad u = 0 \text{ on } \Gamma(\varepsilon), \tag{9.3.37}$$

where $\Gamma(\varepsilon)$ is the boundary of $\Omega(\varepsilon)$.

We shall construct an eigenvalue branch $\lambda(\varepsilon)$ which coincides for $\varepsilon = 0$ with some particular simple eigenvalue, say λ_n, of the base problem. The corresponding eigenfunction (suitably normalized) $u(x,\varepsilon)$ is defined in $\Omega(\varepsilon)$. Differentiating the differential equation in (9.3.37) is permissible at every interior point since such a point remains in the interior for sufficiently small changes in ε. This gives the differential equation $\Delta u_\varepsilon + \lambda u_\varepsilon = -\lambda_\varepsilon u$ in $\Omega(\varepsilon)$.

The question of the boundary condition for u_ε on $\Gamma(\varepsilon)$ is more subtle; it is true that u remains 0 on the changing boundary, but u_ε refers only to the change of u at a fixed point and does not take into account the motion of the boundary. Thus it is the substantial derivative of u with respect to ε that vanishes on $\Gamma(\varepsilon)$, and not u_ε. Following the approach suggested by Joseph and Fosdick [36] and later developed rigorously under the name of *speed method* (see, for instance, Delfour [16]), we shall perform an invertible coordinate transformation which maps each perturbed domain onto the base domain $\Omega(0)$. Let x be the point in $\Omega(\varepsilon)$ that is mapped into the point ξ in $\Omega(0)$; we can then write $\xi = \xi(x,\varepsilon)$ for the mapping that takes $\Omega(\varepsilon)$ into $\Omega(0)$, and $x = x(\xi,\varepsilon)$ for the inverse. An explicit form for the function $\xi(x,\varepsilon)$ will not be needed, as it must play no role in the final answer [the exact eigenvalue $\lambda(\varepsilon)$ is presumably determinate and cannot depend on an artificial domain transformation introduced for mathematical convenience]. With $u(x,\varepsilon)$ the eigenfunction on $\Omega(\varepsilon)$, we see that $u(x(\xi,\varepsilon),\varepsilon) = \overline{u}(\xi,\varepsilon)$ is defined on $\Omega(0)$ for each ε and vanishes when ξ is on $\Gamma(0)$. Thus we can write that

$$\frac{d\overline{u}}{d\varepsilon} = 0, \quad \xi \text{ on } \Gamma(0),$$

and hence

$$\frac{\partial u}{\partial \varepsilon} + \operatorname{grad} u \cdot \frac{dx}{d\varepsilon} = 0, \quad x \text{ on } \Gamma(\varepsilon). \tag{9.3.38}$$

The boundary condition appears to depend on the nature of the transformation $x(\xi,\varepsilon)$, but this is deceptive. We know that $x(\xi,\varepsilon)$ takes $\Omega(0)$ into $\Omega(\varepsilon)$, while $x(\xi,\varepsilon+d\varepsilon)$ takes $\Omega(0)$ into the domain $\Omega(\varepsilon+d\varepsilon)$. The vector dx therefore takes a point in $\Omega(\varepsilon)$ into a point of $\Omega(\varepsilon+d\varepsilon)$. On the boundary, dx takes a point of $\Gamma(\varepsilon)$ into a point of $\Gamma(\varepsilon+d\varepsilon)$; we can conveniently regard this motion as taking place in a direction normal to $\Gamma(\varepsilon)$. Let x be a point on $\Gamma(\varepsilon)$, and let $\hat{\nu}$ be the outward normal to $\Gamma(\varepsilon)$ at that point. This normal intersects $\Gamma(\varepsilon+d\varepsilon)$ at a point $x + \hat{\nu}\,d\nu$, where

$d\nu = d\nu(x, \varepsilon)$ is a number which may be positive or negative. Thus (9.3.38) takes the form

$$u_\varepsilon + \frac{\partial u}{\partial \nu}\frac{d\nu}{d\varepsilon} = 0,$$

and $u_\varepsilon(x, \varepsilon)$ therefore satisfies the BVP

$$\Delta u_\varepsilon + \lambda u_\varepsilon = -\lambda_\varepsilon u, \quad x \in \Omega(\varepsilon); \quad u_\varepsilon = -\frac{\partial u}{\partial \nu}\frac{d\nu}{d\varepsilon} \quad \text{on } \Gamma(\varepsilon).$$

We now set $\varepsilon = 0$ to obtain the equation relating $u_\varepsilon(x, 0)$ and $\lambda_\varepsilon(0)$:

$$\begin{aligned} \Delta u_\varepsilon + \lambda_n u_\varepsilon &= -\,\lambda_\varepsilon(0) e_n(x), \quad x \in \Omega(0); \\ u_\varepsilon &= -\frac{\partial e_n}{\partial \nu}\left(\frac{d\nu}{d\varepsilon}\right)_{\varepsilon=0} \quad \text{on } \Gamma(0). \end{aligned} \tag{9.3.39}$$

Since the homogeneous problem has a nontrivial solution, a solvability condition must be satisfied. Combining the equation $\Delta e_n + \lambda_n e_n = 0$ with (9.3.39) in the usual way, we find that

$$\int_{\Gamma(0)} \left(u_\varepsilon \frac{\partial e_n}{\partial \nu} - e_n \frac{\partial u_\varepsilon}{\partial \nu}\right) dS = \lambda_\varepsilon(0) \int_{\Omega(0)} e_n^2(x)\, dx = \lambda_\varepsilon(0),$$

or, since e_n vanishes on $\Gamma(0)$ and $u_\varepsilon = -(\partial e_n/\partial \nu)(d\nu/d\varepsilon)_{\varepsilon=0}$ on $\Gamma(0)$,

$$\lambda_\varepsilon(0) = -\int_{\Gamma(0)} \left(\frac{\partial e_n}{\partial \nu}\right)^2 \left(\frac{d\nu}{d\varepsilon}\right)_{\varepsilon=0} dS.$$

The first-order change in the eigenvalue λ_n is then given by

$$\lambda(\varepsilon) - \lambda_n \sim \varepsilon \int_{\Gamma(0)} \left(\frac{\partial e_n}{\partial \nu}\right)^2 \left(\frac{d\nu}{d\varepsilon}\right)_{\varepsilon=0} dS. \tag{9.3.40}$$

As a concrete example let us calculate the fundamental eigenvalue of an elliptical membrane with semiaxes $1 - \varepsilon$ and 1, where ε is small. For $\varepsilon = 0$ we have $\lambda_1 = \beta^2$, where β is the first zero of $J_0(x)$, and $e_1(x) = cJ_0(\beta r)$, $r = |x|$, c being a normalization constant chosen so that $2\pi c^2 \int_0^1 r J_0^2(\beta r)\, dr = 1$. We then find from (9.3.40) that

$$\lambda(\varepsilon) \sim \beta^2 - \varepsilon \int_0^{2\pi} c^2 \beta^2 [J_0'(\beta)]^2 (-\cos^2\theta)\, d\theta = \beta^2 + \varepsilon c^2 \beta^2 [J_0'(\beta)]^2 \pi.$$

As another illustration of domain perturbation, we derive Hadamard's formula (see [8]) for the variation of Green's function. Let $\Omega(0)$ be a domain for which Green's function $g(x, \xi; 0)$ is known, and let $\Omega(\varepsilon)$ be a domain for which $g(x, \xi; \varepsilon)$ is sought. For simplicity we deal with the negative Laplacian, but the idea applies to other operators as well. Then $g(x, \xi; \varepsilon)$ satisfies

$$-\Delta g = \delta(x - \xi), \quad x, \xi \in \Omega(\varepsilon); \qquad g = 0 \quad \text{for } x \text{ on } \Gamma(\varepsilon).$$

Proceeding as before, we find that

$$-\Delta g_\varepsilon = 0, \quad x, \xi \in \Omega(\varepsilon); \qquad g_\varepsilon = -\frac{\partial g}{\partial \nu}\frac{d\nu}{d\varepsilon} \quad \text{on } \Gamma(\varepsilon).$$

On setting $\varepsilon = 0$, we see that $g_\varepsilon(x, \xi; 0)$ satisfies

$$-\Delta g_\varepsilon = 0, \quad x, \xi \in \Omega(0); \qquad g_\varepsilon = \frac{\partial g}{\partial \nu}\left(\frac{d\nu}{d\varepsilon}\right)_0 \qquad \text{on } \Gamma(0),$$

from which it follows that

$$g_\varepsilon(\eta, \xi; 0) = \int_{\Gamma_0} \left[\frac{\partial g}{\partial \nu}(x, \xi; 0)\frac{\partial g}{\partial \nu}(x, \eta; 0)\right] \frac{d\nu}{d\varepsilon}(x, 0)\, dS_x. \tag{9.3.41}$$

EXERCISES

9.3.1 Prove that the mapping T defined by (9.3.6) is a contraction on C[0,1] for $|\varepsilon| \leqslant 6$. Hence show that the smallest eigenvalue of

$$-u'' + u = \lambda x^2 u, \quad 0 < x < 1; \qquad u(0) = u(1) = 0$$

must exceed 6.

9.3.2 Compare (9.3.23) and (9.3.30).

9.3.3 Consider steady heat conduction in the strip $= -\infty < x < \infty, |y| < 1$. The boundary temperatures are $u(x, \pm 1) = x$. The strip is filled with a mixture of water and ice. Where the temperature is positive, there is water; where it is negative, there is ice. If the conductivity of water and ice were equal, the interface between the phases would be $x = 0$. By domain perturbation, find an approximate expression for the interface position if the thermal conductivities in water and ice are $1 + \varepsilon$ and 1, respectively.

9.3.4 The geometry is the same as in Exercise 9.3.3, but the boundary temperatures are $u(x, 1) = 1, u(x, -1) = -1$. If the conductivities of water and ice were equal, the interface would be at $y = 0$. For unequal conductivities find the interface by domain perturbation and also by an exact method.

9.4 TECHNIQUES FOR NONLINEAR PROBLEMS

A Branching Problem

Many of the methods introduced in Section 9.3 for linear problems have nonlinear analogs. We shall illustrate their application to concrete examples, beginning with a branching problem:

$$-u'' = \alpha \sin u, \quad 0 < x < 1; \qquad u'(0) = u(1) = 0, \tag{9.4.1}$$

which differs from the buckling problem (9.1.22) for $\phi(x)$ only in the boundary condition at $x = 1$. Every solution ϕ of (9.1.22) with $\alpha > 0$ and $|\phi(0)| < \pi$ generates a solution of (9.4.1) by eliminating the last quarter-wave and rescaling; conversely, every solution of (9.4.1) can be extended into a solution ϕ of (9.1.22). The translation of (9.4.1) into an integral equation is a little simpler than for $\phi(x)$ in (9.1.22) because the latter boundary value problem has a nontrivial solution when $\alpha = 0$.

Problem (9.4.1) has the basic solution $u \equiv 0$, and linearization about $u = 0$ gives the linear BVP

$$-u'' = \alpha u, \quad 0 < x < 1; \qquad u'(0) = u(1) = 0, \tag{9.4.2}$$

whose eigenvalues and normalized eigenfunctions are

$$\alpha_n = \frac{(2n-1)^2\pi^2}{4}, \quad e_n(x) = \sqrt{2}\cos\frac{(2n-1)\pi x}{2}, \quad n = 1, 2, \ldots. \tag{9.4.3}$$

Since 0 is not an eigenvalue, we can construct Green's function $g(x, \xi)$ satisfying

$$-g'' = \delta(x-\xi), \quad 0 < x, \xi < 1; \qquad g'(0, \xi) = g(1, \xi) = 0, \tag{9.4.4}$$

and a straightforward calculation gives

$$g(x, \xi) = 1 - \max(x, \xi). \tag{9.4.5}$$

We can therefore write (9.4.1) as the equivalent integral equation

$$\lambda u(x) = \int_0^1 g(x, \xi)\sin u(\xi)\, d\xi \doteq Au, \quad \lambda = \frac{1}{\alpha}. \tag{9.4.6}$$

Here A is a nonlinear Hammerstein integral operator [see (9.1.32)] whose linearization at $u = 0$ is the linear operator G with kernel $g(x, \xi)$. The equation $Gu = \lambda u$ has eigenvalues $\lambda_n = 1/\alpha_n$ and the same eigenfunctions as (9.4.3). Since A is compact, and the eigenvalues of $G = A'(0)$ are simple, Theorem 9.2.2 guarantees branching from the basic solution at each λ_n. Before calculating the initial shape of these branches by various methods (and at the same time proving the existence independently of Theorem 9.2.2, let us draw some preliminary conclusions from (9.4.1) and (9.4.6).

Multiplication of the differential equation in (9.4.1) by u' and integration from $x = 0$ to $x = 1$ yields

$$[u'(1)]^2 = 2\alpha[1 - \cos u(0)],$$

from which it follows that $\alpha > 0$ unless $u'(1) = 0$ [which in turn implies that (9.4.1) has only the trivial solution]. Therefore, nontrivial solutions of (9.4.1) and (9.4.6) are possible only for $\alpha > 0$ and $\lambda > 0$, respectively. We can say more from (9.4.6) by taking the norm of both sides:

$$|\lambda|\|u\| \leqslant \|G\|\|\sin u\| \leqslant \|G\|\|u\|.$$

Now $\|G\|$ is the largest eigenvalue $\lambda_1 = 4/\pi^2$ of the linear integral operator G. We therefore conclude that (9.4.6) can have nontrivial solutions only if $0 < \lambda \leqslant 4/\pi^2$ [and (9.4.1) only if $\alpha \geqslant \pi^2/4$]. This is a global result and tells us, among other things, that any branch emanating from the basic solution at $\lambda = \lambda_n$ cannot wander beyond the segment $0 < \lambda \leqslant 4/\pi^2$.

We now turn to the branching analysis of (9.4.1). We shall use three methods:

1. The Liapunov-Schmidt method, which deals rigorously with the integral equation (9.4.6) but has computational disadvantages.

2. The Poincaré-Keller method, which exploits the connection between the boundary value problem (9.4.1) and the initial value problem for the same equation.

3. The monotone iteration method, which, unlike the other two, is not based on a local analysis near the branch points, but has the disadvantage that it can be used easily only for the branch of positive (or negative) solutions, that is, the branch emanating from λ_1.

The Liapunov-Schmidt Method

The operator A of (9.4.6) has the property $A0 = 0$ and $A'(0) = G$. In light of (9.1.36), where we set $u_0 = 0$ and $r = Ru$, (9.4.6) becomes

$$\lambda u = Au = Gu + Ru, \quad \lim_{u \to 0} \frac{\|Ru\|}{\|u\|} = 0, \tag{9.4.7}$$

where

$$Gu = \int_0^1 g(x,\xi)u(\xi)\,d\xi, \quad Ru = \int_0^1 g(x,\xi)[\sin u(\xi) - u(\xi)]\,d\xi. \tag{9.4.8}$$

It is useful to further split the remainder Ru into a leading homogeneous term Cu of third degree and a higher-order term Du:

$$\begin{aligned} Ru = Cu + Du = &-\int_0^1 g(x,\xi)\frac{u^3(\xi)}{6}\,d\xi \\ &+ \int_0^1 g(x,\xi)\left(\sin u - u + \frac{u^3}{6}\right)d\xi. \end{aligned} \tag{9.4.9}$$

With a view toward studying branching from the null solution at $\lambda = \lambda_n$, we set

$$\delta = \lambda - \lambda_n, \quad u = ce_n + w, \quad \langle w, e_n \rangle = 0, \tag{9.4.10}$$

so that (9.4.7) becomes

$$Gu - \lambda_n u = \delta u - Ru. \tag{9.4.11}$$

On the left side of (9.4.11) appears the linear self-adjoint integral operator G; λ_n is a simple eigenvalue of G. Therefore, (9.4.11) will be consistent if and only if

$$\langle \delta u - Ru, e_n \rangle = 0.$$

If this solvability condition is satisfied, (9.4.11) has a unique solution orthogonal to e_n. This solution has been called w in (9.4.10) and can be expressed as $T(\delta u - Ru)$, where T is the familiar pseudo inverse given by (9.3.16). Therefore, (9.4.11) is equivalent to the pair of equations $\langle \delta u - Ru, e_n\rangle = 0$, $w = T(\delta u - Ru)$, which, since $Te_n = 0$, can be simplified to

$$\delta c = \langle R(ce_n + w), e_n\rangle, \tag{9.4.12}$$

$$w = T[\delta w - R(ce_n + w)], \tag{9.4.13}$$

which should be compared to (9.3.18) for linear perturbation theory. Our operator R is now nonlinear, and there appears to be no natural perturbation parameter ε; it turns out that c in (9.4.12)–(9.4.13) will play that role. We know that nontrivial solutions branching from the null solution begin with small norm, so that c and w will both be small for small δ; this is quite different from (9.3.18), where c was merely an arbitrary normalization constant (since for a linear problem the eigenfunction is only determined up to a multiplicative constant).

We now turn our attention to the pair of equations (9.4.12)–(9.4.13), where the numbers δ and c and the element w are unknown. A solution with δ, c, w small would represent branching from the null solution near λ_n. To find all such solutions, the procedure (whose details will not be carried out) is as follows. With δ and c fixed and small, it can be shown by contraction methods that (9.4.13) has one and only one small solution $w = \overline{w}(c, \delta)$. This solution is then substituted in (9.4.12) to yield a *single nonlinear equation* in the two real variables c, δ. Of course, none of this is ever carried out exactly. Instead, one obtains from (9.4.13) asymptotic information on $\overline{w}$, which is then used to simplify (9.4.12). In our particular problem it can be shown that for small $|c|, \overline{w}$ is of the order $|c|^3$ uniformly in a δ interval, $|\delta| \leqslant \delta_0$. This tells us, as expected, that near the branch point the solution u of the nonlinear problem is *essentially proportional to* e_n, the eigenfunction of the linearized problem. The order information on w and the specific forms of C and D enable us to write (9.4.12) to the first order as

$$\delta c = \langle C(ce_n), e_n\rangle = c^3\langle Ce_n, e_n\rangle.$$

From the definition of C in (9.4.9) we have

$$\begin{aligned}
\langle Ce_n, e_n\rangle &= -\int_0^1 e_n(x)\,dx \int_0^1 g(x,\xi)\frac{e_n^3(\xi)}{6}\,d\xi \\
&= -\frac{1}{6}\int_0^1 e_n^3(\xi)\,d\xi \int_0^1 g(x,\xi)e_n(x)\,dx \\
&= -\frac{1}{6}\lambda_n \int_0^1 e_n^4(\xi)\,d\xi = -\frac{1}{\pi^2(2n-1)^2},
\end{aligned}$$

which yields

$$\delta = \lambda - \lambda_n \sim -\frac{c^2}{\pi^2(2n-1)^2} \tag{9.4.14}$$

and

$$\alpha - \alpha_n = \frac{1}{\lambda} - \frac{1}{\lambda_n} \sim -\frac{\delta}{\lambda_n^2} = \frac{c^2\pi^2(2n-1)^2}{16}. \tag{9.4.15}$$

Thus the branching diagram (c vs. α) consists of initially parabolic branches opening to the right, as in Figure 9.8.

The Poincaré-Keller Method

Here we work directly with formulation (9.4.1) and a related initial value problem. First let us observe that any nontrivial solution of (9.4.1) must have $u(0) \neq 0$. Along the branch emanating from the trivial solution at $\alpha = \alpha_n$, we expect that the amplitude $u(0)$ will serve as a suitable perturbation parameter. We therefore set

$$u(x) = az(x), \quad z(0) = 1, \tag{9.4.16}$$

and consider the IVP obtained by substituting (9.4.16) into (9.4.1) and temporarily omitting the boundary condition at the right end:

$$z'' + \alpha \frac{\sin az}{a} = 0; \quad z(0) = 1, \quad z'(0) = 0. \tag{9.4.17}$$

The one and only solution of (9.4.17) is denoted by $z(x, \alpha, a)$ and depends analytically on α and a. As $a \to 0$, we have a linear problem whose solution is

$$z(x, \alpha, 0) = \cos \sqrt{\alpha} x. \tag{9.4.18}$$

Any solution of (9.4.17) for $a \neq 0$ that happens to vanish at $x = 1$ will, through (9.4.16), provide a nontrivial solution of (9.4.1). If α and a are chosen randomly in (9.4.17), the corresponding z will not usually vanish at $x = 1$; we are looking for the relationship between α and a that will make

$$0 = z(1, \alpha, a) \doteq b(\alpha, a). \tag{9.4.19}$$

Of course, we already know that $b(\alpha_n, 0) = 0$, but we wish to find solutions with $a \neq 0$, a near 0, α near α_n. The function $b(\alpha, a)$ is not usually explicitly calculable (although in our case it can be expressed in terms of elliptic functions); nevertheless we want to establish the existence and nature of the solutions (α, a) of (9.4.19) near $(\alpha_n, 0)$. The only weapon at our disposal is the implicit function theorem. If $b_\alpha(\alpha_n, 0) \neq 0$, then locally the solution of (9.4.19) can be expressed as $\alpha = \alpha(a)$; if $b_a(\alpha_n, 0) \neq 0$, the solution can be expressed as $a = a(\alpha)$. It is clear from (9.4.17) that $z(x, \alpha, a)$ is an even function of a, so that $z_a(x, \alpha, 0) = 0$ and hence $b_a(\alpha_n, 0) = 0$ [this result could also be obtained by differentiating (9.4.17) with respect to a]. We can calculate $b_\alpha(\alpha_n, 0)$ from (9.4.18):

$$b_\alpha(\alpha_n, 0) = z_\alpha(1, \alpha_n, 0) = -\frac{1}{2\sqrt{\alpha_n}} \sin \sqrt{\alpha_n} = \frac{(-1)^n}{\pi(2n-1)} \neq 0.$$

If the explicit expression (9.4.18) were not available, one could differentiate (9.4.17) with respect to α to obtain the desired result.

The implicit function theorem guarantees the existence of a unique function $\alpha = \alpha_n(a)$, defined near $a = 0$, such that $\alpha_n(0) = \alpha_n$ and $b(\alpha_n(a), a) = 0$. This function $\alpha_n(a)$ generates a solution $az(x, \alpha_n(a), a)$ of (9.4.1) which completely describes the branching at $\alpha = \alpha_n$.

The function $z(x, \alpha_n(a), a)$ satisfies (9.4.17) and also the boundary condition at $x = 1$. To find $\alpha_n(a)$ and z for a near zero, we differentiate (9.4.17) totally with respect to a. To simplify the notation, we let $z(x, \alpha_n(a), a) = Z(x, a)$ and drop the index on $\alpha_n(a)$. Then $Z(x, a)$ satisfies

$$Z'' + \alpha(a)\frac{\sin aZ}{a} = 0; \quad Z(0, a) = 1, \quad Z(1, a) = 0, \quad Z'(0, a) = 0.$$

Taking the a derivative, we find that $Z_a(x, 0) = 0, \alpha_n(0) = 0$. Further differentiation gives, after setting $a = 0$, an equation for $Z_{aa}(x, 0)$:

$$\begin{aligned} Z''_{aa} + \alpha_n Z_{aa} &= \tfrac{1}{3}\alpha_n Z^3(x, 0) - \alpha_{aa}(0)Z(x, 0), \quad 0 < x < 1; \\ Z_{aa}(0, a) &= Z_{aa}(1, a) = 0, \quad Z'_{aa}(0, a) = 0, \end{aligned}$$

where $Z(x, 0) = \cos\sqrt{\alpha_n}\, x$. The solvability condition then gives

$$\alpha_{aa}(0) = \frac{\alpha_n}{3}\frac{\int_0^1 \cos^4\sqrt{\alpha_n}x\,dx}{\int_0^1 \cos^2\sqrt{\alpha_n}x\,dx} = \frac{\alpha_n}{4}.$$

Thus, to the first nonvanishing order, we have

$$\alpha_n(a) - \alpha_n \sim \alpha_{aa}(0)\frac{a^2}{2} = \frac{(2n-1)^2\pi^2}{32}a^2,$$

which is in agreement with (9.4.15) if we take into account the relationship between c and a. In fact, c is the coefficient of the normalized eigenfunction $\sqrt{2}\cos\sqrt{\alpha_n}\, x$. whereas a is the coefficient of $\cos\sqrt{\alpha_n}\, x$; thus $c = a/\sqrt{2}$, as is needed to reconcile the two formulas.

It is worth noting that once we have determined that a is a suitable expansion parameter, and that $\alpha_n(a)$ and $Z(x, a)$ are even functions of a, we can merely substitute series expansions in a^2 in (9.4.17) and calculate the coefficients by equating like powers of a^2.

Monotone Iteration Method

The two preceding methods used for (9.4.1) are essentially local in nature. We have been able to predict that near $\alpha = \alpha_n$ there is branching to the right from the basic solution; there are two nontrivial solutions of small norm for α slightly larger than α_n, and these solutions (*to the first order*) are $\pm a\cos\sqrt{\alpha_n}\, x$, where

$$a = \frac{\sqrt{32}}{(2n-1)\pi}\sqrt{\alpha - \alpha_n}.$$

The functions $\{\cos\sqrt{\alpha_n}\, x\}$, being eigenfunctions of a Sturm-Liouville system, have well-defined nodal properties. The fundamental eigenfunction (corresponding to $n = 1$) is of one sign (and can therefore, of course, be chosen positive); the eigenfunction corresponding to α_n has $n-1$ zeros in $0 < x < 1$. It is interesting to speculate whether these properties carry over to the nonlinear problem. Near the branch point α_n, this is certainly true from the fact that the linearized problem dominates; but as we move along the branch, it is not obvious that the nodal properties are maintained. (For proofs, see Crandall and Rabinowitz [14].) We shall content ourselves with showing that the branching from α_1 provides a *positive solution for all* $\alpha > \alpha_1$ (of course, the "lower" half of the parabolic branch yields a negative solution). The monotone method used is not based on a branching argument. We therefore take α in (9.4.1) as *fixed* and larger than α_1. Our principal result is contained in the following theorem.

Theorem. *For each $\alpha > \alpha_1$, (9.4.1) has one and only one positive solution, that is, there is one and only one solution $u(x, \alpha)$ with the property $u(x, \alpha) > 0, 0 < x < 1$.*

Proof. We now proceed in the rest of the subsection to prove this theorem by applying Theorem 9.1.1.

By adding αu to both sides of (9.4.1), we obtain the equivalent BVP

$$-u'' + \alpha u = \alpha(\sin u + u), \quad 0 < x < 1; \quad u'(0) = u(1) = 0,$$

where $\alpha(\sin u + u)$ is an increasing function of u on the real line. For a given $v(x)$, the solution of the linear BVP

$$-u'' + \alpha u = \alpha(\sin v + v), \quad 0 < x < 1; \quad u'(0) = u(1) = 0 \tag{9.4.20}$$

can be written as

$$u = Tv \doteq \int_0^1 g(x, \xi)[\alpha \sin v(\xi) + \alpha v(\xi)]\, d\xi, \tag{9.4.21}$$

where $g(x, \xi)$ is Green's function satisfying

$$-g'' + \alpha g = \delta(x - \xi), \quad 0 < x, \xi < 1; \quad g'|_{x=0} = g|_{x=1} = 0. \tag{9.4.22}$$

Clearly, T can be regarded as a nonlinear operator on the space $C[0, 1]$ of real-valued continuous functions on $0 \leqslant x \leqslant 1$. Problem (9.4.1) is therefore equivalent to the equation $u = Tu$ or

$$u(x) = \int_0^1 g(x, \xi)[\alpha \sin u(\xi) + \alpha u(\xi)]\, d\xi, \quad 0 < x < 1,$$

so that solutions of (9.4.1) are fixed points of T, and vice versa. Since T is an integral operator, it is hardly surprising that it is compact, but we shall not give a proof.

For $u, v \in C[0, 1]$ we write $u \leqslant v$ if $u(x) \leqslant v(x)$ on $0 \leqslant x \leqslant 1$. We shall prove that T is increasing; that is, if $u \leqslant v$, then $Tu \leqslant Tv$. One proof follows from the fact that $g(x, \xi)$ is positive and that $\alpha \sin u + \alpha u$ is an increasing function of u on the real line. Another proof is based on the following extension of the maximum principle.

Lemma. *If v satisfies the differential inequality*

$$-v'' + \alpha v \geqslant 0, \quad 0 < x < 1; \quad -v'(0) \geqslant 0, \quad v(1) \geqslant 0,$$

then $v(x) \geqslant 0$ on $0 \leqslant x \leqslant 1$, that is, $v \geqslant 0$.

Indeed, if $v(x) < 0$ for some x, then either v has an interior negative minimum or v has a negative minimum at $x = 0$. The differential inequality rules out the first of these possibilities. The second implies that $v(0) < 0$, so that, by the differential inequality, $v''(0) < 0$; these conditions, together with the boundary condition $v'(0) \leqslant 0$, imply that $v(x)$ decreases with x at $x = 0$, hence $x = 0$ cannot be a minimum.

Now let $v \leqslant w$, and set $z = Tw - Tv$. Then, from (9.4.20) and (9.4.21), we find that

$$-z'' + \alpha z = \alpha[w - v + \sin w - \sin v], \quad 0 < x < 1; \quad z'(0) = z(1) = 0.$$

Since $w \geqslant v$, the right side of the differential equation is nonnegative, and the lemma gives $z \geqslant 0$, that is, $Tw \geqslant Tv$.

The iterative sequences (9.1.43) and (9.1.44) of Theorem 9.1.1 are defined in terms of the operator T in (9.4.21) or, equivalently, in terms of the BVP (9.4.20). For instance, $\{u_n\}$ satisfies the equivalent iteration schemes

$$-u_n'' + \alpha u_n = \alpha(\sin u_{n-1} + u_{n-1}), \quad 0 < x < 1; \quad u_n'(0) = u_n(1) = 0,$$

or

$$u_n = Tu_{n-1} = \int_0^1 g(x, \xi)[\alpha \sin u_{n-1}(\xi) + \alpha u_{n-1}(\xi)]\, d\xi,$$

and similarly for the sequence $\{u^n\}$.

As starting elements for iterations we need a lower solution u_0 for $T(u_0 \leqslant Tu_0)$ and an upper solution u^0 for $T(u^0 \geqslant Tu^0)$ with $u_0 \leqslant u^0$. We shall show that these properties follow for elements $u_0 \leqslant u^0$ defined by the inequalities below, which are in terms of the original BVP:

$$-u_0'' \leqslant \alpha \sin u_0, \quad 0 < x < 1; \quad -u_0'(0) \leqslant 0, \quad u_0(1) \leqslant 0, \tag{9.4.23}$$

$$-(u^0)'' \geqslant \alpha \sin u^0, \quad 0 < x < 1; \quad -(u^0)'(0) \geqslant 0, \quad u^0(1) \geqslant 0. \tag{9.4.24}$$

To show that $u_0 \leqslant Tu_0$, we note that $u_1 = Tu_0$, and

$$-u_1'' + \alpha u_1 = \alpha(\sin u_0 + u_0), \quad 0 < x < 1; \quad u_1'(0) = u_1(1) = 0,$$

which, when combined with (9.4.23), gives

$$\begin{aligned} &-(u_1 - u_0)'' + \alpha(u_1 - u_0) \geqslant 0, \quad 0 < x < 1; \\ &-[u_1'(0) - u_0'(0)] \geqslant 0, \quad u_1(1) - u_0(1) \geqslant 0. \end{aligned}$$

Our lemma then shows that $u_1 \geqslant u_0$. A similar proof yields $u^0 \geqslant Tu^0$.

Next we note that $u_0(x) \equiv 0$ and $u^0(x) \equiv \pi$ satisfy (9.4.23) and (9.4.24), respectively. Since T is compact and increasing on $[0, \pi]$, we can apply Theorem 9.1.1 to guarantee the existence of a minimal solution $\underline{u}$ and a maximal solution $\overline{u}$ in the order interval $[0, \pi]$. Thus $0 \leqslant \underline{u}(x) \leqslant \overline{u}(x) \leqslant \pi$, but $\underline{u}$ and $\overline{u}$ might both vanish identically. We must try to find another lower solution $u_0(x)$ which is positive in the interval $0 < x < 1$, so that $\underline{u}(x)$ will be positive in $0 < x < 1$. Physical insight helps; for α slightly above α_1 we expect a positive solution which is nearly proportional to the eigenfunction $e_1(x)$ of the linearized problem. Therefore, we try the lower solution

$$u_0 = \varepsilon e_1, \quad \varepsilon > 0,$$

which clearly satisfies the boundary inequalities in (9.4.23). Moreover, we also have

$$-u_0'' - \alpha \sin u_0 = \varepsilon \alpha_1 e_1 - \alpha \sin \varepsilon e_1 = -\alpha \varepsilon e_1 \left(\frac{\sin \varepsilon e_1}{\varepsilon e_1} - \frac{\alpha_1}{\alpha} \right).$$

With α fixed and larger than α_1, we can choose $\varepsilon > 0$ sufficiently small so that the term in parenthesis is positive and hence $-u_0'' \leqslant \alpha \sin u_0$, as required. Therefore Theorem 9.1.1 now guarantees solutions $\underline{u}(x)$ and $\overline{u}(x)$ satisfying

$$\varepsilon e_1(x) \leqslant \underline{u}(x) \leqslant \overline{u}(x) \leqslant \pi, \quad \alpha > \alpha_1.$$

Strictly speaking, the solution $\underline{u}$ should be denoted as $\underline{u}_\varepsilon$ since it might depend on ε. It is easily shown, however, that in fact this solution is independent of ε for all small positive ε satisfying $(\sin \varepsilon e_1)/\varepsilon e_1 \geqslant \alpha_1/\alpha$. We can therefore legitimately drop the ε subscript. Since e_1 is positive for $0 \leqslant x \leqslant 1$, so is $\underline{u}$, and $\underline{u}$ is the minimal solution in the order interval $[\varepsilon e_1, \pi]$ for every small positive ε. Remarkably, this does not guarantee that $\underline{u}$ is the minimal nontrivial solution in $[0, \pi]$. It is conceivable that there exists a solution ϕ of (9.4.1) which vanishes at some point $0 \leqslant x < 1$ or has a vanishing derivative at $x = 1$. If this were the case, we could not squeeze εe_1 under ϕ no matter how small a positive value of ε were chosen. Fortunately, this situation cannot occur. One can show explicitly that Green's function as defined by (9.4.22) is positive in $0 \leqslant x < 1$ and that $g'(1, \xi) < 0$. It then follows from the equation $\phi = T\phi$ that $\phi > 0$ in $0 \leqslant x < 1$ and that $\phi'(1) < 0$, so that a positive ε can be chosen small enough to satisfy $\varepsilon e_1 \leqslant \phi$. Thus $\underline{u}$ is the minimal nontrivial solution in $[0, \pi]$.

We show that $\underline{u} = \overline{u}$, so that the positive solution is unique. From (9.4.1), satisfied by $\underline{u}$ and $\overline{u}$, we find that

$$-\underline{u}''\overline{u} + \overline{u}''\underline{u} = \alpha(\overline{u} \sin \underline{u} - \underline{u} \sin \overline{u}),$$

and, by Green's theorem and the boundary conditions,

$$0 = \int_0^1 \left[\frac{\sin \underline{u}}{\underline{u}} - \frac{\sin \overline{u}}{\overline{u}} \right] \underline{u}\overline{u} \, dx. \tag{9.4.25}$$

We already know that $\underline{u}(x) \leqslant \overline{u}(x)$ and that $\underline{u}(x) > 0$ on $0 \leqslant x < 1$. Let D be the set on which $\underline{u} < \overline{u}$. Then the integral in (9.4.25) reduces to an integral over D.

Since $(\sin u)/u$ is strictly decreasing on $0 \leqslant u \leqslant \pi$, we have

$$\frac{\sin \underline{u}}{\underline{u}} > \frac{\sin \overline{u}}{\overline{u}} \quad \text{on } D \qquad \text{and} \qquad \overline{u}\underline{u} > 0 \quad \text{on } D.$$

This, however, contradicts the fact that the integral over D vanishes. □

Aspects of Reaction-Diffusion

As illustrated in Section 0.3, there are many settings in which both reaction and diffusion take place. Depending on the nature of the nonlinear reaction term, on the respective size of the various parameters, and on the type of initial and boundary conditions, widely different qualitative behaviors are possible. Examples occur in combustion theory, in gas-solid reactions, in flow through porous rock, in the absorption of contaminants in soils, in oxygen diffusion through tissue, in climate modeling, and in population biology. Among the phenomena encountered are thermal runaway, blow-up, extinction, dead cores, traveling fronts, and transition layers. In recent years some excellent books on various aspects of reaction-diffusion have appeared. Among the most accessible of these are the books by Bebernes and Eberly [7], Buckmaster and Ludford [11], Fife [21], and Grindrod [30].

Although reaction-diffusion problems often occur as systems of equations, we shall only analyze the scalar case of the single nonlinear parabolic equation

$$u_t - D\,\Delta u = f(u). \tag{9.4.26}$$

We begin by studying the ordinary differential equation

$$u_t = f(u) \tag{9.4.27}$$

obtained by setting $D = 0$ in (9.4.26). This nondiffusive case is known as a *lumped parameter* problem in the engineering terminology. We then proceed to the steady-state problem

$$-\Delta u = f(u) \tag{9.4.28}$$

and then piece the results together for the full parabolic problem.

Extinction and Blow-up for First-Order ODEs

Because our dependent variable u is regarded as a concentration, a population, or an absolute temperature, we seek nonnegative solutions of the IVP

$$u_t = f(u), \quad t > 0; \qquad u(0) = u_0 \geqslant 0. \tag{9.4.29}$$

To fix ideas let us think of the evolution of a population $u(t)$ whose net rate of increase (births minus deaths) per unit time is $f(u)$.

Consider first the case where $f(u) > 0$ for all $u > 0$ [if $f(0) = 0$, we take $u_0 > 0$]. We can then divide the differential equation (9.4.29) by $f(u)$ and integrate from time 0 to time t:

$$\int_{u_0}^{u(t)} \frac{dz}{f(z)} = t. \tag{9.4.30}$$

Setting

$$I = \int_{u_0}^{\infty} \frac{dz}{f(z)}, \tag{9.4.31}$$

we see that if I is infinite, then (9.4.30) yields a unique $u(t)$ for each $t > 0$ and $\lim_{t\to\infty} u(t) = 0$. If, however, I is finite, $u(t)$ tends to ∞ as $t \to I-$. The solution blows up in finite time.

Next, suppose that $f(u) > 0, 0 < u < a$, where $a > u_0$, and $f(a) = 0$. An example of this kind is the logistic model (0.3.22). The population experiences positive growth when $u < a$, but the rate of growth decreases to zero as the population tends to a. For times when $u < a$, equation (9.4.30) holds. Setting

$$\int_{u_0}^{a} \frac{dz}{f(z)} = J,$$

we see that if J is infinite (as is the case for the logistic model), (9.4.30) yields a solution $u(t)$ which tends to a as $t \to \infty$. If, on the other hand, J is finite, then $u(t)$ reaches the value a at $t = J$; then $f(a) = 0$ and the behavior for $t > J$ depends on how f is defined for $u > a$. If $f(u) = 0$ for $u \geqslant a$, then $u \equiv a$ for $t \geqslant J$.

Examples.

1. If $f(u) = u^p$, we have blow-up in finite time for $p > 1$, but not for $0 \leqslant p \leqslant 1$. If $f(u) = e^u$ the solution blows up in finite time.

2. If $f(u) = u(a-u)$ and $0 < u_0 < a$, $u(t)$ increases asymptotically to a as $t \to \infty$. If $f(u) = u(a-u)^{1/2}$, u reaches the value a in finite time.

3. Consider an exothermic reaction taking place in a region Ω with the temperature independent of position. In the heat balance we also include a loss term to account for the interaction with the cooler ambient medium, leading to the equation

$$u_t = ke^{\left[\frac{u}{1+\varepsilon u}\right]} - ru, \quad t > 0; \quad u(0) = 0.$$

Here k, ε, and r are positive constants. In the so-called *high-activation energy* approximation, ε is set equal to zero, yielding the simpler model

$$u_t = f(u) \doteq k(e^u - \delta u), \quad t > 0; \quad u(0) = 0, \tag{9.4.32}$$

where $\delta = r/k$. Note that $f(0) > 0$ and that f remains positive for all $u > 0$ if e^u and δu do not intersect. This occurs if and only if $\delta < e$ (losses small relative

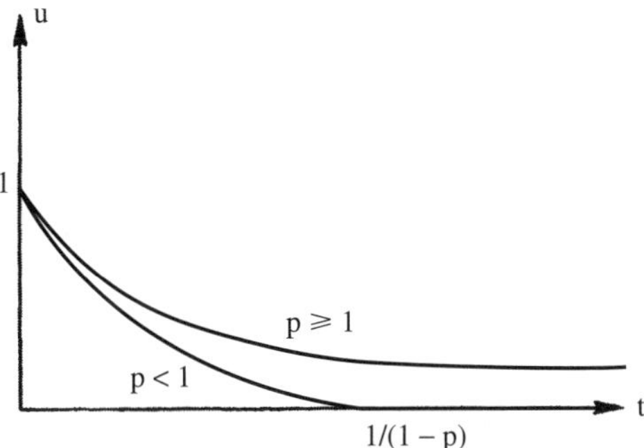

Figure 9.11

to the heat released by the reaction). In that case, $\int_0^\infty dz/f(z)$ is clearly finite and the temperature blows up in finite time. If, however, $\delta > e, e^u$ and δu first intersect at $u = u^* < 1$ and $\int_0^{u^*} dz/f(z) = \infty$, so that $u \to u^*$ as $t \to \infty$.

We now turn to problems where u decreases in time. If u is a concentration, this is an absorption problem; if u is a temperature, it is an endothermic reaction; if u is a population, the death rate exceeds the birth rate. It is convenient to rewrite (9.4.29) as

$$u_t = -f(u), \quad t > 0; \quad u(0) = u_0 > 0,$$

where $f(u) \geqslant 0$ for $u > 0$. It is clear that u decreases monotonically in time; should u reach the value 0 at some finite time T, it is understood that u remains equal to zero for $t \geqslant T$, and we would have *extinction in finite time*. Proceeding as for blow-up problems we divide the differential equation by $f(u)$ and integrate in time to find

$$\int_{u(t)}^{u_0} \frac{dz}{f(z)} = t. \tag{9.4.33}$$

Defining

$$K = \int_0^{u_0} \frac{dz}{f(z)},$$

we see that if $K = \infty$, (9.4.33) has a positive solution which tends to zero as $t \to \infty$. If, however, K is finite, then $u \to 0$ as $t \to K-$ so that $u(t) \equiv 0$ for $t \geqslant K$. This latter case always occurs if $f(0) > 0$, but may or may not occur if $f(0) = 0$.

Example 4. $f(u) = u^p, u_0 = 1$. The equation $u_t = -u^p$ describes the concentration $u(t)$ of a chemical being absorbed at the rate u^p. If $p_1 < p_2$, then $u^{p_1} > u^{p_2}$ for $0 < u < 1$, so that the smaller p is, the stronger the absorption. If $p < 1$, we have extinction at $t = 1/(1-p)$, whereas if $p \geqslant 1, u(t) > 0$ for all t and $\lim_{t\to\infty} u(t) = 0$. (See Figure 9.11.)

Example 5. Consider a spherical (!) lump of sugar dissolving in a cup of hot coffee. In the simplest model, the lump dissolves at a rate proportional to its surface area,

$$V_t = -kS = -k_1 V^{2/3}, \tag{9.4.34}$$

where V and S are the volume and surface area of the spherical lump. In terms of the radius r of the lump, we can rewrite (9.4.34) as

$$r_t = -k, \tag{9.4.35}$$

which remains valid until $t = a/k$ (where a is the initial radius) when $r = 0$ and the sugar has been completely dissolved.

A less naïve model takes into account the fact that there is an upper bound to the amount of sugar that can be dissolved per unit of liquid. Thus V_t is proportional not only to S but also to the difference between the saturation concentration $\overline{C}$ and the actual concentration C of sugar in the liquid. As $\overline{C} \to \infty$, we should recover the earlier model. We are therefore led to the equation

$$r_t = -k\left[1 - \frac{C(t)}{\overline{C}}\right].$$

We have assumed that the liquid, whose constant volume is V_L, is well-stirred so that the concentration of dissolved sugar is the same throughout the liquid. (A more sophisticated model would include both diffusion and the dependence of $\overline{C}$ on the temperature.) In our model,

$$C(t) = \frac{4\pi}{V_L}(a^3 - r^3),$$

so that

$$r_t = -k\left(1 - \frac{C}{\overline{C}}\right) = -k(A + Br^3), \tag{9.4.36}$$

where

$$A = 1 - Ba^3, \qquad B = \frac{4\pi}{3V_L\overline{C}}.$$

Depending on the value of A, we get different behaviors:

(a) If $A > 0$, the coffee can absorb all the sugar. Divide (9.4.36) by $A + Br^3$ and integrate from time 0 to time t to obtain

$$kt = \int_{r(t)}^{a} \frac{dz}{A + Bz^3}.$$

Letting $K = \int_0^a dz/(A + Bz^3)$, we see that $K < \infty$, so the sugar is fully dissolved at $t = K/k$. This is similar to our earlier model corresponding to $B = 0$, leading to $K = a$.

(b) If $A = 0$, $C(t) > 0$ for all t but $\lim_{t\to\infty} C(t) = 0$. The sugar is fully dissolved at $t = \infty$.

(c) If $A < 0$, $kt = \int_r^a dz/(-|A| + Bz^3)$ and the integral is infinite when the lower limit is $r_0 = (|A|/B)^{1/3}$. Thus the sugar is never fully dissolved and

$$\lim_{t\to\infty} r(t) = r_0.$$

The Steady-State Problem

We now turn to the steady-state problem (9.4.28) with vanishing Dirichlet boundary condition:

$$-\Delta u = f(u(x)), x \in \Omega; \quad u = 0, \quad x \in \Gamma, \tag{9.4.37}$$

where Ω is a bounded domain in $\mathbb{R}^n$ and Γ is its boundary.

In many of the applications mentioned at the beginning of the subsection [just before (9.4.26)] the function $u(x)$ must be nonnegative, and we shall be interested in finding *positive solutions*, that is, *nontrivial* solutions which are $\geqslant 0$ on Ω. In view of the boundary condition a positive solution $u(x)$ must take on values between 0 and some positive number u_M. If the diffusion coefficient is itself nonlinear, the differential equation has the form $\operatorname{div}(k(u)\operatorname{grad} u) = f(u)$, which can be reduced to type (9.4.37) by the change of variables $v = \int_0^u k(z)\,dz$.

Positive solutions of (9.4.37) will exist only if f satisfies certain conditions whose nature can be surmised by examining the *much simpler* problem of finding positive numbers u such that

$$lu = f(u), \tag{9.4.38}$$

where $l > 0$. Note that l is a positive linear operator on $\mathbb{R}$, just as $-\Delta$ with vanishing boundary conditions is a positive linear operator on $L_2(\Omega)$. All that is required for (9.4.38) to have positive solution(s) is for the curve $f(u)$ to intersect the straight line lu for $u > 0$. No such solution can exist if f lies either entirely below or entirely above the straight line. Remarkably, the same nonexistence result holds for (9.4.37) if we identify l with the lowest (fundamental) eigenvalue α_1 of

$$-\Delta u = \alpha u, \quad x \in \Omega; \quad u = 0, \quad x \in \Gamma. \tag{9.4.39}$$

We recall that α_1 is simple and that the corresponding eigenfunction $e_1(x)$ can (and will) be chosen to be strictly positive in Ω. (See Exercise 8.3.11.) Moreover, e_1 is normalized so that its maximum is 1. To avoid the multivaluedness associated with linear problems, we shall make the following *nonlinearity assumption*:

The functions $f(z)$ and $\alpha_1 z$ are not identically equal on any interval $(0, z_1)$ with $z_1 > 0$. Note that this condition is automatically satisfied if $f(0) \neq 0$.

We are now in a position to state the following theorem, which is a nonexistence theorem for positive solutions.

Theorem. *Positive solutions of (9.4.37) are possible only if the real-valued function $f(z) - \alpha_1 z$ changes sign for $z > 0$. Moreover, if $f(z) - \alpha_1 z < 0$ for $z > 0$, a solution of (9.4.37) cannot be positive at any point in Ω.*

REMARK. If $f(z) - \alpha_1 z > 0$ for $z > 0$, there may exist solutions that are positive in part of Ω and negative elsewhere in Ω. (See Exercise 9.4.3, part (c), for instance.)

Proof. Write (9.4.37) as

$$-\Delta u - \alpha_1 u = f(u) - \alpha_1 u, \quad x \in \Omega; \qquad u = 0, \quad x \in \Gamma. \tag{9.4.40}$$

The solvability condition for (9.4.40) is

$$\int_\Omega [f(u(x)) - \alpha_1 u(x)] e_1(x)\, dx = 0.$$

Since $e_1(x) > 0$ in Ω, either $f(z) - \alpha_1 z \equiv 0, 0 \leqslant z \leqslant u_M$, or $f(z) - \alpha_1 z$ must change sign in $0 < z < u_M$. The first alternative is impossible by the nonlinearity assumption, so that $f(z) - \alpha_1 z$ must change sign for $z > 0$. Next we want to prove that if $f(z) - \alpha_1 z < 0$ for $z > 0$, then $u(x)$ cannot be positive anywhere in Ω. If $u(x_0) > 0$, there is a domain Ω^* surrounding x_0 for which $u(x) > 0$ with $u = 0$ on the boundary Γ^* of Ω^*. Let (α_1^*, e_1^*) be a fundamental eigenpair for (9.4.39) over Ω^*; since Ω^* is contained in Ω, it follows from the variational characterization of eigenvalues that $\alpha_1^* \geqslant \alpha_1$, and therefore $f(z) - \alpha_1^* z < 0$ for $z > 0$, which, by the first part of the theorem, means that $u(x)$ cannot be a positive solution on Ω^*. We have therefore arrived at a contradiction, and the theorem is proved. □

As we shall see later, the nonexistence of a steady state is associated with blow-up for the evolution equation (9.4.26) when the nonlinearity f (such as λe^u) is rapidly increasing.

Example 6. Consider (9.4.37) with $f(u) = \lambda e^u$ for a fixed $\lambda > 0$. When does the curve λe^u intersect the straight line $\alpha_1 u$ or, equivalently, when does e^u intersect $(\alpha_1/\lambda)u$? Tangency occurs at $u = 1$ when $\alpha_1/\lambda = e$. For greater λ, there is no intersection for any real u, so that (9.4.37) has no solution (positive or otherwise!). If $\lambda \leqslant \alpha_1/e$, solutions are not ruled out and we will see later that a positive solution must exist when $\lambda \leqslant 1/ev_M$. (See Example 8.)

Now that we have excluded some possibilities, when do there actually exist positive solutions? Let us consider first the unforced case: $f(0) = 0$ (see Exercise 9.4.1 for the forced case). We shall establish the existence of a positive solution by a method similar to that used for (9.4.1) and again based on Theorem 9.1.1. To begin we suppose that there exist functions $u_0(x) \leqslant u^0(x)$, continuous on the closure $\overline{\Omega}$ of Ω and satisfying

$$-\Delta u_0 \leqslant f(u_0), \quad x \in \Omega; \qquad u_0 \leqslant 0, \quad x \in \Gamma, \tag{9.4.41}$$

$$-\Delta u^0 \geqslant f(u^0), \quad x \in \Omega; \qquad u^0 \geqslant 0, \quad x \in \Gamma. \tag{9.4.42}$$

Once such elements have been found, we pick M large enough so that

$$f'(u) + M \geqslant 0 \quad \text{for } m_0 \leqslant u \leqslant m^0, \tag{9.4.43}$$

where

$$m_0 = \min_{x \in \overline{\Omega}} u_0(x), \quad m^0 = \max_{x \in \overline{\Omega}} u^0(x).$$

Such a choice of M is possible if f' exists and is bounded on bounded intervals. By adding Mu to both sides of the differential equation, the BVP (9.4.37) becomes

$$-\Delta u + Mu = f(u) + Mu, \quad x \in \Omega; \qquad u = 0, \quad x \in \Gamma, \tag{9.4.44}$$

or, equivalently,

$$u = Tu \doteq \int_{\Omega} g(x, \xi)[f(u(\xi)) + Mu(\xi)]\, d\xi, \tag{9.4.45}$$

where $g(x, \xi)$ is Green's function satisfying

$$-\Delta g + Mg = \delta(x - \xi), \quad x, \xi \in \Omega; \qquad g = 0, \quad x \in \Gamma. \tag{9.4.46}$$

The operator T is a compact operator on the space $C(\overline{\Omega})$ of real-valued continuous functions on $\overline{\Omega}$. Our notation is the same as in the discussion preceding Theorem 9.1.1, with $\overline{\Omega}$ playing the role of R. In view of the definition (9.4.45) of T, the iteration schemes $u^n = Tu^{n-1}$ and $u_n = Tu_{n-1}$ can equally well be expressed in terms of BVPs. For instance, the scheme $u^n = Tu^{n-1}$ is equivalent to

$$-\Delta u^n + Mu^n = f(u^{n-1}) + Mu^{n-1}, \quad x \in \Omega; \qquad u^n = 0, \quad x \in \Gamma. \tag{9.4.47}$$

To use Theorem 9.1.1, it remains to show that $u_0 \leqslant Tu_0$, $u^0 \geqslant Tu^0$, and T is increasing on $[u_0, u^0]$. Let $u_0 \leqslant v \leqslant w \leqslant u^0$, and set $z = Tw - Tv$. Then we find that

$$-\Delta z + Mz = f(w) - f(v) + M(w - v), \quad x \in \Omega; \quad z = 0, \quad x \in \Gamma.$$

The right side of the differential equation is nonnegative by virtue of (9.4.43). By the maximum principle we conclude that $z \geqslant 0$, that is, $Tw \geqslant Tv$, so that T is increasing on $[u_0, u^0]$. Next we show that $u^0 \geqslant Tu^0$ (a similar argument holding for lower solutions). By (9.4.42) and (9.4.47) with $n = 1$, we obtain

$$-\Delta(u^0 - u^1) + M(u^0 - u^1) \geqslant 0, \quad x \in \Omega; \qquad u^0 - u^1 \geqslant 0, \quad x \in \Gamma.$$

The maximum principle again shows that $u^0(x) \geqslant u^1(x)$ in Ω. Since $u^1 = Tu^0$, we have the desired result.

To prove that the minimal and maximal solutions $\underline{u}$ and $\overline{u}$ guaranteed by Theorem 9.1.1 are actually positive in Ω, we must find u_0 satisfying (9.4.41) and $u_0(x) > 0$ in Ω. Of course, we must also find u^0 satisfying (9.4.42) with $u^0 \geqslant u_0$.

For the lower solution we try $u_0 = \varepsilon e_1$, where ε is a positive number and e_1 is the normalized eigenfunction introduced in connection with (9.4.39). We obtain

$$-\Delta u_0 - f(u_0) = \varepsilon e_1 \left[\alpha_1 - \frac{f(\varepsilon e_1)}{\varepsilon e_1} \right]. \tag{9.4.48}$$

For ε small and positive, $f(\varepsilon e_1)/\varepsilon e_1$ is nearly $f'(0)$; therefore, if $f'(0) > \alpha_1$ it will be possible to choose $\varepsilon > 0$ so small that the right side of (9.4.48) will be negative. Since e_1 vanishes on Γ, the boundary inequality in (9.4.41) is clearly satisfied, so that u_0 is a lower solution if $f'(0) > \alpha_1$. Moreover, u_0 is positive in Ω. Let us try to find an upper solution of the same form with ε *so large* that $\alpha_1 - f(\varepsilon e_1)/\varepsilon e_1 \geqslant 0$ in Ω; the controlling factor should be the behavior of $f(z)/z$ for large z, but unfortunately the fact that e_1 vanishes on Γ means that for x near Γ, $f(\varepsilon e_1)/\varepsilon e_1$ will be close to $f'(0)$ even if ε is large. There is a simple remedy if $f(z)$ satisfies the condition

$$\limsup_{z\to\infty} \frac{f(z)}{z} = \beta < \alpha_1. \tag{9.4.49}$$

We then consider a domain Ω^* similar to but slightly larger than Ω, with $\overline{\Omega}$ contained in Ω^*. The fundamental eigenvalue α_1^* of (9.4.39) for the domain Ω^* is slightly smaller than α_1. We can choose Ω^* so that $\beta < \alpha_1^* < \alpha_1$. The corresponding eigenfunction e_1^* is positive on $\overline{\Omega}$ (say, $e_1^* \geqslant \delta$). Trying an upper solution $u^0 = Ae_1^*$, we find that

$$-\Delta u^0 - f(u^0) = Ae_1^* \left[\alpha_1^* - \frac{f(Ae_1^*)}{Ae_1^*}\right],$$

whose right side, in view of (9.4.49), is nonnegative in $\overline{\Omega}$ for A sufficiently large. Simpler upper solutions can be found in some other cases (see Exercise 9.4.2). For instance, if $f(\gamma) = 0$ for some $\gamma > 0$, then $u^0 = \gamma$ is an upper solution.

Assuming that $f(z)$ satisfies (9.4.49) and $f'(0) > \alpha_1$, we have constructed a positive lower solution u_0 and an upper solution u^0. Thus Theorem 9.1.1, guarantees the existence of a positive minimal solution $\underline{u}(x)$ and of a maximal solution $\overline{u}(x)$ in $[\varepsilon e_1, u^0]$. Again one can show that $\underline{u}$ is independent of ε for small ε and that for every positive solution ϕ of (9.4.45) there exists an $\varepsilon > 0$ such that $\varepsilon e_1 \leqslant \phi$. Therefore, $\underline{u}(x)$ is the minimal nontrivial solution in $[0, u^0]$. Clearly, $\underline{u}(x) \leqslant \overline{u}(x)$ for x in Ω; uniqueness follows by the same arguments as were used for (9.4.1) if $f(z)/z$ is strictly decreasing (or if $f'' < 0$).

Our results take on a clearer meaning when we introduce a parameter α in the nonlinear term. Consider the problem

$$-\Delta u = \alpha h(u(x)), \quad x \in \Omega; \qquad u = 0, \quad x \in \Gamma, \tag{9.4.50}$$

where $h(z)$ satisfies

$$h(0) = 0, \quad h'(0) = 1, \quad \limsup_{z\to\infty} \frac{h(z)}{z} = \beta < 1. \tag{9.4.51}$$

Then (9.4.50) has at least one positive solution when

$$\alpha_1 < \alpha < \frac{\alpha_1}{\beta}. \tag{9.4.52}$$

If, furthermore, $h(z)/z$ is strictly decreasing for $z > 0$, the positive solution is unique and there is no positive solution for $\alpha \leqslant \alpha_1$ and $\alpha \geqslant \alpha_1/\beta$ [since then the curves $\alpha_1 z$ and $\alpha h(z)$ do not intersect for $z > 0$].

REMARK. If $\beta = 0$ in (9.4.51), then α_1/β should be interpreted as $+\infty$. Thus, if $h(z)$ is bounded on the positive real axis, a positive solution exists for all $\alpha > \alpha_1$.

For forced problems with $f(0) > 0$, the construction of a lower solution is much simpler indeed: Take $u_0 = 0$; then since $f(0) > 0$, the first iterate u_1 is already positive and we are on our way to constructing an increasing sequence u_n of positive functions. To prove existence of a solution, we must construct a positive upper solution u^0. This question is considered in Exercises 9.4.1, 9.4.2, and 9.4.5, but we now examine two important special cases.

Example 7. Consider the Poisson problem

$$-\Delta v = 1, \quad x \in \Omega; \qquad v = 0, \quad x \in \Gamma. \tag{9.4.52}$$

This is a linear "forced" problem with $f(0) > 0$. The nonlinearity is monotone increasing. The lower solution $v_0 = 0$ generates an increasing sequence v_n of positive functions. To find a positive upper solution, consider (9.4.52) for a circumscribed ball whose radius is a; the solution V_a can be found explicitly (see Exercise 1.3.2):

$$V_a = \frac{a^2 - r^2}{2n}.$$

Note that V_a is positive on Γ and satisfies the differential equation so that it is an upper solution to (9.4.52). Thus (9.4.52) has a solution, easily shown to be unique, satisfying

$$0 < v(x) \leqslant \frac{a^2 - r^2}{2n}. \tag{9.4.53}$$

Therefore, the maximum value of v does not exceed $a^2/2n$.

Example 8. Consider again Example 6 with $f(u) = \lambda e^u$. Clearly, $u_0 = 0$ is a lower solution leading to an increasing sequence of positive iterates. We try to find an upper solution of the form $w = pv$, where p is a positive constant to be determined and v is the solution of (9.4.52). This can be seen to work if

$$\lambda \leqslant pe^{-pv_M}, \tag{9.4.54}$$

where v_M is the maximum of v. Since p is at our disposal we choose it so that (9.4.54) produces the largest range for λ. This is accomplished by taking $p = 1/v_M$, so that for $\lambda \leqslant 1/(v_M e), w = v/v_M$ is an upper solution. For the case of a one-dimensional interval $0 \leqslant x \leqslant 1$, we have $v = x(1-x)/2$ and $v_M = 1/8$. The fundamental eigenvalue of (9.4.39) is $\alpha_1 = \pi^2$. Combining with the results of Example 6, we find that the BVP

$$-u'' = \lambda e^u, \quad 0 < x < 1; \quad u(0) = u(1) = 0$$

has at least one positive solution for $0 < \lambda \leqslant 8/e$ and has no solution for $\lambda > \pi^2/e$.

Steady-State Problems with Absorption

Confining ourselves to power-law absorption, we consider the BVP

$$-\Delta U = -\lambda U^p, \quad x \in \Omega; \quad u(\Gamma) = 1; \quad p > 0. \tag{9.4.55}$$

The problem would be of no interest if the boundary concentration were equal to zero—for then, the unique solution would be $U \equiv 0$. In our case, we are seeking a solution of (9.4.55) which is nonnegative. If we interpret the absorption term to be zero when $U \leqslant 0$, the maximum principle tells us that any solution of (9.4.55) satisfies $0 \leqslant U(x) \leqslant 1$, so that there is no need to impose the nonnegativity requirement on $U(x)$. The PDE implies that $U(x)$ is a convex function. As a simple illustration, take $p = 1$ and Ω the one-dimensional interval $|x| < a$; we then find that $U(x) = (\cosh \sqrt{\lambda}\, x)/(\cosh \sqrt{\lambda}\, a)$, which is strictly positive, convex, and has its minimum at $x = 0$. The fact that the solution depends only on $\sqrt{\lambda}\, x$ is an immediate consequence of (9.4.55): Divide the PDE by λ and introduce new space variables $y = \sqrt{\lambda}\, x$ to remove λ from the PDE. A particular feature of (9.4.55) is that for $p < 1$ and λ sufficiently large, there will exist an interior subregion (known as a *dead core*) where U vanishes. The phenomenon is associated with the non-Lipschitz behavior of the absorption term at $U = 0$ (the derivative of U^p is infinite at $U = 0+$ when $p < 1$).

Uniqueness for (9.4.55) is straightforward. Suppose that U and V are two different solutions. Then $W = U - V$ must be of one sign (which we can take to be positive) on some subdomain $R \subset \Omega$. Since $U = V$ on Γ, we can choose R so that $U = V$ on ∂R (which may, of course, include a portion or all of Γ). The power function being an increasing function, we find that

$$-\Delta W = -\lambda(U^p - V^p) < 0, \quad x \in R; \quad W(\partial R) = 0.$$

The maximum principle then yields $W < 0$ in R, contradicting the assumption that $U > V$ in R.

The proof of existence is more difficult because of the non-Lipschitz behavior when $p < 1$. We can reduce (9.4.55) to a forced problem of type (9.4.37) with zero boundary condition by setting $\Phi = 1 - U$, so that

$$-\Delta \Phi = \lambda(1 - \Phi)^p, \quad x \in \Omega; \quad \Phi(\partial\Omega) = 0. \tag{9.4.56}$$

The functions $\Phi_0 = 0$ and $\Phi^0 = 1$ can then play the respective roles of u_0 and u^0 in (9.4.41)–(9.4.42). If $p \geqslant 1$, f' is bounded on [0, 1] so that the arguments following (9.4.43) carry through and we have proved existence of a solution $\Phi \in [0, 1]$ and hence of a solution U of (9.4.55) in the same order interval. A modified proof is needed when $p < 1$ because f' is unbounded at $\Phi = 1-$. Such a proof was first provided by Amann, but here we only present his results [in the original setting of (9.4.55)]. First, we make the following

Definition. *$\overline{U}(x)$ is said to be an* upper solution *of (9.4.55) if*

$$-\Delta \overline{U} + \lambda(\overline{U})^p \geqslant 0, \quad x \in \Omega; \quad \overline{U}(\Gamma) \geqslant 1. \tag{9.4.57}$$

A lower solution $\overline{U}(x)$ is defined by reversing both inequalities in (9.4.57).

Theorem (Amann). *If we can find a lower solution $\underline{U}$ and an upper solution $\overline{U}$ with $\underline{U} \leqslant \overline{U}$, then there exists a solution of (9.4.55) in the order interval $[\underline{U}, \overline{U}]$.*

REMARK. $\underline{U} = 0$ is a lower solution of (9.4.55) and $\overline{U} = 1$ is an upper solution with $\underline{U} \leqslant \overline{U}$. Therefore, (9.4.55) has a solution $U(x)$ with $0 \leqslant U(x) \leqslant 1$. Since this solution has already been shown to be unique, the bounds on U can be improved by pushing the lower solution upward and the upper solution downward. This idea is used in the comparison theorems which follow.

Let us consider (9.4.55) when a single parameter is changed. For instance, suppose that λ and Ω are fixed but $p_2 > p_1$; then $U^{p_2} \leqslant U^{p_1}$ on [0, 1], so that the absorption is smaller from which we surmise that the solution $U_2(x)$ corresponding to p_2 is larger than the solution $U_1(x)$ corresponding to p_1. The proof is a one-liner:

$$-\Delta U_2 + \lambda U_2^{p_1} \geqslant -\Delta U_2 + \lambda U_2^{p_2} = 0, \quad U_2(\Gamma) = 0,$$

so that U_2 is an upper solution to (9.4.55) formulated with $p = p_1$. Hence $U_1(x) \leqslant U_2(x)$. In a similar way we can prove the following

Theorem (Comparison Theorem).

$$\begin{aligned} &\lambda_2 \geqslant \lambda_1 \quad \textit{implies that } U_2(x) \leqslant U_1(x) \\ &p_2 \geqslant p_1 \quad \textit{implies that } U_2(x) \geqslant U_1(x) \\ &\Omega_2 \supset \Omega_1 \quad \textit{implies that } U_2(x) \leqslant U_1(x) \quad \textit{on } \Omega_1. \end{aligned}$$

To use this theorem we will need some explicit solutions as a basis for comparison. We begin with the one-dimensional half-space problem.

$$-U_{xx} + \lambda U^p = 0, \quad x > 0; \quad U(0) = 1, \quad U(\infty) = 0. \tag{9.4.58}$$

In view of the convexity of U, it follows that U is a decreasing function of x and that $U'(\infty) = 0$.

The special case $p = 1$ gives the elementary solution $U(x) = \exp(-\sqrt{\lambda}\, x)$ which is positive for all $x \geqslant 0$. By comparison we expect the same property for any $p > 1$. In the gas absorption model this means an infinite penetration distance. We shall see below that if $p < 1$, the penetration distance is finite. On setting $y = \sqrt{\lambda}\, x$, the differential equation becomes $-U_{yy} + U^p = 0, y > 0$, with the same boundary conditions. Multiplying by U_y and using the BC at ∞, we find that

$$U_y = -\left(\frac{2}{p+1}\right)^{1/2} U^{(p+1)/2}.$$

As long as $U > 0$ we can divide by $U^{(p+1)/2}$ and integrate from 0 to y to obtain (for $p \neq 1$)

$$U(y) = \left[1 - \frac{y}{P_1}\right]^{2/(1-p)}, \quad P_1 = \frac{\sqrt{2(p+1)}}{1-p}. \tag{9.4.59}$$

For $p > 1$, P_1 is negative and we can rewrite

$$U(y) = \left[1 + \frac{y}{|P_1|}\right]^{2/(p-1)}, \qquad p > 1,$$

which shows that $U(y) > 0$ for all $y \geqslant 0$ and tends to zero as $y \to \infty$. For $p < 1$, P_1 is positive and the expression for U is positive only for $y < P_1$; since U, U', and U'' all vanish at $y = P_1-$, we can extend U to be zero for $y \geqslant P_1$, and still have a classical solution of our differential equation on $(0, \infty)$. In terms of the original variable x, we have

$$U(x) = \begin{cases} \left[1 + \dfrac{\sqrt{\lambda}\,x}{|P_1|}\right]^{-2/(p-1)}, & p > 1, \\ e^{-\sqrt{\lambda}\,x}, & p = 1, \\ \left[1 - \dfrac{\sqrt{\lambda}x}{P_1}\right]_+^{2/(1-p)}, & p < 1, \end{cases} \tag{9.4.60}$$

where $[Z]_+$ stands for the larger of 0 and Z. For $p < 1$ the finite penetration distance is $P_1/\sqrt{\lambda}$. One immediate result of (9.4.60) is that the solution of (9.4.55) cannot have a dead core if $p \geqslant 1$. Indeed, the domain Ω will fit in a half-space and (9.4.60) provides a lower solution to (9.4.55). Since the expression (9.4.60) is positive for $p \geqslant 1$, so must be the solution of (9.4.55).

Our half-space result (9.4.60) is also helpful in establishing a dead core for the slab problem $\Omega : |x| < a$. Let $p < 1$ and choose λ such that $\sqrt{\lambda}\,a > P_1$; let $U(x)$ be the third line of (9.4.60). By piecing together the half-space solutions for $x > -a$ and $x < a$, we see that

$$V(x) = U(x + a) + U(a - x) \tag{9.4.61}$$

is a solution of (9.4.55) for the slab $|x| < a$. Each term has a penetration distance $P_1/\sqrt{\lambda}$ so that $V(x)$ vanishes in the dead core

$$|x| < a - \frac{P_1}{\sqrt{\lambda}}.$$

Our solution has a one-point dead core if $\sqrt{\lambda}\,a = P_1$. If $\sqrt{\lambda}\,a < P_1$, the solution has no dead core and is no longer given by (9.4.61)—in fact, it no longer has an elementary closed form.

Next, we consider the problem of a ball of radius a in $\mathbb{R}^n$. The unique solution of (9.4.55) depends only on $r = |x|$, so that the problem reduces to the BVP for $U(r)$:

$$-\frac{1}{r^{n-1}}\frac{d}{dr}\left(r^{n-1}\frac{dU}{dr}\right) + \lambda U^p = 0, \quad 0 < r < a; \qquad U(a) = 1. \tag{9.4.62}$$

The hidden condition at the singular point $r = 0$ is that U is finite there. We shall not attempt a general solution of the problem. Instead, we look for the possibility of a dead core if $p < 1$ (we already know that no dead core can exist for $p \geqslant 1$).

We begin modestly by seeking a solution of the form Ar^β. Since the Laplacian reduces powers by 2, equality of powers in the two terms of the differential equation requires that

$$\beta - 2 = \beta p \quad \text{or} \quad \beta = \frac{2}{1-p},$$

and to satisfy the BC at $r = a$, we need $A = 1/a^\beta$. Thus our tentative solution is

$$U(r) = \left(\frac{r}{a}\right)^{2/(1-p)}, \tag{9.4.63}$$

which has a one-point dead core at $r = 0$ when $p < 1$. If $p > 1$, (9.4.63) is singular at $r = 0$ and can no longer be the solution of the BVP. When $p < 1$, substitution of (9.4.63) in (9.4.62) shows that we do indeed have a solution for the particular value of λ given by

$$\sqrt{\lambda}\, a = P_n, \quad \text{where } P_n = \left[\frac{2n(1-p) + 4p}{(1-p)^2}\right]^{1/2}, \tag{9.4.64}$$

which is consistent with the slab result ($n = 1$). The Comparison Theorem then predicts that the solution of (9.4.62) has

$$\text{no dead core for } \sqrt{\lambda}\, a < P_n,$$
$$\text{dead core for } \sqrt{\lambda}\, a \geqslant P_n.$$

It can be shown that the dead core has positive measure if $\sqrt{\lambda}\, a > P_n$ and strictly increases with λ (see Friedman and Phillips [25]).

Let us now investigate (9.4.55) for an arbitrary bounded domain Ω in $\mathbb{R}^3$. When $p \geqslant 1$, we have already established that there is no dead core, so we shall consider the case $p < 1$. If λ is small, there will not be a dead core even if $p < 1$. We can quantify this somewhat: Let a_1 be the half-width of the thinnest slab enclosing Ω, let a_2 be the radius of the smallest circular cylinder enclosing Ω, and let a_3 be the radius of the smallest circumscribed sphere. Then our Comparison Theorem guarantees that no dead core exists in Ω if the larger region enclosing Ω has no dead core. Thus Ω has no dead core if

$$\sqrt{\lambda}\, a_1 < P_1, \quad \text{or } \sqrt{\lambda}\, a_2 < P_2, \quad \text{or } \sqrt{\lambda}\, a_3 < P_3. \tag{9.4.65}$$

Which of these gives the largest λ range of nonexistence of a dead core will depend on the shape of Ω. Note that $a_1 \leqslant a_2 \leqslant a_3$, but also $P_1 < P_2 < P_3$, so that the dependence of P_k/a_k on k cannot be predicted in general. Having disposed of nonexistence, what can we say about the existence of a dead core for Ω? Let $x_0 \in \Omega$; since Ω is an open set, x_0 is at a positive distance, say r_0, from the boundary Γ. A ball of radius r_0, with center at x_0, has a dead core if $\sqrt{\lambda}\, r_0 \geqslant P_3$; therefore, Ω

(which encloses this ball) will, by the Comparison Theorem, have a dead core which contains x_0.

We therefore have the following result: Let $x_0 \in \Omega$ and let $r_0 =$ distance from x_0 to Γ, then x_0 is in the dead core for $\sqrt{\lambda} \geqslant P_3/r_0$. Let $r_i =$ radius of the largest inscribed ball in Ω; then a dead core exists for Ω whenever $\sqrt{\lambda} \geqslant P_3/r_i$.

REMARKS

1. Consider (9.4.55) for a bounded, convex domain Ω and let R_Q be a half-space enclosing Ω and touching Γ at the point Q. Since the solution U_Q for the half-space exceeds the solution U_Ω for the domain Ω, the dead core for Ω must be at least at a distance $P_1\lambda^{-1/2}$ from Q. As we vary the point Q on Γ, it is clear that the dead core for Ω must be at least a distance $P_1\lambda^{-1/2}$ from the boundary Γ. On the other hand, points which are at a distance $P_3\lambda^{-1/2}$ from the boundary do belong to the dead core. As $\lambda \to \infty$, we see that the dead core will engulf Ω except for a boundary layer of thickness of order $\lambda^{-1/2}$.

2. For $x_0 \in \Omega$, there is a smallest value of λ for which x_0 belongs to the dead core. Denoting the solution of (9.4.55) by $U(x, \lambda)$, we define, for $p < 1$,

$$\lambda_0 = \inf_{\lambda}\{U(x_0, \lambda) = 0\}, \tag{9.4.66}$$

$$\lambda^* = \inf_{x_0}\{\lambda_0\}. \tag{9.4.67}$$

Thus λ^* is the smallest value of λ for which $U(x, \lambda)$ vanishes somewhere; on the other hand, λ_0 is the smallest value of λ for which $U(x_0, \lambda)$ vanishes. We have already shown that

$$\lambda_0 \leqslant P_3/r_0, \quad \lambda^* \leqslant P_3/r_i. \tag{9.4.67a}$$

The Evolution Problem

We shall study two problems based on (9.4.26). The first leads to possible blow-up in finite time, whereas the second is an absorption problem in which dead cores can form in finite time.

In both problems we will need the notions of upper and lower solutions for parabolic problems. Consider the evolution problem

$$\begin{aligned} &u_t - \Delta u - f(u) = 0, \quad (x,t) \in \Omega \times (0,\infty); \\ &u(x,0) = u_0(x); \qquad u = h(x), \quad (x,t) \in \Gamma \times (0,\infty). \end{aligned} \tag{9.4.68}$$

We say that $\overline{u}(x,t)$ is an upper solution of (9.4.68) if

$$\overline{u}_t - \Delta\overline{u} - f(\overline{u}) \geqslant 0; \qquad \overline{u}(x,0) \geqslant u_0(x); \qquad \overline{u}(\Gamma,t) \geqslant h(x). \tag{9.4.69}$$

Reversal of the three inequalities yields the definition of a lower solution $\underline{u}(x,t)$.

Theorem (Existence). *(Proof omitted). If we can find a lower solution $\underline{u}(x,t)$ and an upper solution $\overline{u}(x,t)$ with $\underline{u}(x,t) \leqslant \overline{u}(x,t)$, then (9.4.68) has a solution $u(x,t)$ satisfying $\underline{u} \leqslant u \leqslant \overline{u}$.*

Problem 1. We consider the forced problem

$$\begin{aligned} &u_t - \Delta u = \lambda e^u, \quad (x,t) \in \Omega \times (0,\infty); \\ &u(x,0) = 0; \qquad u = 0 \quad \text{for } (x,t) \in \Gamma \times (0,\infty). \end{aligned} \tag{9.4.70}$$

We saw earlier in Example 9.4 that when diffusion is absent, the solution blows up in the finite time $I = \int_0^\infty dz/\lambda e^z = 1/\lambda$ [see (9.4.31)]. Thus, there is blow-up in finite time for every $\lambda > 0$, but the time of blow-up depends on λ.

The expected effect of diffusion (the Laplacian term) in (9.4.70) is to mitigate the explosive nature of the reaction term. Indeed, we shall see that blow-up in finite time now occurs only if λ is sufficiently large. Let (α_1, e_1) be defined as in (9.4.39); multiply the PDE in (9.4.70) by $e_1(x)$ and integrate over Ω to obtain

$$\frac{dE}{dt} + \alpha_1 E = \lambda \int_\Omega e^u e_1(x)\,dx, \quad E(0) = 0, \tag{9.4.71}$$

where

$$E(t) = \int_\Omega u(x,t) e_1(x)\,dx,$$

and $e_1(x)$ has been normalized so that its integral over Ω is 1. Obviously, $E(t)$ increases in time. Does it become infinite in finite time?

Jensen's inequality (see Exercise 9.5.3) applied to the convex function e^u enables us to write

$$\int_\Omega e^u e_1(x)\,dx \geqslant e^E,$$

so that (9.4.71) becomes the differential inequality

$$\frac{dE}{dt} \geqslant \lambda(e^E - \beta E), \quad E(0) = 0, \quad \beta = \lambda_1/\lambda. \tag{9.4.72}$$

Initially, $e^E - \beta E$ is positive; it remains positive for all E if $\beta < e$; if $\beta = e$, the line βE is tangent to e^E at $E = 1$; if $\beta > e$, βE first intersects e^E at a value of $E = E^* < 1$.

If $\beta < e$ (that is, $\lambda > \lambda_1/e$) we can divide the differential inequality (9.4.72) by $e^E - \beta E$ and integrate from 0 to t to obtain

$$\int_0^{E(t)} \frac{ds}{e^s - \beta s} \geqslant t. \tag{9.4.73}$$

Since $\int_0^\infty ds/(e^s - \beta s) \doteq I$ is finite, (9.4.73) implies that $E(I)$ must be infinite (which we interpret as blow-up in finite time). On the other hand, if $\beta > e$, then $\int_0^{E*} ds/(e^s - \beta s)$ is infinite and (9.4.73) does not imply blow-up in finite time.

Thus, our only conclusion is that if $\lambda > \lambda_1/e$, we have blow-up in finite time. Note that this is precisely the same condition obtained in Example 6 for nonexistence of a solution to the steady-state problem! When can we guarantee that (9.4.70) has a solution defined for all $t > 0$? We shall use the Existence Theorem with upper and lower solutions defined from (9.4.69) with $f(u) = \lambda e^u, u(x,0) = 0, u(\Gamma, t) = 0$.

Theorem. *The BVP (9.4.70) has a solution for $\lambda \leqslant 1/ev_M$, where v_M is defined in Example 8.*

Proof. Clearly, $\underline{u} \equiv 0$ is a lower solution of (9.4.70). For an upper solution, we try the function $w(x) = v(x)/v_M$ of Example 8. For $\lambda \leqslant 1/ev_M$, w is an upper solution to the steady-state problem and since $w_t = 0$, it also satisfies the differential inequality in (9.4.69) as well as $w(\Gamma, t) = 0$ and $w(x, 0) \geqslant 0$. Thus $w(x)$ is an upper solution of (9.4.70) and hence there exists a solution of (9.4.70) for $\lambda \leqslant 1/ev_M$ and this solution $u(x,t)$ satisfies $0 \leqslant u(x,t) \leqslant v(x)/v_M$. □

Problem 2. We consider the absorption problem

$$\begin{aligned} &u_t - \Delta u = -\lambda u^p, \quad (x,t) \in \Omega \times (0,\infty); \quad p > 0; \\ &u(x,0) = 1, \quad u(\Gamma,t) = 1. \end{aligned} \tag{9.4.74}$$

The corresponding steady state is governed by (9.4.55).

Problem (9.4.74) is of type (9.4.68) with $f(u) = -\lambda u^p, u_0(x) = h(x) = 1$. Clearly, $\underline{u}(x,t) \equiv 0$ is a lower solution and $\overline{u}(x,t) \equiv 1$ is an upper solution. Therefore there exists a solution $u(x,t)$ of (9.4.74) satisfying $0 \leqslant u(x,t) \leqslant 1$. The maximum principle for the heat equation shows that any solution of (9.4.74) satisfies $0 \leqslant u(x,t) \leqslant 1$. Uniqueness is relatively easy to prove (see Exercise 9.4.9). As $t \to \infty$, the solution of (9.4.74) tends to the solution $U(x)$ of the steady-state problem (9.4.55). For the particular initial value in (9.4.74) we can prove that $u(x,t)$ decreases monotonically in time (see Exercise 9.4.9).

The case $p = 1$ (linear absorption) can be solved by separation of variables and gives us some idea of how the solution behaves. The evolution problem is

$$u_t - \Delta u = -\lambda u, \quad x \in \Omega, \quad t > 0; \quad u(x,0) = u(\Gamma,t) = 1, \tag{9.4.75}$$

and the steady-state problem is

$$-\Delta U = -\lambda U, \quad x \in \Omega; \quad U(\Gamma) = 1. \tag{9.4.76}$$

Writing successively $u(x,t) = U + v(x,t), v = e^{-\lambda t}\eta$, we see that $\eta(x,t)$ satisfies the heat equation without absorption:

$$\eta_t - \Delta\eta = 0; \quad \eta(x,0) = 1 - U(x), \quad \eta(\Gamma,t) = 0. \tag{9.4.77}$$

We can then expand the solution in the eigenfunctions of (9.4.39) to obtain

$$\eta(x,t) = \sum_{k=1}^{\infty} A_k e^{-\alpha_k t} e_k(x), \quad A_k = \langle 1 - U, e_k \rangle. \tag{9.4.78}$$

Calculating the A_k from (9.4.76), we find that

$$A_k = \frac{\lambda\langle e_k, 1\rangle}{\alpha_k + \lambda},$$

which yields

$$u(x,t) = U(x) + \lambda e^{-\lambda t} \sum_{k=1}^{\infty} \frac{\langle e_k, 1\rangle}{\alpha_k + \lambda} e^{-\alpha_k t} e_k(x). \tag{9.4.79}$$

Since $U(x) > 0$ in Ω, so is η (by the maximum principle for the heat equation) and hence v. Thus $u(x,t) > U(x)$ for all t. Clearly, $\lim_{t\to\infty} u(x,t) = U(x)$.

Let us now return to the general case of (9.4.74) with p not equal to 1. There is no hope then to find an explicit solution such as (9.4.79), but qualitative properties may carry over. For instance, it is still true that wherever $U(x) > 0$, $u(x,t)$ will be strictly greater than $U(x)$. But what happens if $U(x)$ contains a dead core? This occurs when $p < 1$ and λ is sufficiently large. The following theorem tells us that the time-dependent problem will also have a dead core.

Theorem. *Let $x_0 \in \Omega$ and let $p < 1$. Let $\lambda > \lambda_0$ [see (9.4.66)]. Then $u(x_0,t) = 0$ for*

$$t \geqslant T_0 \doteq \frac{1}{(1-p)(\lambda - \lambda_0)}. \tag{9.4.80}$$

Proof. With $\lambda \geqslant \lambda_0$, we know that $U(x_0, \lambda) = 0$. Our theorem will be proved if we can construct an upper solution $w(x,t)$ of (9.4.74) which vanishes at x_0 for $t \geqslant T_0$. Let us try

$$w(x,t) = U(x, \lambda_0) + z(t), \tag{9.4.81}$$

where $z(t)$ is the solution of $z_t = -\gamma z^p$ with $z(0) = 1$ and γ to be chosen suitably. From Example 4 we know that z vanishes for $t \geqslant 1/(1-p)\gamma$, and hence so does $w(x_0,t)$. Clearly, $w(x,0) \geqslant 1$ and $w(\Gamma,t) \geqslant 1$, so that it remains only to verify the differential inequality (9.4.69) with $f(u) = \lambda u^p$, $u_0(x) = w(x) = 1$. From (9.4.81), we obtain

$$w_t - \Delta w + \lambda w^p = -\gamma z^p - \lambda_0 U^p(x, \lambda_0) + \lambda w^p \geqslant (\lambda - \lambda_0 - \gamma) w^p.$$

It follows that w is an upper solution if $\gamma = \lambda - \lambda_0$. This completes the proof of the theorem. □

REMARKS

1. Since λ_0 is not known explicitly, it may be useful to restate the theorem in terms of a known upper bound for λ_0 such as (9.4.67a): If $\lambda > P_3/r_0$, $u(x_0,t) = 0$ for $t \geqslant [(1-p)(\lambda - P_3/r_0)]^{-1}$.

2. If our interest is only in determining the time T^* of onset of the dead core, we have the following result: $u(x,t) = 0$ for some x when

$$t \geqslant T^* \doteq \frac{1}{(1-p)(\lambda - \lambda^*)}.$$

3. *Extinction in finite time.* Consider (9.4.74) with the boundary condition changed to $u(\Gamma, t) \equiv 0$. Then the steady-state solution is $U(x) \equiv 0$. The question arises as to whether the solution of (9.4.74) becomes identically zero in finite time. Observe that now the solution of $z_t = -\lambda z^p, z(0) = 1$ is an upper solution of (9.4.74) so that $u(x,t) \leqslant z$; if $p < 1, z = 0$ for $t \geqslant [\lambda(1-p)]^{-1}$, so that u vanishes identically for $t \geqslant [\lambda(1-p)]^{-1}$. For a description of how the solution behaves just prior to extinction, see Friedman and Herrero [23].

EXERCISES

9.4.1 Consider the *forced* problem

$$-\Delta u = \alpha h(u), \quad x \in \Omega; \quad u = 0, \quad x \in \Gamma, \tag{9.4.82}$$

with $h(0) > 0$ and $\limsup_{z\to\infty}[h(z)/z] = \beta$. Show that there exists a positive solution for $0 < \alpha < \alpha_1/\beta$. [*Hint*: We can now use $u_0 = 0$ as a lower solution that yields $u_1(x) > 0$ in Ω and hence $\underline{u}(x) > 0$ in Ω (for the unforced problem we cannot use $u_0 = 0$ to construct a positive solution since all iterates vanish identically). Use the same upper solution as for the unforced problem. Under what conditions for h can you obtain uniqueness?]

9.4.2 Consider the unforced problem (9.4.37) with $f'(0) > \alpha_1, f(z)/z$ strictly decreasing, and $f(\gamma)/\gamma = \alpha_1$ for some $\gamma > 0$. Then we have proved that (9.4.37) has one and only one positive solution $u(x)$. Show that $u(x) \geqslant \gamma e_1(x)$. If, furthermore, there is a value δ for which $f(\delta) = 0$, show that $u(x) \leqslant \delta$. In this way obtain bounds on the branch of positive solutions of

$$-\Delta u = \alpha \sin u, \quad x \in \Omega; \quad u = 0, \quad x \in \Gamma,$$
$$-\Delta u = \alpha(u - u^3), \quad x \in \Omega; \quad u = 0, \quad x \in \Gamma,$$
$$-\Delta u = \alpha u - u^3, \quad x \in \Omega; \quad u = 0, \quad x \in \Gamma.$$

In which case does $\|u\| \to \infty$ when $\alpha \to \infty$?

9.4.3 Consider the existence problem for the BVP (9.4.37) without the requirements of positivity.

(a) Show that if $f(z) - \alpha_1 z \neq 0$ for all z, then the BVP has *no* solution.

(b) If $f(z) < \alpha_1 z$ for $z > 0$ and $f(z) > \alpha_1 z$ for $z < 0$, then the BVP has only the trivial solution.

(c) If the inequalities in (b) are reversed, show that nontrivial solutions are possible. [*Hint*: Let $f(z) = \alpha_2 z$.]

9.4.4 Prove the theorem preceding (9.4.40) for the case when the boundary condition is $(\partial u/\partial \nu) + hu = 0$, where h is a positive constant.

9.4.5 If Ω is a unit ball in $\mathbb{R}^3$ and $h(u) = e^u$, the BVP (9.4.82) can be shown to have an infinite number of positive solutions when $\alpha = 2$ (see Joseph and Lundgren [37]). In the one-dimensional case the problem can be solved explicitly, say on the interval $-\frac{1}{2} < x < \frac{1}{2}$. Show that there exists α_c (approximately equal to 3.52) such that there are two positive solutions for $\alpha < \alpha_c$, one for $\alpha = \alpha_c$, and none for $\alpha > \alpha_c$.

9.4.6 Problem (9.4.50) with

$$h(u) = \begin{cases} 0, & 0 \leqslant u \leqslant 1, \\ u - 1, & u \geqslant 1, \end{cases}$$

is a simplified version of a problem that occurs in plasma containment. Show that there exists no positive solution for $\alpha < \alpha_1$. Solve explicitly the one-dimensional problem on the interval $-1 < x < 1$, and show that there exists a positive solution for $\alpha > \alpha_1$, but that the norm of this solution tends to infinity as $\alpha \to \alpha_1+$ (*branching from infinity*).

9.4.7 Consider (9.4.37) when $f(z) < 0$ for $z > \gamma$. Show that *every* positive solution satisfies $u(x) \leqslant \gamma$. Show that if $f(z) = \alpha z(1 - z)$, there is one and only one positive solution for $\alpha > \alpha_1$ and this solution satisfies $u(x) \leqslant 1$. The problem occurs in population studies [Fisher's equation; see (0.3.25)].

9.4.8 Consider again the satellite web coupling problem (4.4.22), repeated here for convenience:

$$-u'' = -\lambda u^4, \quad 0 < x < 1; \quad u(0) = u(1) = 1; \quad \lambda \geqslant 0.$$

(a) Prove that there is at most one positive solution. (Assume that $u_2 > u_1$ on a subdomain, and then use the maximum principle for $u_2 - u_1$ on that subdomain to obtain a contradiction.)

(b) Show that $u_0 = 0$ is a lower solution and $u^0 = 1$ an upper solution.

(c) Show that the iteration scheme $w = Tv$, defined by

$$-w'' + 4\lambda w = -\lambda v^4 + 4\lambda v, \quad 0 < x < 1; \quad w(0) = w(1) = 1,$$

meets the requirements of Theorem 9.1.1, so that a nontrivial positive solution satisfying $0 \leqslant u(x) \leqslant 1$ is guaranteed. Uniqueness follows from (a).

9.4.9 (a) Let $u(x, t)$ and $v(x, t)$ be two solutions of (9.4.74) with $u(x, t) \geqslant v(x, t)$. Set $w = u - v$ and show how the maximum principle leads to $u \equiv v$.

(b) For problem (9.4.74) show that $u_t(x,0) < 0$ in Ω. To this end, begin by letting $v(x,t) = u(x,t+\Delta t)$; now show that v is a lower solution to (9.4.74) and hence that $u(x,t)$ decreases monotonically in time.

9.4.10 (a) Show that the problem

$$-\Delta u = \lambda(u+1), \quad x \in \Omega; \quad u(\partial\Omega) = 0,$$

does not have a positive solution for λ large. Prove existence of a positive solution for λ small. Check your results in the case of the one-dimensional interval $0 < x < 1$.

(b) Consider the problem

$$-\Delta u = \lambda(1+u^m), \quad x \in \Omega; \quad u(\partial\Omega) = 0,$$

and discuss the nonexistence of positive solutions (here $m > 0$ and the behavior will be different for $m < 1$ and $m > 1$).

(c) Discuss finite-time blow-up for the equation

$$u_t - \Delta u = \lambda(1+u^m); \quad u(x,0) = u(\Gamma,t) = 0,$$

when $m > 1$.

9.4.11 Consider the half-space problem (9.4.58) with U^p replaced by $f(U)$, where $f(0) = 0, f(U) > 0$ for $U > 0$. Show that there is finite penetration if and only if I is finite, where

$$I = \int_0^1 \frac{ds}{\sqrt{F(s)}}, \quad F(s) = \int_0^s f(u)\,du.$$

9.4.12 Consider the steady reaction problem

$$-\Delta u = \lambda \exp\left(\frac{u}{1+\varepsilon u}\right), \quad x \in \Omega; \quad u(\Gamma) = 0.$$

(a) Prove existence of a positive solution for all $\lambda > 0, \varepsilon > 0$.

(b) Prove uniqueness for $\varepsilon \geqslant \frac{1}{4}$.

9.4.13 Let H be the Heaviside function and compare the solutions of the problems

(a) $-u'' = \lambda H(1-u), \quad |x| < 1, \quad u(\pm 1) = 0,$

(b) $-u'' = \lambda H(u-1), \quad |x| < 1, \quad u(\pm 1) = 0.$

The second problem is related to the climatology model of Chapter 0.

9.4.14 (a) Let $k(x,y,z)$ be a continuous function on $[a,b] \times [a,b] \times \mathbb{R}$ and consider the nonlinear integral equation (4.4.4) with $f(x) = 0$. Show that the operator T transforms S: $\|u\|_\infty \leqslant C$ into itself if $M_c(b-a) \leqslant c$, where $M_c = \max |k|$ for $x, y \in [a,b]$ and $|z| \leqslant c$. By the Arzela-Ascoli theorem, T is a compact operator and version (b) of the Schauder theorem guarantees the existence of a solution. Show that the nonlinear eigenvalue problem

$$u = \lambda T u$$

has a solution if $|\lambda| \leqslant C/(b-a)M_c$.

(b) Consider the BVP

$$-u'' = \lambda e^u, \quad 0 < x < 1, \quad u(0) = u(1) = 0,$$

and show by (a) that a solution exists if

$$|\lambda| \leqslant 4Ce^{-c}.$$

The optimal choice of C yields $|\lambda| \leqslant 4/e$.

(c) Refine the calculations to show that a solution exists for $|\lambda| \leqslant 8/e$.

9.5 THE STABILITY OF THE STEADY STATE

In the preceding section, as well as in parts of Chapter 8, we have studied whether the solution of an evolution problem tended to a steady state as $t \to \infty$. Of course, the problem becomes more complicated in the presence of multiple steady states or in the absence of any steady state (see the case of the exponential nonlinearity). The situation is very simple for the linear heat equation

$$u_t - \Delta u = q(x), \quad x \in \Omega; \quad u(x,0) = u_0(x); \quad u(\Gamma, t) = h(x),$$

whose solution was shown to tend as $t \to \infty$ to the solution $U(x)$ of

$$-\Delta U = q(x), \quad x \in \Omega; \quad U(x) = h(x), \quad x \in \Gamma.$$

There is another way of looking at the relationship between these two problems. Suppose that the temperature of the body is in its steady state $U(x)$ until some slight disturbance changes the temperature to $U(x) + z(x)$; will the temperature return to the steady state after the disturbance has been turned off? Solving the evolution problem with $u_0(x) = U + z$, we find that its solution $u(x,t)$ tends to U exponentially fast in time. Note that this behavior does not even require z to be small. Thus U is *stable* to both small and large disturbances. This strong property will not, in general, be true for nonlinear problems. The purpose of this section is to study stability questions for which we will find it useful to introduce a somewhat more abstract setting.

In the steady state we study a boundary value problem in the space coordinates (denoted collectively by x); the time-dependent problem is of the initial value type in time and of the boundary value type in space. Although we normally think of the solution as a function $u(x,t)$ in which the variables x and t have more or less equal status, another point of view will be more useful to us. At each time t the solution u is an element or point in a Hilbert space H (admittedly, such an element is a function of the space coordinates, but we prefer to regard this function in toto as a point in Hilbert space); as t changes, the element $u = u(t)$ moves in Hilbert space according to the *evolution equation*

$$\frac{du}{dt} = F(u), \quad t > 0; \quad u|_{t=0} = u_0, \tag{9.5.1}$$

where u_0 is the initial value of u and F is a transformation (generally, nonlinear) of H into itself. The derivatives, if any, with respect to space coordinates are contained in the operator F. Let us illustrate our point of view for the case of source-free heat conduction in a rod of unit length whose initial temperature is $u_0(x)$ and whose ends are kept at 0 temperature. At each time t the temperature (or "state" of the rod) is an element of $H = L_2(0, 1)$; the state u changes in time according to law (9.5.1), where the operator F is defined as

$$Fu \doteq \frac{d^2u}{dx^2}$$

on the subset of H consisting of twice-differentiable functions that vanish at $x = 0$, $x = 1$. In our case F is not defined on all of H, but this is a technicality which does not obscure the general idea. If H is a finite-dimensional space, (9.5.1) reduces to a system of nonlinear ordinary differential equations.

From the evolution point of view, one of the goals is to find $u(t)$ given u_0. This is accomplished through a transformation S_t (known as the *propagator*) which maps H into H so that

$$u(t) = S(t)u_0.$$

For the finite rod problem, S_t is an integral operator whose kernel is Green's function of (8.2.36a), with $l = 1$. For the heat equation on $\mathbb{R}$, S_t is again an integral operator whose kernel is now given by (8.2.27), with $\tau = 0$.

Two remarks should be made at this time. First, with no additional difficulty we could take the operator F to depend explicitly on t, but we shall have no need for this generalization. (When, as in our case, F does not depend explicitly on t, the equation is said to be *autonomous*.) The second remark has to do with equations which, in their classical form, are of higher-order in time. Suppose that we are dealing with the wave equation

$$\frac{\partial^2 u}{\partial t^2} = \frac{\partial^2 u}{\partial x^2} \quad \text{on } 0 < x < 1,$$

for instance. We set $\partial u/\partial t = v$ and consider the Hilbert space of pairs $(u, v) = z$, where u and v are elements of $L_2(0, 1)$. Then the "state" z satisfies an evolution equation of type (9.5.1). We are familiar with this idea for finite-dimensional problems; it is the way in which Newton's laws are reduced to Hamiltonian form.

One of the principal questions related to (9.5.1) is that of the stability of steady states. Let $\tilde{u}$ be an element of H satisfying

$$F(\tilde{u}) = 0.$$

Thus $\tilde{u}$ does not depend on t and is therefore also a solution of (9.5.1) with initial value $\tilde{u}$. We call $\tilde{u}$ a *steady-state solution* or an *equilibrium state*. One is often interested in the effect of small disturbances on an equilibrium state. Does such a disturbance die out, grow, or remain pretty much the same size? The analysis consists of studying (9.5.1) with initial value u_0 near $\tilde{u}$. We shall say that $\tilde{u}$ is *stable* if there exists $\varepsilon > 0$ such that the solution $u(t)$ of (9.5.1) satisfies

$$\lim_{t\to\infty} \|u - \tilde{u}\| = 0 \quad \text{whenever } \|u_0 - \tilde{u}\| \leqslant \varepsilon. \tag{9.5.2}$$

For problems of undamped vibration, (9.5.2) is obviously too stringent a definition of stability; one would be quite satisfied with avoiding resonance (that is, $u - \tilde{u}$ should remain of the same order as $u_0 - \tilde{u}$). However, for the problems we consider, (9.5.2) is adequate. Note that we cannot expect stability for disturbances of arbitrary size. It is only in linear problems that the question of stability is independent of the size of the disturbance.

Let $u(t)$ be a solution of (9.5.1) corresponding to the initial value u_0 close to $\tilde{u}$; setting

$$u = \tilde{u} + h \tag{9.5.3}$$

and substituting in (9.5.1), we find that

$$\frac{dh}{dt} = F(\tilde{u} + h) - F(\tilde{u}) = F_u(\tilde{u})h + r, \tag{9.5.4}$$

where $F_u(\tilde{u})$ is the Fréchet derivative of F at $\tilde{u}$ (that is, the linearization of F at $\tilde{u}$) and r is of higher-order in h: $\lim_{h\to 0}(\|r\|/\|h\|) = 0$. Since the initial value for h is small, it would appear that the behavior of the solution of (9.5.4) is determined by the *linearized equation*

$$\frac{dh}{dt} = F_u(\tilde{u})h. \tag{9.5.5}$$

We shall assume the following *principle of linearized stability*: If all solutions of (9.5.5) decay exponentially in t, so will solutions of (9.5.4) for $h(0)$ sufficiently small; if (9.5.5) has some solution that is exponentially increasing, there will exist a small initial state for which the solution of (9.5.4) will increase rapidly at first (until nonlinearity takes over).

REMARK. This principle is well established for ordinary differential equations and for some classes of partial differential equations. In any event a study of (9.5.4) begins with (9.5.5).

The question is, then, when does (9.5.5) have all solutions exponentially decaying? If we try a solution of the form $h(t) = ke^{\mu t}$, where k is an element of H that is independent of time, we find that

$$F_u(\tilde{u})k = \mu k, \tag{9.5.6}$$

so that k is an eigenvector of the linear operator $F_u(\tilde{u})$ with eigenvalue μ. *If the spectrum of $F_u(\tilde{u})$ is in the left half of the complex μ plane*, all solutions of (9.5.5) decay [and $\tilde{u}$ is a stable solution of (9.5.1)]. Obviously, if $F_u(\tilde{u})$ is compact and self-adjoint, the solution of (9.5.5) can be expanded in terms of the eigenvectors of $F_u(\tilde{u})$ and there is no difficulty in showing that all solutions of (9.5.5) decay.

In many applications the operator F in (9.5.1) also depends on a real parameter λ which can often be controlled. The differential equation in (9.5.1) then becomes

$$\frac{du}{dt} = F(\lambda, u). \tag{9.5.7}$$

Suppose we are given a branch $\tilde{u}(\lambda)$ of steady solutions of (9.5.7) [often, $\tilde{u}(\lambda) \equiv 0$ will be the branch in question], that is, $F(\lambda, \tilde{u}(\lambda)) = 0$. This equilibrium solution will usually be stable for a certain range of λ (which, by redefining λ if necessary, can be taken to be an interval of the form $\lambda < \lambda_0$) and unstable outside this range ($\lambda \geqslant \lambda_0$, say). Let the physical system be in the equilibrium state for some $\lambda < \lambda_0$, and suppose that we slowly increase the value of the parameter λ (which, depending on the application, may be a flow rate, a compressive load, a Reynolds number, etc.). On passing through the *critical value* λ_0, the equilibrium solution becomes unstable and the physical system can no longer follow it. Although there are many possible behaviors when an equilibrium state becomes unstable, we shall consider only some of these.

A new steady-state solution may branch off from $\tilde{u}(\lambda)$ at $\lambda = \lambda_0$, and the physical system may then follow this new equilibrium solution for $\lambda > \lambda_0$. If the original equilibrium solution turns around at $\lambda = \lambda_0$, or if the new branch starts off to the left ($\lambda < \lambda_0$), the physical system may experience a *snap-through* to another steady state not close to the original branch. Another possibility is that a new *time-periodic* solution (initially of small amplitude) will appear at $\lambda = \lambda_0$. Later on we give an example of each of these three cases (which do not exhaust all the possibilities).

The mathematical criterion for a critical value λ_0 is straightforward. For $\lambda < \lambda_0$ the spectrum of $F_u(\lambda, \tilde{u})$ is in the left half of the complex plane, and for $\lambda > \lambda_0$ there is at least one point of the spectrum in the right half of the μ plane. Thus, as λ passes through λ_0, the front of the spectrum of F_u crosses into the right half of the μ plane. Two important cases occur (a) when a simple eigenvalue of F_u crosses through the origin, and (b) when a pair of simple, complex conjugate, eigenvalues cross the imaginary axis. The first of these possibilities is related to the appearance of a new steady state; the second, to a time-periodic solution.

Example 1. Consider the operation of a continuously stirred tank reactor. Replacing u by $u + 1$ in (0.3.10), we find that $u(t)$ satisfies the single nonlinear differential equation

$$\frac{du}{dt} = -u + \frac{1}{\delta}(\beta - u)e^{-\gamma/u+1}, \tag{9.5.8}$$

where δ, β, γ are positive constants, and u is restricted to $0 \leqslant u \leqslant \beta$. The constant δ is the *flow rate*.

The steady-state equation is

$$(\beta - u)e^{-\gamma/u+1} = \delta u. \tag{9.5.9}$$

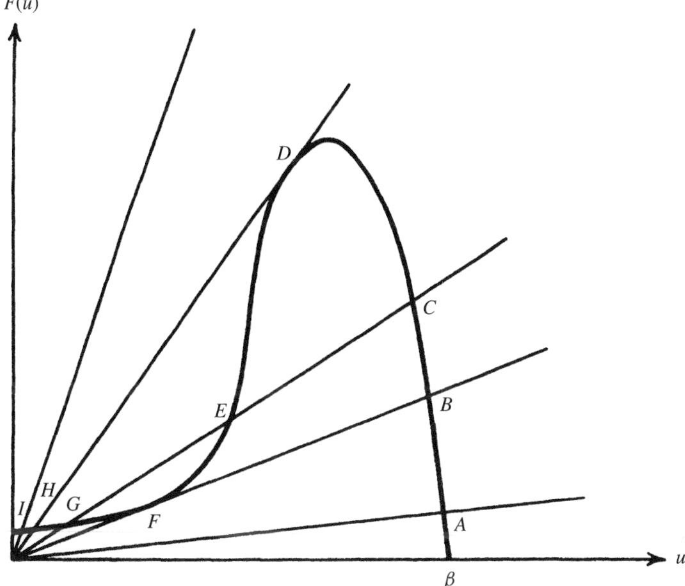

Figure 9.12

The graph of the left side has the general shape of the curve in Figure 9.12; for other values of β and γ the curve may be somewhat flatter. The steady-state solutions are the intersections of the straight line of slope δ and the curve. For δ small there is a single intersection at A. As δ increases we go through the continuous sequence A, B, C, D. If δ is increased above the value δ^* corresponding to D, the only intersection is at a point near H. This corresponds to a violent quenching of the reaction [the ratio of the concentration of the reactant to the feed concentration is given by $1 - (u/\beta)$, so that u near 0 means that very little reaction is taking place, whereas u near β implies nearly complete conversion]. If, however, we proceed in the other direction by decreasing δ from a value corresponding to state I, we pass through the continuous sequence I, H, G, F. On reaching F a further decrease in δ brings a sudden jump to state B with its much higher temperature. In Figure 9.13 we have plotted the steady-state solution as a function of the flow rate δ, that is, we have exhibited the solution set of (9.5.9). It does not appear that a steady state such as E can be reached by continuous increase or decrease in the flow rate; it will come as no surprise that the steady states on the inner loop from D to F are *unstable*, whereas the other steady states are *stable*.

The stability analysis is very simple in our case. Let $\tilde{u}$ be a steady state on the upper part of the curve in Figure 9.13 (say a point such as C, for definiteness). Setting

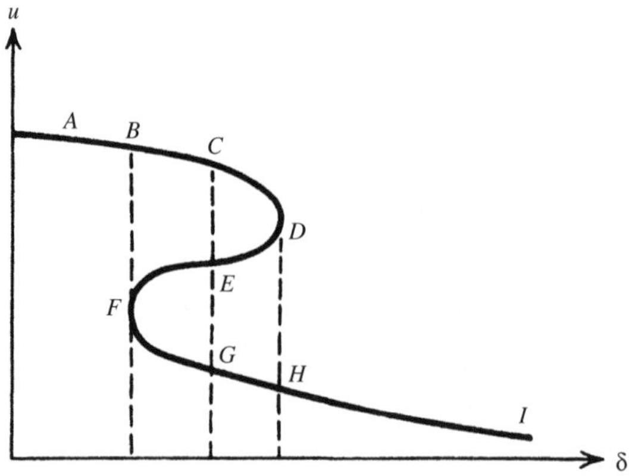

Figure 9.13

$u(t) = \tilde{u} + h(t)$, we find that $h(t)$ satisfies

$$\frac{dh}{dt} = -h + \frac{1}{\delta}[f(\tilde{u} + h) - f(\tilde{u})], \tag{9.5.10}$$

where f is the curve in Figure 9.12 and the initial value of h is taken small in conformity with the assumption of small disturbances from the steady state. Linearization is unnecessary because we can deduce our results directly from the nonlinear equation (9.5.10). At the point C we see from Figure 9.12 that $f(\tilde{u}+h) - f(\tilde{u})$ is negative for small $h > 0$ and positive for small $h < 0$. Thus (9.5.10) shows a tendency to return to equilibrium, and we have stability. Obviously, all that is required for stability is that $-1 + (1/\delta)f'(\tilde{u}) < 0$, and therefore the steady states on the lower part of the curve in Figure 9.13 are also stable (states such as G, H, I). A similar analysis shows that the states on the inner loop (such as E) are unstable.

Example 2. The problem of stability of an equilibrium solution can sometimes be related to the existence or nonexistence of monotone iteration schemes for constructing the equilibrium solution itself. It seems puzzling at first that purely steady considerations could establish dynamic stability, but we can perhaps see the connection by examining the preceding example. Any steady state $\tilde{u}$ of (9.5.8) is a solution of

$$u = f(u), \quad \text{where } f(u) = \frac{1}{\delta}(\beta - u)e^{-\gamma/u+1}. \tag{9.5.11}$$

Such a steady state is stable if for $|u - \tilde{u}|$ small,

$$f(u) - f(\tilde{u}) - (u - \tilde{u}) < 0, \quad u > \tilde{u},$$
$$f(u) - f(\tilde{u}) - (u - \tilde{u}) > 0, \quad u < \tilde{u}.$$

This merely states that the straight line lies above the curve to the right of $\tilde{u}$ and below the curve to the left of $\tilde{u}$. But that is exactly the condition needed to establish a monotone scheme for the solution $\tilde{u}$ of (9.5.11).

Without proof we quote two theorems (see Sattinger [51]) of this type for the evolution equation

$$\frac{du}{dt} - \Delta u - f(u) = 0, \ t > 0, \ x \in \Omega; \ u = c(x), \ x \in \Gamma, \ t > 0; \\ u \text{ given in } \Omega \text{ at } t = 0. \tag{9.5.12}$$

Theorem. *Let $\hat{u}$ be a solution of the steady problem*

$$-\Delta\hat{u} - f(\hat{u}) = 0, \quad x \in \Omega; \quad \hat{u} = c(x), \quad x \in \Gamma, \tag{9.5.13}$$

and let $u_0(x)$ and $u^0(x)$ be lower and upper solutions, respectively, for this steady problem; assume that $u_0 \leqslant u^0$ and that the iteration schemes starting from u_0 and u^0 both converge to $\hat{u}$. Then $\hat{u}$ is a stable solution of (9.5.12), and any solution of (9.5.12) with initial data $v(x)$ satisfying $u_0(x) \leqslant v(x) \leqslant u^0(x)$ tends to $\hat{u}(x)$ as $t \to \infty$.

REMARK. The theorem not only tells us about the stability on the basis of purely steady considerations but also gives an indication of the extent of the stability domain. Recall that a lower solution u_0 satisfies

$$-\Delta u_0 \leqslant f(u_0), \quad x \in \Omega; \quad u_0 \leqslant c(x), \quad x \in \Gamma,$$

and for an upper solution u^0 the inequalities are reversed.

Theorem. *Let $\hat{u}$ be a solution of (9.5.13), and suppose that for each $\varepsilon > 0$ there exists an upper solution $u^0(x, \varepsilon)$ lying* below *$\hat{u}$ with $\hat{u} - u^0 < \varepsilon$ and a lower solution u_0 lying* above *$\hat{u}$ with $u_0 - \hat{u} < \varepsilon$. Then $\hat{u}$ is an* unstable *solution of (9.5.12).*

As illustration of these theorems, let $c(x) = 0$ and $f = \alpha g$, where g satisfies the conditions imposed below (9.4.50) for the existence of a unique positive solution. Then this solution is stable.

If one abandons the fairly severe restrictions imposed on g, one can still prove that supercritical branches are stable and subcritical ones are unstable in the neighborhood of the branch point. Consider, for instance, the equation

$$-\Delta u = \alpha g(u), \quad x \in \Omega; \quad u = 0, \quad x \in \Gamma, \tag{9.5.14}$$

when $g(0) = 0$, $g'(0) = 1$, $g''(u) < 0$ on some interval $-a \leqslant u \leqslant a$. (A typical example would be $g = u - u^2$ or any g whose Maclaurin series begins with $u - cu^2$, $c > 0$.) First we show that the null solution is stable for $\alpha < \alpha_1$ and unstable for $\alpha > \alpha_1$.

We have

$$-\Delta(\varepsilon e_1) - \alpha g(\varepsilon e_1) = \alpha\varepsilon e_1 \left[\frac{\alpha_1}{\alpha} - \frac{g(\varepsilon e_1)}{\varepsilon e_1}\right], \tag{9.5.15}$$

so that for $\alpha < \alpha_1$ and $\varepsilon > 0, \varepsilon e_1$ is an upper solution, whereas $-\varepsilon e_1$ is a lower solution. Thus the null solution (which is the only solution lying between $-\varepsilon e_1$

and εe_1 for all $\varepsilon > 0$) must be *stable*. For $\alpha > \alpha_1$ the situation is reversed: For every small $\varepsilon > 0, \varepsilon e_1$ is a lower solution and $-\varepsilon e_1$ is an upper solution, so that the null solution is unstable. What about the stability of the branch emanating from the simple eigenvalue α_1? In this case the branching looks as in Figure 9.9, at least near the value $\alpha = \alpha_1$ (which corresponds to L). This branch represents a positive solution for values of α slightly larger than α_1 and a negative solution for α just below α_1. The part of the branch for $\alpha > \alpha_1$ is stable for α near α_1; the part of the branch for $\alpha < \alpha_1$ is unstable for α near α_1. Let us prove the first of these statements; if $\alpha > \alpha_1$, then εe_1 with ε small and positive is a lower solution. We shall look for an upper solution of the form Ae_1^*, where e_1^* is defined below (9.4.49). The same calculation gives

$$-\Delta(Ae_1^*) - \alpha g(Ae_1^*) = Ae_1^* \alpha \left[\frac{\alpha_1^*}{\alpha} - \frac{g(Ae_1^*)}{Ae_1^*} \right].$$

By construction α_1^* is only slightly smaller than α_1. The ratio α_1^*/α is therefore only slightly smaller than 1 if α is near α_1. Since $g''(u) < 0$ near $u = 0$, A can be chosen larger than ε, yet sufficiently small so that $g(Ae_1^*)/Ae_1^*$ is smaller than α_1^*/α. One can show that there is a unique positive solution in this alpha range with $u \leqslant Ae_1^*$, so that the iteration schemes starting from εe_1 and Ae_1^* both converge to this positive solution, which must therefore be stable.

Example 3. Hopf Bifurcation. We give a simple example in which the incipient instability of a steady state is accompanied by the appearance of a branch of time-periodic solutions. Consider the pair of nonlinear ordinary differential equations

$$\dot{u}_1 = \lambda u_1 - u_2 - u_1(u_1^2 + u_2^2), \tag{9.5.16a}$$

$$\dot{u}_2 = \lambda u_2 + u_1 - u_2(u_1^2 + u_2^2), \tag{9.5.16b}$$

where the dot indicates differentiation with respect to t. Obviously, $u_1 = u_2 = 0$ is an equilibrium solution for all λ; let us examine its stability. The linearized problem is

$$\left. \begin{aligned} \dot{h}_1 &= \lambda h_1 - h_2 \\ \dot{h}_2 &= \lambda h_2 + h_1 \end{aligned} \right\} \quad \text{or} \quad \dot{h} = Ah. \tag{9.5.17}$$

The spectrum of A consists of the eigenvalues μ_1, μ_2 of the matrix

$$\begin{pmatrix} \lambda & -1 \\ 1 & \lambda \end{pmatrix},$$

from which we conclude that

$$(\lambda - \mu)^2 = -1 \quad \text{or} \quad \mu = \lambda \pm i.$$

Thus, if $\lambda < 0$, the null solution is stable, whereas for $\lambda > 0$ it is unstable. The front of the spectrum (the entire spectrum in this case) crosses the imaginary axis at the conjugate points $\pm i$ at the critical value $\lambda_0 = 0$.

The full nonlinear system can be solved explicitly by introducing polar coordinates $R^2 = u_1^2 + u_2^2$, $\tan\theta = u_2/u_1$. Multiply (9.5.16a) and (9.5.16b) by u_1 and u_2, respectively, and add to obtain

$$\frac{1}{2}\frac{d}{dt}R^2 = \lambda R^2 - R^4. \tag{9.5.18}$$

Multiply (9.5.16b) by u_1 and (9.5.16a) by u_2, and subtract to find

$$\frac{d}{dt}\left(\frac{u_2}{u_1}\right) = 1 + \left(\frac{u_2}{u_1}\right)^2 \quad \text{or} \quad \frac{d}{dt}\tan\theta = 1 + \tan^2\theta.$$

Since straightforward differentiation also gives $(d/dt)\tan\theta = \dot{\theta}(1+\tan^2\theta)$, we conclude that

$$\dot{\theta} = 1. \tag{9.5.19}$$

For $\lambda < 0$ it is clear from (9.5.18) that for every initial condition, R decreases with t and tends to 0 (thus we have global stability of the null solution); and in view of (9.5.19) we proceed at uniform angular velocity along a spiral which converges to the origin. For $\lambda > 0$ the solution of (9.5.18) and (9.5.19) converges to the stable *periodic* solution $R = \sqrt{\lambda}, \dot{\theta} = 1$.

EXERCISES

9.5.1 (a) Consider the IVP

$$\frac{du}{dt} = \lambda u(1-u), \quad u(0) = u_0 \quad \text{with } 0 \leqslant u_0 \leqslant 1.$$

Discuss the stability of the equilibrium solutions $u = 0$ and $u = 1$.

(b) For the elliptic problem

$$-\Delta u = \lambda u(1-u), \quad x \in \Omega; \quad u(\Gamma) = 0,$$

discuss the existence and nonexistence of positive solutions.

(c) For the evolution problem (Fisher's equation)

$$u_t - \Delta u = \lambda u(1-u), \ x \in \Omega, t > 0; \ u(x,0) = u_0(x), \ u(\Gamma, t) = h(x),$$

discuss the stability of the steady-state solutions $u \equiv 0, u \equiv 1$ [corresponding to $h(x) = 0$ and $h(x) = 1$]. Here $0 \leqslant u_0(x) \leqslant 1$.

9.5.2 Consider the system

$$\begin{aligned} \dot{u_1} &= \lambda u_1 - u_2 + u_1(u_1^2 + u_2^2), \\ \dot{u_2} &= \lambda u_2 + u_1 + u_2(u_1^2 + u_2^2). \end{aligned}$$

Discuss possible bifurcations from the equilibrium solution $u_1 = u_2 = 0$. Compare your results with those of Example 3.

9.5.3 *Jensen's inequality*. Let f be convex on $a \leqslant u \leqslant b$. The definition of convexity leads immediately to

$$f(w_1 a + w_2 b) \leqslant w_1 f(a) + w_2 f(b) \quad \text{for all } w_1 \geqslant 0, w_2 \geqslant 0, w_1 + w_2 = 1.$$

(a) Let $a \leqslant u_1 \leqslant u_2 \leqslant \cdots \leqslant u_n \leqslant b$ and let $w_k \geqslant 0,\ k = 1, \ldots, n$, with $w_1 + \cdots + w_n = 1$. Then show that

$$f\left(\sum_{i=1}^{n} w_i u_i\right) \leqslant \sum_{i=1}^{n} w_i f(u_i).$$

(b) Let $w(x)$ be a positive function on Ω with $\int_\Omega w(x)\, dx = 1$. Let $u(x)$ be a function on Ω whose range is in the domain of f. Show that

$$f\left(\int_\Omega w(x) u(x)\, dx\right) \leqslant \int_\Omega w(x) f(u(x))\, dx.$$

9.5.4 For the heat equation on $\mathbb{R}$, show directly that

$$\int_{-\infty}^{\infty} E(x, t; y, 0) E(y, \tau; \xi, 0)\, dy = E(x, t + \tau; \xi, 0),$$

where $E(x, t; \xi, 0) = e^{-(x-\xi)^2/4t}/\sqrt{4\pi t}$. Interpret the result in terms of propagators:

$$S_{t+\tau} = S_t S_\tau.$$

REFERENCES AND ADDITIONAL READING

1. H. Amann, Fixed point equations and nonlinear eigenvalue problems in ordered Banach spaces, *SIAM Rev.*, 18:620, 1976.
2. A. Ambrosetti and G. Prodi, *A Primer of Nonlinear Analysis*, Cambridge University Press, New York, 1993.
3. S. B. Angenent and D. G. Aronson, Optimal asymptotics for solutions to the initial value problem for the porous medium equation, in *Nonlinear Problems in Applied Mathematics* (T. S. Angell, L. Pamela Cook, R. E. Kleinman, and W. E. Olmstead, Eds.), SIAM, Philadelphia, 1996.
4. S. Antman, *Nonlinear Problems of Elasticity*, Springer–Verlag, New York, 1995.
5. R. Aris, *The Mathematical Theory of Diffusion and Reaction in Permeable Catalysts*, Vol. 2, Oxford University Press, New York, 1975.
6. C. Bandle and I. Stakgold, The formation of the dead core in parabolic reaction-diffusion equations, *Trans. Amer. Math. Soc.*, 286, 1984.
7. J. Bebernes and D. Eberly, *Mathematical Problems from Combustion Theory*, Springer–Verlag, New York, 1989.
8. S. Bergman and M. Schiffer, *Kernel Functions and Elliptic Differential Equations in Mathematical Physics*, Academic Press, New York, 1953.

9. F. Bernis, L. A. Peletier, and S. M. Williams, Source type solutions of a fourth-order nonlinear degenerate parabolic equation, *Nonlinear Anal.*, 18, 1992.

10. R. F. Brown, *A Topological Introduction to Nonlinear Analysis*, Birkhäuser, Cambridge, MA, 1993.

11. J. D. Buckmaster and G. S. S. Ludford, *Lectures on Mathematical Combustion*, SIAM, Philadelphia, 1983.

12. S. N. Chow and J. K. Hale, *Methods of Bifurcation Theory*, Springer–Verlag, New York, 1982.

13. D. S. Cohen and H. B. Keller, Some positone problems suggested by nonlinear heat generation, *J. Math. Mech.*, 16:1361, 1967.

14. M. G. Crandall and P. H. Rabinowitz, Bifurcation, perturbation of simple eigenvalues, and linearized stability, *Arch. Ration. Mech. Anal.*, 52:161, 1973.

15. K. Deimling, *Nonlinear Functional Analysis*, Springer–Verlag, New York, 1985.

16. M. C. Delfour, Shape derivatives and differentiability of min-max, in *Shape Optimization and Free Boundaries* (M. C. Delfour, Ed.), Kluwer, Hingham, MA, 1992.

17. J. I. Diaz, *Nonlinear Partial Differential Equations and Free Boundaries*, Vol. I, Pitman, New York, 1985.

18. J. I. Diaz and J. Hernandez, On the existence of a free boundary for a class of reaction diffusion systems, *SIAM J. Math. Anal.*, 15, 1984.

19. P. Drabek and J. Milota, *Methods of Nonlinear Systems: Applications to Differential Equations*, Birkhauser, Cambridge, MA, 2007.

20. P. G. Drazin, *Nonlinear Systems*, Cambridge University Press, New York, 1992.

21. P. Fife, *Mathematical Aspects of Reaction Diffusion Systems*, Springer–Verlag, New York, 1979.

22. J. Franklin, *Methods of Mathematical Economics*, Springer–Verlag, New York, 1980.

23. A. Friedman and M. A. Herrero, Extinction properties of a semilinear elliptic equation, *J. Math. Anal. Appl.*, 124, 1987.

24. A. Friedman and B. McLeod, Blow-up of positive solutions of semilinear heat equations, *Indiana Univ. Math. J.*, 34, 1985.

25. A. Friedman and D. Phillips, The free boundary of a semilinear elliptic equation, *Trans. Amer. Math. Soc.*, 282, 1984.

26. H. Fujita, *Bull. Amer. Math. Soc.*, 75, 1969.

27. Y. Giga and R. V. Kohn, Characterizing blow-up using similarity variables, *Indiana Univ. Math. J.*, 36, 1987.

28. M. Golubitsky and D. G. Schaeffer, *Singularities and Groups in Bifurcation Theory*, Vol. I, Springer–Verlag, New York, 1985.

29. J. Graham-Eagle, On the relation between a nonlinear elliptic equation and its uniform approximation, *J Math. Anal. Appl.*, 177, 1993.

30. P. Grindrod, *Patterns and Waves*, Oxford University Press, New York, 1991.

31. J. Guckenheimer and P. Holmes, *Nonlinear Oscillations, Dynamical Systems, and Bifurcations of Vector Fields*, Springer–Verlag, New York, 1983.

32. J. Hale and H. Koçak, *Dynamics and Bifurcations*, Springer–Verlag, New York, 1991.

33. S. P. Hastings, Some Mathematical Problems from Neurobiology, *Am. Math. Mon.*, 82(9), 1975.

34. F. C. Hoppensteadt and E. M. Izhikevich, *Weakly Connected Neural Networks*, Springer–Verlag, New York, 1997.

35. G. Iooss and D. D. Joseph, *Elementary Stability and Bifurcation Theory*, Springer–Verlag, New York, 1980.

36. D. D. Joseph and R. L. Fosdick, The free surface on a liquid between cylinders rotating at different speeds, part I, *Arch. Ration. Mech. Anal.*, 49:321, 1973.

37. D. D. Joseph and T. S. Lundgren, Quasilinear Dirichlet problems driven by positive sources, *Arch. Ration. Mech. Anal.*, 49:241, 1973.

38. H. B. Keller and J. P. Keener, Perturbed bifurcation theory, *Arch. Ration. Mech. Anal.*, 50:159, 1973.

39. J. B. Keller and S. Antman, *Bifurcation Theory and Nonlinear Eigenvalue Problems*, W. A. Benjamin, New York, 1969.

40. S. Kesavan, *Topics in Functional Analysis and Applications*, Wiley, New York, 1989.

41. W. T. Koiter, Elastic stability and post-buckling behavior, in *Nonlinear Problems* (R. E. Langer, Ed.), University of Wisconsin Press, Madison, WI, 1963.

42. A. A. Lacey and G. C. Wake, Critical initial conditions for spatially-distributed thermal explosions, *J. Austral. Math. Soc., Ser. B.*, 33, 1992.

43. A. C. Lazer and P. J. McKenna, On the number of solutions of a nonlinear Dirichlet problem, *J. Math. Anal. Appl.*, 84, 1981.

44. H. Levine, The role of critical exponents in blow-up theorems, *SIAM Rev.*, 32, 1990.

45. J. D. Logan, *Introduction to Nonlinear Partial Differential Equations*, Wiley, New York, 1997.

46. J. E. Marsden and M. J. Hoffman, *Elementary Classical Analysis*, Freeman, New York, 1993.

47. D. Mitrovic and D. Zubrinic, *Fundamentals of Applied Functional Analysis*, Pitman Monographs and Surveys in Pure and Applied Mathematics, Wiley, New York, 1998.

48. C. V. Pao, *Nonlinear Parabolic and Elliptic Equations*, Plenum, New York, 1992.

49. T. Poston and I. Stewart, *Catastrophe Theory and Its Application*, Pitman, New York, 1978.

50. W. Rudin, *Principles of Mathematical Analysis*, McGraw-Hill, New York, 1976.

51. D. H. Sattinger, *Topics in Stability and Bifurcation Theory*, Lecture Notes in Mathematics 309, Springer-Verlag, New York, 1973.

52. R. Seydel, *From Equilibrium to Chaos*, Elsevier, New York, 1988.

53. J. Smoller, *Shock Waves and Reaction-Diffusion Equations*, Springer–Verlag, New York, 1983.

54. I. Stakgold, Branching of solutions of nonlinear equations, *SIAM Rev.*, 13:289, 1971.

55. I. Stakgold, Diffusion with strong absorption, in *Shape Optimization and Free Boundaries* (M. C. Delfour, Ed.), Kluwer, Hingham, MA, 1992.

56. I. Stakgold, D. D. Joseph, and D. H. Sattinger, *Nonlinear problems in the physical sciences and biology*, Lecture Notes in Mathematics Vol. 322, Springer–Verlag, New York, 1973.

57. I. Stakgold, C. Van der Mee, and S. Vernier-Piro, Traveling waves for gas-solid reactions in porous medium with porosity change, *Dynam. Systems Appl.*, 10(4):589–598, 2001.

58. M. Taylor, *Partial Differential Equations*, III: *Nonlinear Equations*, Springer–Verlag, New York, 1996.

59. J. L. Troutman, *Variational Calculus with Elementary Convexity*, Springer–Verlag, New York, 1983.

60. J. L. Vazquez and L. Véron, Different kinds of singular solutions of nonlinear parabolic equations, in *Nonlinear Problems in Applied Mathematics* (T. S. Angell, L. Pamela Cook, R. E. Kleinman, and W. E. Olmstead, Eds.), SIAM, Philadelphia, 1996.

61. F. B. Weissler, Existence and nonexistence of global solutions for a semilinear heat equation, *Israel J. Math.*, 38, 1981.

62. K. Yosida, *Functional Analysis*, Springer–Verlag, New York, 1980.

63. E. Zeidler, *Nonlinear Functional Analysis and its Applications*, Vol. I, II/A, II/B, III, and IV, Springer–Verlag, New York, 1991.

CHAPTER 10

APPROXIMATION THEORY AND METHODS

In this chapter we develop some additional concepts and tools of nonlinear analysis in Banach spaces and introduce the basic techniques and results of approximation theory in Banach spaces. We also provide a concise summary of the relevant properties of Sobolev and Besov classes of functions as needed for the analysis and approximation of solutions to boundary value problems, and briefly survey some of the main discretization and iterative solution techniques in computational mathematics for constructing and analyzing approximate solutions to linear and nonlinear boundary value problems. This set of topics lies at the intersection of several distinct areas of mathematics, reflecting the surprisingly broad set of mathematical techniques required to understand and make effective use of modern computational methods.

Following the direction of Chapter 9, our goal will be to develop tools needed for analyzing and approximating solutions to nonlinear elliptic equations and systems of such equations, with necessary supporting results for linear equations developed along the way. In keeping with the overall theme of the book, our focus is primarily on the basic mathematical foundational ideas rather than on method description. More to the point: Rather than provide a short survey of many numerical techniques for boundary value problems (which is done more carefully and thoroughly in a number of existing outstanding graduate texts in numerical analysis), our goal here was,

Green's Functions and Boundary Value Problems, Third Edition. By I. Stakgold and M. Holst

instead, to provide the "missing" last chapter that one often wishes were found in either the graduate applied mathematics book or in the graduate numerical analysis book, bridging the two fields with a combination of functional analysis, approximation theory, function space theory, applied harmonic analysis, nonlinear analysis, and (nonlinear) partial differential equations. As only one chapter in an already quite extensive book, we do not have the space for a complete treatment of many of the relevant topics, as can be found for example in [9, 113, 132, 135] and other references in the general area of applied functional analysis. Therefore, our intention is to present a fairly small number of key related topics from these areas that are not easily found in one place, with an emphasis on nonlinear problems. Our experience is that the topics we have selected are among the collection of critical mathematical tools that one finds increasingly necessary to have on hand for doing modern applied analysis and numerical analysis of nonlinear boundary value problems in partial differential equations.

Some of the key *applied analysis* topics to be found here are: further concepts and techniques from linear and nonlinear functional analysis (extending the discussions from Chapters 4 and 5); abstract approximation theory; best and near-best approximation in Banach spaces; a concise survey of Sobolev and Besov spaces and their main properties; and a collection of examples of linear and nonlinear partial differential equations of increasing complexity, illustrating the use of maximum principles, *a priori* energy and L^∞ estimates, ordered Banach space concepts, variational methods, and topological fixed point theorems. Some of the key *numerical analysis* topics to be found here include the basic ideas of finite element construction and approximation theory, fast iterative solvers for discretized linear elliptic equations, and an introduction to methods for solving nonlinear problems.

We note that there are a number of outstanding graduate textbooks and monographs on each of the topics of applied analysis, nonlinear partial differential equations, nonlinear analysis, approximation theory, finite element methods, iterative solution methods, adaptive methods, methods for nonlinear equations, and theoretical numerical analysis. Our approach is to be consistent with previous chapters of the book, listing a fairly small set of central and general books and monographs as references rather than an exhaustive set of specialized literature, and we have tried to list mainly books rather than journal articles except when necessary. We have certainly left out many references that could be (and perhaps should be) in the list.

We now give a more detailed outline of the chapter. In Section 10.1 we briefly survey some standard concepts about nonlinear operators in Banach spaces, including Bochner integration, Gâteaux and Fréchet differentiation, and generalized Taylor expansion of nonlinear maps on general Banach spaces. We also review some topological concepts needed to understand the techniques of nonlinear analysis and approximation theory, such as subspaces, closure of sets, convexity properties of sets and spaces, reflexive spaces, and weakly closed sets. Building on these concepts, we give a brief overview of the main results in variational methods (the calculus of variations). We then give a quick summary of topological fixed point theorems in Banach spaces, followed by a discussion of the Local and Global Inverse Function Theorems along with the Implicit Function Theorem. We finish the section with a brief tutorial on ordered Banach spaces. Our goal is to assemble all of the basic results we need in order to develop the theory of best approximation in Banach spaces in Section 10.2, to understand the basic properties of two particular families of Banach spaces (Sobolev and Besov spaces) studied in Section 10.3, to have a toolkit for analyzing a sequence of linear and nonlinear problems in Section 10.4, and to broadly support the material in the second half of the chapter that focuses primarily on numerical analysis topics (Sections 10.5 to 10.7), as well as to provide some additional analysis foundations for earlier chapters in the book.

In Section 10.2 we develop somewhat more completely the abstract theory of best and near-best linear approximation in Banach and Hilbert spaces encountered earlier in Chapter 4. Our goal is to construct a Banach space version of the best approximation result provided by the Projection Theorem in the case of a Hilbert space. We then develop Petrov-Galerkin methods together with *a priori* error estimates for near-best linear approximation of linear problems, and extend the estimates to a general class of nonlinear problems. We then develop similar *a priori* error estimates for near-best approximation using Galerkin methods applied to several classes of linear problems, as well as to a class of semilinear problems.

Following this, in Section 10.3 we take a closer look at the Sobolev spaces that were briefly encountered in several places earlier in the book (Section 2.6, Example 6 in Section 4.5, and in Section 8.4), and on our way to making sense of fractional order Sobolev spaces, we introduce Besov spaces. These two families of Banach spaces play a central role in the analysis of linear and nonlinear partial differential equations, and also in the modern numerical analysis of finite element and related methods for partial differential equations and integral equations. Although the material in Chapter 8 covering partial differential equations involved only some very basic notions of the (Hilbert) Sobolev space $H^1(\Omega)$, in order to analyze solutions and develop modern approximation techniques for nonlinear partial differential equations, it is important to develop some familiarity with the larger class of (Banach) Sobolev spaces $W^{s,p}(\Omega)$.

In Section 10.4 we subsequently work through three detailed examples from elliptic partial differential equations, where we apply the techniques from the first three sections. In particular, we examine:

(1) A linear coercive problem.
(2) A nonlinear variational problem arising in biophysics.
(3) A nonlinear nonvariational problem arising in astrophysics.

In each case we establish the following three types of results, each of which must be in place in order to establish the result appearing next on the list, with our overall goal being the final result on the list (a near-best approximation result):

(1) *A priori* estimates for any possible solution (energy and/or L^∞ estimates).
(2) Existence of such solutions, uniqueness when possible.
(3) Near-best approximation of such solution(s) in a subspace.

The linear problem involves a fairly straightforward application of the techniques from the first three sections, whereas both of the nonlinear problems have distinct degenerate features, requiring a somewhat more sophisticated use of the techniques from the first three sections. The first nonlinear problem allows for the use of a fairly standard argument involving variational methods, whereas the second nonlinear problem requires the use of more technically sophisticated and delicate fixed point arguments.

In Section 10.5 we then give an overview of approximation methods, focusing mainly on finite element and finite volume methods, including discussions of practical formulation and assembly of linear systems, and *a priori* and *a posteriori* error estimates using our abstract Petrov-Galerkin and Galerkin error estimate frameworks from Section 10.2. We also exploit our new working knowledge of Sobolev and Besov spaces from Section 10.3 to analyze finite element interpolants as a critical component of the error analysis. We briefly discuss finite volume and related methods as types of Petrov-Galerkin methods.

In Section 10.6 we examine closely the problem of fast (optimal or nearly optimal) iterative methods for solving the linear algebraic systems of equations that arise from any of the discretization techniques described in Section 10.5. We give an overview of the theory of stationary iterative methods and the conjugate gradient method for symmetric positive definite problems, and then focus on multigrid and domain decomposition methods. Our goal is again not to give a detailed prescription of how to use these methods for specific boundary value problems (this is done more completely in numerical analysis books), but rather to provide a mathematical foundation for the study of multigrid and domain decomposition-type methods, and give some insight as to how and why these two classes of methods have optimal (linear) or nearly optimal space and time complexity in a very general setting.

In Section 10.7 we consider approximation methods for nonlinear problems based on fixed point and variational techniques, including Newton-type methods and nonlinear multilevel methods. Our focus is on the basic theory of global inexact Newton-type methods and how these are combined with fast iterative methods for linear systems. This is a vast topic, and we only have space to scratch the surface.

10.1 NONLINEAR ANALYSIS TOOLS FOR BANACH SPACES

Let X be a Banach space, which we recall (Chapter 4) means several things: X is a linear (vector) space (algebraically closed) built from a set of vectors X and an associated field of scalars $\mathbb{K}$; X is equipped with a norm satisfying the three norm axioms (see Chapter 4); and X is complete (topologically closed) with respect to this norm. Our interest throughout this section will be primarily a *real* Banach space X, so we will assume that the scalar field $\mathbb{K}$ associated with the vector space structure of X is in fact $\mathbb{R}$ rather than $\mathbb{C}$. Recall that it is convenient when working with sequences in X to use $\mathbb{Z}$ to refer to the (countably infinite) set of all integers, to use $\mathbb{N} = \mathbb{N}_+$ to refer to the subset of positive integers (sometimes called the *natural numbers*), and to use $\mathbb{N}_0$ to refer to the subset of nonnegative integers ($\mathbb{N}$ with zero added to the set).

Recall that when a map $A\colon X \to Y$ is linear, we denote this as $A \in L(X,Y)$. When $A\colon X \to Y$ is both linear and continuous (equivalent to being bounded when A is linear), we use the notation $A \in \mathcal{L}(X,Y)$. The space of bounded linear operators $\mathcal{L}(X,Y)$ is itself a vector space (it is algebraically closed with respect to linear combinations involving the scalar field compatible with the vector space structures of X and Y), and is a normed space with respect to the induced operator norm

$$\|A\|_{\mathcal{L}(X,Y)} = \sup_{0\neq u\in X} \frac{\|Au\|_Y}{\|u\|_X}.$$

A useful fact is that if Y is a Banach space, then so is $\mathcal{L}(X,Y)$, even if X is only a normed space (not complete); this is because Cauchy sequences of operators in $\mathcal{L}(X,Y)$ converge to an element in $\mathcal{L}(X,Y)$ if Cauchy sequences in Y converge to an element in Y. This immediately implies that the space of bounded linear functionals $\mathcal{L}(X,\mathbb{R})$ on a normed vector space is always a Banach space, even when X is not, since the range space $Y = \mathbb{R}$ is complete under the norm $|\cdot|$.

The Banach space of bounded linear functionals on a given normed space X, which we will refer to as the *(topological) dual space of* X and denote as X', is a key object for much of the discussion throughout this chapter. The dual space X' of a normed space X was introduced and studied in some detail in Section 4.8, and as remarked above, it is a Banach space with respect to the induced dual norm

$$\|f\|_{X'} = \sup_{0\neq u\in X} \frac{|f(u)|}{\|u\|_X}.$$

This definition implies the inequality for any member of the dual space

$$|f(u)| \leqslant \|f\|_{X'}\|u\|_X, \qquad \forall u \in X.$$

Let X and Y be Banach spaces, and let X' and Y' be their respective dual spaces. Given a (generally nonlinear) map $F\colon X \to Y'$, we are interested in the following general problem throughout most of this chapter:

$$\text{Find } u \in X \text{ such that } F(u) = 0 \in Y'. \tag{10.1.1}$$

Our definition of dual spaces implies that this problem is equivalent to:

$$\text{Find } u \in X \text{ such that } \langle F(u), v\rangle = 0 \in \mathbb{R}, \qquad \forall v \in Y, \tag{10.1.2}$$

where $\langle F(u), \cdot\rangle$ denotes the action of $F(u) \in Y'$ as a linear functional on Y. We wish both to develop techniques for analyzing solutions to (10.1.1) and to develop approximation theory and (practical) approximation methods for (10.1.1).

To this end, we start with a review of a number of concepts which have been covered in a slightly less formal way in Chapter 9 (in particular, Section 9.1), as well as being touched on in various chapters earlier in the book). Concerning the solution of the operator equation $F(u) = 0$, it is important that the problem be *well-posed*, as introduced in Section 1.2 (sometimes referred to as *well-posed in the Hadamard sense*; see [113]). We say that the problem $F(u) = 0$ is *well-posed* if there is (1) existence, (2) uniqueness, and (3) continuous dependence of the solution on the data of the problem. Recall now that if F is both one-to-one (*injective*) and onto (*surjective*), it is called a *bijection*, in which case the inverse mapping F^{-1} exists, and we would have both existence and uniqueness of solutions to the problem $F(u) = 0$. The queston of continuous dependence of the solution on the data would require more: We would also need *continuity* of F^{-1}.

In the case of linear operators, it is easy to show that boundedness is equivalent to continuity, and that all linear operators on finite-dimensional spaces are bounded. However, this is not true in the general case of nonlinear operators, and a separate notion of continuity is required. A mapping $F\colon D \subset X \to Y$ from a normed space X to a normed space Y is called *continuous* at $u \in D$ if, given $\epsilon > 0$, there exists $\delta = \delta(u, \epsilon) > 0$ such that if $v \in D$ and $\|u - v\|_X < \delta$, then $\|F(u) - F(v)\|_Y < \epsilon$. If F is continuous at each $u \in D$ then F is called continuous on D, and further, if $\delta = \delta(\epsilon)$ then F is called *uniformly continuous* on D.

An equivalent definition of continuity is the more generalizable version based on sequences: We say that $F\colon X \to Y$ is a continuous map from the normed space X to the normed space Y (more generally, they can be metric spaces, or even simply topological spaces) if $\lim_{j\to\infty} u_j = u$ implies that $\lim_{j\to\infty} F(u_j) = F(u)$, where $\{u_j\}$ is a sequence, $u_j \in X$. If both F and F^{-1} are continuous, then F is called a *homeomorphism*. If both F and F^{-1} are differentiable (see the section below for the definition of differentiation of abstract operators in Banach spaces), then F is called a *diffeomorphism*. If both F and F^{-1} are k-times continuously differentiable, then F is called a C^k-*diffeomorphism*. A linear map between two vector spaces is a type of *homomorphism* (structure-preserving map); a linear bijection is called an *isomorphism*.

The following notions of continuity of maps between two Banach spaces are important. They have been encountered previously in Section 4.4 and in Chapter 9; we bring them together here in one place for clarity.

Definition 10.1.1. *Let X and Y be Banach spaces. The mapping $F\colon D \subset X \to Y$ is called* Hölder-continuous *on D with constant γ and exponent p if there exists $\gamma \geqslant 0$ and $p \in (0, 1]$ such that*

$$\|F(u) - F(v)\|_Y \leqslant \gamma \|u - v\|_X^p \quad \forall u, v \in D \subset X.$$

If $p = 1$, then F is called uniformly Lipschitz-continuous *on D, with* Lipschitz constant *γ. If $p = 1$ and $\gamma \in [0, 1)$, then F is called a* contraction *on D.*

Let X and Y be Banach spaces. Recall from Chapter 5 that a *compact operator* $T\colon X \to Y$ is an operator that maps bounded sets in X to relatively compact sets in Y. In other words, for any sequence $\{u_j\} \in X$ such that $\|u_j\|_X < M$ for some finite constant M, the sequence $\{T(u_j)\} \in Y$ contains a subsequence which converges in Y. (Although the discussion in Chapter 5 was restricted to linear operators, this definition also makes sense for a nonlinear operator $T\colon X \to Y$.) Recall that T is *continuous* if $u_j \to u$ as $j \to \infty$ implies that $T(u_j) \to T(u)$ as $j \to \infty$. A related property of a continuous operator is the following: We say that T is *completely continuous* if $u_j \rightharpoonup u$ as $j \to \infty$ implies that $T(u_j) \to T(u)$ as $j \to \infty$. Some useful facts about compact, continuous, and completely continuous operators are:

(1) A compact operator T on a Banach space X is continuous.
(2) A continuous operator T on a Banach space X is completely continuous.
(3) If X is reflexive, then a completely continuous operator T is compact.
(4) Compositions of continuous and compact operators are compact.

Calculus in Banach Spaces

We will need some concepts such as integration, differentiation, and Taylor expansion of general nonlinear maps in Banach spaces, which we now briefly assemble. (See, for example, [113, 188] for more complete discussions.)

Integration in Banach Spaces. Let X be a Banach space and let X' denote its dual space. Consider now the continuous function $u\colon [0,1] \to X$, which we denote by $u \in C^0([0,1];X)$. We would like to extend the notion of integration on the interval $[0,1]$ to Banach space-valued functions such as u, in order to make sense of remainder terms in Taylor expansions of u, as well as for other applications. We will make use of the dual space X' for this purpose.

Definition 10.1.2 (Integration of Banach Space-Valued Functions). *Let X be a Banach space and let $u\colon [0,1] \to X$. If there exists $z \in X$ such that*

$$f(z) = \int_0^1 f(u(t))\, dt, \quad \forall f \in X', \tag{10.1.3}$$

then we define $z \in X$ to be the integral of $u\colon [0,1] \to X$ over the interval $[0,1]$, and write

$$z = \int_0^1 u(t)\, dt. \tag{10.1.4}$$

This definition seems reasonable, since $z \in X$ is determined using only integration of a real-valued function on $[0,1]$, thanks to the action of the dual space. However, we do not actually know if such a $z \in X$ exists satisfying (10.1.3), and whether or not it is unique. The answer for functions $u \in C^0([0,1];X)$ is provided by the following theorem.

Theorem 10.1.1. *Let X be a Banach space, and let $u \in C^0([0,1];X)$. Then there exists a unique $z \in X$ satisfying* (10.1.3). *Moreover,*

$$\|z\|_X = \left\| \int_0^1 u(t)\, dt \right\|_X \leqslant \int_0^1 \|u(t)\|_X\, dt.$$

Furthermore, if $A\colon X \to Y$ is a bounded linear operator from X to a second Banach space Y, then

$$Az = A\left(\int_0^1 u(t)\, dt \right) = \int_0^1 A(u(t))\, dt.$$

Proof. A proof may be found in [113] or [188]. Due to the assumption that the function $u \in C^0([0,1];X)$, the proof relies only on properties of the Riemann integral of real-valued functions on $[0,1]$. □

One can extend the definition of integration to maps $u\colon [a,b] \to X$ in the obvious way, including the case of $a = -\infty$ and/or $b = \infty$ (through the usual limit approach). The usual properties of integration will follow from the same properties of integration on the real line thanks to (10.1.3), such as

$$\int_a^b u(t)\,dt = \int_a^\xi u(t)\,dt + \int_\xi^b u(t)\,dt, \qquad a \leqslant \xi \leqslant b.$$

Finally, we note that the discussion above can be generalized to a class of Banach space-valued functions larger than $C^0([a,b];X)$. In particular, one can define the *Bochner integral* of $u\colon [a,b] \to X$ by interpreting (10.1.3) as the Lebesgue integral on $[a,b]$, and one can show (see [2]) that the analogue of Theorem 10.1.1 holds, as well as the useful familiar properties of the integral. Moreover, the value of the integral is the same as that given in Definition 10.1.2 when both integrals exist. The set of all such functions $u\colon [a,b] \to X$ which are Lebesgue integrable on $[a,b]$ is called the set of *Bochner-integrable functions.*

We can now define the *Bochner spaces* $L^p([a,b];X)$ as generalizations of the spaces $L^p([a,b])$ that we encountered in Chapter 0 and elsewhere in the book. These are normed vector spaces of Bochner integrable functions, with norms defined as

$$\|u\|_{L^p([a,b];X)} = \left(\int_a^b \|u(t)\|_X^p\,dt\right)^{1/p}, \qquad 1 \leqslant p < \infty,$$

$$\|u\|_{L^\infty([a,b];X)} = \operatorname*{ess\,sup}_{t\in(a,b)} \|u(t)\|_X.$$

These can be shown to be Banach spaces for $1 \leqslant p \leqslant \infty$. (See Section 10.3 for an explanation of the *ess sup*, or *essential supremum*, of a set.) One can continue onward and define weak (or distributional) derivatives of Bochner integrable functions with respect to the parameter t by following exactly the same procedure that we used in Section 2.6. This will lead to Sobolev-type classes of Banach space-valued functions (see [2, 70]). Among the most useful examples of Bochner spaces are the solution spaces for time-dependent partial differential equations, where $t \in [a,b]$ represents the time variable and $X = W^{s,p}(\Omega)$ is a Sobolev space defined over a spatial domain $\Omega \subset \mathbb{R}^n$ [see the section on the Sobolev spaces $W^{s,p}(\Omega)$ later in the chapter].

Differentiation in Banach Spaces. The following notions of differentiation of nonlinear operators on Banach spaces are important. They have been previously encountered in Section 9.1 and earlier in the book in a less formal way. We first introduce the following notation for the points in X that can be reached by a convex combination of u and v, where $u, v \in X$, which represents the line of points between u and v:

$$[u,v] = \{(1-\alpha)u + \alpha v \in X,\ \alpha \in [0,1]\}.$$

As a perturbation from u to $v = u + h$, we can write this as

$$[u, u+h] = \{(1-\alpha)u + \alpha(u+h) = u + \alpha h \in X,\ \alpha \in [0,1]\}.$$

Definition 10.1.3. *Let X and Y be Banach spaces, let $F\colon X \to Y$, let $D \subset X$ be an open set, and let $[u, u+h] \subset D$. Then the* Gâteaux- *or* **G**-variation *of F at $u \in D$ in the direction $h \in D$, if it exists, is defined as*

$$F'(u)(h) = \frac{d}{dt} F(u+th)\bigg|_{t=0}. \tag{10.1.5}$$

If the Gâteaux variation produces a bounded linear map at $u \in X$, then the map $F'\colon X \to \mathcal{L}(X,Y)$ acts like a derivative operator.

Definition 10.1.4. *Let X and Y be Banach spaces, let $F\colon X \to Y$, let $D \subset X$ be an open set, and let $[u, u+h] \subset D$. Then the map F is called* Gâteaux- *or* **G**-differentiable *at $u \in D$ in the direction $h \in D$ if there exists a bounded linear operator $F'(u)\colon X \to Y$ such that*

$$\lim_{t\to 0} \frac{1}{t} \|F(u+th) - F(u) - tF'(u)(h)\|_Y = 0.$$

If F is **G**-differentiable for all directions $h \in D$, then F is called **G**-*differentiable* at $u \in D$, and the linear operator $F'(u)$ is called the **G**-*derivative* of F at u. The **G**-derivative can be shown to be unique (see [113]). The *directional derivative* of F at u in the direction h is given by $F'(u)(h)$.

The following stronger notion of differentiation of maps on Banach spaces will often be required throughout the chapter.

Definition 10.1.5. *Let X and Y be Banach spaces, let $F\colon X \to Y$, and let $D \subset X$ be an open set. Then the map F is called* Fréchet- *or* **F**-differentiable *at $u \in D$ if there exists a bounded linear operator $F'(u)\colon X \to Y$ such that:*

$$\lim_{\|h\|_X \to 0} \frac{1}{\|h\|_X} \|F(u+h) - F(x) - F'(u)(h)\|_Y = 0.$$

The bounded linear operator $F'(u)$ is called the **F**-*derivative* of F at u. The **F**-*derivative* of F at u can again be shown to be unique. If the **F**-derivative of F exists at all points $u \in D$, then we say that F is **F**-differentiable on D. If in fact $D = X$, then we simply say that F is **F**-differentiable, and the derivative $F'(\cdot)$ defines a map from X into the space of bounded linear maps, $F'\colon X \to \mathcal{L}(X,Y)$. In this case, we say that $F \in C^1(X;Y)$.

It is clear from the definitions above that the existence of the **F**-derivative implies the existence of the **G**-derivative, in which case it is easy to show that they are identical; otherwise, the **G**-derivative can exist more generally than the **F**-derivative. Similarly, existence of the **G**-derivative implies existence of the **G**-variation. In this case, all three objects are identical. In any case, the formula for computing the variation (10.1.5) is usually the most convenient way to calculate the derivative. Many of the properties of the derivative of smooth functions over domains in $\mathbb{R}^n$ carry over to this abstract setting, including the chain rule: If X, Y, and Z are Banach spaces, and if the maps $F\colon X \to Y$ and $G\colon Y \to Z$ are differentiable, then the derivative of the composition map $H = G \circ F$ also exists, and takes the form

$$H'(u) = (G \circ F)'(u) = G'(F(u)) \circ F'(u),$$

where $H'\colon X \to \mathcal{L}(X, Z)$, $F'\colon X \to \mathcal{L}(X, Y)$, and $G'\colon Y \to \mathcal{L}(Y, Z)$.

Higher order Fréchet (and Gâteaux) derivatives can be defined in the obvious way. We saw that if $F\colon X \to Y$, then $F'\colon X \to \mathcal{L}(X, Y)$, which is a map $F'(u)(\cdot)$ that takes X into the space of bounded linear maps from X into Y. The second derivative will take the form of a map $F''\colon X \to \mathcal{L}(X, \mathcal{L}(X, Y)) = \mathcal{L}(\mathcal{L}(X, X), Y)$, which is a map $F''(u)(v_1, v_2)$ that takes X into the space of bounded bilinear maps (which can be shown to be symmetric) from $X \times X$ into Y. Similarly, the k-th derivative has the form $F^{(k)}\colon X \to \mathcal{L}(\mathcal{L}(X, \dots, X), Y)$, which is a map $F^{(k)}(u)(v_1, \dots, v_k)$ that takes X into the space of bounded k-linear maps from $X \times \cdots \times X$ into Y. If F is k-times **F**-differentiable on all of X, and the map $F^{(k)}\colon X \to \mathcal{L}(\mathcal{L}(X, \dots, X), Y)$ is a continuous map into the space of bounded k-linear maps from $X \times \cdots \times X$ to Y, then in the usual way we say that $F \in C^k(X; Y)$. (See [1, 58, 108] for discussions of general multilinear maps and differentiation.)

REMARK. In the case that $X = \mathbb{R}^n$ and $Y = \mathbb{R}^m$, then (see also Section 9.1) the matrix of all directional derivatives of $F\colon D \subset \mathbb{R}^n \to \mathbb{R}^m$ taken in the coordinate directions is called the *Jacobian matrix*:

$$F'(\mathbf{x}) = \nabla F(\mathbf{x})^T = \left[\frac{\partial F_i(\mathbf{x})}{\partial x_j}\right],$$

where $F(\mathbf{x}) = (F_1(\mathbf{x}), \dots, F_m(\mathbf{x}))^T$ and $\mathbf{x} = (x_1, \dots, x_n)^T$. If $Y = \mathbb{R}^1$, so that $F\colon D \subset \mathbb{R}^n \to \mathbb{R}$ is a linear functional, then $F'(\mathbf{x})$ is the usual *gradient* vector. It is clear from the definitions that the existence of the **G**-derivative implies the existence of the Jacobian matrix (the existence of all partial derivatives of F). In the case that the **F**-derivative exists, the Jacobian matrix is the representation of both the **F**- and **G**-derivatives.

Taylor's Theorem in Banach Spaces. We finish up the discussion on calculus of nonlinear maps in Banach spaces with a result on generalized Taylor expansion that we will need for establishing best approximation-type results for nonlinear equations. We will make use of our earlier definition of the integral of a Banach space-valued function in order to make sense of the Taylor remainder.

Theorem 10.1.2 (Taylor's Theorem in Banach Spaces)**.** *Let X and Y be Banach spaces, and let $F\colon X \to Y$ be a C^k map in an open set $D \subset X$, and let the interval $[u, u + h] \subset D$. Then there exists the Taylor expansion*

$$F(u + h) = F(u) + F'(u)(h) + F''(u)(h, h) + \cdots + F^{(k)}(u)(h)^k + R_{k+1},$$

where $(h)^k = (h, \dots, h)$ *denotes the argument* h *appearing* k *times in the* k*-linear form* $F^{(k)}(u)(h)^k$, *and where the Taylor remainder term* R_{k+1} *has the form:*

$$R_{k+1} = \int_0^1 \frac{(1-t)^k}{k!} F^{(k+1)}(u+th)(h)^{(k+1)}\, dt.$$

Proof. See [188]. □

We have also the following mean value formulas (effectively the Fundamental Theorem of Calculus in Banach spaces), which can be viewed as Theorem 10.1.1 with the lowest-order remainder.

Theorem 10.1.3 (Mean Value Formulas in Banach Spaces)**.** *Let* X *and* Y *be Banach spaces, and let* $F\colon X \to Y$ *be a* C^1 *map in an open set* $D \subset X$, *let* $[u, u+h] \subset D$, *and denote* $v = u + h$. *Then the following hold:*

$$F(u+h) - F(u) = \int_0^1 F'(u+th)h\, dt, \tag{10.1.6}$$

$$F(u+h) - F(u) = F'(u)h + \int_0^1 [F'(u+th) - F'(u)]h\, dt, \tag{10.1.7}$$

$$F(u) - F(v) = \int_0^1 F'(tu + [1-t]v)(u-v)\, dt, \tag{10.1.8}$$

$$\|F(u+h) - F(u)\|_Y \leqslant \sup_{t\in[0,1]} \|F'(u+th)\|_{\mathcal{L}(X,Y)} \|h\|_X, \tag{10.1.9}$$

$$\|F(u) - F(v)\|_Y \leqslant \sup_{t\in[0,1]} \|F'(tu + [1-t]v)\|_{\mathcal{L}(X,Y)} \|u-v\|_X. \tag{10.1.10}$$

Proof. The expression on the right in (10.1.6) is the remainder term arising in Theorem 10.1.2 for $k = 0$. The identity (10.1.7) follows from (10.1.6) simply by adding and substracting $F'(u)h$ from the right side of (10.1.6). Making the substitution $h = v - u$ in (10.1.6) and swapping the roles of u and v gives (10.1.8). The result (10.1.9) follows by taking norms of both sides of (10.1.6) and employing Theorem 10.1.1. The result (10.1.10) follows in a similar way from (10.1.8). □

Note that the two theorems above give meaning to the following representations of Taylor expansion in a Banach space:

$$F(u+h) = F(u) + F'(u)(h) + \mathcal{O}(\|h\|_X^2),$$
$$F(u+h) = F(u) + F'(u)(h) + \frac{1}{2}F''(u)(h,h) + \mathcal{O}(\|h\|_X^3).$$

The following result is one of two key tools that will be needed in Section 10.7 for the analysis of Newton's Method for solving operator equations in Banach spaces.

Lemma 10.1.1 (Linear Approximation)**.** *Let* X *and* Y *be Banach spaces, let* $D \subset X$ *be an open set, let* $F \in C^1(D; Y)$, *and let* $[u, v] \subset D$. *Assume that* F' *is Lipschitz in* D:

$$\|F'(u) - F'(v)\|_{\mathcal{L}(X,Y)} \leqslant \gamma \|u - v\|_X, \quad \forall u, v \in D, \quad \gamma > 0. \tag{10.1.11}$$

Then

$$\|F(v) - [F(u) + F'(u)(v-u)]\|_X \leqslant \frac{\gamma}{2}\|u-v\|_X^2, \quad \forall u, v \in D.$$

Proof. We begin with the remainder formula (10.1.7) from Theorem 10.1.3, taking $h = v - u$ and swapping the roles of u and v, giving

$$F(u) - [F(v) + F'(u)(u-v)] = \int_0^1 [F'(tu + (1-t)v) - F'(u)](u-v)\, dt.$$

We take norms of both sides and use the assumption (10.1.11) to give

$$\begin{aligned} \|F(u) - [F(v) + F'(u)(u-v)]\|_Y &\leqslant \int_0^1 \|F'(tu + (1-t)v) - F'(u)\|_{\mathcal{L}(X,Y)}\, dt \\ &\qquad \cdot \|u-v\|_X \\ &\leqslant \left[\int_0^1 (1-t)\, dt\right] \gamma \|u-v\|_X^2 \\ &= \frac{\gamma}{2}\|u-v\|_X^2. \end{aligned}$$

□

The second key tool we will need later involves using continuity to extend a bounded inverse assumption at a point to an open ball around that point.

Lemma 10.1.2 (Inverse Perturbation Lemma). *Let X be a Banach space and let $F\colon D \subset X \to X$ for an open set $D \subset X$. Assume that there exists an open convex set $U \subset D$ such that $F \in C^1(U; X)$, and a point $u_0 \in U$ such that*

(1) $\|[F'(u_0)]^{-1}\|_{\mathcal{L}(X,X)} \leqslant \beta < \infty$.

(2) $\|F'(u) - F'(v)\|_{\mathcal{L}(X,X)} \leqslant \gamma\|u-v\|_X, \quad \forall u, v \in U \subset D, \quad \gamma > 0.$

Then there exists $\theta \in (0,1)$ sufficiently small so that the open ball $B_\rho(u_0)$ about u_0 of radius $\rho = \theta/[\beta\gamma]$ is contained in U, and

$$\|[F'(u)]^{-1}\|_{\mathcal{L}(X,X)} \leqslant \frac{\beta}{1-\theta} < \infty, \quad \forall u \in B_\rho(u_0) \subset U.$$

Proof. Since U is open and since $u_0 \in U$, it is always possible to find $\theta \in (0,1)$ sufficiently small so that $B_\rho(u_0) \subset U$, where $\rho = \theta/[\beta\gamma]$, with β and γ given and fixed. We now turn to the simple identity

$$F'(u) = F'(u_0)\left\{I + [F'(u_0)]^{-1}[F'(u) - F'(u_0)]\right\}, \tag{10.1.12}$$

which holds for any $u \in B_\rho(u_0)$. This has the form

$$A = B[I - C],$$

where $A = F'(u)$, $B = F'(u_0)$, and $C = [F'(u_0)]^{-1}[F'(u_0) - F'(u)]$. The assumptions (1) and (2) ensure that

$$\begin{aligned}\|C\|_{\mathcal{L}(X,X)} &= \|[F'(u_0)]^{-1}[F'(u_0) - F'(u)]\|_{\mathcal{L}(X,X)} \\ &\leqslant \|[F'(u_0)]^{-1}\|_{\mathcal{L}(X,X)} \, \|[F'(u_0) - F'(u)]\|_{\mathcal{L}(X,X)} \\ &\leqslant \beta\gamma\|u_0 - u\|_X \\ &\leqslant \beta\gamma\rho \\ &= \theta < 1, \quad \forall u \in B_\rho(u) \subset U.\end{aligned}$$

By the Banach Lemma 5.9.1, we have that $I - C$ is invertible, and that

$$\|[I - C]^{-1}\|_{\mathcal{L}(X,X)} \leqslant \frac{1}{1 - \|C\|_{\mathcal{L}(X,X)}} \leqslant \frac{1}{1-\theta}, \quad \forall u \in B_\rho(u) \subset U. \quad (10.1.13)$$

Since B is invertible with norm given by (1), we have that the product of invertible operators $A = B[I - C]$ is invertible with $A^{-1} = [I - C]^{-1}B^{-1}$, and furthermore,

$$\begin{aligned}\|[F'(u)]^{-1}\|_{\mathcal{L}(X,X)} &= \|A^{-1}\|_{\mathcal{L}(X,X)} \\ &= \|[I - C]^{-1}B^{-1}\|_{\mathcal{L}(X,X)} \\ &\leqslant \|[I - C]^{-1}\|_{\mathcal{L}(X,X)} \, \|B^{-1}\|_{\mathcal{L}(X,X)} \\ &\leqslant \frac{\beta}{1-\theta}, \quad \forall u \in B_\rho(u) \subset U.\end{aligned}$$

□

Quadratic Functionals on Hilbert Spaces. We now revisit briefly quadratic functionals on a Hilbert space, which were encountered in Chapters 4 and 8. Let H be a Hilbert space with inner product $(\cdot, \cdot)$, and consider the functional $J\colon H \to \mathbb{R}$, defined in terms of a bounded linear operator $A \in \mathcal{L}(H, H)$ in the following way:

$$J(u) = \frac{1}{2}(Au, u), \quad \forall u \in H.$$

From the definition of the **F**-derivative above, it is easy to see that the **F**-derivative of J at $u \in H$ is a bounded linear functional, or $J'\colon H \to \mathcal{L}(H, \mathbb{R})$. We can calculate the **F**-derivative of J using the expression computing the **G**-variation (10.1.5), or equivalently, we can identify the component of $[J(u + h) - J(u)]$ that is linear in h as follows:

$$\begin{aligned}J(u+h) - J(u) &= \frac{1}{2}(A(u+h), u+h) - \frac{1}{2}(Au, u) \\ &= \frac{1}{2}(Au, u) + \frac{1}{2}(Au, h) + \frac{1}{2}(Ah, u) + \frac{1}{2}(Ah, h) - \frac{1}{2}(Au, u) \\ &= \frac{1}{2}((A + A^T)u, h) + \frac{1}{2}(Ah, h) \\ &= \frac{1}{2}((A + A^T)u, h) + \mathcal{O}(\|h\|^2).\end{aligned}$$

The **F**-*derivative* of $J(\cdot)$ is then

$$\langle J'(u), v\rangle = \frac{1}{2}((A + A^T)u, v), \quad \forall v \in H,$$

where A^T is the adjoint of A with respect to $(\cdot, \cdot)$. It follows from the Riesz Representation Theorem that $\langle J'(u), \cdot\rangle$ can be identified with an element $J'(u) \in H$,

$$\langle J'(u), v\rangle = (J'(u), v), \quad \forall v \in H,$$

called the *gradient* or the **F**-*differential* of J at u, which in this case is given by the expression $J'(u) = \frac{1}{2}(A + A^T)u$.

It is not difficult to see that the second **F**-derivative of J at u can be interpreted as a (symmetric) bilinear form, or $B(\cdot,\cdot) = J''(u)(\cdot,\cdot)\colon H \to \mathcal{L}(H \times H, \mathbb{R})$. To calculate $J''(u)$, we use the expression for computing the **G**-variation (10.1.5), or (again) equivalently, simply identify the component of $[J(u+h) - J(u)]$ that is now *quadratic* in h. From the above, we see that the quadratic term is

$$B(h, h) = \frac{1}{2}(Ah, h).$$

To recover the full symmetric form $B(u, v)$ from $B(h, h)$, we can employ the following standard trick:

$$B(v + w, v + w) = B(v, v) + 2B(v, w) + B(w, w)$$

$$\to \quad B(v, w) = \frac{1}{2}[B(v + w, v + w) - B(v, v) - B(w, w)].$$

This yields:

$$\begin{aligned} J''(u)(v, w) &= \frac{1}{2}[(A(v + w), v + w) - (Av, v) - (Aw, w)] \\ &= \frac{1}{2}[(Av, w) + (Aw, v)] \\ &= \frac{1}{2}[(Av, w) + (w, A^T v)] \\ &= \frac{1}{2}((A + A^T)v, w), \quad \forall v, w \in H. \end{aligned}$$

It now follows from Theorem 8.5.1 that the bilinear form $J''(u)(\cdot,\cdot)$ can be identified with the bounded linear operator $J''(u) \in \mathcal{L}(H, H)$ such that

$$J''(u)(v, w) = (J''(u)v, w) = \frac{1}{2}((A + A^T)v, w),$$

so that $J''(u) = \frac{1}{2}(A + A^T)$. Again, J'' can be computed directly from the definition of the **F**-derivative (or more easily, the **G**-variation), beginning with J'. Finally, note that if A is self-adjoint, then the expressions for J' and J'' simplify to:

$$J'(u) = Au, \qquad \text{and} \qquad J''(u) = A.$$

Consider now the functional $J\colon H \to \mathbb{R}$, defined in terms of a nonlinear operator $F\colon H \to H$ as follows:

$$J(u) = \frac{1}{2}\|F(u)\|^2 = \frac{1}{2}(F(u), F(u)).$$

The following result will be useful in Section 10.7 for globalizing Newton's method for nonlinear problems.

Lemma 10.1.3. *Let X be a Hilbert space, let $F\colon X \to Y$, and let $J\colon X \to \mathbb{R}$ be defined as $J(u) = \frac{1}{2}\|F(u)\|^2$. If $F \in C^1(X;Y)$, then $J \in C^1(X;\mathbb{R})$, and $J'(u) = F'(u)^T F(u)$.*

Proof. We identify the component of $[J(u+h) - J(u)]$ that is linear in h by expanding $F(\cdot)$ in a generalized Taylor series about $u \in H$:

$$\begin{aligned}
J(u+h) - J(u) &= \frac{1}{2}(F(u+h), F(u+h)) - \frac{1}{2}(F(u), F(u)) \\
&= \frac{1}{2}(F(u) + F'(u)h + \cdots, F(u) + F'(u)h + \cdots) \\
&\quad - \frac{1}{2}(F(u), F(u)) \\
&= \frac{1}{2}(F'(u)h, F(u)) + \frac{1}{2}(F(u), F'(u)h) + \mathcal{O}(\|h\|^2) \\
&= (F'(u)^T F(u), h) + \mathcal{O}(\|h\|^2).
\end{aligned}$$

Finally, from the Riesz Representation Theorem we have $J'(u) = F'(u)^T F(u)$. ☐

Subspaces, Algebraic and Topologic Closure, Convexity, and Reflexivity

We now assemble some very basic notions about subspaces, topologically closed subspaces, convex sets, convex spaces, and reflexive spaces, each of which will be important in our discussion of best approximation in Banach spaces.

Subspaces and Closure. Recall that a *subspace* of a Banach (or Hilbert) space X is a self-contained vector space (algebraically closed), and a *closed subspace* is a subspace which is also complete (topologically closed). A useful property of finite-dimensional subspaces is the following.

Theorem 10.1.4. *Let X be a normed vector space, and let $U \subset X$ be a finite-dimensional subspace of X. Then U is closed.*

Proof. Let $u \in X$ and take any sequence $u_j \in U$ such that $u_j \to u$. We know that $\{u_j\}$ is a Cauchy sequence in both X and U. Since U is itself a normed space, and as we discovered in Section 4.8, all finite-dimensional normed spaces are Banach spaces, we thus have that U is complete, and $\{u_j\}$ converges to an element of U. However, nondegeneracy of the norm implies that limits in normed spaces are unique, so the limit $u \in X$ must, in fact, be in U. Therefore, in addition to being algebraically closed as a vector space, U is also topogically closed as a subset of X. □

We now review the concept of orthogonal complement encountered in Chapter 4, and then generalize this concept to a Banach space. Consider first the case of a Hilbert space H: If $U \subset H$ is a subspace, then its complement in H, namely

$$U^{\perp} = \{v \in H \; : \; (v, w)_H = 0, \; \forall w \in U\} \subset H, \tag{10.1.14}$$

is always a closed subspace of H. When U is closed, $U^{\perp}$ is referred to as the *orthogonal complement* of U in H. Orthogonality can in a sense be generalized to a normed space X, through the use of the dual space X'. In particular, we say that a functional $f \in X'$ is orthogonal to a subspace $U \subset X$ if f annihilates all elements of U: $f(v) = 0$, $\forall v \in U \subset X$. We can then consider the generalized orthogonal complement:

$$U^{\perp} = \{f \in X' \; : \; f(w) = 0, \; \forall w \in U\} \subset X', \tag{10.1.15}$$

which is a closed subspace of X'.

Convexity and Reflexivity. A *convex subset* U of a vector space X is a subset $U \subset X$ such that any convex combination of elements taken from U lies again in U:

$$w = \alpha u + (1 - \alpha) v \in U, \qquad \forall u, v \in U, \qquad \alpha \in [0, 1].$$

If the convex combination lies in the interior of U for $u \neq v$ and $\alpha \in (0, 1)$, then U is called a *strictly convex subset.* Note that since a vector space is by definition algebraically closed, the entire vector space forms a convex set, and furthermore, any subspace of a vector space (which is by definition also algebraically closed) also forms a convex set. One of the most important consequences of convexity in Banach spaces concerns convergence. Given a Banach space X, a subset $U \subset X$ is called *closed under weak convergence*, or *weakly closed*, if for all sequences $\{u_j\} \subset U$ such that $u_j \rightharpoonup u$ in X it holds that $u \in U$. (See Section 4.8 for a longer discussion of weakly convergent sequences $u_j \rightharpoonup u$, and see also Exercise 5.7.3.) It can be shown that every weakly closed set U in a Banach space X is closed, but a closed set U in a Banach space X is not necessarily weakly closed; this is because there are at least as many weakly convergent sequences as there are strongly convergent sequences in a Banach space X. Therefore, to require that a set be weakly closed is a *stronger* condition than requiring that a set be closed.

However, there is a particular class of sets which are always weakly closed whenever they are closed, namely *convex sets*. (A similar phenomenon occurs with respect to lower and weak lower semicontinuity of functionals when they have convexity properties; see Theorem 10.1.10.) In order to state this result, we first examine a simple but fundamentally important idea due to Mazur.

Lemma 10.1.4 (Mazur's Lemma). *Let X be a Banach space, and let $u_j \rightharpoonup u \in X$. Then there exists a map $K \colon \mathbb{N} \to \mathbb{N}$ with $K(j) \geqslant j$, together with a sequence*

$\{\alpha_k^{(j)}\}_{k=j}^{K(j)}$, *such that*

$$\sum_{k=j}^{K(j)} \alpha_k^{(j)} = 1, \qquad \alpha_k^{(j)} \geqslant 0,$$

and such that the sequence $\{w_j\}$ *built from the convex combination of terms of* u_j

$$w_j = \sum_{k=j}^{K(j)} \alpha_k^{(j)} u_k,$$

is now strongly convergent to the same limit: $w_j \to u \in X$.

Proof. See [67]. □

An immediate implication of Mazur's lemma is the following.

Theorem 10.1.5. *Let X be a Banach space, and $U \subset X$ be closed and convex. Then U is also weakly closed.*

Proof. Let $u_j \rightharpoonup u \in X$ with $u_j \in U \subset X$. By Mazur's Lemma (Lemma 10.1.4), there exists a second sequence w_j, formed from convex combinations of u_j, such that $w_j \to u \in X$. Since U is convex, $w_j \in U \subset X, \forall j \in \mathbb{N}$. Due to the fact that U is closed (with respect to strong convergence), we must have $u \in U$. However, this implies that U is in fact also weakly closed. □

To state a second important result that follows from Mazur's Lemma, recall that the *convex hull* of a set $U \subset X$, denoted $co(U)$ is the smallest convex set containing U. The *closed convex hull* of a set $U \subset X$, denoted $\overline{co}(U)$ is the smallest closed convex set containing U. Recall that the set $U \subset X$ is *relatively (sequentially) compact* if every sequence in U contains a convergent subsequence; U is *(sequentially) compact* if the limit of the subsequence lies in U. A second important implication of Mazur's lemma concerns compactness properties of convex hulls.

Theorem 10.1.6 (Mazur's Theorem)**.** *Let X be a Banach space. If $U \subset X$ is relatively compact, then $co(U)$ is relatively compact, and $\overline{co}(U)$ is compact.*

Proof. See [188]. □

A *locally convex space* is a topological vector space where each neighborhood of the origin contains a convex neighborhood. All normed spaces are locally convex spaces, since given any neighborhood (open set in X) containing the origin, one can construct a sufficiently small open ball using the norm topology such that the open ball lies in the interior of the given neighborhood. A *strictly convex space* is a normed vector space for which the unit ball is a strictly convex set; this means that given any two points u and v on the boundary of the unit ball in X, the convex combination of u and v meets the boundary of the unit ball only at u and v. More precisely, a Banach space X is strictly convex if and only if $u \neq v$ along with the conditions $\|u\|_X = \|v\|_X = 1$ imply that $\|\alpha u + (1-\alpha)v\|_X < 1$ for all $0 < \alpha < 1$. All inner product spaces are strictly convex, so that a strictly convex normed space has more structure than a normed space, but not as much as an inner product space. Examples of strictly convex Banach spaces include $L^p(\Omega)$ for $1 < p < \infty$, as well as the Sobolev spaces built from these spaces (see Sections 2.6 and 10.3).

A slightly stronger condition is local uniform convexity: A Banach space X is *locally uniformly convex* if for every $\epsilon > 0$ and every $u \in X$ with $\|u\|_X \leqslant 1$ there is a $\delta(\epsilon, u) > 0$ such that for any $v \in X$, $\|v\|_X \leqslant 1$, the condition $\|u + v\| > 2 - \delta$ implies that $\|u - v\| < \epsilon$. If $\delta(\epsilon, u) = \delta(\epsilon)$, then X is called simply *uniformly convex*. This property is essentially that the midpoint of the line between two points in a uniformly convex Banach space X cannot lie near the boundary of the unit ball in X unless the points are close together. It can be shown (see [189]) that

$$\text{Uniform Convexity} \Longrightarrow \text{Local Uniform Convexity} \Longrightarrow \text{Strict Convexity}.$$

Examples of uniformly convex Banach spaces are again $L^p(\Omega)$ for $1 < p < \infty$, as well as the Sobolev spaces built from these spaces (see Sections 2.6 and 10.3).

An important theorem involving convexity of Banach spaces is the following.

Theorem 10.1.7 (Milman-Pettis Theorem). *If X is a uniformly convex Banach space then it is reflexive.*

Proof. A proof may be found in [2, 65]. □

The usual proof that $L^p(\Omega)$ is reflexive for $1 < p < \infty$ is to establish uniform convexity using the *Clarkson inequalities*, and then to appeal to this theorem (see [2]). It is useful to finish this discussion of subspaces and convexity with a result that combines these properties.

Theorem 10.1.8. *Let X be a Banach space, and let $U \subset X$ be a closed subspace of X. Then U is also a Banach space with respect to the norm inherited from X, and*

(1) If X is separable, then so is U.

(2) If X is strictly convex, then so is U.

(3) If X is uniformly convex, then so is U.

(4) If X is reflexive, then so is U.

Proof. That U is complete follows from the closure assumption, and as a subspace, separability is obviously inherited from X. Since convexity is a property of the *norm* on X rather than X itself, convexity (strict and/or uniform) of U follows immediately from these properties of X since U inherits its norm from X. That U is reflexive is substantially more involved; a proof may be found in [183]. □

Variational Methods in Banach Spaces

In the preceding section we assembled some ideas on convexity of norms and spaces, reflexive Banach spaces, strong and weak convergence, and closure with respect to strong and weak convergence, in order to set the stage for our discussion of best approximation in Banach spaces. The main results in best approximation are proved using weak convergence ideas similar to variational methods (direct methods in the calculus of variations); we were first exposed to these ideas in Section 8.4. The material in the preceding section, combined with the material covered in Section 4.8 on dual spaces, reflexivity, and weak convergence, puts us in a position to give a more careful discussion of variational methods. We need access to these methods to establish existence and uniqueness of best approximation in subspaces of a Banach space in Section 10.2, as well as to prove the existence of solutions to continuous and discretized nonlinear partial differential equations in Sections 10.4 and 10.5.

Let X be a Banach space, let U be a subset of X, and let $J\colon U \subset X \to \overline{\mathbb{R}}$ be a (generally nonlinear) functional. The notation $\overline{\mathbb{R}} = \mathbb{R} \cup \{+\infty\} \cup \{-\infty\}$ is used here to allow for functionals J which are potentially unbounded at points in U. We are interested in the following (local) minimization problem:

$$\text{Find } u_0 \in U \subset X \text{ such that } J(u_0) = \inf_{u\in U} J(u), \tag{10.1.16}$$

which we can think about equivalently as:

$$\text{Find } u_0 \in U \subset X \text{ such that } J(u_0) \leqslant J(v), \quad \forall v \in U \subset X. \tag{10.1.17}$$

The functional J is called the *energy* or *objective functional*, and potentially has one or more of the following mathematical properties for a given subset $U \subset X$:

(1) Convex: $J(\alpha u + (1-\alpha)v) \leqslant \alpha J(u) + (1-\alpha)J(v)$, for a convex set $U \subset X$, $\forall u, v \in U$, and any $\alpha \in [0,1]$,

(2) Strictly convex: Convex J with strict inequality for $u \neq v$ and $\alpha \in (0,1)$.

(3) Quasiconvex: The set $U_r = \{u \in U \ : \ J(u) \leqslant R\}$ is convex for all $r \in \mathbb{R}$.

(4) Proper: $\|u_j\|_X \to +\infty \Longrightarrow J(u_j) \to +\infty, \forall u_j \in U \subset X$.

(5) Coercive: $J(u) \geqslant C_0\|u\|_X^2 - C_1, \forall u \in U \subset X$.

(6) Bounded below: $J(u) \geqslant K_0, \forall u \in U \subset X$.

(7) Lower semicontinuity at $u_0 \in X$: $J(u_0) \leqslant \liminf_{j\to\infty} J(u_j)$, $\forall \{u_j\}$ such that $u_j \to u_0 \in X$.

(8) Upper semicontinuity at $u_0 \in X$: $J(u_0) \geqslant \limsup_{j\to\infty} J(u_j)$, $\forall \{u_j\}$ such that $u_j \to u_0 \in X$.

(9) Weak lower semicontinuity at $u_0 \in X$: $J(u_0) \leqslant \liminf_{j\to\infty} J(u_j)$, $\forall \{u_j\}$ such that $u_j \rightharpoonup u_0 \in X$.

One can define continuity, as well as lower and upper semicontinuity, with respect to weak-$*$ convergence as well. The *limit inferior* (or *lim inf*) and *limit superior* (or *lim sup*) appearing in the definitions of lower and upper semicontinuity above are defined as

$$\liminf_{j\to\infty} a_j = \lim_{j\to\infty}\left(\inf_{k\geqslant j} a_k\right), \qquad \limsup_{j\to\infty} a_j = \lim_{j\to\infty}\left(\sup_{k\geqslant j} a_k\right).$$

Note that a functional J is *continuous* if and only if it is both lower and upper semi-continuous. Coercive functionals are always both proper and bounded from below, since

$$-C_1 \leqslant C_0\|u\|_X^2 - C_1 \leqslant J(u) \to +\infty, \quad \text{as } \|u\|_X \to \infty. \tag{10.1.18}$$

In addition, note that weakly lower semicontinuous functions are also lower semi-continuous, but the reverse is not true; there are at least as many weakly convergent sequences as there are strongly convergent sequences in a Banach space. Therefore, weak lower semicontinuity is a *stronger* condition than lower semicontinuity; this is similar to the relationship between weakly closed sets and closed sets. (See the discussion leading up to Lemma 10.1.4.)

One of the simplest and most general results which guarantees the existence of a solution to the minimization problem (10.1.16) is the following, which we can prove using only the concepts we have covered so far.

Theorem 10.1.9. *Let X be a reflexive Banach space, let U be a weakly closed subset of X, and let $J\colon U \subset X \to \overline{\mathbb{R}}$ be a proper, bounded below, and weakly lower semicontinuous functional on U. Then there exists a solution to problem* (10.1.16).

Proof. Since J is bounded from below on U, there exists a greatest lower bound K_0 such that $J(u) \geqslant K_0, \forall u \in U$, and a minimizing sequence $\{u_j\}$ with $u_j \in X$ such that $\lim_{j\to\infty} J(u_j) = K_0$. Since J is proper and $J(u_j) \to K_0$, the sequence u_j must be bounded; otherwise, properness of J would imply $J(u_j) \to +\infty$ as $\|u_j\|_X \to +\infty$. Since $\{u_j\}$ is a bounded sequence in the reflexive Banach space X, there exists $u_0 \in X$ and a weakly convergent subsequence $\{u_{j_k}\}$, which we relabel as $\{u_j\}$ without danger of confusion, such that $u_j \rightharpoonup u_0$. However, since U is weakly closed and $u_j \in U$, we must have that $u_0 \in U \subset X$. Now, since $\{u_j\}$ is a minimizing sequence and J is weakly lower semicontinuous, we have

$$J(u_0) \leqslant \liminf_{j\to\infty} J(u_j) = K_0 \leqslant \inf_{v\in U} J(v),$$

which proves the theorem. □

The weak lower semicontinuity assumption on the functional J can be relaxed to simply lower semicontinuity if J and the subset $U \subset X$ have convexity properties; this is because convex lower semicontinuous functions are weakly lower semicontinuous. This is similar to how adding convexity to a set ensures that a (convex) closed set is also weakly closed (see Theorem 10.1.5).

Theorem 10.1.10. *Let X be a Banach space, let $U \subset X$ be a closed convex subset, and let $J\colon U \subset X \to \overline{\mathbb{R}}$ be convex and lower semicontinuous on U. Then J is weakly lower semicontinuous on U.*

Proof. Given a weakly convergent sequence $u_j \in U \subset X$ with $u_j \rightharpoonup u_0 \in X$, by Mazur's Lemma (Lemma 10.1.4), there exists a sequence $w_j \in U$ built from a convex combination of u_j:

$$w_j = \sum_{k=j}^{K(j)} \alpha_k^{(j)} u_k, \qquad \sum_{k=n}^{J(j)} \alpha_k^{(j)} = 1, \qquad \alpha_k^{(j)} \geqslant 0,$$

such that $w_j \to u_0 \in X$. We now use convexity and lower semicontinuity of J to conclude that

$$\begin{aligned} J(u_0) = J\left(\lim_{j\to\infty} w_j\right) &= J\left(\lim_{j\to\infty} \sum_{k=j}^{K(j)} \alpha_k^{(j)} u_k\right) \leqslant \liminf_{j\to\infty} J\left(\sum_{k=j}^{K(j)} \alpha_k^{(j)} u_k\right) \\ &\leqslant \liminf_{j\to\infty} \left(\sum_{k=j}^{K(j)} \alpha_k^{(j)}\right) J(u_k) = \liminf_{j\to\infty} J(u_j), \end{aligned}$$

which shows J is also weakly lower semicontinuous. □

The following gives a condition to ensure lower semicontinuity of convex functions on a convex set.

Theorem 10.1.11. *Let X be a Banach space, let $U \subset X$ be a closed convex subset, and let $J\colon U \subset X \to \overline{\mathbb{R}}$ be convex on U. If J is* **G***-differentiable on U, then J is weakly lower semicontinuous on U.*

Proof. See [159, 183]. □

The discussion above leads to one of the central results in convex analysis.

Theorem 10.1.12. *Let X be a reflexive Banach space, let U be a closed convex subset of X, and let $J\colon U \subset X \to \overline{\mathbb{R}}$ be a proper, bounded below, convex, and lower semicontinuous functional on U. Then there exists a solution to problem* (10.1.16). *Moreover, if J is strictly convex, then the solution is unique.*

Proof. Since U is a closed convex subset of X, by Theorem 10.1.5 we have that U is also weakly closed. By assumption, J is both proper and bounded from below. Since J is convex and lower semicontinuous, by Theorem 10.1.10 we know J is also weakly lower semicontinuous. Therefore, by Theorem 10.1.9 there exists a solution to problem (10.1.16). Uniqueness in the case of strict convexity is by the following argument. Assume that there are two minimizers u_0 and u_1 of J in $U \subset X$, so that we must have $J(u_0) = J(u_1) = K_0 \leqslant J(v)$, $\forall v \in U \subset X$. Since U is convex, any convex combination $u_\alpha = \alpha u_0 + (1-\alpha)u_1$ remains in U, for $\alpha \in (0,1)$. But strict convexity of J implies that

$$J(u_\alpha) < \alpha J(u_0) + (1-\alpha) J(u_1) = K_0,$$

which contradicts the assumption that u_0 and u_1 were minimizers. □

A similar argument is used later to prove uniqueness of best approximation in a subspace of a Banach space. In fact, we will prove some of the main results on best approximation in Banach spaces by viewing the search for a best approximation as minimizing the particular functional $J(u) = \|u - v\|_X$ over a subspace $U \subset X$, and using some of the variational methods results above to establish the existence of a minimizer of J.

A final result that will be very useful for nonlinear partial differential equations that arise as Euler-Lagrange equations of an energy functional $J\colon X \to \overline{\mathbb{R}}$ is the following. (See Section 10.3 for a review of measurable functions and L^p spaces.)

Theorem 10.1.13 (Tonelli's Theorem). *Let $\Omega \subset \mathbb{R}^n$ be an open set, and for $m \geqslant 1$ let $u\colon \Omega \to \mathbb{R}^m$ be a measurable function. Define the functional $J\colon L^p(\Omega) \to \overline{\mathbb{R}}$ by*

$$J(u) = \int_\Omega F(u(x))\, dx,$$

where $F\colon \mathbb{R}^m \to \overline{R}$ is a continuous function. Then J is weakly lower semicontinuous on $L^p(\Omega)$ for $1 < p < \infty$, and weak-$*$ lower semicontinuous on $L^\infty(\Omega)$, if and only if F is convex.

Proof. A proof may be found in [145]. □

Topological Fixed-Point Theorems

We give a brief review of some standard topological fixed-point theorems in Banach spaces. The Banach Fixed-Point Theorem was presented in Chapter 4, and the Brouwer Theorem, as well as one version of the Schauder Theorem below, were presented and used in Chapter 9; we bring them together here with some related results as a reference and for later use in proving existence of solutions to continuous and discretized nonlinear elliptic equations in Sections 10.4 and 10.5. We will need to recall from Section 4.8 that we defined several distinct notions of *convergence of sequences* in Banach spaces (strong, weak, and weak-* convergence). To state an extended version of the Banach Fixed-Point Theorem below, we need to recall also that we defined several distinct notions of *rate of strong convergence* of sequences [namely Q-linear, Q-superlinear, Q-order(p), and R-order(p)].

Theorem 10.1.14 (Banach Fixed-Point Theorem). *Let U be a closed set in the Banach space X and let $T\colon U \to U$ be a contraction with constant $\alpha \in [0, 1)$, such that $\|T(u) - T(v)\|_X \leqslant \alpha \|u - v\|_X$, $\forall u, v \in U$. Then:*

(1) T has a unique fixed point $u \in U$ such that $u = T(u)$.
(2) For any $u_0 \in U$, the sequence $\{u_j\}$ given by $u_j = T(u_{j-1})$ converges to u.
(3) The convergence rate is Q-linear with constant α, with the following a priori and a posteriori error estimates:

$$\|u - u_j\|_X \leqslant \frac{\alpha^j}{1-\alpha} \|u_1 - u_0\|_X, \qquad \text{(a priori)} \qquad (10.1.19)$$

$$\|u - u_j\|_X \leqslant \frac{\alpha}{1-\alpha} \|u_j - u_{j-1}\|_X. \qquad \text{(a posteriori)} \qquad (10.1.20)$$

Proof. See the proof in Section 4.4 or, for example [108]. In the proof given in Section 4.4, the first part of (1), existence, was established by showing that u_j generated by $u_j = T(u_{j-1})$ was a Cauchy sequence, and U being closed ensured that $u_j \to u \in U \subset X$ for some $u \in U$. The second part of (1), uniqueness, followed in a simple way from the contraction property of T. Conclusion (2), which was that the convergence happened for arbitrary $u_0 \in U$, was immediate in the proof, since the arguments for (1) were for an arbitrary u_0 selected from U. Here, we still must prove conclusion (3) involving the error estimates. These follow from the intermediate steps of the proof in Section 4.4. In particular, it was shown that for $k > j$,

$$\|u_k - u_j\|_X \leqslant \frac{\alpha^j}{1-\alpha}\|u_1 - u_0\|_X. \tag{10.1.21}$$

Taking the limit as $k \to \infty$, employing continuity of the norm, gives (10.1.19). A similar analysis of the finite geometric series that gave rise to (10.1.21) in the proof given in Chapter 4 leads to the alternative bound $k > j$:

$$\|u_k - u_j\|_X \leqslant \frac{\alpha(1-\alpha^{k-j})}{1-\alpha}\|u_j - u_{j-1}\|_X. \tag{10.1.22}$$

Taking the limit as $k \to \infty$ now leads to (10.1.20). □

Theorem 10.1.15 (Brouwer Fixed-Point Theorem). *Let $U \subset \mathbb{R}^n$ be a nonempty, convex, compact set, with $n \geqslant 1$. If $T: U \to U$ is a continuous mapping, then there exists a fixed point $u \in U$ such that $u = T(u)$.*

Proof. See [188]; a short proof can be based on homotopy-invariance of topological degree. An elementary proof based on the No-Differentiable-Retraction Theorem and a simple result on determinants of matrices may be found in [113]. □

The extension of the Brouwer Fixed-Point Theorem to arbitrary (possibly infinite-dimensional) Banach spaces is known as the *Schauder Theorem.*

Theorem 10.1.16 (Schauder Theorem). *Let X be a Banach space, and let $U \subset X$ be a nonempty, convex, compact set. If $T: U \to U$ is a continuous operator, then there exists a fixed point $u \in U$ such that $u = T(u)$.*

Proof. This is a direct extension of the Brouwer Fixed-Point Theorem from $\mathbb{R}^n$ to X; see [188]. The short proof involves a finite-dimensional approximation algorithm and a limiting argument, extending the Brouwer Fixed Point Theorem (itself generally having a more complicated proof) from $\mathbb{R}^n$ to X. □

Theorem 10.1.17 (Schauder Theorem B). *Let X be a Banach space, and let $U \subset X$ be a nonempty, convex, closed, bounded set. If $T: U \to U$ is a compact operator, then there exists a fixed point $u \in U$ such that $u = T(u)$.*

Proof. Since U is bounded, and since T is a compact operator, the set $T(U)$ is relatively compact. Let $V = \overline{co}(T(u))$, the closed convex hull of $T(U)$, which we recall is the smallest closed convex set containing $T(U)$. By Mazur's Theorem (Theorem 10.1.6), V is compact. So we have $T(U) \subset V$, and also by assumption, $T(U) \subset U$. However, since U is convex, and since V is the smallest convex set containing $T(U)$, we must have that $V \subset U$. We now have that $T\colon V \to V$, with V compact and T a continuous map (since compact maps are continuous maps). Therefore, by Theorem 10.1.16, there exists a fixed point $u \in V \subset U$ such that $u = T(u)$. □

The following two results are consequences of Theorem 10.1.17; they do not require identifying a convex subset which is either compact or closed and bounded.

Theorem 10.1.18 (Leray-Schauder/Schaefer Theorem). *Let X be a Banach space, and let $T\colon X \to X$ be a compact operator. If there exists $r > 0$ such that*

$$u = \lambda T(u), \quad \lambda \in (0,1) \quad \textit{implies} \quad \|u\|_X \leqslant r, \tag{10.1.23}$$

then there exists a fixed point $u \in X$ such that $u = T(u)$, and such that $\|u\|_X \leqslant r$.

Proof. See [188]; the proof involves the construction of a compact operator S which acts invariantly on the ball of diameter r which shares its fixed point with T, and then simply invoking Theorem 10.1.17. □

Corollary 10.1.1 (Leray-Schauder/Schaefer Theorem B). *Let X be a Banach space, and let $T\colon X \to X$ be a compact operator. If*

$$\sup_{u \in X} \|T(u)\|_X < +\infty, \tag{10.1.24}$$

then there exists a fixed point $u \in X$ such that $u = T(u)$.

Proof. Condition (10.1.24) implies (10.1.23), so that the result now follows from Theorem 10.1.18. □

In the case of reflexive and separable Banach spaces as introduced in Section 4.8, it is possible to develop a Schauder-like theorem that replaces compactness of the set U in Theorem 10.1.16, or compactness of the operator T in Theorem 10.1.17, with weak sequential continuity of the operator T. Recall that a (possibly nonlinear) map $T\colon X \to Y$ is called *weakly sequentially continuous* if $u_j \rightharpoonup u$ as $j \to \infty$ implies that $T(u_j) \rightharpoonup T(u)$ as $j \to \infty$.

Theorem 10.1.19 (Schauder Theorem C). *Let X be a reflexive, separable Banach space, and let $U \subset X$ be a nonempty, convex, closed, bounded subset. If $T\colon U \to U$ is weakly sequentially continuous, then there exists a fixed point $u \in U$ such that $u = T(u)$.*

Proof. A proof may be found in [188]. □

Note that since every strongly convergent sequence in a Banach space is weakly convergent, but not vice versa, there are generally more weakly convergent sequences than strongly convergent sequences. Hence, weak sequential continuity is a stronger condition than strong sequential continuity. Nevertheless, when compactness is used to obtain strong convergence from weak convergence, we have the following result to ensure that a mapping is weakly sequentially continuous.

Lemma 10.1.5 (Weak Sequential Continuity and Compactness). *Let X and Z be Banach spaces with compact embedding $X \overset{c}{\hookrightarrow} Z$ for embedding operator i. If $T\colon Z \to X$ is continuous, then $F = (T \circ i)\colon X \to X$ is weakly sequentially continuous, and is also compact. Moreover, the map $G = (i \circ T)\colon Z \to Z$ is also compact.*

Proof. Let $u_j \rightharpoonup u$ in X be arbitrary. Compactness of the embedding $X \overset{c}{\hookrightarrow} Z$ gives $u_j \to u$ in Z. Continuity of $T\colon Z \to X$ implies that $T(u_j) \to T(u)$ in X. Since strong convergence in X implies weak convergence in X, we have finally $T(u_j) \rightharpoonup T(u)$ in X, giving that $F = (T \circ i)\colon X \to X$ is weakly sequentially continuous. Moreover, being the composition of continuous and compact maps, both $F = (T \circ i)\colon X \to X$ and $G = (i \circ T)\colon Z \to Z$ are compact. □

See Section 10.3 for more information about continuous and compact embeddings of the form $X \hookrightarrow Z$, where X and Z are two normed spaces.

Inverse and Implicit Function Theorems. The following are three fundamentally important theorems in nonlinear analysis that allow for characterizing local and global invertibility of a nonlinear map in terms of invertibility of the Fréchet derivative operator.

Theorem 10.1.20 (Local Inverse Function Theorem). *Let X and Y be Banach spaces, let $D \subset X$ be an open set, and let $F \in C^1(D; Y)$. If $F'(u) \in \mathcal{L}(X, Y)$ is an isomorphism for each $u \in D$, then there exist open neighborhoods $U \subset D \subset X$ and $V \subset Y$ with $u \in U$ and $F(u) \in V$ such that $F(U) = V$ (F is injective). Moreover, if $G = [F|_U]^{-1}$, then $G \in C^1(V, U)$.*

Proof. The proof is based on the Banach Fixed-Point Theorem (Theorem 10.1.14); see [65, 188]. □

Theorem 10.1.21 (Global Inverse Function Theorem or Hadamard Theorem). *Let X and Y be Banach spaces, and let $F \in C^1(X; Y)$. If $F'(u) \in \mathcal{L}(X, Y)$ is an isomorphism for each $u \in X$, and if there exists a constant $\gamma > 0$ such that*

$$\|[F(u)]^{-1}\|_{\mathcal{L}(X,Y)} \leqslant \gamma, \quad \forall u \in X,$$

then F is a diffeomorphism from X onto Y.

Proof. A proof can be based on showing that F is separately injective and surjective, and then using Theorem 10.1.20. See [65, 188]. □

Theorem 10.1.22 (Implicit Function Theorem). *Let X, Y, and Z be Banach spaces, and let $F: X \times Y \to Z$. Let $(u_0, v_0) \in X \times Y$ be such that*

$$F(u_0, v_0) = 0 \in Z.$$

Let $D \subset X \times Y$ be an open set such that $(u_0, v_0) \in D$, and assume that $F \in C^1(D; Z)$. If $D_2F(u_0, v_0) \in \mathcal{L}(Y, Z)$ (the partial Fréchet derivative of F with respect to the second slot) is an isomorphism, then there exist open neighborhoods $U \subset X$ and $V \subset Y$ with $u_0 \in U$ and $v_0 \in V$ such that for any $u_1 \in U$ there exists a unique $v_1 \in V$ satisfying

$$F(u_1, v_1) = 0 \in Z.$$

Proof. The short proof can be based on the Local Inverse Function Theorem 10.1.20; see [65, 188]. □

We remark that several variations of the Inverse and Implicit Function Theorems may be found, for example, in [111], including the *Approximate Implicit Function Theorem*. These types of results allow for satisfying the assumptions in (for example) Theorem 10.1.22 only approximately (but in a controlled way), yielding essentially the same conclusions. These types of results are particularly useful in the development of homotopy and continuation methods (both analytical and numerical) for nonlinear problems.

Ordered Banach Spaces, Maximum Principles, and Increasing Maps

Let X and Y be Banach spaces, and let X' and Y' be their respective dual spaces. Given a (generally nonlinear) map $F: X \to Y'$, we are again interested in the following general problem:

$$\text{Find } u \in X \text{ such that } F(u) = 0 \in Y'. \tag{10.1.25}$$

As noted earlier, this problem is equivalent to:

$$\text{Find } u \in X \text{ such that } \langle F(u), v \rangle = 0 \in \mathbb{R}, \qquad \forall v \in Y, \tag{10.1.26}$$

where $\langle F(u), \cdot \rangle$ denotes the action of $F(u) \in Y'$ as a linear functional on Y. As stated at the beginning of the chapter, our primary objectives are to develop techniques for analyzing solutions to (10.1.25), and to develop hard results on best and near-best approximation of solutions to general nonlinear operator equations of the form (10.1.25). We are interested in results where $F: X \to Y'$ has no finer structure, and also when F has additional structure of the form

$$F(u) = Au + B(u), \tag{10.1.27}$$

where $B: X \to Y'$ is a (generally nonlinear) operator, and $A: X \to Y'$ is a linear invertible operator. This leads to the equivalent problem:

$$\text{Find } u \in X \text{ such that } Au + B(u) = 0 \in Y'. \tag{10.1.28}$$

If A is invertible, then one can also consider a third equivalent fixed point formulation of (10.1.28):

$$\text{Find } u \in X \text{ such that } u = T(u) \in X, \tag{10.1.29}$$

where $T\colon X \to X$, with $T(u) = -A^{-1}B(u)$. This represents a large class of nonlinear problems of interest, and in most of the problems we will run into, the operators A and B have additional key properties that we can exploit to our advantage, both for analysis of solutions and development of approximation results. In order to be able to state these properties precisely, we must develop some basic concepts and ideas of *ordered Banach spaces*. See [188] and also the appendix of [97] for more extensive treatments.

Ordered Banach Spaces. The notion of order structures on classical function spaces has already been introduced and exploited in Chapter 9 for the material on fixed-point theorems for *order-preserving maps*. These are (generally nonlinear) maps $G\colon X \to Z$ having the property

$$u \leqslant v \Longrightarrow G(u) \leqslant G(v). \tag{10.1.30}$$

Such a map G is called a *monotone increasing map* G, and a map having the second inequality reversed is called a *monotone decreasing map*. In order to make sense of both inequalities in (10.1.30), both X and Z must have order structures. It may be that the order structure we need is not obviously present in X and/or Z, but may be inherited from a third space W which itself is an ordered Banach space, through an embedding $X \hookrightarrow W$ (see Section 10.3 for a discussion of embeddings of Sobolev spaces into other function spaces).

We now define the order structure needed for an abstract ordered Banach space. Recall that a *cone* in a vector space X is a set $C \subset X$ such that $au \in C$, $\forall u \in C$ with $a \in \mathbb{R}$ and $a \geqslant 0$.

Definition 10.1.6. *Let X be a Banach space. An* order cone *in X is a cone $X_+ \subset X$ with the following additional properties:*

(1) $X_+ \subset X$ is nonempty and closed in the topology of X, with $X_+ \neq \{0\}$.
(2) If $u, v \in X_+$ and $a, b \in \mathbb{R}$, $a \geqslant 0$, $b \geqslant 0$, then $au + bv \in X_+$.
(3) If $u \in X_+$ and $-u \in X_+$, then $u = 0$.

The first property states that an order cone X_+ is a nonempty topologically closed subset of X that consists of more than just the vector space zero element of X. The second property says that X_+ satisfies the definition of a cone (take $b = 0$), and, in fact, X_+ is a convex set [take $a \in (0,1)$ and $b = 1 - a$]. The third property ensures that the cone gives an ordering (or rather, a *partial* ordering); it states that other than the apex $0 \in X$ of the cone X_+, both $u \in X$ and $-u \in X$ cannot be in the cone $X_+ \subset X$ at the same time. The order cone $X_+ \subset X$, together with the underlying Banach space X, is referred to as an *ordered Banach space (OBS)*, and is sometimes denoted $\langle X, X_+ \rangle$. One also simply states that X is an OBS when reference to the set X_+ is not specifically required.

We will use the following standard notation for the order structure.

Definition 10.1.7. *Let $\langle X, X_+ \rangle$ be an ordered Banach space, and let $u, v \in X$.*

(1) $u \geqslant v$ *if and only if* $u - v \in X_+$.
(2) $u \leqslant v$ *if and only if* $v - u \in X_+$.
(3) $u > v$ *if and only if* $u \geqslant v$ *and* $u \neq v$.
(4) $u < v$ *if and only if* $u \leqslant v$ *and* $u \neq v$.
(5) $u \ngeqslant v$ *if and only if* $u \geqslant v$ *does not hold.*
(6) $u \nleqslant v$ *if and only if* $u \leqslant v$ *does not hold.*
(7) $u \gg v$ *if and only if* $u - v \in \operatorname{int}(X_+)$.
(8) $u \ll v$ *if and only if* $v - u \in \operatorname{int}(X_+)$.

Basic facts about the order structure on a Banach space X that are easy to establish from Definition 10.1.7 are the following.

Lemma 10.1.6. *Let X be an ordered Banach space. Then for all $u, v, w, \hat{u}, \hat{v} \in X$, and all $a, b \in \mathbb{R}$:*

(1) $u \geqslant u$.
(2) If $u \geqslant v$ *and* $v \geqslant u$, *then* $u = v$.
(3) If $u \geqslant v$ *and* $v \geqslant w$, *then* $u \geqslant w$.
(4) If $u \geqslant v$ *and* $a \geqslant b \geqslant 0$, *then* $au \geqslant bv$.
(5) If $u \geqslant v$ *and* $\hat{u} \geqslant \hat{v}$, *then* $u + \hat{u} \geqslant v + \hat{v}$.
(6) If $u_j \geqslant v_j$, $\forall j \in \mathbb{N}$, *then* $\lim_{j\to\infty} u_j \geqslant \lim_{j\to\infty} v_j$.
(7) If $u \gg v$ *and* $v \gg w$, *then* $u \gg w$.
(8) If $u \gg v$ *and* $a > 0$, *then* $au \gg av$.

Proof. These follow easily from Definition 10.1.7. □

We can now make sense of closed and open intervals in X with respect to the order structure, given any two elements $u, v \in X$ such that $u \leqslant v$:

$$[u, v] = \{w \in X \ : \ u \leqslant w \leqslant v\} \subset X, \tag{10.1.31}$$

$$(u, v) = \{w \in X \ : \ u \ll w \ll v\} \subset X. \tag{10.1.32}$$

In order to draw some conclusions about the properties of these two subsets of X, the order cone $X_+ \subset X$ needs to have some additional properties.

Definition 10.1.8. *Let X be a Banach space, and let $X_+ \subset X$ be an order cone satisfying Definition 10.1.6. Then X_+ may have any of the additional properties:*

(1) Normal cone*: if and only if there exists* $0 < a \in \mathbb{R}$ *such that* $\|v\|_X \leqslant a\|u\|_X$, *whenever* $u, v \in X$ *satisfy* $0 \leqslant v \leqslant u$.
(2) Generating Cone*:* $\operatorname{span}(X_+) = X$.
(3) Total cone*:* $\operatorname{span}(X_+)$ *is dense in* X.

(4) Solid cone*:* $\text{int}(X_+)$ *is nonempty.*

In the above, $\text{span}(X_+) = X_+ - X_+ = \{u - v : u, v \in X_+\}$.

The two most important consequences of these additional properties for our purposes are the following.

Lemma 10.1.7. *Let* $\langle X, X_+ \rangle$ *be an ordered Banach space with a normal order cone* X_+*, and let* $u, v \in X$ *such that* $u \leqslant v$*. Then the closed interval* $[u, v] \subset X$ *is a bounded subset of* X*.*

Proof. We include the short proof (see also [188] or the appendix of [97]). Let $u, v \in X$ such that $u \leqslant v$ be arbitrary, and pick any $w \in [u, v]$. Then we have $u \leqslant w \leqslant v$, or $0 \leqslant w - u \leqslant v - u$. By normality of X_+, there exists $0 < a \in \mathbb{R}$ such that $\|w - u\|_X \leqslant a\|v - u\|_X$. We then have

$$\|w\|_X = \|w-u+u\|_X \leqslant \|w-u\|_X + \|u\|_X \leqslant a\|v-u\|_X + \|u\|_X < \infty. \quad (10.1.33)$$

Since $[u, v]$ was arbitrary, and since inequality (10.1.33) holds for any $w \in [u, v]$, we have that any closed interval $[u, v] \subset X$ is bounded in the norm on X. □

Lemma 10.1.8. *Let* $\langle X, X_+ \rangle$ *be an ordered Banach space with a total order cone* X_+*. Then* X_+ *induces an order cone* $X'_+ \subset X'$ *on the dual space* X' *by*

$$X'_+ = \{f \in X' \ : \ f(u) \geqslant 0, \ \forall u \in X_+\}.$$

Proof. See [188] or the appendix of [97]. □

Monotone Increasing Maps and Maximum Principles. Let X and Z be ordered Banach spaces. Given a map $F\colon X \to Z$ such that $F(u) = Au + B(u)$, where $B\colon X \to Z$ is a nonlinear operator and where $A\colon X \to Z$ is an invertible linear operator, we are interested in solutions to the following mathematically equivalent problems: Find $u \in X$ such that any of the following hold:

$$F(u) = 0 \in Z, \quad (10.1.34)$$

$$Au + B(u) = 0 \in Z, \quad (10.1.35)$$

$$u = T(u) \in X, \quad (10.1.36)$$

where

$$F(u) = Au + B(u), \qquad T(u) = -A^{-1}B(u), \quad (10.1.37)$$

with $T\colon X \to X$. The operators F, A, B, and T may have additional properties. In particular, let $G\colon X \to Z$, where X and Z are ordered Banach spaces, and define the following properties on any subset $U \in X$:

(1) Monotone increasing: $u \leqslant v \Longrightarrow G(u) \leqslant G(v), \ \forall u, v \in U \subset X.$ (10.1.38)

(2) Monotone decreasing: $u \leqslant v \Longrightarrow G(u) \geqslant G(v), \ \forall u, v \in U \subset X.$ (10.1.39)

(3) Maximum principle: $G(u) \leqslant G(v) \Longrightarrow u \leqslant v, \ \forall u, v \in U \subset X.$ (10.1.40)

A map G with one of the first two properties is sometimes called an *order preserving map* (see Chapter 9 for a number of examples). A map G having the third property is often called a *monotone operator*; we will refer to the property as the *maximum principle property* to avoid confusion with the monotone increasing/decreasing properties.

It will be useful to exploit the order structure in X further, and to consider weakening the notion of solution to (10.1.36) to include:

(1) Supersolution: $u_+ \geqslant T(u_+)$, (10.1.41)

(2) Subsolution: $u_- \leqslant T(u_-)$, (10.1.42)

and then to consider the use of sub- and supersolutions to find solutions to (10.1.36). To this end, recall that a map $G\colon X \to Z$ is called *compact* if and only if G is continuous and maps any bounded set in X to a relatively compact set in Z. We need to extend this definition to possibly unbounded sets in X.

Definition 10.1.9. *Let X and Z be Banach spaces, and let $U \subset X$ be a (possibly unbounded) set in X. If $G\colon X \to Z$ is continuous, and if $G(U) \subset Z$ is relatively compact, then we say that G is* image compact, *or* i-compact *on U.*

On bounded sets, compactness and i-compactness are the same condition. On unbounded sets, an i-compact map is clearly compact. However, a compact map is not necessarily i-compact; consider $T = I$, the identity operator, and $X = Z = \mathbb{R}^n$, which is a finite-dimensional set but is at the same time unbounded; as is well-known, G is compact, but it is clearly not i-compact. Therefore, i-compactness is a *stronger* condition than compactness.

We now have the first of two main fixed-point theorems for monotone increasing maps; we have already exploited this first result in Chapter 9.

Theorem 10.1.23 (Fixed Points for Monotone Increasing Maps). *Let X be a real ordered Banach space with order cone X_+. Let $T\colon [u_-, u_+] \subset X \to X$ be a monotone increasing map, and let one of the following two conditions hold:*

(1) X_+ is normal and T is compact.

(2) T is i-compact on $[u_-, u_+]$.

If u_-, u_+ are sub- and supersolutions to $u = T(u)$, then the iterations

$$u_{j+1} = T(u_j), \quad u_0 = u_-,$$
$$\hat{u}_{j+1} = T(\hat{u}_j), \quad \hat{u}_0 = u_+,$$

converge to $u, \hat{u} \in [u_-, u_+]$, respectively, and the following estimate holds:

$$u_- \leqslant u_j \leqslant u \leqslant \hat{u} \leqslant \hat{u}_j \leqslant u_+, \qquad \forall j \in \mathbb{N}.$$

Proof. Note that if X_+ is normal, then the closed interval $[u_0, v_0] \subset X$ is a bounded set in X. If T is also compact, then $T([u_0, v_0]) \subset X$ is relatively compact; hence T is also i-compact with respect to closed intervals generated by a normal order cone.

Therefore, it suffices to prove the theorem for a general order cone that may not be normal, under only the condition that T is i-compact on the (possibly unbounded) set $[u_-, u_+] \subset X$. See [188] for the rest of the argument. □

We are interested in problem (10.1.35), where $A\colon X \to Z$ is invertible and satisfies (10.1.40), and where $B\colon X \to Z$ is monotone decreasing (10.1.39). We have the following simple lemmas which allow for connecting problem (10.1.35) to problem (10.1.36).

Lemma 10.1.9. *Let X and Z be ordered Banach spaces. Let $A\colon X \to Z$ be a linear invertible operator with the maximum principle property* (10.1.40). *Then A^{-1} has the monotone increasing property* (10.1.38).

Proof. Let $u, v \in Y$ such that $u \geqslant v$. Then since $u \geqslant v \Leftrightarrow u - v \geqslant 0$, and since

$$u - v \geqslant 0 \Leftrightarrow A(A^{-1}(u-v)) \geqslant 0 \Rightarrow A^{-1}(u-v) \geqslant 0 \Leftrightarrow A^{-1}u \geqslant A^{-1}v,$$

we have $u \geqslant v \Rightarrow A^{-1}u \geqslant A^{-1}v$, which shows that A^{-1} is monotone increasing. □

Lemma 10.1.10. *Let X and Z be ordered Banach spaces. Let $A\colon X \to Z$ be a linear invertible operator with the maximum principle property* (10.1.40), *and let $B\colon X \to Z$ be monotone decreasing (increasing). Then the operator $T\colon X \to X$ defined by $T(\cdot) = -A^{-1}B(\cdot)$ is monotone increasing (decreasing).*

Proof. Assume that B is monotone decreasing. Then

$$\begin{aligned}
u \geqslant v &\Rightarrow B(u) \leqslant B(v) \\
&\Leftrightarrow B(v) - B(u) \geqslant 0 \\
&\Leftrightarrow A(A^{-1}[B(v) - B(u)]) \geqslant 0 \\
&\Rightarrow A^{-1}[B(v) - B(u)] \geqslant 0 \\
&\Leftrightarrow A^{-1}B(v) - A^{-1}B(u) \geqslant 0 \\
&\Leftrightarrow -A^{-1}B(u) - (-A^{-1}B(v)) \geqslant 0 \\
&\Leftrightarrow T(u) - T(v) \geqslant 0 \\
&\Leftrightarrow T(u) \geqslant T(v),
\end{aligned}$$

which implies that if B is monotone decreasing, then $T = -A^{-1}B$ is monotone increasing. The case of B monotone increasing is similar, resulting in T being monotone decreasing. □

Lemma 10.1.11. *Let X and Z be ordered Banach spaces. Let $A\colon X \to Z$ be a linear invertible operator with maximum principle property* (10.1.40), *let $B\colon X \to Z$, and let $T\colon X \to X$ be defined as $T(u) = -A^{-1}B(u)$. If u_- and u_+ are sub- and supersolutions to* (10.1.35) *then u_- and u_+ are sub- and supersolutions to* (10.1.36).

Proof. Assume that u_+ satisfies $Au_+ + B(u_+) \geqslant 0$. Then

$$Au_+ \geqslant -B(u_+) \Rightarrow u_+ \geqslant -A^{-1}B(u_+) \Leftrightarrow u_+ \geqslant T(u_+).$$

The case of u_- is similar. □

We now have the second of the two main fixed-point theorems for monotone increasing maps; this one exploits the specific structure present in the operators A and B, and will be used in Section 10.4 to establish existence of solutions to a nonlinear elliptic equation.

Theorem 10.1.24 (Method of Sub- and Supersolutions). *Let X and Z be ordered Banach spaces. Let $A\colon X \to Z$ be a linear invertible operator with the maximum principle property* (10.1.40), *and let $B\colon X \to Z$ be compact and monotone decreasing* (10.1.39). *If u_- and u_+ are sub- and supersolutions to* (10.1.35) *in that*

$$Au_- + B(u_-) \leqslant 0, \qquad Au_+ + B(u_+) \geqslant 0,$$

then there exists a solution $u \in [u_-, u_+]$ to (10.1.35) *such that $Au + B(u) = 0$.*

Proof. Due to A being invertible, we can form $T(u) = -A^{-1}B(u)$. Since A satisfies the maximum principle and B is monotone decreasing, by Lemma 10.1.10 we know that T is monotone increasing. Now, since B is compact, we have T is compact since it is the composition of the continuous map A^{-1} and the compact map B. Additionally, by Lemma 10.1.11, the sub- and supersolutions u_- and u_+ of (10.1.35) are also sub- and supersolutions of (10.1.36). Therefore, by Theorem 10.1.23, there exists a solution to $u = T(u)$, and hence to $Au + B(u) = 0$, with $u \in [u_-, u_+]$. □

EXERCISES

10.1.1 Give an example of a function $F\colon \mathbb{R}^2 \to \mathbb{R}$ which has a well-defined gradient $\nabla F\colon \mathbb{R}^2 \to L(\mathbb{R}^2, \mathbb{R})$, but which is not **G**-differentiable in all directions $v \in \mathbb{R}^2$. (*Hint:* The gradient is the **G**-derivative in just the coordinate directions.)

10.1.2 Give an example of a function $F\colon \mathbb{R}^2 \to \mathbb{R}$ which is **G**-differentiable on $\mathbb{R}^2$ but which is not **F**-differentiable on $\mathbb{R}^2$.

10.1.3 Let X be a Hilbert space, and define $J\colon X \to \mathbb{R}$ as

$$J(u) = \frac{1}{2}a(u, u) - f(u),$$

where $a\colon X \times X \to \mathbb{R}$ is a symmetric, bounded, and coercive bilinear form, and where $f\colon X \to \mathbb{R}$ is a bounded linear functional. Prove that J has a unique minimizer $u \in X$ given by the solution of the operator equation $Au = F$, where $F \in X$ arises from $f(\cdot)$ through the Riesz Representation Theorem in Section 4.8 and where $A \in \mathcal{L}(X, X)$ arises from $a(\cdot, \cdot)$ through the Bounded Operator Theorem in Section 8.5.

10.1.4 Show that a compact linear operator on a Banach space is continuous, and in fact is completely continuous.

10.1.5 Prove that the conclusions of Theorem 10.1.9 still hold if two of the assumptions on J, namely that it is proper and bounded below, are replaced by the assumption that the weakly closed subset $U \subset X$ is also bounded.

10.1.6 Prove that the conclusions of Theorem 10.1.12 still hold if two of the assumptions on J, namely that it is proper and bounded below, are replaced by the assumption that the weakly closed subset $U \subset X$ is also bounded.

10.2 BEST AND NEAR-BEST APPROXIMATION IN BANACH SPACES

In this section we consider the problem of finding the best (or possibly near-best, to be made precise below) approximation to a given function in a Banach space from a set of candidate approximations taken from a subspace of the Banach space. This problem was considered briefly in Section 4.6 in the setting of separable Hilbert spaces. Here, the setting will be a more general Banach space X, the function we are interested in will be the (unknown) solution $u \in X$ to a given linear or nonlinear operator equation $F(u) = 0$, and the candidate approximations will form a (typically finite-dimensional) subspace $X_h \subset X$, where h parametrizes the dimension of X_h. We then consider three very specific problems:

(1) Is there a best approximation $u_h \in X_h \subset X$ to $u \in X$? If so, is it unique?

(2) Is there a method for computing the best (or near-best) approximation?

(3) What is the error in such a computable best or near-best approximation?

The answer to the first question (yes) will be provided by the Projection Theorem in the case that X is a Hilbert space, but is also affirmative more generally when X is a Banach space, under some minimal additional conditions on X and on $X_h \subset X$. The answer to the second question will be provided by studying Petrov-Galerkin and Galerkin methods for operator equations in Banach and Hilbert spaces, and the answer to the third question will be uncovered by developing *a priori* error estimates for these two methods, for both linear and nonlinear problems.

Best Approximation in Banach and Hilbert Spaces

Let X be a Banach space. For a given $u \in X$, the error in approximating u by elements in $U \subset X$ is denoted

$$E_U(u) = E_U(u)_X = \inf_{v \in U} \|u - v\|_X. \tag{10.2.1}$$

Of primary interest here is the existence, uniqueness, quality, and computability of a *best approximation* to a given $u \in X$, which is an element $u_0 \in U$ that achieves the infimum in (10.2.1). A *local best approximation* to a given $u \in X$ is an element $u_0 \in U \cap V$ that achieves the infimum in (10.2.1) for some open set V containing u_0.

Best Approximation in Hilbert Spaces. In the case of a Hilbert space H, the existence and uniqueness of a best approximation is given by the following result (a variation of which was presented in Chapter 4), which together with the Riesz Representation Theorem (which is proved using this theorem) are two of the key theorems on Hilbert spaces.

Theorem 10.2.1 (Hilbert Space Projection Theorem). *Let U be a closed subspace of a Hilbert space H. Then for every $u \in H$ we have the decomposition $u = Pu + z$, where $Pu \in U$ and $z \in U^{\perp}$. Moreover, $Pu \in U$ is the unique element in $U \subset H$ closest to $u \in H$:*

$$\|u - Pu\|_X = E_U(u),$$

and characterized by orthogonality of the error $z = u - Pu$ to the subspace U:

$$(u - Pu, w)_H = 0, \quad \forall w \in U.$$

Proof. See Chapter 4, or for example, [57, 116]. □

This theorem provides a solution to the best approximation problem in the setting of a Hilbert space H: The unique best approximation to $u \in H$ in a given (possibly infinite-dimensional) subspace $U \subset H$ is the Hilbert space projection $Pu \in U$, which is characterized by the orthogonality condition for the remainder (or *error*) $z = u - Pu \in U^{\perp}$:

$$(u - Pu, w)_H = 0, \quad \forall w \in U, \tag{10.2.2}$$

or equivalently, $(Pu, w) = (u, w)$, $\forall w \in U$. That the mapping $u \mapsto Pu$ is linear follows from the definitions, and it is not difficult to show that it is both idempotent ($P^2 = P$) and self-adjoint ($P = P^*$), where the Hilbert adjoint of P is the unique linear operator P^* satisfying $(Pu, v)_H = (u, P^* v)_H$, $\forall u, v \in H$. These two properties together characterize an *orthogonal projection* operator. As pointed out in Chapter 4, if H is a separable Hilbert space, then (10.2.2) gives an explicit prescription for the best approximation Pu in terms of basis components for the subspace $U = \text{span}\{\phi_j\}_{j=1}^{\infty}$. In particular, with $Pu = \sum_{j=1}^{\infty} c_j \phi_j$, for some set of basis coefficients $\{c_j\}_{j=1}^{\infty}$, $c_j \in \mathbb{R}$, equation (10.2.2) produces the (countably infinite) matrix system

$$\mathbf{AU} = \mathbf{F}, \tag{10.2.3}$$

with

$$\mathbf{A}_{ij} = (\phi_j, \phi_i)_H, \qquad \mathbf{U}_j = c_j, \qquad \mathbf{F}_i = (u, \phi_i)_H.$$

When the subspace $U \subset H$ is finite-dimensional, $U = \text{span}\{\phi_j\}_{j=1}^{n}$, then the matrix $\mathbf{A}$ is simply an invertible $n \times n$ matrix equation for the expansion coefficients $\{c_j\}_{i=1}^{n}$, yielding the best approximation $Pu = \sum_{j=1}^{n} c_j \phi_j \in U \subset H$ to $u \in H$. The matrix $\mathbf{A}$ is always symmetric positive definite (SPD), due the inner product being a symmetric positive definite bilinear form on $H \times H$. If the basis is (conveniently) orthogonal, or better, orthonormal, then the matrix $\mathbf{A}$ is, in fact, a diagonal matrix. In this case, computing the best approximation is a simple matter of computing the expansion coefficients by projecting the target function onto each basis function (involving evaluating two inner products), and then summing up the linear combination of basis functions using these expansion coefficients.

Best Approximation in Banach Spaces. In the more general setting of a Banach space X, there is no analogue of the Projection Theorem that gives existence and uniqueness of best approximation in a general (possibly infinite-dimensional) closed subspace $U \subset X$. However, it is possible to use the generalized notion of orthogonality, involving the dual space as in (10.1.15), to give the following result.

Theorem 10.2.2. *Let X be a Banach space, and let U be a closed subspace of X. Given an arbitrary $u \in X$, the element $u_0 \in U$ is a best approximation to u if and only if there exists $f \in U^\perp$ such that*

$$\|f\|_{X'} = 1, \qquad \|u - u_0\|_X = f(u).$$

In this case,

$$E_U(u) = \sup_{f \in U^\perp, \|f\|_{X'}=1} f(u).$$

Proof. A proof may be found in [32] or [64]. □

By viewing the problem of best approximation to $u \in X$ for a Banach space X as the problem of minimizing the functional $J \colon X \to \mathbb{R}$ over a subspace $U \subset X$, where J is defined by

$$J(v) = \|u - v\|_X, \tag{10.2.4}$$

then we can make use of the tools we assembled in Section 10.1 on the existence of minimizers of such functionals. We first observe that the norm on a Banach space is a continuous function of its arguments.

Lemma 10.2.1. *Let X be a Banach space with norm $\|\cdot\|_X$. Then the map $u \mapsto \|u\|_X$ is a continuous function of its argument. Similarly, let H be a Hilbert space with inner product $(\cdot,\cdot)_H$. Then the map $\{u, v\} \mapsto (u, v)_H$ is a continuous function of its arguments.*

Proof. Let $u_j \to u \in X$. By the triangle inequality in X, we first note that $\|u\|_X = \|u - u_j + u_j\|_X \leqslant \|u - u_j\|_X + \|u_j\|_X$, which gives

$$|\ \|u\|_X - \|u_j\|_X\ | \leqslant \|u - u_j\|_X \to \infty \text{ as } j \to \infty.$$

The proof for the case of an inner product is similar. (See the exercises at the end of this section.) □

We now observe that J has the following properties.

Lemma 10.2.2. *Let X be a Banach space, let $u \in X$ be given, and let $U \subset X$ be any subspace, including the case $U = X$. Then the functional $J \colon X \to \mathbb{R}$ defined by* (10.2.4) *has the following properties:*

(1) J is continuous (hence lower semicontinuous) on U.

(2) J is convex on U.

(3) J is proper on U.

(4) J is bounded below on U.

Proof. By Lemma 10.2.1, J is continuous. Convexity of J follows from the triangle inequality and the norm property $\|av\|_X = |a|\|v\|_X$, for $\forall a \in \mathbb{R}, \forall v \in U$. That J is proper follows from $\|v\|_X = \|v - u + u\|_X \leqslant \|u - v\|_X + \|u\|_X$, which gives

$$J(v) = \|u - v\|_X \geqslant \|v\|_X - \|u\|_X = C_1\|v\|_X - C_2, \tag{10.2.5}$$

with $C_1 = 1$ and $C_2 = \|u\|_X$. This ensures that $J(v) \to \infty$ as $\|v\|_X \to \infty$. That J is bounded below comes by noting that (10.2.5) ensures $J(v) \geqslant -C_2 > -\infty$. □

Lemma 10.2.2 allows for a very simple existence proof of best approximation in a Banach space, when the subset U is compact.

Theorem 10.2.3. *Let X be a Banach space, and let U be a compact subset of X. Given an arbitrary $u \in X$, there exists a best approximation $u_0 \in U$ to u such that*

$$\|u - u_0\|_X = E_U(u).$$

Proof. For fixed $u \in X$, the distance function $J(\cdot) = \|u - \cdot\|_X : U \to \mathbb{R}$ defines a map $v \mapsto \|u - v\|_X$. By Lemma 10.2.2, this map is a continuous function of $v \in X$. Since it is continuous on the compact set U, it achieves its minimum $E_U(u)$ at a point $u_0 \in U$, yielding a best approximation. □

More generally, when $U \subset X$ is not compact, we can use Lemma 10.2.2 together with the more sophisticated framework we have developed in Section 10.1 to give the following general result.

Theorem 10.2.4. *Let X be a reflexive Banach space, and let $U \subset X$ be a closed convex subset of X. Given an arbitrary $u \in X$, there exists a best approximation $u_0 \in U$ to u such that*

$$\|u - u_0\|_X = E_U(u). \tag{10.2.6}$$

Proof. By Lemma 10.2.2, the functional $J = \|u - u_0\|_X$ is a proper, bounded below, convex, and lower semicontinuous functional on U. Since U is a closed convex subset, by Theorem 10.1.12 there exists a solution to (10.2.6). □

If we consider finite-dimensional subspaces, then we no longer need reflexivity of the enclosing space X, and immediately have the following.

Theorem 10.2.5. *Let X be a Banach space, and let U be a finite-dimensional subspace of X. Given an arbitrary $u \in X$, there exists a best approximation $u_0 \in U$ to u such that*

$$\|u - u_0\|_X = E_U(u).$$

Proof. By Theorem 10.1.4 we know that U is a *closed* subspace of X. Consider now the intersection of $U \subset X$ with the closed ball in X of radius $R = 2\|u\|_X$ about zero:

$$U_R = U \cap B_R(0), \qquad B_R(v) = \{w \in X \ : \ \|v - w\|_X \leqslant R\}.$$

As the intersection of two closed sets, the set U_R is closed. It is also nonempty, since as a subspace U inherits the zero element of X, which lies in both U and $B_R(0)$. Since U_R is closed and bounded, as well as finite-dimensional, it is compact. By Theorem 10.2.3, there exists a best approximation $u_0 \in U_R$ to $u \in X$. Now, for $v \notin U_R$, we know that $\|v\|_X > R = 2\|u\|_X$. However, this implies that

$$\|u-v\|_X \geqslant \|v\|_X - \|u\|_X > 2\|u\|_X - \|u\|_X = \|u\|_X = \|u-0\|_X \geqslant \|u-u_0\|_X,$$

since u_0 is a best approximation in U_R, which contains the element 0. Since one of the inequalities above is strict, there can be no best approximation in U outside U_R, so that u_0 is in fact a best approximation to $u \in X$ from all of U, rather than just from U_R. □

Here is a second distinct proof.

Proof. (Alternative approach.) By definition of $E_U(u)$, we know that there exists a sequence $\{u_j\}$ in U such that $\|u - u_j\|_X \to E_U(u)$. What remains is to prove that $u_j \to u_0$ for some some $u_0 \in U$. Note that

$$\|u_j\|_X = \|u - u + u_j\|_X \leqslant \|u\|_X + \|u - u_j\|_X,$$

therefore, $\{u_j\}$ is a bounded sequence in U. Note that U is a reflexive Banach space, since all finite-dimensional Banach spaces are reflexive, as observed in Section 4.8. By Theorem 4.8.9 there exists a subsequence $\{u_{j_k}\}$ of $\{u_j\}$, which we immediately relabel as $\{u_j\}$ without danger of confusion, and an element $u_0 \in U$, such that $u_j \rightharpoonup u_0 \in U$. However, as noted in Theorem 4.8.6 , weak and strong convergence are equivalent in finite-dimensional Banach spaces, so we have $u_j \to u_0 \in U$. Continuity of the norm implies that $\|u - u_j\|_X \to \|u - u_0\|_X = E_U(u)$. □

To complete the picture, it is important to know when we have a unique best approximation in a Banach space, analogous to the result in the case of a Hilbert space. Adding convexity to a Banach space, which gives it just part of the structure always present in a Hilbert space, is enough to guarantee uniqueness of best approximation.

Theorem 10.2.6. *Let X be a strictly convex Banach space, and let $U \subset X$ be a convex subset of X. Let $u \in X$ be arbitrary. If there exists a best approximation $u_0 \in U$ to u, then u_0 must be unique.*

Proof. Let $u_0, u_1 \in U$ both be distinct best approximations to u, so that $u_0 \neq u_1$ and that $\|u - u_0\|_X = \|u - u_1\|_X = E_U(u)$. Consider now the convex combination $u_\alpha = \alpha u_0 + (1-\alpha)u_1$, for any $\alpha \in (0,1)$. Since U is convex, we know that u_α is also in U. The assumption that $u_0 \neq u_1$ implies $u - u_0 \neq u - u_1$, and the characterization of strict convexity leads to a strict inequality

$$\begin{aligned}
\|u - u_\alpha\|_X &= \|u - \alpha u_0 - (1-\alpha)u_1\|_X \\
&= \|\alpha(u - u_0) + (1-\alpha)(u - u_1)\|_X \\
&< \alpha\|u - u_0\|_X + (1-\alpha)\|u - u_1\|_X \\
&= \alpha E_U(u) + (1-\alpha)E_U(u) \\
&= E_U(u).
\end{aligned}$$

Therefore, $u_\alpha \in U$ is a better approximation than u_0 and u_1, contradicting the assumption that they are both best approximations from U. Therefore, if a best approximation from U exists, it must be unique. □

We finally have the main results we were after, which are analogues of the Hilbert Space Projection Theorem in the more general setting of Banach spaces.

Theorem 10.2.7. *Let X be a strictly convex Banach space, and let $U \subset X$ be a finite-dimensional subspace of X. Given an arbitrary $u \in X$, there exists a unique best approximation $u_0 \in U$ to u such that*

$$\|u - u_0\|_X = E_U(u).$$

Proof. Since U is a finite-dimensional subspace of X, by Theorem 10.2.5, there exists at least one best approximation $u_0 \in U$ to u. Since U is a subspace, it forms a convex subset of X. Furthermore, since X is strictly convex, by Theorem 10.2.6, there is at most one best approximation $u_0 \in U$. Therefore, the $u_0 \in U$ provided by Theorem 10.2.5 is the unique best approximation to $u \in X$. □

More generally, we have the following.

Theorem 10.2.8. *Let X be a reflexive and strictly convex Banach space, and let $U \subset X$ be a closed convex subset of X. Given an arbitrary $u \in X$, there exists a unique best approximation $u_0 \in U$ to u such that*

$$\|u - u_0\|_X = E_U(u). \tag{10.2.7}$$

Proof. By Theorem 10.2.4, the assumptions imply the existence of a best approximation satisfying (10.2.7). By Theorem 10.2.6, the strict convexity assumption on X ensures the best approximation is unique. □

Near-Best Approximation with Petrov-Galerkin Methods

In the previous section, we reviewed the Projection Theorem on Hilbert spaces that guarantees the existence and uniqueness of a best approximation in a closed subspace of a Hilbert space, and then generalized this result to Banach spaces by adding just enough additional structure. In particular, we were able to guarantee existence and uniqueness of a best approximation to an arbitrary element $u \in X$ in a Banach space X from a subspace $U \subset X$ with the following additional structure:

(1) X is a reflexive and strictly convex Banach space.

(2) $U \subset X$ is a closed convex subset (such as a finite-dimensional subspace).

Most of the function spaces of interest to us later in this chapter, namely the L^p spaces for $1 < p < \infty$ and the L^p-based Sobolev spaces, are strictly convex and in fact satisfy the stronger uniform convexity condition due to the Clarkson inequalities (see [2]), which also ensures reflexivity by Theorem 10.1.7. Moreover, we are primarily interested in developing practical approximation techniques whereby we explicitly construct the best (or near-best, see below) approximation, meaning that we will necessarily work only with finite-dimensional subspaces $U \subset X$. Therefore, for our setting of interest, we have existence and uniqueness of best approximation by Theorems 10.2.7 and 10.2.8.

However, we still have a problem; we do not actually have the target function $u \in X$ in our hands, but rather have it only as the solution to a (generally nonlinear) operator equation of the form (10.1.1), or rather (10.1.2). Moreover, unlike the Hilbert space setting, whereby one has an explicit procedure for determining the best approximation using the orthogonality characterization involving the remainder (10.2.2) (which requires being able to explicitly work with the target function $u \in X$), we have only a nonconstructive existence result in Theorem 10.2.7. When u is an unknown solution to a given operator equation, and without an explicit construction as in the case of (separable) Hilbert spaces, how will we explicitly construct the best approximation, or an approximation that we can prove is "close" to the best approximation? The answer is provided by the *Petrov-Galerkin (PG)* and *Galerkin (G)* methods.

We are faced with the following problem:

$$\text{Find } u \in X \text{ such that } \langle F(u), v\rangle = 0 \in \mathbb{R}, \qquad \forall v \in Y, \tag{10.2.8}$$

where X and Y are Banach spaces, and where $F\colon X \to Y'$. Here, $\langle F(u), \cdot\rangle$ represents the action of $F(u)$ as a linear functional on Y, for fixed $u \in X$. A *Petrov-Galerkin (PG) method* looks for an approximation $u_h \approx u$ satisfying the functional problem (10.2.8) in subspaces:

$$\text{Find } u_h \in X_h \subset X \text{ such that } \langle F(u_h), v_h\rangle = 0,\ \forall v_h \in Y_h \subset Y. \tag{10.2.9}$$

A *Galerkin method* is the special case of $Y = X$ and $Y_h = X_h$. Ideally, we would like one or both methods to yield a solution u_h which is the best approximation to the (unknown) $u \in X$ that solves problem (10.2.8). Petrov-Galerkin and Galerkin methods are, in fact, Banach space generalizations of Hilbert space projection in the following sense. If $X = Y = H$ is a Hilbert space, if $\langle F(u), \cdot\rangle = (u, \cdot)_H$, and if we take $X_h = Y_h = U \subset H$ as a closed subspace, then subtracting (10.2.8) and (10.2.9) gives:

$$\text{Find } u_h \in U \subset H \text{ such that } (u - u_h, v)_H = 0, \quad \forall v \in U \subset X. \tag{10.2.10}$$

This is precisely the characterization of the Hilbert space projection in terms of orthogonality of the remainder appearing earlier in (10.2.2), with $u_h = Pu$, and with P the orthogonal projection operator onto $U \subset H$.

Back to the more general problem (10.2.8) and the PG method (10.2.9), let us consider the case $\dim(X_h) = \dim(Y_h) = n < \infty$, with

(1) $X_h = \text{span}\{\phi_1, \ldots, \phi_n\} \subset X$,

(2) $Y_h = \text{span}\{\psi_1, \ldots, \psi_n\} \subset Y$,

for some bases $\{\phi_j\}$ and $\{\psi_j\}$. In this case, similar to our approach yielding (10.2.3) in the case of the Hilbert projection, the problem reduces to determining the appropriate coefficients in the expansion:

$$u_h = \sum_{j=1}^{n} c_j \phi_j. \tag{10.2.11}$$

The PG method (10.2.9) gives n (generally nonlinear) equations for the n coefficients:

$$\text{Find } u_h = \sum_{j=1}^{n} c_j \phi_j \text{ such that } \langle F(u_h), \psi_i \rangle = 0, \quad i = 1, \ldots, n. \tag{10.2.12}$$

For nonlinear algebraic systems of equations, Newton-like methods are often the most robust and efficient of all numerical techniques available; we will examine these methods closely in Section 10.7. For the moment, we will just note that for a PG approximation $u_h = \sum_{j=1}^{n} c_j \phi_j$, an $n \times n$ matrix equation is produced at each iteration of a Newton-type method for solving (10.2.12), which produces the Newton correction $w_h = \sum_{j=1}^{n} d_j \phi_j$ as the solution of a linear system similar to (10.2.3):

$$\mathbf{AW} = \mathbf{B}, \tag{10.2.13}$$

where now

$$\mathbf{A}_{ij} = \langle F'(u_h)\phi_j, \psi_i \rangle, \quad \mathbf{W}_j = d_j, \quad \mathbf{B}_i = -\langle F(u_h), \psi_i \rangle, \quad i, j = 1, \ldots, n.$$

The linear operator $F'(u_h)$ is the Fréchet derivative of F at u_h. While this linear system changes at each Newton iteration for (10.2.12) and must be solved repeatedly, the situation is different if the original operator equation is linear:

$$\langle F(u), v \rangle = a(u, v) - f(v), \quad \forall u \in X, \quad \forall v \in Y.$$

In this case, our problem (10.2.8) and its PG approximation (10.2.9) become:

$$\text{Find } u \in X \text{ such that } a(u, v) = f(v), \qquad \forall v \in Y, \tag{10.2.14}$$

$$\text{Find } u_h \in X_h \subset X \text{ such that } a(u_h, v_h) = f(v_h), \qquad \forall v_h \in Y_h \subset Y. \tag{10.2.15}$$

Here, $a \colon X \times Y \to \mathbb{R}$ is a bilinear form, and $f \colon Y \to \mathbb{R}$ is a linear functional. In this case, the PG system (10.2.12) for the expansion coefficients in (10.2.11) reduces to the purely linear algebraic system (10.2.13), where now

$$\mathbf{A}_{ij} = a(\phi_j, \psi_i), \quad \mathbf{W}_j = c_j, \quad \mathbf{B}_i = f(\psi_i), \quad i, j = 1, \ldots, n.$$

One simply solves (10.2.13) for the coefficients $\{c_j\}$ once and for all; since it is linear, there is no need for a Newton iteration.

Let us now summarize what we have seen in this section:

(1) Petrov-Galerkin (PG) methods are generalizations of Hilbert space projection to Banach spaces, and to (unknown) solutions of operator equations rather than to given (known) target functions;

(2) PG methods reduce to Hilbert space projection in the case that the solution and test spaces are a single Hilbert space, and the nonlinear operator is the inner product on the Hilbert space with a given (known) target function;

(3) PG methods require the solution of nonlinear algebraic systems equations of the form (10.2.12), which can be tackled using standard Newton-type methods;

(4) PG methods for nonlinear problems which employ Newton methods, or PG methods for purely linear problems, require the solution of matrix systems of the form (10.2.13), similar in structure to the *moment matrix* systems (10.2.3) arising in Hilbert space projection.

Regarding the last item in the list, which concerns the linear system (10.2.13), for practical reasons one hopes that:

(1) The cost of storing the matrix A is as close as possible to the optimal $\mathcal{O}(n)$.

(2) The cost of inverting the matrix A is as close as possible to the optimal $\mathcal{O}(n)$.

We examine both of these questions in detail in Section 10.6.

Petrov-Galerkin Error Estimates for Linear Equations. Since it is now clear that we can produce PG approximations in finite-dimensional subspaces by using standard algorithms for solving linear and nonlinear algebraic equations (see Sections 10.6 and 10.7), what remains then is to characterize the *quality* of the PG approximation. Since the results we assembled on the existence of best approximation u_0 in a subspace X_h of a Banach space X did not involve the additional constraint that the approximation u_0 also solve an operator equation in X_h, it is clear that a PG approximation taken from X_h will generally be distinct from the best approximations guaranteed to exist by these theorems. Moreover, since we have shown above that the best approximation is unique in convex subsets $X_h \subset X$, PG approximation will generally not be a best approximation.

The remaining question is then: How good is PG approximation? Ideally, what we would hope to establish is that PG approximation is *near-best*, or *quasi-optimal*, in the sense that

$$\|u - u_h\|_X \leqslant CE_{X_h}(u) = C \inf_{v \in X_h} \|u - v\|_X, \qquad (10.2.16)$$

for some constant $C \geqslant 1$ which is independent of the dimension of the subspace X_h. This would imply that PG approximation is within a constant of being best approximation, and moreover, if the best approximation can be improved by enlarging (or more generally, by changing) the subspace X_h, then the PG approximation will also improve at the same rate.

Let us consider now the error in Petrov-Galerkin approximation in the case of a linear problem, with $X \neq Y$:

$$\text{Find } u \in X \text{ such that } a(u,v) = f(v), \quad \forall v \in Y. \tag{10.2.17}$$

$$\text{Find } u_h \in X_h \subset X \text{ such that } a(u_h,v_h) = f(v_h), \quad \forall v_h \in Y_h \subset Y. \tag{10.2.18}$$

We will need to assume that the following three conditions hold for the bilinear form $a\colon X \times Y \to \mathbb{R}$ and the linear functional $f\colon Y \to \mathbb{R}$ on the Banach spaces X and Y:

$$a(u,v) \leqslant M\|u\|_X\|v\|_Y, \quad M > 0, \ \forall u \in X, \ \forall v \in Y, \tag{10.2.19}$$

$$f(v) \leqslant L\|v\|_Y, \quad L > 0, \ \forall v \in Y, \tag{10.2.20}$$

$$\inf_{u\in X}\sup_{v\in Y} \frac{a(u,v)}{\|u\|_X\|v\|_Y} \geqslant m > 0. \tag{10.2.21}$$

Note that while conditions (10.2.19) and (10.2.20) also hold when restricted to the subspaces X_h and Y_h, the third condition (10.2.21) may in fact not hold when restricted to the subspaces X_h and Y_h. Therefore, we will also need to assume separately that this condition holds on the subspaces X_h and Y_h as a separate fourth (Ladyzhenskaya-Babuška-Brezzi, or LBB) condition:

$$\inf_{u_h\in X_h}\sup_{v_h\in Y_h} \frac{a(u_h,v_h)}{\|u_h\|_X\|v_h\|_Y} \geqslant m_h > 0. \tag{10.2.22}$$

We then have the following standard result on the quality of PG approximation.

Theorem 10.2.9 (Ladyzhenskaya-Babuška-Brezzi Theorem)**.** *Let X and Y be Banach spaces, let $X_h \subset X$ and $Y_h \subset Y$ be subspaces, let $u \in X$ be the solution to the linear problem* (10.2.14), *and let $u_h \in X_h$ be the PG approximation satisfying* (10.2.15). *Suppose that conditions* (10.2.19)–(10.2.22) *hold. Then*

$$\|u-u_h\|_X \leqslant \left(1+\frac{M}{m}\right) E_{X_h}(u). \tag{10.2.23}$$

Proof. Let P_h denote the generalized projection of u onto the unique PG approximation $u_h = P_h u$, and let $\|P_h\|$ denote the induced operator norm on X. Then, using idempotence of P_h we have

$$\begin{aligned}\|u-u_h\|_X &= \|(I-P_h)(u-w_h)\|_X \\ &\leqslant \|I-P_h\| \, \|u-w_h\|_X \\ &\leqslant (1+\|P_h\|) \, \|u-w_h\|_X.\end{aligned} \tag{10.2.24}$$

Using the fact that $a(u_h,v_h) = a(u,v_h)$, $\forall v_h \in Y_h$, one notes:

$$m\|P_h u\|_X = m\|u_h\|_X \leqslant \sup_{v_h\in Y_h} \frac{a(u_h,v_h)}{\|v_h\|_Y} = \sup_{v_h\in Y_h} \frac{a(u,v_h)}{\|v_h\|_Y} \leqslant M\|u\|_X, \tag{10.2.25}$$

giving $\|P_h\| \leqslant M/m$. Employing (10.2.25) in (10.2.24) then gives (10.2.23). □

In the special case that $X = H$ is a Hilbert space, then it was pointed out in [180] that the following result on nontrivial idempotent linear operators can be used to remove the leading "1" in the PG constant in (10.2.23).

Lemma 10.2.3 (Kato,Xu-Zikatanov Lemma). *Let H be a Hilbert space. If $P \in L(H, H)$ satisfies $0 \neq P^2 = P \neq I$, and if $\|\cdot\|$ denotes the induced operator norm on H, then*

$$\|P\| = \|I - P\|.$$

Proof. A proof may be found in [110, 180]. □

To see how Lemma 10.2.3 can be used for this purpose, let P_h denote the generalized projection of u onto the unique PG approximation $u_h = P_h u$, as in the proof of Theorem 10.2.9. The result in Lemma 10.2.3 now gives

$$\begin{aligned} \|u - u_h\|_X &= \|(I - P_h)(u - w_h)\|_X \\ &\leqslant \|I - P_h\| \, \|u - w_h\|_X \\ &= \|P_h\| \, \|u - w_h\|_X. \end{aligned} \tag{10.2.26}$$

Employing (10.2.25) in (10.2.26) now gives the improved constant:

$$\|u - u_h\|_X \leqslant \left(\frac{M}{m}\right) E_{X_h}(u).$$

Petrov-Galerkin Error Estimates for Nonlinear Equations. We now extend the near-best approximation results to a general class of nonlinear problems; we follow the approach in [99], which is based on the ideas of [143, 144].

Let X and Y be Banach spaces. We are concerned with the following nonlinear problem and its Petrov-Galerkin approximation:

$$\text{Find } u \in X \text{ such that } \langle F(u), v\rangle = 0, \qquad \forall v \in Y, \tag{10.2.27}$$

$$\text{Find } u_h \in X_h \subset X \text{ such that } \langle F(u_h), v_h\rangle = 0, \qquad \forall v_h \in Y_h \subset Y, \tag{10.2.28}$$

where $F\colon X \to Y'$, and where $\langle F(u), \cdot\rangle$ represents the action of $F(u)$ as a linear functional on Y, for fixed $u \in X$. A critical basic assumption we again make about the approximation subspaces X_h and Y_h is

$$\dim(X_h) = \dim(Y_h). \tag{10.2.29}$$

We assume that $F \in C^1(X, Y')$, and in order to derive an *a priori* estimate for the error in the PG approximation, we will need to rely on some auxiliary results that characterize "locally good" approximations $G \approx F$. To characterize "local" we recall that the closed ball of radius $\delta > 0$ about the point $u \in X$ is defined as

$$B_\delta(u) = \{w \in X \; : \; \|u - w\|_X \leqslant \delta\} \subset X.$$

We can now state the conditions we need on local approximations.

Assumption 10.2.1. *Assume that $G \in C^1(X, Y')$ and has the following properties:*

(H1) There exists a constant $\delta > 0$ such that G' satisfies

$$\|G'(u) - G'(x)\|_{\mathcal{L}(X,Y')} \leqslant L\|u - x\|_X, \quad \forall x \in B_\delta(u).$$

(H2) $G'(u)$ is an isomorphism from $X \to Y'$, and there exists $K > 0$ such that

$$\left\|[G'(u)]^{-1}\right\|_{\mathcal{L}(Y',X)} \leqslant K.$$

(H3) $\|G(u)\|_{Y'} \leqslant C$, where $C = \min\left\{\frac{\delta}{2K}, \frac{1}{4K^2L}\right\}$.

Properties (H2) and (H3) of Assumption 10.2.1 are stability and consistency conditions, respectively. If G satisfies Assumption 10.2.1, then we have the following lemma from [99], which is a variation of one of the key results in [143].

Lemma 10.2.4 ([99]). *Let G satisfy Assumption 10.2.1. Then there exists a constant $\delta_0 > 0$ and a unique $u_G \in X$ such that $G(u_G) = 0$, and $u_G \in B_{\delta_0}(u)$. Moreover, we have the following* a priori *error estimate:*

$$\|u - u_G\|_X \leqslant 2\left\|[G'(u)]^{-1}\right\|_{\mathcal{L}(Y',X)} \|G(u)\|_{Y'}.$$

Proof. We show existence and uniqueness through the Banach Fixed-Point Theorem (see Section 10.1 and Chapter 4). First define

$$T(x) = x - [G'(u)]^{-1}G(x), \qquad \forall x \in X.$$

This new operator T is well-defined because $G'(u)$ is an isomorphism by (H2) of Assumption 10.2.1. We now show that the assumptions ensure that T is a contraction on X, and that it maps a sufficiently small ball contained in X back into itself. To this end, pick any $x_1, x_2 \in X$. We have by generalized Taylor expansion (see Section 10.1) that

$$\begin{aligned}
T(x_1) - T(x_2) &= (x_1 - x_2) + [G'(u)]^{-1}(G(x_2) - G(x_1)) \\
&= (x_1 - x_2) - [G'(u)]^{-1} \int_0^1 G'(x_1 + t(x_2 - x_1))(x_1 - x_2)\, dt \\
&= [G'(u)]^{-1} \int_0^1 (G'(u) - G'(x_1 + t(x_2 - x_1)))(x_1 - x_2)\, dt.
\end{aligned}$$

Now pick $\delta_0 > 0$ such that $\delta_0 = \min\{\delta, 1/(2LK)\}$. We now show that T is a contraction mapping in the (closed) ball $B_{\delta_0}(u) \subset X$. By (H1) of Assumption 10.2.1, we have

$$\|G'(u) - G'((x_1 + t(x_2 - x_1))\|_{\mathcal{L}(X,Y')} \leqslant L\delta_0, \quad \forall x_1, x_2 \in B_{\delta_0}(u).$$

By (H2) of Assumption 10.2.1 and by the choice of δ_0, we have

$$\|T(x_1) - T(x_2)\|_X \leqslant L\delta_0 \left\|[G'(u)]^{-1}\right\|_{\mathcal{L}(Y',X)} \|x_1 - x_2\|_X \leqslant \frac{1}{2}\|x_1 - x_2\|_X.$$

Therefore, T is a contraction on $B_{\delta_0}(u)$. We now show that T maps the closed subset $B_{\delta_0}(u) \subset X$ into itself. By using the inequality above and (H3) of Assumption 10.2.1, for any $x \in B_{\delta_0}(u)$ we have

$$\begin{aligned}\|T(x) - u\|_X &\leqslant \|T(x) - T(u)\|_X + \|T(u) - u\|_X \\ &\leqslant \frac{1}{2}\|x - u\|_X + \left\|[G'(u)]^{-1}G(u)\right\|_X \\ &\leqslant \frac{1}{2}\delta_0 + KC \leqslant \delta_0.\end{aligned}$$

Therefore, T is a contraction mapping from the closed ball $B_{\delta_0}(u)$ to itself. Thus, there exists a unique $u_G \in B_{\delta_0}(u)$ such that $u_G = T(u_G)$, or in other words, such that $G(u_G) = 0$. Moreover,

$$\|u - u_G\|_X = \|u - T(u_G)\|_X \leqslant 2\left\|[G'(u)]^{-1}\right\|_{\mathcal{L}(Y',X)} \|G(u)\|_{Y'},$$

which completes the proof. □

Lemma 10.2.4 gives an abstract framework for existence, uniqueness, and *a priori* error estimates (giving then continuous dependence) for the approximate method $G(x) = 0$. Based on this lemma, we now construct such an approximate nonlinear operator $G \approx F$ for the Petrov-Galerkin formulation (10.2.28). This turns out to be somewhat nontrivial, since Petrov-Galerkin formulations are built only on the subspaces (X_h, Y_h), whereas the operator $G\colon X \to Y'$ is defined on the pair (X, Y). For each pair (X_h, Y_h), we need to construct an operator $F_h\colon X \to Y'$ such that the weak solution of $F_h(x) = 0$ is equivalent to the solution of (10.2.28). To this end, let us first introduce a bilinear form $a(\cdot,\cdot)\colon X \times Y \to \mathbb{R}$ at $u \in X$:

$$a(x, y) = \langle F'(u)x, y\rangle, \quad \forall x \in X, \quad \forall y \in Y, \tag{10.2.30}$$

which is the linearization of F at u. Denote by M the norm of $a(\cdot,\cdot)$:

$$M = \|a\|_{\mathcal{L}([X,Y],\mathbb{R})} := \sup_{\substack{0\neq x\in X\\ 0\neq y\in Y}} \frac{|a(x,y)|}{\|x\|_X\|y\|_Y} = \|F'(u)\|_{\mathcal{L}(X,Y')}. \tag{10.2.31}$$

This ensures the bound: $a(u, v) \leqslant M\|u\|_X\|v\|_Y$, $\forall u \in X$, $\forall v \in Y$. We assume that "inf-sup" conditions hold for $a(\cdot,\cdot)$: There exists a constant $m > 0$ such that

$$\inf_{x\in X}\sup_{y\in Y} \frac{a(x,y)}{\|x\|_X\|y\|_Y} = \inf_{y\in Y}\sup_{x\in X} \frac{a(x,y)}{\|x\|_X\|y\|_Y} \geqslant m > 0. \tag{10.2.32a}$$

This is equivalent to assuming that $F'(u)$ is an isomorphism from X to Y' with

$$\|[F'(u)]^{-1}\|_{\mathcal{L}(Y',X)} = m^{-1}.$$

In the subspaces (X_h, Y_h), we assume a discrete version of (10.2.32a)

$$\inf_{x\in X_h}\sup_{y\in Y_h}\frac{a(x,y)}{\|x\|_X\|y\|_Y}=\inf_{y\in Y_h}\sup_{x\in X_h}\frac{a(x,y)}{\|x\|_X\|y\|_Y}\geqslant m_h>0. \tag{10.2.32b}$$

The inf-sup condition (10.2.32b) implies the existence of two projectors,

$$\Pi_h^X : X\to X_h \quad\text{and}\quad \Pi_h^Y : Y\to Y_h,$$

defined by

$$a(x-\Pi_h^X x, y_h)=0,\quad \forall y_h\in Y_h,\quad \forall x\in X, \tag{10.2.33}$$
$$a(x_h, y-\Pi_h^Y y)=0,\quad \forall x_h\in X_h,\quad \forall y\in Y. \tag{10.2.34}$$

These operators are stable in the following sense:

$$\|\Pi_h^X\|_{\mathcal{L}(X,X_h)}\leqslant\frac{M}{m_h}\quad\text{and}\quad\|\Pi_h^Y\|_{\mathcal{L}(Y,Y_h)}\leqslant\frac{M}{m_h}. \tag{10.2.35}$$

Consider the projector Π_h^X; the discrete inf-sup condition (10.2.32b) gives

$$m_h\|\Pi_h^X x\|_X\leqslant\sup_{y_h\in Y_h}\frac{a(\Pi_h^X x, y_h)}{\|y_h\|_Y}=\sup_{y_h\in Y_h}\frac{a(x,y_h)}{\|y_h\|_Y}\leqslant M\|x\|_X.$$

Moreover, the discrete inf-sup condition (10.2.32b) guarantees that

$$(\Pi_h^X)^2=\Pi_h^X \text{ and } (\Pi_h^Y)^2=\Pi_h^Y.$$

Now we are ready to define the approximate nonlinear operator $F_h : X\to Y'$:

$$\langle F_h(x), y\rangle := \langle F(x), \Pi_h^Y y\rangle + a(x, y-\Pi_h^Y y),\ \ \forall x\in X,\ y\in Y. \tag{10.2.36}$$

By construction we have that

$$\langle F_h'(x)w, y\rangle := \langle F'(x)w, \Pi_h^Y y\rangle + \langle F'(u)w, y-\Pi_h^Y y\rangle. \tag{10.2.37}$$

In particular, we have $F_h'(u) = F'(u)$. This operator F_h gives rise to another nonlinear problem:

$$\text{Find } w\in X \text{ such that } \langle F_h(w), y\rangle = 0,\quad \forall y\in Y. \tag{10.2.38}$$

The equation (10.2.38) is posed on the entire pair of spaces (X, Y). However, it is not difficult to verify that the solution to (10.2.28) and the zero of (10.2.36) are equivalent:

Lemma 10.2.5 ([143, Lemma 1]). *It holds that $u_h\in X_h$ is a solution of* (10.2.28) *if and only if $u_h\in X$ is a solution of* (10.2.38).

Proof. We include the proof here for completeness. If $u_h\in X_h\subset X$ is a solution to (10.2.28), then

$$\langle F(u_h), v_h\rangle = 0,\quad \forall v_h\in Y_h.$$

Therefore,

$$\langle F(u_h), \Pi_h^Y y\rangle = 0, \quad \forall y \in Y.$$

For the second term in (10.2.36), notice that $u_h \in X_h$, and by the definition of Π_h^Y, we have

$$a(u_h, y - \Pi_h^Y y) = 0, \quad \forall y \in Y.$$

Thus, $u_h \in X$ is a solution to (10.2.38).

Conversely, let $w \in X$ satisfy $\langle F_h(w), y\rangle = 0, \ \forall y \in Y$; that is,

$$\langle F(w), \Pi_h^Y y\rangle + a(w, y - \Pi_h^Y y) = 0, \ \forall y \in Y.$$

By choosing $y = v - \Pi_h^Y v$, we obtain $a(w, v - \Pi_h^Y v) = 0, \quad \forall v \in Y$. By the definition of Π_h^Y and Π_h^X, we then have

$$a(w - \Pi_h^X w, v) = a(w, v - \Pi_h^Y v) = 0, \quad \forall v \in Y.$$

Since the inf-sup condition holds for a, we have $w = \Pi_h^X w \in X_h$. On the other hand, by choosing $y = v_h \in Y_h$, we then have

$$\langle F(w), v_h\rangle = 0, \quad \forall v_h \in Y_h,$$

which implies that $w \in X_h$ is a solution to (10.2.28). □

Lemma 10.2.5 shows that (10.2.38) is a reformulation involving the entire spaces (X, Y), of the original formulation (10.2.28) which was posed only in the subspaces (X_h, Y_h). This result enables us to obtain the well-posedness and an *a priori* error estimate of (10.2.28) by applying Lemma 10.2.4 to F_h. More precisely, we have the following result.

Theorem 10.2.10 ([99]). *Suppose that equation* (10.2.27) *and the discretized form* (10.2.28) *satisfy the inf-sup conditions* (10.2.32a), (10.2.32b), *respectively. Suppose also that F' is Lipschitz continuous at u in a ball of radius $\delta > 0$ about u; that is:*

There exists $\delta > 0$ and L such that for all $w \in B_\delta(u) \subset X$,
$\|F'(u) - F'(w)\|_{\mathcal{L}(X,Y')} \leqslant L\|u - w\|_X$.

Suppose that there exists a subspace $X_0 \subset X_h$ satisfying the approximation condition

$$\inf_{\chi_0 \in X_0} \|u - \chi_0\|_X \leqslant R = M^{-1}\left(1 + \frac{M}{m_h}\right)^{-1} \min\left\{\frac{\delta m}{2}, \frac{m^2}{4L}\right\}. \qquad (10.2.39)$$

Then there exists a constant $\delta_1 > 0$ such that equation (10.2.28) *has a locally unique solution $u_h \in X_h$ in $B_{\delta_1}(u)$. Moreover, we have the* a priori *error estimate:*

$$\|u - u_h\|_X \leqslant \frac{2M}{m}\left(1 + \frac{M}{m_h}\right) E_{X_h}(u). \qquad (10.2.40)$$

Proof. By Lemma 10.2.5, solutions to (10.2.28) and (10.2.38) are equivalent. By choosing $G = F_h$ in Lemma 10.2.4, we only need to verify (H1)–(H3) of Assumption 10.2.1. To this end, note that (10.2.37) implies $F_h'(u) = F'(u)$. Therefore,

$$\|[F'(u)]^{-1}\|_{\mathcal{L}(Y',X)} = m^{-1}$$

from the inf-sup condition (10.2.32a). Then (H2) in Assumption 10.2.1 follows. Again, by (10.2.37) we deduce that for any $w, x \in X$ and $y \in Y$,

$$\langle (F_h'(u) - F_h'(x))w, y\rangle = \langle (F'(u) - F'(x))w, \Pi_h^Y y\rangle.$$

Therefore,

$$\begin{aligned}\|F_h'(u) - F_h'(x)\|_{\mathcal{L}(X,Y')} &\leqslant \|F'(u) - F'(x)\|_{\mathcal{L}(X,Y')} \|\Pi_h^Y\|_{\mathcal{L}(Y,Y_h)}\\ &\leqslant \frac{M}{m_h} \|F'(u) - F'(x)\|_{\mathcal{L}(X,Y')}\\ &\leqslant \frac{ML}{m_h}\|u - x\|_X,\end{aligned}$$

where in the second inequality we used stability (10.2.35) of Π_h^Y. Hence, F_h satisfies (H1) in Assumption 10.2.1. To verify (H3) in Assumption 10.2.1, we have

$$\begin{aligned}\|F_h(u)\|_{Y'} &= \sup_{v\in Y} \frac{\langle F_h(u), v\rangle}{\|v\|_Y} = \sup_{v\in Y} \frac{a(u, v - \Pi_h^Y v)}{\|v\|_Y} = \sup_{v\in Y} \frac{a(u - \Pi_h^X u, v)}{\|v\|_Y}\\ &\leqslant M\|u - \Pi_h^X u\|_X.\end{aligned}$$

By triangle inequality and stability (10.2.35) of Π_h^X, we have

$$\|u - \Pi_h^X u\|_X \leqslant \|u - \chi_h\|_X + \|\Pi_h^X(u - \chi_h)\|_X \leqslant \left(1 + \frac{M}{m_h}\right) \inf_{\chi_h\in X_h} \|u - \chi_h\|_X.$$

Therefore, we obtain

$$\|F_h(u)\|_{Y'} \leqslant M\left(1 + \frac{M}{m_h}\right) \inf_{\chi_h\in X_h} \|u - \chi_h\|_X.$$

Notice that $X_0 \subset X_h$, and by assumption (10.2.39) we have

$$\|F_h(u)\|_{Y'} \leqslant M\left(1 + \frac{M}{m_h}\right) \inf_{\chi_0\in X_0} \|u - \chi_0\|_X \leqslant \min\left\{\frac{\delta m}{2}, \frac{m^2}{4L}\right\}.$$

Hence, (H3) in Assumption 10.2.1 is satisfied. By Lemma 10.2.4, there exists a constant $\delta_1 > 0$ such that (10.2.28) has a locally unique solution $u_h \in X_h$ in $B_{\delta_1}(u)$. Furthermore, we have the following *a priori* error estimate which completes the proof:

$$\begin{aligned}\|u - u_h\|_X &\leqslant 2\left\|[F_h'(u)]^{-1}\right\|_{\mathcal{L}(Y',X)} \|F_h(u)\|_{Y'}\\ &\leqslant \frac{2M}{m}\left(1 + \frac{M}{m_h}\right) \inf_{\chi_h\in X_h} \|u - \chi_h\|_X.\end{aligned}$$

□

REMARK. Theorem 10.2.10 from [99] is similar to [143, Theorem 4]. However, instead of assuming the approximation property

$$\lim_{h \to 0} \inf_{x_h \in X_h} \|u - x_h\|_X = 0$$

as used for the result in [143], here we assume only that there exists some (initial) subspace X_0 that satisfies (10.2.39). This distinction is quite critical for the development of adaptive finite element and other methods, because we cannot (and of course, do not want to) guarantee that $h \to 0$ uniformly; this assumption would require uniform mesh refinement (uniform subspace enrichment). The assumption (10.2.39) is essentially an approximation property of the subspace X_0, since

$$\inf_{\chi_0 \in X_0} \|u - \chi_0\|_X \leqslant \|u - I_0^X u\|_X.$$

In most of the applications we consider, the finite element space X_0 has a certain approximation property, namely that $\|u - I_0^X u\|_X = \mathcal{O}(h_0^\alpha)$ for some $\alpha > 0$, where I_0^X is inclusion or quasi-interpolation. Therefore, the condition (10.2.39) can be satisfied by choosing the mesh size h_0 of the initial triangulation to be sufficiently small.

Near-Best Approximation with Galerkin Methods

We are faced now with the following problem:

$$\text{Find } u \in X \text{ such that } \langle F(u), v \rangle = 0 \in \mathbb{R}, \qquad \forall v \in X, \tag{10.2.41}$$

where X is a Hilbert space, and $F \colon X \to X'$. Here, $\langle F(u), \cdot \rangle$ represents the action of $F(u)$ as a linear functional on X, for fixed $u \in X$. We wish to analyze the error in a Galerkin approximation $u_h \approx u$ satisfying the functional problem (10.2.41) in a subspace:

$$\text{Find } u_h \in X_h \subset X \text{ such that } \langle F(u_h), v_h \rangle = 0, \ \forall v_h \in X_h \subset X. \tag{10.2.42}$$

We now derive error estimates for a Galerkin method applied to problem (10.2.42) in three separate cases:

(1) F is affine, containing a coercive bilinear form (Cea's Lemma);

(2) F is affine, containing a bilinear form satisfying only a Gårding inequality (known as Schatz's trick);

(3) F is semilinear, with a coercive bilinear form and with a monotone Lipschitz nonlinearity.

Galerkin Error Estimates for Linear Coercive Equations. We consider now the special case when (10.2.42) is linear with a coercive bilinear form, meaning that F is affine and has the form $\langle F(u), v\rangle = a(u, v) - f(v)$, so that (10.2.41)–(10.2.42) become

$$\text{Find } u \in X \text{ such that } a(u, v) = f(v), \quad \forall v \in X. \tag{10.2.43}$$

$$\text{Find } u_h \in X_h \subset X \text{ such that } a(u_h, v_h) = f(v_h), \quad \forall v_h \in X_h \subset X. \tag{10.2.44}$$

The conditions (10.2.19)–(10.2.22) now simplify to

$$a(u, v) \leqslant M\|u\|_X\|v\|_X, \quad M > 0,\ \forall u, v \in X, \tag{10.2.45}$$

$$f(v) \leqslant L\|v\|_X, \quad L > 0,\ \forall v \in X, \tag{10.2.46}$$

$$a(u, u) \geqslant m\|u\|_X^2, \quad m > 0,\ \forall u \in X, \tag{10.2.47}$$

with (10.2.47) now capturing both (10.2.21)–(10.2.22). Theorem 10.2.9 now simplifies to

Theorem 10.2.11 (Cea's Lemma). *Let X be a Banach space, let $X_h \subset X$, let $u \in X$ be the solution to the linear problem* (10.2.43)*, and let $u_h \in X_h$ be the Galerkin approximation satisfying* (10.2.44)*. Let conditions* (10.2.45)–(10.2.47) *hold. Then*

$$\|u - u_h\|_X \leqslant \left(\frac{M}{m}\right) E_{X_h}(u). \tag{10.2.48}$$

Proof. We start with the coercivity inequality (10.2.47) and then exploit the Galerkin orthogonality property $a(u - u_h, v_h) = 0, \forall v_h \in X_h$, which yields

$$\begin{aligned} m\|u - u_h\|_X^2 &\leqslant a(u - u_h, u - u_h) \\ &= a(u - u_h, u) \\ &= a(u - u_h, u - w_h) \\ &\leqslant M\|u - u_h\|_X\|u - w_h\|_X, \end{aligned}$$

where at the end we use the boundedness inequality (10.2.45). Assuming first that $\|u - u_h\|_X \neq 0$, then division by $\|u - u_h\|_X$ gives (10.2.48), which we note also holds when $\|u - u_h\|_X = 0$. □

Galerkin Error Estimates and Gårding Inequalities. In a number of problems of interest that we will encounter later, neither the coercivity condition (10.2.47) nor the inf-sup condition (10.2.21) holds for the bilinear form. However, we still wish to use a Petrov-Galerkin-type method, and hope to establish some type of relationship to the best approximation as we have done in the previous sections when one of (10.2.47) or (10.2.21) holds. We begin with a *Gelfand triple* of Hilbert spaces $V \subset H \equiv H' \subset V'$ with continuous embedding, meaning that the *pivot* space H and its dual space H' are identified through the Riesz representation theorem (see Chapter 4), and that the embedding $V \subset H$ is continuous (see Section 10.3). A consequence of this is:

$$\|u\|_H \leqslant C\|u\|_V, \quad \forall u \in V, \tag{10.2.49}$$

where we will assume that the embedding constant $C = 1$ (the norm $\|\cdot\|_V$ can be redefined as necessary). In applications, the embedding $V \subset H$ will often also be compact (see Section 10.3). We are given the following functional problem:

$$\text{Find } u \in V \text{ such that } a(u,v) = f(v), \quad \forall v \in V, \tag{10.2.50}$$

where the bilinear form $a(u,v)\colon V \times V \to \mathbb{R}$ is bounded

$$a(u,v) \leqslant M\|u\|_V\|v\|_V, \quad \forall u,v \in V, \tag{10.2.51}$$

and satisfies a Gårding inequality:

$$m\|u\|_V^2 \leqslant K\|u\|_H^2 + a(u,u), \quad \forall u \in V, \text{ where } m > 0, \tag{10.2.52}$$

and where the linear functional $f(v)\colon V \to \mathbb{R}$ is bounded and thus lies in the dual space V':

$$f(v) \leqslant L\|v\|_V, \quad \forall v \in V.$$

Now, we are interested in the quality of a Galerkin approximation:

$$\text{Find } u_h \in V_h \subset V \text{ such that } a(u_h,v) = a(u,v) = f(v), \quad \forall v \in V_h \subset V. \tag{10.2.53}$$

We will assume that there exists a sequence of approximation subspaces $V_h \subset V$ parametrized by h, with $V_{h_1} \subset V_{h_2}$ when $h_2 < h_1$, and that there exists a sequence $\{a_h\}$, with $\lim_{h\to 0} a_h = 0$, such that

$$\|u - u_h\|_H \leqslant a_h\|u - u_h\|_V, \text{ when } a(u - u_h, v) = 0, \ \forall v \in V_h \subset V. \tag{10.2.54}$$

The assumption (10.2.54) will turn out to be very natural when we examine particular approximation spaces in Section 10.5; it is the assumption that the error in the approximation converges to zero more quickly in the H-norm than in the V-norm. This is easily verified if, for example, V_h are piecewise polynomial approximation spaces (finite element-type spaces; see Section 10.5), under very mild smoothness requirements on the solution u [see Lemmas 2.1 and 2.2 in [179]]. Under these assumptions, we have the following *a priori* error estimate. The result and the main idea for the simple proof we give below go back to Schatz [153] (see also [94, 154, 179]).

Theorem 10.2.12. *Let $V \subset H \subset V'$ be a Gelfand triple of Hilbert spaces with continuous embedding. Assume that* (10.2.50) *is uniquely solvable and that assumptions* (10.2.49), (10.2.51), (10.2.52), *and* (10.2.54) *hold. Then for h sufficiently small, there exists a unique approximation u_h satisfying* (10.2.53), *for which the following quasi-optimal* a priori *error bounds hold:*

$$\|u - u_h\|_V \leqslant CE_{V_h}(u), \tag{10.2.55}$$
$$\|u - u_h\|_H \leqslant Ca_hE_{V_h}(u), \tag{10.2.56}$$

where C is a constant independent of h. If $K \leqslant 0$ in (10.2.52), *then the above holds for all h with $C = M/m$.*

Proof. The following proof follows the idea in [153]. We begin with the Gårding inequality (10.2.52) and then employ (10.2.51):

$$\begin{aligned} m\|u-u_h\|_V^2 - K\|u-u_h\|_H^2 &\leqslant a(u-u_h, u-u_h) \\ &= a(u-u_h, u-v) \\ &\leqslant M\|u-u_h\|_V\|u-v\|_V, \end{aligned} \tag{10.2.57}$$

where we have used Galerkin orthogonality: $a(u-u_h, v) = 0, \ \forall v \in V_h$, to replace u_h with an arbitrary $v \in V_h$. Excluding first the case that $\|u-u_h\|_V = 0$, we divide through by $m\|u-u_h\|_V$ and employ (10.2.49) and (10.2.54), giving $\forall v \in V_h$,

$$\begin{aligned} \left(1 - \frac{Ka_h}{m}\right)\|u-u_h\|_V &\leqslant \|u-u_h\|_V - \frac{K\|u-u_h\|_H^2}{m\|u-u_h\|_V} \\ &\leqslant \frac{M}{m}\|u-v\|_V, \end{aligned} \tag{10.2.58}$$

which we note also holds when $\|u-u_h\|_V = 0$.

Assume first that $K > 0$. Since $\lim_{h\to 0} a_h = 0$, we know that there exists $\overline{h}$ such that $a_h < m/K, \forall h \leqslant \overline{h}$. This implies that $\forall v \in V_h$,

$$\left(1 - \frac{Ka_{\overline{h}}}{m}\right)\|u-u_h\|_V \leqslant \left(1 - \frac{Ka_h}{m}\right)\|u-u_h\|_V \leqslant \frac{M}{m}\|u-v\|_V. \tag{10.2.59}$$

Taking $u = 0$ in (10.2.53) together with $v = 0$ in (10.2.59), with $h \leqslant \overline{h}$, implies that the homogeneous problem

$$\text{Find } u_h \in V_h \text{ such that } a(u_h, v) = 0, \quad \forall\, v \in V_h,$$

has only the trivial solution, so that by the discrete Fredholm alternative a solution u_h to (10.2.53) is unique and therefore exists. Equation (10.2.59) then finally gives (10.2.55) whenever $h \leqslant \overline{h}$, with the choice

$$C = \frac{M}{m\left(1 - \dfrac{Ka_{\overline{h}}}{m}\right)} = \frac{M}{m - Ka_{\overline{h}}}.$$

Assume now that $K \leqslant 0$. Directly from (10.2.58) we can conclude (10.2.55) with $C = M/m$, which is completely independent of h; this then becomes Theorem 10.2.11. Moreover, the continuous and discrete problems are both uniquely solvable due to coercivity (10.2.52), independent of h.

In either case of $K > 0$ or $K \leqslant 0$, the second estimate (10.2.56) now follows immediately from assumption (10.2.54). □

Galerkin Error Estimates for Nonlinear Equations. We now derive an *a priori* error estimate for a Galerkin approximation to an abstract semilinear problem. Let X be a Banach space. We are given the following nonlinear variational problem:

$$\text{Find } u \in X \text{ such that } a(u,v) + \langle b(u), v\rangle = f(v), \quad \forall v \in X, \tag{10.2.60}$$

where the bilinear form $a(u, v)\colon X \times X \to \mathbb{R}$ is bounded,

$$a(u, v) \leqslant M\|u\|_X\|v\|_X, \quad \forall u, v \in X, \tag{10.2.61}$$

and coercive,

$$m\|u\|_X^2 \leqslant a(u, u), \quad \forall u \in X, \text{ where } m > 0, \tag{10.2.62}$$

where the linear functional $f\colon X \to \mathbb{R}$ is bounded and thus lies in the dual space X':

$$f(v) \leqslant L\|v\|_X, \quad \forall v \in X,$$

and where the nonlinear form $\langle b(u), v\rangle\colon X \times X \to \mathbb{R}$ is assumed to be monotone:

$$0 \leqslant \langle b(u) - b(v), u - v\rangle, \quad \forall u, v \in X, \tag{10.2.63}$$

where we have used the notation

$$\langle b(u) - b(v), w\rangle = \langle b(u), w\rangle - \langle b(v), w\rangle.$$

We will assume that $\langle b(u), v\rangle$ is Lipschitz in the following weak sense: If $u \in X$ satisfies (10.2.60), and if $u_h \in X_h$ satisfies (10.2.65) below, then there exists a constant $K > 0$ such that

$$\langle b(u) - b(u_h), u - v_h\rangle \leqslant K\|u - u_h\|_X\|u - v_h\|_X, \qquad \forall v_h \in X_h. \tag{10.2.64}$$

We are interested in the quality of a Galerkin approximation:

$$\text{Find } u_h \in X_h \text{ such that } a(u_h, v_h) + \langle b(u_h), v_h\rangle = f(v_h), \quad \forall v_h \in X_h, \tag{10.2.65}$$

where $X_h \subset X$ or, equivalently,

$$a(u - u_h, v_h) + \langle b(u) - b(u_h), v_h\rangle = 0, \ \forall v_h \in X_h \subset X.$$

Under these assumptions, we have the following *a priori* error estimate for the Galerkin approximation u_h. See [43, 52, 106] for similar types of estimates.

Theorem 10.2.13. *Let X be a Banach space. Assume that* (10.2.60) *and* (10.2.65) *are uniquely solvable and that assumptions* (10.2.61), (10.2.62), (10.2.63), (10.2.64) *hold. Then the approximation u_h satisfying* (10.2.65) *obeys the following quasi-optimal* a priori *error bound:*

$$\|u - u_h\|_X \leqslant \left(\frac{M + K}{m}\right) E_{X_h}(u). \tag{10.2.66}$$

Proof. We begin by subtracting (10.2.65) from (10.2.60) and taking $v = v_h = w_h$ in both equations, where $w_h \in X_h \subset X$, which gives

$$a(u - u_h, w_h) + \langle b(u) - b(u_h), w_h\rangle = 0, \quad \forall w_h \in X_h. \tag{10.2.67}$$

In particular, if $w_h = v_h - u_h \in X_h$ where $v_h \in X_h$, then this implies that

$$\begin{aligned} a(u-u_h, v_h-u_h) &= \langle b(u_h)-b(u), v_h-u_h\rangle \\ &= \langle b(u_h)-b(u), v_h-u\rangle - \langle b(u_h)-b(u), u_h-u\rangle \\ &\leqslant \langle b(u_h)-b(u), v_h-u\rangle, \end{aligned} \tag{10.2.68}$$

where we have employed monotonicity (10.2.63). Beginning now with (10.2.62), we have for arbitrary $v_h \in X_h$ that

$$\begin{aligned} m\|u-u_h\|_X^2 &\leqslant a(u-u_h, u-u_h) \\ &= a(u-u_h, u-v_h) + a(u-u_h, v_h-u_h) \\ &\leqslant a(u-u_h, u-v_h) + \langle b(u_h)-b(u), v_h-u\rangle \\ &\leqslant M\|u-u_h\|_X\|u-v_h\|_X + K\|u-u_h\|_X\|u-v_h\|_X, \end{aligned} \tag{10.2.69}$$

where we have used (10.2.68), (10.2.61), and (10.2.64). Excluding first the special case of $\|u-u_h\|_X = 0$, we divide through by $m\|u-u_h\|_X$, giving

$$\|u-u_h\|_X \leqslant \left(\frac{M+K}{m}\right)\|u-v_h\|_X, \quad \forall v_h \in X_h, \tag{10.2.70}$$

which we note also holds when $\|u-u_h\|_X = 0$, giving (10.2.66). □

EXERCISES

10.2.1 Prove that an inner product on a Hilbert space is always a continuous function of its arguments.

10.2.2 Give a proof of the Hilbert space projection theorem by assuming only the existence of a best approximation.

10.2.3 Use the Hilbert space projection theorem to prove the Riesz Representation Theorem (see Chapter 4).

10.2.4 Prove the analogue of Theorem 10.2.8 in the case of a uniformly convex Banach space X.

10.2.5 Let X be a Banach space, and let $U \subset X$ be a finite-dimensional subspace. Modify the analysis used to prove Theorem 10.2.10 to develop a near-best approximation result for a Petrov-Galerkin approximation to $T(u) = u$, where $T\colon X \to X$. What properties does T need to have to make such a result possible?

10.2.6 Develop an error estimate for a Petrov-Galerkin method ($X \neq Y$) applied to a semilinear problem $Au + B(u) = f \in Y'$, analogous to the result given for a Galerkin method, where $A\colon X \to Y'$ and $B\colon X \to Y'$. Assuming that A gives rise to a bounded bilinear form on $X \times Y$ that satisfies both continuous and discrete inf-sup conditions, what properties must B have in order to establish such a result?

10.3 OVERVIEW OF SOBOLEV AND BESOV SPACES

We now take a closer look at the Sobolev spaces that were encountered briefly earlier in the book (in Section 2.6, Example 6 in Section 4.5, and Section 8.4), and introduce the more general Besov spaces. These two families of Banach spaces play a central role in the analysis of linear and nonlinear partial differential equations, and also in the modern numerical analysis of the finite element method and other methods for partial differential equations and integral equations. Although the material in Chapter 8 covering partial differential equations involved only some very basic notions of the (Hilbert) Sobolev space $H^1(\Omega)$, in order to understand finite element and related approximations, we need to develop more background on the larger class of (Banach) Sobolev spaces $W^{s,p}(\Omega)$. Our main goal here is to assemble a clear and concise survey of the most important properties of these Banach spaces that we make use of in the analysis and approximation of solutions to partial differential equations; complete developments of Sobolev and Besov spaces are given, for example, in [2, 168].

Domain Conditions and Hölder Spaces and Their Properties

Let Ω be an open subset of $\mathbb{R}^n$, possibly unbounded, which we will refer to as a *domain* in $\mathbb{R}^n$. We will denote the boundary of Ω as $\partial\Omega$ and will make some assumptions on the smoothness of the boundary set. There is a standard set of conditions that attempt to characterize the boundary properties of Ω, including (1) *the uniform C^m-regularity condition*; (2) *the strong local Lipschitz condition*; (3) *the uniform cone condition*; (4) *the cone condition*; (5) *the weak cone condition*; and (6) *the segment condition*. These conditions are related in two distinct ways: namely, (1) (for $m \geqslant 2$) $\Longrightarrow$ (2) $\Longrightarrow$ (3) $\Longrightarrow$ (6), and also (3) $\Longrightarrow$ (4) $\Longrightarrow$ (5). See, for example, [2, 82] for a careful discussion of these different properties.

However, in this chapter we are concerned mainly with bounded domains Ω, meaning that there exists $R > 0$ such that $\Omega \subset B_R(0)$, where $B_R(0)$ is the open ball of radius R about zero:

$$B_R(0) = \{x \in \mathbb{R}^n \ : \ \|x\|_{l^2} < R\}.$$

In this case, although condition (1) remains a stronger condition than the other five conditions, conditions (2) to (6) all reduce to the following single condition, which will be satisfied by bounded polyhedral (or smoother) subsets of $\mathbb{R}^n$:

Definition 10.3.1. *Let $\Omega \subset \mathbb{R}^n$ be a bounded domain (an open subset of $\mathbb{R}^n$). We say that Ω has a* Lipschitz *(or* locally Lipschitz*) boundary, and refer to Ω as a* Lipschitz *or* bounded Lipshitz domain, *if to each point $x \in \partial\Omega$ there exists an open set $S_x \subset \mathbb{R}^n$ such that $S_x \cap \partial\Omega$ is the graph of a Lipschitz continuous function.*

In Chapter 2 we introduced the notion of a multi-index as a compact notation for specifying partial differentiation of any order. Recall that a multi-index has the form $\alpha = (\alpha_1, \ldots, \alpha_n)$, $0 \leqslant \alpha_i \in \mathbb{N}_0$, with the following properties and operations defined:

- Order relation: $\alpha \geqslant \beta$ iff $\alpha_i \geqslant \beta_i, \forall i$.
- Magnitude: $|\alpha| \equiv \alpha_1 + \cdots + \alpha_n$.
- Exponentiation: $x^\alpha \equiv x_1^{\alpha_1} \cdots x_n^{\alpha_n}$, for $x \in \mathbb{R}^n$.

We used multi-indices to denote partial differentiation of $u \colon \Omega \to \mathbb{R}$ concisely as

$$D^\alpha u = \frac{\partial^{|\alpha|} u}{\partial x_1^{\alpha_1} \cdots \partial x_n^{\alpha_n}}; \qquad \text{e.g., if } \alpha = (1,2) \Longrightarrow D^\alpha u = \frac{\partial^3 u}{\partial x_1 \, \partial x_2^2}.$$

We now use this compact notation to quickly assemble the classical function spaces that we will need to refer to when discussing the Sobolev spaces. Here we assume that the domain $\Omega \subset \mathbb{R}^n$ is open, not necessarily bounded unless stated explicitly. The classical vector spaces of functions that we have worked with throughout the book so far have been:

$$\begin{aligned} C(\Omega) &= C^0(\Omega) = \{\text{the vector space of continuous functions on } \Omega\}, \\ C^m(\Omega) &= \{u \in C^0(\Omega) \ : \ D^\alpha u \in C^0(\Omega) \text{ for } 0 \leqslant |\alpha| \leqslant m\}, \\ C_0^m(\Omega) &= \{u \in C^m(\Omega) \ : \ \text{supp}(u) \subset\subset \Omega\}, \end{aligned}$$

where the notation "$\text{supp}(u) \subset\subset \Omega$" describing a function with compact support in Ω was explained in Section 2.6. Since $\Omega \subset \mathbb{R}^n$ is an open set, $u \in C^m(\Omega)$ may not be (pointwise) bounded on Ω. Therefore, we introduce $C_B^m(\Omega)$ as the subspace of functions in $C^m(\Omega)$ for which all derivatives up to order m are (pointwise) bounded on Ω, which we can write as

$$C_B^m(\Omega) = \{u \in C^m(\Omega) \ : \ \|u\|_{C^m(\Omega)} < +\infty\},$$

where

$$\|u\|_{C^m(\Omega)} = \max_{0 \leqslant |\alpha| \leqslant m} \sup_{x \in \Omega} |D^\alpha u(x)|.$$

This space can be shown to be complete in this norm, and hence is a Banach space.

Given a function $u \in C^0(\Omega)$ which is bounded and uniformly continuous (u.c.) on Ω, it has a unique continuous extension to the closed set $\overline{\Omega}$. The subspace of $C_B^m(\Omega)$ consisting of functions for which $D^\alpha u$ is bounded and uniformly continuous on Ω for $0 \leqslant |\alpha| \leqslant m$ is denoted

$$C^m(\overline{\Omega}) = \{u \in C_B^m(\Omega) \ : \ D^\alpha u \text{ bounded and u.c. on } \Omega \text{ for } 0 \leqslant |\alpha| \leqslant m\}.$$

This is a closed subspace of $C_B^m(\Omega)$, and hence also a Banach space. It is a proper subset of $C_B^m(\Omega)$; there are functions in $C_B^m(\Omega)$ which are not in $C^m(\overline{\Omega})$ (not uniformly continuous on Ω). However, if Ω is bounded, and satisfies some minimal conditions such as the segment condition (which we know holds, for example, if Ω is a bounded Lipschitz domain), then the two spaces $C_B^m(\Omega)$ and $C^m(\overline{\Omega})$ coincide.

It would have been nice if $C^m(\overline{\Omega})$ were the only Banach spaces that we would have need of, but in the early part of the twentieth century, mathematicians ran into the following fundamental problem when trying to develop a more complete theory of partial differential equations. Let X^0 represent a space of functions defined over $\Omega \subset \mathbb{R}^n$, and let X^m represent the space of functions with all partial derivatives of order m in X^0, so that the norm on X^m can be given by the norm on X^0 through

$$\|u\|_{X^m} = | \{\|D^\alpha u\|_{X^0} \; ; \; 0 \leqslant |\alpha| \leqslant m\} |_q$$

for some $1 \leqslant q \leqslant \infty$, where $|\cdot|_q$ is the discrete norm on l^q (see the next section). In order to develop a satisfying solution theory for partial differential equations, a requirement is that the operator $-\Delta + I \colon X^2 \to X^0$ be a linear isomorophism between the two spaces. Unfortunately, the spaces $C^m(\overline{\Omega})$ above do not have this property. However, the Hölder spaces $C^{m,\lambda}(\overline{\Omega})$ (as well as the $L^p(\Omega)$ spaces encountered later in the chapter) do have this property.

The *Hölder spaces* $C^{m,\lambda}(\overline{\Omega})$ are the subspaces of $C^m(\overline{\Omega})$ consisting of functions u such that $D^\alpha u$ satisfies a Hölder condition with exponent $0 < \lambda \leqslant 1$, for $|\alpha| = m$, or explicitly,

$$|D^\alpha u(x) - D^\alpha u(y)| \leqslant K|x-y|^\lambda, \quad x, y \in \Omega.$$

If we define the *Hölder coefficient* with exponent λ for a function $v \in C^0(\overline{\Omega})$ as

$$|v|_{C^{0,\lambda}(\Omega)} = \sup_{\substack{x,y\in\Omega \\ x\neq y}} \frac{|v(x) - v(y)|}{|x-y|^\lambda},$$

then this leads naturally to the definition of the norm on $C^{m,\lambda}(\overline{\Omega})$ as

$$\|u\|_{C^{m,\lambda}(\overline{\Omega})} = \|u\|_{C^m(\overline{\Omega})} + \max_{|\alpha|=m} |D^\alpha u|_{C^{0,\lambda}(\Omega)}.$$

One can then characterize $C^{m,\lambda}(\overline{\Omega})$ as

$$C^{m,\lambda}(\overline{\Omega}) = \{u \in C^m(\overline{\Omega}) \; : \; \|u\|_{C^{m,\lambda}(\overline{\Omega})} < +\infty\}.$$

The two key theorems that hold when Ω is a bounded domain in $\mathbb{R}^n$ are the *Stone-Weierstrass Theorem* [characterizing when a subset $A \subset C^0(\overline{\Omega})$ is a dense subset] and the *Arzela-Ascoli Theorem* [characterizing when a subset $A \subset C^0(\overline{\Omega})$ is relatively compact], along with embedding theorems (see the next section for definition of an embedding) that relate the spaces $C^m(\overline{\Omega})$ and $C^{m,\lambda}(\overline{\Omega})$.

$L^p(\Omega)$ Spaces and Their Properties

The $L^p(\Omega)$ spaces for $1 \leqslant p \leqslant \infty$ have been encountered repeatedly in earlier chapters; they were seen to be Banach spaces when equipped with the norms

$$\|u\|_p = \|u\|_{L^p(\Omega)} = \left(\int_\Omega |u|^p \, dx\right)^{1/p}, \quad 1 \leqslant p < \infty, \tag{10.3.1}$$

$$\|u\|_\infty = \|u\|_{L^\infty(\Omega)} = \operatorname*{ess\,sup}_{x\in\Omega} |u(x)|. \tag{10.3.2}$$

Some comments are needed for the case $p = \infty$; we say that a function $u \in L^\infty(\Omega)$ if the function is *essentially bounded* on Ω, which means that there exists a constant C such that $|u(x)| \leqslant C$, almost everywhere (in the sense of measure zero) on Ω. The L^∞ norm is simply the greatest lower bound of all such constants C for a given function u and is called the *essential supremum* of u on Ω.

We now define the $L^p(\Omega)$ spaces a little more carefully as follows. Let $\mathcal{M}(\Omega)$ denote the set of Lebesgue-measurable functions defined over a domain $\Omega \subset \mathbb{R}^n$. We then define the space $L^p(\Omega)$, for $1 \leqslant p \leqslant \infty$, by

$$L^p(\Omega) = \{u \in \mathcal{M}(\Omega) \ : \ \|u\|_p < \infty\}.$$

If the domain for the norm is important (for example, is other than Ω), we will include it as an additional subscript: $\|u\|_{p,\Omega}$. Recall now that the case $p = 2$ was special; in this case the L^2 norm is induced by the L^2 inner product

$$(u, v) = (u, v)_0 = \int_\Omega uv\, dx,$$

and it was seen that L^2 subsequently was a Hilbert space. The subscript 0 is sometimes used on the inner product to indicate the relationship between inner products on Sobolev spaces, introduced in the next section.

Below we will often need the discrete norms of summable sequences $\{u_j\}_{j=1}^\infty$, interpreted as members of the Banach spaces "little" l^p; for convenience we will use the following notation for the norms associated with l^p:

$$|u|_p = \left(\sum_{j=1}^\infty |u_j|^p \right)^{1/p}, \qquad 1 \leqslant p < \infty, \tag{10.3.3}$$

$$|u|_\infty = \max_j |u_j|, \tag{10.3.4}$$

with the case of $p = 2$ allowing for the inner product $(u, v)_{l^2} = \sum_{j=1}^\infty u_j v_j$.

Two very basic important inequalities involving the L^p norms are the following (see also Exercise 4.3.7):

(1) Minkowski inequality: If $u, v \in L^p(\Omega)$, then $u + v \in L^p(\Omega)$, and

$$\|u + v\|_p \leqslant \|u\|_p + \|v\|_p, \quad 1 \leqslant p \leqslant \infty.$$

(See Exercise 4.3.7.)

(2) Hölder inequality: If $u \in L^p(\Omega)$ and $v \in L^q(\Omega)$, where $1 \leqslant p, q \leqslant \infty$ and $1/p + 1/q = 1$, then $uv \in L^1(\Omega)$, and

$$\|uv\|_1 \leqslant \|u\|_p \|v\|_q.$$

The *conjugate exponent condition* $1/p + 1/q = 1$ reappears frequently; given an exponent $p \in [1, \infty]$, we will use p' to denote the extended conjugate exponent in the following sense:

$$p' = \begin{cases} \infty, & p = 1, \\ \dfrac{p}{p-1}, & 1 < p < \infty, \\ 1, & p = \infty. \end{cases}$$

The *convolution* of two measurable functions (when it is well-defined) takes the form

$$(u * v)(x) = \int_\Omega u(x - y)v(y)\,dy.$$

It yields a uniformly continuous function when it involves $L^p(\Omega)$ and its conjugate:

$$|(u * v)(x)| \leqslant \|u\|_p \|v\|_{p'}, \quad \forall x \in \Omega,\ \forall u \in L^p(\Omega),\ \forall v \in L^{p'}(\Omega).$$

Some of the basic properties of the Lebesgue integral were summarized in the last section of Chapter 0; we add to these a list of basic facts about the L^p spaces which one typically establishes in a standard graduate course in real analysis (see [2]).

(1) $L^p(\Omega)$ is a Banach space for $1 \leqslant p \leqslant \infty$.
(2) Every Cauchy sequence in $L^p(\Omega)$, $1 \leqslant p \leqslant \infty$, has a subsequence converging pointwise almost everywhere on Ω.
(3) $L^p(\Omega) \subset L^1_{\text{loc}}(\Omega)$ for $1 \leqslant p \leqslant \infty$.
(4) $L^p(\Omega)$ is separable for $1 \leqslant p < \infty$.
(5) $L^p(\Omega)$ is uniformly convex for $1 < p < \infty$.
(6) $L^p(\Omega)$ is reflexive for $1 < p < \infty$.
(7) $C_0(\Omega)$ is dense in $L^p(\Omega)$ for $1 \leqslant p < \infty$.
(8) $C_0^\infty(\Omega)$ is dense in $L^p(\Omega)$ for $1 \leqslant p < \infty$.
(9) $L^p(\Omega)$ is a Banach lattice on Ω for $1 \leqslant p \leqslant \infty$.
(10) If $u \in L^q(\Omega)$ with $1 \leqslant p \leqslant q \leqslant \infty$, then $u \in L^p(\Omega)$ and

$$\|u\|_p \leqslant \text{vol}(\Omega)^{(1/p)-(1/q)} \|u\|_q, \qquad \text{vol}(\Omega) = \int_\Omega 1\,dx.$$

We expand a bit now on some of these properties.

That $L^p(\Omega)$ is a Banach space follows from the fact that it is algebraically closed as a vector space with respect to pointwise (almost everywhere) linear combinations, that the L^p norm satisfies the norm axioms (Section 4.3), and that it can be shown to be complete (topologically closed) with respect to this norm (see [114]). The convergent subsequence property (the second property on the list) is typically established as part of the proof of completeness. The next property follows from the definition of L^1_{loc} given in Section 2.6. Regarding the next two properties on the list, we have discussed separability, uniform convexity, and reflexivity of Banach spaces at length in Section 4.8. In particular, uniform convexity follows from the Clarkson inequalities (see [2]), and uniform convexity implies reflexivity.

The two properties involving *density* refer to the following concept: We say that a Banach space $X \subset Y$ is *dense* in a Banach space Y if given any $u \in Y$ and an arbitrary $\epsilon > 0$, there exists $v \in X$ such that $\|u - v\|_X < \epsilon$. In other words, if $X \subset Y$ is dense in Y, then any function in Y can be approximated arbitrarily well by a function in $X \subset Y$.

Regarding the next property, a *Banach lattice* on Ω is a Banach space $X(\Omega)$ of functions defined over Ω such that if $u \in X(\Omega)$ and $|v(x)| \leqslant |u(x)|$, *a.e.* in Ω, for some measurable function $v(x)$, then also $v \in X(\Omega)$ and $\|v\|_X \leqslant \|u\|_X$. A Banach lattice has what is called the *dominated convergence property* if pointwise convergence of a sequence combined with pointwise domination by a fixed function implies norm convergence, or more explicitly, if for all sequences $\{u_j\}$ converging pointwise in $X(\Omega)$, with $u_j \in X(\Omega)$, and with $|u_j(x)| \leqslant |u(x)|$ *a.e.* for some $u \in X(\Omega)$, then $\lim_{j\to\infty} u_j(x) = 0$ implies that $\lim_{j\to\infty} \|u_j\|_X = 0$.

The last property in the list is an example of a *continuous embedding* of a function space X (in this case, L^q) into a second function space Y (in this case, L^p, with $1 \leqslant p \leqslant q \leqslant \infty$). We examine this idea more carefully after the next section.

Integer Order Sobolev Spaces $W^{m,p}(\Omega)$

In Section 2.6 we defined the *weak* or *distributional* partial derivative of order α of a function $u \in L^1_{\text{loc}}(\Omega)$, for a given multi-index α with $|\alpha| \geqslant 0$, as follows. (Recall that L^1_{loc} was defined in Section 2.6.) We called $D^\alpha u = v_\alpha \in L^1_{\text{loc}}(\Omega)$ the weak partial derivative of order α of $u \in L^1_{\text{loc}}(\Omega)$ if there exists $v_\alpha \in L^1_{\text{loc}}(\Omega)$ such that

$$\int_\Omega u(x)(D^\alpha \phi(x))\, dx = (-1)^{|\alpha|} \int_\Omega v_\alpha(x)\phi(x)\, dx, \qquad \forall \phi \in \mathcal{D}(\Omega), \qquad (10.3.5)$$

with $\mathcal{D}(\Omega)$ the space of test functions as described in Chapter 2. We noted there that $D^\alpha u$ defined in this way is unique (up to sets of measure zero), and if u is sufficiently smooth so that $u \in C^m(\Omega)$ for $m = |\alpha|$, then the classical and weak derivatives are the same object. This led us to introduce the vector spaces of functions which have weak derivatives [in the sense of (10.3.5)] in $L^p(\Omega)$,

$$W^{m,p}(\Omega) = \{u \in L^p(\Omega) \ : \ D^\alpha u \in L^p(\Omega),\ 0 \leqslant |\alpha| \leqslant m\}.$$

with the Banach spaces $L^p(\Omega)$ defined as above. The vector spaces $W^{m,p}(\Omega)$ for $m \geqslant 0$ and $1 \leqslant p \leqslant \infty$ are referred to as *Sobolev spaces*, with the case $m = 0$ coinciding with the Lebesgue spaces $L^p(\Omega)$.

The norm $\|\cdot\|_{m,p}$ on $W^{m,p}(\Omega)$, and the inner product $(\cdot,\cdot)_m$ when $p = 2$ is conveniently written in terms of a seminorm $|\cdot|_{m,p}$ and semi-inner product $[\cdot,\cdot]_m$ as follows:

$$\|u\|_{m,p} = \left(\sum_{k=0}^{m} |u|_{k,p}^p \right)^{1/p}, \qquad |u|_{k,p}^p = \sum_{|\alpha|=k} \|D^\alpha u\|_p^p, \quad 1 \leqslant p < \infty, \quad (10.3.6)$$

$$\|u\|_{m,\infty} = \max_{1 \leqslant k \leqslant m} |u|_{k,\infty}, \qquad |u|_{k,\infty} = \max_{|\alpha|=k} \|D^\alpha u\|_\infty, \quad (10.3.7)$$

$$(u,v)_m = \sum_{k=0}^{m} [u,v]_k, \qquad [u,v]_k = \sum_{|\alpha|=k} (D^\alpha u, D^\alpha v)_0, \quad (10.3.8)$$

with obviously $\|u\|_{0,p} = \|u\|_p$, and also $(u,v)_0 = [u,v]_0 = (u,v)$ recovering the L^2 inner product.

That $|\cdot|_{m,p}$ fails to be a norm and $[\cdot,\cdot]_m$ fails to be an inner product, for $m \geqslant 1$, is due to violating the nondegeneracy property of norms and inner products (Section 4.3) for nonzero constant functions. Again, if the domain for the norm is important (for example, if it differs from Ω), we will use an additional subscript to indicate the domain: $\|\cdot\|_{m,p,\Omega}$.

The space $W^{m,p}$ can now be defined equivalently as

$$W^{m,p}(\Omega) = \{ u \in L^p(\Omega) \; : \; \|u\|_{m,p} < \infty \}.$$

The following related function space is also important:

$$W_0^{m,p}(\Omega) = \{\text{closure of } C_0^\infty(\Omega) \text{ in } W^{m,p}(\Omega) \},$$

which by construction is a Banach space (see the discussion on closure of sets and completion of metric spaces in Chapter 4). As we remarked in Chapter 2, $W^{m,p}(\Omega)$ are Banach spaces for $m \geqslant 0$ and $1 \leqslant p \leqslant \infty$, with algebraic closure as a vector space inherited from that of L^p and the three norm axioms satisfied by the norm (10.3.6), also inherited from those properties of L^p. That $W^{m,p}(\Omega)$ is complete (topologically closed) follows from completeness of $L^p(\Omega)$ and from some simple properties of the distributional derivative; see [2, 70]. In the case $p = 2$, the standard notation $H^m(\Omega) = W^{m,2}(\Omega)$ is used.

Abstract Embedding Operators. One of the properties stated for the L^p spaces in the sections above was an inequality of the form

$$\|u\|_Y \leqslant C \|u\|_X. \quad (10.3.9)$$

This is an example of a *continuous embedding* of a function space X (in this case, L^q) into a second function space Y (in this case, L^p, with $1 \leqslant p \leqslant q \leqslant \infty$). More formally, we say that a normed space X is *continuously embedded* into a normed space Y, and denote this as

$$X \hookrightarrow Y, \quad (10.3.10)$$

if X is a subspace (algebraically closed vector space) of Y, and if the linear map I taking elements of X into Y, referred to as the *embedding operator*, is a continuous map. (Since continuity and boundedness are equivalent for linear operators, this is equivalent to saying that I is a bounded operator.) In other words, if $X \hookrightarrow Y$, and $u \in X$, then also $Iu \in Y$ (which we denote as $u \in Y$), and

$$\|u\|_Y = \|Iu\|_Y \leqslant \|I\|_{\mathcal{L}(X,Y)}\|u\|_X = C\|u\|_X,$$

which is indeed simply an inequality of the form (10.3.9). If the embedding operator I is actually a compact linear operator (see Section 5.7), then we denote this as

$$X \overset{c}{\hookrightarrow} Y, \tag{10.3.11}$$

and refer to it as a *compact embedding*.

The Sobolev Embedding Theorem that we discuss later in the chapter is a collection of embeddings of the form (10.3.10) and (10.3.11), or, equivalently, inequalities of the form (10.3.9), where $X = W^{s,p}(\Omega)$, and Y is one of

$$C_B^j(\Omega), \quad C^j(\overline{\Omega}), \quad C^{j,\lambda}(\overline{\Omega}), \quad L^q(\Omega),$$

for certain ranges of exponents. Recall that the elements of $W^{s,p}(\Omega)$ are distributions, and are not "normal" functions defined everywhere in Ω, but rather, are equivalence classes of functions that are equal only up to sets of measure zero. The interpretation of $W^{s,p}(\Omega) \hookrightarrow C^j(\overline{\Omega})$ is that given $u \in W^{s,p}(\Omega)$, which is an equivalence class of functions, there exists a member of the equivalence class that is also a member of $C^j(\overline{\Omega})$, and (10.3.9) holds. We can alternatively interpret this as: Given an arbitrary $u \in W^{s,p}(\Omega)$, we can redefine u on a subset of points in Ω having measure zero such that the resulting new function $\overline{u} \in C^j(\Omega)$, and (10.3.9) holds. Of course, the function $\overline{u} = u$ in the sense of the norm on $W^{s,p}(\Omega)$, so they remain in the same equivalence class.

Extension Operators for $W^{m,p}(\Omega)$. A useful tool is the idea of an *extension operator*, which (if it exists) allows one to establish a result for Sobolev spaces defined on all of $\mathbb{R}^n$, and then restrict the result to a domain. An example of such an extension is

Definition 10.3.2 (Zero Extensions). *Let u be defined on $\Omega \subset \mathbb{R}^n$. We denote the* zero extension *of u to all of $\mathbb{R}^n$ as*

$$\tilde{u}(x) = \begin{cases} u(x) & \text{if } x \in \Omega, \\ 0 & \text{if } x \in \mathbb{R}^n \setminus \Omega. \end{cases}$$

More generally, following [2] we can define

Definition 10.3.3 (General Extension Operators). *Let Ω be a domain in $\mathbb{R}^n$. A linear operator $E\colon W^{m,p}(\Omega) \to W^{m,p}(\mathbb{R}^n)$ is called a* simple (m,p)-extension operator for the domain Ω *if there exists a constant $C = C(m,p)$ such that for any $u \in W^{m,p}(\Omega)$, it holds that:*

(1) $Eu(x) = u(x)$, *a.e. in Ω;*

(2) $\|Eu\|_{m,p,\mathbb{R}^n} \leqslant C\|u\|_{m,p,\Omega}$.

The operator E *is called a* strong m-extension *if* E *extends a.e.-defined functions in* Ω *to a.e.-defined functions in* $\mathbb{R}^n$*, and if for all* $1 \leqslant p < \infty$ *and all* $0 \leqslant k \leqslant m$*,* E *(restricted to* $W^{k,p}(\Omega)$*) is a simple* (k,p)*-extension. The operator* E *is called a* total extension *if* E *is a strong extension for all* m*.*

If a simple (m,p)-extension operator E exists for a given domain Ω, then that domain automatically inherits various properties of $W^{m,p}(\mathbb{R}^n)$, such as embeddings, without other detailed considerations such as domain regularity. For example, if $W^{m,p}(\mathbb{R}^n) \hookrightarrow X(\mathbb{R}^n)$, with continuous linear embedding operator I, where the space $X(\Omega) = L^q(\Omega)$ or one of the other targets for the Sobolev embedding theorem, then we have

$$\|u\|_{X(\mathbb{R}^n)} \leqslant C_0\|u\|_{m,p,\mathbb{R}^n}. \tag{10.3.12}$$

Having the extension operator E immediately gives the same result for the spaces $W^{m,p}(\Omega)$ and $X(\Omega)$, but with a different constant:

$$\|u\|_{X(\Omega)} \leqslant \|Eu\|_{X(\mathbb{R}^n)} \leqslant C_0\|Eu\|_{m,p,\mathbb{R}^n}, \leqslant C_0 C\|u\|_{m,p,\Omega}.$$

The first inequality holds simply because all the families of norms we are working with have the property that for suitably chosen $p \geqslant 0$,

$$\|u\|_{X(\mathbb{R}^n)} = \left(\|u\|^p_{X(\Omega)} + \|u\|^p_{X(\mathbb{R}^n\backslash\Omega)}\right)^{1/p}.$$

The second inequality holds by the assumed embedding (10.3.12), and the third inequality holds due to E being an extension. In the case of a bounded Lipschitz domain Ω, the existence of a total extension is guaranteed by

Theorem 10.3.1 (Stein Extension Theorem). *Let* Ω *be a domain in* $\mathbb{R}^n$ *satisfying the strong local Lipschitz condition. Then there exists a total extension operator* E *for* Ω*.*

Proof. See [2] for an outline of the proof, or [160] for the full proof. □

Fractional Order Sobolev Spaces $W^{s,p}(\Omega)$

Before we assemble a list of basic important properties of the spaces $W^{m,p}(\Omega)$ and $W^{m,p}_0(\Omega)$, we will define a scale of function spaces that lie *between* two successive Sobolev spaces, such as $X = W^{m,p}(\Omega)$ and $Y = W^{m+1,p}(\Omega)$. In other words, we would like to have a finer notion of differentiability and smoothness that is not restricted to integer order differentiability; our current definition of Sobolev spaces based on weak derivatives involving integer order multi-indices is necessarily restricted to integer order differentiability. Why is this important? Among a number of compelling reasons, it turns out that even to make sense of simple boundary value problems such as the Poisson equation on polygonal domains, we will naturally be led to *fractional order* Sobolev spaces. This will be needed to make sense of the restriction (or *trace*) of a function in $W^{m,p}(\Omega)$ to the boundary $\partial\Omega$ of Ω, which is necessary to impose boundary conditions on the solution to the boundary value problem.

As it turns out, there are a number of distinct ways to construct function spaces that lie between $W^{m,p}(\Omega)$ and $W^{m+1,p}(\Omega)$; we consider several of them below briefly, and then examine how all of these constructions are essentially special cases of two general families of spaces: *Besov spaces* $B^s_{p,q}(\Omega)$ and *Triebel-Lizorkin spaces* $F^s_{p,q}(\Omega)$. In order to understand these types of function spaces, it will be useful to have access to the Fourier transform of a distribution on $\mathbb{R}^n$, which will lead us first to introducing a subspace of the Schwartz distributions $\mathcal{D}'(\Omega)$ introduced in Section 2.6: the space of *tempered distributions* $\mathcal{S}'(\Omega)$. To help the discussion below, we introduce the following notation: Given $s \in \mathbb{R}$, define

$$\begin{aligned} \underline{s} = \lfloor s \rfloor &= \text{ the largest integer } \underline{s} \text{ such that } \underline{s} \leqslant s, \\ \overline{s} = \lceil s \rceil &= \text{ the smallest integer } \overline{s} \text{ such that } \overline{s} \geqslant s. \end{aligned}$$

Given any $s \in \mathbb{R}$, we can then write $s = \underline{s} + \sigma$, where $\sigma \in [0, 1)$.

Fourier Transform and the Space of Tempered Distributions $\mathcal{S}'(\mathbb{R}^n)$. We introduced the Schwartz distributions in Section 2.6 when we defined the notion of weak derivative, but unfortunately, not all Schwartz distributions have well-defined Fourier transforms. Therefore, we need to consider a subspace of the Schwartz distributions known as the space of *tempered distributions*. The approach is to enlarge the space of *test functions* from $\mathcal{D}(\mathbb{R}^n)$ to the *Schwartz space* (of test functions) $\mathcal{S}(\mathbb{R}^n)$, which will produce a smaller space of linear functionals $\mathcal{S}'(\mathbb{R}^n)$ than were present in $\mathcal{D}'(\mathbb{R}^n)$. To enlarge $\mathcal{D}(\mathbb{R}^n)$, we define $\mathcal{S}(\mathbb{R}^n)$ to be the set of all infinitely differentiable functions $\phi\colon \mathbb{R}^n \to \mathbb{R}$ with all derivatives "rapidly decreasing" as $|x| \to \infty$. The "rapid decrease" property is characterized as follows. One defines

$$P_{\alpha,\beta}(\phi) = \sup_{x \in \mathbb{R}^n} |x^\beta D^\alpha \phi(x)|,$$

for any pair of multi-indices α and β. The Schwartz space can then be defined as

$$\mathcal{S} = \mathcal{S}(\mathbb{R}^n) = \{\phi \in C^\infty(\mathbb{R}^n) \ : \ P_{\alpha,\beta}(\phi) < +\infty, \ \forall \alpha, \beta\}.$$

One can show that $P_{\alpha,\beta}(\cdot)$ defines a seminorm, giving $\mathcal{S}$ again the structure of a (locally convex) topological vector space.

The *space of tempered distributions* $\mathcal{S}' = \mathcal{S}'(\mathbb{R}^n)$ is now finally defined as the topological dual space of $\mathcal{S}$. Some useful facts about $\mathcal{S}'$ are:

(1) $f \in \mathcal{S}'$ if and only if $\lim_{j\to\infty} \sup_{x\in\mathbb{R}^n} |x^\beta D^\alpha \phi_j(x)| = 0$ for all α, β implies that $\lim_{j\to\infty} f(\phi_j) = 0$.

(2) If $f \in \mathcal{D}'(\mathbb{R}^n)$ and f has compact support (a property which is defined through the test space), then $f \in \mathcal{S}'$.

(3) If $f \in \mathcal{S}'$, then $f' \in \mathcal{S}'$, where f' is the distributional derivative of f.

(4) If $f \in L^1_{\text{loc}}(\mathbb{R}^n)$ and $f(x) = \mathcal{O}(|x|^r)$ for $r < \infty$, then $f \in \mathcal{S}'$.

(5) If $f \in L^p(\mathbb{R}^n)$, then $f \in \mathcal{S}'$ for any $p \geqslant 1$.
(6) If $f \in \mathcal{S}'$, then $\mathcal{F}f \in \mathcal{S}'$, where $\mathcal{F}$ is the Fourier transform.

It is this last property which will be useful in defining the notion of fractional order differentiation. Recall that the Fourier transform $\mathcal{F}$ and the inverse Fourier transform $\mathcal{F}^{-1}$ are defined for a function $u \in L^1(\mathbb{R}^n)$ as follows:

$$\mathcal{F}(u)(y) = \hat{u}(y) = \frac{1}{(2\pi)^{n/2}} \int_{\mathbb{R}^n} e^{-ix\cdot y} u(x)\, dx, \quad y \in \mathbb{R}^n, \tag{10.3.13}$$

$$\mathcal{F}^{-1}(u)(x) = \check{u}(x) = \frac{1}{(2\pi)^{n/2}} \int_{\mathbb{R}^n} e^{ix\cdot y} u(y)\, dy, \quad x \in \mathbb{R}^n. \tag{10.3.14}$$

That $\hat{u}$ and $\check{u}$ are in L^1 follow from the Hölder inequality and uniform boundedness of e^{iz}. The Fourier transform is a 1-1 map on $\mathcal{S}'(\mathbb{R}^n)$, and if $u \in \mathcal{S}'(\mathbb{R}^n)$, then $u = \mathcal{F}^{-1}(\mathcal{F}(u))$. One of the key results on Fourier transform is

Theorem 10.3.2 (Plancherel's Theorem). *If $u \in L^2(\mathbb{R}^n)$, then $\hat{u}, \check{u} \in L^2(\mathbb{R}^n)$, and*

$$\|u\|_2 = \|\hat{u}\|_2 = \|\check{u}\|_2.$$

Proof. See, for example, [70]. □

Bessel Potential Spaces $H^{s,p}(\Omega)$. Starting with the observation that

$$\widehat{D^\alpha u} = (i\omega)^\alpha \hat{u}, \tag{10.3.15}$$

which follows from integration by parts, where $\hat{u}$ is the Fourier transform of the function $u \in L^2(\mathbb{R}^n)$, Plancherel's Theorem (Theorem 10.3.2) gives

$$\|D^\alpha u\|_2 = \|\omega^\alpha \hat{u}\|_2.$$

This gives a Fourier representation of the integer Sobolev norm

$$\|u\|_{m,2}^2 = \int_{\mathbb{R}^n} \left(\sum_{0\leqslant|\alpha|\leqslant m} \omega^{2\alpha} \right) |\hat{u}(\omega)|^2\, d\omega.$$

Observing that

$$c_1(1+|\omega|^2)^m \leqslant \sum_{0\leqslant|\alpha|\leqslant m} \omega^{2\alpha} \leqslant c_2(1+|\omega|^2)^m, \tag{10.3.16}$$

for constants c_1 and c_2, we can define a norm for $W^{m,2}(\Omega)$ as

$$\|u\|_{m,2}^* = \|(1+|\cdot|^2)^{m/2}\hat{u}\|_{0,2},$$

where $|\cdot|$ is understood to mean the l^2-norm of the integration variable used for the L^2-norm on the right. One can then show that (see [70])

$$c_3\|u\|_{m,2} \leqslant \|u\|_{m,2}^* \leqslant c_4\|u\|_{m,2}$$

for constants c_3 and c_4, so that the two norms are equivalent. On this basis, one defines the fractional norm for all $0 < s \in \mathbb{R}$ as

$$\|u\|_{s,2} = \|(1 + |\cdot|^2)^{s/2}\hat{u}\|_{0,2},$$

and then subsequently the fractional space for all $0 < s \in \mathbb{R}$ as

$$H^s(\mathbb{R}^n) = \{u \in L^2(\mathbb{R}^n) \;:\; \|u\|_{s,2} < +\infty\}.$$

One can show (see [2]) that the norm

$$\|u\|_{m,p}^* = \|(1 + |\cdot|^2)^{m/2}\hat{u}\|_{0,p}$$

is an equivalent norm on $W^{m,p}(\mathbb{R}^n)$, which by similar reasoning leads one to define the *Bessel Potential Spaces*:

$$H^{s,p}(\mathbb{R}^n) = \{u \in L^p(\mathbb{R}^n) \;:\; \|u\|_{s,p} < \infty\},$$

where

$$\|u\|_{s,p} = \|(1 + |\cdot|^2)^{s/2}\hat{u}\|_{0,p}.$$

This reduces to the case above when $p = 2$. The naming of the spaces arises from the fact that the multipliers $(1 + |\cdot|^2)^{s/2}$ are related to Fourier transforms of Bessel functions; see [2]. The spaces $H^{s,p}(\Omega)$ for domains $\Omega \subset \mathbb{R}^n$ are subsequently constructed using extensions such as that provided by Theorem 10.3.1, as outlined above.

Sobolev Spaces $W^{s,p}(\Omega)$. A fractional norm can be defined directly or *intrinsically* in a manner similar to how the Hölder norms are defined; this allows one to start with domains $\Omega \subset \mathbb{R}^n$ without relying on the existence of an extension operator (although as we have seen, this is not a limitation due to, for example, Theorem 10.3.1). For any noninteger real $s > 0$, written as $s = \underline{s} + \sigma$, with $\underline{s} = \lfloor s \rfloor$ and $\sigma \in (0, 1)$, we first define fractional seminorms (for $1 \leqslant p \leqslant \infty$) and semi-inner products (for $p = 2$) as

$$|u|_{s,p}^p = \sum_{|\alpha|=\underline{s}} \int_\Omega \int_\Omega \frac{|D^\alpha u(x) - D^\alpha u(y)|^p}{|x-y|^{n+\sigma p}}\, dx\, dy, \quad 1 \leqslant p < \infty, \tag{10.3.17}$$

$$|u|_{s,\infty} = \max_{|\alpha|=\underline{s}} \operatorname*{ess\,sup}_{\substack{x,y\in\Omega \\ x\neq y}} \frac{|D^\alpha u(x) - D^\alpha u(y)|}{|x-y|^\sigma}\, dx\, dy, \tag{10.3.18}$$

$$[u, v]_s = \sum_{|\alpha|=\underline{s}} \int_\Omega \int_\Omega \frac{(D^\alpha u(x) - D^\alpha u(y))(D^\alpha v(x) - D^\alpha v(y))}{|x-y|^{n+2\sigma}}\, dx\, dy. \tag{10.3.19}$$

With the convention that when $s \in \mathbb{R}$ is an integer, $|u|_{s,p} = 0$, $1 \leqslant p \leqslant \infty$, and $[u, v]_s = 0$, then the fractional norm for $1 \leqslant p < \infty$ and $p = \infty$, and the fractional inner product for $p = 2$, are defined as

$$\|u\|_{s,p} = \left(\|u\|_{\underline{s},p}^p + |u|_{s,p}^p\right)^{1/p}, \quad 1 \leqslant p < \infty, \tag{10.3.20}$$

$$\|u\|_{s,\infty} = \max\left\{\|u\|_{\underline{s},\infty}\,,\ |u|_{s,\infty}\right\}. \tag{10.3.21}$$

$$(u,v)_s = (u,v)_{\underline{s}} + [u,v]_s, \tag{10.3.22}$$

with norms $\|u\|_{m,p}$ and inner products $(u,v)_m$ for integer m defined as in (10.3.6), (10.3.7), and (10.3.8).

This approach gives construction of the Banach spaces $W^{s,p}(\Omega)$ for $1 \leqslant p \leqslant \infty$ and any real $s > 0$ through

$$W^{s,p}(\Omega) = \{u \in L^p(\Omega) \ :\ \|u\|_{s,p} < \infty\},$$

which reduces to the integer definition when s is an integer, and also yields Hilbert spaces $H^s(\Omega)$ for $p = 2$, reducing again to the integer definition when s is an integer. One can show that $W^{s,p}(\mathbb{R}^n) = H^{s,p}(\mathbb{R}^n)$ for m integer and $1 < p < \infty$, and subsequently for domains $\Omega \subset \mathbb{R}^n$ (see [2]) for which there is an extension operator, such as that provided by Theorem 10.3.1.

Besov Spaces $B^s_{p,q}(\Omega)$ and Triebel-Lizorkin Spaces $F^s_{p,q}(\Omega)$

Perhaps the most useful way to organize the distinct constructions of fractional order Sobolev-type function spaces, and understand their relationships, is by first defining two general families of Banach spaces known as *Besov spaces* $B^s_{p,q}(\Omega)$ and *Triebel-Lizorkin spaces* $F^s_{p,q}(\Omega)$. All of the spaces we have described above can be realized as specific instances of these two families, for different choices of the parameters s, p, and q. We note that the discussion below is primarily restricted to the case where the parameters range as $1 \leqslant p, q \leqslant \infty$ (with $p = \infty$ and/or $q = \infty$ always handled in a special way, and sometimes explicitly excluded), $s \geqslant 0$, with $s < 0$ defined later by forming the normed dual spaces of these spaces.

The spaces $B^s_{p,q}(\Omega)$ and $F^s_{p,q}(\Omega)$ can be defined fairly simply when $\Omega = \mathbb{R}^n$, and then restricted to (for example) bounded Lipschitz domains $\Omega \subset \mathbb{R}^n$ using extension operators when they exist for the particular domain class. The definitions of $B^s_{p,q}(\mathbb{R}^n)$ and $F^s_{p,q}(\mathbb{R}^n)$ are through the Fourier transform of a distribution. Given a set of functions $\{\phi_k\}$ forming a partition of unity $\sum_{k=1}^n \phi_k(x) = 1\ \forall x \in \Omega$, and satisfying some other basic assumptions, one defines the following two norms:

$$\|u\|_{B^s_{p,q}(\mathbb{R}^n)} = |\ \|2^{sk}\mathcal{F}^{-1}\phi_k\mathcal{F}u\|_p\ |_q, \tag{10.3.23}$$

$$\|u\|_{F^s_{p,q}(\mathbb{R}^n)} = \|\ |2^{sk}\mathcal{F}^{-1}\phi_k\mathcal{F}u|_q\ \|_p, \tag{10.3.24}$$

which leads to the definitions of the normed vector spaces:

$$B^s_{p,q}(\mathbb{R}^n) = \{u \in \mathcal{S}'(\mathbb{R}^n) \ :\ \|u\|_{B^s_{p,q}(\mathbb{R}^n)} < +\infty\}, \tag{10.3.25}$$

$$F^s_{p,q}(\mathbb{R}^n) = \{u \in \mathcal{S}'(\mathbb{R}^n) \ :\ \|u\|_{F^s_{p,q}(\mathbb{R}^n)} < +\infty\}. \tag{10.3.26}$$

These two general families of function spaces give rise to all of the Banach spaces we have discussed above, for specific choices of the parameters p, q, and s:

(1) *Hölder spaces*: $C^{m,\lambda}(\overline{\Omega}) = B^{m+\lambda}_{\infty,\infty}(\Omega)$.

(2) *Bessel Potential spaces*: $H^{s,p}(\Omega) = B^{s}_{p,2}(\Omega)$.

(3) *Fourier spaces*: $H^{s}(\Omega) = B^{s}_{2,2}(\Omega)$.

(4) *Sobolev spaces*: $W^{s,p}(\Omega) = \begin{cases} F^{s}_{p,2}(\Omega), & s \text{ an integer}, \\ B^{s}_{p,p}(\Omega), & s \text{ not an integer}. \end{cases}$.

Key Embedding and Related Theorems

We now summarize some of the key properties of some of the function spaces above. Allowing Ω to be unbounded would require assuming one of the more complex domain conditions listed above, such as the cone and uniform cone conditions, and then listing a number of different cases of the key theorems based on which domain property holds; see [2, 82]. We are interested here primarily in the analysis and approximation of solutions to boundary value problems posed on Sobolev spaces; therefore, we will be interested primarily in two specific cases:

(1) $\Omega = \mathbb{R}^n$, $n \geqslant 1$.

(2) $\Omega \subset \mathbb{R}^n$, $n \geqslant 1$, an open, bounded, Lipschitz domain.

This will allow us to organize and simplify the results of interest as much as possible. For example, under the assumption that $\Omega \subset \mathbb{R}^n$ is bounded and Lipschitz, Theorem 10.3.1 ensures that there exists a total extension operator E for Ω; the existence of the total extension E allows us to establish results on $\mathbb{R}^n$, and then transfer the result to Ω (see the discussion of extension operators above).

Theorem 10.3.3 (Sobolev Embedding Theorem). *Let $\Omega \subset \mathbb{R}^n$ be a domain satisfying the cone condition, and let $1 \leqslant p < \infty$. Then*

(1) If $sp > n$, then $W^{s+r,p}(\Omega) \hookrightarrow W^{r,q}(\Omega)$, $r \geqslant 0$, $p \leqslant q \leqslant \infty$,
and $W^{s+j,p}(\Omega) \hookrightarrow C^{j}_{B}(\Omega)$, $j \geqslant 0$, j integer.

(2) If $sp = n$, then $W^{s+r,p}(\Omega) \hookrightarrow W^{r,q}(\Omega)$, $r \geqslant 0$, $p \leqslant q < \infty$.

(3) If $sp < n$, then $W^{s+r,p}(\Omega) \hookrightarrow W^{r,q}(\Omega)$, $r \geqslant 0$, $p \leqslant q \leqslant p^* = np/(n - sp)$.

Proof. See [2]. □

Both $\mathbb{R}^n$ and Lipschitz domains Ω in $\mathbb{R}^n$ (open subsets $\Omega \subset \mathbb{R}^n$) satisfy the cone condition, so the embedding theorem holds as stated for the two cases of primary interest. The embeddings are compact under certain additional assumptions.

Theorem 10.3.4 (Rellich-Kondrachov Theorem). *If Ω is a bounded domain, then the embeddings in Theorem 10.3.3 have the following compactness properties:*

(1) If $sp > n$, the embeddings are compact, excluding $q = \infty$.

(2) If $sp = n$, the embeddings are compact.

(3) If $sp < n$, the embeddings are compact, excluding $q = p^* = np/(n - sp)$.

If Ω also satisfies the strong local Lipschitz condition, then the following are compact embeddings:

(1) If $sp > s$, then $W^{s+j,p}(\Omega) \hookrightarrow C^j(\overline{\Omega})$, $j \geqslant 0$, j integer.

(2) If $sp > n \geqslant (s-1)p$, and $0 < \lambda < s - (n/p)$, then $W^{s+j,p}(\Omega) \hookrightarrow C^{j,\lambda}(\overline{\Omega})$, $j \geqslant 0$, j integer.

Proof. See [2]. □

We also have the characterization of traces of functions in $W^{s,p}(\Omega)$ restricted to boundary sets.

Theorem 10.3.5 (Trace Theorem). *Let $\Omega \subset \mathbb{R}^n$ be a Lipschitz domain with boundary denoted as $\Gamma \subset \partial\Omega$, with possibly $\Gamma = \partial\Omega$. Then for a function $u \in W^{s,p}(\Omega)$, with $1 < p < \infty$, a function $g\colon \Gamma \to \mathbb{R}$ is the trace of u on the boundary set Γ if and only if $g \in W^{s-1/p,p}(\Gamma)$. Moreover, there exists a bounded linear operator T, the* trace operator, *such that $g = Tu = u|_\Gamma$, and*

$$\|u\|_{1-1/p,p,\Gamma} \leqslant C_{\text{tr}}\|u\|_{1,p,\Omega}.$$

Proof. This version of the theorem may be found, for example, in [117]. The proof uses the observation that when $\Gamma \subset \partial\Omega$, then

$$\|u\|_{1-1/p,p,\Gamma} \leqslant \|u\|_{1-1/p,p,\partial\Omega},$$

and then works with the trace map on all of $\partial\Omega$. □

Finally, we have the *Poincaré Inequality* which characterizes when the Sobolev seminorm $|\cdot|_{1,p}$ is, in fact, a norm.

Theorem 10.3.6 (Poincaré Inequality). *Let $\Omega \subset \mathbb{R}^n$ be a domain with finite width. [In other words, Ω is bounded between two parallel $(n-1)$-dimensional planes in $\mathbb{R}^n$.] Let $\Gamma \subset \partial\Omega$ have nonzero $(n-1)$-dimensional measure, possibly with $\Gamma = \partial\Omega$. Then for $1 \leqslant p < \infty$,*

$$\|u\|_p \leqslant \rho|u|_{1,p}, \quad \forall u \in W^{1,p}_{0,D}(\Omega) = \{u \in W^{1,p}(\Omega) \ : \ u = 0 \ on \ \Gamma\}.$$

Proof. See [2]. □

Sobolev Spaces on Manifolds. While the L^p and Sobolev spaces $W^{k,p}$ of scalar-valued functions on bounded sets in $\mathbb{R}^n$ are developed in [2], generalizations to manifolds appear in [91], and generalizations to tensor fields appear in [142]. A general construction of real-order Sobolev spaces of sections of vector bundles over closed manifolds may be found in [97]; see also [11, 149, 155].

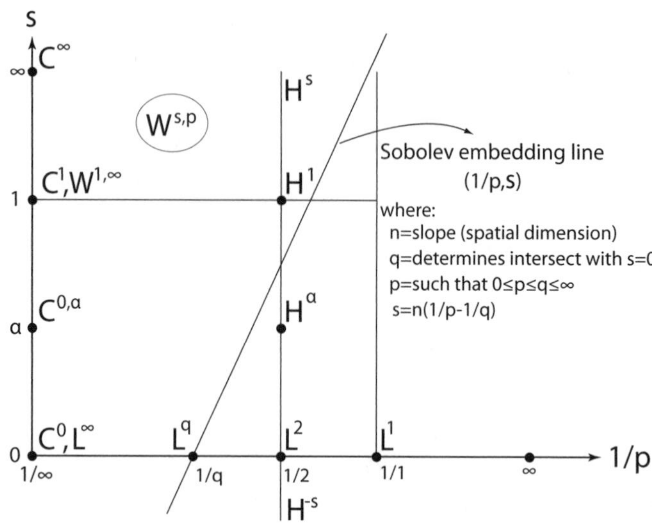

Figure 10.1 DeVore diagram for the $W^{s,p}(\Omega)$ plane within the $B^s_{p,q}(\Omega)$ cube. For a given $q \geqslant 0$, the diagonal Sobolev embedding line is given by the points $(1/p, s)$, where for a given spatial dimension n and a given p such that $0 \leqslant p \leqslant q \leqslant \infty$, the differentiability index of a point on the embedding line is $s = n(1/p - 1/q)$. The slope of the line is then n. The significance of the embedding line is that all function spaces $W^{s,p}(\Omega)$ above the line are continuously embedded in $L^q(\Omega)$. (Note that the $W^{s,p}$ plane intersects the $B^s_{p,q}$ cube only for *noninteger* values of p.)

The DeVore Diagram. The range of q in case (3) of Theorem 10.3.3 is

$$0 \leqslant p \leqslant q \leqslant p^* = \frac{np}{n - sp}.$$

The range of q in cases (1) and (2) of Theorem 10.3.3 are the limiting cases where the denominator $n - sp$ becomes zero or negative. Given n and q, this gives s as a (linear) function of $1/p$:

$$\frac{s}{n} \geqslant \frac{1}{p} - \frac{1}{q} \geqslant 0,$$

which ensures that $W^{s+r,p} \hookrightarrow W^{s,q}$, for $r \geqslant 0$. This can be thought of as an equation for a line in the $(1/p, s)$-plane. The slope of the line is clearly n, with the intersection with the $s = 0$ axis determined by $1/q$.

In Figure 10.1 we display what is commonly called a *DeVore diagram*, which is a graphical way of displaying the relationships between the classical function spaces $C^m(\overline{\Omega})$ and $C^{m,\alpha}(\overline{\Omega})$, the $L^p(\Omega)$ and Sobolev $W^{s,p}(\Omega)$ spaces, and the Besov spaces $B^s_{p,q}(\Omega)$. The diagonal line illustrates the Sobolev Embedding Theorem. For a given $q \geqslant 0$, the diagonal Sobolev embedding line is given by the points $(1/p, s)$, where for a given spatial dimension n and a given p such that $0 \leqslant p \leqslant q \leqslant \infty$, the differentiability index of a point on the embedding line is $s = n(1/p - 1/q)$. The slope of the line is then n. The significance of the embedding line is that all function spaces $W^{s,p}(\Omega)$ above the line are continuously embedded in $L^q(\Omega)$. Note that in the DeVore diagram, the $W^{s,p}(\Omega)$ plane only intersects the $B^s_{p,q}(\Omega)$ cube for *noninteger* values of p.

Positive and Negative Parts of Functions in $W^{1,p}(\Omega)$. We will need the following basic result for working with maximum principles in Sobolev spaces, which is related to the idea of an extension operator. We are given a function $u \in W^{1,p}(\Omega)$, and wish to consider the function $|u|$, as well as the separate positive and negative parts, so that we can write

$$u = u^+ - u^-, \qquad |u| = u^+ + u^-, \tag{10.3.27}$$

with the nonnegative functions u^+ and u^- defined as

$$u^+(x) = \max\{u(x), 0\}, \quad u^-(x) = \max\{-u(x), 0\} = -\min\{u(x), 0\}.$$

It is not completely obvious when these definitions make sense, and in particular, what function spaces $|u|$, u^+, and u^- might belong to. To establish some facts about $|u|$, u^+, and u^-, it is useful to have the following basic result.

Theorem 10.3.7 (Stampacchia Theorem). *Let $F\colon \mathbb{R} \to \mathbb{R}$ be Lipschitz and have the property that $F(0) = 0$. If $\Omega \subset \mathbb{R}^n$ is bounded, if $1 < p < \infty$, and if $u \in W_0^{1,p}(\Omega)$, then $F \circ u \in W_0^{1,p}(\Omega)$.*

Proof. See [113]. □

This result can be used to clarify the situation with regard to $|u|$, u^+, and u^-.

Theorem 10.3.8. *Let $u \in W_0^{1,p}(\Omega)$. Then $|u|, u^+, u^- \in W_0^{1,p}(\Omega)$. Furthermore, if $\alpha, \beta \in \mathbb{R}$, with $\alpha \leqslant 0 \leqslant \beta$, then also $(u-\beta)^+, (u-\alpha)^- \in W_0^{1,p}(\Omega)$. Moreover,*

$$\frac{\partial u^+}{\partial x_k} = \begin{cases} \dfrac{\partial u}{\partial x_k}, & \text{a.e. on } \{u(x) > 0\} \\ 0, & \text{a.e. on } \{u(x) \leqslant 0\}, \end{cases} \qquad \frac{\partial u^-}{\partial x_k} = \begin{cases} 0, & \text{a.e. on } \{u(x) \geqslant 0\} \\ -\dfrac{\partial u}{\partial x_k}, & \text{a.e. on } \{u(x) < 0\}, \end{cases}$$

with $\partial|u|/\partial x_k = \partial u^+/\partial x_k + \partial u^-/\partial x_k$.

Proof. That $|u| \in W_0^{1,p}(\Omega)$ follows from Theorem 10.3.7, since $F(t) = |t|$ is Lipschitz with $F(0) = 0$. That $u^+, u^- \in W_0^{1,p}(\Omega)$ also follows immediately, since solving the two equations (10.3.27) for u^+ and u_- gives

$$u^+ = \frac{|u| + u}{2}, \qquad u^- = \frac{|u| - u}{2}.$$

These are linear combinations of u and $|u|$, both of which are in $W_0^{1,p}(\Omega)$.

Now let $\overline{\phi} = (u-\beta)^+ = \max(u-\beta, 0)$ and $\underline{\phi} = (u-\alpha)^- = \max(\alpha - u, 0)$. Since $\beta \geqslant 0$ and $\alpha \leqslant 0$, we have

$$0 \leqslant \overline{\phi} = (u-\beta)^+ \leqslant u^+,$$
$$0 \leqslant \underline{\phi} = (u-\alpha)^- \leqslant u^-.$$

If $T\colon W^{1,p}(\Omega) \to W^{1-1/p,p}(\partial\Omega)$ is the trace operator (see Theorem 10.3.5),

$$0 \leqslant T\overline{\phi} \leqslant Tu^+ = 0,$$
$$0 \leqslant T\underline{\phi} \leqslant Tu^- = 0,$$

imply that both $\overline{\phi}, \underline{\phi} \in W_0^{1,p}(\Omega)$.

The expressions for the derivatives of u^+ and u^- are derived in [132]. □

We note that the proof of Theorem 10.3.8 in [132] does not go through Theorem 10.3.7, and as a result holds with $W_0^{1,p}(\Omega)$ replaced by $W^{1,p}(\Omega)$.

The primary application of Theorem 10.3.8 will be as a tool for developing *a priori* L^∞ estimates for second-order linear and nonlinear elliptic problems through maximum principles. These results will be of the following form: If $u \in W^{1,p}(\Omega)$ is any solution (which may or may not exist) to the given elliptic problem, then we must also have $u \in W^{1,p}(\Omega) \cap L^\infty(\Omega)$, with u satisfying the following pointwise bounds almost everywhere in Ω:

$$\alpha \leqslant u(x) \leqslant \beta \quad a.e. \text{ in } \Omega. \tag{10.3.28}$$

Obtaining pointwise bounds of the form (10.3.28) in the case of nonlinear problems is quite important; they allow for norm control of nonlinear terms that would otherwise not be possible, or may only be possible in special situations under strong domain and coefficient regularity assumptions. An example along these lines is a useful generalization of Theorem 10.3.7.

Theorem 10.3.9. *Let $F\colon \mathbb{R} \to \mathbb{R}$ be in $C^m(\mathbb{R})$ and have the property that $F(0) = 0$. If $1 < p < \infty$, if $m \geqslant 0$, and if $u \in W^{m,p}(\Omega) \cap L^\infty(\Omega)$, then $F \circ u \in W^{m,p}(\Omega)$, and*

$$\|F(u)\|_{m,p} \leqslant C(F, \|u\|_\infty)(1 + \|u\|_{m,p}),$$

where $C(F, \|u\|_\infty)$ depends on a.e. pointwise values of derivatives of F up to order m, and a.e. pointwise values of u over Ω.

Proof. This is an example of a Gagliardo-Nirenberg-Moser type of estimate; a proof may be found in [166]. □

To finish our discussion, we mention a final useful application, where a quadratic nonlinearity is controlled by the fact that the functions lie in an L^∞ subset of the space $W^{1,p}(\Omega)$; this result would otherwise only hold for sufficiently large p (depending on the spatial dimension; see the section below on the Sobolev Embedding Theorem).

Theorem 10.3.10 (Gradient Product Formula). *Let $\Omega \subset \mathbb{R}^n$ be a bounded domain and let $u, v \in W^{1,p}(\Omega) \cap L^\infty(\Omega)$, $1 \leqslant p \leqslant \infty$. Then $uv \in W^{1,p}(\Omega) \cap L^\infty(\Omega)$, and the following product rule holds: $\nabla(uv) = u\nabla v + v\nabla u$.*

Proof. The proof follows directly by application of Hölder inequalities. □

Closed Intervals in the Ordered Banach Space $W^{s,p}(\Omega)$. Let C_+^∞ be the set of nonnegative smooth (scalar) functions on Ω, where $\Omega \subset \mathbb{R}^n$ is a domain. We can define the following order cone with respect to which the Sobolev spaces $W^{s,p} = W^{s,p}(\Omega)$ are ordered Banach spaces:

$$W_+^{s,p} := \{u \in W^{s,p}(\Omega) \ : \ \langle u, v\rangle \geqslant 0, \quad \forall v \in C_+^\infty \ \}, \tag{10.3.29}$$

where $\langle \cdot, \cdot \rangle$ is the (unique) extension of the L^2 inner product to a bilinear form $W^{s,p} \otimes W^{-s,p'} \to \mathbb{R}$, with $1/p' + 1/p = 1$. The order relation on $W^{s,p}$ is then $u \geqslant w$ iff $u - w \in W_+^{s,p}$. We note that this order cone is normal only for $s = 0$. The ordered Banach spaces $W^{s,p}$, $s \geqslant 0$, $1 \leqslant p \leqslant \infty$, with the order cone defined as in (10.3.29), play a key role in the development of both solution theory and near-best approximation results for nonlinear problems. See Section 10.1, where we review the main properties of ordered Banach spaces.

Given the ordered Banach space $W^{s,p}$ for $s \geqslant 0$ and $p \in [1, \infty]$, and given any two functions $u_-, u_+ \in L^\infty$ such that $u_- \leqslant u_+$ *a.e.* in Ω, we can define the closed interval,

$$[u_-, u_+]_{s,p} := \{u \in W^{s,p} \ : \ u_- \leqslant u \leqslant u_+\} \subset W^{s,p}.$$

We equip $[u_-, u_+]_{s,p}$ with the subspace topology of $W^{s,p}$. We will write $[u_-, u_+]_p$ for $[u_-, u_+]_{0,p}$ when $s = 0$, and simply $[u_-, u_+]$ for $[u_-, u_+]_\infty$ when $s = 0, p = \infty$. When $s = 0$, the $W^{s,p}$ order cone is *normal* for $1 \leqslant p \leqslant \infty$, meaning that closed intervals $[u_-, u_+]_p \subset L^p = W^{0,p}$ are automatically bounded in the metric given by the norm on L^p.

The interval $U = [u_-, u_+]_{s,p} \subset W^{s,p} = Z$ defined using this order structure will play a major role in the development of both solution theory and near-best approximation results in the next section. It will be critically important to establish that U is *convex* (with respect to the vector space structure of Z), *closed* (in the topology of Z), and (when possible) *bounded* (in the metric given by the norm on Z). It will also be important that U be *nonempty* as a subset of Z; this will involve choosing compatible u_- and u_+, meaning simply that $u_- \leqslant u_+$ *a.e.* in Ω. Regarding convexity, closure, and boundedness, we have the following key lemma.

Lemma 10.3.1 (Order Cone Intervals in $W^{s,p}$ [97]). *For $s \geqslant 0$, $1 \leqslant p \leqslant \infty$, the set*

$$U = [u_-, u_+]_{s,p} = \{u \in W^{s,p} \ : \ u_- \leqslant u \leqslant u_+\} \subset W^{s,p}$$

is convex with respect to the vector space structure of $W^{s,p}$ and closed in the topology of $W^{s,p}$. For $s = 0$, $1 \leqslant p \leqslant \infty$, the set U is also bounded with respect to the metric space structure of $L^p = W^{0,p}$.

Proof. That U is convex for $s \geqslant 0$, $1 \leqslant p \leqslant \infty$, follows from the fact that any interval built using order cones is convex. That U is closed in the case of $s = 0$,

$1 \leqslant p \leqslant \infty$, follows from the fact that norm convergence in L^p for $1 \leqslant p \leqslant \infty$ implies pointwise subsequential convergence almost everywhere (see Theorem 3.12 in [150]). That U is bounded when $s = 0$, $1 \leqslant p \leqslant \infty$, follows from the fact that the order cone L^p_+ is normal (see Lemma 10.1.7).

What remains is to show that U is closed in the case of $s > 0$, $1 \leqslant p \leqslant \infty$. The argument is as follows. Let $\{u_k\}_{k=1}^{\infty}$ be a Cauchy sequence in $U \subset W^{s,p} \subset L^p$, with $s > 0$, $1 \leqslant p \leqslant \infty$. From completeness of $W^{s,p}$, there exists $u \in W^{s,p}$ such that $\lim_{k\to\infty} u_k = u \in W^{s,p}$. From the continuous embedding $W^{s,p} \hookrightarrow L^p$ for $s > 0$, we have that

$$\|u_k - u_l\|_p \leqslant C\|u_k - u_l\|_{s,p},$$

so that u_k is also Cauchy in L^p. Moreover, the continuous embedding also implies that u is also the limit of u_k as a sequence in L^p. Since $[u_-, u_+]_{0,p}$ is closed in L^p, we have $u \in [u_-, u_+]_{0,p}$, and so $u \in U = [u_-, u_+]_{s,p} = [u_-, u_+]_{0,p} \cap W^{s,p}$. □

EXERCISES

10.3.1 Let $\Omega \subset \mathbb{R}^n$ be a domain. Prove that $L^p(\Omega)$ is a Banach space for $1 \leqslant p \leqslant \infty$.

10.3.2 Let $\Omega \subset \mathbb{R}^n$ be a domain. Prove that $W^{m,p}(\Omega)$ is a Banach space for $m \in \mathbb{N}_0$ and $1 \leqslant p \leqslant \infty$.

10.3.3 Use integration by parts to prove the identity (10.3.15).

10.3.4 Establish the inequality (10.3.16).

10.3.5 Give a proof of Theorem 10.3.10.

10.3.6 Show that the Dirac distribution $\delta(x) \in H^{-s}(\Omega)$, with $s > n/2$, where n is the spatial dimension, and where $H^{-s}(\Omega) = (H^s_0(\Omega))'$ is the dual space.

10.4 APPLICATIONS TO NONLINEAR ELLIPTIC EQUATIONS

Using the technical machinery we have developed in Sections 10.1 to 10.3, we are now in a position to consider three elliptic partial differential equations of increasing complexity:

(1) A linear second-order coercive elliptic equation.
(2) The nonlinear Poisson-Boltzmann equation arising in biophysics.
(3) The Einstein constraint equations arising in astrophysics.

These problems are similar to some of the problems which we previously encountered in Chapters 8 and 9, and each illustrates different technical difficulties that must be addressed when developing solution and approximation theory.

Our goal is to formulate these problems in the general setting of Sobolev spaces, and then to develop some theoretical results for each problem. Each result will be a necessary piece of the puzzle as we put together *a priori* error estimates for Petrov-Galerkin approximations using finite element and related methods. In particular, the results we will need to develop for each example are:

(1) *A priori* estimates for any weak solutions (energy and/or L^∞ estimates).
(2) Existence of weak solutions; uniqueness when possible.
(3) Near-best (global or local) approximation of weak solutions in a subspace.

The first example, which is linear, will require only energy estimates in order to establish existence, uniqueness, and global near-best approximation. However, the second and third examples are nonlinear, and will require more delicate *a priori* L^∞ estimates in order to make existence and near-best approximation results possible. The second example will allow for the use of simple variational arguments, whereas the third example will require the use of somewhat more sophisticated fixed point arguments.

A General Coercive Linear Elliptic Equation

Let $\Omega \subset \mathbb{R}^n$ be a bounded Lipschitz domain with boundary $\Gamma = \Gamma_D \cup \Gamma_N$, where $\Gamma_D \cap \Gamma_N = \emptyset$. We are concerned with a general second-order linear elliptic equation for an unknown function $\hat{u}\colon \Omega \to \mathbb{R}$, having the form

$$-\nabla \cdot (a\nabla \hat{u}) + b\hat{u} = f \text{ in } \Omega, \tag{10.4.1}$$

$$\hat{u} = g_D \text{ on } \Gamma_D, \tag{10.4.2}$$

$$(a\nabla \hat{u}) \cdot \nu + c\hat{u} = g_N \text{ on } \Gamma_N, \tag{10.4.3}$$

with $b\colon \Omega \to \mathbb{R}$, $f\colon \Omega \to \mathbb{R}$, $g_D\colon \Gamma_D \to \mathbb{R}$, $g_N\colon \Gamma_N \to \mathbb{R}$, $c\colon \Gamma_N \to \mathbb{R}$, with the matrix function $a\colon \Omega \to \mathbb{R}^{n\times n}$, and with the unit normal $\nu\colon \Gamma_N \to \mathbb{R}$. The equation is called *elliptic* if the matrix $a(x) = [a_{ij}(x)]$ is positive definite for all $x \in \Omega$, and *strongly elliptic* if there exists $\lambda > 0$ such that $\sum_{ij} a_{ij}(x)\eta_i\eta_j \geqslant \lambda|\eta|^2$, $\forall x \in \Omega$, $\eta \in \mathbb{R}^n$. The ellipticity conditions will be weakened below to only hold *almost everywhere* (in the sense of measure zero).

If there are discontinuities present in coefficients of equations (10.4.1)–(10.4.3), specifically in the coefficient $a(x)$, then it makes no sense to look for a classical solution. The definition of the weak solution centers on a weaker form of the problem; although the weak solution is defined very generally, one can show (using the Green's integral identities) that the weak solution is exactly the classical solution in the case that the coefficients and the domain are "nice" enough. In addition, finite element methods and variational multigrid methods are constructed from weak formulations of the given elliptic problem. Therefore, we will derive the weak form of problem (10.4.1)–(10.4.3).

Define first the following subspace of $H^1(\Omega)$:

$$H^1_{0,D}(\Omega) = \{u \in H^1(\Omega) \ : \ u = 0 \text{ on } \Gamma_D\}.$$

Note that $H^1_{0,D}(\Omega)$ is a Hilbert space. We now begin by multiplying the strong form equation (10.4.1) by a test function $v \in H^1_{0,D}(\Omega)$ and integrating over Ω to obtain

$$\int_\Omega (-\nabla \cdot (a\nabla \hat{u}) + b\hat{u})\, v\, dx = \int_\Omega fv\, dx,$$

which becomes after integration by parts (which is well-defined in $H^1_{0,D}(\Omega)$ by a density argument),

$$\int_\Omega (a\nabla\hat{u})\cdot\nabla v\,dx - \int_\Gamma v(a\nabla\hat{u})\cdot\nu\,ds + \int_\Omega b\hat{u}v\,dx = \int_\Omega fv\,dx. \tag{10.4.4}$$

Since $v = 0$ on Γ_D, the boundary integral can be reformulated using (10.4.3) as

$$\int_\Gamma v(a\nabla\hat{u})\cdot\nu\,ds = \int_{\Gamma_N} v(a\nabla\hat{u})\cdot\nu\,ds = \int_{\Gamma_N} v(g_N - c\hat{u})\,ds. \tag{10.4.5}$$

If the boundary function g_D is regular enough so that $g_D \in H^{1/2}(\Gamma_D)$, then from Theorem 10.3.5 there exists $w \in H^1(\Omega)$ such that $g_D = \text{tr}\,w$. Employing such a function $w \in H^1(\Omega)$ satisfying $g_D = \text{tr}\,w$, we define the following affine or translated Sobolev space:

$$H^1_{g,D}(\Omega) = \{\hat{u} \in H^1(\Omega)\ :\ u + w,\ u \in H^1_{0,D}(\Omega),\ g_D = \text{tr}\,w\}.$$

It is easily verified that the solution $\hat{u}$ to the problem (10.4.1)–(10.4.3), if one exists, lies in $H^1_{g,D}(\Omega)$, although unfortunately $H^1_{g,D}(\Omega)$ is not a Hilbert space, since it is not a linear space if $g \neq 0$. [If $u, v \in H^1_{g,D}(\Omega)$, then $u + v \notin H^1_{g,D}(\Omega)$.] It is important that the problem be phrased in terms of Hilbert spaces such as $H^1_{0,D}(\Omega)$, in order that certain analysis tools and concepts be applicable. Therefore, we will do some additional transformation of the problem.

So far, we have shown that the solution to the original problem (10.4.1)–(10.4.3) also solves the following problem:

$$\text{Find } \hat{u} \in H^1_{g,D}(\Omega) \text{ such that } \hat{A}(\hat{u}, v) = \hat{F}(v) \quad \forall v \in H^1_{0,D}(\Omega), \tag{10.4.6}$$

where from equations (10.4.4) and (10.4.5), the bilinear form $\hat{A}(\cdot,\cdot)$ and the linear functional $\hat{F}(\cdot)$ are defined as

$$\hat{A}(\hat{u}, v) = \int_\Omega (a\nabla\hat{u}\cdot\nabla v + b\hat{u}v)\,dx + \int_{\Gamma_N} c\hat{u}v\,ds,\quad \hat{F}(v) = \int_\Omega fv\,dx + \int_{\Gamma_N} g_N v\,ds.$$

Since we can write the solution $\hat{u}$ to equation (10.4.6) as $\hat{u} = u + w$ for a fixed w satisfying $g_D = \text{tr}\,w$, we can rewrite the equations completely in terms of u and a new bilinear form $A(\cdot,\cdot)$ and linear functional $F(\cdot)$ as follows:

$$\text{Find } u \in H^1_{0,D}(\Omega) \text{ such that } A(u, v) = F(v) \quad \forall v \in H^1_{0,D}(\Omega), \tag{10.4.7}$$

$$A(u, v) = \int_\Omega a\nabla u\cdot\nabla v + buv\,dx + \int_{\Gamma_N} cuv\,ds, \tag{10.4.8}$$

$$F(v) = \int_\Omega fv\,dx + \int_{\Gamma_N} g_N v\,ds - A(w, v). \tag{10.4.9}$$

Clearly, the "weak" formulation of the problem given by equation (10.4.7) imposes only one order of differentiability on the solution u, and only in the weak sense. It is easily verified that $A\colon H^1_{0,D}(\Omega) \times H^1_{0,D}(\Omega) \to \mathbb{R}$ defines a bilinear form, and $F\colon H^1_{0,D}(\Omega) \to \mathbb{R}$ defines a linear functional.

A Priori Estimates. Although it will be critical in the case of nonlinear problems, it will not be necessary for us to develop *a priori* L^∞ estimates for obtaining pointwise control of solutions in the case of linear problems, in order for us to develop existence, uniqueness, and best approximation results. Therefore, we will focus on developing just a basic *a priori* energy estimate for any solution (if one exists) to (10.4.7).

To make clear what the approach will be, let X be a Hilbert space with norm $\|\cdot\|$, and consider the following abstract problem:

$$\text{Find } u \in X \text{ such that } A(u,v) = F(v) \quad \forall v \in X, \tag{10.4.10}$$

where the bilinear form $A(\cdot,\cdot)$ is bounded and coercive on $X \times X$, and the linear form $F(\cdot)$ is bounded on X, or more explicitly:

$$m\|u\|^2 \leqslant A(u,u), \qquad |A(u,v)| \leqslant M\|u\|\|v\|, \qquad |F(v)| \leqslant L\|v\|, \qquad \forall u,v \in X,$$

where m, M, and L are positive constants. Although we do not yet know whether there is a solution to (10.4.7), we can derive an *a priori* bound (called an *energy estimate*) on the magnitude of the X-norm of any possible solution u in terms of the parameters m, M, and L. To this end, we begin with

$$m\|u\|^2 \leqslant A(u,u) = F(u) \leqslant L\|u\|.$$

Now first assuming that $u \neq 0$, which ensures that $\|u\| \neq 0$ by nondegeneracy of the norm, we have after division

$$\|u\| \leqslant \frac{L}{m}. \tag{10.4.11}$$

Finally, we note that this also holds trivially when $u = 0$. We now apply this simple idea to obtain an energy estimate for our problem. We will need to make the following assumption.

Assumption 10.4.1. *The problem coefficients in* (10.4.1) *to* (10.4.3) *are assumed to satisfy:*

(1) $a_{ij}(x) \leqslant c_1 < \infty$, *a.e. in* Ω, $i,j = 1,\ldots,n$.
(2) $0 \leqslant b(x) \leqslant c_2 < \infty$, *a.e. in* Ω.
(3) $0 \leqslant c(x) \leqslant c_3 < \infty$, *a.e. in* Γ_N.
(4) $g \in H^{1/2}(\Gamma_D)$.
(5) $f \in H^{-1}(\Omega)$.
(6) Strong ellipticity: $\exists \lambda > 0$ *s.t.* $\sum_{ij} a_{ij}(x)\eta_i\eta_j \geqslant \lambda|\eta|^2$, *a.e. in* $\Omega, \forall \eta \in \mathbb{R}^n$.

These very weak assumptions on the coefficients in the problem are sufficient to establish an *a priori* energy estimate.

Theorem 10.4.1 (Energy Estimate)**.** *Under Assumption 10.4.1, if a weak solution to problem* (10.4.1)–(10.4.3) *exists, it depends continuously on the problem data, and the following* a priori *energy estimate holds:*

$$\|u\|_{1,2} \leqslant \frac{\|f\|_{-1} + C_{\mathrm{tr}}\|g_N\|_{2,\Gamma_N} + (n^2c_1 + c_2 + C_{\mathrm{tr}}^2 c_3)\|w\|_{1,2}}{\min\{\lambda/(2\rho^2), \lambda/2\}}.$$

Proof. To begin, it is immediately clear from the linearity of Lebesgue integration that $A(\cdot,\cdot)$ and $F(\cdot)$ define bilinear and linear forms on $H^1_{0,D}(\Omega)$, respectively. If we follow the approach above that leads to the *a priori* bound (10.4.11), we must show that $A(\cdot,\cdot)$ is coercive and that $F(\cdot)$ is bounded; the Dirichlet boundary condition (10.4.2) will also require showing that $A(\cdot,\cdot)$ is bounded.

Consider first the bilinear form $A(\cdot,\cdot)$. Using the notation $D_i = \partial/\partial x_i$, the strong ellipticity assumption and nonnegativity of $b(x)$ and $c(x)$ give

$$\begin{aligned} A(u,u) &= \int_\Omega a\nabla u \cdot \nabla u + bu^2\, dx + \int_{\Gamma_N} cu^2\, dx \\ &= \int_\Omega \sum_{i,j=1}^n a_{ij} D_i u D_j u + bu^2\, dx + \int_{\Gamma_N} cu^2\, dx \\ &\geqslant \lambda \int_\Omega \sum_{i=1}^n |D_i u|^2\, dx = \lambda |u|^2_{1,2}. \end{aligned}$$

Since we have assumed only that $b(x)$ is nonnegative, we must employ the Poincaré inequality (Theorem 10.3.6) to obtain the proper norm of u bounding $A(\cdot,\cdot)$ from below:

$$\begin{aligned} A(u,u) \geqslant \lambda |u|^2_{1,2} &= \lambda\left(\frac{1}{2}|u|^2_{1,2} + \frac{1}{2}|u|^2_{1,2}\right) \geqslant \lambda\left(\frac{1}{2\rho^2}\|u\|_2^2 + \frac{1}{2}|u|^2_{1,2}\right) \\ &\geqslant m\left(\|u\|_2^2 + |u|^2_{1,2}\right) = m\|u\|^2_{1,2}, \end{aligned}$$

where $m = \min\{\lambda/(2\rho^2), \lambda/2\}$. Therefore, $A(\cdot,\cdot)$ is coercive with constant m. We now show that $A(\cdot,\cdot)$ is bounded on $H^1_{0,D}(\Omega)$. By repeated application of Hölder's inequality and Theorem 10.3.5 we have that

$$\begin{aligned} |A(u,v)| &= \left| \int_\Omega a\nabla u \cdot \nabla v + buv\, dx + \int_{\Gamma_N} cuv\, dx \right| \\ &\leqslant \sum_{i,j=1}^n \int_\Omega |a_{ij} D_i u D_j v|\, dx + \int_\Omega |buv|\, dx + \int_{\Gamma_N} |cuv|\, dx \\ &\leqslant \sum_{i,j=1}^n \|a_{ij}\|_\infty \|D_i u\|_2 \|D_j v\|_2 + \|b\|_\infty \|u\|_2 \|v\|_2 \\ &\qquad + \|c\|_{\infty,\Gamma_N} \|u\|_{2,\Gamma_N} \|v\|_{2,\Gamma_N} \\ &\leqslant \sum_{i,j=1}^n \|a_{ij}\|_\infty \|u\|_{1,2} \|v\|_{1,2} + \|b\|_\infty \|u\|_{1,2} \|v\|_{1,2} \\ &\qquad + C_{\text{tr}}^2 \|c\|_\infty \|u\|_{1,2} \|v\|_{1,2} \\ &\leqslant M\|u\|_{1,2}\|v\|_{1,2}, \end{aligned}$$

where $M = n^2 c_1 + c_2 + C_{\text{tr}}^2 c_3 \geqslant \sum_{i,j=1}^n \|a_{ij}\|_\infty + \|b\|_\infty + C_{\text{tr}}^2 \|c\|_\infty$, and where C_{tr} is the constant from Theorem 10.3.5.

Consider now the linear functional $F(\cdot)$. Since $A(\cdot,\cdot)$ is bounded, we have that

$$\begin{aligned}|F(v)| &= \left|\int_\Omega fv\,dx + \int_{\Gamma_N} g_N v - A(w,v)\right| \\ &\leqslant \|f\|_{-1}\|v\|_{1,2} + \|g_N\|_{2,\Gamma_N}\|v\|_{2,\Gamma_N} + M\|w\|_{1,2}\|v\|_{1,2} \\ &\leqslant \|f\|_{-1}\|v\|_{1,2} + \|g_N\|_{2,\Gamma_N} C_{\mathrm{tr}}\|v\|_{1,2} + M\|w\|_{1,2}\|v\|_{1,2} \\ &= L\|v\|_{1,2},\end{aligned}$$

with $L = \|f\|_{-1} + C_{\mathrm{tr}}\|g_N\|_{2,\Gamma_N} + M\|w\|_{1,2}$, where we have used several results, including the Cauchy-Schwarz inequality, Theorem 10.3.5, the definition of $\|\cdot\|_{1,2}$, the fact that $f \in H^{-1}(\Omega)$, and the fact that $w \in H^1(\Omega)$ is fixed. Therefore, $F(\cdot)$ is a bounded linear functional on $H^1_{0,D}(\Omega)$. An *a priori* bound on the magnitude of the solution to the problem (10.4.7)–(10.4.9) in the $H^1_{0,D}(\Omega)$ norm then takes the form of (10.4.11), which is this case becomes

$$\|u\|_{1,2} \leqslant \frac{\|f\|_{-1} + C_{\mathrm{tr}}\|g_N\|_{2,\Gamma_N} + (n^2c_1 + c_2 + C^2_{\mathrm{tr}}c_3)\|w\|_{1,2}}{\min\{\lambda/(2\rho^2), \lambda/2\}}.$$

This bound implies continuous dependence of the solution on the data of the problem, which establishes the theorem. □

Existence and Uniqueness of Solutions. The assumptions made in Theorem 10.4.1 are also sufficient to establish existence and uniqueness, without additional assumptions.

Theorem 10.4.2 (Existence and Uniqueness). *Under Assumption 10.4.1, there exists a unique solution $u \in H^1_{0,D}(\Omega)$ to problem* (10.4.7)–(10.4.9), *and hence a unique weak solution $\hat{u} \in H^1(\Omega)$ to problem* (10.4.1)–(10.4.3).

Proof. Recall from the proof of the energy estimate in Theorem 10.4.1, for the particular weak form PDE problem (10.4.7)–(10.4.9), the bilinear form $A(\cdot,\cdot)$ is coercive with constant m and bounded with constant M, and the linear functional $F(\cdot)$ is bounded with L, where the three constants took the forms

$$\begin{aligned} m = \min\{\lambda/(2\rho^2), \lambda/2\}, \quad M = n^2c_1 + c_2 + C^2_{\mathrm{tr}}c_3, \\ L = \|f\|_{-1} + C_{\mathrm{tr}}\|g_N\|_{2,\Gamma_N} + M\|w\|_{1,2}.\end{aligned}$$

The Hilbert space involved in the arguments was $H^1_{0,D}(\Omega)$, and the various parameters in the three constants arose from the Poincaré inequality, ellipticity assumptions on the operator, and bounds on the PDE coefficients. By the Lax-Milgram Theorem (Theorem 8.5.2), precisely these same three conditions ensure the existence and uniqueness a weak solution $u \in H^1_{0,D}(\Omega)$ to (10.4.7)–(10.4.9). □

Near-Best Approximation with a Galerkin Method. To finish, we have near-best approximation from any subspace X_h. The assumptions we need are precisely those required to establish well-posedness in Theorem 10.4.2, so we need no additional assumptions.

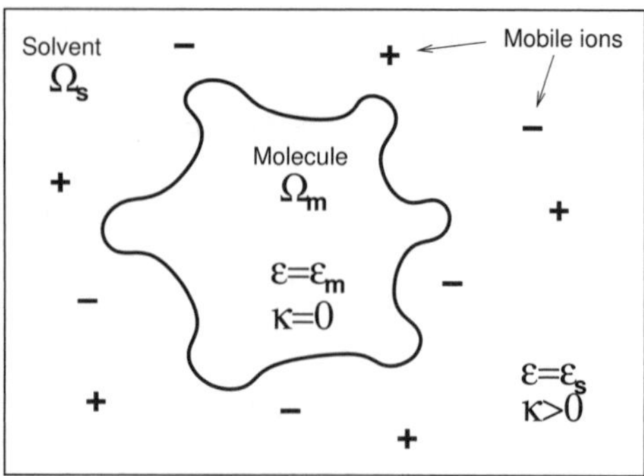

Figure 10.2 Debye-Hückel model of a charged biological structure immersed in a solvent containing mobile ions.

Theorem 10.4.3 (Near-Best Approximation). *Let $X_h \subset H^1_{0,D}(\Omega)$ be a subspace of $H^1_{0,D}(\Omega)$. Then under the assumptions of Theorem 10.4.2, the Galerkin method produces a near-best approximation $u_h \in X_h$ to the solution of problem* (10.4.7)–(10.4.9), *and we have the following* a priori *error estimate:*

$$\|u - u_h\|_{1,2} \leqslant \left(\frac{M}{m}\right) E_{X_h}(u).$$

Proof. The assumptions for Cea's Lemma hold (Theorem 10.2.11), and the result follows. □

The Nonlinear Poisson-Boltzmann Equation

We consider now the nonlinear Poisson-Boltzmann equation arising in biophysics as an example of an important nonlinear problem of wide interest with several features that make both analysis and approximation difficult, including discontinuous coefficients, supercritical nonlinearity, delta function source terms, and infinite domain. More detailed discussions of the equation can be found in [45, 101, 165]. Following the approach in [49, 95], our goal is to develop *a priori* energy estimates, existence and uniqueness results, and a near-best approximation result using Galerkin methods. We will first need to develop *a priori* L^∞ estimates to control the nonlinearity in the problem, in order to make all of the other results possible.

The nonlinear PBE, a second-order nonlinear partial differential equation, is fundamental to Debye-Hückel continuum electrostatic theory [59]. It determines a dimensionless potential around a charged biological structure immersed in a salt solution. (See Figure 10.2.) The PBE arises from Gauss's law, represented by the Poisson equation, relating the electrostatic potential Φ in a dielectric to the charge density ρ:

$$-\nabla \cdot (\varepsilon \nabla \Phi) = \rho.$$

Here ε is the dielectric constant of the medium and is typically piecewise constant; it typically jumps by one or two orders of magnitude at the interface between the charged structure (a biological molecular or membrane) and the solvent (a salt solution). The charge density ρ consists of two components: $\rho = \rho_{\text{macro}} + \rho_{\text{ion}}$. For the macromolecule, the charge density is a sum of δ distributions at N_m point charges

$$\rho_{\text{macro}}(x) = \sum_{i=1}^{N_m} q_i \delta(x - x_i), \qquad q_i = \left(\frac{4\pi e_c^2}{\kappa_B T}\right) z_i,$$

where $\kappa_B > 0$ is the Boltzmann constant, T is the temperature, e_c is the unit of charge, and z_i is the fraction of charge at x_i.

For mobile ions in the solvent, the charge density ρ_{ion} is determined by a Boltzmann distribution that depends on the (unknown) potential, which introduces a nonlinearity into the equation. If the solvent contains N types of ions of valence Z_i and of bulk concentration c_i, then a Boltzmann assumption about the equilibrium distribution of the ions gives

$$\rho_{\text{ion}} = \sum_{i=1}^{N} c_i Z_i e_c \exp\left(-Z_i \frac{e_c \Phi}{\kappa_B T}\right).$$

We now write the PBE for modeling the electrostatic potential of a solvated biological structure. We denote the molecule region by $\Omega_m \subset \mathbb{R}^d$ and consider the solvent region $\Omega_s = \mathbb{R}^d \backslash \bar{\Omega}_m$. We use $\hat{u}$ to denote the dimensionless potential, κ to denote the (strictly positive) Debye-Hückel parameter (which is a function of the ionic strength of the solvent), and $\bar{\kappa}^2 = \epsilon_s \kappa^2$ to denote the modified Debye-Hückel parameter. For a symmetric $1:1$ electrolyte, $N = 2$, $c_i = c_0$, and $Z_i = (-1)^i$, giving the ionic charge density $\rho_{\text{ion}} = -2c_0 e_c \sinh(e_c \Phi/[\kappa_B T])$, the nonlinear Poisson-Boltzmann equation takes the form

$$-\nabla \cdot (\varepsilon \nabla \hat{u}) + \bar{\kappa}^2 \sinh(\hat{u}) = \sum_{i=1}^{N_m} q_i \delta_i \text{ in } \mathbb{R}^d, \qquad \hat{u}(\infty) = 0, \tag{10.4.12}$$

where

$$\varepsilon = \begin{cases} \varepsilon_m & \text{if} \quad x \in \Omega_m, \\ \varepsilon_s & \text{if} \quad x \in \Omega_s, \end{cases} \qquad \text{and} \qquad \bar{\kappa} = \begin{cases} 0 & \text{if} \quad x \in \Omega_m, \\ \sqrt{\varepsilon_s}\, \kappa > 0 & \text{if} \quad x \in \Omega_s. \end{cases}$$

Empirically one has $\varepsilon_m \approx 2$ and $\varepsilon_s \approx 80$. The structure itself (for example, a biological molecule or a membrane) is represented implicitly by ε and $\bar{\kappa}$, as well as explicitly by the N_m point charges $q_i = z_i e_c$ at the positions x_i. The charge positions are located in the strict interior of the molecular region Ω_m. A physically reasonable mathematical assumption is that all charge locations obey the following lower bound on their distance to the solvent region Ω_s for some $\sigma > 0$:

$$|x - x_i| \geqslant \sigma \quad \forall x \in \Omega_s, \quad i = 1, \ldots, N_m. \tag{10.4.13}$$

In some models employing the PBE, there is a third region Ω_l (the Stern layer), a layer between Ω_m and Ω_s. In the presence of a Stern layer, the parameter σ in (10.4.13) increases in value. Our analysis and results below can easily be generalized to this case as well.

Regularization of the Continuous Problem. Following [49, 95, 191, 192], we now introduce a regularized version of the nonlinear PBE for both analysis and discretization purposes. Let $\Omega \subset \mathbb{R}^3$ be convex with boundary $\partial\Omega$, and $\Omega_m \subset \Omega$, with $\partial\Omega_m \cap \partial\Omega = \emptyset$, meaning effectively that $\partial\Omega_m$ is isolated from $\partial\Omega$. One is typically completely free to pick Ω to be a cube, a ball, or some other convenient volume containing the biological structure of interest. The solvent region is chosen as $\Omega_s \cap \Omega$ and will be still denoted Ω_s. On $\partial\Omega$ we choose the boundary condition $\hat{u} = g$, with the function $g \in C^\infty(\partial\Omega)$ defined as

$$g = \left(\frac{e_c^2}{k_B T}\right) \sum_{i=1}^{N_i} \frac{e^{-\kappa|x-x_i|}}{\varepsilon_s |x - x_i|}. \tag{10.4.14}$$

The boundary condition is usually taken to be induced by a known analytical solution to one of several possible simplifications of a linearized version of the problem. Far from the molecule, such analytical solutions provide a highly accurate boundary condition approximation for the general nonlinear PBE on a truncation of $\mathbb{R}^3$. For example, (10.4.14) arises from the use of Green's function for the Helmholtz operator arising from linearizations of the Poisson-Boltzmann operator, where a single constant global dielectric value of ε_s is used to generate the approximate boundary condition. (This is the case of a rod-like molecule approximation; see [101].)

Employing (10.4.14) we obtain the nonlinear PBE on a truncated domain:

$$-\nabla \cdot (\varepsilon \nabla \hat{u}) + \bar{\kappa}^2 \sinh(\hat{u}) = \sum_{i=1}^{N_m} q_i \delta_i \text{ in } \Omega, \qquad \hat{u} = g \text{ on } \partial\Omega. \tag{10.4.15}$$

While this appears to be a boundary value problem for a nonlinear second-order elliptic partial differential equation, the right side contains a linear combination of δ distributions, which individually and together are not in $H^{-1}(\Omega)$. While earlier in the book we assembled techniques for dealing with such distribution sources for linear problems, we will need to decompose the equation in order to use the machinery we developed earlier in the chapter for establishing near-best approximation results. To begin we define G_i and G as

$$-\nabla \cdot (\varepsilon_m \nabla G_i) = q_i \delta_i \text{ in } \mathbb{R}^3, \qquad G_i = \frac{q_i}{\varepsilon_m} \frac{1}{|x - x_i|}, \qquad G = \sum_{i=1}^{N_m} G_i.$$

We now decompose $\hat{u}$ into an unknown u and the known function G:

$$\hat{u} = u + G. \tag{10.4.16}$$

Substituting the decomposition into (10.4.15) gives an equation for the unknown u:

$$-\nabla\cdot(\varepsilon\nabla u)+\bar{\kappa}^2\sinh(u+G)=f_G \text{ in } \Omega, \qquad u=g-G \text{ on } \partial\Omega, \quad (10.4.17)$$

where $f_G=\nabla\cdot((\varepsilon-\varepsilon_m)\nabla G)$. We call (10.4.17) the *regularized Poisson-Boltzmann equation*, or RPBE. The singularities of the δ distributions are transferred to G, which then exhibits degenerate behavior at each $\{x_i\}\subset\Omega_m$. At those points, both $\sinh G(x_i)$ and $\nabla G(x_i)$ exhibit blow-up. However, both of the coefficients $\bar{\kappa}$ and $\varepsilon-\varepsilon_m$ are zero inside Ω_m, where the blow-up behavior arises. Due to this cutoff nature of coefficients, we will be able to show that $f_G\in H^{-1}(\Omega)$, since integration by parts against any test function $v\in H^1_0(\Omega)$ will be finite for any compact domain Ω with reasonably smooth boundary Γ. It is useful to note that away from $\{x_i\}$, the function G is smooth. In particular, we shall make use of the fact $G\in C^\infty(\Omega_s)\cap C^\infty(\Gamma)\cap C^\infty(\partial\Omega)$, giving in particular

$$f_G|_\Gamma\in C^\infty(\Gamma), \qquad (g-G)\in C^\infty(\partial\Omega).$$

A Priori Estimates. We now derive *a priori* L^∞-estimates of the solution of the RPBE. So far we have avoided stating explicitly any smoothness assumptions on the domain Ω, the subdomains Ω_m and Ω_s, and the boundary sets $\partial\Omega$, $\partial\Omega_m$, $\partial\Omega_s$, and $\Gamma=\partial\Omega_m\cap\partial\Omega_s$. We will need to make use of some regularity results for linear interface problems from [152] in order to develop *a priori* L^∞ estimates. In particular, we will make the following assumption.

Assumption 10.4.2. *Let $\Omega\subset\mathbb{R}^n$ be an open bounded convex set in $\mathbb{R}^n$ with Lipschitz boundary $\partial\Omega$. Let $\Omega_m\subset\Omega$ be an open subset of Ω with $C^{1,1}$ boundary $\Gamma=\partial\Omega_m$, which defines the region $\Omega_s=\Omega\setminus\overline{\Omega}_m$, with $\partial\Omega_s=\Gamma\cup\partial\Omega$. We assume that Ω_m lies in the strict interior of Ω, so that $\emptyset=\Gamma\cap\partial\Omega$. We also assume existence of $\sigma>0$ such that* (10.4.13) *holds, so that the charges x_i are isolated to the strict interior of Ω_m.*

This assumption is reasonable, in that we are free to pick the domain Ω as a convex Lipschitz set (such as a convex polyhedron), and the mathematical definitions typically used for Ω_m generate smooth (even C^∞) surfaces.

A possible weak formulation of (10.4.17) is

$$\text{Find } u\in V^e \text{ such that } A(u,v)+(B_G(u),v)=\langle f_G,v\rangle, \quad \forall v\in H^1_0(\Omega), \quad (10.4.18)$$

where

$$A(u,v)=(\varepsilon\nabla u,\nabla v), \qquad (B_G(u),v)=(\bar{\kappa}^2\sinh(u+G),v), \quad (10.4.19)$$

$$\langle f_G,v\rangle=\int_\Omega(\varepsilon-\varepsilon_m)\nabla G\cdot\nabla v, \quad (10.4.20)$$

and where

$$V^e=\{u\in H^1(\Omega)\ :\ e^u,e^{-u}\in L^2(\Omega_s), \text{ and } u=g-G \text{ on } \partial\Omega\}, \quad (10.4.21)$$

which ensures that the terms in (10.4.18) are well-defined for any u from the set V^e, and that u satisfies the appropriate boundary conditions. We cannot apply a standard maximum principle argument directly to (10.4.18) since the source term $f_G \in H^{-1}(\Omega)$, and does not lie in $L^\infty(\Omega)$ as required for use of these techniques. However, we can overcome this problem through further decomposition of the regularized solution u to (10.4.17) into several parts; in particular, we decompose u into $u = u^n + u^l + u^g$, where u^g is the (generalized) *harmonic extension* of u^g of $(g - G)$ into Ω, and where u^l and u^n satisfy

$$-\nabla \cdot (\varepsilon \nabla u^g) = 0 \quad \text{in } \Omega, \quad u^g = g - G, \text{ on } \partial\Omega, \tag{10.4.22}$$

$$-\nabla \cdot (\varepsilon \nabla u^l) = f_G \text{ in } \Omega, \quad u^l = 0 \text{ on } \partial\Omega, \tag{10.4.23}$$

$$-\nabla \cdot (\varepsilon \nabla u^n) + \bar{\kappa}^2 \sinh(u^n + H) = 0 \quad \text{in } \Omega, \quad u^n = 0 \text{ on } \partial\Omega, \tag{10.4.24}$$

where $H = u^l + u^g + G$. The weak forms of (10.4.22)–(10.4.24) are:

$$\text{Find } u^g \in V^g \text{ such that } A(u^g, v) = 0, \ \forall v \in H_0^1(\Omega). \tag{10.4.25}$$

$$\text{Find } u^l \in V^l \text{ such that } A(u^l, v) - \langle f_G, v \rangle = 0, \ \forall v \in H_0^1(\Omega). \tag{10.4.26}$$

$$\text{Find } u^n \in V^n \text{ such that } A(u^n, v) + (B_H(u^n), v) = 0, \ \forall v \in H_0^1(\Omega), \tag{10.4.27}$$

where we define

$$(B_H(u^n), v) = (\bar{\kappa}^2 \sinh(u^n + H), v) \tag{10.4.28}$$

$$= (\bar{\kappa}^2 \sinh(u^n + u^l + u^g + G), v). \tag{10.4.29}$$

The sets

$$V^g = \{u \in H^1(\Omega) \ : \ u|_{\partial\Omega} = g - G\}, \tag{10.4.30}$$

$$V^l = H_0^1(\Omega), \tag{10.4.31}$$

$$V^n = \{u \in H_0^1(\Omega) \ : \ e^u, e^{-u} \in L^2(\Omega_s)\}, \tag{10.4.32}$$

are designed so that the boundary conditions hold and so that all of the terms are well-defined. We first show that there exist unique solutions u^g and u^l to (10.4.25) and (10.4.26), and that, in fact, $u^g, u^l \in L^\infty(\Omega)$.

Lemma 10.4.1 (A Priori L^∞ Estimates: Linear Part [49, 95]). *Let Assumption 10.4.2 hold. There exist unique weak solutions u^g and u^l to* (10.4.25) *and* (10.4.26). *Moreover, it holds that $u^g \in H^1(\Omega) \cap L^\infty(\Omega)$ and $u^l \in H_0^1(\Omega) \cap L^\infty(\Omega)$, and there exist constants $\alpha^g, \beta^g, \alpha^l, \beta^l \in \mathbb{R}$ such that*

$$\alpha^g \leqslant u^g \leqslant \beta^g, \quad \text{a.e. in } \Omega, \tag{10.4.33}$$

$$\alpha^l \leqslant u^l \leqslant \beta^l, \quad \text{a.e. in } \Omega. \tag{10.4.34}$$

Proof. Problem (10.4.25) is simply a particular case of the coercive linear problem (10.4.1)–(10.4.3) studied in this section, with a single inhomogeneous Dirichlet condition (10.4.2) covering the entire boundary. By Theorem 10.4.2, there exists a unique weak solution $u^g \in H^1(\Omega)$ to (10.4.25).

Since $\Delta G = 0$ in Ω_s, integration by parts gives the equivalent representation

$$\langle f_G, v\rangle = ((\varepsilon - \varepsilon_m)\nabla G, \nabla v)_{0,\Omega} = ((\varepsilon_s - \varepsilon_m)\nabla G, \nabla v)_{0,\Omega_s} = \left([\varepsilon]\frac{\partial G}{\partial n_\Gamma}, v\right)_\Gamma,$$

where $[\varepsilon] = \varepsilon_s - \varepsilon_m$ is the jump of ε at the interface. This implies that both $f_G \in H^{-1}(\Omega)$ and $f_G|_\Gamma \in C^\infty(\Gamma)$. In other words, problem (10.4.26) is also a particular case of the coercive linear problem (10.4.1)–(10.4.3) studied in this section, with a single homogeneous Dirichlet condition (10.4.2) covering the entire boundary. By Theorem 10.4.2, there exists a unique weak solution $u^l \in H_0^1(\Omega)$ to (10.4.26).

Note that (10.4.25) is the weak formulation of the classical interface problem

$$-\nabla\cdot(\varepsilon\nabla u^g) = 0 \text{ in } \Omega, \quad [u^g] = 0, \quad \left[\varepsilon\frac{\partial u^g}{\partial n}\right] = 0 \text{ on } \Gamma, \quad \text{and } u^g = g - G \text{ on } \partial\Omega.$$

Similarly, (10.4.26) is the weak formulation of the related interface problem

$$-\nabla\cdot(\varepsilon\nabla u^l) = 0 \text{ in } \Omega \quad [u^l] = 0, \quad \left[\varepsilon\frac{\partial u^l}{\partial n}\right] = f_G \text{ on } \Gamma, \quad \text{and } u^l = 0 \text{ on } \partial\Omega.$$

Since $f_G \in C^\infty(\Gamma)$, $\Gamma \in C^2$, $g - G \in C^\infty(\partial\Omega)$, we have that Assumption 10.4.2, taken together with standard regularity results for classical interface problems (see, for example, [14, 35, 152]), ensures we have $u^g \in H^2(\Omega_m) \cap H^2(\Omega_s) \cap H^1(\Omega)$, and in addition have $u^l \in H^2(\Omega_m) \cap H^2(\Omega_s) \cap H_0^1(\Omega)$, giving $u^g, u^l \in L^\infty(\Omega)$ on the bounded set Ω. Subsequently, (10.4.33) and (10.4.34) hold, with the bounds $\alpha^g = \inf_{x\in\Omega} u^g(x)$, $\beta^g = \sup_{x\in\Omega} u^g(x)$, $\alpha^l = \inf_{x\in\Omega} u^l(x)$, $\beta^l = \sup_{x\in\Omega} u^l(x)$. □

To derive a similar *a priori* L^∞ estimate for the nonlinear part u^n, we first define

$$\alpha' = \arg\max_c \left(\bar{\kappa}^2 \sinh(c + \beta^g + \beta^l + \sup_{x\in\Omega_s} G) \leqslant 0\right), \qquad \alpha^n = \min(\alpha', 0), \tag{10.4.35}$$

$$\beta' = \arg\min_c \left(\bar{\kappa}^2 \sinh(c + \alpha^g + \alpha^l + \inf_{x\in\Omega_s} G) \geqslant 0\right), \qquad \beta^n = \max(\beta', 0), \tag{10.4.36}$$

where $\arg\min_c f(c)$ is defined to be $c \in \mathbb{R}$ which minimizes $f\colon \mathbb{R} \to \mathbb{R}$, and similarly, $\arg\max_c f(c)$ is defined to be the argument which maximizes f.

Lemma 10.4.2 (A Priori L^∞ Estimates: Nonlinear Part [49, 95]). *Let Assumption 10.4.2 hold. Let u^n be any weak solution of* (10.4.27). *If $\alpha^n, \beta^n \in \mathbb{R}$ are as defined in* (10.4.35)–(10.4.36), *then*

$$\alpha^n \leqslant u^n \leqslant \beta^n, \text{ a.e. in } \Omega. \tag{10.4.37}$$

Proof. We first define

$$\overline{\phi} = (u^n - \beta^n)^+ = \max(u^n - \beta^n, 0), \quad \underline{\phi} = (u^n - \alpha^n)^- = \max(\alpha^n - u^n, 0).$$

Since $\beta^n \geqslant 0$ and $\alpha^n \leqslant 0$, by Theorem 10.3.8, we have $\overline{\phi}, \underline{\phi} \in H_0^1(\Omega)$, and we can use them as test functions which are both *pointwise nonnegative* (almost everywhere). Thus for either $\phi = \overline{\phi}$ or for $\phi = -\underline{\phi}$, we have

$$(\varepsilon \nabla u^n, \nabla \phi)_2 + (\bar{\kappa}^2 \sinh(u^n + u^g + u^l + G), \phi)_2 = 0.$$

Note that $\overline{\phi} \geqslant 0$ in Ω and its support set is $\overline{\mathcal{Y}} = \{x \in \bar{\Omega} \mid u^n(x) \geqslant \beta^n\}$. On $\overline{\mathcal{Y}}$, we have

$$\bar{\kappa}^2 \sinh(u^n + u^g + u^l + G) \geqslant \bar{\kappa}^2 \sinh(\beta' + \alpha^g + \alpha^l + \inf_{x \in \Omega_s} G) \geqslant 0.$$

Similarly, $-\underline{\phi} \leqslant 0$ in Ω with support $\underline{\mathcal{Y}} = \{x \in \bar{\Omega} \mid u^n(x) \leqslant \alpha^n\}$. On $\underline{\mathcal{Y}}$, we have

$$\bar{\kappa}^2 \sinh(u^n + u^g + u^l + G) \leqslant \bar{\kappa}^2 \sinh(\alpha' + \beta^g + \beta^l + \sup_{x \in \Omega_s} G) \leqslant 0.$$

Together this implies both

$$0 \geqslant (\varepsilon \nabla u^n, \nabla \overline{\phi})_2 = (\varepsilon \nabla (u^n - \beta^n), \nabla \overline{\phi})_2 = \left[\inf_{x \in \Omega} \varepsilon\right] \|\nabla \overline{\phi}\|_2^2 \geqslant 0,$$

$$0 \geqslant (\varepsilon \nabla u^n, \nabla (-\underline{\phi}))_2 = (\varepsilon \nabla (\alpha^n - u^n), \nabla \underline{\phi})_2 = \left[\inf_{x \in \Omega} \varepsilon\right] \|\nabla \underline{\phi}\|_2^2 \geqslant 0.$$

Using the Poincaré inequality (Theorem 10.3.6) we have finally

$$0 \leqslant \|\phi\|_{1,2} \leqslant (1 + \rho^2)^{1/2} \|\nabla \phi\|_2 \leqslant 0,$$

where $\rho > 0$ from Theorem 10.3.6 depends on the diameter of the domain Ω. This gives $\phi = 0$, for either $\phi = \overline{\phi}$ or $\phi = -\underline{\phi}$. Thus $\alpha^n \leqslant u^n \leqslant \beta^n$ in Ω. □

We finally have *a priori* L^∞ estimates for the full RPBE solution.

Theorem 10.4.4 (A Priori L^∞ Estimates for RPBE [49, 95]). *Let Assumption 10.4.2 hold. Let u be any weak solution of* (10.4.18). *Then*

$$\alpha \leqslant u \leqslant \beta, \quad a.e. \ \textit{in}\ \Omega,$$

where $\alpha = \alpha^g + \alpha^l + \alpha^n$, $\beta = \beta^g + \beta^l + \beta^n$, with α^n, β^n as in Lemma 10.4.2 and with $\alpha^g, \beta^g, \alpha^l, \beta^l$ as in Lemma 10.4.1.

Proof. The proof follows immediately from Lemmas 10.4.1 and 10.4.2. □

Existence and Uniqueness of Solutions. We have shown in Theorem 10.4.4 that any solution $u \in H^1(\Omega)$ to the regularized problem (10.4.17) must lie in the set

$$[\alpha, \beta]_{1,2} := \{u \in H^1(\Omega) \ : \ \alpha \leqslant u \leqslant \beta\} \subset H^1(\Omega),$$

where $\alpha, \beta \in \mathbb{R}$ are as in Theorem 10.4.4. (See the discussion of ordered Banach spaces in Sections 10.1 and 10.3 for more background on order structures on Sobolev spaces.) Since this ensures that $u \in L^\infty(\Omega)$, which subsequently ensures that $e^u \in L^2(\Omega)$, we can replace the set V^e defined in (10.4.32) with the following function space as the set to search for solutions to the RPBE:

$$V = \{u \in [\alpha, \beta]_{1,2} \ : \ u = g - G \text{ on } \partial\Omega\} \subset V^e \subset H^1(\Omega).$$

Our weak formulation of the RPBE now reads:

$$\text{Find } u \in V \text{ s.t. } A(u, v) + (B_G(u), v) + \langle f_G, v\rangle = 0 \quad \forall v \in H^1_0(\Omega), \tag{10.4.38}$$

where A, B_G, and f_G are as in (10.4.19) and (10.4.20). Note that V is not a subspace of $H^1(\Omega)$ since it is not a linear space, due to the inhomogeneous boundary condition requirement. (Consider that if $u, v \in V$, then for any $a \neq 0$ and $b \neq 0$, the linear combination $au + bv = z \notin V$.) However, by Lemma 10.4.1, we have already established existence, uniqueness, and *a priori* L^∞ bounds for u^g and u^l solving (10.4.25) and (10.4.26), leaving only (10.4.27) for the remainder u^n. Therefore, (10.4.38) is mathematically equivalent to

$$\text{Find } u^n \in U \text{ s.t. } A(u^n, v) + (B_H(u^n), v) = 0 \quad \forall v \in H^1_0(\Omega), \tag{10.4.39}$$

where B_H is as in (10.4.28), and where

$$U = \{u \in H^1_0(\Omega) \ : \ u_- \leqslant u \leqslant u_+\} \subset H^1_0(\Omega),$$

where $u_- = \alpha^n$ and $u_+ = \beta^n$ are from Lemma 10.4.2. We now have a weak formulation (10.4.39) that involves looking for a solution in a well-defined subspace U of an ordered Banach space $X = H^1_0(\Omega)$, and are now prepared to establish existence (and uniqueness) of the solution to the RPBE.

We note that existence and uniqueness of solutions to a regularized form of the Poisson-Boltzmann equation was established in the early 1990s in [101], using the same convex analysis argument that we present below. However, *a priori* L^∞ estimates for solutions were not available until the constructions in [49] appeared, so the regularization in [101] was done by artificially replacing delta functions with L^2-approximations. While this yielded an existence and uniqueness result, the regularization was not mathematically equivalent to the nonregularized problem, and therefore did not yield a result for the original nonregularized problem. In [49, 95], the *a priori* L^∞ estimates outlined above were combined with the variational argument from [101], establishing existence and uniqueness of solutions to the RPBE. As the RPBE is mathematically equivalent to the nonregularized problem, this also established existence and uniqueness of solutions to the nonregularized Poisson-Boltzmann equation.

Theorem 10.4.5 (Existence and Uniqueness of RPBE Solutions [49, 95, 101]). *Let Assumption 10.4.2 hold. There exists a unique weak solution $u^n \in U$ to* (10.4.39), *and subsequently there exists a unique weak solution $u \in V$ to the RPBE* (10.4.38).

Proof. We follow the approach in [49, 95, 101]. We begin by defining an energy functional $J\colon U \subset H_0^1(\Omega) \to \overline{\mathbb{R}}$:

$$J(u) = \int_\Omega \frac{\varepsilon}{2}|\nabla u|^2 + \bar{\kappa}^2 \cosh(u + H)\, dx,$$

with $H = u^l + u^g + G$. Note that if u is the solution of the optimization problem, that is,

$$J(u) = \inf_{v \in U} J(v) \leqslant J(v), \ \forall v \in U, \tag{10.4.40}$$

then u is the solution of (10.4.39). To see this, consider that for any $v \in U$ and any $t \in \mathbb{R}$, the function $F(t) = J(u + tv)$ attains its minimum at $t = 0$, and thus we have $F'(0) = 0$. It is easy to see that $J(u)$ is **G**-differentiable at $u \in U$ for any direction $v \in H_0^1(\Omega)$, with

$$\langle DJ(u), v\rangle = A(u, v) + (B_H(u), v),$$

with A and B_H as defined in (10.4.19) and (10.4.28). Therefore, any solution of the optimization problem (10.4.40) is also a solution to the RPBE (10.4.39). We assemble some quick facts about $H_0^1(\Omega)$, $U \subset H_0^1(\Omega)$, and J.

(1) $H_0^1(\Omega)$ is a reflexive Banach space. This is a standard result; see Section 10.3.
(2) U is nonempty, convex, and topologically closed as a subset of $H_0^1(\Omega)$. This holds by Lemma 10.3.1.
(3) J is convex on U: $J(\lambda u + (1 - \lambda)v) \leqslant \lambda J(u) + (1 - \lambda)J(v)$, $\forall u, v \in U$, $\lambda \in (0, 1)$. This follows easily since J is a linear combination of the convex functions x^2 and $\cosh x$.

By Theorem 10.1.12, we can conclude the existence of a solution to (10.4.40), and hence to (10.4.39) and (10.4.38), if we can establish two additional properties of J:

(4) J is lower semicontinuous on U: $J(u) \leqslant \liminf_{j\to\infty} J(u_j)$, $\forall u_j \to u \in U$.
(5) J is coercive on U: $J(u) \geqslant C_0\|u\|_{1,2}^2 - C_1$, $\forall u \in U$.

That J is lower semicontinuous (and in fact, has the stronger property of weak lower semicontinuity), follows from Theorem 10.1.11 since J is both convex and **G**-differentiable on U. That J is coercive follows from $\cosh x \geqslant 0$ and the Poincaré inequality (Theorem 10.3.6),

$$\begin{aligned} J(u) &= \left\|\frac{\varepsilon}{2}|\nabla u|^2\right\|_1 + \left\|\bar{\kappa}^2 \cosh(u + H + G)\right\|_1 \\ &\geqslant \inf_{x\in\Omega} \frac{\varepsilon(x)}{2}|u|_{1,2}^2 \geqslant \inf_{x\in\Omega} \frac{\varepsilon(x)}{2}\left(\frac{1}{2}|u|_{1,2}^2 + \frac{1}{2\rho^2}\|u\|_2^2\right) \\ &\geqslant C_0\|u\|_{1,2}, \end{aligned}$$

with $C_0 = (\inf_{x\in\Omega} \varepsilon(x))\cdot\min\{1/4, 1/(4\rho^2)\}$, where $\rho > 0$ is the Poincaré constant.

All that remains is to show that u is unique. Assume not, and there are two solutions u_1 and u_2 satisfying (10.4.39). Subtracting (10.4.39) for each solution gives

$$A(u_1 - u_2, v) + (B_H(u_1) - B_H(u_2), v) = 0 \quad \forall v \in H_0^1(\Omega).$$

Now take $v = u_1 - u_2$; monotonicity of the nonlinearity defining B_H ensures that $(B_H(u_1) - B_H(u_2), u_1 - u_2) \geqslant 0$, giving

$$0 \geqslant A(u_1 - u_2, u_1 - u_2) \geqslant 2C_0\|u_1 - u_2\|_{1,2} \geqslant 0,$$

where C_0 is as above. This can only hold if $u_1 = u_2$. □

Near-Best Approximation with a Galerkin Method. We now establish near-best approximation from any subspace X_h. Recall that we proved a general abstract near-best approximation result in Section 10.2, namely Theorem 10.2.13, which appears to be appropriate for the nonlinear Poisson-Boltzmann equation. Our problem (10.4.39) has the appropriate form:

$$\text{Find } u \in U \text{ such that } A(u, v) + \langle B_H(u), v\rangle = F(v), \quad \forall v \in H_0^1(\Omega), \tag{10.4.41}$$

with A and B_H as in (10.4.19) and (10.4.28), with the caveat that we restrict our search to $U \subset H_0^1(\Omega)$ to keep the nonlinearity G well-defined. This was justified by establishing *a priori* L^∞ estimates that defined the "box" U, and then we subsequently showed in Theorem 10.4.5 that in fact the box contained a unique solution. A Galerkin approximation can be taken to be:

$$\text{Find } u_h \in X_h \subset H_0^1(\Omega) \text{ such that } A(u_h, v_h) + \langle B_H(u_h), v_h\rangle = F(v_h), \tag{10.4.42}$$

$$\forall v_h \in X_h \subset H_0^1(\Omega). \tag{10.4.43}$$

The bilinear form $A(u, v)\colon H_0^1(\Omega) \times H_0^1(\Omega) \to \mathbb{R}$ defined in (10.4.19) was in fact shown to be both coercive and bounded as part of the proofs of Theorems 10.4.2 and 10.4.5:

$$m\|u\|_{1,2}^2 \leqslant A(u, u), \quad A(u, v) \leqslant M\|u\|_{1,2}\|v\|_{1,2}, \quad \forall u, v \in H_0^1(\Omega). \tag{10.4.44}$$

Recall that $0 < \epsilon_m \leqslant \epsilon_s$; the constants then become simply

$$m = \min\left\{\frac{\epsilon_m}{2\rho^2}, \frac{\epsilon_m}{2}\right\}, \quad M = \epsilon_s, \tag{10.4.45}$$

where ρ is the constant from the Poincaré inequality (Theorem 10.3.6). The nonlinear form $\langle B_H(u), v\rangle\colon U \times H_0^1(\Omega) \to \mathbb{R}$ was easily seen to be monotone:

$$0 \leqslant \langle B_H(u) - B_H(v), u - v\rangle, \quad \forall u, v \in H_0^1(\Omega), \tag{10.4.46}$$

due to the monotonicity of $\sinh(x)$. Regarding the needed Lipschitz property of B_H, we have the following lemma.

Lemma 10.4.3 ([49, 95]). *Let $u \in U$ satisfy* (10.4.41), *let $u_h \in X_h$ satisfy* (10.4.42), *and let u_h be uniformly bounded. Then there exists a constant $K > 0$ such that*

$$\langle B_H(u) - B_H(u_h), u - v_h \rangle \leqslant K \|u - u_h\|_{1,2} \|u - v_h\|_{1,2}, \qquad \forall v_h \in X_h. \quad (10.4.47)$$

Proof. We have shown that the solution u to (10.4.41) satisfies the *a priori* L^∞ bounds in (10.4.37); if similar *a priori* L^∞ bounds hold for u_h solving (10.4.42), then (10.4.47) will follow from a simple mean value argument (see Theorem 10.1.3). In particular, with $(B_H(u), v) = (\bar{\kappa}^2 \sinh(u + H), v)$, by generalized Taylor expansion (Theorem 10.1.2), we have

$$\begin{aligned}
(B_H(u) - B_H(u_h), u - v_h) &= ([\bar{\kappa}^2 \sinh(u + H) - \bar{\kappa}^2 \sinh(u_h + H)], u - v_h) \\
&= (\bar{\kappa}^2 \cosh(\xi u + (1 - \xi) u_h + H)[u - u_h], u - v_h) \\
&\leqslant K \|u - u_h\|_2 \|u - v_h\|_2 \\
&\leqslant K \|u - u_h\|_{1,2} \|u - v_h\|_{1,2},
\end{aligned}$$

where by Hölder's inequality and convexity of $\cosh(x)$, we have

$$K = \|\bar{\kappa}^2\|_{\infty,\Omega_s} (\| \cosh(u + H)\|_{\infty,\Omega_s} + \| \cosh(u_h + H)\|_{\infty,\Omega_s}) < +\infty.$$

□

While discrete *a priori* bounds which guarantee that u_h is uniformly bounded as needed by Lemma 10.4.3 can be shown for certain discretizations such as the finite volume method, it places restrictions on finite element bases and meshes (see the discussion in [49, 95] for more information about when it is possible to satisfy this assumption with piecewise linear finite element spaces X_h).

Theorem 10.4.6 (Near-Best Approximation [49, 95]). *Let Assumption 10.4.2 hold, and let the Galerkin approximation $u_h \in X_h$ satisfying* (10.4.42) *be uniformly bounded. Then it is a near-best approximation to the solution of problem* (10.4.41), *and we have the following* a priori *error bound:*

$$\|u - u_h\|_{1,2} \leqslant \left(\frac{M + K}{m} \right) E_{V_h}(u).$$

Proof. By Lemma 10.4.3, the assumptions for Theorem 10.2.13 hold, so the result follows. □

The Einstein Constraint Equations

Einstein's General Theory of Relativity states that spacetime has the structure of a pseudo-Riemannian 4-manifold $\mathcal{M}$ which, rather than being a static Euclidean space, bends in response to matter and energy density. The theory is represented mathematically by the Einstein Field Equations, a set of 10 second-order partial differential equations published by Einstein in 1915. It remains the most widely accepted theory

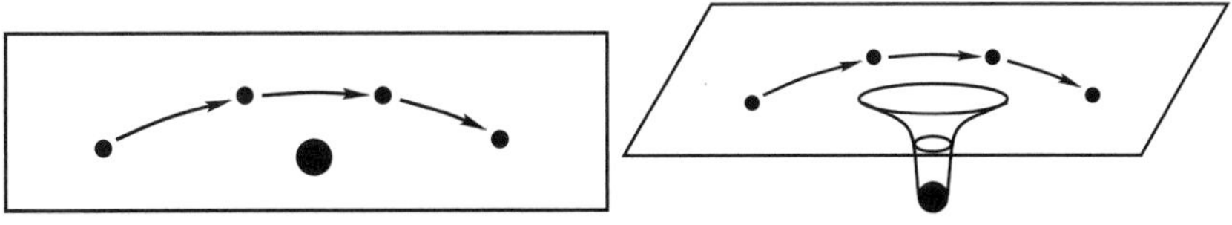

Figure 10.3 Newtonian versus general relativistic theories of gravity.

of the gravitational force field; rather than mass generating an attractive Newtonian field analogous to the electrostatic field generated by a charged body, general relativity postulates that what we experience as gravity is a geometrical effect, a distortion of spacetime caused by matter and energy distribution. The distinctions between the Newtonian and general relativistic theories of gravity are illustrated in Figure 10.3. Although it is not captured by the figure, time is also distorted by matter and energy as described by the Einstein equations. In fact, the time distortion effect is large enough that it must be accounted for in GPS techology, due to the difference in the strength of Earth's gravitational field on the surface of the Earth compared to that in GPS satellite orbit.

The theory predicts that accelerating masses produce gravitational waves of perturbations in the metric tensor of spacetime, which in principle can be detected by early twenty-first century technology. LIGO (Laser Interferometer Gravitational-wave Observatory) is one of several recently constructed gravitational detectors designed to verify this aspect of the theory. The design of LIGO is based on measuring distance changes between mirrors in perpendicular directions as the ripple in the metric tensor propagates through the device. The two L-shaped LIGO observatories (in Washington and Louisiana), with legs at $1.5\,m \times 4\,km$, have phenomenal sensitivity, on the order of 10^{-15} to $10^{-18}m$. Currently, there is tremendous interest in developing highly reliable, high-resolution numerical simulations for the Einstein equations in order to verify the predictions of the theory and also to develop gravitational wave detection science into the ultimate long-range telescope. See [55, 120, 158] for an overview of developments since the field was initiated in the late 1970s.

The Einstein equations have been studied intensively in the mathematics community since the 1950s; see [50]. Our focus here is on a subsystem within the Einstein equations known as the *Einstein constraint equations*; analogous to the situation with the Maxwell system, the Einstein equations are a constrained evolution system in which some of the equations in the system are time independent and represent constraints that are propagated by the remaining evolution equations. The 10 equations for the 10 independent components of the symmetric spacetime metric tensor g_{ab} are the *Einstein Equations*:

$$G_{ab} = \kappa T_{ab}, \quad 0 \leqslant a \leqslant b \leqslant 3, \quad \kappa = 8\pi G/c^4, \tag{10.4.48}$$

where $G_{ab} = R_{ab} - \frac{1}{2}Rg_{ab}$ is the Einstein tensor encoding the deformation of spacetime, and where T_{ab} is the stress-energy tensor that encodes the matter and energy distribution of spacetime. The tensors G_{ab} and T_{ab} are functions of the unknown metric g_{ab}; G_{ab} acts like a quasilinear spacetime differential operator on g_{ab}, whereas T_{ab} is a nonlinear source term in g_{ab}. The Ricci tensor $R_{ab} = R_{acb}{}^{c}$ and the scalar curvature $R = R_a^{\ a}$ are formed from the Riemann (curvature) tensor $R_{abc}{}^{d}$ of the 4-manifold $\mathcal{M}$ by contracting tensor indices; the Riemann tensor arises through the failure of commutativity of covariant differentiation. In the case of calculus in $\mathbb{R}^n$, partial differentiation commutes: $V^a_{\ ,bc} - V^a_{\ ,cb} = 0$, where V^a is a vector field, and where $V^a_{\ ,b} = \partial V^a / \partial x^b$ is a convenient shorthand for partial differentiation. In the case of calculus on a 2-surface or a more general n-manifold, (covariant) partial differentiation does not commute: $V^a_{\ ;bc} - V^a_{\ ;cb} = R^a{}_{dbc} V^d$, where $V^a_{\ ;b} = V^a_{\ ,b} + \Gamma^a_{bc} V^c$, and where Γ^a_{bc} are the Christoffel symbols (connection coefficients) for the surface. What is left over is the Riemann tensor

$$R^a{}_{dbc} = \Gamma^a_{bd,c} - \Gamma^a_{cd,b} + \Gamma^a_{ec}\Gamma^e_{bd} - \Gamma^a_{eb}\Gamma^e_{cd}.$$

Initial value formulations of the Einstein equations (10.4.48) are known to be well-posed (see [90, 174]); various formalisms yield constrained (weakly, strongly, or even symmetric) hyperbolic systems on space-like 3-manifolds $\Omega = \mathcal{S}(t)$ that evolve a Riemannian metric $\hat{h}_{ab}$ and its time derivative forward in time. The manifold and fields $(\Omega, \hat{h}_{ab}, \hat{k}^{ab}, \hat{j}^a, \hat{\rho})$ form an *initial data set* for Einstein's equations if and only if Ω is a three-dimensional smooth manifold, $\hat{h}_{ab}$ is a Riemannian metric on Ω, $\hat{k}^{ab}$ is a symmetric tensor field on Ω, $\hat{j}^a$ and $\hat{\rho}$ are vector and nonnegative scalar fields on Ω, respectively, satisfying a certain energy condition, and the following hold on Ω:

$$\hat{R} + \hat{k}^2 - \hat{k}_{ab}\hat{k}^{ab} - 2\kappa\hat{\rho} = 0, \tag{10.4.49}$$

$$-\hat{\nabla}_a \hat{k}^{ab} + \hat{\nabla}^b \hat{k} + \kappa \hat{j}^b = 0. \tag{10.4.50}$$

Here, $\hat{\nabla}_a$ is the Levi-Civita connection of $\hat{h}_{ab}$, so it satisfies $\hat{\nabla}_a \hat{h}_{bc} = 0$, $\hat{R}$ is the Ricci scalar of the connection $\hat{\nabla}_a$, $\hat{k} = \hat{h}_{ab}\hat{k}^{ab}$ is the trace of $\hat{k}^{ab}$, and $\kappa = 8\pi$ in units where both the gravitational constant and speed of light have value 1. We denote by $\hat{h}^{ab}$ the tensor inverse of $\hat{h}_{ab}$. Tensor indices of hatted quantities are raised and lowered with $\hat{h}^{ab}$ and $\hat{h}_{ab}$, respectively. When (10.4.49)–(10.4.50) hold, the manifold Ω can be embedded as a hypersurface in a four-dimensional manifold corresponding to a solution of the spacetime Einstein field equations, and the pushforward of $\hat{h}^{ab}$ and $\hat{k}^{ab}$ represent the first and second fundamental forms of the embedded hypersurface. This leads to the terminology extrinsic curvature for $\hat{k}^{ab}$, and mean extrinsic curvature for its trace, $\hat{k}$. It suffices for our purposes here to think of Ω as simply a Lipschitz domain in $\mathbb{R}^n$, with any vectors and tensors appearing in the discussion as being simply Cartesian vector and matrix functions of the independent variable ranging over the open set Ω.

We now follow [96, 97] and describe the *conformal method* [182], which consists of finding solutions $\hat{h}_{ab}$, $\hat{k}^{ab}$, $\hat{j}^a$, and $\hat{\rho}$ of (10.4.49)–(10.4.50) using a particular decomposition. To proceed, fix on Ω a Riemannian metric h_{ab} with Levi-Civita connection ∇_a, so it satisfies $\nabla_a h_{bc} = 0$, and has Ricci scalar R. Fix on Ω a symmetric tensor σ^{ab}, trace-free and divergence-free with respect to h_{ab}; that is,

$h_{ab}\sigma^{ab} = 0$ and $\nabla_a \sigma^{ab} = 0$. Also fix on Ω scalar fields τ and ρ, and a vector field j^a, subject to an appropriate energy condition (see [97]). We have denoted by h^{ab} the tensor inverse of h_{ab}, and we use the convention that tensor indices of unhatted quantities are raised and lowered with the tensors h^{ab} and h_{ab}, respectively. Finally, given a smooth vector field w^a on Ω, introduce the conformal Killing operator $\mathcal{L}$ as follows: $(\mathcal{L}w)^{ab} = \nabla^a w^b + \nabla^b w^a - (2/3)\,(\nabla_c w^c)\,h^{ab}$. The conformal method then involves first solving the following equations for a scalar field ϕ and vector field w^a:

$$-\Delta\phi + a_R\phi + a_\tau\,\phi^5 - a_w\,\phi^{-7} - a_\rho\,\phi^{-3} = 0, \tag{10.4.51}$$

$$-\nabla_a(\mathcal{L}w)^{ab} + b^b_\tau\phi^6 + b^b_j = 0, \tag{10.4.52}$$

where we have introduced the Laplace-Beltrami operator $\Delta\phi = h^{ab}\nabla_a\nabla_b\phi$, and the functions $a_R = R/8$, $a_\tau = \tau^2/12$, $a_\rho = \kappa\rho/4$, $b^b_\tau = (2/3)\nabla^b\tau$, $b^b_j = \kappa j^b$, and $a_w = \big[\sigma_{ab} + (\mathcal{L}w)_{ab}\big]\big[\sigma^{ab} + (\mathcal{L}w)^{ab}\big]/8$. One then recovers the tensors $\hat{h}_{ab}$, $\hat{k}^{ab}$, $\hat{j}^a$, and $\hat{\rho}$ through the expressions

$$\hat{h}_{ab} = \phi^4 h_{ab}, \quad \hat{j}^a = \phi^{-10}\,j^a, \quad \hat{\rho} = \phi^{-8}\,\rho, \tag{10.4.53}$$

$$\hat{k}^{ab} = \phi^{-10}\big[\sigma^{ab} + (\mathcal{L}w)^{ab}\big] + \frac{1}{3}\phi^{-4}\tau\,h^{ab}. \tag{10.4.54}$$

A straightforward computation shows that if $\hat{h}_{ab}$, $\hat{k}^{ab}$, $\hat{j}^a$, and $\hat{\rho}$ have the form given in (10.4.53)–(10.4.54), then equations (10.4.49)–(10.4.50) are equivalent to equations (10.4.51)–(10.4.52). Hatted fields represent quantities with physical meaning, except the trace τ of the physical extrinsic curvature $\hat{k}^{ab}$; that is, $\tau = \hat{k}$.

Of interest to us now is the existence, uniqueness, and near-best approximation of solutions to (10.4.51)–(10.4.52). Known results for this problem are summarized in [27, 97]; the equations (10.4.51)–(10.4.52) have been studied almost exclusively in the setting of constant mean extrinsic curvature ($b_\tau = 0$), known as the CMC case. In the CMC case the equations decouple; one then solves the momentum equation first, and uses its solution to solve the Hamiltonian constraint. The CMC case was completely solved over the last 30 years; see [104, 27]. The question of existence of solutions to (10.4.51)–(10.4.52) for nonconstant mean extrinsic curvature ($b_\tau \neq 0$; the "non-CMC case") has remained largely unanswered, with most progress made only in the case that the mean extrinsic curvature is nearly constant (b_τ small; the "near-CMC case"), in the sense that the size of its spatial derivatives is sufficiently small. The near-CMC condition leaves the constraint equations coupled, but ensures that the coupling is weak. In [105], Isenberg and Moncrief established the first existence (and uniqueness) result in the near-CMC case using a contraction argument. Of particular interest here is the "far-from-CMC" solutions, whereby the CMC or near-CMC conditions on the data are not enforced. The first results of this type appeared only recently in [96, 97], followed shortly by [130], based on a combination of topological fixed point arguments and order-preserving maps. We give an overview of this approach below, illustrating the use of many of the techniques we developed in Section 10.1. More detailed proofs may be found in [97] for more general physical situations.

A Priori Estimates. The momentum constraint (10.4.52) is well understood in the case that h_{ab} has no conformal Killing vectors [a vector field v^a is *conformal Killing* iff $(\mathcal{L}v)^{ab} = 0$]. A standard result is the following. Let (Ω, h_{ab}) be a three-dimensional closed C^2 Riemannian manifold, with h_{ab} having no conformal Killing vectors, and let b^a_τ, $b^a_j \in L^p$ with $p \geqslant 2$ and $\phi \in L^\infty$; Then, equation (10.4.52) has a unique solution $w^a \in W^{2,p}$ with

$$c\,\|w\|_{2,p} \leqslant \|\phi\|^6_\infty\,\|b_\tau\|_p + \|b_j\|_p, \tag{10.4.55}$$

where $c > 0$ is a constant. This result is generalized in [97], allowing weaker coefficient differentiability, giving existence of solutions down to $w^a \in W^{1,p}$, with real number $p \geqslant 2$. The proof in [97] is based on Riesz-Schauder theory for compact operators [176].

From inequality (10.4.55) it is not difficult to show that for $p > 3$ the following pointwise estimate holds:

$$a_w \leqslant K_1\,\|\phi\|^{12}_\infty + K_2, \tag{10.4.56}$$

with $K_1 = \frac{1}{2}(c_s c_{\mathcal{L}}/c)^2\|b_\tau\|^2_p$, $K_2 = \frac{1}{4}\|\sigma\|^2_\infty + \frac{1}{2}(c_s c_{\mathcal{L}}/c)^2\|b_j\|^2_p$, where c_s is the constant in the embedding $W^{1,p} \hookrightarrow L^\infty$, and $c_{\mathcal{L}}$ is a bound on the norm of $\mathcal{L}\colon W^{2,p} \to W^{1,p}$. There is no smallness assumption on $\|b_\tau\|_p$, so the near-CMC condition is not required for these results.

Let Ω be closed. The scalar functions ϕ_- and ϕ_+ are called *barriers* (sub- and supersolutions, respectively) iff

$$-\Delta\phi_- + a_R\phi_- + a_\tau\,\phi^5_- - a_w\,\phi^{-7}_- - a_\rho\,\phi^{-3}_- \leqslant 0, \tag{10.4.57}$$

$$-\Delta\phi_+ + a_R\phi_+ + a_\tau\,\phi^5_+ - a_w\,\phi^{-7}_+ - a_\rho\,\phi^{-3}_+ \geqslant 0. \tag{10.4.58}$$

The barriers are *compatible* iff $0 < \phi_- \leqslant \phi_+$, and are *global* iff (10.4.57)–(10.4.58) holds for all w^a solving equation (10.4.52), with source $\phi \in [\phi_-, \phi_+]$. The closed interval

$$[\phi_-, \phi_+] = \{\phi \in L^p : \phi_- \leqslant \phi \leqslant \phi_+ \text{ a.e. in } \Omega\} \tag{10.4.59}$$

is a topologically closed subset of L^p, $1 \leqslant p \leqslant \infty$ (see Theorem 10.3.1). Global supersolutions are difficult to find as a consequence of the nonnegativity of a_w and its estimate (10.4.56), together with the limit (10.4.59). Global supersolution constructions, such as those in [3, 105], rely in a critical way on the near-CMC assumption, which appears as the condition that a suitable norm of $\nabla\tau$ be sufficiently small, or equivalently, that K_1 in (10.4.56) be sufficiently small. Some of the key results in [96, 97] were establishing existence of global supersolutions of the Hamiltonian constraint without the near-CMC assumption. We need the following notation: Given any scalar function $v \in L^\infty$, denote by $v^\wedge = \text{ess sup}_\Omega v$, and $v^\vee = \text{ess inf}_\Omega v$.

Theorem 10.4.7 (Global Supersolution [96, 97]). *Let (Ω, h_{ab}) be a smooth, closed, three-dimensional Riemannian manifold with metric h_{ab} in the positive Yamabe class with no conformal Killing vectors. Let u be a smooth positive solution of the Yamabe problem*

$$-\Delta u + a_R u - u^5 = 0, \tag{10.4.60}$$

and define the constant $k = u^\wedge/u^\vee$. *If the function* τ *is nonconstant and the rescaled matter fields* j^a, ρ, *and traceless transverse tensor* σ^{ab} *are sufficiently small, then*

$$\phi_+ = \epsilon u, \qquad \epsilon = \left[\frac{1}{2K_1 k^{12}}\right]^{\frac{1}{4}} \tag{10.4.61}$$

is a global supersolution of equation (10.4.51).

Proof. Existence of a smooth positive solution u to (10.4.60) is summarized in [118]. Using the notation

$$E(\phi) = -\Delta\phi + a_R\phi + a_\tau\phi^5 - a_w\phi^{-7} - a_\rho\phi^{-3}, \tag{10.4.62}$$

we have to show that $E(\phi_+) \geqslant 0$. Taking $\phi_+ = \epsilon u$, for $\epsilon > 0$ gives the identity $-\Delta\phi_+ + a_R\phi_+ = \epsilon u^5$. We have

$$\begin{aligned}
E(\phi_+) &\geqslant -\Delta\phi_+ + a_R\phi_+ - \frac{K_1(\phi_+^\wedge)^{12} + K_2}{\phi_+^7} - \frac{a_\rho^\wedge}{\phi_+^3} \\
&\geqslant \epsilon\, u^5 - K_1\left[\frac{\phi_+^\wedge}{\phi_+^\vee}\right]^{12}\phi_+^5 - \frac{K_2}{\phi_+^7} - \frac{a_\rho^\wedge}{\phi_+^3} \\
&\geqslant \epsilon u^5\left[1 - K_1\, k^{12}\epsilon^4 - \frac{K_2}{\epsilon^8 u^{12}} - \frac{a_\rho^\wedge}{\epsilon^4 u^8}\right],
\end{aligned}$$

where we have used $\phi_+^\wedge/\phi_+^\vee = u^\wedge/u^\vee = k$. The choice of ϵ made in (10.4.61) is equivalent to $1/2 = 1 - K_1\, k^{12}\epsilon^4$. For this ϵ, impose on the free data σ^{ab}, ρ and j^a, the condition

$$\frac{1}{2} - \frac{K_2}{\epsilon^8(u^\vee)^{12}} - \frac{a_\rho^\wedge}{\epsilon^4(u^\vee)^8} \geqslant 0.$$

Thus for any $K_1 > 0$, we can guarantee $E(\phi_+) \geqslant 0$ for sufficiently small data σ^{ab}, ρ, and j^a. □

Theorem 10.4.7 shows that global supersolutions ϕ_+ can be built without using near-CMC conditions by rescaling solutions to the Yamabe problem (10.4.60); the larger $\|\nabla\tau\|_p$, the smaller the factor ϵ. Existence of the finite positive constant k appearing in Theorem 10.4.7 is related to establishing a Harnack inequality for solutions to the Yamabe problem (see [122]). It remains to construct (again, without near-CMC conditions) a compatible global subsolution satisfying $0 < \phi_- \leqslant \phi_+$.

Theorem 10.4.8 (Global Subsolution [96, 97])**.** *Let the assumptions required for Theorem 10.4.7 hold. If the rescaled matter energy density* ρ *is not identically zero, then there exists a positive global subsolution* ϕ_- *of equation (10.4.51), compatible with the global supersolution in Theorem 10.4.7, so that it satisfies* $0 < \phi_- \leqslant \phi_+$.

Proof. The proof in [96, 97] is based on well-known constructions [51, 129, 137]. □

Existence and Uniqueness of Solutions. We now state some supporting results we need from [97] for solutions of (10.4.51). We state only the results for strong solutions, recovering previous results in [104, 105]. Generalizations allowing weaker differentiability conditions on the coefficients appear in [51, 97, 129].

Theorem 10.4.9 (Hamiltonian Constraint Existence [96, 97]). *Let (Ω, h_{ab}) be a three-dimensional C^2 closed Riemannian manifold. Let the free data τ^2, σ^2, and ρ be in L^p, with $p \geqslant 2$. Let ϕ_- and ϕ_+ be barriers to (10.4.51) for a particular value of the vector $w^a \in W^{1,2p}$. Then, there exists a solution $\phi \in [\phi_-, \phi_+] \cap W^{2,p}$ of the Hamiltonian constraint (10.4.51). Furthermore, if the metric h_{ab} is in the nonnegative Yamabe classes, then ϕ is unique.*

Proof. The proofs in [97] make use of sub- and supersolutions, together with fixed point arguments for monotone increasing maps (see Section 10.1). □

Our main result concerning the coupled constraint system is the following.

Theorem 10.4.10 (Coupled Constraints Existence [96, 97]). *Let (Ω, h_{ab}) be a three-dimensional, smooth, closed Riemannian manifold with metric h_{ab} in the positive Yamabe class with no conformal Killing vectors. Let $p > 3$ and let τ be in $W^{1,p}$. Let σ^2, j^a, and ρ be in L^p and satisfy the assumptions for Theorems 10.4.7 and 10.4.8 to yield a compatible pair of global barriers $0 < \phi_- \leqslant \phi_+$ to the Hamiltonian constraint (10.4.51). Then, there exists a scalar field $\phi \in [\phi_-, \phi_+] \cap W^{2,p}$ and a vector field $w^a \in W^{2,p}$ solving the constraint equations (10.4.51)–(10.4.52).*

Theorem 10.4.10 can be proven using the following topological fixed point result established in [97], based on the Schauder Fixed-Point Theorem. For a review of the Schauder Theorem see Section 10.1. Note that such compactness arguments do not give uniqueness.

Theorem 10.4.11 (Coupled Fixed Point Principle A [96, 97]). *Let X and Y be Banach spaces, and let Z be a Banach space with compact embedding $X \hookrightarrow Z$. Let $U \subset Z$ be a nonempty, convex, closed, bounded subset, and let*

$$S\colon U \to \mathcal{R}(S) \subset Y, \qquad T\colon U \times \mathcal{R}(S) \to U \cap X,$$

be continuous maps. Then there exist $\phi \in U \cap X$ and $w \in \mathcal{R}(S)$ such that

$$\phi = T(\phi, w) \quad \text{and} \quad w = S(\phi). \tag{10.4.63}$$

Proof. The proof will be through a standard variation of the Schauder Fixed-Point Theorem, reviewed as Theorem 10.1.17. The proof is divided into several steps.

Step 1: Construction of a nonempty, convex, closed, bounded subset $U \subset Z$. By assumption we have that $U \subset Z$ is nonempty, convex (involving the vector space structure of Z), closed (involving the topology on Z), and bounded (involving the metric given by the norm on Z).

Step 2: Continuity of a mapping $G\colon U \subset Z \to U \cap X \subset X$. Define the composite operator

$$G := T \circ S\colon U \subset Z \to U \cap X \subset X.$$

The mapping G is continuous, since it is a composition of the continuous operators $S\colon U \subset Z \to \mathcal{R}(S) \subset Y$ and $T\colon U \subset Z \times \mathcal{R}(S) \to U \cap X \subset X$.

Step 3: Compactness of a mapping $F\colon U \subset Z \to U \subset Z$. The compact embedding assumption $X \hookrightarrow Z$ implies that the canonical injection operator $i\colon X \to Z$ is compact. Since the composition of compact and continuous operators is compact, we have the composition $F := i \circ G\colon U \subset Z \to U \subset Z$ is compact.

Step 4: Invoking the Schauder Theorem. Therefore, by a standard variant of the Schauder Theorem (see Theorem 10.1.17), there exists a fixed point $\phi \in U$ such that $\phi = F(\phi) = T(\phi, S(\phi))$. Since $\mathcal{R}(T) = U \cap X$, in fact $\phi \in U \cap X$. We now take $w = S(\phi) \subset \mathcal{R}(S)$ and we have the result. □

The following special case of Theorem 10.4.11 gives some simple sufficient conditions on T to establish the invariance using barriers in an ordered Banach space (for a review of ordered Banach spaces, see Section 10.1).

Theorem 10.4.12 (Coupled Fixed Point Principle B [96, 97]). *Let X and Y be Banach spaces, and let Z be a real ordered Banach space having the compact embedding $X \hookrightarrow Z$. Let $[\phi_-, \phi_+] \subset Z$ be a nonempty interval which is closed in the topology of Z, and set $U = [\phi_-, \phi_+] \cap \overline{B}_M \subset Z$ where $\overline{B}_M$ is the closed ball of finite radius $M > 0$ in Z about the origin. Assume that U is nonempty, and let the maps*

$$S\colon U \to \mathcal{R}(S) \subset Y, \qquad T\colon U \times \mathcal{R}(S) \to U \cap X,$$

be continuous maps. Then there exist $\phi \in U \cap X$ and $w \in \mathcal{R}(S)$ such that

$$\phi = T(\phi, w) \quad \text{and} \quad w = S(\phi).$$

Proof. By choosing the set U as the nonempty intersection of the interval $[\phi_-, \phi_+]$ with a bounded set in Z, we have U bounded in Z. We also have that U is convex with respect to the vector space structure of Z, since it is the intersection of two convex sets $[\phi_-, \phi_+]$ and $\overline{B}_M$. Since U is the intersection of the interval $[\phi_-, \phi_+]$, which by assumption is closed in the topology of Z, with the closed ball $\overline{B}_M$ in Z, U is also closed. In summary, we have that U is nonempty as a subset of Z, closed in the topology of Z, convex with respect to the vector space structure of Z, and bounded with respect to the metric (via normed) space structure of Z. Therefore, the assumptions of Theorem 10.4.11 hold and the result then follows. □

Proof of Theorem 10.4.10. The proof is through Theorem 10.4.12. First, for arbitrary real number $s > 0$, express (10.4.51)–(10.4.52) as

$$L_s\phi + f_s(\phi, w) = 0, \quad (\boldsymbol{L}w)^a + \boldsymbol{f}(\phi)^a = 0, \tag{10.4.64}$$

where $L_s\colon W^{2,p} \to L^p$ and $\boldsymbol{L}\colon W^{2,p} \to L^p$ are defined as $L_s\phi := [-\Delta + s]\phi$, and $(\boldsymbol{L}w)^a := -\nabla_b(\mathcal{L}w)^{ab}$, and where the nonlinear maps $f_s\colon [\phi_-, \phi_+] \times W^{2,p} \to L^p$ and $\boldsymbol{f}\colon [\phi_-, \phi_+] \to L^p$ are defined as

$$\begin{aligned} f_s(\phi, w) &:= [a_R - s]\phi + a_\tau\phi^5 - a_w\phi^{-7} - a_\rho\phi^{-3}, \\ \boldsymbol{f}(\phi)^a &:= b_\tau^a\phi^6 + b_j^a. \end{aligned}$$

Introduce now operators $S\colon [\phi_-, \phi_+] \to W^{2,p}$ and $T\colon [\phi_-, \phi_+] \times W^{2,p} \to W^{2,p}$, which are given by

$$S(\phi)^a := -[\boldsymbol{L}^{-1}\boldsymbol{f}(\phi)]^a, \quad T(\phi, w) := -L_s^{-1} f_s(\phi, w).$$

The mappings S and T are well-defined due to the absence of conformal Killing vectors and by introduction of the positive shift $s > 0$, ensuring that both $\boldsymbol{L}$ and L_s are invertible (see [97]). The equations (10.4.64) have the form (10.4.63) for use of Theorem 10.4.12. We have the Banach spaces $X = W^{2,p}$ and $Y = W^{2,p}$, and the (ordered) Banach space $Z = L^\infty$ with compact embedding $W^{2,p} \hookrightarrow L^\infty$. The compatible barriers form the nonempty, convex, bounded L^∞-interval $U = [\phi_-, \phi_+]$, which we noted earlier is closed in L^p for $1 \leqslant p \leqslant \infty$ (see Theorem 10.3.1). It remains to show that S and T are continuous maps. These properties follow from equation (10.4.55) and from Theorem 10.4.9 with global barriers from Theorems 10.4.7 and 10.4.8, using standard inequalities. Theorem 10.4.10 now follows from Theorem 10.4.12. □

See [97] for generalizations of Theorem 10.4.10 to arbitrary space dimensions and allowing weaker differentiability conditions on the coefficients, establishing existence of nonvacuum, non-CMC weak solutions down to $\phi \in W^{s,p}$, for $(s-1)p > 3$. See also [130] for similar far-from-CMC results in the case of vaccum (without matter, which we needed for our subsolution construction above).

Near-Best Approximation with a Petrov-Galerkin Method. We restrict ourselves to the Hamiltonian constraint, and establish near-best approximation from any subspace X_h. Recall that we proved a general abstract near-best approximation result in Section 10.2, namely Theorem 10.2.13, which appears to be appropriate for the Hamiltonian constraint. The Hamiltonian constraint has the appropriate form:

$$\text{Find } u \in U \text{ such that } A(u, v) + \langle B(u), v\rangle = 0, \;\; \forall v \in X, \tag{10.4.65}$$

where $X \subset H^1(\Omega)$ with the caveat that we restrict our search to $U \subset X$ to keep the nonlinearity B well-defined. This was justified by establishing *a priori* L^∞ estimates (sub- and supersolutions) that defined the "box" U, and then we subsequently showed in Theorem 10.4.9 that, in fact, the box contained a unique solution. A Galerkin approximation can be taken to be:

$$\text{Find } u_h \in X_h \subset X \text{ such that } A(u_h, v_h) + \langle B(u_h), v_h\rangle = 0, \tag{10.4.66}$$
$$\forall v_h \in X_h \subset X. \tag{10.4.67}$$

The bilinear form $A(u, v)\colon X \times X \to \mathbb{R}$ can be shown to be bounded

$$A(u, v) \leqslant M\|u\|_X\|v\|_X, \;\; \forall u, v \in X. \tag{10.4.68}$$

and coercive,

$$m\|u\|_X^2 \leqslant A(u, u), \;\; \forall u \in X, \text{ where } m > 0. \tag{10.4.69}$$

We have shown that the solution u to (10.4.65) satisfies the *a priori* L^∞ bounds in Theorem 10.4.9 of this section; if similar *a priori* L^∞ bounds hold for u_h solving (10.4.66), then as for the other nonlinear example we considered, we have the following result for the Hamiltonian constraint. This result may be found in [94], along with an analogous result for the momentum constraint.

Lemma 10.4.4 ([94]). *Let* $u \in U$ *satisfy* (10.4.65), *let* $u_h \in X_h$ *satisfy* (10.4.66), *and let* u_h *be uniformly bounded. Then there exists a constant* $K > 0$ *such that*

$$\langle B_H(u) - B_H(u_h), u - v_h \rangle \leqslant K\|u - u_h\|_X \|u - v_h\|_X, \qquad \forall v_h \in X_h. \tag{10.4.70}$$

Proof. See [94]. □

Theorem 10.4.13 (Near-Best Approximation [94]). *Let the assumptions for Theorem 10.4.9 hold, and let the Galerkin approximation* $u_h \in X_h$ *satisfying* (10.4.66) *be uniformly bounded. Then it is a near-best approximation to the solution of problem* (10.4.65), *and we have the following* a priori *error bound:*

$$\|u - u_h\|_X \leqslant \left(\frac{M + K}{m}\right) E_{V_h}(u).$$

Proof. By Lemma 10.4.4, the assumptions for Theorem 10.2.13 hold, so the result follows. □

EXERCISES

10.4.1 Consider the linear problem (10.4.1) with no Dirichlet data; in other words, the Robin-Neumann condition (10.4.3) on Γ_N covers the boundary of the entire domain. Without access to the Poincaré inequality, coercivity of the resulting bilinear form $a(\cdot,\cdot)$ arising in the weak formulation will no longer hold. Instead, establish a Gårding inequality of the form

$$m\|u\|_{1,2}^2 \leqslant A(u, v) + K\|u\|_2^2, \quad \forall u \in H^1(\Omega).$$

10.4.2 Establish a generalization of Theorem 10.4.2 using Exercise 10.4.1. You will need to use a Riesz-Schauder (Fredholm Alternative) argument, since you no longer have access to the Lax-Milgram Theorem. The function c appearing in the Robin boundary condition will play a key role in your argument; you may stengthen the assumption on c to

$$0 < c_4 \leqslant c(x) \leqslant c_3 < \infty, \text{ a.e. in } \Gamma_N.$$

(*Hint:* See the analysis of Gårding inequalities in Section 10.2.)

10.4.3 Use the method of sub- and supersolutions (Theorem 10.1.24) to give a different proof of existence of solutions to the Poisson-Boltzmann equation. (*Hint:* Finding sub- and supersolutions will be similar to our construction of *a priori* L^∞ estimates.)

10.4.4 Show that the nonlinearity appearing in the Hamiltonian constraint is actually a convex function when the scalar curvature R appearing in the nonlinearity is nonnegative.

10.4.5 Use a variational method (Theorem 10.1.12) to give an alternative proof of existence of solutions to the Hamiltonian constraint for nonnegative scalar curvature R. (*Hint:* The sub- and supersolutions we have already constructed will provide the nonempty, convex, and closed subset of the solution space in the form of the order interval $[\phi_-, \phi_+]$ for use with a variational argument.)

10.4.6 A type of topological fixed point argument was used in Theorem 10.4.10 to prove the existence of solutions to the coupled Einstein constraints. Is it possible to use a variational argument instead? Is there another approach that could be used for the coupled system?

10.5 FINITE ELEMENT AND RELATED DISCRETIZATION METHODS

At the halfway point of this chapter, let us take stock of where we are. After assembling some additional analysis tools in Section 10.1, we developed in Section 10.2 some results on near-best Petrov-Galerkin approximation of solutions to linear and nonlinear operator equations in Banach spaces. We then gave an overview in Section 10.3 of the particular Banach spaces of interest to us, namely Sobolev and Besov spaces. In Section 10.4 we brought together all of the previous material to analyze a sequence of elliptic problems of increasing complexity, in each case establishing *a priori* estimates (energy and/or L^∞ estimates), existence and uniqueness of solutions, and the existence of a near-best approximation using a Petrov-Galerkin method. In the second half of the chapter we examine some specific discretizations (in this section), the numerical solution of discrete linear problem (in Section 10.6), and the numerical solution of discrete nonlinear problems (in Section 10.7).

The Finite Element Method

We now examine a specific construction of subspaces of Sobolev and Besov spaces for use with Petrov-Galerkin and Galerkin methods, with the goal of approximating solutions to operator equations. The construction is based on piecewise polynomials defined over a disjoint splitting of the spatial domain $\Omega \subset \mathbb{R}^d$ into *elements*, and is known as the *finite element method.* (See Example 6 in Section 8.4 for an introduction involving one spatial dimension.) Some of the most popular classical and modern references include [12, 13, 33, 42, 103, 107, 136, 161, 164]. Some key related references are [9, 10, 44, 141, 167]. In the discussion below, we follow mainly the conventions taken in [52].

A *(conforming) finite element method* can be thought of as a method for constructing *finite element approximation (trial) spaces* $X_h \subset X$ and *test spaces* $Y_h \subset Y$ satisfying three general requirements. The first requirement is on a spatial subdivision

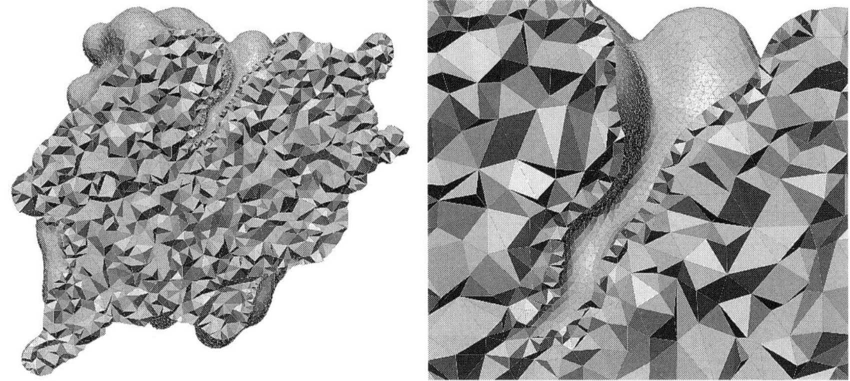

Figure 10.4 Examples of conforming simplex meshes satisfying **FEM1**. These meshes are models of molecular structures, and are constructed using modern computational geometry tools [186, 187].

of the domain Ω that allows for construction of basis functions for X_h and Y_h that will satisfy the other two requirements.

Assumption 10.5.1 (FEM1). *A subdivision $\mathcal{T}_h$ in $\mathbb{R}^d$ is established over $\overline{\Omega} \subset \mathbb{R}^d$ by subdividing $\overline{\Omega}$ into a finite number of subsets K, called* elements, *such that*

(1) $\overline{\Omega} = \cup_{K \in \mathcal{T}_h} K$.

(2) For each $K \in \mathcal{T}_h$, K is closed, $\overset{\circ}{K}$ (the interior of K) is nonempty, and ∂K is Lipschitz continuous.

(3) For distinct $K_1, K_2 \in \mathcal{T}_h$, $\overset{\circ}{K}_1 \cap \overset{\circ}{K}_2 = \emptyset$.

(4) For two distinct polyhedral *elements $K_1, K_2 \in \mathcal{T}_h$, the set $K_1 \cap K_2$ is either empty, or a common p-dimensional subset (vertex, edge, face, etc) in $\Omega \subset \mathbb{R}^d$ with $p < d$.*

If a partitioning $\mathcal{T}_h$ satisfies **FEM1**, it is called a *conforming finite element subdivision* of Ω; it basically ensures that elements meet only at common vertices, edges, or faces. It is possible to relax the *regularity condition* 4) and still develop finite element discretizations; this becomes a *nonconforming finite element subdivision* of Ω. Note that the conditions of **FEM1** are purely topological. If we define

$$h_K = \{ \text{ diameter of smallest circumscribed sphere of } K\}, \tag{10.5.1}$$

$$\rho_K = \{ \text{ diameter of largest inscribed sphere of } K\}. \tag{10.5.2}$$

then it is possible to define geometrical regularity conditions as well:

$$\frac{h_K}{\rho_K} \leqslant \sigma, \quad \forall K \in \mathcal{T}_h, \quad \sigma > 0, \qquad \text{(shape regularity)} \tag{10.5.3}$$

$$\frac{h_{K_1}}{h_{K_2}} \leqslant \nu, \quad \forall K_1, K_2 \in \mathcal{T}_h, \quad \nu > 0. \qquad \text{(quasi-uniformity)} \tag{10.5.4}$$

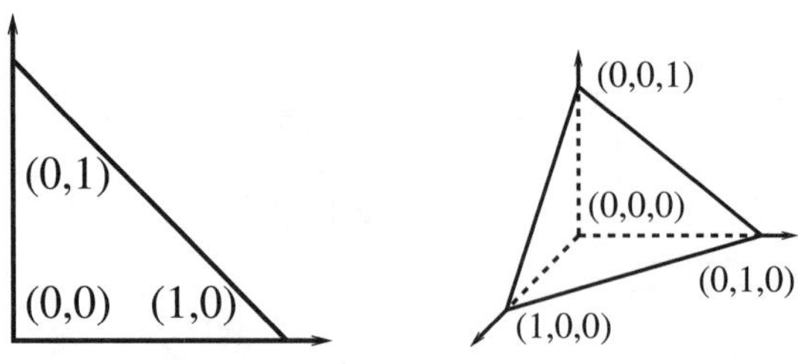

Figure 10.5 Standard reference triangle and tetrahedron.

Shape regularity ensures that the elements K do not become degenerate (flat); this condition may be intentionally relaxed to create a specific type of approximation space (for example, to capture convection phenomena). *Quasi-uniformity* ensures that the elements are all roughly the same size; this condition is intentionally relaxed in *adaptive methods* (see below). When both conditions are satisfied, there is a single parameter h which can be viewed as characterizing the quality of the partition. In this case, the number of nodes n is related to the parameter h as $h = \mathcal{O}(n^{-1/d})$ or $n = \mathcal{O}(h^{-d})$.

The second general requirement is on the approximation (trial) and test spaces X_h and Y_h over $\mathcal{T}_h$.

Assumption 10.5.2 (FEM2). *The restrictions of the approximation space $X_h \subset X$ and the test space $Y_h \subset Y$ to an element $K \in \mathcal{T}_h$, namely*

$$P_K = \{u_h|_K \ : \ u_h \in X_h\}, \qquad Q_K = \{v_h|_K \ : \ v_h \in Y_h\}$$

are finite-dimensional spaces consisting of pure polynomials for each $K \in \mathcal{T}_h$.

The condition **FEM2** may also be relaxed to include much more general functions, although polynomials are often the most convenient, stable, and efficient of the choices. The third general requirement is a locality requirement on a basis representation of X_h and Y_h.

Assumption 10.5.3 (FEM3). *There exist canonical or nodal bases for $X_h \subset X$ and $Y_h \subset Y$ for which the basis functions have the smallest possible (local) support and are easily described. Moreover, dim$(X_h) =$ dim(Y_h).*

Condition **FEM3** ensures the moment matrices (10.2.13) generated by Petrov-Galerkin and Galerkin methods are invertible and highly sparse, leading to efficient storage as well as efficient implementaton of iterative methods for their inversion (see Section 10.6). We can now give a formal definition of a conforming finite element method.

Definition 10.5.1 (Conforming Finite Element Method)**.** *A* conforming finite element method *is a Petrov-Galerkin method for which the finite element triangulation $\mathcal{T}_h$, approximation subspace $X_h \subset X$, and test subspace $Y_h \subset Y$ satisfy:*

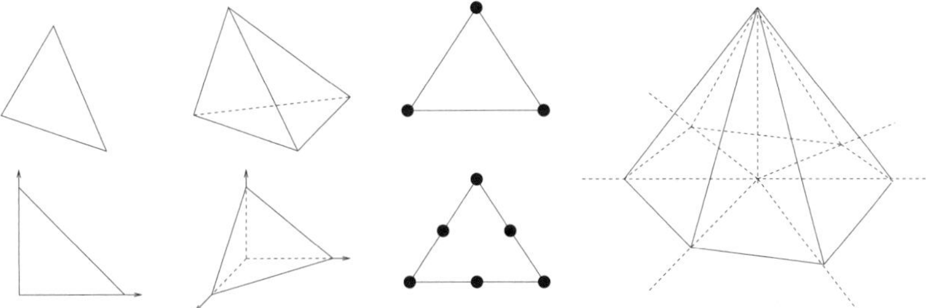

Figure 10.6 *From left to right:* Arbitrary and reference simplex elements in two and three dimensions; nodal placement for C^0 piecewise linear and quadratic elements on triangles; global piecewise linear basis function, built over a triangular mesh, showing the local support structure.

*(1) $\mathcal{T}_h$ is conforming (**FEM1**).*

*(2) For each $K \in \mathcal{T}_h$, P_K and Q_K consist of pure polynomials (**FEM2**).*

*(3) The canonical bases for X_h and Y_h are easy to describe and have local support (**FEM3**).*

Most standard finite element methods can be categorized as *conforming* in the sense of Definition 10.5.1. with possible relaxation of the conditions as described above. The distinction between different methods is in the choice of the polynomial element spaces P_K and Q_K and the imposition of global smoothness conditions between the elements. If the construction is such that $X_h \not\subset X$ and/or $Y_h \not\subset Y$, or if the Petrov-Galerkin method involves approximate forms of the variational problem (by using quadrature to evaluate integrals, using a polyhedron for building $\mathcal{T}_h$ as an approximation to a nonpolygonal Ω, or making other *variational crimes*), then the method is called a *nonconforming finite element method.*

Following [52], a *finite element* in $\mathbb{R}^d$ is the set (K, P, Σ), where

(1) $K \subset \mathbb{R}^d$ is a closed Lipschitz domain with a nonempty interior.

(2) P is a finite-dimensional space of (real-valued) functions over K.

(3) Σ is a finite linearly independent set of functionals $\{\psi_i\}_{i=1}^N$ taken from P' (the dual space of P).

The set Σ is taken to be P*-unisolvent*, meaning that given any c_i, $i = 1, \ldots, N$, there exists a unique $p \in P$ such that

$$\psi(p) = c_i, \quad 1 \leqslant i \leqslant N.$$

This implies the existence of a basis $\{p_i\}_{i=1}^N$ for P such that

$$\psi_j(p_i) = \delta_{ij}, \quad 1 \leqslant j \leqslant N,$$

giving $p = \sum_{i=1}^N \psi_i(p)p_i$, for any $p \in P$, and ensuring that $\dim(P) = \dim(\Sigma)$. The functions p_i are called the *basis functions*, and the functionals ψ_i are called the *degrees of freedom* associated with the *element K*. The terminology *finite element* is often taken to refer to the set K as well as to the entire structure (K, P, Σ).

The most convenient way to construct the spaces P_K and Q_K is often through a *reference or master element* $\tilde{K}$, such as the unit triangle and unit tetrahedron (see Figure 10.5) with vertices in $\mathbb{R}^d$ for $d = 2$ or $d = 3$:

$$\begin{aligned} \tilde{\mathbf{x}}_1 &= (0,0), \\ \tilde{\mathbf{x}}_2 &= (1,0), \\ \tilde{\mathbf{x}}_3 &= (0,1), \end{aligned} \qquad \begin{aligned} \tilde{\mathbf{x}}_1 &= (0,0,0), \\ \tilde{\mathbf{x}}_2 &= (1,0,0), \\ \tilde{\mathbf{x}}_3 &= (0,1,0), \\ \tilde{\mathbf{x}}_4 &= (0,0,1). \end{aligned}$$

With $h_{\tilde{K}}$ and $\rho_{\tilde{K}}$ defined in (10.5.1)–(10.5.2), these shapes satisfy (10.5.3) with σ a small finite constant. On $\tilde{K}$, one can easily define a *reference basis*; for example, in the case of linear functions, the sets of basis functions for $P_{\tilde{K}}$ and $Q_{\tilde{K}}$ on the reference triangle and tetrahedron are easily seen to be

$$\begin{aligned} \tilde{\phi}_1(\tilde{\mathbf{x}}) &= 1 - \tilde{x} - \tilde{y}, \\ \tilde{\phi}_2(\tilde{\mathbf{x}}) &= \tilde{x}, \\ \tilde{\phi}_3(\tilde{\mathbf{x}}) &= \tilde{y}, \end{aligned} \qquad \begin{aligned} \tilde{\phi}_1(\tilde{\mathbf{x}}) &= 1 - \tilde{x} - \tilde{y} - \tilde{z}, \\ \tilde{\phi}_2(\tilde{\mathbf{x}}) &= \tilde{x}, \\ \tilde{\phi}_3(\tilde{\mathbf{x}}) &= \tilde{y}, \\ \tilde{\phi}_4(\tilde{\mathbf{x}}) &= \tilde{z}, \end{aligned}$$

where $\tilde{\mathbf{x}} = (\tilde{x}, \tilde{y})$ when $d = 2$, and where $\tilde{\mathbf{x}} = (\tilde{x}, \tilde{y}, \tilde{z})$ when $d = 3$. These functions have the *Lagrange property*:

$$\tilde{\phi}_i(\tilde{\mathbf{x}}_j) = \delta_{ij}. \tag{10.5.5}$$

Once a basis is given on a reference element, the corresponding basis on an arbitrary element is easy to construct if the elements $K \in \mathcal{T}_h$ represent an *affine equivalent family*. This condition is that for each $K \in \mathcal{T}_h$, there exists an affine map $G_K : \tilde{K} \to K$ such that

$$\mathbf{x} = G_K(\tilde{\mathbf{x}}) = A_K \tilde{\mathbf{x}} + b_K,$$

where $A_K \in \mathbb{R}^{d\times d}$ is an invertible matrix and $b_K \in \mathbb{R}^d$. Such families of elements are all related to the reference element $\tilde{K}$ by rotation, scaling, and translation, as encoded in the affine transformation G_K. Since A_K is invertible, the map G_K is also invertible, giving $G_K^{-1} : K \to \tilde{K}$ of the form

$$\tilde{\mathbf{x}} = G_K^{-1}(\mathbf{x}) = A_K^{-1}\mathbf{x} + A_K^{-1} b_K.$$

The vertices of the element K, as well as any additional interpolation nodes added to define the basis on K, are recovered from those of $\tilde{K}$ through the map $\mathbf{x}_j^K = G_K(\tilde{\mathbf{x}}_j)$, and the basis functions $\{\phi_i^K(\mathbf{x})\}$ on K are defined through G_K as

$$\phi_i^K(\mathbf{x}) = \tilde{\phi}_i \circ G_K^{-1}(\mathbf{x}).$$

Arbitrarily high-order basis functions on each affine-equivalent element $K \in \mathcal{T}_h$ can then be constructed from such a basis on the reference element $\tilde{K}$ in precisely the same way through the map G_K. Function spaces of increasing approximation power can subsequently be built through element subdivision, resulting in highly complex meshes of elements; see Figure 10.7. These meshes may be constructed to satisfy quasi-uniformity (10.5.4) or can be made to intentionally violate this condition for producing adaptive approximation spaces (see below).

We can now consider the subspaces X_h and Y_h that we were after. In order to ensure that $X_h \subset Y$ and $Y_h \subset Y$, it is necessary to impose global continuity. Together with our elementwise construction, we have then

$$X_h = \{u_h \in C^0(\overline{\Omega}) \; : \; u_h|_K \in P_K, \forall K \in \mathcal{T}_h\},$$
$$Y_h = \{v_h \in C^0(\overline{\Omega}) \; : \; v_h|_K \in Q_K, \forall K \in \mathcal{T}_h\}.$$

To allow for homogeneous Dirichlet conditions on some portion of the boundary $\Gamma_D \subset \partial\Omega$, we can define

$$X_{0h,D} = \{u_h \in X_h \; : \; u_h = 0 \text{ on } \Gamma_D\},$$
$$Y_{0h,D} = \{v_h \in Y_h \; : \; v_h = 0 \text{ on } \Gamma_D\},$$

where we drop the subscript D on the spaces when $\Gamma_D = \partial\Omega$. The continuity enforcement allows for the following.

Theorem 10.5.1. *Assume that*

$$P_K, Q_K \subset W^{1,p}(K), \; \forall K \in \mathcal{T}_h, \quad \textit{and} \quad X_h, Y_h \subset C^0(\overline{\Omega}).$$

Then

$$X_h, Y_h \subset W^{1,p}(\Omega) \qquad \textit{and} \qquad X_{0h,D}, Y_{0h,D} \subset W^{1,p}_{0,D}(\Omega).$$

Proof. The following is a minor generalization of the argument in [52] to $W^{1,p}(\Omega)$. Let $v_h \subset X_h \subset L^p(\Omega)$. To conclude that $v_h \in W^{1,p}(\Omega)$, we must find $v_h^i \in L^p(\Omega)$ such that

$$\int_\Omega v_h^i \phi \, dx = -\int_\Omega v_h \partial_i \phi \, dx, \quad \forall \phi \in C_0^\infty(\Omega).$$

We assemble each v_h^i elementwise as $v_h^i|_K = \partial_i(v_h|_K)$. Green's formula ensures that on every element K,

$$\int_K \partial_i(v_h|_K)\phi \, dx = -\int_K v_h|_K \partial_i \phi \, dx + \int_{\partial K} (v_h|_K \cdot \mathbf{n})\phi \, ds,$$

with $\mathbf{n}$ the outward unit normal to ∂K. We now sum up the equation above over all elements K:

$$\int_\Omega \partial_i v_h^i \phi \, dx = -\int_\Omega v_h \partial_i \phi \, dx + \sum_{K \in \mathcal{T}_h} \int_{\partial K} (v_h|_K \cdot \mathbf{n})\phi \, ds.$$

Since the boundary integral above vanishes when it coincides with the boundary of Ω (where $\phi = 0$) and also when it does not (continuity of v_h globally over Ω, combined with the sign change of the unit normals for elements K_1 and K_2 sharing an internal boundary), we are left with

$$\int_\Omega \partial_i v_h^i \phi \, dx = -\int_\Omega v_h \partial_i \phi \, dx, \quad \forall \phi \in C_0^\infty(\Omega),$$

giving that $X_h \subset W^{1,p}(\Omega)$. The same argument works if we start with $v_h \in X_{0h}$, so the theorem follows. □

Therefore, we have constructed subspaces $X_h \subset X$ and $Y_h \subset Y$ with X and Y are $W^{1,p}(\Omega)$ for some p; subspaces of Sobolev classes of higher order are obtained be using higher order bases constructions. As a result, the near-best approximation results from Section 10.2 can be brought to bear; what remains is to use the specific details about how X_h is constructed in order to estimate the best approximation error.

Interpolation and A Priori Error Estimates. Due to property (10.5.5), we can define an *interpolation operator* in the reference element, $\tilde{I}\colon C^0(\tilde{K}) \to P_{\tilde{K}}$, through

$$\tilde{I}\tilde{u}(\tilde{\mathbf{x}}) = \sum_{i=1}^{p} \tilde{u}(\mathbf{x}_i)\tilde{\phi}_i(\tilde{\mathbf{x}}),$$

where p is the number of interpolation points for the given basis $\tilde{\phi}_i$. This is clearly an interpolant of $\tilde{u}$ at the nodes $\mathbf{x}_i$, since $\tilde{I}\tilde{u}(\mathbf{x}_i) = \tilde{u}(\mathbf{x}_i)$. An interpolation operator $I_K\colon C^0(K) \to P_K$ on any affine equivalent $K \in \mathcal{T}_h$ can then be constructed from basis functions constructed on K via the affine map G through $I_K u = \tilde{I}\tilde{u} \circ G_K$, giving

$$I_K u(\mathbf{x}) = \sum_{i=1}^{p} u(\mathbf{x}_i)\phi_i^K(\mathbf{x}).$$

The global interpolant $I_h\colon C^0(\overline{\Omega})) \to X_h$ over all of Ω can then be written as

$$I_h u(\mathbf{x}) = \sum_{i=1}^{n} u(\mathbf{x}_i)\phi_i(\mathbf{x}),$$

where n is the total number of nodes in Ω, with global nodes $\{\mathbf{x}_i\}_{i=1}^n$ shared by elements representing global continuity conditions, and with the *global basis functions* $\{\phi_i\}_{i=1}^n$ associated with each node defined conditionally as

$$\phi_i(\mathbf{x}) = \{\phi_j^K \ : \mathbf{x} \in K \text{ and } \mathbf{x}_i = \mathbf{x}_j^K\}.$$

In other words, to evaluate a global basis function $\phi_i(\mathbf{x})$ at a particular point $\mathbf{x} \in \overline{\Omega}$, one does the following:

(1) Locates the element $K \in \mathcal{T}_h$ containing $\mathbf{x}$. (There may be more than one such K if $\mathbf{x}$ lies on the boundary of K; in that case, pick any K containing $\mathbf{x}$.)

(2) Identifies the local node $\mathbf{x}_j^K$, and subsequently the local basis function ϕ_j^K, with which the global node $\mathbf{x}_i$ is associated.

(3) Sets $\phi_i(\mathbf{x}) = \phi_j^K(\mathbf{x})$.

In the case of piecewise linear basis functions, all nodes lie on vertices of edges of triangles (2D) or tetrahedra (3D), representing shared nodes among all elements with a common vertex. Since a linear function on an edge is completely defined by its values at the two vertices at the endpoints of the edge, the sharing of vertices ensures that piecewise linear functions are globally continuous at all vertices and edges, and subsequently, all faces in the three-dimensional case. This implies that $X_h \subset C^0(\overline{\Omega})$, as needed for Theorem 10.5.1.

Our near-best approximation results for $u_h \in X_h \subset W^{m,p}$ produced by a Petrov-Galerkin method imply that

$$\|u - u_h\|_{m,p} \leqslant C \inf_{v_h \in X_h} \|u - v_h\|_{m,p} \leqslant C\|u - I_h u\|_{m,p},$$

since $I_h u$ can be no better than the best approximation. Therefore, we can use $I_h u$ to give an upper bound on the error in the Petrov-Galerkin approximation. To estimate the interpolation error in an integer Sobolev norm, one notes that

$$\|u - I_h u\|_{m,p}^p = \sum_{K \in \mathcal{T}_h} \|u - I_K u\|_{m,p,K}^p.$$

Hence, we are left with estimating the elementwise seminorms in the sum

$$\|u - I_K u\|_{m,p,K}^p = \sum_{i=0}^{m} |u - I_K u|_{i,p,K}^p.$$

For affine equivalent elements, this can be done in an elegant way through a sequence of preliminary results. The first result characterizes the Jacobian of transformation that arises when mapping Sobolev norms between a reference element and an arbitrary element.

Lemma 10.5.1. *Let $\mathcal{T}_h$ be an affine equivalent family, let $\tilde{K}$ be the associated reference element, and let $K \in \mathcal{T}_h$ be arbitrary. If $u \in W^{m,p}(K)$ for $m \geqslant 0$ and $1 \leqslant p \leqslant \infty$, then $\tilde{u} \in W^{m,p}(\tilde{K})$, and*

$$\frac{1}{C\|A_K^{-1}\|^m |\det(A_K)|^{1/p}} |u|_{m,p,K} \leqslant |\tilde{u}|_{m,p,\tilde{K}} \leqslant \frac{C\|A_K\|^m}{|\det(A_K)|^{1/p}} |u|_{m,p,K}. \quad (10.5.6)$$

Proof. For the proof, see Theorem 3.1.2 in [52]. □

The second result characterizes the geometric relationship (size and shape) between an arbitrary affine equivalent element K and the reference element $\tilde{K}$, in terms of diameters of the circumscribing ($\tilde{h}$, h_K) and inscribing ($\tilde{\rho}$, ρ_K) spheres.

Lemma 10.5.2. *Let $\mathcal{T}_h$ be an affine equivalent family, let $\tilde{K}$ be the associated reference element, and let $K \in \mathcal{T}_h$ be arbitrary. Then*

$$\|A_K\| \leqslant \frac{h_K}{\tilde{\rho}}, \qquad \|A_K^{-1}\| \leqslant \frac{\tilde{h}}{\rho_K}, \quad (10.5.7)$$

where for $B \in \mathbb{R}^{d\times d}$, the norm $\|B\| = \sup_{0\neq x\in\mathbb{R}^d} |Bx|_2/|x|_2$ is the matrix 2-norm.

Proof. We begin with an alternative characterization of the matrix 2-norm:

$$\|A_K\| = \sup_{\substack{\tilde{\mathbf{x}}\in\mathbb{R}^d \\ \tilde{\mathbf{x}}\neq 0}} \frac{|A_K\tilde{\mathbf{x}}|_2}{|\tilde{\mathbf{x}}|_2} = \sup_{\substack{\tilde{\mathbf{x}}\in\mathbb{R}^d \\ \tilde{\mathbf{x}}=\tilde{\rho}}} \frac{|A_K\tilde{\mathbf{x}}|_2}{\tilde{\rho}}.$$

By definition of the inscribing sphere for $\tilde{K}$, for any inscribing sphere of diameter $\tilde{\rho}$ we can find (nonuniquely) two vectors $\tilde{\mathbf{y}}$ and $\tilde{\mathbf{z}}$ lying on the inscribing sphere such that $|\tilde{\mathbf{w}}|_2 = \tilde{\rho}$, where $\tilde{\mathbf{w}} = \tilde{\mathbf{y}} - \tilde{\mathbf{z}}$. We then have

$$|A_K\tilde{\mathbf{w}}|_2 = |A_K(\tilde{\mathbf{y}} - \tilde{\mathbf{z}})|_2 = |A_K\tilde{\mathbf{y}} + b_K - (A_K\tilde{\mathbf{z}} + b_K)|_2 = |\mathbf{y} - \mathbf{z}|_2 = h_K,$$

where h_K is the diameter of the circumscribing sphere for K. This proves the first inequality in (10.5.7). The second inequality follows by starting with the inverse affine transformation from K back to $\tilde{K}$, and then using the same argument. □

The second preliminary result characterizes the approximation properties of polynomial preserving operators in Sobolev norms; here $\mathcal{P}_k(\tilde{K})$ is the set of polynomials of degree k on $\tilde{K}$.

Lemma 10.5.3. *Let $\mathcal{T}_h$ be an affine equivalent family, and let $\tilde{K}$ be the associated reference element. For integers $k \geqslant 0$ and $m \geqslant 0$, and $1 \leqslant p, q \leqslant \infty$, assume that $W^{k+1,p}(\tilde{K})$ and $W^{m,q}(\tilde{K})$ have the embedding relationship*

$$W^{k+1,p}(\tilde{K}) \hookrightarrow W^{m,q}(\tilde{K}).$$

Let $\tilde{I}\colon W^{k+1,p}(\tilde{K}) \to W^{m,q}(\tilde{K})$ satisfy

$$\tilde{I}\tilde{p} = \tilde{p}, \quad \forall \tilde{p} \in \mathcal{P}_k(\tilde{K}).$$

Then there exists a constant $C = C(\tilde{I}, \tilde{K})$ such that $\forall u \in W^{k+1,p}(K)$,

$$|u - I_K u|_{m,p,K} \leqslant C(\text{meas}(\Omega))^{((1/q)-(1/p))} \frac{h_K^{k+1}}{\rho_K^m} |u|_{k+1,p,K}.$$

Proof. The proof involves using Lemmas 10.5.1 and 10.5.2, together with a standard embedding inquality involving L^p and L^q, which can be found, for example, in [2]. See Theorem 3.1.4 in [52] for a complete proof. □

Lemma 10.5.3 now leads to the result we are after, as long as the basis functions used to define our finite element capture all of the polynomials in $\mathcal{P}_k$ as needed for Lemma 10.5.3.

Theorem 10.5.2. *Let $\mathcal{T}_h$ be an affine equivalent family, and let $\tilde{K}$ be the associated reference element. For integers $k \geqslant 0$ and $m \geqslant 0$, and $1 \leqslant p, q \leqslant \infty$, assume that the following embeddings hold:*

$$W^{k+1,p}(\tilde{K}) \hookrightarrow C^s(\tilde{K}), \quad W^{k+1,p}(\tilde{K}) \hookrightarrow W^{m,q}(\tilde{K}),$$

$$\mathcal{P}(\tilde{K}) \subset P_{\tilde{K}} \subset W^{m,q}(\tilde{K}).$$

Then there exists a constant $C = C(\tilde{K}, \tilde{P}, \tilde{\Sigma})$ *such that*

$$|u - I_K u|_{m,p,K} \leqslant C(\text{meas}(\Omega))^{((1/q)-(1/p))} \frac{h_K^{k+1}}{\rho_K^m} |u|_{k+1,p,K}.$$

Proof. See Theorem 3.1.5 in [52]. □

Adaptive Finite Element Methods. The quasi-uniformity condition (10.5.4) ensures that the elements are all roughly the same size; this condition is intentionally relaxed in *adaptive methods*. This allows for minimizing the number of degrees of freedom used in the discretization to achieve a target accuracy in the finite element approximation. These methods are currently the subject of intense research in the numerical analysis community; a recent overview of the work in this area can be found in [134], which contains a large bibliography to work going back to the early 1970s. Recent work in the direction of nonlinear problems can be found in [47, 49, 99, 170].

A posteriori error estimates, or error indicators based on such estimates, are commonly used in conjunction with an adaptive mesh refinement algorithm [15, 16, 171, 172, 179] as part of an *adaptive finite element method*. Common error indicators include weak and strong residual-based indicators [15, 171], indicators based on the solution of local problems [24, 25], and indicators based on the solution of global (but linearized) adjoint or *dual* problems [68, 69]. The challenge for a numerical method is to be as efficient as possible, and *a posteriori* estimates are a basic tool in deciding which parts of the solution require additional attention. While much work on *a posteriori* estimates and indicators has been for linear problems, nonlinear extensions are possible through linearization theorems; see [171, 172]. The typical solve-estimate-refine structure in simplex-based adaptive finite element codes exploiting these *a posteriori* indicators is illustrated in Algorithm 10.5.1.

Algorithm 10.5.1 (Adaptive Finite Element Approximation).

While ($\mathcal{E}(u-u_h)$ is ''large'') do:
- (1) **SOLVE:** Find $u_h \in \bar{u}_h + X_h \subset X$ s.t. $\langle F(u_h), v\rangle = 0, \forall v \in Y_h \subset Y$.
- (2) **ESTIMATE:** $\mathcal{E}(u - u_h)$ over each element.
- (3) **MARK:**
  - (3.1) Initialize two simplex lists: $Q1 = Q2 = \emptyset$.
  - (3.2) Simplices failing error test placed on list $Q1$.
- (4) **REFINE:**
  - (4.1) Bisect simplices in $Q1$, place nonconforming on $Q2$.
  - (4.2) $Q1$ is now empty; set $Q1 = Q2$, $Q2 = \emptyset$.
  - (4.3) If $Q1$ is not empty, go to (4.1).

End While.

The conformity loop (4.1)–(4.3), required to produce a globally "conforming" mesh (described below) at the end of a refinement step, is guaranteed to terminate in a finite number of steps (see [147]), so that the refinements remain local. Element shape is crucial for approximation quality, and there are effectively two known

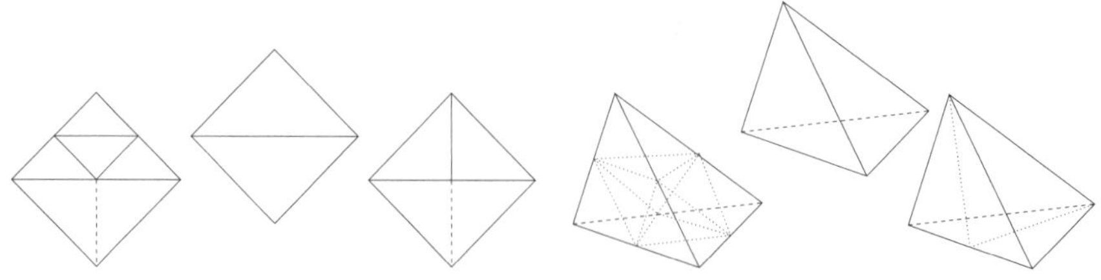

Figure 10.7 Simplex quadrasection supplemented by bisection, versus pure bisection, in two dimensions; octasection versus bisection in three dimensions. See [6, 26, 30, 138, 190].

nondegenerate subdivision algorithms for simplices: *bisection*, whereby each simplex is subdivided into two (in either two or three dimensions), and *octasection* or *quadrasection*, with each simplex subdivided into eight (three dimensions) or four (two dimensions) children. Unlike the two-dimensional case where quadrasection produces four child triangles that are similar to the parent, octasection produces four similar children (the cut corners), and then four additional children which arise from splitting the remaining interior octahedron. There are three distinct ways to split the inner octahedron, and one must choose one specific diagonal of the three for the split to ensure nondegeneracy [30, 138, 190]. Nondegenerate families are also produced in two dimensions if *longest edge bisection* in used [148, 163]; whether longest edge bisection is nondegenerate in three dimensions apparently remains an open question. However, *marked edge bisection*, an algorithm which monitors three successive generations of simplices produced by bisection of an original parent and then chooses a specific bisection edge, was finally shown to be nondegenerate in the 1990s; see [6, 26, 123, 128, 133]. Figure 10.7 shows a single subdivision of a 2-simplex or a 3-simplex using either 4-section (leftmost figure), 8-section (fourth figure from the left), or bisection (third figure from the left and the rightmost figure).

The paired triangle in the 2-simplex case of Figure 10.7 illustrates the nature of conformity and its violation during refinement. A globally conforming simplex mesh is defined as a collection of simplices which meet only at vertices and faces; for example, removing the dotted bisection in the third group from the left in Figure 10.7 produces a nonconforming mesh. Nonconforming simplex meshes create several theoretical as well as practical implementation difficulties; while the queue swapping presented in Algorithm 10.5.1 is a feature specific to FETK [75], an equivalent approach is taken in PLTMG [16] and similar packages [28, 29, 31, 133].

A General Family of Coupled Elliptic Problems. In order to derive an *a posteriori* error estimate for the Petrov-Galerkin approximation of a class of problems, we describe in some detail a general family of coupled nonlinear elliptic problems.

Let $\Omega \subset \mathbb{R}^d$ be a Lipschitz domain. The boundary $(d-1)$-submanifold $\partial\Omega$ is assumed to be formed from two disjoint submanifolds Γ_D and Γ_N satisfying

$$\Gamma_D \cup \Gamma_N = \partial\Omega, \qquad \Gamma_D \cap \Gamma_N = \emptyset. \tag{10.5.8}$$

When convenient in the discussions below, one of the two submanifolds Γ_D or Γ_N may be allowed to shrink to zero measure, leaving the other to cover $\partial\Omega$. We will use standard Cartesian tensor notation below to represent coupled systems of partial differential equations in a concise way; see [126, 157]. For example, we will use the concise notation $V^a_{;b} = \partial V^a / \partial x^b$ to represent partial differentiation of scalars, vectors, and tensors. A product containing repeated indices, such as $V^a W_a$, represents an implied sum (Einstein summation convention), with the sum running from 1 to d unless otherwise noted. For example, the divergence of a vector field will be written $V^a_{;a}$.

Consider now a general second-order elliptic system of (tensor) equations in strong divergence form over Ω:

$$-A^{ia}(x^b, u^j, u^k_{;c})_{;a} + B^i(x^b, u^j, u^k_{;c}) = 0 \text{ in } \Omega, \tag{10.5.9}$$

$$A^{ia}(x^b, u^j, u^k_{;c}) n_a + C^i(x^b, u^j, u^k_{;c}) = 0 \text{ on } \Gamma_N, \tag{10.5.10}$$

$$u^i(x^b) = E^i(x^b) \text{ on } \Gamma_D, \tag{10.5.11}$$

where

$$1 \leqslant a, b, c \leqslant d, \quad 1 \leqslant i, j, k \leqslant n,$$

$$A : \Omega \times \mathbb{R}^n \times \mathbb{R}^{nd} \to \mathbb{R}^{nd}, \quad B : \Omega \times \mathbb{R}^n \times \mathbb{R}^{nd} \to \mathbb{R}^n,$$

$$C : \Gamma_N \times \mathbb{R}^n \times \mathbb{R}^{nd} \to \mathbb{R}^n, \quad E : \Gamma_D \to \mathbb{R}^n.$$

The first term in (10.5.9) represents the divergence of a two-index tensor (an implied sum on the differentiation index a). The divergence-form system (10.5.9)–(10.5.11), together with the boundary conditions, can be viewed as an operator equation of the form

$$F(u) = 0, \quad F : X \to Y', \tag{10.5.12}$$

for some Banach spaces X and Y, where Y' denotes the dual space of Y. The system (10.5.9)–(10.5.11) captures a very general class of coupled systems of elliptic equations. A standard example is the single scalar equation ($n = 1$) with $\Gamma_D = \partial\Omega$ and with

$$A^{ia}(x^b, u^j, u^k_{;c}) = \epsilon \nabla u(x), \quad B^i(x^b, u^j, u^k_{;c}) = -f(x), \quad E^i(x^b) = 0,$$

which leads to simply the Poisson equation with zero Dirichlet conditions:

$$-\nabla \cdot (\epsilon \nabla u) = f, \text{ in } \Omega, \qquad u = 0 \text{ on } \partial\Omega.$$

Other examples of (10.5.9)–(10.5.11) include semi and quasilinear scalar equations and elliptic systems, and coupled systems of such equations. Our interest here is primarily in coupled systems of one or more scalar field equations and one or more d-vector field equations. The unknown n-vector u^i then in general consists of n_s scalars and n_v d-vectors, so that $n = n_s + n_v \cdot d$. To allow the n-component system (10.5.9)–(10.5.11) to be treated notationally as if it were a single n-vector equation, it will be convenient to introduce the following notation for the unknown vector u^i and for a "metric" of the product space of scalar and vector components of u^i:

$$\mathcal{G}_{ij} = \begin{bmatrix} g_{ab}^{(1)} & & 0 \\ & \ddots & \\ 0 & & g_{ab}^{(n_e)} \end{bmatrix}, \qquad u^i = \begin{bmatrix} u^a_{(1)} \\ \vdots \\ u^a_{(n_e)} \end{bmatrix}, \qquad n_e = n_s + n_v. \tag{10.5.13}$$

If $u^a_{(k)}$ is a d-vector we take $g_{ab}^{(k)} = \delta_{ab}$; if $u^a_{(k)}$ is a scalar, we take $g_{ab}^{(k)} = 1$.

Denoting the outward unit normal to $\partial\Omega$ as n_b, recall the Divergence Theorem for a vector field w^b on Ω (see [119]):

$$\int_\Omega w^b{}_{;b}\, dx = \int_{\partial\Omega} w^b n_b\, ds. \tag{10.5.14}$$

Making the choice $w^b = u_{a_1\ldots a_k} v^{a_1\ldots a_k b}$ in (10.5.14) and forming the divergence $w^b{}_{;b}$ by applying the product rule leads to a useful integration-by-parts formula for certain contractions of tensors:

$$\begin{aligned}\int_\Omega u_{a_1\ldots a_k} v^{a_1\ldots a_k b}{}_{;b}\, dx = &\int_{\partial\Omega} u_{a_1\ldots a_k} v^{a_1\ldots a_k b} n_b\, ds \\ &- \int_\Omega v^{a_1\ldots a_k b} u_{a_1\ldots a_k;b}\, dx.\end{aligned} \tag{10.5.15}$$

When $k = 0$ this reduces to the familiar case where u and v are scalars.

The weak form of (10.5.9)–(10.5.11) is obtained by taking the L^2-based duality pairing between a vector v^j (vanishing on Γ_D) lying in a product space of scalars and tensors, and the residual of the tensor system (10.5.9), yielding

$$\int_\Omega \mathcal{G}_{ij} \left(B^i - A^{ia}{}_{;a}\right) v^j\, dx = 0. \tag{10.5.16}$$

Due to the definition of $\mathcal{G}_{ij}$ in (10.5.13), this is simply a sum of integrals of scalars, each of which is a contraction of the type appearing on the left side in (10.5.15). Then using (10.5.15) and (10.5.10) together in (10.5.16), and recalling that $v^i = 0$ on Γ_D satisfying (10.5.8), yields

$$\int_\Omega \mathcal{G}_{ij} A^{ia} v^j{}_{;a}\, dx + \int_\Omega \mathcal{G}_{ij} B^i v^j\, dx + \int_{\Gamma_N} \mathcal{G}_{ij} C^i v^j\, ds = 0. \tag{10.5.17}$$

Equation (10.5.17) leads to a weak formulation of the problem:

$$\text{Find } u \in \bar{u} + X \text{ s.t. } \langle F(u), v\rangle = 0, \quad \forall\, v \in Y, \tag{10.5.18}$$

for suitable Banach spaces of functions X and Y, where the nonlinear weak form $\langle F(\cdot), \cdot\rangle$ can be written as

$$\langle F(u), v\rangle = \int_\Omega \mathcal{G}_{ij}(A^{ia} v^j{}_{;a} + B^i v^j)\, dx + \int_{\Gamma_N} \mathcal{G}_{ij} C^i v^j\, ds. \tag{10.5.19}$$

Under suitable growth conditions on the nonlinearities in F, it can be shown (essentially by applying the Hölder inequality) that there exists p_k, q_k, r_k satisfying $1 < p_k, q_k, r_k < \infty$ such that the choice

$$X = W_{0,D}^{1,r_1} \times \cdots \times W_{0,D}^{1,r_{n_e}}, \quad X = W_{0,D}^{1,q_1} \times \cdots \times W_{0,D}^{1,q_{n_e}}, \tag{10.5.20}$$

$$\frac{1}{p_k} + \frac{1}{q_k} = 1, \quad r_k \geq \min\{p_k, q_k\}, \quad k = 1, \ldots, n_e, \tag{10.5.21}$$

ensures that $\langle F(u), v\rangle$ in (10.5.19) remains finite for all arguments [77].

The affine shift tensor $\bar{u}$ in (10.5.18) represents the essential or Dirichlet part of the boundary condition if there is one; the existence of $\bar{u}$ such that $E = \bar{u}|_{\Gamma_D}$ in the sense of the trace operator is guaranteed by Theorem 10.3.5 as long as E^i in (10.5.11) and Γ_D are smooth enough.

A *Petrov-Galerkin* approximation of the solution to (10.5.18) is the solution to the following subspace problem:

$$\text{Find } u_h \in \bar{u}_h + X_h \subset X \text{ s.t. } \langle F(u_h), v\rangle = 0, \ \ \forall\, v \in Y_h \subset Y, \tag{10.5.22}$$

for some chosen subspaces X_h and Y_h, where $\dim(X_h) = \dim(Y_h) = n$.

Residual A Posteriori Error Estimates. There are several approaches to adaptive error control, although the approaches based on *a posteriori* error estimation are usually the most effective and most general. While most existing work on *a posteriori* estimates has been for linear problems, extensions to the nonlinear case can be made through linearization. For example, consider the nonlinear problem in (10.5.12), which we will write as follows (ignoring the parameters for simplicity):

$$F(u) = 0, \quad F \in C^1(X, Y'), \quad X, Y \text{ Banach spaces}, \tag{10.5.23}$$

and a discretization:

$$F_h(u_h) = 0, \quad F_h \in C^0(X_h, Y_h'), \quad X_h \subset X, \quad Y_h \subset Y. \tag{10.5.24}$$

The nonlinear residual $F(u_h)$ can be used to estimate the error $\|u - u_h\|_X$, through the use of a *linearization theorem* [124, 171]. An example of such a theorem due to Verfürth is the following.

Theorem 10.5.3 ([171]). *Let $u \in X$ be a regular solution of $F(u) = 0$, so that the Gâteaux derivative $F'(u)$ is a linear homeomorphism of X onto Y'. Assume that F' is Lipschitz continuous at u, so that there exists R_0 such that*

$$\gamma = \sup_{u_h \in B(u, R_0)} \frac{\|F'(u) - F'(u_h)\|_{\mathcal{L}(X,Y')}}{\|u - u_h\|_X} < \infty.$$

Let $R = \min\{R_0, \gamma^{-1}\|[F'(u)]^{-1}\|_{\mathcal{L}(Y',X)}, 2\gamma^{-1}\|F'(u)\|_{\mathcal{L}(X,Y')}\}$. Then for all $u_h \in B(u, R)$,

$$C_1\|F(u_h)\|_{Y'} \leqslant \|u - u_h\|_X \leqslant C_2\|F(u_h)\|_{Y'}, \tag{10.5.25}$$

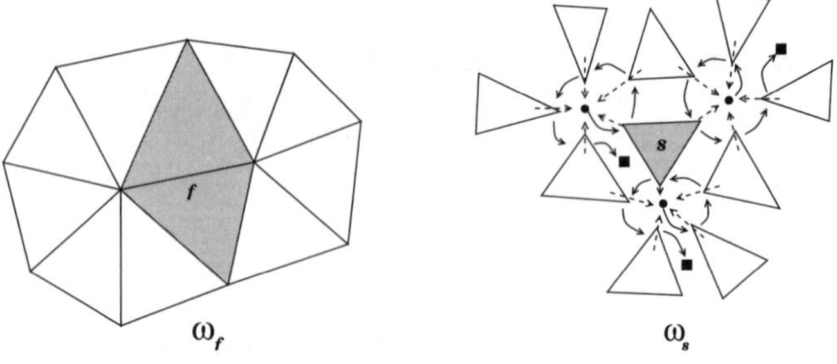

Figure 10.8 The sets ω_f and ω_s.

where $C_1 = \frac{1}{2}\|F'(u)\|^{-1}_{\mathcal{L}(X,Y')}$ *and* $C_2 = 2\|[F'(u)]^{-1}\|_{\mathcal{L}(Y',X)}$.

Proof. See [171]. □

We now derive a strong residual-based *a posteriori* error indicator for general Petrov-Galerkin approximations (10.5.22) to the solutions of general nonlinear elliptic systems of tensors of the form (10.5.9)–(10.5.11). The analysis involves primarily the weak formulation (10.5.18)–(10.5.19). Our derivation follows closely that of Verfürth [171, 172]. The starting point for our residual-based error indicator is the linearization inequality (10.5.25). In our setting of the weak formulation (10.5.18)–(10.5.19), we restrict our discussion here to a single elliptic system for a scalar or a d-vector (i.e., the product space has dimension $n_e = 1$). The linearization inequality then involves standard Sobolev norms for the spaces (10.5.20)–(10.5.21):

$$C_1\|F(u_h)\|_{W^{-1,q}(\Omega)} \leqslant \|u - u_h\|_{W^{1,r}(\Omega)} \leqslant C_2\|F(u_h)\|_{W^{-1,q}(\Omega)}, \tag{10.5.26}$$

for $1/p+1/q = 1$, $r \geqslant \min\{p, q\}$, where $W^{-1,q}(\Omega) = (W^{1,q}(\Omega))'$ denotes the dual space of bounded linear functionals on $W^{1,q}(\Omega)$. The norm of the nonlinear residual $F(\cdot)$ in the dual space of bounded linear functionals on $W^{1,q}(\Omega)$ is defined in the usual way:

$$\|F(u)\|_{W^{-1,q}(\Omega)} = \sup_{0\neq v\in W^{1,q}(\Omega)} \frac{|\langle F(u), v\rangle|}{\|v\|_{W^{1,q}(\Omega)}}. \tag{10.5.27}$$

The numerator is the nonlinear weak form $\langle F(u), v\rangle$ appearing in (10.5.19). In order to derive a bound on the weak form in the numerator we must first introduce quite a bit of notation that we have managed to avoid until now.

To begin, we assume that the domain Ω has been exactly triangulated with a set $\mathcal{S}$ of shape-regular d-simplices (the finite dimension d is arbitrary throughout this discussion). It will be convenient to introduce the following notation:

$$\begin{aligned}
\mathcal{S} &= \text{set of shape-regular simplices triangulating } \Omega, \\
\mathcal{N}(s) &= \text{union of faces in simplex set } s \text{ lying on } \partial_N \Omega, \\
\mathcal{I}(s) &= \text{union of faces in simplex set } s \text{ not in } \mathcal{N}(s), \\
\mathcal{F}(s) &= \mathcal{N}(s) \cup \mathcal{I}(s), \\
\omega_s &= \cup \{ \tilde{s} \in \mathcal{S} \mid s \cap \tilde{s} \neq \emptyset, \text{ where } s \in \mathcal{S} \}, \\
\omega_f &= \cup \{ \tilde{s} \in \mathcal{S} \mid f \cap \tilde{s} \neq \emptyset, \text{ where } f \in \mathcal{F} \}, \\
h_s &= \text{diameter (inscribing sphere) of the simplex } s, \\
h_f &= \text{diameter (inscribing sphere) of the face } f.
\end{aligned}$$

When the argument to one of the face set functions $\mathcal{N}$, $\mathcal{I}$, or $\mathcal{F}$ is, in fact, the entire set of simplices $\mathcal{S}$, we will omit the explicit dependence on $\mathcal{S}$ without danger of confusion. Referring briefly to Figure 10.8 will be convenient. The two darkened triangles on the left in the figure represent the set w_f for the face f shared by the two triangles. The clear triangles on the right represent the set w_s for the darkened triangle s in the center (the set w_s also includes the darkened triangle).

Finally, we will also need some notation to represent discontinuous jumps in function values across faces interior to the triangulation. To begin, for any face $f \in \mathcal{N}$, let n_f denote the unit outward normal; for any face $f \in \mathcal{I}$, take n_f to be an arbitrary (but fixed) choice of one of the two possible face normal orientations. Now, for any $v \in L^2(\Omega)$ such that $v \in C^0(s)\ \forall s \in \mathcal{S}$, define the *jump function*:

$$[v]_f(x) = \lim_{\epsilon \to 0^+} v(x + \epsilon n_f) - \lim_{\epsilon \to 0^-} v(x - \epsilon n_f).$$

We now begin the analysis by splitting the volume and surface integrals in equation (10.5.19) into sums of integrals over the individual elements and faces, and we then employ the divergence theorem (10.5.15) to work backward toward the strong form in each element:

$$\begin{aligned}
\langle F(u), v \rangle &= \int_\Omega \mathcal{G}_{ij} (A^{ia} v^j{}_{;a} + B^i v^j)\, dx + \int_{\partial_N \Omega} \mathcal{G}_{ij} C^i v^j\, ds \\
&= \sum_{s \in \mathcal{S}} \int_s \mathcal{G}_{ij} (A^{ia} v^j{}_{;a} + B^i v^j)\, dx + \sum_{f \in \mathcal{N}} \int_f \mathcal{G}_{ij} C^i v^j\, ds \\
&= \sum_{s \in \mathcal{S}} \int_s \mathcal{G}_{ij} (B^i - A^{ia}{}_{;a}) v^j\, dx + \sum_{s \in \mathcal{S}} \int_{\partial s} \mathcal{G}_{ij} A^{ia} n_a v^j\, ds \\
&\quad + \sum_{f \in \mathcal{N}} \int_f \mathcal{G}_{ij} C^i v^j\, ds.
\end{aligned}$$

Using the fact that (10.5.22) holds for the solution to the discrete problem, we employ the jump function and write

$$\begin{aligned}
\langle F(u_h), v\rangle &= \langle F(u_h), v - v_h\rangle \\
&= \sum_{s\in\mathcal{S}} \int_s \mathcal{G}_{ij}(B^i - A^{ia}{}_{;a})(v^j - v_h^j)\, dx \\
&\quad + \sum_{s\in\mathcal{S}} \int_{\partial s} \mathcal{G}_{ij} A^{ia} n_a (v^j - v_h^j)\, ds \\
&\quad + \sum_{f\in\mathcal{N}} \int_f \mathcal{G}_{ij} C^i (v^j - v_h^j)\, ds \\
&= \sum_{s\in\mathcal{S}} \int_s \mathcal{G}_{ij}(B^i - A^{ia}{}_{;a})(v^j - v_h^j)\, dx \\
&\quad + \sum_{f\in\mathcal{I}} \int_f \mathcal{G}_{ij} \left[A^{ia} n_a\right]_f (v^j - v_h^j)\, ds \\
&\qquad + \sum_{f\in\mathcal{N}} \int_f \mathcal{G}_{ij}(C^i + A^{ia} n_a)(v^j - v_h^j)\, ds \\
&\leqslant \sum_{s\in\mathcal{S}} \left(\|B^i - A^{ia}{}_{;a}\|_{L^p(s)} \|v^j - v_h^j\|_{L^q(s)} \right) \\
&\quad + \sum_{f\in\mathcal{I}} \left(\| \left[A^{ia} n_a\right]_f \|_{L^p(f)} \|v^j - v_h^j\|_{L^q(f)} \right) \\
&\qquad + \sum_{f\in\mathcal{N}} \left(\|C^i + A^{ia} n_a\|_{L^p(f)} \|v^j - v_h^j\|_{L^q(f)} \right),
\end{aligned} \tag{10.5.28}$$

where we have applied the Hölder inequality three times with $1/p + 1/q = 1$.

In order to bound the sums on the right, we will employ a standard tool known as a $W^{1,p}$-quasi-interpolant I_h. An example of such an interpolant is due to Scott and Zhang [156], which we refer to as the SZ-interpolant (see also Clément's interpolant in [53]). Unlike pointwise polynomial interpolation, which is not well-defined for functions in $W^{1,p}(\Omega)$ when the embedding $W^{1,p}(\Omega) \hookrightarrow C^0(\Omega)$ fails, the SZ-interpolant I_h can be constructed quite generally for $W^{1,p}$-functions on shape-regular meshes of 2- and 3-simplices. Moreover, it can be shown to have the following remarkable local approximation properties: For all $v \in W^{1,q}(\Omega)$, it holds that

$$\|v - I_h v\|_{L^q(s)} \leqslant C_s h_s \|v\|_{W^{1,q}(\omega_s)}, \tag{10.5.29}$$

$$\|v - I_h v\|_{L^q(f)} \leqslant C_f h_f^{1-1/q} \|v\|_{W^{1,q}(\omega_f)}. \tag{10.5.30}$$

For the construction of the SZ-interpolant, and for a proof of the approximation inequalities in L^p-spaces for $p \neq 2$, see [156]. A simple construction and the proof of the first inequality may also be found in the appendix of [98].

Employing now the SZ-interpolant by taking $v_h = I_h v$ in equation (10.5.28), using (10.5.29)–(10.5.30), and noting that $1 - 1/q = 1/p$, we have

$$
\begin{aligned}
\langle F(u_h), v\rangle &\leqslant \sum_{s\in\mathcal{S}} C_s h_s \|B^i - A^{ia}{}_{;a}\|_{L^p(s)} \|v^j\|_{W^{1,q}(\omega_s)} \\
&\quad + \sum_{f\in\mathcal{I}} C_f h_f^{1/p} \| \left[A^{ia} n_a\right]_f \|_{L^p(f)} \|v^j\|_{W^{1,q}(\omega_f)} \\
&\quad + \sum_{f\in\mathcal{N}} C_f h_f^{1/p} \|C^i + A^{ia} n_a\|_{L^p(f)} \|v^j\|_{W^{1,q}(\omega_f)} \\
&\leqslant \left(\sum_{s\in\mathcal{S}} C_s^p h_s^p \|B^i - A^{ia}{}_{;a}\|_{L^p(s)}^p \right. \\
&\left. + \sum_{f\in\mathcal{I}} C_f^p h_f \| \left[A^{ia} n_a\right]_f \|_{L^p(f)}^p + \sum_{f\in\mathcal{N}} C_f^p h_f \|C^i + A^{ia} n_a\|_{L^p(f)}^p \right)^{1/p} \\
&\cdot \left(\sum_{s\in\mathcal{S}} \|v^j\|_{W^{1,q}(\omega_s)}^q + \sum_{f\in\mathcal{I}} \|v^j\|_{W^{1,q}(\omega_f)}^q + \sum_{f\in\mathcal{N}} \|v^j\|_{W^{1,q}(\omega_f)}^q \right)^{1/q}, \qquad (10.5.31)
\end{aligned}
$$

where we have used the discrete Hölder inequality to obtain the last inequality.

It is not difficult to show (see [171]) that the simplex shape regularity assumption bounds the number of possible overlaps of the sets ω_s with each other, and also bounds the number of possible overlaps of the sets ω_f with each other. This makes it possible to establish the following two inequalities:

$$
\sum_{s\in\mathcal{S}} \|v^j\|_{W^{1,q}(\omega_s)}^q \leqslant D_s \|v\|_{W^{1,q}(\Omega)}^q, \qquad (10.5.32)
$$

$$
\sum_{f\in\mathcal{F}} \|v^j\|_{W^{1,q}(\omega_f)}^q \leqslant D_f \|v\|_{W^{1,q}(\Omega)}^q, \qquad (10.5.33)
$$

where D_s and D_f depend on the shape regularity constants reflecting these overlap bounds. Therefore, since $\mathcal{I} \subset \mathcal{F}$ and $\mathcal{N} \subset \mathcal{F}$, we employ (10.5.32)–(10.5.33) in (10.5.31), which gives

$$
\begin{aligned}
\langle F(u_h), v\rangle &\leqslant C_5 \|v\|_{W^{1,q}(\Omega)} \cdot \left(\sum_{s\in\mathcal{S}} h_s^p \|B^i - A^{ia}{}_{;a}\|_{L^p(s)}^p \right. \\
&\left. + \sum_{f\in\mathcal{I}} h_f \| \left[A^{ia} n_a\right]_f \|_{L^p(f)}^p + \sum_{f\in\mathcal{N}} h_f \|C^i + A^{ia} n_a\|_{L^p(f)}^p \right)^{1/p},
\end{aligned}
\qquad (10.5.34)
$$

where $C_5 = \max_{\mathcal{S},\mathcal{F}}\{C_s, C_f\} \cdot \max_{\mathcal{S},\mathcal{F}}\{D_s^{1/q}, D_f^{1/q}\}$ depends on the shape regularity of the simplices in $\mathcal{S}$.

We finally use (10.5.34) in (10.5.27) to achieve the upper bound in (10.5.26):

$$\begin{aligned}\|u-u_h\|_{W^{1,r}(\Omega)} &\leqslant C_2\|F(u_h)\|_{W^{-1,q}(\Omega)}\\ &= C_2 \sup_{0\neq v\in W^{1,q}(\Omega)} \frac{|\langle F(u_h),v\rangle|}{\|v\|_{W^{1,q}(\Omega)}}\\ &\leqslant C_2C_5\left(\sum_{s\in\mathcal{S}} h_s^p\|B^i - A^{ia}_{\ ;a}\|^p_{L^p(s)}\right.\\ &\quad+\sum_{f\in\mathcal{I}} h_f\|\left[A^{ia}n_a\right]_f\|^p_{L^p(f)}\\ &\quad\left.+\sum_{f\in\mathcal{N}} h_f\|C^i + A^{ia}n_a\|^p_{L^p(f)}\right)^{1/p}.\end{aligned} \tag{10.5.35}$$

We will make one final transformation that will turn this into a sum of elementwise error indicators that will be easier to work with in an implementation. We only need to account for the interior face integrals (which would otherwise be counted twice) when we combine the sum over the faces into the sum over the elements. This leaves us with the following

Theorem 10.5.4. *Let $u \in W^{1,r}(\Omega)$ be a regular solution of (10.5.9)–(10.5.11), or equivalently of (10.5.18)–(10.5.19). Then under the same assumptions as in Theorem 10.5.3, the following* a posteriori *error estimate holds for a Petrov-Galerkin approximation u_h satisfying (10.5.22):*

$$\|u-u_h\|_{W^{1,r}(\Omega)} \leq C\left(\sum_{s\in\mathcal{S}}\eta_s^p\right)^{1/p}, \tag{10.5.36}$$

where

$$C = 2\cdot\max_{\mathcal{S},\mathcal{F}}\{C_s, C_f\}\cdot\max_{\mathcal{S},\mathcal{F}}\{D_s^{1/q}, D_f^{1/q}\}\cdot\|[F'(u)]^{-1}\|_{\mathcal{L}(W^{-1,q},W^{1,p})},$$

and where the elementwise error indicator η_s is defined as

$$\begin{aligned}\eta_s = &\left(h_s^p\|B^i - A^{ia}_{\ ;a}\|^p_{L^p(s)} + \frac{1}{2}\sum_{f\in\mathcal{I}(s)} h_f\|\left[A^{ia}n_a\right]_f\|^p_{L^p(f)}\right.\\ &\left.+\sum_{f\in\mathcal{N}(s)} h_f\|C^i + A^{ia}n_a\|^p_{L^p(f)}\right)^{1/p}.\end{aligned} \tag{10.5.37}$$

Proof. The proof follows from (10.5.35) and the discussion above. □

The elementwise error indicator in (10.5.37) provides an error bound in the $W^{1,r}$-norm for a general nonlinear elliptic system of the form (10.5.9)–(10.5.11), with

$1/p + 1/q = 1$, $r \geqslant \min\{p, q\}$, which may be more useful than $r = p = q = 2$ in the case of some nonlinear problems; see [171]. It is possible to use a similar analysis to construct lower bounds, dual to (10.5.36); these are useful for performing unrefinement and in accessing the quality of an error-indicator, as well as in proving convergence of adaptive methods; see [134].

Finite Volume Methods

Consider now the second-order linear elliptic partial differential equation for an unknown function $u\colon \Omega \subset \mathbb{R}^d \to \mathbb{R}$, where $d \geqslant 1$:

$$- \nabla \cdot (a \nabla u) + bu = f \text{ in } \Omega \subset \mathbb{R}^d, \qquad u = g \text{ on } \Gamma = \partial\Omega, \tag{10.5.38}$$

with $a\colon \Omega \to \mathbb{R}^{d\times d}$, $b\colon \Omega \to \mathbb{R}$, $f\colon \Omega \to \mathbb{R}$, and $g\colon \Gamma \to \mathbb{R}$, where $\Omega \subset \mathbb{R}^d$ is a Lipschitz domain (bounded open subset of $\mathbb{R}^d$ with a Lipschitz boundary). We assume that the problem is strongly elliptic, in other words, that there exists $\gamma > 0$ such that for almost every $x \in \Omega$, it holds that $y^T a(x) y \geqslant \gamma |y|^2_{l^2}$, $\forall y \in \mathbb{R}^d$. We assume that $a, b \in L^\infty(\Omega)$, $f \in H^{-1}(\Omega)$, $g \in L^2(\Gamma)$, and that $b(x) \geqslant 0$, *a.e.* in Ω. In Section 10.4 we have shown that these assumptions are sufficient to prove that the problem is well-posed in the affine space $w + H^1_0(\Omega)$, where $w \in H^1(\Omega)$ is a fixed function (guaranteed to exist by Theorem 10.3.5) whose trace on Γ is g. The equivalent weak form of the problem in the Hilbert space $H^1_0(\Omega)$ is:

$$\text{Find } u \in H^1_0(\Omega) \text{ such that } A(u,v) = F(v) \quad \forall v \in H^1_0(\Omega), \tag{10.5.39}$$

where

$$A(u,v) = \int_\Omega (a\nabla u \cdot \nabla v + buv)\, dx,$$

$$F(v) = \int_\Omega fv\, dx - A(w,v) = (f,v)_{L^2(\Omega)} - A(w,v),$$

and where $g = \operatorname{tr} w$.

We will also consider the nonlinear (semilinear) equation

$$- \nabla \cdot (a \nabla u) + b(x,u) = f \text{ in } \Omega \subset \mathbb{R}^d, \qquad u = g \text{ on } \Gamma = \partial\Omega, \tag{10.5.40}$$

where we make the same assumptions as made for the problem above, except that we replace the assumption on the coefficient b in (10.5.38) with the following assumption on the (now nonlinear) coefficient function $b\colon \Omega \times \mathbb{R} \to \mathbb{R}$ appearing in (10.5.40):

$$\frac{\partial}{\partial w} b(x,w) \geqslant 0, \qquad \forall x \in \Omega,\ \forall w \in \mathbb{R},$$

which we note is also satisfied by (10.5.38) by defining $b(x,u) = b(x)u$, under the assumption $b(x) \geqslant 0$, $\forall x \in \Omega$. We have shown in Section 10.3 that examples of this type of problem are also well-posed, under subcritical growth conditions on the nonlinearity, or by exploiting maximum principles to derive *a priori* L^∞ bounds on solutions in the critical and supercritical cases.

To perform a finite volume discretization of either (10.5.38) or (10.5.40), we begin by partitioning the domain Ω into the elements or volumes τ^j such that the following list of conditions holds. Note that we are concerned here with the possibility that the functions $\{a, b, f\}$ are piecewise $C^k(\overline{\Omega})$ functions, with k sufficiently large; therefore, we assume that any coefficient discontinuities are regular and can be identified during the discretization process, so that the box boundaries can be aligned with any discontinuities in the coefficients.

- $\Omega \subset \mathbb{R}^d$ is polyhedral, $d = 1, 2$, or 3 (with $d \geqslant 3$ also possible).
- $\Omega \equiv \bigcup_{j=1}^{M} \tau^j$, where the *elements* τ^j are arbitrary hexahedra or tetrahedra.
- $\{a, b, f\}$ have discontinuities along boundaries of the τ^j.
- The union of the $l = 2^d$ corners of all the τ^j form the *nodes* x^i.
- $\{\tau^{j;i}\} \equiv \{\tau^j : x^i \in \tau^j\}$.
- $\tau^{(i)} \equiv \bigcup_j \tau^{j;i} \equiv \{\bigcup_j \tau^j : x^i \in \tau^j\}$.
- Continuity is required of $u(x)$ and of $a\nabla u \cdot \nu$ across interfaces.

We now discuss briefly the finite volume method (or *box method*), a method for discretizing interface problems which yields reliably accurate approximations. This method, in one form or another, has been one of the standard approaches for discretizing two- and three-dimensional interface problems occurring in reactor physics and reservoir simulation since the 1960s [169, 173], and is now one of the most popular methods for discretizing partial differential equations. Reasons for the popularity of these types of methods include enforcement of conservation properties, favorable properties of the resulting algebraic operators, and ease of implementation.

General Formulation of Finite Volume Methods. We begin by integrating (10.5.38) over an arbitrary $\tau^{(i)}$. Note that in many cases the underlying conservation law was originally phrased in integral form. The resulting equation is

$$-\sum_j \int_{\tau^{j;i}} \nabla \cdot (a\nabla u)\, dx + \sum_j \int_{\tau^{j;i}} bu\, dx = \sum_j \int_{\tau^{j;i}} f\, dx.$$

Using the divergence theorem, we can rewrite the first term on the left, yielding

$$-\sum_j \int_{\partial\tau^{j;i}} (a\nabla u) \cdot \nu\, ds + \sum_j \int_{\tau^{j;i}} bu\, dx = \sum_j \int_{\tau^{j;i}} f\, dx,$$

where $\partial\tau^{j;i}$ is the boundary of $\tau^{j;i}$, and ν is the unit normal to the surface of $\tau^{j;i}$.

Note that all interior surface integrals in the first term vanish, since $a\nabla u \cdot \nu$ must be continuous across the interfaces. We are left with

$$-\int_{\partial\tau^{(i)}} (a\nabla u) \cdot \nu\, ds + \sum_j \int_{\tau^{j;i}} bu\, dx = \sum_j \int_{\tau^{j;i}} f\, dx, \tag{10.5.41}$$

where $\partial\tau^{(i)}$ denotes the boundary of $\tau^{(i)}$.

Since this last relationship holds exactly in each $\tau^{(i)}$, we can use (10.5.41) to develop an approximation at the nodes x_i at the "centers" of the $\tau^{(i)}$ by employing quadrature rules and difference formulas. In particular, the volume integrals in the second two terms in (10.5.41) can be approximated with quadrature rules. Similarly, the surface integrals required to evaluate the first term in (10.5.41) can be approximated with quadrature rules, where ∇u is replaced with an approximation. Error estimates can be obtained from difference and quadrature formulas, as in Chapter 6 of [169], or more generally by analyzing the finite volume method as a special Petrov-Galerkin finite element method [23, 46, 71, 86, 112]. We will discuss this a bit further after describing the method in more detail in a simplified setting.

REMARK. This procedure is sometimes referred to as the *integral method* in one dimension, the *box method* in two dimensions, and the *finite volume* method in three dimensions, although it is standard to refer to the method in any dimension as the finite volume or box method.

Finite Volume Methods on Nonuniform Cartesian meshes. We now focus on what is arguably the most important case, $\Omega \subset \mathbb{R}^3$, but we restrict the discussion to the case that the τ^j are hexahedral elements, whose six sides are parallel to the coordinate axes. Our discussion below is valid in the general case described above, but the notation becomes much more complex, and the main ideas become harder to follow. With regard to the notation above, since we are working with $\mathbb{R}^3$, we will define $w = x_1, y = x_2, z = x_3$. We will abuse the notation p_{ijk} to mean both $p(w_i, y_j, z_k)$ as well as (at times) an approximation to $p(w_i, y_j, z_k)$; the context will make it clear which is the case.

By restricting our discussion to elements which are nonuniform Cartesian (or *axiparallel*), the spatial mesh may be characterized by the nodal points $x \in \Omega \subset \mathbb{R}^3$ as the set of points

$$x = (w, y, z) \text{ such that } \begin{cases} w \in \{w_0, w_1, \ldots, w_{I+1}\} \\ y \in \{y_0, y_1, \ldots, y_{J+1}\} \\ z \in \{z_0, z_1, \ldots, z_{K+1}\}. \end{cases}$$

Any such mesh point we denote as $x_{ijk} = (w_i, y_j, z_k)$, and we define the *mesh spacings* as

$$h_i = w_{i+1} - w_i, \qquad h_j = y_{j+1} - y_j, \qquad h_k = z_{k+1} - z_k,$$

which are not required to be equal or uniform.

To each mesh point $x_{ijk} = (w_i, y_j, z_k)$, we associate a closed three-dimensional hexahedral region $\tau^{(ijk)}$ "centered" at x_{ijk}, defined by

$$w \in \left[w_i - \frac{h_{i-1}}{2}, w_i + \frac{h_i}{2}\right], \qquad y \in \left[y_j - \frac{h_{j-1}}{2}, y_j + \frac{h_j}{2}\right],$$

$$z \in \left[y_k - \frac{h_{k-1}}{2}, z_k + \frac{h_k}{2}\right].$$

Integrating (10.5.38) over $\tau^{(ijk)}$ for each meshpoint x_{ijk} and employing the divergence theorem as above yields

$$\int_{\partial\tau^{(ijk)}} (a\nabla u)\cdot\nu\, ds + \int_{\tau^{(ijk)}} bu\, dx = \int_{\tau^{(ijk)}} f\, dx.$$

The volume integrals are now approximated with the quadrature rule:

$$\int_{\tau^{(ijk)}} p\, dx \approx p_{ijk}\left[\frac{(h_{i-1}+h_i)(h_{j-1}+h_j)(h_{k-1}+h_k)}{8}\right].$$

Assuming that the tensor a is diagonal, $a = \operatorname{diag}(a^{(11)}, a^{(22)}, a^{(33)})$, the surface integral then reduces to

$$\int_{\partial\tau^{(ijk)}} \left[a^{(11)}u_w + a^{(22)}u_y + a^{(33)}u_z\right]\cdot\nu\, ds.$$

This integral reduces further to six two-dimensional plane integrals on the six faces of the $\tau^{(ijk)}$, and are approximated by the analogous two-dimensional rule, after approximating the partial derivatives with centered differences. Introducing the notation $p_{i-1/2,j,k} = p(w_i - h_{i-1}/2, y_j, z_k)$ and $p_{i+1/2,j,k} = p(w_i + h_i/2, y_j, z_k)$, the resulting discrete equations for each u_{ijk} are

$$\begin{aligned}(A_{ijk}+B_{ijk})u_{ijk} &- A_{i-1,j,k}u_{i-1,j,k} - A_{i,j-1,k}u_{i,j-1,k} - A_{i,j,k-1}u_{i,j,k-1}\\ &- A_{i+1,j,k}u_{i+1,j,k} - A_{i,j+1,k}u_{i,j+1,k} - A_{i,j,k+1}u_{i,j,k+1}\\ &= F_{ijk}\end{aligned}$$

for $i = 1,\ldots,I,\ j = 1,\ldots,J,\ k = 1,\ldots,K,$ where

$$\begin{aligned}
F_{ijk} &= f_{ijk}\frac{(h_{i-1}+h_i)(h_{j-1}+h_j)(h_{k-1}+h_k)}{8},\\
B_{ijk} &= b_{ijk}\frac{(h_{i-1}+h_i)(h_{j-1}+h_j)(h_{k-1}+h_k)}{8},\\
A_{ijk} &= a^{(11)}_{i-1/2,j,k}\frac{(h_{j-1}+h_j)(h_{k-1}+h_k)}{4h_{i-1}}\\
&\quad + a^{(11)}_{i+1/2,j,k}\frac{(h_{j-1}+h_j)(h_{k-1}+h_k)}{4h_i}\\
&\quad + a^{(22)}_{i,j-1/2,k}\frac{(h_{i-1}+h_i)(h_{k-1}+h_k)}{4h_{j-1}}\\
&\quad + a^{(22)}_{i,j+1/2,k}\frac{(h_{i-1}+h_i)(h_{k-1}+h_k)}{4h_j}\\
&\quad + a^{(33)}_{i,j,k-1/2}\frac{(h_{i-1}+h_i)(h_{j-1}+h_j)}{4h_{k-1}}\\
&\quad + a^{(33)}_{i,j,k+1/2}\frac{(h_{i-1}+h_i)(h_{j-1}+h_j)}{4h_k}
\end{aligned}$$

and where

$$
\begin{aligned}
A_{i-1,j,k} &= -a^{(11)}_{i-1/2,j,k}\frac{(h_{j-1}+h_j)(h_{k-1}+h_k)}{4h_{i-1}},\\
A_{i+1,j,k} &= -a^{(11)}_{i+1/2,j,k}\frac{(h_{j-1}+h_j)(h_{k-1}+h_k)}{4h_{i}},\\
A_{i,j-1,k} &= -a^{(22)}_{i,j-1/2,k}\frac{(h_{i-1}+h_i)(h_{k-1}+h_k)}{4h_{j-1}},\\
A_{i,j+1,k} &= -a^{(22)}_{i,j+1/2,k}\frac{(h_{i-1}+h_i)(h_{k-1}+h_k)}{4h_{j}},\\
A_{i,j,k-1} &= -a^{(33)}_{i,j,k-1/2}\frac{(h_{i-1}+h_i)(h_{j-1}+h_j)}{4h_{k-1}},\\
A_{i,j,k+1} &= -a^{(33)}_{i,j,k+1/2}\frac{(h_{i-1}+h_i)(h_{j-1}+h_j)}{4h_{k}}.
\end{aligned}
$$

Note that since u_{ijk} is known at the six boundaries of the cube: $i=0$, $i=I+1$, $j=0$, $j=J+1$, and $k=0$, $k=K+1$, the equations at points near the boundary can be rewritten to involve only the interior unknowns. For example, for the interior points $u_{1,j,k}$ that are near the $i=0$ boundary, the equation can be written as

$$
\begin{aligned}
(A_{1jk}+B_{1jk})u_{1jk} &- A_{1,j-1,k}u_{1,j-1,k} - A_{1,j,k-1}u_{1,j,k-1}\\
&- A_{1+1,j,k}u_{1+1,j,k} - A_{1,j+1,k}u_{1,j+1,k} - A_{1,j,k+1}u_{1,j,k+1}\\
&= F_{1jk} + A_{0,j,k}u_{0,j,k},
\end{aligned}
$$

where $u_{0,j,k} = g(x_{0,j,k})$ is known due to the boundary condition (10.5.38). After employing the Dirichlet boundary conditions from (10.5.38) in this way for any equation involving a known boundary value, the set of equations above for the approximations u_{ijk} to the solution values $u(x_{ijk})$ at the nodes x_{ijk} can be written together as the single matrix equation

$$
\mathbf{AU} = \mathbf{F}, \tag{10.5.42}
$$

where $\mathbf{A} \in \mathbb{R}^{n\times n}$ and $\mathbf{U}, \mathbf{F} \in \mathbb{R}^n$. As a result of considering the nonuniform Cartesian mesh, we have the opportunity to order the unknowns u_{ijk} in the vector $\mathbf{U}$ in the *lexigraphical* or *natural ordering*. The natural ordering is defined to be the one that varies the indices i, j, k, chronologically, with i varying most rapidly and k varying least rapidly. In this case, the p-th entry of the vector U can be recovered from u_{ijk} using the following simple map:

$$
\begin{gathered}
p = I\times J\times(k-1) + I\times(j-1) + i, \qquad p = 1,\ldots,n = I\times J\times K,\\
i = 1,\ldots,I, \qquad j = 1,\ldots,J, \qquad k = 1,\ldots,K.
\end{gathered} \tag{10.5.43}
$$

The explicit entries of the vectors $U, F \in \mathbb{R}^n$ are then

$$
\mathbf{U}_p = u_{ijk}, \qquad \mathbf{F}_p = f_{ijk},
$$

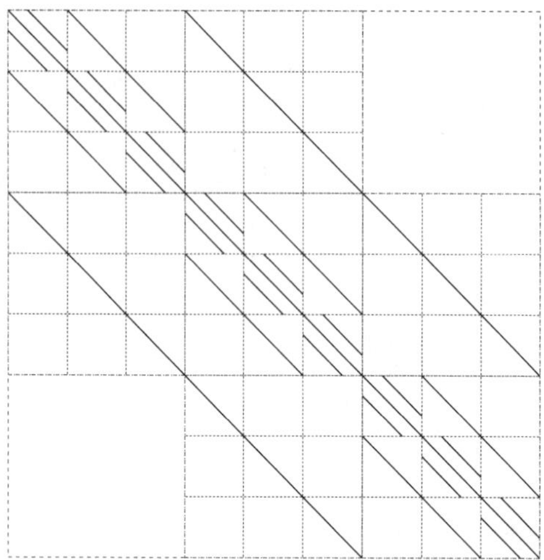

Figure 10.9 Banded matrix structure produced by the finite volume method.

where i, j, k are run through their ranges, and the map $p(i, j, k)$ is as in (10.5.43). If $g \neq 0$ in (10.5.38), then $\mathbf{F}_p = f_{ijk}$ is suitably modified to incorporate information about the boundary function g when x_{ijk} is a point near the boundary Γ. The $\mathbf{A}_{pq}$ entry of $\mathbf{A}$ uses the same map (10.5.43) for two distinct indices $p(i, j, k)$ and $q(i, j, k)$, with most entries zero except for the cases where indices passed to p are the same, or differ only by ± 1 in one or more indices from those passed to q. The result is that the matrix $\mathbf{A}$ will have seven-banded block-tridiagonal form. The banded structure produced in the case of nonuniform Cartesian meshes allows for very efficient implementations of iterative methods for numerical solution of the discrete linear and nonlinear equations; the seven-banded form is depicted in Figure 10.9 for a $3 \times 3 \times 3$ nonuniform Cartesian mesh.

In the case of the nonlinear equation (10.5.40), if we assume that the nonlinear term is *autonomous* in the sense that $b(x, u) = b(u)$, then the derivation is as above, and the resulting system of *nonlinear* algebraic equations is

$$\begin{aligned} A_{ijk}u_{ijk} + B_{ijk}(u_{ijk}) - A_{i-1,j,k}u_{i-1,j,k} - A_{i,j-1,k}u_{i,j-1,k} - A_{i,j,k-1}u_{i,j,k-1} \\ - A_{i+1,j,k}u_{i+1,j,k} - A_{i,j+1,k}u_{i,j+1,k} - A_{i,j,k+1}u_{i,j,k+1} \\ = F_{ijk}, \end{aligned}$$

where

$$b_{ijk}(u_{ijk}) = \left[\frac{(h_{i-1} + h_i)(h_{j-1} + h_j)(h_{k-1} + h_k)}{8} \right] b(u_{ijk}).$$

If the nonlinear term is not autonomous but is at least *separable* in the sense that

$$b(x, u) = \eta(x)\beta(u),$$

then the derivation is the same, except that $\beta(u)$ will replace $b(u)$ above, and an averaging of $\eta(x)$ will multiply the nonlinear term $\beta(u)$.

In either case, the set of equations above for approximations at the nodes x_{ijk} can be written together as the single *nonlinear* algebraic equation:

$$\mathbf{A}\mathbf{U} + \mathbf{B}(\mathbf{U}) = \mathbf{F}, \tag{10.5.44}$$

where $\mathbf{A} \in \mathbb{R}^{n \times n}$, $\mathbf{U}, \mathbf{F} \in \mathbb{R}^n$, and now $\mathbf{B}\colon \mathbb{R}^n \to \mathbb{R}^n$. If natural ordering is used, then again the matrix $\mathbf{A}$ representing the linear part of (10.5.44) is seven-banded and block-tridiagonal.

Properties of Linear and Nonlinear Algebraic Equations. We now wish to establish some basic mathematical properties of linear and nonlinear algebraic systems arising from finite volume discretizations of the linear equation (10.5.38) and of the nonlinear equation (10.5.40). When analyzing linear and nonlinear algebraic equations in isolation from a specific discretization, so that the discussion takes place entirely in the Hilbert space $\langle \mathbb{R}^n, (\cdot,\cdot)_{l^2}\rangle$, we will use a more transparent notation with normal rather than bold fonts, with uppercase letters representing matrices and more general mappings, and with lowercase letters representing vectors. With this notation, we will write the system (10.5.44) as

$$Au + B(u) = f, \tag{10.5.45}$$

with $A \in \mathbb{R}^{n \times n}$, $B\colon \mathbb{R}^n \to \mathbb{R}^n$, $f, u \in \mathbb{R}^n$, and write its simpler linear version as

$$Au = f. \tag{10.5.46}$$

First, we review some background material following [140] and [169].

Recall that an $n \times n$ matrix A is called *reducible* if it can be written as

$$PAP^T = \begin{bmatrix} A_{11} & A_{12} \\ 0 & A_{22} \end{bmatrix},$$

where P is a permutation matrix. If there does not exist a matrix P such that the above holds, then the matrix A is called *irreducible*. The reducibility of a matrix can be determined by examining its *finite directed graph*, which is defined as follows. With the $n \times n$ matrix A are associated the n nodes $N_i, i = 1, \ldots, n$. If the entry a_{ij} of A is nonzero, the directed graph of A contains a directed path *from* N_i *to* N_j. If there exists a directed path *from* any node i *to* any node j, the graph is called *strongly connected*. The following condition is useful.

Theorem 10.5.5. *An $n \times n$ matrix A is irreducible if and only if its directed graph is strongly connected.*

Proof. See [169]. □

Recall that the matrix A is *diagonally dominant* if

$$|a_{ii}| \geqslant \sum_{j=1, j\neq i}^{n} |a_{ij}|, \quad i = 1, \ldots, n. \tag{10.5.47}$$

Further, the matrix A is *irreducibly diagonally dominant* if A is diagonally dominant and irreducible, and if inequality (10.5.47) holds in the strict sense for at least one i. The matrix A is *strictly diagonally dominant* if (10.5.47) holds for all i in the strict sense. The following theorem follows easily from the Gerschgorin Circle Theorem.

Theorem 10.5.6. *If the $n \times n$ matrix A is either strictly or irreducibly diagonally dominant with positive diagonal entries, then A is positive definite.*

Proof. See [169]. □

Recall that a *partial ordering* of the space $\mathcal{L}(\mathbb{R}^n, \mathbb{R}^n)$ of linear operators mapping $\mathbb{R}^n$ into itself may defined in the following way. Let $A \in \mathcal{L}(\mathbb{R}^n, \mathbb{R}^n)$ and $B \in \mathcal{L}(\mathbb{R}^n, \mathbb{R}^n)$. The ordering is defined as

$$A \leqslant B \text{ if and only if } a_{ij} \leqslant b_{ij}, \quad \forall i, j.$$

In particular, we can now write expressions such as $A > 0$, meaning that all the entries of the matrix A are positive.

Two important classes of matrices often arise from the discretization of partial differential equations: *Stieltjes matrices* and *M-matrices*. First note that if A is irreducibly diagonally dominant, with positive diagonal entries and nonpositive off-diagonal entries, it can be shown that $A^{-1} > 0$. Similarly, if A is irreducible and symmetric, with nonpositive off-diagonal entries, then $A^{-1} > 0$ if and only if A is also positive definite. The matrix A is called a *Stieltjes matrix* if A is symmetric positive definite with nonpositive off-diagonal entries. If A is nonsingular, $A^{-1} > 0$, and if A has nonpositive off-diagonal entries, then A is called an *M-matrix*. Clearly, we have that if the matrix A is a Stieltjes matrix then A is also an M-matrix; in fact, a Stieltjes matrix is simply a symmetric M-matrix. As a final remark, the following result will be useful.

Theorem 10.5.7. *If A is an M-matrix, and if D a nonnegative diagonal matrix, then $A + D$ is an M-matrix, and $(A + D)^{-1} \leqslant A^{-1}$.*

Proof. See [140]. □

Consider now the nonlinear algebraic equation

$$F(u) = Au + B(u) = f, \tag{10.5.48}$$

where $A \in \mathcal{L}(\mathbb{R}^n, \mathbb{R}^n)$, and where $B(u)\colon \mathbb{R}^n \to \mathbb{R}^n$ is a nonlinear operator. We will be interested in conditions which guarantee that F is a *homeomorphism* of $\mathbb{R}^n$ onto $\mathbb{R}^n$, meaning that the mapping F is one-to-one and onto, and that both F and F^{-1} are continuous. If this is established for the mapping F, then clearly for any function f, equation (10.5.48) has a unique solution u which depends continuously on f.

We first introduce some notation and then state a useful theorem. A nonlinear operator $B = (b_1, \ldots, b_n)$ is called *diagonal* if the i-th component function b_i is a function only of u_i, where $u = (u_1, \ldots, u_n)$. The composite function F is then called *almost linear* if A is linear and B is diagonal.

The following theorem, which is a finite-dimensional version of the Global Inverse Function Theorem (Theorem 10.1.21) gives a sufficient condition for a nonlinear operator to be (at least) a homeomorphism.

Theorem 10.5.8 (Hadamard Theorem). *If the mapping $F\colon \mathbb{R}^n \to \mathbb{R}^n$ is in $C^1(\mathbb{R}^n)$, and if $\|[F'(x)]^{-1}\| \leqslant \gamma < +\infty$, $\forall x \in \mathbb{R}^n$, then F is a homeomorphism of $\mathbb{R}^n$ onto $\mathbb{R}^n$.*

Proof. See Theorem 10.1.21; a proof may be found in [140]. □

Various properties of systems generated with the finite volume method are discussed at length in [169] for one- and two-dimensional problems. In the three-dimensional case, it is immediately clear from the form of the discrete equations given in (10.5.42) that the resulting matrix is symmetric, with positive diagonal entries and nonpositive off-diagonal entries. It is also immediate from its directed graph in the case of a natural ordering (which is simply the nonuniform Cartesian mesh itself) that the resulting matrix is irreducible. Finally, if at least one Dirichlet point is specified, then strict inequality will hold in the definition of diagonal dominance for at least one equation; hence, the matrix is irreducibly diagonally dominant. This gives the following three-dimensional version of a result [169].

Theorem 10.5.9. *If the matrix A represents the discrete equations* (10.5.42) *generated by a finite volume method discretization applied to the linear problem* (10.5.38), *and at least one Dirichlet point is specified on the boundary, then A is symmetric, irreducibly diagonally dominant, and has positive diagonal entries as well as nonpositive off-diagonal entries. Therefore, A is nonsingular. In addition, A is positive definite, and is therefore a Stieltjes matrix.*

Proof. See the discussion above. □

We now consider the general nonlinear equation

$$Au + B(u) = f, \tag{10.5.49}$$

arising from the finite volume method discretization of (10.5.40) as described earlier using hexahedral nonuniform Cartesian elements. The matrix A clearly has the properties described in Theorem 10.5.9 for the linear case. If we make some simple assumptions about the nonlinear operator $B(u)$, then we have the following result (see [140]), the short proof of which we include for completeness.

Theorem 10.5.10. *If A is an M-matrix, if $B(u)$ is continuously differentiable, and if $B'(u)$ is diagonal and nonnegative for all $u \in \mathbb{R}^n$, then the composite operator $F(u) = Au + B(u)$ is a homeomorphism of $\mathbb{R}^n$ onto $\mathbb{R}^n$.*

Proof. Since B is continuously differentiable, so is the composite operator F, and $F'(u) = A + B'(u)$. For each u, since the linear operator $B'(u)$ is a nonnegative diagonal matrix, we have by Theorem 10.5.7 that the linear operator $F'(u)$ is an M-matrix, and that

$$0 < [F'(u)]^{-1} = [A + B'(u)]^{-1} \leqslant A^{-1}, \quad \forall u \in \mathbb{R}^n.$$

Thus, $\|F'(u)^{-1}\|$ is bounded uniformly. The theorem then follows by the Hadamard Theorem. □

Corollary 10.5.1. *A finite volume discretization of the nonlinear problem* (10.5.40), *using nonuniform Cartesian hexahedral elements, yields a nonlinear algebraic operator F which is a homeomorphism from $\mathbb{R}^n$ onto $\mathbb{R}^n$, so that the discrete nonlinear problem is well-posed.*

Proof. First, it is clear from the preceding section that the matrix A which arises from the discretization is a symmetric M-matrix. Now consider the nonlinear term $B(u)$ in the case of the nonlinear equation (10.5.40). Using the notation for the nonlinear term in (10.5.40) $b(w) = b(x, w) \colon \Omega \times \mathbb{R} \to \mathbb{R}$, we have that $B(u)$ has the form $B(u) = (b(u_1), \ldots, b(u_n))$. Since the nonlinearity $b(\cdot, w)$ has the property that $b(x, w) \colon \Omega \times \mathbb{R} \to \mathbb{R}$, $\partial b(x, w)/\partial w \geqslant 0$, $\forall x \in \Omega$, $\forall w \in \mathbb{R}$, and since we then have that $B'(u) = (b'(u_1), \ldots, b'(u_n))$, we must have $B'(u)$ nonnegative and diagonal for all $u \in \mathbb{R}^n$. The corollary then follows from Theorem 10.5.10. □

REMARK. We note that Corollary 10.5.1 can also be shown using either a discrete convex analysis approach or discrete monotone operator theory, or using *M-function theory* (see [140]); an M-function is the nonlinear extension of an M-matrix.

Error Analysis of the Finite Volume Method. Error estimates for the finite volume method go back at least to the 1960s, and may be found in the books of Varga [169] and Wachspress [173] (referred to as the box method in both books). For two-dimensional analogues of the finite volume methods considered above, the discretization error is $\mathcal{O}(h^2)$, where h is the (uniform) box diameter; this is the same order as low-order finite difference and finite element methods using similar mesh partitionings. Moreover, in the case of uniform meshes, it is now well-known that finite element, box, and finite difference methods are similar or even equivalent; see, for example, [23, 112] for a detailed analysis of why this is the case, and exactly when it happens. In [23], the finite volume method is recast as a Petrov-Galerkin method with piecewise linear trial functions and piecewise constant test functions; this gives access to near-best approximation results of the type appearing in Section 10.2. It was then shown in [23] that in the two-dimensional case, the finite volume method generates approximations comparable in accuracy to finite element methods with piecewise linear basis functions. A related analysis appears in [46].

There are a number of ways to build boxes into an existing simplicial subdivision in order to produce a box method; in [86], a simple condition for the choice of the box vertices was established which ensures that the discrete generalized Laplacian, where $b = 0$ and a is piecewise constant in (10.5.38), is identical for the box method and the piecewise linear finite element method. The three-dimensional case was considered by Kerkhoven [112]; error estimates are derived by considering the finite volume method to be a Petrov-Galerkin discretization, using piecewise linear trial and test functions, of a closely associated "relaxed" problem. The relaxed method is then analyzed, using a modification of Strang's First Lemma [52], as a perturbation to a standard Galerkin discretization. The error induced by moving to the relaxed problem is then analyzed. As a result, error estimates comparable to Galerkin error estimates with piecewise linear trial and test functions are obtained for the finite volume method. More general versions of the box method have since been thoroughly analyzed, including now high-order versions of the method; see [48, 72, 181].

Note one advantage of the finite volume method over finite element methods: The finite volume method applied to problems of the form (10.5.38) and (10.5.40) always produce M-matrices in all spatial dimensions; this is only the case for finite element methods in one dimension, and for a very special class of differential operators and finite element meshes in two dimensions. An M-matrix is essentially a discrete translation of a maximum principle property of the underlying elliptic operator; finite volume methods then in a sense preserve the maximum principle, whereas finite element methods generally do not (except for the special cases mentioned above). This has a large impact on error analysis, particularly for nonlinear scalar equations; *both* continuous and discrete maximum principles are the main tools for obtaining *a priori* pointwise control of any solution to a given elliptic problem, thereby giving *a priori* pointwise control of the pointwise magnitude of the nonlinearity.

In the case of semilinear equations of the type we have considered here, the discretized nonlinear term produced by a finite element discretization is nondiagonal; this is also true in the linear case with a Helmholtz-like term in the operator, which produces a nondiagonal mass matrix. The finite volume method produces only diagonal terms from the Helmholtz-like or nonlinear terms; by *mass lumping* the elements of the mass matrix onto the diagonal this property is regained for finite element methods, which can then recover the discrete maximum principle property. However, some of the other nice features of the finite element method are sacrificed by this "variational crime," such as the multigrid *variational conditions* which we describe later in the chapter when we look at multilevel solvers for discretized elliptic systems.

Discrete Elliptic Operators and Condition Numbers

Consider now the second-order linear elliptic partial differential equation for an unknown function $\hat{u}\colon \Omega \subset \mathbb{R}^d \to \mathbb{R}$, where $d \geqslant 1$:

$$-\nabla \cdot (a\nabla \hat{u}) + b\hat{u} = f \text{ in } \Omega \subset \mathbb{R}^d, \qquad \hat{u} = g \text{ on } \Gamma = \partial\Omega, \tag{10.5.50}$$

with $a\colon \Omega \to \mathbb{R}^{d\times d}$, $b\colon \Omega \to \mathbb{R}$, $f\colon \Omega \to \mathbb{R}$, and $g\colon \Gamma \to \mathbb{R}$, where $\Omega \subset \mathbb{R}^d$ is a Lipschitz domain (bounded open subset of $\mathbb{R}^n$ with a Lipschitz boundary). We assume that the problem is elliptic; in other words, that there exists $\gamma > 0$ such that for almost every $x \in \Omega$, it holds that $y^T a(x) y \geqslant \gamma |y|_{l^2}^2$, $\forall y \in \mathbb{R}^d$. We assume that $a, b \in L^\infty(\Omega)$, $f \in H^{-1}(\Omega)$, $g \in L^2(\Gamma)$, and that $b(x) \geqslant 0$, *a.e.* in Ω. We have shown in Section 10.4 that these assumptions are sufficient to prove that the problem is well-posed in the affine space $w + H_0^1(\Omega)$, where $w \in H^1(\Omega)$ is a fixed function (guaranteed to exist by Theorem 10.3.5) whose trace on Γ is g. The equivalent weak form of the problem in the Hilbert space $H_0^1(\Omega)$, for $\hat{u} = u + w$, is:

$$\text{Find } u \in H_0^1(\Omega) \text{ such that } A(u,v) = F(v) \quad \forall v \in H_0^1(\Omega), \tag{10.5.51}$$

where:

$$A(u,v) = \int_\Omega (a\nabla u \cdot \nabla v + buv)\, dx,$$

$$F(v) = \int_\Omega fv\, dx - A(w,v) = (f,v)_{L^2(\Omega)} - A(w,v),$$

and where $g = \operatorname{tr} w$. In this section we briefly discuss the properties of discrete equations produced by finite element and related discretizations of this problem.

When problem (10.5.50) is discretized with either the finite element method or, for example, the box method, the following matrix problem is generated:

$$\text{Find } u_k \in \mathcal{U}_k \text{ such that } A_k u_k = f_k, \tag{10.5.52}$$

where k represents an index into a hierarchical family of discretizations obtained by mesh refinement. Here A_k is an SPD matrix, and the solution space $\mathcal{U}_k$ is the space of *grid functions*, defined as

$$\mathcal{U}_k = \{u_k \in \mathbb{R}^{n_k} \ : \ u_k(x_k^i) \in \mathbb{R},\ \forall x_k^i \in \Omega_k\},$$

where Ω_k represents the set of nodes x_k^i in Ω. The space $\mathcal{U}_k$ (which is simply $\mathbb{R}^{n_k}$) is a finite-dimensional Hilbert space when equipped with the inner product and norm:

$$(u_k, v_k)_k = h_k^d \sum_{i=1}^{n_k} u_k(x_k^i) v_k(x_k^i), \quad \|u_k\|_k = (u_k, u_k)_k^{1/2}, \quad \forall u_k, v_k \in \mathcal{U}_k. \tag{10.5.53}$$

The A_k-inner product and A_k-norm in $\mathcal{U}_k$ are defined by

$$(u_k, v_k)_{A_k} = (A_k u_k, v_k)_k, \quad \|u_k\|_{A_k} = (u_k, u_k)_{A_k}^{1/2}, \quad \forall u_k, v_k \in \mathcal{U}_k. \tag{10.5.54}$$

It is well-known that for finite element discretization satisfying assumptions (10.5.3) and (10.5.4), the eigenvalues and condition number of the matrix which arises can be bounded by

$$\lambda_{\min}(A_k) \geqslant C_1 h_k^d, \qquad \lambda_{\max}(A_k) \leqslant C_2 h_k^{d-2}, \tag{10.5.55}$$

$$\kappa(A_k) = \frac{\lambda_{\max}(A_k)}{\lambda_{\min}(A_k)} \leqslant \left(\frac{C_2}{C_1}\right) h_k^{-2}, \tag{10.5.56}$$

where a division or multiplication of the matrix (and hence the eigenvalue bounds above) by various powers of h_k is common. For the derivation of these bounds, see for example, [12, 161] for the finite element case.

In the case of the finite element method, the discrete problem can also be interpreted abstractly, which is useful for modern multilevel convergence theories. To begin, the finite element space of C^0-piecewise linear functions defined over the tessellation $\mathcal{T}_k$ at level k is denoted by

$$\mathcal{V}_k = \{u_k \in H_0^1(\Omega) \ : \ u_k|_{\tau_k^i} \in \mathcal{P}_1(\tau_k^i), \ \forall \tau_k^i \in \mathcal{T}_k\},$$

where $\mathcal{P}_1(\tau)$ is the space of polynomials of degree one over τ. The space $\mathcal{V}_k$ is a finite-dimensional Hilbert space when equipped with the inner product and norm in $H_0^1(\Omega)$. If $u_k \in \mathcal{V}_k$, and $\tilde{u}_k \in \mathcal{U}_k$ such that $u_k = \sum_{i=1}^{n_k} \tilde{u}_k(x_k^i)\phi_k^i$ for a Lagrangian basis $\{\phi_k^i\}$, then it can be shown (see [18, 34]) that the discrete norm defined in equation (10.5.53) is equivalent to the L^2-norm in $\mathcal{V}_k$ in the sense that

$$C_1\|u_k\|_{L^2(\Omega)} \leqslant \|\tilde{u}_k\|_k \leqslant C_2\|u_k\|_{L^2(\Omega)}, \quad \forall u_k = \sum_{i=1}^{n_k} \tilde{u}_k(x_k^i)\phi_k^i \in \mathcal{V}_k, \quad \tilde{u}_k \in \mathcal{U}_k.$$

The finite element approximation to the solution of the partial differential equation (10.5.51) has the form:

$$\text{Find } u_k \in \mathcal{V}_k \text{ s.t. } A(u_k, v_k) = (f, v_k)_{L^2(\Omega)} - A(w_k, v_k) \ \forall v_k \in \mathcal{V}_k. \tag{10.5.57}$$

Since $A(\cdot,\cdot)$ is a bilinear form on the finite-dimensional space $\mathcal{V}_k$, by Theorem 8.5.1 there exists a bounded linear operator $A_k \colon \mathcal{V}_k \to \mathcal{V}_k$ such that:

$$(A_k u_k, v_k)_{L^2(\Omega)} = A(u_k, v_k), \qquad \forall u_k, v_k \in \mathcal{V}_k.$$

If we denote the L^2-projection onto $\mathcal{V}_k$ as

$$Q_k \colon L_2(\Omega) \to \mathcal{V}_k, \quad (Q_k f, v_k)_{L^2(\Omega)} = (f, v_k)_{L^2(\Omega)}, \quad \forall f \in L^2(\Omega), v_k \in \mathcal{V}_k,$$

then problem (10.5.57) is equivalent to the abstract problem:

$$\text{Find } u_k \in \mathcal{V}_k \text{ such that } A_k u_k = f_k, \quad \text{where } f_k = Q_k f - A_k w_k. \tag{10.5.58}$$

The operator A_k is an abstract (SPD) operator on $\mathcal{V}_k$, and its representation with respect to the set of finite element basis functions $\{\phi_k^i\}$ is the stiffness matrix with entries $[\mathbf{A}_k]_{ij} = A(\phi_k^j, \phi_k^i)$.

Bounds for the maximum and minimum eigenvalues of the abstract operator A_k can be derived, analogous to those in equation (10.5.56) for the matrices arising in finite element and related discretizations, using a well-known inverse inequality available in the finite element literature. To begin, we first note that if the elements $\mathcal{T}_k$ satisfy the shape-regular and quasi-uniform assumptions (10.5.3) and (10.5.4), then the following *inverse inequality* can be derived for the resulting space $\mathcal{V}_k$ (see [52]):

$$\|u_k\|_{H^1(\Omega)} \leqslant \gamma h_k^{-1}\|u_k\|_{L^2(\Omega)}, \quad \forall u_k \in \mathcal{V}_k. \tag{10.5.59}$$

This inequality can be combined with the usual boundedness and coerciveness conditions on the underlying bilinear form to prove the next result; the bound for $\lambda_{\max}$ is given in [34], and a more careful derivation of both upper and lower bounds on each eigenvalue is given in [177].

Theorem 10.5.11. *The following bounds hold for the abstract discrete operator A_k:*

$$\lambda_{\min}(A_k) \geqslant C_1, \qquad \lambda_{\max}(A_k) \leqslant C_2 h_k^{-2}, \qquad \kappa(A_k) \leqslant \left(\frac{C_2}{C_1}\right) h_k^{-2}.$$

Proof. Given the usual boundedness condition on the bilinear form defined by A_k:

$$A(u_k, v_k) = (A_k u_k, v_k)_{L^2(\Omega)} \leqslant M \|u_k\|_{H^1(\Omega)} \|v_k\|_{H^1(\Omega)}, \quad \forall u_k, v_k \in \mathcal{V}_k,$$

we can combine this with the inverse inequality (10.5.59) to bound the largest eigenvalue of the operator A_k:

$$\begin{aligned}
\lambda_{\max}(A_k) &= \max_{u_k \neq 0} \frac{(A_k u_k, u_k)_{L^2(\Omega)}}{(u_k, u_k)_{L^2(\Omega)}} \leqslant \max_{u_k \neq 0} \frac{M \|u_k\|^2_{H^1(\Omega)}}{\|u_k\|^2_{L^2(\Omega)}} \\
&\leqslant \max_{u_k \neq 0} \frac{M \gamma^2 h_k^{-2} \|u_k\|^2_{L^2(\Omega)}}{\|u_k\|^2_{L^2(\Omega)}} \leqslant M \gamma^2 h_k^{-2} = C_2 h_k^{-2}.
\end{aligned}$$

Similarly, the usual coerciveness condition on the bilinear form defined by A_k is given as

$$A(u_k, u_k) = (A_k u_k, u_k)_{L^2(\Omega)} \geqslant m \|u_k\|^2_{H^1(\Omega)}, \quad \forall u_k \in \mathcal{V}_k,$$

which can be used directly to yield a bound on the smallest eigenvalue of A_k:

$$\begin{aligned}
\lambda_{\min}(A_k) &= \min_{u_k \neq 0} \frac{(A_k u_k, u_k)_{L^2(\Omega)}}{(u_k, u_k)_{L^2(\Omega)}} \geqslant \min_{u_k \neq 0} \frac{m \|u_k\|^2_{H^1(\Omega)}}{\|u_k\|^2_{L^2(\Omega)}} \\
&= \min_{u_k \neq 0} \frac{m(\|u_k\|^2_{L^2(\Omega)} + |u_k|^2_{H^1(\Omega)})}{\|u_k\|^2_{L^2(\Omega)}} \geqslant \min_{u_k \neq 0} \frac{m \|u_k\|^2_{L^2(\Omega)}}{\|u_k\|^2_{L^2(\Omega)}} \geqslant m = C_1.
\end{aligned}$$

The final result follows immediately since

$$\kappa(A_k) = \frac{\lambda_{\max}(A_k)}{\lambda_{\min}(A_k)} \leqslant \left(\frac{C_2}{C_1}\right) h_k^{-2}.$$

□

To conclude, we see that the discrete approximation to problem (10.5.50) can be characterized as the solution to the operator equation:

$$\text{Find } u_k \in \mathcal{H}_k \text{ such that } A_k u_k = f_k, \tag{10.5.60}$$

for some SPD $A_k \in \mathcal{L}(\mathcal{H}_k, \mathcal{H}_k)$ and some finite-dimensional Hilbert space $\mathcal{H}_k$, where $\mathcal{H}_k$ can be interpreted as the grid function space $\mathcal{U}_k$, in which case this is a matrix equation generated by finite element or similar discretization of (10.5.50), or $\mathcal{H}_k$ may be interpreted as the finite element function space $\mathcal{V}_k$, in which case this is an abstract operator equation. In either case, we have seen above that with appropriate assumptions (10.5.3) and (10.5.4) on the underlying mesh $\mathcal{T}_k$, we can relate the maximal and minimal eigenvalues of the resulting matrix or abstract operator A_k to the mesh parameter h_k.

EXERCISES

10.5.1 Use Taylor expansion (Theorem 10.1.2) to give a proof of Theorem 10.5.3.

10.5.2 Give an alternative to Theorem 10.5.3 when $F\colon X \to Y$ has the additional structure

$$F(u) = Au + B(u),$$

where A has the maximum principle property and B is monotone increasing (see Section 10.1).

10.5.3 Use the general residual indicator given by Theorem 10.5.4 to derive a residual indicator for

$$-\nabla \cdot (\epsilon \nabla u) = f \text{ in } \Omega, \quad u = 0 \text{ on } \partial\Omega, \quad \epsilon > 0.$$

10.5.4 Use the general residual indicator given by Theorem 10.5.4 to derive a residual indicator for

$$-\nabla \cdot (\epsilon \nabla u) + bu = f \text{ in } \Omega, \quad \epsilon \nabla u \cdot n = g \text{ on } \partial\Omega, \quad \epsilon, b > 0.$$

10.6 ITERATIVE METHODS FOR DISCRETIZED LINEAR EQUATIONS

In this section we give a survey of classical and modern techniques for iterative solution of linear systems involving matrices arising from any of the discretization techniques considered earlier in this chapter. Our focus will be primarily on fast (optimal or nearly optimal complexity) linear solvers based on multilevel and domain decomposition methods. Our goal here is to develop a basic understanding of the structure of modern optimal and near-optimal complexity methods based on space and/or frequency decompositions, including domain decomposition and multilevel methods. To this end, we first review some basic concepts and tools involving self-adjoint linear operators on a finite-dimensional Hilbert space. The results required for the analysis of linear methods, as well as conjugate gradient methods, are summarized. We then develop carefully the theory of classical linear methods for operator equations. The conjugate gradient method is then considered, and the relationship between the convergence rate of linear methods as preconditioners and the convergence rate of the resulting preconditioned conjugate gradient method is explored in some detail. We then consider linear two-level and multilevel methods as recursive algorithms, and examine various forms of the error propagator that have been key tools for unlocking a complete theoretical understanding of these methods over the last 20 years.

Since our focus has now turned to linear (and in Section 10.7, nonlinear) algebraic systems in finite-dimensional spaces, a brief remark about notation is in order. When we encountered a sequence in a general Banach space X earlier in the chapter, we used a fairly standard notation to denote the sequence, $\{u_j\}_{j=1}^{\infty}$, with j the sequence index. Now that we will be working entirely with sequences in finite-dimensional spaces, it is standard to use a subscript to refer to a particular component of a vector in $\mathbb{R}^n$. Moreover, it will be helpful to use a subscript on a matrix or vector to refer to a particular discrete space when dealing with multiple spaces. Therefore, rather than keep track of three distinct subscripts when we encounter sequences of vectors in multiple discrete spaces, we will place the sequence index as a superscript, for example, $\{u^j\}_{j=1}^{\infty}$. There will be no danger of confusion with the exponentiation operator, as this convention is only used on vectors in a finite-dimensional vector space analogous to $\mathbb{R}^n$. When encountering a sequence of real numbers, such as the coefficients in an expansion of a finite-dimensional basis $\{u^j\}_{j=1}^{n}$, we will continue to denote the sequence using subscripts for the index, such as $\{c_j\}_{j=1}^{n}$. The expression for the expansion would then be $u = \sum_{j=1}^{n} c_j u^j$.

Linear Iterative Methods

When finite element, wavelet, spectral, finite volume, or other standard methods are used to discretize the second-order linear elliptic partial differential equation $Au = f$, a set of linear algebraic equations results, which we denote as

$$A_k u_k = f_k. \tag{10.6.1}$$

The subscript k denotes the discretization level, with larger k corresponding to a more refined mesh, and with an associated mesh parameter h_k representing the diameter of the largest element or volume in the mesh Ω_k. For a self-adjoint strongly elliptic partial differential operator, the matrix A_k produced by finite element and other discretizations is SPD. In this section we are primarily interested in linear iterations for solving the matrix equation (10.6.1) which have the general form

$$u_k^{i+1} = (I - B_k A_k) u_k^i + B_k f_k, \tag{10.6.2}$$

where B_k is an SPD matrix approximating A_k^{-1} in some sense. The classical stationary linear methods fit into this framework, as well as domain decomposition methods and multigrid methods. We will also make use of nonlinear iterations such as the conjugate gradient method, but primarily as a way to improve the performance of an underlying linear iteration.

Linear Operators, Spectral Bounds, and Condition Numbers. We briefly compile some material on self-adjoint linear operators in finite-dimensional spaces which will be used throughout the section. (See Chapters 4 and 5 for a more lengthy and more general exposition.) Let $\mathcal{H}$, $\mathcal{H}_1$, and $\mathcal{H}_2$ be real finite-dimensional Hilbert spaces equipped with the inner product $(\cdot,\cdot)$ inducing the norm $\|\cdot\| = (\cdot,\cdot)^{1/2}$. Since we are concerned only with finite-dimensional spaces, a Hilbert space $\mathcal{H}$ can be thought of as the Euclidean space $\mathbb{R}^n$; however, the preliminary material below and the algorithms we develop are phrased in terms of the unspecified space $\mathcal{H}$, so that the algorithms may be interpreted directly in terms of finite element spaces as well.

If the operator $A\colon \mathcal{H}_1 \to \mathcal{H}_2$ is linear, we denote this as $A \in \mathcal{L}(\mathcal{H}_1, \mathcal{H}_2)$. The *(Hilbert) adjoint* of a linear operator $A \in \mathcal{L}(\mathcal{H}, \mathcal{H})$ with respect to $(\cdot, \cdot)$ is the unique operator A^T satisfying $(Au, v) = (u, A^T v)$, $\forall u, v \in \mathcal{H}$. An operator A is called *self-adjoint* or *symmetric* if $A = A^T$; a self-adjoint operator A is called *positive definite* or simply *positive* if $(Au, u) > 0, \forall u \in \mathcal{H},\ u \neq 0$.

If A is self-adjoint positive definite (SPD) with respect to $(\cdot, \cdot)$, then the bilinear form $A(u, v) = (Au, v)$ defines another inner product on $\mathcal{H}$, which we sometimes denote as $(\cdot, \cdot)_A = A(\cdot, \cdot)$ to emphasize the fact that it is an inner product rather than simply a bilinear form. The A-inner product then induces the A-norm in the usual way: $\|\cdot\|_A = (\cdot, \cdot)_A^{1/2}$. For each inner product the Cauchy-Schwarz inequality holds:

$$|(u, v)| \leqslant (u, u)^{1/2}(v, v)^{1/2}, \qquad |(u, v)_A| \leqslant (u, u)_A^{1/2}(v, v)_A^{1/2}, \qquad \forall u, v \in \mathcal{H}.$$

The adjoint of an operator M with respect to $(\cdot, \cdot)_A$, the *A-adjoint*, is the unique operator M^* satisfying $(Mu, v)_A = (u, M^* v)_A, \forall u, v \in \mathcal{H}$. From this definition it follows that

$$M^* = A^{-1} M^T A. \tag{10.6.3}$$

An operator M is called *A-self-adjoint* if $M = M^*$, and it is called *A-positive* if $(Mu, u)_A > 0, \forall u \in \mathcal{H}, u \neq 0$.

If $N \in \mathcal{L}(\mathcal{H}_1, \mathcal{H}_2)$, then the adjoint satisfies $N^T \in \mathcal{L}(\mathcal{H}_2, \mathcal{H}_1)$ and relates the inner products in $\mathcal{H}_1$ and $\mathcal{H}_2$ as follows:

$$(Nu, v)_{\mathcal{H}_2} = (u, N^T v)_{\mathcal{H}_1}, \quad \forall u \in \mathcal{H}_1, \quad \forall v \in \mathcal{H}_2.$$

Since it is usually clear from the arguments which inner product is involved, we shall drop the subscripts on inner products (and norms) throughout the section, except when necessary to avoid confusion.

For the operator M we denote the eigenvalues satisfying $Mu_i = \lambda_i u_i$ for eigenfunctions $u_i \neq 0$ as $\lambda_i(M)$. The spectral theory for self-adjoint linear operators states that the eigenvalues of the self-adjoint operator M are real and lie in the closed interval $[\lambda_{\min}(M), \lambda_{\max}(M)]$ defined by the Rayleigh quotients:

$$\lambda_{\min}(M) = \min_{u \neq 0} \frac{(Mu, u)}{(u, u)}, \qquad \lambda_{\max}(M) = \max_{u \neq 0} \frac{(Mu, u)}{(u, u)}.$$

Similarly, if an operator M is A-self-adjoint, then the eigenvalues are real and lie in the interval defined by the Rayleigh quotients generated by the A-inner product:

$$\lambda_{\min}(M) = \min_{u \neq 0} \frac{(Mu, u)_A}{(u, u)_A}, \qquad \lambda_{\max}(M) = \max_{u \neq 0} \frac{(Mu, u)_A}{(u, u)_A}.$$

We denote the set of eigenvalues as the spectrum $\sigma(M)$ and the largest of these in absolute value as the spectral radius as $\rho(M) = \max(|\lambda_{\min}(M)|, |\lambda_{\max}(M)|)$. For SPD (or A-SPD) operators M, the eigenvalues of M are real and positive, and the powers M^s for real s are well-defined through the spectral decomposition; see, for example, [89]. Finally, recall that a matrix representing the operator M with respect to any basis for $\mathcal{H}$ has the same eigenvalues as the operator M.

Linear operators on finite-dimensional spaces are bounded, and these bounds define the operator norms induced by the norms $\|\cdot\|$ and $\|\cdot\|_A$:

$$\|M\| = \max_{u\neq 0} \frac{\|Mu\|}{\|u\|}, \qquad \|M\|_A = \max_{u\neq 0} \frac{\|Mu\|_A}{\|u\|_A}.$$

A well-known property is that if M is self-adjoint, then $\rho(M) = \|M\|$. This property can also be shown to hold for A-self-adjoint operators. The following lemma can be found in [7] (as Lemma 4.1), although the proof there is for A-normal matrices rather than A-self-adjoint operators.

Lemma 10.6.1. *If A is SPD and M is A-self-adjoint, then $\|M\|_A = \rho(M)$.*

Proof. We simply note that

$$\|M\|_A = \max_{u\neq 0} \frac{\|Mu\|_A}{\|u\|_A} = \max_{u\neq 0} \frac{(Mu, Mu)_A^{1/2}}{(u,u)_A^{1/2}} = \max_{u\neq 0} \frac{(M^*Mu, u)_A^{1/2}}{(u,u)_A^{1/2}}$$
$$= \lambda_{\max}^{1/2}(M^*M),$$

since M^*M is always A-self-adjoint. Since by assumption M itself is A-self-adjoint, we have that $M^* = M$, which yields $\|M\|_A = \lambda_{\max}^{1/2}(M^*M) = \lambda_{\max}^{1/2}(M^2) = (\max_i[\lambda_i^2(M)])^{1/2} = \max[|\lambda_{\min}(M)|, |\lambda_{\max}(M)|] = \rho(M)$. □

Finally, we define the *A-condition number* of an invertible operator M by extending the standard notion to the A-inner product:

$$\kappa_A(M) = \|M\|_A \|M^{-1}\|_A.$$

In Lemma 10.6.9 we will show that if M is an A-self-adjoint operator, then in fact the following simpler expression holds for the generalized condition number:

$$\kappa_A(M) = \frac{\lambda_{\max}(M)}{\lambda_{\min}(M)}.$$

The Basic Linear Method and Its Error Propagator. Assume that we are faced with the operator equation $Au = f$, where $A \in \mathcal{L}(\mathcal{H}, \mathcal{H})$ is SPD, and we desire the unique solution u. Given a *preconditioner* (an approximate inverse operator) $B \approx A^{-1}$, consider the equivalent *preconditioned system* $BAu = Bf$. The operator B is chosen so that the simple linear iteration

$$u^1 = u^0 - BAu^0 + Bf = (I - BA)u^0 + Bf,$$

which produces an improved approximation u^1 to the true solution u given an initial approximation u^0, has some desired convergence properties. This yields the following basic linear iterative method, which we study in the remainder of this section.

Algorithm 10.6.1 (Basic Linear Method for Solving $Au = f$).

```
Form u^{i+1} from u^i using the affine fixed point iteration:

    u^{i+1} = u^i + B(f - Au^i) = (I - BA)u^i + Bf.
```

Subtracting the iteration equation from the identity $u = u - BAu + Bf$ yields the equation for the error $e^i = u - u^i$ at each iteration:

$$e^{i+1} = (I - BA)e^i = (I - BA)^2 e^{i-1} = \cdots = (I - BA)^{i+1} e^0. \tag{10.6.4}$$

The convergence of Algorithm 10.6.1 is determined completely by the spectral radius of the error propagation operator $E = I - BA$.

Theorem 10.6.1. *The condition $\rho(I - BA) < 1$ is necessary and sufficient for convergence of Algorithm 10.6.1 for an arbitrary initial approximation $u^0 \in \mathcal{H}$.*

Proof. See, for example, [115] or [169]. □

Since $|\lambda|\|u\| = \|\lambda u\| = \|Mu\| \leqslant \|M\| \, \|u\|$ for any norm $\|\cdot\|$, it follows that $\rho(M) \leqslant \|M\|$ for all norms $\|\cdot\|$. Therefore, $\|I - BA\| < 1$ and $\|I - BA\|_A < 1$ are both sufficient conditions for convergence of Algorithm 10.6.1. In fact, it is the norm of the error propagation operator which will bound the reduction of the error at each iteration, which follows from (10.6.4):

$$\|e^{i+1}\|_A \leqslant \|I - BA\|_A \|e^i\|_A \leqslant \|I - BA\|_A^{i+1} \|e^0\|_A. \tag{10.6.5}$$

The spectral radius $\rho(E)$ of the error propagator E is called the *convergence factor* for Algorithm 10.6.1, whereas the norm of the error propagator $\|E\|$ is referred to as the *contraction number* (with respect to the particular choice of norm $\|\cdot\|$).

We now establish some simple properties of the error propagation operator of an abstract linear method. We note that several of these properties are commonly used, especially in the multigrid literature, although the short proofs of the results seem difficult to locate. The particular framework we construct here for analyzing linear methods is based on the work of Xu [178] and the papers referenced therein, on the text by Varga [169], and on [100].

An alternative sufficient condition for convergence of the basic linear method is given in the following lemma, which is similar to *Stein's Theorem* (see [139] or [184]).

Lemma 10.6.2. *If E^* is the A-adjoint of E, and if the operator $I - E^*E$ is A-positive, then $\rho(E) \leqslant \|E\|_A < 1$.*

Proof. By hypothesis, $(A(I - E^*E)u, u) > 0 \; \forall u \in \mathcal{H}$. This then implies that $(AE^*Eu, u) < (Au, u) \; \forall u \in \mathcal{H}$, or $(AEu, Eu) < (Au, u) \; \forall u \in \mathcal{H}$. But this last inequality implies that

$$\rho(E) \leqslant \|E\|_A = \left(\max_{u \neq 0} \frac{(AEu, Eu)}{(Au, u)} \right)^{1/2} < 1.$$

□

We now state three very simple lemmas that we use repeatedly in the following sections.

Lemma 10.6.3. *If A is SPD, then BA is A-self-adjoint if and only if B is self-adjoint.*

Proof. Simply note that $(ABAx, y) = (BAx, Ay) = (Ax, B^T Ay)\ \forall x, y \in \mathcal{H}$. The lemma follows since $BA = B^T A$ if and only if $B = B^T$. □

Lemma 10.6.4. *If A is SPD, then $I - BA$ is A-self-adjoint if and only if B is self-adjoint.*

Proof. Begin by noting that $(A(I - BA)x, y) = (Ax, y) - (ABAx, y) = (Ax, y) - (Ax, (BA)^* y) = (Ax, (I - (BA)^*)y)$, $\forall x, y \in \mathcal{H}$. Therefore, $E^* = I - (BA)^* = I - BA = E$ if and only if $BA = (BA)^*$. But by Lemma 10.6.3, this holds if and only if B is self-adjoint, so the result follows. □

Lemma 10.6.5. *If A and B are SPD, then BA is A-SPD.*

Proof. By Lemma 10.6.3, BA is A-self-adjoint. Since B is SPD, and since $Au \neq 0$ for $u \neq 0$, we have $(ABAu, u) = (BAu, Au) > 0$, $\forall u \neq 0$. Therefore, BA is also A-positive, and the result follows. □

We noted above that the property $\rho(M) = \|M\|$ holds in the case that M is self-adjoint with respect to the inner product inducing the norm $\|\cdot\|$. If B is self-adjoint, the following theorem states that the resulting error propagator $E = I - BA$ has this property with respect to the A-norm.

Theorem 10.6.2. *If A is SPD and B is self-adjoint, then $\|I - BA\|_A = \rho(I - BA)$.*

Proof. By Lemma 10.6.4, $I - BA$ is A-self-adjoint, and by Lemma 10.6.1, the result follows. □

REMARK. Theorem 10.6.2 will be exploited later since $\rho(E)$ is usually much easier to compute numerically than $\|E\|_A$, and since it is the energy norm $\|E\|_A$ of the error propagator E which is typically bounded in various convergence theories for iterative processes.

The following simple lemma, similar to Lemma 10.6.2, will be very useful later.

Lemma 10.6.6. *If A and B are SPD, and if the operator $E = I - BA$ is A-nonnegative, then $\rho(E) = \|E\|_A < 1$.*

Proof. By Lemma 10.6.4, E is A-self-adjoint. By assumption, E is A-nonnegative, so from the discussion earlier in the section we see that E must have real nonnegative eigenvalues. By hypothesis, $(A(I - BA)u, u) \geqslant 0\ \forall u \in \mathcal{H}$, which implies that $(ABAu, u) \leqslant (Au, u)\ \forall u \in \mathcal{H}$. By Lemma 10.6.5, BA is A-SPD, and we have that

$$0 < (ABAu, u) \leqslant (Au, u) \quad \forall u \in \mathcal{H}, \ \ u \neq 0,$$

which implies that $0 < \lambda_i(BA) \leqslant 1 \; \forall \lambda_i \in \sigma(BA)$. Thus, since we also have that $\lambda_i(E) = \lambda_i(I - BA) = 1 - \lambda_i(BA) \; \forall i$, we have

$$\rho(E) = \max_i \lambda_i(E) = 1 - \min_i \lambda_i(BA) < 1.$$

Finally, by Theorem 10.6.2, we have $\|E\|_A = \rho(E) < 1$. □

The following simple lemma relates the contraction number bound to two simple inequalities; it is a standard result which follows directly from the spectral theory of self-adjoint linear operators.

Lemma 10.6.7. *If A is SPD and B is self-adjoint, and $E = I - BA$ is such that*

$$-C_1(Au, u) \leqslant (AEu, u) \leqslant C_2(Au, u), \quad \forall u \in \mathcal{H},$$

for $C_1 \geqslant 0$ and $C_2 \geqslant 0$, then $\rho(E) = \|E\|_A \leqslant \max\{C_1, C_2\}$.

Proof. By Lemma 10.6.4, $E = I - BA$ is A-self-adjoint, and by the spectral theory outlined at the beginning of the earlier section on linear iterative methods, the inequality above simply bounds the most negative and most positive eigenvalues of E with $-C_1$ and C_2, respectively. The result then follows by Theorem 10.6.2. □

Corollary 10.6.1. *If A and B are SPD, then Lemma 10.6.7 holds for some $C_2 < 1$.*

Proof. By Lemma 10.6.5, BA is A-SPD, which implies that the eigenvalues of BA are real and positive by the discussion earlier in the section. By Lemma 10.6.4, $E = I - BA$ is A-self-adjoint, and therefore has real eigenvalues. The eigenvalues of E and BA are related by $\lambda_i(E) = \lambda_i(I - BA) = 1 - \lambda_i(BA) \; \forall i$, and since $\lambda_i(BA) > 0 \; \forall i$, we must have that $\lambda_i(E) < 1 \; \forall i$. Since C_2 in Lemma 10.6.7 bounds the largest positive eigenvalue of E, we have that $C_2 < 1$. □

Convergence Properties of the Linear Method. The generalized condition number κ_A is employed in the following lemma, which states that there is an optimal relaxation parameter for a basic linear method, and gives the best possible convergence estimate for the method employing the optimal parameter. This lemma has appeared many times in the literature in one form or another; see [141].

Lemma 10.6.8. *If A and B are SPD, then*

$$\rho(I - \alpha BA) = \|I - \alpha BA\|_A < 1$$

if and only if $\alpha \in (0, 2/\rho(BA))$. Convergence is optimal (the norm is minimized) when $\alpha = 2/[\lambda_{\min}(BA) + \lambda_{\max}(BA)]$, giving

$$\rho(I - \alpha BA) = \|I - \alpha BA\|_A = 1 - \frac{2}{1 + \kappa_A(BA)} < 1.$$

Proof. Note that $\rho(I - \alpha BA) = \max_\lambda |1 - \alpha\lambda(BA)|$, so that $\rho(I - \alpha BA) < 1$ if and only if $\alpha \in (0, 2/\rho(BA))$, proving the first part of the lemma. We now take $\alpha = 2/[\lambda_{\min}(BA) + \lambda_{\max}(BA)]$, which gives

$$\begin{aligned}
\rho(I - \alpha BA) &= \max_\lambda |1 - \alpha\lambda(BA)| = \max_\lambda (1 - \alpha\lambda(BA)) \\
&= \max_\lambda \left(1 - \frac{2\lambda(BA)}{\lambda_{\min}(BA) + \lambda_{\max}(BA)}\right) \\
&= 1 - \frac{2\lambda_{\min}(BA)}{\lambda_{\min}(BA) + \lambda_{\max}(BA)} \\
&= 1 - \frac{2}{1 + \dfrac{\lambda_{\max}(BA)}{\lambda_{\min}(BA)}}.
\end{aligned}$$

Since BA is A-self-adjoint, by Lemma 10.6.9 we have that the condition number is $\kappa_A(BA) = \lambda_{\max}(BA)/\lambda_{\min}(BA)$, so that if $\alpha = 2/[\lambda_{\min}(BA) + \lambda_{\max}(BA)]$, then

$$\rho(I - \alpha BA) = \|I - \alpha BA\|_A = 1 - \frac{2}{1 + \kappa_A(BA)}.$$

To show that this is optimal, we must solve the mini-max problem: $\min_\alpha[\max_\lambda |1 - \alpha\lambda|]$, where $\alpha \in (0, 2/\lambda_{\max})$. Note that each α defines a polynomial of degree zero in λ, namely $P_o(\lambda) = \alpha$. Therefore, we can rephrase the problem as

$$P_1^{\text{opt}}(\lambda) = \min_{P_o} \left[\max_\lambda |1 - \lambda P_o(\lambda)|\right].$$

It is well-known that the scaled and shifted Chebyshev polynomials give the solution to this "mini-max" problem (see Exercise 10.5.2):

$$P_1^{\text{opt}}(\lambda) = 1 - \lambda P_o^{\text{opt}} = \frac{T_1\left(\dfrac{\lambda_{\max} + \lambda_{\min} - 2\lambda}{\lambda_{\max} - \lambda_{\min}}\right)}{T_1\left(\dfrac{\lambda_{\max} + \lambda_{\min}}{\lambda_{\max} - \lambda_{\min}}\right)}.$$

Since $T_1(x) = x$, we have simply that

$$P_1^{\text{opt}}(\lambda) = \frac{\dfrac{\lambda_{\max} + \lambda_{\min} - 2\lambda}{\lambda_{\max} - \lambda_{\min}}}{\dfrac{\lambda_{\max} + \lambda_{\min}}{\lambda_{\max} - \lambda_{\min}}} = 1 - \lambda\left(\frac{2}{\lambda_{\min} + \lambda_{\max}}\right),$$

showing that, in fact, $\alpha_{\text{opt}} = 2/[\lambda_{\min} + \lambda_{\max}]$. □

Note that if we wish to reduce the initial error $\|e^0\|_A$ by the factor ϵ, then equation (10.6.5) implies that this will be guaranteed if

$$\|E\|_A^{i+1} \leqslant \epsilon.$$

Taking natural logarithms of both sides and solving for i (where we assume that $\epsilon < 1$), we see that the number of iterations required to reach the desired tolerance, as a function of the contraction number, is given by

$$i \geqslant \frac{|\ln \epsilon|}{|\ln \|E\|_A|}. \tag{10.6.6}$$

If the bound on the norm is of the form in Lemma 10.6.8, then to achieve a tolerance of ϵ after i iterations will require that

$$i \geqslant \frac{|\ln \epsilon|}{\left|\ln\left(1 - \dfrac{2}{1+\kappa_A(BA)}\right)\right|} = \frac{|\ln \epsilon|}{\left|\ln\left(\dfrac{\kappa_A(BA)-1}{\kappa_A(BA)+1}\right)\right|}. \tag{10.6.7}$$

Using the approximation

$$\begin{aligned}\ln\left(\frac{a-1}{a+1}\right) &= \ln\left(\frac{1+(-1/a)}{1-(-1/a)}\right) = 2\left[\left(\frac{-1}{a}\right) + \frac{1}{3}\left(\frac{-1}{a}\right)^3 + \frac{1}{5}\left(\frac{-1}{a}\right)^5 + \cdots\right] \\ &< \frac{-2}{a},\end{aligned} \tag{10.6.8}$$

we have $|\ln[(\kappa_A(BA)-1)/(\kappa_A(BA)+1)]| > 2/\kappa_A(BA)$. Thus, we can guarantee (10.6.7) holds by enforcing

$$i \geqslant \frac{1}{2}\kappa_A(BA)|\ln \epsilon| + 1.$$

Therefore, the number of iterations required to reach an error on the order of the tolerance ϵ is then

$$i = \mathcal{O}\left(\kappa_A(BA)|\ln \epsilon|\right).$$

If a single iteration of the method costs $\mathcal{O}(N)$ arithmetic operations, then the overall complexity to solve the problem is $\mathcal{O}(|\ln \|E\|_A|^{-1} N |\ln \epsilon|)$, or $\mathcal{O}(\kappa_A(BA) N |\ln \epsilon|)$. If the quantity $\|E\|_A$ can be bounded by a constant which is less than 1, where the constant is independent of N, or alternatively, if $\kappa_A(BA)$ can be bounded by a constant which is independent of N, then the complexity is near optimal $\mathcal{O}(N|\ln \epsilon|)$.

Note that if E is A-self-adjoint, then we can replace $\|E\|_A$ by $\rho(E)$ in the discussion above. Even when this is not the case, $\rho(E)$ is often used above in place of $\|E\|_A$ to obtain an estimate, and the quantity $R_\infty(E) = -\ln \rho(E)$ is referred to as the *asymptotic convergence rate* (see [169, 184]). In [169], the *average rate of convergence of m iterations* is defined as the quantity $R(E^m) = -\ln(\|E^m\|^{1/m})$, the meaning of which is intuitively clear from equation (10.6.5). Since we have that $\rho(E) = \lim_{m\to\infty} \|E^m\|^{1/m}$ for all bounded linear operators E and norms $\|\cdot\|$ (see [116]), it then follows that $\lim_{m\to\infty} R(E^m) = R_\infty(E)$. While $R_\infty(E)$ is considered the standard measure of convergence of linear iterations (it is called the "convergence rate"; see [184]), this is really an asymptotic measure, and the convergence behavior for the early iterations may be better monitored by using the norm of the propagator E directly in (10.6.6); an example is given in [169], for which $R_\infty(E)$ gives a poor estimate of the number of iterations required.

The Conjugate Gradient Method

Consider now the linear equation $Au = f$ in the space $\mathcal{H}$. The conjugate gradient method was developed by Hestenes and Stiefel [92] for linear systems with symmetric positive definite operators A. It is common to *precondition* the linear system by the SPD *preconditioning operator* $B \approx A^{-1}$, in which case the generalized or preconditioned conjugate gradient method results. Our purpose in this section is to briefly examine the algorithm, its contraction properties, and establish some simple relationships between the contraction number of a basic linear preconditioner and that of the resulting preconditioned conjugate gradient algorithm. These relationships are commonly used, but some of the short proofs seem unavailable.

In [8], a general class of conjugate gradient methods obeying three-term recursions is studied, and it is shown that each instance of the class can be characterized by three operators: an inner product operator X, a preconditioning operator Y, and the system operator Z. As such, these methods are denoted as CG(X,Y,Z). We are interested in the special case that $X = A$, $Y = B$, and $Z = A$, when both B and A are SPD. Choosing the *Omin* [8] algorithm to implement the method CG(A,B,A), the *preconditioned conjugate gradient method* results. In order to present the algorithm, which is more complex than the basic linear method (Algorithm 10.6.1), we will employ some standard notation from the algorithm literature. In particular, we will denote the start of a complex fixed point-type iteration involving multiple steps using the standard notion of a "Do"-loop, where the beginning of the loop, as well as its duration, is denoted with a "Do X" statement, where X represents the conditions for continuing or terminating the loop. The end of the complex iteration will be denoted simply by "End do."

Algorithm 10.6.2 (Preconditioned Conjugate Gradient Algorithm).

Let $u^0 \in \mathcal{H}$ be given.
$r^0 = f - Au^0$, $s^0 = Br^0$, $p^0 = s^0$.
Do $i = 0, 1, \ldots$ until convergence:
  $\alpha_i = (r^i, s^i)/(Ap^i, p^i)$
  $u^{i+1} = u^i + \alpha_i p^i$
  $r^{i+1} = r^i - \alpha_i Ap^i$
  $s^{i+1} = Br^{i+1}$
  $\beta_{i+1} = (r^{i+1}, s^{i+1})/(r^i, s^i)$
  $p^{i+1} = s^{i+1} + \beta_{i+1} p^i$
End do.

If the dimension of $\mathcal{H}$ is n, then the algorithm can be shown to converge in n steps since the preconditioned operator BA is A-SPD [8]. Note that if $B = I$, then this algorithm is exactly the Hestenes and Stiefel algorithm.

Convergence Properties of the Conjugate Gradient Method. Since we wish to understand a little about the convergence properties of the conjugate gradient method and how these will be affected by a linear method representing the preconditioner B, we will briefly review a well-known conjugate gradient contraction bound. To begin, it is not difficult to see that the error at each iteration of Algorithm 10.6.2

can be written as a polynomial in BA times the initial error:

$$e^{i+1} = [I - BAp_i(BA)]e^0,$$

where $p_i \in \mathcal{P}_i$, the space of polynomials of degree i. At each step the energy norm of the error $\|e^{i+1}\|_A = \|u - u^{i+1}\|_A$ is minimized over the *Krylov subspace*:

$$K_{i+1}(BA, Br^0) = \text{span}\{Br^0, (BA)Br^0, (BA)^2Br^0, \ldots, (BA)^iBr^0\}.$$

Therefore,

$$\|e^{i+1}\|_A = \min_{p_i \in \mathcal{P}_i} \|[I - BAp_i(BA)]e^0\|_A.$$

Since BA is A-SPD, the eigenvalues $\lambda_j \in \sigma(BA)$ of BA are real and positive, and the eigenvectors v_j of BA are A-orthonormal. By expanding $e^0 = \sum_{j=1}^n \alpha_j v_j$, we have

$$\begin{aligned}
\|[I - BAp_i(BA)]e^0\|_A^2 &= (A[I - BAp_i(BA)]e^0, [I - BAp_i(BA)]e^0) \\
&= (A[I - BAp_i(BA)] \\
&\qquad \cdot (\sum_{j=1}^n \alpha_j v_j), [I - BAp_i(BA)](\sum_{j=1}^n \alpha_j v_j)) \\
&= (\sum_{j=1}^n [1 - \lambda_j p_i(\lambda_j)]\alpha_j \lambda_j v_j, \sum_{j=1}^n [1 - \lambda_j p_i(\lambda_j)]\alpha_j v_j) \\
&= \sum_{j=1}^n [1 - \lambda_j p_i(\lambda_j)]^2 \alpha_j^2 \lambda_j \\
&\leqslant \max_{\lambda_j \in \sigma(BA)} [1 - \lambda_j p_i(\lambda_j)]^2 \sum_{j=1}^n \alpha_j^2 \lambda_j \\
&= \max_{\lambda_j \in \sigma(BA)} [1 - \lambda_j p_i(\lambda_j)]^2 \sum_{j=1}^n (A\alpha_j v_j, \alpha_j v_j) \\
&= \max_{\lambda_j \in \sigma(BA)} [1 - \lambda_j p_i(\lambda_j)]^2 (A \sum_{j=1}^n \alpha_j v_j, \sum_{j=1}^n \alpha_j v_j) \\
&= \max_{\lambda_j \in \sigma(BA)} [1 - \lambda_j p_i(\lambda_j)]^2 \|e^0\|_A^2.
\end{aligned}$$

Thus, we have that

$$\|e^{i+1}\|_A \leqslant \left(\min_{p_i \in \mathcal{P}_i} \left[\max_{\lambda_j \in \sigma(BA)} |1 - \lambda_j p_i(\lambda_j)| \right] \right) \|e^0\|_A.$$

The scaled and shifted Chebyshev polynomials $T_{i+1}(\lambda)$, extended outside the interval $[-1, 1]$ as in Appendix A of [12], yield a solution to this *mini-max* problem (see

Exercises 10.5.2 and 10.5.3). Using some simple well-known relationships valid for $T_{i+1}(\cdot)$, the following contraction bound is easily derived:

$$\|e^{i+1}\|_A \leqslant 2 \left(\frac{\sqrt{\frac{\lambda_{\max}(BA)}{\lambda_{\min}(BA)}} - 1}{\sqrt{\frac{\lambda_{\max}(BA)}{\lambda_{\min}(BA)}} + 1} \right)^{i+1} \|e^0\|_A = 2\, \delta_{\mathrm{cg}}^{i+1} \, \|e^0\|_A. \tag{10.6.9}$$

The ratio of the extreme eigenvalues of BA appearing in the bound is often mistakenly called the (spectral) condition number $\kappa(BA)$; in fact, since BA is not self-adjoint (it is A-self-adjoint), this ratio is not in general equal to the condition number (this point is discussed in detail in [7]). However, the ratio does yield a condition number in a different norm. The following lemma is a special case of a more general result [7].

Lemma 10.6.9. *If A and B are SPD, then*

$$\kappa_A(BA) = \|BA\|_A \|(BA)^{-1}\|_A = \frac{\lambda_{\max}(BA)}{\lambda_{\min}(BA)}. \tag{10.6.10}$$

Proof. For any A-SPD M, it is easy to show that M^{-1} is also A-SPD, so from the material in the earlier section on linear iterative methods we know that both M and M^{-1} have real, positive eigenvalues. From Lemma 10.6.1 it then holds that

$$\begin{aligned} \|M^{-1}\|_A = \rho(M^{-1}) &= \max_{u \neq 0} \frac{(AM^{-1}u, u)}{(Au, u)} = \max_{u \neq 0} \frac{(AM^{-1/2}u, M^{-1/2}u)}{(AMM^{-1/2}u, M^{-1/2}u)} \\ &= \max_{v \neq 0} \frac{(Av, v)}{(AMv, v)} = \left[\min_{v \neq 0} \frac{(AMv, v)}{(Av, v)} \right]^{-1} = \lambda_{\min}(M)^{-1}. \end{aligned}$$

By Lemma 10.6.5, BA is A-SPD, which together with Lemma 10.6.1 implies that $\|BA\|_A = \rho(BA) = \lambda_{\max}(BA)$. We have then $\|(BA)^{-1}\|_A = \lambda_{\min}(BA)^{-1}$, implying that the A-condition number is given as the ratio of the extreme eigenvalues of BA as in equation (10.6.10). □

More generally, it can be shown that if the operator D is C-normal for some SPD inner product operator C, then the generalized condition number given by the expression $\kappa_C(D) = \|D\|_C \|D^{-1}\|_C$ is equal to the ratio of the extreme eigenvalues of the operator D. A proof of this fact is given in [7], along with a detailed discussion of this and other relationships for more general conjugate gradient methods. The conjugate gradient contraction number δ_{cg} can now be written as

$$\delta_{\mathrm{cg}} = \frac{\sqrt{\kappa_A(BA)} - 1}{\sqrt{\kappa_A(BA)} + 1} = 1 - \frac{2}{1 + \sqrt{\kappa_A(BA)}}.$$

The following lemma is used in the analysis of multigrid and other linear preconditioners (it appears for example in [177]) to bound the condition number of the operator BA in terms of the extreme eigenvalues of the linear preconditioner error propagator $E = I - BA$. We have given our own short proof of this result for completeness.

Lemma 10.6.10. *If A and B are SPD, and $E = I - BA$ is such that*

$$-C_1(Au,u) \leqslant (AEu,u) \leqslant C_2(Au,u), \quad \forall u \in \mathcal{H},$$

for $C_1 \geqslant 0$ and $C_2 \geqslant 0$, then the inequality above must in fact also hold with $C_2 < 1$, and it follows that

$$\kappa_A(BA) \leqslant \frac{1+C_1}{1-C_2}.$$

Proof. First, since A and B are SPD, by Corollary 10.6.1 we have that $C_2 < 1$. Since $(AEu,u) = (A(I-BA)u,u) = (Au,u) - (ABAu,u)$, $\forall u \in \mathcal{H}$, it is immediately clear that

$$-C_1(Au,u) - (Au,u) \leqslant -(ABAu,u) \leqslant C_2(Au,u) - (Au,u), \quad \forall u \in \mathcal{H}.$$

After multiplying by minus 1, we have

$$(1-C_2)(Au,u) \leqslant (ABAu,u) \leqslant (1+C_1)(Au,u), \quad \forall u \in \mathcal{H}.$$

By Lemma 10.6.5, BA is A-SPD, and it follows from the material in the section on linear iterative methods that the eigenvalues of BA are real and positive, and lie in the interval defined by the Rayleigh quotients generated by the A-inner product. From above, we see that the interval is given by $[(1-C_2),(1+C_1)]$, and by Lemma 10.6.9 the result follows. □

The next corollary may be found in [177].

Corollary 10.6.2. *If A and B are SPD, and BA is such that*

$$C_1(Au,u) \leqslant (ABAu,u) \leqslant C_2(Au,u), \quad \forall u \in \mathcal{H},$$

for $C_1 \geqslant 0$ and $C_2 \geqslant 0$, then the above must hold with $C_1 > 0$, and it follows that

$$\kappa_A(BA) \leqslant \frac{C_2}{C_1}.$$

Proof. This follows easily from the argument used in the proof of Lemma 10.6.10. □

The following corollary, which relates the contraction property of a linear method to the condition number of the operator BA, appears without proof in [178].

Corollary 10.6.3. *If A and B are SPD, and $\|I - BA\|_A \leqslant \delta < 1$, then*

$$\kappa_A(BA) \leqslant \frac{1+\delta}{1-\delta}. \tag{10.6.11}$$

Proof. This follows immediately from Lemma 10.6.10 with $\delta = \max\{C_1, C_2\}$. □

Preconditioners and the Acceleration of Linear Methods. We comment briefly on an interesting implication of Lemma 10.6.10, which was pointed out in [177]. It seems that even if a linear method is not convergent, for example if $C_1 > 1$ so that $\rho(E) > 1$, it may still be a good preconditioner. For example, if A and B are SPD, then by Corollary 10.6.1 we always have $C_2 < 1$. If it is the case that $C_2 << 1$, and if $C_1 > 1$ does not become too large, then $\kappa_A(BA)$ will be small and the conjugate gradient method will converge rapidly. A multigrid method (see below) will often diverge when applied to a problem with discontinuous coefficients unless special care is taken. Simply using the conjugate gradient method in conjunction with the multigrid method often yields a convergent (even rapidly convergent) method without employing any of the special techniques that have been developed for these problems; Lemma 10.6.10 gives some insight into this behavior.

The following result from [178] connects the contraction number of the linear method used as the preconditioner to the contraction number of the resulting conjugate gradient method, and it shows that the conjugate gradient method always accelerates a linear method, justifying the terminology "CG acceleration."

Theorem 10.6.3. *If A and B are SPD, and $\|I - BA\|_A \leqslant \delta < 1$, then $\delta_{\mathrm{cg}} < \delta$.*

Proof. An abbreviated proof appears in [178]; we fill in the details here for completeness. Assume that the given linear method has contraction number bounded as $\|I - BA\|_A < \delta$. Now, since the function

$$\frac{\sqrt{\kappa_A(BA)} - 1}{\sqrt{\kappa_A(BA)} + 1}$$

is an increasing function of $\kappa_A(BA)$, we can use the result of Lemma 10.6.10, namely $\kappa_A(BA) \leqslant (1+\delta)/(1-\delta)$, to bound the contraction rate of preconditioned conjugate gradient method as follows:

$$\delta_{\mathrm{cg}} \leqslant \left(\frac{\sqrt{\kappa_A(BA)} - 1}{\sqrt{\kappa_A(BA)} + 1}\right) \leqslant \left(\frac{\sqrt{\dfrac{1+\delta}{1-\delta}} - 1}{\sqrt{\dfrac{1+\delta}{1-\delta}} + 1}\right) \cdot \left(\frac{\sqrt{\dfrac{1+\delta}{1-\delta}} - 1}{\sqrt{\dfrac{1+\delta}{1-\delta}} - 1}\right)$$

$$= \frac{\dfrac{1+\delta}{1-\delta} - 2\sqrt{\dfrac{1+\delta}{1-\delta}} + 1}{\dfrac{1+\delta}{1-\delta} - 1} = \frac{1 - \sqrt{1-\delta^2}}{\delta}.$$

Note that this last term can be rewritten as

$$\delta_{\mathrm{cg}} \leqslant \frac{1 - \sqrt{1-\delta^2}}{\delta} = \delta\left(\frac{1}{\delta^2}[1 - \sqrt{1-\delta^2}]\right).$$

Now, since $0 < \delta < 1$, clearly $1 - \delta^2 < 1$, so that $1 - \delta^2 > (1-\delta^2)^2$. Thus, $\sqrt{1-\delta^2} > 1 - \delta^2$, or $-\sqrt{1-\delta^2} < \delta^2 - 1$, or finally, $1 - \sqrt{1-\delta^2} < \delta^2$. Therefore,

$(1/\delta^2)\left[1-\sqrt{1-\delta^2}\right] < 1$, or

$$\delta_{\mathrm{cg}} \leq \delta \left(\frac{1}{\delta^2} \left[1 - \sqrt{1-\delta^2} \right] \right) < \delta.$$

A more direct proof follows by recalling from Lemma 10.6.8 that the *best* possible contraction of the linear method, when provided with an optimal parameter, is given by

$$\delta_{\mathrm{opt}} = 1 - \frac{2}{1+\kappa_A(BA)},$$

whereas the conjugate gradient contraction is

$$\delta_{\mathrm{cg}} = 1 - \frac{2}{1+\sqrt{\kappa_A(BA)}}.$$

Assuming that $B \neq A^{-1}$, then we always have $\kappa_A(BA) > 1$, so we must have that $\delta_{\mathrm{cg}} < \delta_{\mathrm{opt}} \leqslant \delta$. $\square$

This result implies that it always pays in terms of an improved contraction number to use the conjugate gradient method to accelerate a linear method; the question remains, of course, whether the additional computational labor involved will be amortized by the improvement. This is not clear from the analysis above, and is problem dependent in practice.

Note that if a given linear method requires a parameter α as in Lemma 10.6.8 in order to be competitive, one can simply use the conjugate gradient method as an accelerator for the method without a parameter, avoiding the possibly costly estimation of a good parameter α. Theorem 10.6.3 guarantees that the resulting method will have superior contraction properties, without requiring the parameter estimation. This is exactly why additive multigrid and domain decomposition methods (which we discuss in more detail below) are used almost exclusively as preconditioners for conjugate gradient methods; in contrast to the multiplicative variants, which can be used effectively without a parameter, the additive variants always require a good parameter α to be effective, unless used as preconditioners.

To finish this section, we remark briefly on the complexity of Algorithm 10.6.2. If a tolerance of ϵ is required, then the computational cost to reduce the energy norm of the error below the tolerance can be determined from the expression above for δ_{cg} and from equation (10.6.9). To achieve a tolerance of ϵ after i iterations will require that

$$2\,\delta_{\mathrm{cg}}^{i+1} = 2\left(\frac{\sqrt{\kappa_A(BA)}-1}{\sqrt{\kappa_A(BA)}+1} \right)^{i+1} < \epsilon.$$

Dividing by 2 and taking natural logarithms (and assuming that $\epsilon < 1$) yields

$$i \geqslant \frac{\left| \ln \frac{\epsilon}{2} \right|}{\left| \ln \left(\frac{\sqrt{\kappa_A(BA)}-1}{\sqrt{\kappa_A(BA)}+1} \right) \right|}. \tag{10.6.12}$$

Using (10.6.8) we have $|\ln[(\kappa_A^{1/2}(BA)-1)/(\kappa_A^{1/2}(BA)+1)]| > 2/\kappa_A^{1/2}(BA)$. Thus, we can ensure that (10.6.12) holds by enforcing

$$i \geqslant \frac{1}{2}\kappa_A^{1/2}(BA)\left|\ln\frac{\epsilon}{2}\right| + 1.$$

Therefore, the number of iterations required to reach an error on the order of the tolerance ϵ is

$$i = \mathcal{O}\left(\kappa_A^{1/2}(BA)\left|\ln\frac{\epsilon}{2}\right|\right).$$

If the cost of each iteration is $\mathcal{O}(N)$, which will hold in the case of the sparse matrices generated by standard discretizations of elliptic partial differential equations, then the overall complexity to solve the problem is $\mathcal{O}(\kappa_A^{1/2}(BA)N|\ln[\epsilon/2]|)$. If the preconditioner B is such that $\kappa_A^{1/2}(BA)$ can be bounded independently of the problem size N, then the complexity becomes (near) optimal order $\mathcal{O}(N|\ln[\epsilon/2]|)$.

We make some final remarks regarding the idea of *spectral equivalence.*

Definition 10.6.1. *The SPD operators $A \in \mathcal{L}(\mathcal{H},\mathcal{H})$ and $M \in \mathcal{L}(\mathcal{H},\mathcal{H})$ are called spectrally equivalent if there exist constants $C_1 > 0$ and $C_2 > 0$ such that*

$$C_1(Au,u) \leqslant (Mu,u) \leqslant C_2(Au,u), \quad \forall u \in \mathcal{H}.$$

In other words, A defines an inner product which induces a norm equivalent to the norm induced by the M-inner product. If a given preconditioner B is spectrally equivalent to A^{-1}, then the condition number of the preconditioned operator BA is uniformly bounded.

Lemma 10.6.11. *If the SPD operators B and A^{-1} are spectrally equivalent, then*

$$\kappa_A(BA) \leqslant \frac{C_2}{C_1}.$$

Proof. By hypothesis, we have $C_1(A^{-1}u,u) \leq (Bu,u) \leq C_2(A^{-1}u,u)$, $\forall u \in \mathcal{H}$. But this can be written as

$$\begin{aligned} C_1(A^{-1/2}u, A^{-1/2}u) &\leqslant (A^{1/2}BA^{1/2}A^{-1/2}u, A^{-1/2}u) \\ &\leqslant C_2(A^{-1/2}u, A^{-1/2}u) \end{aligned}$$

or

$$C_1(\tilde{u},\tilde{u}) \leqslant (A^{1/2}BA^{1/2}\tilde{u},\tilde{u}) \leqslant C_2(\tilde{u},\tilde{u}), \quad \forall \tilde{u} \in \mathcal{H}.$$

Now, since $BA = A^{-1/2}(A^{1/2}BA^{1/2})A^{1/2}$, we have that BA is similar to the SPD operator $A^{1/2}BA^{1/2}$. Therefore, the inequality above bounds the extreme eigenvalues of BA, and as a result the lemma follows by Lemma 10.6.9. □

Moreover, if any of the following (equivalent) norm equivalences hold:

$$\begin{aligned} C_1(Au,u) \leqslant &\quad (ABAu,u) &\leqslant C_2(Au,u), \\ C_1(Bu,u) \leqslant &\quad (BABu,u) &\leqslant C_2(Bu,u), \\ C_1(A^{-1}u,u) \leqslant &\quad (Bu,u) &\leqslant C_2(A^{-1}u,u), \\ C_1(B^{-1}u,u) \leqslant &\quad (Au,u) &\leqslant C_2(B^{-1}u,u), \\ C_2^{-1}(Au,u) \leqslant &\quad (B^{-1}u,u) &\leqslant C_1^{-1}(Au,u), \\ C_2^{-1}(Bu,u) \leqslant &\quad (A^{-1}u,u) &\leqslant C_1^{-1}(Bu,u), \end{aligned}$$

then by similar arguments one has

$$\kappa_A(BA) \leqslant \frac{C_2}{C_1}.$$

Of course, since all norms on finite-dimensional spaces are equivalent (which follows from the fact that all linear operators on finite-dimensional spaces are bounded), the idea of spectral equivalence is only important in the case of infinite-dimensional spaces, or when one considers how the equivalence constants behave as one increases the sizes of the spaces. This is exactly the issue in multigrid and domain decomposition theory: As one decreases the mesh size (increases the size of the spaces involved), one would like the quantity $\kappa_A(BA)$ to remain uniformly bounded (in other words, one would like the equivalence constants to remain constant or grow only slowly). A discussion of these ideas appears in [141].

Domain Decomposition Methods

Domain decomposition methods were first proposed by H. A. Schwarz as a theoretical tool for studying elliptic problems on complicated domains, constructed as the union of simple domains. An interesting early reference not often mentioned is [109], containing both analysis and numerical examples and references to the original work by Schwarz. Since the development of parallel computers, domain decomposition methods have become one of the most important practical methods for solving elliptic partial differential equations on modern parallel computers. In this section we briefly describe basic overlapping domain decomposition methods; our discussion here draws much from [66, 100, 178] and the references cited therein.

Given a domain Ω and coarse triangulation by J regions $\{\Omega_k\}$ of mesh size H_k, we refine (several times) to obtain a fine mesh of size h_k. The regions defined by the initial triangulation Ω_k are then extended by δ_k to form the "overlapping subdomains" Ω'_k. Let $\mathcal{V}$ and $\mathcal{V}_0$ denote the finite element spaces associated with the h_k and H_k triangulation of Ω, respectively. Examples of overlapping subdomains constructed in this way over existing coarse simplicial meshes, designed for building piecewise-linear finite element subdomain spaces $\mathcal{V}_k = H_0^1(\Omega'_k) \cap \mathcal{V}$, are shown in Figure 10.10.

To describe overlapping domain decomposition methods, we focus on the following variational problem in $\mathcal{V}$:

$$\text{Find } u \in \mathcal{V} \text{ such that } a(u,v) = f(v), \quad \forall v \in \mathcal{V}, \tag{10.6.13}$$

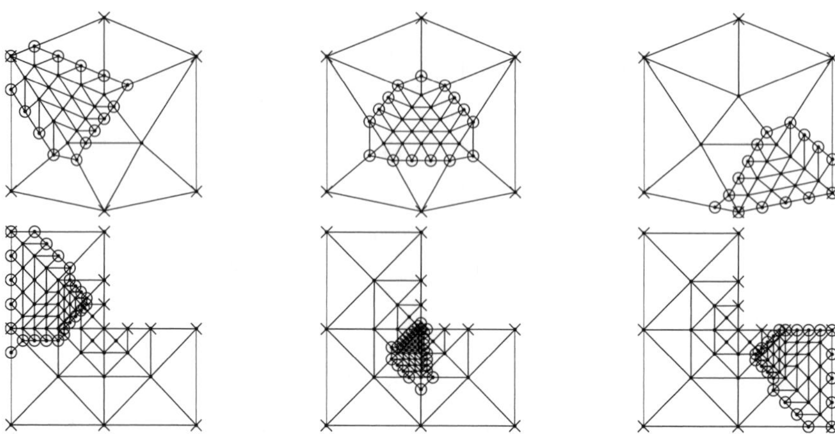

Figure 10.10 Unstructured overlapping subdomain collections for two example domains. The large triangles in the coarse mesh form the nonoverlapping subdomains Ω_k, and the refined regions form the overlapping subdomains Ω'_k. The symbols $\times$ denote nodes lying on the boundary of the global domain Ω, whereas the symbols $\circ$ denote nodes lying on the boundary of a particular subdomain Ω'_k.

where the form $a(\cdot,\cdot)$ is bilinear, symmetric, coercive, and bounded, whereas $f(\cdot)$ is linear and bounded. An overlapping domain decomposition method involves first solving (10.6.13) restricted to each overlapping subdomain Ω'_k:

$$\text{Find } u_k \in \mathcal{V}_k \text{ such that } a(u_k, v_k) = f(v_k), \quad \forall v_k \in \mathcal{V}_k, \tag{10.6.14}$$

and then combining the results to improve an approximation over the entire domain Ω. Since the global problem over Ω was not solved, this procedure must be repeated until it converges to the solution of the global problem (10.6.13). Therefore, overlapping domain decomposition methods can be viewed as iterative methods for solving the variational problem (10.6.13), where each iteration involves approximate projections of the error onto subspaces of $\mathcal{V}$ associated with the overlapping subdomains Ω'_k, which is accomplished by solving the subspace problem (10.6.14).

It is useful to reformulate problems (10.6.13) and (10.6.14) as operator equations in the function spaces defined over Ω and Ω'_k. Let $\mathcal{V}_k = H^1_0(\Omega'_k) \cap \mathcal{V}$, $k = 1, \ldots, J$; it is not difficult to show that $\mathcal{V} = \mathcal{V}_1 + \cdots + \mathcal{V}_J$, where the coarse space $\mathcal{V}_0$ may also be included in the sum. Through the Riesz representation theorem and the Bounded Operator Theorem of Section 4.8, we can associate with the problem above an abstract operator equation $Au = f$, where A is SPD. We denote as A_k the restriction of the operator A to the space $\mathcal{V}_k$, corresponding to (any) discretization of the original problem restricted to the subdomain Ω'_k. Algebraically, it can be shown that $A_k = I_k^T A I_k$, where I_k is the natural inclusion of $\mathcal{V}_k$ into $\mathcal{V}$ and I_k^T is the corresponding projection of $\mathcal{V}$ onto $\mathcal{V}_k$. The property that I_k is the natural inclusion and I_k^T is the corresponding projection holds for both the finite element space $\mathcal{V}_k$ as well as the Euclidean space $\mathbb{R}^{n_k}$. In other words, domain decomposition methods automatically satisfy the so-called *variational condition*:

$$A_k = I_k^T A I_k \tag{10.6.15}$$

in the subspaces $\mathcal{V}_k$, $k \neq 0$, for *any* discretization. Recall that A-orthogonal projection from $\mathcal{V}$ onto $\mathcal{V}_k$ can be written as $P_k = I_k(I_k^T A I_k)^{-1} I_k^T A$, which becomes simply $P_k = I_k A_k^{-1} I_k^T A$ when A_k satisfies the variational condition (10.6.15). If $R_k \approx A_k^{-1}$, we can define the *approximate* A-orthogonal projector from $\mathcal{V}$ onto $\mathcal{V}_k$ as $T_k = I_k R_k I_k^T A$. The case of $R_k = A_k^{-1}$ corresponds to an exact solution of the subdomain problems, giving $T_k = P_k$.

A *multiplicative Schwarz overlapping domain decomposition method*, employing *successive* approximate projections onto the subspaces $\mathcal{V}_k$ and written in terms of the operators A and A_k, has the following form.

Algorithm 10.6.3 (Multiplicative Schwarz Method: Implementation Form).

```
Set u^{i+1} = MS(u^i, f), where u^{i+1} = MS(u^i, f) is defined as:
Do k = 1,...,J
      r_k = I_k^T (f - A u^i)
      e_k = R_k r_k
      u^{i+1} = u^i + I_k e_k
      u^i = u^{i+1}
End do.
```

Note that the first step through the loop in $MS(\cdot,\cdot)$ gives

$$\begin{aligned} u^{i+1} &= u^i + I_1 e_1 \\ &= u^i + I_1 R_1 I_1^T (f - Au^i) \\ &= (I - I_1 R_1 I_1^T A) u^i + I_1 R_1 I_1^T f. \end{aligned}$$

Continuing in this fashion, and by defining $T_k = I_k R_k I_k^T A$, we see that after the full loop in $MS(\cdot,\cdot)$ the solution transforms according to

$$u^{i+1} = (I - T_J)(I - T_{J-1}) \cdots (I - T_1) u^i + Bf,$$

where B is a quite complicated combination of the operators R_k, I_k, I_k^T, and A. By defining $E_k = (I - T_k)(I - T_{k-1}) \cdots (I - T_1)$, we see that $E_k = (I - T_k)E_{k-1}$. Therefore, since $E_{k-1} = I - B_{k-1}A$ for some (implicitly defined) B_{k-1}, we can identify the operators B_k through the recursion $E_k = I - B_k A = (I - T_k)E_{k-1}$, giving

$$\begin{aligned} B_k A &= I - (I - T_k)E_{k-1} = I - (I - B_{k-1}A) + T_k(I - B_{k-1}A) \\ &= B_{k-1}A + T_k - T_k B_{k-1} A = B_{k-1}A + I_k R_k I_k^T A - I_k R_k I_k^T A B_{k-1} A \\ &= \left[B_{k-1} + I_k R_k I_k^T - I_k R_k I_k^T A B_{k-1}\right] A, \end{aligned}$$

so that $B_k = B_{k-1} + I_k R_k I_k^T - I_k R_k I_k^T A B_{k-1}$. But this means that Algorithm 10.6.3 is equivalent to the following.

Algorithm 10.6.4 (Multiplicative Schwarz Method: Operator Form).

Define:
  $u^{i+1} = u^i + B(f - Au^i) = (I - BA)u^i + Bf$,
where the error propagator E is defined by:
  $E = I - BA = (I - T_J)(I - T_{J-1}) \cdots (I - T_1)$,
  $T_k = I_k R_k I_k^T A, \quad k = 1, \ldots, J.$
The implicit operator $B \equiv B_J$ obeys the recursion:
  $B_1 = I_1 R_1 I_1^T, \quad B_k = B_{k-1} + I_k R_k I_k^T - I_k R_k I_k^T A B_{k-1}, \quad k = 2, \ldots, J.$

An *additive Schwarz overlapping domain decomposition method*, employing *simultaneous* approximate projections onto the subspaces $\mathcal{V}_k$, has the form:

Algorithm 10.6.5 (Additive Schwarz Method: Implementation Form).

Set $u^{i+1} = AS(u^i, f)$, where $u^{i+1} = AS(u^i, f)$ is defined as:
$r = f - Au^i$
Do $k = 1, \ldots, J$
  $r_k = I_k^T r$
  $e_k = R_k r_k$
  $u^{i+1} = u^i + I_k e_k$
End do.

Since each loop iteration depends only on the original approximation u^i, we see that the full correction to the solution can be written as the sum

$$u^{i+1} = u^i + B(f - Au^i) = u^i + \sum_{k=1}^{J} I_k R_k I_k^T (f - Au^i),$$

where the preconditioner B has the form $B = \sum_{k=1}^{J} I_k R_k I_k^T$, and the error propagator is $E = I - BA$. Therefore, Algorithm 10.6.5 is equivalent to the following.

Algorithm 10.6.6 (Additive Schwarz Method: Operator Form).

Define:
  $u^{i+1} = u^i + B(f - Au^i) = (I - BA)u^i + Bf$,
where the error propagator E is defined by:
  $E = I - BA = I - \sum_{k=1}^{J} T_k$,
  $T_k = I_k R_k I_k^T A, \quad k = 1, \ldots, J.$
The operator B is defined explicitly as:
  $B = \sum_{k=1}^{J} I_k R_k I_k^T$.

Therefore, the multiplicative and additive domain decomposition methods fit exactly into the framework of a basic linear method (Algorithm 10.6.1) or can be viewed as methods for constructing preconditioners B for use with the conjugate gradient method (Algorithm 10.6.2). If $R_k = A_k^{-1}$, where A_k satisfies the variational condition (10.6.15), then each iteration of the algorithms involves removal of the A-orthogonal projection of the error onto each subspace, either successively (the multiplicative method) or simultaneously (the additive method). If R_k is an approximation to A_k^{-1}, then each step is an approximate A-orthogonal projection.

Multilevel Methods

Multilevel (or *multigrid*) methods are highly efficient numerical techniques for solving the algebraic equations arising from the discretization of partial differential equations. These methods were developed in direct response to the deficiencies of the classical iterations such as the Gauss-Seidel and SOR methods. Some of the early fundamental papers are [18, 40, 84, 162], as well as [17, 19, 185], and a comprehensive analysis of the many different aspects of these methods is given in [85, 178]. The following derivation of two-level and multilevel methods in a recursive operator framework is motivated by some work on finite element-based multilevel and domain decomposition methods, represented, for example, by [38, 66, 100, 178]. Our notation follows the currently established convention for these types of methods; see [100, 178].

Linear Equations in a Nested Sequence of Spaces. In what follows we are concerned with a nested sequence of spaces $\mathcal{H}_1 \subset \mathcal{H}_2 \subset \cdots \subset \mathcal{H}_J \equiv \mathcal{H}$, where $\mathcal{H}_J$ corresponds to the finest or largest space and $\mathcal{H}_1$ the coarsest or smallest. Each space $\mathcal{H}_k$ is taken to be a Hilbert space, equipped with an inner product $(\cdot,\cdot)_k$ which induces the norm $\|\cdot\|_k$. Regarding notation, if $A \in \mathcal{L}(\mathcal{H}_k, \mathcal{H}_k)$, then we denote the operator as A_k. Similarly, if $A \in \mathcal{L}(\mathcal{H}_k, \mathcal{H}_i)$, then we denote the operator as A_k^i. Finally, if $A \in \mathcal{L}(\mathcal{H}_k, \mathcal{H}_k)$ but its operation somehow concerns a specific subspace $\mathcal{H}_i \subset \mathcal{H}_k$, then we denote the operator as $A_{k;i}$. For quantities involving the finest space $\mathcal{H}_J$, we will often leave off the subscripts without danger of confusion.

Now, given such a nested sequence of Hilbert spaces, we assume that associated with each space $\mathcal{H}_k$ is an SPD operator A_k, which defines a second inner product $(\cdot,\cdot)_{A_k} = (A_k\cdot,\cdot)_k$, inducing a second norm $\|\cdot\|_{A_k} = (\cdot,\cdot)_{A_k}^{1/2}$. The spaces $\mathcal{H}_k$ are connected by *prolongation* operators $I_{k-1}^k \in \mathcal{L}(\mathcal{H}_{k-1}, \mathcal{H}_k)$ and *restriction* operators $I_k^{k-1} \in \mathcal{L}(\mathcal{H}_k, \mathcal{H}_{k-1})$, where we assume that the null space of I_{k-1}^k contains only the zero vector, and usually that $I_k^{k-1} = (I_{k-1}^k)^T$, where the (Hilbert) adjoint is with respect to the inner products on the sequence of spaces $\mathcal{H}_k$:

$$(u_k, I_{k-1}^k v_{k-1})_k = ((I_{k-1}^k)^T u_k, v_{k-1})_{k-1}, \quad \forall u_k \in \mathcal{H}_k, \quad \forall v_{k-1} \in \mathcal{H}_{k-1}. \tag{10.6.16}$$

We are given the operator equation $Au = f$ in the finest space $\mathcal{H} \equiv \mathcal{H}_J$, where $A \in \mathcal{L}(\mathcal{H}, \mathcal{H})$ is SPD, and we are interested in iterative algorithms for determining the unique solution u which involves solving problems in the coarser spaces $\mathcal{H}_k$ for $1 \leqslant k < J$. If the equation in $\mathcal{H}$ has arisen from finite element or similar discretization of an elliptic partial differential equation, then operators A_k (and the associated coarse problems $A_k u_k = f_k$) in coarser spaces $\mathcal{H}_k$ for $k < J$ may be defined naturally with the same discretization on a coarser mesh. Alternatively, it is convenient (for theoretical reasons which we discuss later in the chapter) to take the so-called *variational approach* of constructing the coarse operators, where the operators $A_k \in \mathcal{L}(\mathcal{H}_k, \mathcal{H}_k)$ satisfy

$$A_{k-1} = I_k^{k-1} A_k I_{k-1}^k, \qquad I_k^{k-1} = (I_{k-1}^k)^T. \tag{10.6.17}$$

The first condition in (10.6.17) is sometimes referred to as the *Galerkin condition*, whereas the two conditions (10.6.17) together are known as the *variational conditions*, due to the fact that both conditions are satisfied naturally by variational or Galerkin (finite element) discretizations on successively refined meshes. Note that if A_k is SPD, then A_{k-1} produced by (10.6.17) will also be SPD.

In the case that $\mathcal{H}_k = \mathcal{U}_k = \mathbb{R}^{n_k}$, the prolongation operator I_{k-1}^k typically corresponds to d-dimensional interpolation of u_{k-1} to $u_k = I_{k-1}^k u_{k-1}$, where u_{k-1} and u_k are interpreted as grid functions defined over two successively refined (box or finite element) discretizations Ω_{k-1} and Ω_k of the domain $\Omega \subset \mathbb{R}^d$. Since the coarse grid function space has by definition smaller dimension than the fine space, I_{k-1}^k takes the form of a rectangular matrix with more rows than columns. A positive scaling constant $c \in \mathbb{R}$ will appear in the second condition in (10.6.17), which will become $I_k^{k-1} = c(I_{k-1}^k)^T$, due to taking I_k^{k-1} to be the adjoint of I_{k-1}^k with respect to the inner product (10.5.53). This results from $h_k < h_{k-1}$ on two successive spaces, and the subsequent need to scale the corresponding discrete inner product to preserve a discrete notion of volume; this scaling allows for comparing inner products on spaces with different dimensions.

In the case that $\mathcal{H}_k = \mathcal{V}_k$, where $\mathcal{V}_k$ is a finite element subspace, the prolongation corresponds to the natural inclusion of a coarse space function into the fine space, and the restriction corresponds to its natural adjoint operator, which is the L^2-projection of a fine space function onto the coarse space. The variational conditions (10.6.17) then hold for the abstract operators A_k on the spaces $\mathcal{V}_k$, with inclusion and L^2-projection for the prolongation and restriction (see the proof in [85]). In addition, the stiffness matrices representing the abstract operators A_k also satisfy the conditions (10.6.17), where now the prolongation and restriction operators are as in the case of the space $\mathcal{U}_k$. However, we remark that this is true only with *exact evaluation* of the integrals forming the matrix components; the conditions (10.6.17) are violated if quadrature is used. "Algebraic multigrid" are methods based on enforcing (10.6.17) algebraically using a product of sparse matrices; one can develop a strong two-level theory for this class of methods in the case of M-matrices (see, for example, [41, 151]), but it is difficult to develop theoretical results for multilevel versions of these methods.

Many important results have been obtained for multilevel methods in the spaces $\mathcal{H}_k = \mathcal{V}_k$, which rely on certain operator recursions (we point out in particular the papers [36, 38, 177, 178]). Some of these results [38, 178] are "regularity-free" in the sense that they do not require the usual regularity or smoothness assumptions on the solution to the problem, which is important since these are not valid for problems such as those with discontinuous coefficients. As a result, we will develop multilevel algorithms in a recursive form in the abstract spaces $\mathcal{H}_k$.

Two-Level Methods. As we noted earlier, the convergence rate of the classical methods (Gauss-Seidel and similar methods) deteriorate as the mesh size $h_k \to 0$; we examine the reasons for this behavior for a model problem later in this section. However, using the same spectral analysis, one can easily see that the components of the error corresponding to the small eigenvalues of the error propagation operator are

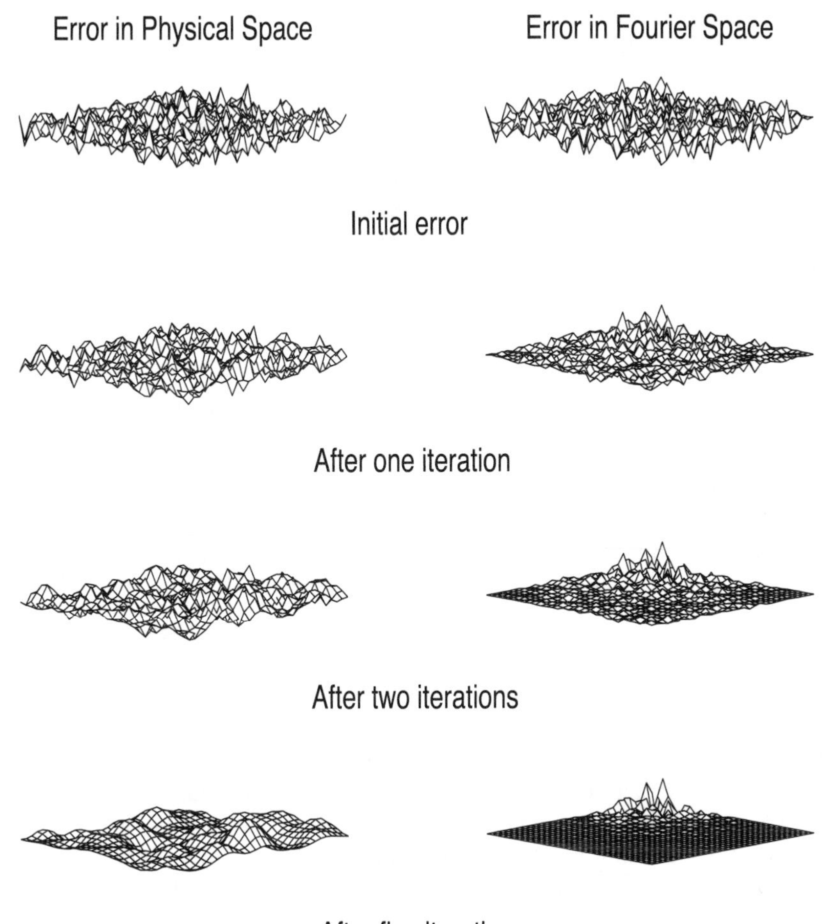

Figure 10.11 Error-smoothing effect of Gauss-Seidel iteration. The error in both physical and Fourier (or frequency) space is shown initially and after one, two, and five iterations. Low-frequency components of the error appear at the rear of the Fourier plots; high-frequency components appear at far left, far right, and in the foreground.

actually being decreased quite effectively even as $h_k \to 0$; these are the rapidly varying or *high-frequency* components in the error. This effect is illustrated graphically in Figure 10.11 for Gauss-Seidel iteration applied to the two-dimensional Poisson equation on the unit square. In the figure, the error in both physical and Fourier (or frequency) space is shown initially and after one, two, and five iterations. In the Fourier space plots, the low-frequency components of the error are found in the rear, whereas the high-frequency components are found to the far left, the far right, and in the foreground. The source function for this example was constructed from a random field (to produce all frequencies in the solution) and the initial guess was taken to be zero.

The observation that classical linear methods are very efficient at reducing the high-frequency modes is the motivation for the multilevel method: A classical linear method can be used to handle the high-frequency components of the error (or to *smooth* the error), and the low-frequency components can be eliminated efficiently on a coarser mesh with fewer unknowns, where the low-frequency modes are well represented.

For the equation $A_k u_k = f_k$ on level k, the smoothing method takes the form of Algorithm 10.6.1 for some operator R_k, the *smoothing operator*, as the approximate inverse of the operator A_k:

$$u_k^{i+1} = u_k^i + R_k(f_k - A_k u_k^i). \tag{10.6.18}$$

In the case of two spaces $\mathcal{H}_k$ and $\mathcal{H}_{k-1}$, the error equation $e_k = A_k^{-1} r_k$ is solved approximately using the coarse space, with the *coarse-level correction operator* $C_k = I_{k-1}^k A_{k-1}^{-1} I_k^{k-1}$ representing the exact solution with A_{k-1}^{-1} in the coarse-level subspace $\mathcal{H}_{k-1}$. The solution is then adjusted by the correction

$$u_k^{i+1} = u_k^i + C_k(f_k - A_k u_k^i). \tag{10.6.19}$$

There are several ways in which these two procedures can be combined.

By viewing multilevel methods as compositions of the simple linear methods (10.6.18) and (10.6.19), a simple yet complete framework for understanding these methods can be constructed. The most important concepts can be discussed with regard to two-level methods and then generalized to more than two levels using an implicit recursive definition of an approximate coarse-level inverse operator.

Consider the case of two nested spaces $\mathcal{H}_{k-1} \subset \mathcal{H}_k$, and the following two-level method:

Algorithm 10.6.7 (Nonsymmetric Two-Level Method).

$v_k = u_k^i + C_k(f_k - A_k u_k^i)$. `[Coarse-level correction]`
$u_k^{i+1} = v_k + R_k(f_k - A_k v_k)$. `[Post-smoothing]`

The coarse-level correction operator has the form $C_k = I_{k-1}^k A_{k-1}^{-1} I_k^{k-1}$, and the smoothing operator is one of the classical iterations. This two-level iteration, a composition of two linear iterations of the form of Algorithm 10.6.1, can itself be written in the form of Algorithm 10.6.1:

$$\begin{aligned} u_k^{i+1} &= v_k + R_k(f_k - A_k v_k) \\ &= u_k^i + C_k(f_k - A_k u_k^i) + R_k f_k - R_k A_k(u_k^i + C_k(f_k - A_k u_k^i)) \\ &= (I - C_k A_k - R_k A_k + R_k A_k C_k A_k) u_k^i + (C_k + R_k - R_k A_k C_k) f_k \\ &= (I - B_k A_k) u_k^i + B_k f_k. \end{aligned}$$

The *two-level operator* B_k, the approximate inverse of A_k which is implicitly defined by the nonsymmetric two-level method, has the form:

$$B_k = C_k + R_k - R_k A_k C_k. \tag{10.6.20}$$

The error propagation operator for the two-level method has the usual form $E_k = I - B_k A_k$, which now can be factored due to the form for B_k above:

$$E_k = I - B_k A_k = (I - R_k A_k)(I - C_k A_k). \tag{10.6.21}$$

In the case that ν post-smoothing iterations are performed in step (2) instead of a single post-smoothing iteration, it is not difficult to show that the error propagation operator takes the altered form

$$I - B_k A_k = (I - R_k A_k)^{\nu}(I - C_k A_k).$$

Now consider a symmetric form of the above two-level method:

Algorithm 10.6.8 (Symmetric Two-Level Method).

$$\begin{aligned} w_k &= u_k^i + R_k^T(f_k - A_k u_k^i). && \texttt{[Pre-smoothing]} \\ v_k &= w_k + C_k(f_k - A_k w_k). && \texttt{[Coarse-level correction]} \\ u_k^{i+1} &= v_k + R_k(f_k - A_k v_k). && \texttt{[Post-smoothing]} \end{aligned}$$

As in the nonsymmetric case, it is a simple task to show that this two-level iteration can be written in the form of Algorithm 10.6.1:

$$u_k^{i+1} = (I - B_k A_k)u_k^i + B_k f_k,$$

where after a simple expansion as for the nonsymmetric method above, the *two-level operator* B_k implicitly defined by the symmetric method can be seen to be

$$B_k = R_k + C_k + R_k^T - R_k A_k C_k - R_k A_k R_k^T - C_k A_k R_k^T + R_k A_k C_k A_k R_k^T.$$

It is easily verified that the factored form of the resulting error propagator E_k^s for the symmetric algorithm is

$$E_k^s = I - B_k A_k = (I - R_k A_k)(I - C_k A_k)(I - R_k^T A_k).$$

Note that the operator $I - B_k A_k$ is A_k-self-adjoint, which by Lemma 10.6.4 is true if and only if B_k is symmetric, implying the symmetry of B_k. The operator B_k constructed by the symmetric two-level iteration is always symmetric if the smoothing operator R_k is symmetric; however, it is also true in the symmetric algorithm above when general nonsymmetric smoothing operators R_k are used, because we use the adjoint R_k^T of the post-smoothing operator R_k as the pre-smoothing operator. The symmetry of B_k is important for use as a preconditioner for the conjugate gradient method, which requires that B_k be symmetric for guarantee of convergence.

REMARK. Note that this alternating technique for producing symmetric operators B_k can be extended to multiple nonsymmetric smoothing iterations, as suggested in [37]. Denote the variable nonsymmetric smoothing operator $R_k^{(i)}$ as

$$R_k^{(j)} = \begin{cases} R_k, & j \text{ odd}, \\ R_k^T, & j \text{ even}. \end{cases}$$

If ν pre-smoothings are performed, alternating between R_k and R_k^T, and ν post-smoothings are performed alternating in the opposite way, then a tedious computation shows that the error propagator has the factored form

$$I - B_k A_k = \left(\prod_{j=\nu}^{1} (I - R_k^{(j)} A_k) \right) (I - C_k A_k) \left(\prod_{j=1}^{\nu} (I - (R_k^{(j)})^T A_k) \right),$$

where we adopt the convention that the first terms indexed by the products appear on the left. It is easy to verify that $I - B_k A_k$ is A_k-self-adjoint, so that B_k is symmetric.

Variational Conditions and A-Orthogonal Projection. Up to this point, we have specified the approximate inverse corresponding to the coarse-level subspace correction only as $C_k = I_{k-1}^k A_{k-1}^{-1} I_k^{k-1}$, for some coarse-level operator A_{k-1}. Consider the case that the variational conditions (10.6.17) are satisfied. The error propagation operator for the coarse-level correction then takes the form

$$I - C_k A_k = I - I_{k-1}^k A_{k-1}^{-1} I_k^{k-1} A_k = I - I_{k-1}^k [(I_{k-1}^k)^T A_k I_{k-1}^k]^{-1} (I_{k-1}^k)^T A_k.$$

This last expression is simply the A_k-orthogonal projector $I - P_{k;k-1}$ onto the complement of the coarse-level subspace, where the unique orthogonal and A_k-orthogonal projectors $Q_{k;k-1}$ and $P_{k;k-1}$ projecting $\mathcal{H}_k$ onto $I_{k-1}^k \mathcal{H}_{k-1}$ can be written as

$$Q_{k;k-1} = I_{k-1}^k [(I_{k-1}^k)^T I_{k-1}^k]^{-1} (I_{k-1}^k)^T,$$
$$P_{k;k-1} = C_k A_k = I_{k-1}^k [(I_{k-1}^k)^T A_k I_{k-1}^k]^{-1} (I_{k-1}^k)^T A_k.$$

In other words, if the variational conditions are satisfied, and the coarse-level equations are solved exactly, then the coarse-level correction projects the error onto the A_k-orthogonal complement of the coarse-level subspace. It is now not surprising that successively refined finite element discretizations satisfy the variational conditions naturally, since they are defined in terms of A_k-orthogonal projections.

Note the following interesting relationship between the symmetric and nonsymmetric two-level methods, which is a consequence of the A_k-orthogonal projection property.

Lemma 10.6.12. *If the variational conditions (10.6.17) hold, then the nonsymmetric and symmetric propagators E_k and E_k^s are related by*

$$\|E_k^s\|_{A_k} = \|E_k\|_{A_k}^2.$$

Proof. Since $I - C_k A_k$ is a projector, we have $(I - C_k A_k)^2 = I - C_k A_k$. It follows that

$$\begin{aligned} E_k^s &= (I - R_k A_k)(I - C_k A_k)(I - R_k^T A_k) \\ &= (I - R_k A_k)(I - C_k A_k)(I - C_k A_k)(I - R_k^T A_k) = E_k E_k^*, \end{aligned}$$

where E_k^* is the A_k-adjoint of E_k. Therefore, the convergence of the symmetric algorithm is related to that of the nonsymmetric algorithm by:

$$\|E_k^s\|_{A_k} = \|E_k E_k^*\|_{A_k} = \|E_k\|_{A_k}^2.$$

□

REMARK. The relationship between the symmetric and nonsymmetric error propagation operators in Lemma 10.6.12 was first pointed out by McCormick in [131], and has been exploited in many papers; see [36, 100, 178]. It allows one to use the symmetric form of the algorithm as may be necessary for use with conjugate gradient methods while exploiting the relationship above to work only with the nonsymmetric error propagator E_k in analysis, which may be easier to analyze.

Multilevel Methods. Consider now the full nested sequence of Hilbert spaces $\mathcal{H}_1 \subset \mathcal{H}_2 \subset \cdots \subset \mathcal{H}_J \equiv \mathcal{H}$. The idea of the multilevel method is to begin with the two-level method, but rather than solve the coarse-level equations exactly, yet another two-level method is used to solve the coarse-level equations approximately, beginning with an initial approximation of zero on the coarse-level. The idea is applied recursively until the cost of solving the coarse system is negligible, or until the coarsest possible level is reached. Two nested simplicial mesh hierarchies for building piecewise-linear finite element spaces in the case $\mathcal{H}_k = \mathcal{V}_k$ are shown in Figure 10.12.

The following is a recursively defined multilevel algorithm which corresponds to the form of the algorithm commonly implemented on a computer. For the system $Au = f$, the algorithm returns the approximate solution u^{i+1} after one iteration of the method applied to the initial approximate u^i.

Algorithm 10.6.9 (Nonsymmetric Multilevel Method: Implementation Form).

Set:
  $u^{i+1} = ML(J, u^i, f)$
where $u_k^{i+1} = ML(k, u_k^i, f_k)$ is defined recursively as:
  If $(k = 1)$ Then:
    $u_1^{i+1} = A_1^{-1} f_1$. [Direct solve]
  Else:
    $v_k = u_k^i + I_{k-1}^k (ML(k-1, 0, I_k^{k-1}(f_k - A_k u_k^i)))$. [Correction]
    $u_k^{i+1} = v_k + R_k(f_k - A_k v_k)$. [Post-smoothing]
  End.

As with the two-level Algorithm 10.6.7, it is a straightforward calculation to write the multilevel Algorithm 10.6.9 in the standard form of Algorithm 10.6.1, where now

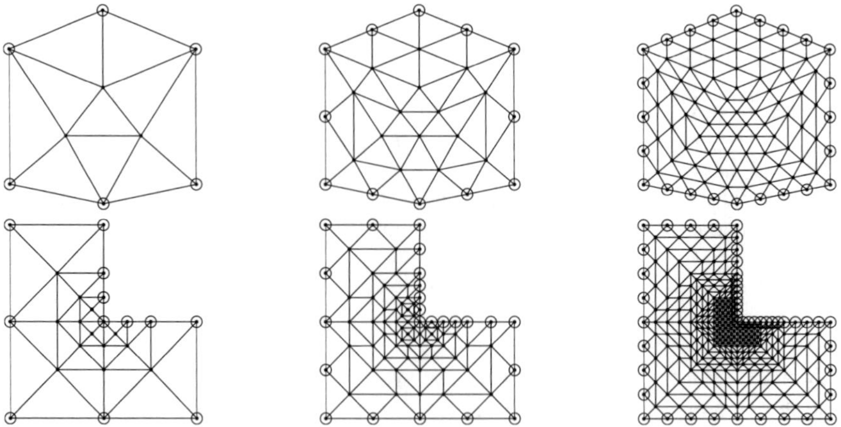

Figure 10.12 Unstructured three-level mesh hierarchies for two example domains. The nested refinements are achieved by successive quadra-section (subdivision into four similar subtriangles). Nested hierarchies of finite element spaces are then built over theses nested triangulations.

the *multilevel operator* $B \equiv B_J$ is defined recursively. To begin, assume that the approximate inverse of A_{k-1} at level $k-1$ implicitly defined by Algorithm 10.6.9 has been explicitly identified and denoted as B_{k-1}. The coarse-level correction step of Algorithm 10.6.9 at level k can then be written as

$$v_k = u_k^i + I_{k-1}^k B_{k-1} I_k^{k-1}(f_k - A_k u_k^i).$$

At level k, Algorithm 10.6.9 can be thought of as the two-level Algorithm 10.6.7, where the two-level operator $C_k = I_{k-1}^k A_{k-1}^{-1} I_k^{k-1}$ has been replaced by the approximation $C_k = I_{k-1}^k B_{k-1} I_k^{k-1}$. From (10.6.20) we see that the expression for the multilevel operator B_k at level k in terms of the operator B_{k-1} at level $k-1$ is given by

$$B_k = I_{k-1}^k B_{k-1} I_k^{k-1} + R_k - R_k A_k I_{k-1}^k B_{k-1} I_k^{k-1}. \tag{10.6.22}$$

We can now state a second multilevel algorithm, which is mathematically equivalent to Algorithm 10.6.9, but which is formulated explicitly in terms of the recursively defined multilevel operators B_k.

Algorithm 10.6.10 (Nonsymmetric Multilevel Method: Operator Form).

Set: $u^{i+1} = u^i + B(f - Au^i)$,
where the operator $B \equiv B_J$ is defined recursively:
  Let $B_1 = A_1^{-1}$, and assume that B_{k-1} has been defined.
  $B_k = I_{k-1}^k B_{k-1} I_k^{k-1} + R_k - R_k A_k I_{k-1}^k B_{k-1} I_k^{k-1}, \quad k = 2, \ldots, J.$

REMARK. Recursive definition of multilevel operators B_k apparently first appeared in [36], although operator recursions for the error propagators $E_k = I - B_k A_k$ appeared earlier in [125]. Many of the results on finite element-based multilevel methods depend on the recursive definition of the multilevel operators B_k.

As was noted for the two-level case, the error propagator at level k can be factored as:

$$E_k = I - B_k A_k = (I - R_k A_k)(I - I_{k-1}^k B_{k-1} I_k^{k-1} A_k). \tag{10.6.23}$$

It can be shown (see [39, 87, 100, 175, 178]) that the multilevel error propagator can actually be factored into a full product.

Lemma 10.6.13. *If variational conditions (10.6.17) hold, the error propagator E of Algorithm 10.6.10 can be factored:*

$$E = I - BA = (I - T_J)(I - T_{J-1}) \cdots (I - T_1), \tag{10.6.24}$$

where

$$T_1 = I_1 A_1^{-1} I_1^T A, \qquad T_k = I_k R_k I_k^T A, \qquad k = 2, \ldots, J,$$

with

$$I_J = I, \quad I_k = I_{J-1}^J I_{J-2}^{J-1} \cdots I_{k+1}^{k+2} I_k^{k+1}, \quad k = 1, \ldots, J-1.$$

Moreover, one has the additional variational condition

$$A_k = I_k^T A I_k. \tag{10.6.25}$$

Proof. Let us begin by expanding the second term in (10.6.23) more fully and then factoring again:

$$\begin{aligned} I - I_{k-1}^k B_{k-1} I_k^{k-1} A_k &= I - I_{k-1}^k (I_{k-2}^{k-1} B_{k-2} I_{k-1}^{k-2} + R_{k-1} \\ &\quad - R_{k-1} A_{k-1} I_{k-2}^{k-1} B_{k-2} I_{k-1}^{k-2}) I_k^{k-1} A_k \\ &= I - I_{k-2}^k B_{k-2} I_k^{k-2} A_k - I_{k-1}^k R_{k-1} I_k^{k-1} A_k \\ &\quad + I_{k-1}^k R_{k-1} (I_k^{k-1} A_k I_{k-1}^k) I_{k-2}^{k-1} B_{k-2} I_k^{k-2} A_k \\ &= I - I_{k-2}^k B_{k-2} I_k^{k-2} A_k - I_{k-1}^k R_{k-1} I_k^{k-1} A_k \\ &\quad + (I_{k-1}^k R_{k-1} I_k^{k-1} A_k)(I_{k-2}^k B_{k-2} I_k^{k-2} A_k) \\ &= (I - I_{k-1}^k R_{k-1} I_k^{k-1} A_k)(I - I_{k-2}^k B_{k-2} I_k^{k-2} A_k), \end{aligned}$$

where we have assumed that the first part of the variational conditions (10.6.17) holds. In general, we have

$$I - I_{k-i}^k B_{k-i} I_k^{k-i} A_k = (I - I_{k-i}^k R_{k-i} I_k^{k-i} A_k)(I - I_{k-i-1}^k B_{k-i-1} I_k^{k-i-1} A_k).$$

Using this result inductively, beginning with $k = J$, the error propagator $E \equiv E_J$ takes the *product form*:

$$E = I - BA = (I - T_J)(I - T_{J-1}) \cdots (I - T_1).$$

The second part of the variational conditions (10.6.17) implies that the T_k are A-self-adjoint and have the form

$$T_1 = I_1 A_1^{-1} I_1^T A, \qquad T_k = I_k R_k I_k^T A, \qquad k = 2, \ldots, J.$$

That (10.6.25) holds follows from the definitions. □

Note that this lemma implies that the multilevel error propagator has precisely the same form as the multiplicative Schwarz domain decomposition error propagator. One can also define an additive version via the sum

$$E = I - BA = T_1 + T_2 + \cdots + T_J, \tag{10.6.26}$$

where B is now an additive preconditioner, again identical in form to the additive Schwarz domain decomposition error propagator. Lemma 10.6.13 made it possible to consider multilevel and domain decomposition methods as particular instances of a general class of *Schwarz methods*, which allowed for the development of a very general convergence theory; see, for example, [66, 87, 93, 178] for more detailed expositions of this convergence theory framework.

The V-Cycle, the W-Cycle, and Nested Iteration. The methods we have just described are standard examples of *multigrid* or *multilevel methods* [85], where we have introduced a few restrictions for convenience, such as equal numbers of pre- and post-smoothings, one coarse space correction per iteration, and pre-smoothing with the adjoint of the post-smoothing operator. These restrictions are unnecessary in practice, but are introduced to make the analysis of the methods somewhat simpler, and to result in a symmetric preconditioner as required for combination with the conjugate gradient method.

The procedure just outlined involving correcting with the coarse space once each iteration is referred to as the *V-cycle* [40]. A similar procedure is the *Variable V-cycle*, whereby the number of smoothing iterations in one cycle is increased as coarser spaces are visited [38]. Another variation is termed the *W-cycle*, in which two coarse space corrections are performed per level at each iteration. More generally, the *p-cycle* would involve p coarse space corrections per level at each iteration for some integer $p \geqslant 1$. The *full multigrid method* [40] or *nested iteration technique* [85] begins with the coarse space, prolongates the solution to a finer space, performs a p-cycle, and repeats the process until a p-cycle is performed on the finest level. The methods can be depicted as in Figure 10.13.

Complexity of Classical, CG, DD, and Multilevel Methods

We compare the complexity of multilevel methods to some classical linear iterations for discrete elliptic equations $Au = f$ on the space U (omitting the subscript k here and below since only one space is involved), where A is an SPD matrix. Our purpose is to explain briefly the motivation for considering the more complex domain decomposition and multilevel methods as essential alternatives to the classical methods.

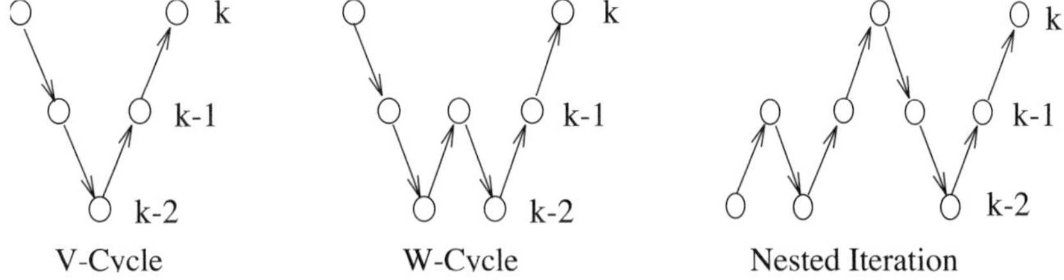

Figure 10.13 The V-cycle, the W-cycle, and nested iteration.

Convergence and Complexity of Classical Methods. Since A is SPD, we may write $A = D - L - L^T$, where D is a diagonal matrix and L a strictly lower-triangular matrix. The Richardson variation of Algorithm 10.6.1 takes λ^{-1} as the approximate inverse $B \approx A^{-1}$ of A, where λ is a bound on the largest eigenvalue of A:

$$u^{i+1} = (I - \lambda^{-1}A)u^i + \lambda^{-1}f. \tag{10.6.27}$$

The Jacobi variation of Algorithm 10.6.1 takes D^{-1} as the approximate inverse B:

$$u^{i+1} = (I - D^{-1}A)u^i + D^{-1}f. \tag{10.6.28}$$

In the Gauss-Seidel variant, the approximate inverse is taken to be $(D-L)^{-1}$, giving

$$u^{i+1} = (I - (D-L)^{-1}A)u^i + (D-L)^{-1}f. \tag{10.6.29}$$

The SOR variant takes the approximate inverse as $\omega(D - \omega L)^{-1}$, giving

$$u^{i+1} = (I - \omega(D - \omega L)^{-1}A)u^i + \omega(D - \omega L)^{-1}f. \tag{10.6.30}$$

When the model problem of the Poisson equation on a uniform mesh is considered, then the eigenvalues of both A and the error propagation matrix $I - BA$ can be determined analytically. This allows for an analysis of the convergence rates of the Richardson, Jacobi, and Gauss-Seidel iterations.

To give an example of the convergence results which are available for these classical methods, first recall that for the real square matrix A, the splitting $A = M - R$ is called a *regular splitting* (see [169]) of A if $R > 0$, M is nonsingular, and $M^{-1} \geqslant 0$. Note that an alternative construction of the Jacobi and Gauss-Seidel methods is through matrix splittings. For example, given the particular matrix splitting $A = M - R = D - (L + U)$, which corresponds to the Jacobi iteration, the resulting iteration can be written in terms of M and R as follows:

$$\begin{aligned} u^{i+1} &= (I - D^{-1}A)u^i + D^{-1}f = (I - M^{-1}(M - R))u^i + M^{-1}f \\ &= M^{-1}Ru^i + M^{-1}f. \end{aligned}$$

Therefore, for a splitting $A = M - R$, the convergence of the resulting linear method is governed completely by the spectral radius of the error propagation matrix, $\rho(M^{-1}R)$. The following standard theorem gives a sufficient condition for convergence of the Jacobi and Gauss-Seidel iterations, which can be considered to be regular splittings of A.

Theorem 10.6.4. *If A is an M-matrix, and M is obtained from A by setting off-diagonal elements of A to zero, then the splitting $A = M - R$ is regular and the corresponding linear iteration defined by the splitting is convergent: $\rho(M^{-1}R) < 1$.*

Proof. This follows from Theorem 10.6.1; see also [169]. □

Given that λ is the largest eigenvalue (or an upper bound on the largest eigenvalue) of A, we remark that Richardson's method is always trivially convergent since each eigenvalue $\lambda_j(E)$ of E is bounded by 1:

$$\lambda_j(E) = \lambda_j(I - BA) = \lambda_j(I - \lambda^{-1}A) = 1 - \lambda^{-1}\lambda_j(A) < 1.$$

However, the following difficulty makes these classical linear methods impractical for large problems. Consider the case of the three-dimensional Poisson's equation on the unit cube with zero Dirichlet boundary conditions, discretized with the box-method on a uniform mesh with m meshpoints in each mesh direction ($n = m^3$) and mesh spacing $h = 1/(m + 1)$. It is well-known that the eigenvalues of the resulting matrix A can be expressed in closed form

$$\lambda_j = \lambda_{\{p,q,r\}} = 6 - 2\cos p\pi h - 2\cos q\pi h - 2\cos r\pi h, \quad p, q, r = 1, \ldots, m.$$

Clearly, the largest eigenvalue of A is $\lambda = 6(1 - \cos m\pi h)$, and the smallest is $\lambda_1 = 6(1 - \cos \pi h)$. It is not difficult to show (see [169] or [184] for the two-dimensional case) that the largest eigenvalue of the Jacobi error propagation matrix $I - D^{-1}A$ is in this case equal to $\cos \pi h$. It is also well-known that for consistently ordered matrices with *Property $\mathcal{A}$* (see [184]), the spectral radius of the Gauss-Seidel error propagation matrix is the square of the Jacobi matrix spectral radius; more generally, the relationship between the Jacobi and Gauss-Seidel spectral radii is given by the *Stein-Rosenberg Theorem* (again see [169], or [184]). An expression for the spectral radius of the SOR error propagation matrix can also be derived; the spectral radii for the classical methods are then:

- Richardson: $\rho(E) = 1 - 6\lambda^{-1}(1 - \cos \pi h) \approx 1 - 3\lambda^{-1}\pi^2 h^2 = 1 - \mathcal{O}(h^2)$
- Jacobi: $\rho(E) = \cos \pi h \approx 1 - \frac{1}{2}\pi^2 h^2 = 1 - \mathcal{O}(h^2)$
- Gauss-Seidel: $\rho(E) = \cos^2 \pi h \approx 1 - \pi^2 h^2 = 1 - \mathcal{O}(h^2)$
- SOR: $\rho(E) \approx 1 - \mathcal{O}(h)$

The same dependence on h is exhibited for one- and two-dimensional problems. Therein lies the problem: As $h \to 0$, then for the classical methods, $\rho(E) \to 1$, so that the methods converge more and more slowly as the problem size is increased.

REMARK. An alternative convergence proof for the Jacobi and Gauss-Seidel iterations follows simply by noting that the matrix $I - E^*E$ is A-positive for both the Jacobi and Gauss-Seidel error propagators E, and by employing Lemma 10.6.2, or the related Stein's Theorem. Stein's Theorem is the basis for the proof of the Ostrowski-Reich SOR convergence theorem (see [139]).

In the case of a uniform $m \times m \times m$ mesh and the standard box-method discretization of Poisson's equation on the unit cube, the resulting algebraic system is of dimension $N = m^3$. It is well-known that the computational complexities of dense, banded, and sparse Gaussian elimination are $\mathcal{O}(N^3)$, $\mathcal{O}(N^{7/3})$, and $\mathcal{O}(N^2)$, respectively, with storage requirements that are also worse than linear (even if the matrix A itself requires only storage linear in N). In order to understand how the iterative methods we have discussed in this chapter compare to direct methods as well as to each other in terms of *complexity*, we must translate their respective known convergence properties for the model problem into a complexity estimate.

Assume now that the discretization error is $\mathcal{O}(h^s)$ for some $s > 0$, which yields a practical linear iteration tolerance of $\epsilon = \mathcal{O}(h^s)$. As remarked earlier, if the mesh is shape-regular and quasi-uniform, then the mesh size h is related to the number of discrete unknowns N through the dimension d of the spatial domain as $h = \mathcal{O}(N^{-1/d})$. Now, for the model problem, we showed above that the spectral radii of the Richardson, Jacobi, and Gauss-Seidel behave as $1 - \mathcal{O}(h^2)$. Since $-\ln(1 - ch^2) \approx ch^2 + \mathcal{O}(h^4)$, we can estimate the number of iterations required to solve the problem to the level of discretization error from (10.6.6) as follows:

$$n \geqslant \frac{|\ln \epsilon|}{|\ln \rho(E)|} = \frac{|\ln h^s|}{|\ln(1 - ch^2)|} \approx \frac{|s \ln h|}{h^2} = \mathcal{O}\left(\frac{|\ln N^{-1/d}|}{N^{-2/d}}\right) = \mathcal{O}(N^{2/d} \ln N).$$

Assuming that the cost of each iteration is $\mathcal{O}(N)$ due to the sparsity of the matrices produced by standard discretization methods, we have that the total computational cost to solve the problem using any of the three methods above for $d = 3$ is $\mathcal{O}(N^{5/3} \ln N)$. A similar model problem analysis can be carried out for other methods.

Convergence and Complexity of Multilevel Methods. Let us now examine the complexity of multilevel methods. Multilevel methods first appeared in the Russian literature in [73]. In his 1961 paper Fedorenko, described a two-level method for solving elliptic equations, and in a second paper from 1964 [74] proved convergence of a multilevel method for Poisson's equation on the square. Many theoretical results have been obtained since these first two papers. In short, what can be proven for multilevel methods under reasonable conditions is that the convergence rate or contraction number (usually, the energy norm of the error propagator E^s) is bounded by a constant below 1, independent of the mesh size and the number of levels, and hence the number of unknowns:

$$\|E^s\|_A \leqslant \delta_J < 1. \tag{10.6.31}$$

In more general situations (such as problems with discontinuous coefficients), analysis yields contraction numbers which decay as the number of levels employed in the method is increased.

If a tolerance of ϵ is required, then the computational cost to reduce the energy norm of the error below the tolerance can be determined from (10.6.6) and (10.6.31):

$$i \geqslant \frac{|\ln \epsilon|}{|\ln \delta_J|} \geqslant \frac{|\ln \epsilon|}{|\ln \|E^s\|_A|}.$$

The discretization error of $\mathcal{O}(h_J^s)$ for some $s > 0$ yields a practical tolerance of $\epsilon = \mathcal{O}(h_J^s)$. As remarked earlier, for a shape-regular and quasi-uniform mesh, the mesh size h_J is related to the number of discrete unknowns n_J through the dimension d of the spatial domain as $n_J = \mathcal{O}(h_J^{-d})$. Assuming that $\delta_J < 1$ independently of J and h_J, we have that the maximum number of iterations i required to reach an error on the order of discretization error is

$$i \geqslant \frac{|\ln \epsilon|}{|\ln \delta_J|} = \mathcal{O}(|\ln h_J|) = \mathcal{O}(|\ln n_J^{-1/d}|) = \mathcal{O}(\ln n_J). \tag{10.6.32}$$

Consider now that the operation count o_J of a single (p-cycle) iteration of Algorithm 10.6.9 with J levels is given by

$$\begin{aligned} o_J &= p o_{J-1} + C n_J = p(p o_{J-2} + C n_{J-1}) + C n_J = \cdots \\ &= p^{J-1} o_1 + C \sum_{k=2}^{J} p^{J-k} n_k, \end{aligned}$$

where we assume that the post-smoothing iteration has cost Cn_k for some constant C independent of the level k, and that the cost of a single coarse-level correction is given by o_{k-1}. Now, assuming that the cost to solve the coarse problem o_1 can be ignored, then it is not difficult to show from the expression for o_J above that the computational cost of each multilevel iteration is $\mathcal{O}(n_J)$ if (and only if) the dimensions of the spaces $\mathcal{H}_k$ satisfy

$$n_{k_1} < \frac{C_1}{p^{k_2-k_1}} n_{k_2}, \quad \forall k_1, k_2, \quad k_1 < k_2 \leqslant J,$$

where C_1 is independent of k. This implies both of the following:

$$n_k < \frac{C_1}{p} n_{k+1}, \qquad n_k < \frac{C_1}{p^{J-k}} n_J, \qquad k = 1, \ldots, J-1.$$

Consider the case of nonuniform Cartesian meshes which are successively refined, so that $h_{k_1} = 2^{k_2-k_1} h_{k_2}$ for $k_1 < k_2$, and in particular $h_{k-1} = 2h_k$. This gives

$$\begin{aligned} n_{k_1} &= C_2 h_{k_1}^{-d} = C_2 (2^{k_2-k_1} h_{k_2})^{-d} = C_2 2^{-d(k_2-k_1)} (C_3 n_{k_2}^{-1/d})^{-d} \\ &= \frac{C_2 C_3^{-d}}{(2^d)^{k_2-k_1}} n_{k_2}. \end{aligned}$$

Therefore, if $2^{d(k_2-k_1)} > p^{k_2-k_1}$, or if $2^d > p$, which is true in two dimensions ($d = 2$) for $p \leqslant 3$, and in three dimensions ($d = 3$) for $p \leqslant 7$, then each multilevel

Table 10.1 Model problem computational complexities of various solvers.

Method	$3D$	$3D$
Dense Gaussian elimination	$\mathcal{O}(N^3)$	$\mathcal{O}(N^3)$
Banded Gaussian elimination	$\mathcal{O}(N^2)$	$\mathcal{O}(N^{2.33})$
Sparse Gaussian elimination	$\mathcal{O}(N^{1.5})$	$\mathcal{O}(N^2)$
Richardson's method	$\mathcal{O}(N^2 \ln N)$	$\mathcal{O}(N^{1.67} \ln N)$
Jacobi iteration	$\mathcal{O}(N^2 \ln N)$	$\mathcal{O}(N^{1.67} \ln N)$
Gauss-Seidel iteration	$\mathcal{O}(N^2 \ln N)$	$\mathcal{O}(N^{1.67} \ln N)$
SOR	$\mathcal{O}(N^{1.5} \ln N)$	$\mathcal{O}(N^{1.33} \ln N)$
Conjugate gradient methods (CG)	$\mathcal{O}(N^{1.5} \ln N)$	$\mathcal{O}(N^{1.33} \ln N)$
Preconditioned CG	$\mathcal{O}(N^{1.25} \ln N)$	$\mathcal{O}(N^{1.17} \ln N)$
Multilevel methods	$\mathcal{O}(N \ln N)$	$\mathcal{O}(N \ln N)$
Nested multilevel methods	$\mathcal{O}(N)$	$\mathcal{O}(N)$
Domain decomposition methods	$\mathcal{O}(N)$	$\mathcal{O}(N)$

iteration has complexity $\mathcal{O}(n_J)$. In particular, one V-cycle ($p = 1$) or W-cycle ($p = 2$) iteration has complexity $\mathcal{O}(n_J)$ for nonuniform Cartesian meshes in two and three dimensions.

If these conditions on the dimensions of the spaces are satisfied, so that each multilevel iteration has cost $\mathcal{O}(n_J)$, then combining this with equation (10.6.32) implies that the overall complexity to solve the problem with a multilevel method is $\mathcal{O}(n_J \ln n_J)$. By using the nested iteration, it is not difficult to show using an inductive argument (see [85]) that the multilevel method improves to optimal order $\mathcal{O}(n_J)$ if $\delta_J < 1$ independent of J and h_J, meaning that the computational cost to solve a problem with n_J pieces of data is Cn_J, for some constant C which does not depend on n_J. Theoretical multilevel studies first appeared in the late 1970s and continuing up through the present have focused on extending the proofs of optimality (or near optimality) to larger classes of problems.

To summarize, the complexities of the methods we have discussed in this chapter plus a few others are given in Table 10.1. The complexities for the conjugate gradient methods applied to the model problem may be found in [12]. The entry for domain decomposition methods is based on the assumption that the complexity of the solver on each subdomain is linear in the number of degrees of freedom in the subdomain (usually requiring the use of a multilevel method), and on the assumption that a global coarse space is solved to prevent the decay of the condition number or contraction constant with the number of subdomains. This table states clearly the motivation for considering the use of multilevel and domain decomposition methods for the numerical solution of elliptic partial differential equations.

EXERCISES

10.6.1 *Derivation of the conjugate gradient method.*

1. The Cayley-Hamilton Theorem states that a square $n \times n$ matrix M satisfies its own characteristic equation:

$$P_n(M) = 0.$$

Using this result, prove that if M is also nonsingular, then the matrix M^{-1} can be written as a matrix polynomial of degree $n-1$ in M, or

$$M^{-1} = Q_{n-1}(M).$$

2. Given an SPD matrix A, show that it defines a new inner product

$$(u, v)_A = (Au, v) = \sum_{i=1}^{n} (Au)_i v_i, \quad \forall u, v \in \mathbb{R}^n,$$

called the *A-inner product*; that is, show that $(u, v)_A$ is a "true" inner product, in that it satisfies the inner product axioms.

3. Recall that the transpose M^T of an $n \times n$ matrix M is defined as

$$M_{ij}^T = M_{ji}.$$

We observed in Section 3.4 that an equivalent characterization of the transpose matrix M^T is that it is the unique *adjoint* operator satisfying

$$(Mu, v) = (u, M^T v), \quad \forall u, v \in \mathbb{R}^n,$$

where $(\cdot, \cdot)$ is the usual Euclidean inner product,

$$(u, v) = \sum_{i=1}^{n} u_i v_i.$$

The A-adjoint of a matrix M, denoted M^*, is defined as the adjoint in the A-inner product; that his, the unique matrix satisfying

$$(AMu, v) = (Au, M^* v), \quad \forall u, v \in \mathbb{R}^n.$$

Show that that an equivalent definition of M^* is

$$M^* = A^{-1} M^T A.$$

4. Consider now the matrix equation

$$Au = f,$$

where A is an $n \times n$ SPD matrix, and u and f are n-vectors. It is common to "precondition" such an equation before attempting a numerical solution, by multiplying by an approximate inverse operator $B \approx A^{-1}$ and then solving the *preconditioned system*:

$$BAu = Bf.$$

If A and B are both SPD, under what conditions is BA also SPD? Show that if A and B are both SPD, then BA is A-SPD (symmetric and positive in the A-inner product).

5. Given an initial guess u^0 for the solution of $BAu = Bf$, we can form the initial residuals

$$r^0 = f - Au^0, \quad s^0 = Br^0 = Bf - BAu^0.$$

Do a simple manipulation to show that the solution u can be written as

$$u = u^0 + Q_{n-1}(BA)s^0,$$

where $Q(\cdot)$ is the matrix polynomial representing $(BA)^{-1}$. In other words, you have established that the solution u lies in a *translated Krylov space*:

$$u \in u^0 + K_{n-1}(BA, s^0),$$

where

$$K_{n-1}(BA, s^0) = \text{span}\{s^0, BAs^0, (BA)^2 s^0, \ldots, (BA)^{n-1} s^0\}.$$

Note that we can view the Krylov spaces as a sequence of expanding subspaces

$$K_0(BA, s^0) \subset K_1(BA, s^0) \subset \cdots \subset K_{n-1}(BA, s^0).$$

6. We will now try to construct an iterative method (the CG method) for finding u. The algorithm determines the best approximation u_k to u in a subspace $K_k(BA, s^0)$ at each step k of the algorithm, by forming

$$u^{k+1} = u^k + \alpha_k p^k,$$

where p^k is such that $p^k \in K_k(BA, s^0)$ at step k, but $p^k \notin K_j(BA, s^0)$ for $j < k$. In addition, we want to enforce minimization of the error in the A-norm,

$$\|e^{k+1}\|_A = \|u - u^{k+1}\|_A,$$

at step k of the algorithm. The next iteration expands the subspace to $K_{k+1}(BA, s^0)$, finds the best approximation in the expanded space, and so on, until the exact solution in $K_{n-1}(BA, s^0)$ is reached.

To realize this algorithm, let us consider how to construct the required vectors p^k in an efficient way. Let $p^0 = s^0$, and consider the construction of an A-orthogonal basis for $K_{n-1}(BA, s^0)$ using the standard Gram-Schmidt procedure:

$$p^{k+1} = BAp^k - \sum_{i=0}^{k} \frac{(BAp^k, p^i)_A}{(p^i, p^i)_A} p^i, \quad k = 0, \ldots, n-2.$$

At each step of the procedure, we will have generated an A-orthogonal (orthogonal in the A-inner product) basis $\{p^0, \ldots, p^k\}$ for $K_k(BA, s^0)$. Now, note that by construction,

$$(p^k, v)_A = 0, \quad \forall v \in K_j(BA, s^0), \ j < k.$$

Using this fact and the fact you established previously that BA is A-self-adjoint, show that the Gram-Schmidt procedure has only three nonzero terms in the sum; namely, for $k = 0, \ldots, n-1$, it holds that

$$p^{k+1} = BAp^k - \frac{(BAp^k, p^k)_A}{(p^k, p^k)_A} p^k - \frac{(BAp^k, p^{k-1})_A}{(p^{k-1}, p^{k-1})_A} p^{k-1}.$$

Thus, there exists an efficient three-term recursion for generating the A-orthogonal basis for the solution space. Note that this three-term recursion is possible due to the fact that we are working with orthogonal (matrix) polynomials!

7. We can nearly write down the CG method now, by attempting to expand the solution in terms of our cheaply generated A-orthogonal basis. However, we need to determine how far to move in each "conjugate" direction p^k at step k after we generate p^k from the recursion. As remarked earlier, we would like to enforce minimization of the quantity

$$\|e^{k+1}\|_A = \|u - u^{k+1}\|_A$$

at step k of the iterative algorithm. It is not difficult to show that this is equivalent to enforcing

$$(e^{k+1}, p^k)_A = 0.$$

Let's assume that we have somehow enforced

$$(e^k, p^i)_A = 0, \quad i < k,$$

at the previous step of the algorithm. We have at our disposal $p^k \in K_k(BA, s^0)$, and let's take our new approximation at step $k+1$ as

$$u^{k+1} = u^k + \alpha_k p^k,$$

for some step length $\alpha_k \in \mathbb{R}$ in the direction p^k. Thus, the error in the new approximation is simply

$$e^{k+1} = e^k + \alpha_k p^k.$$

Show that in order to enforce $(e^{k+1}, p^k)_A = 0$, we must choose α_k to be

$$\alpha_k = \frac{(r^k, p^k)}{(p^k, p^k)_A}.$$

The final algorithm is now as follows.

The Conjugate Gradient Algorithm
Let $u^0 \in \mathcal{H}$ be given.
$r^0 = f - Au^0, \quad s^0 = Br^0, \quad p^0 = s^0.$
Do $k = 0, 1, \ldots$ until convergence:
$\alpha_k = (r^k, p^k)/(p^k, p^k)_A$
$u^{k+1} = u^k + \alpha_k p^k$
$r^{k+1} = r^k - \alpha_k A p^k$
$s^{k+1} = Br^{k+1}$
$\beta_{k+1} = -(BAp^k, p^k)_A/(p^k, p^k)_A$
$\gamma_{k+1} = -(BAp^k, p^{k-1})_A/(p^{k-1}, p^{k-1})_A$
$p^{k+1} = BAp^k + \beta_{k+1} p^k + \gamma_{k+1} p^{k-1}$
End do.

8. Show that equivalent expressions for some of the parameters in CG are:
   (a) $\alpha_k = (r^k, s^k)/(p^k, p^k)_A$
   (b) $\delta_{k+1} = (r^{k+1}, s^{k+1})/(r^k, s^k)$
   (c) $p^{k+1} = s^{k+1} + \delta_{k+1} p^k$

   In other words, the CG algorithm you have derived from first principles in this exercise, using only the idea of orthogonal projection onto an expanding set of subspaces, is mathematically equivalent to Algorithm 10.6.2.

   Remark: The CG algorithm that appears in most textbooks is formulated to employ these equivalent expressions due to the reduction in computational work of each iteration.

10.6.2 *Properties of the conjugate gradient method.*

In this exercise, we will establish some simple properties of the CG method derived in Exercise 10.6.1. (Although this analysis is standard, you will have difficulty finding all of the pieces in one text.)

1. It is not difficult to show that the error in the CG algorithm propagates as

$$e^{k+1} = [I - BAp_k(BA)]e^0,$$

where $p_k \in \mathcal{P}_k$, the space of polynomials of degree k. By construction, we know that this polynomial is such that

$$\|e^{k+1}\|_A = \min_{p_k \in \mathcal{P}_k} \|[I - BAp_k(BA)]e^0\|_A.$$

Now, since BA is A-SPD, we know that it has real positive eigenvalues $\lambda_j \in \sigma(BA)$, and further, that the corresponding eigenvectors v_j of BA are orthonormal. Using the expansion of the initial error

$$e^0 = \sum_{j=1}^{n} \alpha_j v_j,$$

establish the inequality

$$\|e^{k+1}\|_A \leqslant \left(\min_{p_k \in \mathcal{P}_k} \left[\max_{\lambda_j \in \sigma(BA)} |1 - \lambda_j p_k(\lambda_j)| \right] \right) \|e^0\|_A.$$

The polynomial which minimizes the maximum norm above is said to solve a *mini-max problem*.

2. It is well-known in approximation theory that the Chebyshev polynomials

$$T_k(x) = \cos(k \arccos x)$$

solve mini-max problems of the type above, in the sense that they deviate least from zero (in the *max-norm* sense) in the interval $[-1, 1]$, which can be shown to be due to their unique *equi-oscillation* property. (These facts can be found in any introductory numerical analysis text.) If we extend the Chebyshev polynomials outside the interval $[-1, 1]$ in the natural way, it can be shown that shifted and scaled forms of the Chebyshev polynomials solve the mini-max problem above. In particular, the solution is simply

$$1 - \lambda p_k(\lambda) = \tilde{p}_{k+1}(\lambda) = \frac{T_{k+1}\left(\dfrac{\lambda_{\max} + \lambda_{\min} - 2\lambda}{\lambda_{\max} - \lambda_{\min}} \right)}{T_{k+1}\left(\dfrac{\lambda_{\max} + \lambda_{\min}}{\lambda_{\max} - \lambda_{\min}} \right)}.$$

Use an obvious property of the polynomials $T_{k+1}(x)$ to conclude that

$$\|e^{k+1}\|_A \leqslant \left[T_{k+1}\left(\frac{\lambda_{\max} + \lambda_{\min}}{\lambda_{\max} - \lambda_{\min}} \right) \right]^{-1} \|e_0\|_A.$$

3. Use one of the Chebyshev polynomial results given in Exercise 10.6.3 below to refine this inequality to

$$\|e^{k+1}\|_A \leqslant 2\left(\frac{\sqrt{\dfrac{\lambda_{\max}(BA)}{\lambda_{\min}(BA)}}-1}{\sqrt{\dfrac{\lambda_{\max}(BA)}{\lambda_{\min}(BA)}}+1}\right)^{k+1}\|e^0\|_A.$$

Now, recall that the A*-condition number* of the matrix BA is defined just as the normal condition number, except employing the A-norm:

$$\kappa_A(BA) = \|BA\|_A\|(BA)^{-1}\|_A.$$

Since the matrix BA is A-self-adjoint, it can be shown that, in fact,

$$\kappa_A(BA) = \|BA\|_A\|(BA)^{-1}\|_A = \frac{\lambda_{\max}(BA)}{\lambda_{\min}(BA)},$$

so that the error reduction inequality above can be written more simply as

$$\begin{aligned}\|e^{k+1}\|_A &\leqslant 2\left(\frac{\sqrt{\kappa_A(BA)}-1}{\sqrt{\kappa_A(BA)}+1}\right)^{k+1}\|e^0\|_A \\ &= 2\left(1-\frac{2}{1+\sqrt{\kappa_A(BA)}}\right)^{k+1}\|e^0\|_A.\end{aligned}$$

4. Assume that we would like to achieve the following accuracy in our iteration after some number of steps n:

$$\frac{\|e^{n+1}\|_A}{\|e^0\|_A} < \epsilon.$$

Using the approximation

$$\begin{aligned}\ln\left(\frac{a-1}{a+1}\right) &= \ln\left(\frac{1+(-1/a)}{1-(-1/a)}\right) \\ &= 2\left[\left(\frac{-1}{a}\right)+\frac{1}{3}\left(\frac{-1}{a}\right)^3+\frac{1}{5}\left(\frac{-1}{a}\right)^5+\cdots\right] \\ &< \frac{-2}{a},\end{aligned}$$

show that we can achieve this error tolerance if n satisfies

$$n = \mathcal{O}\left(\kappa_A^{1/2}(BA)\left|\ln\frac{\epsilon}{2}\right|\right).$$

5. Many types of matrices have $\mathcal{O}(1)$ nonzeros per row (for example, finite element and other discretizations of ordinary and partial differential equations.) If A is an $n \times n$ matrix, then the cost of one iteration of CG (Algorithm 10.6.2) will be $\mathcal{O}(n)$, as would one iteration of the basic linear method (Algorithm 10.6.1). What is the overall complexity [in terms of n and $\kappa_A(BA)$] to solve the problem to a given tolerance ϵ? If $\kappa_A(BA)$ can be bounded by a constant, independent of the problem size n, what is the complexity? Is this then an optimal method?

10.6.3 *Properties of the Chebyshev polynomials.*

The Chebyshev polynomials are defined as

$$t_n(x) = \cos(n \cos^{-1} x), \qquad n = 0, 1, 2, \ldots.$$

Taking $t_0(x) = 1$, $t_1(x) = x$, it can be shown that the Chebyshev polynomials are an orthogonal family that can be generated by the standard recursion (which holds for any orthogonal polynomial family):

$$t_{n+1}(x) = 2t_1(x)t_n(x) - t_{n-1}(x), \qquad n = 1, 2, 3, \ldots.$$

Prove the following extremely useful relationships:

$$t_k(x) = \frac{1}{2}\left[\left(x + \sqrt{x^2 - 1}\right)^k + \left(x - \sqrt{x^2 - 1}\right)^k\right], \qquad \forall x, \quad (10.6.33)$$

$$t_k\left(\frac{\alpha + 1}{\alpha - 1}\right) > \frac{1}{2}\left(\frac{\sqrt{\alpha} + 1}{\sqrt{\alpha} - 1}\right)^k, \qquad \forall \alpha > 1. \quad (10.6.34)$$

These two results are fundamental in the convergence analysis of the conjugate gradient method in the earlier exercises in the section. [Hint: For the first result, use the fact that $\cos k\theta = (e^{ik\theta} + e^{-ik\theta})/2$. The second result will follow from the first after some algebra.]

10.7 METHODS FOR NONLINEAR EQUATIONS

Building on the material assembled in Section 10.1 on nonlinear equations and calculus in Banach spaces, we now consider some of the classical nonlinear iterations and nonlinear conjugate gradient methods for solving nonlinear equations in finite-dimensional Hilbert spaces. Newton-like methods are then reviewed, including inexact variations and global convergence modifications. We then discuss damped inexact Newton multilevel methods, which involve the coupling of damped Newton methods with linear multilevel methods for approximate solution of the linearized systems. We then combine the damping (or *backtracking*) parameter selection and linear iteration tolerance specification to ensure global superlinear convergence. We also describe nonlinear multilevel methods proposed by Hackbusch and others, which do not involve an outer Newton iteration.

While we only have space to cover a few of the main ideas, our discussion in this section follows closely some of the standard references for nonlinear equations in $\mathbb{R}^n$, such as [78, 140], as well as standard references for generalizations to Banach spaces, such as [108, 188]. For Newton multilevel-type methods, we also follow material from the research monographs [63, 85], as well as the articles [21, 22] and several other references cited in the text.

Standard Methods for Nonlinear Equations in $\mathbb{R}^n$

Let $\mathcal{H}$ be a Hilbert space, endowed with an inner product $(\cdot, \cdot)$ which induces a norm $\|\cdot\|$. Given a map $F\colon \mathcal{H} \to \mathcal{H}$ such that $F(u) = Au + B(u)$, where $B\colon \mathcal{H} \to \mathcal{H}$ is a nonlinear operator and where $A\colon \mathcal{H} \to \mathcal{H}$ is an invertible linear operator, we are interested in solutions to the following mathematically equivalent problems: Find $u \in \mathcal{H}$ such that any of the following hold:

$$F(u) = 0, \tag{10.7.1}$$

$$Au + B(u) = 0, \tag{10.7.2}$$

$$u = T(u), \tag{10.7.3}$$

where

$$F(u) = Au + B(u), \qquad T(u) = -A^{-1}B(u), \tag{10.7.4}$$

with $T\colon \mathcal{H} \to \mathcal{H}$. These three familiar-looking equations also arose at the end of Section 10.1 in our discussions of fixed-point theorems and ordered Banach spaces. In this section, we are interested in iterative algorithms for solving equation (10.7.1) or (10.7.2) in the setting of a finite-dimensional Hilbert space $\mathcal{H}$. We will focus entirely on general iterations of the form

$$u^{i+1} = T(u^i), \tag{10.7.5}$$

where T is as in (10.7.4), or more generally is any mapping which is constructed to have as its fixed point the unique solution u of (10.7.1) and (10.7.2).

The nonlinear extensions of the classical linear methods fit into this framework, as well as the Newton-like methods. Our interest in improved convergence, efficiency, and robustness properties will lead us to damped inexact Newton multilevel methods and nonlinear multilevel methods. We are particularly interested in the nonlinear equations which arise from discretizations of the types of semilinear elliptic partial differential equations we considered in detail in Section 10.4, leading to equations which have the additional structure (10.7.2). It will be useful to consider the following variation of (10.7.2), which obviously can be rewritten in the form of (10.7.2) by suitably redefining the operator B:

$$A_k u_k + B_k(u_k) = f_k. \tag{10.7.6}$$

These types of equations will arise from a box or finite element discretization of the types of semilinear elliptic partial differential equations we encountered in Section 10.4, as discussed in some detail in Section 10.5. The space of grid functions u_k

with values at the nodes of the mesh will be denoted as $\mathcal{U}_k$, and equation (10.7.6) may be interpreted as a nonlinear algebraic equation in the space $\mathcal{U}_k$. Equation (10.7.6) may also be interpreted as an abstract operator equation in the finite element space $\mathcal{V}_k$, as discussed in detail in Sections 10.5 and 10.6. In either case, the operator A_k is necessarily symmetric positive definite (SPD) for the problems and discretization methods we consider, while the form and properties of the nonlinear term $B_k(\cdot)$ depend on the particular problem.

To discuss algorithms for (10.7.6), and in particular multilevel algorithms, we will need a nested sequence of finite-dimensional spaces $\mathcal{H}_1 \subset \mathcal{H}_2 \subset \cdots \mathcal{H}_J \equiv \mathcal{H}$, which are connected by prolongation and restriction operators, as discussed in detail in Section 10.6. We are given the abstract nonlinear problem in the finest space $\mathcal{H}$:

$$\text{Find } u \in \mathcal{H} \text{ such that } Au + B(u) = f, \tag{10.7.7}$$

where $A \in \mathcal{L}(\mathcal{H}, \mathcal{H})$ is SPD and $B(\cdot)\colon \mathcal{H} \to \mathcal{H}$ is a nonlinearity which yields a uniquely solvable problem, and we are interested in iterative algorithms for determining the unique solution u which involves solving problems of the form:

$$\text{Find } u_k \in \mathcal{H}_k \text{ such that } A_k u_k + B_k(u_k) = f_k, \tag{10.7.8}$$

in the coarser spaces $\mathcal{H}_k$ for $1 \leqslant k < J$. When only the finest space $\mathcal{H}$ is employed, we will omit the subscripts on functions, operators, and spaces to simplify the notation.

Nonlinear Extensions of Classical Linear Methods. In this section we review nonlinear conjugate gradient methods and Newton-like methods; we also make a few remarks about extensions of the classical linear methods. We discuss at some length the one-dimensional line search required in the Fletcher-Reeves nonlinear conjugate gradient method, which we will use later for computing a global convergence damping parameter in nonlinear multilevel methods.

In Section 4.8 we discussed three distinct notions of convergence in a Banach or Hilbert space: *strong convergence* (often called simply *convergence*), *weak convergence*, and *weak-** convergence. One result we showed was that strong convergence implies weak convergence (Theorem 4.8.5), although the reverse is generally not true. However, in the setting of practical algorithms for linear and nonlinear equations, which take place in finite-dimensional Banach spaces, these three notions of convergence are all equivalent; therefore, here we are interested simply in *strong convergence* of sequences generated by iterative algorithms. Let X be a Banach space, and for ease of exposition denote the norm on X as $\|\cdot\| = \|\cdot\|_X$. Recall that the sequence $\{u^i\}$ with $u^i \in X$ is said to *converge strongly* to $u \in X$ if $\lim_{i\to\infty} \|u - u^i\| = 0$. In Section 4.8 we also defined several distinct notions of the *rate of (strong) convergence* of a sequence, such as Q-linear, Q-superlinear, Q-order(p), and R-order(p). We are interested primarily here in *fixed point iterations* of the form (10.7.5), where the nonlinear mapping $T(\cdot)$ is the *fixed point mapping*. If $T(\cdot)$ represents some iterative technique for obtaining the solution to a problem, it is important to understand what are necessary or at least sufficient conditions for this iteration to converge to a solution, and what its convergence properties are (such as *rate*). Recall that in Chapter 4 and also in Section 10.1 we examined *contraction operators*, which are maps $T: U \subset X \to U \subset X$ having the property

$$\|T(u) - T(v)\| \leqslant \alpha \|u - v\|, \quad \forall u, v \in U, \quad \alpha \in [0, 1),$$

for some *contraction constant* $\alpha \in [0, 1)$. In Section 4.4 we stated and proved the Banach Fixed-Point Theorem (or Contraction Mapping Theorem). The version of the theorem we gave ensured that a fixed point iteration involving a contraction in a closed subset of a Banach space will converge to a unique fixed point. However, in the proof given in Chapter 4, two results were established as intermediate steps that were not stated as separate conclusions of the theorem, but in fact have importance here; the first is that the convergence rate of the iteration is actually Q-linear, and the second is that the error at each iteration may be bounded by the contraction constant. In Section 10.1 we have restated a version of the Banach Fixed-Point Theorem from Chapter 4, but with these two additional conclusions emphasized.

The classical linear methods discussed in Section 10.6, such as Jacobi and Gauss-Seidel, can be extended in the obvious way to nonlinear algebraic equations of the form (10.7.6). In each case, the method can be viewed as a fixed point iteration of the form (10.7.5). Of course, implementation of these methods, which we refer to as nonlinear Jacobi and nonlinear Gauss-Seidel methods, now requires the solution of a sequence of one-dimensional nonlinear problems for each unknown in one step of the method. A variation that works well, even compared to newer methods, is the nonlinear SOR method. The convergence properties of these types of methods, as well as a myriad of variations and related methods, are discussed in detail in [140]. Note, however, that the same difficulty arising in the linear case also arises here: As the problem size is increased (the mesh size is reduced), these methods converge more and more slowly. As a result, we consider alternative methods, such as nonlinear conjugate gradient methods, Newton-like methods, and nonlinear multilevel methods.

Note that since the one-dimensional problems arising in the nonlinear Jacobi and nonlinear Gauss-Seidel methods are often solved with Newton's method, the methods are also referred to as Jacobi-Newton and Gauss-Seidel-Newton methods, meaning that the Jacobi or Gauss-Seidel iteration is the main or outer iteration, whereas the inner iteration is performed by Newton's method. Momentarily we will consider the other situation: The use of Newton's method as the outer iteration, and a linear iterative method such as multigrid for solution of the linear Jacobian system at each outer Newton iteration. We refer to this method as a Newton multilevel method.

Nonlinear Conjugate Gradient Methods. As we have seen in Sections 10.1 and 10.4, the following minimization problem:

$$\text{Find } u \in \mathcal{H} \text{ such that } J(u) = \min_{v \in \mathcal{H}} J(v), \text{ where } J(u) = \frac{1}{2}(Au, u) + G(u) - (f, u)$$

is equivalent to the associated zero-point problem:

$$\text{Find } u \in \mathcal{H} \text{ such that } F(u) = Au + B(u) - f = 0,$$

where $B(u) = G'(u)$. We assume here that both problems are uniquely solvable. An effective approach for solving the zero-point problem, by exploiting the connection with the minimization problem, is the *Fletcher-Reeves* version [76] of the nonlinear conjugate gradient method, which takes the form:

Algorithm 10.7.1 (Fletcher-Reeves Nonlinear CG Method).

```
Let u^0 ∈ H be given.
r^0 = f − B(u^0) − Au^0,      p^0 = r^0.
Do i = 0,1,... until convergence:
     α_i = (see below)
     u^{i+1} = u^i + α_i p^i
     r^{i+1} = r^i + B(u^i) − B(u^{i+1}) − α_i Ap^i
     β_{i+1} = (r^{i+1}, r^{i+1})/(r^i, r^i)
     p^{i+1} = r^{i+1} + β_{i+1} p^i
End do.
```

The expression for the residual r^{i+1} is from

$$r^{i+1} = -F(u^{i+1}) \tag{10.7.9}$$

$$= f - B(u^{i+1}) - Au^{i+1} \tag{10.7.10}$$

$$= (f - B(u^i) - Au^i) + B(u^i) - B(u^{i+1}) + A(u^i - u^{i+1}) \tag{10.7.11}$$

$$= r^i + B(u^i) - B(u^{i+1}) - \alpha_i Ap^i, \tag{10.7.12}$$

where $u^i - u^{i+1} = \alpha_i p^i$ has been used to obtain the last expresion; this holds by the definition of u^{i+1} in the second step of the "Do Loop" in Algorithm 10.7.1. The directions p^i are computed from the previous direction and the new residual, and the steplength α_i is chosen to minimize the associated functional $J(\cdot)$ in the direction p^i. In other words, α_i is chosen to minimize $J(u^i + \alpha_i p^i)$, which is equivalent to solving the one-dimensional zero-point problem:

$$\frac{dJ(u^i + \alpha_i p^i)}{d\alpha_i} = 0.$$

Given the form of $J(\cdot)$ above, we have that

$$J(u^i + \alpha_i p^i) = \frac{1}{2}(A(u^i + \alpha_i p^i), u^i + \alpha_i p^i) + G(u^i + \alpha_i p^i) - (f, u^i + \alpha_i p^i).$$

A simple differentiation with respect to α_i (and some simplification) gives

$$\frac{dJ(u^i + \alpha_i p^i)}{d\alpha_i} = \alpha_i (Ap^i, p^i) - (r^i, p^i) + (B(u^i + \alpha_i p^i) - B(u^i), p^i),$$

where $r^i = f - B(u^i) - Au^i = -F(u^i)$ is the nonlinear residual. The second derivative with respect to α_i will be useful also, and is easily seen to be

$$\frac{d^2 J(u^i + \alpha_i p^i)}{d\alpha_i^2} = (Ap^i, p^i) + (B'(u^i + \alpha_i p^i)p^i, p^i).$$

Now, Newton's method for solving the zero-point problem for α_i takes the form

$$\alpha_i^{m+1} = \alpha_i^m - \delta^m,$$

where

$$\begin{aligned}\delta^m &= \frac{dJ(u^i + \alpha_i^m p^i)/d\alpha_i}{d^2 J(u^i + \alpha_i^m p^i)/d\alpha_i^2} \\ &= \frac{\alpha_i^m (Ap^i, p^i) - (r^i, p^i) + (B(u^i + \alpha_i^m p^i) - B(u^i), p^i)}{(Ap^i, p^i) + (B'(u^i + \alpha_i^m p^i)p^i, p^i)}.\end{aligned}$$

The quantities (Ap^i, p^i) and (r^i, p^i) can be computed once at the start of each line search for α_i, each requiring an inner product (Ap^i is available from the CG iteration). Each Newton iteration for the new α_i^{m+1} then requires evaluation of the nonlinear term $B(u^i + \alpha_i^m p^i)$ and inner product with p^i, as well as evaluation of the derivative mapping $B'(u^i + \alpha_i p^i)$, application to p^i, followed by inner product with p^i.

In the case that $B(\cdot)$ arises from the discretization of a nonlinear partial differential equation and is of *diagonal form*, meaning that the j-th component function of the vector $B(\cdot)$ is a function of only the j-th component of the vector of nodal values u, or $B_j(u) = B_j(u_j)$, then the resulting Jacobian matrix $B'(\cdot)$ of $B(\cdot)$ is a diagonal matrix. This situation occurs with box-method discretizations or mass-lumped finite element discretizations of semilinear problems. As a result, computing the term $(B'(u^i + \alpha_i p^i)p^i, p^i)$ can be performed with fewer operations than two inner products.

The total cost for each Newton iteration (beyond the first) is then evaluation of $B(\cdot)$ and $B'(\cdot)$, and something less than three inner products. Therefore, the line search can be performed fairly inexpensively in certain situations. If alternative methods are used to solve the one-dimensional problem defining α_i, then evaluation of the Jacobian matrix can be avoided altogether, although as we remarked earlier, the Jacobian matrix is cheaply computable in the particular applications we are interested in here.

Note that if the nonlinear term $B(\cdot)$ is absent, then the zero-point problem is linear and the associated energy functional is quadratic:

$$F(u) = Au - f = 0, \qquad J(u) = \frac{1}{2}(Au, u) - (f, u).$$

In this case, the Fletcher-Reeves CG algorithm reduces to exactly the Hestenes-Stiefel [92] linear conjugate gradient algorithm (Algorithm 10.6.2 with the preconditioner $B = I$). The exact solution to the linear problem $Au = f$, as well as to the associated minimization problem, can be reached in no more than n_k steps (in exact arithmetic, that is), where n_k is the dimension of the space $\mathcal{H}$ (see, for example, [140]). The calculation of the steplength α_i no longer requires the iterative solution of a one-dimensional minimization problem with Newton's method, since

$$\frac{dJ(u^i + \alpha_i p^i)}{d\alpha_i} = \alpha_i (Ap^i, p^i) - (r^i, p^i) = 0$$

yields an explicit expression for the α_i which minimizes the functional J in the direction p^i:

$$\alpha_i = \frac{(r^i, p^i)}{(Ap^i, p^i)}.$$

See Exercise 10.6.1 for a guided derivation of the linear conjugate gradient method from first principles.

Newton's Method. We now consider one of the most powerful techniques for solving nonlinear problems: Newton's method. A classic reference for much of the following material is [78]. Given the nonlinear map $F\colon D \subset \mathcal{H} \to \mathcal{H}$ for some finite-dimensional Hilbert space $\mathcal{H}$, where $F \in C^2(\mathcal{H})$, we can derive Newton's method by starting with the generalized Taylor expansion (Theorem 10.1.2):

$$F(u+h) = F(u) + F'(u)h + \mathcal{O}(\|h\|^2). \tag{10.7.13}$$

One wants to find $u \in D \subset \mathcal{H}$ such that $F(u) = 0$, but have only an initial approximation $u^0 \approx u$. If the Taylor expansion could be used to determine h such that $F(u^0 + h) = 0$, then the problem would be solved by taking $u = u^0 + h$. Although the Taylor expansion is an infinite series in h, we can solve approximately for h by truncating the series after the first two terms, leaving

$$0 = F(u^0 + h) = F(u^0) + F'(u^0)h. \tag{10.7.14}$$

Writing this as an iteration leads to

$$\begin{aligned} F'(u^i)h^i &= -F(u^i) \\ u^{i+1} &= u^i + h^i. \end{aligned}$$

In other words, the Newton iteration is simply the fixed point iteration

$$u^{i+1} = T(u^i) = u^i - F'(u^i)^{-1}F(u^i). \tag{10.7.15}$$

By viewing the Newton iteration as a fixed point iteration, a very general convergence theorem can be proven in a general Banach space X.

Theorem 10.7.1 (Newton Kantorovich Theorem). *Let X be a Banach space, let $D \subset X$ be an open set, and let $F \in C^1(D; X)$. If there exists $u^0 \in D$ and an open ball $B_\rho(u^0) \subset D$ of radius $\rho > 0$ about u^0 such that*

(1) $F'(u^0)$ is nonsingular, with $\|F'(u^0)^{-1}\|_{\mathcal{L}(X,X)} \leqslant \beta$,
(2) $\|u^1 - u^0\|_X = \|F'(u^0)^{-1}F(u^0)\|_X \leqslant \alpha$,
(3) $\|F'(u) - F'(v)\|_X \leqslant \gamma\|u - v\|_X$, $\forall u, v \in B_\rho(u^0)$,
(4) $\alpha\beta\gamma < \frac{1}{2}$, and $\rho \leqslant [1 - \sqrt{1 - 2\alpha\beta\gamma}]/[\beta\gamma]$,

then the Newton iterates produced by (10.7.15) converge strongly at a q-linear rate to a unique $u^* \in B_\rho(u^0) \subset D$.

Proof. See, for example [108, 140, 188]. □

If one assumes the existence of the solution $F(u^*) = 0$, then theorems such as the following one (see also [111]) give an improved rate of convergence.

Theorem 10.7.2 (Quadratic Convergence of Newton's Method). *Let X be a Banach space, let $D \subset X$ be an open set, and let $F \in C^1(D; X)$. If there exists $u^* \in D$ and an open ball $B_\rho(u^*) \subset D$ of radius $\rho > 0$ about u^* such that*

(1) $F(u^*) = 0$,
(2) $F'(u^*)$ *is nonsingular, with* $\|F'(u^*)^{-1}\|_{\mathcal{L}(X,X)} \leqslant \beta$,
(3) $\|F'(u) - F'(v)\|_X \leqslant \gamma\|u - v\|_X$, $\forall u, v \in B_\rho(u^*)$,
(4) $\rho\beta\gamma < \frac{2}{3}$,

then for any $u^0 \in B_\rho(u^*) \subset D$, the Newton iterates produced by (10.7.15) are well-defined and remain in $B_\rho(u^*)$, and converge strongly at a q-quadratic rate to $u^* \in B_\rho(u^0) \subset D$.

Proof. Since by assumption the ball $B_\rho(u^*)$ is already contained in D, we can take $\theta = \rho\beta\gamma < 2/3$ in the Inverse Perturbation Lemma (Lemma 10.1.2), to extend the bound on the inverse of F' in assumption (2) to all of $B_\rho(u^*)$:

$$\|[F'(u)]^{-1}\|_{\mathcal{L}(X,Y)} \leqslant \frac{\beta}{1 - \rho\beta\gamma}, \quad \forall u \in B_\rho(u^*). \tag{10.7.16}$$

We now consider the behavior of the error in the Newton iteration:

$$\begin{aligned} u^{n+1} - u^* &= -[F'(u^n)]^{-1}F(u^n) - u^* \\ &= [F'(u^n)]^{-1}[F(u^*) - F(u^n) - F'(u^n)u^*] \\ &= [F'(u^n)]^{-1}[F(u^n + h) - \{F(u^n) + F'(u^n)(u^n + h)\}], \end{aligned} \tag{10.7.17}$$

where we have defined $h = u^* - u^n$ and used the fact that $F(u^*) = 0$. Taking norms of both sides of (10.7.17) and employing (10.7.16) and the Linear Approximation Lemma (Lemma 10.1.1), gives

$$\begin{aligned} \|u^{n+1} - u^*\|_X &= \|[F'(u^n)]^{-1}[F(u^n + h) - \{F(u^n) + F'(u^n)(u^n + h)\}]\| \\ &= \|[F'(u^n)]^{-1}\|_{\mathcal{L}(X,X)} \\ &\quad \cdot \|[F(u^n + h) - \{F(u^n) + F'(u^n)(u^n + h)\}]\|_{\mathcal{L}(X,X)} \\ &\leqslant \frac{\beta\gamma}{2(1 - \rho\beta\gamma)}\|u^* - u^n\|_X^2. \end{aligned}$$

Since $\|u^* - u^n\|_X \leqslant \rho$ and

$$0 < \frac{\beta\gamma}{2(1 - \rho\beta\gamma)} \leqslant \frac{1}{\rho}\left(\frac{\rho\beta\gamma}{2(1 - \rho\beta\gamma)}\right) \leqslant \frac{1}{\rho}\left(\frac{1}{2} \cdot \frac{2}{3} \cdot 3\right) \leqslant \frac{1}{\rho},$$

we have $\|u^{n+1} - u^*\|_X \leqslant \|u^* - u^n\|_X \leqslant \rho$, giving $u^{n+1} \in B_\rho(u^*)$, which completes the proof. □

There are several variations of the standard Newton iteration (10.7.15) commonly used for nonlinear algebraic equations which we mention briefly. A *quasi*-Newton method refers to a method which uses an approximation to the true Jacobian matrix for solving the Newton equations. A *truncated*-Newton method uses the true Jacobian matrix in the Newton iteration, but solves the Jacobian system only approximately, using an iterative linear solver in which the iteration is stopped early or *truncated*. These types of methods are referred to collectively as *Inexact* or *approximate* Newton methods, where in the most general case an approximate Newton direction is produced in some unspecified fashion. It can be shown that the convergence behavior of these inexact Newton methods is similar to the standard Newton's method, and theorems similar to (10.7.1) can be established (see [108] and the discussions below).

Global Inexact Newton Iteration

For our purposes here, the inexact Newton approach will be of interest, for the following reasons. First, in the case of semilinear partial differential equations which consist of a leading linear term plus a nonlinear term which does not depend on derivatives of the solution, the nonlinear algebraic equations generated often have the form

$$F(u) = Au + B(u) - f = 0.$$

The matrix A is SPD, and the nonlinear term $B(\cdot)$ is often simple, and in fact is often of *diagonal form*, meaning that the j-th component of the vector function $B(u)$ is a function of only the j-th entry of the vector u, or $B_j(u) = B_j(u_j)$; this occurs, for example, in the case of a box-method discretization, or a mass-lumped finite element discretization of semilinear equations. Further, it is often the case that the derivative $B'(\cdot)$ of the nonlinear term $B(\cdot)$, which will be a diagonal matrix due to the fact that $B(\cdot)$ is of diagonal form, can be computed (and applied to a vector) at low expense. If this is the case, then the true Jacobian matrix is available at low cost:

$$F'(u) = A + B'(u).$$

A second reason for our interest in the inexact Newton approach is that the efficient multilevel methods described in Section 10.6 for the linearized semilinear equations can be used effectively for the Jacobian systems; this is because the Jacobian $F'(u)$ is essentially the linearized semilinear operator, where only the diagonal Helmholtz-like term $B'(\cdot)$ changes from one Newton iteration to the next.

Regarding the assumptions on the function $F(\cdot)$ and the Jacobian $F'(\cdot)$ appearing in Theorem 10.7.1, although they may seem unnatural at first glance, they are essentially the minimal conditions necessary to show that the Newton iteration, viewed as a fixed point iteration, is a *contraction*, so that a contraction argument may be employed (see [108]). Since a contraction argument is used, no assumptions on the existence or uniqueness of a solution are required. A disadvantage of proving Newton convergence through a contraction argument is that only Q-linear convergence is shown. This can be improved to R-quadratic through the idea of *majorization* [108].

If additional assumptions are made, such as the existence of a unique solution, then Q-quadratic convergence can be shown; see [108, 140].

Global Newton Convergence Through Damping. As noted in the preceding section, Newton-like methods converge if the initial approximation is "close" to the solution; different convergence theorems require different notions of closeness. If the initial approximation is close enough to the solution, then superlinear or Q-order(p) convergence occurs. However, the fact that these theorems require a good initial approximation is also indicated in practice: it is well known that Newton's method will converge slowly or fail to converge at all if the initial approximation is not good enough.

On the other hand, methods such as those used for unconstrained minimization can be considered to be "globally" convergent methods, although their convergence rates are often extremely poor. One approach to improving the robustness of a Newton iteration without losing the favorable convergence properties close to the solution is to combine the iteration with a global minimization method. In other words, we can attempt to force global convergence of Newton's method by requiring that

$$\|F(u^{i+1})\| < \|F(u^i)\|,$$

meaning that we require a decrease in the value of the function at each iteration. But this is exactly what global minimization methods, such as the nonlinear conjugate gradient method, attempt to achieve: progress toward the solution at each step.

More formally, we wish to define a minimization problem, such that the solution of the zero-point problem we are interested in also solves the associated minimization problem. Let us define the following two problems:

Problem 1: Find $u \in \mathcal{H}$ such that $F(u) = 0$.
Problem 2: Find $u \in \mathcal{H}$ such that $J(u) = \min_{v \in \mathcal{H}} J(v)$.

We assume that Problem 2 has been defined so that the unique solution to Problem 1 is also the unique solution to Problem 2; note that in general there may not exist a *natural* functional $J(\cdot)$ for a given $F(\cdot)$, although we will see in a moment that it is always possible to construct an appropriate functional $J(\cdot)$.

A *descent direction* for the functional $J(\cdot)$ at the point u is any direction v such that the directional derivative of $J(\cdot)$ at u in the direction v is negative, or $J'(u)(v) = (J'(u), v) < 0$. If v is a descent direction, then it is not difficult to show that there exists some $\lambda > 0$ such that

$$J(u + \lambda v) < J(u). \tag{10.7.18}$$

This follows from generalized Taylor expansion (Theorem 10.1.2), since

$$J(u + \lambda v) = J(u) + \lambda (J'(u), v) + \mathcal{O}(\lambda^2).$$

If λ is sufficiently small and $(J'(u), v) < 0$ holds (v is a descent direction), then clearly $J(u + \lambda v) < J(u)$. In other words, if a descent direction can be found at the current solution u^i, then an improved solution u^{i+1} can be found for some steplength in the descent direction v, that is, by performing a one-dimensional line search for λ until (10.7.18) is satisfied.

Therefore, if we can show that the Newton direction is a descent direction, then performing a one-dimensional line search in the Newton direction will always guarantee progress toward the solution. In the case that we define the functional as

$$J(u) = \frac{1}{2}\|F(u)\|^2 = \frac{1}{2}(F(u), F(u)),$$

we can show that the Newton direction is a descent direction. While the following result is easy to show for $\mathcal{H} = \mathbb{R}^n$, we showed more generally in Lemma 10.1.3 that it is also true in the general case of an arbitrary Hilbert space when $\|\cdot\| = (\cdot,\cdot)^{1/2}$:

$$J'(u) = F'(u)^T F(u).$$

Now, the Newton direction at u is simply $v = -F'(u)^{-1}F(u)$, so if $F(u) \neq 0$, then

$$(J'(u), v) = -(F'(u)^T F(u), F'(u)^{-1}F(u)) = -(F(u), F(u)) < 0.$$

Therefore, the Newton direction is always a descent direction for this particular choice of $J(\cdot)$, and by the introduction of the damping parameter λ, the Newton iteration can be made globally convergent in the sense described above.

Damped Inexact Newton Multilevel Methods. Given the problem of n_k nonlinear algebraic equations and n_k unknowns

$$F(u) = Au + B(u) - f = 0,$$

for which we desire the solution u, the ideal algorithm for this problem is one that (1) always converges, and (2) has optimal complexity, which in this case means $\mathcal{O}(n_k)$.

As we have just seen, Newton's method can be made essentially globally convergent with the introduction of a damping parameter. In addition, close to the root, Newton's method has at least superlinear convergence properties. If a method with linear convergence properties is used to solve the Jacobian systems at each Newton iteration, and the complexity of the linear solver is the dominant cost of each Newton iteration, then the complexity properties of the linear method will determine the complexity of the resulting Newton iteration asymptotically, as long as the number of Newton iterations does not grow with the size of the discretization. This last property can in fact be shown for Newton iterations; see [4].

We have discussed in detail in Section 10.6 the convergence and complexity properties of multilevel methods; in many situations they can be shown to have optimal complexity, and in many others this behavior can be demonstrated empirically. With an efficient inexact solver such as a multilevel method for the early damped iterations, employing a more stringent tolerance for the later iterations as the root is approached, a very efficient yet robust nonlinear iteration should result. Following [21, 22], here we combine the robust damped Newton methods with the fast linear multilevel solvers developed in Section 10.6 for inexact solution of the Jacobian systems.

The conditions for linear solver tolerance to ensure superlinear convergence have been given in [60, 61]. Guidelines for choosing damping parameters to ensure global convergence and yet allow for superlinear convergence have been established in [21]. Combination with linear multilevel iterative methods for the semiconductor problem has been considered in [22], along with questions of complexity. We outline the basic algorithm below, specializing it to the particular form of a nonlinear problem of interest. We then give some results on damping and inexactness tolerance selection strategies.

We restrict our discussion here to the following nonlinear problem, which has arisen, for example, from the discretization of a nonlinear elliptic problem:

$$F(u) = Au + B(u) - f = 0.$$

The derivative has the form

$$F'(u) = A + B'(u).$$

The damped inexact Newton iteration for this problem takes the form:

Algorithm 10.7.2 (Damped Inexact Newton Method).

$$[A + B'(u^i)]\, v^i = f - Au^i - B(u^i). \quad \texttt{[Inexact solve]}$$
$$u^{i+1} = u^i + \lambda_i v^i. \quad \texttt{[Correction]}$$

We can employ the linear multilevel methods of Section 10.6 in step (1) of Algorithm 10.7.2. A convergence analysis of the undamped method is given in [85]. A detailed convergence analysis of the damped method is given in [22]. Below, we outline what guidelines exist for selection of the damping parameters and the linear iteration tolerance.

Note that due to the special form of the nonlinear operator, the damping step can be implemented in a surprisingly efficient manner. During the one-dimensional line search for the parameter λ_i, we continually check for satisfaction of the inequality

$$\|F(u^i + \lambda_i v^i)\| < \|F(u^i)\|.$$

The term on the right is available from the previous Newton iteration. The term on the left, although it might appear to involve computing the full nonlinear residual, in fact can avoid the operator-vector product contributed by the linear term. Simply note that

$$\begin{aligned} F(u^i + \lambda_i v^i) &= A[u^i + \lambda_i v^i] + B(u^i + \lambda_i v^i) - f \\ &= [Au^i - f] + \lambda_i [Av^i] + B(u^i + \lambda_i v^i). \end{aligned}$$

The term $[Au^i - f]$ is available from the previous Newton iteration, and $[Av^i]$ needs to be computed only once at each Newton step. Computing $F(u^i + \lambda_i v^i)$ for each damping step beyond the first requires only the operation $[Au^i - f] + \lambda_i [Av^i]$ for the new damping parameter λ_i, and evaluation of the nonlinear term at the new damped solution, $B(u^i + \lambda_i v^i)$.

Local and Global Superlinear Convergence. Quasi-Newton methods are studied in [61], and a "characterization" theorem is established for the sequence of approximate Jacobian systems. This theorem establishes sufficient conditions on the sequence $\{B_i\}$, where $B_i \approx F'$, to ensure superlinear convergence of a quasi-Newton method. An interesting result which they obtained is that the "consistency" condition is not required, meaning that the sequence $\{B_i\}$ need not converge to the true Jacobian $F'(\cdot)$ at the root of the equation $F(u) = 0$, and superlinear convergence can still be obtained.

In [61], a characterization theorem shows essentially that the full or true Newton step must be approached, asymptotically, in both length and direction, to attain superlinear convergence in a quasi-Newton iteration.

Inexact Newton methods are studied directly in [60]. Their motivation is the use of iterative solution methods for approximate solution of the true Jacobian systems. They establish conditions on the accuracy of the inexact Jacobian solution at each Newton iteration which will ensure superlinear convergence. The inexact Newton method is analyzed in the form

$$F'(u^i)v^i = -F(u^i) + r^i, \qquad \frac{\|r^i\|}{\|F(u^i)\|} \leqslant \eta_i,$$
$$u^{i+1} = u^i + v^i.$$

In other words, the quantity r^i, which is simply the residual of the Jacobian linear system, indicates the inexactness allowed in the approximate solution of the linear system, and is exactly what one would monitor in a linear iterative solver. It is established that if the *forcing sequence* $\eta_i < 1$ for all i, then the method above is locally convergent. Their main result is the following theorem.

Theorem 10.7.3 (Dembo-Eisenstat-Steihaug). *Assume that there exists a unique u^* such that $F(u^*) = 0$, that $F(\cdot)$ is continuously differentiable in a neighborhood of u^*, that $F'(u^*)$ is nonsingular, and that the inexact Newton iterates $\{u^i\}$ converge to u^*. Then:*

(1) The convergence rate is superlinear if $\lim_{i\to\infty} \eta_i = 0$.

(2) The convergence rate is Q-order at least $1 + p$ if $F'(u^*)$ is Hölder continuous with exponent p, and

$$\eta_i = \mathcal{O}(\|F(u^i)\|^p), \ \textit{as } i \to \infty.$$

(3) The convergence rate is R-order at least $1 + p$ if $F'(u^*)$ is Hölder continuous with exponent p, and if $\{\eta_i\} \to 0$ with R-order at least $1 + p$.

Proof. See [60]. □

As a result of this theorem, they suggest the tolerance rule:

$$\eta_i = \min\left\{\frac{1}{2}, C\|F(u^i)\|^p\right\}, \quad 0 < p \leqslant 1, \tag{10.7.19}$$

which guarantees Q-order convergence of at least $1 + p$; a similar criterion is

$$\eta_i = \min \left\{ \frac{1}{i}, \|F(u^i)\|^p \right\}, \quad 0 < p \leqslant 1. \tag{10.7.20}$$

Necessary and Sufficient Conditions for Inexact Descent. Note the following subtle point regarding the combination of inexact Newton methods and damping procedures for obtaining global convergence properties: Only the *exact* Newton direction is guaranteed to be a descent direction. Once inexactness is introduced into the Newton direction, there is no guarantee that damping will achieve global convergence in the sense outlined above. However, the following simple result gives a necessary and sufficient condition on the tolerance of the Jacobian system solution for the inexact Newton direction to be a descent direction.

Theorem 10.7.4. *The inexact Newton method (Algorithm 10.7.2) for $F(u) = 0$ will generate a descent direction v at the point u if and only if the residual of the Jacobian system $r = F'(u)v + F(u)$ satisfies*

$$(F(u), r) < (F(u), F(u)).$$

Proof. (See, for example, [101].) We remarked earlier that an equivalent minimization problem (appropriate for Newton's method) to associate with the zero point problem $F(u) = 0$ is given by $\min_{u \in \mathcal{H}} J(u)$, where $J(u) = (F(u), F(u))/2$. We also noted that the derivative of $J(u)$ can be written as $J'(u) = F'(u)^T F(u)$. Now, the direction v is a descent direction for $J(u)$ if and only if $(J'(u), v) < 0$. The exact Newton direction is $v = -F'(u)^{-1} F(u)$, and as shown earlier is always a descent direction. Consider now the inexact direction satisfying

$$F'(u)v = -F(u) + r \qquad \text{or} \qquad v = F'(u)^{-1}[r - F(u)].$$

This inexact direction is a descent direction if and only if:

$$\begin{aligned}
(J'(u), v) &= (F'(u)^T F(u), F'(u)^{-1}[r - F(u)]) \\
&= (F(u), r - F(u)) \\
&= (F(u), r) - (F(u), F(u)) \\
&< 0,
\end{aligned}$$

which is true if and only if the residual of the Jacobian system r satisfies

$$(F(u), r) < (F(u), F(u)).$$

□

This leads to the following very simple sufficient condition for descent.

Corollary 10.7.1. *The inexact Newton method (Algorithm 10.7.2) for* $F(u) = 0$ *yields a descent direction* v *at the point* u *if the residual of the Jacobian system* $r = F'(u)v + F(u)$ *satisfies*

$$\|r\| < \|F(u)\|.$$

Proof. (See, for example, [101].) From the proof of Theorem 10.7.4 we have

$$(J'(u), v) = (F(u), r) - (F(u), F(u)) \leqslant \|F(u)\|\|r\| - \|F(u)\|^2,$$

where we have employed the Cauchy-Schwarz inequality. Therefore, if we have $\|r\| < \|F(u)\|$, then the rightmost term is clearly negative [unless $F(u) = 0$], so that v is a descent direction. □

Note that most stopping criteria for the Newton iteration involve evaluating $F(\cdot)$ at the previous Newton iterate u^i. The quantity $F(u^i)$ will have been computed during the computation of the previous Newton iterate u^i, and the tolerance for u^{i+1} which guarantees descent requires that $(F(u^i), r) < (F(u^i), F(u^i))$ by Theorem 10.7.4. This involves only the inner product of r and $F(u^i)$, so that enforcing this tolerance requires only an additional inner product during the Jacobian linear system solve, which for n_k unknowns introduces an additional n_k multiplications and n_k additions. In fact, a scheme may be employed in which only a residual tolerance requirement for superlinear convergence is checked until an iteration is reached in which it is satisfied. At this point, the descent direction tolerance requirement can be checked, and additional iterations will proceed with this descent stopping criterion until it too is satisfied. If the linear solver reduces the norm of the residual monotonically (such as any of the linear methods of Section 10.6), then the first stopping criterion need not be checked again.

In other words, this adaptive Jacobian system stopping criterion, enforcing a tolerance on the residual for local superlinear convergence *and* ensuring a descent direction at each Newton iteration, can be implemented at the same computational cost as a simple check on the norm of the residual of the Jacobian system.

Alternatively, the sufficient condition given in Corollary 10.7.1 may be employed at no additional cost, since only the norm of the residual needs to be computed, which is also what is required to ensure superlinear convergence using Theorem 10.7.3.

Global Superlinear Convergence. In [21], an analysis of inexact Newton methods is performed, where a damping parameter has been introduced. Their goal was to establish selection strategies for both the linear solve tolerance and the damping parameters at each Newton iteration, in an attempt to achieve global superlinear convergence of the damped inexact Newton iteration. It was established, similar to the result in [61], that the Jacobian system solve tolerance must converge to zero (exact solve in the limit), and the damping parameters must converge to 1 (the full Newton step in the limit), for superlinear convergence to be achieved. There are several technical assumptions on the function $F(\cdot)$ and the Jacobian $F'(\cdot)$ in their paper; we summarize one of their main results in the following theorem, as it applies to the inexact Newton framework we have constructed in this section.

Theorem 10.7.5 (Bank and Rose). *Suppose that $F\colon D \subset \mathcal{H} \to \mathcal{H}$ is a homeomorphism on $\mathcal{H}$. Assume also that $F(\cdot)$ is differentiable on closed bounded sets D, that $F'(u)$ is nonsingular and uniformly Lipschitz continuous on such sets D, and that the closed level set*

$$S_o = \{u \mid \|F(u)\| \leqslant \|F(u^0)\|\}$$

is a bounded set. Suppose now that the forcing and damping parameters η_i and λ_i satisfy

$$\eta_i \leqslant C\|F(x^i)\|^p, \quad \eta_i \leqslant \eta_0, \quad \eta_0 \in (0,1),$$

$$\lambda_i = \frac{1}{1 + K_i\|F(x^i)\|}, \quad 0 \leqslant K_i \leqslant K_0, \quad \textit{so that} \quad \lambda_i \leqslant 1.$$

Then there exists $u^* \in \mathcal{H}$ such that $F(u^*) = 0$, and with any $u^0 \in \mathcal{H}$, the sequence generated by the damped inexact Newton method

$$F'(u^i)v^i = -F(u^i) + r^i, \qquad \frac{\|r^i\|}{\|F(u^i)\|} \leqslant \eta_i, \tag{10.7.21}$$

$$u^{i+1} = u^i + \lambda_i v^i, \tag{10.7.22}$$

converges to $u^* \in S_0 \subset \mathcal{H}$. In addition, on the set S_0, the sequence $\{u^i\}$ converges to u^* at rate Q-order at least $1 + p$.

Proof. See [22]. □

Note that by forcing $\eta_i \leqslant \eta_0 < 1$, it happens that the residual of the Jacobian system in Theorem 10.7.5 satisfies $\|r^i\| \leqslant \eta_i\|F(u^i)\| \leqslant \|F(u^i)\|$, which by Corollary 10.7.1 always ensures that the inexact Newton direction produced by their algorithm is a descent direction. The sequence $\{K_i\}$ is then selected so that each parameter is larger than a certain quantity [inequality 2.14 in [22]], which is a guarantee that an appropriate steplength for actual descent is achieved, without line search. We remark that there is also a weaker convergence result in [22] which essentially states that the convergence rate of the damped inexact Newton method above is R-linear or Q-order$(1 + p)$ on certain sets which are slightly more general than the set S_0. The parameter selection strategy suggested in [22] based on Theorem 10.7.5 is referred to as *Algorithm Global*. The idea of the algorithm is to avoid the typical searching strategies required for other global methods by employing the sequence K_i above.

Backtracking for Sufficient Descent. One of the standard choices for backtracking in Algorithm 10.7.2 to ensure global convergence is to successively reduce the size of the damping parameter λ_i at step i of the Newton iteration according to

$$\lambda_i = \frac{1}{2^k}, \quad k = 0, 1, 2, \ldots,$$

where k is incrementally increased from $k = 0$ (giving the full Newton step with $\lambda_i = 1$) to a sufficiently large number until descent (10.7.18) occurs. However, consider the following example from [78]. Let $F\colon \mathbb{R} \to \mathbb{R}$ be given as $F(u) = u$, and take $\lambda_i = 1/2^{i+1}$, with $u^0 = 2$. The Newton direction at each step remains constant at $v^i = u^i$, $\forall i$, which generates the sequence $\{2, 1, 3/4, 27/32, \ldots\}$, converging to approximately 0.58, yet the solution to $F(u) = 0$ in this case is $u = 0$. The failure of convergence is due to the damping; it is a result of $\lambda_i \to 0$ while $v^i \to v \neq 0$.

To avoid this problem of stalling during the damping procedure, one can enforce a stronger *sufficient descent condition*. By analyzing a linear model of F (see [78]), one can show if $F \in C^1(\mathcal{H})$, then for a fixed $\mu \in (0, 1)$, the following condition can always be satisfied for $\lambda_i \in (0, 1]$ sufficiently small:

$$\|F(u^i + \lambda_i v^i)\| \leqslant (1 - \lambda_i \mu)\|F(u^i)\|. \tag{10.7.23}$$

The result of enforcing this condition is that if $\lambda_i \not\to 0$, then descent cannot stall unless $\|F(u^i)\| \to 0$.

We now describe a globally convergent inexact Newton algorithm that is fairly easy to understand and implement, motivated by the simple necessary and sufficient descent conditions established in the preceding section, as well as the stronger sufficient descent condition described above.

Algorithm 10.7.3 (Damped Inexact Newton method).

Do:

$$F'(u^i)v^i = -F(u^i) + r^i, \quad \text{TEST}(r^i) = \mathit{TRUE}, \qquad \texttt{[Inexact solve]}$$
$$u^{i+1} = u^i + \lambda_i v^i, \qquad \texttt{[Correction]}$$

where parameters λ_i and $\text{TEST}(r^i)$ are defined as:

(1) $\text{TEST}(r^i)$:
- If: $\|r^i\| \leqslant C\|F(u^i)\|^{p+1}$, $C > 0$, $p > 0$,
- And: $(F(u^i), r^i) < (F(u^i), F(u^i))$,
- Then: $\text{TEST} \equiv \mathit{TRUE}$;
- Else: $\text{TEST} \equiv \mathit{FALSE}$.

(2) For fixed $\mu \in (0, 1)$, find λ_i by line search so that:
$\|F(u^i + \lambda_i v^i)\| \leqslant (1 - \lambda_i \mu)\|F(u^i)\|$.
Always possible if $\text{TEST}(r^i) = \mathit{TRUE}$.
Full inexact Newton step $\lambda = 1$ always tried first.

An alternative $\text{TEST}(r^i)$ is as follows:

(1′) $\text{TEST}(r^i)$:
- If: $\|r^i\| \leqslant C\|F(u^i)\|^{p+1}$, $C > 0$, $p > 0$,
- And: $\|r^i\| < \|F(u^i)\|$,
- Then: $\text{TEST} \equiv \mathit{TRUE}$;
- Else: $\text{TEST} \equiv \mathit{FALSE}$.

In Algorithm 10.7.3, the damping parameters λ_i selected in (2) ensure the enforcement of the stronger sufficient descent condition described above, to avoid having the backtracking procedure stall before reaching the solution. The second condition in (1) is the necessary and sufficient condition for the inexact Newton direction to be a descent direction, established in Theorem 10.7.4. The second condition in (1′) of Algorithm 10.7.3 is the weaker sufficient condition established in Corollary 10.7.1. Note that in early iterations when Q-order$(1 + p)$ for $p > 0$ is not to be expected, just satisfying one of the descent conditions is (necessary and) sufficient for progress toward the solution. The condition $\eta_i < 1$ in Theorem 10.7.5 implies that the inexact Newton directions produced by the algorithm are, by Corollary 10.7.1, descent directions. Algorithm 10.7.3 decouples the descent and superlinear convergence conditions and would allow for the use of only the weakest possible test of $(F(u^i), r^i) < (F(u^i), F(u^i))$ far from the solution, ensuring progress toward the solution with the least amount of work per Newton step.

Note also that the Q-order$(1 + p)$ condition

$$\|r^i\| \leqslant C\|F(u^i)\|^{p+1}$$

does *not* guarantee a descent direction, so that it is indeed important to satisfy the descent condition separately. The Q-order$(1 + p)$ condition *will* impose descent if

$$C\|F(u^i)\|^{p+1} < \|F(u^i)\|,$$

which does not always hold. If one is close to the solution, so that $\|F(u^i)\| < 1$, and if $C \leqslant 1$, then the Q-order$(1 + p)$ condition will imply descent. By this last comment, we see that if $\|F(u^i)\| < 1$ and $C \leq 1$, then the full inexact Newton step is a descent direction, and since we attempt this step first, we see that the algorithm reduces to the algorithm studied in [60] near the solution; therefore, Theorem 10.7.3 applies to Algorithm 10.7.3 near the solution without modification.

Nonlinear Multilevel Methods.

Nonlinear multilevel methods were developed originally in [40, 83]. These methods attempt to avoid Newton linearization by accelerating nonlinear relaxation methods with multiple coarse problems. We are again concerned with the problem

$$F(u) = Au + B(u) - f = 0.$$

Let us introduce the notation $M(\cdot) = A+B(\cdot)$, which yields the equivalent problem:

$$M(u) = f.$$

While there is no direct analogue of the linear error equation in the case of a nonlinear operator $M(\cdot)$, a modified equation for e^i can be used. Given an approximation u^i to the true solution u at iteration i, the equations

$$r^i = f - M(u^i), \qquad M(u) = M(u^i + e^i) = f,$$

where r^i and e^i are the residual and error, give rise to the expressions

$$u^i = M^{-1}(f - r^i), \qquad e^i = M^{-1}(f) - u^i,$$

which together give an expression for the error:

$$e^i = (u^i + e^i) - u^i = M^{-1}(f) - M^{-1}(f - r^i).$$

This expression can be used to develop two- and multiple-level methods as in the linear case.

Nonlinear Two-Level Methods. Consider now the case of two nested finite-dimensional spaces $\mathcal{H}_{k-1} \subset \mathcal{H}_k$, where $\mathcal{H}_k$ is the fine space and $\mathcal{H}_{k-1}$ is a lower-dimensional coarse space, connected by a prolongation operator $I_{k-1}^k : \mathcal{H}_{k-1} \to \mathcal{H}_k$ and a restriction operator $I_k^{k-1} : \mathcal{H}_k \to \mathcal{H}_{k-1}$. These spaces may, for example, correspond to either the finite element spaces $\mathcal{V}_k$ or the grid function spaces $\mathcal{U}_k$ arising from the discretization of a nonlinear elliptic problem on two successively refined meshes, as discussed above.

Assuming that the error can be smoothed efficiently as in the linear case, then the error equation can be solved in the coarser space. If the solution is transferred to the coarse space as $u_{k-1}^i = I_k^{k-1} u_k^i$, then the coarse space source function can be formed as $f_{k-1} = M_{k-1}(u_{k-1}^i)$. Transferring the residual r_k to the coarse space as $r_{k-1}^i = I_k^{k-1} r_k^i$, the error equation can then be solved in the coarse space as

$$e_{k-1}^i = I_k^{k-1} u_k^i - M_{k-1}^{-1}(M_{k-1}(I_k^{k-1} u_k^i) - I_k^{k-1} r_k^i).$$

The solution is corrected as

$$\begin{aligned} u_k^{i+1} &= u_k^i + I_{k-1}^k e_{k-1}^i \\ &= u_k^i + I_{k-1}^k [I_k^{k-1} u_k^i - M_{k-1}^{-1}(M_{k-1}(I_k^{k-1} u_k^i) - I_k^{k-1}[f_k - M_k(u_k^i)])] \\ &= K_k(u_k^i, f_k). \end{aligned}$$

Therefore, the nonlinear coarse space correction can be viewed as a fixed point iteration.

The algorithm implementing the nonlinear error equation is known as the *full approximation scheme* [40] or the *nonlinear multigrid method* [85]. The two-level version of this iteration can be formulated as:

Algorithm 10.7.4 (Nonlinear Two-Level Method).

$v_k = K_k(u_k^i, f_k)$. `[Correction]`
$u_k^{i+1} = S_k(v_k, f_k)$. `[Post-smoothing]`

Algorithm 10.7.4 will require a nonlinear relaxation operator $S_k(\cdot)$ in step (2), and restriction and prolongation operators as in the linear case, as well as the solution of the nonlinear coarse space equations, to apply the mapping $K_k(\cdot)$ in step (1).

Nonlinear Multilevel Methods. We consider now a nested sequence of finite-dimensional spaces $\mathcal{H}_1 \subset \mathcal{H}_2 \subset \cdots \subset \mathcal{H}_J \equiv \mathcal{H}$, where $\mathcal{H}_J$ is the finest space and $\mathcal{H}_1$ the coarsest space, each space being connected to the others via prolongation and restriction operators, as discussed above.

The *multi*-level version of Algorithm 10.7.4 would employ another two-level method to solve the coarse space problem in step (1), and can be described recursively as follows:

Algorithm 10.7.5 (Nonlinear Multilevel Method).

Do:

$u^{i+1} = NML(J, u^i, f)$.

where $u_k^{\text{NEW}} = NML(k, u_k^{\text{OLD}}, f_k)$ is defined recursively:

If $(k = 1)$ Then:
$u_1^{\text{NEW}} = M_1^{-1}(f_1)$. [Direct solve]
Else:
$r_{k-1} = I_k^{k-1}(f_k - M_k(u_k^{\text{OLD}}))$, [Restrict residual]
$u_{k-1} = I_k^{k-1} u_k^{\text{OLD}}$ [Restrict solution]
$f_{k-1} = M_{k-1}(u_{k-1}) - r_{k-1}$ [Coarse source]
$w_{k-1} = u_{k-1} - NML(k-1, u_{k-1}, f_{k-1})$ [Coarse solution]
$w_k = I_{k-1}^k w_{k-1}$ [Coarse correction]
$\lambda =$ (see below) [Damping parameter]
$v_k = u_k^{\text{OLD}} + \lambda w_k$ [Correction]
$u_k^{\text{NEW}} = S_k(v_k, f_k)$. [Post-smoothing]
End.

The practical aspects of this algorithm and variations are discussed in [40]. A convergence theory has been discussed in [85] and in the sequence of papers [88, 146].

Damping Parameter. Note that we have introduced a damping parameter λ in the coarse space correction step of Algorithm 10.7.5, analogous to the damped inexact Newton multilevel method discussed earlier. In fact, without this damping parameter, the algorithm fails for difficult problems such as those with exponential or rapid nonlinearities (this is also true for the Newton iteration without damping).

To explain how the damping parameter is chosen, we refer back to the earlier discussion of nonlinear conjugate gradient methods. We begin with the following energy functional:

$$J_k(u_k) = \frac{1}{2}(A_k u_k, u_k)_k + B_k(u_k) - (f_k, u_k)_k.$$

As we have seen, the resulting minimization problem:

$$\text{Find } u_k \in \mathcal{H}_k \text{ such that } J_k(u_k) = \min_{v_k \in \mathcal{H}_k} J_k(v_k)$$

is equivalent to the associated zero-point problem:

$$\text{Find } u_k \in \mathcal{H}_k \text{ such that } F_k(u_k) = A_k u_k + B_k(u_k) - f_k = 0,$$

where $B_k(u_k) = G_k'(u_k)$. In other words, $F_k(\cdot)$ is a gradient mapping of the associated energy functional $J_k(\cdot)$, where we assume that both problems above are uniquely solvable.

In [88] it is shown [with suitable conditions on the nonlinear term $B_k(\cdot)$ and satisfaction of a nonlinear form of the variational conditions] that the prolongated coarse space correction $w_k = I_{k-1}^k w_{k-1}$ is a descent direction for the functional $J_k(\cdot)$. Therefore, there exists some $\lambda > 0$ such that

$$J_k(u_k + \lambda w_k) < J_k(u_k).$$

Minimization of $J_k(\cdot)$ along the descent direction w_k is equivalent to solving the following one-dimensional problem:

$$\frac{dJ(u_k + \lambda w_k)}{d\lambda} = 0.$$

As in the discussion of the nonlinear conjugate gradient method, the one-dimensional problem can be solved with Newton's method:

$$\lambda^{m+1} = \lambda^m - \frac{\lambda^m (A_k w_k, w_k)_k - (r_k, w_k)_k + (B_k(u_k + \lambda^m w_k) - B_k(u_k), w_k)_k}{(A_k w_k, w_k)_k + (B_k'(u_k + \lambda^m w_k) w_k, w_k)_k}.$$

Now, recall that the "direction" from the coarse space correction has the form $w_k = I_{k-1}^k w_{k-1}$. Defining the quantities

$$\begin{aligned}
A_1 &= \lambda^m (A_k I_{k-1}^k w_{k-1}, I_{k-1}^k w_{k-1})_k, \\
A_2 &= (r_k, I_{k-1}^k w_{k-1})_k, \\
A_3 &= (B_k(u_k + \lambda^m I_{k-1}^k w_{k-1}) - B_k(u_k), I_{k-1}^k w_{k-1})_k,
\end{aligned}$$

the Newton correction for λ then takes the form

$$\frac{A_1 - A_2 + A_3}{(A_k I_{k-1}^k w_{k-1}, I_{k-1}^k w_{k-1})_k + (B_k'(u_k + \lambda^m I_{k-1}^k w_{k-1}) I_{k-1}^k w_{k-1}, I_{k-1}^k w_{k-1})_k}.$$

If the linear variational conditions are satisfied:

$$A_{k-1} = I_k^{k-1} A_k I_{k-1}^k, \qquad I_k^{k-1} = (I_{k-1}^k)^T, \tag{10.7.24}$$

and we define the quantities

$$\begin{aligned}
B_1 &= \lambda^m (A_{k-1} w_{k-1}, w_{k-1})_{k-1}, \\
B_2 &= (r_{k-1}, w_{k-1})_{k-1}, \\
B_3 &= (I_k^{k-1}(B_k(u_k + \lambda^m I_{k-1}^k w_{k-1}) - B_k(u_k)), w_{k-1})_{k-1},
\end{aligned}$$

then this expression becomes

$$\frac{B_1 - B_2 + B_3}{(A_{k-1} w_{k-1}, w_{k-1})_{k-1} + (I_k^{k-1} B_k'(u_k + \lambda^m I_{k-1}^k w_{k-1}) I_{k-1}^k w_{k-1}, w_{k-1})_{k-1}}.$$

It can be shown [88] that as in the linear case, a conforming finite element discretization of the nonlinear elliptic problem we are considering, on two successively refined meshes, satisfies the following so-called *nonlinear variational conditions*:

$$A_{k-1} + B_{k-1}(\cdot) = I_k^{k-1} A_k I_{k-1}^k + I_k^{k-1} B_k(I_{k-1}^k \cdot), \qquad I_k^{k-1} = (I_{k-1}^k)^T. \tag{10.7.25}$$

As in the linear case, these conditions are usually required [88] to show theoretical convergence results about nonlinear multilevel methods. Unfortunately, unlike the linear case, there does not appear to be a way to enforce these conditions algebraically [at least for the strictly nonlinear term $B_k(\cdot)$] in an efficient way. Therefore, if we employ discretization methods other than finite element methods, or cannot approximate the integrals accurately (such as if discontinuities occur within elements on coarser levels) for assembling the discrete nonlinear system, then the variational conditions will be violated. With the algebraic approach, we will have to be satisfied with violation of the nonlinear variational conditions, at least for the strictly nonlinear term $B_k(\cdot)$, in the case of the nonlinear multilevel method.

In [88] an expression is given for λ in an attempt to avoid solving the one-dimensional minimization problem. Certain norm estimates are required in their expression for λ, which depends on the particular nonlinearity; therefore, the full line search approach may be more robust, although more costly. There is an interesting result regarding the damping parameter in the linear case, first noticed in [88]. If the nonlinear term $B(\cdot)$ is absent, the zero-point problem is linear and the associated energy functional is quadratic:

$$F_k(u_k) = A_k u_k - f_k = 0, \qquad J_k(u_k) = \frac{1}{2}(A_k u_k, u_k)_k - (f_k, u_k)_k.$$

As in the conjugate gradient algorithm, the calculation of the steplength λ no longer requires the iterative solution of a one-dimensional minimization problem with Newton's method, since

$$\frac{dJ(u_k + \lambda w_k)}{d\lambda} = \lambda(A_k w_k, w_k)_k - (r_k, w_k)_k = 0$$

yields an explicit expression for λ which minimizes the functional $J_k(\cdot)$ in the direction w_k:

$$\lambda = \frac{(r_k, w_k)_k}{(A_k w_k, w_k)_k}.$$

Since $w_k = I_{k-1}^k w_{k-1}$, we have that

$$\begin{aligned} \lambda &= \frac{(r_k, w_k)_k}{(A_k w_k, w_k)_k} \\ &= \frac{(r_k, I_{k-1}^k w_{k-1})_k}{(A_k I_{k-1}^k w_{k-1}, I_{k-1}^k w_{k-1})_k} \\ &= \frac{((I_{k-1}^k)^T r_k, w_{k-1})_{k-1}}{((I_{k-1}^k)^T A_k I_{k-1}^k w_{k-1}, w_{k-1})_{k-1}}. \end{aligned}$$

Therefore, if the variational conditions (10.7.24) are satisfied, the damping parameter can be computed cheaply with only coarse space quantities:

$$\lambda = \frac{(I_k^{k-1} r_k, w_{k-1})_{k-1}}{(I_k^{k-1} A_k I_{k-1}^k w_{k-1}, w_{k-1})_{k-1}} = \frac{(r_{k-1}, w_{k-1})_{k-1}}{(A_{k-1} w_{k-1}, w_{k-1})_{k-1}}.$$

Note that in the two-level case, $w_{k-1} = A_{k-1}^{-1} r_{k-1}$, so that $\lambda = 1$ always holds. Otherwise, numerical experiments show that $\lambda \geqslant 1$, and it is argued [88] that this is always the case. Adding the parameter λ to the linear multilevel algorithms of Section 10.6 guarantees that the associated functional $J_k(\cdot)$ is minimized along the direction defined by the coarse space correction. A simple numerical example in [88] illustrates that, in fact, the convergence rate of the linear algorithm can be improved to a surprising degree by employing the damping parameter.

Stopping Criteria for Nonlinear Iterations.

As in a linear iteration, there are several quantities which can be monitored during a nonlinear iteration to determine whether a sufficiently accurate approximation u^{i+1} to the true solution u^* has been obtained. Possible choices, with respect to any norm $\|\cdot\|$, include:

(1) Nonlinear residual:	$\|F(u^{i+1})\|$	$\leqslant FTOL$
(2) Relative residual:	$\|F(u^{i+1})\|/\|F(u^0)\|$	$\leqslant RFTOL$
(3) Iterate change:	$\|u^{i+1} - u^i\|$	$\leqslant UTOL$
(4) Relative change:	$\|u^{i+1} - u^i\|/\|u^{i+1}\|$	$\leqslant RUTOL$
(5) Contraction estimate:	$\|u^{i+1} - u^i\|/\|u^i - u^{i-1}\|$	$\leqslant CTOL.$

We also mention a sixth option, which attempts to obtain an error estimate from the Contraction Mapping Theorem (Theorem 10.1.14) by estimating the contraction constant α of the nonlinear fixed point mapping $T(\cdot)$ associated with the iteration. The constant is estimated as follows:

$$\alpha = \frac{\|u^{i+1} - u^i\|}{\|u^i - u^{i-1}\|} = \frac{\|T(u^i) - T(u^{i-1})\|}{\|u^i - u^{i-1}\|},$$

and the Contraction Mapping Theorem gives the error estimate-based criterion:

$$\text{(6) Error estimate: } \|u^* - u^{i+1}\| \leqslant \frac{\alpha}{1-\alpha}\|u^{i+1} - u^i\| \leqslant ETOL.$$

There are certain difficulties with employing any of these conditions alone. For example, if the iteration has temporarily stalled, then criteria (3) and (4) would prematurely halt the iteration. On the other hand, if the scaling of the function $F(\cdot)$ is such that $\|F(\cdot)\|$ is always very small, then criterion (1) could halt the iteration early. Criterion (2) attempts to alleviate this problem in much the same way as a relative stopping criterion in the linear case. However, if the initial approximation u^0 is such that $\|F(u^0)\|$ is extremely large, then (3) could falsely indicate that a good approximation has been reached. Criterion (5) cannot be used to halt the iteration alone, as it gives no information about the quality of the approximation; it would be useful in a Newton iteration to detect when the region of fast convergence has been entered.

Criterion (6) may be the most reliable stand-alone criterion, although it depends on the accuracy of the contraction number estimate. If the contraction number is constant (linear convergence) over many iterations or goes to zero monotonically (superlinear convergence), then this should be reliable; otherwise, the contraction estimate may have no bearing on the true contraction constant for the mapping $T(\cdot)$, and the error estimate may be poor.

Several dual criteria have been proposed in the literature. For example, the combination of (4) and (5) was suggested in [20], since (4) attempts to detect if convergence has been reached, whereas (5) attempts to ensure that (4) has not been satisfied simply due to stalling of the iteration. In [62], the combination of (4) and (1) is suggested, where (1) attempts to prevent halting on (4) due to stalling. The idea of scaling the components of u^{i+1} in (1) and $F(u^{i+1})$ in (2) is also recommended in [62], along with use of the maximum norm $\|\cdot\|_\infty$. In [78], other combinations are suggested [with an optimization orientation, some combinations involving the associated functional $J(\cdot)$].

EXERCISES

10.7.1 Let X and Y be Banach spaces and let $F \in C^2(X, Y)$. Use only the mean value theorem (Theorem 10.1.3) to derive the following Taylor-series expansion with integral remainder:

$$F(u+h) = F(u) + F'(u)h + \int_0^1 (1-t)F''(u+th)(h,h)\,dt.$$

[*Hint:* Expand $F'(u+h)$ using one of the formulas from Theorem 10.1.3, and then differentiate with respect to h using the chain rule.]

10.7.2 Find $J'(u)$ (a row vector function), $\nabla J(u)$ (a column vector function), and $\nabla^2 J(u)$ (the symmetric Hessian matrix of J) for the following functions of n variables.

(1) $J(u) = (1/2)u^T Au - u^T f$, where $A \in \mathbb{R}^{n\times n}$.
(2) $J(u) = (1/2)u^T Au - u^T f$, where $A \in \mathbb{R}^{n\times n}$, *and also* $A = A^T$.
(3) $J(u) = (1/2)u^T A^T Au - u^T Af$, where $A \in \mathbb{R}^{m\times n}$, and $f \in \mathbb{R}^m$.
(4) $J(u) = \|u\|_{l^2} = \left(\sum_{i=1}^n u_i^2\right)^{1/2}$.

[*Hint:* Do not use the information that you are working with the particular normed space $\mathbb{R}^n$; just think of $\mathbb{R}^n$ as an arbitrary Hilbert space H, and compute the derivatives using the convenient expression for the **G**-variation in (10.1.5).]

10.7.3 In [78], the sufficient descent condition (10.7.23) is derived by requiring the reduction in $\|F(u)\|_X$ be no worse than μ times the reduction in the *linear model* of F given by Taylor expansion $F(u^i + h) = F(u^i) + F'(u^i)h + \mathcal{O}(\|h\|_X^2)$. Setting $w = u^i + h$, we can write the expansion as a linear model $L^i(w)$ plus a remainder:

$$F(w) = L^i(w) + \mathcal{O}(\|h\|_X^2),$$

where $L^i(w) = F(u^i) + F'(u^i)(w - u^i)$. Prove that condition (10.7.23) is equivalent to

$$\frac{\|F(u^i)\|_X - \|F(u^i + \lambda_i v^i)\|_X}{\|L^i(u^i)\|_X - \|L^i(u^i + \lambda_i v^i)\|_X} \geqslant \mu.$$

10.7.4 Prove that Newton's method converges Q-linearly by using the Banach Fixed-Point Theorem. If you assume the existence of a solution, then you need to simply give conditions on F and F' which guarantee that the fixed point operator defined by the Newton iteration is a contraction on a sufficiently small ball around the solution. Can you construct a proof using the Banach Fixed-Point Theorem that also gives existence of the solution, without assuming it *a priori*? Can you recover something faster than Q-linear convergence?

10.7.5 *For Fun:* Construct a Newton iteration for computing the reciprocal of a positive real number without performing division. (This has been a standard algorithm for doing division in computer arithmetic units, together with a lookup table of good initial approximations.)

REFERENCES AND ADDITIONAL READING

1. R. Abraham, J. E. Marsden, and T. Ratiu, *Manifolds, Tensor Analysis, and Applications*, Springer-Verlag, New York, 1988.
2. R. A. Adams and J. F. Fornier, *Sobolev Spaces*, 2nd ed., Academic Press, San Diego, CA, 2003.
3. P. Allen, A. Clausen, and J. Isenberg, Near-constant mean curvature solutions of the Einstein constraint equations with non-negative Yamabe metrics, 2007, available as arXiv:0710.0725 [gr-qc].
4. E. L. Allgower, K. Böhmer, F. A. Potra, and W. C. Rheinboldt, A mesh-independence principle for operator equations and their discretizations, *SIAM J. Numer. Anal.*, 23(1):160–169, 1986.
5. D. Arnold, R. Falk, and R. Winther, Finite element exterior calculus: From hodge theory to numerical stability, *Bull. Amer. Math. Soc. (N.S.)*, 47:281–354, 2010.
6. D. Arnold, A. Mukherjee, and L. Pouly, Locally adapted tetrahedral meshes using bisection, *SIAM J. Sci. Statist. Comput.*, 22(2):431–448, 1997.
7. S. Ashby, M. Holst, T. Manteuffel, and P. Saylor, The role of the inner product in stopping criteria for conjugate gradient iterations, *BIT*, 41(1):26–53, 2001.
8. S. F. Ashby, T. A. Manteuffel, and P. E. Saylor, A taxonomy for conjugate gradient methods, *SIAM J. Numer. Anal.*, 27(6):1542–1568, 1990.
9. K. Atkinson and W. Han, *Theoretical Numerical Analysis*, Springer Verlag, New York, 2001.
10. J. Aubin, *Approximation of Elliptic Boundary-Value Problems*, Wiley, New York, 1972.
11. T. Aubin, *Nonlinear Analysis on Manifolds: Monge-Ampère Equations*, Springer-Verlag, New York, 1982.
12. O. Axelsson and V. Barker, *Finite Element Solution of Boundary Value Problems*, Academic Press, Orlando, FL, 1984.
13. A. K. Aziz, *The Mathematical Foundations of the Finite Element Method with Applications to Partial Differential Equations*, Academic Press, New York, 1972.

14. I. Babuška, The finite element method for elliptic equations with discontinuous coefficients, *Computing*, 5(3):207–213, 1970.

15. I. Babuška and W. Rheinboldt, Error estimates for adaptive finite element computations, *SIAM J. Numer. Anal.*, 15:736–754, 1978.

16. R. E. Bank, *PLTMG: A Software Package for Solving Elliptic Partial Differential Equations, Users' Guide 8.0*, Software, Environments and Tools, Vol. 5, SIAM, Philadelphia, 1998.

17. R. E. Bank and T. F. Dupont, Analysis of a two-level scheme for solving finite element equations, Tech. Rep. CNA–159, Center for Numerical Analysis, University of Texas at Austin, 1980.

18. R. E. Bank and T. F. Dupont, An optimal order process for solving finite element equations, *Math. Comp.*, 36(153):35–51, 1981.

19. R. E. Bank, T. F. Dupont, and H. Yserentant, The hierarchical basis multigrid method, *Numer. Math.*, 52:427–458, 1988.

20. R. E. Bank and D. J. Rose, Parameter selection for Newton-like methods applicable to nonlinear partial differential equations, *SIAM J. Numer. Anal.*, 17(6):806–822, 1980.

21. R. E. Bank and D. J. Rose, Global Approximate Newton Methods, *Numer. Math.*, 37:279–295, 1981.

22. R. E. Bank and D. J. Rose, Analysis of a multilevel iterative method for nonlinear finite element equations, *Math. Comp.*, 39(160):453–465, 1982.

23. R. E. Bank and D. J. Rose, Some error estimates for the box method, *SIAM J. Numer. Anal.*, 24(4):777–787, 1987.

24. R. E. Bank and R. K. Smith, A posteriori error estimates based on hierarchical bases, *SIAM J. Numer. Anal.*, 30(4):921–935, 1993.

25. R. E. Bank and A. Weiser, Some a posteriori error estimators for elliptic partial differential equations, *Math. Comp.*, 44(170):283–301, 1985.

26. E. Bänsch, Local mesh refinement in 2 and 3 dimensions, *Impact Comput. Sci. Engr.*, 3:181–191, 1991.

27. R. Bartnik and J. Isenberg, The constraint equations, in *The Einstein Equations and Large Scale Behavior of Gravitational Fields* (P. Chruściel and H. Friedrich, Eds.), pp. 1–38, Birkhäuser, Berlin, 2004.

28. P. Bastian, K. Birken, K. Johannsen, S. Lang, N. Neuss, H. Rentz-Reichert, and C. Wieners, UG - A flexible software toolbox for solving partial differential equations, in *Computing and Visualization in Science*, pp. 27–40, 1997.

29. R. Beck, B. Erdmann, and R. Roitzsch, KASKADE 3.0: An object-oriented adaptive finite element code, Tech. Rep. TR95–4, Konrad-Zuse-Zentrum for Informationstechnik, Berlin, 1995.

30. J. Bey, Tetrahedral grid refinement, *Computing*, 55(4):355–378, 1995.

31. J. Bey, Adaptive grid manager: AGM3D manual, Tech. Rep. 50, SFB 382, Mathematics Institute, University of Tubingen, Tubingen, Germany, 1996.

32. D. Braess, *Nonlinear Approximation Theory*, Springer-Verlag, Berlin, 1980.

33. D. Braess, *Finite Elements*, Cambridge University Press, Cambridge, MA, 1997.

34. D. Braess and W. Hackbusch, A new convergence proof for the multigrid method including the V-cycle, *SIAM J. Numer. Anal.*, 20(5):967–975, 1983.

35. J. Bramble and J. King, A finite element method for interface problems in domains with smooth boundaries and interfaces, *Adv. Comput. Math.*, 6(1):109–138, 1996.

36. J. H. Bramble and J. E. Pasciak, New convergence estimates for multigrid algorithms, *Math. Comp.*, 49(180):311–329, 1987.

37. J. H. Bramble and J. E. Pasciak, The analysis of smoothers for multigrid algorithms, *Math. Comp.*, 58(198):467–488, 1992.

38. J. H. Bramble, J. E. Pasciak, J. Wang, and J. Xu, Convergence estimates for multigrid algorithms without regularity assumptions, *Math. Comp.*, 57:23–45, 1991.

39. J. H. Bramble, J. E. Pasciak, J. Wang, and J. Xu, Convergence estimates for product iterative methods with applications to domain decomposition and multigrid, *Math. Comp.*, 57:1–21, 1991.

40. A. Brandt, Multi-level adaptive solutions to boundary-value problems, *Math. Comp.*, 31:333–390, 1977.

41. A. Brandt, Algebraic multigrid theory: The symmetric case, *Appl. Math. Comput.*, 19:23–56, 1986.

42. S. C. Brenner and L. R. Scott, *The Mathematical Theory of Finite Element Methods*, 2nd ed., Springer-Verlag, New York, 2002.

43. F. Brezzi, Mathematical theory of finite elements, in *State-of-the-Art Surveys on Finite Element Technology* (A. K. Noor and W. D. Pilkey, Eds.), pp. 1–25, The American Society of Mechanical Engineers, New York, NY, 1985.

44. F. Brezzi and M. Fortin, *Mixed and Hybrid Finite Element Methods*, Springer-Verlag, New York, 1991.

45. J. M. Briggs and J. A. McCammon, Computation unravels mysteries of molecular biophysics, *Comput. Phys.*, 6(3):238–243, 1990.

46. Z. Cai, J. Mandel, and S. F. McCormick, The finite volume element method for diffusion equations on general triangulations, *SIAM J. Numer. Anal.*, 28:392–402, 1991.

47. C. Carstensen, Convergence of adaptive FEM for a class of degenerate convex minimization problem, *Preprint*, 2006.

48. L. Chen, A new class of high order finite volume methods for second order elliptic equations, *SIAM J. Numer. Anal.*, 47(6):4021–4043, 2010.

49. L. Chen, M. Holst, and J. Xu, The finite element approximation of the nonlinear Poisson-Boltzmann Equation, *SIAM J. Numer. Anal.*, 45(6):2298–2320, 2007, available as arXiv:1001.1350 [math.NA].

50. Y. Choquet-Bruhat, Sur l'intégration des équations de la relativité générale, *J. Rational Mech. Anal.*, 5:951–966, 1956.

51. Y. Choquet-Bruhat, Einstein constraints on compact n-dimensional manifolds, *Class. Quantum Grav.*, 21:S127–S151, 2004.

52. P. G. Ciarlet, *The Finite Element Method for Elliptic Problems*, North-Holland, New York, 1978.

53. P. Clément, Approximation by finite element functions using local regularization, *RAIRO Anal. Numer.*, 2:77–84, 1975.

54. A. Cohen, *Numerical Analysis of Wavelet Methods*, North-Holland, New York, 2003.

55. G. Cook and S. Teukolsky, Numerical relativity: Challenges for computational science, in *Acta Numerica*, Vol. 8 (A. Iserles, Ed.), pp. 1–44, Cambridge University Press, New York, 1999.

56. I. Daubechies, *Ten Lectures on Wavelets*, SIAM, Philadelphia, 1992.

57. P. J. Davis, *Interpolation and Approximation*, Dover, New York, 1963.

58. L. Debnath and P. Mikusinński, *Introduction to Hilbert Spaces with Applications*, Academic Press, San Diego, CA, 1990.

59. P. Debye and E. Hückel, Zur Theorie der Elektrolyte: I. Gefrierpunktserniedrigung und verwandte Erscheinungen, *Phys. Z.*, 24(9):185–206, 1923.

60. R. S. Dembo, S. C. Eisenstat, and T. Steihaug, Inexact Newton Methods, *SIAM J. Numer. Anal.*, 19(2):400–408, 1982.

61. J. E. Dennis, Jr. and J. J. Moré, Quasi-Newton methods, motivation and theory, *SIAM Rev.*, 19(1):46–89, 1977.

62. J. E. Dennis, Jr. and R. B. Schnabel, *Numerical Methods for Unconstrained Optimization and Nonlinear Equations*, Prentice-Hall, Englewood Cliffs, NJ, 1983.

63. P. Deuflhard, *Newton Methods for Nonlinear Problems*, Springer-Verlag, Berlin, 2004.

64. R. A. DeVore and G. G. Lorentz, *Constructive Approximation*, Springer-Verlag, New York, 1993.

65. P. Drabek and J. Milota, *Methods of Nonlinear Analysis: Applications to Differential Equations*, Birkhäuser, Berlin, 2000.

66. M. Dryja and O. B. Widlund, Towards a unified theory of domain decomposition algorithms for elliptic problems, in *Third International Symposium on Domain Decomposition Methods for Partial Differential Equations* (T. F. Chan, R. Glowinski, J. Périaux, and O. B. Widlund, Eds.), pp. 3–21, SIAM, Philadelphia, 1989.

67. I. Ekeland and R. Temam, *Convex Analysis and Variational Problems*, North-Holland, New York, 1976.

68. D. Estep, A posteriori error bounds and global error control for approximations of ordinary differential equations, *SIAM J. Numer. Anal.*, 32:1–48, 1995.

69. D. Estep, M. Holst, and M. Larson, Generalized Green's functions and the effective domain of influence, *SIAM J. Sci. Comput.*, 26(4):1314–1339, 2005.

70. L. C. Evans, *Partial Differential Equations*, Vol. 19 of *Graduate Studies in Mathematics*, American Mathematical Society, Providence, RI, 2010.

71. R. Eymard, T. Gallouët, and R. Herbin, Convergence of finite volume schemes for semilinear convection diffusion equations, *Numer. Math.*, 82(1):91–116, 1999.

72. R. Eymard, T. Gallouët, and R. Herbin, Finite volume methods, in *Handbook of numerical analysis, Vol. VII*, pp. 713–1020, North-Holland, Amsterdam, 2000.

73. R. P. Fedorenko, A relaxation method for solving elliptic difference equations, *USSR Comput. Math. Math. Phys.*, 1(5):1092–1096, 1961.

74. R. P. Fedorenko, The speed of convergence of one iterative process, *USSR Comput. Math. Math. Phys.*, 4(3):227–235, 1964.

75. FETK, The Finite Element ToolKit, `http://www.FETK.org`.

76. R. Fletcher and C. Reeves, Function minimization by conjugate gradients, *Comput. J.*, 7:149–154, 1964.

77. S. Fucik and A. Kufner, *Nonlinear Differential Equations*, Elsevier Scientific, New York, 1980.

78. P. E. Gill, W. Murray, and M. H. Wright, *Practical Optimization*, Academic Press, London and New York, 1981.

79. M. S. Gockenbach, *Partial Differential Equations: Analytical and Numerical Methods*, SIAM, Philadelphia, 2002.

80. M. Griebel, Multilevel algorithms considered as iterative methods on indefinite systems, in *Proceedings of the Second Copper Mountain Conference on Iterative Methods* (T. Manteuffel, Ed.), 1992.

81. M. Griebel and M. A. Schweitzer, A particle-partition of unity method for the solution of elliptic, parabolic, and hyperbolic PDEs, *SIAM J. Sci. Statist. Comput.*, 22(3):853–890, 2000.

82. P. Grisvard, *Elliptic Problems in Nonsmooth Domains*, Pitman, Marshfield, MA, 1985.

83. W. Hackbusch, On the fast solutions of nonlinear elliptic equations, *Numer. Math.*, 32:83–95, 1979.

84. W. Hackbusch, Multi-grid convergence theory, in *Multigrid Methods: Proceedings of Köln-Porz Conference on Multigrid Methods, Lecture notes in Mathematics 960* (W. Hackbusch and U. Trottenberg, Eds.), Springer-Verlag, Berlin, Germany, 1982.

85. W. Hackbusch, *Multi-grid Methods and Applications*, Springer-Verlag, Berlin, 1985.

86. W. Hackbusch, On first and second order box schemes, *Computing*, 41(4):277–296, 1989.

87. W. Hackbusch, *Iterative Solution of Large Sparse Systems of Equations*, Springer-Verlag, Berlin, 1994.

88. W. Hackbusch and A. Reusken, Analysis of a damped nonlinear multilevel method, *Numer. Math.*, 55:225–246, 1989.

89. P. R. Halmos, *Finite-Dimensional Vector Spaces*, Springer-Verlag, Berlin, 1958.

90. S. W. Hawking and G. F. R. Ellis, *The Large Scale Structure of Space-Time*, Cambridge University Press, Cambridge, UK, 1973.

91. E. Hebey, *Sobolev Spaces on Riemannian Manifolds*, Vol. 1635 of *Lecture Notes in Mathematics*, Springer-Verlag, Berlin, New York, 1996.

92. M. R. Hestenes and E. Stiefel, Methods of conjugate gradients for solving linear systems, *J. Res. of NBS*, 49:409–435, 1952.

93. M. Holst, An algebraic Schwarz theory, Tech. Rep. CRPC-94-12, Applied Mathematics and CRPC, California Institute of Technology, 1994.

94. M. Holst, Adaptive numerical treatment of elliptic systems on manifolds, *Adv. Comput. Math.*, 15(1–4):139–191, 2001, available as arXiv:1001.1367 [math.NA].

95. M. Holst, J. McCammon, Z. Yu, Y. Zhou, and Y. Zhu, Adaptive finite element modeling techniques for the Poisson-Boltzmann equation, to appear in Comm. Comput. Phys. Available as arXiv:1009.6034 [math.NA].

96. M. Holst, G. Nagy, and G. Tsogtgerel, Far-from-constant mean curvature solutions of Einstein's constraint equations with positive Yamabe metrics, *Phys. Rev. Lett.*, 100(16):161101.1–161101.4, 2008, available as arXiv:0802.1031 [gr-qc].

97. M. Holst, G. Nagy, and G. Tsogtgerel, Rough solutions of the Einstein constraints on closed manifolds without near-CMC conditions, *Comm. Math. Phys.*, 288(2):547–613, 2009, available as arXiv:0712.0798 [gr-qc].

98. M. Holst and E. Titi, Determining projections and functionals for weak solutions of the Navier-Stokes equations, in *Recent Developments in Optimization Theory and Nonlinear Analysis*, Vol. 204 of *Contemporary Mathematics* (Y. Censor and S. Reich, Eds.), American Mathematical Society, Providence, RI, 1997, available as arXiv:1001.1357 [math.AP].

99. M. Holst, G. Tsogtgerel, and Y. Zhu, Local convergence of adaptive methods for nonlinear partial differential equations, submitted for publication. Available as arXiv:1001.1382 [math.NA].

100. M. Holst and S. Vandewalle, Schwarz methods: To symmetrize or not to symmetrize, *SIAM J. Numer. Anal.*, 34(2):699–722, 1997, available as arXiv:1001.1362 [math.NA].

101. M. J. Holst, *The Poisson-Boltzmann Equation: Analysis and Multilevel Numerical Solution*, Ph.D. dissertation, University of Illinois at Urbana-Champaign, 1994.

102. G. Hsiao and W. Wendland, *Boundary Integral Equations*, Vol. 164 of *Applied Mathematical Sciences Series*, Springer-Verlag, Berlin, 2008.

103. T. J. R. Hughes, *The Finite Element Method*, Dover, New York, 2000.

104. J. Isenberg, Constant mean curvature solution of the Einstein constraint equations on closed manifold, *Class. Quantum Grav.*, 12:2249–2274, 1995.

105. J. Isenberg and V. Moncrief, A set of nonconstant mean curvature solution of the Einstein constraint equations on closed manifolds, *Class. Quantum Grav.*, 13:1819–1847, 1996.

106. J. W. Jerome and T. Kerkhoven, A finite element approximation theory for the drift diffusion semiconductor model, *SIAM J. Numer. Anal.*, 28(2):403–422, 1991.

107. C. Johnson, *Numerical Solution of Partial Differential Equations by the Finite Element Method*, Cambridge University Press, Cambridge, UK, 1987.

108. L. V. Kantorovich and G. P. Akilov, *Functional Analysis*, Pergamon Press, New York, 1982.

109. L. V. Kantorovich and V. I. Krylov, *Approximate Methods of Higher Analysis*, P. Noordhoff, Groningen, The Netherlands, 1958.

110. T. Kato, *Perturbation Theory for Linear Operators*, Springer-Verlag, Berlin, 1980.

111. H. B. Keller, *Numerical Methods in Bifurcation Problems*, Tata Institute of Fundamental Research, Bombay, India, 1987.

112. T. Kerkhoven, Piecewise linear Petrov-Galerkin analysis of the box-method, *SIAM J. Numer. Anal.*, 33(5):1864–1884, 1996.

113. S. Kesavan, *Topics in Functional Analysis and Applications*, Wiley, New York, 1989.

114. A. N. Kolmogorov and S. V. Fomin, *Introductory Real Analysis*, Dover, New York, 1970.

115. R. Kress, *Linear Integral Equations*, Springer-Verlag, Berlin, 1989.

116. E. Kreyszig, *Introductory Functional Analysis with Applications*, Wiley, New York, 1990.

117. A. Kufner, O. John, and S. Fucik, *Function Spaces*, Noordhoff International, Leyden, The Netherlands, 1977.

118. J. Lee and T. Parker, The Yamabe problem, *Bull. Amer. Math. Soc.*, 17(1):37–91, 1987.

119. J. M. Lee, *Riemannian Manifolds*, Springer-Verlag, New York, 1997.

120. L. Lehner, Numerical relativity: A review, *Class. Quantum Grav.*, 18:R25–R86, 2001.

121. R. J. LeVeque, *Finite Difference Methods for Ordinary and Partial Differential Equations: Steady-State and Time-Dependent Problems*, SIAM, Philadelphia, 2007.

122. Y. Y. Li and L. Zhang, A Harnack type inequality for the Yamabe equation in low dimensions, *Calc. Var.*, 20:133–151, 2004.

123. A. Liu and B. Joe, Quality local refinement of tetrahedral meshes based on bisection, *SIAM J. Sci. Statist. Comput.*, 16(6):1269–1291, 1995.

124. J. Liu and W. Rheinboldt, A posteriori finite element error estimators for indefinite elliptic boundary value problems, *Numer. Funct. Anal. Optim.*, 15(3):335–356, 1994.

125. J.-F. Maitre and F. Musy, Multigrid methods: Convergence theory in a variational framework, *SIAM J. Numer. Anal.*, 21(4):657–671, 1984.

126. J. E. Marsden and T. J. R. Hughes, *Mathematical Foundations of Elasticity*, Dover, New York, 1994.

127. R. Mattheij, W. Rienstra, and J. ten Thije Boonkkamp, *Partial Differential Equations: Modeling, Analysis, Computation*, SIAM, Philadelphia, 2005.

128. J. Maubach, Local bisection refinement for N-simplicial grids generated by relection, *SIAM J. Sci. Statist. Comput.*, 16(1):210–277, 1995.

129. D. Maxwell, Rough solutions of the Einstein constraint equations on compact manifolds, *J. Hyp. Differential Equations*, 2(2):521–546, 2005.

130. D. Maxwell, A class of solutions of the vacuum Einstein constraint equations with freely specified mean curvature, 2008, available as arXiv:0804.0874 [gr-qc].

131. S. F. McCormick, Multigrid methods for variational problems: Further results, *SIAM J. Numer. Anal.*, 21(2):255–263, 1984.

132. D. Mitrović and D. Zubrinić, *Fundamentals of Applied Functional Analysis*, Pitman Monographs and Surveys in Pure and Applied Mathematics, Longman Scientific & Tecnical, Wiley, New York, 1998.

133. A. Mukherjee, *An Adaptive Finite Element Code for Elliptic Boundary Value Problems in Three Dimensions with Applications in Numerical Relativity*, Ph.D. dissertation, Department of Mathematics, The Pennsylvania State University, 1996.

134. R. Nochetto, K. Siebert, and A. Veeser, Theory of adaptive finite element methods: An introduction, in *Multiscale, Nonlinear and Adaptive Approximation* (R. DeVore and A. Kunoth, Eds.), pp. 409–542, Springer, 2009, dedicated to Wolfgang Dahmen on the Occasion of His 60th Birthday.

135. J. T. Oden and L. F. Demkowicz, *Applied Functional Analysis*, CRC Series in Computational Mechanics and Applied Analysis, CRC Press, Boca Raton, FL, 1996.

136. J. T. Oden and J. N. Reddy, *An Introduction to The Mathematical Theory of Finite Elements*, Wiley, New York, 1976.

137. N. O'Murchadha and J. York, Jr., Existence and uniqueness of solutions of the Hamiltonian constraint of general relativity on compact manifolds, *J. Math. Phys.*, 14(11):1551–1557, 1973.

138. E. G. Ong, Uniform refinement of a tetrahedron, Tech. Rep. CAM 91-01, Department of Mathematics, UCLA, 1991.

139. J. M. Ortega, *Numerical Analysis: A Second Course*, Academic Press, New York, 1972.

140. J. M. Ortega and W. C. Rheinboldt, *Iterative Solution of Nonlinear Equations in Several Variables*, Academic Press, New York, 1970.

141. P. Oswald, *Multilevel Finite Element Approximation*, B. G. Teubner, Stuttgart, Germany, 1994.

142. R. Palais, *Foundations of Global Non-linear Analysis*, W. A. Benjamin, New York, 1968.

143. J. Pousin and J. Rappaz, Consistency, stability, a priori and a posteriori errors for Petrov-Galerkin methods applied to nonlinear problems, *Numer. Math.*, 69(2):213–231, 1994.

144. J. Rappaz, Numerical approximation of PDEs and Clément's interpolation, *Partial Differential Equations Func. Anal.*, 168:237–250, 2006.

145. M. Renardy and R. C. Rogers, *An Introduction to Partial Differential Equations*, Springer-Verlag, New York, 1993.

146. A. Reusken, Convergence of the multilevel full approximation scheme including the V-cycle, *Numer. Math.*, 53:663–686, 1988.

147. M. Rivara, Algorithms for refining triangular grids suitable for adaptive and multigrid techniques, *Internat. J. Numer. Methods Engrg.*, 20:745–756, 1984.

148. I. Rosenberg and F. Stenger, A lower bound on the angles of triangles constructed by bisecting the longest side, *Math. Comp.*, 29:390–395, 1975.

149. S. Rosenberg, *The Laplacian on a Riemannian Manifold*, Cambridge University Press, Cambridge, UK, 1997.

150. W. Rudin, *Real and Complex Analysis*, McGraw-Hill, New York, 1987.

151. J. W. Ruge and K. Stüben, Algebraic multigrid (AMG), in *Multigrid Methods*, Vol. 3 of *Frontiers in Applied Mathematics* (S. F. McCormick, Ed.), pp. 73–130, SIAM, Philadelphia, 1987.

152. G. Savare, Regularity results for elliptic equations in Lipschitz domains, *J. of Func. Anal.*, 152(1):176–201, 1998.

153. A. H. Schatz, An oberservation concerning Ritz-Galerkin methods with indefinite bilinear forms, *Math. Comp.*, 28(128):959–962, 1974.

154. A. H. Schatz and J. Wang, Some new error estimates for Ritz-Galerkin methods with minimal regularity assumptions, *Math. Comp.*, 62:445–475, 2000.

155. G. Schwarz, *Hodge Decomposition: A Method for Solving Boundary Value Problems*, Springer-Verlag, New York, 1991.

156. L. R. Scott and S. Zhang, Finite element interpolation of nonsmooth functions satisfying boundary conditions, *Math. Comp.*, 54(190):483–493, 1990.

157. L. A. Segel, *Mathematics Applied to Continuum Mechanics*, Dover, New York, 1977.

158. E. Seidel, New developments in numerical relativity, *Helv. Phys. Acta*, 69:454–471, 1996.

159. R. E. Showalter, *Hilbert Space Methods for Partial Differential Equations*, Pitman, Marshfield, MA, 1979.

160. E. M. Stein, *Singular Integrals and Differentiability Properties of Functions*, Princeton Mathematical Series, No. 30, Princeton University Press, Princeton, N.J., 1970.

161. G. Strang and G. Fix, *An Analysis of the Finite Element Method*, Prentice-Hall, Englewood Cliffs, NJ, 1973.

162. K. Stüben and U. Trottenberg, Multigrid methods: Fundamental algorithms, model problem analysis and applications, in *Multigrid Methods: Proceedings of Köln-Porz Conference on Multigrid Methods, Lecture notes in Mathematics 960* (W. Hackbusch and U. Trottenberg, Eds.), Springer-Verlag, Berlin, Germany, 1982.

163. M. Stynes, On faster convergence of the bisection method for all triangles, *Math. Comp.*, 35:1195–1201, 1980.

164. B. Szabó and I. Babuška, *Finite Element Analysis*, Wiley, New York, 1991.

165. C. Tanford, *Physical Chemistry of Macromolecules*, John Wiley & Sons, New York, NY, 1961.

166. M. E. Taylor, *Partial Differential Equations*, Vol. III, Springer-Verlag, New York, 1996.

167. V. Thomee, *Galerkin Finite Element Methods for Parabolic Problems*, Springer-Verlag, New York, 1997.

168. H. Triebel, *Interpolation Theory, Function Spaces, and Differential Operators*, North-Holland, Amsterdam, 1978.

169. R. S. Varga, *Matrix Iterative Analysis*, Prentice-Hall, Englewood Cliffs, NJ, 1962.

170. A. Veeser, Convergent adaptive finite elements for the nonlinear Laplacian, *Numer. Math.*, 92:743–770, 2002.

171. R. Verfürth, A posteriori error estimates for nonlinear problems: Finite element discretizations of elliptic equations, *Math. Comp.*, 62(206):445–475, 1994.

172. R. Verfürth, *A Review of* A Posteriori *Error Estimation and Adaptive Mesh-Refinement Techniques*, Wiley, New York, 1996.

173. E. L. Wachspress, *Iterative Solution of Elliptic Systems and Applications to the Neutron Diffusion Equations of Reactor Physics*, Prentice-Hall, Englewood Cliffs, NJ, 1966.

174. R. M. Wald, *General Relativity*, University of Chicago Press, Chicago, IL, 1984.

175. J. Wang, Convergence analysis without regularity assumptions for multigrid algorithms based on SOR smoothing, *SIAM J. Numer. Anal.*, 29(4):987–1001, 1992.

176. J. Wloka, *Partial Differential Equations*, Cambridge University Press, Cambridge, MA, 1992.

177. J. Xu, *Theory of Multilevel Methods*, Ph.D. dissertation, Department of Mathematics, Penn State University, 1989, technical Report AM 48.

178. J. Xu, Iterative methods by space decomposition and subspace correction, *SIAM Rev.*, 34(4):581–613, 1992.

179. J. Xu and A. Zhou, Local and parallel finite element algorithms based on two-grid discretizations, *Math. Comp.*, 69:881–909, 2000.

180. J. Xu and L. Zikatanov, Some observations on Babuška and Brezzi theories, *Numer. Math.*, 94:195–202, 2003.

181. J. Xu and Q. Zou, Analysis of linear and quadratic simplicial finite volume methods for elliptic equations, *Numer. Math.*, 111(3):469–492, 2009.

182. J. W. York, Jr., Kinematics and dynamics of general relativity, in *Sources of Gravitational Radiation* (L. L. Smarr, Ed.), pp. 83–126, Cambridge University Press, Cambridge, MA, 1979.

183. K. Yosida, *Functional Analysis*, Springer-Verlag, Berlin, 1980.

184. D. M. Young, *Iterative Solution of Large Linear Systems*, Academic Press, New York, 1971.

185. H. Yserentant, On the multi-level splitting of finite element spaces, *Numer. Math.*, 49:379–412, 1986.

186. Z. Yu, M. Holst, Y. Cheng, and J. McCammon, Feature-preserving adaptive mesh generation for molecular shape modeling and simulation, *J. of Mol. Graph. Model.*, 26:1370–1380, 2008.

187. Z. Yu, M. Holst, and J. McCammon, High-fidelity geometric modeling for biomedical applications, *Finite Elem. Anal. Des.*, 44(11):715–723, 2008.

188. E. Zeidler, *Nonlinear Functional Analysis and Its Applications*, Vol. I: Fixed Point Theorems, Springer-Verlag, New York, 1991.

189. E. Zeidler, *Nonlinear Functional Analysis and Its Applications*, Vol. III: Variational Methods and Optimization, Springer-Verlag, New York, 1991.

190. S. Zhang, *Multi-level Iterative Techniques*, Ph.D. dissertation, Department of Mathematics, Pennsylvania State University, 1988.

191. Y. Zhou, M. Holst, and J. McCammon, Nonlinear elastic modeling of macromolecular conformational change induced by electrostatic forces, *J. Math. Anal. Appl.*, 340(1):135–164, 2008, available as arXiv:1001.1371 [math.AP].

192. Z. Zhou, P. Payne, M. Vasquez, N. Kuhn, and M. Levitt, Finite-difference solution of the Poisson-Boltzmann equation: Complete elimination of self-energy, *J. Comput. Chem.*, 11(11):1344–1351, 1996.

193. O. C. Zienkiewicz and R. L. Taylor, *The Finite Element Method*, 5th ed., Vol. 1: The Basis, Butterworth-Heinemann, 2000.

194. O. C. Zienkiewicz and R. L. Taylor, *The Finite Element Method*, 5th ed., Vol. 2: Solid Mechanics, Butterworth-Heinemann, 2000.

195. O. C. Zienkiewicz and R. L. Taylor, *The Finite Element Method*, 5th ed., Vol. 3: Fluid Dynamics, Butterworth-Heinemann, 2000.

INDEX

PURE AND APPLIED MATHEMATICS

A Wiley Series of Texts, Monographs, and Tracts

Founded by RICHARD COURANT

Editors Emeriti: MYRON B. ALLEN III, DAVID A. COX, PETER HILTON, HARRY HOCHSTADT, PETER LAX, JOHN TOLAND

ADÁMEK, HERRLICH, and STRECKER—Abstract and Concrete Catetories
ADAMOWICZ and ZBIERSKI—Logic of Mathematics
AINSWORTH and ODEN—A Posteriori Error Estimation in Finite Element Analysis
AKIVIS and GOLDBERG—Conformal Differential Geometry and Its Generalizations
ALLEN and ISAACSON—Numerical Analysis for Applied Science
*ARTIN—Geometric Algebra
ATKINSON, HAN, and STEWART—Numerical Solution of Ordinary Differential Equations
AUBIN—Applied Functional Analysis, Second Edition
AZIZOV and IOKHVIDOV—Linear Operators in Spaces with an Indefinite Metric
BASENER—Topology and Its Applications
BERG—The Fourier-Analytic Proof of Quadratic Reciprocity
BERKOVITZ—Convexity and Optimization in $\mathbb{R}^n$
BERMAN, NEUMANN, and STERN—Nonnegative Matrices in Dynamic Systems
BOYARINTSEV—Methods of Solving Singular Systems of Ordinary Differential Equations
BRIDGER—Real Analysis: A Constructive Approach
BURK—Lebesgue Measure and Integration: An Introduction
*CARTER—Finite Groups of Lie Type
CASTILLO, COBO, JUBETE, and PRUNEDA—Orthogonal Sets and Polar Methods in Linear Algebra: Applications to Matrix Calculations, Systems of Equations, Inequalities, and Linear Programming
CASTILLO, CONEJO, PEDREGAL, GARCIÁ, and ALGUACIL—Building and Solving Mathematical Programming Models in Engineering and Science
CHATELIN—Eigenvalues of Matrices
CLARK—Mathematical Bioeconomics: The Mathematics of Conservation, Third Edition
COX—Galois Theory
†COX—Primes of the Form $x^2 + ny^2$: Fermat, Class Field Theory, and Complex Multiplication
*CURTIS and REINER—Representation Theory of Finite Groups and Associative Algebras
*CURTIS and REINER—Methods of Representation Theory: With Applications to Finite Groups and Orders, Volume I
CURTIS and REINER—Methods of Representation Theory: With Applications to Finite Groups and Orders, Volume II
DINCULEANU—Vector Integration and Stochastic Integration in Banach Spaces
*DUNFORD and SCHWARTZ—Linear Operators
Part 1—General Theory
Part 2—Spectral Theory, Self Adjoint Operators in Hilbert Space
Part 3—Spectral Operators
FARINA and RINALDI—Positive Linear Systems: Theory and Applications

*Now available in a lower priced paperback edition in the Wiley Classics Library.
†Now available in paperback.

FATICONI—The Mathematics of Infinity: A Guide to Great Ideas
FOLLAND—Real Analysis: Modern Techniques and Their Applications
FRÖLICHER and KRIEGL—Linear Spaces and Differentiation Theory
GARDINER—Teichmüller Theory and Quadratic Differentials
GILBERT and NICHOLSON—Modern Algebra with Applications, Second Edition
*GRIFFITHS and HARRIS—Principles of Algebraic Geometry
GRILLET—Algebra
GROVE—Groups and Characters
GUSTAFSSON, KREISS and OLIGER—Time Dependent Problems and Difference Methods
HANNA and ROWLAND—Fourier Series, Transforms, and Boundary Value Problems, Second Edition
*HENRICI—Applied and Computational Complex Analysis
Volume 1, Power Series—Integration—Conformal Mapping—Location of Zeros
Volume 2, Special Functions—Integral Transforms—Asymptotics—Continued Fractions
Volume 3, Discrete Fourier Analysis, Cauchy Integrals, Construction of Conformal Maps, Univalent Functions
*HILTON and WU—A Course in Modern Algebra
*HOCHSTADT—Integral Equations
JOST—Two-Dimensional Geometric Variational Procedures
KHAMSI and KIRK—An Introduction to Metric Spaces and Fixed Point Theory
*KOBAYASHI and NOMIZU—Foundations of Differential Geometry, Volume I
*KOBAYASHI and NOMIZU—Foundations of Differential Geometry, Volume II
KOSHY—Fibonacci and Lucas Numbers with Applications
LAX—Functional Analysis
LAX—Linear Algebra and Its Applications, Second Edition
LOGAN—An Introduction to Nonlinear Partial Differential Equations, Second Edition
LOGAN and WOLESENSKY—Mathematical Methods in Biology
MARKLEY—Principles of Differential Equations
MORRISON—Functional Analysis: An Introduction to Banach Space Theory
NAYFEH—Perturbation Methods
NAYFEH and MOOK—Nonlinear Oscillations
O'LEARY—Revolutions of Geometry
O'NEIL—Beginning Partial Differential Equations, Second Edition
PANDEY—The Hilbert Transform of Schwartz Distributions and Applications
PETKOV—Geometry of Reflecting Rays and Inverse Spectral Problems
*PRENTER—Splines and Variational Methods
PROMISLOW—A First Course in Functional Analysis
RAO—Measure Theory and Integration
RASSIAS and SIMSA—Finite Sums Decompositions in Mathematical Analysis
RENELT—Elliptic Systems and Quasiconformal Mappings
RIVLIN—Chebyshev Polynomials: From Approximation Theory to Algebra and Number Theory, Second Edition
ROCKAFELLAR—Network Flows and Monotropic Optimization
ROITMAN—Introduction to Modern Set Theory
ROSSI—Theorems, Corollaries, Lemmas, and Methods of Proof
*RUDIN—Fourier Analysis on Groups
SENDOV—The Averaged Moduli of Smoothness: Applications in Numerical Methods and Approximations
SENDOV and POPOV—The Averaged Moduli of Smoothness

*Now available in a lower priced paperback edition in the Wiley Classics Library.
†Now available in paperback.

SEWELL—The Numerical Solution of Ordinary and Partial Differential Equations, Second Edition
SEWELL—Computational Methods of Linear Algebra, Second Edition
SHICK—Topology: Point-Set and Geometric
SHISKOWSKI and FRINKLE—Principles of Linear Algebra With *Maple*™
*SIEGEL—Topics in Complex Function Theory
Volume 1—Elliptic Functions and Uniformization Theory
Volume 2—Automorphic Functions and Abelian Integrals
Volume 3—Abelian Functions and Modular Functions of Several Variables
SMITH and ROMANOWSKA—Post-Modern Algebra
ŠOLÍN–Partial Differential Equations and the Finite Element Method
STADE—Fourier Analysis
STAHL—Introduction to Topology and Geometry
STAKGOLD and HOLST—Green's Functions and Boundary Value Problems, Third Edition
STANOYEVITCH—Introduction to Numerical Ordinary and Partial Differential Equations Using MATLAB®
*STOKER—Differential Geometry
*STOKER—Nonlinear Vibrations in Mechanical and Electrical Systems
*STOKER—Water Waves: The Mathematical Theory with Applications
WATKINS—Fundamentals of Matrix Computations, Third Edition
WESSELING—An Introduction to Multigrid Methods
†WHITHAM—Linear and Nonlinear Waves
ZAUDERER—Partial Differential Equations of Applied Mathematics, Third Edition

*Now available in a lower priced paperback edition in the Wiley Classics Library.
†Now available in paperback.